BASIC TECHNICAL MATHEMATICS

OTHER ADDISON-WESLEY TITLES OF RELATED INTEREST

- *Basic Technical Mathematics with Calculus,* Seventh Edition, by Allyn J. Washington
- *Basic Technical Mathematics with Calculus, Metric Version,* Seventh Edition, by Allyn J. Washington
- *Introduction to Technical Mathematics,* Fourth Edition, by Allyn J. Washington and Mario F. Triola
- *Technical Calculus with Analytic Geometry,* Third Edition, by Allyn J. Washington

BASIC TECHNICAL MATHEMATICS

SEVENTH EDITION

ALLYN J. WASHINGTON

Dutchess Community College

ADDISON-WESLEY

An imprint of Addison Wesley Longman, Inc.

Reading, Massachusetts • Menlo Park, California • New York • Harlow, England
Don Mills, Ontario • Sydney • Mexico City • Madrid • Amsterdam

Publisher: Jason A. Jordan

Acquisitions Editor: Jennifer Crum

Assistant Editor: Adam Hamel

Editorial Project Manager: Ruth Berry

Managing Editor: Ron Hampton

Text Design and Cover Art Direction: Susan Carsten

Production Services: Greg Hubit Bookworks

Production Coordinator: Jane DePasquale

Prepress Services Buyer: Caroline Fell

Marketing Manager: Craig Bleyer

Marketing Coordinators: Laura Rogers and Elan Hanson

Manufacturing Coordinator: Evelyn Beaton

Composition: The Beacon Group

ABOUT THE COVER

Millennial Cityscape is a three-dimensional rendering of a computer motherboard created using KPT Bryce software with enhanced textures added in Adobe Photoshop.

LIBRARY OF CONGRESS CATALOGING-IN-PUBLICATION DATA

Washington, Allyn J.
  Basic technical mathematics / Allyn J. Washington.—7th ed.
    p. cm.
  Includes index.
  ISBN 0-201-35664-3
  1. Mathematics. I. Title.
QA39.2.W37 1999
512′.1—dc21   98-48201
                                              CIP

7 8 9 10—CW—03

CONTENTS

PREFACE

SCOPE OF THE BOOK

This book is intended primarily for students in technical and pre-engineering technology programs or other programs for which coverage of basic mathematics is required.

Chapters 1 through 20 provide the necessary background in algebra and trigonometry for analytic geometry and calculus courses, and Chapters 1 through 21 provide the necessary background for calculus courses. There is an integrated treatment of mathematical topics, primarily algebra and trigonometry, which are necessary for a sound mathematical background for the technician. Numerous applications from many fields of technology are included, primarily to indicate where and how mathematical techniques are used. However, it is not necessary that the student have a specific knowledge of the technical area from which any given problem is taken.

Most students using this text will have a background including some algebra and geometry. However, the material is presented in adequate detail for those whose background is deficient to some extent in these areas. The material presented here is sufficient for two or three semesters.

One of the primary reasons for the arrangement of topics in this text is to present material in an order that allows a student to take courses concurrently in allied technical areas, such as physics and electricity. These allied courses normally require a student to know certain mathematical topics by certain definite times; yet the traditional order of topics in mathematics makes it difficult to attain this coverage without loss of continuity. However, the material in this book can be rearranged to fit any appropriate sequence of topics. Another feature of this text is that certain topics that are traditionally included primarily for mathematical completeness have been covered only briefly or have been omitted.

The approach used here is basically an intuitive one. It is not unduly rigorous mathematically, although all appropriate terms and concepts are introduced as needed and given an intuitive or algebraic foundation. The book's aim is to help the student develop a feeling for mathematical methods, and is not simply to provide a collection of formulas. The text emphasizes that it is essential for the student to have a sound background in algebra and trigonometry in order to understand and succeed in any subsequent work in mathematics.

NEW FEATURES

The seventh edition of *Basic Technical Mathematics* includes all the basic features of the earlier editions. However, many sections have been rewritten to some degree to include additional or revised explanatory material, examples, and exercises. Some sections have been extensively revised. Specifically, among the new features of this edition are the following:

THE GRAPHING CALCULATOR

The coverage of the many operations that can be performed with a graphing calculator has been greatly expanded in this edition. The graphing calculator is used throughout the text in examples and exercises to help reinforce and develop many topics. There are over 140 all-new graphing calculator screens in this edition. The coverage starts in Section 1-3, where it is used for calculational purposes, and its use for graphing starts in Section 3-5. Additional notes on its use are found in Appendix C.

Appendix C presents 14 graphing calculator programs that show how the calculator's use can be expanded even more. Also, a *Graphing Calculator Lab Manual* showing detailed procedures is available as a supplement.

REVISED COVERAGE

Chapter 22 on statistics has been significantly revised. Section 22-1 on frequency distributions has been rewritten. Section 22-4 is a new section on normal distributions, and Section 22-5 is a new section on statistical process control.

WORD PROBLEMS

There are over 70 examples throughout the text that show the complete solutions of word problems. These are now clearly noted by the phrase *solving a word problem* placed to the left of the example. There are also about 700 exercises in which word problems are to be solved.

WRITING EXERCISES

As in the previous edition, there is one specific writing exercise included at the end of each chapter. These exercises give the student practice in writing explanations of problem solutions. Also, there are over 140 additional exercises throughout the book (at least five in each chapter) that require at least a sentence or two of explanation along with the answer. These are now specifically noted by the (W) symbol placed to the left of the exercise number. A special *Index of Writing Exercises* is included in the back of the book.

EXERCISES

There are over 1000 new exercises, including over 200 that illustrate technical applications. Most of these have replaced exercises from the previous edition. There are about 7300 exercises total in this edition.

FIGURES

There are over 140 new figures in this edition, and all of the over 140 calculator screens are new. There are now over 900 figures in the text, in addition to those in the answer section.

ADDITIONAL FEATURES

PAGE LAYOUT
Special attention has been given to the page layout. Nearly all examples are started and completed on the same page (there are only two exceptions, and each of these is presented on facing pages). Also, all figures are shown immediately adjacent to the material in which they are discussed.

SPECIAL EXPLANATORY COMMENTS
Throughout the book, special explanatory comments in color have been used in the examples to emphasize and clarify certain important points. Arrows are often used to indicate clearly the part of the example to which reference is made.

PROBLEM-SOLVING TECHNIQUES
Techniques and procedures that summarize the approaches to solving many types of problems have been clearly outlined in color-shaded boxes.

IMPORTANT FORMULAS
Throughout the book, important formulas are set off and displayed so that they can be easily located and used.

SUBHEADS AND KEY TERMS
Many sections include subheads to indicate where the discussion of a new topic starts within the section, and other key terms are noted in the margin for emphasis and easy reference. Also, where appropriate, chapter equations from earlier material are shown in the margin.

SPECIAL CAUTION AND NOTE INDICATORS

CAUTION ▶

NOTE ▶

Two special margin indicators (as shown at the left) are used. The caution indicator is used to identify points with which students commonly make errors or which they tend to have more difficulty in handling. The note indicator is used to point out text material that is of particular importance in developing or understanding the topic under discussion.

CHAPTER INTRODUCTIONS
Each chapter is introduced by identifying the topics to be developed and some of the important areas of technical applications. A particular type of application is illustrated through a drawing at the beginning of each chapter, and in Chapters 2 through 22 a problem related to this application is solved in an example later in the chapter.

CHAPTER EQUATIONS, REVIEW EXERCISES, AND PRACTICE TEST
At the end of each chapter, all important equations are listed together for easy reference. Each chapter is also followed by a set of review exercises that covers all of the material of the chapter. Following the chapter equations and review exercises is a chapter practice test that students can use to check their understanding of the material. Solutions to all practice test problems are given in the back of the book.

APPLICATIONS AND UNITS OF MEASUREMENT

The examples and exercises illustrate the application of mathematics to all fields of technology. Many relate to modern technology such as computer design, computer-assisted design (CAD), electronics, solar energy, lasers, fiber optics, holography, the environment, and space technology. A special *Index of Applications* for the exercises is included near the end of the book.

SUPPLEMENTARY TOPICS

In response to the needs of certain programs, two additional sections of text material are included after Chapter 22. The topics covered are *Gaussian elimination* and *rotation of axes.*

EXAMPLES

There are about 1000 worked examples in this edition. Of these, over 200 illustrate technical applications.

ANSWERS TO EXERCISES

The answers to all the odd-numbered exercises (except the end-of-chapter writing exercises) are given at the back of the book. The *Student's Solutions Manual* contains solutions for every other odd-numbered section exercise and the *Instructor's Solutions Manual* contains solutions for every section exercise. The answers to all exercises are given in the *Answer Book.*

FLEXIBILITY OF MATERIAL COVERAGE

The order of material coverage can be changed in many places, and certain sections may be omitted without loss of continuity of coverage. Users of earlier editions have indicated the successful use of numerous variations in coverage. Any changes will depend on the type of course and completeness required. Several of the possible variations in coverage are discussed in the *Answer Book.*

SUPPLEMENTS

Supplements to this text include an *Instructor's Solutions Manual* with detailed solutions to every section exercise, as well as a *Student's Solutions Manual* with detailed solutions for every other odd-numbered section exercise. An *Answer Book* for instructors includes answers to all exercises (including review exercises), along with information on how to use the text. A *Printed Test Bank* contains two short-answer test forms for every chapter. A *Graphing Calculator Lab Manual,* by Robert Seaver of Lorain County Community College, presents detailed instructions on the use of graphing calculators. Addison Wesley Longman's test-generating software—*TestGen-EQ* with *QuizMaster-EQ*—is also available. This software provides a computerized test bank of algorithmically defined problems organized by chapter, and it allows instructors to create and print tests in a variety of formats. Using QuizMaster-EQ, instructors can choose to create quizzes for students to take on-line; Quiz-Master automatically grades the quizzes and can generate a number of grading reports. Instructors may obtain copies of any of these supplements by contacting their Addison Wesley Longman sales consultant.

A Web site is also available for this text at www.technicalmath.com. The site contains collaborative activities, chapter openers with Web links, chapter quizzes, a glossary, and lists of supplements for students and instructors.

QUESTIONS/COMMENTS/INFORMATION

We welcome your comments about this text. You may write to the publisher or author at:

Addison Wesley
Mathematics Marketing
75 Arlington Street, Suite 300
Boston, MA 02116

You may also contact us via e-mail at math@awl.com to leave comments, suggestions, or questions for the publisher or the author.

ACKNOWLEDGMENTS

The author gratefully acknowledges the contributions of the following reviewers. Their detailed comments and many suggestions were of great assistance in preparing this seventh edition.

Kathleen M. Acks
Maui Community College

Carl W. Anderson
Johnson County Community College

Ginny Anson
Northeastern Iowa Community College

Russell Baker
Howard Community College

Irving Bansfield
Algonquin College

Tony Biles
College of North Atlantic

Wayne Braith
St. Cloud State University

Sid Britton
New Brunswick Community College

James Cassidy
Suffolk Community College

Michael Chen
British Columbia Institute of Technology

Stephen T. Corbin
J. Sargeant Reynolds Community College

Cynthia L. Coulter
Catawba Valley Community College

D. Demedash
Red River College

Chris J. Diorietes
Fayetteville Technical Community College

Kodwo Ewusi
Delaware Technical and Community College

Lionel Geller
Dawson College

Parviz Ghavami
Texas State Technical College

Glen Goodale
Dawson College

Carole E. Goodson
University of Houston

Bob Hamel
Sault College

Mark S. Harris
Monroe Community College, Harlingen Campus

Mary Ann Hovis
Lima Technical College

Malvern Huseman
Northern Alberta Institute of Technology

Marilyn Jacobi
Gateway Community-Technical College

Joseph Jordan
John Tyler Community College

Larry Juse
Northern Alberta Institute of Technology

James Lai
Nova Scotia Institute of Technology

Colin Lawrence
British Columbia Institute of Technology

Pamela Lowry
Lawrence Technological University

Joe MacPherson
University College of Cape Breton

Asad Maham
Radio College of Canada

Paul Maini
Suffolk County Community College

Marcel Maupin
Oklahoma State University, Oklahoma City

Ron McPhee
St. Lawrence College

Carol A. McVey
Florence Darlington Technical College

Blair Mennie
Algonquin College

E. Kurt Mobley
Augusta Technical Institute

Gharib Mohamed
Northern Alberta Institute of Technology

Don Nevin
Vermont Technical College

Tom Nichols
North Central Technical College

Robert Opel
Waukesha County Technical College

Barry Parsons
*Southern Alberta Institute of
Technology*

Derek Randall
Camosun College

Gar Randall
College of North Atlantic

Scott P. Randby
University of Akron

Don Reichman
Mercer County Community College

Louise Routledge
*British Columbia Institute of
Technology*

Radha Shrinivas
*St. Louis Community College,
Forest Park*

Jack Sontrop
Loyalist College

Chris Sparks
*Southern Alberta Institute of
Technology*

Thomas J. Stark
*Cincinnati State Technical and
Community College*

David Sumerel
Piedmont Technical College

Richard P. Truchon
*New Hampshire Community
Technical College*

Sue Vallery
Sir Sandford Fleming College

Frank Walton
Lethbridge Community College

Richard Watkins
Tidewater Community College

Robert Woods
Broome Community College

I again wish to thank the members of the mathematics department of Dutchess Community College who made many suggestions for the earlier editions.

Special thanks go to Bob Martin of Tarrant County Junior College for preparing the *Answer Book,* the *Student's Solutions Manual,* and the *Instructor's Solutions Manual.* Special thanks also to Robert Seaver of Lorain County Community College for preparing the *Graphing Calculator Lab Manual,* and to Thomas Stark of Cincinnati State Technical and Community College for the *RISERS* approach to solving word problems in Appendix A.

My thanks and gratitude go to Jim Bryant who drew all of the chapter-opener drawings, and to Martha Ghent and Jeff Suzuki for assisting in the tedious task of reading proofs.

I gratefully acknowledge the cooperation and support of my editor, Jennifer Crum, and production editor, Greg Hubit, whose valuable assistance was greatly appreciated. I also wish to thank Ron Doleman, Publisher for Addison Wesley Longman's Canadian College Editorial group, for his valuable assistance.

Also of great assistance during the production of this edition were Ron Hampton, Ruth Berry, Meredith Nightingale, Jane DePasquale, Susan Carsten, Adam Hamel, and Lorie Reilly of the Addison Wesley Longman staff. Finally, special mention is due my wife, Millie, for her help with checking answers, as well as for her patience and support all through the preparation of this edition, and all earlier editions.

A.J.W.

BASIC TECHNICAL MATHEMATICS

1

BASIC ALGEBRAIC OPERATIONS

Mathematics has played a most important role in the development and understanding of the great advances in technology and science. This has resulted in a continually increased use of mathematics by technicians in all fields.

With the mathematics developed in this text, we can solve many kinds of applied problems. Consider the scene shown at the right. A few of the technical areas of application are

◆ surveying and construction (the bridge)
◆ physics, mechanical design, machine design, and drafting (the ship)
◆ electronics (communications and circuitry)
◆ architecture (the buildings)

as well as many others. In support of these technologies, many other technologies would also be used. Lasers would be used in surveying, and computers would certainly be used in many ways.

In this text there are applications from technical areas related to those noted above and many others. These technologies include (but are not limited to) aeronautical, automotive, business, environmental, heat and air conditioning, medical, meteorology, petroleum, product design, solar, space, and wastewater. To solve the applied problems in this text will require a knowledge of the mathematics presented, but will *not* require any prior knowledge of the field of application.

Although we cannot solve the more advanced types of problems that arise, we can form a foundation for the more advanced mathematics that is used to solve such problems. Therefore, a real understanding of the mathematics given in this text will be of great value to you in your future work.

A thorough understanding of algebra is essential in the study of any of the fields of mathematics. It is very important for you to learn and understand the basic concepts developed in this text; otherwise, the result will be a weak foundation in mathematics and in the various technical fields where mathematics is applied. Development of this understanding will require a serious commitment on your part. The author sincerely wishes you the best success.

We begin our study of mathematics by reviewing some of the basic concepts and operations that deal with numbers and symbols. With these we shall be able to develop the topics in algebra necessary for progress into other fields of mathematics, such as trigonometry and calculus.

1-1 NUMBERS

The way we represent numbers has been evolving for thousands of years. The first numbers used were for counting objects, and stand for whole quantities. Today, *these numbers are called* **positive integers** (*or* **natural numbers**). The positive integers are represented by the symbols 1, 2, 3, 4, and so on.

It is also necessary to have numbers that can represent parts of certain quantities. *The name* **positive rational number** *is given to any number that we can represent by the division of one positive integer by another. Numbers that cannot be written as the division of one integer by another are termed* **irrational.**

Irrational numbers were discussed by the Greek mathematician Pythagoras in about 540 B.C.

■**EXAMPLE 1**   The numbers 5 and 19 are positive integers. They are also rational numbers since they may be written as $\frac{5}{1}$ and $\frac{19}{1}$. Normally we do not write the 1's in the denominators.

The numbers $\frac{5}{8}$, $\frac{11}{3}$, and $\frac{106}{17}$ are positive rational numbers, since the numerator and denominator of each are positive integers.

The numbers $\sqrt{2}$ and π are irrational. It is not possible to find two integers that represent these numbers if one of the integers is divided by the other. For example, $\frac{22}{7}$ is not *exactly* equal to π; it is an *approximation*.

The number $\frac{2}{\sqrt{3}}$ is irrational. The numerator is an integer, but the denominator is irrational and $\frac{2}{\sqrt{3}}$ cannot be written as one integer divided by another. ■

THE REAL NUMBER SYSTEM

It is also necessary to have **negative numbers** in order to have a numerical answer to problems such as $5 - 8$. *Thus,* $-1, -2, -3$, *and so on are the* **negative integers.** *The number* **zero** *is an integer, but it is neither positive nor negative. This means that the* **integers** *are the numbers* $\ldots, -3, -2, -1, 0, 1, 2, 3$, *and so on.*

The integers, the rational numbers, and the irrational numbers, which include all such numbers that are zero, positive, or negative, make up what we call the **real number system.** We shall use real numbers throughout this text, with one important exception. In Chapter 12 we shall be using **imaginary numbers,** *which is the name given to square roots of negative numbers.* (The symbol j is used to designate $\sqrt{-1}$, which is not part of the real number system.) However, until Chapter 12, when we discuss operations on imaginary numbers in detail, it will be necessary only to recognize them if they occur.

■**EXAMPLE 2**   The number 7 is an integer. It is also rational since $7 = \frac{7}{1}$, and it is a real number since the real numbers include all the rational numbers.

The number 3π is irrational, and it is real since the real numbers include all the irrational numbers.

The numbers $\sqrt{-10}$ and $-\sqrt{-7}$ are imaginary numbers.

The number $\frac{-3}{7}$ is rational and real. The number $-\sqrt{7}$ is irrational and real.

The number $\frac{\pi}{6}$ is irrational and real. The number $\frac{\sqrt{-3}}{2}$ is imaginary. ■

Fractions were used by early Egyptians and Babylonians. They were used for calculations that involved parts of measurements, property, and possessions.

A **fraction** *may contain any number or symbol representing a number in its numerator or in its denominator. Therefore, a fraction may be a number that is rational, irrational, or imaginary.*

■**EXAMPLE 3**   The numbers $\frac{2}{7}$ and $\frac{-3}{2}$ are fractions, and they are rational.

The numbers $\frac{\sqrt{2}}{9}$ and $\frac{6}{\pi}$ are fractions, but they are not rational numbers. It is not possible to express either as one integer divided by another integer.

The number $\frac{\sqrt{-3}}{2}$ is a fraction, and it is an imaginary number. ■

THE NUMBER LINE

Real numbers may be represented by points on a line. We draw a horizontal line and designate some point on it by *O*, which we call the **origin** (see Fig. 1-1). The integer *zero* is located at this point. Equal intervals are marked to the right of the origin, and the positive integers are placed at these positions. The other positive rational numbers are located between the integers.

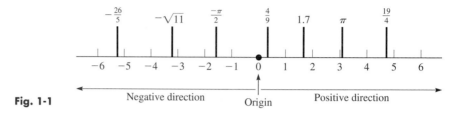

Fig. 1-1

Now we can give a meaning of direction to negative numbers. By starting at the origin and proceeding to the left, *defined as the* **negative direction,** we locate all the negative numbers. *As shown in Fig. 1-1, the positive numbers are to the right of the origin and the negative numbers are to the left of the origin.* Representing numbers in this way will be especially useful when we study graphical methods.

It will not be proved here, but the rational numbers do not take up all the positions on the line; the remaining points represent irrational numbers.

We next define another important concept of a number. *The* **absolute value** *of a positive number is the number itself, and the absolute value of a negative number is the corresponding positive number.* On the number line we may interpret the absolute value of a number as the distance between the origin and the number. The absolute value is denoted by writing the number between vertical lines, as shown in the following example.

■**EXAMPLE 4**   The absolute value of 6 is 6, and the absolute value of -7 is 7. We write these as $|6| = 6$ and $|-7| = 7$. See Fig. 1-2.

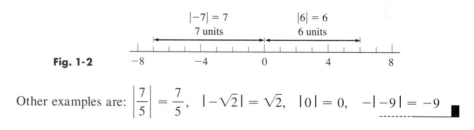

Fig. 1-2

Other examples are: $\left|\dfrac{7}{5}\right| = \dfrac{7}{5},\quad |-\sqrt{2}| = \sqrt{2},\quad |0| = 0,\quad -|-9| = -9$ ■

On the number line, *if a first number is to the right of a second number, then the first number is said to be* **greater than** *the second. If the first number is to the left of the second, it is* **less than** *the second number.* The symbol $>$ is used to designate "is greater than," and the symbol $<$ is used to designate "is less than." These are called **signs of inequality.** See Fig. 1-3.

The symbols $=$, $<$, and $>$ were introduced by English mathematicians in the late 1500s.

■**EXAMPLE 5**

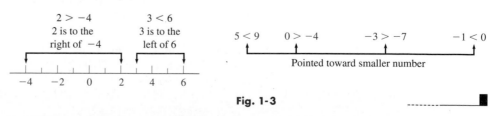

Fig. 1-3

Every number, except zero, has a **reciprocal.** *The reciprocal of a number is 1 divided by the number.*

EXAMPLE 6 The reciprocal of 7 is $\frac{1}{7}$. The reciprocal of $\frac{2}{3}$ is

$$\frac{1}{\frac{2}{3}} = 1 \times \frac{3}{2} = \frac{3}{2} \qquad \text{Invert and multiply (from arithmetic).}$$

The reciprocal of π is $1/\pi$. The reciprocal of -5 is $-\frac{1}{5}$. Note that the negative sign is retained in the reciprocal of a negative number. -------------------■

For reference, see Appendix B for units of measurement and the symbols used for them.

In applications, *numbers that represent a measurement and are written with units of measurement are called* **denominate numbers.** The next example illustrates the use of units and the symbols that represent them.

EXAMPLE 7 To show that a wire is 10 feet long, we write the length as 10 ft.

To show that the speed of a rocket is 1500 meters per second, we write the speed as 1500 m/s. (Note the use of s for second. We use s rather than sec.)

To show that the area of a computer chip is 0.75 square inch, we write this as 0.75 in.^{2}. (We will not use sq in.)

To show that the volume of a container is 500 cubic centimeters, we write the volume as 500 cm^3. (We will not use cu cm or cc.) -------------------■

LITERAL NUMBERS

Until now we have used numbers in their explicit form. However, it is usually more convenient to state definitions and operations on numbers in a general form. *To do this we represent the numbers by letters, referred to as* **literal numbers.**

For example, we can say, "If a is to the right of b on the number line, then $a > b$." This is more convenient than saying, "If a first number is to the right of a second number on the number line, then the first number is greater than the second number." The statement "the reciprocal of a number n is $1/n$" is another example of using a letter to stand for a number in general.

In an algebraic discussion, certain literal numbers may take on any allowable value, whereas other literal numbers represent the same number throughout the discussion. *Those literal numbers that may vary in a given problem are called* **variables,** *and those that are held fixed are called* **constants.**

Common usage normally designates letters near the end of the alphabet as variables and letters near the beginning of the alphabet as constants. Any exceptions to this usage would be specifically noted.

EXAMPLE 8 (a) The resistance of an electric resistor is R. The current I in the resistor equals the voltage V divided by R, written as $I = V/R$. For this resistor, I and V may take on various values, and R is fixed. This means I and V are variables and R is a constant. For *another* resistor, the value of R may differ.

(b) The fixed cost for a calculator manufacturer to operate a certain plant is b dollars per day, and it costs a dollars to produce each calculator. The total daily cost C to produce n calculators is

$$C = an + b$$

Here, C and n are variables, and a and b are constants. For *another* plant, the values of a and b would probably differ. -------------------■

EXERCISES *1-1*

In Exercises 1–4, designate each of the given numbers as being an integer, rational, irrational, real, or imaginary. (More than one designation may be correct.)

1. $3, -\pi$

2. $\dfrac{5}{4}, \sqrt{-4}$

3. $-\sqrt{-6}, \dfrac{\sqrt{7}}{3}$

4. $-\dfrac{7}{3}, \dfrac{\pi}{6}$

In Exercises 5–8, find the absolute value of each number.

5. $3, \dfrac{7}{2}$

6. $-4, \sqrt{2}$

7. $-\dfrac{6}{7}, -\sqrt{3}$

8. $-\dfrac{\pi}{2}, -\dfrac{19}{4}$

In Exercises 9–16, insert the correct sign of inequality (> or <) between the given pairs of numbers.

9. 6   8

10. 7   5

11. π -1

12. -4 0

13. -4 $-|-3|$

14. $-\sqrt{2}$ -1.42

15. $-\dfrac{1}{3}$ $-\dfrac{1}{2}$

16. -0.6 0.2

In Exercises 17–20, find the reciprocal of each number.

17. $3, -\dfrac{1}{3}$

18. $\dfrac{1}{6}, -\dfrac{4}{\sqrt{3}}$

19. $-\dfrac{5}{\pi}, x$

20. $-\dfrac{8}{3}, \dfrac{y}{b}$

In Exercises 21–24, locate each number on a number line as in Fig. 1-1.

21. $2.5, -\dfrac{1}{2}$

22. $\sqrt{3}, -\dfrac{12}{5}$

23. $-\dfrac{\sqrt{2}}{2}, 2\pi$

24. $\dfrac{123}{19}, -\dfrac{\pi}{6}$

In Exercises 25–40, solve the given problems. Refer to Appendix B for units of measurement and their symbols.

25. List the following numbers in numerical order, starting with the smallest: $-1, 9, \pi, \sqrt{5}, |-8|, -|-3|, -18$.

26. List the following numbers in numerical order, starting with the smallest: $\frac{1}{5}, -\sqrt{10}, -|-6|, -4, 0.25, |-\pi|$.

27. If a and b are positive integers and $b > a$, what type of number is represented by
(a) $b - a$, (b) $a - b$, (c) $\dfrac{b - a}{b + a}$?

28. If a and b represent positive integers, what kind of number is represented by (a) $a + b$, (b) a/b, (c) $a \times b$?

29. For any positive or negative integer: (a) Is its absolute value always an integer? (b) Is its reciprocal always a rational number?

30. For any positive or negative rational number: (a) Is its absolute value always a rational number? (b) Is its reciprocal always a rational number?

31. Describe the location of a number x on the number line when (a) $x > 0$, (b) $x < -4$.

32. Describe the location of a number x on the number line when (a) $|x| < 1$, (b) $|x| > 2$.

(W) 33. For a number $x > 1$, describe the location on the number line of the reciprocal of x.

(W) 34. For a number $x < 0$, describe the location on the number line of the number with a value of $|x|$.

35. The heat loss L through a certain type of insulation of thickness t is given by $L = a/t$, where a has a fixed value for this type of insulation. Identify the variables and constants.

36. A sensitive gauge measures the total weight w of a container and the water that forms in it as vapor condenses. It is found that $w = c\sqrt{0.1t + 1}$, where c is the weight of the container and t is the time of condensation. Identify the variables and constants.

37. The memory of a certain computer has a bits in each byte. Express the number N of bits in n kilobytes in an equation. (A *bit* is a single digit, and bits are grouped in *bytes* in order to represent special characters. Generally there are 8 bits per byte. If necessary, see Appendix B for the meaning of *kilo*.)

38. A piece y inches long is cut from a board x feet long. Give an equation for the length L, in inches, of the remaining piece.

(W) 39. In a laboratory report, a student wrote "$-20°C > -30°C$." Is this statement correct? Explain.

40. After 5 s, the pressure on a valve is less than 60 lb/in.^{2} (pounds per square inch). Using t to represent time and p to represent pressure, this statement can be written "for $t > 5$ s, $p < 60$ lb/in.^{2}." In this way, write the statement "when the current I in a circuit is less than 4 A, the voltage V is greater than 12 V."

1-2 FUNDAMENTAL OPERATIONS OF ALGEBRA

In performing operations with numbers, we know that certain basic laws are valid. *These are called the* **fundamental laws of algebra.**

THE COMMUTATIVE AND ASSOCIATIVE LAWS

For example, if two numbers are added, it does not matter in which order they are added. (For example, $5 + 3 = 8$ and $3 + 5 = 8$, or $5 + 3 = 3 + 5$.) This statement, generalized and assumed correct for all possible combinations of numbers to be added, is called the **commutative law** for addition. The law states that *the sum of two numbers is the same, regardless of the order in which they are added.* We make no attempt to prove this law in general but accept its validity.

In the same way, we have the **associative law** for addition, which states that *the sum of three or more numbers is the same, regardless of the way in which they are grouped for addition.* For example, $3 + (5 + 6) = (3 + 5) + 6$.

The laws just stated for addition are also true for multiplication. Therefore, *the product of two numbers is the same, regardless of the order in which they are multiplied,* and *the product of three or more numbers is the same, regardless of the way in which they are grouped for multiplication.* For example, $2 \times 5 = 5 \times 2$, and $5 \times (4 \times 2) = (5 \times 4) \times 2$.

THE DISTRIBUTIVE LAW

Another very important law is the **distributive law.** It states that *the product of one number and the sum of two or more other numbers is equal to the sum of the products of the first number and each of the other numbers of the sum.* For example,

$$4(3 + 5) = 4 \times 3 + 4 \times 5$$

In practice these laws are used intuitively, except perhaps for the distributive law. However, it is necessary to state them and to accept them so that we may use them as a basis for later results.

Not all operations are commutative and associative. For example, division is not commutative, since the order of division of two numbers does matter. For example, $\frac{6}{5} \neq \frac{5}{6}$ ($\neq$ is read "does not equal").

Using literal numbers, the fundamental laws of algebra are as follows:

Commutative law of addition: $a + b = b + a$

Associative law of addition: $a + (b + c) = (a + b) + c$

Commutative law of multiplication: $ab = ba$

Associative law of multiplication: $a(bc) = (ab)c$

Distributive law: $a(b + c) = ab + ac$

Each of these laws is an example of an *identity,* in that the expression to the left of the $=$ sign equals the expression to the right for any value of each of *a, b,* and *c*.

Operations on Positive and Negative Numbers

When using the basic operations (addition, subtraction, multiplication, division) on positive and negative numbers, we determine the result to be either positive or negative according to the following rules. From Section 1-1 we recall that *a positive number is preceded by no sign.* Therefore, in using these rules we indicate the "sign" of a positive number by simply writing the number itself.

Addition of two numbers of the same sign. *Add their absolute values and assign the sum their common sign.*

■EXAMPLE 1 (a) $2 + 6 = 8$ the sum of two positive numbers is positive

(b) $-2 + (-6) = -(2 + 6) = -8$ the sum of two negative numbers is negative

The negative number -6 is placed in parentheses since it is also preceded by a plus sign showing addition. It is not necessary to place the -2 in parentheses. ■

Addition of two numbers of different signs. *Subtract the number of smaller absolute value from the number of larger absolute value, and assign to the result the sign of the number of larger absolute value.*

■EXAMPLE 2

(a) $2 + (-6) = -(6 - 2) = -4$ the negative 6 has the larger absolute value

(b) $-6 + 2 = -(6 - 2) = -4$

(c) $6 + (-2) = 6 - 2 = 4$ the positive 6 has the larger absolute value

(d) $-2 + 6 = 6 - 2 = 4$

the subtraction of absolute values ■

Subtraction of one number from another. *Change the sign of the number being subtracted and change the subtraction to addition. Perform the addition.*

■EXAMPLE 3 (a) $2 - 6 = 2 + (-6) = -(6 - 2) = -4$

Note that after changing the subtraction to addition, and changing the sign of 6 to make it -6, we have precisely the same illustration as Example 2(a).

(b) $-2 - 6 = -2 + (-6) = -(2 + 6) = -8$

Note that after changing the subtraction to addition, and changing the sign of 6 to make it -6, we have precisely the same illustration as Example 1(b).

(c) $-a - (-a) = -a + a = 0$

SUBTRACTION OF A NEGATIVE NUMBER

This shows that subtracting a number from itself results in zero, even if the number is negative. Therefore, *subtracting a negative number is equivalent to adding a positive number of the same absolute value.* ■

Multiplication and division of two numbers. *The product (or quotient) of two numbers of the same sign is positive. The product (or quotient) of two numbers of different signs is negative.*

■EXAMPLE 4

(a) $3(12) = 3 \times 12 = 36$ $\dfrac{12}{3} = 4$ result is positive if both numbers are positive

(b) $-3(-12) = 3 \times 12 = 36$ $\dfrac{-12}{-3} = 4$ result is positive if both numbers are negative

(c) $3(-12) = -(3 \times 12) = -36$ $\dfrac{-12}{3} = -\dfrac{12}{3} = -4$ result is negative if one number is positive and the other is negative

(d) $-3(12) = -(3 \times 12) = -36$ $\dfrac{12}{-3} = -\dfrac{12}{3} = -4$ ■

Order of Operations

For an expression in which there is a combination of operations, we must perform them in the proper order. Often it is clear by the grouping of numbers as to the proper order of performing the operations. Numbers are grouped by symbols such as **parentheses,** ( ), and the **bar,** ____, between the numerator and the denominator of a fraction. However, if the order of some operations is not defined by specific groupings, we use the following order of operations:

Order of Operations

  1. *Operations within specific groupings are done first.*

  2. *Perform multiplications and divisions (from left to right).*

  3. *Then perform additions and subtractions (from left to right).*

■**EXAMPLE 5**   **(a)** $20 \div (2 + 3)$ is evaluated by first adding $2 + 3$ and then dividing. The grouping of $2 + 3$ is clearly shown by the parentheses. Therefore, we have $20 \div (2 + 3) = 20 \div 5 = 4$.

  (b) $20 \div 2 + 3$ is evaluated by first dividing 20 by 2 and then adding. No specific grouping is shown, and therefore the division is done before the addition. This means $20 \div 2 + 3 = 10 + 3 = 13$.

CAUTION▶   **(c)** $16 - 2 \times 3$ is evaluated by ***first multiplying*** 2 ***by*** 3 and then subtracting. We *do* **not** *first subtract* 2 *from* 16. Therefore, $16 - 2 \times 3 = 16 - 6 = 10$.

  (d) $16 \div 2 \times 4$ is evaluated by first dividing 16 by 2 and then multiplying. From left to right, the division occurs first. Therefore, $16 \div 2 \times 4 = 8 \times 4 = 32$.

  (e) $16 \div (2 \times 4)$ is evaluted by first multiplying 2 by 4 and then dividing. The grouping of 2×4 is clearly shown by parentheses. Therefore, this means that $16 \div (2 \times 4) = 16 \div 8 = 2$.   ∎

When evaluating expressions, it is generally more convenient to change the operations and numbers so that the result is found by the addition and subtraction of positive numbers. When this is done, we must remember that

$$a + (-b) = a - b \tag{1-1}$$

$$a - (-b) = a + b \tag{1-2}$$

■**EXAMPLE 6**   **(a)** $7 + (-3) - 6 = 7 - 3 - 6 = 4 - 6 = -2$ using Eq. (1-1)

(b) $\dfrac{18}{-6} + 5 - (-2) = -3 + 5 + 2 = 2 + 2 = 4$ using Eq. (1-2)

(c) $2(-3) - 2(-4) + \dfrac{25}{-5} = -6 - (-8) + (-5) = -6 + 8 - 5 = -3$

(d) $\dfrac{-12}{2 - 8} + \dfrac{5 - 1}{2(-1)} = \dfrac{-12}{-6} + \dfrac{4}{-2} = 2 + (-2) = 2 - 2 = 0$

In illustrations (b) and (c) we see that the multiplications and divisions were done before the additions and subtractions. In (d) we see that the groupings $(2 - 8$ and $5 - 1)$ were evaluated first. Then we did the divisions and, finally, the addition. ∎

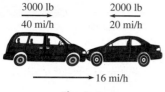

3000 lb → 40 mi/h

2000 lb ← 20 mi/h

→ 16 mi/h

Fig. 1-4

EXAMPLE 7 A 3000-lb van going at 40 mi/h ran head-on into a 2000-lb car going at 20 mi/h. An insurance investigator determined the velocity of the vehicles immediately after the collision from the following calculation. See Fig. 1-4.

$$\frac{3000(40) + (2000)(-20)}{3000 + 2000} = \frac{120,000 + (-40,000)}{3000 + 2000} = \frac{120,000 - 40,000}{5000}$$

$$= \frac{80,000}{5000} = 16 \text{ mi/h}$$

The numerator and the denominator must be evaluated before the division is performed. The multiplications in the numerator are performed first, followed by the addition in the denominator and the subtraction in the numerator. ▪

Operations with Zero

Since operations with zero tend to cause some difficulty, we will show them here.

If a is a real number, the operations of addition, subtraction, multiplication, and division with zero are as follows:

$$a + 0 = a$$

$$a - 0 = a \qquad 0 - a = -a$$

$$a \times 0 = 0$$

$$0 \div a = \frac{0}{a} = 0 \qquad \text{if} \qquad a \neq 0 \qquad (\neq \text{ means "is not equal to"})$$

EXAMPLE 8 (a) $5 + 0 = 5$ (b) $-6 - 0 = -6$ (c) $0 - 4 = -4$
(d) $\frac{0}{6} = 0$ (e) $\frac{0}{-3} = 0$ (f) $\frac{5 \times 0}{7} = \frac{0}{7} = 0$ ▪

Note that there is no result defined for division by zero. To understand the reason for this, consider the results for $\frac{6}{2}$ and $\frac{6}{0}$.

$$\frac{6}{2} = 3 \qquad \text{since} \qquad 2 \times 3 = 6$$

If $\frac{6}{0} = b$, then $0 \times b = 6$. This cannot be true because $0 \times b = 0$ for any value of b. Thus,

division by zero is undefined

(The special case of $\frac{0}{0}$ is termed *indeterminate*. If $\frac{0}{0} = b$, then $0 = 0 \times b$, which is true for any value of b. Therefore, no specific value of b can be determined.)

EXAMPLE 9

$\frac{2}{5} \div 0$ is undefined $\frac{8}{0}$ is undefined $\left(\frac{7 \times 0}{0 \times 6} \text{ is indeterminate}\right)$ ▪

The operations with zero will not cause any difficulty if we remember to

CAUTION ▶   *never divide by 0*

Division by zero is the only undefined basic operation. All the other operations with zero may be performed as for any other number.

EXERCISES *1-2*

In Exercises 1–32, evaluate each of the given expressions by performing the indicated operations.

1. $8 + (-4)$ **2.** $-4 + (-7)$ **3.** $-3 + 9$

4. $18 - 21$ **5.** $-19 - (-16)$ **6.** $8 - (-4)$

7. $8(-3)$ **8.** $-9(3)$ **9.** $-7(-5)$

10. $\dfrac{-9}{3}$ **11.** $\dfrac{-60}{-3}$ **12.** $\dfrac{28}{-7}$

13. $-2(4)(-5)$ **14.** $3(-4)(6)$ **15.** $\dfrac{2(-5)}{10}$

16. $\dfrac{-64}{2(-4)}$ **17.** $9 - 0$ **18.** $\dfrac{0}{-6}$

19. $\dfrac{17}{0}$ **20.** $\dfrac{2 - (-5)}{0}$

21. $8 - 3(-4)$ **22.** $20 + 8 \div 4$

23. $3 - 2(6) + \left|\dfrac{8}{2}\right|$ **24.** $0 - (-6)(-8) + (-10)$

25. $\dfrac{3(-6)(-2)}{0 - 4}$ **26.** $\dfrac{7 - |-5|}{-1(-2)}$

27. $\dfrac{24}{3 + (-5)} - 4(-9)$ **28.** $\dfrac{-18}{3} - \dfrac{4 - 6}{-1}$

29. $-7 - \dfrac{-14}{2} - 3(2)$ **30.** $-7(-3) + \dfrac{6}{-3} - (-9)$

31. $\dfrac{3(-9) - 2(-3)}{3 - 10}$ **32.** $\dfrac{2(-7) - 4(-2)}{-9 - (-9)}$

In Exercises 33–40, determine which of the fundamental laws of algebra is demonstrated.

33. $6(7) = 7(6)$ **34.** $6 + 8 = 8 + 6$

35. $6(3 + 1) = 6(3) + 6(1)$ **36.** $4(5 \times 7) = (4 \times 5)(7)$

37. $3 + (5 + 9) = (3 + 5) + 9$

38. $8(3 - 2) = 8(3) - 8(2)$

39. $(2 \times 3) \times 9 = 2 \times (3 \times 9)$

40. $(3 \times 6) \times 7 = 7 \times (3 \times 6)$

In Exercises 41–44, for numbers a and b, determine which of the following expressions equals the given expression.
(a) $a + b$, (b) $a - b$, (c) $b - a$, (d) $-a - b$

41. $-a + (-b)$ **42.** $b - (-a)$

43. $-b - (-a)$ **44.** $-a - (-b)$

In Exercises 45–52, answer the given questions.

45. (a) What is the sign of the product of an even number of negative numbers? (b) What is the sign of the product of an odd number of negative numbers?

(W) 46. Is subtraction commutative? Explain.

47. The daily high temperature (in °C) in the Falkland Islands in the southern Atlantic Ocean during the first week in July were recorded as 7, 3, −2, −3, −1, 4, and 6. What was the average daily temperature for the week? (Divide the algebraic sum of the readings by the number of readings.)

48. The electric current was measured in a given ac circuit at equal intervals as 0.7 mA, −0.2 mA, −0.9 mA, and −0.6 mA. What was the change in the current between (a) the first two readings, (b) the middle two readings, and (c) the last two readings?

49. One oil well drilling rig drills 100 m deep the first day and 200 m deeper the second day. A second rig drills 200 m deep the first day and 100 m deeper the second day. In showing that the total depth drilled by each rig was the same, state what fundamental law of algebra is illustrated.

50. An electronics dealer averaged 18 min each when selling 5 computers, and 5 min each when selling 18 cellular phones. In showing that the total time in selling the computers equals the total time in selling the phones, what fundamental law of algebra is illustrated?

51. Each of three delivery trucks is loaded with 50 cases of regular soda and 40 cases of diet soda. Set up the expression for the total number of cases on all three trucks. What fundamental law of algebra is illustrated?

52. A jet travels 600 mi/h relative to the air. The wind is blowing at 50 mi/h. If the jet travels with the wind for 3 h, set up the expression for the distance traveled. What fundamental law of algebra is illustrated?

1-3 CALCULATORS AND APPROXIMATE NUMBERS

Electronic scientific calculators were first developed in the early 1970s, and graphing calculators were introduced in the late 1980s.

In the remainder of the book it is assumed that you will do most of your calculations on a calculator. A *graphing calculator* can be used for all of these calculations and many other operations. Although a *scientific calculator* can perform the calculations, it cannot perform many of the other required operations we will cover. Therefore, in this text we will restrict our coverage of calculator use to the graphing calculator.

A brief discussion of the use of a graphing calculator appears in Appendix C, and sample calculator screens for various operations will appear throughout the book. However, there are many different models of graphing calculators, and *the notation and screen appearance for many operations will differ from one model to another.* If you are not familiar with the use of your calculator, you should *practice using it and review its manual* to be sure of the order in which the keys are used. It will take only a little practice to be able to use many of the basic operations. However, some of the specialized operations may require a review of the manual to be certain of the required key sequences. Following is an example of a basic calculation done on a graphing calculator.

■**EXAMPLE 1**   Calculate the value of $38.3 - 12.9(-3.58)$.

The numbers are entered as follows. The calculator will perform the multiplication first, following the order of operations shown in the previous section. The sign of -3.58 is entered using the $(-)$ key, before the 3.58 is entered. The display on the calculator screen is shown in Fig. 1-5.

$$38.3 \boxed{-} 12.9 \boxed{\times} \boxed{(-)} 3.58 \quad \text{ENTER}$$

This means that $38.3 - 12.9(-3.58) = 84.482$.

Note in the display that the negative sign of -3.58 is smaller and a little higher to distinguish it from the minus sign for subtraction. Also note the * shown for multiplication; the asterisk is the standard computer symbol for multiplication.

Looking back into Section 1-2, we see that *the minus sign is used in two different ways:* (1) to indicate subtraction and (2) to designate a negative number. This is clearly shown on a graphing calculator since there is a key for each purpose. The $\boxed{-}$ key is used for subtraction, and the $\boxed{(-)}$ key is used before a number to make it negative.

We will first use a graphing calculator for the purpose of graphing in Section 3-5. Before then we will show some calculational uses of a graphing calculator. Throughout the book we will show many uses of a graphing calculator, although *key sequences will generally not be shown,* except as they may appear in the display screens, *since they may vary from one model to another.*

Approximate Numbers and Significant Digits

We must consider the accuracy of the numbers used in calculations, as the final result should not be written with any more accuracy than is proper. For example, if the numbers in Example 1 are *approximate,* the result should be written as 84.5, not 84.482. We will see the reason for this later in this section.

Most numbers in technical and scientific work are **approximate numbers,** having been determined by some measurement. Certain other numbers are **exact numbers,** having been determined by a definition or a counting process.

■**EXAMPLE 2**   If a voltage shown on a voltmeter is read as 116 V, the 116 is approximate. Another voltmeter may show the voltage as 115.7 V. However, this voltage cannot be determined *exactly.*

If a computer prints out the number of names on a list as 97, this 97 is exact. We know it is not 96 or 98. Since 97 was found by precise counting, it is exact.

By definition, 60 s = 1 min, and the 60 and the 1 are exact.

All calculator screens shown with text material are for a TI-83. They are intended only as an illustration of a calculator screen for the particular operation. Screens for other models may differ.

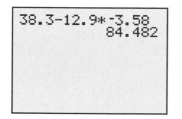

```
38.3-12.9*-3.58
            84.482
```

Fig. 1-5

Some calculator keys on different models are labeled differently. For example, on some models the EXE key is equivalent to the ENTER key.

SIGNIFICANT DIGITS

An approximate number may have to include some zeros to properly locate the decimal point. *Except for these zeros, all other digits are called* **significant digits.** The next example illustrates how we determine the number of significant digits.

■**EXAMPLE 3**   All numbers in this example are assumed to be approximate.
34.7 has three significant digits.
0.039 has two significant digits. The zeros properly locate the decimal point.
706.1 has four significant digits. The zero is not used for the location of the decimal point. It shows the number of tens in 706.1.

CAUTION▶

5.90 has three significant digits. *The zero is not necessary as a placeholder* and should not be written unless it is significant.
8900 has two significant digits, unless information is known about the number that makes one or both zeros significant. Without such information, we assume the zeros are placeholders for proper location of the decimal point.
Other approximate numbers with the number of significant digits are:
0.0005 (one), 960,000 (two), 0.0709 (three), 1.070 (four), 700.00 (five)

From Example 3 we see that *all nonzero digits are significant. Also, zeros not used as placeholders (for location of the decimal point) are significant.*

ACCURACY AND PRECISION

In calculations with approximate numbers, the number of significant digits and the position of the decimal point are important. *The* **accuracy** *of a number refers to the number of significant digits it has,* whereas *the* **precision** *of a number refers to the decimal position of the last significant digit.*

■**EXAMPLE 4**   An electric current is measured as 0.31 A on one ammeter and as 0.312 A on another ammeter. Here 0.312 is more precise since its last digit represents thousandths and 0.31 is expressed only to hundredths. Also, 0.312 is more accurate since it has three significant digits and 0.31 has only two.
A concrete driveway is 230 ft long and 0.4 ft thick. Here 230 is more accurate (two significant digits) and 0.4 is more precise (expressed to tenths).

ROUNDING OFF

The last significant digit of an approximate number is not completely accurate. It has usually been determined by estimation or *rounding off.* However, it is in error at most by one-half of a unit in its place value.

■**EXAMPLE 5**   When we write the voltage in Example 2 as 115.7 V, we are saying that the voltage is at least 115.65 V and no more than 115.75 V. Any value between these two, rounded off to tenths, would be expressed as 115.7 V.
In changing the fraction 2/3 to the approximate decimal value 0.667, we are saying that the value is between 0.6665 and 0.6675.

On graphing calculators it is possible to set the number of decimal places (to the right of the decimal point) to which results will be rounded off.

To **round off** *a number to a specified number of significant digits, discard all digits to the right of the last significant digit (replace them with zeros if needed to place the decimal point). If the first digit discarded is 5 or more, increase the last significant digit by 1 (round up). If the first digit discarded is less than 5, do not change the last significant digit (round down).*
Note that if the only digit discarded is a 5, then rounding up, rounding down, or rounding to the nearest even number are all equally appropriate. Rounding up is probably most commonly used, although rounding to the nearest even number is used in many references.

■EXAMPLE 6 70,360 rounded off to three significant digits is 70,400. Here 3 is the third significant digit and the next digit is 6. Since 6 > 5, we add 1 to 3 and the result, 4, becomes the third significant digit of the approximation. The 6 is then replaced with a zero in order to keep the decimal point in the proper position.

70,430 rounded off to three significant digits, or to the nearest hundred, is 70,400. Here the 3 is replaced with a zero.

187.35 rounded off to four significant digits, or to tenths, is 187.4.

187.349 rounded off to four significant digits is 187.3.

71,500 rounded off to two significant digits is 72,000.

71,499 rounded off to two significant digits is 71,000.

Operations with Approximate Numbers

NOTE▶ When performing operations on approximate numbers, *we must not express the re-sult to an accuracy or precision that is not valid.* The next two examples show how results might be written with incorrect accuracy.

16.3   ft
0.927 ft
─────────
17.227 ft

■EXAMPLE 7 A pipe is made in two sections. One is measured as 16.3 ft long and the other as 0.927 ft long. What is the total length of the two sections together?

It appears that we simply add the numbers as shown at the left. However, 16.3 is precise only to tenths, which means it might have been as small as 16.25 ft or as large as 16.35 ft. Considering only the precision of 16.3, the total length is 17.177 ft or 17.277 ft, or between these values. If rounded off to two significant digits, these values agree, and if rounded off to tenths they differ by 0.1 (17.2 and 17.3). When rounded off to hundredths, they do not agree, since the third digit differs (17.18 and 17.28). Thus, these values differ significantly when rounded off to a precision be-yond tenths, the precision of 16.3. The 0.927 does not change this precision, since it is rounded off to thousandths. We then conclude that the total length should be rounded off to *tenths,* the precision of 16.3. Therefore, the total length should be written as 17.2 ft.

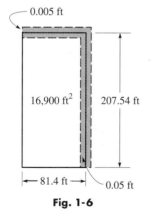

0.005 ft

16,900 ft^2 207.54 ft

81.4 ft   0.05 ft

Fig. 1-6

■EXAMPLE 8 We find the area of the rectangular piece of land in Fig. 1-6 by multiplying the length, 207.54 ft, by the width, 81.4 ft. Using a calculator, we find that $(207.54)(81.4) = 16,893.756$. This apparently means the area is 16,893.756 ft^2.

However, *the area should not be expressed with this accuracy.* Since the length and width are both approximate, we have

$(207.535 \text{ ft})(81.35 \text{ ft}) = 16,882.97225 \text{ ft}^2$ least possible area

$(207.545 \text{ ft})(81.45 \text{ ft}) = 16,904.54025 \text{ ft}^2$ greatest possible area

These values agree when rounded off to three significant digits (16,900 ft^2), but do not agree when rounded off to a greater accuracy. Therefore, we conclude the result is accurate only to *three* significant digits, which means the area is 16,900 ft^2. Note that the width is accurate to *three* significant digits and the length to five significant digits.

Following are the rules used in expressing the result when we perform basic operations on approximate numbers. They are based on reasoning similar to that shown in Examples 7 and 8.

Operations with Approximate Numbers

1. *When approximate numbers are added or subtracted, the result is expressed with the precision of the least precise number.*

2. *When approximate numbers are multiplied or divided, the result is expressed with the accuracy of the least accurate number.*

3. *When the root of an approximate number is found, the result is expressed with the accuracy of the number.*

CAUTION ▶

We should always express the result of a calculation with the proper accuracy or precision. **Using a calculator, it is necessary to round off the result if additional digits are displayed.** *Therefore, it is necessary to note the accuracy or precision of the numbers being used.*

> When using a calculator, round off only the final result.

■**EXAMPLE 9**   Find the sum of the approximate numbers 73.2, 8.0627, and 93.57.

Showing the addition in the standard way, and using a calculator, we have

$$
\begin{array}{r}
73.2 \quad \longleftarrow \text{ least precise number (expressed to tenths)}\\
8.0627 \\
\underline{93.57} \\
174.8327 \quad \longleftarrow \text{ final display must be rounded to tenths}
\end{array}
$$

Therefore, the sum of these approximate numbers is 174.8.   ■

> When rounding off a number, it may seem difficult to discard the extra digits. However, if you keep those digits, you show a number with too great an accuracy, and it is incorrect to do so.

■**EXAMPLE 10**   In finding the product of the approximate numbers 2.4832 and 30.5 on a calculator, the final display shows 75.7376. However, since 30.5 has only three significant digits, the product is 75.7.

In Example 1 we calculated that $38.3 - 12.9(-3.58) = 84.482$. We know that $38.3 - 12.9(-3.58) = 38.3 + 46.182 = 84.482$. If these numbers are approximate, we must round off the result to tenths, which means the sum is 84.5. We see that *where there is a combination of operations, the final operation determines how the final result is to be rounded off.*   ■

NOTE ▶

If an exact number is used in a calculation, there is no limit to the number of decimal places it may take on. *The accuracy of the result is limited only by the approximate numbers involved.*

■**EXAMPLE 11**   Using the exact number 600 and the approximate number 2.7, we express the result to tenths if the numbers are added or subtracted. If they are multiplied or divided, we express the result to two significant digits. Since 600 is exact, the accuracy of the result depends only on the approximate number 2.7.

$$
\begin{array}{ll}
600 + 2.7 = 602.7 & 600 - 2.7 = 597.3 \\
600 \times 2.7 = 1600 & 600 \div 2.7 = 220
\end{array}
$$ ■

A note regarding the equal sign ($=$) is in order. We will use it for its defined meaning of "equals exactly" and when the result is an approximate number that has been properly rounded off. Although $\sqrt{27.8} \approx 5.27$, where $\approx$ means "equals approximately," we write $\sqrt{27.8} = 5.27$, since 5.27 has been properly rounded off.

ESTIMATING RESULTS   You should *make a rough estimate* of the result when using a calculator. An estimation may prevent accepting an incorrect result after using an incorrect calculator sequence, particularly if the calculator result is far from the estimated value.

EXAMPLE 12   In Example 1 we found that

$$38.3 - 12.9(-3.58) = 84.482 \qquad \text{using exact numbers}$$

When using the calculator, if we forgot to make 3.58 negative, the display would be -7.882, or if we incorrectly entered 38.3 as 83.3, the display would be 129.482.

However, if we estimate the result as

$$40 - 10(-4) = 80$$

we know that a result of -7.882 or 129.482 cannot be correct.

In making the estimation, we can often use one-significant-digit approximations. If the calculator result is far from the estimation, we should do the calculation again.

EXERCISES *1-3*

In Exercises 1–4, determine whether the given numbers are approximate or exact.

1. A car with 8 cylinders travels at 55 mi/h.

2. A computer chip 0.002 mm thick is priced at $7.50.

3. A cube of gold 1 cm on an edge weighs 19.3 g.

4. A calculator has 50 keys, and its battery lasted for 50 h.

In Exercises 5–10, determine the number of significant digits in each of the given approximate numbers.

5. 107; 3004   **6.** 3600; 730   **7.** 6.80; 6.08

8. 0.8735; 0.0075   **9.** 3000; 3000.1   **10.** 1.00; 0.01

In Exercises 11–16, determine which of the pair of approximate numbers is (a) more precise and (b) more accurate.

11. 30.8; 0.01   **12.** 0.041; 7.673   **13.** 0.1; 78.0

14. 7040; 0.004   **15.** 7000; 0.004   **16.** 50.060; 8.914

In Exercises 17–24, round off the given approximate numbers (a) to three significant digits and (b) to two significant digits.

17. 4.936   **18.** 80.53   **19.** 50,893   **20.** 31,490

21. 9549   **22.** 30.96   **23.** 0.9499   **24.** 0.9999

In Exercises 25–36, assume that all numbers are approximate. (a) Perform the indicated operations on a calculator and (b) estimate the result and compare with the calculator result.

25. $3.8 + 0.154 + 47.26$

26. $12.78 + 1.0495 - 1.633$

27. $3.64(17.06)$

28. $0.49 \div 827$

29. $0.0350 - \dfrac{0.0450}{1.909}$

30. $\dfrac{0.3275}{1.096 \times 0.50085}$

31. $\dfrac{0.26(-0.4095)}{50.75(0.937)}$

32. $\dfrac{326.0}{2.060(3894) - 4008}$

33. $\dfrac{23.962 \times 0.01537}{10.965 - 8.249}$

34. $\dfrac{0.69378 + 0.04997}{257.4 \times 3.216}$

35. $\dfrac{3872}{503.1} - \dfrac{2.056 \times 309.6}{395.2}$

36. $\dfrac{1}{0.5926} + \dfrac{3.6957}{2.935 - 1.054}$

In Exercises 37–40, perform the indicated operations. The first number is approximate, and the second number is exact.

37. $0.9788 + 14.9$ **38.** $17.311 - 22.98$

39. $3.142(65)$ **40.** $8.62 \div 1728$

In Exercises 41–44, answer the given questions. Refer to Appendix B for units of measurement and their symbols.

41. The manual for a heart monitor lists the frequency of the ultrasound wave as 2.75 MHz. What are the least possible and the greatest possible frequencies?

42. An automobile manufacturer states that the gas tank on a certain car holds approximately 82 L. What are the least possible and greatest possible capacities of this tank?

(W) **43.** A flash of lightning struck a tower 3.25 mi from a person. The thunder was heard 15 s later. The person calculated the speed of sound and reported it as 1144 ft/s. What is wrong with this conclusion?

(W) **44.** A student reports the electric current in a certain experiment as 0.02 A and later notes that the current is 0.023 A. The student states the change in current is 0.003 A. What is wrong with this conclusion?

In Exercises 45–56, perform the indicated calculations on a calculator.

45. Calculate: (a) $2.2 + 3.8 \times 4.5$ (b) $(2.2 + 3.8) \times 4.5$
(Note the use of parentheses for grouping.)

46. Calculate: (a) $6.03 \div 2.25 + 1.77$ (b) $6.03 \div (2.25 + 1.77)$
(Note the use of parentheses for grouping.)

47. (a) Show that π is not exactly equal to 3.1416.
(b) Show that π is not exactly equal to 22/7.

48. (a) What is the calculator display for $2 \div 0$?
(b) What is the calculator display for $0 \div 0$?

49. At some point in the decimal equivalent of a rational number, some sequence of digits will start repeating endlessly. An irrational number never has an endlessly repeating sequence of digits. Find the decimal equivalents of (a) 8/33 and (b) π. Note the repetition for 8/33 and that no such repetition occurs for π.

(W) 50. Following Exercise 49, show that the decimal equivalent of the fraction 124/990 indicates that it is rational. Why is the last digit different?

51. Three adjacent lots have road frontages of 75.4 ft, 39.66 ft, and 81 ft, respectively. What is the total frontage of these lots?

52. Two jets flew at 938 km/h and 1450 km/h, respectively. How much faster was the second jet?

53. If 1 K of computer memory has 1024 bytes, how many bytes are there in 256 K of memory? (All numbers are exact.)

54. The power (in watts) developed in an electric circuit is the product of the current (in amperes) and the voltage. What is the power developed in a circuit in which the current is 0.0125 A and the voltage is 12.68 V?

55. The percent of alcohol in a certain car engine coolant is found by performing the calculation $\dfrac{100(40.63 + 52.96)}{105.30 + 52.96}$. Find this percent of alcohol. The number 100 is exact.

56. The evaporation rate (in gal/day) of a wastewater holding pond is found by calculating $145(1.05 + \frac{1}{236})$. Determine this rate.

$1\text{-}4$ EXPONENTS

In this section we shall introduce some basic terminology and notation that are important to the algebraic expressions developed in later sections.

We often have a number multiplied by itself several times. Rather than writing the number over and over, we use the notation a^n, where a is the number and n is the number of times it appears in the product. *In the expression a^n, the number a is called the* **base,** *and the number n is called the* **exponent;** *in words, a^n is read as the* **"nth power of a."**

■EXAMPLE 1   **(a)** $4 \times 4 \times 4 \times 4 \times 4 = 4^5$ (the fifth power of 4)

(b) $(-2)(-2)(-2)(-2) = (-2)^4$ (the fourth power of -2)

(c) $a \times a = a^2$ (the second power of a, called "a squared")

(d) $(\frac{1}{5})(\frac{1}{5})(\frac{1}{5}) = (\frac{1}{5})^3$ (the third power of $\frac{1}{5}$, called "$\frac{1}{5}$ cubed")

The basic operations with exponents will now be stated symbolically. We first state them for positive integers as exponents. Therefore, if m and n are positive integers, we have the following important operations for exponents.

Two forms are shown for Eqs. (1-4) in order that the resulting exponent is a positive integer. We consider negative and zero exponents after the next three examples.

$$a^m \times a^n = a^{m+n} \tag{1-3}$$

$$\frac{a^m}{a^n} = a^{m-n} \ (m > n, a \neq 0), \qquad \frac{a^m}{a^n} = \frac{1}{a^{n-m}} \ (m < n, a \neq 0) \tag{1-4}$$

$$(a^m)^n = a^{mn} \tag{1-5}$$

$$(ab)^n = a^n b^n, \qquad \left(\frac{a}{b}\right)^n = \frac{a^n}{b^n} \ (b \neq 0) \tag{1-6}$$

EXAMPLE 2   Applying Eq. (1-3), we have

add exponents

$$a^3 \times a^5 = a^{3+5} = a^8$$

We see that this result is correct since we can also write

(3 factors of a)(5 factors of a) ——————— 8 factors of a

$$a^3 \times a^5 = (a \times a \times a)(a \times a \times a \times a \times a) = a^8$$

Applying the first form of Eqs. (1-4), we have

$$5 > 3$$

$$\frac{a^5}{a^3} = a^{5-3} = a^2, \qquad \frac{a^5}{a^3} = \frac{\cancel{a} \times \cancel{a} \times \cancel{a} \times a \times a}{\cancel{a} \times \cancel{a} \times \cancel{a}} = a^2$$

Applying the second form of Eqs. (1-4), we have

$$\frac{a^3}{a^5} = \frac{1}{a^{5-3}} = \frac{1}{a^2}, \qquad \frac{a^3}{a^5} = \frac{\cancel{a} \times \cancel{a} \times \cancel{a}}{\cancel{a} \times \cancel{a} \times \cancel{a} \times a \times a} = \frac{1}{a^2}$$

$$5 > 3$$

EXAMPLE 3   Applying Eq. (1-5), we have

multiply exponents

$$(a^5)^3 = a^{5(3)} = a^{15}, \qquad (a^5)^3 = (a^5)(a^5)(a^5) = a^{5+5+5} = a^{15}$$

Applying the first form of Eqs. (1-6), we have

$$(ab)^3 = a^3 b^3, \qquad (ab)^3 = (ab)(ab)(ab) = a^3 b^3$$

Applying the second form of Eqs. (1-6), we have

$$\left(\frac{a}{b}\right)^3 = \frac{a^3}{b^3}, \qquad \left(\frac{a}{b}\right)^3 = \left(\frac{a}{b}\right)\left(\frac{a}{b}\right)\left(\frac{a}{b}\right) = \frac{a^3}{b^3}$$

CAUTION▶   When an expression involves a product or a quotient of different bases, *only exponents of the same base may be combined.* Consider the following example.

EXAMPLE 4   Other illustrations using Eqs. (1-3) to (1-6) are as follows:

(a) $(-x^2)^3 = [(-1)x^2]^3 = (-1)^3(x^2)^3 = -x^6$

exponent of 1 ⌐   add exponents of a

(b) $ax^2(ax)^3 = ax^2(a^3 x^3) = a^4 x^5$ ◀ add exponents of x

(c) $\dfrac{(3 \times 2)^4}{(3 \times 5)^3} = \dfrac{3^4 2^4}{3^3 5^3} = \dfrac{3 \times 2^4}{5^3}$ **(d)** $\dfrac{(ry^3)^2}{r(y^2)^4} = \dfrac{r^2 y^6}{ry^8} = \dfrac{r}{y^2}$

CAUTION▶   In illustration (b), note that ax^2 *means a times the square of x and does* **not** *mean* $a^2 x^2$, whereas $(ax)^3$ *does* mean $a^3 x^3$.

EXAMPLE 5 In the analysis of the deflection of a beam, the expression that follows is simplified as shown.

$$\frac{1}{2}\left(\frac{PL}{4EI}\right)\left(\frac{2}{3}\right)\left(\frac{L}{2}\right)^2 = \frac{1}{2}\left(\frac{PL}{4EI}\right)\left(\frac{2}{3}\right)\left(\frac{L^2}{2^2}\right) = \frac{\overset{1}{\cancel{2}}PL(L^2)}{\underset{1}{\cancel{2}}(3)(4)(4)EI} = \frac{PL^3}{48EI}$$

L is the length of the beam, and P is the force applied to it. E and I are constants related to the beam. In *simplifying* this expression, we combined exponents of L and divided out the 2 that was in the numerator and in the denominator. $\blacksquare$

Zero and Negative Exponents

We developed Eqs. (1-3) to (1-6) by using positive integers as exponents. A positive integer is generally, but not always, preferred in the final result. We now show how zero and negative integers are used as exponents.

In Eqs. (1-4), if $n = m$, we would have $a^m/a^m = a^{m-m} = a^0$. Also, $a^m/a^m = 1$, since any nonzero quantity divided by itself equals 1. Therefore, for Eqs. (1-4) to hold when $m = n$, we have

$$\boxed{a^0 = 1 \qquad (a \neq 0)} \tag{1-7}$$

Equation (1-7) gives the definition of zero as an exponent. Since a has not been specified, this equation states that *any nonzero algebraic expression raised to the zero power is* 1. Also, the other laws of exponents are valid for this definition.

EXAMPLE 6 (a) $5^0 = 1$ (b) $(2x)^0 = 1$ (c) $(ax + b)^0 = 1$

(d) $(a^2 b^0 c)^2 = a^4 b^0 c^2 = a^4 c^2$ (e) $2t^0 = 2(1) = 2$

$\qquad\qquad\quad b^0 = 1$

CAUTION ▶ We note in illustration (e) that *only t is raised to the zero power*. If the quantity $2t$ were raised to the zero power, it would be written as $(2t)^0$. $\blacksquare$

If we apply the first form of Eqs. (1-4) to the case where $n > m$, the resulting exponent is negative. This leads to the definition of a negative exponent.

EXAMPLE 7 Applying both forms of Eqs. (1-4) to a^2/a^7, we have

$$\frac{a^2}{a^7} = a^{2-7} = a^{-5} \qquad \text{and} \qquad \frac{a^2}{a^7} = \frac{1}{a^{7-2}} = \frac{1}{a^5}$$

If these results are to be consistent, then $a^{-5} = \dfrac{1}{a^5}$. $\blacksquare$

Following the reasoning of Example 7, if we define

$$\boxed{a^{-n} = \frac{1}{a^n} \qquad (a \neq 0)} \tag{1-8}$$

then all of the laws of exponents will hold for negative integers.

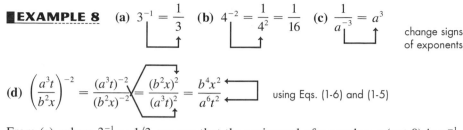

EXAMPLE 8   (a) $3^{-1} = \dfrac{1}{3}$ (b) $4^{-2} = \dfrac{1}{4^2} = \dfrac{1}{16}$ (c) $\dfrac{1}{a^{-3}} = a^3$

change signs
of exponents

(d) $\left(\dfrac{a^3 t}{b^2 x}\right)^{-2} = \dfrac{(a^3 t)^{-2}}{(b^2 x)^{-2}} = \dfrac{(b^2 x)^2}{(a^3 t)^2} = \dfrac{b^4 x^2}{a^6 t^2}$ using Eqs. (1-6) and (1-5)

From (a), where $3^{-1} = 1/3$, we see that the reciprocal of a number x (not 0) is x^{-1}.

Order of Operations

In Section 1-2 we saw that it is necessary to follow a particular order of operations when performing the basic operations on numbers. Since raising a number to a power is a form of multiplication, this operation is performed before additions and subtractions. In fact, it is performed before multiplications and divisions.

> **Order of Operations**
> 1. *Operations within specific groupings*
> 2. *Powers*
> 3. *Multiplications and divisions (from left to right)*
> 4. *Additions and subtractions (from left to right)*

EXAMPLE 9 $8 - (-1)^2 - 2(-3)^2 = 8 - 1 - 2(9)$
$$= 8 - 1 - 18 = -11$$

CAUTION ▶

Since there were no specific groupings, we first squared -1 and -3. Next we found the product $2(9)$ in the last term. Finally, the subtractions were performed. Note carefully that *we did not change the sign of -1 before we squared it.*

Evaluating Algebraic Expressions

An algebraic expression is **evaluated** *by* **substituting** *given values of the literal numbers in the expression and calculating the result.* On a calculator the $\boxed{x^2}$ key is used to square numbers, and the $\boxed{\wedge}$ or $\boxed{x^y}$ key is used for other powers.

On many calculators there is a specific key or key sequence to evaluate x^3.

```
16*4.2²
           282.24
```

Fig. 1-7

EXAMPLE 10   The distance (in ft) that an object falls in 4.2 s is found by substituting 4.2 for t in the expression $16.0t^2$. We show this as

$t = 4.2$ s ⟵——————— substituting

$16.0(4.2)^2 = 280$ ft   estimation ⟶ $20(4)^2 = 320$

The result is rounded off to two significant digits (the accuracy of t). The calculator will square the 4.2 before multiplying. See Fig. 1-7.

As we stated in Section 1-3, in evaluating an expression on a calculator, we should also estimate its value as in Example 10. For the estimate we note that *a negative number raised to an even power gives a positive value, and a negative number raised to an odd power gives a negative value.*

EXAMPLE 11   Using the meaning of a power of a number, we have

$$(-2)^2 = (-2)(-2) = 4, \qquad (-2)^3 = (-2)(-2)(-2) = 4(-2) = -8$$
$$(-2)^4 = 16, \quad (-2)^5 = -32, \qquad (-2)^6 = 64, \qquad (-2)^7 = \underline{-128} \quad \blacksquare$$

EXAMPLE 12   A wire made of a special alloy has an electric resistance R (in Ω) given by $R = a + 0.0115T^3$, where T (in °C) is the temperature (between -4°C and 4°C). Find R for $a = 0.838$ Ω and $T = -2.87$°C.

Substituting these values, we have

$$R = 0.838 + 0.0115(-2.87)^3 \qquad \text{estimation:}$$
$$= 0.566 \ \Omega \qquad\qquad\qquad 0.8 + 0.01(-3)^3 = 0.8 + 0.01(-27) = 0.53$$

Note in the estimation that $(-3)^3 = -27$.   --------- $\blacksquare$

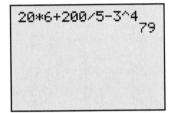

Fig. 1-8

Graphing calculators generally use computer symbols in the display for some of the operations to be performed. These symbols are as follows:

Multiplication: $*$ Division: $/$ Powers: $\wedge$

Therefore, to calculate the value of $20 \times 6 + 200/5 - 3^4$, we use the key sequence

20 $\boxed{\times}$ 6 $\boxed{+}$ 200 $\boxed{\div}$ 5 $\boxed{-}$ 3 $\boxed{\wedge}$ 4

CAUTION ▶ with the result of 79 shown in the display of Fig. 1-8. ***Note carefully that 200 is divided only by 5.*** If it were divided by $5 - 3^4$, then we would use parentheses and show the expression to be evaluated as $20 \times 6 + 200/(5 - 3^4)$.

EXERCISES *1-4*

In Exercises 1–48, simplify the given expressions. Express results with positive exponents only.

1. $x^3 x^4$ **2.** $y^2 y^7$ **3.** $2b^4 b^2$

4. $3k(k^5)$ **5.** $\dfrac{m^5}{m^3}$ **6.** $\dfrac{x^6}{x}$

7. $\dfrac{n^5}{n^9}$ **8.** $\dfrac{s}{s^4}$ **9.** $(a^2)^4$

10. $(x^8)^3$ **11.** $(t^5)^4$ **12.** $(n^3)^7$

13. $(2n)^3$ **14.** $(ax)^5$ **15.** $(ax^4)^2$

16. $(3a^2)^3$ **17.** $\left(\dfrac{2}{b}\right)^3$ **18.** $\left(\dfrac{x}{y}\right)^7$

19. $\left(\dfrac{x^2}{2}\right)^4$ **20.** $\left(\dfrac{3}{n^3}\right)^3$ **21.** 7^0

22. $(8a)^0$ **23.** $-3x^0$ **24.** $6v^0$

25. 6^{-1} **26.** -10^{-3} **27.** $\dfrac{1}{R^{-2}}$

28. $\dfrac{1}{t^{-5}}$ **29.** $(-t^2)^7$ **30.** $(-y^3)^5$

31. $(2x^2)^6$ **32.** $-(-c^4)^4$ **33.** $(4xa^{-2})^0$

34. $3(LC^{-1})^0$ **35.** $-b^5 b^{-3}$ **36.** $2c^4 c^{-7}$

37. $\dfrac{2a^4}{(2a)^4}$ **38.** $\dfrac{x^2 x^3}{(x^2)^3}$ **39.** $\dfrac{(n^2)^4}{(n^4)^2}$

40. $\dfrac{(3t)^{-1}}{3t^0}$ **41.** $(5^0 x^2 a^{-1})^{-1}$ **42.** $(3m^{-2}n^4)^{-2}$

43. $\left(\dfrac{4a}{x}\right)^{-3}$ **44.** $\left(\dfrac{2b^2}{y^5}\right)^{-2}$ **45.** $(-8gs^3)^2$

46. $ax^2(-a^2x)^2$ **47.** $\dfrac{15a^2n^5}{3an^6}$ **48.** $\dfrac{(ab^2)^3}{a^2b^8}$

In Exercises 49–56, evaluate the given expressions. In Exercises 51–56, all numbers are approximate.

49. $7(-4) - (-5)^2$ **50.** $6 + (-2)^5 - (-2)(8)$

51. $-(-26.5)^2 - (-9.85)^3$ **52.** $-0.711^2 - (-0.809)^6$

53. $\dfrac{3.07(-1.86)}{(-1.86)^4 + 1.596}$ **54.** $\dfrac{15.66^2 - (-4.017)^4}{1.044(-3.68)}$

55. $2.38(-60.7)^2 - 2540/1.17^3 + 0.806^5(26.1^3 - 9.88^4)$

56. $0.513(-2.778) - (-3.67)^3 + 0.889^4/(1.89 - 1.09^2)$

In Exercises 57–60, perform the indicated operations.

57. In designing a cam for a pump, the expression $\pi\left(\dfrac{r}{2}\right)^3\left(\dfrac{4}{3\pi r^2}\right)$ is used. Simplify this expression.

58. For a certain integrated electric circuit, it is necessary to simplify the expression $\dfrac{gM}{2\pi f C (2\pi f M)^2}$. Perform this simplification.

59. In order to find the electric power (in W) consumed by an electric light, the expression $i^2 R$ must be evaluated, where i is the current (in A) and R is the resistance (in Ω). Find the power consumed if $i = 0.525$ A and $R = 250\ \Omega$.

60. In designing a building, it was determined that the forces acting on an I beam would deflect the beam an amount (in cm), given by $\dfrac{x(1000 - 20x^2 + x^3)}{1850}$, where x is the distance (in m) from one end of the beam. Find the deflection for $x = 6.85$ m. (The 1000 and 20 are exact.)

1-5 SCIENTIFIC NOTATION

In technical and scientific work we often encounter numbers that are either very large or very small. Such numbers are illustrated in the next example.

Television was invented in the 1920s and first used commercially in the 1940s.

EXAMPLE 1 Television signals travel at about 30,000,000,000 cm/s. The mass of the earth is about 6,600,000,000,000,000,000,000 tons. A typical individual fiber in a fiber-optic communications cable has a diameter of 0.000005 m. Some X rays have a wavelength of about 0.000000095 cm.

The use of fiber optics was developed in the 1950s.

Writing numbers like those in Example 1 is inconvenient in ordinary notation. Also, calculators and computers require a more efficient way of expressing such numbers in order to work with them. Therefore, a convenient and useful notation, called *scientific notation,* is used to represent such numbers.

X rays were discovered by Roentgen in 1895.

A number in **scientific notation** *is expressed as the product of a number greater than or equal to 1 and less than 10, and a power of 10, and is written as*

$$P \times 10^k$$

where $1 \le P < 10$ and k is an integer. (The symbol $\le$ means "is less than or equal to.") See the following example.

EXAMPLE 2 **(a)** $340{,}000 = 3.4(100{,}000) = 3.4 \times 10^5$

(b) $0.000503 = \dfrac{5.03}{10{,}000} = \dfrac{5.03}{10^4} = 5.03 \times 10^{-4}$ between 1 and 10

(c) $6.82 = 6.82(1) = 6.82 \times 10^0$

From Example 2 we can establish a way of changing numbers from ordinary notation to scientific notation. *The decimal point is moved so that only one nonzero digit is to its left. The number of places moved is the value of k, which is positive if the decimal point is moved to the left and negative if moved to the right.*

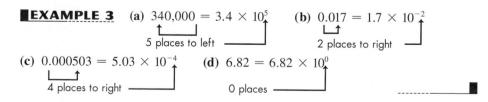

EXAMPLE 3 **(a)** $340{,}000 = 3.4 \times 10^5$ **(b)** $0.017 = 1.7 \times 10^{-2}$
5 places to left   2 places to right

(c) $0.000503 = 5.03 \times 10^{-4}$ **(d)** $6.82 = 6.82 \times 10^0$
4 places to right   0 places

To change a number from scientific notation to ordinary notation, we reverse the procedure used in Example 3. This is shown in the next example.

EXAMPLE 4   To change 5.83×10^6 to ordinary notation, we move the decimal point 6 places to the right. Additional zeros must be included to properly locate the decimal point. This means we write

$$5.83 \times 10^6 = 5,830,000$$

6 places to right

To change 8.06×10^{-3} to ordinary notation, we must move the decimal point 3 places to the left. Again, additional zeros must be included. Therefore,

$$8.06 \times 10^{-3} = 0.00806$$

3 places to left

As seen in these examples, scientific notation is an important application of the use of positive and negative exponents. Also, its importance is shown by the metric system use of prefixes to denote powers of 10, as shown in Appendix B.

Scientific notation also provides a practical way to handle calculations with very large or very small numbers. First, all numbers are expressed in scientific notation. Then the calculation can be performed on numbers between 1 and 10, using the laws of exponents to find the power of 10 in the final result.

EXAMPLE 5   In designing a computer, it was determined that it would be able to process 803,000 bits of data in 0.00000525 s. (See Exercise 37 of Exercises 1-1 for a brief note on computer data.) The rate of processing the data is

$$5 - (-6) = 11$$

$$\frac{803,000}{0.00000525} = \frac{8.03 \times 10^5}{5.25 \times 10^{-6}} = \left(\frac{8.03}{5.25}\right) \times 10^{11} = 1.53 \times 10^{11} \text{ bits/s}$$

As shown, it is proper to leave the result *(rounded off)* in scientific notation. This method is useful when using a calculator and then estimating the result. In this case the estimate is $(8 \times 10^5) \div (5 \times 10^{-6}) = 1.6 \times 10^{11}$.

Another advantage of scientific notation is that the precise number of significant digits of a number can be shown directly, even when the final significant digit is 0.

EXAMPLE 6   In evaluating the product $(750,000,000,000)(0.00644)$, and if we know that 750,000,000,000 has *three* significant digits, we can show the significant digits in the product as

$$(7.50 \times 10^{11})(6.44 \times 10^{-3}) = 48.3 \times 10^8$$

If the answer is to be given in scientific notation, we should express it as the product of a number between 1 and 10 and a power of 10. This means we should rewrite it as

$$48.3 \times 10^8 = (4.83 \times 10)(10^8)$$

not between 1 and 10 $= 4.83 \times 10^9$

We can enter numbers in scientific notation on a calculator, as well as have results given automatically in scientific notation. See the following example.

As for large numbers, consider these. A *googol* has been defined as 1 followed by 100 zeros. This means it can be written as 10^{100}. A *googolplex* is 10^{googol}.

The number 7.50×10^{11} cannot be entered on a scientific calculator in the form 750000000000. It could be entered on a graphing calculator.

Fig. 1-9

■EXAMPLE 7 The wavelength λ (in m) of the light in a red laser beam can be found from the following calculation. Note the significant digits in the numerator.

$$\lambda = \frac{3{,}000{,}000}{4{,}740{,}000{,}000{,}000} = \frac{3.00 \times 10^6}{4.74 \times 10^{12}} = 6.33 \times 10^{-7} \text{ m}$$

The key sequence is 3 $\boxed{\text{EE}}$ 6 $\boxed{\div}$ 4.74 $\boxed{\text{EE}}$ 12 $\boxed{\text{ENTER}}$. See Fig. 1-9. ■

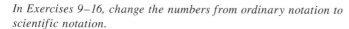

EXERCISES *1-5*

In Exercises 1–8, change the numbers from scientific notation to ordinary notation.

1. 4.5×10^4 **2.** 6.8×10^7 **3.** 2.01×10^{-3}

4. 9.61×10^{-5} **5.** 3.23×10^0 **6.** 8.40×10^0

7. 1.86×10 **8.** 1×10^{-1}

In Exercises 9–16, change the numbers from ordinary notation to scientific notation.

9. 40,000   **10.** 560,000   **11.** 0.0087

12. 0.7   **13.** 6   **14.** 1.09

15. 0.063   **16.** 0.0000908

In Exercises 17–20, perform the indicated calculations using a calculator and by first expressing all numbers in scientific notation.

17. 28,000(2,000,000,000)   **18.** 50,000(0.006)

19. $\dfrac{88{,}000}{0.0004}$ **20.** $\dfrac{0.00003}{6{,}000{,}000}$

In Exercises 21–28, perform the indicated calculations using a calculator. All numbers are approximate.

21. 1280(865,000)(43.8)

22. 0.0000659(0.00486)(3,190,000,000)

23. $\dfrac{0.0732(6710)}{0.00134(0.0231)}$ **24.** $\dfrac{0.00452}{2430(97{,}100)}$

25. $(3.642 \times 10^{-8})(2.736 \times 10^5)$ **26.** $\dfrac{9.368 \times 10^{-12}}{4.651 \times 10^4}$

27. $\dfrac{1.000 \times 10^7}{(3.10 \times 10^{-5})(1.07 \times 10^{14})}$ **28.** $\dfrac{7.3009 \times 10^{-2}}{5.9843(2.5036 \times 10^{-20})}$

In Exercises 29–36, change numbers in ordinary notation to scientific notation or change numbers in scientific notation to ordinary notation. See Appendix B for an explanation of symbols used.

29. The power plant at Grand Coulee Dam produces 6,500,000 kW of power.

30. The maximum pressure exerted by the human heart is 16,000 Pa.

31. A fiber-optic system requires 0.000003 W of power.

32. A red blood cell measures 0.0075 mm across.

33. To attain an energy density of that in some laser beams, an object would have to be heated to about $10^{30}\,°$C.

34. Among the stars nearest the earth, Centaurus A is about 4.07×10^{13} km away.

35. The power of a radio signal from Galileo, the space probe to Jupiter, is 1.6×10^{-12} W.

36. The electrical force between two electrons is about 2.4×10^{-43} times the gravitational force between them.

In Exercises 37–40, perform the indicated calculations.

37. A computer can do an addition in 7.5×10^{-15} s. How long does it take to perform 5.6×10^6 additions?

38. Uranium is used in nuclear reactors to generate electricity. About 0.000000039% of the uranium disintegrates each day. How much of 0.085 mg of uranium disintegrates in a day?

39. A TV signal travels at 1.86×10^5 mi/s for a total of 4.57×10^4 mi from the station transmitter to a satellite and then to a receiver dish. How long does it take the signal to go from the transmitter to the dish?

40. At $0°$C, the refrigerant Freon is a vapor at a pressure of $P = 1.378 \times 10^5$ Pa. If the volume of vapor is $V = 7.865 \times 10^3$ cm^3, find the value of PV.

In Exercises 41–44, perform the indicated calculations by first expressing all numbers in scientific notation.

41. Hudson Bay has an area of about 730,000 km^2. What is the area of Hudson Bay in square centimeters?

42. The rate of energy radiation (in W) from an object is found by evaluating the expression kT^4, where T is the thermodynamic temperature. Find this value for the human body, for which $k = 0.000000057$ W/K^4 and $T = 303$ K.

43. In a microwave receiver circuit, the resistance R of a wire 1 m long is given by $R = k/d^2$, where d is the diameter of the wire. Find R if $k = 0.00000002196$ $\Omega \cdot$m^2 and $d = 0.00007998$ m.

44. The closest the earth gets to the sun is 91,400,000 mi, and it takes the light from the sun 491 s to reach the earth. What is the speed of light? Compare this with the speed of the TV signal in Exercise 39.

1-6 ROOTS AND RADICALS

A problem often encountered is: What number multiplied by itself n times gives another specified number? For example, what number squared is 9? We can see that either 3 or -3 is a proper answer. *We call either 3 or -3 a* **square root** *of 9*, since $3^2 = 9$ or $(-3)^2 = 9$.

In order to have a general notation for the square root and have it represent *one* number, *we define the* **principal square root** *of a to be positive if a is positive and represent it by* $\sqrt{a}$. This means $\sqrt{9} = 3$ and not -3.

The general notation for the **principal nth root** *of a is* $\sqrt[n]{a}$. (When $n = 2$, we do not put the 2 for n.) *The* $\sqrt{\ }$ *sign is called a* **radical sign.**

■**EXAMPLE 1**   (a) $\sqrt{2}$ (the square root of 2)   (b) $\sqrt[3]{2}$ (the cube root of 2)

(c) $\sqrt[4]{2}$ (the fourth root of 2)   (d) $\sqrt[7]{6}$ (the seventh root of 6)   ----------■

In order to have a single defined value for all roots (not just square roots) and to consider only real number roots, *we define the* **principal nth root** *of a to be positive if a is positive and to be negative if a is negative and n is odd.* (If a is negative and n is even, the roots are not real.)

■**EXAMPLE 2**   (a) $\sqrt{169} = 13$ ($\sqrt{169} \neq -13$)   (b) $-\sqrt{64} = -8$

(c) $\sqrt[3]{27} = 3$

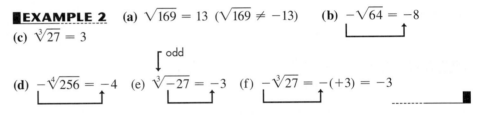

(d) $-\sqrt[4]{256} = -4$ (e) $\sqrt[3]{-27} = -3$ (f) $-\sqrt[3]{27} = -(+3) = -3$ ----------■

Another property of square roots is developed by noting illustrations such as $\sqrt{36} = \sqrt{4 \times 9} = \sqrt{4} \times \sqrt{9} = 2 \times 3 = 6$. In general, this property states that *the square root of a product of positive numbers is the product of their square roots.*

$$\sqrt{ab} = \sqrt{a}\sqrt{b} \quad (a \text{ and } b \text{ positive real numbers})$$ **(1-9)**

This property is used in simplifying radicals. It is most useful if either a or b is a **perfect square**, *which is the square of a rational number.*

■**EXAMPLE 3**   (a) $\sqrt{8} = \sqrt{(4)(2)} = \sqrt{4}\sqrt{2} = 2\sqrt{2}$

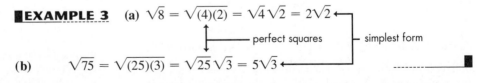

(b) $\sqrt{75} = \sqrt{(25)(3)} = \sqrt{25}\sqrt{3} = 5\sqrt{3}$ ----------■

In order to represent the square root of a number *exactly,* we use Eq. (1-9) to write it in simplest form. However, a decimal *approximation* is often acceptable, and we use the $\boxed{\sqrt{x}}$ key on a calculator. (We will show another way of finding the root of a number in Chapter 11.)

Fig. 1-10

EXAMPLE 4 After reaching its greatest height, the time (in s) for a rocket to fall h ft is found by evaluating $0.25\sqrt{h}$. Find the time for the rocket to fall 1260 ft.

The calculator evaluation is shown in Fig. 1-10. From the display we see that the rocket takes 8.9 s to fall 1260 ft. The result is rounded off to two significant digits, the accuracy of 0.25 (an approximate number).

In simplifying a radical, *all operations under a radical sign must be done before finding the root.* The horizontal bar groups the numbers under it.

CAUTION ▶

EXAMPLE 5 (a) $\sqrt{16 + 9} = \sqrt{25} = 5$ first perform the addition $16 + 9$

However, $\sqrt{16 + 9}$ is *not* $\sqrt{16} + \sqrt{9} = 4 + 3 = 7$

(b) $\sqrt{2^2 + 6^2} = \sqrt{4 + 36} = \sqrt{40} = \sqrt{4}\sqrt{10} = 2\sqrt{10}$, but $\sqrt{2^2 + 6^2}$ is *not* $\sqrt{2^2} + \sqrt{6^2} = 2 + 6 = 8$

A more detailed discussion of radicals and imaginary numbers is found in Chapters 11 and 12.

In defining the principal square root, we did not define the square root of a negative number. However, in Section 1-1 we defined the square root of a negative number to be an **imaginary number.** More generally, *the even root of a negative number is an imaginary number, and the odd root of a negative number is a negative real number.*

EXAMPLE 6

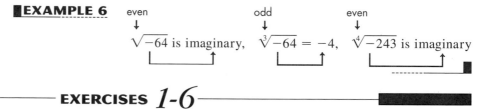

EXERCISES *1-6*

In Exercises 1–28, simplify the given expressions. In each of 1–16, the result is an integer.

1. $\sqrt{25}$ **2.** $\sqrt{81}$ **3.** $-\sqrt{121}$ **4.** $-\sqrt{36}$

5. $-\sqrt{49}$ **6.** $\sqrt{225}$ **7.** $\sqrt{400}$ **8.** $-\sqrt{900}$

9. $\sqrt[3]{125}$ **10.** $\sqrt[4]{16}$ **11.** $\sqrt[3]{-216}$ **12.** $\sqrt[5]{-32}$

13. $(\sqrt{5})^2$ **14.** $(\sqrt{19})^2$ **15.** $(\sqrt[3]{31})^3$ **16.** $(\sqrt[4]{53})^4$

17. $(-\sqrt{18})^2$ **18.** $-\sqrt{32}$ **19.** $\sqrt{12}$ **20.** $\sqrt{50}$

21. $2\sqrt{84}$ **22.** $4\sqrt{108}$ **23.** $\dfrac{7^2\sqrt{81}}{3^2\sqrt{49}}$ **24.** $\dfrac{2^5\sqrt[5]{243}}{3\sqrt{144}}$

25. $\sqrt{36 + 64}$ **26.** $\sqrt{25 + 144}$

27. $\sqrt{3^2 + 9^2}$ **28.** $\sqrt{8^2 - 4^2}$

In Exercises 29–36, find the value of each square root by use of a calculator. Each number is approximate.

29. $\sqrt{85.4}$ **30.** $\sqrt{3762}$ **31.** $\sqrt{0.4729}$ **32.** $\sqrt{0.0627}$

33. (a) $\sqrt{1296 + 2304}$ (b) $\sqrt{1296} + \sqrt{2304}$

34. (a) $\sqrt{10.6276 + 2.1609}$ (b) $\sqrt{10.6276} + \sqrt{2.1609}$

35. (a) $\sqrt{0.0429^2 - 0.0183^2}$ (b) $\sqrt{0.0429^2} - \sqrt{0.0183^2}$

36. (a) $\sqrt{3.625^2 + 0.614^2}$ (b) $\sqrt{3.625^2} + \sqrt{0.614^2}$

In Exercises 37–44, solve the given problems.

37. The velocity (in ft/s) of water flowing from an open valve in the side of a tank is $\sqrt{64.4h}$, where h is the height (in ft) of the top of the water from the valve. Find the velocity for $h = 2.75$ ft.

38. The resistance in an amplifier circuit is found by evaluating $\sqrt{Z^2 - X^2}$. Find the resistance for $Z = 5.362$ Ω and $X = 2.875$ Ω.

39. The speed (in m/s) of sound in seawater is found by evaluating $\sqrt{B/d}$ for $B = 2.18 \times 10^9$ Pa and $d = 1.03 \times 10^3$ kg/m³. Find this speed, which is important in locating underwater objects using sonar.

40. The time (in s) for an object to fall h feet is given by the expression $\sqrt{h}/4.0$. How long does it take a person to fall 55 ft from a fifth-floor window into a net while escaping a fire?

41. A TV screen is 15.2 in. wide and 11.4 in. high. The length of a diagonal (the dimension used to describe it—from one corner to the opposite corner) is found by evaluating $\sqrt{w^2 + h^2}$, where w is the width and h is the height. Find the diagonal.

42. A car costs \$22,000 new and is worth \$15,000 two years later. The annual rate of depreciation is found by evaluating $100(1 - \sqrt{V/C})$, where C is the cost and V is the value after two years. At what rate did the car depreciate? (100 and 1 are exact.)

(W) 43. Is it always true that $\sqrt{a^2} = a$? Explain.

44. For what values of x is (a) $x > \sqrt{x}$, (b) $x = \sqrt{x}$, (c) $x < \sqrt{x}$?

1-7 ADDITION AND SUBTRACTION OF ALGEBRAIC EXPRESSIONS

Since we use letters to represent numbers, we conclude that all operations valid for numbers are also valid for literal numbers. In this section we discuss the methods for combining literal numbers.

Addition, subtraction, multiplication, division, and taking of roots are known as **algebraic operations.** *Any combination of numbers and literal symbols that results from algebraic operations is known as an* **algebraic expression.**

When an algebraic expression consists of several parts connected by plus signs and minus signs, each part (along with its sign) is known as a **term** *of the expression. If a given expression is made up of the product of a number of quantities, each of these quantities, or any product of them, is called a* **factor** *of the expression.*

CAUTION ▶   It is important to ***distinguish clearly between terms and factors*** since some operations that are valid for terms are not valid for factors, and conversely.

■ EXAMPLE 1 $3xy + 6x^2 - 7x\sqrt{y}$ is an algebraic expression with three terms: $3xy$, $6x^2$, and $-7x\sqrt{y}$.

The first term, $3xy$, has individual factors of 3, x, and y. Also, any product of these factors is also a factor of $3xy$. Thus, additional expressions that are factors of $3xy$ are $3x$, $3y$, xy, and $3xy$ itself.   ----------■

■ EXAMPLE 2 $7x(y^2 + x) - \dfrac{x + y}{6x}$ is an algebraic expression with terms $7x(y^2 + x)$ and $-\dfrac{x + y}{6x}$.

The term $7x(y^2 + x)$ has individual factors of 7, x, and $(y^2 + x)$, as well as products of these factors. The factor $y^2 + x$ has two terms, y^2 and x.

The numerator of the term $-\dfrac{x + y}{6x}$ has two terms, and the denominator has factors of 2, 3, and x. The negative sign can be treated as a factor of -1.   ----------■

An algebraic expression containing only one term is called a **monomial.** *An expression containing two terms is a* **binomial,** *and one containing three terms is a* **trinomial.** *Any expression containing two or more terms is called a* **multinomial.** Therefore, any binomial or trinomial is also a multinomial.

In any given term, the numbers and literal symbols multiplying any given factor constitute the **coefficient** *of that factor. The product of all the numbers in explicit form is known as the* **numerical coefficient** *of the term. All terms that differ at most in their numerical coefficients are known as* **similar** *or* **like** *terms. That is,* similar terms have the same variables with the same exponents.

■ EXAMPLE 3   **(a)** $7x^3\sqrt{y}$ is a monomial. It has a numerical coefficient of 7. The coefficient of $\sqrt{y}$ is $7x^3$, and the coefficient of x^3 is $7\sqrt{y}$.

(b) The expression $3a + 2b - 5a$ is a trinomial. The terms $3a$ and $-5a$ are similar since the literal parts are the same.   ----------■

like terms

unlike other terms due to factor of *a*

EXAMPLE 4 (a) $4 \times 2b + 81b - 6ab$

is a multinomial of three terms (a trinomial). The first term has a numerical coefficient of 8 ($= 4 \times 2$), the second has a numerical coefficient of 81, and the third has a numerical coefficient of -6 (the sign is attached to the numerical coefficient). The first two terms are similar, since they differ only in numerical coefficients. The third term is not similar to the others, for it has the factor *a*.

(b) The commutative law tells us that $x^2y^3 = y^3x^2$. Therefore, the two terms of the expression $3x^2y^3 + 5y^3x^2$ are similar.

Some calculators can display algebraic expressions and perform algebraic operations.

In adding and subtracting algebraic expressions, we combine similar (or like) terms into a single term. The final **simplified** expression will contain only terms that are not similar.

EXAMPLE 5 (a) $3x + 2x - 5y = 5x - 5y$

Since there are two similar terms in the original expression, they are added together so that the simplified result has two unlike terms.

(b) $6a^2 - 7a + 8ax$ cannot be simplified since there are no like terms.

(c) $6a + 5c + 2a - c = 6a + 2a + 5c - c$ commutative law

$\qquad\qquad\qquad\quad = 8a + 4c$ add like terms

In writing algebraic expressions, it is often necessary to group certain terms together. For this purpose we use **symbols of grouping.** In this text we use **parentheses,** ( ), **brackets,** [ ], and **braces,** { }. The **bar,** which is used with radicals and fractions, also groups terms. In earlier sections we used parentheses and the bar for grouping.

CAUTION ▶ When adding and subtracting algebraic expressions, it is often necessary to remove symbols of grouping. To do so we must *change the sign of **every term** within the symbols if the grouping is preceded by a minus sign. If the symbols of grouping are preceded by a plus sign, each term within the symbols retains its original sign.* This is a result of the distributive law, $a(b + c) = ab + ac$.

EXAMPLE 6 (a) $2(a + 2x) = 2a + 2(2x)$ use distributive law

$\qquad\qquad\qquad\quad = 2a + 4x$

(b) $-(+a - 3c) = (-1)(+a - 3c)$ treat $-$ sign as -1

$\qquad\qquad\quad = (-1)(+a) + (-1)(-3c)$

$\qquad\qquad\quad = -a + 3c$ note change of signs

Normally, the $+a$ would be written simply as *a*.

In Example 6(a) we see that $2(a + 2x) = 2a + 4x$. This is true for any values of *a* and *x* and is therefore an identity (see p. 6). In fact, an expression and any proper form to which it may be changed form an identity when they are shown as equal to each other.

EXAMPLE 7

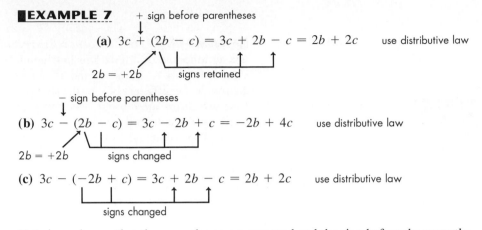

(a) $3c + (2b - c) = 3c + 2b - c = 2b + 2c$ use distributive law

(b) $3c - (2b - c) = 3c - 2b + c = -2b + 4c$ use distributive law

(c) $3c - (-2b + c) = 3c + 2b - c = 2b + 2c$ use distributive law

Note in each case that the parentheses are removed and the sign before the parentheses is also removed.

EXAMPLE 8   (a) $-(2x - 3c) + (c - x) = -2x + 3c + c - x = 4c - 3x$

(b) $4(t - 3 - 2t^2) - (6t + t^2 - 4) = 4t - 12 - 8t^2 - 6t - t^2 + 4$

$$= -9t^2 - 2t - 8$$

EXAMPLE 9   In designing a certain machine part, it is necessary to perform the following simplification.

$$16(8 - x) - 2(8x - x^2) - (64 - 16x + x^2) = 128 - 16x - 16x + 2x^2 - 64 + 16x - x^2$$
$$= 64 - 16x + x^2$$

NOTE ▶   It is fairly common to have expressions in which more than one symbol of grouping is to be removed in the simplification. Normally, *when several symbols of grouping are to be removed, it is more convenient to remove the innermost symbols first.* This is illustrated in the following example.

EXAMPLE 10   (a) $3ax - [ax - (5s - 2ax)] = 3ax - [ax - 5s + 2ax]$
$$= 3ax - ax + 5s - 2ax \qquad \text{remove parentheses}$$
$$= 5s \qquad \qquad \text{remove brackets}$$

(b) $3a^2b - \{[a - (2a^2b - a)] + 2b\} = 3a^2b - \{[a - 2a^2b + a] + 2b\}$
$$= 3a^2b - \{a - 2a^2b + a + 2b\} \qquad \text{remove parentheses}$$
$$= 3a^2b - a + 2a^2b - a - 2b \qquad \text{remove brackets}$$
$$= 5a^2b - 2a - 2b \qquad \text{remove braces}$$

Calculators use only parentheses for grouping symbols, and we often need to use one set of parentheses within another set. These are called **nested parentheses.** In the next example, note that the innermost parentheses are removed first.

EXAMPLE 11 $2 - (3x - 2(5 - (7 - x))) = 2 - (3x - 2(5 - 7 + x))$
$$= 2 - (3x - 10 + 14 - 2x)$$
$$= 2 - 3x + 10 - 14 + 2x$$
$$= -x - 2$$

CAUTION ▶ One of the most common errors made by beginning students is changing the sign of only the first term when removing symbols of grouping preceded by a minus sign. *Remember,* **if the symbols are preceded by a minus sign, we must change the sign of** all *terms.*

───────────────────── **EXERCISES** *1-7* ─────────────────────

In the following exercises, simplify the given algebraic expressions.

1. $5x + 7x - 4x$

2. $6t - 3t - 4t$

3. $2y - y + 4x$

4. $4c + d - 6c$

5. $2a - 2c - 2 + 3c - a$

6. $x - 2y + 3x - y + z$

7. $a^2b - a^2b^2 - 2a^2b$

8. $xy^2 - 3x^2y^2 + 2xy^2$

9. $s + (4 + 3s)$

10. $5 + (3 - 4n + p)$

11. $v - (4 - 5x + 2v)$

12. $2a - (b - a)$

13. $2 - 3 - (4 - 5a)$

14. $\sqrt{A} + (h - 2\sqrt{A}) - 3\sqrt{A}$

15. $(a - 3) + (5 - 6a)$

16. $(4x - y) - (-2x - 4y)$

17. $-(t - 2u) + (3u - t)$

18. $2(x - 2y) + (5x - y)$

19. $3(2r + s) - (-5s - r)$

20. $3(a - b) - 2(a - 2b)$

21. $-7(6 - 3j) - 2(j + 4)$

22. $-(5t + a^2) - 2(3a^2 - 2st)$

23. $-[(6 - n) - (2n - 3)]$

24. $-[(a - b) - (b - a)]$

25. $2[4 - (t^2 - 5)]$

26. $3[-3 - (a - 4)]$

27. $-2[-x - 2a - (a - x)]$

28. $-2[-3(x - 2y) + 4y]$

29. $a\sqrt{LC} - [3 - (a\sqrt{LC} + 4)]$

30. $9v - [6 - (v - 4) + 4v]$

31. $8c - \{5 - [2 - (3 + 4c)]\}$

32. $7y - \{y - [2y - (x - y)]\}$

33. $5p - (q - 2p) - [3q - (p - q)]$

34. $-(4 - x) - [(5x - 7) - (6x + 2)]$

35. $-2\{-(4 - x^2) - [3 + (4 - x^2)]\}$

36. $-\{-[-(x - 2a) - b] - a\}$

37. $5V^2 - (6 - (2V^2 + 3))$

38. $-2x + 2((2x - 1) - 5)$

39. $-(3t - (7 + 2t - (5t - 6)))$

40. $a^2 - 2(x - 5 - (7 - 2(a^2 - 2x) - 3x))$

41. In determining the size of a V belt to be used with an engine, the expression $3D - (D - d)$ is used. Simplify this expression.

42. When finding the current in a transistor circuit, the expression $i_1 - (2 - 3i_2) + i_2$ is used. Simplify this expression. (The numbers below the i's are *subscripts*. Different subscripts denote different variables.)

43. Water leaked into a gasoline storage tank at an oil refinery. Finding the pressure in the tank leads to the expression $-4(b - c) - 3(a - b)$. Simplify this expression.

44. Research on a plastic building material leads to $[(B + \frac{4}{3}\alpha) + 2(B - \frac{2}{3}\alpha)] - [(B + \frac{4}{3}\alpha) - (B - \frac{2}{3}\alpha)]$. Simplify this expression.

1-8 MULTIPLICATION OF ALGEBRAIC EXPRESSIONS

To find the product of two or more monomials, we use the laws of exponents as given in Section 1-4 and the laws for multiplying signed numbers as stated in Section 1-2. We first multiply the numerical coefficients to find the numerical coefficient of the product. Then we multiply the literal numbers, remembering that *the exponents may be combined only if the base is the same.*

■EXAMPLE 1 (a) $3c^5(-4c^2) = -12c^7$ multiply numerical coefficients and add exponents of c

(b) $(-2b^2y^3)(-9aby^5) = 18ab^3y^8$ add exponents of same base

(c) $2xy(-6cx^2)(3xcy^2) = -36c^2x^4y^3$

If a product contains a monomial that is raised to a power, *we must first raise it to the indicated power* before proceeding with the multiplication.

EXAMPLE 2   (a) $3(2a^2x)^3(-ax) = 3(8a^6x^3)(-ax) = -24a^7x^4$

(b) $2s^3(-st^4)^2(4s^2t) = 2s^3(s^2t^8)(4s^2t) = 8s^7t^9$

We find the product of a monomial and a multinomial by using the distributive law, which states that we *multiply each term of the multinomial by the monomial.* We must be careful to assign the correct sign to each term of the result, using the rules for multiplication of signed numbers.

EXAMPLE 3   (a) $2ax(3ax^2 - 4yz) = 2ax(3ax^2) + (2ax)(-4yz) = 6a^2x^3 - 8axyz$

(b) $5cy^2(-7cx - ac) = (5cy^2)(-7cx) + (5cy^2)(-ac) = -35c^2xy^2 - 5ac^2y^2$

It is generally not necessary to write out the middle step as it appears in the preceding example. We write the answer directly. For example, Example 3(a) would appear as $2ax(3ax^2 - 4yz) = 6a^2x^3 - 8axyz.$

We find the product of two or more multinomials by using the distributive law and the laws of exponents. The result is that *we multiply each term of one multinomial by each term of the other, and add the results.*

EXAMPLE 4 $(x - 2)(x + 3) = x(x) + x(3) + (-2)(x) + (-2)(3)$

$= x^2 + 3x - 2x - 6 = x^2 + x - 6$

Finding the power of an algebraic expression is equivalent to using the expression as a factor the number of times indicated by the exponent. It is often convenient to write the power of an algebraic expression in this form before multiplying.

two factors

EXAMPLE 5   (a) $(x + 5)^2 = (x + 5)(x + 5) = x^2 + 5x + 5x + 25$

$= x^2 + 10x + 25$

(b) $(2a - b)^3 = (2a - b)(2a - b)(2a - b)$ the exponent 3 indicates three factors

$= (2a - b)(4a^2 - 4ab + b^2)$

$= 8a^3 - 8a^2b + 2ab^2 - 4a^2b + 4ab^2 - b^3$

$= 8a^3 - 12a^2b + 6ab^2 - b^3$

We should note in illustration (a) that

CAUTION ▶ $(x + 5)^2$ **is not** *equal to* $x^2 + 25$

since the term $10x$ is not included. We must follow the proper procedure and not simply square each of the terms within the parentheses.

EXAMPLE 6 An expression used with a lens of a certain telescope is simplified as shown.

$$a(a + b)^2 + a^3 - (a + b)(2a^2 - s^2)$$
$$= a(a + b)(a + b) + a^3 - (2a^3 - as^2 + 2a^2b - bs^2)$$
$$= a(a^2 + ab + ab + b^2) + a^3 - 2a^3 + as^2 - 2a^2b + bs^2$$
$$= a^3 + a^2b + a^2b + ab^2 - a^3 + as^2 - 2a^2b + bs^2$$
$$= ab^2 + as^2 + bs^2$$

EXERCISES *1-8*

In the following exercises, perform the indicated multiplications.

1. $(a^2)(ax)$

2. $(2xy)(x^2y^3)$

3. $-ac^2(acx^3)$

4. $-2s^2(-4cs)^2$

5. $(2ax^2)^2(-2ax)$

6. $6pq^3(3pq^2)^2$

7. $a(-a^2x)^3(-2a)$

8. $-2m^2(-3mn)(m^2n)^2$

9. $i^2(R + r)$

10. $2x(p - q)$

11. $-3s(s^2 - 5t)$

12. $-3b(2b^2 - b)$

13. $5m(m^2n + 3mn)$

14. $a^2bc(2ac - 3a^2h)$

15. $3x(-x - y + 2)$

16. $b^2x^2(x^2 - 2x + 1)$

17. $ab^2c^4(ac - bc - ab)$

18. $-4c^2(-9gc - 2c + g^2)$

19. $ax(cx^2)(x + y^3)$

20. $-2(-3st^3)(3s - 4t)$

21. $(x - 3)(x + 5)$

22. $(a + 7)(a + 1)$

23. $(x + 5)(2x - 1)$

24. $(4t_1 + t_2)(2t_1 - 3t_2)$

25. $(2a - b)(3a - 2b)$

26. $(4w^2 - 3)(3w^2 - 1)$

27. $(2s + 7t)(3s - 5t)$

28. $(5p - 2q)(p + 8q)$

29. $(x^2 - 1)(2x + 5)$

30. $(3y^2 + 2)(2y - 9)$

31. $(x^2 - 2x)(x + 4)$

32. $(2ab^2 - 5t)(-ab^2 - 6t)$

33. $(x + 1)(x^2 - 3x + 2)$

34. $(2x + 3)(x^2 - x - 5)$

35. $(4x - x^3)(2 + x - x^2)$

36. $(5a - 3c)(a^2 + ac - c^2)$

37. $2(a + 1)(a - 9)$

38. $-5(y - 3)(y + 6)$

39. $2L(L + 1)(L - 4)$

40. $ax(x + 4)(7 - x^2)$

41. $(2x - 5)^2$

42. $(x - 3)^2$

43. $(x_1 + 3x_2)^2$

44. $(2m + 1)^2$

45. $(xyz - 2)^2$

46. $(b - 2x^2)^2$

47. $2(x + 8)^2$

48. $3(3R + 4)^2$

49. $(2 + x)(3 - x)(x - 1)$

50. $(3x - c^2)^3$

51. $3x(x + 2)^2(2x - 1)$

52. $[(x - 2)^2(x + 2)]^2$

53. In a particular computer design containing n circuit elements, n^2 switches are needed. Find the expression for the number of switches needed for $n + 100$ circuit elements.

54. An expression found in chemical thermodynamics is $p(c - 1) + 2 - c(p - 1)$. Multiply and simplify.

55. In finding the maximum power in part of a microwave transmitter circuit, the expression $(R_1 + R_2)^2 - 2R_2(R_1 + R_2)$ is used. Multiply and simplify.

56. In determining the deflection of a certain steel beam, the expression $27x^2 - 24(x - 6)^2 - (x - 12)^3$ is used. Multiply and simplify.

1-9 DIVISION OF ALGEBRAIC EXPRESSIONS

To find the quotient of one monomial divided by another, we use the laws of exponents and the laws for dividing signed numbers. Again, the *exponents may be combined only if the base is the same.*

EXAMPLE 1 (a) $\dfrac{3c^7}{c^2} = 3c^{7-2} = 3c^5$ (b) $\dfrac{16x^3y^5}{4xy^2} = \dfrac{16}{4}(x^{3-1})(y^{5-2}) = 4x^2y^3$

(c) $\dfrac{-6a^2xy^2}{2axy^4} = -\left(\dfrac{6}{2}\right)\dfrac{a^{2-1}x^{1-1}}{y^{4-2}} = -\dfrac{3a}{y^2}$

As shown in illustration (c), we use only positive exponents in the final result unless there are specific instructions otherwise.

From arithmetic we may show how a multinomial is to be divided by a monomial. When adding fractions, say $\frac{2}{7}$ and $\frac{3}{7}$, we have

$$\frac{2}{7} + \frac{3}{7} = \frac{2+3}{7}$$

Looking at this from *right to left*, we see that *the quotient of a multinomial divided by a monomial is found by dividing each term of the multinomial by the monomial and adding the results.* This can be shown as

$$\frac{a+b}{c} = \frac{a}{c} + \frac{b}{c}$$

CAUTION ▶   Be careful: Although $\dfrac{a+b}{c} = \dfrac{a}{c} + \dfrac{b}{c}$, we must note that $\dfrac{c}{a+b}$ *is not* $\dfrac{c}{a} + \dfrac{c}{b}$.

■EXAMPLE 2   (a) $\dfrac{4a^2 + 8a}{2a} = \dfrac{4a^2}{2a} + \dfrac{8a}{2a} = 2a + 4$ each term of numerator divided by denominator

(b) $\dfrac{4x^3y - 8x^3y^2 + 2x^2y}{2x^2y} = \dfrac{4x^3y}{2x^2y} - \dfrac{8x^3y^2}{2x^2y} + \dfrac{2x^2y}{2x^2y}$

$$= 2x - 4xy + 1$$

(c) $\dfrac{a^3bc^4 - 6abc + 9a^2b^3c - 3}{3ab^2c^3} = \dfrac{a^2c}{3b} - \dfrac{2}{bc^2} + \dfrac{3ab}{c^2} - \dfrac{1}{ab^2c^3}$

We usually do not write the middle step shown in illustrations (a) and (b). The divisions are done by inspection, and the result appears as in illustration (c).   ■

■EXAMPLE 3   The expression $\dfrac{2p + v^2d + 2ydg}{2dg}$ is used when analyzing the operation of an irrigation pump. Performing the indicated division, we have

$$\frac{2p + v^2d + 2ydg}{2dg} = \frac{p}{dg} + \frac{v^2}{2g} + y$$ ■

If each term in an algebraic sum is a number or is of the form ax^n, where n is a nonnegative integer, we call the expression a **polynomial** *in x. The greatest value of the exponent n that appears is the* **degree** *of the polynomial.*

A multinomial and a polynomial may differ in that a polynomial cannot have terms like $\sqrt{x}$ or $1/x^2$, but a multinomial may have such terms. Also, a polynomial may have only one term, whereas a multinomial has at least two terms.

■EXAMPLE 4   (a) $3 + 2x^2 - x^3$ is a polynomial of degree 3.
(b) $x^4 - 3x^2 - \sqrt{x}$ is not a polynomial.

(c) $4x^5$ is a polynomial of degree 5.   (d) $\dfrac{1}{4x^5}$ is not a polynomial.

Expressions (a) and (b) are also multinomials since each has more than one term. Expression (b) is not a polynomial due to the $\sqrt{x}$ term. Expression (c) is a polynomial since the exponent is a positive integer. It is also a monomial. Expression (d) is not a polynomial since it can be written as $\frac{1}{4}x^{-5}$, and when written in the form ax^n, n is not positive.   ■

Division of One Polynomial by Another

To divide one polynomial by another, first arrange the dividend (the polynomial to be divided) and the divisor in descending powers of the variable. Then divide the first term of the dividend by the first term of the divisor. The result is the first term of the quotient. Next, multiply the entire divisor by the first term of the quotient and subtract the product from the dividend. Divide the first term of this difference by the first term of the divisor. This gives the second term of the quotient. Multiply this term by the entire divisor and subtract the product from the first difference. Repeat this process until the remainder is zero or a term of lower degree than the divisor. This process is similar to long division of numbers.

EXAMPLE 5   Perform the division $(6x^2 + x - 2) \div (2x - 1)$.
$\left(\text{This division can also be indicated in the fractional form } \dfrac{6x^2 + x - 2}{2x - 1}. \right)$

We set up the division as we would for long division in arithmetic. Then, following the procedure outlined above, we have the following:

$$
\begin{array}{r}
3x + 2 \quad\quad \frac{6x^2}{2x} \\
2x - 1 \overline{)\, 6x^2 + x - 2} \\
\end{array}
$$

CAUTION   *subtract* $6x^2 - 3x \longleftarrow 3x(2x - 1)$

$6x^2 - 6x^2 = 0 \longrightarrow 4x - 2$
$x - (-3x) = 4x \longrightarrow 4x - 2$ *subtract*
$\underline{}$
0

The remainder is zero and the quotient is $3x + 2$. Note that when we subtracted $-3x$ from x, we obtained $4x$.

EXAMPLE 6   Perform the division $\dfrac{4x^3 + 6x^2 + 1}{2x - 1}$. Since there is no x-term in the dividend, we should leave space for any x-terms that might arise.

$$
\begin{array}{r}
2x^2 + 4x + 2 \\
\end{array}
$$

divisor $2x - 1 \overline{)\, 4x^3 + 6x^2 \quad\quad + 1}$ dividend
$\underline{4x^3 - 2x^2 \quad\quad\quad}$ subtract
$6x^2 - (-2x^2) = 8x^2 \longrightarrow 8x^2 \quad\quad + 1$
$\underline{8x^2 - 4x \quad\quad}$ subtract
$0 - (-4x) = 4x \quad\quad \longrightarrow 4x + 1$
$\underline{4x - 2}$ subtract
3 remainder

The quotient in this case is written as $2x^2 + 4x + 2 + \dfrac{3}{2x - 1}$.

EXERCISES *1-9*

In the following exercises, perform the indicated divisions.

1. $\dfrac{8x^3y^2}{-2xy}$ **2.** $\dfrac{-18b^7c^3}{bc^2}$ **3.** $\dfrac{-16r^3t^5}{-4r^5t}$ **4.** $\dfrac{51mn^5}{17m^2n^2}$

5. $\dfrac{(15x^2)(4bx)(2y)}{30bxy}$ **6.** $\dfrac{(5st)(8s^2t^3)}{10s^3t^2}$ **7.** $\dfrac{6(ax)^2}{-ax^2}$

8. $\dfrac{12a^2b}{(3ab^2)^2}$ **9.** $\dfrac{a^2x + 4xy}{x}$ **10.** $\dfrac{2m^2n - 6mn}{2m}$

11. $\dfrac{3rst - 6r^2st^2}{3rs}$ **12.** $\dfrac{-5a^2n - 10an^2}{5an}$

13. $\dfrac{4pq^3 + 8p^2q^2 - 16pq^5}{4pq^2}$ **14.** $\dfrac{a^2x_1x_2^2 + ax_1^3 - ax_1}{ax_1}$

15. $\dfrac{2\pi fL - \pi fR^2}{\pi fR}$ **16.** $\dfrac{2(ab)^4 - a^3b^4}{3(ab)^3}$

17. $\dfrac{3ab^2 - 6ab^3 + 9a^2b^2}{9a^2b^2}$ **18.** $\dfrac{2x^2y^2 + 8xy - 12x^2y^4}{2x^2y^2}$

19. $\dfrac{x^{n+2} + ax^n}{x^n}$ **20.** $\dfrac{3a(x + y)b^2 - (x + y)}{a(x + y)}$

21. $(2x^2 + 7x + 3) \div (x + 3)$

22. $(3x^2 - 11x - 4) \div (x - 4)$

23. $\dfrac{x^2 - 3x + 2}{x - 2}$ **24.** $\dfrac{2x^2 - 5x - 7}{x + 1}$

25. $\dfrac{x - 14x^2 + 8x^3}{2x - 3}$ **26.** $\dfrac{6x^2 + 6 + 7x}{2x + 1}$

27. $(4x^2 + 23x + 15) \div (4x + 3)$

28. $(6x^2 - 20x + 16) \div (3x - 4)$

29. $\dfrac{x^3 + 3x^2 - 4x - 12}{x + 2}$ **30.** $\dfrac{3x^3 + 19x^2 + 16x - 20}{3x - 2}$

31. $\dfrac{2x^4 + 4x^3 + 2}{x^2 - 1}$ **32.** $\dfrac{2x^3 - 3x^2 + 8x - 2}{x^2 - x + 2}$

33. $\dfrac{x^3 + 8}{x + 2}$ **34.** $\dfrac{D^3 - 1}{D - 1}$

35. $\dfrac{x^2 - 2xy + y^2}{x - y}$ **36.** $\dfrac{3r^2 - 5rR + 2R^2}{r - 3R}$

37. In the optical theory dealing with lasers, the following expression arises: $\dfrac{8A^5 + 4A^3\mu^2E^2 - A\mu^4E^4}{8A^4}$. Perform the indicated division. (μ is the Greek letter mu.)

38. In finding the total resistance of the resistors shown in Fig. 1-11, the expression $\dfrac{6R_1 + 6R_2 + R_1R_2}{6R_1R_2}$ is used. Perform the indicated division.

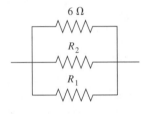

Fig. 1-11

39. A computer model shows that the temperature change T in a certain freezing unit is found by using the expression $\dfrac{3T^3 - 8T^2 + 8}{T - 2}$. Perform the indicated division.

40. In analyzing the displacement of a certain valve, the expression $\dfrac{s^2 - 2s - 2}{s^4 + 4}$ is used. Find the reciprocal of this expression and perform the indicated division.

1-10 SOLVING EQUATIONS

In this section we show how algebraic operations are used in solving equations. In the following sections we show some of the important applications of equations.

   An **equation** *is an algebraic statement that two algebraic expressions are equal.* Any value of the literal number representing the **unknown** that produces equality when **substituted** in the equation is said to **satisfy** the equation.

▮**EXAMPLE 1**   The equation $3x - 5 = x + 1$ is true only if $x = 3$. For $x = 3$ we have $4 = 4$; for $x = 2$ we have $1 = 3$, which is not correct.

   This equation is valid for only one value of the unknown. *An equation valid only for certain values of the unknown is a* **conditional equation.** In this section, nearly all equations we solve will be conditional equations that are satisfied by only one value of the unknown.

We have previously noted *identities* on pages 6 and 27.

Equations can be solved on many calculators. However, an estimate of the answer may be required in order to find the solution.

Although we may multiply both sides of an equation by zero, this produces $0 = 0$, which is not useful in finding the solution.

The word *algebra* comes from Arabic and means "a restoration." It refers to the fact that when a number has been added to one side of an equation, the same number must be added to the other side to maintain equality.

EXAMPLE 2   (a) The equation $x^2 - 4 = (x - 2)(x + 2)$ is true for all values of x. For example, if $x = 3$, we have $9 - 4 = (3 - 2)(3 + 2)$, or $5 = 5$. If $x = -1$, we have $-3 = -3$. *An equation valid for all values of the unknown is an* **identity.**

(b) The equation $x + 5 = x + 1$ is not true for any value of x. For any value of x we try, we find that the left side is 4 greater than the right side. *Such an equation is called a* **contradiction.**

To **solve** *an equation we find the values of the unknown that satisfy it.* There is one basic rule to follow when solving an equation:

Perform the same operation on both sides of the equation.

We do this to isolate the unknown and thus to find its values.

By performing the same operation on both sides of an equation, the two sides remain equal. Thus,

we may add the same number to both sides, subtract the same number from both sides, multiply both sides by the same number, or divide both sides by the same number (not zero).

EXAMPLE 3   For each of the following equations, we may isolate x, and thereby solve the equation, by performing the indicated operation.

$x - 3 = 12$	$x + 3 = 12$	$\dfrac{x}{3} = 12$	$3x = 12$
add 3 to both sides	subtract 3 from both sides	multiply both sides by 3	divide both sides by 3
$x - 3 + 3 = 12 + 3$	$x + 3 - 3 = 12 - 3$	$3\left(\dfrac{x}{3}\right) = 3(12)$	$\dfrac{3x}{3} = \dfrac{12}{3}$
$x = 15$	$x = 9$	$x = 36$	$x = 4$

Each solution should be checked by substitution in the original equation.

The solution of an equation generally requires a combination of the basic operations. The following examples illustrate the solution of such equations.

EXAMPLE 4   Solve the equation $2t - 7 = 9$.

We are to perform basic operations to both sides of the equation to finally isolate t on one side. The steps to be followed are suggested by the form of the equation, and in this case are as follows:

$$2t - 7 = 9 \qquad \text{original equation}$$
$$2t - 7 + 7 = 9 + 7 \qquad \text{add 7 to both sides}$$
$$2t = 16 \qquad \text{combine like terms}$$
$$\frac{2t}{2} = \frac{16}{2} \qquad \text{divide both sides by 2}$$
$$t = 8 \qquad \text{simplify}$$

NOTE ▶ Therefore, we conclude that $t = 8$. Checking *in the original equation*, we have

$$2(8) - 7 \overset{?}{=} 9, \qquad 16 - 7 \overset{?}{=} 9, \qquad \text{or} \qquad 9 = 9$$

Therefore, the solution checks.

With simpler numbers, the step of adding or subtracting a term, or multiplying or dividing by a factor, is usually done by inspection and not actually written down.

EXAMPLE 5   Solve the equation $3n + 4 = n - 6$.

$$2n + 4 = -6 \qquad \text{n subtracted from both sides—by inspection}$$
$$2n = -10 \qquad \text{4 subtracted from both sides—by inspection}$$
$$n = -5 \qquad \text{both sides divided by 2—by inspection}$$

Checking *in the original equation,* we have $-11 = -11$.   ■

EXAMPLE 6   Solve the equation $x - 7 = 3x - (6x - 8)$.

$$x - 7 = 3x - 6x + 8 \qquad \text{parentheses removed}$$
$$x - 7 = -3x + 8 \qquad \text{x-terms combined on right}$$
$$4x - 7 = 8 \qquad \text{$3x$ added to both sides}$$
$$4x = 15 \qquad \text{7 added to both sides}$$
$$x = \tfrac{15}{4} \qquad \text{both sides divided by 4}$$

Checking in the original equation, we obtain (after simplifying) $-\tfrac{13}{4} = -\tfrac{13}{4}$.   ■

NOTE ▶   Note that we ***always check in the** **original equation.*** This is done since errors may have been made in finding the later equations.

From these examples we see that the following steps are used in solving the basic equations of this section.

Procedure for Solving Equations

1. *Remove grouping symbols (distributive law).*

2. *Combine any like terms on each side (also after step 3).*

3. *Perform the same operations on both sides until x = result is obtained.*

4. *Check the solution in the original equation.*

NOTE ▶   If an equation contains numbers not easily combined by inspection, the best procedure is to *first solve for the unknown and then perform the calculation.*

EXAMPLE 7   When finding the current i (in A) in a certain radio circuit, the following equation and solution are used.

$$0.0595 - 0.525i - 8.85(i + 0.00316) = 0$$
$$0.0595 - 0.525i - 8.85i - 8.85(0.00316) = 0 \qquad \text{note how the above}$$
$$\qquad\qquad\qquad\qquad\qquad\qquad\qquad\qquad\qquad\;\; \text{procedure is followed}$$
$$(-0.525 - 8.85)i = 8.85(0.00316) - 0.0595$$
$$i = \frac{8.85(0.00316) - 0.0595}{-0.525 - 8.85} \qquad \text{evaluate}$$
$$= 0.00336 \text{ A}$$

NOTE ▶   In checking, if we use the result above we find the left side of the original equation to be 3.4×10^{-5}, and if we use the unrounded calculator value we get zero. Although the *result* should be rounded off, *no rounding off should be done before the final calculation.* An inaccurate result might otherwise be obtained.   ■

Except for the third illustration of Example 3, we have not solved an equation that contained a fraction. We now consider one type of equation with fractions.

RATIO AND PROPORTION

The quotient a/b is also called the **ratio** *of a to b. An equation stating that two ratios are equal is called a* **proportion.** Since a proportion is an equation, if one of the numbers is unknown, we can solve for its value as with any equation.

■**EXAMPLE 8**   The ratio of 2 to 3 is the quotient 2/3. The ratio of 4 to 6 is 4/6. Since these ratios are equal, we have the proportion

$$\frac{2}{3} = \frac{4}{6}$$

If the ratio of x to 8 equals the ratio of 3 to 4, we have the proportion

$$\frac{x}{8} = \frac{3}{4}$$

We can solve this equation by multiplying both sides by 8. This gives

$$8\left(\frac{x}{8}\right) = 8\left(\frac{3}{4}\right) \qquad \text{or} \qquad x = 6$$

Since 6/8 = 3/4, the solution checks.

■**EXAMPLE 9**   The ratio of electric current I (in A) to the voltage V across a resistor is constant. If $I = 1.52$ A for $V = 60.0$ V, find I for $V = 82.0$ V.
Since the ratio of I to V is constant, we have the following solution.

$$\frac{1.52}{60.0} = \frac{I}{82.0}$$

$$82.0\left(\frac{1.52}{60.0}\right) = 82.0\left(\frac{I}{82.0}\right)$$

$$2.08 \text{ A} = I$$

$$I = 2.08 \text{ A}$$

Checking in the original equation, we have 0.0253 A/V = 0.0254 A/V. Since the value of I is rounded up, the solution checks.

We will use the terms *ratio* and *proportion* (particularly ratio) when we study trigonometry in Chapter 4. A more detailed discussion of ratio and proportion is found in Chapter 18. Also, a general method of solving equations with fractions is given in Chapter 6.

━━━━ **EXERCISES** *1-10* ━━━━

In Exercises 1–32, solve the given equations.

1. $x - 2 = 7$

2. $x - 4 = 1$

3. $x + 5 = 4$

4. $s + 6 = -3$

5. $\dfrac{t}{2} = 5$

6. $\dfrac{x}{4} = -2$

7. $4x = -20$

8. $2x = 12$

9. $3t + 5 = -4$

10. $5D - 2 = 13$

11. $5 - 2y = 3$

12. $8 - 5t = 18$

13. $3x + 7 = x$

14. $6 + 4L = 5 - 3L$

15. $2(s - 4) = s$

16. $3(n - 2) = -n$

17. $6 - (r - 4) = 2r$

18. $5 - (x + 2) = 5x$

19. $2(x - 3) - 5x = 7$

20. $4(x + 7) - x = -7$

21. $x - 5(x - 2) = 2$ **22.** $5x - 2(x - 5) = 4x$

23. $7 - 3(1 - 2p) = 4 + 2p$

24. $3 - 6(2 - 3t) = t - 5$

In Exercises 25–32, all numbers are approximate.

25. $5.8 - 0.3(x - 6.0) = 0.5x$

26. $1.9t = 0.5(4.0 - t) - 0.8$

27. $0.15 - 0.24(C - 0.50) = 0.63$

28. $27.5(5.17 - 1.44x) = 73.4$

29. $\dfrac{x}{2.0} = \dfrac{17}{6.0}$ **30.** $\dfrac{3.0}{7.0} = \dfrac{x}{42}$

31. $\dfrac{165}{223} = \dfrac{13V}{15}$ **32.** $\dfrac{276x}{17.0} = \dfrac{1360}{46.4}$

In Exercises 33–40, solve the given problems.

33. In finding the rate (in mi/h) at which a polluted stream flows, the equation $15(5.5 + v) = 24(5.5 - v)$ is used. Find the rate v.

34. To find the voltage V in a circuit in a TV remote-control unit, the equation $1.12V - 0.67(10.5 - V) = 0$ is used. Find V.

35. In blending two gasolines of different octanes, in order to find the number n of gallons of one octane needed, the equation $0.14n + 0.06(2000 - n) = 0.09(2000)$ is used. Find n, given

that 0.06 and 0.09 are exact and the first zero of 2000 is significant.

36. In order to find the distance x such that the weights are balanced on the lever shown in Fig. 1-12, the equation $210(3x) = 55.3x + 38.5(8.25 - 3x)$ must be solved. Find x. (3 is exact.)

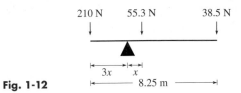

Fig. 1-12

37. A capsule contains two medications in the ratio of 5 to 2. If there are 150 mg of the first medication in the capsule, how many milligrams of the second are there?

38. A person 1.8 m tall is photographed with a 35-mm camera, and the film image is 20 mm. Under the same conditions, how tall is a person whose film image is 16 mm?

W 39. In solving the equation $2(x - 3) + 1 = 2x - 5$, what conclusion can be made?

W 40. In solving the equation $7 - (2 - x) = x + 2$, what conclusion can be made?

1-11 FORMULAS AND LITERAL EQUATIONS

One of the most important applications of equations occurs in the use of formulas in geometry, physics, engineering, and many other fields. *A **formula** is an equation that expresses a rule and uses letters to represent certain quantities.* For example, one of the most famous formulas in the history of science is Einstein's equivalence of mass and energy, $E = mc^2$. The symbol E represents energy, as does mc^2. The formula states that the equivalent energy E of a mass m is found by multiplying m by the square of c, the speed of light.

Often it is necessary to solve a formula for a particular symbol that appears in it. We do this in the same way as we solve any equation. *We isolate the desired symbol by use of the basic algebraic operations.*

Einstein published his first paper on relativity in 1905.

■**EXAMPLE 1**   In the formula for the equivalent energy E of a mass m, $E = mc^2$, solve for m.

$$E = mc^2 \qquad \text{original equation}$$

$$\frac{E}{c^2} = m \qquad \text{divide both sides by } c^2$$

$$m = \frac{E}{c^2}$$

Since the symbol for which we are solving is usually shown on the left of the equal sign, the sides were switched, as shown.   ----------∎

The subscript $_0$ makes v_0 a different literal symbol from v. (We have used subscripts in a few of the earlier exercises.)

EXAMPLE 2   A formula relating acceleration a, velocity v, initial velocity v_0, and time t is $v = v_0 + at$. Solve for t.

$$v - v_0 = at \qquad \text{v_0 subtracted from both sides}$$

$$t = \frac{v - v_0}{a} \qquad \text{both sides divided by a and then sides switched}$$

NOTE ▶   As we can see from Examples 1 and 2, we can solve for the indicated literal number just as we solved for the unknown in the previous section. That is, *we perform the basic algebraic operations on the various literal numbers that appear in the same way we perform them on explicit numbers.* Other illustrations are shown in the following examples.

EXAMPLE 3   In the study of the forces on a certain beam, the equation

$$W = \frac{L(wL + 2P)}{8}$$

is used. Solve for P.

Be careful! Just as subscripts can indicate different literal numbers, a capital letter and the same letter in lowercase are different literal numbers. In this example W and w are different. Also see Exercises 23 and 36 for this section.

$$8W = \frac{8L(wL + 2P)}{8} \qquad \text{multiply both sides by 8}$$

$$8W = L(wL + 2P) \qquad \text{simplify right side}$$

$$8W = wL^2 + 2LP \qquad \text{remove parentheses}$$

$$8W - wL^2 = 2LP \qquad \text{subtract wL^2 from both sides}$$

$$P = \frac{8W - wL^2}{2L} \qquad \text{divide both sides by $2L$ and switch sides}$$

EXAMPLE 4   The effect of temperature is important when measurements must be made with great accuracy. The volume V of a special precision container at temperature T in terms of the volume at temperature T_0 is given by

$$V = V_0[1 + b(T - T_0)]$$

where b depends on the material of which the container is made. Solve for T.

Since we are to solve for T, we must isolate the term containing T. This can be done by first removing the grouping symbols and then isolating the term with T.

$$V = V_0[1 + b(T - T_0)] \qquad \text{original equation}$$

$$V = V_0[1 + bT - bT_0] \qquad \text{remove parentheses}$$

$$V = V_0 + bTV_0 - bT_0V_0 \qquad \text{remove brackets}$$

$$V - V_0 + bT_0V_0 = bTV_0 \qquad \text{subtract V_0 and add bT_0V_0 to both sides}$$

$$T = \frac{V - V_0 + bT_0V_0}{bV_0} \qquad \text{divide both sides by bV_0 and switch sides}$$

NOTE ▶   If we wish to determine the value of any literal number in an expression for which we know values of the other literal numbers, we should *first solve for the required symbol and then substitute the given values.* This is illustrated in the following example.

EXAMPLE 5   The electric resistance R (in Ω) of a resistor changes with the temperature T (in °C) according to $R = R_0 + R_0 \alpha T$, where R_0 is the resistance at 0°C. For a given resistor, $R_0 = 712$ Ω and $\alpha = 0.00455/$°C. Find the value of T for $R = 825$ Ω.

We first solve for T and then substitute the given values.

$$R = R_0 + R_0 \alpha T$$

$$R - R_0 = R_0 \alpha T$$

$$T = \frac{R - R_0}{\alpha R_0}$$

Now substituting, we have

$$T = \frac{825 - 712}{(0.00455)(712)}$$

$$= 34.9°C \quad \text{rounded off}$$

estimation:

$$\frac{800 - 700}{0.005(700)} = \frac{1}{0.035} = 30$$

EXERCISES *1-11*

In Exercises 1–8, solve for the given letter.

1. $ax = b$, for x

2. $cy + d = 0$, for y

3. $4n + 1 = 4m$, for n

4. $bt - 3 = a$, for t

5. $ax + 6 = 2ax - c$, for x

6. $s - 6n^2 = 3s + 4$, for s

7. $\frac{1}{2}t - (4 - a) = 2a$, for t

8. $7 - (p - \frac{1}{3}x) = 3p$, for x

In Exercises 9–32, each of the given formulas arises in the technical or scientific area of study shown. Solve for the indicated letter.

9. $\theta = kA + \lambda$, for λ (robotics)

10. $W = S_d T - Q$, for Q (air conditioning)

11. $E = IR$, for R (electricity)

12. $Q = SLd^2$, for L (machine design)

13. $P = 2\pi Tf$, for T (mechanics)

14. $PV = nRT$, for T (chemistry)

15. $p = p_a + dgh$, for h (hydrodynamics)

16. $2Q = 2I + A + S$, for I (nuclear physics)

17. $A = \dfrac{Rt}{PV}$, for t (jet engine design)

18. $u = -\dfrac{eL}{2m}$, for L (spectroscopy)

19. $ct^2 = 0.3t - ac$, for a (medical technology)

20. $FL = P_1 L - P_1 d + P_2 L$, for d (construction)

21. $C_0^2 = C_1^2(1 + 2V)$, for V (electronics)

22. $A_1 = A(M + 1)$, for M (photography)

23. $P = n(p - c)$, for p (economics)

24. $T = 3(T_2 - T_1)$, for T_1 (oil drilling)

25. $Q_1 = P(Q_2 - Q_1)$, for Q_2 (refrigeration)

26. $p - p_a = dg(y_2 - y_1)$, for y_2 (pressure gauges)

27. $N = N_1 T - N_2(1 - T)$, for N_1 (machine design)

28. $t_a = t_c + (1 - h)t_m$, for h (computer access time)

29. $L = \pi(r_1 + r_2) + 2x_1 + x_2$, for r_1 (pulleys)

30. $I = \dfrac{VR_2 + VR_1(1 + \mu)}{R_1 R_2}$, for μ (electronics)

31. $P = \dfrac{V_1(V_2 - V_1)}{gJ}$, for V_2 (jet engine power)

32. $W = T(S_1 - S_2) - Q$, for S_2 (refrigeration)

In Exercises 33–36, find the indicated values.

33. The pressure p (in kPa) at a depth h (in m) below the surface of water is given by the formula $p = p_0 + kh$, where p_0 is the atmospheric pressure. Find h for $p = 205$ kPa, $p_0 = 101$ kPa, and $k = 9.80$ kPa/m.

34. A formula used in determining the total transmitted power P_t in an AM radio signal is $P_t = P_c(1 + 0.500m^2)$. Find P_c if $P_t = 680$ W and $m = 0.925$.

35. A formula relating the Fahrenheit temperature F and the Celsius temperature C is $F = \frac{9}{5}C + 32$. Find the Celsius temperature that corresponds to 90.2°F.

36. In forestry, a formula used to determine the volume V of a log is $V = \frac{1}{2}L(B + b)$, where L is the length of the log and B and b are the areas of the ends. Find b (in ft^2) if $V = 38.6$ ft^3, $L = 16.1$ ft, and $B = 2.63$ ft^2. See Fig. 1-13.

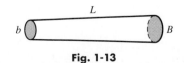

Fig. 1-13

1-12 APPLIED WORD PROBLEMS

Mathematics allows us to solve many kinds of applied problems. Some of these problems are given in equation or formula form and can be solved directly. However, in practice it is often necessary to first set up equations by using known formulas and the given conditions. Such problems are at first word problems, and it is necessary to translate them into mathematical terms for solution.

Usually the most difficult part in solving a word problem is identifying the information that leads to the equation. Finding this information requires that you carefully read the problem, being sure that you understand all the terms and expressions that are used. Following is a general approach that you should use.

See Appendix A, page A-3, for a variation to the method outlined in these steps. You might find it helpful. Also, some suggested study aids are briefly discussed in Appendix A.

Procedure for Solving Word Problems

1. *Read the statement of the problem.* First read it quickly for a general overview. Then *reread* slowly and carefully, *listing the information* given.

2. *Clearly identify the unknown quantities* and then *assign an appropriate letter to represent one of them,* stating this choice clearly.

3. *Specify the other unknown quantities* in terms of the one in step 2.

4. *If possible, make a sketch* using the known and unknown quantities.

5. *Analyze the statement* of the problem and *write the necessary equation.* This is often the most difficult step because **some of the information may be implied and not explicitly stated.** Again, a very careful reading of the statement is necessary.

6. *Solve the equation,* clearly stating the solution.

7. *Check the solution with the original statement* of the problem.

CAUTION ▶

Carefully read the following examples. Note how the above steps are followed.

■**EXAMPLE 1**   A 17-lb beam is supported at each end. The supporting force at one end is 3 lb more than at the other end. Find the forces.

Since the force at each end is required, we write

Be sure to carefully identify your choice for the unknown. In most problems there is really a choice. Using the word *let* clearly shows that a specific choice has been made.

$$\text{let } F = \text{the smaller force} \qquad \text{step 2}$$

as a way of establishing the unknown for the equation. Any appropriate letter could be used, and we could have let it represent the larger force.

Also, since the other force is 3 lb more, we write

The statement after "let *x* (or some other appropriate letter) =" should be clear. It should completely define the chosen unknown.

$$F + 3 = \text{the larger force} \qquad \text{step 3}$$

We now draw the sketch in Fig. 1-14.   step 4

Since the forces at each end of the beam support the weight of the beam, we have the equation

$$F + (F + 3) = 17 \qquad \text{step 5}$$

This equation can now be solved: $2F = 14$

$$F = 7 \text{ lb} \qquad \text{step 6}$$

F 17 lb $F + 3$

Fig. 1-14

Thus, the smaller force is 7 lb and the larger force is 10 lb. This checks with the original statement of the problem.   step 7

EXAMPLE 2   In designing an electric circuit, it is found that 34 resistors with a total resistance of 56 Ω are required. Two different resistances, 1.5 Ω and 2.0 Ω, are used. How many of each are in the circuit?

Since we want to find the number of each resistance, we

$$\text{let } x = \text{number of 1.5-}\Omega \text{ resistors}$$

Also, since there are 34 resistors in all,

$$34 - x = \text{number of 2.0-}\Omega \text{ resistors}$$

"Let x = 1.5-Ω resistors" is incomplete. We want to find out how many of them there are.

See Appendix A, page A-3, for a "sketch" that might be used with this example.

We also know that the total resistance of all resistors is 56 Ω. This means that

1.5 Ω   number   2.0 Ω

each   each   number

$$1.5x \qquad\qquad + \quad 2.0(34 - x) = 56 \ \leftarrow \text{total resistance of all resistors}$$

total resistance   total resistance

of 1.5-Ω resistors   of 2.0-Ω resistors

$$1.5x + 68 - 2.0x = 56$$
$$-0.5x = -12$$
$$x = 24$$

Therefore, there are 24 1.5-Ω resistors and 10 2.0-Ω resistors. The total resistance of these is $24(1.5) + 10(2.0) = 36 + 20 = 56$ Ω. We see that this checks with the statement of the problem.

EXAMPLE 3   A medical researcher finds that a given sample of an experimental drug can be divided into 4 more slides with 5 mg each than with 6 mg each. How many slides with 5 mg each can be made up?

We are asked to find the number of slides with 5 mg, and therefore we

$$\text{let } x = \text{number of slides with 5 mg}$$

Since the sample may be divided into 4 more slides with 5 mg each than of 6 mg each, we know that

$$x - 4 = \text{number of slides with 6 mg}$$

Since *it is the same sample that is to be divided,* the total mass of the drug on each type is the same. This means

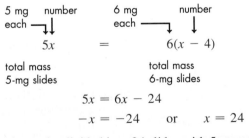

$$5x = 6x - 24$$
$$-x = -24 \qquad \text{or} \qquad x = 24$$

Therefore, the sample can be divided into 24 slides with 5 mg each, or 20 slides with 6 mg each. Since the total mass, 120 mg, is the same for each set of slides, the solution checks with the statement of the problem.

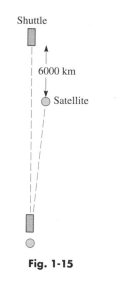

Fig. 1-15

A maneuver similar to the one in this example was used to repair the Hubble space telescope in 1993.

"Let x = methanol" is incomplete. We want to find out the volume (in L) that is to be added.

■**EXAMPLE 4**   A space shuttle maneuvers so that it may "capture" an already orbiting satellite that is 6000 km ahead. If the satellite is moving at 27,000 km/h and the shuttle is moving at 29,500 km/h, how long will it take the shuttle to reach the satellite? (All digits shown are significant.)

First, we let t = the time for the shuttle to reach the satellite. Then, using the fact that the shuttle must go 6000 km farther in the same time, we draw the sketch in Fig. 1-15. Next we use the formula *distance = rate × time* ($d = rt$). This leads to the following equation and solution.

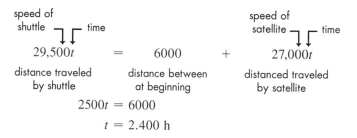

$$2500t = 6000$$
$$t = 2.400 \text{ h}$$

This means that it will take the shuttle 2.400 h to reach the satellite. In 2.400 h the shuttle will travel 70,800 km and the satellite will travel 64,800 km. We see that the solution checks with the statement of the problem.   ■

■**EXAMPLE 5**   A refinery has 7600 L of a gasoline and methanol blend that is 5.00% methanol. How much pure methanol must be added to this blend so that the resulting blend is 10.0% methanol? (All data have three significant digits.)

First let x = the number of liters of methanol to be added. We want the total volume of methanol to be 10.0% of the volume of the final blend, which is the volume in the original blend plus the volume that is added. See Fig. 1-16. This leads to the following equation and solution.

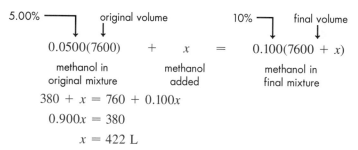

$$380 + x = 760 + 0.100x$$
$$0.900x = 380$$
$$x = 422 \text{ L}$$

Therefore, 422 L of methanol are to be added. The result checks (to three significant digits) since there would be 802 L of methanol of a total final volume of 8020 L.

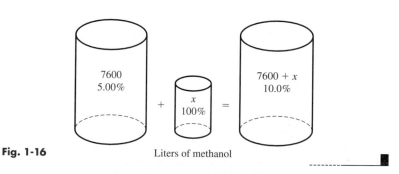

Fig. 1-16   Liters of methanol   ■

EXERCISES *1-12*

Solve the following problems by first setting up an appropriate equation. Assume all data are accurate to two significant digits unless greater accuracy is given.

1. Two computer software programs cost $390 together. If one costs $114 more than the other, what is the cost of each?

2. The flow of one stream into a lake is 1700 ft^3/s more than the flow of a second stream. In one hour 1.98×10^7 ft^3 flow into the lake from the two streams. What is the flow rate of each?

3. Approximately 4.5 million wrecked cars are recycled in two consecutive years. There were 700,000 more recycled the second year than the first year. How many are recycled each year?

4. During a promotion a business found its Web site was accessed as often on the first day as on the second and third days combined. On the first day the number was four times that of the third day, and on the second day it was 4000 more than on the third day. How many times was it accessed each day?

5. A developer purchased 70 acres of land for $900,000. If part cost $20,000 per acre and the remainder cost $10,000 per acre, how many acres did the developer buy at each price?

6. A vial contains 2000 mg, which is to be used for two dosages. One patient is to be administered 660 mg more than another. How much should be administered to each?

7. In the design of a bridge, an engineer determines that four fewer 18-m girders are needed for the span length than 15-m girders. How many 18-m girders are needed?

8. A fuel oil storage depot had an 8-week supply on hand. However, cold weather caused the supply to be used in 6 weeks when 5000 gal extra were used each week. How many gallons were in the original supply?

9. The sum of three electric currents that come together at a point in an integrated circuit is zero. If the second current is double the first, and the third current is 9.2 μA more than the first, what are the currents? (The sign of a current indicates the direction of flow.)

10. A trucking firm uses one fleet of trucks on round-trip delivery routes of 8 h, and a second fleet, with five more trucks than the first, is used on round-trip delivery routes of 6 h. Budget allotments allow 198 h of delivery time in a week. How many trucks are in each fleet?

11. A primary natural gas pipeline feeds into three smaller pipelines, each of which is 2.6 km longer than the main pipeline. If the total length of the four pipelines is 35.4 km, what is the length of each section of the line?

12. Ten 6.0-V and 12-V batteries have a total voltage of 84 V. How many of each type are there?

13. If $8000 is invested at 7.00%, how much must be invested at 9.00% to have total interest of $1550?

14. Two stock investments cost $15,000. One stock then had a 40% gain and the other a 10% loss. If the net profit is $2000, how much was invested in each stock?

15. One lap at the Indianapolis Speedway is 2.50 mi. In a race, a car stalls and then starts 30.0 s after a second car. The first car travels at 260 ft/s, and the second car travels at 240 ft/s. How long does it take the first car to overtake the second, and which car will be ahead after eight laps?

16. The supersonic jet Concorde made a trip averaging 100 mi/h less than the speed of sound for 1.00 h and averaging 400 mi/h more than the speed of sound for 3.00 h. If the trip covered 3980 mi, what is the speed of sound?

17. Trains at each end of the 50.0-km long Eurotunnel under the English Channel start at the same time into the tunnel. Find their speeds if the train from France travels 8.0 km/h faster than the train from England and they pass in 17.0 min. See Fig. 1-17.

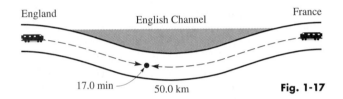

England   English Channel   France

17.0 min   50.0 km   **Fig. 1-17**

18. An executive traveling at 60.0 mi/h would arrive 10.0 min early for an appointment, whereas at a speed of 45.0 mi/h the executive would arrive 5.0 min early. How much time is there until the appointment?

19. A ski lift takes a skier up a slope at 50 m/min. The skier then skis down the slope at 150 m/min. If one round-trip takes 24 min, how long is the slope?

20. A computer chip manufacturer produces two types of chips. In testing a total of 6100 chips of both types, 0.50% of one type and 0.80% of the other type were defective. If a total of 38 defective chips were found, how many of each type were tested?

21. Two gasoline distributors, A and B, are 228 mi apart on Interstate 80. A charges $0.85 per gallon and B charges $0.80 per gallon. Each charges 0.05¢ per gallon per mile for delivery. Where on Interstate 80 is the cost to the customer the same?

22. An outboard engine uses a gasoline-oil fuel mixture in the ratio of 15 to 1. How much gasoline must be mixed with a gasoline-oil mixture, which is 75% gasoline, to make 8.0 L of the mixture for the outboard engine?

23. How many grams of solder that is 15% tin must be mixed with 90 g of solder that is 45% tin in order to have solder that is 25% tin?

24. By weight, a certain roadbed material is 75% crushed rock, and another is 30% crushed rock. How many tons of each must be mixed in order to have 250 tons of material that is 50% crushed rock?

CHAPTER EQUATIONS

Commutative law of addition: $a + b = b + a$

Commutative law of multiplication: $ab = ba$

Associative law of addition: $a + (b + c) = (a + b) + c$

Associative law of multiplication: $a(bc) = (ab)c$

Distributive law: $a(b + c) = ab + ac$

$a + (-b) = a - b$ (1-1)

$a^m \times a^n = a^{m+n}$ (1-3)

$\dfrac{a^m}{a^n} = a^{m-n}$ $(m > n, a \neq 0)$,

$a - (-b) = a + b$ (1-2)

$\dfrac{a^m}{a^n} = \dfrac{1}{a^{n-m}}$ $(m < n, a \neq 0)$ (1-4)

$(a^m)^n = a^{mn}$ (1-5)

$(ab)^n = a^n b^n,$ $\left(\dfrac{a}{b}\right)^n = \dfrac{a^n}{b^n}$ $(b \neq 0)$ (1-6)

$a^0 = 1$ $(a \neq 0)$ (1-7)

$a^{-n} = \dfrac{1}{a^n}$ $(a \neq 0)$ (1-8)

$\sqrt{ab} = \sqrt{a}\sqrt{b}$ (a and b positive real numbers)   (1-9)

REVIEW EXERCISES

In Exercises 1–12, evaluate the given expressions.

1. $(-2) + (-5) - 3$

2. $6 - 8 - (-4)$

3. $\dfrac{(-5)(6)(-4)}{(-2)(3)}$

4. $\dfrac{(-9)(-12)(-4)}{24}$

5. $-5 - |2(-6)| + \dfrac{-15}{3}$

6. $3 - 5(-2) - \dfrac{12}{-4}$

7. $\dfrac{18}{3 - 5} - (-4)^2$

8. $-(-3)^2 - \dfrac{-8}{(-2) - |-4|}$

9. $\sqrt{16} - \sqrt{64}$

10. $-\sqrt{144} + \sqrt{49}$

11. $(\sqrt{7})^2 - \sqrt[3]{8}$

12. $-\sqrt[4]{16} + (\sqrt{6})^2$

In Exercises 13–20, simplify the given expressions. Where appropriate, express results with positive exponents only.

13. $(-2rt^2)^2$

14. $(3a^0 b^{-2})^3$

15. $\dfrac{18m^3 n^4 t}{-3mn^5 t^3}$

16. $\dfrac{15p^4 q^2 r}{5pq^5 r}$

17. $\dfrac{-16s^{-2}(st^2)}{-2st^{-1}}$

18. $\dfrac{-35x^{-1}y(x^2 y)}{5xy^{-1}}$

19. $\sqrt{45}$

20. $\sqrt{9 + 36}$

In Exercises 21–24, for each number (a) determine the number of significant digits, and (b) round off each to two significant digits.

21. 8840

22. 21,450

23. 9.040

24. 0.700

In Exercises 25–28, evaluate the given expressions. All numbers are approximate.

25. $37.3 - 16.92(1.067)^2$

26. $\dfrac{8.896 \times 10^{-12}}{3.5954 + 6.0449}$

27. $\dfrac{\sqrt{0.1958 + 2.844}}{3.142(65)^2}$

28. $\dfrac{1}{0.03568} + \dfrac{37,466}{29.63^2}$

In Exercises 29–60, perform the indicated operations.

29. $a - 3ab - 2a + ab$

30. $xy - y - 5y - 4xy$

31. $6LC - (3 - LC)$

32. $-(2x - b) - 3(-x - 5b)$

33. $(2x - 1)(x + 5)$

34. $(x - 4y)(2x + y)$

35. $(x + 8)^2$

36. $(2x + 3y)^2$

37. $\dfrac{2h^3 k^2 - 6h^4 k^5}{2h^2 k}$

38. $\dfrac{4a^2 x^3 - 8ax^4}{2ax^2}$

39. $4R - [2r - (3R - 4r)]$

40. $3b - [2b + 3a - (2a - 3b)] + 4a$

41. $2xy - \{3z - [5xy - (7z - 6xy)]\}$

42. $x^2 + 3b + [(b - y) - 3(2b - y + z)]$

43. $(2x + 1)(x^2 - x - 3)$

44. $(x - 3)(2x^2 - 3x + 1)$

45. $-3y(x - 4y)^2$

46. $-s(4s - 3t)^2$

47. $3p[(q - p) - 2p(1 - 3q)]$

48. $3x[2y - r - 4(s - 2r)]$

49. $\dfrac{12p^3q^2 - 4p^4q + 6pq^5}{2p^4q}$

50. $\dfrac{27s^3t^2 - 18s^4t + 9s^2t}{9s^2t}$

51. $(2x^2 + 7x - 30) \div (x + 6)$

52. $(4x^2 + 15x - 21) \div (2x + 7)$

53. $\dfrac{3x^3 - 7x^2 + 11x - 3}{3x - 1}$

54. $\dfrac{w^3 - 4w^2 + 7w - 12}{w - 3}$

55. $\dfrac{4x^4 + 10x^3 + 18x - 1}{x + 3}$

56. $\dfrac{8x^3 - 14x + 3}{2x + 3}$

57. $-3\{(r + s - t) - 2[(3r - 2s) - (t - 2s)]\}$

58. $(1 - 2x)(x - 3) - (x + 4)(4 - 3x)$

59. $\dfrac{2y^3 + 9y^2 - 7y + 5}{2y - 1}$

60. $\dfrac{6x^2 + 5xy - 4y^2}{2x - y}$

In Exercises 61–72, solve the given equations.

61. $3x + 1 = x - 8$

62. $4y - 3 = 5y + 7$

63. $\dfrac{5x}{7} = \dfrac{3}{2}$

64. $\dfrac{3}{4} = \dfrac{2A}{9}$

65. $6x - 5 = 3(x - 4)$

66. $-2(-4 - y) = 3y$

67. $2s + 4(3 - s) = 6$

68. $-(4 + v) = 2(2v - 5)$

69. $3t - 2(7 - t) = 5(2t + 1)$

70. $6 - 3x - (8 - x) = x - 2(2 - x)$

71. $2.7 + 2.0(2.1x - 3.4) = 0.1$

72. $0.250(6.721 - 2.44x) = 2.08$

In Exercises 73–80, change numbers in ordinary notation to scientific notation or change numbers in scientific notation to ordinary notation. (See Appendix B for an explanation of the symbols that are used.)

73. The escape velocity (the velocity required to leave the earth's gravitational field) is about 25,000 mi/h.

74. When pictures of the surface of Mars were transmitted to earth from the Pathfinder mission in 1997, Mars was about 192,000,000 km from earth.

75. Police radar has a frequency of 1.02×10^9 Hz.

76. The World Trade Center has nearly 10^7 ft^2 of office space.

77. A biological cell has a surface area of 0.0000012 cm^2.

78. An optical coating on glass to reduce reflections is about 0.00000015 m thick.

79. The maximum safe level of radiation in the air of a home due to radon gas is 1.5×10^{-1} Bq/L. (Bq is the symbol for Bequerel, the metric unit of radioactivity, where 1 Bq = 1 decay/s.)

80. A computer can execute an instruction in 6.3×10^{-10} s.

In Exercises 81–96, solve for the indicated letter. Where noted, the given formula arises in the technical or scientific area of study.

81. $3s + 2 = 5a$, for s

82. $5 - 7t = 6b$, for t

83. $3(4 - x) = 8 + 2n$, for x

84. $6 - 3b = 5(7 + 2v)$, for v

85. $R = n^2Z$, for Z (electricity)

86. $R = \dfrac{2GM}{c^2}$, for G (astronomy: black holes)

87. $I = P + Prt$, for t (business)

88. $V = IR + Ir$, for R (electricity)

89. $m = dV(1 - e)$, for e (solar heating)

90. $mu = (m + M)v$, for M (physics: momentum)

91. $N_1 = T(N_2 - N_3) + N_3$, for N_2 (mechanics, gears)

92. $2(J + 1) = \dfrac{f}{B}$, for J (spectroscopy)

93. $R = \dfrac{A(T_2 - T_1)}{H}$, for T_2 (thermal resistance)

94. $Z^2\left(1 - \dfrac{\lambda}{2a}\right) = k$, for λ (radar design)

95. $d = kx^2[3(a + b) - x]$, for a (mechanics: beams)

96. $V = V_0[1 + 3a(T_2 - T_1)]$, for T_2 (thermal expansion)

In Exercises 97–100, perform the indicated calculations.

97. The CN Tower in Toronto is 0.553 km high. The Sears Tower in Chicago is 443 m high. How much higher is the CN Tower than the Sears Tower?

98. The time (in s) it takes a computer to check n memory cells is found by evaluating $(n/2650)^2$. Find the time to check 48 cells.

99. The combined electric resistance of two parallel resistors is found by evaluating the expression $\dfrac{R_1R_2}{R_1 + R_2}$. Evalute this for $R_1 = 0.0275\ \Omega$ and $R_2 = 0.0590\ \Omega$.

100. The distance (in m) from the earth for which the gravitational force of the earth on a spacecraft equals the gravitational force of the sun on it is found by evaluating $1.5 \times 10^{11}\sqrt{m/M}$, where m and M are the masses of the earth and sun, respectively. Find this distance for $m = 5.98 \times 10^{24}$ kg and $M = 1.99 \times 10^{30}$ kg.

In Exercises 101–104, simplify the given expressions.

101. An analysis of the electric potential of a conductor includes the expression $2V(r - a) - V(b - a)$. Simplify this expression.

102. In finding the value of an annuity, the expression $(Ai - R)(1 + i)^2$ is used. Multiply out this expression.

103. A computer analysis of the velocity of a link in an industrial robot leads to the expression $4(t + h) - 2(t + h)^2$. Simplify this expression.

104. When analyzing the motion of a communications satellite, the expression $\dfrac{k^2r - 2h^2k + h^2rv^2}{k^2r}$ is used. Perform the indicated division.

In Exercises 105–116, solve the given problems. All data are accurate to two significant digits unless more are given.

105. One computer hard disk has 1.5 times the memory of another hard disk. If their combined memory is 10.5 gigabytes, what is the memory of each?

106. Three chemical reactions each produce oxygen. If the first produces twice that of the second, the third produces twice that of the first, and the combined total is 560 cm³, what volume is produced by each?

107. The voltage across a resistor equals the current times the resistance. In a microprocessor circuit, one resistor is 1200 Ω greater than another. The sum of the voltages across them is 12.0 mV. Find the resistances if the current is 2.4 μA in each.

108. A air sample contains 4.0 ppm (parts per million) of two pollutants. The concentration of one is four times the other. What are the concentrations?

109. A 5.0-ft piece of tubing weighs 1.7 lb. What is the weight of a 27-ft piece of the same tubing?

110. The fuel for a two-cycle motorboat engine is a mixture of gasoline and oil in the ratio of 15 to 1. How many liters of each are in 6.6 L of mixture?

111. A ship enters the Red Sea from the Suez Canal, moving south at 17.4 mi/h. Two hours later a second ship enters the Red Sea at the southern end, moving north at 21.8 mi/h. If the Red Sea is 1390 mi long, when will the ships pass?

112. A helicopter used in fighting a forest fire travels at 105 mi/h from the fire to a pond and 70 mi/h with water from the pond to the fire. If a round-trip takes 30 min, how long does it take from the pond to the fire? See Fig. 1-18.

113. One grade of oil has 0.50% of an additive, and a higher grade has 0.75% of the additive. How many liters of each must be used to have 1000 L of a mixture with 0.65% of the additive?

114. Fifty pounds of a cement-sand mixture is 40% sand. How many pounds of sand must be added for the resulting mixture to be 60% sand?

115. An architect plans to have 25% of the floor area of a house in ceramic tile. In all but the kitchen and entry, there are 2200 ft² of floor area, 15% of which is tile. What area can be planned for the kitchen and entry if each has an all-tile floor?

116. A *karat* equals 1/24 part of gold in an alloy (e.g., 9-karat gold is 9/24 gold). How many grams of 9-karat gold must be mixed with 18-karat gold to get 200 g of 14-karat gold?

Writing Exercise

117. A person was asked to explain how to evaluate the expression $\dfrac{1.70}{0.0246(0.0309)}$, where the numbers are approximate. The explanation was: "Divide 1.70 by 0.0246, and then multiply by 0.0309. The calculator display is the answer." Write a short paragraph stating what is wrong with the explanation. (What is the correct result?)

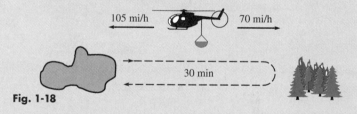

105 mi/h   70 mi/h

30 min

Fig. 1-18

PRACTICE TEST

In Problems 1–5, evaluate the given expressions. In Problems 3 and 5, the numbers are approximate.

1. $\sqrt{9 + 16}$ **2.** $\dfrac{(7)(-3)(-2)}{(-6)(0)}$ **3.** $\dfrac{3.372 \times 10^{-3}}{7.526 \times 10^{12}}$

4. $\dfrac{(+6)(-2) - 3(-1)}{5 - 2}$ **5.** $\dfrac{346.4 - 23.5}{287.7} - \dfrac{0.944^3}{(3.46)(0.109)}$

In Problems 6–12, perform the indicated operations and simplify. When exponents are used, use only positive exponents in the result.

6. $(2a^0 b^{-2} c^3)^{-3}$ **7.** $(2x + 3)^2$ **8.** $3m^2(am - 2m^3)$

9. $\dfrac{8a^3 x^2 - 4a^2 x^4}{-2ax^2}$ **10.** $\dfrac{6x^2 - 13x + 7}{2x - 1}$

11. $(2x - 3)(x + 7)$ **12.** $3x - [4x - (3 - 2x)]$

13. Solve for y: $5y - 2(y - 4) = 7$

14. Solve for x: $3(x - 3) = x - (2 - 3d)$

15. Express 0.0000036 in scientific notation.

16. List the numbers -3, $|-4|$, $-\pi$, $\sqrt{2}$, and 0.3 in numerical order.

17. What fundamental law is illustrated by $3(5 + 8) = 3(5) + 3(8)$?

18. (a) How many significant digits are in the number 3.0450? (b) Round it off to two significant digits.

19. If P dollars is deposited in a bank that compounds interest n times a year, the value of the account after t years is found by evaluating $P(1 + i/n)^{nt}$, where i is the annual interest rate. Find the value of an account for which $P = \$1000$, $i = 5\%$, $n = 2$, and $t = 3$ years (values are exact).

20. In finding the illuminance from a light source, the expression $8(100 - x)^2 + x^2$ is used. Simplify this expression.

21. The equation $L = L_0[1 + \alpha(t_2 - t_1)]$ is used when studying thermal expansion. Solve for t_2.

22. An alloy weighing 20 lb is 30% copper. How many pounds of another alloy, which is 80% copper, must be added for the final alloy to be 60% copper?

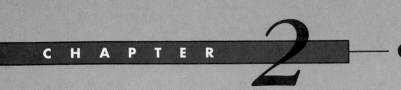
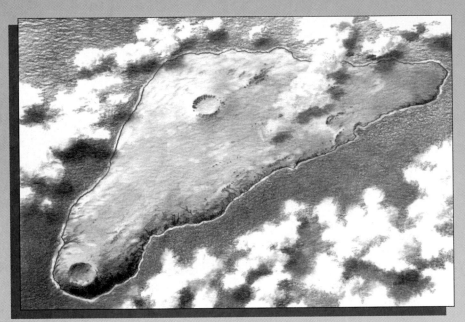

In Section 2-5 we see how to find an excellent approximation of the area of an irregular geometric figure, such as an island. Above is a view of Easter Island, a remote Pacific island, noted for its mysterious stone statues.

Many applied problems in technology and science involve the shape and size of objects. Since geometry deals with shape and size, the topics and methods of geometry are important in many areas of technology. These include architecture, construction, instrumentation, surveying, mechanical design, product design of all possible types, surveying and civil engineering, as well as many other areas of engineering.

The study of geometry includes the properties and measurements of angles, lines, and surfaces and the basic figures they form. In this chapter we review the more important methods and formulas for calculating the important geometric measures, such as area and volume. Numerous technical applications are shown.

Geometric figures and concepts are also basic to the development of many other areas of mathematics, such as graphing and trigonometry. We will start our study of graphs in Chapter 3 and trigonometry in Chapter 4.

2-1 LINES AND ANGLES

In establishing the properties and measures of the basic geometric figures, it is not possible to define every geometric term and prove every statement that we use. There are certain words and concepts that must be used without definition. In general, *the meanings of the geometric terms* **point, line,** *and* **plane** *are accepted without actually being defined.* These terms give us a starting point for definitions of other useful geometric terms.

*The amount of rotation of a **ray** (or **half-line**) about its endpoint is called an* **angle.** A ray is that part of a line (the word *line* means *straight line*) to one side of a fixed point on the line. *The fixed point is the* **vertex** *of the angle. One complete rotation of a ray is an angle with a measure of 360* **degrees,** *written as 360°.* Some special types of angles are as follows:

Name of angle	Measure of angle
Right angle	*90°*
Straight angle	*180°*
Acute angle	*Between 0° and 90°*
Obtuse angle	*Between 90° and 180°*

The use of 360 comes from the ancient Babylonians and their number system based on 60 rather than 10, as we use today. However, the specific reason for the choice of 360 is not known. (One theory is that 360 is divisible by many smaller numbers and is close to the number of days in a year.)

■**EXAMPLE 1**   Figure 2-1(a) shows a right angle (marked as ⌐). The vertex of the angle is point *B*, and the ray is the half-line *BA*. Figure 2-1(b) shows a straight angle. Figure 2-1(c) shows an acute angle, denoted as ∠*E* (or ∠*DEF* or ∠*FED*). In Fig. 2-1(d), ∠*G* is an obtuse angle.

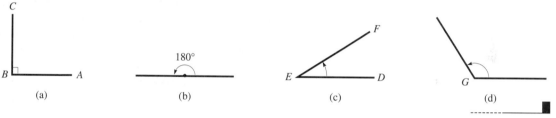

Fig. 2-1   (a)   (b)   (c)   (d)

If two lines intersect such that the angle between them is a right angle, the lines are **perpendicular.** *Lines in the same plane that do not intersect are* **parallel.** These are illustrated in the following example.

■**EXAMPLE 2**   In Fig. 2-2(a), lines *AC* and *DE* are perpendicular (which is shown as $AC \perp DE$) since they meet in a right angle (again, shown as ⌐) at *B*. In Fig. 2-2(b), lines *AB* and *CD* are drawn so that they do not meet, even if extended. Therefore, these lines are parallel (which can be shown as $AB \parallel CD$).

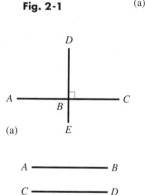

(a)

(b)

Fig. 2-2

If the sum of the measures of two angles is 180°, then the angles are called **supplementary angles.** *Each angle is the* **supplement** *of the other. If the sum of the measures of two angles is 90°, the angles are called* **complementary angles.** *Each is the* **complement** *of the other.*

■**EXAMPLE 3**   (a) In Fig. 2-3(a), ∠*BAC* = 55°, and in Fig. 2-3(b), ∠*DEF* = 125°. Since 55° + 125° = 180°, ∠*BAC* and ∠*DEF* are supplementary angles.

(b) In Fig. 2-4, we see that ∠*POQ* is a right angle, or ∠*POQ* = 90°. Since ∠*POR* + ∠*ROQ* = ∠*POQ* = 90°, ∠*POR* is the complement of ∠*ROQ* (or ∠*ROQ* is the complement of ∠*POR*).

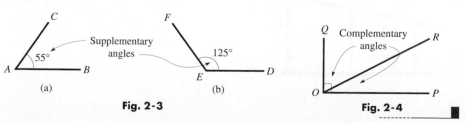

(a)   (b)   **Fig. 2-3**   **Fig. 2-4**

In many geometric figures, it is necessary to refer to certain specific pairs of angles. *Two angles that have a common vertex and a side common between them are known as* **adjacent angles.** *If two lines cross to form equal angles on opposite sides of the point of intersection, which is the common vertex, these equal angles are called* **vertical angles.** These are illustrated in the following example.

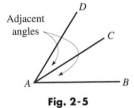

Adjacent angles

Fig. 2-5

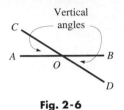

Vertical angles

Fig. 2-6

EXAMPLE 4   (a) In Fig. 2-5, $\angle BAC$ and $\angle CAD$ have a common vertex at A and the common side AC between them so that $\angle BAC$ and $\angle CAD$ are adjacent angles.

(b) In Fig. 2-6, lines AB and CD intersect at point O. Here $\angle AOC$ and $\angle BOD$ are vertical angles, and they are equal. Also, $\angle BOC$ and $\angle AOD$ are vertical angles and are equal.

We should also be able to identify *the sides of an angle which are adjacent to the angle.* In Fig. 2-5, sides AB and AC are adjacent to $\angle BAC$, and in Fig. 2-6, sides OB and OD are adjacent to angle $\angle BOD$. Identifying sides adjacent and opposite an angle in a triangle is important in trigonometry.

In a plane, *if a line crosses two or more parallel or nonparallel lines, it is called a* **transversal.** In Fig. 2-7, $AB \parallel CD$, and the transversal of these two parallel lines is the line EF.

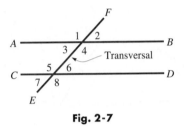

Fig. 2-7

When a transversal crosses a pair of parallel lines, certain pairs of equal angles result. In Fig. 2-7, the **corresponding angles** are equal (that is, $\angle 1 = \angle 5$, $\angle 2 = \angle 6$, $\angle 3 = \angle 7$, and $\angle 4 = \angle 8$). Also, the **alternate-interior angles** are equal ($\angle 3 = \angle 6$ and $\angle 4 = \angle 5$) and the **alternate-exterior angles** are equal ($\angle 1 = \angle 8$ and $\angle 2 = \angle 7$).

NOTE ▸

When working with geometric figures, it is important that you *readily identify angles of equal measure* such as vertical angles and alternate-interior angles. Proper identification of these angles is often the key to finding the other sides and angles of a geometric figure.

When more than two parallel lines are crossed by *two* transversals, such as is shown in Fig. 2-8, *the segments of the transversals between the same two parallel lines are called* **corresponding segments.** A useful theorem is that *the ratios of corresponding segments of the transversals are equal.* In Fig. 2-8, this means that

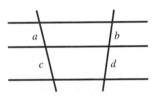

Fig. 2-8

$$\frac{a}{b} = \frac{c}{d} \qquad\qquad \textbf{(2-1)}$$

EXAMPLE 5   In Fig. 2-9, part of the beam structure within a building is shown. The vertical beams are parallel. From the distances between beams that are shown, determine the distance x between the middle and right vertical beams.

Using Eq. (2-1), we have

$$\frac{6.50}{5.65} = \frac{x}{7.75}$$

$$x = \frac{6.50(7.75)}{5.65}$$

$$= 8.92 \text{ ft} \qquad \text{rounded off}$$

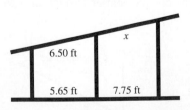

Fig. 2-9

EXERCISES *2-1*

In Exercises 1–8, identify the indicated angles and sides in Fig. 2-10. In Exercises 5 and 6, also evaluate the indicated angles.

1. Two acute angles
2. Two right angles
3. The straight angle
4. The obtuse angle
5. If $\angle CBD = 65°$, find its complement.
6. If $\angle CBD = 65°$, find its supplement.
7. The sides adjacent to $\angle DBC$
8. The angle adjacent to $\angle DBC$

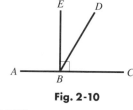

Fig. 2-10

In Exercises 9 and 10, use Fig. 2-11. In Exercises 11 and 12, use Fig. 2-12. Determine the indicated angles.

9. $\angle AOB$ 10. $\angle AOC$ 11. $\angle 3$ 12. $\angle 4$

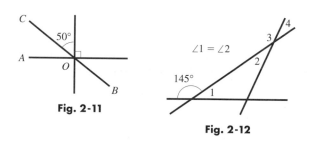

Fig. 2-11

Fig. 2-12

In Exercises 13–16, use Fig. 2-13. In Exercises 17–20, use Fig. 2-14. Determine the indicated angles.

13. $\angle 1$ 14. $\angle 2$ 15. $\angle 3$ 16. $\angle 4$
17. $\angle FCE$ 18. $\angle ECD$ 19. $\angle BCE$ 20. $\angle BFC$

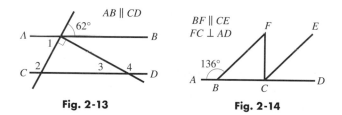

Fig. 2-13 **Fig. 2-14**

In Exercises 21–24, determine the indicated segments in Fig. 2-15.

21. a 22. b 23. c 24. d

In Exercises 25–28, solve the given problems.

25. A steam pipe is connected in sections AB, BC, and CD, as shown in Fig. 2-16. What is the angle between sections BC and CD if $AB \parallel CD$?

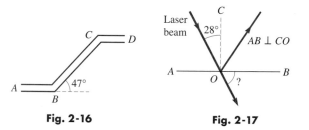

Fig. 2-16 **Fig. 2-17**

26. A laser beam striking a surface is partly reflected, and the remainder of the beam passes straight through the surface, as shown in Fig. 2-17. Find the angle between the surface and the part that passes through.

27. Find the distance on Dundas St. W between Dufferin St. and Ossington Ave. in Toronto, as shown in Fig. 2-18. The north–south streets are parallel.

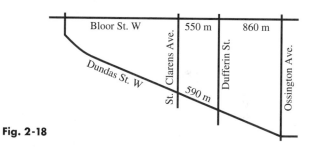

Fig. 2-18

28. An electric circuit board has equally spaced parallel wires with connections at points A, B, and C, as shown in Fig. 2-19. How far is A from C, if $BC = 2.15$ cm?

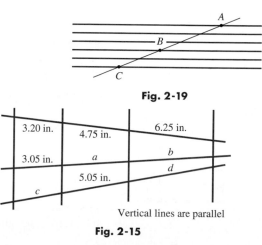

Fig. 2-19

Fig. 2-15

2-2 TRIANGLES

When part of a plane is bounded and closed by straight-line segments, it is called a **polygon.** In general, polygons are named according to the number of sides they have. *A* **triangle** *has three sides, a* **quadrilateral** *has four sides, a* **pentagon** *has five sides, a* **hexagon** *has six sides, and so on.* The polygons of greatest importance are the triangle and the quadrilateral. Therefore, in this section we review the properties of triangles, and in the following section we will consider the quadrilateral. Many of the properties of triangles are important in the study of trigonometry, which we will start in Chapter 4.

Types and Properties of Triangles

There are several important types of triangles. In a **scalene triangle,** no two sides are equal in length. In an **isosceles triangle,** two of the sides are equal in length, and the two *base angles* (the angles opposite the equal sides) are equal. In an **equilateral triangle,** the three sides are equal in length, and each of the three angles is 60°.

One of the most important triangles in scientific and technical applications is the **right triangle.** *In a right triangle, one of the angles is a right angle. The side opposite the right angle is called the* **hypotenuse,** *and the other two sides are called* **legs.** Each of these triangles is illustrated in the following example.

∎EXAMPLE 1   Figure 2-20(a) shows a scalene triangle. We see that each side is of a different length. Figure 2-20(b) shows an isosceles triangle with two equal sides of 2 in. and equal base angles of 40°. Figure 2-20(c) shows an equilateral triangle, each side of which is 5 cm and each interior angle of which is 60°. Figure 2-20(d) shows a right triangle. The hypotenuse is side *AB*.

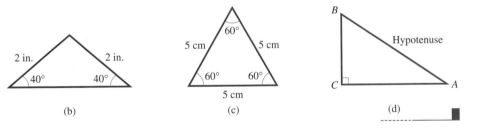

Fig. 2-20   (a)   (b)   (c)   (d)

One very important property of a triangle is that

the sum of the measures of the three angles of a triangle is 180°.

In the next example, we show this property by using material from Section 2-1.

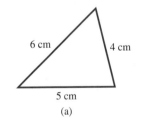

Fig. 2-21

∎EXAMPLE 2   In Fig. 2-21, since $\angle 1$, $\angle 2$, and $\angle 3$ constitute a straight angle,

$$\angle 1 + \angle 2 + \angle 3 = 180°$$

Also, by noting alternate interior angles, we see that $\angle 1 = \angle 4$ and $\angle 3 = \angle 5$. Therefore, by substitution we have

$$\angle 4 + \angle 2 + \angle 5 = 180°$$

Therefore, if two of the angles of a triangle are known, the third may be found by subtracting the sum of the first two from 180°.

SOLVING A WORD PROBLEM

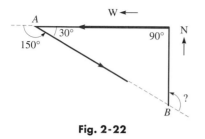

Fig. 2-22

█EXAMPLE 3   An airplane is flying north and then makes a 90° turn to the west. Later it makes another left turn of 150°. What is the angle of a third left turn that will cause the plane to again fly north? See Fig. 2-22.

From Fig. 2-22, we see that the interior angle of the triangle at *A* is the supplement of 150°, or 30°. Since the sum of the measures of the interior angles of the triangle is 180°, the interior angle at *B* is

$$\angle B = 180° - (90° + 30°) = 60°$$

The required angle is the supplement of 60°, which is 120°.   ---------█

A line segment drawn from a vertex of a triangle to the *midpoint* of the opposite side is called a **median** of the triangle. A basic property of a triangle is that *the three medians meet at a single point, called the* **centroid** *of the triangle.* See Fig. 2-23. Also, *the three* **angle bisectors** (lines from the vertices that divide the angles in half) *meet at a common point.* See Fig. 2-24.

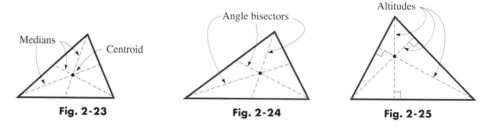

Fig. 2-23   **Fig. 2-24**   **Fig. 2-25**

An **altitude** *(or* **height***) of a triangle is the line segment drawn from a vertex perpendicular to the opposite side (or its extension), which is called the* **base** *of the triangle. The three altitudes of a triangle meet at a common point.* See Fig. 2-25. The three common points of the medians, angle bisectors, and altitudes are generally not the same point for a given triangle.

Perimeter and Area of a Triangle

We now consider two of the most basic measures of a plane geometric figure. The first of these is its **perimeter,** *which is the total distance around it.* In the following example we find the perimeter of a triangle.

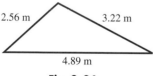

Fig. 2-26

█EXAMPLE 4   Find the perimeter *p* of a triangle with sides 2.56 m, 3.22 m, and 4.89 m. See Fig. 2-26.

Using the definition of perimeter, for this triangle we have

$$p = 2.56 + 3.22 + 4.89 = 10.67 \text{ m}$$

Therefore, the distance around the triangle is 10.67 m. We express the results to hundredths since each side is given to hundredths.   ---------█

The second important measure of a geometric figure is its **area.** Although the concept of area is primarily intuitive, it is easily defined and calculated for the basic geometric figures. *Area gives a measure of the surface of the figure,* just as perimeter gives the measure of the distance around it. The formula for the area of a triangle is given on the next page.

The area A of a triangle of base b and altitude h is

$$A = \tfrac{1}{2}bh \qquad\qquad \text{(2-2)}$$

The following example illustrates the use of Eq. (2-2).

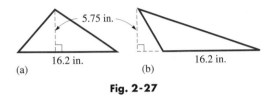

(a)   16.2 in.   (b)   16.2 in.

5.75 in.

Fig. 2-27

■**EXAMPLE 5**   Find the areas of the triangles in Fig. 2-27(a) and Fig. 2-27(b).

Even though the triangles are of different shapes, we see that the base b of each triangle is 16.2 in. and that the altitude h of each is 5.75 in. Therefore, the area of each triangle is

$$A = \tfrac{1}{2}bh = \tfrac{1}{2}(16.2)(5.75) = 46.6 \text{ in.}^2 \quad\text{---------}■$$

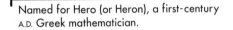

Named for Hero (or Heron), a first-century A.D. Greek mathematician.

Another formula for the area of a triangle that is particularly useful when we have *a triangle with three known sides and no right angle* is **Hero's formula,** which is given in Eq. (2-3).

$$A = \sqrt{s(s-a)(s-b)(s-c)}, \qquad\qquad \text{(2-3)}$$
$$\text{where } s = \tfrac{1}{2}(a+b+c)$$

In Eq. (2-3), a, b, and c are the lengths of the sides and s is one-half of the perimeter.

■**EXAMPLE 6**   A surveyor measures the three sides of a triangular parcel of land between two intersecting straight roads to be 206 ft, 293 ft, and 187 ft, as shown in Fig. 2-28. Find the area of this parcel.

In order to use Eq. (2-3), we first find s.

$$s = \tfrac{1}{2}(206 + 293 + 187) = \tfrac{1}{2}(686) = 343 \text{ ft}$$

Now, substituting in Eq. (2-3), we have

$$A = \sqrt{343(343 - 206)(343 - 293)(343 - 187)} = 19{,}100 \text{ ft}^2$$

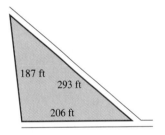

187 ft
293 ft
206 ft

Fig. 2-28

The result has been rounded off to three significant digits. In using a calculator, the value of s is stored in memory and then used to find A. It is not necessary to write down anything except the final result.   ---------■

The Pythagorean Theorem

As we have noted, one of the most important geometric figures in technical applications is the right triangle. A very important property of a right triangle is given by the **Pythagorean theorem,** which states that

*in a **right triangle,** the square of the length of the hypotenuse equals the sum of the squares of the lengths of the other two sides.*

If c is the length of the hypotenuse and a and b are the lengths of the other two sides, the Pythagorean theorem is

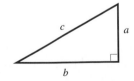

Named for the Greek mathematician Pythagoras (sixth-century B.C.).

c a

b

Fig. 2-29

$$c^2 = a^2 + b^2 \qquad\qquad \text{(2-4)}$$

See Fig. 2-29.

SOLVING A WORD PROBLEM

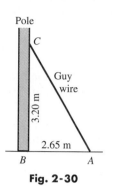

Pole

Fig. 2-30

Calculators can be programmed to perform specific calculations. See Appendix C for a graphing calculator program PYTHAGTH. It can be used to find a side of a right triangle, given the other two sides.

EXAMPLE 7 A pole is perpendicular to the level ground around it. A guy wire is attached 3.20 m up the pole and at a point on the ground, 2.65 m from the pole. How long is the guy wire?

From the given information, we sketch the pole and guy wire as shown in Fig. 2-30. Using the Pythagorean theorem and then substituting, we have

$$AC^2 = AB^2 + BC^2$$
$$= 2.65^2 + 3.20^2$$
$$AC = \sqrt{2.65^2 + 3.20^2} = 4.15 \text{ m}$$

The guy wire is 4.15 m long. In using the calculator, parentheses are used to group $2.65^2 + 3.20^2$.

Similar Triangles

The perimeter and area of a triangle are measures of its *size*. We now consider the shape of triangles.

Two triangles are **similar** *if they have the same shape (but not necessarily the same size).* There are two very important properties of similar triangles.

Properties of Similar Triangles

1. *The corresponding angles of similar triangles are equal.*
2. *The corresponding sides of similar triangles are proportional.*

NOTE ▸

If it can be shown that either of these properties is true for two triangles, we may then conclude that the triangles are similar. This means that *if one property is true, then the other is also true*. In two triangles that are similar, *the* **corresponding sides** *are the sides, one in each triangle, which are between the same pair of equal corresponding angles.* These properties are illustrated in the following example.

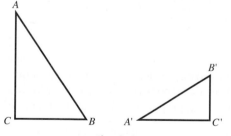

Fig. 2-31

EXAMPLE 8 In Fig. 2-31, a pair of similar triangles are shown. They are similar even though the corresponding parts are not in the same position relative to the page. Using standard symbols, we can write $\triangle ABC \sim \triangle A'B'C'$, where $\triangle$ means "triangle" and $\sim$ means "is similar to."

The pairs of corresponding angles are A and A', B and B', and C and C'. This means $A = A'$, $B = B'$, and $C = C'$.

The pairs of corresponding sides are AB and $A'B'$, BC and $B'C'$, and AC and $A'C'$. In order to show that these corresponding sides are proportional, we write

$$\frac{AB}{A'B'} = \frac{BC}{B'C'} = \frac{AC}{A'C'} \longleftarrow \text{ sides of } \triangle ABC$$
$$\longleftarrow \text{ sides of } \triangle A'B'C'$$

If we know that two triangles are similar, we can use the two basic properties of similar triangles to find the unknown parts of one triangle from the known parts of the other triangle. The following example illustrates how this is done in a practical application.

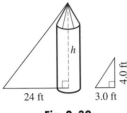

Fig. 2-32

■**EXAMPLE 9**   On level ground a silo casts a shadow 24 ft long. At the same time, a pole 4.0 ft high casts a shadow 3.0 ft long. How tall is the silo? See Fig. 2-32.

The rays of the sun are essentially parallel. The two triangles in Fig. 2-32 are similar since *each has a right angle and the angles at the tops are equal.* The other angles must be equal since the sum of the angles is 180°. The lengths of the hypotenuses are of no importance in this problem, so we use only the other sides in stating the ratios of corresponding sides. Denoting the height of the silo as h, we have

$$\frac{h}{4.0} = \frac{24}{3.0}, \qquad h = 32 \text{ ft}$$

We conclude that the silo is 32 ft high.

One of the most practical uses of similar geometric figures is that of **scale drawings**. Maps, charts, blueprints, and most drawings that appear in books are familiar examples of scale drawings. Actually, there have been many scale drawings used in this book already.

In any scale drawing, all distances are drawn a certain ratio of the distances they represent, and all angles equal the angles they represent. Consider the following example.

■**EXAMPLE 10**   In drawing a map of the area shown in Fig. 2-33, a scale of 1 cm = 200 km is used. In measuring the distance between Chicago and Toronto on the map, we find it to be 3.5 cm. The actual distance x between Chicago and Toronto is found from the proportion

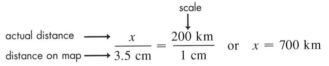

If we did not have the scale but knew that the distance between Chicago and Toronto is 700 km, then by measuring distances on the map between Chicago and Toronto (3.5 cm) and between Toronto and Philadelphia (2.7 cm), we could find the distance between Toronto and Philadelphia. It is found from the following proportion, determined by use of similar triangles:

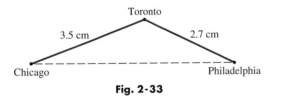

Fig. 2-33

$$\frac{700 \text{ km}}{3.5 \text{ cm}} = \frac{y}{2.7 \text{ cm}}$$

$$y = \frac{2.7(700)}{3.5} = 540 \text{ km}$$

Similarity requires *equal* angles and *proportional* sides. *If the corresponding angles and the corresponding sides of two triangles are equal, the two triangles are* **congruent.** As a result of this definition, the areas and perimeters of congruent triangles are also equal. Informally, we can say that similar triangles have the same shape, whereas congruent triangles have the same shape and same size.

EXAMPLE 11 A right triangle with legs of 2 in. and 4 in. is congruent to any other right triangle with legs of 2 in. and 4 in. However, it is similar to any right triangle with legs of 5 in. and 10 in., since the corresponding sides are proportional. See Fig. 2-34.

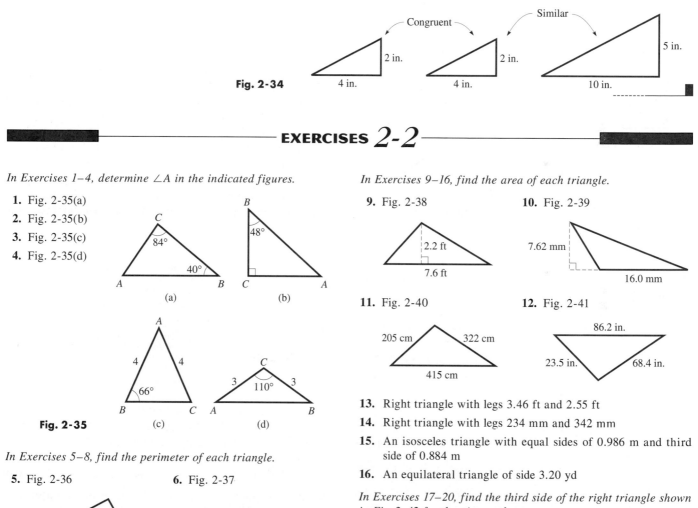

Fig. 2-34

EXERCISES *2-2*

In Exercises 1–4, determine ∠A in the indicated figures.

1. Fig. 2-35(a)
2. Fig. 2-35(b)
3. Fig. 2-35(c)
4. Fig. 2-35(d)

Fig. 2-35

In Exercises 5–8, find the perimeter of each triangle.

5. Fig. 2-36   **6.** Fig. 2-37

7. An equilateral triangle of side 21.5 cm

8. An isosceles triangle with equal sides of 2.45 in., and third side of 3.22 in.

In Exercises 9–16, find the area of each triangle.

9. Fig. 2-38   **10.** Fig. 2-39

11. Fig. 2-40   **12.** Fig. 2-41

13. Right triangle with legs 3.46 ft and 2.55 ft

14. Right triangle with legs 234 mm and 342 mm

15. An isosceles triangle with equal sides of 0.986 m and third side of 0.884 m

16. An equilateral triangle of side 3.20 yd

In Exercises 17–20, find the third side of the right triangle shown in Fig. 2-42 for the given values.

17. $a = 13.8$ ft, $b = 22.7$ ft
18. $a = 2.48$ m, $b = 1.45$ m
19. $a = 17.5$ cm, $c = 55.1$ cm
20. $b = 0.474$ in., $c = 0.836$ in.

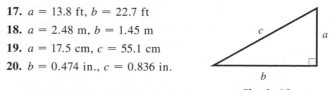

Fig. 2-42

In Exercises 21–24, use the right triangle in Fig. 2-43.

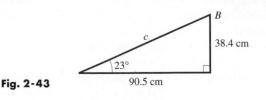

Fig. 2-43

21. Find $\angle B$.

22. Find side c.

23. Find the perimeter

24. Find the area.

In Exercises 25–40, solve the given problems.

25. In Fig. 2-44, show that $\triangle MKL \sim \triangle MNO$.

26. In Fig. 2-45, show that $\triangle ACB \sim \triangle ADC$.

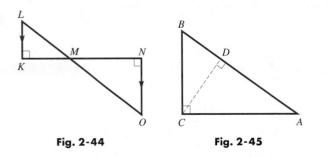

Fig. 2-44   **Fig. 2-45**

27. In Fig. 2-44, if $KN = 15$, $MN = 9$, and $MO = 12$, find LM.

28. In Fig. 2-45, if $AD = 9$ and $AC = 12$, find AB.

29. A tooth on a saw is in the shape of an isosceles triangle. If the angle at the point is 38°, find the two base angles.

30. A transmitting tower is supported by a wire that makes an angle of 52° with the level ground. What is the angle between the tower and the wire?

31. Find the area of a triangular patio with sides of 17.5 ft, 18.7 ft, and 19.5 ft.

32. The Bermuda Triangle is sometimes defined as an equilateral triangle 1600 km on a side, with vertices in Bermuda, Puerto Rico, and the Florida coast. Assuming it is flat, what is its approximate area?

33. The sail of a sailboat is in the shape of a right triangle with sides of 8.0 ft, 15 ft, and 17 ft. What is the area of the sail?

34. Three straight city streets enclose a right-triangular city block. If the shorter sides are 260 m and 320 m, find (a) the area of the block and (b) the distance around the block.

35. An observer is 550 m from the launch pad of a rocket. After the rocket has ascended 750 m, how far is it from the observer?

36. The base of a 20.0-ft ladder is 8.0 ft from a wall. How far up on the wall does the ladder reach?

37. A rectangular room is 18 ft long, 12 ft wide, and 8.0 ft high. What is the length of the longest diagonal from one corner to another corner of the room?

38. On a blueprint, a hallway is 45.6 cm long. The scale is 1.2 cm = 1.0 m. How long is the hallway?

39. Two parallel guy wires are attached to a vertical pole 4.5 m and 5.4 m above the ground. They are secured on the level ground at points 1.2 m apart. How long are the guy wires?

40. To find the width ED of a river, a surveyor places markers at A, B, C, and D, as shown in Fig. 2-46. The markers are placed such that $AB \parallel ED$, $BC = 50.0$ ft, $DC = 312$ ft, and $AB = 80.0$ ft. How wide is the river?

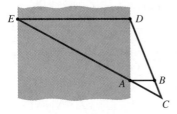

Fig. 2-46

2-3 QUADRILATERALS

In this section we consider another important type of polygon. *A **quadrilateral** is a closed plane figure that has four sides,* and these four sides form four interior angles. A general quadrilateral is shown in Fig. 2-47.

A **diagonal** *of a polygon is a straight line segment joining any two nonadjacent vertices.* The dashed line is one of two possible diagonals of the quadrilateral shown in Fig. 2-48.

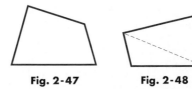

Fig. 2-47   **Fig. 2-48**

Types of Quadrilaterals

A **parallelogram** *is a quadrilateral in which opposite sides are parallel.* In a parallelogram, opposite sides are equal and opposite angles are equal. *A* **rhombus** *is a parallelogram with four equal sides.*

A **rectangle** *is a parallelogram in which intersecting sides are perpendicular,* which means that all four interior angles are right angles. In a rectangle, the longer side is usually called the **length** and the shorter side is called the **width.** *A* **square** *is a rectangle with four equal sides.*

A **trapezoid** *is a quadrilateral in which two sides are parallel.* The parallel sides are called the **bases** of the trapezoid.

■EXAMPLE 1 A parallelogram is shown in Fig. 2-49(a). Opposite sides a are equal in length, as are opposite sides b. A rhombus with equal sides s is shown in Fig. 2-49(b). A rectangle is shown in Fig. 2-49(c). The length is labeled l, and the width is labeled w. A square with equal sides s is shown in Fig. 2-49(d). A trapezoid with bases b_1 and b_2 is shown in Fig. 2-49(e).

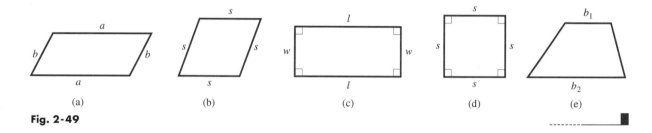

(a)   (b)   (c)   (d)   (e)

Fig. 2-49

Perimeter and Area of a Quadrilateral

Since the perimeter of a polygon is the distance around it, *the perimeter of a quadrilateral is the sum of the lengths of its four sides.* Consider the next example.

SOLVING A WORD PROBLEM

■EXAMPLE 2 An architect designs a room with a rectangular window 36 in. high and 21 in. wide, with another window above in the shape of an equilateral triangle, 21 in. on a side. See Fig. 2-50. How much molding is needed for these windows?

The length of molding is the sum of the perimeters of the windows. For the rectangular window, the opposite sides are equal, which means the perimeter is twice the length l plus twice the width w. For the equilateral triangle, the perimeter is three times the side s. Therefore, the length L of molding is

$$L = 2l + 2w + 3s$$
$$= 2(36) + 2(21) + 3(21)$$
$$= 177 \text{ in.}$$

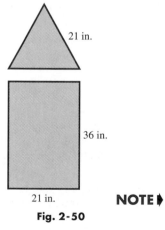

21 in.

36 in.

21 in.   **NOTE▶**

Fig. 2-50

We could write down formulas for the perimeters of the different kinds of triangles and quadrilaterals. However, if we *remember the meaning of perimeter as being the total distance around a geometric figure,* such formulas are not necessary.

For the areas of the square, rectangle, parallelogram, and trapezoid, we have the following formulas.

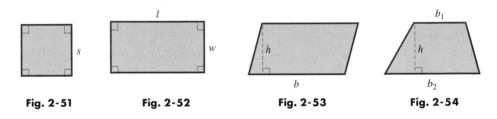

$A = s^2$	Square of side s (Fig. 2-51)	**(2-5)**
$A = lw$	Rectangle of length l and width w (Fig. 2-52)	**(2-6)**
$A = bh$	Parallelogram of base b and height h (Fig. 2-53)	**(2-7)**
$A = \frac{1}{2}h(b_1 + b_2)$	Trapezoid of bases b_1 and b_2, and height h (Fig. 2-54)	**(2-8)**

Fig. 2-51   **Fig. 2-52**   **Fig. 2-53**   **Fig. 2-54**

Since a rectangle, a square, and a rhombus are special types of parallelograms, the area of these figures can be found from Eq. (2-7). The area of a trapezoid is of importance when we find areas of irregular geometric figures in Section 2-5.

■**EXAMPLE 3**   A city park is designed with lawn areas in the shape of a right triangle, a parallelogram, and a trapezoid, as shown in Fig. 2-55, with walkways between them. Find the area of each section of lawn and the total lawn area.

$$A_1 = \tfrac{1}{2}bh = \tfrac{1}{2}(72)(45) = 1600 \text{ ft}^2 \qquad A_2 = bh = (72)(45) = 3200 \text{ ft}^2$$
$$A_3 = \tfrac{1}{2}h(b_1 + b_2) = \tfrac{1}{2}(45)(72 + 35) = 2400 \text{ ft}^2$$

The total lawn area is about 7200 ft^2.

Following is an example of another word problem involving a quadrilateral. In this example it is necessary to follow the procedure on page 41 in order to first set up the equation which leads to the solution.

■**EXAMPLE 4**   The length of a rectangular computer chip is 2.0 mm longer than its width. Find the dimensions of the chip if its perimeter is 26.4 mm.

Since the dimensions, the length and the width, are required, we let w = the width of the chip. Since the length is 2.0 mm more than the width, we know that $w + 2.0$ = the length of the chip. See Fig. 2-56.

Since the perimeter of a rectangle is twice the length plus twice the width, we have the equation

$$2(w + 2.0) + 2w = 26.4$$

since the perimeter is given as 26.4 mm. This is the equation we need.

Solving this equation, we have

$$2w + 4.0 + 2w = 26.4$$
$$4w = 22.4$$
$$w = 5.6 \text{ mm} \quad \text{and} \quad w + 2.0 = 7.6 \text{ mm}$$

Therefore, the length is 7.6 mm and the width is 5.6 mm. These values check with the statements of the original problem.

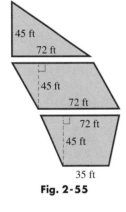

45 ft

72 ft

45 ft

72 ft

72 ft

45 ft

35 ft

Fig. 2-55

SOLVING A WORD PROBLEM

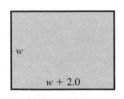

w

$w + 2.0$

Fig. 2-56

EXERCISES *2-3*

In Exercises 1–8, find the perimeter of each figure.

1. Square: side of 0.65 m
2. Rhombus: side of 2.46 ft
3. Rectangle: $l = 46.5$ in., $w = 37.4$ in.
4. Rectangle: $l = 14.2$ cm, $w = 12.6$ cm
5. The parallelogram in Fig. 2-57
6. The parallelogram in Fig. 2-58
7. The trapezoid in Fig. 2-59
8. The trapezoid in Fig. 2-60

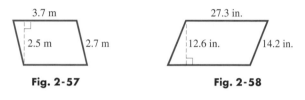

Fig. 2-57   **Fig. 2-58**

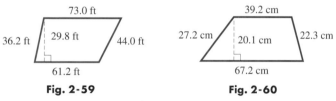

Fig. 2-59   **Fig. 2-60**

In Exercises 9–16, find the area of each figure.

9. Square: $s = 2.7$ mm   10. Square: $s = 15.6$ ft
11. Rectangle: $l = 46.5$ in., $w = 37.4$ in.
12. Rectangle: $l = 14.2$ cm, $w = 12.6$ cm
13. The parallelogram in Fig. 2-57
14. The parallelogram in Fig. 2-58
15. The trapezoid in Fig. 2-59
16. The trapezoid in Fig. 2-60

In Exercises 17–20, set up a formula for the indicated perimeter or area. (Do not include dashed lines.)

17. The perimeter of the figure in Fig. 2-61 (a parallelogram and a square attached)
18. The perimeter of the figure in Fig. 2-62 (two trapezoids attached)
19. The area of the figure in Fig. 2-61
20. The area of the figure in Fig. 2-62

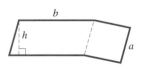

Fig. 2-61   **Fig. 2-62**

In Exercises 21–28, solve the given problems.

21. A machine part is in the shape of a square with equilateral triangles attached to two sides (see Fig. 2-63). Find the perimeter of the machine part.

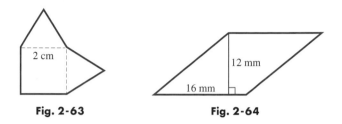

Fig. 2-63   **Fig. 2-64**

22. Part of an electric circuit is wired in the configuration of a rhombus and one of its altitudes as shown in Fig. 2-64. What is the length of wire in this part of the circuit?

23. A beam support in a building is in the shape of a parallelogram, as shown in Fig. 2-65. Find the area of the side of the beam shown.

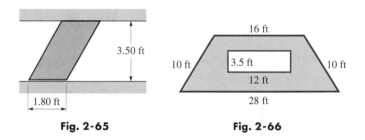

Fig. 2-65   **Fig. 2-66**

24. Each of two walls (with rectangular windows) of an A-frame house has the shape of a trapezoid as shown in Fig. 2-66. If a gallon of paint covers 320 ft^2, how much paint is required to paint these walls? (All data are accurate to two significant digits.)

25. A walkway 3.0 m wide is constructed along the outside edge of a square courtyard. If the perimeter of the courtyard is 320 m, what is the perimeter of the square formed by the outer edge of the walkway?

26. An architect designs a rectangular window such that the width of the window is 18 in. less than the height. If the perimeter of the window is 180 in., what are its dimensions?

27. A designer plans the top of a rectangular workbench to be four times as long as it is wide, and then determines that if the width is 2.5 ft greater and the length is 4.7 ft less, it would be a square. What are its dimensions?

28. A rectangular security area is enclosed on one side by a wall, and the other sides are fenced. The length of the wall is twice the width of the area. The total cost of building the wall and fence is $13,200. If the wall costs $50.00/m and the fence costs $5.00/m, find the dimensions of the area.

2-4 CIRCLES

The next geometric figure we consider is the circle. *All points on a* **circle** *are at the same distance from a fixed point, the* **center** *of the circle. The distance from the center to a point on the circle is the* **radius** *of the circle. The distance between two points on the circle on a line through the center is the* **diameter.** Therefore, the diameter d is twice the radius r, or $d = 2r$. See Fig. 2-67.

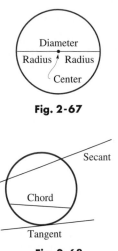

Fig. 2-67

There are also certain special types of lines associated with a circle. *A* **chord** *is a line segment having its endpoints on the circle. A* **tangent** *is a line that touches (does not pass through) the circle at one point. A* **secant** *is a line that passes through two points of the circle.* See Fig. 2-68.

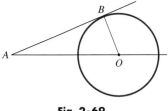

Fig. 2-68

An important property of a tangent is that *a tangent to a circle is perpendicular to the radius drawn to the point of contact.* This is illustrated in the next example.

EXAMPLE 1 In Fig. 2-69, O is the center of the circle, and AB is tangent at B. If $\angle OAB = 25°$, find $\angle AOB$.

Since the center is O, OB is a radius of the circle. A tangent is perpendicular to a radius at the point of tangency, which means $\angle ABO = 90°$ so that

$$\angle OAB + \angle OBA = 25° + 90° = 115°$$

Since the sum of the angles of a triangle is 180°, we have

$$\angle AOB = 180° - 115° = 65°$$

Fig. 2-69

Circumference and Area of a Circle

The perimeter of a circle is called the **circumference.** The formulas for the circumference and area of a circle are as follows:

$c = 2\pi r$	Circumference of a circle of radius r	**(2-9)**
$A = \pi r^2$	Area of a circle of radius r	**(2-10)**

The symbol π (a Greek letter), which we use as a number, was first used in this way as a number in the 1700s.

Here, π equals approximately 3.1416. In using a calculator, π can be entered by using the $\boxed{\pi}$ key.

EXAMPLE 2 A circular oil spill has a diameter of 2.4 km. This oil spill is to be enclosed within a length of special flexible tubing. What is the area of the spill, and how long must the tubing be? See Fig. 2-70.

We find the area by using Eq. (2-10). Since $d = 2r$, $r = d/2 = 1.2$ km. Therefore, the area is

$$A = \pi r^2 = \pi(1.2)^2$$
$$= 4.5 \text{ km}^2$$

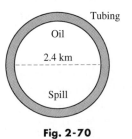

Fig. 2-70

The length of the tubing needed to enclose the oil spill is the circumference of the circle. Therefore,

$$c = 2\pi r = 2\pi(1.2) \qquad \text{note that } c = \pi d$$
$$= 7.5 \text{ km}$$

Results have been rounded off to two significant digits, the accuracy of d.

Many applied problems involve a combination of geometric figures. The following example illustrates one such combination.

SOLVING A WORD PROBLEM

EXAMPLE 3 A machine part is a square of side 3.25 in. with a quarter circle removed (see Fig. 2-71). Find the perimeter and the area of one side of the part.

Setting up a formula for the perimeter, we add the two sides of length s to *one-fourth of the circumference of a circle with radius s.* For the area, we *subtract the area of one-fourth of a circle from the area of the square.* This gives

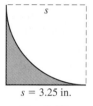

$s = 3.25$ in.

Fig. 2-71

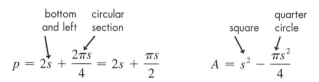

$$p = 2s + \frac{2\pi s}{4} = 2s + \frac{\pi s}{2} \qquad A = s^2 - \frac{\pi s^2}{4}$$

where s is the side of the square and the radius of the circle. Evaluating, we have

$$p = 2(3.25) + \frac{\pi(3.25)}{2} = 11.6 \text{ in.}$$

$$A = 3.25^2 - \frac{\pi(3.25)^2}{4} = 2.27 \text{ in.}^2$$

Circular Arcs and Angles

*An **arc** is part of a circle,* and *an angle formed at the center by two radii is a **central angle**.* The measure of an arc is the same as the central angle between the ends of the radii that define the arc. *A **sector** of a circle is the region bounded by two radii and the arc they intercept.* A **segment** *of a circle is the region bounded by a chord and its arc.* (There are two possible segments for a given chord. The smaller region is a *minor segment,* and the larger region is a *major segment.*). These are illustrated in the following example.

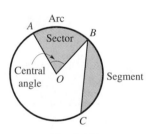

Fig. 2-72

EXAMPLE 4 In Fig. 2-72, a sector of the circle is between radii OA and OB and arc AB (which is denoted by $\overset{\frown}{AB}$). If the measure of the central angle at O between the radii is 70°, the measure of $\overset{\frown}{AB}$ is also 70°.

In Fig. 2-72, a segment of the circle is between chord BC and arc BC ($\overset{\frown}{BC}$). ∎

*An **inscribed angle** of an arc is one for which the endpoints of the arc are points on the sides of the angle and for which the vertex is a point (not an endpoint) of the arc.* An important property of a circle is that *the measure of an inscribed angle is one-half of its intercepted arc.*

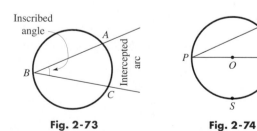

Fig. 2-73   **Fig. 2-74**

EXAMPLE 5 (a) In the circle shown in Fig. 2-73, $\angle ABC$ is inscribed in $\overset{\frown}{ABC}$, and it intercepts $\overset{\frown}{AC}$. If $\overset{\frown}{AC} = 60°$, then $\angle ABC = 30°$.

(b) In the circle shown in Fig. 2-74, PQ is a diameter, and $\angle PRQ$ is inscribed in the semicircular $\overset{\frown}{PRQ}$. Since $\overset{\frown}{PSQ} = 180°$, $\angle PRQ = 90°$. From this we conclude that *an angle inscribed in a semicircle is a right angle.* ∎

Radian Measure of an Angle

To this point we have measured all angles in degrees. There is another measure of an angle, the *radian*, which is defined in terms of an arc of a circle. We will find it of importance when we study trigonometry.

If a central angle of a circle intercepts an arc equal in length to the radius of the circle, the measure of the central angle is defined as one **radian.** See Fig. 2-75. Since the radius can be marked off along the circumference 2π times (about 6.283 times), we see that 2π rad $= 360°$ (where rad is the symbol for radian). Therefore,

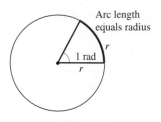

Arc length equals radius

Fig. 2-75

$$\boxed{\pi \text{ rad} = 180°}$$

(2-11)

is a basic relationship between radians and degrees.

■**EXAMPLE 6**   **(a)** If we divide each side of Eq. (2-11) by π, we get

$$1 \text{ rad} = 57.3°$$

where the result has been rounded off.

(b) To change an angle of 118.2° to radian measure, we have

$$118.2° = 118.2° \left(\frac{\pi \text{ rad}}{180°} \right) = 2.06 \text{ rad}$$

By multiplying 118.2° by π rad/180°, the unit of measurement that remains is rad, since the degrees "cancel." We will review radian measure again when we study trigonometry.

EXERCISES *2-4*

In Exercises 1–4, refer to the circle with center at O in Fig. 2-76. Identify the following.

1. (a) A secant line
   (b) A tangent line

2. (a) Two chords
   (b) An inscribed angle

3. (a) Two perpendicular lines
   (b) An isosceles triangle

4. (a) A segment
   (b) A sector with an acute central angle

Fig. 2-76

In Exercises 5–8, find the circumference of the circle with the given radius or diameter.

5. $r = 2.75$ ft

6. $r = 0.563$ m

7. $d = 23.1$ mm

8. $d = 8.2$ in.

In Exercises 9–12, find the area of the circle with the given radius or diameter.

9. $r = 0.0952$ yd

10. $r = 45.8$ cm

11. $d = 2.33$ m

12. $d = 12.56$ ft

In Exercises 13–16, refer to Fig. 2-77, where AB is a diameter, TB is a tangent line at B, and $\angle ABC = 65°$. Determine the indicated angles.

13. $\angle CBT$

14. $\angle BCT$

15. $\angle CAB$

16. $\angle BTC$

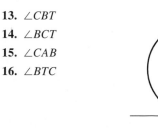

Fig. 2-77

In Exercises 17–20, refer to Fig. 2-78. Determine the indicated arcs and angles.

17. $\overset{\frown}{BC}$

18. $\overset{\frown}{AB}$

19. $\angle ABC$

20. $\angle ACB$

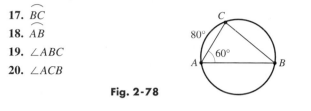

Fig. 2-78

In Exercises 21–24, change the given angles to radian measure.

21. 22.5°   **22.** 60.0°   **23.** 125.2°   **24.** 323.0°

In Exercises 25–28, find a formula for the indicated perimeter or area.

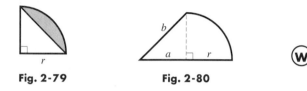

Fig. 2-79   **Fig. 2-80**

25. The perimeter of the quarter-circle in Fig. 2-79

26. The perimeter of the figure in Fig. 2-80. A quarter-circle is attached to a triangle

27. The area of the segment of the quarter-circle in Fig. 2-79

28. The area of the figure in Fig. 2-80

In Exercises 29–36, solve the given problems.

29. The radius of the earth's equator is 3960 mi. What is the circumference?

30. As a ball bearing rolls along a straight track, it makes 11.0 revolutions while traveling a distance of 109 mm. Find its radius.

31. Using a tape measure, the circumference of a tree is found to be 112 in. What is the diameter of the tree (assuming a circular cross section)?

32. What is the area of the largest circle that can be cut from a rectangular plate 21.2 cm by 15.8 cm?

33. A patio is designed with semicircular areas attached to a square, as shown in Fig. 2-81. Find the area of the patio.

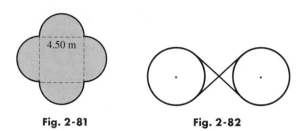

Fig. 2-81   **Fig. 2-82**

34. Find the length of the pulley belt shown in Fig. 2-82, if the belt crosses at right angles. The radius of each pulley wheel is 5.50 in.

(W) 35. The velocity of an object moving in a circular path is directed tangent to the circle in which it is moving. A stone on a string moves in a vertical circle, and the string breaks after 5.5 revolutions. If the string was initially in a vertical position, in what direction does it move after the string breaks? Explain.

36. Part of a circular gear with 24 teeth is shown in Fig. 2-83. Find the indicated angle.

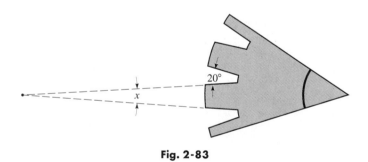

Fig. 2-83

2-5 MEASUREMENT OF IRREGULAR AREAS

To this point the figures for which we have found areas are well defined, and the areas can be found by direct use of a specific formula. In practice, however, it may be necessary to find the area of a figure with an irregular perimeter or one for which there is no specific formula. In this section we show two methods of finding a very good *approximation* of such an area. These methods are particularly useful in technical areas such as surveying, architecture, and mechanical design.

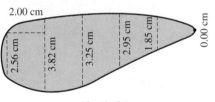

Fig. 2-84

The Trapezoidal Rule

The first method is based on dividing the required area into trapezoids with equal heights. Considering the area shown in Fig. 2-84, we draw parallel lines at n equal intervals between the edges of the area. We then join the ends of these parallel line segments to form adjacent trapezoids. The sum of the areas of the trapezoids gives a good approximation to the required area.

Calling the lengths of the parallel lines $y_0, y_1, y_2, \ldots, y_n$ and the height of each trapezoid h (the distance between the parallel lines), the total area A is the sum of areas of all the trapezoids. This gives us

$$A = \underset{\substack{\text{first} \\ \text{trapezoid}}}{\frac{h}{2}(y_0 + y_1)} + \underset{\substack{\text{second} \\ \text{trapezoid}}}{\frac{h}{2}(y_1 + y_2)} + \underset{\substack{\text{third} \\ \text{trapezoid}}}{\frac{h}{2}(y_2 + y_3)} + \cdots + \underset{\substack{\text{next-to-last} \\ \text{trapezoid}}}{\frac{h}{2}(y_{n-2} + y_{n-1})} + \underset{\substack{\text{last} \\ \text{trapezoid}}}{\frac{h}{2}(y_{n-1} + y_n)}$$

$$= \frac{h}{2}(y_0 + y_1 + y_1 + y_2 + y_2 + y_3 + \cdots + y_{n-2} + y_{n-1} + y_{n-1} + y_n)$$

Therefore, the approximate area is

$$A = \frac{h}{2}(y_0 + 2y_1 + 2y_2 + \cdots + 2y_{n-1} + y_n) \qquad \textbf{(2-12)}$$

Equation (2-12) is known as the **trapezoidal rule.** The following examples illustrate its use.

■EXAMPLE 1 A plate cam for opening and closing a valve is shown in Fig. 2-85. Widths of the face of the cam are shown at 2.00-cm intervals. Find the area of the face of the cam.

From the figure we see that

$$y_0 = 2.56 \text{ cm}, \ y_1 = 3.82 \text{ cm}, \ y_2 = 3.25 \text{ cm}, \ y_3 = 2.95 \text{ cm},$$
$$y_4 = 1.85 \text{ cm}, \ y_5 = 0.00 \text{ cm}$$

(In making such measurements, often a y-value at one end (or both ends) is zero. In such a case, the end "trapezoid" is actually a triangle.) From the given information in this example, $h = 2.00$ cm. Therefore, using the trapezoidal rule, Eq. (2-12), we have

$$A = \frac{2.00}{2}[2.56 + 2(3.82) + 2(3.25) + 2(2.95) + 2(1.85) + 0.00]$$
$$= 26.3 \text{ cm}^2$$

The area of the face of the cam is approximately 26.3 cm^2. ------------■

Fig. 2-85

When approximating the area with trapezoids, we omit small parts of the area for some trapezoids and include small extra areas for other trapezoids. The omitted areas often approximate the extra areas, which makes the approximation better. Also, the use of smaller intervals improves the approximation since the total omitted area or total extra area is smaller.

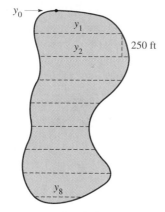

Fig. 2-86

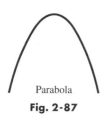

Parabola
Fig. 2-87

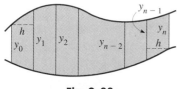

Fig. 2-88

Named for the English mathematician Thomas Simpson (1710–1761).

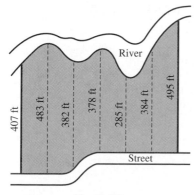

Fig. 2-89

■**EXAMPLE 2**   A surveyor measures the width of a small lake at 250-ft intervals from one end of the lake, as shown in Fig. 2-86. The widths found are as follows:

Distance from one end (ft)	0	250	500	750	1000	1250	1500	1750	2000
Width (ft)	0	940	920	890	740	550	770	960	220

From the table we see that $y_0 = 0$ ft, $y_1 = 940$ ft, ..., $y_8 = 220$ ft, and $h = 250$ ft. Therefore, using the trapezoidal rule, the approximate area of the lake is found as follows:

$$A = \frac{250}{2}[0 + 2(940) + 2(920) + 2(890) + 2(740) + 2(550) + 2(770) + 2(960) + 220]$$

$$= 1{,}500{,}000 \text{ ft}^2$$

Simpson's Rule

For the second method of measuring an irregular area, we also draw parallel lines at equal intervals between the edges of the area. We then join the ends of these parallel lines with curved *arcs*. This takes into account the fact that the perimeters of most figures are curved. The arcs used in this method are not arcs of a circle, but arcs of a *parabola*. A parabola is shown in Fig. 2-87 and is discussed in detail in Chapter 21. (Examples of parabolas are (1) the path of a ball that has been thrown and (2) the cross section of a microwave "dish.")

The development of this method requires advanced mathematics. Therefore, we will simply state the formula to be used. It might be noted that the form of the equation is similar to that of the trapezoidal rule.

The approximate area of the geometric figure shown in Fig. 2-88 is given by

$$A = \frac{h}{3}(y_0 + 4y_1 + 2y_2 + 4y_3 + \cdots + 2y_{n-2} + 4y_{n-1} + y_n) \qquad \text{(2-13)}$$

Equation (2-13) is known as **Simpson's rule.** *In using Eq. (2-13),* *the number n of intervals of width h must be even.*

■**EXAMPLE 3**   A parking lot is proposed for a riverfront area in a town. The town engineer measured the widths of the area at 100 ft (three sig. digits) intervals, as shown in Fig. 2-89. Find the area available for parking.

First we see that there are six intervals, which means Eq. (2-13) may be used. With $y_0 = 407$ ft, $y_1 = 483$ ft, ..., $y_6 = 495$ ft, and $h = 100$ ft, we have

$$A = \frac{100}{3}[407 + 4(483) + 2(382) + 4(378) + 2(285) + 4(384) + 495]$$

$$= 241{,}000 \text{ ft}^2$$

For most areas, Simpson's rule gives a somewhat better approximation than the trapezoidal rule. The accuracy of Simpson's rule is also usually improved by using smaller intervals.

See the chapter introduction.

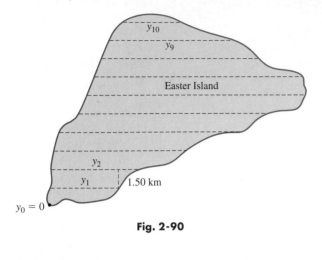

Fig. 2-90

EXAMPLE 4   From an aerial photograph, a cartographer determines the width of Easter Island at 1.50-km intervals as shown in Fig. 2-90. The widths found are as follows:

Distance from south end (km)	0	1.50	3.00	4.50	6.00	7.50	9.00	10.5	12.0	13.5	15.0
Width (km)	0	4.8	5.7	10.5	15.2	18.5	18.8	17.9	11.3	8.8	3.1

Since there are ten intervals, Simpson's rule may be used. From the table we have the following values: $y_0 = 0$, $y_1 = 4.5$, $y_2 = 5.4, \ldots, y_9 = 8.3$, $y_{10} = 2.6$, and $h = 1.5$. Using Simpson's rule, the cartographer would approximate the area of Easter Island as follows:

$$A = \frac{1.50}{3}(0 + 4(4.8) + 2(5.7) + 4(10.5) + 2(15.2) + 4(18.5)$$
$$+ 2(18.8) + 4(17.9) + 2(11.3) + 4(8.8) + 3.1) = 174 \text{ km}^2$$

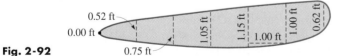

EXERCISES *2-5*

In Exercises 1–12, calculate the indicated areas. All data are accurate to at least two significant digits.

1. The widths of a kidney-shaped swimming pool were measured at 2.0-m intervals as shown in Fig. 2-91. Calculate the surface area of the pool, using the trapezoidal rule.

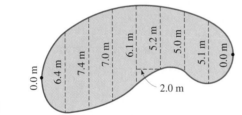

Fig. 2-91

2. Calculate the surface area of the swimming pool in Fig. 2-91, using Simpson's rule.

3. The widths of a cross section of an airplane wing are measured at 1.00-ft intervals as shown in Fig. 2-92. Calculate the area of the cross section, using Simpson's rule.

Fig. 2-92

4. Calculate the area of the cross section of the airplane wing in Fig. 2-92, using the trapezoidal rule.

5. Using aerial photography, the widths of an area burned by a forest fire were measured at 0.5-mi intervals as shown in the following table.

Distance (mi)	0.0	0.5	1.0	1.5	2.0	2.5	3.0	3.5	4.0
Width (mi)	0.6	2.2	4.7	3.1	3.6	1.6	2.2	1.5	0.8

Determine the area burned by the fire by using the trapezoidal rule.

6. Find the area burned by the forest fire of Exercise 5, using Simpson's rule.

7. A cartographer measured the width of Kruger National Park (and game reserve) in South Africa at 6.0-mm intervals on a map, as shown in Fig. 2-93. The widths are shown in the list that follows. Find the area of the park if the scale of the map is 1.0 mm = 6.0 km.

$y_0 = 7$ mm, $y_1 = 15$ mm
$y_2 = 7$ mm, $y_3 = 11$ mm
$y_4 = 13$ mm, $y_5 = 10$ mm
$y_6 = 9$ mm, $y_7 = 12$ mm
$y_8 = 8$ mm, $y_9 = 3$ mm

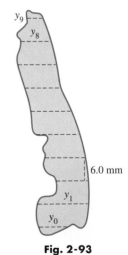

Fig. 2-93

8. The widths of an oval-shaped floor were measured at 1.5-m intervals, as shown in the following table.

Distance (m)	0.0	1.5	3.0	4.5	6.0	7.5	9.0	10.5	12.0
Width (m)	0.0	5.0	7.2	8.3	8.6	8.3	7.2	5.0	0.0

Find the area of the floor by using Simpson's rule.

9. The widths of the baseball playing area in Boston's Fenway Park at 45-ft intervals are shown in Fig. 2-94. Find the playing area using the trapezoidal rule.

10. Find the playing area of Fenway Park (see Exercise 9) by Simpson's rule.

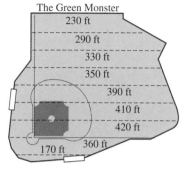

Fig. 2-94

11. Soundings taken across a river channel give the following values of distance from one shore with the corresponding depth of the channel.

Distance (ft)	0	50	100	150	200	250	300	350	400	450	500
Depth (ft)	5	12	17	21	22	25	26	16	10	8	0

Find the area of the channel using Simpson's rule.

12. The widths of a bell crank are measured at 2.0-in. intervals, as shown in Fig. 2-95. Find the area of the bell crank if the two connector holes are each 2.50 in. in diameter.

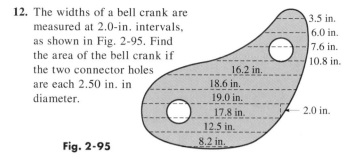

Fig. 2-95

In Exercises 13–16, calculate the area of the circle by the indicated method.

The lengths of parallel chords of a circle that are 0.250 in. apart are given in the following table. The diameter of the circle is 2.000 in. The distance shown is the distance from one end of a diameter.

Distance (in.)	0.000	0.250	0.500	0.750	1.000	1.250	1.500	1.750	2.000
Length (in.)	0.000	1.323	1.732	1.936	2.000	1.936	1.732	1.323	0.000

Using the formula $A = \pi r^2$, the area of the circle is 3.14 in.^{2}.

W 13. Find the area of the circle using the trapezoidal rule and only the values of distance of 0.000 in., 0.500 in., 1.000 in., 1.500 in., and 2.000 in. with the corresponding values of the chord lengths. Explain why the value found is less than 3.14 in.^{2}.

W 14. Find the area of the circle using the trapezoidal rule and all values in the table. Explain why the value found is closer to 3.14 in.^{2} than the value found in Exercise 13.

W 15. Find the area of the circle using Simpson's rule and the same table values as in Exercise 13. Explain why the value found is closer to 3.14 in.^{2} than the value found in Exercise 13.

W 16. Find the area of the circle using Simpson's rule and all values in the table. Explain why the value found is closer to 3.14 in.^{2} than the value found in Exercise 15.

2-6 SOLID GEOMETRIC FIGURES

We now review the formulas for the *volume* and *surface area* of some basic solid geometric figures. Just as area is a measure of the surface of a plane geometric figure, **volume** is a measure of the space occupied by a solid geometric figure.

One of the most common solid figures is the **rectangular solid.** This figure has six sides (**faces**), and opposite sides are rectangles. All intersecting sides are perpendicular to each other. The **bases** of the rectangular solid are the top and bottom faces. A **cube** is a rectangular solid with all six faces being equal squares.

A **right circular cylinder** is *generated* by rotating a rectangle about one of its sides. Each **base** is a circle, and the *cylindrical surface* is perpendicular to each of the bases. The **height** is one side of the rectangle, and the **radius** of the base is the other side.

A **right circular cone** is generated by rotating a right triangle about one of its legs. The **base** is a circle, and the **slant height** is the hypotenuse of the right triangle. The **height** is one leg of the right triangle, and the **radius** of the base is the other leg.

The bases of a **right prism** are equal and parallel polygons, and the sides are rectangles. The **height** of a prism is the perpendicular distance between bases. The base of a **pyramid** is a polygon, and the other faces, the **lateral faces**, are triangles that meet at a common point, the **vertex**. A **regular pyramid** has congruent triangles for its lateral faces.

A **sphere** is generated by rotating a circle about a diameter. The **radius** is a line segment joining the center and a point on the sphere. The **diameter** is a line segment through the center and having its endpoints on the sphere.

In the following formulas, *V* represents the *volume,* *A* represents the *total surface area,* *S* represents the *lateral surface area* (bases not included), *B* represents the *area of the base,* and *p* represents the *perimeter of the base.*

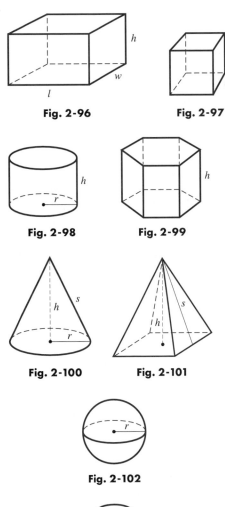

Fig. 2-96   **Fig. 2-97**

Fig. 2-98   **Fig. 2-99**

Fig. 2-100   **Fig. 2-101**

Fig. 2-102

Fig. 2-103

$V = lwh$ Rectangular solid (Fig. 2-96)		**(2-14)**
$A = 2lw + 2lh + 2wh$		**(2-15)**
$V = e^3$ Cube (Fig. 2-97)		**(2-16)**
$A = 6e^2$		**(2-17)**
$V = \pi r^2 h$ Right circular cylinder (Fig. 2-98)		**(2-18)**
$A = 2\pi r^2 + 2\pi rh$		**(2-19)**
$S = 2\pi rh$		**(2-20)**
$V = Bh$ Right prism (Fig. 2-99)		**(2-21)**
$S = ph$		**(2-22)**
$V = \frac{1}{3}\pi r^2 h$ Right circular cone (Fig. 2-100)		**(2-23)**
$A = \pi r^2 + \pi rs$		**(2-24)**
$S = \pi rs$		**(2-25)**
$V = \frac{1}{3}Bh$ Regular pyramid (Fig. 2-101)		**(2-26)**
$S = \frac{1}{2}ps$		**(2-27)**
$V = \frac{4}{3}\pi r^3$ Sphere (Fig. 2-102)		**(2-28)**
$A = 4\pi r^2$		**(2-29)**

Equation (2-21) is valid for any prism, and Eq. (2-26) is valid for any pyramid. There are other types of cylinders and cones, but we shall restrict our attention to right circular cylinders and right circular cones, and we will often use "cylinder" or "cone" when referring to them.

The **frustum** of a cone or pyramid is the solid figure that remains after the top is cut off by a plane parallel to the base. Figure 2-103 shows the frustum of a cone.

EXAMPLE 1 What volume of concrete is needed for a driveway 25.0 m long, 2.75 m wide, and 0.100 m thick?

The driveway is a rectangular solid for which $l = 25.0$ m, $w = 2.75$ m, and $h = 0.100$ m. Using Eq. (2-14), we have

$$V = (25.0)(2.75)(0.100) = 6.88 \text{ m}^3$$

SOLVING A WORD PROBLEM

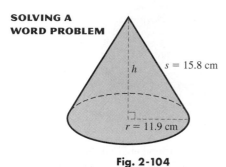

Fig. 2-104

EXAMPLE 2 Calculate the volume of a right circular cone for which the radius $r = 11.9$ cm and the *slant height $s = 15.8$ cm*. See Fig. 2-104.

To find the volume using Eq. (2-23), we need the radius and height of the cone. Therefore, we must first find the height. As noted, the radius and height are the legs of a right triangle. and the slant height is the hypotenuse. To find the height we *use the Pythagorean theorem.*

$$s^2 = r^2 + h^2 \quad \text{Pythagorean theorem}$$
$$h^2 = s^2 - r^2 \quad \text{solve for } h$$
$$h = \sqrt{s^2 - r^2}$$
$$= \sqrt{15.8^2 - 11.9^2} = 10.4 \text{ cm}$$

Now, calculating the volume (in using a calculator it is not necessary to record the value of h), we have

$$V = \tfrac{1}{3}\pi r^2 h \qquad \text{Eq. (2-23)}$$
$$= \tfrac{1}{3}\pi (11.9^2)(10.4) \qquad \text{substituting}$$
$$= 1540 \text{ cm}^3$$

SOLVING A WORD PROBLEM

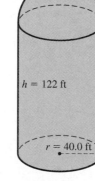

Fig. 2-105

EXAMPLE 3 A grain storage building is in the shape of a cylinder surmounted by a hemisphere *(half a sphere)*. See Fig. 2-105. Find the volume of grain that can be stored if the height of the cylinder is 122 ft and its radius is 40.0 ft.

The total volume of the structure is the volume of the cylinder plus the volume of the hemisphere. By the construction we see that the radius of the hemisphere is the same as the radius of the cylinder. Therefore,

$$\overset{\text{cylinder}}{} \overset{\text{hemisphere}}{}$$
$$V = \pi r^2 h + \tfrac{1}{2}(\tfrac{4}{3}\pi r^3) = \pi r^2 h + \tfrac{2}{3}\pi r^3$$
$$= \pi (40.0)^2 (122) + \tfrac{2}{3}\pi (40.0)^3$$
$$= 747{,}000 \text{ ft}^3$$

EXERCISES *2-6*

In Exercises 1–16, find the volume or area of each solid figure for the given values. See Figs. 2-96 to 2-102.

1. Volume of cube: $e = 7.15$ ft

2. Volume of right circular cylinder: $r = 23.5$ cm, $h = 48.4$ cm

3. Total surface area of right circular cylinder: $r = 6.89$ m, $h = 2.33$ m

4. Area of sphere: $r = 67$ in.

5. Volume of sphere: $r = 0.877$ yd

6. Volume of right circular cone: $r = 25.1$ mm, $h = 5.66$ mm

7. Lateral area of right circular cone: $r = 78.0$ cm, $s = 83.8$ cm

8. Lateral area of regular pyramid: $p = 3.45$ ft, $s = 2.72$ ft

9. Volume of regular pyramid: square base of side 16 in., $h = 13$ in.

10. Volume of right prism: square base of side 29.0 cm, $h = 11.2$ cm

11. Lateral area of regular prism: equilateral triangle base of side 1.092 m, $h = 1.025$ m

12. Lateral area of right circular cylinder: diameter $= 25.0$ ft, $h = 34.7$ ft

13. Volume of hemisphere: diameter $= 0.83$ yd

14. Volume of regular pyramid: square base of side 22.4 m, $s = 14.2$ m

15. Total surface area of right circular cone: $r = 0.339$ cm, $h = 0.274$ cm

16. Total surface area of pyramid: all faces and base are equilateral triangles of side 3.67 in.

In Exercises 17–28, find the indicated areas and volumes.

17. A rectangular box is to be used to store radioactive materials. The inside of the box is 12.0 in. long, 9.50 in. wide, and 8.75 in. deep. What is the area of sheet lead that must be used to line the inside of the box?

18. A swimming pool is 50.0 ft wide, 78.0 ft long, 3.50 ft deep at one end, and 8.50 ft deep at the other end. How many cubic feet of water can it hold? (The slope on the bottom is constant.)

19. The Alaskan oil pipeline is 750 mi long and has a diameter of 4.0 ft. What is the maximum volume of the pipeline?

20. A glass prism used in the study of optics has a right triangular base. The legs of the triangle are 3.00 cm and 4.00 cm. The prism is 8.50 cm high. What is the total surface area of the prism? See Fig. 2-106.

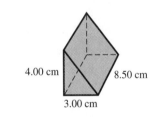

4.00 cm
8.50 cm

Fig. 2-106

3.00 cm

21. The Great Pyramid of Egypt has a square base approximately 250 yd on a side. The height of the pyramid is about 160 yd. What is its volume? See Fig. 2-107.

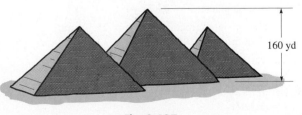

160 yd

Fig. 2-107

22. A paper cup is in the shape of a cone with radius 1.80 in. and height 3.50 in. What is the surface area of the cup?

23. *Spaceship Earth* (shown in Fig. 2-108) at Epcot Center in Florida is a sphere of 165 ft in diameter. What is the volume of *Spaceship Earth?*

165 ft

Fig. 2-108

24. The diameter of a spherical balloon is 12 cm. It is then further inflated until its diameter is 24 cm. By how much was the surface area increased?

25. A special wedge in the shape of a regular pyramid has a square base 16.0 mm on a side. The height of the wedge is 40.0 mm. What is the total surface area of the wedge?

26. What is the area of a paper label that is to cover the lateral surface of a cylindrical can 3.00 in. in diameter and 4.25 in. high? The ends of the label will overlap 0.25 in. when the label is placed on the can.

27. A dipstick is made to measure the volume remaining in the conical container shown in Fig. 2-109. How far below the full mark (at the top of the container) on the stick should the mark for half-full be placed?

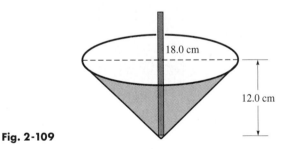

18.0 cm

12.0 cm

Fig. 2-109

28. The side view of a rivet is shown in Fig. 2-110. It is a conical part on a cylindrical part. Find the volume of the rivet.

0.625 in.
2.75 in.

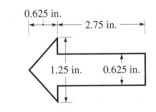

1.25 in.
0.625 in.

Fig. 2-110

CHAPTER EQUATIONS

Line segments	Fig. 2-8	$\dfrac{a}{b} = \dfrac{c}{d}$	(2-1)
Triangle		$A = \frac{1}{2}bh$	(2-2)
Hero's formula		$A = \sqrt{s(s-a)(s-b)(s-c)},$ where $s = \frac{1}{2}(a+b+c)$	(2-3)
Pythagorean theorem	Fig. 2-29	$c^2 = a^2 + b^2$	(2-4)
Square	Fig. 2-51	$A = s^2$	(2-5)
Rectangle	Fig. 2-52	$A = lw$	(2-6)
Parallelogram	Fig. 2-53	$A = bh$	(2-7)
Trapezoid	Fig. 2-54	$A = \frac{1}{2}h(b_1 + b_2)$	(2-8)
Circle		$c = 2\pi r$	(2-9)
		$A = \pi r^2$	(2-10)
Radians	Fig. 2-75	$\pi \text{ rad} = 180°$	(2-11)
Trapezoidal rule	Fig. 2-84	$A = \dfrac{h}{2}(y_0 + 2y_1 + 2y_2 + \cdots + 2y_{n-1} + y_n)$	(2-12)
Simpson's rule	Fig. 2-88	$A = \dfrac{h}{3}(y_0 + 4y_1 + 2y_2 + 4y_3 + \cdots + 2y_{n-2} + 4y_{n-1} + y_n)$	(2-13)
Rectangular solid	Fig. 2-96	$V = lwh$	(2-14)
		$A = 2lw + 2lh + 2wh$	(2-15)
Cube	Fig. 2-97	$V = e^3$	(2-16)
		$A = 6e^2$	(2-17)
Right circular cylinder	Fig. 2-98	$V = \pi r^2 h$	(2-18)
		$A = 2\pi r^2 + 2\pi rh$	(2-19)
		$S = 2\pi rh$	(2-20)
Right prism	Fig. 2-99	$V = Bh$	(2-21)
		$S = ph$	(2-22)
Right circular cone	Fig. 2-100	$V = \frac{1}{3}\pi r^2 h$	(2-23)
		$A = \pi r^2 + \pi rs$	(2-24)
		$S = \pi rs$	(2-25)
Regular pyramid	Fig. 2-101	$V = \frac{1}{3}Bh$	(2-26)
		$S = \frac{1}{2}ps$	(2-27)
Sphere	Fig. 2-102	$V = \frac{4}{3}\pi r^3$	(2-28)
		$A = 4\pi r^2$	(2-29)

REVIEW EXERCISES

In Exercises 1–4, use Fig. 2-111. Determine the indicated angles.

1. $\angle CGE$ **2.** $\angle EGF$ **3.** $\angle DGH$ **4.** $\angle EGI$

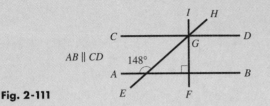

$AB \parallel CD$ 148°

Fig. 2-111

In Exercises 5–12, find the indicated sides of the right triangle shown in Fig. 2-112.

5. $a = 9$, $b = 40$, $c = ?$

6. $a = 14$, $b = 48$, $c = ?$

7. $a = 40$, $c = 58$, $b = ?$

8. $b = 56$, $c = 65$, $a = ?$

9. $a = 6.30$, $b = 3.80$, $c = ?$

10. $a = 126$, $b = 251$, $c = ?$

11. $b = 29.3$, $c = 36.1$, $a = ?$

12. $a = 0.782$, $c = 0.885$, $b = ?$

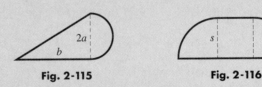

Fig. 2-112

In Exercises 13–20, find the perimeter or area of the indicated figure.

13. Perimeter: equilateral triangle of side 8.5 mm

14. Perimeter: rhombus of side 15.2 in.

15. Area: triangle, $b = 3.25$ ft, $h = 1.88$ ft

16. Area: triangle of sides 17.5 cm, 13.8 cm, 11.9 cm

17. Circumference of circle: $d = 98.4$ mm

18. Perimeter: rectangle, $l = 2.98$ yd, $w = 1.86$ yd

19. Area: trapezoid, $b_1 = 67.2$ in., $b_2 = 83.8$ in., $h = 34.2$ in.

20. Area: circle, $d = 32.8$ m

In Exercises 21–24, find the volume of the indicated solid geometric figure.

21. Prism: base is right triangle with legs 26.0 cm and 34.0 cm, height is 14.0 cm

22. Cylinder: base radius 36.0 in., height 24.0 in.

23. Pyramid: base area 3850 ft², height 125 ft

24. Sphere: diameter 22.1 mm

In Exercises 25–28, find the surface area of the indicated solid geometric figure.

25. Total area of cube of edge 5.20 m

26. Total area of cylinder: base diameter 1.20 ft, height 5.80 ft

27. Lateral area of cone: base radius 18.2 in., height 11.5 in.

28. Total area of sphere: diameter 0.884 m

In Exercises 29–32, use Fig. 2-113. Line CT is tangent to the circle with center at O. Find the indicated angles.

29. $\angle BTA$

30. $\angle TAB$

31. $\angle BTC$

32. $\angle ABT$

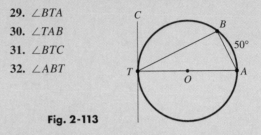

Fig. 2-113

In Exercises 33–36, use Fig. 2-114. Given that $AB = 4$, $BC = 4$, $CD = 6$, and $\angle ADC = 53°$, find the indicated angle and lengths.

33. $\angle ABE$

34. AD

35. BE

36. AE

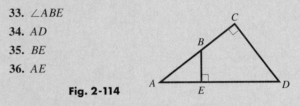

Fig. 2-114

In Exercises 37–40, find the formulas for the indicated perimeters and areas.

37. Perimeter of Fig. 2-115 (a right triangle and semicircle attached)

38. Perimeter of Fig. 2-116 (a square with a quarter circle at each end)

39. Area of Fig. 2-115   **40.** Area of Fig. 2-116

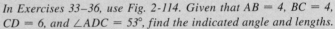

Fig. 2-115   **Fig. 2-116**

In Exercises 41–44, answer the given questions and explain your reasoning.

41. Is a square also a rectangle, a parallelogram, and a rhombus?

42. If the measures of two angles of one triangle equal the measures of the measures of two angles of a second triangle, are the two triangles similar?

(W) 43. If the dimensions of a plane geometric figure are each multiplied by n, by how much is the area multiplied? Explain, using a circle to illustrate.

(W) 44. If the dimensions of a solid geometric figure are each multiplied by n, by how much is the volume multiplied? Explain, using a cube to illustrate.

In Exercises 45–64, solve the given problems.

45. A ramp for the disabled is designed so that it rises 1.2 m over a horizontal distance of 7.8 m. How long is the ramp?

46. An airplane is 2100 ft directly above one end of a 9500-ft runway. How far is the plane from the glide-slope indicator on the ground at the other end of the runway?

47. A radio transmitting tower is supported by guy wires. The tower and three parallel guy wires are shown in Fig. 2-117. Find the distance *AB* along the tower.

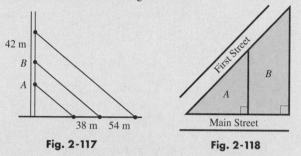

Fig. 2-117   **Fig. 2-118**

48. Find the areas of lots *A* and *B* in Fig. 2-118. *A* has a frontage on Main St. of 140 ft, and *B* has a frontage on Main St. of 84 ft. The boundary between lots is 120 ft.

49. Find the area of the side of the building shown in Fig. 2-119 (a triangle over a rectangle).

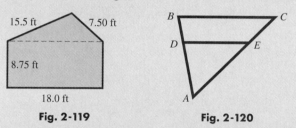

Fig. 2-119   **Fig. 2-120**

50. The metal support in the form of △*ABC* shown in Fig. 2-120 is strengthened by brace *DE*, which is parallel to *BC*. How long is the brace if *AB* = 24 in., *AD* = 16 in., and *BC* = 33 in.?

51. A typical scale for an aerial photograph is 1/18450. In an 8.00-in.-by-10.0-in. photograph with this scale, what is the longest distance (in mi) between two locations in the photograph?

52. For a hydraulic press, the mechanical advantage is the ratio of the large piston area to the small piston area. Find the mechanical advantage if the pistons have diameters of 3.10 cm and 2.25 cm.

53. The diameter of the earth is 7920 mi, and a satellite is in orbit at an altitude of 210 mi. How far does the satellite travel in one rotation about the earth?

54. The roof of the Louisiana Superdome in New Orleans is supported by a circular steel tension ring 651 m in circumference. Find the area covered by the roof.

55. A rectangular piece of wallboard is 8.0 ft long and 4.0 ft wide. Two 1.0-ft diameter holes are cut out for heating ducts. What is the area of the remaining piece?

56. The diameter of the sun is 1.38×10^6 km, the diameter of the earth is 1.27×10^4 km, and the distance from the earth to the sun (center to center) is 1.50×10^8 km. What is the distance from the center of the earth to the end of the shadow due to the rays from the sun?

57. Using aerial photography, the width of an oil spill is measured at 250-m intervals, as shown in Fig. 2-121. Using Simpson's rule, find the area of the oil spill.

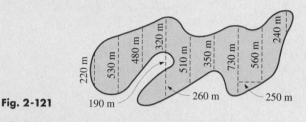

Fig. 2-121

58. To build a highway, it is necessary to cut through a hill. A surveyor measured the cross-sectional areas at 250-ft intervals through the cut as shown in the following table. Using the trapezoidal rule, determine the volume of soil to be removed.

Dist. (ft)	0	250	500	750	1000	1250	1500	1750
Area (ft²)	560	1780	4650	6730	5600	6280	2260	230

59. The Hubble space telescope is within a cylinder 4.3 m in diameter and 13 m long. What is the volume within this cylinder?

60. A horizontal cross section of a concrete bridge pier is a regular hexagon (six sides, all equal in length, and all internal angles are equal), each side of which is 2.50 m long. If the height of the pier is 6.75 m, what is the volume of concrete in the pier?

61. Two persons are talking to each other on cellular phones. One is 2.4 mi from the relay tower, and the other is 3.7 mi from the tower. If the angle between their signals at the tower is 90.0°, how far apart are the two persons?

62. A railroad track 1000.00 ft long expands 0.20 ft (2.4 in.) during the afternoon (due to an increase in temperature of about 30°F). Assuming that the track cannot move at either end and that the increase in length causes a bend straight up in the middle of the track, how high is the top of the bend?

63. A hot-water tank is in the shape of a right circular cylinder surmounted by a hemisphere. The total height of the tank is 6.75 ft, and the diameter of the base is 2.50 ft. How many gallons does the tank hold? (7.48 gal = 1.00 ft³)

64. A tent is in the shape of a regular pyramid surmounted on a cube. If the edge of the cube is 2.50 m and the total height of the tent is 3.25 m, find the area of the material used in making the tent (not including any floor area).

Writing Exercise

65. The Pentagon, headquarters of the U.S. Department of Defense, is one of the world's largest office buildings. It is a regular pentagon (five sides, all equal in length, and all internal angles are equal) 921 ft on a side, with a diagonal of length 1490 ft. Using these data, draw a sketch and write one or two paragraphs to explain how to find the area covered within the outside perimeter of the Pentagon. (What is the area?)

PRACTICE TEST

1. In Fig. 2-122, determine ∠1.

2. In Fig. 2-122, determine ∠2.

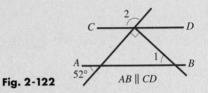

Fig. 2-122 $AB \parallel CD$

3. A tree is 8.0 ft high and casts a shadow 10.0 ft long. At the same time, a telephone pole casts a shadow 25.0 ft long. How tall is the pole?

4. Find the area of a triangle with sides of 2.46 cm, 3.65 cm, and 4.07 cm.

5. What is the diagonal distance along the floor between corners of a rectangular room 12.5 ft wide and 17.0 ft long?

6. Find the surface area of a tennis ball whose circumference is 21.0 cm.

7. Find the volume of a right circular cone of radius 2.08 m and height 1.78 m.

8. In Fig. 2-123, find ∠1.

9. In Fig. 2-123, find ∠2.

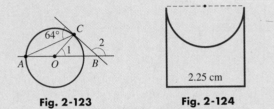

Fig. 2-123   **Fig. 2-124**

10. In Fig. 2-124, find the perimeter of the figure shown. It is a square with a semicircle removed.

11. In Fig. 2-124, find the area of the figure shown.

12. The width of a marshy area is measured at 50-ft intervals, with the results shown in the following table. Using the trapezoidal rule, find the area of the marsh. (All data accurate to two or more significant digits.)

Distance (ft)	0	50	100	150	200	250	300
Width (ft)	0	90	145	260	205	110	20

3 FUNCTIONS AND GRAPHS

In technology and science, as well as in everyday life, we see that one quantity depends on one or more other quantities. Plant growth depends on sunlight and rainfall; traffic flow depends on roadway design; the sales tax on an item depends on the cost of the item; the time required to access an Internet site depends on the speed at which a computer processes data. These are but a few of the innumerable possible examples.

Determining how one quantity depends on other quantities is one of the primary goals of science. A rule that relates such quantities is of great importance and usefulness in science and technology. In mathematics such a rule is called a *function*, and we will start this chapter with a discussion of functions.

A way of actually seeing how one quantity depends on another is by means of a *graph*. Basic methods of drawing and using graphs are also taken up in this chapter.

The electric power produced in a circuit depends on the resistance R in the circuit. In Section 3-4 we draw a graph of
$P = \dfrac{100R}{(0.50 + R)^2}$ to see this type of relationship.

3-1 INTRODUCTION TO FUNCTIONS

In most of the formulas of Chapter 1, one quantity was given in terms of one or more other quantities. It is obvious, then, that the various quantities are related by means of the formula. One important method of finding such formulas is through scientific observation and experimentation.

If we were to perform an experiment to determine whether or not a relationship exists between the distance an object drops and the time it falls, observation of the results would indicate (approximately, at least) that $s = 16t^2$, where s is the distance in feet and t is the time in seconds. We would therefore see that distance and time for a falling object are related.

A similar study of the pressure and the volume of a gas at constant temperature would show that as pressure increases, volume decreases according to the formula $PV = k$, where k is a constant. Electrical measurements of current and voltage with respect to a particular resistor would show that $V = kI$, where V is the voltage, I is the current, and k is a constant.

The Italian scientist Galileo (1564–1642) is generally credited with first using the *experimental method* by which controlled experiments are used to study natural phenomena. Until he showed that the distance an object falls in a given time does not depend on its weight, the concept of Aristotle (384–332 B.C.) that heavier objects fall faster than lighter ones was accepted.

Considerations such as these lead us to one of the most important and basic concepts in mathematics.

DEFINITION OF A FUNCTION

> *Whenever a relationship exists between two variables such that for every value of the first, there is only one corresponding value of the second, we say that the second variable is a **function** of the first variable.*
>
> *The first variable is called the **independent variable,** and the second variable is called the **dependent variable.***

Using controlled experiments, Galileo wanted to find mathematical formulas to describe natural phenomena.

The first variable is termed *independent* since permissible values can be assigned to it arbitrarily, and the second variable is termed *dependent* since its value is determined by the choice of the independent variable. *Values of the independent variable and dependent variable are to be real numbers.* Therefore, there may be restrictions on their possible values. This is discussed in the following section.

EXAMPLE 1   In the equation $y = 2x$, we see that y is a function of x, since for each value of x there is only one value of y. For example, if we substitute $x = 3$, we get $y = 6$ and no other value. By arbitrarily assigning values to x and then substituting, we see that the values of y we obtain *depend* on the values chosen for x. Therefore, x is the independent variable and y is the dependent variable. ▬

EXAMPLE 2   The power P developed in a certain resistor by a current I is given by $P = 4I^2$. Here P is a function of I. The dependent variable is P, and the independent variable is I. ▬

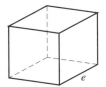

Fig. 3-1

EXAMPLE 3   Figure 3-1 shows a cube of edge e. In order to express the volume V as a function of the edge e, we recall from Chapter 2 that $V = e^3$. Here V is a function of e, since for each value of e there is only one value of V. The dependent variable is V, and the independent variable is e.

If the equation relating the volume and the edge of a cube is written as $e = \sqrt[3]{V}$—that is, if the edge is expressed in terms of the volume—e is a function of V. In this case, e is the dependent variable, and V is the independent variable. ▬

There are many ways to express functions. Formulas, tables, charts, and graphs can also define functions. Functions will be of importance throughout the book, and we will use a number of different types of functions in later chapters.

Functional Notation

For convenience of notation, the phrase

"function of x" is written as $f(x)$.

This means that "y is a function of x" may be written as $y = f(x)$. Here f denotes *dependence* and does not represent a quantity or a variable. Therefore, it also follows that $f(x)$ *does not mean f times x.*

CAUTION ▶

EXAMPLE 4   If $y = 6x^3 - 5x$, we say that y is a function of x. This function is $6x^3 - 5x$. It is also common to write such a function as $f(x) = 6x^3 - 5x$. However, y and $f(x)$ represent the same expression, $6x^3 - 5x$. Using y, the quantities are shown, and using $f(x)$, the functional dependence is shown. ▬

A graphing calculator can be used to evaluate a function in several ways. One is to directly substitute the value into the function as shown in Fig. 3-2(a). A second is to enter the function as Y_1 and evaluate as shown in Fig. 3-2(b). A third way, which is very useful when many values are to be used, is to enter the function as Y_1 and use the *table* feature as shown in Fig. 3-2(c).

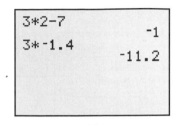

(a)

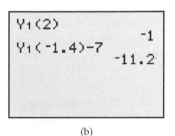

(b)

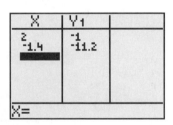

(c)

Fig. 3-2

One of the most important uses of functional notation is to designate the value of the function for a particular value of the independent variable. That is,

the value of the function $f(x)$ when $x = a$ is written as $f(a)$.

This is illustrated in the next example.

■**EXAMPLE 5**   For a function $f(x)$, the value of $f(x)$ for $x = 2$ may be expressed as $f(2)$. Thus, substituting 2 for x in $f(x) = 3x - 7$, we have

$$f(2) = 3(2) - 7 = -1 \qquad \text{substitute 2 for } x$$

The value of $f(x)$ for $x = -1.4$ is

$$f(-1.4) = 3(-1.4) - 7 = -11.2 \qquad \text{substitute } -1.4 \text{ for } x \qquad ■$$

We must note that *whatever number a represents, to find $f(a)$, we substitute a for x in $f(x)$.* This is true even if a is a literal number, as in the following examples.

■**EXAMPLE 6**   If $g(t) = \dfrac{t^2}{2t + 1}$, to find $g(a^3)$ we substitute a^3 for t in $g(t)$.

$$g(a^3) = \frac{(a^3)^2}{2a^3 + 1} = \frac{a^6}{2a^3 + 1}$$

For the same function, $g(3) = \dfrac{3^2}{2(3) + 1} = \dfrac{9}{7}$. In both cases, to obtain the value of $g(t)$, we simply substitute the value within the parentheses for t. ────■

■**EXAMPLE 7**   The electric resistance R of a particular resistor as a function of the temperature T is given by $R = 10.0 + 0.10T + 0.001T^2$. If a given temperature T is increased by 10°C, what is the value of R for the increased temperature as a function of the temperature T?

We are to determine R for a temperature of $T + 10$. Since

$$f(T) = 10.0 + 0.10T + 0.001T^2$$

then

$$f(T + 10) = 10.0 + 0.10(T + 10) + 0.001(T + 10)^2 \qquad \text{substitute } T + 10 \text{ for } T$$
$$= 10.0 + 0.10T + 1.0 + 0.001T^2 + 0.02T + 0.1$$
$$= 11.1 + 0.12T + 0.001T^2 \qquad\qquad ■$$

At times we need to define more than one function of the independent variable. We then use different symbols to denote the different functions. For example, $f(x)$ and $g(x)$ may represent different functions, such as $f(x) = 5x - 3$ and $g(x) = ax^2 + x$ (where a is a constant).

■**EXAMPLE 8**   For $f(x) = 5x - 3$ and $g(x) = ax^2 + x$, we have

$$f(-4) = 5(-4) - 3 = -23 \qquad \text{substitute } -4 \text{ for } x \text{ in } f(x)$$

$$g(-4) = a(-4)^2 + (-4) = 16a - 4 \qquad \text{substitute } -4 \text{ for } x \text{ in } g(x) \qquad ■$$

A function may be looked upon as a set of instructions. These instructions tell us how to obtain the value of the dependent variable for a particular value of the independent variable, even if the instructions are expressed in literal symbols.

■**EXAMPLE 9**   The function $f(x) = x^2 - 3x$ tells us to "square the value of the independent variable, multiply the value of the independent variable by 3, and subtract the second result from the first." An analogy would be a computer that was programmed so that when a number was entered into the program, it would square the number, then multiply the number by 3, and finally subtract the second result from the first. This is represented in diagram form in Fig. 3-3.

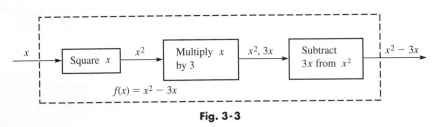

Fig. 3-3

NOTE ▶   The functions $f(t) = t^2 - 3t$ and $f(n) = n^2 - 3n$ are the same as the function $f(x) = x^2 - 3x$, since the operations performed on the independent variable are the same. *Although different literal symbols appear, this does not change the function.*

In later chapters we will use a number of special functions, and these special functions are represented by particular symbols. For example, in trigonometry we shall come across the "sine of the angle θ," where the sine is a function of θ. This is designated by $\sin \theta$.

EXERCISES *3-1*

In Exercises 1–8, determine the appropriate functions.

1. Express the area A of a circle as a function of (a) its radius r and (b) its diameter d.

2. Express the circumference c of a circle as a function of (a) its radius r and (b) its diameter d.

3. Express the volume V of a sphere as a function of its diameter d.

4. Express the edge e of a cube as a function of its surface area A.

5. Express the area A of a rectangle of width 5 as a function of its length l.

6. Express the volume V of a right circular cone of height 8 as a function of the radius r of the base.

7. Express the area A of a square as a function of its side s; express the side s of a square as a function of its area A.

8. Express the perimeter p of a square as a function of its side s; express the side s of a square as a function of its perimeter p.

In Exercises 9–20, evaluate the given functions.

9. $f(x) = 2x + 1$; find $f(1)$ and $f(-1)$.

10. $f(x) = 5x - 9$; find $f(2)$ and $f(-2)$.

11. $f(x) = 5 - 3x$; find $f(-2)$ and $f(0.4)$.

12. $f(T) = 7.2 - 2.5T$; find $f(2.6)$ and $f(-4)$.

13. $\phi(x) = \dfrac{6 - x^2}{2x}$; find $\phi(1)$ and $\phi(-2)$.

14. $H(q) = \dfrac{8}{q} + 2\sqrt{q}$; find $H(4)$ and $H(0.16)$.

15. $g(t) = at^2 - a^2t$; find $g(-\frac{1}{2})$ and $g(a)$.

16. $s(y) = 6\sqrt{y + 1} - 3$; find $s(8)$ and $s(a^2)$.

17. $K(s) = 3s^2 - s + 6$; find $K(-s)$ and $K(2s)$.

18. $T(t) = 5t + 7$; find $T(-2t)$ and $T(t + 1)$.

19. $f(x) = 2x + 4$; find $f(3x) - 3f(x)$.

20. $f(x) = 2x^2 + 1$; find $f(x + 2) - [f(x) + 2]$.

In Exercises 21–24, evaluate the given functions. The values of the independent variable are approximate.

21. Given $f(x) = 5x^2 - 3x$, find $f(3.86)$ and $f(-6.92)$.

22. Given $g(t) = \sqrt{t + 1.0604} - 6t^3$, find $g(0.9261)$.

23. Given $F(H) = \dfrac{2H^2}{H + 0.03685}$, find $F(-0.08466)$.

24. Given $f(x) = \dfrac{x^4 - 2.0965}{6x}$, find $f(1.9654)$.

(W) *In Exercises 25–32, state the instructions of the function in words as in Example 9.*

25. $f(x) = x^2 + 2$ **26.** $f(x) = 2x - 6$

27. $g(y) = 6y - y^3$ **28.** $\phi(s) = 8 - 5s + s^2$

29. $R(r) = 3(2r + 5) - 1$ **30.** $f(z) = \dfrac{4z}{5 - z}$

31. $p(t) = \dfrac{2t - 3}{t + 2}$ **32.** $Y(y) = 2 + \dfrac{5y}{2(y - 3)}$

In Exercises 33–36, write the equation as given by the statement. Then write the indicated function using functional notation.

33. y is equal to the square of x.

34. s is equal to the square root of $t + 2$.

35. The area A of the Bering Glacier in Alaska, given that its present area is 8430 km^2 and that it is melting at the rate of $140t$ km^2, where t is the time in centuries.

36. The electrical resistance R of a certain ammeter, in which the resistance of the coil is R_c, is the product of 10 and R_c divided by the sum of 10 and R_c.

In Exercises 37–40, solve the given problems.

37. A demolition ball is used to tear down a building. Its distance s (in m) above the ground as a function of the time t (in s) after it is dropped is $s = 17.5 - 4.9t^2$. Since $s = f(t)$, find $f(1.2)$.

38. The change C (in in.) in the length of a 100-ft steel bridge girder from its length at 40°F, as a function of the temperature T, is given by $C = 0.014(T - 40)$. Since $C = f(T)$, find $f(15)$.

39. The stopping distance d (in ft) of a car going v mi/h is given by $d = v + 0.05v^2$. Since $d = f(v)$, find $f(30)$, $f(2v)$, and $f(60)$, using both $f(v)$ and $f(2v)$.

40. The electric power P (in W) dissipated in a resistor of resistance R (in Ω) is given by the function $P = \dfrac{200R}{(100 + R)^2}$. Since $P = f(R)$, find $f(R + 10)$.

$3\text{-}2$ MORE ABOUT FUNCTIONS

Domain and Range

As we mentioned in the previous section, using only real numbers may result in restrictions as to the permissible values of the independent and dependent variables. *The complete set of possible values of the independent variable is called the* **domain** *of the function,* and *the complete set of all possible resulting values of the dependent variable is called the* **range** *of the function.* Therefore, using real numbers in the domain and range of a function,

NOTE ▶   *values that lead to division by zero or to imaginary values may not be included.*

▐**EXAMPLE 1**   The function $f(x) = x^2 + 2$ is defined for all real values of x. This means its domain is written as *all real numbers*. However, since x^2 is never negative, $x^2 + 2$ is never less than 2. We then write the range as *all real numbers* $f(x) \geq 2$, where the symbol $\geq$ means "is greater than or equal to."

The function $f(t) = \dfrac{1}{t + 2}$ is not defined for $t = -2$, for this value would require division by zero. Also, no matter how large t becomes, $f(t)$ will never exactly equal zero. Therefore, the domain of this function is *all real numbers except* -2, and the range is *all real numbers except 0*. ┄┄┄┄┄┄┄┄ ▌

■**EXAMPLE 2**   The function $g(s) = \sqrt{3 - s}$ is not defined for real numbers greater than 3, since such values make $3 - s$ negative and would result in imaginary values for $g(s)$. This means that the domain of this function is *all real numbers* $s \leq 3$, where the symbol $\leq$ means "is less than or equal to."

Also, since $\sqrt{3 - s}$ means the principal square root of $3 - s$ (see Section 1-6), we know that $g(s)$ cannot be negative. This tells us that the range of the function is *all real numbers* $g(s) \geq 0$.

In Examples 1 and 2 we determined the domain of each function by looking for those values of the independent variable that cannot be used. The range of each was found through an inspection of the function. When possible, this is the procedure we use, although it is often necessary to use more advanced methods to find the range of a function. Even in the second illustration of Example 1, we had to take a special look at the function to find the range. For this reason, until we develop other methods, we will look only for the domain of some functions.

Later in this chapter we will see that the graphing calculator is useful in finding the range of a function.

■**EXAMPLE 3**   Find the domain of the function $f(x) = 16\sqrt{x} + \dfrac{1}{x}$.

From the term $16\sqrt{x}$ we see that x must be greater than or equal to zero in order to have real values. The term $\dfrac{1}{x}$ indicates that x cannot be zero, because of division by zero. Thus, putting these together, the domain is *all real numbers* $x > 0$.

As for the range, it is *all real numbers* $f(x) \geq 12$. More advanced methods are needed to determine this.

We have seen that the domains of some functions are restricted to particular values. It can also happen that the domain of a function is restricted by definition or by practical considerations in an application. Consider the illustrations in the following example.

■**EXAMPLE 4**   A function defined as

$$f(x) = x^2 + 4 \quad \text{for } x > 2$$

has a domain restricted to real numbers greater than 2 by definition. Thus, $f(5) = 29$, but $f(1)$ is not defined, since 1 is not in the domain. Also, the range is all real numbers greater than 8.

The height h (in m) of a certain projectile as a function of the time t (in s) is found using the equation

$$h = 20t - 4.9t^2$$

Generally, negative values of time do not have meaning in such an application. This leads us to state the domain as values of $t \geq 0$. Of course, the projectile will not continue in flight indefinitely, and there is some upper limit on the value of t. These restrictions are not usually stated unless there is a particular reason that affects the solution.

The following example illustrates a function that is defined differently for different intervals of the domain.

EXAMPLE 5 In a certain electric circuit, the current i (in mA) is a function of the time t (in s), which means $i = f(t)$. The function is

$$f(t) = \begin{cases} 8 - 2t & \text{for } 0 \le t \le 4 \text{ s} \\ 0 & \text{for } t > 4 \text{ s} \end{cases}$$

Since negative values of t are not usually meaningful, $f(t)$ is not defined for $t < 0$. Find the current for $t = 3$ s, $t = 6$ s, and $t = -1$ s.

We are to find $f(3)$, $f(6)$, and $f(-1)$, and we see that values of this function are determined differently depending on the value of t. Since 3 is between 0 and 4,

$$f(3) = 8 - 2(3) = 2 \quad \text{or} \quad i = 2 \text{ mA}$$

Since 6 is greater than 4, $f(6) = 0$, or $i = 0$ mA. We see that $i = 0$ mA for all values of t that are 4 or greater.

Since $f(t)$ is not defined for $t < 0$, $f(-1)$ is not defined. ▬▬▬▬▬■

Functions from Verbal Statements

In the previous section we wrote functions in mathematical form from given statements and by using geometric information. It is often necessary to determine a mathematical function from a given statement. The function is formed using methods similar to those we used in establishing equations from statements in Chapter 1. In the following examples we set up such functions.

SOLVING A WORD PROBLEM

EXAMPLE 6 The fixed cost for a company to operate a certain plant is $3000 per day. It also costs $4 for each unit produced in the plant. Express the daily cost C of operating the plant as a function of the number n of units produced.

The daily total cost C equals the fixed cost of $3000 plus the cost of producing n units. Since the cost of producing one unit is $4, the cost of producing n units is $4n$. Thus, the total cost C, where $C = f(n)$, is

$$C = 3000 + 4n$$

Here we know that the domain is all values of $n \ge 0$, with some upper limit on n based on the production capacity of the plant. ▬▬▬▬▬■

SOLVING A WORD PROBLEM

EXAMPLE 7 A metallurgist melts and mixes m grams of solder that is 40% tin with n grams of another solder that is 20% tin to get a final solder mixture that contains 200 g of tin. Express n as a function of m. See Fig. 3-4.

The statement leads to the following equation.

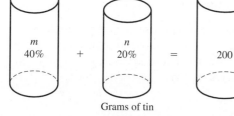

Grams of tin

Fig. 3-4

tin in first solder		tin in second solder		total amount of tin
$0.40m$	$+$	$0.20n$	$=$	200

Since we want $n = f(m)$, we now solve for n.

$$0.20n = 200 - 0.40m$$
$$n = 1000 - 2m$$

This is the required function. Since neither m nor n can be negative, the domain is all values $0 \le m \le 500$ g, which means that m is greater than or equal to 0 g and less than or equal to 500 g. The range is all values $0 \le n \le 1000$ g. ▬▬▬▬▬■

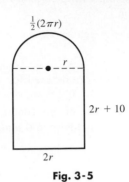

$\frac{1}{2}(2\pi r)$

r

$2r + 10$

$2r$

Fig. 3-5

EXAMPLE 8   An architect designs a window such that it has the shape of a rectangle with a semicircle on top, as shown in Fig. 3-5. The base of the window is 10 cm less than the height of the rectangular part. Express the perimeter p of the window as a function of the radius r of the circular part.

   We know that the perimeter is the distance around the window. Since the top part is a semicircle and the circumference of a circle is $2\pi r$, the length of the top circular part is $\frac{1}{2}(2\pi r)$. This also tells us that the dashed line, and therefore the base of the window, is $2r$. Finally, the fact that the base is 10 cm less than the height of the rectangular part tells us that each vertical part is $2r + 10$. This means that the perimeter p, where $p = f(r)$, is

$$p = \tfrac{1}{2}(2\pi r) + 2r + 2(2r + 10)$$
$$= \pi r + 2r + 4r + 20$$
$$= \pi r + 6r + 20$$

We see that the required function is $p = \pi r + 6r + 20$. Since the radius cannot be negative and there would be no window if $r = 0$, the domain of the function is all values $0 < r \le R$, where R is a maximum possible value of r determined by design considerations.

   In the definition of a function, it was stipulated that any value for the independent variable must yield only a single value of the dependent variable. This requirement is stressed in more advanced courses, and we will use it again in Chapter 20. *If a value of the independent variable yields one or more values of the dependent variable, the relationship is called a* **relation** *instead of a function.* A relation involves two variables related so that values of the second variable can be determined from values of the first variable. A function is a relation in which each value of the first variable yields only one value of the second. A function is therefore a special type of relation. However, there are relations that are not functions.

RELATION

EXAMPLE 9   For $y^2 = 4x^2$, if $x = 2$, then y can be either 4 or -4. Since a value of x yields more than one single value for y, we see that $y^2 = 4x^2$ is a relation, not a function.

------ EXERCISES *3-2* ------

In Exercises 1–8, determine the domain and range of the given functions. In Exercises 7 and 8, explain your answers.

1. $f(x) = x + 5$

2. $g(u) = 3 - u^2$

3. $G(R) = \dfrac{3.2}{R}$

4. $F(r) = \sqrt{r + 4}$

5. $f(s) = \dfrac{2}{s^2}$

6. $T(t) = 2t^4 + t^2 - 1$

(W) 7. $H(h) = 2h + \sqrt{h} + 1$

(W) 8. $f(x) = \dfrac{6}{\sqrt{2 - x}}$

In Exercises 9–12, determine the domain of the given functions.

9. $Y(y) = \dfrac{y + 1}{\sqrt{y - 2}}$

10. $f(n) = \dfrac{n}{6 - 2n}$

11. $f(D) = \dfrac{D}{D - 2} + \dfrac{4}{D + 4} - \dfrac{D - 3}{D - 6}$

12. $g(x) = \dfrac{\sqrt{x - 2}}{x - 3}$

In Exercises 13–16, evaluate the indicated functions.

$$F(t) = 3t - t^2 \quad \text{for } t \le 2 \qquad h(s) = \begin{cases} 2s & \text{for } s < -1 \\ s + 1 & \text{for } s \ge -1 \end{cases}$$

$$f(x) = \begin{cases} x + 1 & \text{for } x < 1 \\ \sqrt{x + 3} & \text{for } x \ge 1 \end{cases} \qquad g(x) = \begin{cases} \dfrac{1}{x} & \text{for } x \ne 0 \\ 0 & \text{for } x = 0 \end{cases}$$

13. Find $F(2)$ and $F(3)$.

14. Find $h(-8)$ and $h\left(-\dfrac{1}{2}\right)$.

15. Find $f(1)$ and $f\left(-\dfrac{1}{4}\right)$.

16. Find $g\left(\dfrac{1}{5}\right)$ and $g(0)$.

In Exercises 17–28, determine the appropriate functions.

17. A motorist travels at 40 mi/h for 2 h and then at 55 mi/h for t hours. Express the distance d traveled as a function of t.

18. Express the cost C of insulating a cylindrical water tank of height 2 m as a function of its radius r, if the cost of insulation is \$3 per square meter.

19. A rocket burns up at the rate of 2 tons/min after falling out of orbit into the atmosphere. If the rocket weighed 5500 tons before reentry, express its weight w as a function of the time t, in minutes, of reentry.

20. A computer part costs \$3 to produce and distribute. Express the profit p made by selling 100 of these parts as a function of the price of c dollars each.

21. Upon ascending, a weather balloon ices up at the rate of 0.5 kg/m after reaching an altitude of 1000 m. If the mass of the balloon below 1000 m is 110 kg, express its mass m as a function of its altitude h if $h > 1000$ m.

22. A chemist adds x liters of a solution that is 50% alcohol to 100 L of a solution that is 70% alcohol. Express the number n of liters of alcohol in the final solution as a function of x.

23. A company installs underground cable at a cost of \$500 for the first 50 ft (or up to 50 ft) and \$5 for each foot thereafter. Express the cost C as a function of the length l of underground cable if $l > 50$ ft.

24. The *mechanical advantage* of an inclined plane is the ratio of the length of the plane to its height. Express the mechanical advantage M of a plane of length 8 m as a function of its height h.

25. The capacities (in L) of two oil-storage tanks are x and y. The tanks are initially full; 1200 L is removed from them by taking 10% of the contents of the first tank and 40% of the contents of the second tank. (a) Express y as a function of x. (b) Find $f(400)$.

26. A special buoy 36 in. in diameter is floating in a lake and is more than half above the water. Express the circumference c of the circle of intersection of the buoy and water as a function of the depth d to which the buoy sinks.

27. In studying the electric current that is induced in wire rotating through a magnetic field, a piece of wire 60 cm long is cut into two pieces. One of these is bent into a circle and the other into a square. Express the total area A of the two figures as a function of the perimeter p of the square.

28. The cross section of an air-conditioning duct is in the shape of a square with semicircles on each side. See Fig. 3-6. Express the area A of this cross section as a function of the diameter d (in cm) of the circular part.

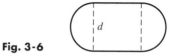

Fig. 3-6

In Exercises 29–36, solve the given problems.

29. A computer program displays a circular image of radius 6 in. If the radius is decreased by x in., express the area of the image as a function of x. What are the domain and range of $A = f(x)$?

30. A helicopter 120 m from a person takes off vertically. Express the distance d from the person to the helicopter as a function of the height h of the helicopter. What are the domain and the range of $d = f(h)$? See Fig. 3-7.

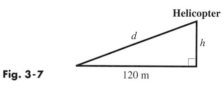

Fig. 3-7

31. A truck travels 300 km in t hours. Express the average speed s of the truck as a function of t. What are the domain and range of $s = f(t)$?

32. A rectangular grazing range with an area of 8 mi^2 is to be fenced. Express the length l of the field as a function of its width w. What are the domain and range of $l = f(w)$?

(W) 33. The resonant frequency f (in Hz) in a certain electric circuit as a function of the capacitance is $f = \dfrac{1}{2\pi\sqrt{C}}$. Describe the domain of this function.

34. A jet is traveling directly between Calgary, Alberta, and Portland, Oregon, which are 550 mi apart. If the jet is x mi from Calgary and y mi from Portland, find the domain of $y = f(x)$.

35. Express the mass m of the weather balloon in Exercise 21 as a function of any height h in the same manner as the function in Example 5 (and Exercises 13–16) was expressed.

36. Express the cost C of installing any length l of the underground cable in Exercise 23 in the same manner as the function in Example 5 (and Exercises 13–16) was represented.

3-3 RECTANGULAR COORDINATES

One of the most valuable ways of representing a function is by graphical representation. By using graphs we are able to obtain a "picture" of the function, and by using this picture we can learn a great deal about the function.

To make a graphical representation of a function, we recall from Chapter 1 that numbers can be represented by points on a line. For a function, we have values of the independent variable as well as the corresponding values of the dependent variable. Therefore, it is necessary to use two different lines to represent the values from each of these sets of numbers. We do this by placing the lines perpendicular to each other.

We place one line horizontally and label it the **x-axis.** The values of the independent variable are normally placed on this axis. *The other line is placed vertically and labeled the* **y-axis.** Normally the y-axis is used for values of the dependent variable. *The point of intersection is called the* **origin.** This is the **rectangular coordinate system.**

Rectangular (Cartesian) coordinates were developed by the French mathematician Descartes (1596–1650).

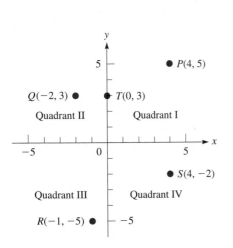

Fig. 3-8

On the x-axis, positive values are to the right of the origin, and negative values are to the left of the origin. On the y-axis, positive values are above the origin, and negative values are below it. *The four parts into which the plane is divided are called* **quadrants,** which are numbered as in Fig. 3-8.

A point P on the plane is designated by the pair of numbers (x, y), where x is the value of the independent variable and y is the corresponding value of the dependent variable. *The x-value, called the* **abscissa,** *is the perpendicular distance of P from the y-axis. The y-value, called the* **ordinate,** *is the perpendicular distance of P from the x-axis.* The values x and y together, written as (x, y), are the **coordinates** of the point P. Note carefully that *the x-value is always written first, and the y-value is written second.*

NOTE ▶

■ EXAMPLE 1 Locate the points $A(2, 1)$ and $B(-4, -3)$ on the rectangular coordinate system.

The coordinates $(2, 1)$ for A mean that the point is 2 units to the *right* of the y-axis and 1 unit *above* the y-axis, as shown in Fig. 3-9. The coordinates $(-4, -3)$ for B mean that the point is 4 units to the *left* of the y-axis and 3 units *below* the x-axis, as shown.

The abscissa of point A is 2, and the ordinate of A is 1. For point B, its abscissa is -4, and its ordinate is -3. ■

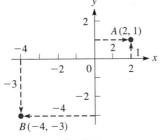

Fig. 3-9

■ EXAMPLE 2 The positions of points $P(4, 5)$, $Q(-2, 3)$, $R(-1, -5)$, $S(4, -2)$, and $T(0, 3)$ are shown in Fig. 3-8 above. We see that this representation allows for *one point for any pair of values* (x, y). Also note that the point $T(0, 3)$ is on the y-axis. Any such point that is on either axis is not *in* any of the four quadrants. ■

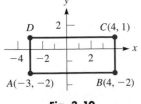

Fig. 3-10

■ EXAMPLE 3 Three vertices of the rectangle in Fig. 3-10 are $A(-3, -2)$, $B(4, -2)$, and $C(4, 1)$. What is the fourth vertex?

We use the fact that opposite sides of a rectangle are equal and parallel to find the solution. Since both vertices of the base AB of the rectangle have a y-coordinate of -2, the base is parallel to the x-axis. Therefore, the top of the rectangle must also be parallel to the x-axis. Thus, the vertices of the top must both have a y-coordinate of 1, since one of them has a y-coordinate of 1. In the same way, the x-coordinates of the left side must both be -3. Therefore, the fourth vertex is $D(-3, 1)$. ■

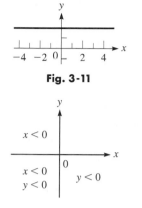

Fig. 3-11

EXAMPLE 4   Where are all points whose ordinates are 2?

Since the ordinate is the *y*-value, we can see that the question could be stated as: "Where are all points for which $y = 2$?" Since all such points are 2 units above the *x*-axis, the answer can be stated as "on a line 2 units above the *x*-axis." See Fig. 3-11. ∎

EXAMPLE 5   Where are all points (x, y) for which $x < 0$ and $y < 0$?

Noting that $x < 0$ means "*x* is less than zero," or "*x* is negative," and that $y < 0$ means the same for *y*, we want to determine where both *x* and *y* are negative. Our answer is "in the third quadrant," since both coordinates are negative for all points in the third quadrant, and this is the only quadrant for which this is true. See Fig. 3-12. ∎

Fig. 3-12

EXERCISES *3-3*

In Exercises 1 and 2, determine (at least approximately) the coordinates of the points specified in Fig. 3-13.

1. *A, B, C*

2. *D, E, F*

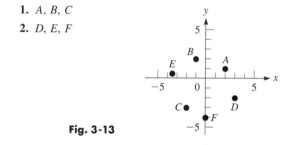

Fig. 3-13

In Exercises 3 and 4, plot the given points.

3. $A(2, 7)$, $B(-1, -2)$, $C(-4, 2)$

4. $A(3, \frac{1}{2})$, $B(-6, 0)$, $C(-\frac{5}{2}, -5)$

In Exercises 5–8, plot the given points and then join these points, in the order given, by straight-line segments. Name the geometric figure formed.

5. $A(-1, 4)$, $B(3, 4)$, $C(1, -2)$

6. $A(0, 3)$, $B(0, -1)$, $C(4, -1)$

7. $A(-2, -1)$, $B(3, -1)$, $C(3, 5)$, $D(-2, 5)$

8. $A(-5, -2)$, $B(4, -2)$, $C(6, 3)$, $D(-3, 3)$

In Exercises 9–12, find the indicated coordinates.

9. Three vertices of a rectangle are $(5, 2)$, $(-1, 2)$, and $(-1, 4)$. What are the coordinates of the fourth vertex?

10. Two vertices of an equilateral triangle are $(7, 1)$ and $(2, 1)$. What is the abscissa of the third vertex?

11. *P* is the point $(3, 2)$. Locate point *Q* such that the *x*-axis is the perpendicular bisector of the line segment joining *P* and *Q*.

12. *P* is the point $(-4, 1)$. Locate point *Q* such that the line segment joining *P* and *Q* is bisected by the origin.

In Exercises 13–28, answer the given questions.

13. Where are all the points whose abscissas are 1?

14. Where are all the points whose ordinates are -3?

15. Where are all points such that $y = 3$?

16. Where are all points such that $x = -2$?

17. Where are all the points whose abscissas equal their ordinates?

18. Where are all the points whose abscissas equal the negative of their ordinates?

19. What is the abscissa of all points on the *y*-axis?

20. What is the ordinate of all points on the *x*-axis?

21. Where are all the points for which $x > 0$?

22. Where are all the points for which $y < 0$?

23. Where are all the points for which $x < -1$?

24. Where are all the points for which $y > 4$?

25. In which quadrants is the ratio y/x positive?

26. In which quadrants is the ratio y/x negative?

27. Find the distance (a) between $(3, -2)$ and $(-5, -2)$ and (b) between $(3, -2)$ and $(3, 4)$.

(W) 28. Find the distance between $(-5, -2)$ and $(3, 4)$. Explain your method. (Hint: See Exercise 27.)

3-4 THE GRAPH OF A FUNCTION

Now that we have introduced the concepts of a function and the rectangular coordinate system, we are in a position to determine the graph of a function. In this way we shall obtain a visual representation of a function.

The graph of a function is the set of all points whose coordinates (x, y) satisfy the functional relationship $y = f(x)$. Since $y = f(x)$, we can write the coordinates of the points on the graph as $(x, f(x))$. Writing the coordinates in this manner tells us exactly how to find them. *We assume a certain value for x and then find the value of the function of x. These two numbers are the coordinates of the point.*

Since there is no limit to the possible number of points that can be chosen, we normally select a few values of x, obtain the corresponding values of the function, plot these points, and then join them. Therefore, we use the following basic procedure in plotting a graph.

As to just what values of x to choose and how many to choose, with a little experience you will usually be able to tell if you have enough points to plot an accurate graph.

> **Procedure for Plotting the Graph of a Function**
> 1. *Let x take on several values and calculate the corresponding values of y.*
> 2. *Tabulate these values, arranging the table so that **values of x are increasing.***
> 3. *Plot the points and join them from left to right by a **smooth curve** (not short straight-line segments).*

■**EXAMPLE 1**   Graph the function $f(x) = 3x - 5$.

For purposes of graphing, we let $y = f(x)$, or $y = 3x - 5$. We then let x take on various values and determine the corresponding values of y. Note that once we choose a given value of x, we have no choice about the corresponding y-value, as it is determined by evaluating the function. If $x = 0$, we find that $y = -5$. This means that the point $(0, -5)$ is on the graph of the function $3x - 5$. Choosing another value of x—for example, 1—we find that $y = -2$. This means that the point $(1, -2)$ is on the graph of the function $3x - 5$. Continuing to choose a few other values of x, we tabulate the results, as shown in Fig. 3-14. It is best to arrange the table so that the values of x increase; then there is no doubt how they are to be connected, for they are then connected in the order shown. Finally, we connect the points in Fig. 3-14 and see that the graph of the function $3x - 5$ is a straight line.

The table of values for a graph can be seen on a graphing calculator using the *table* feature. Below, in Fig. 3-15, is the calculator display of the table for Example 1.

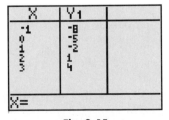

Fig. 3-15

Select values of x and calculate corresponding values of y

$f(x) = 3x - 5$
$f(-1) = 3(-1) - 5 = -8$
$f(0) = 3(0) - 5 = -5$
$f(1) = 3(1) - 5 = -2$
$f(2) = 3(2) - 5 = 1$
$f(3) = 3(3) - 5 = 4$

Tabulate with values of x increasing

x	y
-1	-8
0	-5
1	-2
2	1
3	4

Plot points and join from left to right with smooth curve

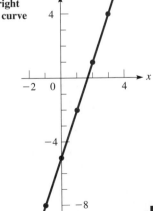

Fig. 3-14

EXAMPLE 2   Graph the function $f(x) = 2x^2 - 4$.

First, we let $y = 2x^2 - 4$ and tabulate the values as shown in Fig. 3-16. In determining the values in the table, we must take particular care to obtain the correct values of y for negative values of x. ***Mistakes are relatively common when dealing with negative numbers.*** We must carefully use the laws for signed numbers. For example, if $x = -2$, we have $y = 2(-2)^2 - 4 = 2(4) - 4 = 8 - 4 = 4$. Once the values are obtained, we plot and connect the points with a smooth curve, as shown.

CAUTION ▶

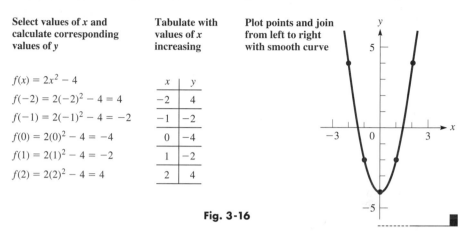

Select values of x and calculate corresponding values of y		Tabulate with values of x increasing		Plot points and join from left to right with smooth curve

$f(x) = 2x^2 - 4$

$f(-2) = 2(-2)^2 - 4 = 4$

$f(-1) = 2(-1)^2 - 4 = -2$

$f(0) = 2(0)^2 - 4 = -4$

$f(1) = 2(1)^2 - 4 = -2$

$f(2) = 2(2)^2 - 4 = 4$

x	y
-2	4
-1	-2
0	-4
1	-2
2	4

Fig. 3-16

When graphing a function, there are some special points about which we should be careful. These include the following:

RESTRICTIONS ON A GRAPH

1. Since the graphs of most common functions are smooth, any irregularities in the graph should be carefully checked. In these cases it usually helps to take values of x between those values where the question arises.

2. The domain of the function may not include all values of x (remember, *division by zero is not defined, and only real values of the variables are permissible*).

3. In applications, we must be careful to use values that are meaningful. As we have seen, negative values for many quantities, such as time, are not generally meaningful.

The following examples illustrate these points.

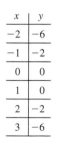

x	y
-2	-6
-1	-2
0	0
1	0
2	-2
3	-6

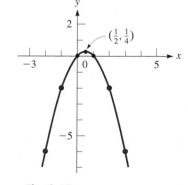

Fig. 3-17

EXAMPLE 3   Graph the function $y = x - x^2$.

First we determine the values in the table, as shown with Fig. 3-17. Again, we must be careful when dealing with negative values of x. For the value $x = -1$, we have $y = (-1) - (-1)^2 = -1 - 1 = -2$. Once all values in the table have been found and plotted, we note that $y = 0$ *for both* $x = 0$ *and* $x = 1$. The question arises—*what happens between these values?* Trying $x = \frac{1}{2}$, we find that $y = \frac{1}{4}$. Using this point completes the information needed to complete an accurate graph. Also note that in plotting these graphs we do not stop the graph with the last points determined but indicate that the curve continues by drawing the graph past these points.

EXAMPLE 4   Graph the function $y = 1 + \dfrac{1}{x}$.

CAUTION ▶

In finding the points on this graph, as shown in Fig. 3-18, we note that y is not defined for $x = 0$, due to division by zero. Thus, $x = 0$ is not in the domain, and *we must be careful not to have any part of the curve cross the y-axis* ($x = 0$). Although we cannot let $x = 0$, we can choose other values of x between -1 and 1 that are close to zero. In doing so, we find that as x gets closer to zero, the points get closer and closer to the y-axis, although they do not reach or touch it. In this case the y-axis is called an **asymptote** of the curve.   ⬛

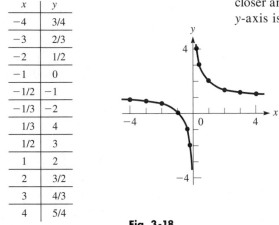

x	y
-4	3/4
-3	2/3
-2	1/2
-1	0
-1/2	-1
-1/3	-2
1/3	4
1/2	3
1	2
2	3/2
3	4/3
4	5/4

Fig. 3-18

x	y
-1	0
0	1
1	1.4
2	1.7
3	2
4	2.2
5	2.4
6	2.6
7	2.8
8	3

Fig. 3-19

EXAMPLE 5   Graph the function $y = \sqrt{x + 1}$.

NOTE ▶

When finding the points for the graph, we may not let x take on any value less than -1, for *all such values would lead to imaginary values for y* and are not in the domain. Also, since we have the positive square root indicated, the range consists of all values of y that are positive or zero ($y \geq 0$). See Fig. 3-19. Note that the graph starts at $(-1, 0)$.   ⬛

See the chapter introduction.

The unit of power, the watt (W), is named for James Watt (1736–1819), a British engineer.

EXAMPLE 6   The electric power P (in W) delivered by a certain battery as a function of the resistance R (in Ω) in the circuit is given by $P = \dfrac{100R}{(0.50 + R)^2}$. Plot P as a function of R.

Since negative values for the resistance have no physical significance, we should not plot any values of P for negative values of R. The following table is obtained.

R (Ω)	0	0.25	0.50	1.0	2.0	3.0	4.0	5.0	10.0
P (W)	0.0	44.4	50.0	44.4	32.0	24.5	19.8	16.5	9.1

The values 0.25 and 0.50 are used for R when it is found that P is less for $R = 2$ than for $R = 1$. The sharp change in direction at $R = 1$ should be checked by using these additional points to better see how the curve changes near $R = 1$ and thereby obtain a smoother curve. See Fig. 3-20.

Also note that the scale on the P-axis is different from that on the R-axis. This reflects the different magnitudes and ranges of values used for each of the variables. Different scales are normally used in such cases.

We can make various conclusions from the graph. For example, we see that the maximum power of 50 W occurs for $R = 0.5 \ \Omega$. Also, P decreases as R increases beyond 0.5 Ω. We will consider further the information that can be read from a graph in the next section.   ⬛

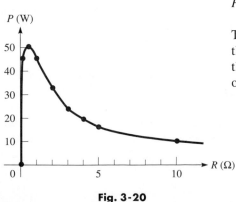

Fig. 3-20

The following example illustrates the graph of a function that is defined differently for different intervals of the domain.

EXAMPLE 7 Graph the function $f(x) = \begin{cases} 2x + 1 & \text{for } x \leq 1 \\ 6 - x^2 & \text{for } x > 1 \end{cases}$

First, we let $y = f(x)$ and then tabulate the necessary values. In evaluating $f(x)$ we must be careful to use the proper part of the definition. In order to see where to start the curve for $x > 1$, we evaluate $6 - x^2$ for $x = 1$, but

NOTE ▶ we must realize that ***the curve does not include this point*** $(1, 5)$ and starts immediately to its right. To show that this is not part of the curve, we draw it as an open circle. See Fig. 3-21. Such a function, with a "break" in it, is called *discontinuous*. --------- ∎

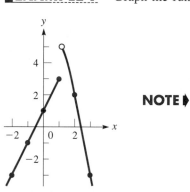

$f(-2) = 2(-2) + 1 = -3$

$f(-1) = 2(-1) + 1 = -1$

$f(0) = 2(0) + 1 = 1$

$f(1) = 2(1) + 1 = 3$

$f(2) = 6 - 2^2 = 2$

$f(3) = 6 - 3^2 = -3$

x	y
-2	-3
-1	-1
0	1
1	3
2	2
3	-3

Fig. 3-21

In Section 3-1, we defined a function such that it has only one value of the dependent variable for each value of the independent variable. At the end of Section 3-2, we stated that a relation may have more than one such value of the dependent variable. The following example illustrates how to use the graph to determine whether or not a relation is a function.

x	y
-2	± 4
-1	± 2
0	0
1	± 2
2	± 4

Two values of y for $x = 2$

EXAMPLE 8 In Example 9 of Section 3-2, we noted that $y^2 = 4x^2$ is a relation, but not a function. Since $y = 4$ or $y = -4$, normally written as $y = \pm 4$, for $x = 2$, we have two values of y for $x = 2$. Therefore, it is not a function. Making the table as shown at the left, we then draw the graph in Fig. 3-22. When making the table, we also note that there are two values of y for every value of x, except $x = 0$.

If any vertical line that crosses the x-axis in the domain intersects the graph in more than one point, it is the graph of a relation that is not a function. Any such vertical line in any of the previous graphs of this section would intersect the graph in only one point. This shows that they are graphs of functions. --------- ∎

Fig. 3-22

Functions of a particular type have graphs of a specific basic shape, and many have been named. Two examples of this are the *straight line* (Example 1) and the *parabola* (Examples 2 and 3). We consider the straight line again in Chapter 5 and the parabola in Chapter 7. Other types of graphs are found in many of the later chapters, with a detailed analysis of some of them in Chapter 21.

The use of graphs is extensive in mathematics and in applications. For example, in the next section we use graphs to solve equations. Numerous applications can be noted for most types of curves. For example, the parabola has applications in microwave dish design, suspension bridge design, and the path (trajectory) of a baseball. We will note many more applications of graphs throughout the book.

EXERCISES $3\text{-}4$

In Exercises 1–32, graph the given functions.

1. $y = 3x$
2. $y = -2x$
3. $y = 2x - 4$
4. $y = 3x + 5$
5. $y = 7 - 2x$
6. $y = 5 - 3x$
7. $y = \frac{1}{2}x - 2$
8. $y = 6 - \frac{1}{3}x$
9. $y = x^2$
10. $y = -2x^2$
11. $y = 3 - x^2$
12. $y = x^2 - 3$
13. $y = \frac{1}{2}x^2 + 2$
14. $y = 2x^2 + 1$
15. $y = x^2 + 2x$
16. $h = 20t - 5t^2$
17. $y = x^2 - 3x + 1$
18. $y = 2 + 3x + x^2$
19. $V = e^3$
20. $y = -2x^3$
21. $y = x^3 - x^2$
22. $y = 3x - x^3$
23. $y = x^4 - 4x^2$
24. $y = x^3 - x^4$
25. $P = \dfrac{1}{V}$
26. $y = \dfrac{2}{x + 2}$
27. $y = \dfrac{4}{x^2}$
28. $p = \dfrac{1}{n^2 + 0.5}$
29. $y = \sqrt{x}$
30. $y = \sqrt{4 - x}$
31. $y = \sqrt{16 - x^2}$
32. $y = \sqrt{x^2 - 16}$

In Exercises 33–52, graph the indicated functions.

33. In blending gasoline, the number of gallons n of 85 octane gas to be blended with m gallons of 92 octane gas is given by the equation $n = 0.40m$. Plot n as a function of m.

34. The length L (in cm) of a pulley belt is 12 cm longer than the circumference of one of the pulley wheels. Express L as a function of the radius r of the wheel and plot L as a function of r.

35. For a certain model of truck, its resale value V (in dollars) as a function of its mileage m is $V = 50,000 - 0.2m$. Plot V as a function of m for $m \leq 100,000$ mi.

36. The resistance R (in Ω) of a resistor as a function of the temperature T (in °C) is given by $R = 250(1 + 0.0032T)$. Plot R as a function of T.

37. The consumption of fuel c (in L/h) of a certain engine is determined as a function of the number r of r/min of the engine, to be $c = 0.011r + 4.0$. This formula is valid from 500 r/min to 3000 r/min. Plot c as a function of r. (r is the symbol for revolution.)

38. The profit P (in dollars) a retailer makes in selling 50 cordless phones is given by $P = 50(p - 30)$, where p is the selling price. Plot P as a function of p for $p \geq \$20$ and $p \leq \$100$.

39. The rate H (in W) at which heat is developed in the filament of an electric light bulb as a function of the electric current I (in A) is $H = 240I^2$. Plot H as a function of I.

40. The total annual fraction f of energy supplied by solar energy to a home as a function of the area A (in m^2) of the solar collector is $f = 0.065\sqrt{A}$. Plot f as a function of A.

41. The maximum speed v (in mi/h) at which a car can safely travel around a circular turn of radius r (in ft) is given by $r = 0.42v^2$. Plot r as a function of v.

42. The height h (in m) of a rocket as a function of the time t (in s) is given by the function $h = 1500t - 4.9t^2$. Plot h as a function of t, assuming level terrain.

43. A formula used to determine the number N of board feet of lumber that can be cut from a 4-ft section of a log of diameter d (in in.) is $N = 0.22d^2 - 0.71d$. Plot N as a function of d for values of d from 10 in. to 40 in.

44. A copper electrode with a mass of 25.0 g is placed in a solution of copper sulfate. An electric current is passed through the solution, and 1.6 g of copper is deposited on the electrode each hour. Express the total mass m on the electrode as a function of the time t and plot the graph.

45. A land developer is considering several options of dividing a large tract into building lots, many of which would have perimeters of 200 m. For these, the minimum width would be 30 m and the maximum width would be 70 m. Express the areas A of these lots as a function of their widths w and plot the graph.

46. The distance p (in m) from a camera with a 50-mm lens to the object being photographed is a function of the magnification m of the camera given by $p = \dfrac{0.05(1 + m)}{m}$. Plot the graph for positive values of m up to 0.50.

47. A measure of the light beam that can be passed through an optic fiber is its numerical aperture N. For a particular optic fiber, N is a function of the index of refraction n of the glass in the fiber, given by $N = \sqrt{n^2 - 1.69}$. Plot the graph for $n \leq 2.00$.

48. The force F (in N) exerted by a cam on the arm of a robot is a function of the distance x shown in Fig. 3-23. The function is $F = x^4 - 12x^3 + 46x^2 - 60x + 25$. Plot the graph. (Note that x varies from 1 cm to 5 cm.)

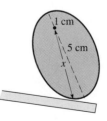

Fig. 3-23

W **49.** Plot the graphs of $y = x$ and $y = |x|$ on the same coordinate system. Explain why the graphs differ.

W **50.** Plot the graphs of $y = 2 - x$ and $y = |2 - x|$ on the same coordinate system. Explain why the graphs differ.

51. Plot the graph of $f(x) = \begin{cases} 3 - x & \text{for } x < 1 \\ x^2 + 1 & \text{for } x \geq 1 \end{cases}$

52. Plot the graph of $f(x) = \begin{cases} \dfrac{1}{x - 1} & \text{for } x < 0 \\ \sqrt{x + 1} & \text{for } x \geq 0 \end{cases}$

In Exercises 53–56, determine whether or not the indicated graph is that of a relation which is a function.

53. Fig. 3-24(a)   **54.** Fig. 3-24(b)

55. Fig. 3-24(c)   **56.** Fig. 3-24(d)

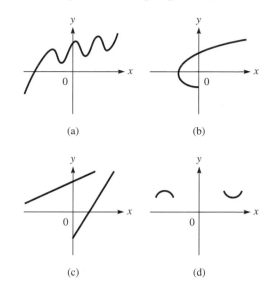

(a)   (b)

(c)   (d)

Fig. 3-24

3-5 GRAPHS ON THE GRAPHING CALCULATOR

We have discussed the use of a graphing calculator for calculations, and in earlier sections of this chapter noted its use (in the margin) for evaluating a function and finding a table of values for graphing. In this section we will see that a graphing calculator can display a graph quickly and easily.

As we pointed out in Chapter 1, *you must be familiar with the sequence of keys to be used* on your calculator. For example, on some calculators the *graph* key is used after the function has been entered, whereas on others it puts the calculator into the graphing mode. A brief discussion of the features of graphing calculators is found in Appendix C. The manual for a particular model should be used for a detailed coverage of its features.

> See Appendix C for a list of graphing calculator features and examples of calculator use (with page references) that are included in this text.

> The specific detail shown in any nongraphic display depends on the model. A graph with the same *window* settings should appear about the same on all models.

■**EXAMPLE 1**   To graph the function $y = 2x + 8$, we first display $Y_1=$, and then enter the $2x + 8$. The display for this is shown in Fig. 3-25(a).

Next we use the *window* (or *range*) feature to set the part of the domain and the range that will be seen in the *viewing window*. For this function we set

Xmin = −6, Xmax = 2, Xscl = 1, Ymin = −2, Ymax = 10, Yscl = 1

in order to get a good view of the graph. The display for the *window* settings is shown in Fig. 3-25(b).

We then display the graph, using the *graph* (or *exe*) key. The display showing the graph of $y = 2x + 8$ is shown in Fig. 3-25(c).

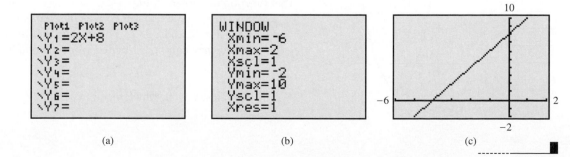

Fig. 3-25   (a)   (b)   (c)

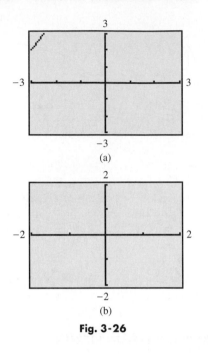

Fig. 3-26

In later examples we generally will not show the *window* settings in the text. They will be shown outside the display, as in Fig. 3-26.

In Example 1, we noted that the *window* feature sets the intervals of the domain and range of the function, which is seen in the viewing window. Unless the settings are appropriate for the given function, the viewing window may not display a good view of the graph. Consider the following example.

■ EXAMPLE 2   In Example 1, the *window* settings were chosen to give a good view of the graph of $y = 2x + 8$. However, if we had chosen settings of Xmin = −3, Xmax = 3, Ymin = −3, and Ymax = 3, we get the view shown in Fig. 3-26(a). We see that very little of the graph can be seen. It is possible that no part of the graph will be seen, such as is shown in Fig. 3-26(b), where Xmin = −2, Xmax = 2, Ymin = −2, and Ymax = 2. As we see, care must be used in choosing these settings. This includes the scale settings (Xscl and Yscl), which should be chosen so as to give several (not too many or too few) markers on each axis.

Since these settings are easily changed, at first it is usually best to choose intervals between the min and max values that are larger than is probably necessary to get the general location of the curve. These intervals can then be reduced to obtain the view that is needed.

Also note that the calculator constructs the graph just as we did in the previous section—by plotting points (square dots called *pixels*). It just does it a lot faster. ■

Solving Equations Graphically

An equation can be solved by use of a graph. Most of the time the solution will be approximate, but with a graphing calculator it is possible to get good accuracy in the result. The procedure used is as follows:

Procedure for Solving an Equation Graphically

1. *Collect all terms on one side of the equal sign.* This gives us the equation $f(x) = 0$.

2. *Set $y = f(x)$ and graph this function.*

3. *Find the points where the graph crosses the x-axis. These points are called the **x-intercepts** of the graph.* At these points, $y = 0$.

4. *The values of x for which $y = 0$ are the solutions of the equation. (These values are called the **zeros** of the function $f(x)$.)*

The next example is done without the graphing calculator to illustrate the method.

■ EXAMPLE 3   Graphically solve the equation $x^2 - 2x = 1$.

Following the above procedure, we first rewrite the equation as $x^2 - 2x - 1 = 0$ and then set $y = x^2 - 2x - 1$. Next we plot the graph of this function as shown in Fig. 3-27.

For this curve, we see that it crosses the x-axis between $x = -1$ and $x = 0$ and between $x = 2$ and $x = 3$. This means that there are *two* solutions of the equation, and they can be estimated from the graph as

$$x = -0.5 \quad \text{and} \quad x = 2.5$$

For these values, $y = 0$, which means $x^2 - 2x - 1 = 0$, which in turn means $x^2 - 2x = 1$. ■

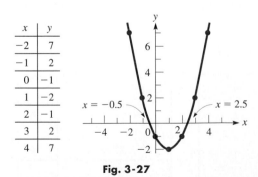

x	y
−2	7
−1	2
0	−1
1	−2
2	−1
3	2
4	7

Fig. 3-27

We now show the use of a graphing calculator in solving equations graphically. In the next example we solve the equation of Example 3.

EXAMPLE 4   Using a graphing calculator, solve the equation $x^2 - 2x = 1$.

Again, we rewrite the equation as $x^2 - 2x - 1 = 0$ and set $y = x^2 - 2x - 1$. Then knowing the approximate solution from Example 3, we have the display shown in Fig. 3-28(a). The scale on each axis is 0.5.

To get more accurate values of x for which $y = 0$, we use the *trace* feature. To find the y-value closest to zero, we move the *cursor* (a blinking pixel) with the arrow keys. In this way we have the display shown in Fig. 3-28(b), and we find the solutions to be approximately $x = -0.4$ and $x = 2.4$. We should note that the pixel for which y is closest to zero generally will not have a y-coordinate of *exactly zero*.

Much more accurate values can be found by using the *zoom* feature, with which any region of the screen can be greatly magnified. Fig. 3-28(c) shows the screen when *zoom* is used near $x = 2.4$. The cursor shows that the solution is about $x = 2.415$. In the same way, the other solution is about $x = -0.415$.

Checking these solutions *in the original equation,* we get $1.002225 = 1$ for each. This shows that they check, since we know that they are approximate.

Depending on the calculator model, the equation and cursor coordinates may be displayed when using *trace*.

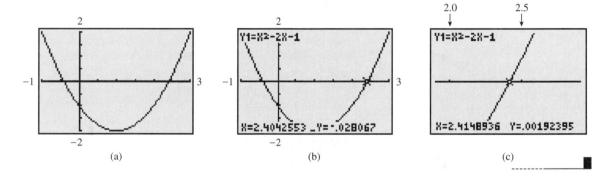

Fig. 3-28   (a)   (b)   (c)

Many calculators have a *zero* (or *root*) feature. The equation in Example 5 can be solved using this feature by finding the zeros on the graph of the function $y = 2x^3 + 3x^2 - 3x$.

Many calculators have an *intersect* feature. The equation in Example 5 can be solved using this feature by finding the point where the curves $y = x^2(2x + 3)$ and $y = 3x$ cross.

EXAMPLE 5   Solve the equation $t^2(2t + 3) = 3t$ graphically.

Collecting all terms on the left leads to the equation $2t^3 + 3t^2 - 3t = 0$, and we then let $y = 2t^3 + 3t^2 - 3t$. On a calculator we graph $y = 2x^3 + 3x^2 - 3x$ (letting $t = x$) as shown in Fig. 3-29(a). The *window* settings are usually the *default* settings.

This gives us a general idea of the graph, but to get better accuracy we change the settings to those shown in Fig. 3-29(b).

Now using the *trace* feature, we get the solutions of $t = -2.2$, $t = 0$, and $t = 0.7$. Here, $t = 0$ is exact (which the calculator may show with $x = 0$, $y = 0$, and no other decimal positions shown). The other solutions are approximate. Checking these solutions, we find that they check (and also that $t = 0$ is an exact solution).

More accurate values can be obtained by using the *zoom* feature.

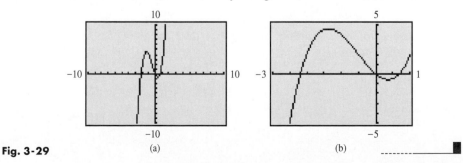

Fig. 3-29   (a)   (b)

SOLVING A WORD PROBLEM

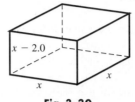

Fig. 3-30

EXAMPLE 6   A rectangular box whose volume is 30 cm^3 (two significant digits) is made with a square base and a height that is 2.0 cm less than the length of a side of the base. Find the dimensions of the box by first setting up the necessary equation and then solving it graphically.

Let x = the length of a side of the square base (see Fig. 3-30); the height is then $x - 2.0$. This means that the volume is $(x)(x)(x - 2.0)$, or $x^2(x - 2.0)$. Since the volume is 30 cm^3, we have the equation

$$x^2(x - 2.0) = 30$$

To solve it graphically, we first rewrite the equation as $x^3 - 2.0x^2 - 30 = 0$ and then set $y = x^3 - 2.0x^2 - 30$. The graphing calculator view is shown in Fig. 3-31. We use only positive values of x since negative values have no meaning. Using the *trace* feature (or *zero* feature or *intersect* feature), we find that $x = 3.9$ is the approximate solution. Therefore, the dimensions are 3.9 cm, 3.9 cm, and 1.9 cm. Checking, these dimensions give a volume of 29 cm^3. The difference from 30 cm^3 is due to the use of the approximate value of x. A better check can be made by using the unrounded calculator value for x.

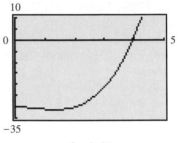

Fig. 3-31

Finding the Range of a Function

In Section 3-2 we stated that advanced methods are often necessary to find the range of a function and that we would see that a graphing calculator is useful for this purpose. As with solving equations, we can get very good approximations of the range for most functions by using a graphing calculator.

The method is simply to graph the function and see for what values of y there is a point on the graph. These values of y give us the range of the function.

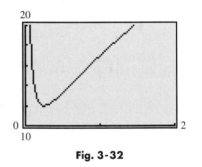

Fig. 3-32

EXAMPLE 7   Graphically find the range of the function $y = 16\sqrt{x} + \dfrac{1}{x}$.

If we start by using the default *window* settings of Xmin = -10, Xmax = 10, Ymin = -10, Ymax = 10, we see no part of the graph. Looking back to the function, we note that x cannot be zero because that would require dividing by zero in the $1/x$ term. Also, $x > 0$ since $\sqrt{x}$ is not defined for $x < 0$. Thus, we should try Xmin = 0 and Ymin = 0.

We also can easily see that $y = 17$ for $x = 1$. This means we should try a value more than 17 for Ymax. Using Ymax = 20, after two or three *window* settings, we use the settings shown in Fig. 3-32. The display shows the range to be all real numbers greater than about 12. (Actually, the range is all real numbers $y \geq 12$.)

─────────────────────── EXERCISES $3\text{-}5$ ───────────────────────

In Exercises 1–16, use a graphing calculator to solve the given equations to the nearest 0.1.

1. $7x - 5 = 0$

2. $8x + 3 = 0$

3. $x^2 - 41 = 0$

4. $2x^2 = 1$

5. $x^2 + x - 5 = 0$

6. $x^2 - 2x - 4 = 0$

7. $2x^2 - x = 7$

8. $w(w - 4) = 9$

9. $x^3 = 4x$

10. $x^3 - 3 = 3x$

11. $x^4 - 2x = 0$

12. $5x = 2x^5$

13. $\sqrt{5R + 2} = 3$

14. $\sqrt{x} + 3x = 7$

15. $\dfrac{1}{x^2 + 1} = 0$

16. $x - 2 = \dfrac{1}{x}$

In Exercises 17–24, use a graphing calculator to find the range of the given functions. (The functions of Exercises 21–24 are the same as the functions of Exercises 9–12 of Section 3-2.)

17. $y = \dfrac{4}{x^2 - 4}$

18. $y = \dfrac{x + 1}{x^2}$

19. $y = \dfrac{x^2}{x + 1}$

20. $y = \dfrac{x}{x^2 - 4}$

21. $Y(y) = \dfrac{y + 1}{\sqrt{y - 2}}$

22. $f(n) = \dfrac{n}{6 - 2n}$

23. $f(D) = \dfrac{D}{D - 2} + \dfrac{4}{D + 4} - \dfrac{D - 3}{D - 6}$

24. $g(x) = \dfrac{\sqrt{x - 2}}{x - 3}$

In Exercises 25–32, solve the indicated equations graphically. Assume all data are accurate to two significant digits unless greater accuracy is given.

25. In an electric circuit, the current i (in A) as a function of voltage v is given by $i = 0.01v - 0.06$. Find v for $i = 0$.

26. For tax purposes, a corporation assumes that one of its computers depreciates according to the equation $V = 90{,}000 - 12{,}000t$, where V is the value (in dollars) of the computer after t years. According to this formula, when will the computer be fully depreciated (no value)?

27. Two cubical coolers together hold 40.0 L (40,000 cm³). If the inside edge of one is 5.00 cm greater than the inside edge of the other, what is the inside edge of each?

28. The pressure loss P (in lb/in.² per 100 ft) in a fire hose is given by $P = 0.00021Q^2 + 0.013Q$, where Q is the rate of flow (in gal/min). Find Q if $P = 1.2$ lb/in.².

29. The length of a rectangular solar panel is 12 cm more than its width. If its area is 520 cm², find its dimensions.

30. The height h (in ft) of a rocket as a function of the time t (in s) of flight is given by $h = 50 + 280t - 16t^2$. Determine when the rocket is at ground level.

31. In finding the illumination at a point x feet from one of two light sources that are 100 ft apart, it is necessary to solve the equation $9x^3 - 2400x^2 + 240{,}000x - 8{,}000{,}000 = 0$. Find x.

32. A rectangular storage bin is to be made from a rectangular piece of sheet metal 12 in. by 10 in., by cutting out equal corners of side x and bending up the sides. See Fig. 3-33. Find x if the storage bin is to hold 90 in.³.

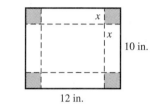

Fig. 3-33

In Exercises 33–36, solve the given problems.

33. Solve the equation of Example 4 to the nearest 0.0001 by using an appropriate feature of a graphing calculator.

34. Solve the equation of Example 5 to the nearest 0.001 by using an appropriate feature of a graphing calculator.

35. The cutting speed s (in ft/min) of a saw in cutting a particular type of metal piece is given by $s = \sqrt{t - 4t^2}$, where t is the time in seconds. What is the maximum cutting speed in this operation (to two significant digits)? (Hint: Find the range.)

(W) 36. Referring to Exercise 32, explain how to determine the maximum possible capacity for a storage bin constructed in this way. What is the maximum possible capacity (to three significant digits)?

3-6 GRAPHS OF FUNCTIONS DEFINED BY TABLES OF DATA

In Section 3-4 we showed how to construct the graph of a function defined by a formula or an equation. However, as we stated in Section 3-1, there are other ways to express functions. One of the most important ways to show the relationship between variables is by using a table of values found by observation or experimentation.

Often the data from an experiment indicate that the variables could have a formula that relates them, although the formula may not be known. Data from experiments in the various fields of science and technology generally have variables that are related in such a way. In this case, when we are plotting the graph, the points should be connected by a smooth curve.

Statistical data often give values that are taken for certain intervals or are averaged over specified intervals, and there is no meaning to the intervals *between* points on the graph. On these graphs, the points should be connected by straight-line segments, where this is done only to make the points stand out better and make the graph easier to read. Example 1 illustrates this type of graph.

EXAMPLE 1   The electric energy usage (in kW·h) for a certain all-electric house for each month of a year is shown in the following table. Plot these data.

Month	Jan	Feb	Mar	Apr	May	Jun
Energy usage	2626	3090	2542	1875	1207	892

Month	July	Aug	Sep	Oct	Nov	Dec
Energy usage	637	722	825	1437	1825	2427

Using statistical plotting features, Fig. 3-35 shows a graphing calculator display of the data of Example 1.

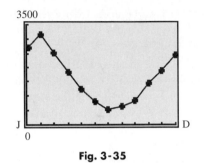

Fig. 3-35

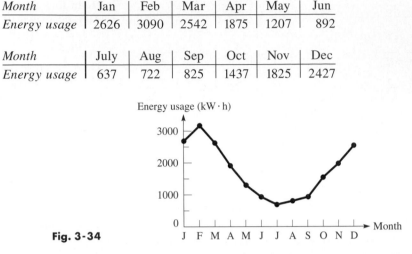

Fig. 3-34

We know that there is no meaning to the intervals between the months, since we have the total number of kilowatt-hours for each month. Therefore, we use straight-line segments, but only to make the points stand out better. See Fig. 3-34.

In the next example, we illustrate data that give a graph in which the points are connected by a smooth curve and not by straight-line segments.

EXAMPLE 2   Steam in a boiler was heated to 150°C. Its temperature was recorded each minute, giving the values in the following table. Plot the graph.

The Celsius degree is named for the Swedish astronomer Andres Celsius (1701–1744). He designated 100° as the freezing point of water and 0° as the boiling point. These were later reversed.

Time (minutes)	0.0	1.0	2.0	3.0	4.0	5.0
Temperature (°C)	150.0	142.8	138.5	135.2	132.7	130.8

Since the temperature changes in a continuous way, there is meaning to the values in the intervals between points. Therefore, these points are joined by a smooth curve, as in Fig. 3-36. Also note that most of the vertical scale was used for the required values, with the indicated break in the scale between 0 and 130.

In Chapter 22 we see how to find an equation that approximates the function relating the variables in a table of values.

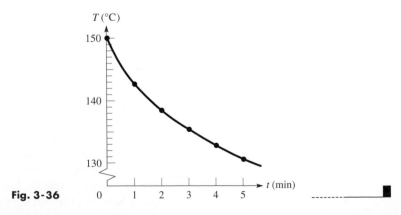

Fig. 3-36

If the graph relating two variables is known, values can be obtained directly from the graph. In finding such a value through the inspection of the graph, we are *reading the graph.*

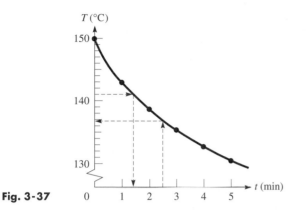

Fig. 3-37

■**EXAMPLE 3**   For the cooling steam in Example 2, we can estimate values of one variable for given values of the other variable.

If we want to know the temperature after 2.5 min, we estimate 0.5 of the interval between 2 and 3 on the *t*-axis and mark this point. See Fig. 3-37. Then we draw a vertical line from this point to the curve. From the point where it intersects the curve, we draw a horizontal line to the *T*-axis. We now estimate that the line crosses at $T = 146.7°$C. Obviously, the number of tenths is a rough estimate.

In the same way, if we want to determine how long the steam took to cool down to $141.0°$C, we go from 141.0 marker on the *T*-axis to the curve and then to the *t*-axis. This crosses at about $t = 1.4$ min.

Linear Interpolation

In Example 3 we see that we can estimate values from a graph. However, unless a very accurate graph is drawn with expanded scales for both variables, only very approximate values can be found. There is a method, called *linear interpolation,* which uses the table itself to get more accurate results.

Linear interpolation *assumes that if a particular value of one variable lies between two of those listed in the table, then the corresponding value of the other variable is at the same proportional distance between the listed values.* On the graph, linear interpolation assumes that two points defined in the table are connected by a straight line. Although this is generally not correct, it is a good approximation if the values in the table are sufficiently close together.

Before the use of electronic calculators, interpolation was used extensively in mathematics textbooks for finding values from mathematics tables. It is still of use when using scientific and technical tables.

■**EXAMPLE 4**   For the cooling steam in Example 2, we can use interpolation to find its temperature after 1.4 min. Since 1.4 min is $\frac{4}{10}$ of the way from 1.0 min to 2.0 min, we shall assume that the value of T we want is $\frac{4}{10}$ of the way between 142.8 and 138.5, the values of T for 1.0 min and 2.0 min, respectively. The difference between these values is 4.3, and $\frac{4}{10}$ of 4.3 is 1.7 (rounded off to tenths). Subtracting (the values of T are decreasing) 1.7 from 142.8, we obtain 141.1. Thus, the required value of T is about $141.1°$C. (Note that this agrees well with the result in Example 3.)

Another method of indicating the interpolation is shown in Fig. 3-38. From the figure we have the proportion

$$\frac{0.4}{1.0} = \frac{x}{4.3}$$

$$x = 1.7 \quad \text{rounded off}$$

Therefore,

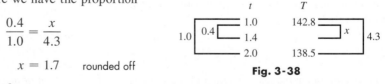

Fig. 3-38

$$142.8 - 1.7 = 141.1°C$$

is the required value of T. If the values of T had been increasing, we would have added 1.7 to the value of T for 1.0 min.

EXERCISES $3\text{-}6$

In Exercises 1–8, represent the data graphically.

1. The diesel fuel production (in 1000's of gallons) at a certain refinery during an 8-week period was as follows:

Week	1	2	3	4	5	6	7	8
Production	765	780	840	850	880	840	760	820

2. The *exchange rate* for the number of Canadian dollars equal to one United States dollar for 1984–1998 is as follows:

Year	1984	1986	1988	1990	1992	1994	1996	1998
No. Can. dollars	1.30	1.39	1.23	1.17	1.21	1.38	1.37	1.48

3. The amount of material necessary to make a cylindrical gallon container depends on the diameter, as shown in this table.

Diameter (in.)	3.0	4.0	5.0	6.0	7.0	8.0	9.0
Material (in.^{2})	322	256	224	211	209	216	230

4. An oil burner propels air that has been heated to 90°C. The temperature then drops as the distance from the burner increases, as shown in the following table.

Distance (m)	0.0	1.0	2.0	3.0	4.0	5.0	6.0
Temperature (°C)	90	84	76	66	54	46	41

5. A changing electric current in a coil of wire will induce a voltage in a nearby coil. Important in the design of transformers, the effect is called *mutual inductance*. For two coils, the mutual inductance (in H) as a function of the distance between them is given in the following table.

Distance (cm)	0.0	2.0	4.0	6.0	8.0	10.0	12.0
M. ind. (H)	0.77	0.75	0.61	0.49	0.38	0.25	0.17

6. The temperatures felt by the body as a result of the *wind chill factor* for an outside temperature of 30°F (as determined by the National Weather Service) are given in the following table.

Wind speed (mi/h)	5	10	15	20	25	30	35	40	45
Temp. felt (°F)	27	16	9	4	1	−2	−4	−5	−6

7. The time required for a sum of money to double in value, when compounded annually, is given as a function of the interest rate in the following table.

Rate (%)	4	5	6	7	8	9	10
Time (years)	17.7	14.2	11.9	10.2	9.0	8.0	7.3

8. The torque T of an engine, as a function of the frequency f of rotation, was measured as follows.

f(r/min)	500	1000	1500	2000	2500	3000	3500
T (ft·lb)	175	90	62	45	34	31	27

In Exercises 9 and 10, use the graph in Fig. 3-36, which relates the temperature of cooling steam and the time. Find the indicated values by reading the graph.

9. (a) For $t = 4.3$ min, find T. (b) For $T = 145.0°C$, find t.

10. (a) For $t = 1.8$ min, find T. (b) For $T = 133.5°C$, find t.

In Exercises 11 and 12, use the following table, which gives the voltage produced by a certain thermocouple as a function of the temperature of the thermocouple. Plot the graph. Find the indicated values by reading the graph.

Temperature (°C)	0	10	20	30	40	50
Voltage (volts)	0.0	2.9	5.9	9.0	12.3	15.8

11. (a) For $T = 26°C$, find V. (b) For $V = 13.5$ V, find T.

12. (a) For $T = 32°C$, find V. (b) For $V = 4.8$ V, find T.

In Exercises 13–16, find the indicated values by means of linear interpolation.

13. In Exercise 5, find the inductance for $d = 9.2$ cm.

14. In Exercise 6, find the temperature for $s = 22$ mi/h.

15. In Exercise 7, find the rate for $t = 10.0$ years.

16. In Exercise 8, find the torque for $t = 2300$ r/min.

In Exercises 17–20, use the following table that gives the rate R of discharge from a tank of water as a function of the height H of water in the tank. For Exercises 17 and 18, plot the graph and find the values from the graph. For Exercises 19 and 20, find the indicated values by linear interpolation.

Height (ft)	0	1.0	2.0	4.0	6.0	8.0	12
Rate (ft^3/s)	0	10	15	22	27	31	35

17. (a) For $R = 20$ ft^3/s, find H. (b) For $H = 2.5$ ft, find R.

18. (a) For $R = 34$ ft^3/s, find H. (b) For $H = 6.4$ ft, find R.

19. Find R for $H = 1.7$ ft.   **20.** Find H for $R = 25$ ft^3/s.

In Exercises 21–24, use the following table, which gives the fraction (as a decimal) of the total heating load of a certain system that will be supplied by a solar collector of area A (in m^2). Find the indicated values by linear interpolation.

f	0.22	0.30	0.37	0.44	0.50	0.56	0.61
A (m^2)	20	30	40	50	60	70	80

21. For $A = 36$ m^2, find f.   **22.** For $A = 52$ m^2, find f.

23. For $f = 0.59$, find A.   **24.** For $f = 0.27$, find A.

In Exercises 25–28, a method of finding values beyond those given is considered. By using a straight-line segment to extend a graph beyond the last known point, we can estimate values from the extension of the graph. The method is known as **linear extrapolation.** *Use this method to estimate the required values from the given graphs.*

25. Using Fig. 3-37, estimate T for $t = 5.3$ min.

26. Using the graph for Exer. 11 and 12, estimate V for $T = 55°C$.

27. Using the graph for Exer. 17–20, estimate R for $H = 13$ ft.

28. Using the graph for Exer. 17–20, estimate R for $H = 16$ ft.

REVIEW EXERCISES

In Exercises 1–4, determine the appropriate function.

1. The radius of circular water wave increases at the rate of 2 m/s. Express the area of the circle as a function of the time t (in s).

2. A conical sheet metal hood is to cover an area 6 m in diameter. Find the total surface area A of the hood as a function of its height h.

3. One computer printer prints at the rate of 2000 lines/min for x minutes, and a second printer prints at the rate of 1800 lines/min for y minutes. Together they print 50,000 lines. Find y as a function of x.

4. Fencing around a rectangular storage depot area costs twice as much along the front as along the other three sides. The back costs $10 per foot. Express the cost C of the fencing as a function of the width w if the length (along the front) is 20 ft longer than the width.

In Exercises 5–12, evaluate the given functions.

5. $f(x) = 7x - 5$; find $f(3)$ and $f(-6)$.

6. $g(I) = 8 - 3I$; find $g(\frac{1}{6})$ and $g(-4)$.

7. $H(h) = \sqrt{1 - 2h}$; find $H(-4)$ and $H(2h)$.

8. $\phi(v) = \dfrac{3v - 2}{v + 1}$; find $\phi(-2)$ and $\phi(v + 1)$.

9. $f(x) = 3x^2 - 2x + 4$; find $f(x + h) - f(x)$.

10. $F(x) = x^3 + 2x^2 - 3x$; find $F(3 + h) - F(3)$.

11. $f(x) = 3 - 2x$; find $f(2x) - 2f(x)$.

12. $f(x) = 1 - x^2$; find $[f(x)]^2 - f(x^2)$.

In Exercises 13–16, evaluate the given functions. Values of the independent variable are approximate.

13. $f(x) = 8.07 - 2x$; find $f(5.87)$ and $f(-4.29)$.

14. $g(x) = 7x - x^2$; find $g(45.81)$ and $g(-21.85)$.

15. $G(S) = \dfrac{S - 0.087629}{3.0125S}$; find $G(0.17427)$ and $G(0.053206)$.

16. $h(t) = \dfrac{t^2 - 4t}{t^3 + 564}$; find $h(8.91)$ and $h(-4.91)$.

In Exercises 17–20, determine the domain and the range of the given functions.

17. $f(x) = x^4 + 1$

18. $G(z) = \dfrac{4}{z^3}$

19. $g(t) = \dfrac{2}{\sqrt{t + 4}}$

20. $F(y) = 1 - 2\sqrt{y}$

In Exercises 21–32, plot the graphs of the given functions. Check these graphs by using a graphing calculator.

21. $y = 4x + 2$

22. $y = 5x - 10$

23. $y = 4x - x^2$

24. $y = x^2 - 8x - 5$

25. $y = 3 - x - 2x^2$

26. $y = 6 + 4x + x^2$

27. $y = x^3 - 6x$

28. $V = 3 - 0.5s^3$

29. $y = 2 - x^4$

30. $y = x^4 - 4x$

31. $y = \dfrac{x}{x + 1}$

32. $Z = \sqrt{25 - 2R^2}$

In Exercises 33–40, use a graphing calculator to solve the given equations to the nearest 0.1.

33. $7x - 3 = 0$

34. $3x + 11 = 0$

35. $x^2 + 1 = 6x$

36. $3t - 2 = t^2$

37. $x^3 - x^2 = 2 - x$

38. $5 - x^3 = 2x^2$

39. $\dfrac{1}{x} = 2x$

40. $\sqrt{x} = 2x - 1$

In Exercises 41–44, use a graphing calculator to find the range of the given function.

41. $y = x^4 - 5x^2$

42. $y = x\sqrt{4 - x^2}$

43. $A = w + \dfrac{2}{w}$

44. $y = 2x + \dfrac{3}{\sqrt{x}}$

In Exercises 45–52, answer the given questions.

(W) 45. Explain how $A(a, b)$ and $B(b, a)$ may be in different quadrants.

46. Determine the distance from the origin to the point (a, b).

47. Two vertices of an equilateral triangle are $(0, 0)$ and $(2, 0)$. What is the third vertex?

48. The points $(1, 2)$ and $(1, -3)$ are two adjacent vertices of a square. Find the other vertices.

49. An equation used in electronics with a transformer antenna is $I = 12.5\sqrt{1 + 0.5m^2}$. For $I = f(m)$, find $f(0.55)$.

50. The percent p of wood lost in cutting it into boards 1.5 in. thick due to the thickness t (in in.) of the saw blade is $p = \dfrac{100t}{t + 1.5}$. Find p if $t = 0.4$ in. That is, since $p = f(t)$, find $f(0.4)$.

51. The angle A (in degrees) of a robot arm with the horizontal as a function of time t (for 0.0 s to 6.0 s) is given by $A = 8.0 + 12t^2 - 2.0t^3$. What is the greatest value of A to the nearest 0.1°? See Fig. 3-39. (Hint: Find the range.)

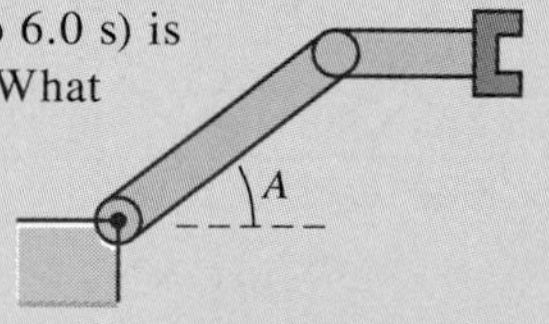

Fig. 3-39

52. The electric power P (in W) produced by a certain battery is $P = \dfrac{24R}{R^2 + 1.40R + 0.49}$, where R is the resistance (in Ω) in the circuit. What is the maximum power produced? (Hint: Find the range.) See Example 6 on page 90.

In Exercises 53–64, plot the graphs of the indicated functions.

53. When El Niño, a Pacific Ocean current, moves east and warms the water off South America, weather patterns in many parts of the world change significantly. Special buoys along the equator in the Pacific Ocean send data via satellite to monitoring stations. If the temperature T (in °C) at one of these buoys is $T = 28.0 + 0.15t$, where t is the time in weeks between Jan. 1 and Aug. 1 (30 weeks), plot the graph of $T = f(t)$.

54. There are 500 L of oil in a tank that has the capacity of 100,000 L. It is filled at the rate of 7000 L/h. Determine the function relating the number of liters N and the time t while the tank is being filled. Plot N as a function of t.

55. For a given temperature, five times the Fahrenheit reading F less nine times the Celsius C reading is 160. Plot the graph of $C = f(F)$.

56. A company buys a new copier for $1000 and determines that it costs $10 per day to use it (for paper, toner, etc.). Plot the total cost C to operate the copier as a function of the number n of days of use.

57. For a certain laser device, the laser output power P (in mW) is negligible if the drive current i is less than 80 mA. From 80 mA to 140 mA, $P = 1.5 \times 10^{-6}i^3 - 0.77$. Plot the graph of $P = f(i)$.

58. It is determined that a good approximation for the cost C (in cents/mi) of operating a certain car at a constant speed v (in mi/h) is given by $C = 0.025v^2 - 1.4v + 35$. Plot C as a function of v for $v = 10$ mi/h to $v = 60$ mi/h.

59. A medical researcher exposed a virus culture to an experimental vaccine. It was observed that the number of live cells N in the culture as a function of the time t (in h) after exposure was given by $N = \dfrac{1000}{\sqrt{t+1}}$. Plot the graph of $N = f(t)$.

60. The electric field E (in V/m) from a certain electric charge is given by $E = 25/r^2$, where r is the distance (in m) from the charge. Plot the graph of $E = f(r)$ for values of r up to 10 cm.

61. To draw the approximate shape of an irregular shoreline, a surveyor measured the distances d from a straight wall to the shoreline at 20-ft intervals along the wall, as shown in the following table. Plot the graph of distance d as a function of the distance D along the wall.

D (ft)	0	20	40	60	80	100	120	140	160
d (ft)	15	32	56	33	29	47	68	31	52

62. The percent p of a computer network that is in use during a particular loading cycle as a function of the time t (in s) is given in the following table. Plot the graph of $p = f(t)$.

t (s)	0.0	0.2	0.4	0.6	0.8	1.0	1.2	1.4	1.6
P (%)	0	45	85	90	85	85	60	10	0

63. In an experiment measuring the pressure p (in kPa) at a given depth d (in m) of seawater, the results in the following table were found. Plot the graph of $p = f(d)$ and from the graph determine $f(10)$.

d (m)	0.0	3.0	6.0	9.0	12	15
p (kPa)	101	131	161	193	225	256

64. The vertical sag s (in ft) at the middle of an 800-ft power line as a function of the temperature T (in °F) is given in the following table. See Fig. 3-40. For the function $s = f(T)$, find $f(47)$ by linear interpolation.

T (°F)	0	20	40	60
s (ft)	10.2	10.6	11.1	11.7

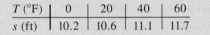

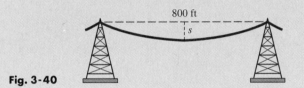

Fig. 3-40

In Exercises 65–72, solve the indicated equations graphically.

65. A person 250 mi from home starts toward home and travels at 60 mi/h for the first 2.0 h and then slows down to 40 mi/h for the rest of the trip. How long does it take the person to be 70 mi from home?

66. One industrial cleaner contains 30% of a certain solvent, and another contains 10% of the solvent. To get a mixture containing 50 gal of the solvent, 120 gal of the first cleaner is used. How much of the second must be used?

67. The solubility s (in kg/m^3 of water) of a certain type of fertilizer is given by $s = 135 + 4.9T + 0.19T^2$, where T is the temperature (in °C). Find T for $s = 500$ kg/m^3.

68. A 2.00-L (2000-cm^3) metal container is to be made in the shape of a right circular cylinder. Express the total area A of metal necessary as a function of the radius r of the base. Then find A for $r = 6.00$ cm, 7.00 cm, and 8.00 cm.

69. In an oil pipeline, the velocity v (in ft/s) of the oil as a function of the distance x (in ft) from the wall of the pipe is given by $v = 7.6x - 2.1x^2$. Find x for $v = 5.6$ ft/s. The diameter of the pipe is 3.50 ft.

70. One ball bearing is 1.00 mm more in radius and has twice the volume of another ball bearing. What is the radius of each?

71. A computer, using data from a refrigeration plant, estimates that in the event of a power failure, the temperature (in °C) in the freezers would be given by $T = \dfrac{4t^2}{t + 2} - 20$, where t is the number of hours after the power failure. How long would it take for the temperature to reach 0°C?

72. Two electrical resistors in parallel (see Fig. 3-41) have a combined resistance R_T given by $R_T = \dfrac{R_1 R_2}{R_1 + R_2}$. If $R_2 = R_1 + 2.0$, express R_T as a function of R_1 and find R_1 if $R_T = 6.0\ \Omega$.

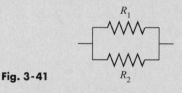

R_1

R_2

Fig. 3-41

Writing Exercise

73. In one or two paragraphs, explain how you would solve the following problem using a graphing calculator: The inner surface area A of a 250.0-cm³ cylindrical cup, as a function of the radius r of the base is $A = \pi r^2 + \dfrac{500.0}{r}$. Find r if $A = 175.0$ cm². (What is the answer?) (See if you can derive the formula.)

PRACTICE TEST

1. Given $f(x) = 2x - x^2 + \dfrac{8}{x}$, find $f(-4)$ and $f(2.385)$.

2. A rocket has a mass of 2000 Mg at liftoff. If the first-stage engines burn fuel at the rate of 10 Mg/s, find the mass m of the rocket as a function of the time t (in s) while the first-stage engines operate.

3. Plot the graph of the function $f(x) = 4 - 2x$.

4. Use a graphing calculator to solve the equation $2x^2 - 3 = 3x$ to the nearest 0.1.

5. Plot the graph of the function $y = \sqrt{4 + 2x}$.

6. Locate all points (x, y) for which $x < 0$ and $y = 0$.

7. Find the domain and the range of the function $f(x) = \sqrt{6 - x}$.

8. A window has the shape of a semicircle over a square, as shown in Fig. 3-42. Express the area of the window as a function of the radius of the circular part.

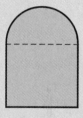

Fig. 3-42

9. The voltage V and current i (in mA) for a certain electrical experiment were measured as shown in the following table. Plot the graph of $i = f(V)$ and from the graph find $f(45.0)$.

Voltage (V)	10.0	20.0	30.0	40.0	50.0	60.0
Current (mA)	145	188	220	255	285	315

10. From the table in Problem 9, find the voltage for $i = 200$ mA.

CHAPTER 4

THE TRIGONOMETRIC FUNCTIONS

Using trigonometry it often is possible to calculate distances that may not be directly measurable. In Section 4-5 we show how we can measure the distance from the Goodyear blimp to a point on the football field.

Many applied problems in science and technology require the use of triangles, especially right triangles, for their solution. Included among these are problems in air navigation, surveying, the motion of rockets and missiles, structural design, and optics. Problems involving forces acting on objects and the measurement of distances between various parts of the solar system and universe can also be solved. Even certain types of electric circuits can be analyzed by the use of triangles.

In **trigonometry** we develop methods for measuring the sides and angles of triangles as well as solving related applied problems. Because of the great number of applications it has in many areas, trigonometry is considered one of the most practical and relevant branches of mathematics.

In this chapter we introduce the basic trigonometric functions and show a number of applications of right triangles from many areas of science and technology. In later chapters we will use the trigonometric functions with other types of triangles and their applications.

Also in later chapters we will see how the trigonometric functions are applied without reference to triangles. Such applications are found in electronics, mechanical vibrations, acoustics, optics, and other fields.

4-1 ANGLES

In the review of geometry in Chapter 2, we stated a basic definition of an *angle.* In this section we extend this definition and give other important definitions related to angles.

An **angle** *is generated by rotating a ray about its fixed endpoint from an* **initial position** *to a* **terminal position.** *The initial position is called the* **initial side** *of the angle, the terminal position is called the* **terminal side,** *and the fixed endpoint is the* **vertex.** The angle itself is the amount of rotation from the initial side to the terminal side.

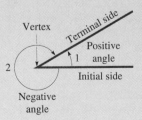

Fig. 4-1

*If the rotation of the terminal side from the initial side is **counterclockwise**, the angle is defined as **positive**. If the rotation is **clockwise**, the angle is **negative**.* In Fig. 4-1, ∠1 is positive and ∠2 is negative.

There are many symbols used to designate angles. Among the most widely used are certain Greek letters such as θ (theta), ϕ (phi), α (alpha), and β (beta). Capital letters representing the vertex (e.g., ∠A or simply A) and other literal symbols, such as x and y, are also used commonly.

In Chapter 2 we introduced two measurements of an angle. These are the *degree* and the *radian*. Since degrees and radians are both used on calculators and computers, we will briefly review the relationship between them in this section. However, we will not make use of radians until Chapter 8.

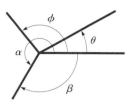

Fig. 4-2

From Section 2-1 we recall that *a **degree** is* 1/360 *of one complete rotation.* In Fig. 4-2, ∠θ = 30°, ∠ϕ = 140°, ∠α = 240°, and ∠β = −120°. Note that β is drawn in a clockwise direction to show that it is negative. The other angles are drawn in a counterclockwise direction to show that they are positive angles.

In Chapter 2 we used degrees and decimal parts of a degree. Most calculators use degrees in this decimal form. Another traditional way is to divide a degree into 60 equal parts called **minutes;** each minute is divided into 60 equal parts called **seconds.** The symbols ′ and ″ are used to designate minutes and seconds, respectively.

In Fig. 4-2 we note that angles α and β have the same initial and terminal sides. *Such angles are called* **coterminal angles.** An understanding of coterminal angles is important in certain concepts of trigonometry.

■EXAMPLE 1   Determine the values of two angles that are coterminal with an angle of 145.6°.

Since there are 360° in a complete rotation, we can find one coterminal angle by considering the angle that is 360° larger than the given angle. This gives us an angle of 505.6°. Another method of finding a coterminal angle is to subtract 145.6° from 360° and then consider the resulting angle to be negative. This means that the original angle and the derived angle would make up one complete rotation, when put together. This method leads us to the angle of −214.4° (see Fig. 4-3). These methods could be employed repeatedly to find other coterminal angles.   --------■

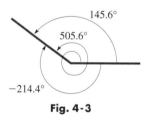

Fig. 4-3

Angle Conversions

NOTE▶

Although we will use only degrees as a measure of an angle in this chapter, when using a calculator we must *be careful to have the calculator set in the correct mode.* Therefore, to be sure that the calculator is using degrees, set the *mode* feature to degrees. In later chapters we will have use for a setting of radians. We can change from one measure to the other by using a calculator feature or using the definition of a radian in Section 2-4 that π rad = 180°.

Before the extensive use of calculators it was common to use degrees and minutes in tables, whereas calculators use degrees and decimal parts of a degree. Changing from one form to another can be done directly on a calculator by use of the *dms (degree-minute-second)* feature. The following examples illustrate angle conversions by using the definitions of the different measures of an angle and by using the appropriate calculator features.

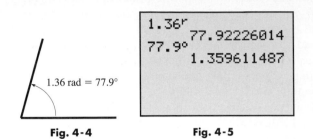

Fig. 4-4   **Fig. 4-5**

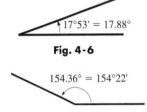

Fig. 4-6

$154.36° = 154°22'$

Fig. 4-7

▌**EXAMPLE 2**   Express 1.36 rad in degrees.

We know that π rad $= 180°$, which means 1 rad $= 180°/\pi$. Therefore,

$$1.36 \text{ rad} = 1.36\left(\frac{180°}{\pi}\right) = 77.9° \qquad \text{to nearest } 0.1°$$

This angle is shown in Fig. 4-4. We again note that degrees and radians are simply two different ways of measuring an angle.

In Fig. 4-5, a graphing calculator display shows the conversions of 1.36 rad to degrees (calculator in degree mode) and 77.9° to radians (calculator in radian mode).

▌**EXAMPLE 3**   To change $17°53'$ to decimal form, we use the fact that $1° = 60'$. Therefore, $53' = \left(\frac{53}{60}\right)° = 0.88°$ to nearest $0.01°$

This means that $17°53' = 17.88°$. This angle is shown in Fig. 4-6.

To change $154.36°$ to an angle measured to the nearest minute, we have

$$0.36° = 0.36(60') = 22'$$

This means that $154.36° = 154°22'$. See Fig. 4-7.

Standard Position of an Angle

If the initial side of the angle is the positive x-axis and the vertex is the origin, the angle is said to be in **standard position.** The angle is then determined by the position of the terminal side. *If the terminal side is in the first quadrant, the angle is called a* **first-quadrant angle.** Similar terms are used when the terminal side is in the other quadrants. *If the terminal side coincides with one of the axes, the angle is a* **quadrantal angle.** For an angle in standard position, the terminal side can be determined if we know any point, except the origin, on the terminal side.

▌**EXAMPLE 4**   A standard position angle of 60° is a first-quadrant angle with its terminal side 60° from the *x*-axis. See Fig. 4-8(a).

A second-quadrant angle of 130° is shown in Fig. 4-8(b).

A third-quadrant angle of 225° is shown in Fig. 4-8(c).

A fourth-quadrant angle of 340° is shown in Fig. 4-8(d).

A standard position angle of −120° is shown in Fig. 4-8(e). Since the terminal side is in the third quadrant, it is a third-quadrant angle.

A standard position angle of 90° is a quadrantal angle since its terminal side is the positive *y*-axis. See Fig. 4-8(f).

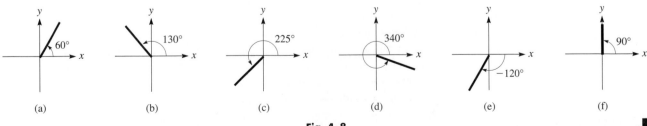

(a)   (b)   (c)   (d)   (e)   (f)

Fig. 4-8

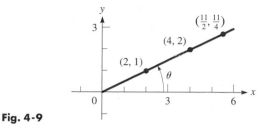

Fig. 4-9

EXAMPLE 5 In Fig. 4-9, θ is in standard position, and the terminal side is uniquely determined by knowing that it passes through the point (2, 1). The same terminal side passes through (4, 2) and $(\frac{11}{2}, \frac{11}{4})$, among other points. Knowing that the terminal side passes through any one of these points makes it possible to determine the terminal side of the angle. ∎

━━━━━━━━━ EXERCISES *4-1* ━━━━━━━━━

In Exercises 1–4, draw the given angles.

1. 60°, 120°, −90°   **2.** 330°, −150°, 450°

3. 50°, −360°, −30°   **4.** 45°, 245°, −250°

In Exercises 5–12, determine one positive and one negative coterminal angle for each angle given.

5. 45°   **6.** 73°

7. −150°   **8.** 162°

9. 70°30′   **10.** 153°47′

11. 278.1°   **12.** −197.6°

In Exercises 13–16, by means of the definition of a radian, change the given angles in radians to equal angles expressed in degrees to the nearest 0.01°.

13. 0.265 rad   **14.** 0.838 rad

15. 1.447 rad   **16.** 3.642 rad

In Exercises 17–20, use a calculator conversion sequence to change the given angles in radians to equal angles expressed in degrees to the nearest 0.01°.

17. 0.329 rad   **18.** 2.089 rad

19. 4.110 rad   **20.** 0.067 rad

In Exercises 21–24, use a calculator conversion sequence to change the given angles to equal angles expressed in radians to three significant digits.

21. 56.0°   **22.** 137.4°

23. 284.8°   **24.** −17.5°

In Exercises 25–28, change the given angles to equal angles expressed to the nearest minute.

25. 47.50°   **26.** 315.80°

27. −5.62°   **28.** 142.87°

In Exercises 29–32, change the given angles to equal angles expressed in decimal form to the nearest 0.01°.

29. 15°12′   **30.** 157°39′

31. 301°16′   **32.** −4°47′

In Exercises 33–40, draw angles in standard position such that the terminal side passes through the given point.

33. (4, 2)   **34.** (−3, 8)

35. (−3, −5)   **36.** (6, −1)

37. (−7, 5)   **38.** (−4, −2)

39. (2, −5)   **40.** (1, 6)

In Exercises 41 and 42, change the given angles to equal angles expressed in decimal form to the nearest 0.001°. In Exercises 43 and 44, change the given angles to equal angles expressed to the nearest second.

41. 21°42′36″   **42.** 7°16′23″

43. 86.274°   **44.** 57.019°

4-2 DEFINING THE TRIGONOMETRIC FUNCTIONS

In Chapter 2 we reviewed many of the basic geometric figures and their properties. Important to the definitions and development in this section are the right triangle, the Pythagorean theorem, and the properties of similar triangles. We now briefly review similar triangles and their properties.

As stated in Section 2-2, *two triangles are* **similar** *if they have the same shape (but not necessarily the same size).* Similar triangles have the following important properties.

Properties of Similar Triangles

1. *Corresponding angles are equal.*

2. *Corresponding sides are proportional.*

The **corresponding sides** *are the sides, one in each triangle, that are between the same pair of equal* **corresponding angles.**

■**EXAMPLE 1**   In Fig. 4-10 the triangles are similar and are lettered so that corresponding sides and angles have the same letters. That is, angles A_1 and A_2, angles B_1 and B_2, and angles C_1 and C_2 are pairs of corresponding angles. The pairs of corresponding sides are a_1 and a_2, b_1 and b_2, and c_1 and c_2. From the properties of similar triangles, we know that the corresponding angles are equal, or

$$\angle A_1 = \angle A_2, \quad \angle B_1 = \angle B_2, \quad \angle C_1 = \angle C_2$$

Also, the corresponding sides are proportional, which we can show as

$$\frac{a_1}{a_2} = \frac{b_1}{b_2}, \quad \frac{a_1}{a_2} = \frac{c_1}{c_2}, \quad \frac{b_1}{b_2} = \frac{c_1}{c_2}$$

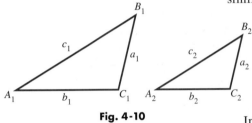

Fig. 4-10

In Example 1, if we multiply both sides of

$$\frac{a_1}{a_2} = \frac{b_1}{b_2} \quad \text{by} \quad \frac{a_2}{b_1}, \quad \text{we get} \quad \frac{a_1}{a_2}\left(\frac{a_2}{b_1}\right) = \frac{b_1}{b_2}\left(\frac{a_2}{b_1}\right)$$

which when simplified gives

$$\frac{a_1}{b_1} = \frac{a_2}{b_2}$$

This shows us that *when two triangles are similar*

> *the ratio of one side to another side in one triangle is the same as the ratio of the corresponding sides in the other triangle.*

Using this we now proceed to the definitions of the trigonometric functions.

We now place an angle θ in standard position and drop perpendicular lines from points on the terminal side to the *x*-axis, as shown in Fig. 4-11. In doing this we set up similar triangles, each with one vertex at the origin and one side along the *x*-axis.

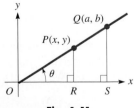

Fig. 4-11

■**EXAMPLE 2**   In Fig. 4-11 we can see that triangles *ORP* and *OSQ* are similar since their corresponding angles are equal (each has the same angle at *O*, a right angle, and therefore equal angles at *P* and *Q*). This means that ratios of the lengths of corresponding sides are equal. For example,

$$\frac{RP}{OR} = \frac{SQ}{OS} \quad \text{which is the same as} \quad \frac{y}{x} = \frac{b}{a}$$

For any position (except at the origin) of *Q* on the terminal side of θ, the ratio b/a of its ordinate to its abscissa will equal y/x.

For any angle θ in standard position, six different ratios may be set up. Because of the property of similar triangles, any given ratio has the same value for any point on the terminal side that is chosen. For a different angle, with a different terminal side, the ratio will have a different value. Thus, the values of the ratios depend on the position of the terminal side, which means that *the values of the ratios depend on the size of the angle, and there is only one value for each ratio for a given angle.* Recalling the meaning of a function, we see that *the ratios are functions of the angle. These functions are called the* **trigonometric functions,** and they are defined as follows (see Fig. 4-12):

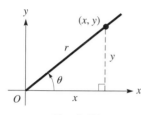

Fig. 4-12

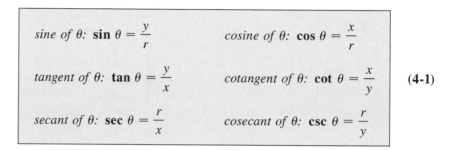

$$
\begin{aligned}
&\textit{sine of } \theta:\ \mathbf{sin}\ \theta = \frac{y}{r} && \textit{cosine of } \theta:\ \mathbf{cos}\ \theta = \frac{x}{r} \\[2mm]
&\textit{tangent of } \theta:\ \mathbf{tan}\ \theta = \frac{y}{x} && \textit{cotangent of } \theta:\ \mathbf{cot}\ \theta = \frac{x}{y} \\[2mm]
&\textit{secant of } \theta:\ \mathbf{sec}\ \theta = \frac{r}{x} && \textit{cosecant of } \theta:\ \mathbf{csc}\ \theta = \frac{r}{y}
\end{aligned}
\qquad \textbf{(4-1)}
$$

Here, *the distance r from the origin to the point is called the* **radius vector.** Also note that we have used the abbreviations that are used most of the time when working with these functions.

In this chapter we shall restrict our attention to the trigonometric functions of acute angles (angles between $0°$ and $90°$). However, *the definitions in Eqs. (4-1) are general and may be used for angles of any magnitude.* Discussion of the trigonometric functions of angles in general, along with other important properties, is found in Chapters 8 and 20.

We should note that a given function is not defined if the denominator is zero. The denominator is zero in $\tan \theta$ and $\sec \theta$ for $x = 0$, and in $\cot \theta$ and $\csc \theta$ for $y = 0$. In all cases we will assume that $r > 0$. If $r = 0$, there would be no terminal side and therefore no angle. These restrictions affect the domain of these functions. We will discuss the domains and ranges of the trigonometric functions in Chapter 10, when we consider the graphs of these functions.

Evaluating the Trigonometric Functions

When evaluating the trigonometric functions we use the definitions in Eqs. (4-1). We also often use the *Pythagorean theorem,* which we discussed in Section 2-2. For reference, we restate it here. For the right triangle in Fig. 4-13, with hypotenuse c and legs a and b, we have

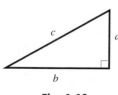

Fig. 4-13

$$
\boxed{c^2 = a^2 + b^2}
\qquad \textbf{(4-2)}
$$

Following are examples of evaluating the trigonometric functions of an angle when a point on the terminal side of the angle is given or can be found.

EXAMPLE 3   Find the values of the trigonometric functions of the angle θ with its terminal side passing through the point $(3, 4)$.

By placing the angle in standard position, as shown in Fig. 4-14, and drawing the terminal side through $(3, 4)$, we find by use of the Pythagorean theorem that

$$r = \sqrt{3^2 + 4^2} = \sqrt{25} = 5$$

Using the values $x = 3$, $y = 4$, and $r = 5$, we find that

$$\sin \theta = \frac{4}{5} \qquad \cos \theta = \frac{3}{5} \qquad \tan \theta = \frac{4}{3}$$

$$\cot \theta = \frac{3}{4} \qquad \sec \theta = \frac{5}{3} \qquad \csc \theta = \frac{5}{4}$$

We have left each of these results in the form of a fraction, which is considered to be an *exact form* in that there has been no approximation made. In writing decimal values, we find that $\tan \theta = 1.333$ and $\sec \theta = 1.667$, where these values have been rounded off and are therefore *approximate*.   ---------■

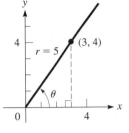

Fig. 4-14

EXAMPLE 4   Find the values of the trigonometric functions of the angle whose terminal side passes through $(7.27, 4.49)$. The coordinates are approximate.

We show the angle and the given point in Fig. 4-15. From the Pythagorean theorem, we have

$$r = \sqrt{7.27^2 + 4.49^2} = 8.545$$

(Here we show a rounded-off value of r. It is not actually necessary to record the value of r since its value can be stored in the memory of a calculator. The reason for recording it here is to show the values used in the calculation of each of the trigonometric functions.) Therefore, we have the following values:

$$\sin \theta = \frac{4.49}{8.545} = 0.525 \qquad \cos \theta = \frac{7.27}{8.545} = 0.851$$

$$\tan \theta = \frac{4.49}{7.27} = 0.618 \qquad \cot \theta = \frac{7.27}{4.49} = 1.62$$

$$\sec \theta = \frac{8.545}{7.27} = 1.18 \qquad \csc \theta = \frac{8.545}{4.49} = 1.90$$

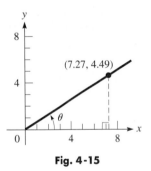

Fig. 4-15

Since the coordinates are approximate, the results are rounded off.

The display in a graphing calculator window is shown in Fig. 4-16. (The key sequences are indicated by the display.) The first result shown is the value of r, and the calculator stores this result under *ans*. In order to evaluate all of the trigonometric functions, we then store the value of r as the value of x. (It is not necessary to enter *ans*. The first key to use for line 3 is *sto*.) This is shown in the third and fourth lines of the display. To find $\sin \theta$ we continue with the fifth line of the display, and the result is shown on the sixth line. To get values of the other functions, we can use the value of r stored as x (here, x is chosen for convenience of calculator use—it is *not* the same as the x-coordinate of the point.)   ---------■

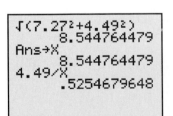

Fig. 4-16

In Example 4, we expressed the result as sin θ = 0.525. A common error is to omit the angle and give the value as sin = 0.525. This is a meaningless expression, for *we must show the angle* for which we have the value of a function.

CAUTION ▶

If one of the trigonometric functions is known, it is possible to find the values of the other functions. The following example illustrates the method.

■EXAMPLE 5 If we know that sin θ = 3/7 and that θ is a first-quadrant angle, we know the ratio of the ordinate to the radius vector (y to r) is 3 to 7. Therefore, the point on the terminal side for which y = 3 can be found by use of the Pythagorean theorem. The x-value for this point is

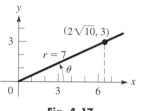

Fig. 4-17

$$x = \sqrt{7^2 - 3^2} = \sqrt{49 - 9} = \sqrt{40} = 2\sqrt{10}$$

Therefore, the point $(2\sqrt{10}, 3)$ is on the terminal side, as shown in Fig. 4-17.

Therefore, using the values $x = 2\sqrt{10}$, y = 3, and r = 7, we have the other trigonometric functions of θ. They are

$$\cos \theta = \frac{2\sqrt{10}}{7}, \quad \tan \theta = \frac{3}{2\sqrt{10}}, \quad \cot \theta = \frac{2\sqrt{10}}{3}, \quad \sec \theta = \frac{7}{2\sqrt{10}}, \quad \csc \theta = \frac{7}{3}$$

These values are *exact*. *Approximate* decimal values found on a calculator are

$$\cos \theta = 0.9035, \quad \tan \theta = 0.4743, \quad \cot \theta = 2.108$$
$$\sec \theta = 1.107, \quad \csc \theta = 2.333$$

--- **EXERCISES** *4-2* ---

In Exercises 1–16, find values of the trigonometric functions of the angle (in standard position) whose terminal side passes through the given points. For Exercises 1–12, give answers in exact form. For Exercises 13–16, the coordinates are approximate.

1. (6, 8)   **2.** (5, 12)   **3.** (15, 8)   **4.** (24, 7)

5. (9, 40)   **6.** (16, 30)   **7.** (1, $\sqrt{15}$)   **8.** ($\sqrt{3}$, 2)

9. (1, 1)   **10.** (6, 5)   **11.** (5, 2)   **12.** (1, $\frac{1}{2}$)

13. (3.25, 5.15)   **14.** (0.687, 0.943)

15. (0.08623, 0.01327)   **16.** (37.65, 21.87)

In Exercises 17–24, find the values of the indicated functions. In Exercises 17–20, give answers in exact form. In Exercises 21–24, the values are approximate.

17. Given cos θ = 12/13, find sin θ and cot θ.

18. Given sin θ = 1/2, find cos θ and csc θ.

19. Given tan θ = 1, find sin θ and sec θ.

20. Given sec θ = 4/3, find tan θ and cos θ.

21. Given sin θ = 0.750, find cot θ and csc θ.

22. Given cos θ = 0.326, find sin θ and tan θ.

23. Given cot θ = 0.254, find cos θ and tan θ.

24. Given csc θ = 1.20, find sec θ and cos θ.

In Exercises 25–28, each point listed is on the terminal side of an angle. Show that each of the indicated functions is the same for each of the points.

25. (3, 4), (6, 8), (4.5, 6), sin θ and tan θ

26. (5, 12), (15, 36), (7.5, 18), cos θ and cot θ

27. (2, 1), (4, 2), (8, 4), tan θ and sec θ

28. (3, 2), (6, 4), (9, 6), csc θ and cos θ

In Exercises 29–32, answer the given questions.

29. If tan θ = 3/4, what is the value of $\sin^2 \theta + \cos^2 \theta$? [$\sin^2 \theta = (\sin \theta)^2$]

30. What is x if (2, 5) and (7, x) are on the same terminal side?

31. From the definitions of the trigonometric functions, it can be seen that csc θ is the reciprocal of sin θ. What function is the reciprocal of cos θ?

(W) 32. Refer to the definitions of the trigonometric functions in Eqs. (4-1). Is the quotient of one of the functions divided by cos θ equal to tan θ? Explain.

4-3 VALUES OF THE TRIGONOMETRIC FUNCTIONS

We have been able to calculate the trigonometric functions if we knew one point on the terminal side of the angle. However, in practice it is more common to know the angle in degrees, for example, and to be required to find the functions of this angle. Therefore, we must be able to determine the trigonometric functions of angles in degrees.

One way to determine the functions of a given angle is to make a scale drawing. That is, we draw the angle in standard position using a protractor and then measure the lengths of the values of x, y, and r for some point on the terminal side. By using the proper ratios we may determine the functions of this angle.

We may also use certain geometric facts to determine the functions of some particular angles. The following two examples illustrate this procedure.

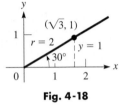

Fig. 4-18

■ EXAMPLE 1   From geometry we find that the side opposite a 30° angle in a right triangle is one-half of the hypotenuse. Using this fact and letting $y = 1$ and $r = 2$ (see Fig. 4-18), we calculate that $x = \sqrt{2^2 - 1^2} = \sqrt{3}$ from the Pythagorean theorem. Therefore, with $x = \sqrt{3}$, $y = 1$, and $r = 2$, we have

$$\sin 30° = \frac{1}{2} \qquad \cos 30° = \frac{\sqrt{3}}{2} \qquad \tan 30° = \frac{1}{\sqrt{3}}$$

In a similar way we may determine the values of the functions of 60° to be

$$\sin 60° = \frac{\sqrt{3}}{2} \qquad \cos 60° = \frac{1}{2} \qquad \tan 60° = \sqrt{3}$$

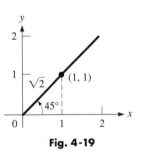

Fig. 4-19

■ EXAMPLE 2   Find sin 45°, cos 45°, and tan 45°.

From geometry we know that in an isosceles right triangle the angles are 45°, 45°, and 90°. We know that the sides are in proportion 1, 1, $\sqrt{2}$, respectively. Putting the 45° angle in standard position, we find $x = 1$, $y = 1$, and $r = \sqrt{2}$ (see Fig. 4-19). From this we determine

$$\sin 45° = \frac{1}{\sqrt{2}} \qquad \cos 45° = \frac{1}{\sqrt{2}} \qquad \tan 45° = 1$$

As in Example 1, we have given the exact values. Decimal approximations are given in the table that follows.

Summarizing the results for 30°, 45°, and 60°, we have:

θ	30°	45°	60°	(decimal approximations) 30°	45°	60°
$\sin \theta$	$\dfrac{1}{2}$	$\dfrac{1}{\sqrt{2}}$	$\dfrac{\sqrt{3}}{2}$	0.500	0.707	0.866
$\cos \theta$	$\dfrac{\sqrt{3}}{2}$	$\dfrac{1}{\sqrt{2}}$	$\dfrac{1}{2}$	0.866	0.707	0.500
$\tan \theta$	$\dfrac{1}{\sqrt{3}}$	1	$\sqrt{3}$	0.577	1.000	1.732

NOTE ▶   It is helpful to be familiar with these values, as they are used in later sections.

The scale-drawing method for finding values of the trigonometric functions gives only approximate results, and geometric methods work only for a limited number of angles. However, it is possible to find these values to any required degree of accuracy through more advanced mathematical methods (using calculus and what are known as *power series*).

The values of the trigonometric functions $\sin \theta$, $\cos \theta$, and $\tan \theta$ are programmed into graphing calculators. For all of our work in the remainder of this chapter, *be sure that your calculator is set for **degrees*** (not radians). The following examples illustrate the use of a calculator in finding trigonometric values.

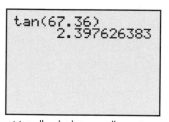

Not all calculators will require parentheses around 67.36.

Fig. 4-20

EXAMPLE 3 Using a graphing calculator to find the value of tan 67.36°, we first enter the function and then the angle, just as we have written it. The resulting display is shown in Fig. 4-20.

Therefore, we see that tan 67.36° = 2.397626383. ∎

Not only are we able to find values of the trigonometric functions if we know the angle, but we can also find the angle if we know that value of a function. In doing this, we are actually using another important type of mathematical function, an **inverse trigonometric function.** They are discussed in detail in Chapter 20. For the purpose of using a calculator at this point, it is sufficient to recognize and understand the notation that is used.

INVERSE TRIGONOMETRIC FUNCTIONS

The notation for "the angle whose sine is x" is $\sin^{-1} x$. This is called the *inverse sine function*. Equivalent meanings are given to $\cos^{-1} x$ (the angle whose cosine is x) and $\tan^{-1} x$ (the angle whose tangent is x). (The -1 used with a *function* indicates an *inverse function* and is *not* a negative exponent.) On a calculator the $\sin^{-1}$ key is used to find the angle when the sine of that angle is known. The following example illustrates the use of the equivalent $\cos^{-1}$ key.

CAUTION ▶

Another notation that is used for $\sin^{-1} x$ is arcsin x.

Fig. 4-21

EXAMPLE 4 If $\cos \theta = 0.3527$, which means that $\theta = \cos^{-1} 0.3527$ (θ is the angle whose cosine is 0.3527), we can use a graphing calculator to find θ. The display for this is shown in Fig. 4-21.

Therefore, we see that $\theta = 69.35°$ (rounded off). ∎

When using the trigonometric functions, the angle is often *approximate*. Angles of 2.3°, 92.3°, and 182.3° are angles with equal accuracy, which shows that *the accuracy of an angle does not depend on the number of digits shown*. The measurement of an angle and the accuracy of its trigonometric functions are shown in the following table.

Angles and Accuracy of Trigonometric Functions

Measurement of Angle to Nearest	Accuracy of Trigonometric Function
1°	2 significant digits
0.1° or 10′	3 significant digits
0.01° or 1′	4 significant digits

We rounded off the result in Example 4 according to this table.

It is generally possible to set up the solution of a problem in terms of the sine, cosine, or tangent. However, there are times when a value of the cotangent, secant, or cosecant is given or is needed. We will now show how values of these functions are found using a calculator.

RECIPROCAL FUNCTIONS

From the definitions of the trigonometric functions, we see that $\csc \theta = r/y$ and $\sin \theta = y/r$. This means that *the value of* $\csc \theta$ *is the reciprocal of the value of* $\sin \theta$. Therefore, by use of the reciprocal key, x^{-1} (here, the -1 *is* an exponent) of a calculator, the values of these functions can be found. (Remember, by the definition of a reciprocal, $x^{-1} = 1/x$).

NOTE▶

▌EXAMPLE 5 To find the value of sec 27.82°, we use the fact that

$$\sec 27.82° = \frac{1}{\cos 27.82°} \qquad \text{or} \qquad \sec 27.82° = (\cos 27.82°)^{-1}$$

Therefore, we are to find the reciprocal of the value of cos 27.82°. This value can be found using either the first two lines, or the third and fourth lines, of the calculator display shown in Fig. 4-22.

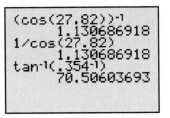

From the display, we see that sec 27.82° = 1.131, with the results rounded off according to the table at the bottom of page 113. ------------▇

Fig. 4-22

▌EXAMPLE 6 To find the value of θ if $\cot \theta = 0.354$, we use the fact that

$$\tan \theta = \frac{1}{\cot \theta} = \frac{1}{0.354}$$

The value found by using a graphing calculator is shown in the fifth and sixth lines of Fig. 4-22.

Therefore, $\theta = 70.5°$ (rounded off). ------------▇

In the following example, we see how to find the value of one function if we know the value of another function of the same angle.

▌EXAMPLE 7 Find $\sin \theta$ if $\sec \theta = 2.504$.

Since the value of $\sec \theta$ is known, we know that $\cos \theta = 2.504^{-1}$ (or $1/2.504$). This in turn tells us that $\theta = \cos^{-1}(2.504^{-1})$. Since we are to find the value of $\sin \theta$, we can see that

$$\sin \theta = \sin(\cos^{-1}(2.504^{-1}))$$

Therefore, we have the calculator display shown in Fig. 4-23.

This means that $\sin \theta = 0.9168$ (rounded off). ------------▇

Fig. 4-23

Calculators have been used extensively since the 1970s. Until then the values of the trigonometric functions were generally found by the use of tables (where linear interpolation was used to find values to one more place than was shown in the table). Many standard sources with tables are still available with a precision to at least 10′ or 0.1° (these tables give values with a precision to about five decimal places). However, a calculator is much easier to use than a table, and it can give values to a much greater accuracy. Therefore, we will not use tables in this text.

The following example illustrates the use of the value of a trigonometric function in an applied problem. We consider various types of applications later in the chapter.

EXAMPLE 8 When a rocket is launched, its horizontal velocity v_x is related to the velocity v with which it is fired by the equation $v_x = v \cos \theta$ (which means $v(\cos \theta)$, but does *not* mean $\cos \theta \, v$, which is the same as $\cos (\theta \, v)$). Here, θ is the angle between the horizontal and the direction in which it is fired (see Fig. 4-24). Find v_x if $v = 1250$ m/s and $\theta = 36.0°$.

Substituting the given values of v and θ in $v_x = v \cos \theta$, we have

$$v_x = 1250 \cos 36.0°$$
$$= 1010 \text{ m/s}$$

Therefore, the horizontal velocity is 1010 m/s.

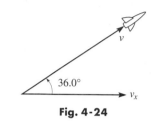

Fig. 4-24

EXERCISES 4-3

In Exercises 1–4, use a protractor to draw the given angle. Measure off 10 units (centimeters are convenient) along the radius vector. Then measure the corresponding values of x and y. From these values determine the trigonometric functions of the angle.

1. 40°   **2.** 75°   **3.** 15°   **4.** 53°

In Exercises 5–20, find the values of the trigonometric functions. Round off results according to the table following Example 4.

5. sin 22.4°   **6.** cos 72.5°

7. tan 57.6°   **8.** sin 36.0°

9. cos 15.71°   **10.** tan 8.653°

11. sin 84°   **12.** cos 47°

13. cot 67.78°   **14.** csc 22.81°

15. sec 50.4°   **16.** cot 41.8°

17. csc 49.3°   **18.** sec 7.8°

19. cot 85.96°   **20.** csc 76.30°

In Exercises 21–36, find θ for each of the given trigonometric functions. Round off results according to the table following Example 4.

21. $\cos \theta = 0.3261$ **22.** $\tan \theta = 2.470$

23. $\sin \theta = 0.9114$ **24.** $\cos \theta = 0.0427$

25. $\tan \theta = 0.207$ **26.** $\sin \theta = 0.109$

27. $\cos \theta = 0.65007$ **28.** $\tan \theta = 5.7706$

29. $\csc \theta = 1.245$ **30.** $\sec \theta = 2.045$

31. $\cot \theta = 0.1443$ **32.** $\csc \theta = 1.012$

33. $\sec \theta = 3.65$ **34.** $\cot \theta = 2.08$

35. $\csc \theta = 3.262$ **36.** $\cot \theta = 0.1519$

In Exercises 37 and 38, use a calculator to verify the given relationships. [$\sin^2 \theta = (\sin \theta)^2$]

37. $\dfrac{\sin 43.7°}{\cos 43.7°} = \tan 43.7°$ **38.** $\sin^2 77.5° + \cos^2 77.5° = 1$

W *In Exercises 39 and 40, explain why the given statements are true for an acute angle θ.*

39. $\sin \theta$ is always between 0 and 1.

40. $\tan \theta$ can equal any positive real number.

In Exercises 41–44, find the values of the indicated trigonometric functions.

41. Find $\sin \theta$, given $\tan \theta = 1.936$.

42. Find $\cos \theta$, given $\sin \theta = 0.6725$.

43. Find $\tan \theta$, given $\sec \theta = 1.3698$.

44. Find $\csc \theta$, given $\cos \theta = 0.1063$.

In Exercises 45–48, solve the given problems.

45. The sound produced by a jet engine was measured at a distance of 100 m in all directions. The loudness d of the sound (in decibels) was found to be $d = 70.0 + 30.0 \cos \theta$, where the 0° line was directed in front of the engine. Calculate d for $\theta = 54.5°$.

46. A brace is used in the structure shown in Fig. 4-25. Its length is $l = a(\sec \theta + \csc \theta)$. Find l if $a = 28.0$ cm and $\theta = 34.5°$.

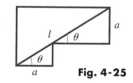

Fig. 4-25

47. The current i in a certain electric circuit is given by the equation $i = I \sin(\theta - 5.0°)$. Find i for $I = 1.80$ A and $\theta = 46.3°$.

48. A submarine dives such that the horizontal distance h and vertical distance v are related by $v = h \tan \theta$. Here θ is the angle of the dive, as shown in Fig. 4-26. Find θ if $h = 2.35$ mi and $v = 1.52$ mi.

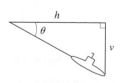

Fig. 4-26

4-4 THE RIGHT TRIANGLE

From geometry we know that a triangle, by definition, consists of three sides and has three angles. If one side and any other two of these six parts of the triangle are known, it is possible to determine the other three parts. One of the three known parts must be a side, for if we know only the three angles, we can conclude only that an entire set of similar triangles has those particular angles.

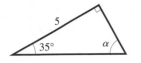

Fig. 4-27

■EXAMPLE 1 Assume that one side and two angles are known, such as the side of 5 and the angles of 35° and 90° in the triangle in Fig. 4-27. Then we may determine the third angle α by the fact that the sum of the angles of a triangle is always 180°. Of all possible similar triangles having the three angles of 35°, 90°, and 55° (which is α), we have the one with the particular side of 5 between angles of 35° and 90°. Only one triangle with these parts is possible (in the sense that all triangles with the given parts are *congruent* and have equal corresponding angles and sides).

--------■

To **solve a triangle** *means that, when we are given three parts of a triangle (at least one a side), we are to find the other three parts.* In this section we are going to demonstrate the method of solving a right triangle. *Since one angle of the triangle will be 90°, it is necessary to know one side and one other part.* Also, we know that the sum of the three angles of a triangle is 180°, and this in turn tells us that *the sum of the other two angles, both acute, is* 90°. *Any two acute angles whose sum is 90° are said to be* **complementary.**

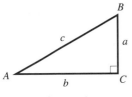

Fig. 4-28

For consistency, when we are labeling the parts of the right triangle *we shall use the letters A and B to denote the acute angles and C to denote the right angle. The letters a, b, and c will denote the sides opposite these angles, respectively. Thus, side c is the hypotenuse of the right triangle.* See Fig. 4-28.

In solving right triangles we shall find it convenient to express the trigonometric functions of the acute angles in terms of the sides. By placing the vertex of angle A at the origin and the vertex of right angle C on the positive x-axis, as shown in Fig. 4-29, we have the following ratios for angle A in terms of the sides of the triangle.

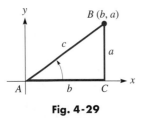

Fig. 4-29

$$\sin A = \frac{a}{c} \qquad \cos A = \frac{b}{c} \qquad \tan A = \frac{a}{b}$$

$$\cot A = \frac{b}{a} \qquad \sec A = \frac{c}{b} \qquad \csc A = \frac{c}{a}$$

(4-3)

If we should place the vertex of B at the origin, instead of the vertex of angle A, we would obtain the following ratios for the functions of angle B (see Fig. 4-30):

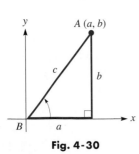

Fig. 4-30

$$\sin B = \frac{b}{c} \qquad \cos B = \frac{a}{c} \qquad \tan B = \frac{b}{a}$$

$$\cot B = \frac{a}{b} \qquad \sec B = \frac{c}{a} \qquad \csc B = \frac{c}{b}$$

(4-4)

Equations (4-3) and (4-4) show that we may generalize our definitions of the trigonometric functions of an acute angle of a right triangle (we have chosen ∠A in Fig. 4-31) to be as follows:

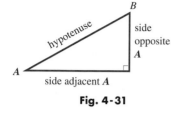

Fig. 4-31

$$\sin A = \frac{\text{side opposite } A}{\text{hypotenuse}} \qquad \cos A = \frac{\text{side adjacent } A}{\text{hypotenuse}}$$

$$\tan A = \frac{\text{side opposite } A}{\text{side adjacent } A} \qquad \cot A = \frac{\text{side adjacent } A}{\text{side opposite } A} \qquad \text{(4-5)}$$

$$\sec A = \frac{\text{hypotenuse}}{\text{side adjacent } A} \qquad \csc A = \frac{\text{hypotenuse}}{\text{side opposite } A}$$

Which side is adjacent or opposite depends on the angle being considered. In Fig. 4-31 the side opposite A is adjacent to B, and the side adjacent to A is opposite B.

Using the definitions in this form, we can solve right triangles without placing the angle in standard position. The angle need only be a part of any right triangle.

We note from the above discussion that $\sin A = \cos B$, $\tan A = \cot B$, and $\sec A = \csc B$. From this we conclude that *cofunctions of acute complementary angles are equal.* The sine function and cosine function are cofunctions, the tangent function and cotangent function are cofunctions, and the secant function and cosecant functions are cofunctions. This property of the values of the trigonometric functions is illustrated in the following example.

EXAMPLE 2   Given $a = 4$, $b = 7$, and $c = \sqrt{65}$, find $\sin A$, $\cos A$, and $\tan A$ (see Fig. 4-32) in exact form and in approximate decimal form (to three significant digits).

$$\sin A = \frac{\text{side opposite angle } A}{\text{hypotenuse}} = \frac{4}{\sqrt{65}} = 0.496$$

$$\cos A = \frac{\text{side adjacent angle } A}{\text{hypotenuse}} = \frac{7}{\sqrt{65}} = 0.868$$

$$\tan A = \frac{\text{side opposite angle } A}{\text{side adjacent angle } A} = \frac{4}{7} = 0.571 \qquad \text{■}$$

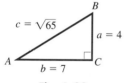

Fig. 4-32

EXAMPLE 3   Finding $\sin B$, $\cos B$, and $\tan B$ for the triangle in Fig. 4-32, we have

$$\sin B = \frac{\text{side opposite angle } B}{\text{hypotenuse}} = \frac{7}{\sqrt{65}} = 0.868$$

$$\cos B = \frac{\text{side adjacent angle } B}{\text{hypotenuse}} = \frac{4}{\sqrt{65}} = 0.496$$

$$\tan B = \frac{\text{side opposite angle } B}{\text{side adjacent angle } B} = \frac{7}{4} = 1.75$$

In Fig. 4-32 we note that A and B are complementary angles. Comparing with values found in Example 2, we see that $\sin A = \cos B$ and $\cos A = \sin B$. 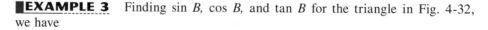 ■

We are now ready to solve right triangles. The method is given and illustrated on the next page.

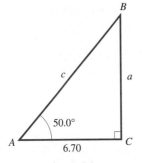

Fig. 4-33

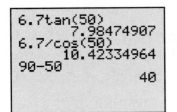

Fig. 4-34

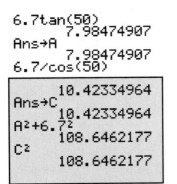

Fig. 4-35

Procedure for Solving a Right Triangle

1. *Sketch a right triangle and label the known and unknown sides and angles.*

2. *Express each of the three unknown parts in terms of the known parts and solve for the unknown parts.*

3. *Check the results.* The sum of the angles should be 180°. If only one side is given, check the computed side with the Pythagorean theorem. If two sides are given, check the angles and computed side by using appropriate trigonometric functions.

■**EXAMPLE 4**   Solve the right triangle with $A = 50.0°$ and $b = 6.70$.

We first sketch the right triangle shown in Fig. 4-33. (In making the sketch, we should be careful to follow proper labeling of the triangle as outlined on the previous page.) We then express unknown side a in terms of known side b and known angle A and solve for a. We will then do the same for unknown side c and unknown angle B.

Finding side a, we know that $\tan A = \dfrac{a}{b}$, which means that $a = b \tan A$. Thus,

$$a = 6.70 \tan 50.0° = 7.98 \qquad \text{lines 1 \& 2 of calculator display in Fig. 4-34}$$

Next, solving for side c, we have $\cos A = \dfrac{b}{c}$, which means $c = \dfrac{b}{\cos A}$.

$$c = \frac{6.70}{\cos 50.0°} = 10.4 \qquad \text{lines 3 \& 4 of calculator display in Fig. 4-34}$$

Now solving for B, we know that $A + B = 90°$, or

$$B = 90° - A$$
$$= 90° - 50.0° = 40.0° \qquad \text{lines 5 \& 6 of calculator display in Fig. 4-34}$$

Therefore, $a = 7.98$, $c = 10.4$, and $B = 40.0°$.

Checking the angles: $A + B + C = 50.0° + 40.0° + 90° = 180°$
Checking the sides: $10.4^2 = 108.16$
$$7.98^2 + 6.70^2 = 108.57$$

Since the computed values were rounded off, the values 108.16 and 108.57 show that the values for sides a and c check.

As we calculate the values of the unknown parts, if we store each in the calculator memory, we can get a better check of the solution. In Fig. 4-35 we see the solution for each side, its storage in memory, and the check of these values. The lines above the display are those that are replaced as the solution proceeds. From this calculator display we see that the values for sides a and c check very accurately.   ■

NOTE ▶   In finding the unknown parts, we first expressed them in terms of the known parts. We do this because *it is best to use given values in calculations.* If we use one computed value to find another computed value, any error in the first would be carried to the value of the second. For example, in Example 4, if we were to find the value of c by using the value of a, any error in a would cause c to be in error as well.

See Appendix C for a graphing calculator program SLVRTTRI. It can be used to solve a right triangle, given the two legs.

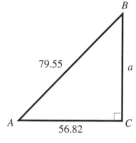

Fig. 4-36

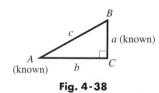

Fig. 4-37

As we noted earlier, the symbol $\approx$ means "equals approximately."

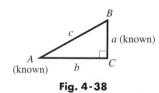

Fig. 4-38

We should also point out that, by inspection, we can make a rough check on the sides and angles of any triangle.

The longest side is always opposite the largest angle, and the shortest side is always opposite the smallest angle.

In a right triangle, *the hypotenuse is always the longest side.* We see that this is true for the sides and angles for the triangle in Example 4, where c is the longest side (opposite the 90° angle), and b is the shortest side and is opposite the angle of 40°.

EXAMPLE 5   Solve the right triangle with $b = 56.82$ and $c = 79.55$.

We sketch the right triangle as shown in Fig. 4-36. Since two sides are given, we will use the Pythagorean theorem to find the third side a. Also, we will use the cosine to find $\angle A$.

Since $c^2 = a^2 + b^2$, $a^2 = c^2 - b^2$. Therefore,

$$a = \sqrt{c^2 - b^2} = \sqrt{79.55^2 - 56.82^2}$$

$$= 55.67 \qquad \text{lines 1 \& 2 shown in Fig. 4-37}$$

Since $\cos A = \dfrac{b}{c}$, we have

$$\cos A = \frac{56.82}{79.55}, \qquad A = \cos^{-1}\left(\frac{56.82}{79.55}\right) = 44.42° \qquad \text{lines 3–5 shown in Fig. 4-37}$$

It is not necessary to actually calculate the ratio 56.82/79.55. In the same way, we find the value of angle B.

$$\sin B = \frac{56.82}{79.55}, \qquad B = \sin^{-1}\left(\frac{56.82}{79.55}\right) = 45.58° \qquad \text{lines 6–8 shown in Fig. 4-37}$$

Although we used a different function, we did use exactly the same ratio to find B as we used to find A. Therefore, many texts would find B from the fact that $A + B = 90°$, or $B = 90° - A = 90° - 44.42° = 45.58°$. This is also acceptable since any possible error should be discovered when the solution is checked.

We have now found that

$$a = 55.67 \qquad A = 44.42° \qquad B = 45.58°$$

Checking the sides and angles, we first note that side a is the shortest side and is opposite the smallest angle, $\angle A$. Also, the hypotenuse is the longest side. Next, using the sine function (we could use the cosine or tangent) to check the sides, we have

$$\sin 44.42° = \frac{55.67}{79.55}, \text{ or } 0.6999 \approx 0.6998 \qquad \sin 45.58° = \frac{56.82}{79.55}, \text{ or } 0.7142 \approx 0.7143$$

This shows that the values check. As we noted at the end of Example 4, we would get a more accurate check if we save the calculator values as they are found and use them for the check.

EXAMPLE 6   Given that $\angle A$ and side a are known, express the unknown parts of the right triangle in terms of A and a.

We sketch a right triangle as shown in Fig. 4-38, and set up the required expressions as follows:

Since $\dfrac{a}{b} = \tan A$, we have $b = \dfrac{a}{\tan A}$. Since $\dfrac{a}{c} = \sin A$, we have $c = \dfrac{a}{\sin A}$.

Since A is known, $B = 90° - A$.

EXERCISES *4-4*

In Exercises 1–4, draw appropriate figures and verify through observation that only one triangle may contain the given parts (that is, any other which may be drawn will be congruent).

1. A 60° angle included between sides of 3 in. and 6 in.

2. A side of 4 in. included between angles of 40° and 50°

3. A right triangle with a hypotenuse of 5 cm and a leg of 3 cm

4. A right triangle with a 70° angle between the hypotenuse and a leg of 5 cm

In Exercises 5–24, solve the right triangles with the given parts. Round off results. Refer to Fig. 4-39.

5. $A = 77.8°$, $a = 6700$

6. $A = 18.4°$, $c = 0.0897$

7. $a = 150$, $c = 345$

8. $a = 93.2$, $c = 124$

9. $B = 32.1°$, $c = 23.8$

10. $B = 64.3°$, $b = 0.652$

11. $b = 82$, $c = 88$

12. $a = 5920$, $b = 4110$

13. $A = 32.10°$, $c = 56.85$

14. $B = 12.60°$, $c = 18.42$

15. $a = 56.73$, $b = 44.09$

16. $a = 9.908$, $c = 12.63$

17. $B = 37.5°$, $a = 0.862$

18. $A = 52°$, $b = 8.4$

19. $B = 74.18°$, $b = 1.849$

20. $A = 51.36°$, $a = 369.2$

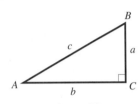

Fig. 4-39

21. $a = 591.87$, $b = 264.93$

22. $b = 2.9507$, $c = 5.0864$

23. $A = 12.975°$, $b = 14.592$

24. $B = 84.942°$, $a = 7413.5$

In Exercises 25–28, find the part of the triangle labeled either x or A in the indicated figure.

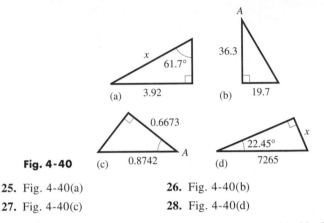

Fig. 4-40

25. Fig. 4-40(a)

26. Fig. 4-40(b)

27. Fig. 4-40(c)

28. Fig. 4-40(d)

In Exercises 29–32, refer to Fig. 4-39. In Exercises 29–31, the listed parts are assumed known. Express the other parts in terms of these known parts.

29. A, c

30. a, b

31. B, a

(W) 32. In Fig. 4-39, is there any combination of two given parts (not including $\angle C$) that does not give a unique solution of the triangle? Explain.

4-5 APPLICATIONS OF RIGHT TRIANGLES

SOLVING A WORD PROBLEM

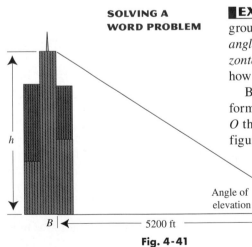

Angle of elevation 16°

5200 ft

Fig. 4-41

■**EXAMPLE 1**   The Sears Tower in Chicago can be seen from a point on the ground known to be 5200 ft from the base of the tower. The **angle of elevation** (*the angle between the horizontal and the line of sight, when the object is **above** the horizontal*) from the observer to the top of the tower is 16°. Based on this information, how high is the Sears Tower?

By drawing an appropriate figure, as shown in Fig. 4-41, we show the given information and that which is required. Here we have let h be the height of the tower, O the position of the observer, and B the point at the base of Sears Tower. From the figure we see that

$$\frac{h}{5200} = \tan 16° \qquad \frac{\text{required opposite side}}{\text{given adjacent side}} = \text{tangent of given angle}$$

$$h = 5200 \tan 16°$$

$$= 1500 \text{ ft}$$

Here we have rounded off the result since the data are good only to two significant digits. (The Sears Tower was completed in 1974 and is the tallest building in the world. Its height is actually 1454 ft.)

See the chapter introduction.

SOLVING A WORD PROBLEM

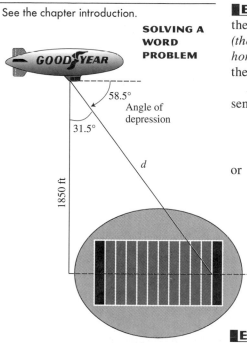

Fig. 4-42

SOLVING A WORD PROBLEM

Fig. 4-43

SOLVING A WORD PROBLEM

█EXAMPLE 2 The Goodyear blimp is 1850 ft above the ground and south of the Rose Bowl in California during a Super Bowl game. The **angle of depression** *(the angle between the horizontal and the line of sight, when the object is **below** the horizontal)* of the north goal line from the blimp is 58.5°. How far is the observer in the blimp from the goal line?

Again we sketch an appropriate figure as shown in Fig. 4-42. Here we let *d* represent the required distance. From the figure we see that

$$\frac{1850}{d} = \cos 31.5° \qquad \frac{\text{given adjacent side}}{\text{required hypotenuse}} = \text{cosine of known angle}$$

or

$$d = \frac{1850}{\cos 31.5°}$$
$$= 2170 \text{ ft}$$

Here we have rounded off the result to three significant digits, the accuracy of the given information. (The Super Bowl games in 1977, 1980, 1987, and 1993 were played in the Rose Bowl.)

█EXAMPLE 3 A missile is launched at an angle of 26.55° with respect to the horizontal. If it travels in a straight line over level terrain for 2.000 min and its average speed is 6355 km/h, what is its altitude at this time?

In Fig. 4-43 we let *h* represent the altitude of the missile after 2.000 min (altitude is measured on a perpendicular line). Also, we determine that in this time the missile has flown 211.8 km in a direct line from the launching site. This is found from the fact that it travels at 6355 km/h for $\frac{1}{30.00}$ h (2.000 min), and distance = speed × time. We therefore have $(6355 \text{ km/h})\left(\frac{1}{30.00} \text{ h}\right) = 211.8$ km. This means

$$\frac{h}{211.8} = \sin 26.55° \qquad \frac{\text{required opposite side}}{\text{known hypotenuse}} = \text{sine of given angle}$$
$$h = 211.8(\sin 26.55°)$$
$$= 94.67 \text{ km}$$

█EXAMPLE 4 A driver coming to an intersection sees the word STOP in the roadway. From the measurements shown in Fig. 4-44, find the angle θ that the letters make at the driver's eye.

From the figure we know sides *BS* and *BE* in triangle *BES,* and sides *BT* and *BE* in triangle *BET.* This means we can find ∠*TEB* and ∠*SEB* by use of the tangent. We then find θ from the fact that $\theta = \angle TEB - \angle SEB.$

$$\tan\angle TEB = \frac{18.0}{1.20}, \qquad \angle TEB = 86.2°$$

$$\tan\angle SEB = \frac{15.0}{1.20}, \qquad \angle SEB = 85.4°$$

$$\theta = 86.2° - 85.4° = 0.8°$$

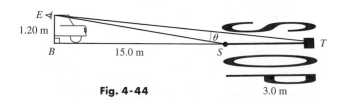

Fig. 4-44

Lasers were first produced in the late 1950s.

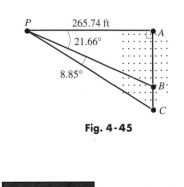

Fig. 4-45

EXAMPLE 5 Using lasers, a surveyor makes the measurements shown in Fig. 4-45, where points B and C are in a marsh. Find the distance between B and C.

Since the distance $BC = AC - AB$, BC is found by finding AC and AB and subtracting.

$$\frac{AB}{265.74} = \tan 21.66° \quad \text{or} \quad AB = 265.74 \tan 21.66°$$

$$\frac{AC}{265.74} = \tan(21.66° + 8.85°) \quad \text{or} \quad AC = 265.74 \tan 30.51°$$

$$BC = AC - AB = 265.74 \tan 30.51° - 265.74 \tan 21.66°$$
$$= 51.06 \text{ ft}$$

EXERCISES $4\text{-}5$

In the following exercises, solve the given problems. Sketch an appropriate figure unless the figure is given.

1. A straight 120-ft culvert is built down a hillside that makes an angle of 54.0° with the horizontal. Find the height of the hill.

2. A point near the top of the Leaning Tower of Pisa is 50.5 m from a point at the base of the tower (measured along the tower). This top point is also directly above a point on the ground 4.25 m from the same base point. What angle does the tower make with the ground? See Fig. 4-46.

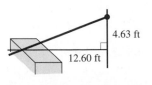

Fig. 4-46

3. A tree has a shadow 22.8 ft long when the angle of elevation of the sun is 62.6°. How tall is the tree?

4. The straight arm of a robot is 1.25 m long and makes an angle of 13.0° above a horizontal conveyor belt. How high above the belt is the end of the arm? See Fig. 4-47.

Fig. 4-47

5. The headlights of an automobile are set such that the beam drops 2.00 in. for each 25.0 ft in front of the car. What is the angle between the beam and the road?

6. A bullet was fired such that it just grazed the top of a table. It entered a wall, which is 12.60 ft from the graze point in the table, at a point 4.63 ft above the tabletop. At what angle was the bullet fired above the horizontal? See Fig. 4-48.

Fig. 4-48

7. A robot is on the surface of Mars. The angle of depression from a camera in the robot to a rock on the surface of Mars is 13.33°. The camera is 196.0 cm above the surface. How far from the camera is the rock?

8. From a small boat 2500 ft downstream from the Horseshoe Falls (Canadian) at Niagara Falls, the angle of elevation of the top of the Falls is 4.0°. How high are the Falls? See Fig. 4-49.

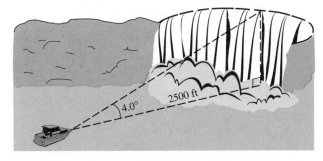

Fig. 4-49

9. In designing a new building, a doorway is 2.65 ft above the ground. A ramp for the disabled, at an angle of 6.0° with the ground, is to be built to the doorway. How long will the ramp be?

10. On a test flight, during the landing of the space shuttle, the ship was 325 ft above the end of the landing strip. It then came in on a constant angle of 7.5° with the landing strip. How far from the end of the landing strip did it first touch ground?

11. A rectangular piece of plywood 4.00 ft by 8.00 ft is cut from one corner to an opposite corner. What are the angles between edges of the resulting pieces?

12. A guardrail is to be constructed around the top of a circular observation tower. The diameter of the observation area is 12.3 m. If the railing is constructed with 30 equal straight sections, what should be the length of each section?

13. The angle of inclination of a road is often expressed as *percent grade,* which is the vertical rise divided by the horizontal run (expressed as a percent). See Fig. 4-50. A 6.0% grade corresponds to a road that rises 6.0 ft for every 100 ft along the horizontal. Find the angle of inclination that corresponds to a 6.0% grade.

Fig. 4-50
Horizontal run
Rise

14. A tabletop is in the shape of a regular octagon (8 sides). What is the greatest distance across the table if one edge of the octagon is 0.750 m?

15. To get a good view of a person in front of a teller's window, it is determined that a surveillance camera at a bank should be directed at a point 15.5 ft to the right and 6.75 ft below the camera. See Fig. 4-51. At what angle of depression should the camera be directed?

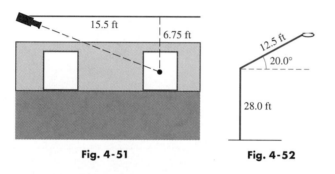

Fig. 4-51
15.5 ft
6.75 ft

12.5 ft
20.0°
28.0 ft

Fig. 4-52

16. A street light is designed as shown in Fig. 4-52. How high above the street is the light?

17. A straight driveway is 85.0 ft long, and the top is 12.0 ft above the bottom. What angle does it make with the horizontal?

18. Part of the Tower Bridge in London is a drawbridge. This part of the bridge is 76.0 m long. When each half is raised, the distance between them is 8.0 m. What angle does each half make with the horizontal? See Fig. 4-53.

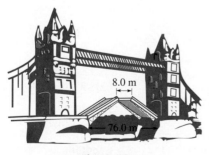

8.0 m
76.0 m

Fig. 4-53

19. A square wire loop is rotating in the magnetic field between two poles of a magnet in order to induce an electric current. The axis of rotation passes through the center of the loop and is midway between the poles, as shown in the side view in Fig. 4-54. How far is the edge of the loop from either pole if the side of the square is 7.30 cm and the poles are 7.66 cm apart when the angle between the loop and the vertical is 78.0°?

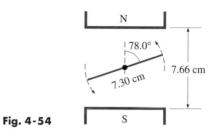

N
78.0°
7.66 cm
7.30 cm
S

Fig. 4-54

20. From a space probe circling Io, one of Jupiter's moons, at an altitude of 552 km, it was observed that the angle of depression of the horizon was 39.7°. What is the radius of Io?

21. A manufacturing plant is designed to be in the shape of a regular pentagon with 92.5 ft on each side. A security fence surrounds the building to form a circle, and each corner of the building is to be 25.0 ft from the closest point on the fence. How much fencing is required?

22. A surveyor on the New York City side of the Hudson River wishes to find the height of a cliff *(palisade)* on the New Jersey side. Figure 4-55 shows the measurements which were made. How high is the cliff? (In the figure, the triangle containing the height *h* is vertical and perpendicular to the river.)

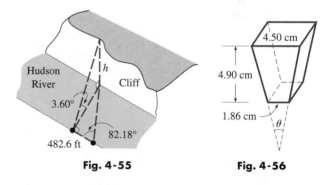

Hudson River
Cliff
h
3.60°
82.18°
482.6 ft

Fig. 4-55

4.50 cm
4.90 cm
1.86 cm
θ

Fig. 4-56

23. Find the angle θ in the taper shown in Fig. 4-56. (The front face is an isosceles trapezoid.)

24. What is the circumference of the Arctic Circle (latitude 66°32′N)? The radius of the earth is 3960 mi.

25. A stairway 1.0 m wide goes from the bottom of a cylindrical storage tank to the top at a point halfway around the tank. The handrail on the outside of the stairway makes an angle of 31.8° with the horizontal, and the radius of the tank is 11.8 m. Find the length of the handrail. See Fig. 4-57.

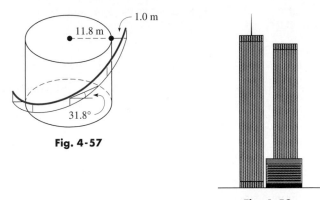

Fig. 4-57

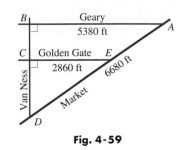

Fig. 4-58

26. An antenna is on the top of the World Trade Center. From a point on the river 7800 ft from the Center, the angles of elevation of the top and bottom of the antenna are 12.1° and 9.9°, respectively. How tall is the antenna? (Disregard the small part of the antenna near the base that cannot be seen.) The World Trade Center is shown in Fig. 4-58.

27. Some of the streets of San Francisco are shown in Fig. 4-59. The distances between intersections *A* and *B*, *A* and *D*, and *C* and *E* are shown. How far is it between intersections *C* and *D*?

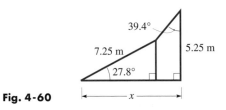

Fig. 4-59

28. A supporting girder structure is shown in Fig. 4-60. Find the length *x*.

Fig. 4-60

CHAPTER EQUATIONS

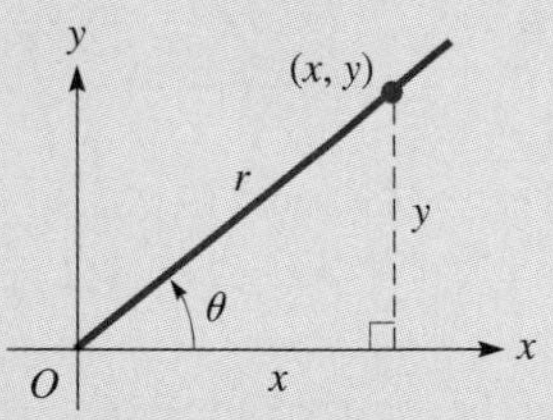

$$\sin \theta = \frac{y}{r} \qquad \cos \theta = \frac{x}{r}$$

$$\tan \theta = \frac{y}{x} \qquad \cot \theta = \frac{x}{y} \qquad (4\text{-}1)$$

$$\sec \theta = \frac{r}{x} \qquad \csc \theta = \frac{r}{y}$$

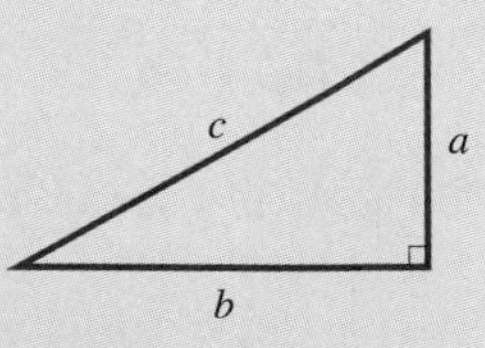

Pythagorean theorem

$$c^2 = a^2 + b^2 \qquad (4\text{-}2)$$

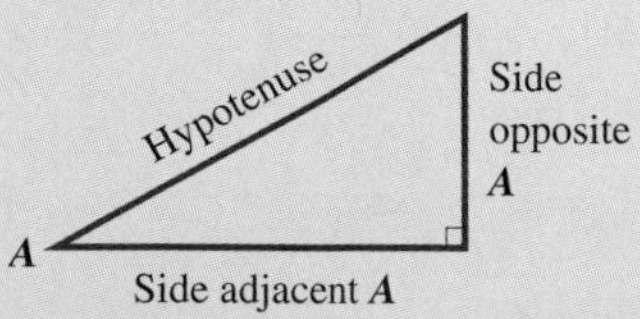

$$\sin A = \frac{\text{side opposite } A}{\text{hypotenuse}} \qquad \cos A = \frac{\text{side adjacent } A}{\text{hypotenuse}}$$

$$\tan A = \frac{\text{side opposite } A}{\text{side adjacent } A} \qquad \cot A = \frac{\text{side adjacent } A}{\text{side opposite } A} \qquad (4\text{-}5)$$

$$\sec A = \frac{\text{hypotenuse}}{\text{side adjacent } A} \qquad \csc A = \frac{\text{hypotenuse}}{\text{side opposite } A}$$

REVIEW EXERCISES

In Exercises 1–4, find the smallest positive angle and the smallest negative angle (numerically) coterminal with but not equal to the given angle.

1. $17.0°$ **2.** $248.3°$ **3.** $-217.5°$ **4.** $-7.6°$

In Exercises 5–8, express the given angles in decimal form.

5. $31°54'$ **6.** $174°45'$ **7.** $38°6'$ **8.** $321°27'$

In Exercises 9–12, express the given angles to the nearest minute.

9. $17.5°$ **10.** $65.4°$ **11.** $49.7°$ **12.** $126.25°$

In Exercises 13–16, determine the trigonometric functions of the angles (in standard position) whose terminal side passes through the given points. Give answers in exact form.

13. $(24,7)$ **14.** $(5,4)$ **15.** $(4,4)$ **16.** $(1.2,0.5)$

In Exercises 17–20, find the indicated trigonometric functions. Give answers in decimal form, rounded off to three significant digits.

17. Given $\sin \theta = \frac{5}{13}$, find $\cos \theta$ and $\cot \theta$.

18. Given $\cos \theta = \frac{3}{8}$, find $\sin \theta$ and $\tan \theta$.

19. Given $\tan \theta = 2$, find $\cos \theta$ and $\csc \theta$.

20. Given $\cot \theta = 4$, find $\sin \theta$ and $\sec \theta$.

In Exercises 21–28, find the values of the trigonometric functions. Round off results.

21. $\sin 72.1°$ **22.** $\cos 40.3°$

23. $\tan 61.64°$ **24.** $\sin 49.09°$

25. $\sec 18.4°$ **26.** $\csc 82.4°$

27. $(\cot 7.06°)(\sin 7.06°) - \cos 7.06°$

28. $(\sec 79.36°)(\sin 79.36°) - \tan 79.36°$

In Exercises 29–36, find θ for each of the given trigonometric functions. Round off results.

29. $\cos \theta = 0.950$ **30.** $\sin \theta = 0.63052$

31. $\tan \theta = 1.574$ **32.** $\cos \theta = 0.1345$

33. $\csc \theta = 4.713$ **34.** $\cot \theta = 0.7561$

35. $\sec \theta = 2.54$ **36.** $\csc \theta = 1.92$

In Exercises 37–48, solve the right triangles with the given parts. Refer to Fig. 4-61.

37. $A = 17.0°, b = 6.00$

38. $B = 68.1°, a = 1080$

39. $a = 81.0, b = 64.5$

40. $a = 1.06, c = 3.82$

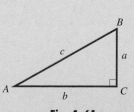

Fig. 4-61

41. $A = 37.5°, a = 12.0$ **42.** $B = 15.7°, c = 12.6$

43. $b = 6.508, c = 7.642$ **44.** $a = 72.14, b = 14.37$

45. $A = 49.67°, c = 0.8253$ **46.** $B = 4.38°, b = 5.682$

47. $a = 11.652, c = 15.483$ **48.** $a = 724.39, b = 852.44$

In Exercises 49–76, solve the given applied problems.

49. The voltage e at any instant in a coil of wire that is turning in a magnetic field is given by $e = E \cos \alpha$, where E is the maximum voltage and α is the angle the coil makes with the field. Find the acute angle α if $e = 56.9$ V and $E = 339$ V.

50. A formula for the area of a quadrilateral is $A = \frac{1}{2}d_1 d_2 \sin \theta$, where d_1 and d_2 are the lengths of the diagonals and θ is the angle between them. Find the area of a four-sided carpet remnant with diagonals 3.25 ft and 4.38 ft and $\theta = 72.0°$.

51. For a car rounding a curve, the road should be banked at an angle θ according to the equation $\tan \theta = \dfrac{v^2}{gr}$. Here, v is the speed of the car and r is the radius of the curve in the road. See Fig. 4-62. Find θ for $v = 80.7$ ft/s (55.0 mi/h), $g = 32.2$ ft/s^2, and $r = 950$ ft.

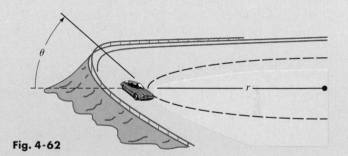

Fig. 4-62

52. The *apparent power S* in an electric circuit in which the power is P and the impedance phase angle is θ is given by $S = P \sec \theta$. Given $P = 12.0$ V $\cdot$ A and $\theta = 29.4°$, find S.

53. A surveyor measures two sides and the included angle of a triangular tract of land to be $a = 31.96$ m, $b = 47.25$ m, and $C = 64.09°$. (a) Show that a formula for the area A of the tract is $A = \frac{1}{2}ab \sin C$. (b) Find the area of the tract.

54. A water channel has the cross section of an isosceles trapezoid. See Fig. 4-63. (a) Show that a formula for the area of the cross section is $A = bh + h^2 \cot \theta$. (b) Find A if $b = 12.6$ ft, $h = 4.75$ ft, and $\theta = 37.2°$.

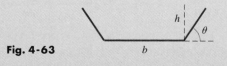

Fig. 4-63

55. In tracking an airplane on radar, it is found that the plane is 27.5 km on a direct line from the control tower, with an angle of elevation of 10.3°. What is the altitude of the plane?

56. A straight emergency chute for an airplane is 16.0 ft long. In being tested, the end of the chute is 8.5 ft above the ground. What angle does the chute make with the ground?

57. The window of a house is shaded as shown in Fig. 4-64. What percent of the window is shaded when the angle of elevation θ of the sun is 65°?

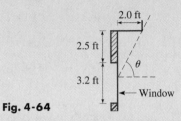

Fig. **4-64**

58. The windshield on an automobile is inclined 42.5° with respect to the horizontal. Assuming that the windshield is flat and rectangular, what is its area if it is 4.80 ft wide and the bottom is 1.50 ft in front of the top?

59. A water slide at an amusement park is 85 ft long and is inclined at an angle of 52° with the horizontal. How high is the top of the slide above the water level?

60. The distance from the ground level to the underside of a cloud is called the *ceiling*. See Fig. 4-65. A ground observer 950 m from a searchlight aimed vertically notes that the angle of elevation of the spot of light on a cloud is 76°. What is the ceiling?

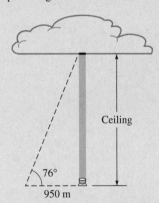

Fig. **4-65**

61. The vertical cross section of an attic room in a house is shown in Fig. 4-66. Find the distance *d* across the floor.

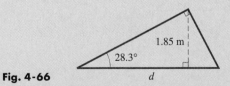

Fig. **4-66**

62. The impedance *Z* and resistance *R* in an alternating-current circuit may be represented by letting the impedance be the hypotenuse of a right triangle and the resistance be the side adjacent to the phase angle θ. If $R = 1750 \ \Omega$ and $\theta = 17.38°$, find *Z*.

63. A typical aqueduct built by the Romans dropped on average at an angle of about 0.03° to allow gravity to move the water from the source to the city. For such an aqueduct of 65 km in length, how much higher was the source than the city?

64. A Coast Guard boat 2.75 km from a straight beach can travel at 37.5 km/h. By traveling along a line that is at 69.0° with the beach, how long will it take it to reach the beach? See Fig. 4-67.

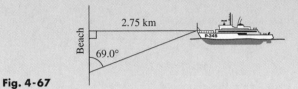

Fig. **4-67**

65. In the structural support shown in Fig. 4-68, find *x*.

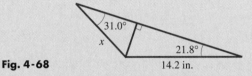

Fig. **4-68**

66. A person standing on a level plain hears the sound of a plane, looks in the direction of the sound, but the plane is not there (familiar?). When the sound was heard, it was coming from a point at an angle of elevation of 25°, and the plane was traveling at 450 mi/h (660 ft/s) at a constant altitude of 2800 ft along a straight line. If the plane later passes directly over the person, at what angle of elevation should the person have looked directly to see the plane when the sound was heard? (The speed of sound is 1130 ft/s.) See Fig. 4-69.

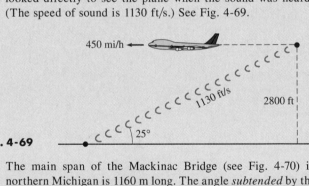

Fig. **4-69**

67. The main span of the Mackinac Bridge (see Fig. 4-70) in northern Michigan is 1160 m long. The angle *subtended* by the span at the eye of an observer in a helicopter is 2.2°. Show that the distance calculated from the helicopter to the span is about the same if the line of sight is perpendicular to the end or to the middle of the span.

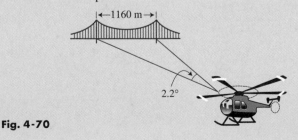

Fig. **4-70**

68. Each side piece of the trellis shown in Fig. 4-71 makes an angle of 80.0° with the ground. Find the length of each side piece and the area covered by the trellis.

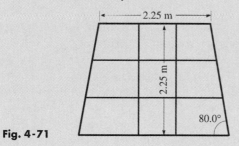

Fig. 4-71

W 69. The surface of a soccer ball consists of 20 regular hexagons (6 sides) interlocked around 12 regular pentagons (5 sides). See Fig. 4-72. (a) If the side of each hexagon and pentagon is 45.0 mm, what is the surface area of the soccer ball? (b) Find the surface area, given that the diameter of the ball is 222 mm. (c) Assuming that the given values are accurate, account for the difference in the values found in parts (a) and (b).

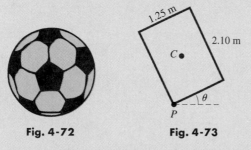

Fig. 4-72   **Fig. 4-73**

70. Through what angle θ must the crate shown in Fig. 4-73 be tipped in order that its center of gravity C is directly above the pivot point P?

71. A laser beam is transmitted with a "width" of 0.00200°. What is the diameter of a spot of the beam on an object 52,500 km distant? See Fig. 4-74.

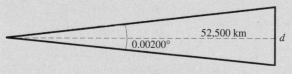

Fig. 4-74

72. Find the gear angle θ in Fig. 4-75 if $t = 0.180$ in.

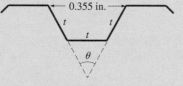

Fig. 4-75

73. A hang glider is directly above the shore of a lake. An observer on a hill is 375 m along a straight line from the shore. From the observer, the angle of elevation of the hang glider is 42.0°, and the angle of depression of the shore is 25.0°. How far above the shore is the hang glider?

74. A ground observer sights a weather balloon to the east at an angle of elevation of 15.0°. A second observer 2.35 mi to the east of the first also sights the balloon to the east at an angle of elevation of 24.0°. How high is the balloon? See Fig. 4-76.

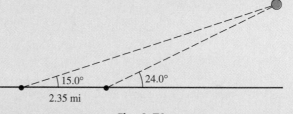

Fig. 4-76

75. A uniform strip of wood 5.0 cm wide frames a trapezoidal window as shown in Fig. 4-77. Find the left dimension l of the outside of the frame.

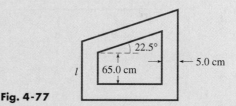

Fig. 4-77

76. A crop-dusting plane flies over a level field at a height of 25 ft. If the dust leaves the plane through a 30° angle and hits the ground after the plane travels 75 ft, how wide a strip is dusted? See Fig. 4-78.

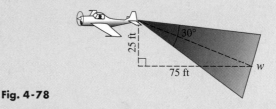

Fig. 4-78

Writing Exercise

77. Two students were discussing a problem in which a distant object was seen through an angle of 2.3°. One found the length of the object (perpendicular to the line of sight) using the tangent of the angle, and the other had the same answer using the sine of the angle. Write a paragraph explaining how this is possible.

PRACTICE TEST

1. Express 37°39′ in decimal form.

2. Find θ to the nearest 0.01° if $\cos\theta = 0.3726$.

3. A ship's captain, desiring to travel due south, discovers due to an improperly functioning instrument, the ship has gone 22.62 km in a direction 4.05° east of south. How far from its course (to the east) is the ship?

4. Find $\tan\theta$ in fractional form if $\sin\theta = \frac{2}{3}$.

5. Find $\csc\theta$ if $\tan\theta = 1.294$.

6. Solve the right triangle in Fig. 4-79 if $A = 37.4°$ and $b = 52.8$.

7. Solve the right triangle in Fig. 4-79 if $a = 2.49$ and $c = 3.88$.

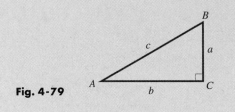

Fig. 4-79

8. In finding the wavelength λ (the Greek lambda) of light, the equation $\lambda = d\sin\theta$ is used. Find λ if $d = 30.05$ μm and $\theta = 1.167°$. (μ is the prefix for 10^{-6}.)

9. Determine the trigonometric functions of an angle in standard position if its terminal side passes through $(5, 2)$. Give answers in exact and decimal forms.

10. A surveyor sights two points directly ahead. Both are at an elevation 18.525 m lower than the observation point. How far apart are the points if the angles of depression are 13.500° and 21.375°, respectively? See Fig. 4-80.

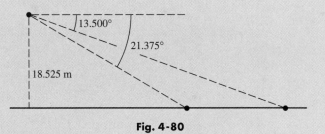

Fig. 4-80

CHAPTER 5

SYSTEMS OF LINEAR EQUATIONS; DETERMINANTS

The solution of many technical and scientific problems requires that we deal with several quantities that are related in a number of ways. This can lead to more than one equation relating these quantities. For example, when determining the price at which to sell a computer, we must consider (among many others) the time required for research, development, and production of the computer, as well as the costs of its various components.

Two or more equations that relate variables are found in nearly all fields of science and technology. These include aeronautics, transportation, the analysis of forces acting on a structure, the electric currents in different parts of a circuit, the components of a chemical compound, and medical dosages. These technical problems often require solutions that satisfy all equations at the same time.

In this chapter we restrict our attention to *linear* equations (variables occur only to the first power). We will consider *systems* of two equations with two unknowns and systems of three equations with three unknowns. Systems with other kinds of equations and systems with more unknowns are taken up in Chapters 14 and 16.

In designing an industrial robot, the forces acting on each link must be carefully analyzed. In Section 5-6 we see how a system of equations is solved to find these forces.

5-1 LINEAR EQUATIONS

In general, *an equation is termed* **linear** *in a given set of variables if each term contains only one variable, to the first power, or is a constant.*

EXAMPLE 1 $5x - t + 6 = 0$ is linear in x and t, but $5x^2 - t + 6 = 0$ is not linear, due to the presence of x^2.

The equation $4x + y = 8$ is linear in x and y, but $4xy + y = 8$ is not, due to the presence of xy.

The equation $x - 6y + z - 4w = 7$ is linear in x, y, z, and w, but the equation $x - \frac{6}{y} + z - 4w = 7$ is not, due to the presence of $\frac{6}{y}$, where y appears in the denominator.

An equation that can be written in the form

$$ax + b = 0 \qquad \text{(5-1)}$$

is known as a **linear equation in one unknown.** Here, a and b are constants. We have already discussed the solution to this type of equation in Section 1-10. In general, *the* **solution,** *or* **root,** *of the equation is* $x = -b/a$. Also, it is noted that the solution of the linear equation is the same as the *zero* of the **linear function** $f(x) = ax + b$.

■**EXAMPLE 2** The equation $2x + 7 = 0$ is a linear equation of the form of Eq. (5-1), with $a = 2$ and $b = 7$.

The solution to the linear equation $2x + 7 = 0$, or the zero of the linear function $f(x) = 2x + 7$, is $-7/2$. From Chapter 3 we recall that the *zero* of a function $f(x)$ is the value of x for which $f(x) = 0$. ----------■

As we noted in the chapter introduction, there are a great many applied problems that involve more than one unknown quantity. In the next example, two specific illustrations are given.

Named for the German physicist Gustav Kirchhoff (1824–1887).

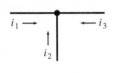

Fig. 5-1

■**EXAMPLE 3** **(a)** A basic law of direct-current electricity, known as *Kirchhoff's first law,* may be stated as "The algebraic sum of the currents entering any junction in a circuit is zero." If three wires are joined at a junction, this law leads to the linear equation

$$i_1 + i_2 + i_3 = 0$$

where i_1, i_2, and i_3 are the currents in each of the wires. (Either one or two of these currents must have a negative sign, showing that it is acually leaving the junction.) See Fig. 5-1.

(b) When determining two forces F_1 and F_2 acting on a beam, we might encounter an equation such as

$$2F_1 + 4F_2 = 200 \qquad \text{----------}■$$

An equation that can be written in the form

$$ax + by = c \qquad \text{(5-2)}$$

NOTE ▶

is known as a **linear equation in two unknowns.** In Chapter 3 we considered many equations that can be written in this form. We found that for each value of x, there is a corresponding value for y. Each of these pairs of numbers is a **solution** to the equation, although we did not call it that at the time. *A solution is any set of numbers, one for each variable, that satisfies the equation.* When we represent the solutions in the form of a graph, we see that the graph of any linear equation in two unknowns is a straight line. Also, graphs of linear equations in one unknown, those for which $a = 0$ or $b = 0$, are also straight lines. Thus, we see the significance of the name *linear.* In the next section we shall further consider the graph of the linear equation.

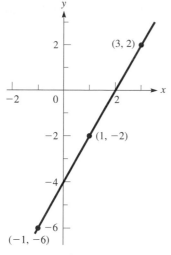

Fig. 5-2

EXAMPLE 4 The equation $2x - y - 4 = 0$ is a linear equation in two unknowns, x and y, since we can write it in the form of Eq. (5-2) as $2x - y = 4$. Following the methods of Chapter 3, to graph this equation we would write it in the more convenient form $y = 2x - 4$. We see from Fig. 5-2 that the graph is a straight line.

The coordinates of any point on the line give us a solution of this equation. For example, the point $(1, -2)$ is on the line. This means that $x = 1$, $y = -2$ is a solution of the equation. We can show that these values are a solution by substituting in the equation $2x - y - 4 = 0$. This gives us

$$2(1) - (-2) - 4 = 0, \qquad 2 + 2 - 4 = 0, \qquad 0 = 0$$

Since we have equality, $x = 1$, $y = -2$ is a solution. In the same way we can show that $x = 3$, $y = 2$ is a solution.

If we are given the value of one variable, we find the value of the other variable, which gives a solution by substitution. For example, for the equation $2x - y - 4 = 0$, if $x = -1$, we have

$$2(-1) - y - 4 = 0$$
$$-2 - y - 4 = 0$$
$$y = -6$$

Therefore, $x = -1$, $y = -6$ is a solution, and the point $(-1, -6)$ is on the graph.

Two linear equations, each containing the same two unknowns,

$$a_1x + b_1y = c_1$$
$$a_2x + b_2y = c_2$$

(5-3)

are said to form a **system of simultaneous linear equations.** *A* **solution of the system** *is any pair of values* (x, y) *that satisfies both equations.* Methods of finding the solutions to such systems are the principal concern of this chapter.

EXAMPLE 5 The perimeter of the piece of computer paper shown in Fig. 5-3 is 104 cm. The length is 4 cm more than the width.

From these two statements, we can set up two equations in the two unknown quantities, the length and the width. Letting $l =$ the length and $w =$ the width, we have

$$2l + 2w = 104$$
$$l - w = 4$$

as a system of simultaneous linear equations. The solution of this system is $l = 28$ cm and $w = 24$ cm. These values satisfy both equations since

$$2(28) + 2(24) = 104$$

and

$$28 - 24 = 4$$

This is the only pair of values that satisfies *both* equations. Methods for finding such solutions are taken up in later sections of this chapter.

Fig. 5-3

— EXERCISES *5-1* —

In Exercises 1–4, determine whether or not the given pairs of values are solutions of the given linear equations in two unknowns.

1. $2x + 3y = 9$; $(3, 1)$, $(5, \frac{1}{3})$

2. $5x + 2y = 1$; $(2, -4)$, $(1, -2)$

3. $-3x + 5y = 13$; $(-1, 2)$, $(4, 5)$

4. $x - 4y = 10$; $(2, -2)$, $(2, 2)$

In Exercises 5–8, for each given value of x, determine the value of y that gives a solution to the given linear equations in two unknowns.

5. $5x - y = 6$; $x = 1$, $x = -2$

6. $2x + 7y = 8$; $x = -3$, $x = 2$

7. $x - 4y = 2$; $x = 3$, $x = -0.4$

8. $3x - 2y = 9$; $x = \frac{2}{3}$, $x = -3$

In Exercises 9–16, determine whether or not the given pair of values is a solution of the given system of simultaneous linear equations.

9. $x - y = 5$ $x = 4$, $y = -1$
  $2x + y = 7$

10. $2x + y = 8$ $x = -1$, $y = 10$
  $3x - y = -13$

11. $A + 5B = -7$ $A = -2$, $B = 1$
  $3A - 4B = -4$

12. $-3x + y = 1$ $x = \frac{1}{3}$, $y = 2$
  $6x - 3y = -4$

13. $2x - 5y = 0$ $x = \frac{1}{2}$, $y = -\frac{1}{5}$
  $4x + 10y = 4$

14. $6i_1 + i_2 = 5$ $i_1 = 1$, $i_2 = -1$
  $3i_1 - 4i_2 = -1$

15. $3x - 2y = 2.2$ $x = 0.6$, $y = -0.2$
  $5x + y = 2.8$

16. $x - 7y = -3.2$ $x = -1.1$, $y = 0.3$
  $2x + y = 2.5$

In Exercises 17–20, answer the given questions.

17. In planning a search pattern from an aircraft carrier, a pilot plans to fly at p mi/h relative to a wind that is blowing at w mi/h. Traveling with the wind, the ground speed would be 300 mi/h, and against the wind the ground speed would be 220 mi/h. This leads to two equations:

$$p + w = 300$$
$$p - w = 220$$

Are the speeds 260 mi/h and 40 mi/h?

18. The electric resistance R of a certain resistor is a function of the temperature T given by the equation $R = aT + b$, where a and b are constants. If $R = 1200 \ \Omega$ when $T = 10.0°C$ and $R = 1280 \ \Omega$ when $T = 50.0°C$, we can find the constants a and b by substituting and obtaining the equations

$$1200 = 10.0a + b$$
$$1280 = 50.0a + b$$

Are the constants $a = 4.00 \ \Omega/°C$ and $b = 1160 \ \Omega$?

19. The forces acting on part of a structure are shown in Fig. 5-4. An analysis of the forces leads to the equations

$$0.80F_1 + 0.50F_2 = 50$$
$$0.60F_1 - 0.87F_2 = 12$$

Are the forces 45 N and 28 N?

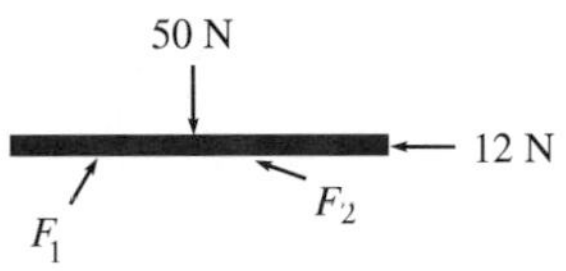

Fig. 5-4

20. Using the data that fuel consumption for transportation contributes a percent p_1 of pollution that is 16% less than the percent p_2 of all other sources combined, the equations

$$p_1 + p_2 = 100$$
$$p_2 - p_1 = 16$$

can be set up. Are the percents $p_1 = 58$ and $p_2 = 42$?

5-2 GRAPHS OF LINEAR FUNCTIONS

As we mentioned in the previous section, the main topic of this chapter is finding solutions of systems of linear equations. Our first method of solving a system of equations will be a graphical one. Therefore, before taking up this method in the next section, we shall first develop additional ways of graphing a linear equation. We will then be able to analyze and check the graph of a linear equation quickly, often by inspection.

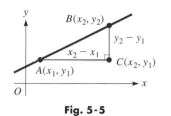

Fig. 5-5

Consider the line that passes through points A, with coordinates (x_1, y_1), and B, with coordinates (x_2, y_2), in Fig. 5-5. Point C is horizontal from A and vertical from B. Thus, C has coordinates (x_2, y_1), and there is a right angle at C. One way of measuring the steepness of this line is to find the ratio of the vertical distance to the horizontal distance between two points. Therefore, *we define the* **slope** *of the line through two points as the difference in the y-coordinates divided by the difference in the x-coordinates.* For points A and B, the slope m is

SLOPE

$$m = \frac{y_2 - y_1}{x_2 - x_1}$$ (5-4)

The slope is often referred to as the *rise* (vertical change) over the *run* (horizontal change). Note that the slope of a vertical line, for which $x_2 = x_1$, is undefined (for $x_2 = x_1$, the denominator of Eq. (5-4) is zero).

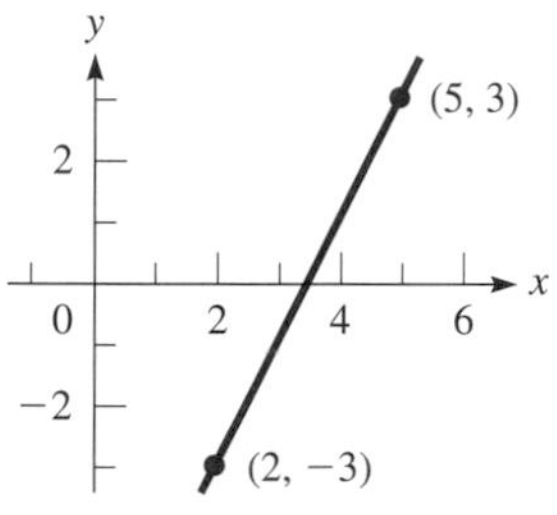

■EXAMPLE 1   Find the slope of the line through the points $(2, -3)$ and $(5, 3)$.
    In Fig. 5-6, we draw the line through the two given points. By taking $(5, 3)$ as (x_2, y_2), then (x_1, y_1) is $(2, -3)$. We may choose either point as (x_2, y_2), but *once the choice is made the order must be maintained.* Using Eq. (5-4), the slope is

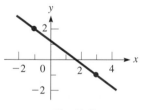

Fig. 5-6

$$m = \frac{3 - (-3)}{5 - 2} = \frac{6}{3} = 2$$

The rise is 2 units for each unit (of run) it moves from left to right.   ----------■

■EXAMPLE 2   Find the slope of the line through $(-1, 2)$ and $(3, -1)$.
    In Fig. 5-7, we draw the line through these two points. By taking (x_2, y_2) as $(3, -1)$ and (x_1, y_1) as $(-1, 2)$, the slope is

$$m = \frac{-1 - 2}{3 - (-1)} = \frac{-3}{3 + 1} = -\frac{3}{4}$$

The line *falls* 3 units for each 4 units it moves from left to right.   ----------■

Fig. 5-7

We note in Example 1 that *as x increases, y increases and that slope is **positive.*** In Example 2, *as x increases, y decreases and that slope is **negative.*** Also, *the larger the absolute value of the slope, the more nearly vertical is the line.*

■EXAMPLE 3   For each of the following lines shown in Fig. 5-8, we show the difference in the y-coordinates and in the x-coordinates between two points.
    In Fig. 5-8(a), a line with a slope of 5 is shown. It rises sharply.
    In Fig. 5-8(b), a line with a slope of $\frac{1}{2}$ is shown. It rises slowly.
    In Fig. 5-8(c), a line with a slope of -5 is shown. It falls sharply.
    In Fig. 5-8(d), a line with a slope of $-\frac{1}{2}$ is shown. It falls slowly.

Fig. 5-8

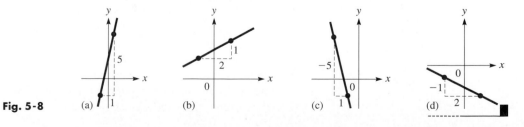

(a)   (b)   (c)   (d)

Slope-Intercept Form of the Equation of a Straight Line

We now show how the slope is related to the equation of a straight line. In Fig. 5-9, if we have two points, $(0, b)$ and a general point (x, y), the slope is

$$m = \frac{y - b}{x - 0}$$

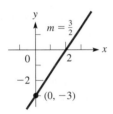

Fig. 5-9

Simplifying this, we have $mx = y - b$, or

$$\boxed{y = mx + b} \qquad (5\text{-}5)$$

In Eq. (5-5), m is the slope, and b is the y-coordinate of the point where it crosses the y-axis. *This point is the **y-intercept** of the line,* and its coordinates are $(0, b)$. *Equation (5-5) is the **slope-intercept** form of the equation of a straight line. The coefficient of x is the slope, and the constant is the ordinate of the y-intercept.* The point $(0, b)$ and simply b are both referred to as the y-intercept. Since the x-coordinate of the y-intercept is 0, this should cause no confusion.

NOTE ▶

◼ EXAMPLE 4   Find the slope and the y-intercept of the line $y = \frac{3}{2}x - 3$.

Since the equation is written as y as a function of x, it is in the form of Eq. (5-5). Therefore, since the coefficient of x is $\frac{3}{2}$, the slope of the line is $\frac{3}{2}$. Also, since we can write the equation as

$$y = \overset{\text{slope}}{\frac{3}{2}}x + \overset{\text{y-intercept ordinate}}{(-3)}$$

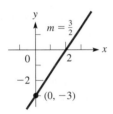

Fig. 5-10

we see that the constant is -3, which means the y-intercept is the point $(0, -3)$. The line is shown in Fig. 5-10.   ◼

◼ EXAMPLE 5   Find the slope and the y-intercept of the line $2x + 3y = 4$.

Here, *we must first write the equation in the slope-intercept form.* Solving for y gives us

$$y = \overset{\text{slope}}{-\frac{2}{3}}x + \overset{\text{y-intercept ordinate}}{\frac{4}{3}}$$

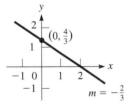

Fig. 5-11

Therefore, the slope is $-\frac{2}{3}$ and the y-intercept is the point $(0, \frac{4}{3})$. See Fig. 5-11.   ◼

◼ EXAMPLE 6   In analyzing an electric current, two of the currents, i_1 and i_2, were related by $i_2 = 2i_1 + 1$. Graph this equation using the slope-intercept form.

Since the equation is solved for i_2, we treat i_1 as the independent variable and i_2 as the dependent variable. This means that the slope of the line is 2. Also, since the b-term is 1, we can see that the intercept is $(0, 1)$.

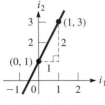

Fig. 5-12

We can use this information to sketch the line, as shown in Fig. 5-12. Since the slope is 2, we know that i_2 increases 2 units for each unit of increase of i_1. Thus, starting at the i_2-intercept $(0, 1)$, if i_1 increases by 1, i_2 increases by 2, and we are at the point $(1, 3)$. The line must pass through $(1, 3)$, as well as through $(0, 1)$. Therefore, we draw the line through these points. (Negative values may be used for electric currents since the sign shows the direction of flow.)   ◼

Sketching Lines by Intercepts

Another way of sketching the graph of a straight line is to find two points on the line and then draw the line through these points. Two points that are easily determined are those where the line crosses the *y*-axis and the *x*-axis. We already know that the point where it crosses the *y*-axis is the *y*-intercept. In the same way, *the point where it crosses the x-axis is called the* **x-intercept,** and the coordinates of the *x*-intercept are $(a, 0)$. These points are easily found because in each case one of the coordinates is zero. By setting $x = 0$ and $y = 0$, in turn, and determining the corresponding value of the other unknown, we obtain the coordinates of the intercepts. A third point should be found as a check. This method is sufficient unless the line passes through the origin. Then both intercepts are at the origin and one more point must be determined, or we must use the slope-intercept method. Example 7 illustrates how a line is sketched by finding its intercepts.

NOTE ▶

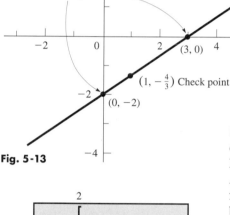

Fig. 5-13

■**EXAMPLE 7**   Sketch the graph of $2x - 3y = 6$ by finding its intercepts and one checkpoint (see Fig. 5-13).

First, we let $x = 0$. This gives us $-3y = 6$, or $y = -2$. Thus, the point $(0, -2)$ is on the graph. Next, let $y = 0$, and this gives $2x = 6$, or $x = 3$. Thus, the point $(3, 0)$ is on the graph. The point $(0, -2)$ is the *y*-intercept, and $(3, 0)$ is the *x*-intercept. These two points are sufficient to sketch the line, but we should find another point as a check. Choosing $x = 1$, we find that $y = -\frac{4}{3}$. This means the point $(1, -\frac{4}{3})$ should be on the line. From Fig. 5-13 we see that it is on the line. --------■

A graphing calculator can be used to graph a linear equation, just as we did in Section 3-5. However, even when using a graphing calculator we must first solve the equation for *y* (in order to enter it into the calculator). In doing so, we will often have the equation in slope-intercept form, or easily be able to write it in this form. Also, both intercepts can often be found by inspection. Making appropriate choices for the *window* settings can be helped by noting the slope and the *y*-intercept, or both intercepts.

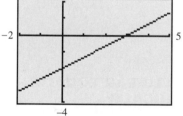

Fig. 5-14

■**EXAMPLE 8**   **(a)** Use a graphing calculator to display the graph of the equation in Example 7.

First, we solve the equation $2x - 3y = 6$ for *y*. This gives us $y = \frac{2}{3}x - 2$. Therefore on the calculator we enter $y_1 = 2x/3 - 2$ and use the *window* settings shown in Fig. 5-14. We can see that the graph is the same as that in Fig. 5-13.

(b) Use a graphing calculator to display the graph of the equation $4x + 5y = 100$.

Solving for *y*, we get $y = -\frac{4}{5}x + 20$, and therefore in the calculator we enter the function $y_1 = -4x/5 + 20$. We also note immediately that the *y*-intercept is $(0, 20)$, which means we must be careful in choosing the *window* settings. We can also determine that the *x*-intercept is $(25, 0)$. Therefore, we have the settings and graph shown in Fig. 5-15. --------■

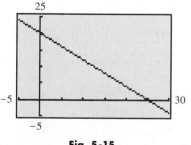

Fig. 5-15

Further details and discussion of slope and the graphs of linear equations are found in Chapter 21.

EXERCISES *5-2*

In Exercises 1–8, find the slope of the line that passes through the given points.

1. $(1, 0)$, $(3, 8)$
2. $(3, 1)$, $(2, 7)$
3. $(-1, 2)$, $(-4, 17)$
4. $(-1, -2)$, $(2, 10)$
5. $(5, -3)$, $(-2, -5)$
6. $(3, -4)$, $(-7, 1)$
7. $(0.4, 0.5)$, $(-0.2, 0.2)$
8. $(-2.8, 3.4)$, $(1.2, 4.2)$

In Exercises 9–16, sketch the line with the given slope and y-intercept.

9. $m = 2$, $(0, -1)$
10. $m = 3$, $(0, 1)$
11. $m = -3$, $(0, 2)$
12. $m = -4$, $(0, -2)$
13. $m = \frac{1}{2}$, $(0, 0)$
14. $m = \frac{2}{3}$, $(0, -1)$
15. $m = -9$, $(0, 20)$
16. $m = -0.3$, $(0, -1.4)$

In Exercises 17–24, find the slope and the y-intercept of the line with the given equation and sketch the graph using the slope and the y-intercept. A graphing calculator can be used to check your graph.

17. $y = -2x + 1$
18. $y = -4x$
19. $y = x + 4$
20. $y = \frac{4}{5}x + 2$
21. $5x - 2y = 40$
22. $6x - 2y = 7$
23. $2x + 6y = 3$
24. $10x + 3y = 30$

In Exercises 25–32, find the x-intercept and the y-intercept of the line with the given equation. Sketch the line using the intercepts. A graphing calculator can be used to check the graph.

25. $x + 2y = 4$
26. $3x + y = 3$
27. $4x - 3y = 12$
28. $x - 5y = 5$

29. $y = 3x + 6$
30. $y = -2x - 4$
31. $y = -12x + 30$
32. $y = 0.25x + 4.5$

In Exercises 33–36, sketch the indicated lines.

33. The diameter of the large end, d (in in.), of a certain type of machine tool can be found from the equation $d = 0.2l + 1.2$, where l is the length of the tool. Sketch d as a function of l, for values of l to 10 in. See Fig. 5-16.

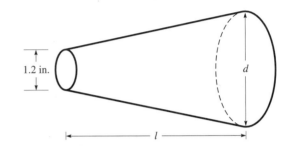

Fig. 5-16

34. In testing an anticholesterol drug, it was found that each gram of drug administered reduced a person's blood cholesterol level by 2 units. Set up the function relating the cholesterol level C as a function of the dosage d for a person whose cholesterol level is 310 before taking the drug. Sketch the graph.

35. In mixing 85 octane gasoline and 93 octane gasoline to produce 91 octane gasoline, the equation $0.85x + 0.93y = 910$ is used. Sketch the graph.

36. Two electric currents, I_1 and I_2 (in mA), in part of a circuit in a computer are related by the equation $4I_1 - 5I_2 = 2$. Sketch I_2 as a function of I_1. These currents can be negative.

5-3 SOLVING SYSTEMS OF TWO LINEAR EQUATIONS IN TWO UNKNOWNS GRAPHICALLY

Since a solution of a system of simultaneous linear equations in two unknowns is any pair of values (x, y) that satisfies *both* equations, graphically *the solution would be the coordinates of the point of intersection of the two lines.* This must be the case, for the coordinates of this point constitute the only pair of values to satisfy *both* equations. (In some special cases there may be no solution; in others there may be many solutions. See Examples 5 and 6.)

Therefore, when we solve two simultaneous linear equations in two unknowns graphically, *we must graph each line and determine the point of intersection.* This may, of course, lead to *approximate results* if the lines cross at points not used to determine the graph.

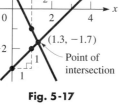

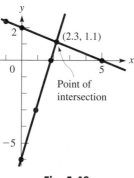

Fig. 5-17

Fig. 5-18

█EXAMPLE 1 Solve the system of equations

$$y = x - 3$$
$$y = -2x + 1$$

Since each of the equations is in slope-intercept form, we see that $m = 1$ and $b = -3$ for the first line and that $m = -2$ and $b = 1$ for the second line. Using these values we sketch the lines, as shown in Fig. 5-17.

From the figure we see that *the lines cross at about the point* $(1.3, -1.7)$. This means that the solution is approximately

$$x = 1.3 \qquad y = -1.7$$

(The exact solution is $x = \frac{4}{3}$, $y = -\frac{5}{3}$.)

NOTE▶ In checking the solution, be careful to substitute the values in *both* equations. Making these substitutions gives us

$$-1.7 \overset{?}{=} 1.3 - 3 \qquad \text{and} \qquad -1.7 \overset{?}{=} -2(1.3) + 1$$
$$= -1.7 \qquad\qquad\qquad \approx -1.6$$

These values show that the solution checks. (The point $(1.3, -1.7)$ is *on* the first line and *almost on* the second line. The difference in values when checking the values for the second line is due to the fact that the solution is *approximate*.) ▬

█EXAMPLE 2 Solve the system of equations

$$2x + 5y = 10$$
$$3x - y = 6$$

We could write each equation in slope-intercept form in order to sketch the lines. Also, we could use the form in which they are written to find the intercepts. Choosing to find the intercepts and draw lines through them, we let $y = 0$; then $x = 0$. Therefore, we find that the intercepts of the first line are the points $(5, 0)$ and $(0, 2)$. A third point is $(-1, \frac{12}{5})$. The intercepts of the second line are $(2, 0)$ and $(0, -6)$. A third point is $(1, -3)$. Plotting these points and drawing the proper straight lines, we see that the lines cross at about $(2.3, 1.1)$. [The exact values are $(\frac{40}{17}, \frac{18}{17})$.] The solution of the system of equations is approximately $x = 2.3$, $y = 1.1$ (see Fig. 5-18).

Checking, we have

$$2(2.3) + 5(1.1) \overset{?}{=} 10 \qquad \text{and} \qquad 3(2.3) - 1.1 \overset{?}{=} 6$$
$$10.1 \approx 10 \qquad\qquad\qquad 5.8 \approx 6$$

This verifies that the solution is correct to the accuracy obtainable from the graph. ▬

As we have seen, it is difficult to get a good approximation of the coordinates of the point of intersection without being very careful in sketching both lines. The graphing calculator can be used to find this point with much greater accuracy than is possible by hand-sketching the lines. Once we locate the point of intersection, the features of the calculator allow us to get the required accuracy. This is illustrated in the following example.

EXAMPLE 3 Using a graphing calculator, we graph the lines of Example 1 in Fig. 5-19(a) and those of Example 2 in Fig. 5-19(b).

In Fig. 5-19(a), we let $y_1 = x - 3$ and $y_2 = -2x + 1$. Using the *trace* and *zoom* features, we find (to the nearest 0.001) that $x = 1.333$, $y = -1.667$.

In Fig. 5-19(b), we let $y_1 = -2x/5 + 2$ and $y_2 = 3x - 6$. Using the *trace* and *zoom* features, we find (to the nearest 0.001) that $x = 2.353$, $y = 1.059$.

Many calculators have an *intersect* feature. These solutions can be found using this feature (which may require an estimate of the point of intersection).

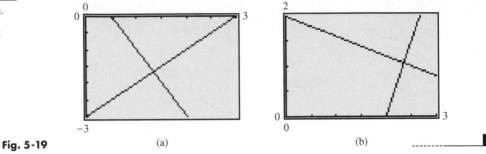

Fig. 5-19   (a)   (b)

Word Problems Involving Two Linear Equations

Linear equations in two unknowns are often useful in solving word problems. Just as in Section 1-12, *we must read the statement carefully in order to identify the unknowns and the information for setting up the equations.* The following example illustrates the method.

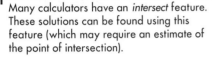

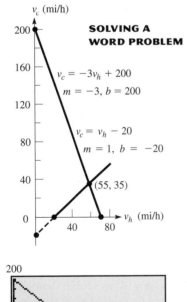

SOLVING A
WORD PROBLEM

$v_c = -3v_h + 200$
$m = -3, b = 200$

$v_c = v_h - 20$
$m = 1, b = -20$

$(55, 35)$

(a)

EXAMPLE 4 A driver traveled for 1.5 h at a constant speed along a highway. Then, through a construction zone, the driver reduced the car's speed by 20 mi/h for 30 min. If 100 mi were covered in the 2.0 h, what were the two speeds?

First, we let $v_h =$ the highway speed and $v_c =$ the speed in the construction zone. Two equations are found by using

1. distance = rate × time (for units, mi = $(\frac{\text{mi}}{\text{h}})$h), and
2. the fact that "the driver reduced the car's speed by 20 mi/h for 30 min."

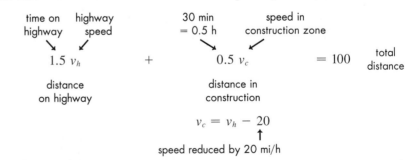

$$v_c = v_h - 20$$
↑
speed reduced by 20 mi/h

Using v_h as the independent variable and v_c as the dependent variable, the sketch is shown in Fig. 5-20(a). With $v_c = y$ and $v_h = x$, we have the calculator display shown in Fig. 5-20(b) with the indicated settings, and $y_1 = -3x + 200$, $y_2 = x - 20$.

We see that the point of intersection is $(55, 35)$, which means that the solution is $v_c = 55$ mi/h and $v_c = 35$ mi/h. Checking *in the statement of the problem*, we have $(1.5 \text{ h})(55 \text{ mi/h}) + (0.5 \text{ h})(35 \text{ mi/h}) = 100$ mi.

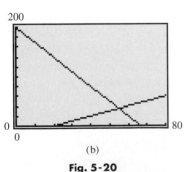

(b)

Fig. 5-20

Inconsistent and Dependent Systems

The lines of each system in the previous examples intersect in a single point, and each system has *one* solution. Such systems are called *consistent* and *independent*. Most systems that we will encounter will have just one solution. However, as we will now show, *not all systems have just one such solution.* The following examples illustrate a system that has *no solution* and a system that has an *unlimited number of solutions.*

■EXAMPLE 5   Solve the system of equations

$$x = 2y + 6$$
$$6y = 3x - 6$$

Writing each of these equations in slope-intercept form (Eq. 5-5), we have for the first equation

$$y = \tfrac{1}{2}x - 3$$

For the second equation we have

$$y = \tfrac{1}{2}x - 1$$

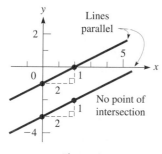

Fig. 5-21

From these we see that each line has a slope of $\tfrac{1}{2}$ and that the y-intercepts are $(0, -3)$ and $(0, -1)$. Therefore, we know that the y-intercepts are different, but the *slopes are the same*. Since the slope indicates that each line rises $\tfrac{1}{2}$ unit for y for each unit x increases, *the lines are parallel and do not intersect,* as is shown in Fig. 5-21. This means that ***there are no solutions*** for this system of equations. *Such a system is called* **inconsistent.**   ---------■

■EXAMPLE 6   Solve the system of equations

$$x - 3y = 9$$
$$-2x + 6y = -18$$

We find that the intercepts and a third point for the first line are $(9, 0)$, $(0, -3)$, and $(3, -2)$. In determining the intercepts for the second line, we find that they are $(9, 0)$ and $(0, -3)$, which are also the intercepts of the first line. As a check we find that the point $(3, -2)$ also satisfies the equation of the second line. This means that the two lines are really the same line. Another check is to write each of these equations in slope-intercept form. This gives us

$$y = \tfrac{1}{3}x - 3$$

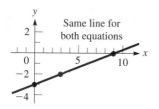

Fig. 5-22

for each equation. See Fig. 5-22.

Since the lines are the same, *the coordinates of any point on this common line constitute a solution of the system.* Since ***no unique solution can be determined,*** *the system is called* **dependent.**   ---------■

We will encounter inconsistent and dependent systems in the sections that follow. We will show that there are also algebraic ways of showing whether a given system is consistent, inconsistent, or dependent.

EXERCISES 5-3

In Exercises 1–16, solve each system of equations by sketching the graphs. Use the slope and the y-intercept or both intercepts. Estimate each result to the nearest 0.1 if necessary.

1. $y = -x + 4$
$y = x - 2$

2. $y = \frac{1}{2}x - 1$
$y = -x + 8$

3. $y = 2x - 6$
$y = -\frac{1}{3}x + 1$

4. $y = \frac{1}{2}x - 4$
$y = 2x + 2$

5. $3x + 2y = 6$
$x - 3y = 3$

6. $4R - 3V = -8$
$6R + V = 6$

7. $2x - 5y = 10$
$3x + 4y = -12$

8. $-5x + 3y = 15$
$2x + 7y = 14$

9. $s - 4t = 8$
$2s = t + 4$

10. $y = 4x - 6$
$y = 2x + 4$

11. $y = -x + 3$
$y = -2x + 3$

12. $p - 6 = 6v$
$v = 3 - 3p$

13. $x - 4y = 6$
$2y = x + 4$

14. $x + y = 3$
$3x - 2y = 14$

15. $-2r_1 + 2r_2 = 7$
$4r_1 - 2r_2 = 1$

16. $2x - 3y = -5$
$3x + 2y = 12$

In Exercises 17–28, solve each system of equations to the nearest 0.1 for each variable by using a graphing calculator.

17. $x = 4y + 2$
$3y = 2x + 3$

18. $1.2x - 2.4y = 4.8$
$3.0x = -2.0y + 7.2$

19. $4.0x - 3.5y = 1.5$
$0.7y + 0.1x = 0.7$

20. $5F - 2T = 7$
$3F + 4T = 8$

21. $x - 5y = 10$
$2x - 10y = 20$

22. $18x - 3y = 7$
$2y = 1 + 12x$

23. $1.9v = 3.2t$
$1.2t - 2.6v = 6$

24. $4x - y = 3$
$2x + 3y = 0$

25. $5x = y + 3$
$4x = 2y - 3$

26. $0.75u + 0.67v = 5.9$
$2.1u - 3.9v = 4.8$

27. $3x = 8y + 12$
$-6x + 16y = 6$

28. $y = 6x + 2$
$12x - 2y = -4$

In Exercises 29–32, graphically solve the given problems to the stated accuracy.

29. Chains support a crate as shown in Fig. 5-23. The equations relating tensions T_1 and T_2 are given below. Determine the tensions to the nearest 1 N from the graph.

$0.8T_1 - 0.6T_2 = 12$
$0.6T_1 + 0.8T_2 = 68$

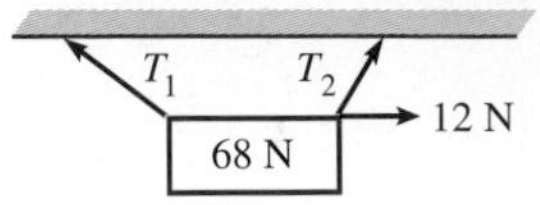

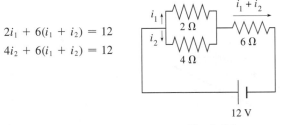

Fig. 5-23

30. The equations relating the currents i_1 and i_2 shown in Fig. 5-24 are given below. Find the currents to the nearest 0.1 A.

$2i_1 + 6(i_1 + i_2) = 12$
$4i_2 + 6(i_1 + i_2) = 12$

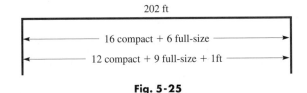

Fig. 5-24

31. An architect designing a parking lot has a row 202 ft wide to divide into spaces for compact cars and full-size cars. The architect determines that 16 compact car spaces and 6 full-size car spaces use the width, or that 12 compact car spaces and 9 full-size car spaces use all but 1 ft of the width. What are the widths (to the nearest 0.1 ft) of the spaces being planned? See Fig. 5-25.

202 ft

16 compact + 6 full-size

12 compact + 9 full-size + 1ft

Fig. 5-25

32. A total of 42 tons of two types of ore is to be loaded into a smelter. The first type contains 6.0% copper, and the second contains 2.4% copper. Find the necessary amounts of each ore (to the nearest 1 ton) to produce 2 tons of copper.

5-4 **SOLVING SYSTEMS OF TWO LINEAR EQUATIONS IN TWO UNKNOWNS ALGEBRAICALLY**

The graphical method of solving two linear equations is good for obtaining a "picture" of the solution. One problem is that graphical methods usually give *approximate* results. If exact solutions are required, we turn to other methods. In this section we present two algebraic methods of solution.

Solution by Substitution

The first method involves the *elimination of one variable by* **substitution.** An outline of the method is as follows:

> ### Solution of Two Linear Equations by Substitution
>
> 1. *Solve one equation for one of the unknowns.*
> 2. **Substitute** *this solution into the* **other** *equation. At this point we have a linear equation in one unknown.*
> 3. *Solve the resulting equation for the value of the unknown it contains.*
> 4. *Substitute this value into the equation of step 1 and solve for the other unknown.*
> 5. *Check the values in* **both original equations.**

The following examples illustrate the method.

EXAMPLE 1   Use the method of elimination by substitution to solve the system of equations

$$x - 3y = 6$$
$$2x + 3y = 3$$

The first step is to solve one of the equations for one of the unknowns. The choice of which equation and which unknown depends on ease of algebraic manipulation. In this system, it is somewhat easier to solve the first equation for x. Therefore, performing this operation we have

Step 1
$$x = 3y + 6 \qquad\qquad\qquad \textbf{(A1)}$$

We then substitute this expression into the second equation in place of x, giving

in second equation, x replaced by $3y + 6$

Step 2
$$2(\overline{3y + 6}) + 3y = 3$$

Solving this equation for y, we obtain

Step 3
$$6y + 12 + 3y = 3$$
$$9y = -9$$
$$y = -1$$

We now put the value $y = -1$ into the first of the original equations. Since we have already solved this equation for x in terms of y, Eq. (A1), we obtain

Step 4
$$x = 3(-1) + 6 = 3$$

Step 5   Therefore, the solution of the system is $x = 3$, $y = -1$. As a check, we substitute these values into each of the original equations. We obtain $3 - 3(-1) = 6$ and $2(3) + 3(-1) = 3$, which verifies the solution.

EXAMPLE 2   Use the method of elimination by substitution to solve the system of equations

$$-5x + 2y = -4$$
$$10x + 6y = 3$$

It makes little difference which equation or which unknown is chosen. Therefore, choosing to solve the first equation for y, we obtain

$$2y = 5x - 4$$
$$y = \frac{5x - 4}{2} \tag{B1}$$

Substituting this expression into the second equation, we have

$$10x + 6\left(\frac{5x - 4}{2}\right) = 3 \qquad \text{in second equation, } y \text{ replaced by } \frac{5x - 4}{2}$$

We now proceed to solve this equation for x.

$$10x + 3(5x - 4) = 3$$
$$10x + 15x - 12 = 3$$
$$25x = 15$$
$$x = \frac{3}{5}$$

Substituting this value into the expression for y, Eq. (B1), we obtain

$$y = \frac{5(3/5) - 4}{2} = \frac{3 - 4}{2} = -\frac{1}{2}$$

Therefore, the solution of this system is $x = \frac{3}{5}$, $y = -\frac{1}{2}$. Substituting these values in both original equations shows that the solution checks. ---------■

Solution by Addition or Subtraction

The method of elimination by substitution is useful if one equation can easily be solved for one of its unknowns. However, the numerical coefficients often make this method somewhat cumbersome. So we now present another algebraic method of solving a system of linear equations.

The second method is the *elimination of a variable by means of* **addition or subtraction.** The basis of this method is to

> Some prefer always to use addition of the terms of the resulting equations. This method is appropriate as it avoids possible errors that may be caused when subtracting. However, it may require that all signs of one equation be changed.

multiply (if necessary) all the terms of each equation by a constant chosen so that the coefficients of one of the unknowns will be numerically the same in both equations. (*Numerically* means we disregard the signs for this step.)

If these numerically equal coefficients are *opposite in sign,* we **add** the terms on each side of the resulting equations. If these numerically equal coefficients have the *same sign,* we **subtract** the terms on each side of one equation from the terms of the other equation. After either *adding* or *subtracting* we have a simple linear equation in one unknown, which we then solve for the unknown. We then substitute this value into *either* one of the original equations to find the value of the other unknown.

Some graphing calculators have a specific feature for solving simultaneous linear equations. It is necessary only to enter the coefficients and constants to get the solution.

■EXAMPLE 3   Use the method of elimination by addition or subtraction to solve the system of equations

$$x - 3y = 6$$
$$2x + 3y = 3$$

We look at the coefficients to determine the best way to eliminate one of the unknowns. Since the coefficients of the *y*-terms are *numerically the same and opposite in sign,* we may immediately *add* terms of the two equations together to eliminate *y*. Adding the terms of the left sides and adding terms of the right sides, we obtain

$$x + 2x - 3y + 3y = 6 + 3$$
$$3x = 9$$
$$x = 3$$

Substituting this value into the first equation, we obtain

$$3 - 3y = 6$$
$$-3y = 3$$
$$y = -1$$

The solution $x = 3$, $y = -1$ agrees with the results obtained for the same problem illustrated in Example 1 of this section.   ■

■EXAMPLE 4   Use addition or subtraction to solve the system of equations

$$3x - 2y = 4$$
$$x + 3y = 2$$

CAUTION▶

Looking at the coefficients of *x* and *y*, we see that we must multiply the second equation by 3 to make the coefficients of *x* the same. To make the coefficients of *y* numerically the same, we must multiply the first equation by 3 and the second equation by 2. Thus, the best method is to multiply the second equation by 3 and eliminate *x*. Doing this (be careful to multiply the terms on **both** sides: *a common error is to forget to multiply the value on the* **right**), the coefficients of *x* have the *same sign*. Therefore, we *subtract* terms of the second equation from those of the first equation.

$$3x - 2y = 4$$
$$\underline{3x + 9y = 6} \qquad \text{each term of second equation multiplied by 3}$$

$$3x - 3x = 0 \quad\rceil \quad -11y = -2 \qquad \text{subtract}$$
$$-2y - (+9y) = -11y \quad\rceil \quad y = \frac{2}{11} \qquad 4 - 6 = -2$$

In order to find the value of *x*, we substitute $y = \frac{2}{11}$ into one of the original equations. Choosing the second equation (its form is somewhat simpler), we have

$$x + 3\left(\frac{2}{11}\right) = 2$$
$$11x + 6 = 22 \qquad \text{multiply each term by 11}$$
$$x = \frac{16}{11}$$

We arrive at the solution $x = \frac{16}{11}$, $y = \frac{2}{11}$. Substituting these values into both of the original equations shows that the solution checks.   ■

EXAMPLE 5   As we noted in Example 4, we can solve the system of equations by first multiplying the first equation by 3 and the second equation by 2, thereby eliminating y. Doing this, we have

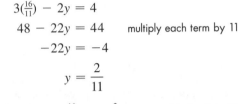

$$9x - 6y = 12 \qquad \text{each term of first equation multiplied by 3}$$
$$2x + 6y = 4 \qquad \text{each term of second equation multiplied by 2}$$
$$11x = 16 \qquad \text{add}$$
$$9x + 2x = 11x$$
$$x = \frac{16}{11} \qquad 12 + 4 = 16$$
$$-6y + 6y = 0$$

At this point we can find the value of y by substituting $x = \frac{16}{11}$ into one of the original equations, or we could eliminate x as is done in Example 4. Substitution in the first of the original equations gives us

$$3(\tfrac{16}{11}) - 2y = 4$$
$$48 - 22y = 44 \qquad \text{multiply each term by 11}$$
$$-22y = -4$$
$$y = \frac{2}{11}$$

Therefore, the solution is $x = \frac{16}{11}$, $y = \frac{2}{11}$, as we obtained in Example 4.

A calculator solution of this system is shown in Fig. 5-26, where $y_1 = 3x/2 - 2$ and $y_2 = -x/3 + 2/3$. The point of intersection is $(1.455, 0.182)$. This solution is the same as the algebraic solutions since $\frac{16}{11} = 1.455$ and $\frac{2}{11} = 0.182$. ◼

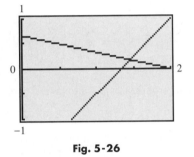

Fig. 5-26

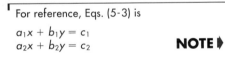

For reference, Eqs. (5-3) is

$$a_1x + b_1y = c_1$$
$$a_2x + b_2y = c_2$$

NOTE ▶

The best form in which to have the equations of the system for solution by addition or subtraction is the form shown in Examples 3 and 4, which is the same form as shown in Eqs. (5-3). That is, the x-term and the y-term in each equation is on the left and the constant is on the right. *If the equations are not written in this form, both should be rewritten in this form before proceeding with the solution.*

EXAMPLE 6   In solving the system of equations

$$4x = 2y + 3$$
$$-y + 2x - 2 = 0$$

we should first write the equations in the form noted just above. Doing this, we have

$$4x - 2y = 3$$
$$2x - y = 2$$

When we multiply the second equation by 2 and subtract, we obtain

$$4x - 2y = 3$$
$$4x - 2y = 4$$
$$4x - 4x = 0 \longrightarrow 0 = -1 \longleftarrow 3 - 4 = -1$$
$$-2y - (-2y) = 0$$

Since we know 0 does not equal -1, we conclude that there is no solution. When we obtain a result of $0 = a$ $(a \neq 0)$, the system of equations is *inconsistent*. As we discussed in the previous section, the lines that represent the equations are parallel. This is shown in the calculator display in Fig. 5-27, where $y_1 = 2x - 3/2$, $y_2 = 2x - 2$. ◼

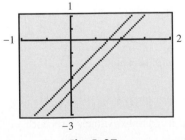

Fig. 5-27

NOTE ▶ In Example 6 we showed that a result of $0 = a$ ($a \neq 0$) indicates the system of equations is inconsistent. If we obtain the result $0 = 0$, the system is *dependent*. As shown in the previous section, this means that there is an unlimited number of solutions and the lines that represent the equations are the same line.

The following example gives another complete illustration of solving a word problem by first setting up the proper equations.

SOLVING A WORD PROBLEM

■**EXAMPLE 7** By weight, one alloy is 70% copper and 30% zinc. Another alloy is 40% copper and 60% zinc. How many grams of each of these are required to make 300 g of an alloy that is 60% copper and 40% zinc?

Let A = the required number of grams of the first alloy and B = the required number of grams of the second alloy. Our equations are determined from:

1. The total weight of the final alloy is 300 g: $A + B = 300$.

2. The final alloy will have 180 g of copper (60% of 300 g), and this comes from 70% of A (0.70A) and 40% of B (0.40B): $0.70A + 0.40B = 180$.

> The second equation is based on the amount of copper. We could have used the amount of zinc, which would have led to the equation $0.30A + 0.60B = 0.40(300)$. Since we need only two equations, we may use any two of these three equations to find the solution.

These two equations can now be solved simultaneously.

$$A + B = 300 \qquad \text{sum of weights is 300 g}$$

copper $\longrightarrow$ $0.70A + 0.40B = 180 \longleftarrow$ 60% of 300 g

70% weight of first alloy   40% weight of second alloy

$$4A + 4B = 1200 \qquad \text{multiply each term of first equation by 4}$$

$$\underline{7A + 4B = 1800} \qquad \text{multiply each term of second equation by 10}$$

$$3A \quad = 600 \qquad \text{subtract first equation from second equation}$$

$$A = 200 \text{ g}$$

$$B = 100 \text{ g} \qquad \text{by substituting into first equation}$$

Checking with the statement of the problem, using the percentages of zinc, we have $0.30(200) + 0.60(100) = 0.40(300)$, or 60 g + 60 g = 120 g. We use zinc here since we used the equation for the percentages of copper. ■

EXERCISES $5\text{-}4$

In Exercises 1–12, solve the given systems of equations by the method of elimination by substitution.

1. $x = y + 3$
$x - 2y = 5$

2. $x = 2y + 1$
$2x - 3y = 4$

3. $p = V - 4$
$V + p = 10$

4. $y = 2x + 10$
$2x + y = -2$

5. $x + y = -5$
$2x - y = 2$

6. $3x + y = 1$
$3x - 2y = 16$

7. $2x + 3y = 7$
$6x - y = 1$

8. $2s + 2t = 1$
$4s - 2t = 17$

9. $3x + 2y = 7$
$2y = 9x + 11$

10. $3A + 3B = -1$
$5A = -6B - 1$

11. $0.4p - 0.3n = 0.6$
$0.2p + 0.4n = -0.5$

12. $6.0x + 4.8y = -8.4$
$4.8x - 6.5y = -7.8$

In Exercises 13–24, solve the given systems of equations by the method of elimination by addition or subtraction.

13. $x + 2y = 5$
$x - 2y = 1$

14. $x + 3y = 7$
$2x + 3y = 5$

15. $2x - 3y = 4$
$2x + y = -4$

16. $R - 4r = 17$
$3R + 4r = 3$

17. $2x + 3y = 8$
$x = 2y - 3$

18. $3x - y = 3$
$4x = 3y + 14$

19. $v + 2t = 7$
$2v + 4t = 9$

20. $3x - y = 5$
$-9x + 3y = -15$

21. $2x - 3y - 4 = 0$
$3x + 2 = 2y$

22. $3i_1 + 5 = -4i_2$
$3i_2 = 5i_1 - 2$

23. $0.3R = 0.7Z + 0.4$
$0.5Z = 0.7 - 0.2R$

24. $2.50x + 2.25y = 4.00$
$3.75x - 6.75y = 3.25$

In Exercises 25–32, solve the given systems of equations by either method of this section.

25. $2x - y = 5$
$6x + 2y = -5$

26. $3x + 2y = 4$
$6x - 6y = 13$

27. $6x + 3y + 4 = 0$
$5y = -9x - 6$

28. $1 + 6q = 5p$
$3p - 4q = 7$

29. $3x - 6y = 15$
$4x - 8y = 20$

30. $2x + 6y = -3$
$-6x - 18y = 5$

31. $1.2V + 10.8 = -8.4C$
$3.6C + 4.8V + 13.2 = 0$

32. $0.66x + 0.66y = -0.77$
$0.33x - 1.32y = 1.43$

In Exercises 33–36, solve the given systems of equations by an appropriate algebraic method.

33. Find the voltages V_1 and V_2 of the batteries shown in Fig. 5-28. The terminals are aligned in the same direction in Fig. 5-28(a) and in opposite directions in Fig. 5-28(b).

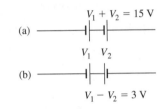

Fig. 5-28

34. A spring of length L is stretched x cm for each newton of weight hung from it. Weights of 3 N and then 5 N are hung from the spring, leading to the equations

$$L + 3x = 18$$
$$L + 5x = 22$$

Solve for L and x.

35. Two grades of gasoline are mixed to make a blend with 1.50% of a special additive. Combining x liters of a grade with 1.80% of the additive to y liters of a grade with 1.00% of the additive gives 10,000 L of the blend. The equations relating x and y are

$$x + y = 10{,}000$$
$$0.0180x + 0.0100y = 0.0150(10{,}000)$$

Find x and y (to three significant digits).

36. A 6.0% solution and a 15.0% solution of a drug are added to 200 mL of a 20.0% solution to make 1200 mL of a 12.0% solution for a proper dosage. The equations relating the number of milliliters of the added solutions are

$$x + y + 200 = 1200$$
$$0.060x + 0.150y + 0.200(200) = 0.120(1200)$$

Find x and y (to three significant digits).

In Exercises 37–44, set up appropriate systems of two linear equations and solve the systems algebraically. All data are accurate to at least two significant digits.

37. In a test of a heat-seeking rocket, a first rocket is launched at 2000 ft/s, and the heat-seeking rocket is launched along the same flight path 12 s later at a speed of 3200 ft/s. Find the times t_1 and t_2 of flight of the rockets until the heat-seeking rocket destroys the first rocket.

38. The *torque* of a force is the product of the force and the perpendicular distance from a specified point. If a lever is supported at only one point and is in balance, the sum of the torques (about the support) of forces acting on one side of the support must equal the sum of the torques of the forces acting on the other side. Find the forces F_1 and F_2 that are in the positions shown in Fig. 5-29(a) and then move to the positions in Fig. 5-29(b). The lever weighs 20 N and is in balance in each case.

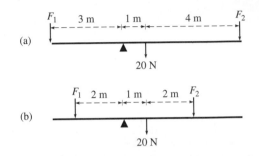

Fig. 5-29

39. A small isolated farm uses a windmill and a gas generator for power. During a 10-day period they produced 3010 kW·h of power, with the windmill operating at 45.0% of capacity and the generator at capacity. During the following 10-day period they produced 2900 kW·h with the windmill at 72.0% of capacity and the generator down 60 h for repairs (at capacity otherwise). What is the capacity (in kW) of each?

40. An underwater (but near the surface) explosion is detected by sonar on a ship 30 s before it is heard on the deck. If sound travels at 5000 ft/s in water and 1100 ft/s in air, how far is the ship from the explosion? See Fig. 5-30.

Fig. 5-30

41. As a 40-ft pulley belt makes one revolution, one of the two pulley wheels makes one more revolution than the other. Another wheel of half the radius replaces the smaller wheel and makes six more revolutions than the larger wheel for one revolution of the belt. Find the circumferences of the wheels.

42. In mixing a weed-killing chemical, a 40% solution of the chemical is mixed with an 85% solution to get 20 L of a 60% solution. How much of each solution is needed?

W) 43. What conclusion can you draw from a sales report that states that "sales this month were $8000 more than last month, which means that total sales for both months are $4000 more than twice the sales last month"?

W) 44. For an electric circuit, a report stated that current i_1 is twice current i_2 and that twice the sum of the two currents less 6 times i_2 is 6 mA. Explain your conclusion about the values of the currents found from this report.

$5\text{-}5$ SOLVING SYSTEMS OF TWO LINEAR EQUATIONS IN TWO UNKNOWNS BY DETERMINANTS

Consider two linear equations in two unknowns, as given in Eqs. (5-3):

$$
\begin{aligned}
a_1x + b_1y &= c_1 \\
a_2x + b_2y &= c_2
\end{aligned}
\tag{5-3}
$$

If we multiply the first of these equations by b_2 and the second by b_1, we obtain

$$
\begin{aligned}
a_1b_2x + b_1b_2y &= c_1b_2 \\
a_2b_1x + b_2b_1y &= c_2b_1
\end{aligned}
\tag{5-6}
$$

We see that the coefficients of y are the same. Thus, subtracting the second equation from the first, we can solve for x. The solution can be shown to be

$$
x = \frac{c_1b_2 - c_2b_1}{a_1b_2 - a_2b_1}
\tag{5-7}
$$

In the same manner, we may show that

$$
y = \frac{a_1c_2 - a_2c_1}{a_1b_2 - a_2b_1}
\tag{5-8}
$$

The expression $a_1b_2 - a_2b_1$, which appears in each of the denominators of Eqs. (5-7) and (5-8), is an example of a special kind of expression called a *determinant of the second order*. The determinant $a_1b_2 - a_2b_1$ is denoted by

$$
\begin{vmatrix} a_1 & b_1 \\ a_2 & b_2 \end{vmatrix}
$$

Therefore, by definition, *a* **determinant of the second order** *is*

$$
\begin{vmatrix} a_1 & b_1 \\ a_2 & b_2 \end{vmatrix} = a_1b_2 - a_2b_1
\tag{5-9}
$$

The numbers a_1 and b_1 are called the **elements** *of the first* **row** *of the determinant. The numbers a_1 and a_2 are the elements of the first* **column** *of the determinant.* In the same manner, the numbers a_2 and b_2 are the elements of the second row, and the numbers b_1 and b_2 are the elements of the second column. *The numbers a_1 and b_2 are the elements of the* **principal diagonal,** *and the numbers a_2 and b_1 are the elements of the* **secondary diagonal.** Thus, one way of stating the definition indicated in Eq. (5-9) is that *the value of a determinant of the second order is found by taking the product of the elements of the principal diagonal and subtracting the product of the elements of the secondary diagonal.*

A diagram that is often helpful for remembering the expansion of a second-order determinant is shown in Fig. 5-31. The following examples illustrate how we carry out the evaluation of determinants.

Determinants were invented by the German mathematician Gottfried Wilhelm Leibniz (1646–1716).

SECOND-ORDER DETERMINANT

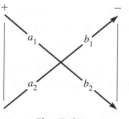

Fig. 5-31

EXAMPLE 1 $\begin{vmatrix} -5 & 8 \\ 3 & 7 \end{vmatrix} = (-5)(7) - 3(8) = -35 - 24 = -59$

EXAMPLE 2 (a) $\begin{vmatrix} 4 & 6 \\ 3 & 17 \end{vmatrix} = 4(17) - (3)(6) = 68 - 18 = 50$

(b) $\begin{vmatrix} 4 & 6 \\ -3 & 17 \end{vmatrix} = 4(17) - (-3)(6) = 68 + 18 = 86$

(c) $\begin{vmatrix} 3.6 & 6.1 \\ -3.2 & -17.2 \end{vmatrix} = 3.6(-17.2) - (-3.2)(6.1) = -42.4$

Note the signs of the terms being combined.

We note that the numerators and denominators of Eqs. (5-7) and (5-8) may be written as determinants. The numerators of the equations are

$$\begin{vmatrix} c_1 & b_1 \\ c_2 & b_2 \end{vmatrix} \quad \text{and} \quad \begin{vmatrix} a_1 & c_1 \\ a_2 & c_2 \end{vmatrix}$$

Therefore, the solutions for x and y of the system of equations

$$\begin{aligned} a_1 x + b_1 y &= c_1 \\ a_2 x + b_2 y &= c_2 \end{aligned} \tag{5-3}$$

may be written directly in terms of determinants, without algebraic operations, as

NOTE CAREFULLY THE LOCATION OF c_1 AND c_2

$$x = \frac{\begin{vmatrix} c_1 & b_1 \\ c_2 & b_2 \end{vmatrix}}{\begin{vmatrix} a_1 & b_1 \\ a_2 & b_2 \end{vmatrix}} \quad \text{and} \quad y = \frac{\begin{vmatrix} a_1 & c_1 \\ a_2 & c_2 \end{vmatrix}}{\begin{vmatrix} a_1 & b_1 \\ a_2 & b_2 \end{vmatrix}} \tag{5-10}$$

For this reason, determinants provide a quick and easy method of solution of systems of equations. Again, *the denominator of each of Eqs. (5-10) is the same.*

The determinant of the denominator is made up of the coefficients of x and y. Also, the determinant of the numerator of the solution for x is obtained from the determinant of the denominator by **replacing the column of a's by the column of c's.** *The determinant of the numerator of the solution for y is obtained from the determinant of the denominator by* **replacing the column of b's by the column of c's.**

This result is referred to as **Cramer's rule.** In using Cramer's rule we must be sure that the equations are written in the form of Eqs. (5-3) before setting up the determinants.

The following examples illustrate the method of solving systems of equations by determinants.

Named for the Swiss mathematician Gabriel Cramer (1704–1752).

■**EXAMPLE 3**   Solve the following system of equations by determinants:

$$2x + y = 1$$
$$5x - 2y = -11$$

We first note that the equations are in the proper form of Eqs. (5-3) for solution by determinants. Next, we set up the determinant for the denominator, which consists of the four coefficients in the system, written as shown. It is

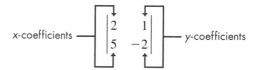

For finding *x,* the determinant in the numerator is obtained from this determinant by replacing the first column by the constants that appear on the right sides of the equations. Thus, *the numerator for the solution for x is*

For finding *y,* the determinant in the numerator is obtained from the determinant of the denominator by replacing the second column by the constants that appear on the right sides of the equations. Thus, *the numerator for the solution for y is*

Now we set up the solutions for *x* and *y* using the determinants above.

Determinants can be evaluated on a
graphing calculator. This is shown on
page 158.

$$x = \frac{\begin{vmatrix} 1 & 1 \\ -11 & -2 \end{vmatrix}}{\begin{vmatrix} 2 & 1 \\ 5 & -2 \end{vmatrix}} = \frac{1(-2) - (-11)(1)}{2(-2) - (5)(1)} = \frac{-2 + 11}{-4 - 5} = \frac{9}{-9} = -1$$

$$y = \frac{\begin{vmatrix} 2 & 1 \\ 5 & -11 \end{vmatrix}}{\begin{vmatrix} 2 & 1 \\ 5 & -2 \end{vmatrix}} = \frac{2(-11) - (5)(1)}{-9} = \frac{-22 - 5}{-9} = 3$$

Therefore, the solution to the system of equations is $x = -1$, $y = 3$.
Substituting these values into the equations, we have

$$2(-1) + 3 \stackrel{?}{=} 1 \quad \text{and} \quad 5(-1) - 2(3) \stackrel{?}{=} -11$$
$$1 = 1 \qquad\qquad\qquad -11 = -11$$

which shows that they check.

NOTE▶   Since the same determinant appears in each denominator, *it needs to be evaluated only once.* This means that three determinants are to be evaluated in order to solve the system. ---------■

EXAMPLE 4   Solve the following system of equations by determinants. All numbers are approximate.

$$5.3x + 7.2y = 4.5$$
$$3.2x - 6.9y = 5.7$$

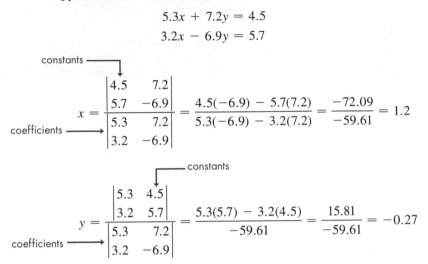

constants

$$x = \frac{\begin{vmatrix} 4.5 & 7.2 \\ 5.7 & -6.9 \end{vmatrix}}{\begin{vmatrix} 5.3 & 7.2 \\ 3.2 & -6.9 \end{vmatrix}} = \frac{4.5(-6.9) - 5.7(7.2)}{5.3(-6.9) - 3.2(7.2)} = \frac{-72.09}{-59.61} = 1.2$$

coefficients

constants

$$y = \frac{\begin{vmatrix} 5.3 & 4.5 \\ 3.2 & 5.7 \end{vmatrix}}{\begin{vmatrix} 5.3 & 7.2 \\ 3.2 & -6.9 \end{vmatrix}} = \frac{5.3(5.7) - 3.2(4.5)}{-59.61} = \frac{15.81}{-59.61} = -0.27$$

coefficients

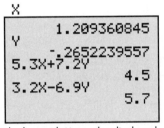

The line with X may be displaced off the screen.

Fig. 5-32

The calculations can be done completely on the calculator using the following procedure: (1) Evaluate the denominator and store this value; (2) divide the value of each numerator by the value of the denominator, storing the values of x and y; (3) use the stored values of x and y for checking the solution; (4) round off results (if the numbers are approximate).

Using this procedure, we store the value of the denominator D ($D = -59.61$) in order to calculate x and y, which are stored as X and Y. The check of the solution is shown in the calculator display shown in Fig. 5-32. ▪

SOLVING A WORD PROBLEM

EXAMPLE 5   Two investments totaling $18,000 yield an annual income of $700. If the first investment has an interest rate of 5.5% and the second a rate of 3.0%, what is the value of each investment?

Let $x = $ the value of the first investment and $y = $ the value of the second investment. We know that the total of the two investments is $18,000. This leads to the equation $x + y = 18,000$. The first investment yields $0.055x$ dollars annually, and the second yields $0.030y$ dollars annually. This leads to the equation $0.055x + 0.030y = 700$. These two equations are then solved simultaneously.

$$x + y = 18,000 \qquad \text{sum of investments}$$
$$0.055x + 0.030y = 700 \longleftarrow \text{income}$$

5.5% value   3.0% value

$$x = \frac{\begin{vmatrix} 18,000 & 1 \\ 700 & 0.030 \end{vmatrix}}{\begin{vmatrix} 1 & 1 \\ 0.055 & 0.030 \end{vmatrix}} = \frac{540 - 700}{0.030 - 0.055} = \frac{-160}{-0.025} = 6400$$

The value of y can be found most easily by substituting this value of x into the first equation, $y = 18,000 - x = 18,000 - 6400 = 11,600$.

Therefore, the values invested are $6400 and $11,600, respectively. Checking, we see that the total income is $6400(0.055) + $11,600(0.030) = 700. This agrees with the statement of the problem. ▪

CAUTION▶

The equations must be in the form of Eqs. (5-3) before the determinants are set up. The specific positions of the values in the determinants are based on that form of writing the system. If either unknown is missing from an equation, a zero must be placed in the proper position. Also, from Example 4 we see that determinants are easier to use than other algebraic methods when the coefficients are decimals.

NOTE▶

If the determinant of the denominator is zero, we do not have a unique solution since this would require division by zero. If the determinant of the denominator is zero and that of the numerator is not zero, the system is *inconsistent*. If the determinants of both numerator and denominator are zero, the system is *dependent*.

EXERCISES 5-5

In Exercises 1–12, evaluate the given determinants.

1. $\begin{vmatrix} 2 & 4 \\ 3 & 1 \end{vmatrix}$ **2.** $\begin{vmatrix} -1 & 3 \\ 2 & 6 \end{vmatrix}$ **3.** $\begin{vmatrix} 3 & -5 \\ 7 & -2 \end{vmatrix}$

4. $\begin{vmatrix} -4 & 7 \\ 1 & -3 \end{vmatrix}$ **5.** $\begin{vmatrix} 8 & -10 \\ 0 & 4 \end{vmatrix}$ **6.** $\begin{vmatrix} -4 & -3 \\ -8 & -6 \end{vmatrix}$

7. $\begin{vmatrix} -2 & 11 \\ -7 & -8 \end{vmatrix}$ **8.** $\begin{vmatrix} -6 & 12 \\ -15 & 3 \end{vmatrix}$ **9.** $\begin{vmatrix} 0.7 & -1.3 \\ 0.1 & 1.1 \end{vmatrix}$

10. $\begin{vmatrix} 0.20 & -0.05 \\ 0.28 & 0.09 \end{vmatrix}$ **11.** $\begin{vmatrix} 16 & -8 \\ 42 & -15 \end{vmatrix}$ **12.** $\begin{vmatrix} 43 & -7 \\ -81 & 16 \end{vmatrix}$

In Exercises 13–24, solve the given systems of equations by determinants. (These are the same as those for Exercises 13–24 of Section 5-4.)

13. $x + 2y = 5$
$x - 2y = 1$

14. $x + 3y = 7$
$2x + 3y = 5$

15. $2x - 3y = 4$
$2x + y = -4$

16. $R - 4r = 17$
$3R + 4r = 3$

17. $2x + 3y = 8$
$x = 2y - 3$

18. $3x - y = 3$
$4x = 3y + 14$

19. $v + 2t = 7$
$2v + 4t = 9$

20. $3x - y = 5$
$-9x + 3y = -15$

21. $2x - 3y - 4 = 0$
$3x + 2 - 2y$

22. $3i_1 + 5 = -4i_2$
$3i_2 = 5i_1 - 2$

23. $0.3R = 0.7Z + 0.4$
$0.5Z = 0.7 - 0.2R$

24. $2.50x + 2.25y = 4.00$
$3.75x - 6.75y = 3.25$

In Exercises 25–32, solve the given systems of equations by determinants. All numbers are approximate.

25. $2.1x - 1.0y = 5.2$
$5.8x + 1.6y = -5.4$

26. $3.3x + 1.9y = 4.2$
$5.4x - 6.4y = 13.2$

27. $5.5R = -9.0t - 6.8$
$6.0t + 3.8R + 4.0 = 0$

28. $12 + 66y = 53x$
$32x - 39y = 71$

29. $301x - 529y = 1520$
$385x - 741y = 2540$

30. $0.25d + 0.63n = -0.37$
$-0.61d - 1.80n = 0.55$

31. $1.2y + 10.8 = -8.4x$
$3.5x + 4.8y + 12.9 = 0$

32. $6541x + 4397y = -7732$
$3309x - 8755y = 7622$

In Exercises 33–36, solve the given systems of equations by determinants. All numbers are accurate to at least two significant digits.

33. The forces acting on a link of an industrial robot are shown in Fig. 5-33. The equations for finding forces F_1 and F_2 are

$F_1 + F_2 = 21$
$2F_1 = 5F_2$

Find F_1 and F_2.

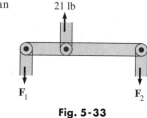

Fig. 5-33

34. The area of a quadrilateral is

$$A = \frac{1}{2}\left(\begin{vmatrix} x_0 & x_1 \\ y_0 & y_1 \end{vmatrix} + \begin{vmatrix} x_1 & x_2 \\ y_1 & y_2 \end{vmatrix} + \begin{vmatrix} x_2 & x_3 \\ y_2 & y_3 \end{vmatrix} + \begin{vmatrix} x_3 & x_0 \\ y_3 & y_0 \end{vmatrix} \right)$$

where (x_0, y_0), (x_1, y_1), (x_2, y_2), and (x_3, y_3) are the rectangular coordinates of the vertices of the quadrilateral, listed counterclockwise. (This *surveyor's formula* can be generalized to find the area of any polygon.)

A surveyor records the locations of the vertices of a quadrilateral building lot on a rectangular coordinate system as $(12.79, 0.00)$, $(67.21, 12.30)$, $(53.05, 47.12)$, and $(10.09, 53.11)$, where distances are in meters. Find the area of the lot.

35. An airplane begins a flight with a total of 36.0 gal of fuel stored in two separate wing tanks. During the flight, 25.0% of the fuel in one tank is used, and in the other tank 37.5% of the fuel is used. If the total fuel used is 11.2 gal, the amounts x and y used from each tank can be found by solving the system of equations

$x + y = 36.0$
$0.250x + 0.375y = 11.2$

Find x and y.

36. A certain amount of fuel contains 150,000 Btu of potential heat. Part is burned at 80% efficiency, and the remainder is burned at 70% efficiency such that the total amount of heat actually delivered is 114,000 Btu. Find the amounts x and y burned with each efficiency by solving the following equations.

$x + y = 150,000$
$0.80x + 0.70y = 114,000$

In Exercises 37–44, set up appropriate systems of two linear equations in two unknowns and then solve the systems by determinants. All numbers are accurate to at least two significant digits.

37. A boat carrying illegal drugs leaves a port and travels at 42 mi/h. A Coast Guard cutter leaves the port 24 min later and travels at 50 mi/h in pursuit of the boat. Find the times each has traveled when the cutter overtakes the boat with drugs. See Fig. 5-34.

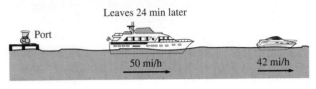

Leaves 24 min later

Port

50 mi/h   42 mi/h

Fig. 5-34

38. In framing the wall of a house, each horizontal plate is 3 ft 3 in. longer than each vertical stud. Find the lengths of each plate and each of the 9 studs if a total of 89 ft of lumber is used. See Fig. 5-35.

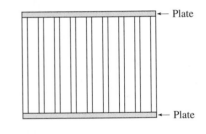

← Plate

Fig. 5-35

← Plate

39. A machinery sales representative receives a fixed salary plus a sales commission each month. If $6200 is earned on sales of $70,000 in one month and $4700 is earned on sales of $45,000 in the following month, what is the fixed salary and the commission percent?

40. Two types of electromechanical carburetors are being assembled and tested. Each of the first type requires 15 min of assembly time and 2 min of testing time. Each of the second type requires 12 min of assembly time and 3 min of testing time. If 222 min of assembly time and 45 min of testing time are available, how many of each type can be assembled and tested if all the time is used?

41. A pharmacist is mixing a 3.0% saline solution and an 8.0% saline solution to get 2.0 L of a 6.0% solution. How much of each solution is needed?

42. The velocity of sound in steel is 15,900 ft/s faster than the velocity of sound in air. One end of a long steel bar is struck and an instrument at the other end measures the time it takes for the sound to reach it. The sound in the bar takes 0.0120 s, and the sound in the air takes 0.180 s. What are the velocities of sound in air and in steel?

43. In an experiment, a variable voltage V is in a circuit with a fixed voltage V_0 and a resistance R. The voltage V is related to the current i in the circuit by $V = Ri - V_0$. If $V = 5.8$ V for $i = 2.0$ A and $V = 24.7$ V for $i = 6.2$ A, find V as a function of i.

44. A shipment of 320 cellular phone and radar detectors was destroyed due to a truck accident. On the insurance claim the shipper stated that each phone was worth $110, each detector was worth $160, and their total value was $40,700. How many of each were in the shipment?

$5\text{-}6$ SOLVING SYSTEMS OF THREE LINEAR EQUATIONS IN THREE UNKNOWNS ALGEBRAICALLY

Many technical and scientific problems involve the solution of systems of linear equations that have three, four, and occasionally even more unknowns. In this section we discuss the solution of systems with three unknowns, and in Chapter 16 we will show how systems with even more unknowns are solved.

Solving such systems algebraically or by determinants is very similar to solving systems of two linear equations in two unknowns. However, the numerical work can be very lengthy, especially when dealing with more than three unknowns. In this section we will note how a calculator can be used to evaluate a determinant without finding all the individual products (the calculator does that). Graphical solutions are not used, since a linear equation in three unknowns represents a plane in space. We shall, however, briefly show graphical interpretations of systems of three linear equations in three unknowns at the end of this section.

A system of three linear equations in three unknowns written in the form

$$\begin{aligned} a_1x + b_1y + c_1z &= d_1 \\ a_2x + b_2y + c_2z &= d_2 \\ a_3x + b_3y + c_3z &= d_3 \end{aligned}$$

(5-11)

has as its solution the set of values x, y, and z that satisfy all three equations simultaneously. The method of solution involves multiplying *two* of the equations by the proper numbers to eliminate *one* of the unknowns between these equations. We then repeat this process, using a ***different pair*** of the original equations, being sure that we eliminate the same unknown as we did between the first pair of equations. At this point we have two linear equations in two unknowns that can be solved by any of the methods previously discussed.

EXAMPLE 1 Solve the following system of equations.

(1)	$4x + y + 3z = 1$	
(2)	$2x - 2y + 6z = 11$	
(3)	$-6x + 3y + 12z = -4$	
(4)	$8x + 2y + 6z = 2$	(1) multiplied by 2
	$2x - 2y + 6z = 11$	(2)
(5)	$10x \quad\quad + 12z = 13$	adding
(6)	$12x + 3y + 9z = 3$	(1) multiplied by 3
	$-6x + 3y + 12z = -4$	(3)
(7)	$18x \quad\quad - 3z = 7$	subtracting
	$10x + 12z = 13$	(5)
(8)	$72x - 12z = 28$	(7) multiplied by 4
(9)	$82x \quad\quad = 41$	adding
(10)	$x = \frac{1}{2}$	
(11)	$18(\frac{1}{2}) - 3z = 7$	substituting (10) in (7)
(12)	$-3z = -2$	
(13)	$z = \frac{2}{3}$	
(14)	$4(\frac{1}{2}) + y + 3(\frac{2}{3}) = 1$	substituting (13) and (10) in (1)
(15)	$2 + y + 2 = 1$	
(16)	$y = -3$	

We can choose to first eliminate any one of the three unknowns. We have chosen y.

At this point we could have used Eqs. (1) and (2) or Eqs. (2) and (3), but we must set them up to eliminate y.

Some graphing calculators have a specific feature for solving simultaneous linear equations. It is necessary only to enter the coefficients and constants to get the solution.

Thus, the solution is $x = \frac{1}{2}$, $y = -3$, $z = \frac{2}{3}$. Substituting in the equations, we have

$$4(\tfrac{1}{2}) + (-3) + 3(\tfrac{2}{3}) \stackrel{?}{=} 1 \qquad 2(\tfrac{1}{2}) - 2(-3) + 6(\tfrac{2}{3}) \stackrel{?}{=} 11 \qquad -6(\tfrac{1}{2}) + 3(-3) + 12(\tfrac{2}{3}) \stackrel{?}{=} -4$$

$$1 = 1 \qquad\qquad\qquad 11 = 11 \qquad\qquad\qquad -4 = -4$$

We see that the solution checks. If we had first eliminated x or z, we would have completed the solution in a similar manner.

See the chapter introduction.

EXAMPLE 2   An industrial robot and the forces acting on its main link are shown in Fig. 5-36. An analysis of the forces leads to the following equations relating the forces. Determine the forces.

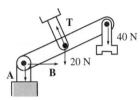

Fig. 5-36

(1)	$A + 60 = 0.8T$			
(2)	$B = 0.6T$			
(3)	$8A + 6B + 80 = 5T$			
(4)	$5A \qquad - 4T = -300$		(1) multiplied by 5 and rewritten	
(5)	$5B - 3T = 0$		(2) multiplied by 5 and rewritten	
(6)	$8A + 6B - 5T = -80$		(3) rewritten	
(7)	$40A \qquad - 32T = -2400$		(4) multiplied by 8	
(8)	$40A + 30B - 25T = -400$		(6) multiplied by 5	
(9)	$-30B - 7T = -2000$		subtracting	
(10)	$30B - 18T = 0$		(5) multiplied by 6	
(11)	$-25T = -2000$		adding	
(12)	$T = 80$ N			
(13)	$A + 60 = 64$		(12) substituted in (1)	
(14)	$A = 4$ N			
(15)	$B = 48$ N		(12) substituted in (2)	

Therefore, the forces are $A = 4$ N, $B = 48$ N, and $T = 80$ N. The solution checks when substituted into the original equations. Note that we could have started the solution by substituting Eq. (2) into Eq. (3), thereby first eliminating B. ∎

SOLVING A WORD PROBLEM

EXAMPLE 3   Three voltages, e_1, e_2, and e_3, where e_3 is three times e_1, are in series with the same polarity (see Fig. 5-37(a)) and have a total voltage of 85 mV. If e_2 is reversed in polarity (see Fig. 5-37(b)), the voltage is 35 mV. Find the voltages.

Since the voltages are in series with the same polarity, $e_1 + e_2 + e_3 = 85$. Then, since e_3 is three times e_1, we have $e_3 = 3e_1$. Then, with the reversed polarity of e_2, we have $e_1 - e_2 + e_3 = 35$. Writing these equations in standard form, we have the following solution.

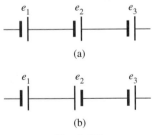

(a)

(b)

Fig. 5-37

(1)	$e_1 + e_2 + e_3 = 85$	
(2)	$3e_1 \qquad - e_3 = 0$	rewriting second equation
(3)	$e_1 - e_2 + e_3 = 35$	
(4)	$2e_1 \qquad + 2e_3 = 120$	adding (1) and (3)
(5)	$6e_1 \qquad - 2e_3 = 0$	(2) multiplied by 2
(6)	$8e_1 \qquad = 120$	adding
(7)	$e_1 = 15$ mV	
(8)	$3(15) - e_3 = 0$	substituting (7) in (2)
(9)	$e_3 = 45$ mV	
(10)	$15 + e_2 + 45 = 85$	substituting (7) and (9) in (1)
(11)	$e_2 = 25$ mV	

Therefore, the three voltages are 15 mV, 25 mV, and 45 mV.

Checking the solution, the sum of the three voltages is 85 mV, e_3 is three times e_1, and the sum of e_1 and e_3 less e_2 is 35 mV. ∎

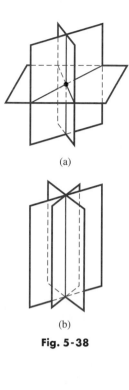

(a)

(b)

Fig. 5-38

The systems of equations we have solved in this section have had unique solutions. However, linear systems with more than two unknowns may also have an unlimited number of solutions or may be inconsistent. After eliminating unknowns, if we obtain $0 = 0$, there is an unlimited number of solutions. If we obtain $0 = a$ ($a \neq 0$), the system is inconsistent and there is no solution. (See Exercises 25–28 of this section.)

In the introduction to this section we noted that a linear equation in three unknowns represents a plane in space. For a system of three linear equations in three unknowns, if the planes intersect at a point, there is a unique solution (Fig. 5-38(a)). If the three planes intersect in a line, there is an unlimited number of solutions (Fig. 5-38(b)). If the planes do not have a common intersection, the system is inconsistent. The planes can be parallel (Fig. 5-39(a)), two of the planes can be parallel (Fig. 5-39(b)), or they can intersect in pairs in three parallel lines (Fig. 5-32(c)). If any plane is coincident with another plane, the system is inconsistent or there is an unlimited number of solutions, depending on whether or not any of the planes are parallel.

Fig. 5-39

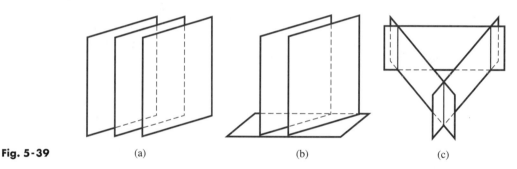

(a)   (b)   (c)

For systems of equations with more than three unknowns, the solution is found in a manner similar to that used with three unknowns. For example, with four unknowns one of the unknowns is eliminated between three different pairs of equations. The result is three equations in the remaining three unknowns. The solution then follows the procedure used with three unknowns.

EXERCISES 5-6

In Exercises 1–16, solve the given systems of equations.

1. $x + y + z = 2$
$x - z = 1$
$x + y = 1$

2. $x + y - z = -3$
$x + z = 2$
$2x - y + 2z = 3$

3. $2x + 3y + z = 2$
$-x + 2y + 3z = -1$
$-3x - 3y + z = 0$

4. $2x + y - z = 4$
$4x - 3y - 2z = -2$
$8x - 2y - 3z = 3$

5. $5l + 6w - 3h = 6$
$4l - 7w - 2h = -3$
$3l + w - 7h = 1$

6. $3r + s - t = 2$
$r - 2s + t = 0$
$4r - s + t = 3$

7. $2x - 2y + 3z = 5$
$2x + y - 2z = -1$
$4x - y - 3z = 0$

8. $2u + 2v + 3w = 0$
$3u + v + 4w = 21$
$-u - 3v + 7w = 15$

9. $3x - 7y + 3z = 6$
$3x + 3y + 6z = 1$
$5x - 5y + 2z = 5$

10. $8x + y + z = 1$
$7x - 2y + 9z = -3$
$4x - 6y + 8z = -5$

11. $p + 2q + 2r = 0$
$2p + 6q - 3r = -1$
$4p - 3q + 6r = -8$

12. $3x + 3y + z = 6$
$2x + 2y - z = 9$
$4x + 2y - 3z = 16$

13. $2x + 3y - 5z = 7$
$4x - 3y - 2z = 1$
$8x - y + 4z = 3$

14. $2i_1 - 4i_2 - 4i_3 = 3$
$3i_1 + 8i_2 + 2i_3 = -11$
$4i_1 + 6i_2 - i_3 = -8$

15. $r - s - 3t - u = 1$
$2r + 4s - 2u = 2$
$3r + 4s - 2t = 0$
$r + 2t - 3u = 3$

16. $3x + 2y - 4z + 2t = 3$
$5x - 3y - 5z + 6t = 8$
$2x - y + 3z - 2t = 1$
$-2x + 3y + 2z - 3t = -2$

In Exercises 17–20, solve the given systems of equations by determinants. All numbers are accurate to at least two significant digits.

17. A medical supply company has 1150 worker-hours for production, maintenance, and inspection. Using this and other factors, the number of hours used for each operation, *P, M,* and *I,* respectively, is found by solving the following system of equations.

$$P + M + I = 1150$$
$$P = 4I - 100$$
$$P = 6M + 50$$

18. Three oil pumps fill three different tanks. The pumping rates of the pumps (in L/h) are r_1, r_2, and r_3, respectively. Because of malfunctions, they do not operate at capacity each time. Their rates can be found by solving the following system of equations.

$$r_1 + r_2 + r_3 = 14{,}000$$
$$r_1 + 2r_2 \quad\;\; = 13{,}000$$
$$3r_1 + 3r_2 + 2r_3 = 36{,}000$$

19. The forces acting on a certain girder, as shown in Fig. 5-40, can be found by solving the following system of equations.

$$0.707F_1 - 0.800F_2 \qquad\quad = 0$$
$$0.707F_1 + 0.600F_2 - \quad F_3 = 10.0$$
$$3.00\; F_2 - 3.00F_3 = 20.0$$

Find the forces, in newtons.

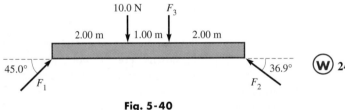

Fig. 5-40

20. In applying Kirchhoff's laws (e.g., see Beiser, *Modern Technical Physics,* 6th ed., p. 550) to the electric circuit shown in Fig. 5-41, the following equations are found. Determine the indicated currents I_A, I_B, and I_c (in A).

$$I_A + I_B + I_C = 0$$
$$4I_A - 10I_B \qquad\;\; = 3$$
$$-10I_B + 5I_C = 6$$

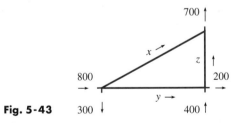

Fig. 5-41

In Exercises 21–24, set up systems of three linear equations in three unknowns and solve for the indicated quantities. All numbers are accurate to at least two significant digits.

21. The angle θ between two links of a robot arm is given by $\theta = at^3 + bt^2 + ct$, where t is the time during an 11.8-s cycle. If $\theta = 19.0°$ for $t = 1.00$ s, $\theta = 30.9°$ for $t = 3.00$ s, and $\theta = 19.8°$ for $t = 5.00$ s, find the equation $\theta = f(t)$. See Fig. 5-42.

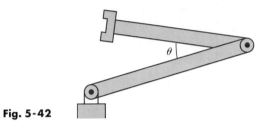

Fig. 5-42

22. Three computer line printers print 18,500 lines when operating together for 2 min. They print 17,500 lines if they operate for 3 min, 2 min, and 1 min, respectively. The first and third printers print 6000 lines when they operate together for 1 min. What is the rate (in lines/minute) at which each prints? (All data are accurate to three significant digits, although times are shown with only one digit.)

23. By weight, one fertilizer is 20% potassium, 30% nitrogen, and 50% phosphorus. A second fertilizer has percents of 10, 20, and 70, respectively, and a third fertilizer has percents of 0, 30, and 70, respectively. How much of each must be mixed to get 200 lb of fertilizer with percents of 12, 25, and 63, respectively?

(W) 24. The average traffic flow (number of vehicles) from noon until 1 P.M. in a certain section of one-way streets in a city is shown in Fig. 5-43. Explain why an analysis of the flow through these intersections is not sufficient to obtain unique values for *x, y,* and *z.*

	700 ↑
800 →	*x* ↗ ... *z* ↑ ... 200 →
	y →

Fig. 5-43   300 ↓   400 ↑

In Exercises 25–28, show that the given systems of equations have either an unlimited number of solutions or no solution. If there is an unlimited number of solutions, find one of them.

25.
$$x - 2y - 3z = 2$$
$$x - 4y - 13z = 14$$
$$-3x + 5y + 4z = 0$$

26.
$$x - 2y - 3z = 2$$
$$x - 4y - 13z = 14$$
$$-3x + 5y + 4z = 2$$

27.
$$3x + 3y - 2z = 2$$
$$2x - y + z = 1$$
$$x - 5y + 4z = -3$$

28.
$$3x + y - z = -3$$
$$x + y - 3z = -5$$
$$-5x - 2y + 3z = -7$$

5-7 SOLVING SYSTEMS OF THREE LINEAR EQUATIONS IN THREE UNKNOWNS BY DETERMINANTS

Just as systems in two linear equations in two unknowns can be solved by determinants, so can systems of three linear equations in three unknowns. The system

$$
\begin{aligned}
a_1x + b_1y + c_1z &= d_1 \\
a_2x + b_2y + c_2z &= d_2 \\
a_3x + b_3y + c_3z &= d_3
\end{aligned}
\qquad \text{(5-11)}
$$

can be solved in general terms by the method of elimination by addition or subtraction. This leads to the following solutions for x, y, and z.

$$
x = \frac{d_1b_2c_3 + d_3b_1c_2 + d_2b_3c_1 - d_3b_2c_1 - d_1b_3c_2 - d_2b_1c_3}{a_1b_2c_3 + a_3b_1c_2 + a_2b_3c_1 - a_3b_2c_1 - a_1b_3c_2 - a_2b_1c_3}
$$

$$
y = \frac{a_1d_2c_3 + a_3d_1c_2 + a_2d_3c_1 - a_3d_2c_1 - a_1d_3c_2 - a_2d_1c_3}{a_1b_2c_3 + a_3b_1c_2 + a_2b_3c_1 - a_3b_2c_1 - a_1b_3c_2 - a_2b_1c_3} \qquad \text{(5-12)}
$$

$$
z = \frac{a_1b_2d_3 + a_3b_1d_2 + a_2b_3d_1 - a_3b_2d_1 - a_1b_3d_2 - a_2b_1d_3}{a_1b_2c_3 + a_3b_1c_2 + a_2b_3c_1 - a_3b_2c_1 - a_1b_3c_2 - a_2b_1c_3}
$$

The expressions that appear in the numerators and denominators of Eqs. (5-12) are examples of a **determinant of the third order.** *This determinant is defined by*

$$
\begin{vmatrix}
a_1 & b_1 & c_1 \\
a_2 & b_2 & c_2 \\
a_3 & b_3 & c_3
\end{vmatrix} = a_1b_2c_3 + a_3b_1c_2 + a_2b_3c_1 - a_3b_2c_1 - a_1b_3c_2 - a_2b_1c_3 \qquad \text{(5-13)}
$$

The elements, rows, columns, and diagonals of a third-order determinant are defined just as are those of a second-order determinant. For example, the principal diagonal is made up of the elements a_1, b_2, and c_3.

Probably the easiest way of remembering the method of finding the value of a third-order determinant is as follows: *Rewrite the first two columns to the right of the determinant. The products of the elements of the principal diagonal and the two parallel diagonals to the right of it are then added. The products of the elements of the secondary diagonal and the two parallel diagonals to the right of it are subtracted from the first sum. The algebraic sum of these six products gives the value of the determinant.* These products are indicated in Fig. 5-44.

This method is used only for third-order determinants. It does not work for determinants of order higher than three.

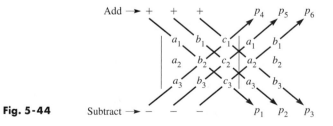

Fig. 5-44

Examples 1 and 2 illustrate this method of evaluating third-order determinants. (Other methods are shown in Chapter 16.)

Determinants can be evaluated on a calculator using the *matrix* feature. Figure 5-45 shows two windows for the determinant in Example 2. The first shows the window for entering the numbers in the *matrix* (the array of numbers—see page 426). The first four lines of the second window show the matrix displayed, and the last two lines show the evaluation of the determinant.

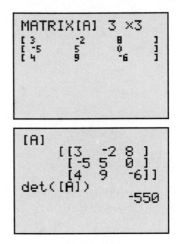

Fig. 5-45

EXAMPLE 1

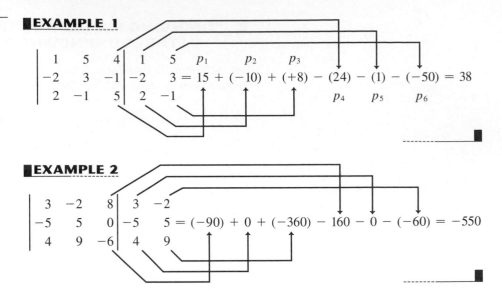

$$\begin{vmatrix} 1 & 5 & 4 \\ -2 & 3 & -1 \\ 2 & -1 & 5 \end{vmatrix} \begin{matrix} 1 & 5 \\ -2 & 3 \\ 2 & -1 \end{matrix} = 15 + (-10) + (+8) - (24) - (1) - (-50) = 38$$

EXAMPLE 2

$$\begin{vmatrix} 3 & -2 & 8 \\ -5 & 5 & 0 \\ 4 & 9 & -6 \end{vmatrix} \begin{matrix} 3 & -2 \\ -5 & 5 \\ 4 & 9 \end{matrix} = (-90) + 0 + (-360) - 160 - 0 - (-60) = -550$$

Inspection of Eqs. (5-12) reveals that the numerators of these solutions may also be written in terms of determinants. Thus, we may write the general solution to a system of three equations in three unknowns as

$$x = \frac{\begin{vmatrix} d_1 & b_1 & c_1 \\ d_2 & b_2 & c_2 \\ d_3 & b_3 & c_3 \end{vmatrix}}{\begin{vmatrix} a_1 & b_1 & c_1 \\ a_2 & b_2 & c_2 \\ a_3 & b_3 & c_3 \end{vmatrix}} \qquad y = \frac{\begin{vmatrix} a_1 & d_1 & c_1 \\ a_2 & d_2 & c_2 \\ a_3 & d_3 & c_3 \end{vmatrix}}{\begin{vmatrix} a_1 & b_1 & c_1 \\ a_2 & b_2 & c_2 \\ a_3 & b_3 & c_3 \end{vmatrix}} \qquad z = \frac{\begin{vmatrix} a_1 & b_1 & d_1 \\ a_2 & b_2 & d_2 \\ a_3 & b_3 & d_3 \end{vmatrix}}{\begin{vmatrix} a_1 & b_1 & c_1 \\ a_2 & b_2 & c_2 \\ a_3 & b_3 & c_3 \end{vmatrix}} \tag{5-14}$$

If the determinant of the denominator is not zero, there is a unique solution to the system of equations. (If all determinants are zero, there is an unlimited number of solutions. If the determinant of the denominator is zero and any of the determinants of the numerators is not zero, the system is inconsistent, and there is no solution.)

An analysis of Eqs. (5-14) shows that the situation is precisely the same as it was when we were using determinants to solve systems of two linear equations. That is, the determinants in the denominators in the expressions for *x, y,* and *z* are the same. They consist of elements that are the coefficients of the unknowns. The determinant of the numerator of the solution for *x* is the same as that of the denominator, except that the column of *d*'s replaces the column of *a*'s. The determinant in the numerator of the solution for *y* is the same as that of the denominator, except that the column of *d*'s replaces the column of *b*'s. The determinant of the numerator of the solution for *z* is the same as the determinant of the denominator, except that the column of *d*'s replaces the column of *c*'s. To summarize:

CRAMER'S RULE

NOTE ▶

The determinant in the denominator of each is made up of the coefficients of x, y, and z. The determinants in the numerators are the same as that in the denominator, except that **the column of d's replaces the column of coefficients of the unknown for which we are solving.**

This, again, is **Cramer's rule.** Remember, the equations must be written in the standard form shown in Eqs. (5-11) before the determinants are formed.

EXAMPLE 3 Solve the following system by determinants.

$$3x + 2y - 5z = -1$$
$$2x - 3y - z = 11$$
$$5x - 2y + 7z = 9$$

Figure 5-46(a) shows the calculator display for the solutions for x and y in Example 3. Here, det[A] and det[B] are the determinants for the numerators for x and y, respectively, and det[D] is the determinant for the denominator.

Figure 5-46(b) shows the solution for z and the use of the *fraction* feature in changing the result for z into a fraction.

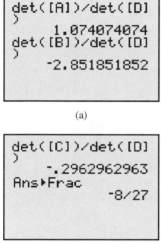

(a)

(b)

Fig. 5-46

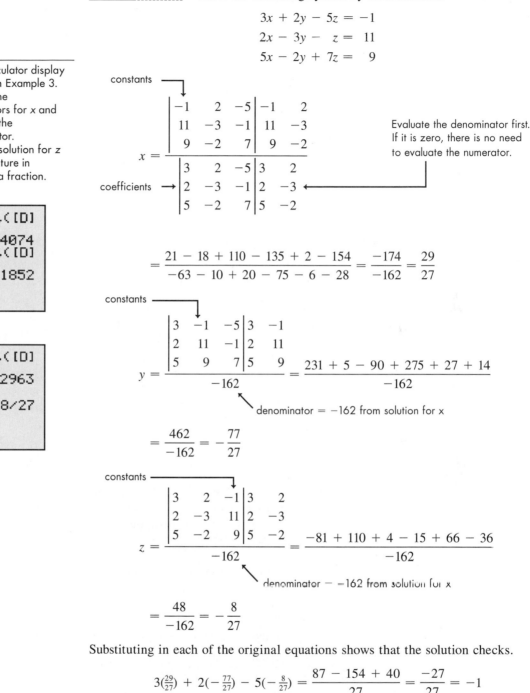

constants ⟶

Evaluate the denominator first. If it is zero, there is no need to evaluate the numerator.

$$x = \frac{\begin{vmatrix} -1 & 2 & -5 \\ 11 & -3 & -1 \\ 9 & -2 & 7 \end{vmatrix} \begin{matrix} -1 & 2 \\ 11 & -3 \\ 9 & -2 \end{matrix}}{\begin{vmatrix} 3 & 2 & -5 \\ 2 & -3 & -1 \\ 5 & -2 & 7 \end{vmatrix} \begin{matrix} 3 & 2 \\ 2 & -3 \\ 5 & -2 \end{matrix}}$$

coefficients ⟶

$$= \frac{21 - 18 + 110 - 135 + 2 - 154}{-63 - 10 + 20 - 75 - 6 - 28} = \frac{-174}{-162} = \frac{29}{27}$$

constants ⟶

$$y = \frac{\begin{vmatrix} 3 & -1 & -5 \\ 2 & 11 & -1 \\ 5 & 9 & 7 \end{vmatrix} \begin{matrix} 3 & -1 \\ 2 & 11 \\ 5 & 9 \end{matrix}}{-162} = \frac{231 + 5 - 90 + 275 + 27 + 14}{-162}$$

denominator $= -162$ from solution for x

$$= \frac{462}{-162} = -\frac{77}{27}$$

constants ⟶

$$z = \frac{\begin{vmatrix} 3 & 2 & -1 \\ 2 & -3 & 11 \\ 5 & -2 & 9 \end{vmatrix} \begin{matrix} 3 & 2 \\ 2 & -3 \\ 5 & -2 \end{matrix}}{-162} = \frac{-81 + 110 + 4 - 15 + 66 - 36}{-162}$$

denominator $= -162$ from solution for x

$$= \frac{48}{-162} = -\frac{8}{27}$$

Substituting in each of the original equations shows that the solution checks.

$$3(\tfrac{29}{27}) + 2(-\tfrac{77}{27}) - 5(-\tfrac{8}{27}) = \frac{87 - 154 + 40}{27} = \frac{-27}{27} = -1$$

$$2(\tfrac{29}{27}) - 3(-\tfrac{77}{27}) - (-\tfrac{8}{27}) = \frac{58 + 231 + 8}{27} = \frac{297}{27} = 11$$

$$5(\tfrac{29}{27}) - 2(-\tfrac{77}{27}) + 7(-\tfrac{8}{27}) = \frac{145 + 154 - 56}{27} = \frac{243}{27} = 9$$

After the values of x and y were determined, we could have evaluated z by substituting the values of x and y into one of the original equations. ∎

SOLVING A WORD PROBLEM

EXAMPLE 4   An 8.0% solution, an 11% solution, and an 18% solution of nitric acid are to be mixed to get 150 mL of a 12% solution. If the volume of acid from the 8.0% solution equals half the volume of acid from the other two solutions, how much of each is needed?

Let x = volume of 8.0% solution needed, y = volume of 11% solution needed, and z = volume of 18% solution needed.

The fact that the sum of the volumes of the three solutions is 150 mL leads to the equation $x + y + z = 150$. Since there are $0.080x$ mL of pure acid from the first solution, $0.11y$ mL from the second solution, and $0.18z$ mL from the third solution, and $0.12(150)$ mL in the final solution, we are led to the equation $0.080x + 0.11y + 0.18z = 18$. Finally, using the last stated condition, we have the equation $0.080x = 0.5(0.11y + 0.18z)$. These equations are then written in the form of Eqs. (5-11) and solved.

$$
\begin{aligned}
x + \quad y + \quad z &= 150 & &\text{sum of volumes}\\
0.080x + \quad 0.11y + \quad 0.18z &= 18 & &\text{volumes of pure acid}\\
\text{acid in 8.0\% solution} \quad 0.080x \qquad\qquad &= 0.055y + 0.090z & &\text{one-half of acid in others}
\end{aligned}
$$

$$
\begin{aligned}
x + \quad y \quad\quad z &= 150 & &\text{standard form of}\\
0.080x + \quad 0.11y + \quad 0.18z &= 18 & &\text{Eqs. (5-11) by}\\
0.080x - 0.055y - 0.090z &= 0 & &\text{rewriting third equation}
\end{aligned}
$$

$$
x = \frac{\begin{vmatrix} 150 & 1 & 1 \\ 18 & 0.11 & 0.18 \\ 0 & -0.055 & -0.090 \end{vmatrix} \begin{matrix} 150 & 1 \\ 18 & 0.11 \\ 0 & -0.055 \end{matrix}}{\begin{vmatrix} 1 & 1 & 1 \\ 0.080 & 0.11 & 0.18 \\ 0.080 & -0.055 & -0.090 \end{vmatrix} \begin{matrix} 1 & 1 \\ 0.080 & 0.11 \\ 0.080 & -0.055 \end{matrix}}
$$

$$
= \frac{-1.485 + 0 - 0.990 - 0 + 1.485 + 1.620}{-0.0099 + 0.0144 - 0.0044 - 0.0088 + 0.0099 + 0.0072} = \frac{0.630}{0.0084} = 75
$$

$$
y = \frac{\begin{vmatrix} 1 & 150 & 1 \\ 0.080 & 18 & 0.18 \\ 0.080 & 0 & -0.090 \end{vmatrix} \begin{matrix} 1 & 150 \\ 0.080 & 18 \\ 0.080 & 0 \end{matrix}}{0.0084}
$$

$$
= \frac{-1.620 + 2.160 + 0 - 1.440 - 0 + 1.080}{0.0084} = \frac{0.180}{0.0084} = 21
$$

$$
z = \frac{\begin{vmatrix} 1 & 1 & 150 \\ 0.080 & 0.11 & 18 \\ 0.080 & -0.055 & 0 \end{vmatrix} \begin{matrix} 1 & 1 \\ 0.080 & 0.11 \\ 0.080 & -0.055 \end{matrix}}{0.0084}
$$

The value of z can also be found by substituting x = 75 and y = 21 into the first equation.

$$
= \frac{0 + 1.440 - 0.660 - 1.320 + 0.990 - 0}{0.0084} = \frac{0.450}{0.0084} = 54
$$

Therefore, 75 mL of the 8.0% solution, 21 mL of the 11% solution, and 54 mL of the 18% solution are required to make the 12% solution. Results have been rounded off to two significant digits, the accuracy of the data. Checking with the statement of the problem, we see that these volumes total 150 mL.

As we previously noted, the calculations can be done completely on the calculator using either the *matrix* feature (as illustrated with Examples 2 and 3), or using the method outlined after Example 4 on page 150. Briefly, using this method for three equations is: (1) evaluate and store the value of the denominator; (2) divide the value of each numerator by the value of the denominator, storing the values of *x, y,* and *z;* (3) use the stored values of *x, y,* and *z* to check the solution.

EXERCISES 5-7

In Exercises 1–12, evaluate the given third-order determinants.

1. $\begin{vmatrix} 5 & 4 & -1 \\ -2 & -6 & 8 \\ 7 & 1 & 1 \end{vmatrix}$
2. $\begin{vmatrix} -7 & 0 & 0 \\ 2 & 4 & 5 \\ 1 & 4 & 2 \end{vmatrix}$

3. $\begin{vmatrix} 8 & 9 & -6 \\ -3 & 7 & 2 \\ 4 & -2 & 5 \end{vmatrix}$
4. $\begin{vmatrix} -2 & 4 & -1 \\ 5 & -1 & 4 \\ 4 & -8 & 2 \end{vmatrix}$

5. $\begin{vmatrix} -3 & -4 & -8 \\ 5 & -1 & 0 \\ 2 & 10 & -1 \end{vmatrix}$
6. $\begin{vmatrix} 10 & 2 & -7 \\ -2 & -3 & 6 \\ 6 & 5 & -2 \end{vmatrix}$

7. $\begin{vmatrix} 4 & -3 & -11 \\ -9 & 2 & -2 \\ 0 & 1 & -5 \end{vmatrix}$
8. $\begin{vmatrix} 9 & -2 & 0 \\ -1 & 3 & -6 \\ -4 & -6 & -2 \end{vmatrix}$

9. $\begin{vmatrix} 5 & 4 & -5 \\ -3 & 2 & -1 \\ 7 & 1 & 3 \end{vmatrix}$
10. $\begin{vmatrix} 20 & 0 & -15 \\ -4 & 30 & 1 \\ 6 & -1 & 40 \end{vmatrix}$

11. $\begin{vmatrix} 0.1 & -0.2 & 0 \\ -0.5 & 1 & 0.4 \\ -2 & 0.8 & 2 \end{vmatrix}$
12. $\begin{vmatrix} 0.2 & -0.5 & -0.4 \\ 1.2 & 0.3 & 0.2 \\ -0.5 & 0.1 & -0.4 \end{vmatrix}$

In Exercises 13–28, solve the given systems of equations by use of determinants. (Exercises 15–26 are the same as Exercises 1–12 of Section 5-6.)

13. $2x + 3y + z = 4$
$3x - z = -3$
$x - 2y + 2z = -5$

14. $4x + y + z = 2$
$2x - y - z = 4$
$3y + z = 2$

15. $x + y + z = 2$
$x - z = 1$
$x + y = 1$

16. $x + y - z = -3$
$x + z = 2$
$2x - y + 2z = 3$

17. $2x + 3y + z = 2$
$-x + 2y + 3z = -1$
$-3x - 3y + z = 0$

18. $2x + y - z = 4$
$4x - 3y - 2z = -2$
$8x - 2y - 3z = 3$

19. $5l + 6w - 3h = 6$
$4l - 7w - 2h = -3$
$3l + w - 7h = 1$

20. $3r + s - t = 2$
$r - 2s + t = 0$
$4r - s + t = 3$

21. $2x - 2y + 3z = 5$
$2x + y - 2z = -1$
$4x - y - 3z = 0$

22. $2u + 2v + 3w = 0$
$3u + v + 4w = 21$
$-u - 3v + 7w = 15$

23. $3x - 7y + 3z = 6$
$3x + 3y + 6z = 1$
$5x - 5y + 2z = 5$

24. $8x + y + z = 1$
$7x - 2y + 9z = -3$
$4x - 6y + 8z = -5$

25. $p + 2q + 2r = 0$
$2p + 6q - 3r = -1$
$4p - 3q + 6r = -8$

26. $3x + 3y + z = 6$
$2x + 2y - z = 9$
$4x + 2y - 3z = 16$

27. $3.0x + 4.5y - 7.5z = 10.5$
$4.8x - 3.6y - 2.4z = 1.2$
$4.0x - 0.5y + 2.0z = 1.5$

28. $26L - 52M - 52N = 39$
$45L + 96M + 40N = -80$
$55L + 62M - 11N = -48$

In Exercises 29–32, solve the given problems by determinants. In Exercises 31 and 32, set up appropriate systems of equations. All numbers are accurate to at least two significant digits.

29. In analyzing the forces on the bell-crank mechanism shown in Fig. 5-47, the following equations are obtained. Find the forces.

$$A \quad - 0.60F = 80$$
$$B - 0.80F = 0$$
$$6.0A \quad - 10F = 0$$

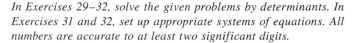

Fig. 5-47   80 N

30. Find the indicated electric currents (in A) in the circuit shown in Fig. 5-48 from the following equations.

$$i_A + i_B + i_C = 0$$
$$-8.2i_B + 10i_C = 0$$
$$4.3i_A - 8.2i_B = 6.5$$

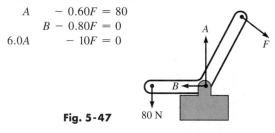

Fig. 5-48

31. A person spent 1.10 h in a car going to an airport, 1.95 h flying in a jet, and 0.520 h in a taxi to reach the final destination. The jet's speed averaged 12.0 times that of the car, which averaged 15.0 mi/h more than the taxi. What was the average speed of each if the trip covered 1140 mi?

32. An intravenous aqueous solution is made from three mixtures to get 500 mL with 6.0% of one medication, 8.0% of a second medication, and 86% water. The percents in the mixtures are, respectively, 5.0, 20, 75 (first), 0, 5.0, 95 (second), and 10, 5.0, 85 (third). How much of each is used?

CHAPTER EQUATIONS

Linear equation in one unknown	$ax + b = 0$	(5-1)
Linear equation in two unknowns	$ax + by = c$	(5-2)
System of two linear equations	$\begin{aligned} a_1x + b_1y &= c_1 \\ a_2x + b_2y &= c_2 \end{aligned}$	(5-3)
Definition of slope	$m = \dfrac{y_2 - y_1}{x_2 - x_1}$	(5-4)
Slope-intercept form	$y = mx + b$	(5-5)
Second-order determinant	$\begin{vmatrix} a_1 & b_1 \\ a_2 & b_2 \end{vmatrix} = a_1b_2 - a_2b_1$	(5-9)
Cramer's rule	$x = \dfrac{\begin{vmatrix} c_1 & b_1 \\ c_2 & b_2 \end{vmatrix}}{\begin{vmatrix} a_1 & b_1 \\ a_2 & b_2 \end{vmatrix}} \quad \text{and} \quad y = \dfrac{\begin{vmatrix} a_1 & c_1 \\ a_2 & c_2 \end{vmatrix}}{\begin{vmatrix} a_1 & b_1 \\ a_2 & b_2 \end{vmatrix}}$	(5-10)
System of three linear equations	$\begin{aligned} a_1x + b_1y + c_1z &= d_1 \\ a_2x + b_2y + c_2z &= d_2 \\ a_3x + b_3y + c_3z &= d_3 \end{aligned}$	(5-11)
Third-order determinant	$\begin{vmatrix} a_1 & b_1 & c_1 \\ a_2 & b_2 & c_2 \\ a_3 & b_3 & c_3 \end{vmatrix} = a_1b_2c_3 + a_3b_1c_2 + a_2b_3c_1 - a_3b_2c_1 - a_1b_3c_2 - a_2b_1c_3$	(5-13)
Cramer's rule	$x = \dfrac{\begin{vmatrix} d_1 & b_1 & c_1 \\ d_2 & b_2 & c_2 \\ d_3 & b_3 & c_3 \end{vmatrix}}{\begin{vmatrix} a_1 & b_1 & c_1 \\ a_2 & b_2 & c_2 \\ a_3 & b_3 & c_3 \end{vmatrix}} \quad y = \dfrac{\begin{vmatrix} a_1 & d_1 & c_1 \\ a_2 & d_2 & c_2 \\ a_3 & d_3 & c_3 \end{vmatrix}}{\begin{vmatrix} a_1 & b_1 & c_1 \\ a_2 & b_2 & c_2 \\ a_3 & b_3 & c_3 \end{vmatrix}} \quad z = \dfrac{\begin{vmatrix} a_1 & b_1 & d_1 \\ a_2 & b_2 & d_2 \\ a_3 & b_3 & d_3 \end{vmatrix}}{\begin{vmatrix} a_1 & b_1 & c_1 \\ a_2 & b_2 & c_2 \\ a_3 & b_3 & c_3 \end{vmatrix}}$	(5-14)

REVIEW EXERCISES

In Exercises 1–4, evaluate the given determinants.

1. $\begin{vmatrix} -2 & 5 \\ 3 & 1 \end{vmatrix}$ **2.** $\begin{vmatrix} 4 & 0 \\ -2 & -6 \end{vmatrix}$

3. $\begin{vmatrix} -18 & -33 \\ -21 & 44 \end{vmatrix}$ **4.** $\begin{vmatrix} 0.91 & -1.2 \\ 0.73 & -5.0 \end{vmatrix}$

In Exercises 5–8, find the slopes of the lines that pass through the given points.

5. $(2, 0), (4, -8)$ **6.** $(-1, -5), (-4, 4)$

7. $(4, -2), (-3, -4)$ **8.** $(-6, \frac{1}{2}), (1, -\frac{7}{2})$

In Exercises 9–12, find the slopes and y-intercepts of the lines with the given equations. Sketch the graphs.

9. $y = -2x + 4$ **10.** $y = \frac{2}{3}x - 3$

11. $8x - 2y = 5$ **12.** $3x = 8 + 3y$

In Exercises 13–20, solve the given systems of equations graphically.

13. $\begin{aligned} y &= 2x - 4 \\ y &= -\tfrac{3}{2}x + 3 \end{aligned}$ **14.** $\begin{aligned} y &= -3x + 3 \\ y &= 2x - 6 \end{aligned}$

15. $\begin{aligned} 4x - y &= 6 \\ 3x + 2y &= 12 \end{aligned}$ **16.** $\begin{aligned} 2x - 5y &= 10 \\ 3x + y &= 6 \end{aligned}$

17. $7x = 2y + 14$
$y = -4x + 4$

18. $5x = 15 - 3y$
$y = 6x - 12$

19. $3M + 4N = 6$
$2M - 3N = 2$

20. $5x + 2y = 5$
$2x - 4y = 3$

In Exercises 21–30, solve the given systems of equations by using an appropriate algebraic method.

21. $x + 2y = 5$
$x + 3y = 7$

22. $2x - y = 7$
$x + y = 2$

23. $4x + 3y = -4$
$y = 2x - 3$

24. $x = -3y - 2$
$-2x - 9y = 2$

25. $3i + 4v = 6$
$9i + 8v = 11$

26. $3x - 6y = 5$
$7x + 2y = 4$

27. $7x = 2y - 6$
$7y = 12 - 4x$

28. $3R = 8 - 5I$
$6I = 8R + 11$

29. $0.9x - 1.1y = 0.4$
$0.6x - 0.3y = 0.5$

30. $0.42x - 0.56y = 1.26$
$0.98x - 1.40y = -0.28$

In Exercises 31–40, solve the given systems of equations by determinants. (These are the same as for Exercises 21–30.)

31. $x + 2y = 5$
$x + 3y = 7$

32. $2x - y = 7$
$x + y = 2$

33. $4x + 3y = -4$
$y = 2x - 3$

34. $x = -3y - 2$
$-2x - 9y = 2$

35. $3i + 4v = 6$
$9i + 8v = 11$

36. $3x - 6y = 5$
$7x + 2y = 4$

37. $7x = 2y - 6$
$7y = 12 - 4x$

38. $3R = 8 - 5I$
$6I = 8R + 11$

39. $0.9x - 1.1y = 0.4$
$0.6x - 0.3y = 0.5$

40. $0.42x - 0.56y = 1.26$
$0.98x - 1.40y = -0.28$

(W) *In Exercises 41–44, explain your answers. In Exercises 41–43, choose an exercise from among Exercises 31–40 that you think is most easily solved by the indicated method.*

41. Substitution **42.** Addition or subtraction **43.** Determinants

44. Considering the slope m and the y-intercept b, explain how to determine if there is (a) a unique solution, (b) an inconsistent solution, or (c) a dependent solution.

In Exercises 45–48, evaluate the given determinants.

45. $\begin{vmatrix} 4 & -1 & 8 \\ -1 & 6 & -2 \\ 2 & 1 & -1 \end{vmatrix}$

46. $\begin{vmatrix} -5 & 0 & -5 \\ 2 & 3 & -1 \\ -3 & 2 & 2 \end{vmatrix}$

47. $\begin{vmatrix} -2.2 & -4.1 & 7.0 \\ 1.2 & 6.4 & -3.5 \\ -7.2 & 2.4 & -1.0 \end{vmatrix}$

48. $\begin{vmatrix} 30 & 22 & -12 \\ 0 & -34 & 44 \\ 35 & -41 & -27 \end{vmatrix}$

In Exercises 49–54, solve the given systems of equations algebraically. In Exercises 53 and 54, the numbers are approximate.

49. $2x + y + z = 4$
$x - 2y - z = 3$
$3x + 3y - 2z = 1$

50. $x + 2y + z = 2$
$3x - 6y + 2z = 2$
$2x - z = 8$

51. $2r + s + 2t = 8$
$3r - 2s - 4t = 5$
$-2r + 3s + 4t = -3$

52. $2u + 2v - w = -2$
$4u - 3v + 2w = -2$
$8u - 4v - 3w = 13$

53. $3.6x + 5.2y - z = -2.2$
$3.2x - 4.8y + 3.9z = 8.1$
$6.4x + 4.1y + 2.3z = 5.1$

54. $32t + 24u + 63v = 32$
$42t - 31u + 19v = 132$
$48t + 12u + 11v = 0$

In Exercises 55–60, solve the given systems of equations by determinants. In Exercises 59 and 60, the numbers are approximate. (These systems are the same as for Exercises 49–54.)

55. $2x + y + z = 4$
$x - 2y - z = 3$
$3x + 3y - 2z = 1$

56. $x + 2y + z = 2$
$3x - 6y + 2z = 2$
$2x - z = 8$

57. $2r + s + 2t = 8$
$3r - 2s - 4t = 5$
$-2r + 3s + 4t = -3$

58. $2u + 2v - w = -2$
$4u - 3v + 2w = -2$
$8u - 4v - 3w = 13$

59. $3.6x + 5.2y - z = -2.2$
$3.2x - 4.8y + 3.9z = 8.1$
$6.4x + 4.1y + 2.3z = 5.1$

60. $32t + 24u + 63v = 32$
$42t - 31u + 19v = 132$
$48t + 12u + 11v = 0$

In Exercises 61–64, solve for x. Here we see that we can solve an equation in which the unknown is an element of a determinant.

61. $\begin{vmatrix} 2 & 5 \\ 1 & x \end{vmatrix} = 3$

62. $\begin{vmatrix} -1 & x \\ 3 & 4 \end{vmatrix} = 7$

63. $\begin{vmatrix} x & 1 & 2 \\ 0 & -1 & 3 \\ -2 & 2 & 1 \end{vmatrix} = 5$

64. $\begin{vmatrix} 1 & 2 & -1 \\ -2 & 3 & x \\ -1 & 2 & -2 \end{vmatrix} = -3$

In Exercises 65–68, let $1/x = u$ and $1/y = v$. Solve for u and v and then solve for x and y. In this way we see how to solve systems of equations involving reciprocals.

65. $\dfrac{1}{x} - \dfrac{1}{y} = \dfrac{1}{2}$
$\dfrac{1}{x} + \dfrac{1}{y} = \dfrac{1}{4}$

66. $\dfrac{1}{x} + \dfrac{1}{y} = 3$
$\dfrac{2}{x} + \dfrac{1}{y} = 1$

67. $\dfrac{2}{x} + \dfrac{3}{y} = 3$
$\dfrac{5}{x} - \dfrac{6}{y} = 3$

68. $\dfrac{3}{x} - \dfrac{2}{y} = 4$
$\dfrac{2}{x} + \dfrac{4}{y} = 1$

In Exercises 69 and 70, determine the value of k that makes the system dependent. In Exercises 71 and 72, determine the value of k that makes the system inconsistent.

69. $3x - ky = 6$
$x + 2y = 2$

70. $5x + 20y = 15$
$2x + ky = 6$

71. $kx - 2y = 5$
$4x + 6y = 1$

72. $2x - 5y = 7$
$kx + 10y = 2$

In Exercises 73 and 74, solve the given systems of equations by any appropriate method. All numbers in 73 are accurate to two significant digits, and in 74 they are accurate to three significant digits.

73. A 20-ft crane arm with a supporting cable and with a 100-lb box suspended from its end has forces acting on it as shown in Fig. 5-49. Find the forces (in lb) from the following equations.

$$F_1 + 2.0F_2 = 280$$
$$0.87F_1 - F_3 = 0$$
$$3.0F_1 - 4.0F_2 = 600$$

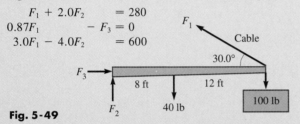

Fig. 5-49

74. In applying Kirchhoff's laws (see Exercise 20 of Section 5-6) to the electric circuit shown in Fig. 5-50, the following equations result. Find the indicated currents (in A).

$$i_1 + i_2 + i_3 = 0$$
$$5.20i_1 - 3.25i_2 = 8.33 - 6.45$$
$$3.25i_2 - 2.62i_3 = 6.45 - 9.80$$

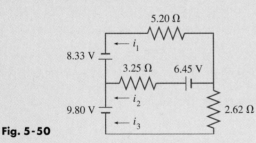

Fig. 5-50

In Exercises 75–84, set up systems of equations and solve by any appropriate method. All numbers are accurate to at least two significant digits.

75. A computer analysis showed that the temperature T of the ocean water within 1000 m of a nuclear-plant discharge pipe was given by $T = \dfrac{a}{x + 100} + b$, where x is the distance from the pipe and a and b are constants. If $T = 14°C$ for $x = 0$ and $T = 10°C$ for $x = 900$ m, find a and b.

(W) 76. In measuring the angles of a triangular parcel, a surveyor noted that one of the angles equaled the sum of the other two angles. What conclusion can be drawn?

77. A satellite is to be launched from a space shuttle. It is calculated that the satellite's speed will be 24,200 km/h if launched directly ahead of the shuttle, or 21,400 km/h if launched directly to the rear of the shuttle. What is the speed of the shuttle and the launching speed of the satellite relative to the shuttle? See Fig. 5-51.

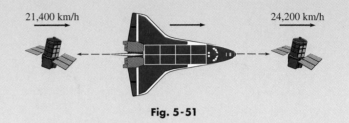

Fig. 5-51

78. The velocity v of sound is a function of the temperature T according to the function $v = aT + b$, where a and b are constants. If $v = 337.5$ m/s for $T = 10.0°C$ and $v = 346.6$ m/s for $T = 25.0°C$, find v as a function of T.

79. The power (in W) dissipated in an electric resistance (in Ω) equals the resistance times the square of the current (in A). If 1.0 A flows through resistance R_1 and 3.0 A flows through resistance R_2, the total power dissipated is 14.0 W. If 3.0 A flows through R_1 and 1.0 A flows through R_2, the total power dissipated is 6.0 W. Find R_1 and R_2.

80. Twelve equal rectangular ceiling panels are placed as shown in Fig. 5-52. If each panel is 6.0 in. longer than it is wide and a total of 132.0 ft of edge and middle strips is used, what are the dimensions of the room?

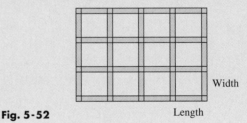

Fig. 5-52

81. The weight of a lever may be considered to be at its center. A 20-ft lever of weight w is balanced on a fulcrum 8 ft from one end by a load L at that end. If a load of $4L$ is placed at that end, it requires a 20-lb weight at the other end to balance the lever. What is the load initial L and the weight w of the lever? (See Exercise 38 of Section 5-4.) See Fig. 5-53.

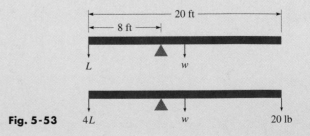

Fig. 5-53

82. Two fuel mixtures, one of 2.0% oil and 98.0% gasoline and another of 8.0% oil and 92.0% gasoline, are to be used to make 10.0 L of a fuel that is 4.0% oil and 96.0% gasoline for use in a chain saw. How much of each mixture is needed?

83. Three computer programs *A, B,* and *C,* require a total of 140 MB (megabytes) of hard-disk memory. If three other programs, two requiring the same memory as *B* and one the same as *C,* are added to a disk with *A, B,* and *C,* a total of 236 MB are required. If three other programs, one requiring the same memory as *A* and two the same memory as *C,* are added to a disk with *A, B,* and *C,* a total of 304 MB are required. How much memory is required for each of *A, B,* and *C?*

84. One ampere of electric current is passed through a solution of sulfuric acid, silver nitrate, and cupric sulfate, releasing hydrogen gas, silver, and copper. A total mass of 1.750 g is released. The mass of silver deposited is 3.40 times the mass of copper deposited, and the mass of copper and 70.0 times the mass of hydrogen combined equals the mass of silver deposited less 0.037 g. How much of each is released?

Writing Exercise

85. Write one or two paragraphs giving reasons for choosing a particular method of solving the following problem. If a first pump is used for 2.2 h and a second pump is used for 2.7 h, 1100 ft^3 can be removed from a wastewater holding tank. If the first pump is used for 1.4 h and the second for 2.5 h, 840 ft^3 can be removed. How much can each pump remove in 1.0 h? (What is the result to two significant digits?)

PRACTICE TEST

1. Find the slope of the line through $(2, -5)$ and $(-1, 4)$.

2. Solve by substitution:
$$x + 2y = 5$$
$$4y = 3 - 2x$$

3. Solve by determinants:
$$3x - 2y = 4$$
$$2x + 5y = -1$$

4. By finding the slope and *y*-intercept, sketch the graph of $2x + y = 4$

5. The perimeter of a rectangular ranch is 24 km, and the length is 6.0 km more than the width. Set up equations relating the length *l* and the width *w* and then solve for *l* and *w*.

6. Solve the following system of equations graphically. Determine the values of *x* and *y* to the nearest 0.1.
$$2x - 3y = 6$$
$$4x + y = 4$$

7. By volume, one alloy is 60% copper, 30% zinc, and 10% nickel. A second alloy has percents of 50, 30, and 20, respectively. A third alloy is 30% copper and 70% nickel. How much of each metal is needed to make 100 cm^3 of a resulting alloy with percents of 40, 15, and 45, respectively?

8. Solve for *y* by determinants:
$$3x + 2y - z = 4$$
$$2x - y + 3z = -2$$
$$x \qquad + 4z = 5$$

6

FACTORING AND FRACTIONS

The basic algebraic operations we introduced and reviewed in Chapter 1 have been sufficient for use in developing other topics to this point. However, material that we shall encounter throughout the rest of the book will require additional algebraic methods beyond those we have already developed. Therefore, in this chapter we consider additional algebraic operations with products, quotients, and fractions. In later chapters these operations in turn will allow us to develop other topics that have scientific and technical applications.

Although the primary purpose of this chapter is to introduce and develop additional algebraic methods, there are many technical applications that use these operations. The important applications in optics are indicated by the picture of the Hubble space telescope at the left. Other areas of applications include electronics and mechanical design.

Great advances in our knowledge of the universe have been made through the use of the Hubble space telescope (shown above) since the mid-1990s. In Section 6-8 we illustrate the use of algebraic operations in the study of planetary motion and in the design of lenses and reflectors used in telescopes.

6-1 SPECIAL PRODUCTS

In working with algebraic expressions, we use certain types of products so often that we should be very familiar with them. We show these here in general form.

$$a(x + y) = ax + ay \qquad \text{(6-1)}$$
$$(x + y)(x - y) = x^2 - y^2 \qquad \text{(6-2)}$$
$$(x + y)^2 = x^2 + 2xy + y^2 \qquad \text{(6-3)}$$
$$(x - y)^2 = x^2 - 2xy + y^2 \qquad \text{(6-4)}$$
$$(x + a)(x + b) = x^2 + (a + b)x + ab \qquad \text{(6-5)}$$
$$(ax + b)(cx + d) = acx^2 + (ad + bc)x + bd \qquad \text{(6-6)}$$

We recognize Eq. (6-1) as the very important *distributive law*. Equations (6-2) to (6-6) are found by using the distributive law along with the associative and commutative laws and the operations on positive and negative numbers that were developed in Chapter 1.

Equations (6-1) to (6-6) are identities. (See page 35.) Each is valid for all sets of values of the literal numbers.

Equations (6-1) to (6-6) are important **special products** *and should be known* **thoroughly.** When these products are quickly recognized, they allow us to perform many multiplications easily, often by inspection. In using them we must realize that any of the literal numbers may represent an expression that in turn represents a number.

EXAMPLE 1 (a) Using Eq. (6-1) in the following product, we have

$$6(3r + 2s) = 6(3r) + 6(2s) = 18r + 12s$$

(b) Using Eq. (6-2), we have

DIFFERENCE OF SQUARES

$$(3r + 2s)(3r - 2s) = (3r)^2 - (2s)^2 = 9r^2 - 4s^2$$

sum of 3r and 2s   difference of 3r and 2s   difference of squares

Using Eq. (6-1) in (a), we have $a = 6$. In (a) and (b), $3r = x$ and $2s = y$. ∎

EXAMPLE 2 Using Eqs. (6-3) and (6-4) in the following products, we have

(a) $(5a + 2)^2 = (5a)^2 + 2(5a)(2) + 2^2 = 25a^2 + 20a + 4$

square   twice product   square

(b) $(5a - 2)^2 = (5a^2) - 2(5a)(2) + 2^2 = 25a^2 - 20a + 4$

In these illustrations we let $x = 5a$ and $y = 2$. *It should be emphasized that*

CAUTION ▶ $(5a + 2)^2$ is **not** $(5a)^2 + 2^2$, or $25a^2 + 4$

We must be careful to follow the form of Eqs. (6-3) and (6-4) properly and include the middle term, $20a$. (See Example 5 of Section 1-8 on page 30.) ∎

EXAMPLE 3 Using Eqs. (6-5) and (6-6) in the following products, we have

(a) $(x + 5)(x - 3) = x^2 + [5 + (-3)]x + (5)(-3) = x^2 + 2x - 15$

(b) $(4x + 5)(2x - 3) = (4x)(2x) + [(4)(-3) + (5)(2)]x + (5)(-3)$
$$= 8x^2 - 2x - 15$$ ∎

Generally, when we use these special products, we find the middle term mentally and write down the result directly, as shown in the next example.

EXAMPLE 4

(a) $(y - 5)(y + 5) = y^2 - 25$ no middle term—Eq. (6-2)

(b) $(3x - 2)^2 = 9x^2 - 12x + 4$ middle term $= 2(3x)(-2)$—Eq. (6-4)

(c) $(x^2 - 4)(x^2 + 7) = x^4 + 3x^2 - 28$ middle term $= (7 - 4)x^2$—Eq. (6-5)
$(x^2)^2 = x^4$ ∎

At times it is necessary to use more than one of the special products to simplify an algebraic expression. When this happens, it may be necessary to write down an intermediate step. This is illustrated in the following examples.

EXAMPLE 5   (a) When analyzing the forces on a certain type of beam, the expression $Fa(L - a)(L + a)$ occurs. In expanding this expression, we first multiply $L - a$ by $L + a$ by use of Eq. (6-2). The expansion is completed by using Eq. (6-1), the distributive law.

$$Fa(L - a)(L + a) = Fa(L^2 - a^2) = FaL^2 - Fa^3$$

(b) The electrical power delivered to the resistor R in Fig. (6-1) is $R(i_1 + i_2)^2$. Here, i_1 and i_2 are electric currents. To expand this expression we first perform the square by use of Eq. (6-3) and then complete the expansion by use of Eq. (6-1).

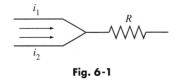

Fig. 6-1

$$R(i_1 + i_2)^2 = R(i_1^2 + 2i_1i_2 + i_2^2) = Ri_1^2 + 2Ri_1i_2 + Ri_2^2$$

EXAMPLE 6   In determining the product $(x + y - 2)^2$, we may group the quantity $(x + y)$ in an intermediate step. This leads to

$$(x + y - 2)^2 = [(x + y) - 2]^2 = (x + y)^2 - 2(x + y)(2) + 2^2$$
$$= x^2 + 2xy + y^2 - 4x - 4y + 4$$

In this example we used Eqs. (6-3) and (6-4).

Special Products Involving Cubes

There are four other special products that occur less frequently. However, they are sufficiently important that they should be readily recognized. They are shown in Eqs. (6-7) to (6-10).

$$(x + y)^3 = x^3 + 3x^2y + 3xy^2 + y^3 \tag{6-7}$$
$$(x - y)^3 = x^3 - 3x^2y + 3xy^2 - y^3 \tag{6-8}$$
$$(x + y)(x^2 - xy + y^2) = x^3 + y^3 \tag{6-9}$$
$$(x - y)(x^2 + xy + y^2) = x^3 - y^3 \tag{6-10}$$

The following examples illustrate the use of Eqs. (6-7) to (6-10).

EXAMPLE 7   (a) $(x + 4)^3 = x^3 + 3(x^2)(4) + 3(x)(4^2) + 4^3$ Eq. (6-7)
$$= x^3 + 12x^2 + 48x + 64$$

(b) $(2x - 5)^3 = (2x)^3 - 3(2x)^2(5) + 3(2x)(5^2) - 5^3$ Eq. (6-8)
$$= 8x^3 - 60x^2 + 150x - 125$$

EXAMPLE 8 (a) $(x + 3)(x^2 - 3x + 9) = x^3 + 3^3$ Eq. (6-9)

$$= x^3 + 27$$

(b) $(x - 2)(x^2 + 2x + 4) = x^3 - 2^3$ Eq. (6-10)

$$= x^3 - 8$$

EXERCISES 6-1

In Exercises 1–36, find the indicated products directly by inspection. It should not be necessary to write down intermediate steps (except possibly when using Eq. (6-6)).

1. $40(x - y)$
2. $2x(a - 3)$
3. $2x^2(x - 4)$
4. $3a^2(2a + 7)$
5. $(y + 6)(y - 6)$
6. $(s + 2t)(s - 2t)$
7. $(3v - 2)(3v + 2)$
8. $(ab - c)(ab + c)$
9. $(4x - 5y)(4x + 5y)$
10. $(7s + 2t)(7s - 2t)$
11. $(12 + 5ab)(12 - 5ab)$
12. $(2xy - 11)(2xy + 11)$
13. $(5f + 4)^2$
14. $(i_1 + 3)^2$
15. $(2x + 7)^2$
16. $(5a + 2b)^2$
17. $(L^2 - 1)^2$
18. $(b^2 - 6)^2$
19. $(4a + 7xy)^2$
20. $(3x + 10y)^2$
21. $(4x - 2y)^2$
22. $(a - 5p)^2$
23. $(6s - t)^2$
24. $(3p - 4q)^2$
25. $(x + 1)(x + 5)$
26. $(y - 8)(y + 5)$
27. $(3 + C^2)(6 + C^2)$
28. $(1 - e^3)(7 - e^3)$
29. $(3x - 1)(2x + 5)$
30. $(2x - 7)(2x + 1)$
31. $(4x - 5)(5x + 1)$
32. $(2y - 1)(3y - 1)$
33. $(5v - 3)(4v + 5)$
34. $(7s + 6)(2s + 5)$
35. $(3x + 7y)(2x - 9y)$
36. $(8x - y)(3x + 4y)$

Use the special products of this section to determine the products of Exercises 37–60. You may need to write down one or two intermediate steps.

37. $2(x - 2)(x + 2)$
38. $5(n - 5)(n + 5)$
39. $2a(2a - 1)(2a + 1)$
40. $4c(2c - 3)(2c + 3)$
41. $6a(x + 2b)^2$
42. $7r(5r + 2b)^2$
43. $5n^2(2n + 5)^2$
44. $8p(p - 7)^2$
45. $[(2R + 3r)(2R - 3r)]^2$
46. $[(6t - y)(6t + y)]^2$
47. $(x + y + 1)^2$
48. $(x + 2 + 3y)^2$
49. $(3 - x - y)^2$
50. $2(x - y + 1)^2$
51. $(5 - t)^3$
52. $(2s + 3)^3$
53. $(2x + 5t)^3$
54. $(x - 5y)^3$
55. $(w + h - 1)(w + h + 1)$
56. $(2a - c + 2)(2a - c - 2)$

57. $(x + 2)(x^2 - 2x + 4)$
58. $(s - 3)(s^2 + 3s + 9)$
59. $(4 - 3x)(16 + 12x + 9x^2)$
60. $(2x + 3a)(4x^2 - 6ax + 9a^2)$

Use the special products of this section to determine the products in Exercises 61–68. Each comes from the technical area indicated.

61. $P_1(P_0c + G)$ (computers)
62. $h^2L(L + 1)$ (spectroscopy)
63. $4(p + DA)^2$ (photography)
64. $(2J + 3)(2J - 1)$ (lasers)
65. $\frac{1}{2}\pi(R + r)(R - r)$ (architecture)
66. $w(1 - h)(4 - h^2)$ (hydrodynamics)
67. $\frac{L}{6}(x - a)^3$ (mechanics: beams)
68. $(1 - z)^2(1 + z)$ (motion: gyroscope)

In Exercises 69–72, solve the given problems.

69. The length of a piece of rectangular floor tile is 3 cm more than twice the side x of a second square piece of tile. The width of the rectangular piece is 3 cm less than twice the side of the square piece. Find the area of the rectangular piece in terms of x (in expanded form).

70. The radius of a circular oil spill is r. It then increases in radius by 40 m before being contained. Find the area of the oil spill at the time it is contained in terms of r (in expanded form). See Fig. 6-2.

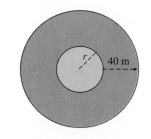

Fig. 6-2

71. Verify Eq. (6-8) by multiplication. Then show by substitution that the same value is obtained for each side for the values $x = -2$ and $y = 3$.

(W) 72. Verify Eqs. (6-9) and (6-10) by multiplication. Then explain why we may refer to Eqs. (6-1) to (6-10) as *identities*.

$6\text{-}2$ FACTORING: COMMON FACTOR AND DIFFERENCE OF SQUARES

At times we want to determine which expressions can be multiplied together to equal a given algebraic expression. We know from Section 1-7 that when an algebraic expression is the product of two or more quantities, each of these quantities is a *factor* of the expression. Therefore, *determining these factors, which is essentially reversing the process of finding a product, is called* **factoring.**

In our work on factoring we shall consider only the factoring of polynomials (see Section 1-9) that have integers as coefficients for all terms. Also, all factors will have integral coefficients. *A polynomial or a factor is called* **prime** *if it contains no factors other than +1 or −1 and plus or minus itself.* Also, we say that *an expression is* **factored completely** *if it is expressed as a product of its prime factors.*

EXAMPLE 1 When we factor the expression $12x + 6x^2$ as

$$12x + 6x^2 = 2(6x + 3x^2)$$

we see that it has not been factored completely. The factor $6x + 3x^2$ is not prime, for it may be factored as

$$6x + 3x^2 = 3x(2 + x)$$

Therefore, the expression $12x + 6x^2$ is factored completely as

$$12x + 6x^2 = 6x(2 + x)$$

The factors x and $2 + x$ are prime. We can factor the numerical coefficient, 6, as $2(3)$, but it is standard not to write numerical coefficients in factored form. ∎

NOTE▶ To factor expressions easily, we must be familiar with algebraic multiplication, particularly the special products of the preceding section. *The solution of factoring problems is heavily dependent on the recognition of special products.* The special products also provide methods of checking answers and deciding whether or not a given factor is prime.

Common Monomial Factors

Often an expression contains a monomial that is common to each term of the expression. Therefore, *the first step in factoring any expression should be to factor out any* **common monomial factor** *that may exist.* To do this, we note the common factor by inspection and then use the reverse of the distributive law, Eq. (6-1), to show the factored form. The following examples illustrate factoring a common monomial factor out of an expression.

For reference, Eq. (6-1) is
$a(x + y) = ax + ay.$

EXAMPLE 2 In factoring $6x - 2y$, we note each term contains a factor of 2.

$$6x - 2y = 2(3x) - 2y = 2(3x - y)$$

Here, 2 is the common monomial factor, and $2(3x - y)$ is the required factored form of $6x - 2y$. Once the common factor has been identified, it is not actually necessary to write a term like $6x$ as $2(3x)$. The result can be written directly.

NOTE▶ *We check our result by multiplication.* Here, $2(3x - y) = 6x - 2y$, which is the original expression. ∎

In Example 2 we determined the common factor of 2 by inspection. This is normally the way in which a common factor is found. Once the common factor has been found, the other factor can be determined by dividing the original expression by the common factor.

EXAMPLE 3 Factor: $4ax^2 + 2ax$.

The numerical factor 2 and the literal factors a and x are common to each term. Therefore, the common monomial factor of $4ax^2 + 2ax$ is $2ax$. This means that

$$4ax^2 + 2ax = 2ax(2x + 1)$$

Note the presence of the 1 in the factored form. When we divide $4ax^2 + 2ax$ by $2ax$, we get

$$\frac{4ax^2 + 2ax}{2ax} = \frac{4ax^2}{2ax} + \frac{2ax}{2ax} = 2x + 1$$

Although it is a common error to omit the 1,

CAUTION ▶   *we must include the 1 in the factor 2x + 1.*

Without the 1, when the factored form is multiplied out, we would not obtain the proper expression.

Usually, the division shown in this example is done by inspection. However, we show it here to emphasize the actual operation that is being performed when we factor out a common factor.

EXAMPLE 4 Factor: $6a^5x^2 - 9a^3x^3 + 3a^3x^2$.

After inspecting each term, we determine that each contains a factor of 3, a^3, and x^2. Thus, the common monomial factor is $3a^3x^2$. This means that

$$6a^5x^2 - 9a^3x^3 + 3a^3x^2 = 3a^3x^2(2a^2 - 3x + 1)$$

In these examples, we note that factoring an expression does not actually change the expression, although it does change the *form* of the expression. In equating the expression to its factored form, we write an *identity*.

It is often necessary to use factoring when solving an equation. This is illustrated in the following example.

FM radio was developed in the early 1930s.

EXAMPLE 5 An equation used in the analysis of FM reception is $R_F = \alpha(2R_A + R_F)$. Solve for R_F.

The steps in the solution are as follows:

$$R_F = \alpha(2R_A + R_F) \qquad \text{original equation}$$
$$R_F = 2\alpha R_A + \alpha R_F \qquad \text{use distributive law}$$
$$R_F - \alpha R_F = 2\alpha R_A \qquad \text{subtract } \alpha R_F \text{ from both sides}$$
$$R_F(1 - \alpha) = 2\alpha R_A \qquad \text{factor out } R_F \text{ on left}$$
$$R_F = \frac{2\alpha R_A}{1 - \alpha} \qquad \text{divide both sides by } 1 - \alpha$$

We see that we collected both terms containing R_F on the left in order that we could factor and thereby solve for R_F.

Factoring the Difference of Two Squares

For reference, Eq. (6-2) is $(x + y)(x - y) = x^2 - y^2$.

Another important form for factoring is based on the special product of Eq. (6-2). In Eq. (6-2) we see that the product of the sum and difference of two numbers results in the difference between the squares of the two numbers. Therefore, *factoring the difference of two squares gives factors that are the sum and the difference of the numbers.*

EXAMPLE 6   In factoring $x^2 - 16$, we note that x^2 is the square of x and that 16 is the square of 4. Therefore,

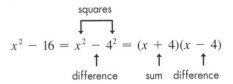

Usually in factoring an expression of this type, we do not actually write out the middle step as shown.

EXAMPLE 7   (a) Since $4x^2$ is the square of $2x$ and 9 is the square of 3, we may factor $4x^2 - 9$ as

$$4x^2 - 9 = (2x)^2 - 3^2 = (2x + 3)(2x - 3)$$

(b) In the same way,

$$(y - 3)^2 - 16x^4 = (y - 3 + 4x^2)(y - 3 - 4x^2)$$

where we note that $16x^4 = (4x^2)^2$.

Complete Factoring

NOTE▶

As indicated previously, *if it is possible to factor out a common monomial factor, this factoring should be done first.* We should then inspect the resulting factors to see if more factoring can be done. It is possible, for example, that the resulting factor is a difference of squares. Thus, complete factoring often requires more than one step. Be sure to include all prime factors in writing the result.

EXAMPLE 8   (a) In factoring $20x^2 - 45$, we note a common factor of 5 in each term. Therefore, $20x^2 - 45 = 5(4x^2 - 9)$. However, the factor $4x^2 - 9$ itself is the difference of squares. Therefore, $20x^2 - 45$ is completely factored as

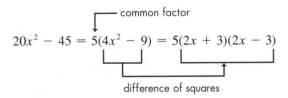

(b) In factoring $x^4 - y^4$, we note that we have the difference of two squares. Therefore, $x^4 - y^4 = (x^2 + y^2)(x^2 - y^2)$. However, the factor $x^2 - y^2$ is also the difference of squares. This means that

$$x^4 - y^4 = (x^2 + y^2)(x^2 - y^2) = (x^2 + y^2)(x + y)(x - y)$$

CAUTION▶   *The factor $x^2 + y^2$ is prime.* It is **not** equal to $(x + y)^2$. (See Example 2 of Section 6-1.)

Factoring by Grouping

The terms in a polynomial can sometimes be grouped such that the polynomial can then be factored by the methods of this section. The following example illustrates this method of *factoring by grouping*.

EXAMPLE 9   Factor: $2x - 2y + ax - ay$.

We see that there is no common factor to all four terms, but that each of the first two terms contains a factor of 2, and each of the third and fourth terms contains a factor of a. Grouping terms this way and then factoring each group, we have

$$2x - 2y + ax - ay = (2x - 2y) + (ax - ay)$$
$$= 2(x - y) + a(x - y)$$

NOTE ▶   Each of the two terms we now have contains a ***common binomial factor*** of $x - y$. Since this is a common factor, we have

$$2(x - y) + a(x - y) = (x - y)(2 + a)$$

This means that

$$2x - 2y + ax - ay = (x - y)(2 + a)$$

where the expression on the right is the factored form of the polynomial. ■

The general method of factoring by grouping can be used with several types of groupings. We will discuss another type in the following section.

EXERCISES 6-2

In Exercises 1–40, factor the given expressions completely.

1. $6x + 6y$ **2.** $3a - 3b$ **3.** $5a - 5$

4. $2x^2 + 2$ **5.** $3x^2 - 9x$ **6.** $4s^2 + 20s$

7. $7b^2h - 28b$ **8.** $5a^2 - 20ax$

9. $12n^2 + 6n$ **10.** $18p^3 - 3p^2$

11. $2x + 4y - 8z$ **12.** $10a - 5b + 15c$

13. $3ab^2 - 6ab + 12ab^3$ **14.** $4pq - 14q^2 - 16pq^2$

15. $12pq^2 - 8pq - 28pq^3$ **16.** $27u^2b - 24ab - 9a$

17. $2a^2 - 2b^2 + 4c^2 - 6d^2$ **18.** $5a + 10ax - 5ay + 20az$

19. $x^2 - 4$ **20.** $r^2 - 25$ **21.** $100 - 9A^2$

22. $49 - Z^4$ **23.** $36a^4 - 1$ **24.** $81z^2 - 1$

25. $81s^2 - 25t^2$ **26.** $36s^2 - 121t^2$

27. $144n^2 - 169p^4$ **28.** $36a^2b^2 - 169c^2$

29. $(x + y)^2 - 9$ **30.** $(a - b)^2 - 1$

31. $2x^2 - 8$ **32.** $5a^2 - 125$ **33.** $3x^2 - 27z^2$

34. $4x^2 - 100y^2$ **35.** $2(I - 3)^2 - 8$

36. $a(x + 2)^2 - ay^2$ **37.** $x^4 - 16$

38. $y^4 - 81$ **39.** $x^8 - 1$ **40.** $2x^4 - 8y^4$

In Exercises 41–44, solve for the indicated letter.

41. $2a - b = ab + 3$, for a **42.** $n(x + 1) = 5 - x$, for x

43. $3 - 2s = 2(3 - st)$, for s

44. $k(2 - y) = y(2k - 1)$, for y

In Exercises 45–52, factor the given expressions by grouping as illustrated in Example 9.

45. $3x - 3y + bx - by$ **46.** $am + an + cn + cm$

47. $a^2 + ax - ab - bx$ **48.** $2y - y^2 - 6y^4 + 12y^3$

49. $x^3 + 3x^2 - 4x - 12$ **50.** $S^3 - 5S^2 - S + 5$

51. $x^2 - y^2 + x - y$ **52.** $4p^2 - q^2 + 2p + q$

In Exercises 53–56, factor the given expressions. In Exercises 57–60, solve for the indicated letter. Each comes from the technical area indicated.

53. $2\pi rh + 2\pi r^2$ (container surface area)

54. $4d^2D^2 - 4d^3D - d^4$ (machine design)

55. $aD_1^2 - aD_2^2$ (surveying)

56. $rR^2 - r^3$ (pipeline flow)

57. $i_3 = (1 + a)i_1 - ai_2$, for a (electricity)

58. $nV + n_1v = n_1V$, for n_1 (acoustics)

59. $ER = AtT_0 - AtT_1$, for t
(energy conservation)

60. $R = kT_2^4 - kT_1^4$ (Solve for k and factor the resulting denominator.)   (energy: radiation)

$6\text{-}3$ FACTORING TRINOMIALS

In the previous section we introduced the concept of factoring and considered factoring based on special products of Eqs. (6-1) and (6-2). We now note that the special products formed from Eqs. (6-3) to (6-6) all result in trinomial (three-term) polynomials. Thus, trinomials of the types formed by these products are important expressions to be factored, and this section is devoted to them.

When factoring an expression based on Eq. (6-5), we start with the expression on the right and then find the factors that are at the left. Therefore, by writing Eq. (6-5) with sides reversed, we have

---
For reference, Eq. (6-5) is
$(x + a)(x + b) = x^2 + (a + b)x + ab$.

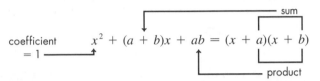

$$\text{coefficient} = 1 \qquad x^2 + (a + b)x + ab = (x + a)(x + b)$$

Factoring a trinomial for which the coefficient of x^2 is 1.

We are to find integers a and b, and they are found by noting that

1. *the coefficient of x^2 is 1,*

2. *the final constant is the product of the constants a and b in the factors, and*

CAUTION ▶

3. *the coefficient of x is the sum of a and b.*

As in Section 6-2 we shall consider only factors in which all terms have integral coefficients.

▌EXAMPLE 1   In factoring $x^2 + 3x + 2$, we set it up as

$$x^2 + \underset{\text{sum}}{\overset{\uparrow}{3x}} + \underset{\text{product}}{\overset{\uparrow}{2}} = (x\boxed{}) \qquad (x\boxed{})$$
$$\text{integers}$$

The constant 2 tells us that the product of the required integers is 2. Thus, the only possibilities are 2 and 1 (or 1 and 2). The plus sign before the 2 indicates that the sign before the 1 and 2 in the factors must be the same, either plus or minus. Since the coefficient of x, 3, is the sum of the integers, the plus sign before the 3 tells us that both signs are positive. Therefore,

$$x^2 + 3x + 2 = (x + 2)(x + 1)$$

In factoring $x^2 - 3x + 2$, the analysis is the same until we note that the middle term is negative. This tells us that both integers are negative in this case. Therefore,

$$x^2 - 3x + 2 = (x - 2)(x - 1)$$

For a trinomial containing x^2 and 2 to be factorable, the middle term must be $+3x$ or $-3x$. No other combination of integers gives the proper middle term. Therefore, the expression

$$x^2 + 4x + 2$$

cannot be factored. The integers would have to be 2 and 1, but the middle term would not be $4x$.   --------▐

EXAMPLE 2   **(a)** In order to factor $x^2 + 7x - 8$, *we must find two integers whose product is −8 and whose sum is +7.* The possible factors of −8 are

$$-8 \text{ and } +1 \qquad 8 \text{ and } -1 \qquad -4 \text{ and } +2 \qquad +4 \text{ and } -2$$

Inspecting these, we see that only +8 and −1 have the sum of +7. Therefore,

$$x^2 + 7x - 8 = (x + 8)(x - 1)$$

In choosing the correct values for the integers, it is usually fairly easy to find a pair for which the product is the final term. However, choosing the pair of integers that correctly fits the middle term is the step that often is not done properly. *Special attention must be given to choosing the integers so that the expansion of the resulting factors has the correct* **middle term** *of the original expression.*
   (b) In the same way, we have

$$x^2 - x - 12 = (x - 4)(x + 3)$$

since −4 and +3 is the only pair of integers whose product is −12 and whose sum is −1.
   (c) Also,

$$x^2 - 5xy + 6y^2 = (x - 3y)(x - 2y)$$

since −3 and −2 is the only pair of integers whose product is +6 and whose sum is −5. Here we find second terms of each factor with a product of $6y^2$ and sum of −5xy, which means that each second term must have a factor of y, as we have shown above.   ∎

For reference, Eqs. (6-3) and (6-4) are
$(x + y)^2 = x^2 + 2xy + y^2$
$(x - y)^2 = x^2 - 2xy + y^2$

In factoring a trinomial in which the first and third terms are perfect squares, we may find that the expression fits the form of Eq. (6-3) or Eq. (6-4), as well as the form of Eq. (6-5). The following example illustrates this case.

EXAMPLE 3   **(a)** To factor $x^2 + 10x + 25$, we must find two integers whose product is +25 and whose sum is +10. Since $5^2 = 25$, we note that this expression may fit the form of Eq. (6-3). This can be the case only if the first and third terms are perfect squares. Since the sum of +5 and +5 is +10, we have

$$x^2 + 10x + 25 = (x + 5)(x + 5)$$

or

$$x^2 + 10x + 25 = (x + 5)^2$$

(b) To factor $A^4 - 20A^2 + 100$, we note that $A^4 = (A^2)^2$ and $100 = 10^2$. This means that this expression may fit the form of Eq. (6-4). Since the sum of −10 and −10 is −20, we have

$$A^4 - 20A^2 + 100 = (A^2 - 10)(A^2 - 10)$$
$$= (A^2 - 10)^2$$

(c) Just because the first and third terms are perfect squares, we must realize that the expression may be factored but does *not* fit the form of either Eq. (6-3) or (6-4). One example, using x^2 and 100 for the first and third term, is

$$x^2 + 29x + 100 = (x + 25)(x + 4) \qquad\qquad ∎$$

Factoring General Trinomials

For reference, Eq. (6-6) is
$(ax + b)(cx + d) = acx^2 + (ad + bc)x + bd.$

Factoring expressions based on the special product of Eq. (6-6) often requires some trial and error. However, the amount of trial and error can be kept to a minimum with a careful analysis of the coefficients of x^2 and the constant. Rewriting Eq. (6-6) with sides reversed, we have

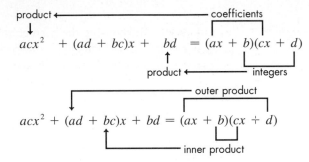

This diagram shows us that

1. *the coefficient of x^2 is the product of the coefficients a and c in the factors,*
2. *the final constant is the product of the constants b and d in the factors, and*

Factoring a trinomial for which the coefficient of x^2 is any positive integer.

CAUTION ▶

3. ***the coefficient of x is the sum of the inner and outer products.***

In finding the factors, we must try possible combinations of a, b, c, and d that give the proper inner and outer products for the middle term of the given expression.

■**EXAMPLE 4**   When factoring $2x^2 + 11x + 5$, we take the factors of 2 to be $+2$ and $+1$ (we will use only positive coefficients a and c when the coefficient of x^2 is positive). We now set up the factoring as

$$2x^2 + 11x + 5 = (2x\boxed{})(x\boxed{})$$

NOTE ▶

Since the product of the integers to be found is $+5$, only integers of the same sign need be considered. Also since ***the sum of the outer and inner products is +11,*** the integers are positive. The factors of $+5$ are $+1$ and $+5$, and -1 and -5, which means that $+1$ and $+5$ is the only possible pair. Now, trying the factors

$$(2x + 5)(x + 1)$$
$$+5x$$
$$+2x$$
$$+2x + 5x = +7x$$

we see that $7x$ is not the correct middle term.
   Next, trying

$$(2x + 1)(x + 5)$$
$$+x$$
$$+10x$$
$$+x + 10x = +11x$$

we have the correct sum of $+11x$. Therefore,

$$2x^2 + 11x + 5 = (2x + 1)(x + 5)$$

According to this analysis, the expression $2x^2 + 10x + 5$ is not factorable, but the following expression is:

$$2x^2 + 7x + 5 = (2x + 5)(x + 1)$$

■EXAMPLE 5 In factoring $4x^2 + 4x - 3$, the coefficient 4 in $4x^2$ shows that the possible coefficients of x in the factors are 4 and 1, or 2 and 2. The 3 shows that

NOTE ▶ the only possible constants in the factors are 1 and 3, and *minus sign with the 3 tells us that these integers have **different signs.*** This gives us the following possible combinations of factors, along with the resulting sum of the outer and inner products:

$$(4x - 3)(x + 1): 4x - 3x = +x$$
$$(4x - 1)(x + 3): 12x - x = +11x$$
$$(2x + 3)(2x - 1): -2x + 6x = +4x$$
$$(2x - 3)(2x + 1): 2x - 6x = -4x$$

We see that the factors that have the correct middle term of $+4x$ are $(2x + 3)(2x - 1)$. This means that

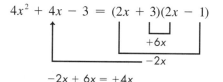

$$-2x + 6x = +4x$$

Expressing the result with the factors reversed is an equally correct answer.

Another hint that is given by the coefficients of the original expression is that *the plus sign with the 4x tells us that the larger of the outer and inner products must be positive.* Also, of course when we have the correct sum of the outer and inner terms, we need not try any other combinations of factors.

■EXAMPLE 6 $6s^2 + 19st - 20t^2 = (6s - 5t)(s + 4t)$

$$+24st - 5st = +19st$$

There are numerous possibilities for the combinations of 6 and 20. However, we

CAUTION ▶ must remember to **check carefully that the middle term** *of the expression is the* **proper result of the factors** *we have chosen.*

■EXAMPLE 7 **(a)** In factoring $9x^2 - 6x + 1$, we note that $9x^2$ is the square of $3x$ and 1 is the square of 1. Therefore, we recognize that this expression might fit the perfect square form of Eq. (6-4). This leads us to factor it tentatively as

$$9x^2 - 6x + 1 = (3x - 1)^2 \qquad \text{check middle term: } 2(3x)(-1) = -6x$$

However, before we can be certain that this is correct, we must check to see if the middle term of the expansion of $(3x - 1)^2$ is $-6x$, which is what it must be to fit the form of Eq. (6-4). When we expand $(3x - 1)^2$, we find that the middle term is $-6x$ and therefore that the factorization is correct.

(b) In the same way, we have

$$36x^2 + 84xy + 49y^2 = (6x + 7y)^2 \qquad \text{check middle term: } 2(6x)(7y) = 84xy$$

NOTE ▶ since the middle term of the expansion of $(6x + 7y)^2$ is $84xy$. *We must be careful to include the factors of y in the second terms of the factors.*

COMMON MONOMIAL FACTORS

As pointed out in Section 6-2, we must be careful to see that we have factored an expression completely. *We look for common monomial factors first* and then check each resulting factor. This check of each factor should be made each time we complete a step in factoring.

■**EXAMPLE 8**   When factoring $2x^2 + 6x - 8$, we first note the common monomial factor of 2. This leads to

$$2x^2 + 6x - 8 = 2(x^2 + 3x - 4)$$

We now notice that $x^2 + 3x - 4$ is also factorable. Therefore,

$$2x^2 + 6x - 8 = 2(x + 4)(x - 1)$$

Now each factor is prime.

Having noted the common factor of 2 prevents our having to check factors of 2 and 8. If we had not noted the common factor, we might have arrived at factorizations of

$$(2x + 8)(x - 1)   \text{or}   (2x - 2)(x + 4)$$

CAUTION ▶

(possibly after a number of trials). Each is correct as far as it goes, but also each of these is *not complete*. Since $2x + 8 = 2(x + 4)$, or $2x - 2 = 2(x - 1)$, we can arrive at the proper result shown above. In this way we would have

$$2x^2 + 6x - 8 = (2x + 8)(x - 1)$$
$$= 2(x + 4)(x - 1)$$

or

$$2x^2 + 6x - 8 = (2x - 2)(x + 4)$$
$$= 2(x - 1)(x + 4)$$

Although these factorizations are correct, *it is better to factor out the common factor first*. By doing so, the number of possible factoring combinations is greatly reduced, and the factoring can be done more easily.

Liquid-fuel rockets were designed in the United States in the 1920s but were developed by German engineers. They were first used in the 1940s during the Second World War.

■**EXAMPLE 9**   A study of the path of a certain rocket leads to the expression $16t^2 + 240t - 1600$, where t is the time of flight. Factor this expression.

An inspection shows that there is a common factor of 16. (This might be found by noting successive factors of 2 or 4.) Factoring out 16 leads to

$$16t^2 + 240t - 1600 = 16(t^2 + 15t - 100)$$
$$= 16(t + 20)(t - 5)$$

Here, factors of 100 need to be checked for sums equal to 15. This might take a little time, but it is much simpler than looking for factors of 16 and 1600 with sums equal to 240.

Factoring by Grouping

In the previous section, we introduced the method of factoring by grouping. The following examples use this method for factoring (1) a trinomial and (2) an expression that can be written as the difference of squares.

EXAMPLE 10 Factor the trinomial $6x^2 + 7x - 20$ by grouping.

To factor the trinomial $ax^2 + bx + c$ by grouping, we first find two numbers whose product is ac and whose sum is b. For $6x^2 + 7x - 20$, this means we want two numbers with a product of -120 and a sum of 7. Trying products of numbers with different signs and a sum of 7, we find the numbers are -8 and 15. We write $6x^2 + 7x - 20$ with x-terms having coefficients of -8 and 15 and then complete the factorization by grouping, as follows.

$$6x^2 + 7x - 20 = 6x^2 - 8x + 15x - 20$$
$$= (6x^2 - 8x) + (15x - 20) \qquad \text{group first two terms and last two terms}$$
$$= 2x(3x - 4) + 5(3x - 4) \qquad \text{find common factor of each group and}$$
$$= (3x - 4)(2x + 5) \qquad \text{note common factor of } 3x - 4$$

Multiplication verifies that these are the correct factors. ▄

EXAMPLE 11 Factor: $x^2 - 4xy + 4y^2 - 9$.

We see that the first three terms of this expression represent $(x - 2y)^2$. Thus, grouping these terms, we have the following solution:

$$x^2 - 4xy + 4y^2 - 9 = (x^2 - 4xy + 4y^2) - 9 \qquad \text{group terms}$$
$$= (x - 2y)^2 - 9 \qquad \text{factor grouping (note difference of squares)}$$
$$= [(x - 2y) + 3][(x - 2y) - 3] \qquad \text{factor difference of squares}$$
$$= (x - 2y + 3)(x - 2y - 3)$$

Note that not all groupings work. If we had seen the combination $4y^2 - 9$, which is factorable, and then grouped the first two terms and the last two terms, this would not have led to the factorization. ▄

━━━━━━━━━━ EXERCISES 6-3 ━━━━━━━━━━

In Exercises 1–48, factor the given expressions completely.

1. $x^2 + 5x + 4$
2. $x^2 - 5x - 6$
3. $s^2 - s - 42$
4. $a^2 + 14a - 32$
5. $t^2 + 5t - 24$
6. $r^2 - 11r + 18$
7. $x^2 + 2x + 1$
8. $D^2 + 8D + 16$
9. $x^2 - 4xy + 4y^2$
10. $b^2 - 12bc + 36c^2$
11. $3x^2 - 5x - 2$
12. $2n^2 - 13n - 7$
13. $3y^2 - 8y - 3$
14. $5x^2 + 9x - 2$
15. $2s^2 + 13s + 11$
16. $7y^2 - 12y + 5$
17. $3f^4 - 16f^2 + 5$
18. $5R^4 - 3R^2 - 2$
19. $2t^2 + 7t - 15$
20. $3n^2 - 20n + 20$
21. $3t^2 - 7tu + 4u^2$
22. $3x^2 + xy - 14y^2$
23. $4x^2 - 3x - 7$
24. $2z^2 + 13z - 5$
25. $9x^2 + 7xy - 2y^2$
26. $4r^2 + 11rs - 3s^2$
27. $4m^2 + 20m + 25$
28. $16q^2 + 24q + 9$
29. $4x^2 - 12x + 9$
30. $a^2c^2 - 2ac + 1$
31. $9t^2 - 15t + 4$
32. $6t^4 + t^2 - 12$
33. $8b^6 + 31b^3 - 4$
34. $12n^2 + 8n - 15$

35. $4p^2 - 25pq + 6q^2$
36. $12x^2 + 4xy - 5y^2$
37. $12x^2 + 47xy - 4y^2$
38. $8r^2 - 14rs - 9s^2$
39. $2x^2 - 14x + 12$
40. $6y^2 - 33y - 18$
41. $4x^2 + 14x - 8$
42. $12B^2 + 22BH - 4H^2$
43. $ax^3 + 4a^2x^2 - 12a^3x$
44. $6x^4 - 13x^3 + 5x^2$
45. $a^2 + 2ab + b^2 - 4$
46. $x^2 - 6xy + 9y^2 - 4z^2$
47. $25a^2 - 25x^2 - 10xy - y^2$
48. $r^2 - s^2 + 2st - t^2$

In Exercises 49–56, factor the given expressions completely. Each is from the technical area indicated.

49. $4s^2 + 16s + 12$ (electricity)
50. $3e^2 + 18e - 1560$ (fuel efficiency)
51. $200n^2 - 2100n - 3600$ (biology)
52. $bT^2 - 40bT + 400b$ (thermodynamics)
53. $wx^4 - 5wLx^3 + 6wL^2x^2$ (beam design)
54. $1 - 2r^2 + r^4$ (lasers)
55. $3Adu^2 - 4Aduv + Adv^2$ (water power)
56. $k^2A^2 + 2k\lambda A + \lambda^2 - \alpha^2$ (robotics)

$6\text{-}4$ THE SUM AND DIFFERENCE OF CUBES

In Section 6-2 we saw that the difference of squares is factorable but that the sum of squares is prime. We now turn our attention to the sum of cubes and the difference of cubes. By writing Eqs. (6-9) and (6-10) with sides reversed, we have

$$x^3 + y^3 = (x + y)(x^2 - xy + y^2) \qquad \text{(6-9)}$$
$$x^3 - y^3 = (x - y)(x^2 + xy + y^2) \qquad \text{(6-10)}$$

We see that both the sum of cubes and the difference of cubes are factorable. In Eqs. (6-9) and (6-10), neither of the expressions $x^2 - xy + y^2$ or $x^2 + xy + y^2$ is factorable. Both expressions are prime.

■EXAMPLE 1

(a) $x^3 + 8 = x^3 + 2^3$

$= (x + 2)[(x)^2 - 2x + 2^2]$

$= (x + 2)(x^2 - 2x + 4)$

(b) $x^3 - 1 = x^3 - 1^3$

$= (x - 1)[(x)^2 + (1)(x) + 1^2]$

$= (x - 1)(x^2 + x + 1)$ ■

■EXAMPLE 2 $8 - 27x^3 = 2^3 - (3x)^3$ $8 = 2^3$ and $27x^3 = (3x)^3$

$= (2 - 3x)[2^2 + 2(3x) + (3x)^2]$

$= (2 - 3x)(4 + 6x + 9x^2)$ ■

As we have stated in the previous sections on factoring, *we should start any factoring process by first checking for any common monomial factor* that might be present in each term of the expression. Also, we should check that our result has been factored **completely.**

■EXAMPLE 3   In factoring $ax^5 - ax^2$, we first note that each term has a common factor of ax^2. This is factored out to get $ax^2(x^3 - 1)$. However, the expression is not completely factored since $1 = 1^3$, which means that $x^3 - 1$ is the difference of cubes. We complete the factoring by the use of Eq. (6-10). Therefore,

$$ax^5 - ax^2 = ax^2(x^3 - 1)$$
$$= ax^2(x - 1)(x^2 + x + 1) \qquad ■$$

■EXAMPLE 4   The volume of material used to make a steel bearing with a hollow core is given by $\dfrac{4}{3}\pi R^3 - \dfrac{4}{3}\pi r^3$. Factor this expression.

$$\frac{4}{3}\pi R^3 - \frac{4}{3}\pi r^3 = \frac{4}{3}\pi(R^3 - r^3) \qquad \text{common factor of } \frac{4}{3}\pi$$

$$= \frac{4}{3}\pi(R - r)(R^2 + Rr + r^2) \qquad \text{using Eq. (6-10)} \qquad ■$$

In our study of factoring, we have seen that we should

first factor out any common monomial factor

and then see if the remaining expression can be further factored as one of the following types:

1. *Difference of squares*
2. *Factorable trinomial*
3. *Sum or difference of cubes*
4. *Factorable by grouping*

Remember, an expression should be factored *completely.*

EXERCISES 6-4

In Exercises 1–24, factor the given expressions completely.

1. $x^3 + 1$
2. $R^3 + 27$
3. $8 - t^3$
4. $8r^3 - 1$
5. $27x^3 - 8a^3$
6. $64x^3 + 125$
7. $2x^3 + 16$
8. $3y^3 - 81$
9. $6A^4 + 6A$
10. $8s^3 - 8$
11. $6x^3y - 6x^3y^4$
12. $12a^3 + 96a^3b^3$
13. $x^6y^3 + x^3y^6$
14. $16r^3 - 432$
15. $3a^6 - 3a^2$
16. $x^6 - 81y^2$
17. $0.001R^3 - 0.064r^3$
18. $0.027x^3 + 0.125$
19. $27L^6 + 216L^3$
20. $a^3s^5 - 8000a^3s^2$
21. $(a + b)^3 - 64$
22. $125 + (2x + y)^3$
23. $64 + x^6$
24. $a^6 - b^6$

In Exercises 25 and 26, perform the indicated operations.

25. Perform the division $(x^5 - y^5) \div (x - y)$. Noting the result, determine the quotient $(x^7 - y^7) \div (x - y)$ without dividing. From these results, factor $x^5 - y^5$ and $x^7 - y^7$.

26. Perform the division $(x^5 + y^5) \div (x + y)$. Noting the result, determine the quotient $(x^7 + y^7) \div (x + y)$ without dividing. From these results, factor $x^5 + y^5$ and $x^7 + y^7$.

In Exercises 27–32, factor the given expressions completely. Each is from the technical area indicated.

27. $2x^3 + 250$ (computer image)
28. $kT^3 - kT_0^3$ (thermodynamics)
29. $D^4 - d^3D$ (machine design)
30. $pb^3 + p(8a^3)$ (business)
31. $QH^4 + Q^4H$ (thermodynamics)
32. $(h + 2t)^3 - h^3$ (container design)

6-5 EQUIVALENT FRACTIONS

When we deal with algebraic expressions, we must be able to work effectively with fractions. Since algebraic expressions are representations of numbers, the basic operations on fractions from arithmetic form the basis of our algebraic operations. In this section we demonstrate a very important property of fractions, and in the following two sections we establish the basic algebraic operations with fractions.

FUNDAMENTAL PRINCIPLE OF FRACTIONS

This important property of fractions, often referred to as the **fundamental principle of fractions,** is that *the value of a fraction is unchanged if both numerator and denominator are multiplied or divided by the same number, provided this number is not zero.* Two fractions are said to be **equivalent** if one can be obtained from the other by use of the fundamental principle.

▌**EXAMPLE 1**   If we multiply the numerator and the denominator of the fraction $\frac{6}{8}$ by 2, we obtain the equivalent fraction $\frac{12}{16}$. If we divide the numerator and the denominator of $\frac{6}{8}$ by 2, we obtain the equivalent fraction $\frac{3}{4}$. Therefore, the fractions $\frac{6}{8}, \frac{3}{4}$, and $\frac{12}{16}$ are equivalent.

EXAMPLE 2   We may write

$$\frac{ax}{2} = \frac{3a^2x}{6a}$$

since the fraction on the right is obtained from the fraction on the left by multiplying the numerator and the denominator by $3a$. Therefore, the fractions are equivalent.

Simplest Form, or Lowest Terms, of a Fraction

One of the most important operations to be performed on a fraction is that of reducing it to its **simplest form,** or **lowest terms.**

> *A fraction is said to be in its simplest form if the numerator and the denominator have no common integral factors other than $+1$ or -1.*

NOTE ▶ In *reducing* a fraction to its simplest form, we use the fundamental principle of fractions by *dividing* both the numerator and the denominator by all factors that are common to each. (It will be assumed throughout this text that if any of the literal symbols were to be evaluated, numerical values would be restricted so that none of the denominators would be equal to zero. Thereby, we avoid the undefined operation of division by zero.)

EXAMPLE 3   In order to reduce the fraction

$$\frac{16ab^3c^2}{24ab^2c^5}$$

to its lowest terms, we note that both the numerator and the denominator contain the factor $8ab^2c^2$. Therefore, we may write

$$\frac{16ab^3c^2}{24ab^2c^5} = \frac{2b(8ab^2c^2)}{3c^3(8ab^2c^2)} = \frac{2b}{3c^3} \quad \text{common factor}$$

Here we divided out the common factor. The resulting fraction is in lowest terms, since there are no common factors in the numerator and the denominator other than $+1$ or -1.

CANCELLATION   In simplifying fractions, we must be very careful in performing the basic step of reducing the fraction to its simplest form. That is,

> **divide** *both the numerator and the denominator* **by the common factor.**

CAUTION ▶ *This process is called* **cancellation.** However, it is a very common error to remove *any* expression that appears in both the numerator and the denominator. If a *term* is removed in this way, it is an incorrect application of the cancellation process. We must always *cancel only factors* that are in both the numerator and the denominator. (Remember—*terms* are separated by $+$ and $-$ signs.) In the following example we illustrate this very common error in the simplification of fractions. Follow it very carefully.

EXAMPLE 4   When simplifying the expression

$$\frac{x^2(x-2)}{x^2-4}$$ a term, but not a factor, of the denominator

CAUTION ▶   many students would "cancel" the x^2 from the numerator and the denominator. This is *incorrect*, since x^2 *is a term only* of the denominator.

   In order to simplify the above fraction properly, we should factor the denominator. We obtain

$$\frac{\overset{1}{x^2(\cancel{x-2})}}{\underset{1}{(\cancel{x-2})(x+2)}} = \frac{x^2}{x+2}$$

Here, the common *factor* $x-2$ has been divided out. ∎

   The following examples illustrate the proper simplification of fractions.

EXAMPLE 5   **(a)** $\dfrac{2a}{2ax} = \dfrac{1}{x}$ ⎯⎯ $2a$ is a factor of the numerator and the denominator

We divide out the common factor of $2a$.

   (b) $\dfrac{2a}{2a+x}$ $2a$ is a term, but not a factor, of the denominator

CAUTION ▶   *This cannot be reduced,* since *there are no common **factors** in the numerator and the denominator.* ∎

EXAMPLE 6 $\dfrac{2x^2+8x}{x+4} = \dfrac{2x\overset{1}{(\cancel{x+4})}}{\underset{1}{\cancel{x+4}}} = \dfrac{2x}{1}$

$$= 2x$$

   The numerator and the denominator were each divided by $x+4$ after we factored the numerator. The only remaining factor in the denominator is 1, and it is generally not written in the final result. Another way of writing the denominator is $1(x+4)$, which shows the ***factor*** of 1 more clearly. ∎

EXAMPLE 7 $\dfrac{x^2-4x+4}{x^2-4} = \dfrac{(x-2)\overset{1}{(\cancel{x-2})}}{(x+2)\underset{1}{(\cancel{x-2})}}$

$$= \frac{x-2}{x+2}$$ x is a term, but not a factor

CAUTION ▶   Here the numerator and the denominator have each been *factored first and then the common factor $x-2$ has been divided out.* In the final form, neither the x's nor the 2's may be canceled, since they are not common *factors*. ∎

EXAMPLE 8   In the mathematical analysis of the vibrations in a certain mechanical system, the following expression and simplification are used:

$$\frac{8s + 12}{4s^2 + 26s + 30} = \frac{4(2s + 3)}{2(2s^2 + 13s + 15)} = \frac{\overset{2}{\cancel{4}}\overset{1}{\cancel{(2s + 3)}}}{\underset{1}{\cancel{2}}\underset{1}{\cancel{(2s + 3)}}(s + 5)}$$

$$= \frac{2}{s + 5}$$

In the third fraction we see that the factors common to both the numerator and the denominator are 2 and $(2s + 3)$.

Factors That Differ Only in Sign

In simplifying fractions we must be able to distinguish between factors that differ only in *sign*. Since $-(y - x) = -y + x = x - y$, we have

$$\boxed{x - y = -(y - x)}$$ (6-11)

CAUTION ▶   Here we see that *factors $x - y$ and $y - x$ differ only in sign.* The following examples illustrate the simplification of fractions where a change of signs is necessary.

EXAMPLE 9 $\dfrac{x^2 - 1}{1 - x} = \dfrac{(x - 1)(x + 1)}{-(x - 1)} = \dfrac{x + 1}{-1} = -(x + 1)$

CAUTION ▶   In the second fraction, *we replaced $1 - x$ with the equal expression $-(x - 1)$.* In the third fraction, the common factor $x - 1$ was divided out. Finally, we expressed the result in the more convenient form by dividing $x + 1$ by -1, which makes the quantity $x + 1$ negative. This also means that another correct answer can be written as $-x - 1$.

EXAMPLE 10

$$\frac{2x^4 - 128x}{20 + 7x - 3x^2} = \frac{2x(x^3 - 64)}{(4 - x)(5 + 3x)} = \frac{2x(x - 4)(x^2 + 4x + 16)}{-(x - 4)(3x + 5)}$$

$$= -\frac{2x(x^2 + 4x + 16)}{3x + 5}$$

Again, the factor $4 - x$ has been replaced by the equal expression $-(x - 4)$. This allows us to recognize the common factor of $x - 4$.

We also note that the order of the terms of the factor $5 + 3x$ was changed in writing the third fraction. This was done only to write the terms in the more standard form with the x-term first. However, since both terms are *positive,* it is simply an application of the commutative law of addition, and the factor itself is not actually changed.

EXERCISES $6\text{-}5$

In Exercises 1–8, multiply the numerator and the denominator of each fraction by the given factor and obtain an equivalent fraction.

1. $\dfrac{2}{3}$ (by 7)

2. $\dfrac{7}{5}$ (by 9)

3. $\dfrac{ax}{y}$ (by 2x)

4. $\dfrac{2x^2y}{3n}$ (by $2xn^2$)

5. $\dfrac{2}{x+3}$ (by $x-2$)

6. $\dfrac{7}{a-1}$ (by $a+2$)

7. $\dfrac{a(x-y)}{x-2y}$ (by $x+y$)

8. $\dfrac{x-1}{x+1}$ (by $x-1$)

In Exercises 9–16, divide the numerator and the denominator of each fraction by the given factor and obtain an equivalent fraction.

9. $\dfrac{28}{44}$ (by 4)

10. $\dfrac{25}{65}$ (by 5)

11. $\dfrac{4x^2y}{8xy^2}$ (by $2x$)

12. $\dfrac{6a^3b^2}{9a^5b^4}$ (by $3a^2b^2$)

13. $\dfrac{2(x-1)}{(x-1)(x+1)}$ (by $x-1$)

14. $\dfrac{(x+5)(x-3)}{3(x+5)}$ (by $x+5$)

15. $\dfrac{s^2-3s-10}{2s^2+3s-2}$ (by $s+2$)

16. $\dfrac{6x^2+13x-5}{6x^3-2x^2}$ (by $3x-1$)

In Exercises 17–52, reduce each fraction to simplest form.

17. $\dfrac{2a}{8a}$

18. $\dfrac{6x}{15x}$

19. $\dfrac{18x^2y}{24xy}$

20. $\dfrac{2a^2xy}{6axyz^2}$

21. $\dfrac{a+b}{5a^2+5ab}$

22. $\dfrac{t-a}{t^2-a^2}$

23. $\dfrac{6a-4b}{4a-2b}$

24. $\dfrac{5r-20s}{10r-5s}$

25. $\dfrac{4x^2+1}{4x^2-1}$

26. $\dfrac{x^2-y^2}{x^2+y^2}$

27. $\dfrac{3x^2-6x}{x-2}$

28. $\dfrac{10x^2+15x}{2x+3}$

29. $\dfrac{2y+3}{4y^3+6y^2}$

30. $\dfrac{3t-6}{4t^3-8t^2}$

31. $\dfrac{x^2-8x+16}{x^2-16}$

32. $\dfrac{4a^2+12ab+9b^2}{4a^2+6ab}$

33. $\dfrac{2w^4+5w^2-3}{w^4+11w^2+24}$

34. $\dfrac{3y^3+7y^2+4y}{y^2+5y+4}$

35. $\dfrac{5x^2-6x-8}{x^3+x^2-6x}$

36. $\dfrac{4r^2-8rs-5s^2}{6r^2-17rs+5s^2}$

37. $\dfrac{x^4-16}{x+2}$

38. $\dfrac{3+x(4+x)}{3+x}$

39. $\dfrac{t+4}{(2t+9)t+4}$

40. $\dfrac{8A^5+8A^4+2A^3}{4A+2}$

41. $\dfrac{(x-1)(3+x)}{(3-x)(1-x)}$

42. $\dfrac{(2x-1)(x+6)}{(x-3)(1-2x)}$

43. $\dfrac{y-x}{2x-2y}$

44. $\dfrac{x^2-y^2}{y-x}$

45. $\dfrac{2x^2-9x+4}{4x-x^2}$

46. $\dfrac{3a^2-13a-10}{5+4a-a^2}$

47. $\dfrac{(x+5)(x-2)(x+2)(3-x)}{(2-x)(5-x)(3+x)(2+x)}$

48. $\dfrac{(2x-3)(3-x)(x-7)(3x+1)}{(3x+2)(3-2x)(x-3)(7+x)}$

49. $\dfrac{x^3+y^3}{2x+2y}$

50. $\dfrac{w^3-8}{w^2+2w+4}$

51. $\dfrac{6x^2+2x}{27x^3+1}$

52. $\dfrac{3a^3-24}{a^2-4a+4}$

Ⓦ In Exercises 53–56, after finding the simplest form of each fraction, explain why it cannot be simplified more.

53. (a) $\dfrac{x^2(x+2)}{x^2+4}$ (b) $\dfrac{x^4+4x^2}{x^4-16}$

54. (a) $\dfrac{2x+3}{2x+6}$ (b) $\dfrac{2(x+6)}{2x+6}$

55. (a) $\dfrac{x^2-x-2}{x^2-x}$ (b) $\dfrac{x^2-x-2}{x^2+x}$

56. (a) $\dfrac{x^3-x}{1-x}$ (b) $\dfrac{2x^2+4x}{2x^2+4}$

In Exercises 57–60, reduce each fraction to simplest form. Each is from the indicated area of application.

57. $\dfrac{mu^2-mv^2}{mu-mv}$ (nuclear energy)

58. $\dfrac{16(t^2-2tt_0+t_0^2)(t-t_0-3)}{3t-3t_0}$ (rocket motion)

59. $\dfrac{E^2R^2-E^2r^2}{(R^2+2Rr+r^2)^2}$ (electricity)

60. $\dfrac{r_0^3-r_i^3}{r_0^2-r_i^2}$ (machine design)

$6\text{-}6$ MULTIPLICATION AND DIVISION OF FRACTIONS

From arithmetic we recall that *the product of two fractions is a fraction whose numerator is the product of the numerators and whose denominator is the product of the denominators of the given fractions.* Also, we recall that *we can find the quotient of two fractions by inverting the divisor and proceeding as in multiplication.* Symbolically, multiplication of fractions is indicated by

MULTIPLICATION OF FRACTIONS

$$\frac{a}{b} \times \frac{c}{d} = \frac{ac}{bd}$$

and division is indicated by

DIVISION OF FRACTIONS

$$\frac{a}{b} \div \frac{c}{d} = \frac{\dfrac{a}{b}}{\dfrac{c}{d}} = \frac{a}{b} \times \frac{d}{c} = \frac{ad}{bc}$$

The rule for division may be verified by use of the fundamental principle of fractions. By multiplying the numerator and the denominator of the fraction

$$\frac{\dfrac{a}{b}}{\dfrac{c}{d}} \quad \text{by} \quad \frac{d}{c} \quad \text{we obtain} \quad \frac{\dfrac{a}{b} \times \dfrac{d}{c}}{\dfrac{c}{d} \times \dfrac{d}{c}} = \frac{\dfrac{ad}{bc}}{1} = \frac{ad}{bc}$$

The following three examples illustrate the multiplication of fractions.

EXAMPLE 1 (a) $\dfrac{3}{5} \times \dfrac{2}{7} = \dfrac{(3)(2)}{(5)(7)} = \dfrac{6}{35}$ ← multiply numerators
← multiply denominators

(b) $\dfrac{3a}{5b} \times \dfrac{15b^2}{a} = \dfrac{(3a)(15b^2)}{(5b)(a)}$

$\qquad = \dfrac{45ab^2}{5ab} = \dfrac{9b}{1}$

$\qquad = 9b$

In illustration (b), we divided out the common factor of $5ab$ to reduce the resulting fraction to its lowest terms.

When multiplying fractions we shall usually want to express the final result in simplest form, which is generally its most useful form. Since all factors in the numerators and all factors in the denominators are to be multiplied, we should

NOTE ▶ *first only **indicate** the multiplication, but not actually perform it, and then factor the numerator and the denominator of the result.*

By *indicate* we mean to write the factors of the numerator side by side and do the same for the factors of the denominator. If we were to multiply out the numerator and the denominator before factoring, it is very possible that we would be unable to factor the result and therefore would not be able to simplify it. The following example illustrates this point.

EXAMPLE 2   In performing the multiplication

$$\frac{3(x-y)}{(x-y)^2} \times \frac{(x^2-y^2)}{6x+9y}$$

if we multiplied out the numerators and the denominators before performing any factoring, we would have to simplify the fraction

$$\frac{3x^3 - 3x^2y - 3xy^2 + 3y^3}{6x^3 - 3x^2y - 12xy^2 + 9y^3}$$

It is possible to factor the resulting numerator and denominator, but finding any common factors this way is very difficult. If we first indicate the multiplications, but do not actually perform them, and then factor completely, we have

$$\frac{3(x-y)}{(x-y)^2} \times \frac{(x^2-y^2)}{6x+9y} = \frac{3(x-y)(x^2-y^2)}{(x-y)^2(6x+9y)} = \frac{3(x-y)(x+y)(x-y)}{(x-y)^2(3)(2x+3y)}$$

$$= \frac{3(x-y)^2(x+y)}{3(x-y)^2(2x+3y)}$$

$$= \frac{x+y}{2x+3y}$$

The common factor $3(x-y)^2$ is readily recognized using this procedure.

EXAMPLE 3

$$\frac{2x-4}{4x+12} \times \frac{2x^2+x-15}{3x-1} = \frac{2(x-2)(2x-5)(x+3)}{4(x+3)(3x-1)} \quad \begin{array}{l}\leftarrow \text{multiplications} \\ \leftarrow \text{indicated}\end{array}$$

$$= \frac{(x-2)(2x-5)}{2(3x-1)}$$

It is possible to factor and indicate the product of the factors, showing only a single step, as we have done here.

Here the common factor is $2(x+3)$. It is permissible to multiply out the final form of the numerator and the denominator, but it is often preferable to leave the numerator and the denominator in factored form, as indicated.

The following examples illustrate the division of fractions.

EXAMPLE 4

multiply

(a) $\dfrac{6x}{7} \div \dfrac{5}{3} = \dfrac{6x}{7} \times \dfrac{3}{5} = \dfrac{18x}{35}$

invert

multiply

(b) $\dfrac{\frac{3a^2}{5c}}{\frac{2c^2}{a}} = \dfrac{3a^2}{5c} \times \dfrac{a}{2c^2} = \dfrac{3a^3}{10c^3}$

invert

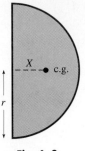

Fig. 6-3

EXAMPLE 5   When finding the center of gravity (c.g.) of a uniform flat semi-circular metal plate, the equation $X = \dfrac{4\pi r^3}{3} \div \left(\dfrac{\pi r^2}{2} \times 2\pi \right)$ is derived. Simplify the right side of this equation to find X as a function of r in simplest form. See Fig. 6-3.

The parentheses indicate that we should perform the multiplication first.

$$X = \frac{4\pi r^3}{3} \div \left(\frac{\pi r^2}{2} \times 2\pi \right) = \frac{4\pi r^3}{3} \div \left(\frac{2\pi^2 r^2}{2} \right)$$

$$= \frac{4\pi r^3}{3} \div (\pi^2 r^2) = \frac{4\pi r^3}{3} \times \frac{1}{\pi^2 r^2}$$

$$= \frac{4\pi r^3}{3\pi^2 r^2} = \frac{4r}{3\pi} \qquad \text{divide out the common factor of } \pi r^2$$

This is the exact solution. Approximately, $X = 0.424r$.

EXAMPLE 6

$$\frac{x + y}{3} \div \frac{2x + 2y}{6x + 15y} = \frac{x + y}{3} \times \frac{6x + 15y}{2x + 2y} = \frac{(x + y)(3)(2x + 5y)}{3(2)(x + y)} \qquad \text{indicate multiplication}$$

$$\text{invert}$$

$$= \frac{2x + 5y}{2} \qquad \text{simplify}$$

EXAMPLE 7

$$\frac{\dfrac{4 - x^2}{x^2 - 3x + 2}}{\dfrac{x + 2}{x^2 - 9}} = \frac{4 - x^2}{x^2 - 3x + 2} \times \frac{x^2 - 9}{x + 2} \qquad \text{invert}$$

$$= \frac{(2 - x)(2 + x)(x - 3)(x + 3)}{(x - 2)(x - 1)(x + 2)} \qquad \text{factor and indicate multiplications}$$

$$= \frac{-(x - 2)(x + 2)(x - 3)(x + 3)}{(x - 2)(x - 1)(x + 2)} \qquad \begin{array}{l}\text{replace } (2 - x) \text{ with} \\ -(x - 2) \text{ and } (2 + x) \\ \text{with } (x + 2)\end{array}$$

$$= -\frac{(x - 3)(x + 3)}{x - 1} \quad \text{or} \quad \frac{(x - 3)(x + 3)}{1 - x} \qquad \text{simplify}$$

Note the use of Eq. (6-11) in the simplification and in expressing an alternate form of the result. The factor $2 - x$ was replaced by its equivalent $-(x - 2)$, and then $x - 1$ was replaced by $-(1 - x)$.

EXERCISES 6-6

In Exercises 1–36, simplify the given expressions involving the indicated multiplications and divisions.

1. $\dfrac{3}{8} \times \dfrac{2}{7}$

2. $\dfrac{11}{5} \times \dfrac{13}{33}$

3. $\dfrac{4x}{3y} \times \dfrac{9y^2}{2}$

4. $\dfrac{18sy^3}{ax^2} \times \dfrac{(ax)^2}{3s}$

5. $\dfrac{2}{9} \div \dfrac{4}{7}$

6. $\dfrac{5}{16} \div \dfrac{25}{13}$

7. $\dfrac{xy}{az} \div \dfrac{bz}{ay}$

8. $\dfrac{sr^2}{2t} \div \dfrac{st}{4}$

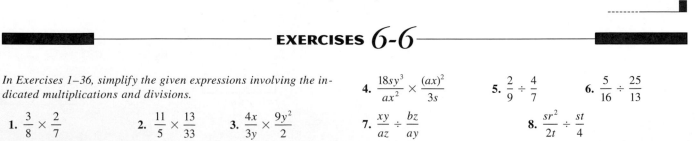

9. $\dfrac{4x + 12}{5} \times \dfrac{15t}{3x + 9}$

10. $\dfrac{y^2 + 2y}{6z} \times \dfrac{z^3}{y^2 - 4}$

11. $\dfrac{u^2 - v^2}{u + 2v}(3u + 6v)$

12. $(x - y)\dfrac{x + 2y}{x^2 - y^2}$

13. $\dfrac{2a + 8}{15} \div \dfrac{a^2 + 8a + 16}{25}$

14. $\dfrac{a^2 - a}{3a + 9} \div \dfrac{a^2 - 2a + 1}{a^2 - 9}$

15. $\dfrac{x^4 - 9}{x^2} \div (x^2 + 3)^2$

16. $\dfrac{9x^2 - 16}{x + 1} \div (4 - 3x)$

17. $\dfrac{3ax^2 - 9ax}{10x^2 + 5x} \times \dfrac{2x^2 + x}{a^2x - 3a^2}$

18. $\dfrac{2x^2 - 18}{x^3 - 25x} \times \dfrac{3x - 15}{2x^2 + 6x}$

19. $\dfrac{x^4 - 1}{8x + 16} \times \dfrac{2x^2 - 8x}{x^3 + x}$

20. $\dfrac{2x^2 - 4x - 6}{x^2 - 3x} \times \dfrac{x^3 - 4x^2}{4x^2 - 4x - 8}$

21. $\dfrac{ax + x^2}{2b - cx} \div \dfrac{a^2 + 2ax + x^2}{2bx - cx^2}$

22. $\dfrac{s^4 - 11s^2 + 28}{s^2 + 3} \div \dfrac{s^2 - 4}{2s^2 + 3}$

23. $\dfrac{35a + 25}{12a + 33} \div \dfrac{28a + 20}{36a + 99}$

24. $\dfrac{2a^3 + a^2}{2b^3 + b^2} \div \dfrac{2ab + a}{2ab + b}$

25. $\dfrac{x^2 - 6x + 5}{4x^2 - 17x - 15} \times \dfrac{6x + 21}{2x^2 + 5x - 7}$

26. $\dfrac{n^2 + 5n}{3n^2 + 8n - 4} \times \dfrac{2n^2 - 8}{n^3 + 2n^2 - 15n}$

27. $\dfrac{7x^2 + 27x - 4}{6x^2 + x - 15} \div \dfrac{4x^2 + 17x + 4}{8x^2 - 10x - 3}$

28. $\dfrac{4x^3 - 9x}{8x^2 + 10x - 3} \div \dfrac{2x^3 - 3x^2}{8x^2 + 18x - 5}$

29. $\dfrac{7x^2}{3a} \div \left(\dfrac{a}{x} \times \dfrac{a^2x}{x^2}\right)$

30. $\left(\dfrac{3u}{8v^2} \div \dfrac{9u^2}{2w^2}\right) \times \dfrac{2u^4}{15vw}$

31. $\left(\dfrac{4t^2 - 1}{t - 5} \div \dfrac{2t + 1}{2t}\right) \times \dfrac{2t^2 - 50}{4t^2 + 4t + 1}$

32. $\dfrac{2x^2 - 5x - 3}{x - 4} \div \left(\dfrac{x - 3}{x^2 - 16} \times \dfrac{1}{3 - x}\right)$

33. $\dfrac{x^3 - y^3}{2x^2 - 2y^2} \times \dfrac{x^2 + 2xy + y^2}{x^2 + xy + y^2}$

34. $\dfrac{2M^2 + 4M + 2}{6M - 6} \div \dfrac{5M + 5}{M^2 - 1}$

35. $\dfrac{ax + bx + ay + by}{p - q} \times \dfrac{3p^2 + 4pq - 7q^2}{a + b}$

36. $\dfrac{x^4 + x^5 - 1 - x}{x - 1} \div \dfrac{x + 1}{x}$

In Exercises 37–40, simplify the given expressions. The technical application of each is indicated.

37. $\dfrac{d}{2} \div \dfrac{v_1d + v_2d}{4v_1v_2}$ (average velocity)

38. $\dfrac{w\pi rbt}{2btg}\left(\dfrac{2r}{12\pi}\right)\left(\dfrac{144v^2}{r^2}\right)$ (stress on a rotating hoop)

39. $\dfrac{2\pi}{\lambda}\left(\dfrac{a + b}{2ab}\right)\left(\dfrac{ab\lambda}{2a + 2b}\right)$ (optics)

40. $(p_1 - p_2) \div \left(\dfrac{\pi a^4 p_1 - \pi a^4 p_2}{8lu}\right)$ (hydrodynamics)

6-7 ADDITION AND SUBTRACTION OF FRACTIONS

From arithmetic we recall that *the sum of a set of fractions that all have the same denominator is the sum of the numerators divided by the common denominator.* Since algebraic expressions represent numbers, this fact is also true in algebra. Addition and subtraction of such fractions are illustrated in the following example.

EXAMPLE 1   (a) $\dfrac{5}{9} + \dfrac{2}{9} - \dfrac{4}{9} = \dfrac{5 + 2 - 4}{9}$ ← sum of numerators

← same denominators

$= \dfrac{3}{9} = \dfrac{1}{3}$ ← final result in lowest terms

use parentheses to show subtraction of both terms

(b) $\dfrac{b}{ax} + \dfrac{1}{ax} - \dfrac{2b - 1}{ax} = \dfrac{b + 1 - (2b - 1)}{ax} = \dfrac{b + 1 - 2b + 1}{ax}$

$= \dfrac{2 - b}{ax}$

Lowest Common Denominator

If the fractions to be combined do not all have the same denominator, we must first change each to an equivalent fraction so that the resulting fractions do have the same denominator. Normally, the denominator that is most convenient and useful is the **lowest common denominator** (abbreviated as **LCD**). *This is the product of all the prime factors that appear in the denominators, with each factor raised to the **highest power** to which it appears **in any one** of the denominators.* This means that the lowest common denominator is the *simplest* algebraic expression into which all given denominators will divide exactly. Following is the procedure for finding the lowest common denominator of a set of fractions.

Procedure for Finding the Lowest Common Denominator

1. *Factor each denominator into its prime factors.*

2. *For each different prime factor that appears, note the highest power to which it is raised in any one of the denominators.*

3. *Form the product of all the different prime factors, each raised to the power found in step 2. This product is the lowest common denominator.*

The two examples that follow illustrate the method of finding the lowest common denominator.

EXAMPLE 2   Find the lowest common denominator of the fractions

$$\frac{3}{4a^2b} \qquad \frac{5}{6ab^3} \qquad \frac{1}{4ab^2}$$

We now express each denominator in terms of powers of its prime factors.

highest powers   already seen to be highest power of 2

$$4a^2b = 2^2a^2b \qquad 6ab^3 = 2 \times 3 \times ab^3 \qquad 4ab^2 = 2^2ab^2$$

The prime factors to be considered are 2, 3, a, and b. The largest exponent of 2 that appears is 2. Therefore, 2^2 is a factor of the lowest common denominator.

CAUTION ▶   *What matters is that **the highest power of 2 that appears is 2,** not the fact that 2 appears in all three denominators with a total of five factors.*

The largest exponent of 3 that appears is 1 (understood in the second denominator). Therefore, 3 is a factor of the lowest common denominator. The largest exponent of a that appears is 2, and the largest exponent of b that appears is 3. Thus, a^2 and b^3 are factors of the lowest common denominator. Therefore, the lowest common denominator of the fractions is

$$2^2 \times 3 \times a^2b^3 = 12a^2b^3$$

This is the simplest expression into which *each* of the denominators above will divide exactly.

EXAMPLE 3 Find the LCD of the following fractions:

$$\frac{x-4}{x^2-2x+1} \qquad \frac{1}{x^2-1} \qquad \frac{x+3}{x^2-x}$$

Factoring each of the denominators, we find that the fractions are

$$\frac{x-4}{(x-1)^2} \qquad \frac{1}{(x-1)(x+1)} \qquad \frac{x+3}{x(x-1)}$$

The factor $(x-1)$ appears in all the denominators. It is squared in the first fraction and appears only to the first power in the other two fractions. Thus, we must have $(x-1)^2$ as a factor in the LCD. We do not need a higher power of $(x-1)$ since, as far as this factor is concerned, each denominator will divide into it evenly. Next, the second denominator has a factor of $(x+1)$. Therefore, the LCD must also have a factor of $(x+1)$; otherwise, the second denominator would not divide into it exactly. Finally, the third denominator shows that a factor of x is also needed. The LCD is therefore $x(x+1)(x-1)^2$. All three denominators will divide exactly into this expression, and there is no simpler expression for which this is true. ∎

Addition and Subtraction of Fractions

Once we have found the lowest common denominator for the fractions, we multiply the numerator and the denominator of each fraction by the proper quantity to make the resulting denominator in each case the lowest common denominator. After this step, it is necessary only to add the numerators, place this result over the common denominator, and simplify.

EXAMPLE 4 Combine: $\dfrac{2}{3r^2} + \dfrac{4}{rs^3} - \dfrac{5}{3s}$.

By looking at the denominators, we see that the factors necessary in the lowest common denominator are 3, r, and s. The 3 appears only to the first power, the largest exponent of r is 2, and the largest exponent of s is 3. Therefore, the lowest common denominator is $3r^2s^3$. We now wish to write each fraction with this quantity as the denominator. Since the denominator of the first fraction already contains factors of 3 and r^2, **it is necessary to introduce the factor of s^3.** In other words, we **NOTE▶** must multiply the numerator and the denominator of this fraction by s^3. For similar reasons, we must multiply the numerators and the denominators of the second and third fractions by $3r$ and r^2s^2, respectively. This leads to

$$\frac{2}{3r^2} + \frac{4}{rs^3} - \frac{5}{3s} = \frac{2(s^3)}{(3r^2)(s^3)} + \frac{4(3r)}{(rs^3)(3r)} - \frac{5(r^2s^2)}{(3s)(r^2s^2)} \qquad \text{change to equivalent fractions with LCD}$$

factors needed in each

$$= \frac{2s^3}{3r^2s^3} + \frac{12r}{3r^2s^3} - \frac{5r^2s^2}{3r^2s^3}$$

$$= \frac{2s^3 + 12r - 5r^2s^2}{3r^2s^3} \qquad \text{combine numerators over LCD} \quad ∎$$

EXAMPLE 5

$$\frac{a}{x-1} + \frac{a}{x+1} = \frac{a(x+1)}{(x-1)(x+1)} + \frac{a(x-1)}{(x+1)(x-1)}$$ change to equivalent fractions with LCD

factors needed

$$= \frac{ax + a + ax - a}{(x+1)(x-1)}$$ combine numerators over LCD

$$= \frac{2ax}{(x+1)(x-1)}$$ simplify

When we multiply each fraction by the quantity required to obtain the proper denominator, we do not actually have to write the common denominator under each numerator. Placing all the products that appear in the numerators over the common denominator is sufficient. Hence, the illustration in this example would appear as

$$\frac{a}{x-1} + \frac{a}{x+1} = \frac{a(x+1) + a(x-1)}{(x-1)(x+1)} = \frac{ax + a + ax - a}{(x-1)(x+1)}$$

$$= \frac{2ax}{(x-1)(x+1)}$$

EXAMPLE 6   The following expression is found in the analysis of the dynamics of missile firing. The indicated addition is performed as shown.

$$\frac{1}{s} - \frac{1}{s+4} + \frac{8}{s^2 + 8s + 16} = \frac{1}{s} - \frac{1}{s+4} + \frac{8}{(s+4)^2}$$ factor third denominator

$$= \frac{1(s+4)^2 - 1(s)(s+4) + 8s}{s(s+4)^2}$$ LCD has one factor of s and two factors of (s + 4)

$$= \frac{s^2 + 8s + 16 - s^2 - 4s + 8s}{s(s+4)^2}$$ expand terms of numerator

$$= \frac{12s + 16}{s(s+4)^2} = \frac{4(3s+4)}{s(s+4)^2}$$ simplify/factor

We factored the numerator in the final result to see whether or not there were any factors common to the numerator and the denominator. Since there are none, either form of the result is acceptable.

EXAMPLE 7

$$\frac{3x}{x^2 - x - 12} - \frac{x-1}{x^2 - 8x + 16} - \frac{6-x}{2x-8} = \frac{3x}{(x-4)(x+3)} - \frac{x-1}{(x-4)^2} - \frac{6-x}{2(x-4)}$$ factor denominators

$$= \frac{3x(2)(x-4) - (x-1)(2)(x+3) - (6-x)(x-4)(x+3)}{2(x-4)^2(x+3)}$$ change to equivalent fraction with LCD

$$= \frac{6x^2 - 24x - 2x^2 - 4x + 6 + x^3 - 7x^2 - 6x + 72}{2(x-4)^2(x+3)}$$ expand in numerator

$$= \frac{x^3 - 3x^2 - 34x + 78}{2(x-4)^2(x+3)}$$ simplify

In doing this kind of problem, many errors may arise in the use of the minus sign.

CAUTION▶   Remember, *if a minus sign precedes an expression, the **signs of all terms must be changed*** before they can be combined with other terms.

Complex Fractions

A **complex fraction** *is one in which the numerator, the denominator, or both the numerator and the denominator contain fractions.* The following examples illustrate the simplification of complex fractions.

EXAMPLE 8

The original complex fraction can be written as a division as follows:
$$\frac{2}{x} \div \left(1 - \frac{4}{x}\right)$$

$$\frac{\dfrac{2}{x}}{1 - \dfrac{4}{x}} = \frac{\dfrac{2}{x}}{\dfrac{x-4}{x}} \qquad \text{first perform subtraction in denominator}$$

$$= \frac{2}{x} \times \frac{x}{x-4} = \frac{2\cancel{x}}{\cancel{x}(x-4)} \qquad \text{invert divisor and multiply}$$

$$= \frac{2}{x-4} \qquad \text{simplify}$$

EXAMPLE 9

$$\frac{1 - \dfrac{2}{x}}{\dfrac{1}{x} + \dfrac{2}{x^2 + 4x}} = \frac{1 - \dfrac{2}{x}}{\dfrac{1}{x} + \dfrac{2}{x(x+4)}} = \frac{\dfrac{x-2}{x}}{\dfrac{x+4+2}{x(x+4)}} \qquad \text{perform subtraction and addition}$$

$$= \frac{x-2}{x} \times \frac{x(x+4)}{x+6} \qquad \text{invert divisor and multiply}$$

$$= \frac{(x-2)(\cancel{x})(x+4)}{\cancel{x}(x+6)} \qquad \text{indicate multiplication}$$

$$= \frac{(x-2)(x+4)}{x+6} \qquad \text{simplify}$$

EXERCISES 6-7

In Exercises 1–44, perform the indicated operations and simplify.

1. $\dfrac{3}{5} + \dfrac{6}{5}$

2. $\dfrac{2}{13} + \dfrac{6}{13}$

3. $\dfrac{1}{x} + \dfrac{7}{x}$

4. $\dfrac{2}{a} + \dfrac{3}{a}$

5. $\dfrac{1}{2} + \dfrac{3}{4}$

6. $\dfrac{5}{9} - \dfrac{1}{3}$

7. $\dfrac{3}{4x} + \dfrac{7a}{4}$

8. $\dfrac{t-3}{a} - \dfrac{t}{2a}$

9. $\dfrac{a}{x} - \dfrac{b}{x^2}$

10. $\dfrac{2}{s^2} + \dfrac{3}{s}$

11. $\dfrac{6}{5x^3} + \dfrac{a}{25x}$

12. $\dfrac{a}{6y} - \dfrac{2b}{3y^4}$

13. $\dfrac{2}{5a} + \dfrac{1}{a} - \dfrac{a}{10}$

14. $\dfrac{1}{2A} - \dfrac{6}{B} - \dfrac{9}{4C}$

15. $\dfrac{x+1}{x} - \dfrac{x-3}{y} - \dfrac{2-x}{xy}$

16. $5 + \dfrac{1-x}{2} - \dfrac{3+x}{4}$

17. $\dfrac{3}{2x-1} + \dfrac{1}{4x-2}$

18. $\dfrac{5}{6y+3} - \dfrac{a}{8y+4}$

19. $\dfrac{4}{x(x+1)} - \dfrac{3}{2x}$

20. $\dfrac{3}{ax+ay} - \dfrac{1}{a^2}$

21. $\dfrac{s}{2s-6} + \dfrac{1}{4} - \dfrac{3s}{4s-12}$

22. $\dfrac{2}{x+2} - \dfrac{3-x}{x^2+2x} + \dfrac{1}{x}$

23. $\dfrac{3R}{R^2-9} - \dfrac{2}{3R+9}$

24. $\dfrac{2}{x^2+4x+4} - \dfrac{3}{x+2}$

25. $\dfrac{3}{x^2-8x+16} - \dfrac{2}{4-x}$

26. $\dfrac{1}{a^2-1} - \dfrac{2}{3-3a}$

27. $\dfrac{3}{x^2 - 11x + 30} - \dfrac{2}{x^2 - 25}$

28. $\dfrac{x - 1}{2x^3 - 4x^2} + \dfrac{5}{x - 2}$

29. $\dfrac{x - 1}{3x^2 - 13x + 4} - \dfrac{3x + 1}{4 - x}$

30. $\dfrac{x}{4x^2 - 12x + 5} + \dfrac{2x - 1}{4x^2 - 4x - 15}$

31. $\dfrac{t}{t^2 - t - 6} - \dfrac{2t}{t^2 + 6t + 9} + \dfrac{t}{t^2 - 9}$

32. $\dfrac{5}{2x^3 - 3x^2 + x} - \dfrac{x}{x^4 - x^2} + \dfrac{2 - x}{2x^2 + x - 1}$

33. $\dfrac{1}{w^3 + 1} + \dfrac{1}{w + 1} - 2$

34. $\dfrac{2}{8 - x^3} + \dfrac{1}{x^2 - x - 2}$

35. $\dfrac{\dfrac{1}{x}}{1 - \dfrac{1}{x}}$

36. $\dfrac{x - \dfrac{1}{x}}{1 - \dfrac{1}{x}}$

37. $\dfrac{\dfrac{x}{y} - \dfrac{y}{x}}{1 + \dfrac{y}{x}}$

38. $\dfrac{\dfrac{V^2 - 9}{V}}{\dfrac{1}{V} - \dfrac{1}{3}}$

39. $\dfrac{2 - \dfrac{1}{x} - \dfrac{2}{x + 1}}{\dfrac{1}{x^2 + 2x + 1} - 1}$

40. $\dfrac{\dfrac{2}{a} - \dfrac{1}{4} - \dfrac{3}{4a - 4b}}{\dfrac{1}{4a^2 - 4b^2} - \dfrac{2}{b}}$

41. $\dfrac{\dfrac{3}{x} + \dfrac{1}{x^2 + x}}{\dfrac{1}{x + 1} - \dfrac{1}{x - 1}}$

42. $\dfrac{\dfrac{1}{2x} - \dfrac{1}{4x^2 - 2x}}{\dfrac{6}{4x^2 - 1} - \dfrac{2}{2x + 1}}$

43. $\dfrac{\dfrac{r + s}{r - s} - \dfrac{r - s}{r + s}}{1 + \dfrac{r - s}{r + s}}$

44. $\dfrac{\dfrac{1}{u - v} + \dfrac{1}{2u + 2v}}{\dfrac{2u}{2u^2 - 3uv + v^2} + \dfrac{2}{2u - v}}$

The expression $f(x + h) - f(x)$ is frequently used in the study of calculus. (If necessary, refer to Section 3-1 for a review of functional notation.) In Exercises 45–48, determine and then simplify this expression for the given functions.

45. $f(x) = \dfrac{x}{x + 1}$

46. $f(x) = \dfrac{3}{2x - 1}$

47. $f(x) = \dfrac{1}{x^2}$

48. $f(x) = \dfrac{2}{x^2 + 4}$

In Exercises 49–55, simplify the given expressions. In Exercise 56, answer the given question.

49. Using the definitions of the trigonometric functions given in Section 4-2, find an expression that is equivalent to $(\tan \theta)(\cot \theta) + (\sin \theta)^2 - \cos \theta$, in terms of x, y, and r.

50. Using the definitions of the trigonometric functions given in Section 4-2, find an expression that is equivalent to $\sec \theta - (\cot \theta)^2 + \csc \theta$, in terms of x, y, and r.

51. If $f(x) = 2x - x^2$, find $f\left(\dfrac{1}{a}\right)$.

52. If $f(x) = x^2 + x$, find $f\left(a + \dfrac{1}{a}\right)$.

53. If $f(x) = x - \dfrac{2}{x}$, find $f(a + 1)$.

54. If $f(x) = \dfrac{x + 1}{2} + \dfrac{3}{x}$, find $f(2a)$.

55. The sum of two numbers a and b is divided by the sum of their reciprocals. Simplify the expression for this quotient.

(W) 56. When adding fractions, explain why it is better to find the lowest common denominator rather than any denominator that is common to the fractions.

In Exercises 57–64, perform the indicated operations. Each expression occurs in the indicated area of application.

57. $\dfrac{3}{4\pi} - \dfrac{3H_0}{4\pi H}$ (transistor theory)

58. $1 + \dfrac{9}{128T} - \dfrac{27P}{64T^3}$ (thermodynamics)

59. $\dfrac{2n^2 - n - 4}{2n^2 + 2n - 4} + \dfrac{1}{n - 1}$ (optics)

60. $\dfrac{b}{x^2 + y^2} - \dfrac{2bx^2}{x^4 + 2x^2y^2 + y^4}$ (magnetic field)

61. $\left(\dfrac{3Px}{2L^2}\right)^2 + \left(\dfrac{P}{2L}\right)^2$ (force on a weld)

62. $\dfrac{a}{b^2h} + \dfrac{c}{bh^2} - \dfrac{1}{6bh}$ (strength of materials)

63. $\dfrac{\dfrac{L}{C} + \dfrac{R}{sC}}{sL + R + \dfrac{1}{sC}}$ (electricity)

64. $\dfrac{\dfrac{m}{c}}{1 - \dfrac{p^2}{c^2}}$ (airfoil design)

$6\text{-}8$ EQUATIONS INVOLVING FRACTIONS

Many important equations in science and technology have fractions in them. Although the solution of these equations will still involve the use of the basic operations stated in Section 1-10, an additional procedure can be used to eliminate the fractions and thereby help lead to the solution. The method is to

NOTE ▶   *multiply each term of the equation by the lowest common denominator.*

The resulting equation will not involve fractions and can be solved by methods previously discussed. The following examples illustrate how to solve equations involving fractions.

EXAMPLE 1   Solve for x: $\dfrac{x}{12} - \dfrac{1}{8} = \dfrac{x+2}{6}$.

We first note that the lowest common denominator of the terms of the equation is 24. Therefore, we multiply each term by 24. This gives

$$\frac{24(x)}{12} - \frac{24(1)}{8} = \frac{24(x+2)}{6} \qquad \text{each term multiplied by LCD}$$

We reduce each term to its lowest terms and solve the resulting equation.

$$2x - 3 = 4(x+2) \qquad \text{each term reduced}$$
$$2x - 3 = 4x + 8$$
$$-2x = 11$$
$$x = -\frac{11}{2}$$

When we check this solution in the original equation, we obtain $-\frac{7}{12}$ on each side of the equal sign. Therefore, the solution is correct.   ▬

EXAMPLE 2   Solve for x: $\dfrac{x}{2} - \dfrac{1}{b^2} = \dfrac{x}{2b}$.

We first determine that the lowest common denominator of the terms of the equation is $2b^2$. We then multiply each term by $2b^2$ and continue with the solution.

$$\frac{2b^2(x)}{2} - \frac{2b^2(1)}{b^2} = \frac{2b^2(x)}{2b} \qquad \text{each term multiplied by LCD}$$
$$b^2 x - 2 = bx \qquad \text{each term reduced}$$
$$b^2 x - bx = 2$$
$$x(b^2 - b) = 2 \qquad \text{factor}$$
$$x = \frac{2}{b^2 - b}$$

Note the use of factoring in arriving at the final result. Checking shows that each side of the original equation is equal to $\dfrac{1}{b^2(b-1)}$. (The check here is somewhat lengthy.)   ▬

See the chapter introduction.

The first telescope was invented by Lippershay, a Dutch lens maker, in about 1608.

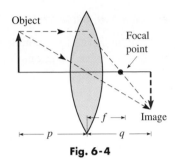

Object

Focal point

f Image

p q

Fig. 6-4

For reference, Eq. (6-11) is $x - y = -(y - x)$.

See the chapter introduction.

In 1609 the Italian scientist Galileo (1564–1642) learned of the invention of the telescope and developed it for astronomical observations. Among his first discoveries were the four largest moons of the planet Jupiter.

■**EXAMPLE 3**   An equation relating the focal length f of a lens with the object distance p and the image distance q is given below. See Fig. 6-4. Solve for q.

$$f = \frac{pq}{p + q} \qquad \text{given equation}$$

Since the only denominator is $p + q$, the LCD is also $p + q$. By first multiplying each term by $p + q$, the solution is completed as follows:

$$f(p + q) = \frac{pq(p + q)}{p + q} \qquad \text{each term multiplied by LCD}$$

$$fp + fq = pq \qquad \text{reduce term on right}$$

$$fq - pq = -fp$$

$$q(f - p) = -fp \qquad \text{factor}$$

$$q = \frac{-fp}{f - p} \qquad \text{divide by } f - p$$

$$= -\frac{fp}{-(p - f)} \qquad \text{use Eq. (6-11)}$$

$$= \frac{fp}{p - f}$$

The last form is preferred since there is no minus sign before the fraction. However, either form of the result is correct.   ◼

■**EXAMPLE 4**   When developing the equations that describe the motion of the planets, the equation

$$\frac{1}{2}v^2 - \frac{GM}{r} = -\frac{GM}{2a}$$

is found. Solve for M.

We first determine that the lowest common denominator of the terms of the equation is $2ar$. Multiplying each term by $2ar$ and proceeding, we have

$$\frac{2ar(v^2)}{2} - \frac{2ar(GM)}{r} = -\frac{2ar(GM)}{2a} \qquad \text{each term multiplied by LCD}$$

$$arv^2 - 2aGM = -rGM \qquad \text{each term reduced}$$

$$rGM - 2aGM = -arv^2$$

$$M(rG - 2aG) = -arv^2 \qquad \text{factor}$$

$$M = -\frac{arv^2}{rG - 2aG}$$

$$= \frac{arv^2}{2aG - rG}$$

The second form of the result is obtained by using Eq. (6-11). Again, note the use of factoring to arrive at the final result.   ◼

EXAMPLE 5 Solve for x: $\dfrac{2}{x+1} - \dfrac{1}{x} = -\dfrac{2}{x^2+x}$.

Multiplying each term by the lowest common denominator $x(x+1)$, we have

$$\frac{2(x)(x+1)}{x+1} - \frac{x(x+1)}{x} = -\frac{2x(x+1)}{x(x+1)}$$

Now, simplifying each fraction, we have

$$2x - (x+1) = -2$$

We now complete the solution.

$$2x - x - 1 = -2$$
$$x = -1$$

Checking this solution *in the original equation,* we see that we have zero in the denominators of the first and third terms of the equation. Since division by zero is undefined (see Section 1-2), $x = -1$ cannot be a solution. *Thus there is* **no solution** *to this equation.* This example points out clearly why it is necessary to check solutions in the original equation. It also shows that *whenever we multiply each term by a common denominator that* **contains the unknown,** *it is possible to obtain a value that is not a solution of the original equation. Such a value is termed an* **extraneous solution.** Only certain equations will lead to extraneous solutions, but we must be careful to identify them when they occur. ▬

CAUTION ▶

EXTRANEOUS SOLUTIONS

A number of stated problems give rise to equations involving fractions. The following example illustrates the solution of such a problem.

SOLVING A WORD PROBLEM

EXAMPLE 6 An industrial firm uses a computer system that processes and prints out its data for an average day in 20 h. In order to process the data more rapidly and to handle increased future computer needs, the firm plans to add new components to the system. One set of new components can process the data in 12 h, without the present system. How long would it take the new system, a combination of the present system and the new components, to process the data?

First, we let x = the number of hours for the new system to process the data. Next, we know that it takes the present system 20 h to do it. This means that it processes $\frac{1}{20}$ of the data in one hour, or $\frac{1}{20}x$ of the data in x hours. In the same way, the new components can process $\frac{1}{12}x$ of the data in x hours. When x hours have passed, the new system will have processed all of the data. Therefore,

<table>
<tr><td align="center">part of data processed
by present system</td><td></td><td align="center">part of data processed
by new components</td><td></td><td></td><td></td></tr>
<tr><td align="center">$\dfrac{x}{20}$</td><td align="center">$+$</td><td align="center">$\dfrac{x}{12}$</td><td align="center">$=$</td><td align="center">1</td><td align="center">one complete processing
(all of data)</td></tr>
</table>

$$\frac{60x}{20} + \frac{60x}{12} = 60(1) \qquad \text{each term multiplied by LCD of 60}$$

$$3x + 5x = 60 \qquad \text{each term reduced}$$

$$8x = 60$$

$$x = \frac{60}{8} = 7.5 \text{ h}$$

Therefore, the new system should take about 7.5 h to process the data. ▬

The first large-scale electronic computer was the ENIAC. It was constructed at the Univ. of Pennsylvania in the mid-1940s and used until 1955. It had 18,000 vacuum tubes and occupied 15,000 ft^2 of floor space.

EXERCISES 6-8

In Exercises 1–28, solve the given equations and check the results.

1. $\dfrac{x}{2} + 6 = 2x$

2. $\dfrac{x}{5} + 2 = \dfrac{15 + x}{10}$

3. $\dfrac{x}{6} - \dfrac{1}{2} = \dfrac{x}{3}$

4. $\dfrac{3N}{8} - \dfrac{3}{4} = \dfrac{N - 4}{2}$

5. $\dfrac{1}{2} - \dfrac{t - 5}{6} = \dfrac{3}{4}$

6. $\dfrac{2x - 7}{3} + 5 = \dfrac{1}{5}$

7. $\dfrac{3x}{7} - \dfrac{5}{21} = \dfrac{2 - x}{14}$

8. $\dfrac{x - 3}{12} - \dfrac{2}{3} = \dfrac{1 - 3x}{2}$

9. $\dfrac{3}{x} + 2 = \dfrac{5}{3}$

10. $\dfrac{1}{2y} - \dfrac{1}{2} = 4$

11. $3 - \dfrac{x - 2}{5x} = \dfrac{1}{5}$

12. $\dfrac{1}{2R} - \dfrac{1}{3} = \dfrac{2}{3R}$

13. $\dfrac{2y}{y - 1} = 5$

14. $\dfrac{x}{2x - 3} = 4$

15. $\dfrac{2}{s} = \dfrac{3}{s - 1}$

16. $\dfrac{5}{n + 2} = \dfrac{3}{2n}$

17. $\dfrac{5}{2x + 4} + \dfrac{3}{x + 2} = 2$

18. $\dfrac{3}{4x - 6} + \dfrac{1}{4} = \dfrac{5}{2x - 3}$

19. $\dfrac{2}{Z - 5} - \dfrac{3}{10 - 2Z} = 3$

20. $\dfrac{4}{4 - x} + 2 - \dfrac{2}{12 - 3x} = \dfrac{1}{3}$

21. $\dfrac{1}{x} + \dfrac{3}{2x} = \dfrac{2}{x + 1}$

22. $\dfrac{3}{t + 3} - \dfrac{1}{t} = \dfrac{5}{2t + 6}$

23. $\dfrac{7}{y} = \dfrac{3}{y - 4} + \dfrac{7}{2y^2 - 8y}$

24. $\dfrac{1}{2x + 3} = \dfrac{5}{2x} - \dfrac{4}{2x^2 + 3x}$

25. $\dfrac{1}{x^2 - x} - \dfrac{1}{x} = \dfrac{1}{x - 1}$

26. $\dfrac{2}{x^2 - 1} - \dfrac{2}{x + 1} = \dfrac{1}{x - 1}$

27. $\dfrac{2}{B^2 - 4} - \dfrac{1}{B - 2} = \dfrac{1}{2B + 4}$

28. $\dfrac{2}{2x^2 + 5x - 3} - \dfrac{1}{4x - 2} + \dfrac{3}{2x + 6} = 0$

In Exercises 29–44, solve for the indicated letter. In Exercises 33–44, each of the given formulas arises in the technical or scientific area of study listed.

29. $2 - \dfrac{1}{b} + \dfrac{3}{c} = 0$, for c

30. $\dfrac{2}{3} - \dfrac{h}{x} = \dfrac{1}{6x}$, for x

31. $\dfrac{t - 3}{b} - \dfrac{t}{2b - 1} = \dfrac{1}{2}$, for t

32. $\dfrac{1}{a^2 + 2a} - \dfrac{y}{2a} = \dfrac{2y}{a + 2}$, for y

33. $\dfrac{s - s_0}{t} = \dfrac{v + v_0}{2}$, for v (velocity of object)

34. $S = \dfrac{P}{A} + \dfrac{Mc}{I}$, for P (machine design)

35. $V_0 = \dfrac{V_r A}{1 + \beta A}$, for β (electricity)

36. $K = \dfrac{ax}{x + b}$, for x (medicine)

37. $z = \dfrac{1}{g_m} - \dfrac{jX}{g_m R}$, for R (FM transmission)

38. $A = \dfrac{1}{2} wp - \dfrac{1}{2} w^2 - \dfrac{\pi}{8} w^2$, for p (architecture)

39. $P = \dfrac{RT}{V - b} - \dfrac{a}{V^2}$, for T (thermodynamics)

40. $\dfrac{1}{x} + \dfrac{1}{nx} = \dfrac{1}{f}$, for n (photography)

41. $\dfrac{1}{R_1} = \dfrac{N_2^2}{N_1^2 R_2} + \dfrac{N_3^2}{N_1^2 R_3}$, for R_1 (transformer resistance)

42. $D = \dfrac{wx^4}{24EI} - \dfrac{wLx^3}{6EI} + \dfrac{wL^2x^2}{4EI}$, for w (beam design)

43. $\dfrac{1}{f} = (n - 1)\left(\dfrac{1}{R_1} + \dfrac{1}{R_2}\right)$, for R_1 (optics)

44. $P = \dfrac{\dfrac{1}{1 + i}}{1 - \dfrac{1}{1 + i}}$, for i (business)

In Exercises 45–52, set up appropriate equations and solve the given stated problems. All numbers are accurate to at least two significant digits.

45. One pipe can fill a certain oil storage tank in 4.0 h, and a second pipe can fill it in 6.0 h. How long will it take to fill the tank if both pipes operate together?

46. One company determines that it will take its crew 450 h to clean up a chemical dump site, and a second company determines that it will take its crew 600 h to clean up the site. How long will it take the two crews working together?

47. One automatic packaging machine can package 100 boxes of machine parts in 12 min, and a second machine can do it in 10 min. A newer model machine can do it in 8.0 min. How long would it take the three machines working together?

48. A painting crew can paint a structure in 12 h, or the crew can paint it in 7.2 h when working with a second crew. How long would it take the second crew to do the job if working alone?

49. A jet takes the same time to travel 2580 km with the wind as it does to travel 1800 km against the wind. If its speed relative to the air is 450 km/h, what is the speed of the wind?

50. An engineer travels from Aberdeen, Scotland, to the Montrose oil field in the North Sea on a ship that averages 28 km/h. After spending 6.0 h at the field, the engineer returns to Aberdeen in a helicopter that averages 140 km/h. If the total trip takes 15.0 h, how far is the Montrose oil field from Aberdeen?

51. The current through each of the resistances R_1 and R_2 in Fig. 6-5 equals the voltage V divided by the resistance. The sum of the currents equals the current i in the rest of the circuit. Find the voltage if $i = 1.2$ A, $R_1 = 2.7$ Ω, and $R_2 = 6.0$ Ω.

52. A fox, pursued by a greyhound, has a start of 60 leaps. He makes 9 leaps while the greyhound makes but 6; but, 3 leaps of the greyhoud are equivalent to 7 of the fox. How many leaps must the greyhound make to overcome the fox? (Copied from Davies, Charles. *Elementary Algebra,* New York; A.S. Barnes & Burr, 1852.) (Hint: Let the unit of distance be one fox leap.)

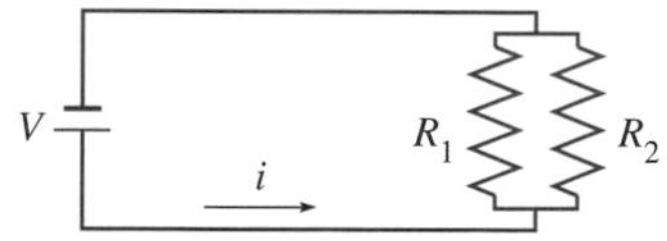

Fig. 6-5

CHAPTER EQUATIONS

$$a(x + y) = ax + ay \tag{6-1}$$
$$(x + y)(x - y) = x^2 - y^2 \tag{6-2}$$
$$(x + y)^2 = x^2 + 2xy + y^2 \tag{6-3}$$
$$(x - y)^2 = x^2 - 2xy + y^2 \tag{6-4}$$
$$(x + a)(x + b) = x^2 + (a + b)x + ab \tag{6-5}$$
$$(ax + b)(cx + d) = acx^2 + (ad + bc)x + bd \tag{6-6}$$
$$(x + y)^3 = x^3 + 3x^2y + 3xy^2 + y^3 \tag{6-7}$$
$$(x - y)^3 = x^3 - 3x^2y + 3xy^2 - y^3 \tag{6-8}$$
$$(x + y)(x^2 - xy + y^2) = x^3 + y^3 \tag{6-9}$$
$$(x - y)(x^2 + xy + y^2) = x^3 - y^3 \tag{6-10}$$
$$x - y = -(y - x) \tag{6-11}$$

REVIEW EXERCISES

In Exercises 1–12, find the products by inspection. No intermediate steps should be necessary.

1. $3a(4x + 5a)$
2. $-7xy(4x^2 - 7y)$
3. $(2a + 7b)(2a - 7b)$
4. $(x - 4z)(x + 4z)$
5. $(2a + 1)^2$
6. $(4x - 3y)^2$
7. $(b - 4)(b + 7)$
8. $(y - 5)(y - 7)$
9. $(2x + 5)(x - 9)$
10. $(4ax - 3)(5ax + 7)$
11. $(2c^a + d)(2c^a - d)$
12. $(3s^n - 2t)(3s^n + 2t)$

In Exercises 13–44, factor the given expressions completely.

13. $3s + 9t$
14. $7x - 28y$
15. $a^2x^2 + a^2$
16. $3ax - 6ax^4 - 9a$
17. $W^2 - 144$
18. $900 - n^2$
19. $16(x + 2)^2 - t^4$
20. $25s^4 - 36t^2$
21. $9t^2 - 6t + 1$
22. $4x^2 - 12x + 9$
23. $25t^2 + 10t + 1$
24. $4x^2 + 36xy + 81y^2$
25. $x^2 + x - 56$
26. $x^2 - 4x - 45$
27. $t^4 - 5t^2 - 36$
28. $N^4 - 11N^2 + 10$

29. $2x^2 - x - 36$
30. $5x^2 + 2x - 3$
31. $4x^2 - 4x - 35$
32. $9x^2 + 7x - 16$
33. $10b^2 + 23b - 5$
34. $12x^2 - 7xy - 12y^2$
35. $4x^2 - 64y^2$
36. $4a^2x^2 + 26a^2x + 36a^2$
37. $250 - 16y^6$
38. $a^4 + 64a$
39. $8x^3 + 27$
40. $R^3 - 125r^3$
41. $ab^2 - 3b^2 + a - 3$
42. $axy - ay + ax - a$
43. $nx + 5n - x^2 + 25$
44. $ty - 4t + y^2 - 16$

In Exercises 45–68, perform the indicated operations and express results in simplest form.

45. $\dfrac{48ax^3y^6}{9a^3xy^6}$

46. $\dfrac{-39r^2s^4t^8}{52rs^5t}$

47. $\dfrac{6x^2 - 7x - 3}{4x^2 - 8x + 3}$

48. $\dfrac{p^4 - 4p^2 - 4}{p^4 - p^2 - 12}$

49. $\dfrac{4x + 4y}{35x^2} \times \dfrac{28x}{x^2 - y^2}$

50. $\left(\dfrac{6x - 3}{x^2}\right)\left(\dfrac{4x^2 - 12x}{12x - 6}\right)$

51. $\dfrac{18 - 6x}{x^2 - 6x + 9} \div \dfrac{x^2 - 2x - 15}{x^2 - 9}$

52. $\dfrac{6x^2 - xy - y^2}{2x^2 + xy - y^2} \div \dfrac{4x^2 - 16y^2}{x^2 + 3xy + 2y^2}$

53. $\dfrac{\dfrac{3x}{7x^2 + 13x - 2}}{\dfrac{6x^2}{x^2 + 4x + 4}}$

54. $\dfrac{\dfrac{3x - 3y}{2x^2 + 3xy - 2y^2}}{\dfrac{3x^2 - 3y^2}{x^2 + 4xy + 4y^2}}$

55. $\dfrac{x + \dfrac{1}{x} + 1}{x^2 - \dfrac{1}{x}}$

56. $\dfrac{\dfrac{4}{y} - 4y}{2 - \dfrac{2}{y}}$

57. $\dfrac{4}{9x} - \dfrac{5}{12x^2}$

58. $\dfrac{3}{10a^2} + \dfrac{1}{4a^3}$

59. $\dfrac{6}{x} - \dfrac{7}{2x} + \dfrac{3}{xy}$

60. $\dfrac{T}{T^2 + 2} - \dfrac{1}{T^3 + 2T}$

61. $\dfrac{a + 1}{a + 2} - \dfrac{a + 3}{a}$

62. $\dfrac{y}{y + 2} - \dfrac{1}{y^2 + 2y}$

63. $\dfrac{2x}{x^2 + 2x - 3} - \dfrac{1}{x^2 + 3x}$

64. $\dfrac{x}{4x^2 + 4x - 3} - \dfrac{3}{4x^2 - 9}$

65. $\dfrac{3x}{2x^2 - 2} - \dfrac{2}{4x^2 - 5x + 1}$

66. $\dfrac{2x - 1}{4 - x} + \dfrac{x + 2}{5x - 20}$

67. $\dfrac{3x}{x^2 + 2x - 3} - \dfrac{2}{x^2 + 3x} + \dfrac{x}{x - 1}$

68. $\dfrac{3}{y^4 - 2y^3 - 8y^2} + \dfrac{y - 1}{y^2 + 2y} - \dfrac{y - 3}{y^2 - 4y}$

In Exercises 69–76, solve the given equations.

69. $\dfrac{x}{2} - 3 = \dfrac{x - 10}{4}$

70. $\dfrac{2x}{c} - \dfrac{1}{2c} = \dfrac{3}{c} - x$, for x

71. $\dfrac{2}{t} - \dfrac{1}{at} = 2 + \dfrac{a}{t}$, for t

72. $\dfrac{3}{a^2 y} - \dfrac{1}{ay} = \dfrac{9}{a}$, for y

73. $\dfrac{2x}{x^2 - 3x} - \dfrac{3}{x} = \dfrac{1}{2x - 6}$

74. $\dfrac{3}{x^2 + 3x} - \dfrac{1}{x} = \dfrac{1}{x + 3}$

75. Given $f(x) = \dfrac{1}{x + 2}$, solve for x if $f(x + 2) = 2f(x)$.

76. Given $f(x) = \dfrac{x}{x + 1}$, solve for x if $3f(x) + f\left(\dfrac{1}{x}\right) = 2$.

In Exercises 77–92, perform the given operations. Where indicated, the expression is found in the stated technical area.

77. Show that $xy = \dfrac{1}{4}[(x + y)^2 - (x - y)^2]$.

78. Show that $x^2 + y^2 = \dfrac{1}{2}[(x + y)^2 + (x - y)^2]$.

79. Multiply: $2zS(S + 1)$.   (solid-state physics)

80. Expand: $[2b + (n - 1)\lambda]^2$.   (optics)

81. Factor: $\pi r_1^2 l - \pi r_2^2 l$.   (jet plane fuel supply)

82. Factor: $cT_2 - cT_1 + RT_2 - RT_1$.   (pipeline flow)

83. Factor: $4s^3 + 56s^2 + 96s$.   (electricity)

84. Factor: $9600t + 8400t^2 - 1200t^3$.   (solar energy)

85. Simplify and express in factored form: $(2R - r)^2 - (r^2 + R^2)$ (aircraft radar)

86. Express in factored form: $2R(R + r) - (R + r)^2$. (electricity: power)

87. Expand and simplify: $(n + 1)^3(2n + 1)^3$. (fluid flow in pipes)

88. Expand and simplify: $2(e_1 - e_2)^2 + 2(e_2 - e_3)^2$. (mechanical design)

89. Expand and simplify: $10a(T - t) + a(T - t)^2$. (instrumentation)

90. Expand the third term and then factor by grouping: $pa^2 + (1 - p)b^2 - [pa + (1 - p)b]^2$. (nuclear physics)

91. A metal cube of edge x is heated and each edge increases by 4 mm. Express the increase in volume in factored form.

92. Express the difference in volumes of two ball bearings of radii r mm and 3 mm in factored form. ($r > 3$mm)

In Exercises 93–104, perform the indicated operations and simplify the given expressions. Each expression is from the indicated technical area of application.

93. $\left(\dfrac{2wtv^2}{Dg}\right)\left(\dfrac{b\pi^2 D^2}{n^2}\right)\left(\dfrac{6}{bt^2}\right)$ (machine design)

94. $\dfrac{m}{c} \div \left[1 - \left(\dfrac{p}{c}\right)^2\right]$ (airfoil design)

95. $\dfrac{\dfrac{\pi ka}{2}(R^4 - r^4)}{\pi ka(R^2 - r^2)}$ (flywheel rotation)

96. $\dfrac{V}{kp} - \dfrac{RT}{k^2 p^2}$ (electric motors)

97. $1 - \dfrac{d^2}{2} + \dfrac{d^4}{24} - \dfrac{d^6}{120}$ (aircraft emergency locator transmitter)

98. $\dfrac{wx^2}{2T_0} + \dfrac{kx^4}{12T_0}$ (bridge design)

99. $\dfrac{N + n}{2} + \dfrac{(N - n)^2}{4\pi^2 C}$ (machine design)

100. $\dfrac{Am}{k} - \dfrac{g}{2}\left(\dfrac{m}{k}\right)^2 + \dfrac{AML}{k}$ (rocket fuel)

101. $1 - \dfrac{3a}{4r} - \dfrac{a^3}{4r^3}$ (hydrodynamics)

102. $\dfrac{1}{F} + \dfrac{1}{f} - \dfrac{d}{fF}$ (optics)

103. $\dfrac{\dfrac{u^2}{2g} - x}{\dfrac{1}{2gc^2} - \dfrac{u^2}{2g} + x}$ (mechanism design)

104. $\dfrac{V}{\dfrac{1}{2R} + \dfrac{1}{2R + 2}}$ (electricity)

In Exercises 105–112, solve for the indicated letter. Each equation is from the indicated technical area of application.

105. $W = mgh_2 - mgh_1$, for m (work done on object)

106. $R_1(V_2 - V_1) + R_2 V_2 = 3R_1 R_2$, for V_2 (electricity)

107. $R = \dfrac{wL}{H(w + L)}$, for L (architecture)

108. $\dfrac{q_2 - q_1}{d} = \dfrac{f + q_1}{D}$, for q_1 (photography)

109. $s^2 + \dfrac{cs}{m} + \dfrac{kL^2}{mb^2} = 0$, for c (mechanical vibrations)

110. $I = \dfrac{A}{x^2} + \dfrac{B}{(10 - x)^2}$, for A (optics)

111. $V = \dfrac{i}{sC} + \dfrac{V_0}{s}$, for C (electricity)

112. $C = \dfrac{k}{n(1 - k)}$, for k (reinforced concrete design)

In Exercises 113–120, set up appropriate equations and solve the given stated problems. All numbers are accurate to at least two significant digits.

113. If a certain car's lights are left on, the battery will be dead in 4.0 h. If only the radio is left on, the battery will be dead in 24 h. How long will the battery last if both the lights and the radio are left on?

114. Two pumps are being used to fight a fire. One pumps 5000 gal in 20 min, and the other pumps 5000 gal in 25 min. How long will it take the two pumps together to pump 5000 gal?

115. One computer can solve a certain problem in 3.0 s. With the aid of a second computer, the problem is solved in 1.0 s. How long would the second computer take to solve the problem alone?

116. An auto mechanic can do a certain motor job in 3.0 h, and with an assistant he can do it in 2.1 h. How long would it take the assistant to do the job alone?

117. The *relative density* of an object may be defined as its weight in air w_a, divided by the difference of its weight in air and its weight when submerged in water, w_w. For a lead weight, $w_a = 1.097 \, w_w$. Find the relative density of lead.

118. A car travels halfway to its destination at 80.0 km/h and the remainder of the distance at 60.0 km/h. What is the average speed of the car for the trip?

119. For electric resistors in parallel, the reciprocal of the combined resistance equals the sum of the reciprocals of the individual resistances. For three resistors of 12 Ω, R ohms, and $2R$ ohms in parallel, the combined resistance is 6.0 Ω. Find R.

120. An ambulance averaged 36 mi/h going to an accident and 48 mi/h on its return to the hospital. If the total time for the round-trip was 40 min, including 5 min at the accident scene, how far from the hospital was the accident?

Writing Exercise

121. An architecture student encounters the fraction

$$\frac{2r^2 + 5r - 3}{2r^2 + 7r + 3}$$

Simplify this fraction, and write a paragraph to describe your procedure. When you "cancel," explain what basic operation is being performed.

PRACTICE TEST

1. Find the product: $2x(2x - 3)^2$.

2. The following equation is used in electricity.

Solve for R_1: $\dfrac{1}{R} = \dfrac{1}{R_1 + r} + \dfrac{1}{R_2}$.

3. Reduce to simplest form: $\dfrac{2x^2 + 5x - 3}{2x^2 + 12x + 18}$.

4. Factor: $4x^2 - 16y^2$.

In Problems 5–7, perform the indicated operations and simplify.

5. $\dfrac{3}{4x^2} - \dfrac{2}{x^2 - x} - \dfrac{x}{2x - 2}$

6. $\dfrac{x^2 + x}{2 - x} \div \dfrac{x^2}{x^2 - 4x + 4}$

7. $\dfrac{1 - \dfrac{3}{2x + 2}}{\dfrac{x}{5} - \dfrac{1}{2}}$

8. If one riveter can do a job in 12 days, and a second riveter can do it in 16 days, how long would it take for them to do it together?

QUADRATIC EQUATIONS

In Section 7-3 we see how the design of a swimming pool and the surrounding pool area involves the solution of a quadratic equation.

The solution of basic equations was introduced in Chapter 1. Then in Chapter 5, we extended the solution of equations to systems of linear equations. There are many other types of equations that we shall discuss in this chapter and in later chapters. In this chapter we consider the important *quadratic equation.* To develop methods of solution of the quadratic equation, we use factoring and the algebraic operations with fractions discussed in Chapter 6, as well as graphical methods.

There are many applications of quadratic equations. These include projectile motion, electric circuits, architecture, mechanical systems, forces on structures, and product design.

In the first three sections of this chapter, we present algebraic methods of solving quadratic equations. In Section 7-4 we discuss graphical solutions, including the use of the graphing calculator.

7-1 QUADRATIC EQUATIONS; SOLUTION BY FACTORING

Given that a, b, and c are constants (a ≠ 0), the equation

$$ax^2 + bx + c = 0 \qquad\qquad (7\text{-}1)$$

is called the **general quadratic equation in x.** The left side of Eq. (7-1) is a polynomial function of degree 2. *This function, $f(x) = ax^2 + bx + c$, is known as the* **quadratic function.** Any equation that can be simplified and then written in the form of Eq. (7-1) is a quadratic equation in one unknown.

Among the applications of quadratic equations and functions we have the following examples: In finding the time t of flight of a projectile, we have the equation $s_0 + v_0t - 16t^2 = 0$; in analyzing the electric current i in a circuit, the function $f(i) = Ei - Ri^2$ is found; and in determining the forces at a distance x along a beam, the function $f(x) = ax^2 + bLx + cL^2$ is used.

Since it is the x^2-term in Eq. (7-1) that distinguishes the quadratic equation from other types of equations, the equation is not quadratic if $a = 0$. However, either b or c (or both) may be zero, and the equation is still quadratic. No power of x higher than the second may be present in a quadratic equation. Also, we should be able to properly identify a quadratic equation even when it does not initially appear in the form of Eq. (7-1). The following two examples illustrate how we may recognize quadratic equations.

■EXAMPLE 1   The following are quadratic equations.

$$\underset{\underset{a=1}{\uparrow}}{x^2} - \underset{\underset{b=-4}{\uparrow}}{4x} - \underset{\underset{c=-5}{\uparrow}}{5} = 0$$

To show this equation in the form of Eq. (7-1), it can be written as $1x^2 + (-4)x + (-5) = 0$.

$$\underset{\underset{a=3}{\uparrow}}{3x^2} - \underset{\underset{c=-6}{\uparrow}}{6} = 0$$

Since there is no x-term, $b = 0$.

$$\underset{\underset{a=2}{\uparrow}}{2x^2} + \underset{\underset{b=7}{\uparrow}}{7x} = 0$$

Since no constant appears, $c = 0$.

$(a - 3)x^2 - ax + 7 = 0$ The constants in Eq. (7-1) may include literal expressions. In this case, $a - 3$ takes the place of a, $-a$ takes the place of b, and $c = 7$.

$4x^2 - 2x = x^2$ After all nonzero terms have been collected on the left side, the equation becomes $3x^2 - 2x = 0$.

$(x + 1)^2 = 4$ Expanding the left side and collecting all nonzero terms on the left, we have $x^2 + 2x - 3 = 0$. ■

> If either (or both) the first power term or the constant is missing, the quadratic is called *incomplete*.

■EXAMPLE 2   The following are not quadratic equations.

$bx - 6 = 0$ There is no x^2-term.

$x^3 - x^2 - 5 = 0$ There should be no term of degree higher than 2. Thus there can be no x^3-term in a quadratic equation.

$x^2 + x - 7 = x^2$ When terms are collected, there will be no x^2-term. ■

Solutions of a Quadratic Equation

> Until the 1600s most mathematicians did not accept negative, irrational, or imaginary roots of an equation. It was also generally accepted that an equation had only one root.

From our previous work with equations, we recall that *the* **solution** *of an equation consists of all numbers* (**roots**) *which, when substituted in the equation, produce equality.* There are *two* such roots for a quadratic equation. Occasionally these roots turn out to be equal, as is shown in Example 3, and only one number is actually a solution. Also, due to the presence of the x^2-term, the roots may be imaginary numbers. If the roots are imaginary, at this point all we wish to do is to recognize them when they occur.

EXAMPLE 3 (a) The quadratic equation

$$3x^2 - 7x + 2 = 0$$

has roots $x = 1/3$ and $x = 2$. This can be seen by substituting in the equation.

$$3\left(\frac{1}{3}\right)^2 - 7\left(\frac{1}{3}\right) + 2 = 3\left(\frac{1}{9}\right) - \frac{7}{3} + 2 = \frac{1}{3} - \frac{7}{3} + 2 = \frac{0}{3} = 0$$

$$3(2)^2 - 7(2) + 2 = 3(4) - 14 + 2 = 12 - 14 + 2 = 0$$

(b) The quadratic equation

$$4x^2 - 4x + 1 = 0$$

has the **double root** *(both roots are the same)* of $x = \frac{1}{2}$. This can be seen to be a solution by substitution:

$$4\left(\frac{1}{2}\right)^2 - 4\left(\frac{1}{2}\right) + 1 = 4\left(\frac{1}{4}\right) - 2 + 1 = 1 - 2 + 1 = 0$$

(c) The quadratic equation

$$x^2 + 9 = 0$$

has the imaginary roots of $\sqrt{-3}$ and $-\sqrt{-3}$. ────────■

In this section we shall deal only with quadratic equations whose quadratic expression is factorable. Therefore, all roots will be rational. Using the fact that

NOTE▶
a product is zero if any of its factors is zero

we have the following steps in solving a quadratic equation.

Procedure for Solving a Quadratic Equation by Factoring
1. *Collect all terms on the left and simplify* (to the form of Eq. (7-1)).
2. *Factor the quadratic expression.*
3. *Set each factor equal to zero.*
4. *Solve the resulting linear equations.*
5. *Check the solutions in the original equation.*

The solutions of the resulting linear equations make up the solution of the quadratic equation.

EXAMPLE 4 $x^2 - x - 12 = 0$

$$(x - 4)(x + 3) = 0 \qquad \text{factor}$$

$$x - 4 = 0 \qquad x + 3 = 0 \qquad \text{set each factor equal to zero}$$

$$x = 4 \qquad x = -3 \qquad \text{solve}$$

The roots are $x = 4$ and $x = -3$. We can check them in the original equation by substitution. Therefore, we have

$$(4)^2 - (4) - 12 \overset{?}{=} 0 \qquad (-3)^2 - (-3) - 12 \overset{?}{=} 0$$

$$0 = 0 \qquad\qquad\qquad 0 = 0$$

Both roots satisfy the original equation. ────────■

EXAMPLE 5 $2x^2 + 7x - 4 = 0$

$$(2x - 1)(x + 4) = 0 \qquad \text{factor}$$

$$2x - 1 = 0 \quad \text{or} \quad x = \frac{1}{2} \qquad \begin{array}{l}\text{set each factor equal}\\\text{to zero and solve}\end{array}$$

$$x + 4 = 0 \quad \text{or} \quad x = -4$$

Therefore, the roots are $x = \frac{1}{2}$ and $x = -4$. These roots can be checked by the same procedure used in Example 4.

EXAMPLE 6 $x^2 + 4 = 4x$ equation not in form of Eq. (7-1)

$$x^2 - 4x + 4 = 0 \qquad \text{subtract } 4x \text{ from both sides}$$

$$(x - 2)^2 = 0 \qquad \text{factor}$$

$$x - 2 = 0 \quad \text{or} \quad x = 2 \qquad \text{solve}$$

Since $(x - 2)^2 = (x - 2)(x - 2)$, both factors are the same. This means there is a double root of $x = 2$. Substitution shows that $x = 2$ satisfies the original equation.

It is essential for the quadratic expression on the left to be equal to zero (on the right), because if a product equals a nonzero number, it is probable that neither factor will give us a correct root. Again, the first step must be to write the equation in the form of Eq. (7-1).

EQUATIONS WITH FRACTIONS   A number of equations involving fractions lead to quadratic equations after the fractions are eliminated. The following two examples, the second being a stated problem, illustrate the process of solving such equations with fractions.

EXAMPLE 7   Solve for x: $\dfrac{1}{x} + 3 = \dfrac{2}{x + 2}$.

$$\frac{x(x + 2)}{x} + 3x(x + 2) = \frac{2x(x + 2)}{x + 2} \qquad \begin{array}{l}\text{multiply each term by the LCD,}\\x(x + 2)\end{array}$$

$$x + 2 + 3x^2 + 6x = 2x \qquad \text{reduce each term}$$

$$3x^2 + 5x + 2 = 0 \qquad \text{collect terms on left}$$

$$(3x + 2)(x + 1) = 0 \qquad \text{factor}$$

$$3x + 2 = 0 \quad \text{or} \quad x = -\tfrac{2}{3} \qquad \text{set each factor equal to zero and solve}$$

$$x + 1 = 0 \quad \text{or} \quad x = -1$$

Checking in the original equation, we have

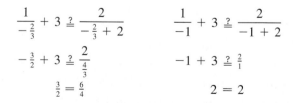

We see that the roots check. Remember, if either value gives division by zero, the root is extraneous, and must be excluded from the solution.

SOLVING A WORD PROBLEM

■EXAMPLE 8 A lumber truck travels 60 mi from a sawmill to a lumber camp and then back in 7 h travel time. If the truck averages 5 mi/h less on the return trip than on the trip to the camp, find its average speed to the camp. See Fig. 7-1.

Let v = the average speed (in mi/h) of the truck going to the camp. This means that the average speed of the return trip was $v - 5$ mi/h.

We also know that $d = vt$ (distance equals speed times time), which tells us that $t = d/v$. Thus, the time for each part of the trip is the distance divided by the speed. This leads to the equation

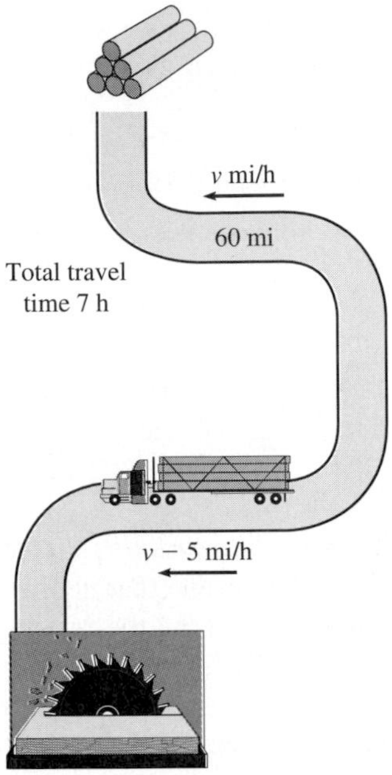

time to camp	time from camp	total time

$$\frac{60}{v} + \frac{60}{v - 5} = 7$$

$$60(v - 5) + 60v = 7v(v - 5) \qquad \text{multiply each term by } v(v - 5)$$

$$7v^2 - 155v + 300 = 0 \qquad \text{collect terms on the left}$$

$$(7v - 15)(v - 20) = 0 \qquad \text{factor}$$

$$7v - 15 = 0 \quad \text{or} \quad v = \frac{15}{7} \qquad \text{set each factor equal to zero and solve}$$

$$v - 20 = 0 \quad \text{or} \quad v = 20$$

The factors lead to two possible solutions, but only one has meaning for this problem. The value $v = \frac{15}{7}$ cannot be the solution, since the return speed of 5 mi/h less would be negative. Therefore, the solution is $v = 20$ mi/h, which means that the return speed was 15 mi/h. Thus, the trip to the camp took 3 h, and the return trip took 4 h, which therefore shows that the solution checks. ------------■

v mi/h

60 mi

Total travel time 7 h

v − 5 mi/h

Fig. 7-1

EXERCISES *7-1*

Ⓦ *In Exercises 1–8, determine whether or not the given equations are quadratic. If the resulting form is quadratic, identify a, b, and c, with a > 0. Otherwise, explain why the resulting form is not quadratic.*

1. $x^2 + 5 = 8x$

2. $5x^2 = 9 - x$

3. $x(x - 2) = 4$

4. $(3x - 2)^2 = 2$

5. $x^2 = (x + 2)^2$

6. $x(2x^2 + 5) = 7 + 2x^2$

7. $x(x^2 + x - 1) = x^3$

8. $(x - 7)^2 = (2x + 3)^2$

In Exercises 9–44, solve the given quadratic equations by factoring.

9. $x^2 - 4 = 0$

10. $x^2 - 400 = 0$

11. $4y^2 - 9 = 0$

12. $x^2 - 0.16 = 0$

13. $x^2 - 8x - 9 = 0$

14. $s^2 + s - 6 = 0$

15. $R^2 - 7R + 12 = 0$

16. $x^2 - 11x + 30 = 0$

17. $x^2 = -2x$

18. $x^2 = 7x$

19. $27m^2 = 3$

20. $5p^2 = 80$

21. $3x^2 - 13x + 4 = 0$

22. $7x^2 + 3x - 4 = 0$

23. $A^2 + 8A + 16 = 0$

24. $4x^2 - 20x + 25 = 0$

25. $6x^2 = 13x - 6$

26. $6z^2 = 6 + 5z$

27. $4x(x + 1) = 3$

28. $9t^2 = 9 - t(43 + t)$

29. $x^2 - x - 1 = 1$

30. $2x^2 - 7x + 6 = 3$

31. $x^2 - 4b^2 = 0$

32. $a^2x^2 - 1 = 0$

33. $40x - 16x^2 = 0$

34. $15L = 20L^2$

35. $8s^2 + 16s = 90$

36. $18t^2 - 48t + 32 = 0$

37. $(x + 2)^3 = x^3 + 8$

38. $V(V^2 - 4) = V^2(V - 1)$

39. $(x + a)^2 - b^2 = 0$

40. $x^2(a^2 + 2ab + b^2) - x(a + b) = 0$

41. The bending moment M of a simply supported beam of length L with a uniform load of w kg/m at a distance x from one end is $M = \frac{1}{2}wLx - \frac{1}{2}wx^2$. For what values of x is $M = 0$?

42. The mass m (in Mg) of the fuel supply in the first-stage booster of a rocket is $m = 135 - 6t - t^2$, where t is the time (in s) after launch. When does the booster run out of fuel?

43. The power P (in MW) produced between midnight and noon by a nuclear power plant is $P = 4h^2 - 48h + 744$, where h is the hour of the day. At what time is the power 664 MW?

44. In determining the speed s (in mi/h) of a car while studying its fuel economy, the equation $s^2 - 16s = 3072$ is used. Find s.

In Exercises 45–48, solve the given equations involving fractions.

45. $\dfrac{1}{x-3} + \dfrac{4}{x} = 2$

46. $2 - \dfrac{1}{x} = \dfrac{3}{x+2}$

47. $\dfrac{1}{2x} - \dfrac{3}{4} = \dfrac{1}{2x+3}$

48. $\dfrac{x}{2} + \dfrac{1}{x-3} = 3$

In Exercises 49–52, set up the appropriate quadratic equations and solve.

49. The spring constant k is the force F divided by the amount x it stretches ($k = F/x$). See Fig. 7-2(a). For two springs in series (see Fig. 7-2(b)), the reciprocal of the spring constant k_c for the combination equals the sum of the reciprocals of the individual spring constants. Find the spring constants for each of two springs in series if $k_c = 2$ N/cm and one spring constant is 3 N/cm more than the other.

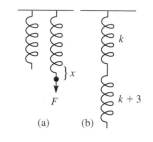

Fig. 7-2   (a)   (b)

50. The reciprocal of the combined resistance R of two resistances R_1 and R_2 connected in parallel (see Fig. 7-3(a)) is equal to the sum of the reciprocals of the individual resistances. If the two resistances are connected in series (see Fig. 7-3(b)), their combined resistance is the sum of their individual resistances. If two resistances connected in parallel have a combined resistance of 3.0 Ω and the same two resistances have a combined resistance of 16 Ω when connected in series, what are the resistances?

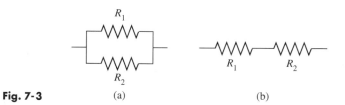

Fig. 7-3   (a)   (b)

51. A jet flew 1200 mi with a tailwind of 50 mi/h. The tailwind then changed to 20 mi/h for the remaining 570 mi of the flight. If the total time of the flight was 3.0 h, find the speed of the jet relative to the air.

52. A rectangular solar panel is 20 cm by 30 cm. By adding the same amount to each dimension, the area is doubled. How much is added?

7-2 COMPLETING THE SQUARE

Many quadratic equations cannot be solved by factoring. This is true of most quadratic equations that arise in applications. Therefore, we now develop a method that can be used to solve any quadratic equation. *The method is called **completing the square**.* In the next section we shall use completing the square to develop a general formula that can be used to solve any quadratic equation.

In the first example that follows we show the solution of a type of quadratic equation that arises while using the method of completing the square. In the examples that follow it, the method itself is used and described.

The method of completing the square is used later in Section 7-4 and also in Chapter 21.

■**EXAMPLE 1**   In solving $x^2 = 16$, we may write it as $x^2 - 16 = 0$ and complete the solution by factoring. This gives us $x = 4$ and $x = -4$. Therefore, we see that the principal square root of 16 and its negative both satisfy the original equation. Thus, we may solve $x^2 = 16$ by equating x to the principal square root of 16 and to its negative. The roots of $x^2 = 16$ are 4 and -4.

We may solve $(x - 3)^2 = 16$ in a similar way by equating $x - 3$ to 4 and to -4. Thus,

$$x - 3 = 4 \quad \text{or} \quad x - 3 = -4$$

Solving these equations, we obtain the roots 7 and -1.

We may solve $(x - 3)^2 = 17$ in the same way. Thus,

$$x - 3 = \sqrt{17} \quad \text{or} \quad x - 3 = -\sqrt{17}$$

The roots are therefore $3 + \sqrt{17}$ and $3 - \sqrt{17}$. Decimal approximations of these roots are 7.123 and -1.123.

■ EXAMPLE 2 We wish to find the roots of the quadratic equation

$$x^2 - 6x - 8 = 0$$

First, we note that the left side is not factorable. However, we can see that

NOTE ▶ $\quad x^2 - 6x \quad$ *is part of the special product* $\quad (x - 3)^2 = x^2 - 6x + 9$

and that this special product is a *perfect square*. By adding 9 to $x^2 - 6x$, we would have $(x - 3)^2$. Therefore, we rewrite the original equation as

$$x^2 - 6x = 8$$

NOTE ▶ and then **add 9** *to* **both** *sides* of the equation. The result is

$$x^2 - 6x + 9 = 17$$

The left side of this equation may be rewritten, giving

$$(x - 3)^2 = 17$$

Now, as in the third illustration of Example 1, we have

$$x - 3 = \pm\sqrt{17}$$

The $\pm$ sign means that $x - 3 = \sqrt{17}$ or $x - 3 = -\sqrt{17}$.
By adding 3 to each side, we obtain

$$x = 3 \pm \sqrt{17}$$

which means $x = 3 + \sqrt{17}$ and $x = 3 - \sqrt{17}$ are the two roots of the equation.
Therefore, we see that by creating an expression that is a perfect square and then using the principal square root and its negative, we were finally able to solve the equation as two linear equations. $\quad\quad$ ------------ ■

How we determine the number to be added to complete the square is based on the special products in Eqs. (6-3) and (6-4). We rewrite these as

$$(x + a)^2 = x^2 + 2ax + a^2 \tag{7-2}$$
$$(x - a)^2 = x^2 - 2ax + a^2 \tag{7-3}$$

We must be certain that the coefficient of the x^2-term is 1 before we start to complete the square. The coefficient of x in each case is numerically $2a$, and the number

CAUTION ▶ added to complete the square is a^2. Thus, *if we* **take half the coefficient of the** **x-term and square this result,** *we have the number that completes the square.* In our example, the numerical coefficient of the x-term was 6, and 9 was added to complete the square. Therefore, the procedure for solving a quadratic equation by completing the square is as follows:

Solving a Quadratic Equation by Completing the Square

1. *Divide each side by a (the coefficient of x^2).*
2. *Rewrite the equation with the constant on the right side.*
3. *Complete the square: Add the square of one-half of the coefficient of x to both sides.*
4. *Write the left side as a square and simplify the right side.*
5. *Equate the square root of the left side to the principal square root of the right side and to its negative.*
6. *Solve the two resulting linear equations.*

The following example illustrates the step-by-step procedure for solving a quadratic equation by completing the square.

EXAMPLE 3 Solve the following quadratic equation by completing the square.

$$2x^2 + 16x - 9 = 0$$

1. Divide each side by 2 to make the coefficient of x^2 equal to 1.

$$x^2 + 8x - \frac{9}{2} = 0$$

2. Put the constant on the right by adding $\frac{9}{2}$ to both sides.

$$x^2 + 8x = \frac{9}{2}$$

CAUTION ▶   **3.** The coefficient of the x-term is 8. Therefore, divide 8 by 2 to get 4. Now square 4 to get 16, and add 16 to both sides.

$$\frac{1}{2}(8) = 4; \ 4^2 = 16$$

$$x^2 + 8x + 16 = \frac{9}{2} + 16 = \frac{41}{2}$$

4. Write the left side as $(x + 4)^2$.

$$(x + 4)^2 = \frac{41}{2} \qquad \text{take the square root of each side}$$

5. Equate $x + 4$ to the principal square root of $\frac{41}{2}$ and its negative.

$$x + 4 = \pm\sqrt{\tfrac{41}{2}}$$

6. Solve for x.

$$x = -4 \pm \sqrt{\tfrac{41}{2}}$$

Therefore the roots are $-4 + \sqrt{\tfrac{41}{2}}$ and $-4 - \sqrt{\tfrac{41}{2}}$. Using a calculator to approximate these roots, we get 0.5277 and −8.528. ∎

EXERCISES 7-2

In Exercises 1–8, solve the given quadratic equations by finding appropriate square roots as in Example 1.

1. $x^2 = 25$

2. $x^2 = 100$

3. $x^2 = 7$

4. $x^2 = 15$

5. $(x - 2)^2 = 25$

6. $(x + 2)^2 = 100$

7. $(x + 3)^2 = 7$

8. $(x - 4)^2 = 10$

In Exercises 9–28, solve the given quadratic equations by completing the square. Exercises 9–12 and 15–18 may be checked by factoring.

9. $x^2 + 2x - 8 = 0$

10. $x^2 - x - 6 = 0$

11. $D^2 + 3D + 2 = 0$

12. $t^2 + 5t - 6 = 0$

13. $x^2 - 4x + 2 = 0$

14. $R^2 + 10R - 4 = 0$

15. $v^2 + 2v - 15 = 0$

16. $x^2 - 8x + 12 = 0$

17. $2s^2 + 5s = 3$

18. $4x^2 + x = 3$

19. $3y^2 = 3y + 2$

20. $3x^2 = 3 - 4x$

21. $2y^2 - y - 2 = 0$

22. $9v^2 - 6v - 2 = 0$

23. $5T^2 - 10T + 4 = 0$

24. $4V^2 + 9 = 12V$

25. $9x^2 + 6x + 1 = 0$

26. $2x^2 - 3x + 2a = 0$

27. $x^2 + 2bx + c = 0$

28. $px^2 + qx + r = 0$

$7\text{-}3$ THE QUADRATIC FORMULA

We shall now use the method of completing the square to derive a general formula that may be used for the solution of any quadratic equation.

Consider Eq. (7-1), the general quadratic equation:

$$ax^2 + bx + c = 0 \qquad (a \neq 0)$$

When we divide through by a, we obtain

$$x^2 + \frac{b}{a}x + \frac{c}{a} = 0$$

Subtracting c/a from each side, we have

$$x^2 + \frac{b}{a}x = -\frac{c}{a}$$

Half of b/a is $b/2a$, which squared is $b^2/4a^2$. Adding $b^2/4a^2$ to each side gives us

$$x^2 + \frac{b}{a}x + \frac{b^2}{4a^2} = -\frac{c}{a} + \frac{b^2}{4a^2}$$

Writing the left side as a perfect square, and combining fractions on the right side, we have

$$\left(x + \frac{b}{2a}\right)^2 = \frac{b^2 - 4ac}{4a^2}$$

Equating $x + \dfrac{b}{2a}$ to the principal square root of the right side and its negative,

$$x + \frac{b}{2a} = \frac{\pm\sqrt{b^2 - 4ac}}{2a}$$

When we subtract $b/2a$ from each side and simplify the resulting expression, we obtain the **quadratic formula:**

QUADRATIC FORMULA

$$\boxed{x = \frac{-b \pm \sqrt{b^2 - 4ac}}{2a}} \qquad (7\text{-}4)$$

To solve a quadratic equation by using the quadratic formula, we need only write the equation in the standard form of Eq. (7-1), identify a, b, and c, and substitute these numbers directly into the formula.

■EXAMPLE 1   Solve: $x^2 - 5x + 6 = 0$.

$$a = 1 \quad b = -5 \quad c = 6$$

Here, using the indicated values of a, b, and c in the quadratic formula, we have

$$x = \frac{-(-5) \pm \sqrt{(-5)^2 - 4(1)(6)}}{2(1)} = \frac{5 \pm \sqrt{25 - 24}}{2} = \frac{5 \pm 1}{2}$$

$$x = \frac{5 + 1}{2} = 3 \quad \text{or} \quad x = \frac{5 - 1}{2} = 2$$

The roots $x = 3$ and $x = 2$ check when substituted in the original equation.   ■

EXAMPLE 2 Solve: $2x^2 - 7x - 5 = 0$.

$$a = 2 \quad b = -7 \quad c = -5$$

Substituting the values for a, b, and c in the quadratic formula, we have

$$x = \frac{-(-7) \pm \sqrt{(-7)^2 - 4(2)(-5)}}{2(2)} = \frac{7 \pm \sqrt{49 + 40}}{4} = \frac{7 \pm \sqrt{89}}{4}$$

$$x = \frac{7 + \sqrt{89}}{4} = 4.108 \quad \text{or} \quad x = \frac{7 - \sqrt{89}}{4} = -0.6085$$

> See Appendix C for a graphing calculator program QUADFORM. It can be used to find the roots of a quadratic equation.

The exact roots are $x = \dfrac{7 \pm \sqrt{89}}{4}$ (this form is often used when the roots are irrational). Approximate decimal values are $x = 4.108$ and $x = -0.6085$. ∎

EXAMPLE 3 Solve: $9x^2 + 24x + 16 = 0$.
In this example, $a = 9$, $b = 24$, and $c = 16$. Thus,

$$x = \frac{-24 \pm \sqrt{24^2 - 4(9)(16)}}{2(9)} = \frac{-24 \pm \sqrt{576 - 576}}{18} = \frac{-24 \pm 0}{18} = -\frac{4}{3}$$

> Many calculators have an equation-solving feature. Also, some calculators have a specific feature for solving higher-degree equations.

Here both roots are $-\frac{4}{3}$, and we write the result as $x = -\frac{4}{3}$ and $x = -\frac{4}{3}$. We will get a double root when $b^2 = 4ac$, as in this case. ∎

EXAMPLE 4 Solve: $3x^2 - 5x + 4 = 0$.
In this example, $a = 3$, $b = -5$, and $c = 4$. Therefore,

$$x = \frac{-(-5) \pm \sqrt{(-5)^2 - 4(3)(4)}}{2(3)} = \frac{5 \pm \sqrt{25 - 48}}{6} = \frac{5 \pm \sqrt{-23}}{6}$$

We see that the roots contain imaginary numbers. This happens if $b^2 < 4ac$. ∎

We can generalize on the results of these examples as to the character of the roots of a quadratic equation. This is done by noting the value of $b^2 - 4ac$, which is called the **discriminant**. If a, b, and c are rational numbers (see Section 1-1), we have the following:

> If $b^2 - 4ac$ is positive and a perfect square, $ax^2 + bx + c$ is factorable.

Character of the Roots of a Quadratic Equation

1. If $b^2 - 4ac$ is positive and a perfect square (see Section 1-6), the roots are real, rational, and unequal. (See Example 1, where $b^2 - 4ac = 1$.)

2. If $b^2 - 4ac$ is positive but not a perfect square, the roots are real, irrational, and unequal. (See Example 2, where $b^2 - 4ac = 89$.)

3. If $b^2 - 4ac = 0$, the roots are real, rational, and equal. (See Example 3, where $b^2 - 4ac = 0$.)

4. If $b^2 - 4ac < 0$, the roots contain imaginary numbers and are unequal. (See Example 4, where $b^2 - 4ac = -23$.)

We can use the value of $b^2 - 4ac$ to help in checking the roots or in finding the character of the roots without having to solve the equation completely.

EXAMPLE 5   Solve: $2x^2 = 4x + 3$.

First, we must put the equation in the proper form with zero on the right:

$$2x^2 - 4x - 3 = 0$$

Now we identify $a = 2$, $b = -4$, and $c = -3$, which leads to the solution

$$x = \frac{-(-4) \pm \sqrt{(-4)^2 - 4(2)(-3)}}{2(2)} = \frac{4 \pm \sqrt{16 + 24}}{4}$$

$$= \frac{4 \pm \sqrt{40}}{4} = \frac{4 \pm 2\sqrt{10}}{4} = \frac{2(2 \pm \sqrt{10})}{4} = \frac{2 \pm \sqrt{10}}{2}$$

Approximate decimal results are $x = 2.581$ and $x = -0.581$. The radical form of the answer is found by simplifying radicals as shown in Section 1-6.

Here, $b^2 - 4ac = 40$, and the roots are real, irrational, and unequal. ∎

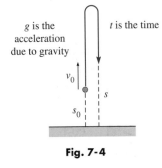

g is the acceleration due to gravity

t is the time

v_0

s

s_0

Fig. 7-4

EXAMPLE 6   An equation used in the analysis of projectile motion (see Fig. 7-4) is $s = s_0 + v_0 t - \frac{1}{2}gt^2$. Solve for t.

$$gt^2 - 2v_0 t - 2(s_0 - s) = 0 \qquad \text{multiply by } -2, \text{ put in form of Eq. (7-1)}$$

In this form we see that $a = g$, $b = -2v_0$, and $c = -2(s_0 - s)$.

$$t = \frac{-(-2v_0) \pm \sqrt{(-2v_0)^2 - 4g(-2)(s_0 - s)}}{2g} = \frac{2v_0 \pm \sqrt{4(v_0^2 + 2gs_0 - 2gs)}}{2g}$$

$$= \frac{2v_0 \pm 2\sqrt{v_0^2 + 2gs_0 - 2gs}}{2g} = \frac{v_0 \pm \sqrt{v_0^2 + 2gs_0 - 2gs}}{g}$$

We see that we can solve quadratic equations which have literal coefficients. ∎

SOLVING A WORD PROBLEM

See the chapter introduction.

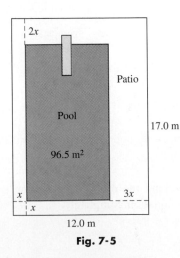

2x

Patio

Pool

96.5 m²

17.0 m

x

x

3x

12.0 m

Fig. 7-5

EXAMPLE 7   A rectangular area 17.0 m long and 12.0 m wide is to be used for a patio with a rectangular pool. One end and one side of the patio (the area around the pool for chairs, sunning, etc.) are to be the same width. The other end with the diving board is to be twice as wide, and the other side is to be three times as wide as the end without the diving board. If the area of the pool is to be 96.5 m², what are the widths of the patio ends and sides, and dimensions of the pool? See Fig. 7-5.

First, we let $x =$ the width of the narrow end and side of the patio. This means that the other end is $2x$ in width and the other side is $3x$ in width. Since the area of the pool is 96.5 m², we find x as follows:

pool length · pool width · pool area

$$(17.0 - 3x)(12.0 - 4x) = 96.5$$

$$204 - 68.0x - 36.0x + 12x^2 = 96.5$$

$$12x^2 - 104.0x + 107.5 = 0$$

$$x = \frac{-(-104.0) \pm \sqrt{(-104.0)^2 - 4(12)(107.5)}}{2(12)} = \frac{104.0 \pm \sqrt{5656}}{24}$$

Evaluating the roots, we get $x = 7.5$ m and $x = 1.2$ m. The value $x = 7.5$ m cannot be the required result since the width of the patio would be greater than the width of the entire area. For the value $x = 1.2$ m, the pool would have a length of 12.4 m and a width of 7.2 m. These give an area of 96.5 m², which means it checks. The widths of the patio area are then 1.2 m, 1.2 m, 2.4 m, and 3.6 m. ∎

CAUTION ▶

It must be emphasized that, in using the quadratic formula, the entire expression $-b \pm \sqrt{b^2 - 4ac}$ is divided by $2a$. *It is a relatively common error to divide only the radical* $\sqrt{b^2 - 4ac}$.

The quadratic formula provides a quick general method for solving quadratic equations. Proper recognition and substitution of the coefficients a, b, and c is all that is required to complete the solution, regardless of the nature of the roots.

EXERCISES $7\text{-}3$

In Exercises 1–32, solve the given quadratic equations, using the quadratic formula. Exercises 1–4 are the same as Exercises 9–12 of Section 7-2.

1. $x^2 + 2x - 8 = 0$ **2.** $x^2 - x - 6 = 0$

3. $D^2 + 3D + 2 = 0$ **4.** $t^2 + 5t - 6 = 0$

5. $x^2 - 4x + 2 = 0$ **6.** $x^2 + 10x - 4 = 0$

7. $v^2 + 2v - 15 = 0$ **8.** $V^2 - 8V + 12 = 0$

9. $2s^2 + 5s = 3$ **10.** $4x^2 + x = 3$

11. $3y^2 = 3y + 2$ **12.** $3x^2 = 3 - 4x$

13. $2y^2 - y - 2 = 0$ **14.** $9v^2 - 6v - 2 = 0$

15. $30y^2 + 23y - 40 = 0$ **16.** $40x^2 - 62x - 63 = 0$

17. $2t^2 + 10t = -15$ **18.** $2d(d - 2) = -7$

19. $s^2 = 9 + s(1 - 2s)$ **20.** $6r^2 = 6r + 1$

21. $4x^2 = 9$ **22.** $6x = x^2$

23. $15 + 4z = 32z^2$ **24.** $4x^2 - 12x = 7$

25. $x^2 - 0.20x - 0.40 = 0$ **26.** $3.2x^2 = 2.5x + 7.6$

27. $0.29Z^2 - 0.18 = 0.63Z$ **28.** $12.5x^2 + 13.2x = 15.5$

29. $x^2 + 2cx - 1 = 0$ **30.** $x^2 - 7x + (6 + a) = 0$

31. $b^2x^2 - (b + 1)x + (1 - a) = 0$

32. $c^2x^2 - x - 1 = x^2$

In Exercises 33–36, solve the given quadratic equations. All numbers are accurate to at least two significant digits.

33. In machine design, in finding the outside diameter D_0 of a hollow shaft, the equation $D_0^2 - DD_0 - 0.25D^2 = 0$ is used. Solve for D_0 if $D = 3.625$ cm.

34. A missile is fired vertically into the air. The distance s (in ft) above the ground as a function of time t (in s) is given by $s = 300 + 500t - 16t^2$. (a) When will the missile hit the ground? (b) When will the missile be 1000 ft above the ground?

35. In calculating the current in an electric circuit with an inductance L, a resistance R, and a capacitance C, it is necessary to solve the equation $Lm^2 + Rm + 1/C = 0$. Solve for m in terms of L, R, and C. See Fig. 7-6.

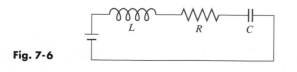

Fig. 7-6

36. In finding the radius r of a circular arch of height h and span b, an architect used the formula
$$r = \frac{b^2 + 4h^2}{8h}.$$ Solve for h.

In Exercises 37–40, without solving the given equations, determine the character of the roots.

37. $2x^2 - 7x = -8$ **38.** $3x^2 + 19x = 14$

39. $3.6t^2 + 2.1 = 7.7t$ **40.** $0.45s^2 + 0.33 = 0.12s$

In Exercises 41–44, set up appropriate equations and solve the word problems. All numbers are accurate to at least two significant digits.

41. The length of a tennis court is 12.8 m more than its width. If the area of a tennis court is 262 m², what are its dimensions? See Fig. 7-7.

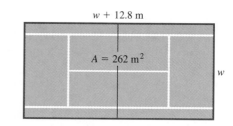

w + 12.8 m

$A = 262$ m²

w

Fig. 7-7

42. Two circular oil spills are tangent to each other. If the distance between centers is 800 m and they cover a combined area of 1.02×10^6 m², what is the radius of each? See Fig. 7-8.

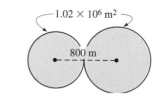

1.02×10^6 m²

800 m

Fig. 7-8

43. In remodeling a house, an architect finds that by adding the same amount to each dimension of a 12-ft by 16-ft rectangular room the area would be increased by 80 ft². How much must be added to each dimension?

44. Two pipes together drain a wastewater holding tank in 6.00 h. If used alone to empty the tank, one takes 2.00 h longer than the other. How long does each take to empty the tank if used alone?

$7\text{-}4$ THE GRAPH OF THE QUADRATIC FUNCTION

We have developed the basic algebraic methods of solving a quadratic equation. We shall now discuss the graph of the quadratic function and show the graphical solution of a quadratic equation, using the methods developed in Chapter 3.

In Section 7-1 we noted that $ax^2 + bx + c$ is the quadratic function. As in Chapter 3, by letting $y = ax^2 + bx + c$ we can graph this function. The next example briefly reviews the graph of a quadratic function done in this way.

▌**EXAMPLE 1**   Graph the function $f(x) = x^2 + 2x - 3$.

First, we let $y = x^2 + 2x - 3$. We can then set up a table of values and graph the function as shown in Fig. 7-9, or we can display its graph on a graphing calculator as shown in Fig. 7-10.

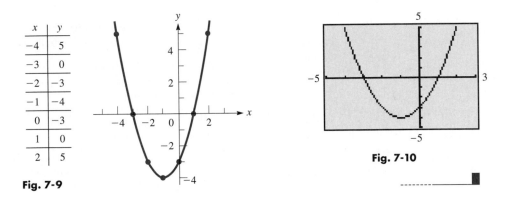

x	y
-4	5
-3	0
-2	-3
-1	-4
0	-3
1	0
2	5

Fig. 7-9

Fig. 7-10

The graph of any quadratic function $y = ax^2 + bx + c$ will have the same basic shape as that shown in Fig. 7-9, and it is called a **parabola.** (In Section 3-4 we briefly noted that graphs in Examples 2 and 3 were parabolas.) A parabola defined by a quadratic function can open upward (as in Example 1) or downward. The location of the parabola and how it opens depend on the values of a, b, and c.

In Example 1 the parabola has a minimum point at $(-1, -4)$ and the curve opens

NOTE ▶ upward. *All parabolas have an* **extreme point** *of this type. If $a > 0$, the parabola has a* **minimum point** *and it opens* **upward.** *If $a < 0$, the parabola has a* **maximum point** *and it opens* **downward.** *The extreme point of the parabola is also known as its* **vertex.**

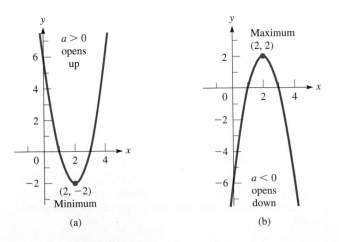

(a)

(b)

Fig. 7-11

▌**EXAMPLE 2**   The graph of $y = 2x^2 - 8x + 6$ is shown in Fig. 7-11(a). For this parabola, $a = 2$ $(a > 0)$ and it opens upward. The vertex (a minimum point) is $(2, -2)$.

The graph of $y = -2x^2 + 8x - 6$ is shown in Fig. 7-11(b). For this parabola, $a = -2$ $(a < 0)$ and it opens downward. The vertex (a maximum point) is $(2, 2)$.

We can sketch the graph of a parabola by using its basic shape and knowing the location of two or three points, including the vertex. Even when using a graphing calculator, we can get a check on the graph by knowing the vertex and how the parabola opens.

In order to find the coordinates of the extreme point we start with the quadratic function

$$y = ax^2 + bx + c$$

Then, we factor a from the two terms containing x, obtaining

$$y = a\left(x^2 + \frac{b}{a}x\right) + c$$

Now, completing the square of the terms within parentheses, we have

$$y = a\left(x^2 + \frac{b}{a}x + \frac{b^2}{4a^2}\right) + c - \frac{b^2}{4a}$$

$$= a\left(x + \frac{b}{2a}\right)^2 + c - \frac{b^2}{4a}$$

We now look at the factor $\left(x + \dfrac{b}{2a}\right)^2$. If $x = -b/2a$, the term is zero. If x is any other value, $\left(x + \dfrac{b}{2a}\right)^2$ is positive. Thus, if $a > 0$, the value of y increases from the minimum value where $x = -b/2a$, and if $a < 0$, the value of y decreases from the maximum value where $x = -b/2a$. *This means that $x = -b/2a$ is the x-coordinate of the vertex (extreme point). The y-coordinate is found by substituting this x-value in the function.*

Another easily found point is the y-intercept. As with a linear equation, we find the y-intercept where $x = 0$. For $y = ax^2 + bx + c$, if $x = 0$, then $y = c$. *This means that the point $(0, c)$ is the y-intercept.*

■EXAMPLE 3 For the graph of the function $y = 2x^2 - 8x + 6$, find the vertex and y-intercept and sketch the graph. (This function is also used in Example 2.)

First, $a = 2$ and $b = -8$. This means that the x-coordinate of the vertex is

$$\frac{-b}{2a} = \frac{-(-8)}{2(2)} = \frac{8}{4} = 2$$

and the y-coordinate is

$$y = 2(2^2) - 8(2) + 6 = -2$$

Thus, the vertex is $(2, -2)$. Since $a > 0$, it is a minimum point.

Since $c = 6$, the y-intercept is $(0, 6)$.

We can use the minimum point $(2, -2)$ and the y-intercept $(0, 6)$, along with the fact that the graph is a parabola, to get an approximate sketch of the graph. Noting that a parabola increases (or decreases) away from the vertex in the same way on each side of it (it is *symmetric* to a vertical line through the vertex), we sketch the graph in Fig. 7-12. We see that it is the same graph as that shown in Fig. 7-11(a). ■

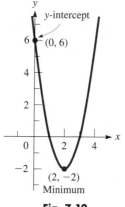

Fig. 7-12

If we are not using a graphing calculator, we may need one or two additional points to get a good sketch of a parabola. This would be true if the y-intercept is close to the vertex. Two points we can find are the x-intercepts, if the parabola crosses the x-axis (one point if the vertex is on the x-axis). They are found by setting $y = 0$ and solving the quadratic equation $ax^2 + bx + c = 0$. Also, we may simply find one or two points other than the vertex and the y-intercept. Sketching a parabola in this way is shown in the following two examples.

■EXAMPLE 4   Sketch the graph of $y = -x^2 + x + 6$.

We first note that $a = -1$ and $b = 1$. Therefore, the x-coordinate of the maximum point ($a < 0$) is $-\frac{1}{2(-1)} = \frac{1}{2}$. The y-coordinate is $-(\frac{1}{2})^2 + \frac{1}{2} + 6 = \frac{25}{4}$. This means that the maximum point is $(\frac{1}{2}, \frac{25}{4})$.

The y-intercept is $(0, 6)$.

Using these points in Fig. 7-13, we see that they are close together and do not give a good idea of how wide the parabola opens. Therefore, setting $y = 0$, we solve the equation

$$-x^2 + x + 6 = 0 \qquad \text{multiply each term by } -1$$

or

$$x^2 - x - 6 = 0$$

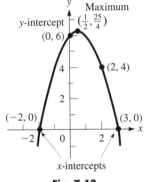

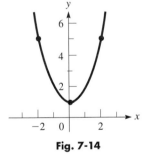

y-intercept $(\frac{1}{2}, \frac{25}{4})$ Maximum
$(0, 6)$
$(2, 4)$
$(-2, 0)$
$(3, 0)$
x-intercepts

Fig. 7-13

This equation is factorable. Thus,

$$(x - 3)(x + 2) = 0, -2$$
$$x = 3, -2$$

This means that the x-intercepts are $(3, 0)$ and $(-2, 0)$, as shown in Fig. 7-13.

Also, rather than finding the x-intercepts, we can let $x = 2$ (or some value to the right of the vertex) and then use the point $(2, 4)$.   ━━━━■

■EXAMPLE 5   Sketch the graph of $y = x^2 + 1$.

Since there is no x-term, $b = 0$. This means that the x-coordinate of the minimum point ($a > 0$) is 0 and that the minimum point and the y-intercept are both $(0, 1)$. We know that the graph opens upward, since $a > 0$, which in turn means that it does not cross the x-axis. Now, letting $x = 2$ and $x = -2$, we find the points $(2, 5)$ and $(-2, 5)$ on the graph, which is shown in Fig. 7-14.   ━━━━■

Fig. 7-14

We can see that the domain of the quadratic function is all x. Knowing that the graph has either a maximum point or a minimum point, we can see that the range of the quadratic function must be restricted to certain values of y. In Example 3, the range is $f(x) \geq -2$; in Example 4, the range is $f(x) \leq \frac{25}{4}$; and in Example 5, the range is $f(x) \geq 1$.

Solving Quadratic Equations Graphically

In Chapter 3 we showed how an equation can be solved graphically. Following that method, to solve the equation $ax^2 + bx + c = 0$, we let $y = ax^2 + bx + c$ and graph the function. The roots of the equation are the x-coordinates of the points for which $y = 0$ (the x-intercepts). The following examples illustrate solving quadratic equations graphically.

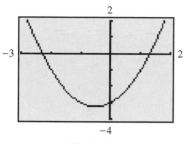

Fig. 7-15

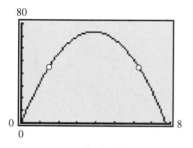

Fig. 7-16

EXAMPLE 6 Solve the equation $3x = x(2 - x) + 3$ graphically.

First, collect all terms on the left of the equal sign. This gives us $x^2 + x - 3 = 0$. We then let $y = x^2 + x - 3$ and graph this function. The minimum point is $(-\frac{1}{2}, -\frac{13}{4})$, and the y-intercept is $(0, -3)$.

Using the graphing calculator, these points help us choose the *window* settings shown for the graph of the function in Fig. 7-15. Now using the *trace* and *zoom* features (or the *intersect* feature), we find the roots to be

$$x = -2.30 \quad \text{and} \quad x = 1.30$$

These values (or the stored calculator values) check when substituted in the original equation.

EXAMPLE 7 A projectile is fired vertically upward from the ground with a velocity of 38 m/s. Its distance above the ground is given by $s = -4.9t^2 + 38t$, where s is the distance (in m) and t is the time (in s). Graph the function and from the graph determine (1) when the projectile will hit the ground and (2) how long it takes to reach 45 m above the ground.

Since $c = 0$, the s-intercept is at the origin. Using $-b/2a$, we find the maximum point at about $(3.9, 74)$. These points help us choose (we use x for t, and y for s, on the calculator) the *window* settings for the graph shown in Fig. 7-16.

Using the *trace* and *zoom* features (or the *intersect* feature), we can now determine the required answers.

1. The projectile will hit the ground for $s = 0$, which is shown at the right t-intercept. Thus, it will take about 7.8 s to hit the ground.

2. We find the time for the projectile to get 45 m above the ground by finding the values of t for $s = 45$ m. We see that there are *two* points for which $s = 45$ m. The values of t at these points tell us that the times are $t = 1.5$ s and $t = 6.3$ s.

From the maximum point, we can also conclude that the maximum height reached by the projectile is about 74 m. This can also be seen on the graph.

EXERCISES *7-4*

In Exercises 1–8, sketch the graph of each parabola by using only the vertex and the y intercept. Check the graph using a graphing calculator.

1. $y = x^2 - 6x + 5$ **2.** $y = -x^2 - 4x - 3$

3. $y = -3x^2 + 10x - 4$ **4.** $y = 2x^2 + 8x - 5$

5. $y = x^2 - 4x$ **6.** $y = -2x^2 - 5x$

7. $y = -2x^2 - 4x - 3$ **8.** $y = x^2 - 3x + 4$

In Exercises 9–12, sketch the graph of each parabola by using the vertex, the y-intercept, and the x-intercepts. Check the graph using a graphing calculator.

9. $y = x^2 - 4$ **10.** $y = x^2 + 3x$

11. $y = -2x^2 - 6x + 8$ **12.** $y = -3x^2 + 12x - 5$

In Exercises 13–16, sketch the graph of each parabola by using the vertex, the y-intercept, and two other points, not including the x-intercepts. Check the graph using a graphing calculator.

13. $y = 2x^2 + 3$ **14.** $y = x^2 + 2x + 2$

15. $y = -2x^2 - 2x - 6$ **16.** $y = -3x^2 - x$

In Exercises 17–24, use a graphing calculator to solve the given equations. If there are no real roots, state this as the answer.

17. $2x^2 - 3 = 0$ **18.** $5 - x^2 = 0$

19. $-3x^2 + 11x - 5 = 0$ **20.** $2x^2 = 7x + 4$

21. $x(2x - 1) = -3$ **22.** $2x - 5 = x^2$

23. $6x^2 = 18 - 7x$ **24.** $3x^2 - 25 = 20x$

(W) *In Exercises 25–28, use a graphing calculator to graph all three parabolas on the same coordinate system. State the changes in the graphs in parts (b) and (c) from the graph in part (a).*

25. (a) $y = x^2$ (b) $y = 3x^2$ (c) $y = \frac{1}{3}x^2$

26. (a) $y = x^2$ (b) $y = (x - 3)^2$ (c) $y = (x + 3)^2$

27. (a) $y = x^2$ (b) $y = -x^2$ (c) $y = 3x - x^2$

28. (a) $y = x^2$ (b) $y = x^2 + 3$ (c) $y = x^2 - 3$

In Exercises 29–36, solve the given applied problems.

29. An equipment company determines that the area A (in m^2) covered by a rectangular tarpaulin is given by $A = w(8 - w)$, where w is the width of the tarpaulin, and its perimeter is 16 m. Sketch the graph of A as a function of w.

30. Under specified conditions, the pressure loss L (in lb/in.^{2} per 100 ft), in the flow of water through a water line in which the flow is q gal/min, is given by $L = 0.0002q^2 + 0.005q$. Sketch the graph of L as a function of q, for $q < 100$ gal/min.

31. When analyzing the power P (in W) dissipated in an electric circuit, the equation $P = 50i - 3i^2$ results. Here i is the current (in A). Sketch the graph of $P = f(i)$.

32. Tests show that the power P (in hp) of an automobile engine as a function of r (in r/min—rpm) is given by $P = -5.0 \times 10^{-6}r^2 + 0.050r - 45$ ($1500 < r < 6000$ r/min). Sketch the graph of P vs. r and find the maximum power that is produced.

33. A missile is fired vertically upward such that its distance s (in ft) above the ground is given by $s = 150 + 250t - 16t^2$, where t is the time (in s). Sketch the graph and then determine from the graph (a) when the missile will hit the ground (to 0.1 s), (b) how high it will go (to 3 significant digits), and (c) how long it will take to reach a height of 800 ft (to 0.1 s).

34. In a certain electric circuit, the resistance R (in Ω) that gives resonance is found by solving the equation $25R = 3(R^2 + 4)$. Solve this equation graphically (to 0.1 Ω).

35. A security fence is to be built around a rectangular parking area of 20,000 ft^2. If the front side of the fence costs \$20/ft and the other three sides cost \$10/ft, solve graphically for the dimensions (to 1 ft) of the parking area if the fence is to cost \$7500. See Fig. 7-17.

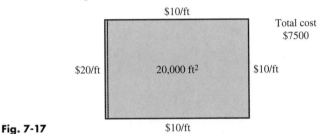

Fig. 7-17

36. An airplane pilot could decrease the time t (in h) needed to travel the 630 mi from Ottawa to Milwaukee by 20 min if the plane's speed v is increased by 40 mi/h. Set up the appropriate equation and solve graphically for v (to two significant digits).

CHAPTER EQUATIONS

Quadratic equation	$ax^2 + bx + c = 0$	(7-1)
Quadratic formula	$x = \dfrac{-b \pm \sqrt{b^2 - 4ac}}{2a}$	(7-4)

REVIEW EXERCISES

In Exercises 1–12, solve the given quadratic equations by factoring.

1. $x^2 + 3x - 4 = 0$

2. $x^2 + 3x - 10 = 0$

3. $x^2 - 10x + 16 = 0$

4. $P^2 - 6P - 27 = 0$

5. $3x^2 + 11x = 4$

6. $6y^2 = 11y - 3$

7. $6t^2 = 13t - 5$

8. $3x^2 + 5x + 2 = 0$

9. $6s^2 = 25s$

10. $6n^2 - 23n - 35 = 0$

11. $4x^2 - 8x = 21$

12. $6x^2 = 8 - 47x$

In Exercises 13–24, solve the given quadratic equations by using the quadratic formula.

13. $x^2 - x - 110 = 0$

14. $x^2 + 3x - 18 = 0$

15. $x^2 + 2x - 5 = 0$

16. $D^2 - 7D - 1 = 0$

17. $2x^2 - x = 36$

18. $3x^2 + x = 14$

19. $4s^2 - 3s - 2 = 0$

20. $5x^2 + 7x - 2 = 0$

21. $2.1x^2 + 2.3x + 5.5 = 0$

22. $0.30R^2 - 0.42R = 0.15$

23. $6x^2 = 9 - 4x$

24. $24x^2 = 25x + 20$

In Exercises 25–36, solve the given quadratic equations by any appropriate algebraic method.

25. $x^2 + 4x - 4 = 0$ **26.** $x^2 + 3x + 1 = 0$

27. $3x^2 + 8x + 2 = 0$ **28.** $3p^2 = 28 - 5p$

29. $4v^2 = v + 5$ **30.** $6x^2 - x + 2 = 0$

31. $2C^2 + 3C + 7 = 0$ **32.** $4y^2 - 5y = 8$

33. $a^2x^2 + 2ax + 2 = 0$ **34.** $16r^2 - 8r + 1 = 0$

35. $ax^2 = a^2 - 3x$ **36.** $2bx = x^2 - 3b$

In Exercises 37–40, solve the given quadratic equations by completing the square.

37. $x^2 - x - 30 = 0$ **38.** $x^2 - 2x - 5 = 0$

39. $2x^2 - x - 4 = 0$ **40.** $4x^2 - 8x - 3 = 0$

In Exercises 41–44, solve the given equations.

41. $\dfrac{x-4}{x-1} = \dfrac{2}{x}$ **42.** $\dfrac{V-1}{3} = \dfrac{5}{V} + 1$

43. $\dfrac{x^2 - 3x}{x - 3} = \dfrac{x^2}{x + 2}$ **44.** $\dfrac{x-2}{x-5} = \dfrac{15}{x^2 - 5x}$

In Exercises 45–48, sketch the graphs of the given functions by using the vertex, the y-intercept, and one or two other points.

45. $y = 2x^2 - x - 1$ **46.** $y = -4x^2 - 1$

47. $y = x - 3x^2$ **48.** $y = 2x^2 + 8x - 10$

In Exercises 49–52, solve the given equations by using a graphing calculator. If there are no real roots, state this as the answer.

49. $2x^2 + x - 4 = 0$ **50.** $-4x^2 - x - 1 = 0$

51. $3x^2 = -x - 2$ **52.** $x(15x - 12) = 8$

In Exercises 53–60, solve the given quadratic equations by any appropriate method. All numbers are accurate to at least two significant digits.

53. In a natural gas pipeline, the velocity v (in m/s) of the gas as a function of the distance x (in cm) from the wall of the pipe is given by $v = 5.2x - x^2$. Determine x for $v = 4.8$ m/s.

54. At an altitude h (in ft) above sea level, the boiling point of water is lower by T °F than the boiling point at sea level, which is 212°F. The difference can be approximated by solving the equation $T^2 + 520T - h = 0$. What is the boiling point in Boulder, Colorado (altitude 5300 ft)?

55. The height h of an object ejected at an angle θ from a vehicle moving with velocity v is given by $h = vt \sin \theta - 16t^2$, where t is the time of flight. Find t (to 0.1 s) if $v = 44$ ft/s, $\theta = 65°$, and $h = 18$ ft.

56. In studying the emission of light, in order to determine the angle at which the intensity is a given value, the equation $\sin^2 A - 4 \sin A + 1 = 0$ must be solved. Find angle A (to 0.1°). $(\sin^2 A = (\sin A)^2.)$

57. A computer analysis shows that the number n of electronic components a company should produce for supply to equal demand is found by solving
$$\frac{n^2}{500{,}000} = 144 - \frac{n}{500}. \text{ Find } n.$$

58. To determine the resistances of two resistors that are to be in parallel in an electric circuit, it is necessary to solve the equation $\dfrac{20}{R} + \dfrac{20}{R + 10} = \dfrac{1}{5}$. Find R (to nearest 1 Ω).

59. In designing a cylindrical container, the formula $A = 2\pi r^2 + 2\pi rh$ (Eq. (2-19)) is used. Solve for r.

60. In determining the number of bytes b that can be stored on a hard disk, the equation $b = kr(R - r)$ is used. Solve for r.

In Exercises 61–72, set up the necessary equation where appropriate and solve the given problems. All numbers are accurate to at least two significant digits.

61. In testing the effects of a drug, the percent of the drug in the blood was given by $p = 0.090t - 0.015t^2$, where t is the time (in h) after the drug was administered. Sketch the graph of $p = f(t)$.

62. In an electric circuit, the voltage V as a function of the time t (in min) is given by $V = 9.8 - 9.2t + 2.3t^2$. Sketch the graph of $V = f(t)$, for $t \le 5$ min.

63. By adding the same amount to its length and its width, a developer increased the area of a rectangular lot by 3000 m² to make it 80 m by 100 m. What were the original dimensions of the lot? See Fig. 7-18.

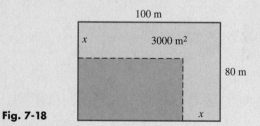

Fig. 7-18

64. A machinery pedestal is made of two concrete cubes, one on top of the other. The pedestal is 8.00 ft high and contains 152 ft³ of concrete. Find the edge of each cube.

65. Concrete contracts as it dries. If the volume of a cubical concrete block is 29 cm³ less and each edge is 0.10 cm less after drying, what was the original length of an edge of the block?

66. A military jet flies directly over and at right angles to the straight course of a commercial jet. The military jet is flying at 200 mi/h faster than four times the speed of the commercial jet. How fast is each going if they are 2050 mi apart (on a direct line) after 1 h?

67. The width of a rectangular TV screen is 5.7 in. more than its height. If the diagonal is 27.0 in., find the dimensions of the screen.

(W) **68.** Find the exact value of x that is defined in terms of the *continuing fraction* at the right (the pattern continues endlessly). Explain your method of solution.

$$x = 2 + \cfrac{1}{2 + \cfrac{1}{2 + \cfrac{1}{2 + \cdots}}}$$

69. An electric utility company is placing utility poles along a road. It is determined that five fewer poles per kilometer would be necessary if the distance between poles were increased by 10 m. How many poles are being placed each kilometer?

70. An architect is designing a Norman window (a semicircular part over a rectangular part) as shown in Fig. 7-19. If the area of the window is to be 16.0 ft^2 and the height of the rectangular part is 4.00 ft, find the radius of the circular part.

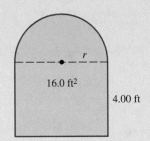

Fig. 7-19

16.0 ft^2

r

4.00 ft

71. A testing station found p parts per million (ppm) of sulfur dioxide in the air as a function of the hour h of the day to be $p = 0.00174(10 + 24h - h^2)$. Sketch the graph of $p = f(h)$ and, from the graph, find the time when $p = 0.200$ ppm.

72. A compact disc (CD) is made such that it is 53.0 mm from the edge of the center hole to the edge of the disc. Find the radius of the hole if 1.36% of the disc is removed in making the hole. See Fig. 7-20.

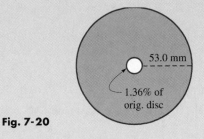

53.0 mm

1.36% of orig. disc

Fig. 7-20

Writing Exercise

73. An electronics student is asked to solve the equation $\dfrac{1}{R_T} = \dfrac{1}{R} + \dfrac{1}{R + 1}$ for R. Write one or two paragraphs explaining your procedure for solving this equation. (What is the solution?)

PRACTICE TEST

In Problems 1–4, algebraically solve for x.

1. $x^2 - 3x - 5 = 0$

2. $2x^2 = 9x - 4$

3. $\dfrac{3}{x} - \dfrac{2}{x + 2} = 1$

4. $2x^2 - x = 6 - 2x(3 - x)$

5. Sketch the graph of $y = 2x^2 + 8x + 5$ using the extreme point and the y-intercept.

6. In electricity the formula $P = EI - RI^2$ is used. Solve for I in terms of E, P, and R.

7. Solve by completing the square: $x^2 - 6x - 9 = 0$.

8. The perimeter of a rectangular window is 8.4 m, and its area is 3.8 m^2. Find its dimensions.

8 TRIGONOMETRIC FUNCTIONS OF ANY ANGLE

When we were dealing with the trigonometric functions in Chapter 4, we restricted ourselves to right triangles and functions of acute angles measured in degrees. Since we did define the functions in general, we can use these same definitions for finding the functions of any possible angle.

In this chapter we will evaluate the trigonometric functions of angles of any magnitude. Also, we shall develop the use of radian measure, as well as use angles measured in degrees.

There are numerous applications of triangles that are not right triangles and of radian measure. Many of the triangle applications in areas such as navigation and forces on structures are given in the next chapter. Applications of radian measure in areas such as mechanical vibrations, electric currents, and rotational motion are given in this chapter and in Chapter 10.

Cruise ships generally measure distances traveled in nautical miles. In Section 8-4 we will see how a nautical mile is defined as a distance along an arc of the earth's surface.

8-1 SIGNS OF THE TRIGONOMETRIC FUNCTIONS

We recall the definitions of the trigonometric functions that were given in Section 4-2. *Here the point (x, y) is a point on the terminal side of angle θ, and r is the radius vector.* See Fig. 8-1.

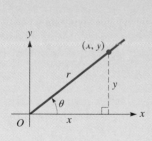

Fig. 8-1

$$\sin \theta = \frac{y}{r} \qquad \cos \theta = \frac{x}{r} \qquad \tan \theta = \frac{y}{x}$$

$$\cot \theta = \frac{x}{y} \qquad \sec \theta = \frac{r}{x} \qquad \csc \theta = \frac{r}{y}$$

(8-1)

As we have noted, these definitions are valid for an angle of any magnitude. In this section we determine the *signs* of the trigonometric functions of angles with terminal sides in each of the four quadrants.

We see that we can find the functions if we know the values of the coordinates (x, y) on the terminal side of θ and the radius vector r. Of course, if either x or y is zero in the denominator, the function is undefined, and we will consider this further in the next section. *Remembering that r is always considered to be positive, we can see that the various functions will vary in sign in each of the quadrants, depending on the signs of x and y.*

	Quadrant			
	I	II	III	IV
sin θ	+	+	−	−

If the terminal side of the angle is in the first or second quadrant, the value of sin θ will be positive, but if the terminal side is in the third or fourth quadrant, sin θ is negative. This is because *the sign of* sin θ *depends on the sign of the y-coordinate* of the point on the terminal side, and y is positive if the point is above the x-axis, and y is negative if this point is below the x-axis. See Fig. 8-2.

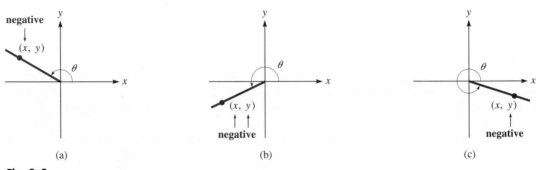

(a)   (b)   (c)

Fig. 8-2

■EXAMPLE 1   The value of sin 20° is positive, since the terminal side of 20° is in the first quadrant. The value of sin 160° is positive, since the terminal side of 160° is in the second quadrant. The values of sin 200° and sin 340° are negative, since the terminal sides of these angles are in the third and fourth quadrants, respectively.   --------▬

	Quadrant			
	I	II	III	IV
tan θ	+	−	+	−

The sign of tan θ *depends on the ratio of y to x.* In the first quadrant both x and y are positive, and therefore the ratio y/x is positive. In the third quadrant both x and y are negative, and the ratio y/x is positive. In the second quadrant x is negative and y is positive, and in the fourth quadrant x is positive and y is negative. Therefore, in these quadrants, the ratio y/x is negative. See Fig. 8-2.

■EXAMPLE 2   The values of tan 20° and tan 200° are positive, since the terminal sides of these angles are in the first and third quadrants, respectively. The values of tan 160° and tan 340° are negative, since the terminal sides of these angles are in the second and fourth quadrants, respectively.   --------▬

	Quadrant			
	I	II	III	IV
cos θ	+	−	−	+

The sign of cos θ *depends on the sign of x.* Since x is positive in the first and fourth quadrants, cos θ is positive in these quadrants. Since x is negative in the second and third quadrants, cos θ is negative in these quadrants. See Fig. 8-2.

■EXAMPLE 3   The values of cos 20° and cos 340° are positive, since the terminal sides of these angles are in the first and fourth quadrants, respectively. The values of cos 160° and cos 200° are negative, since the terminal sides of these angles are in the second and third quadrants, respectively.   --------▬

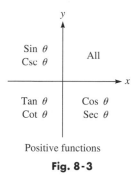

Sin θ
Csc θ | All

Tan θ
Cot θ | Cos θ
Sec θ

Positive functions

Fig. 8-3

Since csc θ is defined in terms of y and r, as is sin θ, the sign of csc θ is the same as that of sin θ. For the same reason, cot θ has the same sign as tan θ, and sec θ has the same sign as cos θ. Therefore:

All functions of first-quadrant angles are positive. Sin θ and csc θ are positive for second-quadrant angles. Tan θ and cot θ are positive for third-quadrant angles. Cos θ and sec θ are positive for fourth-quadrant angles. All others are negative.

This is shown in Fig. 8-3.

This discussion does not include the *quadrantal angles,* those angles with terminal sides on one of the axes. They will be discussed in the next section.

EXAMPLE 4 The following functions have *positive* values: sin 150°, sin (−200°), cos 8°, cos 300°, cos (−40°), tan 220°, tan (−100°), cot 260°, cot (−310°), sec 280°, sec (−37°), csc 140°, and csc (−190°).

EXAMPLE 5 The following functions have *negative* values: sin 190°, sin 325°, cos 95°, cos (−120°), tan 172°, tan 295°, cot 105°, cot (−60°), sec 135°, sec (−135°), csc 240°, and csc 355°.

A calculator will always give the correct sign for a trigonometric function of a given angle. However, as we will see in the next section, a calculator will *not* always give the required angle for a given value of a function.

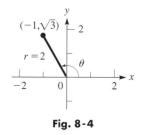

Fig. 8-4

EXAMPLE 6 Determine the trigonometric functions of θ if the terminal side of θ passes through $(-1,\sqrt{3})$. See Fig. 8-4.

We know that $x = -1$, $y = \sqrt{3}$, and from the Pythagorean theorem we find that $r = 2$. Therefore, the trigonometric functions of θ are

$$\sin \theta = \frac{\sqrt{3}}{2} = 0.8660 \qquad \cos \theta = -\frac{1}{2} = -0.5000 \quad \tan \theta = -\sqrt{3} = -1.732$$

$$\cot \theta = -\frac{1}{\sqrt{3}} = -0.5774 \quad \sec \theta = -2 = -2.000 \qquad \csc \theta = \frac{2}{\sqrt{3}} = 1.155$$

The point $(-1,\sqrt{3})$ is in the second quadrant, and the signs of the functions of θ are those of a second-quadrant angle.

EXERCISES $8\text{-}1$

In Exercises 1–8, determine the sign of the given trigonometric functions.

1. sin 36°, cos 120° **2.** tan 320°, sec 185°

3. csc 98°, cot 82° **4.** cos 260°, csc 290°

5. sec 150°, tan 220° **6.** sin 335°, cot 265°

7. cos 348°, csc 238° **8.** cot 110°, sec 309°

9. tan 460°, sin (−110°) **10.** csc (−200°), cos 550°

11. cot (−2°), cos 710° **12.** sin 539°, tan (−480°)

In Exercises 13–20, find the trigonometric functions of θ if the terminal side of θ passes through the given point.

13. (2, 1) **14.** (−1, 1)

15. (−2, −3) **16.** (4, −3)

17. (−5, 12) **18.** (−3, −4)

19. (50, −20) **20.** (9, 10)

In Exercises 21–32, determine the quadrant in which the terminal side of θ lies, subject to both given conditions.

21. $\sin \theta > 0$, $\cos \theta < 0$ **22.** $\tan \theta > 0$, $\cos \theta < 0$

23. $\sec \theta < 0$, $\cot \theta < 0$ **24.** $\cos \theta > 0$, $\csc \theta < 0$

25. $\csc \theta < 0$, $\tan \theta < 0$ **26.** $\sec \theta > 0$, $\csc \theta > 0$

27. $\sin \theta < 0$, $\tan \theta > 0$ **28.** $\cot \theta < 0$, $\sin \theta < 0$

29. $\tan \theta < 0$, $\cos \theta > 0$ **30.** $\sec \theta > 0$, $\csc \theta < 0$

31. $\sin \theta > 0$, $\cot \theta < 0$ **32.** $\tan \theta > 0$, $\csc \theta < 0$

$8\text{-}2$ TRIGONOMETRIC FUNCTIONS OF ANY ANGLE

The trigonometric functions of acute angles (angles less than 90°) were discussed in Section 4-3, and in the previous section we determined the signs of the trigonometric functions in each of the four quadrants. In this section we show how we can find the trigonometric functions of an angle of any magnitude. This information will be very important in Chapter 9, when we discuss vectors and oblique triangles, and in Chapter 10, when we graph the trigonometric functions. Even *a calculator will not always give the required angle for a given value of a function.*

CAUTION ▶

Any angle in standard position is coterminal with some positive angle less than 360°. Since the terminal sides of coterminal angles are the same, the trigonometric functions of coterminal angles are the same. Therefore, we need consider only the problem of finding the values of the trigonometric functions of positive angles less than 360°.

Fig. 8-5

■**EXAMPLE 1**   The following pairs of angles are coterminal.

$$390° \quad \text{and} \quad 30° \qquad -60° \quad \text{and} \quad 300°$$
$$900° \quad \text{and} \quad 180° \qquad -150° \quad \text{and} \quad 210°$$

From this we conclude that the trigonometric functions of both angles in each of these pairs are equal. That is, for example,

$$\sin 390° = \sin 30° \quad \text{and} \quad \tan(-150°) = \tan 210°$$

See Fig. 8-5.   ◼

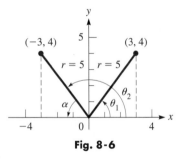

Fig. 8-6

Considering the definitions of the functions, we see that the values of the functions depend only on the values of *x, y,* and *r*. The absolute value of a function of a second-quadrant angle is equal to the value of the same function of a first-quadrant angle. For example, considering angles θ_1 and θ_2 in Fig. 8-6, for angle θ_2 with terminal side passing through $(-3, 4)$, $\tan \theta_2 = -4/3$, or $|\tan \theta_2| = 4/3$. For angle θ_1, with terminal side passing through $(3, 4)$, $\tan \theta_1 = 4/3$, and we see that $|\tan \theta_2| = \tan \theta_1$. Triangles containing angles θ_1 and α are congruent, which means $\theta_1 = \alpha$. Knowing that the absolute value of a function of θ_2 equals the same function of θ_1 means that

$$|F(\theta_2)| = |F(\theta_1)| = |F(\alpha)| \tag{8-2}$$

where F represents any of the trigonometric functions.

REFERENCE ANGLE

The angle labeled α is called the **reference angle.** *The reference angle of a given angle is the acute angle formed by the terminal side of the angle and the x-axis.*

Using Eq. (8-2) and the fact that $\alpha = 180° - \theta_2$, we may conclude that the value of any trigonometric function of any second-quadrant angle is found from

$$\boxed{F(\theta_2) = \pm F(180° - \theta_2) = \pm F(\alpha)} \tag{8-3}$$

NOTE ▶

The *sign* used depends on whether the *function* is positive or negative in the second quadrant.

■ EXAMPLE 2   In Fig. 8-6, the trigonometric functions of θ are as follows:

$$\sin \theta_2 = \sin(180° - \theta_2) = \sin \alpha = \sin \theta_1 = \tfrac{4}{5} = 0.8000$$

$$\cos \theta_2 = -\cos \theta_1 = -\tfrac{3}{5} = -0.6000, \quad \tan \theta_2 = -\tfrac{4}{3} = -1.333$$

$$\cot \theta_2 = -\tfrac{3}{4} = -0.7500, \quad \sec \theta_2 = -\tfrac{5}{3} = -1.667, \quad \csc \theta_2 = \tfrac{5}{4} = 1.250 \quad ■$$

In the same way, we derive the formulas for trigonometric functions of any third- or fourth-quadrant angle. In Fig. 8-7 the reference angle α is found by subtracting 180° from θ_3 and the functions of α and θ_1 are numerically equal. In Fig. 8-8 the reference angle α is found by subtracting θ_4 from 360°.

$$F(\theta_3) = \pm F(\theta_3 - 180°) = \pm F(\alpha) \qquad \text{(8-4)}$$

$$F(\theta_4) = \pm F(360° - \theta_4) = \pm F(\alpha) \qquad \text{(8-5)}$$

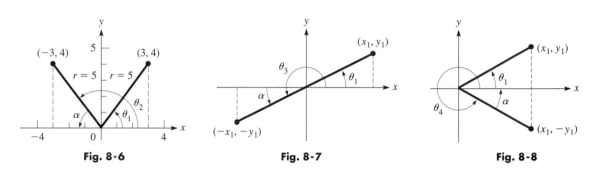

Fig. 8-6   Fig. 8-7   Fig. 8-8

■ EXAMPLE 3   If $\theta_3 = 210°$, the trigonometric functions of θ_3 are found by using Eq. (8-4) as follows. See Fig. 8-9.

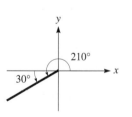

Fig. 8-9

same function   reference angle

$$\sin 210° = -\sin(210° - 180°) = -\sin 30° = -\frac{1}{2} = -0.5000$$

quadrant III | sin θ, csc θ − in quadrant III

$$\csc 210° = -\csc 30° = -2.000$$

$$\cos 210° = -\cos 30° = -0.8660 \qquad \sec 210° = -\sec 30° = -1.155$$

cos θ, sec θ − in quadrant III

$$\tan 210° = +\tan 30° = +0.5774 \qquad \cot 210° = +\cot 30° = +1.732$$

tan θ, cot θ + in quadrant III

Here we see that the reference angle is $210° - 180° = 30°$ and that we can express the value of a function of 210° in terms of the same function of 30°. We must be careful to attach the correct sign to the result. ---------■

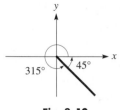

Fig. 8-10

■EXAMPLE 4   If $\theta_4 = 315°$, the trigonometric functions of θ_4 are found by using Eq. (8-5) as follows. See Fig. 8-10.

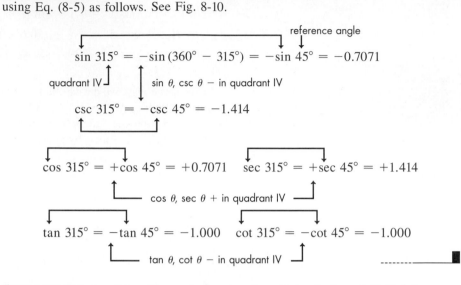

reference angle

$$\sin 315° = -\sin(360° - 315°) = -\sin 45° = -0.7071$$

quadrant IV ⅃   sin θ, csc θ − in quadrant IV

$$\csc 315° = -\csc 45° = -1.414$$

$$\cos 315° = +\cos 45° = +0.7071 \qquad \sec 315° = +\sec 45° = +1.414$$

cos θ, sec θ + in quadrant IV

$$\tan 315° = -\tan 45° = -1.000 \qquad \cot 315° = -\cot 45° = -1.000$$

tan θ, cot θ − in quadrant IV

■EXAMPLE 5   Other illustrations using Eqs. (8-3), (8-4), and (8-5) follow:

same function   reference angle

$$\sin 160° = +\sin(180° - 160°) = \sin 20° = 0.3420$$
$$\tan 110° = -\tan(180° - 110°) = -\tan 70° = -2.747$$
$$\cos 225° = -\cos(225° - 180°) = -\cos 45° = -0.7071$$
$$\cot 260° = +\cot(260° - 180°) = \cot 80° = 0.1763$$
$$\sec 304° = +\sec(360° - 304°) = \sec 56° = 1.788$$
$$\sin 357° = -\sin(360° - 357°) = -\sin 3° = -0.0523$$

↑   ⅃
determines   proper sign for function in quadrant
quadrant

Fig. 8-11

A calculator can directly evaluate functions like those in Examples 3, 4, and 5. The function and angle are entered, and the calculator gives the value, with the proper sign. The reciprocal key is used to find cot θ, sec θ, and csc θ, as shown in Section 4-3. A calculator display for tan 110° and sec 304° of Example 5 is shown in Fig. 8-11.

In most examples we will round off values to four significant digits (as we did in Examples 3, 4, and 5). However, *if the angle is approximate, we must use the guidelines in Section 4-3 for rounding off values.*

NOTE ▶

■EXAMPLE 6   A formula for finding the area of a triangle, knowing sides a and b and the included $\angle C$, is $A = \frac{1}{2}ab \sin C$. A surveyor uses this formula to find the area of a triangular tract of land for which $a = 173.2$ m, $b = 156.3$ m, and $C = 112.51°$. See Fig. 8-12.

To find the area, we substitute into the formula, which gives us

$$A = \frac{1}{2}(173.2)(156.3)\sin 112.51°$$
$$= 12{,}500 \text{ m}^2 \qquad \text{rounded to four significant digits}$$

When these numbers are entered into the calculator, the calculator automatically uses a positive value for sin 112.51°.

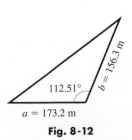

Fig. 8-12

Knowing how to use the reference angle is important when using a calculator, because *if we have the value of a function and want to find the angle,*

CAUTION ▶   ***the calculator will not necessarily give us directly the required angle.***

It will give us an angle we can use, but whether or not it is the required angle for the problem will depend on the problem being solved.

When a value of a trigonometric function is entered into a calculator, it is programmed to give the angle as follows:

For positive values of a function, the calculator displays positive acute angles.
For negative values of $\sin \theta$ *and* $\tan \theta$, *the calculator displays negative acute angles for* $\sin^{-1}$ *and* $\tan^{-1}$.
For negative values of $\cos \theta$, *the calculator displays angles between* $90°$ *and* $180°$ *for* $\cos^{-1}$.

The reason that the calculator displays these values is shown in Chapter 20, when the inverse trigonometric functions are discussed in detail.

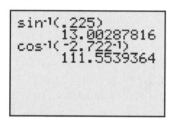

Fig. 8-13

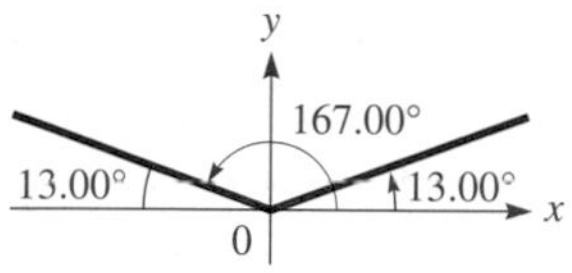

■EXAMPLE 7   For $\sin \theta = 0.2250$, we see from the first two lines of the calculator display shown in Fig. 8-13 that $\theta = 13.00°$ (rounded off).

This result is correct, but we must remember that

$$\sin (180° - 13.00°) = \sin 167.00° = 0.2250$$

also. If we need only an acute angle, $\theta = 13.00°$ is correct. However, if a second-quadrant angle is required, we see that $\theta = 167.00°$ is the angle (see Fig. 8-14). These values can be checked by finding $\sin 13.00°$ and $\sin 167.00°$. ⎯⎯⎯■

Fig. 8-14

■EXAMPLE 8   For $\sec \theta = -2.722$ and $0° \le \theta < 360°$ (this means θ may equal $0°$ or be between $0°$ and $360°$), we see from the third and fourth lines of the calculator display in Fig. 8-13 that $\theta = 111.55°$ (rounded off).

The angle $111.55°$ is the second-quadrant angle, but $\sec \theta < 0$ in the third quadrant as well. The reference angle is $\alpha = 180° - 111.55° = 68.45°$, and the third-quadrant angle is $180° + 68.45° = 248.45°$. Therefore, the two angles between $0°$ and $360°$ for which $\sec \theta = -2.722$ are $111.55°$ and $248.45°$ (see Fig. 8-15). These angles can be checked by finding $\sec 111.55°$ and $\sec 248.45°$.   ⎯⎯⎯■

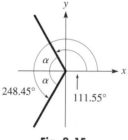

Fig. 8-15

■EXAMPLE 9   Given that $\tan \theta = 2.050$ and $\cos \theta < 0$, find the angle θ for $0° \le \theta < 360°$.

Since $\tan \theta$ is positive and $\cos \theta$ is negative, θ must be a third-quadrant angle. A calculator will display an angle of $64.00°$ (rounded off) for $\tan \theta = 2.050$. However, since we need a third-quadrant angle, *we must add* $64.00°$ *to* $180°$. Thus, the required angle is $244.00°$ (see Fig. 8-16). Check by finding $\tan 244.00°$.

If we are given that $\tan \theta = -2.050$ and $\cos \theta < 0$, the calculator will display an angle of $-64.00°$ for $\tan \theta = -2.050$. We would then have to *recognize that the reference angle is* $64.00°$ *and subtract it from* $180°$ *to get* $116.00°$, the required second-quadrant angle. This can be checked by finding $\tan 116.00°$. ⎯⎯⎯■

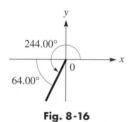

Fig. 8-16

NOTE ▶   We see that the calculator gives the reference angle (disregarding any minus signs) in all cases except when $\cos \theta$ is negative. To avoid confusion from the angle displayed by the calculator, *a good procedure is to find the reference angle first.* Then it can be used to determine the angle required by the problem.

We can find the reference angle by entering the absolute value of the function. The displayed angle will be the reference angle. Then the required angle θ is found by using the reference angle as desribed earlier, and as shown in Eqs. (8-6) for θ in the given quadrant. The angle should be checked as indicated in the earlier examples.

$$\begin{aligned}\theta &= \alpha & \text{(first quadrant)}\\ \theta &= 180° - \alpha & \text{(second quadrant)}\\ \theta &= 180° + \alpha & \text{(third quadrant)}\\ \theta &= 360° - \alpha & \text{(fourth quadrant)}\end{aligned}$$

(8-6)

EXAMPLE 10   Given that $\cos \theta = -0.1298$, find θ for $0° \le \theta < 360°$.

Since $\cos \theta$ is negative, θ is either a second-quadrant angle or a third-quadrant angle. Using 0.1298, the calculator tells us that the reference angle is 82.54°.

To get the required second-quadrant angle, we subtract 82.54° from 180° and obtain 97.46°. To get the required third-quadrant angle, we add 82.54° to 180° to obtain 262.54°. See Fig. 8-17.

If we use -0.1298, the calculator displays the required second-quadrant angle of 97.46°. However, to get the third-quadrant angle we must then subtract 97.46° from 180° to get the reference angle of 82.54°. The reference angle is then added to 180° to obtain the result of 262.54°. Also note that it is better to store the reference angle in memory rather than recalculate it. ∎

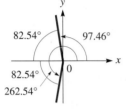

Fig. 8-17

Using Eqs. (8-3) through (8-5) we may find the value of any function when the terminal side of the angle lies *in* one of the quadrants. We now consider *the angle for which the terminal side is along one of the axes, a* **quadrantal angle.** Using the definitions of the functions (recalling that $r > 0$), we obtain the following values.

QUADRANTAL ANGLES

θ	$\sin \theta$	$\cos \theta$	$\tan \theta$	$\cot \theta$	$\sec \theta$	$\csc \theta$
0°	0.000	1.000	0.000	undef.	1.000	undef.
90°	1.000	0.000	undef.	0.000	undef.	1.000
180°	0.000	−1.000	0.000	undef.	−1.000	undef.
270°	−1.000	0.000	undef.	0.000	undef.	−1.000
360°	Same as the functions of 0° (same terminal side)					

The values in the table may be verified by referring to the figures in Fig. 8-18.

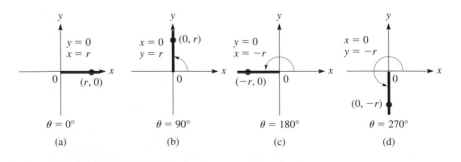

Fig. 8-18   (a)   (b)   (c)   (d)

EXAMPLE 11 Since $\sin \theta = y/r$, by looking at Fig. 8-18(a) we can see that $\sin 0° = 0/r = 0$.

Since $\tan \theta = y/x$, from Fig. 8-18(b) we see that $\tan 90° = r/0$, which is undefined due to the division by zero. If we use a calculator to find $\tan 90°$, the display would indicate an error (due to division by zero).

Since $\cos \theta = x/r$, from Fig. 8-18(c) we see that $\cos 180° = -r/r = -1$.

Since $\cot \theta = x/y$, from Fig. 8-18(d) we see that $\cot 270° = 0/-r = 0$. ∎

EXERCISES *8-2*

In Exercises 1–6, express the given trigonometric function in terms of the same function of a positive acute angle.

1. $\sin 160°$, $\cos 220°$
2. $\tan 91°$, $\sec 345°$
3. $\tan 105°$, $\csc 302°$
4. $\cos 190°$, $\cot 290°$
5. $\cos 400°$, $\tan (-400°)$
6. $\tan 920°$, $\csc (-550°)$

In Exercises 7–36, the given angles are approximate. In Exercises 7–12, find the values of the given trigonometric functions by finding the reference angle and attaching the proper sign.

7. $\sin 195°$
8. $\tan 311°$
9. $\cos 106.3°$
10. $\sin 103.4°$
11. $\sec 328.33°$
12. $\cot 136.53°$

In Exercises 13–18, find the values of the given trigonometric functions directly from a calculator.

13. $\tan 152.4°$
14. $\cos 341.4°$
15. $\sin 310.36°$
16. $\tan 242.68°$
17. $\csc 194.82°$
18. $\sec 261.08°$

In Exercises 19–32, find θ for $0° \le \theta < 360°$.

19. $\sin \theta = -0.8480$
20. $\tan \theta = -1.830$
21. $\cos \theta = 0.4003$
22. $\sin \theta = 0.6374$
23. $\cot \theta = -0.212$
24. $\csc \theta = -1.09$
25. $\sin \theta = 0.870$, $\cos \theta < 0$
26. $\tan \theta = 0.932$, $\sin \theta < 0$
27. $\cos \theta = -0.12$, $\tan \theta > 0$
28. $\sin \theta = -0.192$, $\tan \theta < 0$
29. $\tan \theta = -1.366$, $\cos \theta > 0$
30. $\cos \theta = 0.5726$, $\sin \theta < 0$
31. $\sec \theta = 2.047$, $\cot \theta < 0$
32. $\cot \theta = -0.3256$, $\csc \theta > 0$

In Exercises 33–36, determine the function that satisfies the given conditions.

33. Find $\tan \theta$ when $\sin \theta = -0.5736$ and $\cos \theta > 0$.
34. Find $\sin \theta$ when $\cos \theta = 0.422$ and $\tan \theta < 0$.
35. Find $\cos \theta$ when $\tan \theta = -0.809$ and $\csc \theta > 0$.
36. Find $\cot \theta$ when $\sec \theta = 1.122$ and $\sin \theta < 0$.

In Exercises 37–40, insert the proper sign, $>$ or $<$ or $=$, between the given expressions.

37. $\sin 90°$ $2 \sin 45°$
38. $\cos 360°$ $2 \cos 180°$
39. $\tan 180°$ $\tan 0°$
40. $\sin 270°$ $3 \sin 90°$

In Exercises 41–44, evaluate the given expressions.

41. The current i in an alternating-current circuit is given by $i = i_m \sin \theta$, where i_m is the maximum current in the circuit. Find i if $i_m = 0.0259$ A and $\theta = 495.2°$.

42. A force F is related to force F_x directed along the x-axis by $F = F_x \sec \theta$, where θ is the standard position angle for F. Find F if $F_x = -29.2$ N and $\theta = 127.6°$. See Fig. 8-19.

Fig. 8-19

43. For the slider mechanism shown in Fig. 8-20, $y \sin \alpha = x \sin \beta$. Find y if $x = 6.78$ in., $\alpha = 31.3°$, and $\beta = 104.7°$.

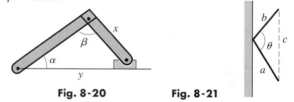

Fig. 8-20   **Fig. 8-21**

44. A laser follows the path shown in Fig. 8-21. The angle θ is related to the distances a, b, and c by $2ab \cos \theta = a^2 + b^2 - c^2$. Find θ if $a = 15.3$ cm, $b = 12.9$ cm, and $c = 24.5$ cm.

In Exercises 45–48, the trigonometric functions of negative angles are considered. In Exercises 46–48, use Eqs. (8-7).

(W) 45. Using the definitions of the trigonometric functions, explain why $\sin(-\theta) = -\sin \theta$. See Fig. 8-22. Also verify the remaining equations derived in Eqs. (8-7).

$$\sin (-\theta) = -\sin \theta$$
$$\cos (-\theta) = \cos \theta$$
$$\tan (-\theta) = -\tan \theta$$
$$\cot (-\theta) = -\cot \theta \quad \textbf{(8-7)}$$
$$\sec (-\theta) = \sec \theta$$
$$\csc (-\theta) = -\csc \theta$$

Fig. 8-22

46. Find (a) $\sin (-60°)$ and   (b) $\cos (-156°)$.
47. Find (a) $\tan (-100°)$ and   (b) $\cot (-215°)$.
48. Find (a) $\sec (-310°)$ and   (b) $\csc (-35°)$.

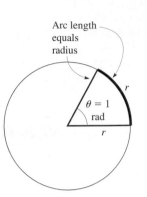

Arc length equals radius

$\theta = 1$ rad

Fig. 8-23

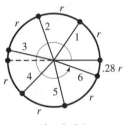

Fig. 8-24

8-3 RADIANS

For many problems in which trigonometric functions are used, particularly those involving the solution of triangles, degree measurements of angles are convenient and quite sufficient. However, division of a circle into 360 equal parts is by definition, and it is arbitrary and artificial (see the margin comment page 49).

In numerous other types of applications and in more theoretical discussions, the *radian* is a more meaningful measure of an angle. We defined the radian in Chapter 2 and reviewed it briefly in Chapter 4. In this section we discuss the radian in detail and start by reviewing its definition.

> A **radian** *is the measure of an angle with its vertex at the center of a circle and with an intercepted arc on the circle equal in length to the radius of the circle.* See Fig. 8-23.

Since the circumference of any circle in terms of its radius is given by $c = 2\pi r$, the ratio of the circumference to the radius is 2π. This means that the radius may be laid off 2π (about 6.28) times along the circumference, regardless of the length of the radius. Therefore, we see that radian measure is independent of the radius of the circle. The definition of a radian is based on an important property of a circle and is therefore a more natural measure of an angle. In Fig. 8-24 the numbers on each of the radii indicate the number of radians in the angle measured in standard position. The circular arrow shows an angle of 6 radians.

Since the radius may be laid off 2π times along the circumference, it follows that there are 2π radians in one complete rotation. Also, there are 360° in one complete rotation. Therefore, 360° is *equivalent* to 2π radians. It then follows that the relation between degrees and radians is 2π rad = 360°, or

CONVERTING ANGLES

$$\pi \text{ rad} = 180° \tag{8-8}$$

DEGREES TO RADIANS

$$1° = \frac{\pi}{180} \text{ rad} = 0.01745 \text{ rad} \tag{8-9}$$

RADIANS TO DEGREES

$$1 \text{ rad} = \frac{180°}{\pi} = 57.30° \tag{8-10}$$

We see from Eqs. (8-8) through (8-10) that we convert angle measurements from degrees to radians or from radians to degrees as follows:

Procedure for Converting Angle Measurements

1. *To convert an angle measured in degrees to the same angle measured in radians,* **multiply the number of degrees by $\pi/180°$.**

2. *To convert an angle measured in radians to the same angle measured in degrees,* **multiply the number of radians by $180°/\pi$.**

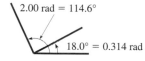

2.00 rad = 114.6°

18.0° = 0.314 rad

Fig. 8-25

EXAMPLE 1

converting degrees to radians

(a) $18.0° = \left(\dfrac{\pi}{180°}\right)(18.0°) = \dfrac{\pi}{10.0} = \dfrac{3.14}{10.0} = 0.314$ rad

degrees cancel   (See Fig. 8-25.)

converting radians to degrees

(b) 2.00 rad $= \left(\dfrac{180°}{\pi}\right)(2.00) = \dfrac{360°}{3.14} = 114.6°$ (See Fig. 8-25.)

In multiplying by $\pi/180°$ or $180°/\pi$, we are actually only multiplying by 1 because π rad $= 180°$. Although the unit of measurement is different, *the angle is the same.* See Appendix B on unit conversions.

Due to the nature of the definition of the radian, it is very common to express radians in terms of π. Expressing angles in terms of π is illustrated in the following example.

EXAMPLE 2

converting degrees to radians

(a) $30° = \left(\dfrac{\pi}{180°}\right)(30°) = \dfrac{\pi}{6}$ rad

(b) $45° = \left(\dfrac{\pi}{180°}\right)(45°) = \dfrac{\pi}{4}$ rad   (See Fig. 8-26.)

converting radians to degrees

(c) $\dfrac{\pi}{2}$ rad $= \left(\dfrac{180°}{\pi}\right)\left(\dfrac{\pi}{2}\right) = 90°$

(d) $\dfrac{3\pi}{4}$ rad $= \left(\dfrac{180°}{\pi}\right)\left(\dfrac{3\pi}{4}\right) = 135°$ (See Fig. 8-26.)

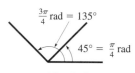

$\frac{3\pi}{4}$ rad = 135°

45° = $\frac{\pi}{4}$ rad

Fig. 8-26

We wish now to make a very important point. Since π is a special way of writing the number (slightly greater than 3) that is the ratio of the circumference of a circle to its diameter, it is the ratio of one distance to another. Thus, radians really have no units, and *radian measure amounts to measuring angles in terms of real numbers.* It is this property of radians that makes them useful in many situations. Therefore, when radians are being used, it is customary that no units are indicated for the angle.

NOTE ▶

CAUTION ▶

When no units are indicated, the radian is understood to be the unit of angle measurement.

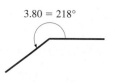

3.80 = 218°

Fig. 8-27

EXAMPLE 3   (a) $60° = \left(\dfrac{\pi}{180°}\right)(60.0°) = \dfrac{\pi}{3.00} = 1.05$

no units indicates radian measure

(b) $3.80 = \left(\dfrac{180°}{\pi}\right)(3.80) = 218°$

Since no units are indicated for 1.05 and 3.80 in this example, they are known to be in radian measure. Here, we must know that 3.80 is an angle measure. See Fig. 8-27.

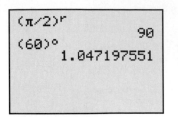

Fig. 8-28

Using the *angle* feature, a calculator can be used directly to change an angle expressed in degrees to an angle expressed in radians, or from an angle expressed in radians to an angle expressed in degrees (as we pointed out in Chapter 4 on page 106). In Fig. 8-28, we show the calculator display for changing $\pi/2$ rad to degrees (calculator in degree mode), and for changing $60°$ to radians (calculator in radian mode).

We can use a calculator to find the value of a function of an angle in radians. If the calculator is in radian mode, it then uses values in radians directly and will *consider any angle entered to be in radians*. The mode can be changed as needed, but

CAUTION ▶   *always be careful to have your calculator in the proper mode.*

Check the setting in the *mode* feature. If you are working in degrees, use the degree mode, but if you are working in radians, use the radian mode.

▌EXAMPLE 4   **(a)** To find the value of sin 0.7538, put the calculator in radian mode (note that no units are shown with 0.7538), and the value is found as shown in the first two lines of the calculator display in Fig. 8-29. Therefore,

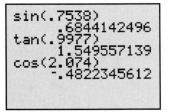

Fig. 8-29

no units indicates radian measure

$$\sin 0.7538 = 0.6844$$

(b) From the third and fourth lines of the display in Fig. 8-29, we see that

$$\tan 0.9977 = 1.550$$

(c) From the last two lines of the display in Fig. 8-29, we see that

$$\cos 2.074 = -0.4822$$

In each case the calculator was in radian mode.   ■

In the following application, the resulting angle is a unitless number, and it is therefore in radian measure.

▌EXAMPLE 5   The velocity v of an object undergoing simple harmonic motion at the end of a spring is given by

$$v = A \sqrt{\frac{k}{m}} \cos \sqrt{\frac{k}{m}}\, t \qquad \text{the angle is } \left(\sqrt{\frac{k}{m}} \right)(t)$$

Here, m is the mass of the object (in g), k is a constant depending on the spring, A is the maximum distance the object moves, and t is the time (in s). Find the velocity (in cm/s) after 0.100 s of a 36.0-g object at the end of a spring for which $k = 400$ g/s², if $A = 5.00$ cm.

Substituting, we have

$$v = 5.00 \sqrt{\frac{400}{36.0}} \cos \sqrt{\frac{400}{36.0}}\, (0.100)$$

Using calculator memory for $\sqrt{\frac{400}{36.0}}$, and with the calculator in radian mode, we have

$$v = 15.7 \text{ cm/s} \qquad\qquad ■$$

For certain special situations, we may need to know a reference angle in radians. In order to determine the proper quadrant, we should remember that $\frac{1}{2}\pi = 90°$, $\pi = 180°$, $\frac{3}{2}\pi = 270°$, and $2\pi = 360°$. These are shown in Table 8-1 along with the approximate decimal values for angles in radians. See Fig. 8-30.

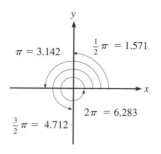

$\pi = 3.142$ $\frac{1}{2}\pi = 1.571$

$\frac{3}{2}\pi = 4.712$ $2\pi = 6.283$

Fig. 8-30

Table 8-1 Quadrantal Angles

Degrees	Radians	Radians (decimal)
90°	$\frac{1}{2}\pi$	1.571
180°	π	3.142
270°	$\frac{3}{2}\pi$	4.712
360°	2π	6.283

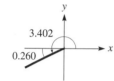

3.402

0.260

Fig. 8-31

EXAMPLE 6 An angle of 3.402 is greater than 3.142 but less than 4.712. Thus, it is a third-quadrant angle, and the reference angle is $3.402 - \pi = 0.260$. The π key can be used. See Fig. 8-31.

An angle of 5.210 is between 4.712 and 6.283. Therefore, it is in the fourth quadrant and the reference angle is $2\pi - 5.210 = 1.073$. ∎

EXAMPLE 7 Express θ in radians, such that $\cos \theta = 0.8829$ and $0 \leq \theta < 2\pi$.

We are to find θ in radians for the given value of the $\cos \theta$. Also, since θ is restricted to values between 0 and 2π, we must find a first-quadrant angle and a fourth-quadrant angle ($\cos \theta$ is positive in the first and fourth quadrants). With the calculator in radian mode, we find that

$$\cos 0.4888 = 0.8829$$

Therefore, for the fourth-quadrant angle,

$$\cos (2\pi - 0.4888) = \cos 5.794$$

This means

$$\theta = 0.4888 \quad \text{or} \quad \theta = 5.794$$

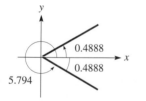

0.4888

0.4888

5.794

Fig. 8-32

See Fig. 8-32. ∎

When one first encounters radian measure,

CAUTION▶ *expressions such as* $\sin 1$ *and* $\sin \theta = 1$ *are often confused.*

The first is equivalent to $\sin 57.30°$, since $57.30° = 1$ (radian). The second means that θ is the angle for which the sine is 1. Since we know that $\sin 90° = 1$, we then can say that $\theta = 90°$ or $\theta = \pi/2$. The following example gives additional illustrations of evaluating expressions involving radians.

EXAMPLE 8 (a) $\sin \dfrac{\pi}{3} = \dfrac{\sqrt{3}}{2}$ since $\dfrac{\pi}{3} = 60°$.

(b) $\sin 0.6050 = 0.5688$. (0.6050 rad = 34.66°.)

(c) $\tan \theta = 1.709$ means that $\theta = 59.67°$ (smallest positive θ).

(d) Since $59.67° = 1.041$, we can state that $\tan 1.041 = 1.708$. ∎

── EXERCISES *8-3* ──

In Exercises 1–8, express the given angle measurements in radian measure in terms of π.

1. 15°, 150°   **2.** 12°, 225°

3. 75°, 330°   **4.** 36°, 315°

5. 210°, 270°   **6.** 240°, 300°

7. 160°, 260°   **8.** 66°, 350°

In Exercises 9–16, the given numbers express angle measure. Express the measure of each angle in terms of degrees.

9. $\dfrac{2\pi}{5}, \dfrac{3\pi}{2}$ **10.** $\dfrac{3\pi}{10}, \dfrac{5\pi}{6}$

11. $\dfrac{\pi}{18}, \dfrac{7\pi}{4}$ **12.** $\dfrac{7\pi}{15}, \dfrac{4\pi}{3}$

13. $\dfrac{17\pi}{18}, \dfrac{5\pi}{3}$ **14.** $\dfrac{11\pi}{36}, \dfrac{5\pi}{4}$

15. $\dfrac{\pi}{12}, \dfrac{3\pi}{20}$ **16.** $\dfrac{7\pi}{30}, \dfrac{4\pi}{15}$

In Exercises 17–24, express the given angles in radian measure. Round off results to the number of significant digits in the given angle.

17. 23.0°   **18.** 54.3°

19. 252°   **20.** 104°

21. 333.5°   **22.** 168.7°

23. 178.5°   **24.** 86.1°

In Exercises 25–32, the given numbers express angle measure. Express the measure of each angle in terms of degrees, with the same accuracy as in the given value.

25. 0.750   **26.** 0.240

27. 3.407   **28.** 1.703

29. 2.45   **30.** 34.4

31. 16.42   **32.** 100.0

In Exercises 33–40, evaluate the given trigonometric functions by first changing the radian measure to degree measure. Round off results to four significant digits.

33. $\sin \dfrac{\pi}{4}$ **34.** $\cos \dfrac{\pi}{6}$

35. $\tan \dfrac{5\pi}{12}$ **36.** $\sin \dfrac{7\pi}{18}$

37. $\cos \dfrac{5\pi}{6}$ **38.** $\tan \dfrac{4\pi}{3}$

39. sec 4.5920   **40.** cot 3.2732

In Exercises 41–48, evaluate the given trigonometric functions directly, without first changing to degree measure.

41. tan 0.7359   **42.** cos 0.9308

43. sin 4.24   **44.** tan 3.47

45. sec 2.07   **46.** sin 2.34

47. cot 4.86   **48.** csc 6.19

In Exercises 49–56, find θ to four significant digits for 0 ≤ θ < 2π.

49. sin θ = 0.3090   **50.** cos θ = −0.9135

51. tan θ = −0.2126   **52.** sin θ = −0.0436

53. cos θ = 0.6742   **54.** cot θ = 1.860

55. sec θ = −1.307   **56.** csc θ = 3.940

In Exercises 57–60, evaluate the given expressions.

57. A flat plate of weight W oscillates as shown in Fig. 8-33. Its potential energy V is given by $V = \frac{1}{2}Wb\theta^2$, where θ is measured in radians. Find V if $W = 8.75$ lb, $b = 0.75$ ft, and $\theta = 5.5°$.

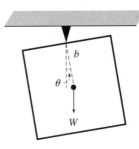

Fig. 8-33

W 58. The charge q (in C) on a capacitor as a function of time is $q = A \sin \omega t$. If t is measured in seconds, in what units is ω measured? Explain.

59. The height h of a rocket launched 1200 m from an observer is found to be $h = 1200 \tan \dfrac{5t}{3t + 10}$ for $t < 10$ s, where t is the time after launch. Find h for $t = 8.0$ s.

60. The electric intensity I (in W/m²) from the two radio antennas shown in Fig. 8-34 is a function of the angle θ given by $I = 0.023 \cos^2(\pi \sin \theta)$. Find I for $\theta = 40.0°$. ($\cos^2 \alpha = (\cos \alpha)^2$.)

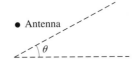

Fig. 8-34   ● Antenna

8-4 APPLICATIONS OF RADIAN MEASURE

Radian measure has numerous applications in mathematics and technology, some of which were indicated in the last four exercises of the previous section. In this section we illustrate a number of additional applications.

Arc Length

From geometry we know that *the length of an arc on a circle is proportional to the central angle* and that the length of arc of a complete circle is the circumference. Letting *s* stand for the length of arc, we may state that $s = 2\pi r$ for a complete circle. Since 2π is the central angle (in radians) of the complete circle, *we have for the length of arc*

Fig. 8-35

$$\boxed{s = \theta r \quad (\theta \text{ in radians})} \tag{8-11}$$

for any circular arc with central angle θ. If we know the central angle in radians and the radius of a circle, we can find the length of a circular arc directly by using Eq. (8-11). See Fig. 8-35.

EXAMPLE 1   In Fig. 8-36, $\theta = \pi/6$ and $r = 3.00$ in. Therefore,

$$s = \left(\frac{\pi}{6}\right)(3.00) = \frac{\pi}{2.00} = 1.57 \text{ in.}$$

where the θ in radians is indicated. ∎

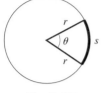

Fig. 8-36

Among the important applications of arc length are distances on the earth's surface. For most purposes the earth may be regarded as a sphere (the diameter at the equator is slightly greater than the distance between the poles). A *great circle* of the earth (or any sphere) is the circle of intersection of the surface of the sphere and a plane that passes through the center.

The equator is a great circle, and is designated as 0° *latitude*. Other *parallels of latitude* are parallel to the equator, with diameters decreasing to zero at the poles, which are 90° N and 90° S. See Fig. 8-37.

Meridians of longitude are half great circles between the poles. The *prime meridian* through Greenwich, England, is designated as 0°, with meridians to 180° measured east and west from Greenwich. Positions on the surface of the earth are designated by longitude and latitude.

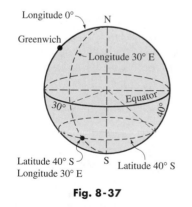

Fig. 8-37

See the chapter introduction.

EXAMPLE 2   The traditional definition of a *nautical mile* is the length of arc along a great circle of the earth for a central angle of 1′. The modern international definition is a distance of 1852 m. What measurement of the earth's radius does this definition use?

Here, $\theta = 1' = (1/60)°$ and $s = 1852$ m. Solving for *r*, we have

$$r = \frac{s}{\theta} = \frac{1852}{\left(\frac{1}{60}\right)°\left(\frac{\pi}{180°}\right)} = 6.367 \times 10^6 \text{ m} = 6367 \text{ km}$$

Historically, the fact that the earth is not a perfect sphere has led to many variations in the distance used for a nautical mile. ∎

Fig. 8-38

Area of a Sector of a Circle

Another application of radians is finding the area of a sector of a circle (see Fig. 8-38). We recall from geometry that areas of sectors of circles are proportional to their central angles. The area of a circle is $A = \pi r^2$, which can be written as $A = \frac{1}{2}(2\pi)r^2$. Since the angle for a complete circle is 2π, *the area of any sector of a circle in terms of the radius and central angle (in radians) is*

$$A = \frac{1}{2}\theta r^2 \quad (\theta \text{ in radians}) \tag{8-12}$$

EXAMPLE 3 (a) The area of a sector of a circle with central angle 218° and a radius of 5.25 in. (see Fig. 8-39(a)) is

$$A = \frac{1}{2}(218)\underbrace{\left(\frac{\pi}{180}\right)}_{\theta \text{ in radians}}(5.25)^2 = 52.4 \text{ in.}^2$$

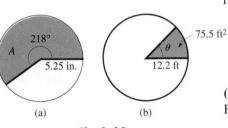

(a)   (b)

Fig. 8-39

(b) Given that the area of a sector is 75.5 ft² and the radius is 12.2 ft (see Fig. 8-39(b)), we find the central angle by solving for θ and then substituting.

$$\theta = \frac{2A}{r^2} = \frac{2(75.5)}{(12.2)^2} = 1.01 \overset{\text{no units indicates radian measure}}{}$$

This means that the central angle is 1.01 rad, or 57.9°.   ∎

CAUTION ▶ We should note again that the equations in this section require that the angle θ *is expressed in radians.* A common error is to use θ in degrees.

Angular Velocity

The average velocity of a moving object is defined by $v = s/t$, where v is the average velocity, s is the distance traveled, and t is the elapsed time. For an object moving in a circular path with constant speed, the distance traveled is the length of arc through which it moves. Therefore, if we divide both sides of Eq. (8-11) by t, we obtain

$$\frac{s}{t} = \frac{\theta r}{t} = \frac{\theta}{t}r$$

where θ/t is called the *angular velocity* and is designated by ω. Therefore,

$$v = \omega r \tag{8-13}$$

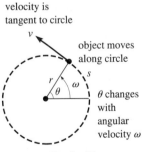

velocity is tangent to circle

object moves along circle

θ changes with angular velocity ω

Fig. 8-40

Equation (8-13) expresses the relationship between the **linear velocity** *v and the* **angular velocity** *ω of an object moving around a circle of radius r.* See Fig. 8-40. In the figure, v is shown directed tangent to the circle, for that is its direction for the position shown. The direction of v changes constantly.

The units for ω are radians per unit of time. In this way the formula can be used directly. However, in practice, ω is often given in revolutions per minute or in some similar unit. In these cases it is necessary to convert the units of ω to radians per unit of time before substituting in Eq. (8-13).

EXAMPLE 4   A person on a hang glider is moving in a horizontal circular arc of radius 90.0 m with an angular velocity of 0.125 rad/s. The person's linear velocity is

$$v = (0.125)(90.0) = 11.3 \text{ m/s}$$

(Remember that radians are numbers and are not included in the final set of units.) This means that the person is moving along the circumference of the arc at 11.3 m/s (40.7 km/h).

SOLVING A WORD PROBLEM

The first U.S. communications satellite was launched in July 1962.

EXAMPLE 5   A communications satellite remains at an altitude of 22,320 mi above a point on the equator. If the radius of the earth is 3960 mi, what is the velocity of the satellite?

In order for the satellite to remain over a point on the equator, it must rotate exactly once each day around the center of the earth (and it must remain at an altitude of 22,320 mi). Since there are 2π radians in each revolution, the angular velocity is

$$\omega = \frac{1 \text{ r}}{1 \text{ day}} = \frac{2\pi \text{ rad}}{24 \text{ h}} = 0.2618 \text{ rad/h}$$

The radius of the circle through which the satellite moves is its altitude plus the radius of the earth, or $22,320 + 3960 = 26,280$ mi. Thus, the velocity is

$$v = 0.2618(26,280) = 6880 \text{ mi/h}$$

SOLVING A WORD PROBLEM

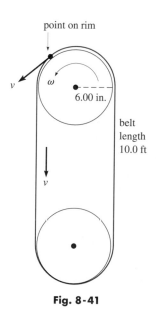

point on rim

6.00 in.

belt
length
10.0 ft

Fig. 8-41

EXAMPLE 6   A pulley belt 10.0 ft long takes 2.00 s to make one complete revolution. The radius of the pulley is 6.00 in. What is the angular velocity (in revolutions per minute) of a point on the rim of the pulley? See Fig. 8-41.

Since the linear velocity of a point on the rim of the pulley is the same as the velocity of the belt, $v = 10.0/2.00 = 5.00$ ft/s. The radius of the pulley is $r = 6.00$ in. $= 0.500$ ft, and we can find ω by substituting into Eq. (8-13). This gives us

$$v = \omega r$$
$$5.00 = \omega(0.500)$$
$$\omega = 10.0 \text{ rad/s} \qquad \text{multiply by 60 s/1 min}$$
$$= 600 \text{ rad/min} \qquad \text{multiply by 1 r/}2\pi \text{ rad}$$
$$= 95.5 \text{ r/min} \qquad \text{r is the symbol for revolution}$$

As shown in Appendix B, the change of units can be handled algebraically as

$$10.0\frac{\text{rad}}{\text{s}} \times 60\frac{\text{s}}{\text{min}} = 600\frac{\text{rad}}{\text{min}}$$

$$\frac{600 \text{ rad/min}}{2\pi \text{ rad/r}} = 600\frac{\text{rad}}{\text{min}} \times \frac{1}{2\pi}\frac{\text{r}}{\text{rad}} = 95.5 \text{ r/min}$$

EXAMPLE 7   The current at any time in a certain alternating-current electric circuit is given by $i = I \sin 120\pi t$, where I is the maximum current and t is the time in seconds. Given that $I = 0.0685$ A, find i for $t = 0.00500$ s.

Substituting, with the calculator in radian mode, we get

$$i = 0.0685 \sin(120\pi)(0.00500)$$
$$= 0.0651 \text{ A}$$

EXERCISES 8-4

In Exercises 1–12, for an arc length s, area of sector A, and central angle θ of a circle of radius r, find the indicated quantity for the given values.

1. $r = 3.30$ in., $\theta = \pi/3$, $s = ?$
2. $r = 21.2$ cm, $\theta = 2.65$, $s = ?$
3. $s = 1010$ mm, $\theta = 136.0°$, $r = ?$
4. $s = 0.3456$ ft, $\theta = 73.61°$, $r = ?$
5. $s = 0.3913$ mi, $r = 0.9449$ mi, $\theta = ?$
6. $s = 3.19$ m, $r = 2.29$ m, $\theta = ?$
7. $r = 4.9$ cm, $\theta = 3.6$, $A = ?$
8. $r = 46.3$ in., $\theta = 2\pi/5$, $A = ?$
9. $A = 0.0119$ ft^2, $\theta = 326.0°$, $r = ?$
10. $A = 1200$ mm^2, $\theta = 17°$, $r = ?$
11. $A = 16.5$ m^2, $r = 4.02$ m, $\theta = ?$
12. $A = 67.8$ mi^2, $r = 67.8$ mi, $\theta = ?$

In Exercises 13–48, solve the given problems.

13. While playing, the left spool of a VCR turns through 820°. For this part of the tape, it is 3.30 cm from the center of the spool to the tape. What length of tape is played?

(W) 14. The latitude of Miami is 26° N, and the latitude of the north end of the Panama Canal is 9° N. Both are at a longitude of 80° W. What is the distance between Miami and the Canal? Explain how the angle used in the solution is found. The radius of the earth is 3960 mi.

15. Of the estimated natural gas reserves in North America, 4.59×10^9 m^3 are in the United States, 2.66×10^9 m^3 are in Canada, and 1.99×10^9 m^3 are in Mexico. In making a *circle graph* (circular sectors represent percentages of the whole—a *pie chart*) with a radius of 4.00 cm for these data, what are the central angle and area of the sector that represents Canada's reserves?

16. A section of sidewalk is a circular sector of radius 3.00 ft and central angle 50.6°. What is the area of this section of sidewalk?

17. When between 12:00 noon and 1:00 P.M. are the minute and hour hands of a clock 180° apart?

18. A cam is in the shape of a circular sector, as shown in Fig. 8-42. What is the perimeter of the cam?

Fig. 8-42   165.58°   1.875 in.

19. A lawn sprinkler can water up to a distance of 65.0 ft. It turns through an angle of 115.0°. What area can it water?

20. A spotlight beam sweeps through a horizontal angle of 75.0°. If the range of the spotlight is 250 ft, what area can it cover?

21. If a car makes a U-turn in 6.0 s, what is its average angular velocity in the turn?

22. The roller on a computer printer makes 2200 r/min. What is its angular velocity?

23. What is the floor area of the hallway shown in Fig. 8-43? The outside and inside of the hallway are circular arcs.

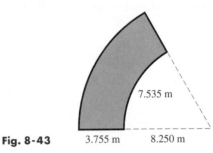

Fig. 8-43   3.755 m   8.250 m

7.535 m

24. The arm of a car windshield wiper is 12.75 in. long and is attached at the middle of a 15.00 in. blade. (Assume that the arm and blade are in line.) What area of the windshield is cleaned by the wiper if it swings through 110.0° arcs?

25. Part of a railroad track follows a circular arc with a central angle of 28.0°. If the radius of the arc of the inner rail is 93.67 ft and the rails are 4.71 ft apart, how much longer is the outer rail than the inner rail?

26. A wrecking ball is dropped as shown in Fig. 8-44. Its velocity at the bottom of its swing is $v = \sqrt{2gh}$, where g is the acceleration due to gravity. What is its angular velocity at the bottom if $g = 9.80$ m/s^2 and $h = 4.80$ m?

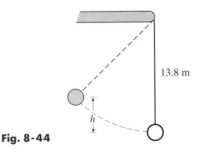

13.8 m

h

Fig. 8-44

27. Part of a security fence is built 2.50 m from a cylindrical storage tank 11.2 m in diameter. What is the area between the tank and this part of the fence if the central angle of the fence is 75.5°? See Fig. 8-45.

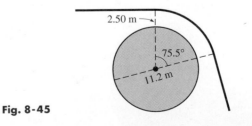

2.50 m

75.5°

11.2 m

Fig. 8-45

28. Through what angle does the drum in Fig. 8-46 turn in order to lower the crate 18.5 ft?

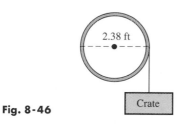

Fig. 8-46

2.38 ft

Crate

29. A section of road follows a circular arc with a central angle of 15.6°. The radius of the inside of the curve is 285.0 m, and the road is 15.2 m wide. What is the volume of the concrete in the road if it is 0.305 m thick?

30. The propeller of the motor on a motorboat is rotating at 130 rad/s. What is the linear velocity of a point on the tip of a blade if it is 22.5 cm long?

31. A storm causes a pilot to follow a circular-arc route, with a central angle of 12.8°, from city A to city B rather than the straight-line route of 185.0 km. How much farther does the plane fly due to the storm?

32. An interstate route exit is a circular arc 330 m long with a central angle of 79.4°. What is the radius of curvature of the exit?

33. A special vehicle for traveling on glacial ice has tires that are 12.0 ft in diameter. If the vehicle travels at 3.5 mi/h, what is the angular velocity of the tire in revolutions per minute?

34. The sweep second hand of a watch is 15.0 mm long. What is the linear velocity of the tip?

35. A computer diskette has a diameter of 3.50 in. and rotates at 360.0 r/min. What is the linear velocity of a point on the outer edge?

36. A rotating circular restaurant at the top of a hotel has a diameter of 25.0 m. If it completes one revolution in 30.0 min, what is the velocity of its outer surface?

37. Two streets meet at an angle of 82.0°. What is the length of the piece of curved curbing at the intersection if it is constructed along the arc of a circle 15.0 ft in radius? See Fig. 8-47.

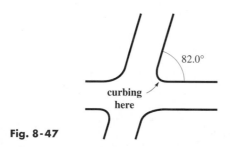

82.0°

curbing here

Fig. 8-47

38. An ammeter needle is deflected 52.00° by a current of 0.2500 A. The needle is 3.750 in. long, and a circular scale is used. How long is the scale for a maximum current of 1.500 A?

39. A drill bit $\frac{3}{8}$ in. in diameter rotates at 1200 r/min. What is the linear velocity of a point on its circumference?

40. A helicopter blade is 2.75 m long and is rotating at 420 r/min. What is the linear velocity of the tip of the blade?

41. A waterwheel used to generate electricity has paddles 3.75 m long. The speed of the end of a paddle is one-fourth that of the water. If the water is flowing at the rate of 6.50 m/s, what is the angular velocity of the waterwheel?

42. A jet is traveling westward with the sun directly overhead (the jet is on a line between the sun and the center of the earth). How fast must the jet fly in order to keep the sun directly overhead? (Assume that the earth's radius is 3960 mi, the altitude of the jet is low, and the earth rotates about its axis once in 24.0 h.)

43. A 1500-kW wind turbine (windmill) rotates at 40.0 r/min. What is the linear velocity of a point on the end of a blade, if the blade is 35 ft long (from the center of rotation)?

44. What is the linear velocity of a point in Charleston, South Carolina, which is at a latitude of 32°46′ N? The radius of the earth is 3960 mi.

45. Through what total angle does the drive shaft of a car rotate in one second when the tachometer reads 2400 r/min?

46. A patio is in the shape of a circular sector with a central angle of 160.0°. It is enclosed by a railing of which the circular part is 11.6 m long. What is the area of the patio?

47. An oil storage tank 4.25 m long has a flat bottom as shown in Fig. 8-48. The radius of the circular part is 1.10 m. What volume of oil does the tank hold?

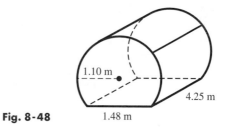

1.10 m

4.25 m

Fig. 8-48   1.48 m

48. Two equal beams of light illuminate the area shown in Fig. 8-49. What area is lit by both beams?

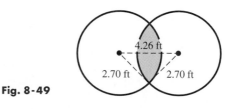

4.26 ft

2.70 ft   2.70 ft

Fig. 8-49

In Exercises 49–52, another use of radians is illustrated.

(W) 49. Use a calculator (in radian mode) to evaluate the ratios $(\sin \theta)/\theta$ and $(\tan \theta)/\theta$ for $\theta = 0.1$, 0.01, 0.001, and 0.0001. From these values explain why it is possible to say that

$$\sin \theta = \tan \theta = \theta \qquad (8\text{-}14)$$

approximately for very small angles.

50. Using Eq. (8-14), evaluate $\tan 0.001°$. Compare with a calculator value.

51. An astronomer observes that a star 12.5 light-years away moves through an angle of $0.2''$ in one year. Assuming it moved in a straight line perpendicular to the initial line of observation, how many miles did the star move? (1 light-year = 5.88×10^{12} mi.) Use Eq. (8-14).

52. In calculating a back line of a lot, a surveyor discovers an error of $0.05°$ in an angle measurement. If the lot is 136.0 m deep, by how much is the back line calculation in error? See Fig. 8-50. Use Eq. (8-14).

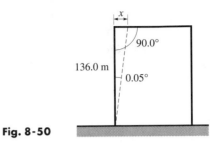

Fig. 8-50

CHAPTER EQUATIONS

α is reference angle

$$\sin \theta = \frac{y}{r} \qquad \cos \theta = \frac{x}{r} \qquad \tan \theta = \frac{y}{x}$$

$$\cot \theta = \frac{x}{y} \qquad \sec \theta = \frac{r}{x} \qquad \csc \theta = \frac{r}{y} \qquad (8\text{-}1)$$

$$F(\theta_2) = \pm F(180° - \theta_2) = \pm F(\alpha) \qquad (8\text{-}3)$$

$$F(\theta_3) = \pm F(\theta_3 - 180°) = \pm F(\alpha) \qquad (8\text{-}4)$$

$$F(\theta_4) = \pm F(360° - \theta_4) = \pm F(\alpha) \qquad (8\text{-}5)$$

$$\theta = \alpha \qquad \text{(first quadrant)}$$

$$\theta = 180° - \alpha \qquad \text{(second quadrant)}$$

$$\theta = 180° + \alpha \qquad \text{(third quadrant)} \qquad (8\text{-}6)$$

$$\theta = 360° - \alpha \qquad \text{(fourth quadrant)}$$

$$\pi \text{ rad} = 180° \qquad (8\text{-}8)$$

Radian-degree conversions

$$1° = \frac{\pi}{180} \text{ rad} = 0.01745 \text{ rad} \qquad (8\text{-}9)$$

$$1 \text{ rad} = \frac{180°}{\pi} = 57.30° \qquad (8\text{-}10)$$

Circular arc length

$$s = \theta r \qquad (\theta \text{ in radians}) \qquad (8\text{-}11)$$

Circular sector area

$$A = \frac{1}{2}\theta r^2 \qquad (\theta \text{ in radians}) \qquad (8\text{-}12)$$

Linear and angular velocity

$$v = \omega r \qquad (8\text{-}13)$$

REVIEW EXERCISES

In Exercises 1–4, find the trigonometric functions of θ. The terminal side of θ passes through the given point.

1. (6, 8)　　　　　　　**2.** (−12, 5)

3. (7, −2)　　　　　　**4.** (−2, −3)

In Exercises 5–8, express the given trigonometric functions in terms of the same function of a positive acute angle.

5. cos 132°, tan 194°　　　**6.** sin 243°, cot 318°

7. sin 289°, sec (−15°)　　**8.** cos 103°, csc (−100°)

In Exercises 9–12, express the given angle measurements in terms of π.

9. 40°, 153°　　　　　　**10.** 22.5°, 324°

11. 48°, 202.5°　　　　　**12.** 27°, −162°

In Exercises 13–20, the given numbers represent angle measure. Express the measure of each angle in degrees.

13. $\dfrac{7\pi}{5}, \dfrac{13\pi}{18}$　　　　**14.** $\dfrac{3\pi}{8}, \dfrac{7\pi}{20}$

15. $\dfrac{\pi}{15}, \dfrac{11\pi}{6}$　　　　**16.** $\dfrac{17\pi}{10}, \dfrac{5\pi}{4}$

17. 0.560　　　　　　　**18.** 1.354

19. 3.607　　　　　　　**20.** 14.5

In Exercises 21–28, express the given angles in radians (not in terms of π).

21. 102°　　　　　　　**22.** 305°

23. 20.25°　　　　　　**24.** 148.38°

25. 262.05°　　　　　**26.** 18.72°

27. 136.2°　　　　　　**28.** 385.4°

In Exercises 29–48, determine the values of the given trigonometric functions directly on a calculator. The angles are approximate. Express answers to Exercises 41–44 to four significant digits.

29. cos 245.5°　　　　　**30.** sin 141.3°

31. cot 295°　　　　　　**32.** tan 184°

33. csc 247.82°　　　　**34.** sec 96.17°

35. sin 205.24°　　　　**36.** cos 326.72°

37. tan 301.4°　　　　　**38.** sin 103.9°

39. tan 256.42°　　　　**40.** cos 162.32°

41. $\sin \dfrac{9\pi}{5}$　　　　　　**42.** $\sec \dfrac{5\pi}{8}$

43. $\cos \dfrac{7\pi}{6}$　　　　　　**44.** $\tan \dfrac{23\pi}{12}$

45. sin 0.5906　　　　　**46.** tan 0.8035

47. csc 2.153　　　　　**48.** cot 5.190

In Exercises 49–52, find θ in degrees for 0° ≤ θ < 360°.

49. tan θ = 0.1817　　　**50.** sin θ = −0.9323

51. cos θ = −0.4730　　**52.** cot θ = 1.196

In Exercises 53–56, find θ in radians for 0 ≤ θ < 2π.

53. cos θ = 0.8387　　　**54.** sin θ = 0.1045

55. sin θ = −0.8650　　**56.** tan θ = 2.840

In Exercises 57–60, find θ in degrees for 0° ≤ θ < 360°.

57. cos θ = −0.7222, sin θ < 0

58. tan θ = −1.683, cos θ < 0

59. cot θ = 0.4291, cos θ < 0

60. sin θ = 0.2626, tan θ < 0

In Exercises 61–64, for an arc of length s, area of sector A, and central angle θ of circle of radius r, find the indicated quantity for the given values.

61. s = 20.3 in., θ = 107.5°, r = ?

62. s = 584 ft, r = 106 ft, θ = ?

63. A = 265 mm², r = 12.8 mm, θ = ?

64. A = 0.908 km², θ = 234.5°, r = ?

In Exercises 65–80, solve the given problems.

65. The instantaneous power p (in W) input to a resistor in an alternating-current circuit is $p = p_m \sin^2 377t$, where p_m is the maximum power input and t is the time (in s). Find p for $p_m = 0.120$ W and t = 2.00 ms. $(\sin^2 \theta = (\sin \theta)^2.)$

66. The horizontal distance x through which a pendulum moves is given by $x = a(\theta + \sin \theta)$, where a is a constant and θ is the angle between the vertical and the pendulum. Find x for a = 45.0 cm and θ = 0.175.

67. A sector gear with a pitch radius of 8.25 in. and a 6.60-in. arc of contact is shown in Fig. 8-51. What is the sector angle θ?

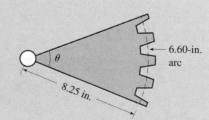

← 6.60-in. arc

θ

8.25 in.

Fig. 8-51

68. Two pulleys have radii of 10.0 in. and 6.00 in., and their centers are 40.0 in. apart. If the pulley belt is uncrossed, what must be the length of the belt?

(W) 69. The longitude of Anchorage, Alaska, is 150° W, and the longitude of St. Petersburg, Russia, is 30° E. Both cities are at a latitude of 60° N. (a) Find the great circle distance (see page 235) from Anchorage to St. Petersburg over the north pole. (b) Find the distance between them along the 60° N latitude arc. The radius of the earth is 3960 mi. What do the results show?

70. A piece of circular filter paper 15.0 cm in diameter is folded such that its effective filtering area is the same as that of a sector with central angle of 220°. What is the filtering area?

71. To produce an electric current, a circular loop of wire of diameter 25.0 cm is rotating about its diameter at 60.0 r/s in a magnetic field. What is the greatest linear velocity of any point on the loop?

72. Find the area of the decorative glass panel shown in Fig. 8-52. The panel is made of two equal circular sectors and an isosceles triangle.

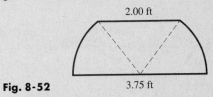

2.00 ft

Fig. 8-52   3.75 ft

73. A circular hood is to be used over a piece of machinery. It is to be made from a circular piece of sheet metal 3.25 ft in radius. A hole 0.75 ft in radius and a sector of central angle 80.0° are to be removed to make the hood. What is the area of the top of the hood?

74. The chain on a chain saw is driven by a sprocket 7.50 cm in diameter. If the chain is 108 cm long and makes one revolution in 0.250 s, what is the angular velocity (in r/s) of the sprocket?

75. An *ultracentrifuge,* used to observe the sedimentation of particles such as proteins, may rotate as fast as 80,000 r/min. If it rotates at this rate and is 7.20 cm in diameter, what is the linear velocity of a particle at the outer edge?

76. A computer is programmed to shade in a sector of a pie chart 2.44 cm in radius. If the perimeter of the shaded sector is 7.32 cm, what is the central angle (in degrees) of the sector? See Fig. 8-53.

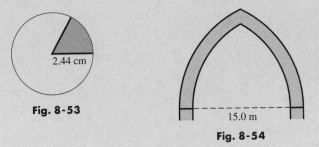

2.44 cm

Fig. 8-53

15.0 m

Fig. 8-54

77. A Gothic arch, commonly used in medieval European structures, is formed by two circular arcs. In one type, each arc is one-sixth of a circle, with the center of each at the base on the end of the other arc. See Fig. 8-54. Therefore, the width of the arch equals the radius of each arc. For such an arch, find the area of the opening if the width is 15.0 m.

78. The Trans-Alaska Pipeline was assembled in sections 40.0 ft long and 4.00 ft in diameter. If the depth of the oil in one horizontal section is 1.00 ft, what is the volume of oil in this section?

79. A laser beam is transmitted with a "width" of 0.0008° and makes a circular spot of radius 2.50 km on a distant object. How far is the object from the source of the laser beam? Use Eq. (8-14).

80. The planet Venus subtends an angle of 15″ to an observer on earth. If the distance between Venus and earth is 1.04×10^8 mi, what is the diameter of Venus? Use Eq. (8-14).

Writing Exercise

81. Write a paragraph explaining how you determine the units for the result of the following problem: An astronaut in a spacecraft circles the moon once each 1.95 h. If the altitude of the spacecraft is constant at 70.0 mi, what is its velocity? The radius of the moon is 1080 mi. (What is the answer?)

PRACTICE TEST

1. Change 150° to radians in terms of π.

2. Express sin 205° in terms of the sine of a positive acute angle. Do not evaluate.

3. Find sin θ and sec θ if θ is in standard position and the terminal side passes through (−9, 12).

4. An airplane propeller blade is 2.80 ft long and rotates at 2200 r/min. What is the linear velocity of a point on the tip of the blade?

5. Given that 3.572 is the measure of an angle, express the angle in degrees.

6. If tan θ = 0.2396, find θ, in degrees, for $0° \leq \theta < 360°$.

7. If cos θ = −0.8244 and csc θ < 0, find θ in radians for $0 \leq \theta < 2\pi$.

8. The floor of a sunroom is in the shape of a circular sector of arc length 32.0 ft and radius 8.50 ft. What is the area of the floor?

CHAPTER 9

VECTORS AND OBLIQUE TRIANGLES

In the applications considered to this point we have dealt with only the magnitudes of the various quantities used. In this chapter we shall study *vectors,* for which we must indicate the *direction* of a given quantity as well as its magnitude. In order to specify the direction we use the trigonometric functions of any required angle, as developed in Chapter 8. Vectors are of great importance in many fields of science and technology, including physics, engineering, structural design, and navigation.

After developing the basic concepts associated with vectors, we then study methods of solving triangles that are not right triangles (*oblique* triangles). In using such triangles we must be able to use the trigonometric functions of oblique angles. As with right triangles, the applications of oblique triangles are found in many fields of science and technology.

The wind must be considered to find the proper heading for an aircraft. In Section 9-5 we use vectors and an oblique triangle to show how this may be done.

9-1 INTRODUCTION TO VECTORS

SCALARS

A great many quantities with which we deal may be described by specifying their magnitudes. Generally, one can describe lengths of objects, areas, time intervals, monetary amounts, temperatures, and numerous other quantities by specifying a number: the magnitude of the quantity. *Such quantities are known as* **scalar** *quantities.*

VECTORS

Many other quantities are fully described only when both their magnitude and direction are specified. Such quantities are known as **vectors.** Examples of vectors are velocity and force. The following example illustrates a vector quantity and the difference between scalars and vectors.

EXAMPLE 1 A jet is traveling at 600 mi/h. From this statement alone we know only the *speed* of the jet. *Speed is a scalar quantity,* and it tells us only the *magnitude* of the rate. Knowing only the speed of the jet, we know the rate at which it is moving, but we do not know where it is headed.

If we add the phrase "in a direction 10° south of west" to the sentence above about the jet, we specify the direction of travel as well as the speed. We then know the *velocity* of the jet; that is, we know the *direction* of travel as well as the *magnitude* of the rate at which it is traveling. *Velocity is a vector quantity.* Knowing the velocity of the jet, we know where it is headed and the rate at which it is moving.

Let us analyze an example of the action of two vectors. Consider a boat moving in a river. For purposes of this example, we shall assume that the boat is driven by a motor that can move it at 8 mi/h in still water. We shall assume the current is moving downstream at 6 mi/h. We immediately see that the movement of the boat depends on the direction in which it is headed. If the boat heads downstream, it can travel at 14 mi/h, for the current is going 6 mi/h and the boat moves at 8 mi/h with respect to the water. If the boat heads upstream, however, it moves at the rate of only 2 mi/h, since the action of the boat and that of the river are counter to each other. If the boat heads directly across the river, the point it reaches on the other side will not be directly opposite the point from which it started. We can see that this is so because we know that as the boat heads across the river, the river is moving the boat downstream *at the same time.*

This last case should be investigated further. Assume that the river is 0.4 mi wide where the boat is crossing. It then takes 0.05 h (0.4 mi ÷ 8 mi/h = 0.05 h) to cross. In 0.05 h the river will carry the boat 0.3 mi (0.05 h × 6 mi/h = 0.3 mi) downstream. Therefore, when the boat reaches the other side it will be 0.3 mi downstream. From the Pythagorean theorem, we find that the boat traveled 0.5 mi from its starting point to its finishing point.

$$d^2 = 0.4^2 + 0.3^2 = 0.25$$
$$d = 0.5 \text{ mi}$$

Since this 0.5 mi was traveled in 0.05 h, the magnitude of the velocity of the boat was actually

$$v = \frac{d}{t} = \frac{0.5 \text{ mi}}{0.05 \text{ h}} = 10 \text{ mi/h}$$

Also, we see that the direction of this velocity can be represented along a line that makes an angle θ with the line directed directly across the river as shown in Fig. 9-1. We can find this angle by noting that

$$\tan \theta = \frac{0.3 \text{ mi}}{0.4 \text{ mi}} = 0.75$$
$$\theta = \tan^{-1} 0.75 = 37°$$

Therefore, when headed directly across the stream, the velocity of the boat is 10 mi/h at an angle of 37° downstream from a line directed directly across the stream.

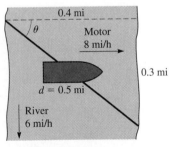

Fig. 9-1

Addition of Vectors

We have just seen two velocity vectors being *added*. Note that these vectors are not added the way numbers are added. We must take into account their directions as well as their magnitudes. Reasoning along these lines, let us now define the sum of two vectors.

We will represent a vector quantity by a letter printed in **boldface** type. The same letter in *italic* (lightface) type represents the magnitude only. Thus, **A** is a vector of magnitude *A*. In handwriting, one usually places an arrow over the letter to represent a vector, such as $\vec{A}$.

Let **A** and **B** represent vectors directed from *O* to *P* and *P* to *Q*, respectively (see Fig. 9-2). *The vector sum* **A** + **B** *is the vector* **R**, *from the* **initial point** *O to the* **terminal point** *Q. Here, vector* **R** *is called the* **resultant**. *In general, a resultant is a single vector that is the vector sum of any number of other vectors.*

There are two common methods of adding vectors by means of a diagram. The first is illustrated in Fig. 9-3. To add **B** to **A**, we shift **B** parallel to itself until its tail touches the head of **A**. *The vector sum* **A** + **B** *is the resultant vector* **R**, *which is drawn from the tail of* **A** *to the head of* **B**. In using this method, we can move a vector for addition as long as we *keep its magnitude and direction unchanged*. (Since the magnitude and direction specify a vector, two vectors in different *locations* are considered the same if they have the same magnitude and direction.) When using a diagram to add vectors, it must be drawn with reasonable accuracy.

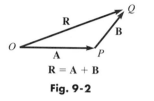

Fig. 9-2

CAUTION ▶

POLYGON METHOD

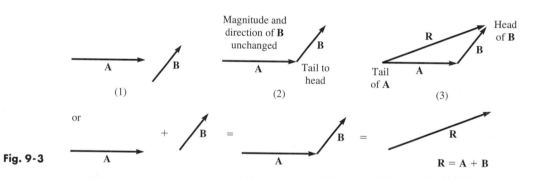

Fig. 9-3

Three or more vectors are added in the same general manner. We place the initial point of the second vector at the terminal point of the first vector, the initial point of the third vector at the terminal point of the second vector, and so on. The resultant is the vector from the initial point of the first vector to the terminal point of the last vector. The order in which they are added does not matter.

■EXAMPLE 2 The addition of vectors **A**, **B**, and **C** is shown in Fig. 9-4.

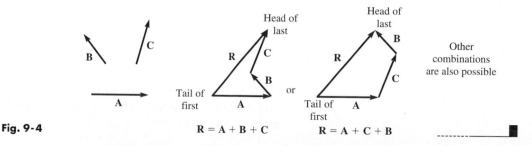

Fig. 9-4

PARALLELOGRAM METHOD   Another method that is convenient when two vectors are being added is to *let the two vectors being added be the sides of a parallelogram. The resultant is then the diagonal of the parallelogram.* The initial point of the resultant is the *common initial point of the two vectors being added.* In using this method the vectors are first placed tail to tail. This method is illustrated in the following example.

■**EXAMPLE 3**   The addition of vectors **A** and **B** is shown in Fig. 9-5.

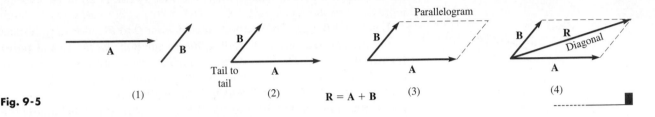

Fig. 9-5

SCALAR MULTIPLE OF VECTOR   If vector **C** is in the same direction as vector **A,** and **C** has a magnitude *n* times that of **A,** then **C** = *n***A,** where *the vector n**A** is called the* **scalar multiple** *of vector* **A.** This means that 2**A** is a vector that is twice as long as **A** but is *in the same direction.* Note carefully that only the magnitudes of **A** and 2**A** are different, and their directions are the same. The addition of scalar multiples of vectors is illustrated in the following example.

■**EXAMPLE 4**   For vectors **A** and **B** in Fig. 9-6, find vector 3**A** + 2**B**.

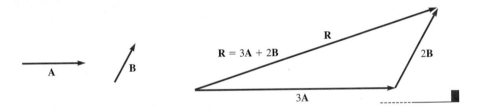

Fig. 9-6

SUBTRACTION OF VECTORS   Vector **B** may be subtracted from vector **A** by reversing the direction of **B** and proceeding as in vector addition. Thus, **A** − **B** = **A** + (−**B**), where *the minus sign indicates that vector* −**B** *has the opposite direction of vector* **B.** Vector subtraction is illustrated in the following example.

NOTE ▶

■**EXAMPLE 5**   For vectors **A** and **B** in Fig. 9-7, find vector 2**A** − **B**.

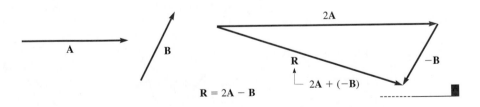

Fig. 9-7

Among the most important applications of vectors is that of the forces acting on a structure or on an object. The next example shows the addition of forces by using the parallelogram method.

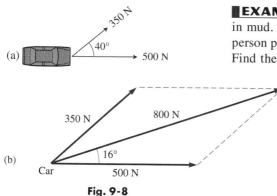

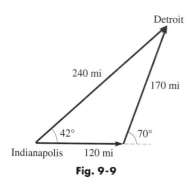

Fig. 9-8

EXAMPLE 6 Two persons pull horizontally on ropes attached to a car mired in mud. One person pulls with a force of 500 N directly to the right, and the other person pulls with a force of 350 N at 40° from the first force, as shown in Fig. 9-8(a). Find the resultant force on the car.

We make a scale drawing of the forces as shown in Fig. 9-8(b), measuring the magnitudes of the forces with a ruler and the angles with a protractor. (The scale drawing of the forces is made larger and with a different scale than that in Fig. 9-8(a) in order to get better accuracy.) We then complete the parallelogram and draw in the diagonal that represents the resultant force. Finally, we find that the resultant force is about 800 N, and that it acts at an angle of about 16° from the first force.

Two other important vector quantities are *velocity* and *displacement*. Velocity as a vector is illustrated in Example 1. *The* **displacement** *of an object is the change in its position. Displacement is given by the distance from a reference point and the angle from a reference direction.* The following example illustrates the difference between *distance* and *displacement*.

EXAMPLE 7 A jet travels due east from Indianapolis for 120 mi and then turns 70° north of east and travels another 170 mi to Detroit. Find the displacement of Detroit from Indianapolis.

We make a scale drawing in Fig. 9-9 to show the route taken by the jet. Measuring distances with a ruler and angles with a protractor, we find that Detroit is about 240 mi from Indianapolis, at an angle of about 42° north of east. By giving both the magnitude and *the direction,* we have given the displacement.

If the jet returned directly from Detroit to Indianapolis, its *displacement* from Indianapolis would be *zero,* although it traveled a *distance* of 530 mi.

Fig. 9-9

EXERCISES 9-1

(W) In Exercises 1–4, determine whether a scalar or a vector is described in (a) and (b). Explain your answers.

1. (a) A person traveled 300 km to the southwest.
(b) A person traveled 300 km.

2. (a) A small-craft warning reports winds of 25 mi/h.
(b) A small-craft warning reports winds out of the north at 25 mi/h.

3. (a) An arm of an industrial robot pushes with a 10-lb force downward on a part.
(b) A part is being pushed with a 10-lb force by an arm of an industrial robot.

4. (a) A ballistics test shows that a bullet hit a wall at a speed of 400 ft/s.
(b) A ballistics test shows that a bullet hit a wall at a speed of 400 ft/s perpendicular to the wall.

In Exercises 5–8, add the given vectors by drawing the appropriate resultant. Use the parallelogram method in Exercises 7 and 8.

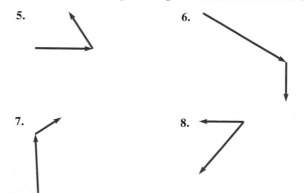

In Exercises 9–28, find the indicated vector sums and differences with the given vectors by means of diagrams. (You might find graph paper to be helpful.)

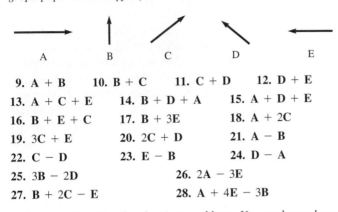

9. **A + B**   10. **B + C**   11. **C + D**   12. **D + E**

13. **A + C + E**   14. **B + D + A**   15. **A + D + E**

16. **B + E + C**   17. **B + 3E**   18. **A + 2C**

19. **3C + E**   20. **2C + D**   21. **A − B**

22. **C − D**   23. **E − B**   24. **D − A**

25. **3B − 2D**   26. **2A − 3E**

27. **B + 2C − E**   28. **A + 4E − 3B**

In Exercises 29–36, solve the given problems. Use a ruler and protractor as in Examples 6 and 7.

29. Two forces that act on an airplane wing are called the *lift* and the *drag*. Find the resultant of these forces acting on the airplane wing in Fig. 9-10.

Fig. 9-10

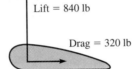

Lift = 840 lb

Drag = 320 lb

30. Two electric charges create an electric field intensity, a vector quantity, at a given point. The field intensity is 30 kN/C to the right and 60 kN/C at an angle of 45° above the horizontal to the right. Find the resultant electric field intensity at this point.

31. A rocket takes off, moving at 2000 ft/s horizontally and 1500 ft/s vertically. Find its resultant velocity.

32. A small plane travels at 120 mi/h in still air. It is headed due south in a wind of 30 mi/h from the northeast. What is the resultant velocity of the plane?

33. A driver takes the wrong road at an intersection and travels 4 mi north, then 6 mi east, and finally 10 mi to the southeast to reach the home of a friend. What is the displacement of the friend's home from the intersection?

34. A ship travels 20 km in a direction of 30° south of east and then turns due south for another 40 km. What is the ship's displacement from its initial position?

35. Three ropes hold a helium-filled balloon in place, but two of the ropes break. The remaining rope holds the balloon with a tension of 510 N at an angle of 80° with the ground due to a wind. The weight (a vertical force) of the balloon and contents is 400 N, and the upward buoyant force is 900 N. The wind creates a horizontal force of 90 N on the balloon. What is the resultant force on the balloon?

36. A crate weighing 100 N is suspended by two ropes. The force in one rope is 70 N and is directed to the left at an angle of 60° above the horizontal. What must be the other force in order that the resultant force (including the weight) on the crate is zero?

$9\text{-}2$ COMPONENTS OF VECTORS

Adding vectors by means of diagrams is very useful in developing an understanding of vector quantities. However, unless the diagrams are drawn with great care and accuracy, the results we can obtain are quite approximate. Therefore, it is necessary to develop other methods to obtain results of sufficient accuracy.

In this section we show how a given vector can be considered to be the sum of two other vectors, with any required degree of accuracy. In the following section we show how this will allow us to add vectors in order to obtain the sum of vectors with the required accuracy in the result.

Two vectors which, when added together, have a resultant equal to the original vector are called **components** *of the original vector.* In the illustration of the boat in Section 9-1, the velocities of 8 mi/h across the river and 6 mi/h downstream are components of the 10 mi/h vector directed at the angle θ.

In practice, there are certain components of a vector that are of particular importance. If a vector is placed with its initial point at the origin of a rectangular coordinate system and its direction is indicated by an angle in standard position, we may find its *x*- and *y*-**components.** *These components are vectors directed along the coordinate axes and that, when added together, have a resultant equal to the given vector.* The initial points of these components are at the origin, and the terminal points are located at the points where perpendicular lines from the terminal point of the given vector cross the axes. *Finding these component vectors is called* **resolving** *the vector into its components.*

RESOLVING A VECTOR INTO COMPONENTS

■EXAMPLE 1   Find the x- and y-components of the vector **A** shown in Fig. 9-11. The magnitude of **A** is 7.25.

From the figure we see that A_x, the magnitude of the x-component $\mathbf{A}_x$, is related to **A** by

$$\frac{A_x}{A} = \cos 62.0°$$

$$A_x = A \cos 62.0°$$

In the same way, A_y, the magnitude of the y-component $\mathbf{A}_y$, is related to **A** ($\mathbf{A}_y$ could be placed along the vertical dashed line) by

$$\frac{A_y}{A} = \sin 62.0°$$

$$A_y = A \sin 62.0°$$

From these relations, knowing that $A = 7.25$, we have

$$A_x = 7.25 \cos 62.0° = 3.40$$
$$A_y = 7.25 \sin 62.0° = 6.40$$

This means that the x-component is directed along the x-axis to the right and has a magnitude of 3.40. Also, the y-component is directed along the y-axis upward and its magnitude is 6.40. These two component vectors can replace vector **A,** since the effect they have is the same as **A.**   ----------■

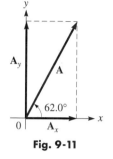

Fig. 9-11

■EXAMPLE 2   Resolve a vector 14.4 units long and directed at an angle of 126.0° into its x- and y-components. See Fig. 9-12.

Placing the initial point of the vector at the origin and putting the angle in standard position, we see that the vector directed along the x-axis, $\mathbf{V}_x$, is related to the vector **V** of magnitude V by

$$V_x = V \cos 126.0°$$

(with labels: "magnitude of vector" pointing to V, and "standard position angle" pointing to $126.0°$)

or in terms of the reference angle by

$$V_x = -V \cos 54.0°$$

(with labels: "reference angle" pointing to $54.0°$, and "directed along negative x-axis" pointing to the minus sign)

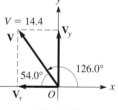

Fig. 9-12

since $\cos 126.0° = -\cos 54.0°$. We see that the minus sign shows that the x-component is directed in the negative direction, that is, to the left.

Since the vector directed along the y-axis, $\mathbf{V}_y$, could also be placed along the vertical dashed line, it is related to the vector **V** by

$$V_y = V \sin 126.0° = V \sin 54.0°$$

Thus, the vectors $\mathbf{V}_x$ and $\mathbf{V}_y$ have the magnitudes

$$V_x = 14.4 \cos 126.0° = -8.46 \qquad V_y = 14.4 \sin 126.0° = 11.6$$

Therefore, we have resolved the given vector into two components: one, directed along the negative x-axis, of magnitude 8.46, and the other, directed along the positive y-axis, of magnitude 11.6.

From Examples 1 and 2, we can see that the steps used in finding the *x*- and *y*-components of a vector are as follows:

Steps Used in Finding the x- AND y-COMPONENTS OF A VECTOR

1. *Place vector* **A** *such that θ is in standard position.*

2. *Calculate* A_x *and* A_y *from* $A_x = A \cos \theta$ *and* $A_y = A \sin \theta$. We may use the reference angle if we note the direction of the component.

3. *Check the components* to see if each is in the correct direction and has a magnitude that is proper for the reference angle.

EXAMPLE 3 Resolve vector **A,** of magnitude 375.4 and direction $\theta = 205.32°$, into its *x*- and *y*-components. See Fig. 9-13.

By placing **A** such that θ is in standard position, we see that

$$A_x = A \cos 205.32° = 375.4 \cos 205.32° = -339.3$$

and

$$A_y = A \sin 205.32° = 375.4 \sin 205.32° = -160.5$$

directed along negative axis

The angle 205.32° places the vector in the third quadrant, and we see that each of the components is directed along the negative axis. This must be the case for a third-quadrant angle. Also, the reference angle is 25.32°, and we see that the magnitude of A_x is greater than the magnitude of A_y, which must be true for a reference angle that is less than 45°.

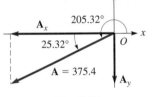

Fig. 9-13

EXAMPLE 4 The tension **T** in a cable supporting the sign shown in Fig. 9-14(a) is 85.0 lb. If the cable makes an angle of 53.5° with the horizontal, find the horizontal and vertical components of the tension.

The tension in the cable is the force that the cable exerts on the sign. Showing the tension in Fig. 9-14(b), we see that

$$T_y = T \sin 53.5° = 85.0 \sin 53.5°$$
$$= 68.3 \text{ lb}$$
$$T_x = T \cos 53.5° = 85.0 \cos 53.5°$$
$$= 50.6 \text{ lb}$$

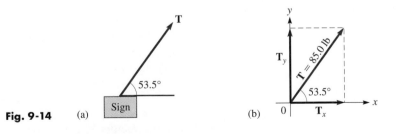

Fig. 9-14   (a)   (b)

The angle 53.5° is the reference angle, and $T_y > T_x$. This must be true for a reference angle greater than 45°.

— EXERCISES 9-2 —

In Exercises 1–4, find the horizontal and vertical components of the vectors shown in the given figures. In each the magnitude of the vector is 750.

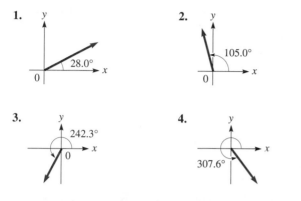

1.

2.

105.0°

3.

242.3°

4.

307.6°

In Exercises 5–16, find the x- and y- components of the given vectors by use of the trigonometric functions.

5. Magnitude 8.60, $\theta = 68.0°$

6. Magnitude 9750, $\theta = 243.0°$

7. Magnitude 76.8, $\theta = 145.0°$

8. Magnitude 0.0998, $\theta = 296.0°$

9. Magnitude 9.04, $\theta = 283.3°$

10. Magnitude 16,400, $\theta = 156.5°$

11. Magnitude 2.65, $\theta = 197.3°$

12. Magnitude 67.8, $\theta = 22.5°$

13. Magnitude 0.8734, $\theta = 157.83°$

14. Magnitude 509.4, $\theta = 221.87°$

15. Magnitude 89,760, $\theta = 7.84°$

16. Magnitude 1.806, $\theta = 301.83°$

In Exercises 17–24, find the required horizontal and vertical components of the given vectors.

17. A nuclear submarine approaches the surface of the ocean at 25.0 km/h and at an angle of 17.3° with the surface. What are the components of its velocity? See Fig. 9-15.

17.3°

25.0 km/h

Fig. 9-15

18. Water is flowing downhill at 18.0 ft/s through a pipe that is at an angle of 66.4° with the horizontal. What are the components of its velocity?

19. The tension in a rope attached to a boat is 55.0 lb. The rope is attached to the boat 12.0 ft below the level at which it is being drawn in. At the point where there are 36.0 ft of rope out, what force tends to bring the boat toward the wharf, and what force tends to raise the boat? See Fig. 9-16.

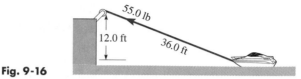

55.0 lb

12.0 ft

36.0 ft

Fig. 9-16

20. A car is being unloaded from a ship. It is supported by a cable from a crane and guided into position by a horizontal rope. If the tension in the cable is 2790 lb and the cable makes an angle of 3.5° with the vertical, what are the weight W of the car and the tension T in the rope? (The weight of the cable is negligible to that of the car.) See Fig. 9-17.

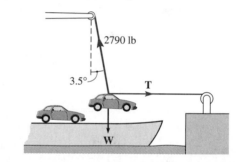

2790 lb

3.5°

T

W

Fig. 9-17

21. A jet is 145 km at a position 37.5° north of east of Tahiti. What are the components of the jet's displacement from Tahiti?

22. The end of a robot arm is 3.50 ft on a line 78.6° above the horizontal from the point where it does a weld. What are the components of the displacement from the end of the robot arm to the welding point?

23. At one point the *Pioneer* space probe was entering the gravitational field of Jupiter at an angle of 2.55° below the horizontal with a velocity of 18,550 mi/h. What were the components of its velocity?

24. Two upward forces are acting on a bolt. One force of 60.5 lb acts at an angle of 82.4° above the horizontal, and the other force of 37.2 lb acts at an angle of 50.5° below the first force. What is the total upward force on the bolt? See Fig. 9-18.

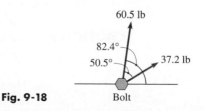

60.5 lb

82.4°

50.5°

37.2 lb

Fig. 9-18

Bolt

9-3 VECTOR ADDITION BY COMPONENTS

Now that we have developed the meaning of the components of a vector, we are able to add vectors to any degree of required accuracy. To do this we use the components of the vector, the Pythagorean theorem, and the tangent of the standard position angle of the resultant. In the following example, two vectors at right angles are added.

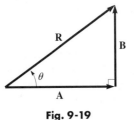

Fig. 9-19

■EXAMPLE 1   Add vectors **A** and **B**, with $A = 14.5$ and $B = 9.10$. The vectors are at right angles, as shown in Fig. 9-19.

We can find the magnitude R of the resultant vector **R** by use of the Pythagorean theorem. This leads to

$$R = \sqrt{A^2 + B^2} = \sqrt{(14.5)^2 + (9.10)^2}$$
$$= 17.1$$

We shall now determine the direction of the resultant vector **R** by specifying its direction as the angle θ in Fig. 9-19, that is, the angle that **R** makes with vector **A**. Therefore, we have

$$\tan \theta = \frac{B}{A} = \frac{9.10}{14.5}$$
$$\theta = 32.1°$$

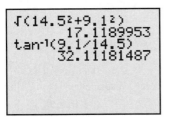

Fig. 9-20

Calculation of the values of R and θ are shown in the calculator window in Fig. 9-20. As we showed in Chapter 4, it is not necessary to record or store the value of the quotient 9.10/14.5, as we can calculate the value of θ directly on the calculator as $\tan^{-1}(9.10/14.5)$.

Therefore, we see that **R** is a vector of magnitude $R = 17.1$ and in a direction 32.1° from vector **A.**

We note that Fig. 9-19 shows vectors **A** and **B** as horizontal and vertical, respectively. This means they are the horizontal and vertical components of the resultant vector **R**. However, we would find the resultant of any two vectors *at right angles* in the same way.   --------▬

If vectors are to be added and they are not at right angles, we first place each vector with its tail at the origin. Next, we resolve each vector into its x- and y-components. We then add all the x-components and add all the y-components to determine the x- and y-components of the resultant. Then, by using the Pythagorean theorem we find the magnitude of the resultant, and by use of the tangent we find the angle that gives us the direction of the resultant.

CAUTION ▶   *Remember, a vector is not completely specified unless both its magnitude and its direction are specified.* A common error is to determine the magnitude, but not to find the angle θ that is used to define its direction.

The following examples illustrate the addition of vectors by first finding the components of the given vectors.

EXAMPLE 2   Find the resultant of two vectors **A** and **B** such that $A = 1200$, $\theta_A = 270.0°$, $B = 1750$, and $\theta_B = 115.0°$.

We first place the vectors on a coordinate system with the tail of each at the origin as shown in Fig. 9-21(a). We then resolve each vector into its x- and y-components, as shown in Fig. 9-21(b) and as calculated below. (Note that **A** is vertical and has no horizontal component.) Next, the components are combined, as in Fig. 9-21(c) and as calculated. Finally, the magnitude of the resultant and the angle θ (to determine the direction), as shown in Fig. 9-21(d), are calculated.

Fig. 9-21

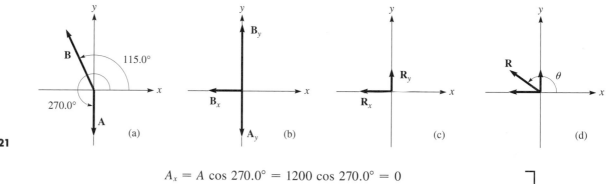

$$A_x = A \cos 270.0° = 1200 \cos 270.0° = 0$$
$$B_x = B \cos 115.0° = 1750 \cos 115.0° = -739.6$$
$$A_y = A \sin 270.0° = 1200 \sin 270.0° = -1200$$
$$B_y = B \sin 115.0° = 1750 \sin 115.0° = 1586$$

Fig. 9-21(b)

$$R_x = A_x + B_x = 0 - 739.6 = -739.6$$
$$R_y = A_y + B_y = -1200 + 1586 = 386$$

Fig. 9-21(c)

$$R = \sqrt{R_x^2 + R_y^2} = \sqrt{(-739.6)^2 + 386^2} = 834$$
$$\tan \theta = \frac{R_y}{R_x} = \frac{386}{-739.6} \qquad \theta = 152.4° \longleftarrow 180° - 27.6°$$

Fig. 9-21(d)

Thus, the resultant has a magnitude of 834 and is directed at a standard-position angle of 152.4°. In finding θ from a calculator,

CAUTION ▶   *the calculator display shows an angle of* $-27.6°$*. However, we know θ is a second-quadrant angle, since* R_x *is negative and* R_y *is positive.*

Therefore, we must use 27.6° as a reference angle. For this reason, it is usually advisable to *find the reference angle first* by disregarding the signs of R_x and R_y when finding θ. Thus,

$$\tan \theta_{ref} = \left| \frac{R_y}{R_x} \right| = \frac{386}{739.6} \qquad \theta_{ref} = 27.6°$$

The values shown in this example have been rounded off. In using the calculator, R_x and R_y are each calculated in one step and stored for the calculation of R and θ, and we will show these steps in the next example. However, here we wished to show the individual steps and results to more clearly show the method.

When we found R_y we saw that we were able to do a vector addition as a scalar addition. Also, since the magnitude of the components and the resultant are much smaller than either of the original vectors, this vector addition would have been difficult to do accurately by means of a diagram.

EXAMPLE 3 Find the resultant **R** of the two vectors shown in Fig. 9-22(a), **A** of magnitude 8.075 and standard position angle of 57.26° and **B** of magnitude 5.437 and standard-position angle of 322.15°.

In Fig. 9-22(b) we show the components of vectors **A** and **B,** and then in Fig. 9-22(c) we show the resultant and its components.

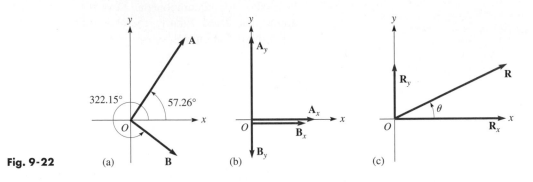

Fig. 9-22 (a)   (b)   (c)

Vector	Magnitude	Angle	x-component		y-component	
A	8.075	57.26°	$A_x = 8.075 \cos 57.26°$	$= 4.367$	$A_y = 8.075 \sin 57.26°$	$= 6.792$
B	5.437	322.15°	$B_x = 5.437 \cos 322.15°$	$= 4.293$	$B_y = 5.437 \sin 322.15°$	$= -3.336$
R			$R_x = A_x + B_x$	$= 8.660$	$R_y = A_y + B_y$	$= 3.456$

$$R = \sqrt{R_x^2 + R_y^2} = \sqrt{(8.660)^2 + (3.456)^2} = 9.324$$

$$\theta = \tan^{-1}\frac{R_y}{R_x} = \tan^{-1}\left(\frac{3.456}{8.660}\right) = 21.76° \quad \longleftarrow \quad \text{don't forget the direction}$$

```
8.075cos(57.26)+
5.437cos(322.15)
        8.660346589
Ans→C
        8.660346589
8.075sin(57.26)+
5.437sin(322.15)
        3.456028938
Ans→D
        3.456028938
√(C²+D²)
        9.324469908
tan⁻¹(D/C)
        21.75514218
```

Fig. 9-23

The resultant vector is 9.324 units long and is directed at a standard-position angle of 21.76°, as shown in Fig. 9-22(c). We know that the resultant is in the first quadrant since both R_x and R_y are positive.

In the table above, we have shown rounded-off values for each result. However, when using a calculator it is necessary only to calculate R_x and R_y in one step each, store these values, and use them to calculate R and θ. This is shown in the calculator window in Fig. 9-23. The lines shown above the window are those that are displaced in proceeding with the solution.

In the calculator solution $C = R_x$ and $D = R_y$. In finding this solution we see that both R_x and R_y are positive, which means that θ is a first-quadrant angle. ∎

Some general formulas can be derived from the previous examples. For a given vector **A,** directed at an angle θ, of magnitude A, and with components A_x and A_y, we have the following relations:

$$A_x = A \cos \theta \quad A_y = A \sin \theta \tag{9-1}$$

$$A = \sqrt{A_x^2 + A_y^2} \tag{9-2}$$

$$\theta_{\text{ref}} = \tan^{-1}\frac{|A_y|}{|A_x|} \tag{9-3}$$

The value of θ is found by using the reference angle from Eq. (9-3) and the quadrant in which the resultant lies.

From the previous examples we see that the following procedure is used for adding vectors.

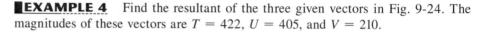

> ## Procedure for Adding Vectors By Components
> 1. *Add the x-components of the given vectors to obtain R_x.*
> 2. *Add the y-components of the given vectors to obtain R_y.*
> 3. *Find the magnitude of the resultant* **R**. *Use Eq. (9-2) in the form*
> $$R = \sqrt{R_x^2 + R_y^2}$$
> 4. *Find the standard-position angle θ for the resultant* **R**. *First find the reference angle θ_{ref} for the resultant* **R** *by using Eq. (9-3) in the form*
> $$\theta_{ref} = \tan^{-1}\frac{|R_y|}{|R_x|}$$

Some calculators have a specific feature for adding vectors.

See Appendix C for a graphing calculator program ADDVCTR. It can be used to add vectors.

■**EXAMPLE 4**   Find the resultant of the three given vectors in Fig. 9-24. The magnitudes of these vectors are $T = 422$, $U = 405$, and $V = 210$.

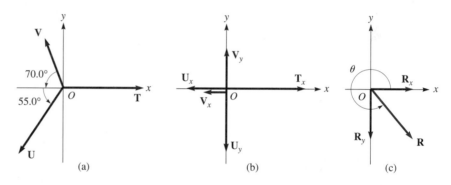

Fig. 9-24   (a)   (b)   (c)

We could change the given angles to standard-position angles. However, we will use the given angles, *being careful to give the proper sign to each component*. In the following table we show the x- and y-components of the given vectors, and the sums of these components give us the components of **R**.

```
422cos(0)-405cos
(55)-210cos(70)
        117.8773132
Ans→A
        117.8773132
422sin(0)-405sin
(55)+210sin(70)

      -134.4211276
Ans→B
      -134.4211276
√(A²+B²)
        178.7850679
tan⁻¹(abs(B)/A)
        48.75165091
```

Fig. 9-25

Vector	Magnitude	Ref. Angle	x-component	y-component
T	422	0°	$422 \cos 0° = 422.0$	$422 \sin 0° = 0.0$
U	405	55.0°	$-405 \cos 55.0° = -232.3$	$-405 \sin 55.0° = -331.8$
V	210	70.0°	$-210 \cos 70.0° = -71.8$	$+210 \sin 70.0° = 197.3$
R			117.9	−134.5

Note the variation due to rounding off

In the calculator solution shown in Fig. 9-25, A = R_x, B = R_y, and abs means absolute value. From this display we have

$$R = 179 \quad \text{and} \quad \theta_{ref} = 48.8°$$

Since R_x is positive and R_y is negative, we know that θ is a fourth-quadrant angle. Therefore, to find θ we subtracted θ_{ref} from 360°. This means

$$\theta = 180° - 48.8° = 311.2°$$

EXERCISES 9-3

*In Exercises 1–4, vectors **A** and **B** are at right angles. Find the magnitude and direction (the angle from vector **A**) of the resultant.*

1. $A = 14.7$
$B = 19.2$

2. $A = 592$
$B = 195$

3. $A = 3.086$
$B = 7.143$

4. $A = 1734$
$B = 3297$

In Exercises 5–12, with the given sets of components, find R and θ.

5. $R_x = 5.18, R_y = 8.56$

6. $R_x = 89.6, R_y = -52.0$

7. $R_x = -0.982, R_y = 2.56$

8. $R_x = -729, R_y = -209$

9. $R_x = -646, R_y = 2030$

10. $R_x = -31.2, R_y = -41.2$

11. $R_x = 0.6941, R_y = -1.246$

12. $R_x = 7.627, R_y = -6.353$

In Exercises 13–28, add the given vectors by using the trigonometric functions and the Pythagorean theorem.

13. $A = 18.0, \theta_A = 0.0°$
$B = 12.0, \theta_B = 27.0°$

14. $A = 154, \theta_A = 90.0°$
$B = 128, \theta_B = 43.0°$

15. $C = 56.0, \theta_C = 76.0°$
$D = 12.0, \theta_D = 160.0°$

16. $A = 6.89, \theta_A = 123.0°$
$B = 29.0, \theta_B = 260.0°$

17. $A = 9.821, \theta_A = 34.27°$
$B = 17.45, \theta_B = 752.50°$

18. $E = 1.653, \theta_E = 36.37°$
$F = 0.9807, \theta_F = 253.06°$

19. $A = 12.653, \theta_A = 98.472°$
$B = 15.147, \theta_B = 332.092°$

20. $L = 121.36, \theta_L = 292.362°$
$M = 112.98, \theta_M = 197.892°$

21. $A = 21.9, \theta_A = 236.2°$
$B = 96.7, \theta_B = 11.5°$
$C = 62.9, \theta_C = 143.4°$

22. $A = 6300, \theta_A = 189.6°$
$B = 1760, \theta_B = 320.1°$
$C = 3240, \theta_C = 75.4°$

23. $U = 0.364, \theta_U = 175.7°$
$V = 0.596, \theta_V = 319.5°$
$W = 0.129, \theta_W = 100.6°$

24. $A = 6.4, \theta_A = 126°$
$B = 5.9, \theta_B = 238°$
$C = 3.2, \theta_C = 72°$

25. The vectors shown in Fig. 9-26

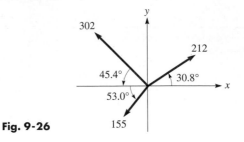

Fig. 9-26

26. The vectors shown in Fig. 9-27

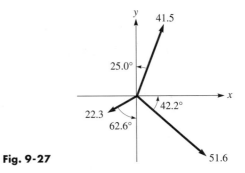

Fig. 9-27

27. Three forces of 3200 lb, 1300 lb, and 2100 lb act on a bolt as shown in Fig. 9-28. Find the resultant force.

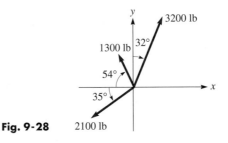

Fig. 9-28

28. A naval crusier on maneuvers travels 54.0 km at 18.7° west of north, then turns and travels 64.5 km at 15.6° south of east, and finally turns to travel 72.4 km at 38.1° east of south. Find its displacement from its original position. See Fig. 9-29.

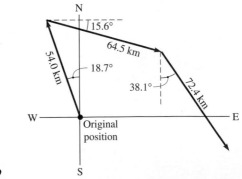

Fig. 9-29

9-4 APPLICATIONS OF VECTORS

In Section 9-1 we introduced the important vector quantities of force, velocity, and displacement, and we found vector sums by use of diagrams. Now we can use the method of Section 9-3 to find sums of these kinds of vectors and others and to use them in various types of applications.

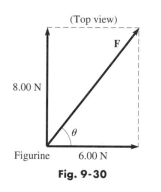

(Top view)

F

8.00 N

θ

Figurine   6.00 N

Fig. 9-30

■**EXAMPLE 1**   In centering a figurine on a table, two persons apply forces on it. These forces are at right angles and have magnitudes of 6.00 N and 8.00 N. The angle between their lines of action is 90.0°. What is the resultant of these forces on the figurine?

By means of an appropriate diagram (Fig. 9-30), we may better visualize the actual situation. We note that a good choice of axes (unless specified, it is often convenient to choose the *x*- and *y*-axes to fit the problem) is to have the *x*-axis in the direction of the 6.00-N force and the *y*-axis in the direction of the 8.00-N force. (This is possible since the angle between them is 90°.) With this choice we note that the two given forces will be the *x*- and *y*-components of the resultant. Therefore, we arrive at the following results:

$$F_x = 6.00 \text{ N}, \quad F_y = 8.00 \text{ N}$$
$$F = \sqrt{(6.00)^2 + (8.00)^2} = 10.0 \text{ N}$$
$$\theta = \tan^{-1}\frac{F_y}{F_x} = \tan^{-1}\frac{8.00}{6.00}$$
$$= 53.1°$$

We would state that the resultant has a magnitude of 10.0 N and acts at an angle of 53.1° from the 6.00-N force.   ------■

The metric unit of force, the newton (N), is named for the great English mathematician and physicist Sir Isaac Newton (1642–1727). His name will appear on other pages of this text, as some of his many accomplishments are noted.

SOLVING A WORD PROBLEM

■**EXAMPLE 2**   A ship sails 32.50 mi due east and then turns 41.25° north of east. After sailing another 16.18 mi, where is it with reference to the starting point?

In this problem we are to find the resultant displacement of the ship from the two given displacements. The problem is diagramed in Fig. 9-31, where the first displacement is labeled vector **A** and the second as vector **B.**

Since east corresponds to the positive *x*-direction, we see that the *x*-component of the resultant is $\mathbf{A} + \mathbf{B}_x$ and the *y*-component of the resultant is $\mathbf{B}_y$. Therefore, we have the following results:

$$R_x = A + B_x = 32.50 + 16.18 \cos 41.25°$$
$$= 32.50 + 12.16$$
$$= 44.66 \text{ mi}$$
$$R_y = 16.18 \sin 41.25° = 10.67 \text{ mi}$$
$$R = \sqrt{(44.66)^2 + (10.67)^2} = 45.92 \text{ mi}$$
$$\theta = \tan^{-1}\frac{10.67}{44.66} = 13.44°$$

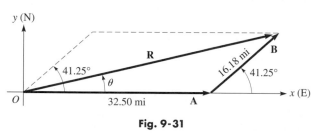

y (N)

41.25°

R

θ

16.18 mi

B

41.25°

O 32.50 mi   A x (E)

Fig. 9-31

Therefore, the ship is 45.92 mi from the starting point, in a direction 13.44° north of east.   ------■

An aircraft's *heading* is the direction in which it is pointed. Its *air speed* is the speed at which it travels through the air surrounding it. Due to the wind, the heading and air speed, and its actual direction and speed relative to the ground, will differ.

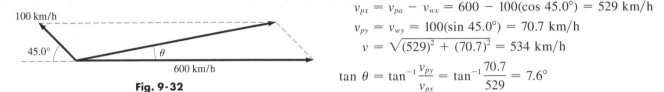

Fig. 9-32

EXAMPLE 3   An airplane headed due east is in a wind blowing from the southeast. What is the resultant velocity of the plane with respect to the surface of the earth if the velocity of the plane with respect to the air is 600 km/h and that of the wind is 100 km/h? See Fig. 9-32.

Let $\mathbf{v}_{px}$ be the velocity of the plane in the x-direction (east), $\mathbf{v}_{py}$ the velocity of the plane in the y-direction, $\mathbf{v}_{wx}$ the x-component of the velocity of the wind, $\mathbf{v}_{wy}$ the y-component of the velocity of the wind, and $\mathbf{v}_{pa}$ the velocity of the plane with respect to the air. Therefore,

$$v_{px} = v_{pa} - v_{wx} = 600 - 100(\cos 45.0°) = 529 \text{ km/h}$$

$$v_{py} = v_{wy} = 100(\sin 45.0°) = 70.7 \text{ km/h}$$

$$v = \sqrt{(529)^2 + (70.7)^2} = 534 \text{ km/h}$$

$$\tan \theta = \tan^{-1} \frac{v_{py}}{v_{px}} = \tan^{-1} \frac{70.7}{529} = 7.6°$$

We have determined that the plane is traveling 534 km/h and is flying in a direction 7.6° north of east. From this we observe that a plane does not necessarily head in the direction of its destination. --------■

NOTE ▶

As we have seen, an important vector quantity is the force acting on an object. One of the most important applications of vectors involves forces that are in **equilibrium.** *For an object to be in equilibrium, the net force acting on it in any direction must be zero.* This condition is satisfied if the sum of the x-components of the force is zero and the sum of the y-components of the force is also zero. The following two examples illustrate forces in equilibrium.

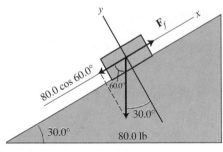

Fig. 9-33

Newton's *third law of motion* states that when an object exerts a force on another object, the second object exerts on the first object a force of the same magnitude but in the opposite direction. The force exerted by the plank on the block is an illustration of this law. (Sir Isaac Newton, again. See page 257).

EXAMPLE 4   A cement block is resting on a straight inclined plank that makes an angle of 30.0° with the horizontal. If the block weighs 80.0 lb, what is the force of friction between the block and the plank?

The weight of the cement block is the force exerted on the block due to gravity. Therefore, the weight is directed vertically downward. The frictional force tends to oppose the motion of the block and is directed upward along the plank. The frictional force must be sufficient to counterbalance that component of the weight of the block that is directed down the plank for the block to be at rest (not moving). The plank itself "holds up" that component of the weight that is perpendicular to the plank. A convenient set of coordinates (see Fig. 9-33) is one with the origin at the center of the block, and with the x-axis directed up the plank and the y-axis perpendicular to the plank. The magnitude of the frictional force $\mathbf{F}_f$ is given by

$$F_f = 80.0 \cos 60.0° = 40.0 \text{ lb} \qquad \text{component of weight down plank}$$
equals frictional force

We have used the 60.0° angle since it is the reference angle. We could have expressed the frictional force as $F_f = 80.0 \sin 30.0°$.

Here, we have assumed that the block is small enough that we may calculate all forces as though they act at the center of the block (although we know that the frictional force acts at the surface between the block and the plank). --------■

SOLVING A WORD PROBLEM

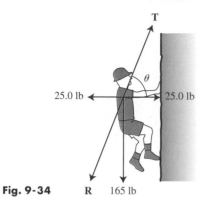

Fig. 9-34   R   165 lb

SOLVING A WORD PROBLEM

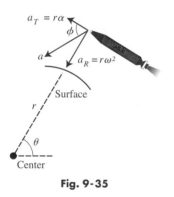

Fig. 9-35

■**EXAMPLE 5**   A 165-lb mountain climber suspended by a rope pushes on the side of a cliff with a horizontal force of 25.0 lb. What is the tension **T** in the rope if the climber is in equilibrium? See Fig. 9-34.

For the climber to be in equilibrium, the tension in the rope must be equal and opposite to the resultant of the climber's weight and the force against the cliff. This means that the magnitude of the *x*-component of the tension is 25.0 lb (the reaction force of the cliff—another illustration of Newton's third law—see Example 4) and the magnitude of the *y*-component is 165 lb (see Fig. 9-34). Therefore,

$$T = \sqrt{25.0^2 + 165^2} = 167 \text{ lb}$$

$$\theta = \tan^{-1}\frac{165}{25.0} = 81.4°$$

■**EXAMPLE 6**   For a spacecraft moving in a circular path around the earth, the tangential component $\mathbf{a}_T$ and the centripetal component $\mathbf{a}_R$ of its acceleration are given by the expressions shown in Fig. 9-35. The radius of the circle through which it is moving (from the center of the earth to the spacecraft) is *r*, its angular velocity is ω, and its angular acceleration is α (the rate at which ω is changing).

While going into orbit, at one point a spacecraft is moving in a circular path 230 km above the surface of the earth. At this point $r = 6.60 \times 10^6$ m, $\omega = 1.10 \times 10^{-3}$ rad/s, and $\alpha = 0.420 \times 10^{-6}$ rad/s². Calculate the magnitude of the resultant acceleration and the angle it makes with the tangential component.

$$a_R = r\omega^2 = 6.60 \times 10^6 (1.10 \times 10^{-3})^2 = 7.99 \text{ m/s}^2$$
$$a_T = r\alpha = 6.60 \times 10^6 (0.420 \times 10^{-6}) = 2.77 \text{ m/s}^2$$

Since a tangent line to a circle is perpendicular to the radius at the point of tangency, $\mathbf{a}_T$ is perpendicular to $\mathbf{a}_R$. Thus,

$$a = \sqrt{a_T^2 + a_R^2} = 8.45 \text{ m/s}^2$$

$$\phi = \tan^{-1}\frac{a_R}{a_T} = \tan^{-1}\frac{7.99}{2.77} = 70.9°$$

The value of $a = 8.45$ m/s² is found directly on a calculator, without first rounding off the values of a_R and a_T. If we use the rounded-off values of 7.99 m/s² and 2.77 m/s² we would get 8.46 m/s².

EXERCISES *9-4*

In Exercises 1–24, solve the given problems.

1. Two hockey players strike the puck at the same time, hitting it with horizontal forces of 5.75 lb and 3.25 lb that are perpendicular to each other. Find the resultant of these forces.

2. To straighten a small tree, two horizontal ropes perpendicular to each other are attached to the tree. If the tensions in the ropes are 18.5 lb and 23.0 lb, what is the resultant force on the tree?

3. In lifting a heavy piece of equipment from the mud, a cable from a crane exerts a vertical force of 6500 N, and a cable from a truck exerts a force of 8300 N at 10.0° above the horizontal. Find the resultant of these forces.

4. At a point in the plane, two electric charges create an electric field (a vector quantity) of 25.9 kN/C at 10.8° above the horizontal to the right and 12.6 kN/C at 83.4° below the horizontal to the right. Find the resultant electric field.

5. A motorboat leaves a dock and travels 1580 ft due west, then turns 35.0° to the south and travels another 1640 ft to a second dock. What is the displacement of the second dock from the first dock?

6. Toronto is 650 km at 19.0° north of east from Chicago. Cincinnati is 390 km at 48.0° south of east from Chicago. What is the displacement of Cincinnati from Toronto?

7. From a fixed point, a surveyor locates a pole at 215.6 ft due east and a building corner at 358.2 ft at 37.72° north of east. What is the displacement of the building from the pole?

8. A rocket is launched with a vertical component of velocity of 2840 km/h and a horizontal component of velocity of 1520 km/h. What is its resultant velocity?

9. A storm front is moving east at 22.0 km/h and south at 12.5 km/h. Find the resultant velocity of the front.

10. In an accident, a truck with momentum (a vector quantity) of 22,100 kg·m/s strikes a car with momentum of 17,800 kg·m/s from the rear. The angle between their directions of motion is 25.0°. What is the resultant momentum?

11. In an automobile safety test, a shoulder and seat belt exerts a force of 95.0 lb directly backward and a force of 83.0 lb backward at an angle of 20.0° below the horizontal on a dummy. If the belt holds the dummy from moving farther forward, what force did the dummy exert on the belt? See Fig. 9-36.

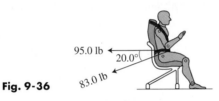

95.0 lb 20.0°

83.0 lb

Fig. 9-36

12. Two perpendicular forces act on a ring at the end of a chain that passes over a pulley and holds an automobile engine. If the forces have the values shown in Fig. 9-37, what is the weight of the engine?

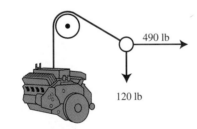

490 lb

120 lb

Fig. 9-37

13. A plane flies at 550 km/h into a head wind of 60 km/h at 78° with the direction of the plane. Find the resultant velocity of the plane with respect to the ground. See Fig. 9-38.

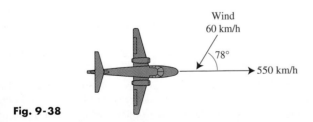

Wind
60 km/h

78°

550 km/h

Fig. 9-38

14. A ship's navigator determines that the ship is moving through the water at 17.5 mi/h with a heading of 26.3° north of east, but that the ship is actually moving at 19.3 mi/h in a direction of 33.7° north of east. What is the velocity of the current?

15. A space shuttle is moving in orbit at 18,250 mi/h. A satellite is launched to the rear at 120 mi/h at an angle of 5.20° from the direction of the shuttle. Find the velocity of the shuttle.

16. A block of ice slides down a (frictionless) ramp with an acceleration of 5.3 m/s². If the ramp makes an angle of 32.7° with the horizontal, find g, the acceleration due to gravity. See Fig. 9-39.

5.3 m/s² g

32.7°

Fig. 9-39

17. While starting up, a circular saw blade 8.20 in. in diameter is rotating at 212 rad/min and has an angular acceleration of 318 rad/min². What is the acceleration of the tip of one of the teeth? (See Example 6.)

18. A boat travels across a river, reaching the opposite bank at a point directly opposite that from which it left. If the boat travels 6.00 km/h in still water and the current of the river flows at 3.00 km/h, what was the velocity of the boat in the water?

19. In searching for a boat lost at sea, a Coast Guard cutter leaves a port and travels 75.0 mi due east. It then turns north of east and travels another 75.0 mi, and finally turns another 65.0° toward the west and travels another 75.0 mi. What is its displacement from the port?

20. A car is held stationary on a ramp by two forces. One is the force of 480 lb by the brakes, which hold it from rolling down the ramp. The other is a reaction force by the ramp of 2250 lb, perpendicular to the ramp. This force keeps the car from going through the ramp. See Fig. 9-40. What is the weight of the car, and at what angle with the horizontal is the ramp inclined?

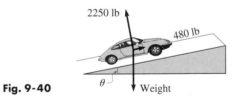

2250 lb

480 lb

θ Weight

Fig. 9-40

21. A plane is moving at 75.0 m/s, and a package with weather instruments is ejected horizontally from the plane at 15.0 m/s, perpendicular to the direction of the plane. If the vertical velocity v_v (in m/s), as a function of time t (in s) of fall, is given by $v_v = 9.80t$, what is the velocity of the package after 2.00 s (before its parachute opens)?

22. A flat rectangular barge, 48.0 m long and 20.0 m wide, is headed directly across a stream at 4.5 km/h. The stream flows at 3.8 km/h. What is the velocity, relative to the riverbed, of a person walking diagonally across the barge at 5.0 km/h while facing the opposite upstream bank?

23. In Fig. 9-41, a long, straight conductor perpendicular to the plane of the paper carries an electric current *i*. A bar magnet having poles of strength *m* lies in the plane of the paper. The vectors $\mathbf{H}_i$, $\mathbf{H}_N$, and $\mathbf{H}_S$ represent the components of the magnetic intensity $\mathbf{H}$ due to the current and to the N and S poles of the magnet, respectively. The magnitudes of the components of $\mathbf{H}$ are given by

$$H_i = \frac{1}{2\pi}\frac{i}{a} \qquad H_N = \frac{1}{4\pi}\frac{m}{b^2} \qquad H_S = \frac{1}{4\pi}\frac{m}{c^2}$$

Given that $a = 0.300$ m, $b = 0.400$ m, $c = 0.300$ m, the length of the magnet is 0.500 m, $i = 4.00$ A, and $m = 2.00$ A·m, calculate the resultant magnetic intensity $\mathbf{H}$. The component $\mathbf{H}_i$ is parallel to the magnet.

Fig. 9-41

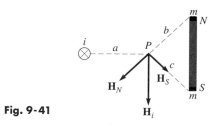

24. Solve the problem of Exercise 23 if $\mathbf{H}_i$ is directed away from the magnet, making an angle of 10.0° with the direction of the magnet.

9-5 OBLIQUE TRIANGLES, THE LAW OF SINES

To this point we have limited our study of triangle solution to right triangles. However, *many triangles that require solution do not contain a right angle. Such triangles are called* **oblique triangles.** We now discuss solutions of oblique triangles.

In Section 4-4 we stated that we need to know three parts, at least one of them a side, to solve a triangle. There are four possible such combinations of parts, and these combinations are as follows:

Case 1. Two angles and one side
Case 2. Two sides and the angle opposite one of them
Case 3. Two sides and the included angle
Case 4. Three sides

There are several ways in which oblique triangles may be solved, but we shall restrict our attention to the two most useful methods, the **law of sines** and the **law of cosines.** In this section we shall discuss the law of sines and show that it may be used to solve Case 1 and Case 2.

Let *ABC* be an oblique triangle with sides *a*, *b*, and *c* opposite angles *A*, *B*, and *C*, respectively. By drawing a perpendicular *h* from *B* to side *b*, or its extension, we see from Fig. 9-42(a) that

$$h = c \sin A \quad \text{or} \quad h = a \sin C \tag{9-4}$$

and from Fig. 9-42(b)

$$h = c \sin A \quad \text{or} \quad h = a \sin (180° - C) = a \sin C \tag{9-5}$$

We see that the results are precisely the same in Eqs. (9-4) and (9-5). Setting the results for *h* equal to each other, we have

$$c \sin A = a \sin C \quad \text{or} \quad \frac{a}{\sin A} = \frac{c}{\sin C} \tag{9-6}$$

By dropping a perpendicular from *A* to *a*, we also derive the result

$$c \sin B = b \sin C \quad \text{or} \quad \frac{b}{\sin B} = \frac{c}{\sin C} \tag{9-7}$$

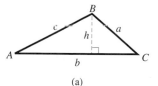

(a)

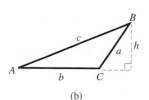

(b)

Fig. 9-42

Combining Eqs. (9-6) and (9-7), *for any triangle with sides a, b, and c, opposite angles A, B, and C, respectively,* such as the one shown in Fig. 9-43, *we have the* **law of sines:**

LAW OF SINES

$$\frac{a}{\sin A} = \frac{b}{\sin B} = \frac{c}{\sin C} \qquad (9\text{-}8)$$

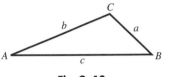

Fig. 9-43

Another form of the law of sines can be obtained by equating the reciprocals of each of the fractions in Eq. (9-8). The law of sines is a statement of proportionality between the sides of a triangle and the sines of the angles opposite them. We should note that there are actually three equations combined in Eq. (9-8). Of these, we use the one with three known parts of the triangle and we find the fourth part. In finding the complete solution of a triangle, it may be necessary to use two of the three equations.

Case 1: Two Angles and One Side

Now we see how the law of sines is used in the solution of a triangle in which two angles and one side are known. If two angles are known, the third may be found from the fact that the sum of the angles in a triangle is 180°. At this point we must be able to find the ratio between the given side and the sine of the angle opposite it. Then, by use of the law of sines, we may find the other two sides.

■**EXAMPLE 1**   Given $c = 6.00$, $A = 60.0°$, and $B = 40.0°$, find a, b, and C.
First, we can see that

$$C = 180.0° - (60.0° + 40.0°) = 80.0°$$

We now know side c and angle C, which allows us to use Eq. (9-8). Therefore, using the equation relating a, A, c, and C, we have

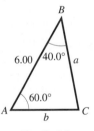

Fig. 9-44

$$\frac{a}{\sin 60.0°} = \frac{6.00}{\sin 80.0°} \quad \text{or} \quad a = \frac{6.00 \sin 60.0°}{\sin 80.0°} = 5.28$$

Now, using the equation relating b, B, c, and C, we have

$$\frac{b}{\sin 40.0°} = \frac{6.00}{\sin 80.0°} \quad \text{or} \quad b = \frac{6.00 \sin 40.0°}{\sin 80.0°} = 3.92$$

```
180-(60+40)
            80
6sin(60)/sin(80)
    5.276311449
6sin(40)/sin(80)
    3.916221868
```

Fig. 9-45

Thus, $a = 5.28$, $b = 3.92$, and $C = 80.0°$. See Fig. 9-44. We could also have used the form of Eq. (9-8) relating a, A, b, and B in order to find b, but any error in calculating a would make b in error as well. Of course, any error in calculating C would make both a and b in error.
    The calculator solution is shown in Fig. 9-45. ---------■

EXAMPLE 2   Solve the triangle with the following given parts: $a = 63.71$, $A = 56.29°$, and $B = 97.06°$. See Fig. 9-46.

From the figure, we see that we are to find angle C and sides b and c. We first determine angle C.

$$C = 180° - (A + B) = 180° - (56.29° + 97.06°)$$
$$= 26.65°$$

We will now use the ratio $a/\sin A$ to find sides b and c, using Eq. (9-8) (the law of sines).

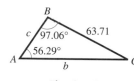

Fig. 9-46

$$\underset{B}{\underset{\uparrow}{\frac{b}{\sin 97.06°}}} = \underset{A}{\underset{\uparrow}{\frac{\overset{a}{\overset{\downarrow}{63.71}}}{\sin 56.29°}}} \quad \text{or} \quad b = \frac{63.71 \sin 97.06°}{\sin 56.29°} = 76.01$$

and

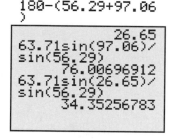

Fig. 9-47

$$\underset{C}{\underset{\uparrow}{\frac{c}{\sin 26.65°}}} = \underset{A}{\underset{\uparrow}{\frac{\overset{a}{\overset{\downarrow}{63.71}}}{\sin 56.29°}}} \quad \text{or} \quad c = \frac{63.71 \sin 26.65°}{\sin 56.29°} = 34.35$$

Thus, $b = 76.01$, $c = 34.35$, and $C = 26.65°$. See Fig. 9-46.
The calculator solution is shown in Fig. 9-47.

If the given information is appropriate, the law of sines may be used to solve applied problems. The following example illustrates the use of the law of sines in such a problem.

SOLVING A WORD PROBLEM

EXAMPLE 3   Two observers A and B sight a helicopter due east. The observers are 3540 ft apart, and the angles of elevation they each measure to the helicopter are 32.0° and 44.0°, respectively. How far is observer A from the helicopter? See Fig. 9-48.

Letting H represent the position of the helicopter, we see that angle B within the triangle ABH is $180° - 44.0° = 136.0°$. This means that the angle at H within the triangle is

$$H = 180° - (32.0° + 136.0°) = 12.0°$$

Now, using the law of sines to find required side AH, we have

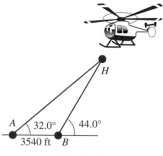

Fig. 9-48

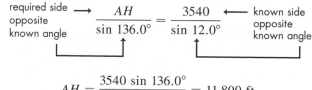

or

$$AH = \frac{3540 \sin 136.0°}{\sin 12.0°} = 11{,}800 \text{ ft}$$

Thus, observer A is about 11,800 ft from the helicopter.

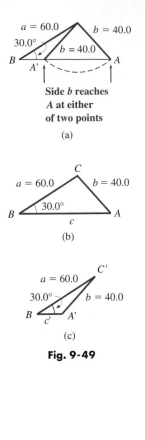

Side b reaches
A at either
of two points

(a)

(b)

(c)

Fig. 9-49

Case 2: Two Sides and the Angle Opposite One of Them

For a triangle in which we know two sides and the angle opposite one of the given sides, the solution will be either *one triangle,* or *two triangles,* or even possibly *no triangle.* In the following examples we illustrate how each of these results is possible.

▌**EXAMPLE 4**   Solve the triangle with the following given parts: $a = 60.0$, $b = 40.0$, and $B = 30.0°$.

By making a good scale drawing, Fig. 9-49(a), we see that the angle opposite a may be at either position A or A'. Both positions of this angle satisfy the given parts. Therefore, there are two triangles that result. Using the law of sines, we solve the case in which A, opposite side a, is an acute angle.

$$\frac{60.0}{\sin A} = \frac{40.0}{\sin 30.0°} \quad \text{or} \quad \sin A = \frac{60.0 \sin 30.0°}{40.0}$$

$$A = \sin^{-1}\left(\frac{60.0 \sin 30.0°}{40.0}\right) = 48.6°$$

$$C = 180° - (30.0° + 48.6°) = 101.4°$$

Therefore, $A = 48.6°$ and $C = 101.4°$. Using the law of sines again to find c, we have

$$\frac{c}{\sin 101.4°} = \frac{40.0}{\sin 30.0°}$$

$$c = \frac{40.0 \sin 101.4°}{\sin 30.0°} = 78.4$$

Thus we have $A = 48.6°$, $C = 101.4°$, and $c = 78.4$. See Fig. 9-49(b).

The other solution is the case in which A', opposite side a, is an obtuse angle. Here we have

$$A' = 180° - A = 180° - 48.6°$$
$$= 131.4°$$
$$C' = 180° - (30.0° + 131.4°)$$
$$= 18.6°$$

Using the law of sines to find c', we have

$$\frac{c'}{\sin 18.6°} = \frac{40.0}{\sin 30.0°}$$

$$c' = \frac{40.0 \sin 18.6°}{\sin 30.0°} = 25.5$$

This means that the second solution is $A' = 131.4°$, $C' = 18.6°$, and $c' = 25.5$. See Fig. 9-49(c).

The complete sequence for the calculator solution is shown in Fig. 9-50. The upper window shows the completion of the solution for A, C, and c. The lower window shows the solution for A', C', and c'. ----------▪

```
sin-1(60sin(30)/4
0)
        48.59037789
Ans→A
        48.59037789
180-(30+A)
        101.4096221
Ans→C
        101.4096221
40sin(C)/sin(30)
        78.41903734
```

```
180-A
        131.4096221
180-(30+Ans)
        18.59037789
40sin(Ans)/sin(3
0)
        25.50401112
```

Fig. 9-50

■**EXAMPLE 5** In Example 4, if $b > 60.0$, only one solution would result. In this case, side b would intercept side c at A. It also intercepts the extension of side c, but this would require that angle B not be included in the triangle (see Fig. 9-51). Thus, only one solution may result if $b > a$.

In Example 4, there would be *no solution* if side b were not at least 30.0. If this were the case, side b would not be long enough to even touch side c. It can be seen that b must at least equal $a \sin B$. If it is just equal to $a \sin B$, there is *one solution*, a right triangle. See Fig. 9-52.

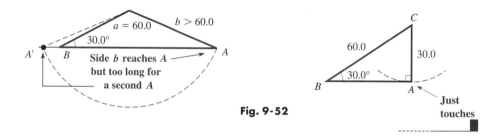

Fig. 9-51

Fig. 9-52

AMBIGUOUS CASE

Summarizing the results for Case 2 as illustrated in Examples 4 and 5, we make the following conclusions. Given sides a and b and angle A (assuming here that a and A ($A < 90°$) are corresponding parts), we have the following summary of solutions for Case 2.

CAUTION▶

> ### Summary of Solutions:
> ### Two Sides and the Angle Opposite One of Them
> 1. **No solution** *if $a < b \sin A$.* See Fig. 9-53(a).
> 2. **A right triangle solution** *if $a = b \sin A$.* See Fig. 9-53(b).
> 3. **Two solutions** *if $b \sin A < a < b$.* See Fig. 9-53(c).
> 4. **One solution** *if $a > b$.* See Fig. 9-53(d).

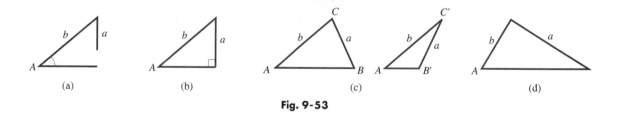

(a)   (b)   (c)   (d)

Fig. 9-53

NOTE▶

We note that *in order to have two solutions, we must know two sides and the angle opposite one of the sides, and the shorter side must be opposite the known angle.*

If there is *no solution*, the calculator will indicate an *error*. If the solution is a *right triangle*, the calculator will show an angle of *exactly 90°* (no extra decimal digits will be displayed.)

For the reason that two solutions may result from it, Case 2 is called the **ambiguous case.** The following example illustrates Case 2 in an applied problem.

SOLVING A WORD PROBLEM

See the chapter introduction.

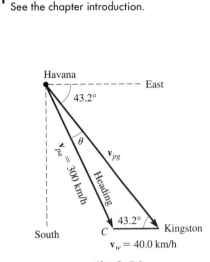

Fig. 9-54

█ EXAMPLE 6 Kingston, Jamaica is 43.2° south of east of Havana, Cuba. What should be the heading of a plane from Havana to Kingston if the wind is from the west at 40.0 km/h and the plane's speed with respect to the air is 300 km/h?

The heading should be set so that the resultant of the plane's velocity with respect to the air $\mathbf{v}_{pa}$ and the velocity of the wind $\mathbf{v}_w$ will be in the direction from Havana to Kingston. This means that the resultant velocity $\mathbf{v}_{pg}$ of the plane with respect to the ground must be at an angle of 43.2° south of east from Havana.

Using the given information we draw the vector triangle shown in Fig. 9-54. In the triangle, we know that the angle at Kingston is 43.2° by noting the alternate interior angles (see page 50). By finding θ, the required heading can be found. There can be only one solution, since $v_{pa} > v_w$. Using the law of sines, we have

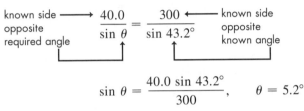

$$\sin \theta = \frac{40.0 \sin 43.2°}{300}, \qquad \theta = 5.2°$$

Therefore, the heading should be 43.2° + 5.2° = 48.4° south of east. Compare this example with Example 3 on page 258.

If we attempt to use the law of sines for the solution of Case 3 or Case 4, we find that we do not have sufficient information to complete one of the ratios. These cases can, however, be solved by the law of cosines, which we shall consider in the next section.

█ EXAMPLE 7 Given the three sides of a triangle $a = 5$, $b = 6$, $c = 7$, we would set up the ratios

$$\frac{5}{\sin A} = \frac{6}{\sin B} = \frac{7}{\sin C}$$

However, since there is no way to determine a complete ratio from these equations, we cannot find the solution of the triangle in this manner.

──────────── **EXERCISES** *9-5* ────────────

In Exercises 1–20, solve the triangles with the given parts.

1. $a = 45.7$, $A = 65.0°$, $B = 49.0°$

2. $b = 3.07$, $A = 26.0°$, $C = 120.0°$

3. $c = 4380$, $A = 37.4°$, $B = 34.6°$

4. $a = 93.2$, $B = 17.9°$, $C = 82.6°$

5. $a = 4.601$, $b = 3.107$, $A = 18.23°$

6. $b = 3.625$, $c = 2.946$, $B = 69.37°$

7. $b = 77.51$, $c = 36.42$, $B = 20.73°$

8. $a = 150.4$, $c = 250.9$, $C = 76.43°$

9. $b = 0.0742$, $B = 51.0°$, $C = 3.4°$

10. $c = 729$, $B = 121.0°$, $C = 44.2°$

11. $a = 63.8$, $B = 58.4°$, $C = 22.2°$

12. $a = 13.0$, $A = 55.2°$, $B = 67.5°$

13. $b = 4384$, $B = 47.43°$, $C = 64.56°$

14. $b = 283.2$, $B = 13.79°$, $C = 76.38°$

15. $a = 5.240$, $b = 4.446$, $B = 48.13°$

16. $a = 89.45$, $c = 37.36$, $C = 15.62°$

17. $b = 2880$, $c = 3650$, $B = 31.4°$

18. $a = 0.841$, $b = 0.965$, $A = 57.1°$

19. $a = 45.0$, $b = 126$, $A = 64.8°$

20. $a = 20$, $c = 10$, $C = 30°$

In Exercises 21–32, use the law of sines to solve the given problems.

21. The Pentagon (headquarters of the U.S. Department of Defense) is one of the world's largest office buildings in the world. It is a regular pentagon (five sides), 921 ft on a side. Find the greatest straight-line distance from one point on the outside of the building to another outside point (the length of a diagonal).

22. Two ropes hold a 175-lb crate as shown in Fig. 9-55. Find the tensions T_1 and T_2 in the ropes. (*Hint:* Move vectors so that they are tail to head to form a triangle. The vector sum $T_1 + T_2$ must equal 175 lb for equilibrium. See page 258.)

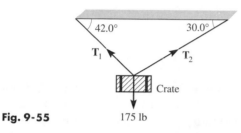

Fig. 9-55   175 lb

23. Find the tension **T** in the left guy wire attached to the top of the tower shown in Fig. 9-56. (*Hint:* The horizontal components of the tensions must be equal and opposite for equilibrium. See page 258. Thus, move the tension vectors tail to head to form a triangle with a vertical resultant. This resultant equals the upward force at the top of the tower for equilibrium. This last force is not shown and does not have to be calculated.)

Fig. 9-56

24. Find the distance from Atlanta to Raleigh, North Carolina, from Fig. 9-57.

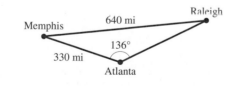

Fig. 9-57

25. Find the distance between Gravois Ave. and Jefferson Ave. along Arsenal St. in St. Louis, from Fig. 9-58.

Fig. 9-58

26. When an airplane is landing at an 8250-ft runway, the angles of depression to the ends of the runway are 10.0° and 13.5°. How far is the plane from the near end of the runway?

27. Find the total length of the path of the laser beam that is shown in Fig. 9-59.

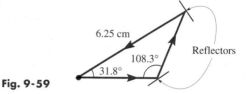

Fig. 9-59

28. In widening a highway, it is necessary for a construction crew to cut into the bank along the highway. The present angle of elevation of the straight slope of the bank is 23.0°, and the new angle is to be 38.5°, leaving the top of the slope at its present position. If the slope of the present bank is 220 ft long, how far horizontally into the bank at its base must they dig?

29. A communications satellite is directly above the extension of a line between receiving towers A and B. It is determined from radio signals that the angle of elevation of the satellite from tower A is 89.2°, and the angle of elevation from tower B is 86.5°. See Fig. 9-60. If A and B are 1290 km apart, how far is the satellite from A? (Neglect the curvature of the Earth.)

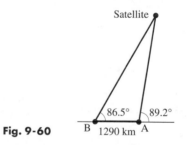

Fig. 9-60   B 1290 km A

30. Point *P* on the mechanism shown in Fig. 9-61 is driven back and forth horizontally. If the minimum value of angle θ is 32.0°, what is the distance between extreme positions of *P*? What is the maximum possible value of angle θ?

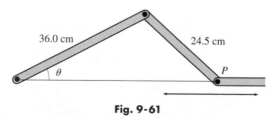

Fig. 9-61

31. A boat owner wishes to cross a river 2.60 km wide and go directly to a point on the opposite side 1.75 km downstream. The boat goes 8.00 km/h in still water, and the stream flows at 3.50 km/h. What should the boat's heading be?

32. A triangular support was measured to have a side of 25.3 in., a second side of 14.0 in., and an angle of 36.5° opposite the second side. Find the length of the third side.

$9\text{-}6$ THE LAW OF COSINES

As we noted at the end of the preceding section, the law of sines cannot be used to solve a triangle if the only information given is that of Case 3 (two sides and the included angle) or Case 4 (three sides). Therefore, it is necessary to develop another method of finding at least one more part of the triangle. Therefore, in this section we develop the *law of cosines,* by which we can solve a triangle for Case 3 and Case 4. After obtaining another part of the triangle by the law of cosines, we can then use the law of sines to complete the solution. We do this because the law of sines generally provides a simpler method of solution than does the law of cosines.

Consider any oblique triangle, for example, either of the triangles shown in Fig. 9-62. For each of these triangles we see that $h/b = \sin A$, or $h = b \sin A$. Also, by using the Pythagorean theorem, we obtain $a^2 = h^2 + x^2$ for each triangle. Thus (with $(\sin A)^2 = \sin^2 A$), we have

$$a^2 = b^2 \sin^2 A + x^2 \tag{9-9}$$

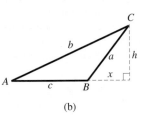

(a)

(b)

Fig. 9-62

In Fig. 9-62(a), we see that $(c - x)/b = \cos A$, or $c - x = b \cos A$. Solving for x, we have $x = c - b \cos A$. In Fig. 9-62(b), we have $c + x = b \cos A$, and solving for x, we have $x = b \cos A - c$. Substituting these relations into Eq. (9-9), we obtain

$$a^2 = b^2 \sin^2 A + (c - b \cos A)^2$$

and

$$a^2 = b^2 \sin^2 A + (b \cos A - c)^2$$

$$\tag{9-10}$$

respectively. When expanded, these give

$$a^2 = b^2 \sin^2 A + b^2 \cos^2 A + c^2 - 2bc \cos A$$
$$= b^2(\sin^2 A + \cos^2 A) + c^2 - 2bc \cos A \tag{9-11}$$

Recalling the definitions of the trigonometric functions, we know that $\sin \theta = y/r$ and $\cos \theta = x/r$. Thus, $\sin^2 \theta + \cos^2 \theta = (y^2 + x^2)/r^2$. However, $x^2 + y^2 = r^2$, which means that

$$\sin^2 \theta + \cos^2 \theta = 1 \tag{9-12}$$

This equation is valid for any angle θ, since we have made no assumptions as to any of the properties of θ. Therefore, by substituting Eq. (9-12) into Eq. (9-11), we arrive at the **law of cosines.**

LAW OF COSINES

$$a^2 = b^2 + c^2 - 2bc \cos A \tag{9-13}$$

Using the method above, we may also show that

$$b^2 = a^2 + c^2 - 2ac \cos B$$

and

$$c^2 = a^2 + b^2 - 2ab \cos C$$

Case 3: Two Sides and the Included Angle

If we know two sides and the included angle of a triangle, we see by the forms of the law of cosines that we may directly solve for the side opposite the given angle. Thus, as we noted at the beginning of this section, we may complete the solution by using the law of sines.

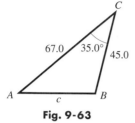

Fig. 9-63

■EXAMPLE 1   Solve the triangle with $a = 45.0$, $b = 67.0$, and $C = 35.0°$. See Fig. 9-63.

Since angle C is known, we first solve for side c, using the law of cosines in the form

$$c^2 = a^2 + b^2 - 2ab \cos C$$

Substituting, we have

unknown side opposite known angle

$$c^2 = 45.0^2 + 67.0^2 - 2(45.0)(67.0)\cos 35.0°$$

known sides

$$c = \sqrt{45.0^2 + 67.0^2 - 2(45.0)(67.0)\cos 35.0°} = 39.7$$

From the law of sines, we now have

$$\frac{45.0}{\sin A} = \frac{67.0}{\sin B} = \frac{39.7}{\sin 35.0°} \quad \begin{matrix} \leftarrow \text{sides} \\ \leftarrow \text{opposite} \\ \text{angles} \end{matrix}$$

$$\sin A = \frac{45.0 \sin 35.0°}{39.7}, \qquad A = 40.6°$$

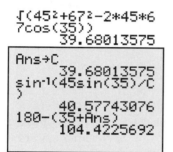

Fig. 9-64

The calculator solution is shown in Fig. 9-64. As shown, we found that $B = 104.4°$ using the fact that the sum of the angles is 180°, rather than using the relation from the law of sines.

Therefore, we have found that $c = 39.7$, $A = 40.6°$, and $B = 104.4°$.   ∎

■EXAMPLE 2   Solve the triangle with $a = 0.1762$, $c = 0.5034$, and $B = 129.20°$. See Fig. 9-65.

Again, the given parts are two sides and the included angle, or Case 3. Since B is the known angle, we use the form of the law of cosines that includes angle B. This means that we find b first.

$$b^2 = a^2 + c^2 - 2ac \cos B$$
$$= 0.1762^2 + 0.5034^2 - 2(0.1762)(0.5034)\cos 129.20°$$
$$b = 0.6297$$

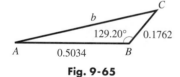

Fig. 9-65

From the law of sines, we have

$$\frac{0.1762}{\sin A} = \frac{0.6297}{\sin 129.20°} = \frac{0.5034}{\sin C}$$

$$\sin A = \frac{0.1762 \sin 129.20°}{0.6297}, \qquad A = 12.52°$$

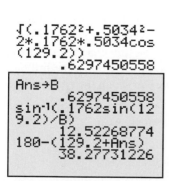

Fig. 9-66

As shown in Fig. 9-66, we complete the solution by using the fact that the sum of the angles is 180° to find $C = 38.28°$.

Thus, $b = 0.6297$, $A = 12.52°$, and $C = 38.28°$.   ∎

Case 4: Three Sides

CAUTION ▶

Given the three sides of a triangle, we may solve for the angle opposite any of the sides using the law of cosines. In solving for an angle, *the best procedure is to find the largest angle first.* This avoids the ambiguous case if we switch to the law of sines and the triangle has an obtuse angle. *The largest angle is always opposite the longest side.* Another procedure is to use the law of cosines to find two angles.

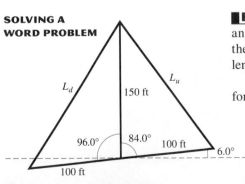

Fig. 9-68

■**EXAMPLE 3**   Solve the triangle given the three sides $a = 49.33$, $b = 21.61$, and $c = 42.57$. See Fig. 9-67.

Since the longest side is $a = 49.33$, we first solve for angle A.

$$a^2 = b^2 + c^2 - 2bc \cos A$$

$$\cos A = \frac{b^2 + c^2 - a^2}{2bc} = \frac{21.61^2 + 42.57^2 - 49.33^2}{2(21.61)(42.57)}$$

$$A = 94.81°$$

From the law of sines, we now have

$$\frac{49.33}{\sin A} = \frac{21.61}{\sin B} = \frac{42.57}{\sin C}$$

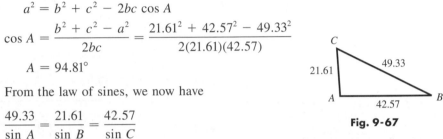

Fig. 9-67

The complete calculator solution is shown in Fig. 9-68. Therefore,

$$B = 25.88° \quad \text{and} \quad C = 59.31°$$ --------

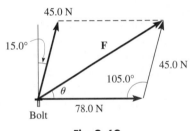

Fig. 9-69

■**EXAMPLE 4**   Two forces are acting on a bolt. One is a 78.0-N force acting horizontally to the right, and the other is a force of 45.0 N acting upward to the right, 15.0° from the vertical. Find the resultant force **F**. See Fig. 9-69.

Moving the 45.0-N vector to the right, and using the lower triangle with the 105.0° angle, we find the magnitude of **F** as

$$F = \sqrt{78.0^2 + 45.0^2 - 2(78.0)(45.0)\cos 105.0°}$$
$$= 99.6 \text{ N}$$

To find θ, we use the law of sines:

$$\frac{45.0}{\sin \theta} = \frac{99.6}{\sin 105.0°}, \quad \sin \theta = \frac{45.0 \sin 105.0°}{99.6}$$

This gives us $\theta = 25.9°$.

We can also solve this problem using vector components. --------

SOLVING A WORD PROBLEM

Fig. 9-70

■**EXAMPLE 5**   A vertical radio antenna is to be built on a hill that makes an angle of 6.0° with the horizontal. Guy wires are to be attached at a point 150 ft up the antenna and at points 100 ft from the base of the antenna. What will be the lengths of the guy wires positioned directly up and directly down the hill?

Making an appropriate figure, as in Fig. 9-70, we can set up the equations needed for the solution.

$$L_u^2 = 100^2 + 150^2 - 2(100)(150)\cos 84.0°$$
$$L_u = 171 \text{ ft}$$
$$L_d^2 = 100^2 + 150^2 - 2(100)(150)\cos 96.0°$$
$$L_d = 189 \text{ ft}$$ --------

Following is a summary of solving an oblique triangle by use of the law of sines or the law of cosines.

> **Solving Oblique Triangles**
> **Case 1: Two Angles and One Side**
> Find the unknown angle by subtracting the sum of the known angles from 180°. Use the **law of sines** to find the unknown sides.
> **Case 2: Two Sides and the Angle Opposite One of Them**
> Use the known side and the known angle opposite it to find the angle opposite the other known side. Find the third angle from the fact that the sum of the angles is 180°. Use the **law of sines** to find the third side. **CAUTION:** There may be two solutions. See page 265 for a summary of Case 2 and the *ambiguous case.*
> **Case 3: Two Sides and the Included Angle**
> Find the third side by using the **law of cosines.** Find a second angle by using the **law of sines.** Complete the solution using the fact that the sum of the angles is 180°.
> **Case 4: Three Sides**
> Find the *largest angle* (opposite the longest side) by using the **law of cosines.** Find a second angle by using the **law of sines.** Complete the solution by using the fact that the sum of the angles is 180°.

Other variations in finding the solutions can be used. For example, after finding the third side in Case 3 or finding the largest angle in Case 4, the solution can be completed by using the law of sines. All angles in Case 4 can be found by using the law of cosines. The methods shown above are those which are normally used.

EXERCISES 9-6

In Exercises 1–20, solve the triangles with the given parts.

1. $a = 6.00$, $b = 7.56$, $C = 54.0°$
2. $b = 87.3$, $c = 34.0$, $A = 130.0°$
3. $a = 4530$, $b = 924$, $C = 98.0°$
4. $a = 0.0845$, $c = 0.116$, $B = 85.0°$
5. $a = 39.53$, $b = 45.22$, $c = 67.15$
6. $a = 23.31$, $b = 27.26$, $c = 29.17$
7. $a = 385.4$, $b = 467.7$, $c = 800.9$
8. $a = 0.2433$, $b = 0.2635$, $c = 0.1538$
9. $a = 320$, $b = 847$, $C = 158.0°$
10. $b = 18.3$, $c = 27.1$, $A = 58.7°$
11. $a = 21.4$, $c = 4.28$, $B = 86.3°$
12. $a = 11.3$, $b = 5.10$, $C = 77.6°$
13. $b = 103.7$, $c = 159.1$, $C = 104.67°$
14. $a = 49.32$, $b = 54.55$, $B = 114.36°$
15. $a = 0.4937$, $b = 0.5956$, $c = 0.6398$
16. $a = 69.72$, $b = 49.30$, $c = 56.29$
17. $a = 723$, $b = 598$, $c = 158$
18. $a = 1.78$, $b = 6.04$, $c = 4.80$
19. $a = 15$, $A = 15°$, $B = 140°$
20. $a = 17$, $b = 24$, $c = 37$

In Exercises 21–32, use the law of cosines to solve the given problems.

21. A nuclear submarine leaves its base and travels at 23.5 mi/h. For 2.00 h it travels along a course 32.1° north of west. It then turns an additional 21.5° north of west and travels for another 1.00 h. How far from its base is it?

22. The robot arm shown in Fig. 9-71 places packages on a conveyor belt. What is the distance x?

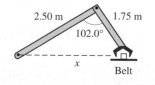

Fig. 9-71

23. In order to get around an obstruction, an oil pipeline is constructed in two straight sections, one 3.756 km long and the other 4.675 km long, with an angle of 168.85° between the sections where they are joined. How much more pipeline was necessary due to the obstruction?

24. In a baseball field the four bases are at the vertices of a square 90.0 ft on a side. The pitching rubber is 60.5 ft from home plate. See Fig. 9-72. How far is it from the pitching rubber to first base?

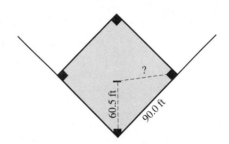

Fig. 9-72

25. A plane leaves an airport and travels 624 km due east. It then turns toward the north and travels another 326 km. It then turns again less than 180° and travels another 846 km directly back to the airport. Through what angles did it turn?

26. The apparent depth of an object submerged in water is less than its actual depth. A coin is actually 5.00 in. from an observer's eye just above the surface, but it appears to be only 4.25 in. The real light ray from the coin makes an angle with the surface that is 8.1° greater than the angle the apparent ray makes. How much deeper is the coin than it appears to be? See Fig. 9-73.

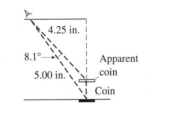

Fig. 9-73

27. A nut is in the shape of a regular hexagon (six sides). If each side is 9.53 mm, what opening on a wrench is necessary to tighten the nut? See Fig. 9-74.

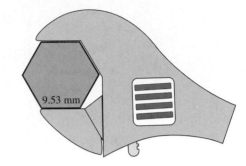

Fig. 9-74

28. Two ropes support a 78.3-lb crate from above. The tensions in the ropes are 50.6 lb and 37.5 lb. What is the angle between the ropes? (See Exercise 22 of Section 9-5.)

29. A ferryboat travels at 11.5 km/h with respect to the water. Because of the river current, it is traveling at 12.7 km/h with respect to the land in the direction of its destination. If the ferryboat's heading is 23.6° from the direction of its destination, what is the velocity of the current?

30. The airline distance from Denver to Dallas is 660 mi. It is 800 mi from Denver to St. Louis and 550 mi from Dallas to St. Louis. Find the angle between the routes from Dallas.

31. An air traffic controller sights two planes that are due east from the control tower and headed toward each other. One is 15.8 mi from the tower at an angle of elevation of 26.4°, and the other is 32.7 mi from the tower at an angle of elevation of 12.4°. How far apart are the planes?

W 32. A triangular machine part has sides of 5 cm and 8 cm. Explain why the law of sines, or the law of cosines, is used to start the solution of the triangle if the third known part is (a) the third side, (b) the angle opposite the 5-cm side, or (c) the angle between the 5-cm and 8-cm sides.

CHAPTER EQUATIONS

Vector components	$A_x = A \cos \theta \qquad A_y = A \sin \theta$	**(9-1)**
	$A = \sqrt{A_x^2 + A_y^2}$	**(9-2)**
	$\theta_{\text{ref}} = \tan^{-1} \dfrac{\lvert A_y \rvert}{\lvert A_x \rvert}$	**(9-3)**
Law of sines	$\dfrac{a}{\sin A} = \dfrac{b}{\sin B} = \dfrac{c}{\sin C}$	**(9-8)**
Law of cosines	$a^2 = b^2 + c^2 - 2bc \cos A$	**(9-13)**
	$b^2 = a^2 + c^2 - 2ac \cos B$	
	$c^2 = a^2 + b^2 - 2ab \cos C$	

REVIEW EXERCISES

In Exercises 1–4, find the x- and y-components of the given vectors by use of the trigonometric functions.

1. $A = 65.0$, $\theta_A = 28.0°$
2. $A = 8.05$, $\theta_A = 149.0°$
3. $A = 0.9204$, $\theta_A = 215.59°$
4. $A = 657.1$, $\theta_A = 343.74°$

*In Exercises 5–8, vectors **A** and **B** are at right angles. Find the magnitude and direction of the resultant.*

5. $A = 327$
   $B = 505$
6. $A = 68$
   $B = 29$
7. $A = 4964$
   $B = 3298$
8. $A = 26.52$
   $B = 89.86$

In Exercises 9–16, add the given vectors by use of the trigonometric functions and the Pythagorean theorem.

9. $A = 780$, $\theta_A = 28.0°$
   $B = 346$, $\theta_B = 320.0°$
10. $J = 0.0120$, $\theta_J = 370.5°$
    $K = 0.00781$, $\theta_K = 260.0°$
11. $A = 22.51$, $\theta_A = 130.16°$
    $B = 7.604$, $\theta_B = 200.09°$
12. $A = 18,760$, $\theta_A = 110.43°$
    $B = 4835$, $\theta_B = 350.20°$
13. $Y = 51.33$, $\theta_Y = 12.25°$
    $Z = 42.61$, $\theta_Z = 291.77°$
14. $A = 70.31$, $\theta_A = 122.54°$
    $B = 30.29$, $\theta_B = 214.82°$
15. $A = 75.0$, $\theta_A = 15.0°$
    $B = 26.5$, $\theta_B = 192.4°$
    $C = 54.8$, $\theta_C = 344.7°$
16. $S = 8120$, $\theta_S = 141.9°$
    $T = 1540$, $\theta_T = 165.2°$
    $U = 3470$, $\theta_U = 296.0°$

In Exercises 17–36, solve the triangles with the given parts.

17. $A = 48.0°$, $B = 68.0°$, $a = 14.5$
18. $A = 132.0°$, $b = 7.50$, $C = 32.0°$
19. $a = 22.8$, $B = 33.5°$, $C = 125.3°$
20. $A = 71.0°$, $B = 48.5°$, $c = 8.42$
21. $A = 17.85°$, $B = 154.16°$, $c = 7863$
22. $u = 1.985$, $b = 4.189$, $c = 3.652$
23. $b = 76.07$, $c = 40.53$, $B = 110.09°$
24. $A = 77.06°$, $a = 12.07$, $c = 5.104$
25. $b = 14.5$, $c = 13.0$, $C = 56.6°$
26. $B = 40.6°$, $b = 7.00$, $c = 18.0$
27. $a = 186$, $B = 130.0°$, $c = 106$
28. $b = 750$, $c = 1100$, $A = 56°$
29. $a = 7.86$, $b = 2.45$, $C = 22.0°$
30. $a = 0.208$, $c = 0.697$, $B = 105.4°$
31. $A = 67.16°$, $B = 96.84°$, $c = 532.9$
32. $A = 43.12°$, $a = 7.893$, $b = 4.113$

33. $a = 17$, $b = 12$, $c = 25$
34. $a = 9064$, $b = 9953$, $c = 1106$
35. $a = 5.30$, $b = 8.75$, $c = 12.5$
36. $a = 47.4$, $b = 40.0$, $c = 45.5$

In Exercises 37–60, solve the given problems.

37. For any triangle ABC show that
$$\frac{a^2 + b^2 + c^2}{2abc} = \frac{\cos A}{a} + \frac{\cos B}{b} + \frac{\cos C}{c}$$

Ⓦ 38. In solving a triangle for Case 3 (two sides and the included angle), explain what type of solution is obtained if the included angle is a right angle.

39. An architect determines the two acute angles and one of the legs of a right triangular wall panel. Show that the area A_t is
$$A_t = \frac{a^2 \sin B}{2 \sin A}$$

40. A surveyor determines the three angles and one side of a triangular tract of land. (a) Show that the area A_t can be found from $A_t = \dfrac{a^2 \sin B \sin C}{2 \sin A}$. (b) For a right triangle, show that this agrees with the formula in Exercise 39.

41. Find the horizontal and vertical components of the force shown in Fig. 9-75.

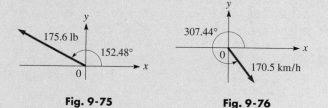

Fig. 9-75   **Fig. 9-76**

42. Find the horizontal and vertical components of the velocity shown in Fig. 9-76.

43. In a ballistics test, a bullet was fired into a block of wood with a velocity of 2200 ft/s and at an angle of 71.3° with the surface of the block. What was the component of the velocity perpendicular to the surface?

44. A storm cloud is moving at 15 mi/h from the northwest. A television tower is 60° south of east of the cloud. What is the component of the cloud's velocity toward the tower?

45. During a 3.00-minute period after taking off, the supersonic jet Concorde traveled at 480 km/h at an angle of 24.0° above the horizontal. What was its gain in altitude during this period?

46. In Fig. 9-77 force **F** represents the total surface tension force around the circumference on the liquid in the capillary tube. The vertical component of **F** holds up the liquid in the tube above the liquid surface outside the tube. What is the vertical component of **F**?

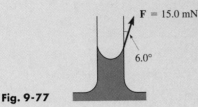

Fig. 9-77

47. A helium-filled balloon rises vertically at 3.5 m/s as the wind carries it horizontally at 5.0 m/s. What is the resultant velocity of the balloon?

48. A crater on the moon is 150 mi in diameter. If the distance to the moon (to each side of the crater) from the earth is 240,000 mi, what angle is subtended by the crater at an observer's position on the earth?

49. In Fig. 9-78 a damper mechanism in an air-conditioning system is shown. If $\theta = 27.5°$ when the spring is at its shortest and longest lengths, what are these lengths?

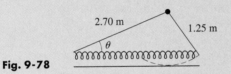

Fig. 9-78

50. A bullet is fired from the ground of a level field at an angle of 39.0° above the horizontal. It travels in a straight line at 2200 ft/s for 0.20 s when it strikes a target. The sound of the strike is recorded 0.32 s later on the ground. If sound travels at 1130 ft/s, where is the recording device located?

51. Two satellites are being observed at the same observing station. One is 22,500 mi from the station, and the other is 18,700 mi away. The angle between their lines of observation is 105.4°. How far apart are the satellites?

52. Find the side x in the truss in Fig. 9-79.

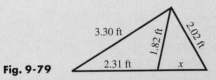

Fig. 9-79

53. The angle of depression of a fire noticed west of a fire tower is 6.2°. The angle of depression of a pond, also west of the tower, is 13.5°. If it is known that the tower is 2.25 km from the pond on a direct line to the pond, how far is the fire from the pond?

54. A surveyor wishes to find the distance between two points between which there is a security-restricted area. The surveyor measures the distance from each of these points to a third point and finds them to be 226.73 m and 185.12 m. If the angle between the lines of sight from the third point to the other points is 126.724°, how far apart are the two points?

55. Atlanta is 290 mi and 51.0° south of east from Nashville. The pilot of an airplane due north of Atlanta radios Nashville and finds the plane is on a line 10.5° south of east from Nashville. How far is the plane from Nashville?

56. In going around a storm, a plane flies 125 mi south, then 140 mi at 30.0° south of west, and finally 225 mi at 15.0° north of west. What is the displacement of the plane from its original position?

57. A sailboat is headed due north, and its sail is set perpendicular to the wind, which is from the south of west. The component of the force of the wind in the direction of the heading is 480 N, and the component perpendicular to the heading (the *drift* component) is 650 N. What is the force exerted by the wind, and what is the direction of the wind? See Fig. 9-80.

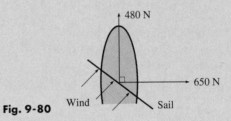

Fig. 9-80

58. Boston is 650 km and 21.0° south of west from Halifax, Nova Scotia. Radio signals locate a ship 10.5° east of south from Halifax and 5.6° north of east from Boston. How far is the ship from each city?

59. One end of a 1450-ft bridge is sighted from a distance of 3250 ft. The angle between the lines of sight of the ends of the bridge is 25.2°. From these data, how far is the observer from the other end of the bridge?

60. A plane is traveling horizontally at 1250 ft/s. A missile is fired horizontally from it 30.0° from the direction in which the plane is traveling. If the missile leaves the plane at 2040 ft/s, what is its velocity 10.0 s later if the vertical component is given by $v_V = -32.0t$ (in ft/s)?

Writing Exercise

61. A laser experiment uses a prism with a triangular base that has sides of 2.00 cm, 3.00 cm, and 4.50 cm. Write two or three paragraphs explaining how to find the angles of the triangle.

PRACTICE TEST

In all triangle solutions, sides a, b, c, are opposite angles A, B, C, respectively.

1. By use of a diagram, find the vector sum $2\mathbf{A} + \mathbf{B}$ for the given vectors.

2. For the triangle in which $a = 22.5$, $B = 78.6°$, and $c = 30.9$, find b.

3. A surveyor locates a tree 36.50 m to the northeast of a set position. The tree is 21.38 m north of a utility pole. What is the displacement of the utility pole from the set position?

4. For the triangle in which $A = 18.9°$, $B = 104.2°$, and $a = 426$, find c.

5. Solve the triangle in which $a = 9.84$, $b = 3.29$, and $c = 8.44$.

6. Find the horizontal and vertical components of a vector of magnitude 871 that is directed at a standard-position angle of 284.3°.

7. A ship leaves a port and travels due west. At a certain point it turns 31.5° north of west and travels an additional 42.0 mi to a point 63.0 mi on a direct line from the port. How far from the port is the point where the ship turned?

8. Find the sum of the vectors for which $A = 449$, $\theta_A = 74.2°$, $B = 285$, and $\theta_B = 208.9°$. Use the trigonometric functions and the Pythagorean theorem.

9. Solve the triangle for which $a = 22.3$, $b = 29.6$, and $A = 36.5°$.

10 GRAPHS OF THE TRIGONOMETRIC FUNCTIONS

One of the clearest ways of showing the properties of the trigonometric functions is by means of their graphs. In addition, their graphs are very valuable in numerous areas of application that do not involve solving triangles. These important applications are found in electronics, communications, optics, acoustics, mechanical vibrations, and in many areas of physics.

These graphs are particularly useful in applications that involve any type of wave motion and periodic values, which repeat on a regular basis. Filtering electronic signals in communications, mixing musical sounds on a tape in a recording studio, studying the seasonal temperatures of an area, and analyzing ocean waves and tides illustrate some of the many applications of this type of periodic motion.

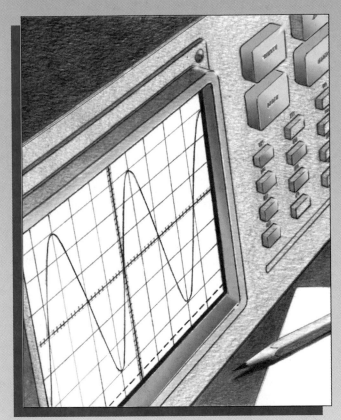

In Section 10-6 we show the resulting curve when an oscilloscope is used to combine and display electric signals.

10-1 GRAPHS OF $y = a \sin x$ AND $y = a \cos x$

The graphs of the trigonometric functions are constructed on the rectangular coordinate system. In plotting and sketching the trigonometric functions, *it is normal to express the angle in radians. By using radians, x and the trigonometric function of x are expressed as real numbers.*

Therefore, in order that we can plot and sketch the graphs of the trigonometric functions,

NOTE ▶   *it is necessary to be able to readily use angles expressed in radians.*

If necessary, review Section 8-3 on radian measure of angles for this purpose.

In this section, the graphs of the sine and cosine functions are shown. We begin by making a table of values of x and y for the function $y = \sin x$, where we are using x and y in the standard way as the *independent variable* and *dependent variable*. We plot the points to obtain the graph in Fig. 10-1.

x	0	$\frac{\pi}{6}$	$\frac{\pi}{3}$	$\frac{\pi}{2}$	$\frac{2\pi}{3}$	$\frac{5\pi}{6}$	π	$\frac{7\pi}{6}$	$\frac{4\pi}{3}$	$\frac{3\pi}{2}$	$\frac{5\pi}{3}$	$\frac{11\pi}{6}$	2π
y	0	0.5	0.87	1	0.87	0.5	0	-0.5	-0.87	-1	-0.87	-0.5	0

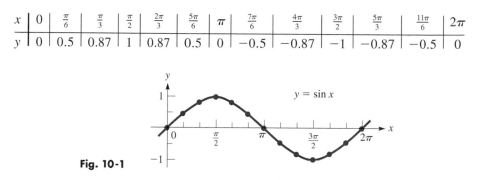

Fig. 10-1

The graph of $y = \cos x$ may be drawn in the same way. The next table gives values for plotting the graph of $y = \cos x$, and the graph is shown in Fig. 10-2.

x	0	$\frac{\pi}{6}$	$\frac{\pi}{3}$	$\frac{\pi}{2}$	$\frac{2\pi}{3}$	$\frac{5\pi}{6}$	π	$\frac{7\pi}{6}$	$\frac{4\pi}{3}$	$\frac{3\pi}{2}$	$\frac{5\pi}{3}$	$\frac{11\pi}{6}$	2π
y	1	0.87	0.5	0	-0.5	-0.87	-1	-0.87	-0.5	0	0.5	0.87	1

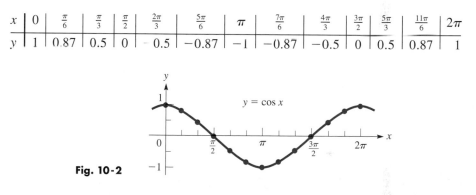

Fig. 10-2

The graphs are continued beyond the values shown in the tables to indicate that *they continue indefinitely in each direction.* To show this more clearly, in Figs. 10-3 and 10-4, we show the graphs of $y = \sin x$ and $y = \cos x$ from $x = -10$ to $x = 10$. (Note that $2\pi \approx 6.3$ for the graphs in Figs. 10-1 and 10-2.)

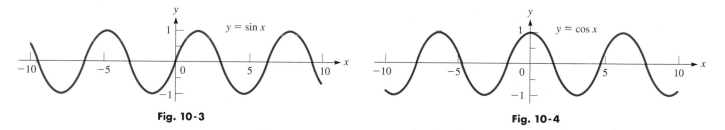

Fig. 10-3

Fig. 10-4

From these tables and graphs, it can be seen that *the graphs of y = sin x and y = cos x are of exactly the same shape (called* **sinusoidal***), with the cosine curve displaced π/2 units to the left of the sine curve.* The shape of these curves should be recognized readily, with special note as to the points at which they cross the axes. This information will be especially valuable in *sketching* similar curves, since the basic sinusoidal shape remains the same. It will not be necessary to plot numerous points every time we wish to sketch such a curve.

AMPLITUDE

To obtain the graph of $y = a \sin x$, we note that all the y-values obtained for the graph of $y = \sin x$ are to be multiplied by the number a. In this case the greatest value of the sine function is $|a|$. *The number $|a|$ is called the **amplitude** of the curve and represents the greatest y-value of the curve.* Also, the curve will have no value less than $-|a|$. This is true for $y = a \cos x$ as well as for $y = a \sin x$.

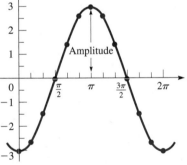

Fig. 10-5

■**EXAMPLE 1** Plot the graph of $y = 2 \sin x$.

Since $a = 2$, the amplitude of this curve is $|2| = 2$. This means that the maximum value of y is 2 and the minimum value is $y = -2$. The table of values follows, and the curve is shown in Fig. 10-5.

x	0	$\frac{\pi}{6}$	$\frac{\pi}{3}$	$\frac{\pi}{2}$	$\frac{2\pi}{3}$	$\frac{5\pi}{6}$	π
y	0	1	1.73	2	1.73	1	0

x	$\frac{7\pi}{6}$	$\frac{4\pi}{3}$	$\frac{3\pi}{2}$	$\frac{5\pi}{3}$	$\frac{11\pi}{6}$	2π
y	-1	-1.73	-2	-1.73	-1	0

■**EXAMPLE 2** Plot the graph of $y = -3 \cos x$.

In this case $a = -3$, and this means that the amplitude is $|-3| = 3$. Therefore, the maximum value of y is 3, and the minimum value of y is -3. The table of values follows, and the curve is shown in Fig. 10-6.

x	0	$\frac{\pi}{6}$	$\frac{\pi}{3}$	$\frac{\pi}{2}$	$\frac{2\pi}{3}$	$\frac{5\pi}{6}$	π
y	-3	-2.6	-1.5	0	1.5	2.6	3

x	$\frac{7\pi}{6}$	$\frac{4\pi}{3}$	$\frac{3\pi}{2}$	$\frac{5\pi}{3}$	$\frac{11\pi}{6}$	2π
y	2.6	1.5	0	-1.5	-2.6	-3

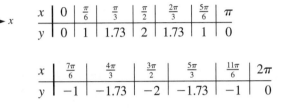

Fig. 10-6

TABLE 10-1

	$x = 0, \pi, 2\pi$	$\frac{\pi}{2}, \frac{3\pi}{2}$
$y = a \sin x$	zeros	max. or min.
$y = a \cos x$	max. or min.	zeros

Note from Example 2 that *the effect of the negative sign before the number a is to **invert** the curve about the x-axis.* The effect of the number a can also be seen readily from these examples.

From the previous examples we see that the function $y = a \sin x$ has zeros for $x = 0, \pi, 2\pi$ and that it has its maximum or minimum values for $x = \pi/2, 3\pi/2$. The function $y = a \cos x$ has its zeros for $x = \pi/2, 3\pi/2$ and its maximum or minimum values for $x = 0, \pi, 2\pi$. This is summarized in Table 10-1. Therefore, by knowing the general shape of the sine curve, where it has its zeros, and what its amplitude is, *we can rapidly **sketch** curves of the form $y = a \sin x$ and $y = a \cos x$.*

Since the graphs of $y = a \sin x$ and $y = a \cos x$ can extend indefinitely to the right and to the left, we see that the domain of each is all real numbers. We should note that the key values of $x = 0, \pi/2, \pi, 3\pi/2$, and 2π are those only for x from 0 to 2π. Corresponding values ($x = 5\pi/2, 3\pi$, and their negatives) could also be used. Also from the graphs we can readily see that the range of these functions is $-|a| \le f(x) \le |a|$.

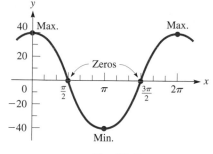

Fig. 10-7

EXAMPLE 3 Sketch the graph of $y = 40 \cos x$.

First, we set up a table of values for the points where the curve has its zeros, maximum points, and minimum points.

x	0	$\frac{\pi}{2}$	π	$\frac{3\pi}{2}$	2π
y	40	0	-40	0	40
	max.		min.		max.

Now, we plot these points and join them, knowing the basic sinusoidal shape of the curve. See Fig. 10-7.

The graphs of $y = a \sin x$ and $y = a \cos x$ can be displayed easily on a graphing calculator. In the next example we will see that the calculator displays the features of the curve we should expect from our previous discussion.

EXAMPLE 4 Display the graph of $y = -2 \sin x$ on a graphing calculator.

Using *radian* mode on the calculator, Fig. 10-8(a) shows the calculator view for the key values in the following table.

x	0	$\frac{\pi}{2}$	π	$\frac{3\pi}{2}$	2π
y	0	-2	0	2	0
		min.		max.	

We see the amplitude of 2 and the effect of the negative sign in inverting the curve, as expected. Fig. 10-8(b) shows the calculator graph for $x = -10$ to $x = 10$.

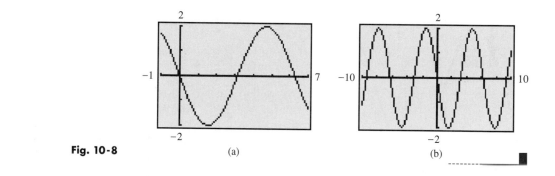

Fig. 10-8   (a)   (b)

EXERCISES *10-1*

In Exercises 1–4, complete the following table for the given functions and then plot the resulting graphs.

x	$-\pi$	$-\frac{3\pi}{4}$	$-\frac{\pi}{2}$	$-\frac{\pi}{4}$	0	$\frac{\pi}{4}$	$\frac{\pi}{2}$	$\frac{3\pi}{4}$	π
y									

x	$\frac{5\pi}{4}$	$\frac{3\pi}{2}$	$\frac{7\pi}{4}$	2π	$\frac{9\pi}{4}$	$\frac{5\pi}{2}$	$\frac{11\pi}{4}$	3π
y								

1. $y = \sin x$

2. $y = \cos x$

3. $y = 3 \cos x$

4. $y = -4 \sin x$

In Exercises 5–20, sketch the graphs of the given functions. Check each using a graphing calculator.

5. $y = 3 \sin x$

6. $y = 5 \sin x$

7. $y = \frac{5}{2} \sin x$

8. $y = 35 \sin x$

9. $y = 2 \cos x$

10. $y = 3 \cos x$

11. $y = 0.8 \cos x$

12. $y = \frac{3}{2} \cos x$

13. $y = -\sin x$

14. $y = -3 \sin x$

15. $y = -1.5 \sin x$

16. $y = -0.2 \sin x$

17. $y = -\cos x$

18. $y = -8 \cos x$

19. $y = -50 \cos x$

20. $y = -0.4 \cos x$

Although units of π are convenient, we must remember that π is only a number. Numbers that are not multiples of π may be used. In Exercises 21–24, plot the indicated graphs by finding the values of y that correspond to values of x of 0, 1, 2, 3, 4, 5, 6, and 7 on a calculator. (Remember, the numbers 0, 1, 2, and so on represent radian measure.)

21. $y = \sin x$ **22.** $y = -3 \sin x$

23. $y = \cos x$ **24.** $y = 2 \cos x$

In Exercises 25–28, the graph of a function of the form $y = a \sin x$ or $y = a \cos x$ is shown. Determine the specific function for each.

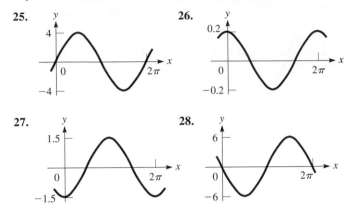

10-2 GRAPHS OF $y = a \sin bx$ AND $y = a \cos bx$

In graphing the function $y = \sin x$, we see that the values of y repeat every 2π units of x. This is because $\sin x = \sin(x + 2\pi) = \sin(x + 4\pi)$, and so forth. For any function F, we say that it has a *period P* if $F(x) = F(x + P)$. For functions that are periodic, such as the sine and the cosine, *the **period** is the x-distance between a point and the next corresponding point for which the value of y repeats.*

PERIOD OF A FUNCTION

Let us now plot the curve $y = \sin 2x$. This means that we choose a value for x, multiply this value by 2, and find the sine of the result. This leads to the following table of values for this function.

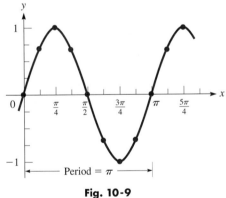

Fig. 10-9

x	0	$\frac{\pi}{8}$	$\frac{\pi}{4}$	$\frac{3\pi}{8}$	$\frac{\pi}{2}$	$\frac{5\pi}{8}$	$\frac{3\pi}{4}$	$\frac{7\pi}{8}$	π	$\frac{9\pi}{8}$	$\frac{5\pi}{4}$
$2x$	0	$\frac{\pi}{4}$	$\frac{\pi}{2}$	$\frac{3\pi}{4}$	π	$\frac{5\pi}{4}$	$\frac{3\pi}{2}$	$\frac{7\pi}{4}$	2π	$\frac{9\pi}{4}$	$\frac{5\pi}{2}$
y	0	0.7	1	0.7	0	-0.7	-1	-0.7	0	0.7	1

Plotting these points, we have the curve shown in Fig. 10-9.

From the table and Fig. 10-9, we see that $y = \sin 2x$ repeats after π units of x. The effect of the 2 is that the period of $y = \sin 2x$ is half the period of the curve of $y = \sin x$. We then conclude that if the period of a function $F(x)$ is P, then the period of $F(bx)$ is P/b. Since each of the functions $\sin x$ and $\cos x$ has a period of 2π,

NOTE ▶ *each of the functions $\sin bx$ and $\cos bx$ has a period of $2\pi/b$.*

■ **EXAMPLE 1**   **(a)** The period of $\sin \overset{\overset{b}{\downarrow}}{3}x$ is $\dfrac{2\pi}{3}$, which means that the curve of the function $y = \sin 3x$ will repeat every $\frac{2\pi}{3}$ (approximately 2.09) units of x.

(b) The period of $\cos 4x$ is $\dfrac{2\pi}{4} = \dfrac{\pi}{2}$.

(c) The period of $\sin \frac{1}{2}x$ is $\dfrac{2\pi}{\frac{1}{2}} = 4\pi$. In this case we see that the period is longer than that of the basic sine curve.

■

EXAMPLE 2   (a) The period of $\sin \pi x$ is $2\pi/\pi = 2$. That is, the curve of the function $\sin \pi x$ repeats every 2 units.

(b) The period of $\cos 3\pi x$ is $\dfrac{2\pi}{3\pi} = \dfrac{2}{3}$.

(c) The period of $\sin \dfrac{\pi}{4} x$ is $\dfrac{2\pi}{\frac{\pi}{4}} = 8$.

(d) The period of $\sin 3x$ is $\dfrac{2\pi}{3} = 2.09$, and the period of $\sin \pi x$ is $\frac{2\pi}{\pi} = 2$. We can see that these two periods differ only slightly since π is only slightly larger than 3.

Combining the value of the period with the value of the amplitude from Section 10-1, we conclude that *the functions $y = a$ sin bx and $y = a$ cos bx each has an amplitude of |a| and a period of $2\pi/b$.* These properties are very useful in sketching these functions.

EXAMPLE 3   Sketch the graph of $y = 3 \sin 4x$ for $0 \le x \le \pi$.

Since $a = 3$, we can immediately see that the amplitude is 3. Also, from the $4x$ we see that the period is $2\pi/4 = \pi/2$. Therefore, we know that $y = 0$ for $x = 0$ and also for $x = \pi/2$. Since we know that this sine function is zero halfway between $x = 0$ and $x = \pi/2$, we find that $y = 0$ for $x = \pi/4$. The graph of the sine function reaches its maximum or minimum values halfway between the zeros. This tells us that $y = 3$ for $x = \pi/8$, and $y = -3$ for $x = 3\pi/8$. A table for these important values follows. We should also note that the values of x that are listed are those for which $4x = 0$, $\pi/2$, π, $3\pi/2$, 2π, and so on.

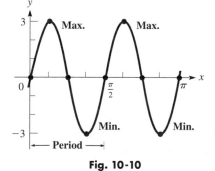

Fig. 10-10

x	0	$\frac{\pi}{8}$	$\frac{\pi}{4}$	$\frac{3\pi}{8}$	$\frac{\pi}{2}$	$\frac{5\pi}{8}$	$\frac{3\pi}{4}$	$\frac{7\pi}{8}$	π
y	0	3	0	-3	0	3	0	-3	0

Using the values from the table and the fact that we know the curve is sinusoidal in form, we sketch the graph of this function in Fig. 10-10. We see again that knowing the key values and the basic shape of the curve allows us to *sketch* the graph of the curve quickly and easily.

We see from Example 3 that *an important distance in sketching a sine curve or a cosine curve is one-fourth of the period.* For $y = a$ sin bx, it is one-fourth of the period from the origin to the first value of x where y is at its maximum (or minimum) value. Then we proceed another one-fourth period to a zero, another one-fourth period to the next minimum (or maximum) value, another to the next zero (this is where the period is completed), and so on. Thus,

NOTE ▶   *by finding one-fourth of the period, we can easily find the important values for sketching the curve.*

Similarly, one-fourth of the period is used in sketching the graph of $y = a$ cos bx. For this function, its maximum (or minimum) occurs for $y = 0$. At the following one-fourth period values, there is a zero, a minimum (or maximum), a zero, and a maximum (or minimum) at the start of the next period.

On the next page we summarize the important values for sketching the graphs of $y = a$ sin bx and $y = a$ cos bx.

> **Important Values for Sketching** $y = a \sin bx$ **and** $y = a \cos bx$
> 1. *The amplitude:* $|a|$
> 2. *The period:* $2\pi/b$
> 3. *Values of the function for each one-fourth period*

■**EXAMPLE 4**   Sketch the graph of $y = -2 \cos 3x$ for $0 \le x \le 2\pi$.

We note that the amplitude is 2 and the period is $\frac{2\pi}{3}$. This means that one-fourth of the period is $\frac{1}{4} \times \frac{2\pi}{3} = \frac{\pi}{6}$. Since the cosine curve is at a maximum or minimum for $x = 0$, we find that $y = -2$ for $x = 0$ (the negative value is due to the minus sign before the function), which means it is a minimum point. The curve then has a zero at $x = \frac{\pi}{6}$, a maximum value of 2 at $x = 2(\frac{\pi}{6}) = \frac{\pi}{3}$, a zero at $x = 3(\frac{\pi}{6}) = \frac{\pi}{2}$, and its next value of -2 at $x = 4(\frac{\pi}{6}) = \frac{2\pi}{3}$, and so on. Therefore, we have the following table.

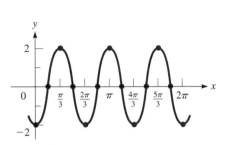

x	0	$\frac{\pi}{6}$	$\frac{\pi}{3}$	$\frac{\pi}{2}$	$\frac{2\pi}{3}$	$\frac{5\pi}{6}$	π	$\frac{7\pi}{6}$	$\frac{4\pi}{3}$	$\frac{3\pi}{2}$	$\frac{5\pi}{3}$	$\frac{11\pi}{6}$	2π
y	-2	0	2	0	-2	0	2	0	-2	0	2	0	-2

Fig. 10-11

Using this table and the sinusoidal shape of the cosine curve, we sketch the function of Fig. 10-11. ■

■**EXAMPLE 5**   A generator produces a voltage $V = 200 \cos 50\pi t$, where t is the time in seconds (50π has units of rad/s; thus, $50\pi t$ is an angle in radians). Use a graphing calculator to display the graph of V as a function of t for $0 \le t \le 0.06$ s.

The amplitude is 200 V and the period is $2\pi/(50\pi) = 0.04$ s. Since the period is not in terms of π, it is more convenient to use decimal units for t rather than to use units in terms of π as in the previous graphs. Thus, we have the following table of values:

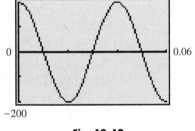

t (seconds)	0	0.01	0.02	0.03	0.04	0.05	0.06
V (volts)	200	0	-200	0	200	0	-200

For the graphing calculator, we use x for t and y for V. This means we graph the function $y_1 = 200 \cos 50\pi x$, as shown in Fig. 10-12. From the amplitude of 200 V and the above table, we choose the *window* values shown in Fig. 10-12. We do not consider negative values of t, for they have no real meaning in this problem. ■

Fig. 10-12

─────────── EXERCISES *10-2* ───────────

In Exercises 1–20, find the period of each function.

1. $y = 2 \sin 6x$
2. $y = 4 \sin 2x$
3. $y = 3 \cos 8x$
4. $y = 28 \cos 10x$
5. $y = -2 \sin 12x$
6. $y = -\sin 5x$
7. $y = -\cos 16x$
8. $y = -4 \cos 2x$
9. $y = 520 \sin 2\pi x$
10. $y = 2 \sin 3\pi x$

11. $y = 3 \cos 4\pi x$
12. $y = 4 \cos 10\pi x$
13. $y = 3 \sin \frac{1}{3}x$
14. $y = -2 \sin \frac{2}{5}x$
15. $y = -\frac{1}{2} \cos \frac{2}{3}x$
16. $y = \frac{1}{3} \cos \frac{1}{4}x$
17. $y = 0.4 \sin \dfrac{2\pi x}{3}$
18. $y = 1.5 \cos \dfrac{\pi x}{10}$
19. $y = 3.3 \cos \pi^2 x$
20. $y = 2.5 \sin \dfrac{2x}{\pi}$

In Exercises 21–40, sketch the graphs of the given functions. Check each using a graphing calculator. (These are the same functions as in Exercises 1–20.)

21. $y = 2 \sin 6x$

22. $y = 4 \sin 2x$

23. $y = 3 \cos 8x$

24. $y = 28 \cos 10x$

25. $y = -2 \sin 12x$

26. $y = -\sin 5x$

27. $y = -\cos 16x$

28. $y = -4 \cos 2x$

29. $y = 520 \sin 2\pi x$

30. $y = 2 \sin 3\pi x$

31. $y = 3 \cos 4\pi x$

32. $y = 4 \cos 10\pi x$

33. $y = 3 \sin \frac{1}{3}x$

34. $y = -2 \sin \frac{2}{5}x$

35. $y = -\frac{1}{2} \cos \frac{2}{3}x$

36. $y = \frac{1}{3} \cos \frac{1}{4}x$

37. $y = 0.4 \sin \dfrac{2\pi x}{3}$

38. $y = 1.5 \cos \dfrac{\pi x}{10}$

39. $y = 3.3 \cos \pi^2 x$

40. $y = 2.5 \sin \dfrac{2x}{\pi}$

In Exercises 41–44, the period is given for a function of the form $y = \sin bx$. Write the function corresponding to the given period.

41. $\dfrac{\pi}{3}$

42. $\dfrac{2\pi}{5}$

43. 2

44. 6

In Exercises 45–48, sketch the indicated graphs.

45. The standard electric voltage in a 60-Hz alternating-current circuit is given by $V = 170 \sin 120\pi t$, where t is the time in seconds. Sketch the graph of V as a function of t for $0 \le t \le 0.05$ s.

46. To tune the instruments of an orchestra before a concert, an A note is struck on a piano. The piano wire vibrates with a displacement y (in mm) given by $y = 3.20 \cos 880\pi t$, where t is in seconds. Sketch the graph of y vs. t for $0 \le t \le 0.01$ s.

47. The velocity v (in in./s) of a piston is $v = 450 \cos 3600t$, where t is in seconds. Sketch the graph of v vs. t for $0 \le t \le 0.006$ s.

48. The displacement y (in m) of the end of a robot arm for welding is $y = 12.75 \sin 0.419t$, where t is in seconds. Sketch the graph of y as a function of t for $0 \le t \le 15$ s.

In Exercises 49–52, the graph of a function of the form $y = a \sin bx$ or $y = a \cos bx$ is shown. Determine the specific function for each.

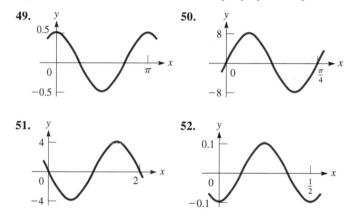

49.

50.

51.

52.

$10\text{-}3$ GRAPHS OF $y = a \sin(bx + c)$ AND $y = a \cos(bx + c)$

Another important quantity in graphing the sine and cosine functions is the *phase angle*. In the function $y = a \sin(bx + c)$, c *represents the* **phase angle**. Its meaning is illustrated in the following example.

■EXAMPLE 1 Sketch the graph of $y = \sin(2x + \frac{\pi}{4})$.

Note that $c = \pi/4$. Therefore, in order to obtain values for the table, we assume a value for x, multiply it by 2, add $\pi/4$ to this value, and then find the sine of the result. The values that are shown are those for which $2x + \pi/4 = 0$, $\pi/4$, $\pi/2$, $3\pi/4$, π, and so on, which are the important values for $y = \sin 2x$.

x	$-\frac{\pi}{8}$	0	$\frac{\pi}{8}$	$\frac{\pi}{4}$	$\frac{3\pi}{8}$	$\frac{\pi}{2}$	$\frac{5\pi}{8}$	$\frac{3\pi}{4}$	$\frac{7\pi}{8}$	π
y	0	0.7	1	0.7	0	-0.7	-1	-0.7	0	0.7

When we solve $2x + \pi/4 = 0$, we get $x = -\pi/8$, and this gives $y = \sin 0 = 0$. The other values for y are found in the same way. The graph is shown in Fig. 10-13. ■

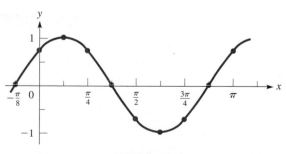

Fig. 10-13

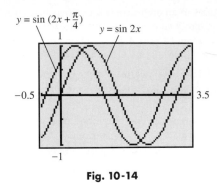

Fig. 10-14

See Appendix C for a graphing calculator program SINECURV. It displays the graphs of $y = \sin x$, $y = 2 \sin x$, $y = \sin 2x$, and $y = 2 \sin (2x - \pi/3)$.

Carefully note the difference between $y = \sin (bx + c)$ and $y = \sin bx + c$. Writing $\sin (bx + c)$ means to find the sine of the quantity of $bx + c$, whereas $\sin bx + c$ means to find the sine of bx and then add the value c.

We see from Example 1 that *the graph of $y = \sin (2x + \frac{\pi}{4})$ is precisely the same as the graph of $y = \sin 2x$, except that it is **shifted** $\pi/8$ units to the left*. In Fig. 10-14 a graphing calculator view shows the graphs of $y = \sin 2x$ and $y = \sin (2x + \frac{\pi}{4})$. We see that the shapes are the same and that the graph of $y = \sin (2x + \frac{\pi}{4})$ is about 0.4 unit ($\pi/8 \approx 0.39$) to the left of the graph of $y = \sin 2x$.

In general, the effect of c in the equation $y = a \sin (bx + c)$ is to shift the curve of $y = a \sin bx$ to the left if $c > 0$, or shift the curve to the right if $c < 0$. The amount of this shift is given by $-c/b$. Due to its importance in sketching curves, *the quantity $-c/b$ is called the* **displacement** (*or* **phase shift**).

We can see the reason that the displacement is $-c/b$ by noting corresponding points on the graphs of $y = \sin bx$ and $y = \sin (bx + c)$. For $y = \sin bx$, when $x = 0$, then $y = 0$. For $y = \sin (bx + c)$, when $x = -c/b$, then $y = 0$. The point $(-c/b, 0)$ on the graph of $y = \sin (bx + c)$ is $-c/b$ units to the left of the point $(0, 0)$ on the graph of $y = \sin x$. In Fig. 10-14, $-c/b = -\pi/8$.

Therefore, we use the displacement combined with the amplitude and the period along with the other information from the previous sections to sketch curves of the functions $y = a \sin (bx + c)$ and $y = a \cos (bx + c)$, where $b > 0$.

Important Quantities to Determine for Sketching Graphs of
$y = a \sin(bx + c)$ **and** $y = a \cos(bx + c)$

$$\text{Amplitude} = |a|$$
$$\text{Period} = \frac{2\pi}{b} \qquad\qquad\qquad (10\text{-}1)$$
$$\text{Displacement} = -\frac{c}{b}$$

By use of these quantities and the one-fourth period distance, the graphs of the sine and cosine functions can be readily sketched. A general illustration of the graph of $y = a \sin (bx + c)$ is shown in Fig. 10-15. Note that

CAUTION ▶

the displacement is **negative** *(to the left) for $c > 0$*   (Fig. 10-15(a))
the displacement is **positive** *(to the right) for $c < 0$*   (Fig. 10-15(b))

Note that we can find the displacement for the graphs of $y = a \sin (bx + c)$ by solving $bx + c = 0$ for x. We see that $x = -c/b$.

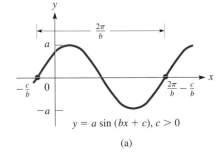

(a)

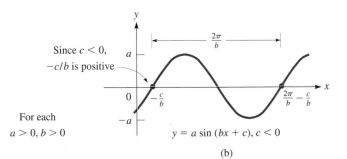

(b)

Fig. 10-15

EXAMPLE 2   Sketch the graph of $y = 2 \sin(3x - \pi)$.

First, we note that $a = 2$, $b = 3$, and $c = -\pi$. Therefore, the amplitude is 2, the period is $2\pi/3$, and the displacement is $-(-\pi/3) = \pi/3$. (We can also get the displacement from $3x - \pi = 0$, $x = \pi/3$.)

We see that the curve "starts" at $x = \pi/3$ and starts repeating $2\pi/3$ units to the right of this point. Be sure to grasp this point well. *The period tells us the number of units along the x-axis between such corresponding points.* One-fourth of the period is $\frac{1}{4}(\frac{2\pi}{3}) = \frac{\pi}{6}$.

Important values are at $\frac{\pi}{3}$, $\frac{\pi}{3} + \frac{\pi}{6} = \frac{\pi}{2}$, $\frac{\pi}{3} + 2(\frac{\pi}{6}) = \frac{2\pi}{3}$, and so on. We now make the table of important values and sketch the graph shown in Fig. 10-16.

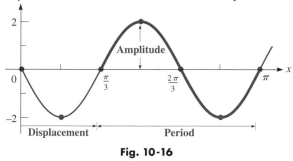

Fig. 10-16

x	0	$\frac{\pi}{6}$	$\frac{\pi}{3}$	$\frac{\pi}{2}$	$\frac{2\pi}{3}$	$\frac{5\pi}{6}$	π
y	0	-2	0	2	0	-2	0

We note that since the period is $2\pi/3$, the curve passes through the origin. ∎

EXAMPLE 3   Sketch the graph of the function $y = -\cos(2x + \frac{\pi}{6})$.

First we determine that

(1) the amplitude is 1

(2) the period is $\frac{2\pi}{2} = \pi$

(3) the displacement is $-\frac{\pi}{6} \div 2 = -\frac{\pi}{12}$

We now make a table of important values, noting that the curve starts repeating π units to the right of $-\frac{\pi}{12}$.

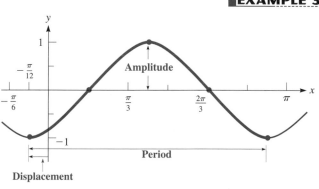

Fig. 10-17

x	$-\frac{\pi}{12}$	$\frac{\pi}{6}$	$\frac{5\pi}{12}$	$\frac{2\pi}{3}$	$\frac{11\pi}{12}$
y	-1	0	1	0	-1

From this table we sketch the graph in Fig. 10-17. ∎

Each of the heavy portions of the graphs in Figs. 10-16 and 10-17 is called a *cycle* of the curve. *A **cycle** is any section of the graph that includes exactly one period.*

EXAMPLE 4   View the graph of $y = 2 \cos(\frac{1}{2}x - \frac{\pi}{6})$ on a graphing calculator. From the values $a = 2$, $b = 1/2$, and $c = -\pi/6$, we determine that

(1) the amplitude is 2

(2) the period is $2\pi \div \frac{1}{2} = 4\pi$

(3) the displacement is $-(-\frac{\pi}{6}) \div \frac{1}{2} = \frac{\pi}{3}$

We now make a table of important values.

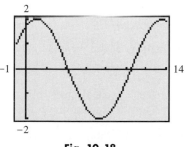

Fig. 10-18

x	$\frac{\pi}{3}$	$\frac{4\pi}{3}$	$\frac{7\pi}{3}$	$\frac{10\pi}{3}$	$\frac{13\pi}{3}$
y	2	0	-2	0	2

This table helps us choose the values for the *window* settings in Fig. 10-18. We choose Xmin $= -1$ in order to start to the left of the y-axis and Xmax $= 14$ since $13\pi/3 \approx 13.6$. Also, we choose Ymin $= -2$ and Ymax $= 2$ since the amplitude is 2. We see that the graph in Fig. 10-18 is a little more than one cycle. ∎

The following example illustrates the use of the graph of a trigonometric function in an applied problem.

■EXAMPLE 5   The cross section of a certain water wave is $y = 0.7 \sin\left(\frac{\pi}{2}x + \frac{\pi}{4}\right)$, where x and y are measured in feet. Display two cycles of y vs. x on a graphing calculator.

From the values $a = 0.7$ ft, $b = \pi/2$ ft^{-1} (this means 1/ft, or per foot), and $c = \pi/4$, we can find the amplitude, period, and displacement:

(1) amplitude $= 0.7$ ft

(2) period $= \dfrac{2\pi}{\frac{\pi}{2}} = 4$ ft

(3) displacement $= -\dfrac{\frac{\pi}{4}}{\frac{\pi}{2}} = -0.5$ ft

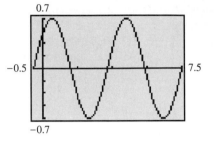

Fig. 10-19

Using these values, we choose the following values for the *window* settings.

(1) Xmin $= -0.5$ (the displacement is -0.5 ft)

(2) Xmax $= 7.5$ (the period is 4 ft, and we want two periods, starting at $x = -0.5$)($-0.5 + 8 = 7.5$)

(3) Ymin $= -0.7$, Ymax $= 0.7$ (the amplitude is 0.7 ft)

The graphing calculator view is shown in Fig. 10-19. The negative values of x have the significance of giving points to the wave to the left of the origin. (When *time* is used, no actual physical meaning is generally given to negative values of t.) ■

EXERCISES 10-3

In Exercises 1–24, determine the amplitude, period, and displacement for each function. Then sketch the graphs of the functions. Check each using a graphing calculator.

1. $y = \sin\left(x - \dfrac{\pi}{6}\right)$

2. $y = 3\sin\left(x + \dfrac{\pi}{4}\right)$

3. $y = \cos\left(x + \dfrac{\pi}{6}\right)$

4. $y = 2\cos\left(x - \dfrac{\pi}{8}\right)$

5. $y = 2\sin\left(2x + \dfrac{\pi}{2}\right)$

6. $y = -\sin\left(3x - \dfrac{\pi}{2}\right)$

7. $y = -\cos(2x - \pi)$

8. $y = 4\cos\left(3x + \dfrac{\pi}{3}\right)$

9. $y = \dfrac{1}{2}\sin\left(\dfrac{1}{2}x - \dfrac{\pi}{4}\right)$

10. $y = 2\sin\left(\dfrac{1}{4}x + \dfrac{\pi}{2}\right)$

11. $y = 3\cos\left(\dfrac{1}{3}x + \dfrac{\pi}{3}\right)$

12. $y = \dfrac{1}{3}\cos\left(\dfrac{1}{2}x - \dfrac{\pi}{8}\right)$

13. $y = \sin\left(\pi x + \dfrac{\pi}{8}\right)$

14. $y = -2\sin(2\pi x - \pi)$

15. $y = \dfrac{3}{4}\cos\left(4\pi x - \dfrac{\pi}{5}\right)$

16. $y = 25\cos\left(3\pi x + \dfrac{\pi}{2}\right)$

17. $y = -0.6\sin(2\pi x - 1)$

18. $y = 1.8\sin\left(\pi x + \dfrac{1}{3}\right)$

19. $y = 40\cos(3\pi x + 2)$

20. $y = 3\cos(6\pi x - 1)$

21. $y = \sin(\pi^2 x - \pi)$

22. $y = -\dfrac{1}{2}\sin\left(2x - \dfrac{1}{\pi}\right)$

23. $y = -\dfrac{3}{2}\cos\left(\pi x + \dfrac{\pi^2}{6}\right)$

24. $y = \pi\cos\left(\dfrac{1}{\pi}x + \dfrac{1}{3}\right)$

In Exercises 25 and 26, sketch the indicated curves. In Exercises 27 and 28, use a graphing calculator to view the indicated curves.

25. A wave traveling in a string may be represented by the equation $y = A\sin 2\pi\left(\dfrac{t}{T} - \dfrac{x}{\lambda}\right)$. Here, A is the amplitude, t is the time the wave has traveled, x is the distance from the origin, T is the time required for the wave to travel one *wavelength* λ (the Greek letter lambda). Sketch three cycles of the wave for which $A = 2.00$ cm, $T = 0.100$ s, $\lambda = 20.0$ cm, and $x = 5.00$ cm.

26. The electric current i (in μA) in a certain circuit is given by $i = 3.8\cos 2\pi(t + 0.20)$, where t is the time in seconds. Sketch three cycles of this function.

27. A certain satellite circles the earth such that its distance y, in miles north or south (altitude is not considered) from the equator, is $y = 4500 \cos{(0.025t - 0.25)}$, where t is the time (in min) after launch. View two cycles of the graph.

28. In performing a test on a patient, a medical technician used an ultrasonic signal given by the equation $I = A \sin{(\omega t + \theta)}$. View two cycles of the graph of I vs. t if $A = 5$ nW/m^2, $\omega = 2 \times 10^5$ rad/s, and $\theta = 0.4$.

Ⓦ *In Exercises 29–32, give the specific form of the equation by evaluating a, b, and c through an inspection of the given curve. Explain how a, b, and c are found.*

29. $y = a \sin{(bx + c)}$
   Fig. 10-20

30. $y = a \cos{(bx + c)}$
   Fig. 10-20

31. $y = a \cos{(bx + c)}$
   Fig. 10-21

32. $y = a \sin{(bx + c)}$
   Fig. 10-21

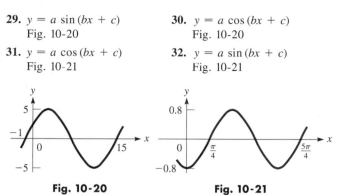

Fig. 10-20   **Fig. 10-21**

10-4   GRAPHS OF $y = \tan x$, $y = \cot x$, $y = \sec x$, $y = \csc x$

In this section we briefly consider the graphs of the other trigonometric functions. We show the basic form of each curve, and then from these we are able to sketch other curves for these functions.

Considering the values of the trigonometric functions we found in Chapter 8, we set up the following table for $y = \tan x$. The graph is shown in Fig. 10-22.

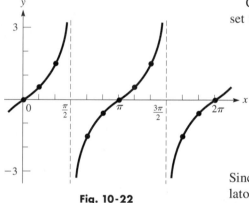

Fig. 10-22

x	0	$\frac{\pi}{6}$	$\frac{\pi}{3}$	$\frac{\pi}{2}$	$\frac{2\pi}{3}$	$\frac{5\pi}{6}$	π
y	0	0.6	1.7	*	-1.7	-0.6	0

x	$\frac{7\pi}{6}$	$\frac{4\pi}{3}$	$\frac{3\pi}{2}$	$\frac{5\pi}{3}$	$\frac{11\pi}{6}$	2π
y	0.6	1.7	*	-1.7	-0.6	0

*Undefined.

NOTE▶

Since the curve is not defined for $x = \pi/2$, $x = 3\pi/2$, and so forth, we use a calculator and find that the value of $\tan x$ becomes very large as x gets closer to $\pi/2$, although *there is no point on the curve for* $x = \pi/2$. For example, if $x = 1.55$, $\tan x = 48.08$ ($\pi/2 \approx 1.57$). We note that *the period of the tangent curve is π.* This differs from the period of the sine and cosine functions

By knowing the values of $\sin x$, $\cos x$, and $\tan x$, we can find the necessary values of $\csc x$, $\sec x$, and $\cot x$. This is due to the reciprocal relationships among the functions that we showed in Section 4-3. We show these relationships by

$$\csc x = \frac{1}{\sin x} \qquad \sec x = \frac{1}{\cos x} \qquad \cot x = \frac{1}{\tan x} \qquad \textbf{(10-2)}$$

Therefore, to graph $y = \cot x$, $y = \sec x$, and $y = \csc x$, we can obtain the necessary values from the corresponding reciprocal function. On the following page we show the graphs of these functions, as well as a more extensive graph of $y = \tan x$.

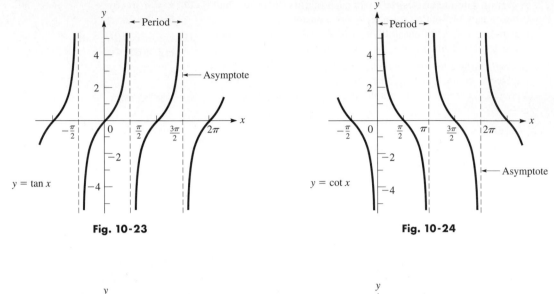

Fig. 10-23   Fig. 10-24

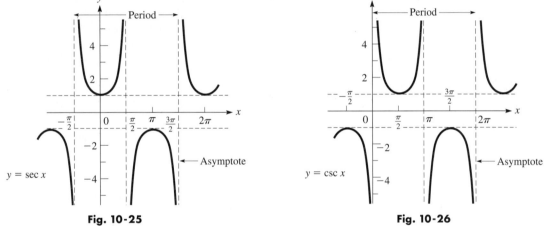

Fig. 10-25   Fig. 10-26

We see from these graphs that the period of $y = \tan x$ and $y = \cot x$ is π and that the period of $y = \sec x$ and $y = \csc x$ is 2π. *The vertical dashed lines in these figures are* **asymptotes** (see Sections 3-4 and 21-6). The curves *approach* these lines, but they never actually touch them.

The functions are not defined for the values of x for which the curve has asymptotes. This means that the domains do not include these values of x. Thus, we see that the domains of $y = \tan x$ and $y = \sec x$ include all real numbers, except the values $x = -\pi/2$, $\pi/2$, $3\pi/2$, and so on. The domain of $y = \cot x$ and $y = \csc x$ include all real numbers except $x = -\pi$, 0, π, 2π, and so on.

From the graphs we see that the ranges of $y = \tan x$ and $y = \cot x$ are all real numbers, but that the ranges of $y = \sec x$ and $y = \csc x$ do not include the real numbers between -1 and 1.

To sketch functions such as $y = a \sec x$, we may first sketch $y = \sec x$ and then multiply the y-values by a. *Here a is not an amplitude,* since the ranges of these functions are not limited in the same way they are for the sine and cosine functions.

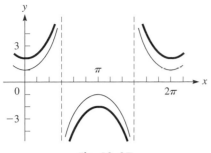

Fig. 10-27

■EXAMPLE 1   Sketch the graph of $y = 2 \sec x$.

First, we sketch in $y = \sec x$, shown as the light curve in Fig. 10-27. Then we multiply the y-values of this secant function by 2. Although we can only estimate these values and do this approximately, a reasonable graph can be sketched this way. The desired curve is shown in color in Fig. 10-27.   ----------■

Using a graphing calculator we can display the graphs of these functions more easily and more accurately than by sketching them. By knowing the general shape and period of the function, the values for the *window* settings can be determined without having to reset them too often.

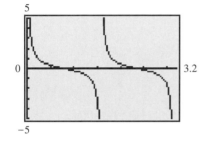

Fig. 10-28

■EXAMPLE 2   View at least two cycles of the graph of $y = 0.5 \cot 2x$ on a graphing calculator.

Since the period of $y = \cot x$ is π, the period of $y = \cot 2x$ is $\pi/2$. Therefore, we choose the *window* settings as follows:

Xmin = 0   ($x = 0$ is one asymptote of the curve)

Xmax = 3.2   ($\pi \approx 3.14$; the period is $\pi/2$; two periods is π)

Ymin = -5, Ymax = 5   (the range is all x; this shows enough of the curve)

We must remember to enter the function as $y_1 = 0.5 \, (\tan 2x)^{-1}$, since $\cot x = (\tan x)^{-1}$. The graphing calculator view is shown in Fig. 10-28. We can view many more cycles of the curve with appropriate *window* settings.   ----------■

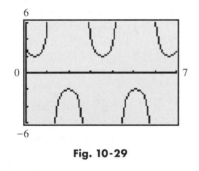

Fig. 10-29

■EXAMPLE 3   View at least two periods of the graph of $y = 2 \sec (2x - \frac{\pi}{4})$ on a graphing calculator.

Since the period of $\sec x$ is 2π, the period of $\sec (2x - \frac{\pi}{4})$ is $2\pi/2 = \pi$. Recalling that $\sec x = (\cos x)^{-1}$, the curve will have the same displacement as $y = \cos (2x - \frac{\pi}{4})$. This displacement is $-\frac{-\pi/4}{2} = \frac{\pi}{8}$. Therefore, we choose the following *window* settings.

Xmin = 0   (the displacement is positive)

Xmax = 7   (displacement = $\pi/8$; period = π; $\pi/8 + 2\pi = 17\pi/8 \approx 6.7$)

Ymin = -6, Ymax = 6   (there is no curve between $y = -2$ and $y = 2$)

With $y_1 = 2(\cos (2x - \pi/4))^{-1}$, Fig. 10-29 shows the calculator view.   ----------■

━━━━━━━━ EXERCISES *10-4* ━━━━━━━━

In Exercises 1–4, fill in the following table for each function and plot the graph from these points.

x	$-\frac{\pi}{2}$	$-\frac{\pi}{3}$	$-\frac{\pi}{4}$	$-\frac{\pi}{6}$	0	$\frac{\pi}{6}$	$\frac{\pi}{4}$	$\frac{\pi}{3}$	$\frac{\pi}{2}$	$\frac{2\pi}{3}$	$\frac{3\pi}{4}$	$\frac{5\pi}{6}$	π
y													

1. $y = \tan x$ **2.** $y = \cot x$ **3.** $y = \sec x$ **4.** $y = \csc x$

In Exercises 5–12, sketch the graphs of the given functions by use of the basic curve forms (Figs. 10-23, 10-24, 10-25, and 10-26). See Example 1.

5. $y = 2 \tan x$ **6.** $y = 3 \cot x$

7. $y = \frac{1}{2} \sec x$ **8.** $y = \frac{3}{2} \csc x$

9. $y = -2 \cot x$ **10.** $y = -0.1 \tan x$

11. $y = -3 \csc x$ **12.** $y = -\frac{1}{2} \sec x$

In Exercises 13–20, view at least two cycles of the graphs of the given functions on a graphing calculator.

13. $y = \tan 2x$ **14.** $y = 2 \cot 3x$

15. $y = \frac{1}{2} \sec 3x$ **16.** $y = 4 \csc 2x$

17. $y = 2 \cot \left(2x + \dfrac{\pi}{6} \right)$ **18.** $y = \tan \left(3x - \dfrac{\pi}{2} \right)$

19. $y = 18 \csc \left(3x - \dfrac{\pi}{3} \right)$ **20.** $y = 3 \sec \left(2x + \dfrac{\pi}{4} \right)$

In Exercises 21–24, sketch the appropriate graphs. Check each on a graphing calculator.

21. A drafting student draws a circle through the three vertices of a right triangle. The hypotenuse of the triangle is the diameter d of the circle, and from Fig. 10-30, we see that $d = a \sec \theta$. Sketch the graph of d as a function of θ for $a = 3.00$ in.

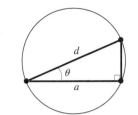

Fig. 10-30

22. Near Antarctica, an iceberg with a vertical face 200 m high is seen from a small boat. At a distance x from the iceberg, the angle of elevation θ of the top of the iceberg can be found from the equation $x = 200 \cot \theta$. Sketch x as a function of θ.

23. A mechanism with two springs is shown in Fig. 10-31, where point A is restricted to move horizontally. From the law of sines we see that $b = (a \sin B) \csc A$. Sketch the graph of b as a function of A for $a = 4.00$ cm and $B = \pi/4$.

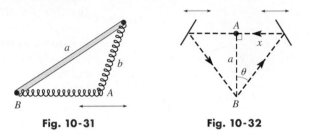

Fig. 10-31   **Fig. 10-32**

24. In a laser experiment, two mirrors move horizontally in equal and opposite distances from point A. The laser path from and to point B is shown in Fig. 10-32. From the figure we see that $x = a \tan \theta$. Sketch the graph of $x = f(\theta)$ for $a = 5.00$ cm.

$10\text{-}5$ APPLICATIONS OF THE TRIGONOMETRIC GRAPHS

In this section we introduce an important physical concept and indicate some of the technical applications.

In Section 8-4 we discussed the velocity of an object moving in a circular path. When this object moves with constant velocity, its *projection* on a diameter moves with what is known as **simple harmonic motion.** For example, consider an object moving around a circle in a plane parallel to rays of light. The movement of the object's shadow on a wall (perpendicular to the light) is simple harmonic motion. The object could be the end of a spoke of a rotating wheel.

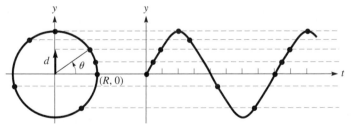

Fig. 10-33

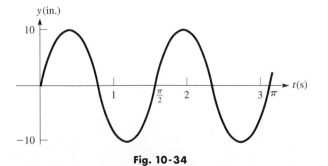

Fig. 10-34

■EXAMPLE 1   In Fig. 10-33, assume a particle starts at the end of the radius at $(R, 0)$, and moves counterclockwise around the circle with constant angular velocity ω. *The displacement of the projection on the y-axis is d and is given by $d = R \sin \theta$.* The displacement is shown for a few different positions of the end of the radius.

Since $\theta/t = \omega$, or $\theta = \omega t$, we have

$$\boxed{d = R \sin \omega t} \qquad \textbf{(10-3)}$$

as the equation for the displacement of this projection, with time t as the independent variable.

For the case where $R = 10.0$ in. and $\omega = 4.00$ rad/s, we have

$$d = 10.0 \sin 4.00t$$

By sketching or viewing the graph of this function, we can find the displacement d of the projection for a given time t. The graph is shown in Fig. 10-34.   ■

In Example 1, note that *time is the independent variable.* This is motion for which the object (the end of the projection) remains at the same horizontal position ($x = 0$) and moves only vertically according to a sinusoidal function. In the previous sections, we dealt with functions in which y is a sinusoidal function of the horizontal displacement x. Think of a water wave. At *one point* of the wave, the motion is only vertical and sinusoidal with time. At *one given time,* a picture would indicate a sinusoidal movement from one horizontal position to the next.

EXAMPLE 2   A windmill is used to pump water. The radius of the blade is 2.5 m, and it is moving with constant angular velocity. If the vertical displacement of the end of the blade is timed from the point it is at an angle of 45° ($\pi/4$ rad) from the horizontal [see Fig. 10-35(a)], the displacement d is given by

$$d = 2.5 \sin\left(\omega t + \frac{\pi}{4}\right)$$

If the blade makes an angle of 90° ($\pi/2$ rad) when $t = 0$ (see Fig. 10-35(b)), the displacement d is given by

$$d = 2.5 \sin\left(\omega t + \frac{\pi}{2}\right)$$

or $d = 2.5 \cos \omega t$

If timing started at the first maximum for the displacement, the resulting curve for the displacement would be that of the cosine function.

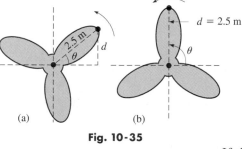

(a)   (b)

Fig. 10-35

Other examples of simple harmonic motion are (1) the movement of a pendulum bob through its arc (a very close approximation to simple harmonic motion), (2) the motion of an object "bobbing" in water, (3) the movement of the end of a vibrating rod (which we hear as sound), and (4) the displacement of a weight moving up and down on a spring. Other phenomena that give rise to equations like those for simple harmonic motion are found in the fields of optics, sound, and electricity. The equations for such phenomena have the same mathematical form because they result from vibratory movement or motion in a circle.

EXAMPLE 3   A very important use of the trigonometric curves arises in the study of alternating current, which is caused by the motion of a wire passing through a magnetic field. If the wire is moving in a circular path, with angular velocity ω, the current i in the wire at time t is given by an equation of the form

$$i = I_m \sin(\omega t + \alpha)$$

where I_m is the maximum current attainable and α is the phase angle.

The current may be represented by a sinusoidal wave. Given that $I_m = 6.00$ A, $\omega = 120\pi$ rad/s, and $\alpha = \pi/6$, we have the equation

$$i = 6.00 \sin\left(120\pi t + \frac{\pi}{6}\right)$$

From this equation we see that the amplitude is 6.00 A, the period is $\frac{1}{60}$ s, and the displacement is $-\frac{1}{720}$ s. From these values we draw the graph as shown in Fig. 10-36. Since the current takes on both positive and negative values, we conclude that it moves alternately in one direction and then the other.

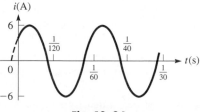

Fig. 10-36

Named for the German physicist Heinrich
Hertz (1857–1894).

It is a common practice to express the rate of rotation in terms of *the* **frequency**
f, the number of cycles per second, rather than directly in terms of the angular
velocity ω, the number of radians per second. *The unit for frequency is the*
hertz (Hz), *and* 1 Hz = 1 cycle/s. Since there are 2π rad in one cycle, we have

$$\omega = 2\pi f \qquad\qquad (10\text{-}4)$$

It is the frequency *f* that is referred to in electric current, on radio stations, for musical tones, and so on.

■EXAMPLE 4   For the electric current in Example 3, $\omega = 120\pi$ rad/s. The
corresponding frequency *f* is

$$f = \frac{120\pi}{2\pi} = 60 \text{ Hz}$$

This means that 120π rad/s corresponds to 60 cycles/s. This is the standard frequency used for alternating current.

EXERCISES *10-5*

A graphing calculator may be used in the following exercises.

*In Exercises 1 and 2, sketch two cycles of the curve of the projec-
tion of Example 1 as a function of time for the given values.*

1. $R = 2.40$ cm, $\omega = 2.00$ rad/s

2. $R = 1.80$ ft, $f = 0.250$ Hz

*In Exercises 3 and 4, a point on a cam is 8.30 cm from the center of
rotation. The cam is rotating with a constant angular velocity, and
the vertical displacement d = 8.30 cm for t = 0 s. See Fig. 10-37.
Sketch two cycles of d as a function of t for the given values.*

3. $f = 3.20$ Hz

4. $\omega = 3.20$ rad/s

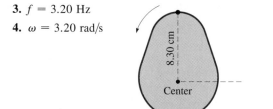

Center

Fig. 10-37

In Exercises 5 and 6, a satellite is orbiting the earth such that its
displacement D north of the equator (or south if D < 0) is given by
D = A sin $(\omega t + \alpha)$. *Sketch two cycles of D as a function of t for*
the given values.

5. $A = 500$ mi, $\omega = 3.60$ rad/h, $\alpha = 0$

6. $A = 850$ km, $f = 1.6 \times 10^{-4}$ Hz, $\alpha = \pi/3$

*In Exercises 7 and 8, for an alternating-current circuit in which
the voltage e is given by e = E* sin$(\omega t + \alpha)$, *sketch two cycles of
the voltage as a function of time for the given values.*

7. $E = 170$ V, $f = 60.0$ Hz, $\alpha = -\pi/3$

8. $E = 80$ V, $\omega = 377$ rad/s, $\alpha = \pi/2$

*In Exercises 9 and 10, refer to the wave in the string described in
Exercise 25 of Section 10-3. For a point on the string, the displace-
ment y is given by $y = A \sin 2\pi\left(\dfrac{t}{T} - \dfrac{x}{\lambda}\right)$. We see that each point
on the string moves with simple harmonic motion. Sketch two cy-
cles of y as a function of t for the given values.*

9. $A = 3.20$ cm, $T = 0.050$ s, $\lambda = 40.0$ cm, $x = 5.00$ cm

10. $A = 1.45$ in., $T = 0.250$ s, $\lambda = 24.0$ in., $x = 20.0$ in.

*In Exercises 11 and 12, the air pressure within a plastic container
changes above and below the external atmospheric pressure by
$p = p_0 \sin 2\pi ft$. Sketch two cycles of $p = f(t)$ for the given values.*

11. $p_0 = 2.80$ lb/in.^{2}, $f = 2.30$ Hz

12. $p_0 = 45.0$ kPa, $f = 0.450$ Hz

In Exercises 13–16, sketch the required curves.

13. The vertical displacement *y* of a point at the end of a propeller
blade of a small boat is $y = 14.0 \sin 40.0\pi t$. Sketch two cycles
of *y* (in cm) as a function of *t* (in s).

(W) 14. The rotating beacon of a parked police car is 12 m from a
straight wall. (a) Sketch the graph of the length *L* of the light
beam, where $L = 12 \sec \pi t$, for $0 \le t \le 2.0$ s. (b) Which
part(s) of the graph show meaningful values? Explain.

15. Sketch two cycles of the radio signal
$e = 0.014 \cos(2\pi ft + \pi/4)$ (*e* in volts, *f* in hertz, and *t* in seconds) for a station broadcasting with $f = 950$ kHz ("95" on
the AM radio dial).

16. Sketch two cycles of the acoustical intensity *I* of the sound
wave for which $I = A \cos(2\pi ft - \alpha)$, given that *t* is in seconds, $A = 0.027$ W/cm^2, $f = 240$ Hz, and $\alpha = 0.80$.

10-6 COMPOSITE TRIGONOMETRIC CURVES

Many applications involve functions that in themselves are a combination of two or more simpler functions. In this section we discuss methods by which the curve of such a function can be found by combining values from the simpler functions.

■**EXAMPLE 1**   Sketch the graph of $y = 2 + \sin 2x$.

This function is the sum of the simpler functions $y_1 = 2$ and $y_2 = \sin 2x$. We may find values for y by adding 2 to each important value of $y_2 = \sin 2x$.

For $y_2 = \sin 2x$, the amplitude is 1 and the period is $2\pi/2 = \pi$. Therefore, we obtain the values in the following table and sketch the graph in Fig. 10-38.

x	0	$\frac{\pi}{4}$	$\frac{\pi}{2}$	$\frac{3\pi}{4}$	π
$\sin 2x$	0	1	0	-1	0
$2 + \sin 2x$	2	3	2	1	2

y

3
2
1

0   $\frac{\pi}{4}$ $\frac{\pi}{2}$ $\frac{3\pi}{4}$ π x
−1

Fig. 10-38

Addition of Ordinates

Another way to sketch the resulting graph is to *first sketch the two simpler curves and then add the y-values graphically. This method is called* **addition of ordinates** and is illustrated in the following example.

■**EXAMPLE 2**   Sketch the graph of $y = 2 \cos x + \sin 2x$.

On the same set of coordinate axes, we sketch the curves $y = 2 \cos x$ and $y = \sin 2x$. These are shown as dashed and solid light curves in Fig. 10-39. For various values of x, we determine the distance above or below the x-axis of each curve and add these distances, noting that those above the axis are positive and those below the axis are negative. We thereby *graphically* ***add*** *the y-values* of these two curves for these values of x to obtain the points on the resulting curve, shown in color in Fig. 10-39.

For example, for the x-value at A we add the two lengths shown (side-by-side for clarity) to get the length for y. At B we see that both lengths are negative, and the value for y is the sum of these two negative values. At C one length is positive and one is negative, and we must subtract the lower length from the upper one to get the length for y.

We add (or subtract) these lengths for enough x-values to get a proper curve. Some points are easily found. Where one curve crosses the x-axis, its value is zero, and the resulting curve has its point on the other curve for this value of x. Here, $\sin 2x$ is zero at $x = \pi/2$, π, and so forth. For these values, the points on the resulting curve lie on the curve of $2 \cos x$.

We should also add the values where each curve is at its maximum or its minimum. In this case, $\sin 2x = 1$ at $x = \pi/4$, and the two values should be added here to get a point on the resulting curve.

At $x = 5\pi/4$, *we must take extra care in combining values, since*

$\sin 2x$ is **positive** and $2 \cos x$ is **negative.**

Reasonable care and accuracy are needed to sketch a proper resulting curve. The graph is shown in Fig. 10-39.

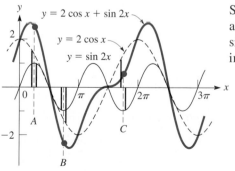

y

$y = 2 \cos x + \sin 2x$
$y = 2 \cos x$
$y = \sin 2x$

2

0   π 2π 3π x

−2

A B C

Fig. 10-39

We have shown how a fairly complex curve can be sketched graphically. However, it is expected that a graphing calculator (or a computer *grapher*) will be used to view most graphs, particularly ones that are difficult to sketch. The graphing calculator can display such curves much more easily and with much greater accuracy. We can use information about the amplitude, period, and displacement in choosing values for the *window* feature on the calculator.

■**EXAMPLE 3**   Use a graphing calculator to display the graph of $y = \frac{x}{2} - \cos x$.

Here, we note that the curve is a combination of the straight line $y = x/2$ and the trigonometric curve $y = \cos x$. There are several good choices for the *window* settings, depending on how much of the curve is to be viewed. To see a little more than one period of $\cos x$, we can make the following choices:

Xmin $= -1$ (to start to the left of the y-axis)

Xmax $= 7$ (the period of $\cos x$ is $2\pi \approx 6.3$)

Ymin $= -2$ (the line passes through $(0, 0)$; the amplitude of $y = \cos x$ is 1)

Ymax $= 4$ (the slope of the line is $1/2$)

The graphing calculator view of the curve is shown in Fig. 10-40(a). The graph of $y = \frac{x}{2} - \cos x$, $y = \frac{x}{2}$, and $y = -\cos x$ is shown in Fig. 10-40(b).

Fig. 10-40   (a)

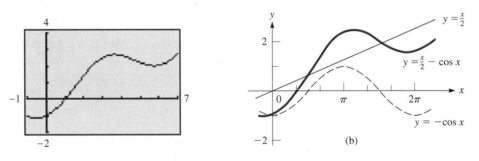

(b)

NOTE ▶   The reason for showing $y = -\cos x$, and not $y = \cos x$, is that if *addition of ordinates* were being used, *it is much easier to add graphic values than to subtract them*. In using the method of addition of ordinates, we could *add* the ordinates of $y = x/2$ and $y = -\cos x$ to get the resulting curve of $y = \frac{x}{2} - \cos x$. ----------■

■**EXAMPLE 4**   View the graph of $y = \cos \pi x - 2 \sin 2x$ on a graphing calculator.

The combination of $y = \cos \pi x$ and $y = 2 \sin 2x$ leads to the following choices for the *window* settings:

Xmin $= -1$ (to start to the left of the y-axis)

Xmax $= 7$ (the periods are 2 and π; this shows at least two periods of each)

Ymin $= -3$, Ymax $= 3$ (the sum of the amplitudes is 3)

There are many possible choices for Xmin and Xmax to get a good view of the graph on a calculator. However, since the sum of the amplitudes is 3, we know that the curve cannot be below $y = -3$ or above $y = 3$.

The graphing calculator view is shown in Fig. 10-41.

This graph can be constructed by using addition of ordinates, although it is difficult to do very accurately. ----------■

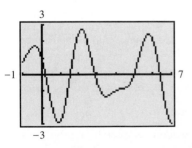

Fig. 10-41

Lissajous Figures

An important application of trigonometric curves is made when they are added at *right angles.* The methods for doing this are shown in the following examples.

EXAMPLE 5   Plot the graph for which the values of x and y are given by the equations $y = \sin 2\pi t$ and $x = 2 \cos \pi t$. *Equations given in this form, x and y in terms of a third variable, are called* **parametric equations.**

Since both x and y are in terms of t, by assuming values of t we find corresponding values of x and y and use these values to plot the graph. Since the periods of $\sin 2\pi t$ and $2 \cos \pi t$ are $t = 1$ and $t = 2$, respectively, we will use values of $t = 0$, 1/4, 1/2, 3/4, 1, and so on. These give us convenient values of 0, $\pi/4$, $\pi/2$, $3\pi/4$, π, and so on to use in the table. We plot the points in Fig. 10-42.

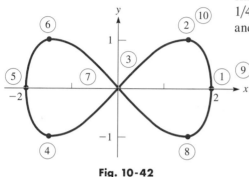

Fig. 10-42

Named for the French physicist Jules Lissajous (1822–1880).

t	0	$\frac{1}{4}$	$\frac{1}{2}$	$\frac{3}{4}$	1	$\frac{5}{4}$	$\frac{3}{2}$	$\frac{7}{4}$	2	$\frac{9}{4}$
x	2	1.4	0	−1.4	−2	−1.4	0	1.4	2	1.4
y	0	1	0	−1	0	1	0	−1	0	1
Point number	1	2	3	4	5	6	7	8	9	10

Since x and y are trigonometric functions of a third variable t and since the x- and y-axes are at right angles, values of x and y obtained in this manner result in a combination of two trigonometric curves at right angles. *Figures obtained in this manner are called* **Lissajous figures.** Note that the Lissajous figure in Fig. 10-42 *is not a function* since there are *two* values of y for each value of x (except $x = -2$, 0, 2) in the domain.

In practice, Lissajous figures can be shown by applying different voltages to an *oscilloscope* and displaying the electric signals on a screen similar to that on a television set.

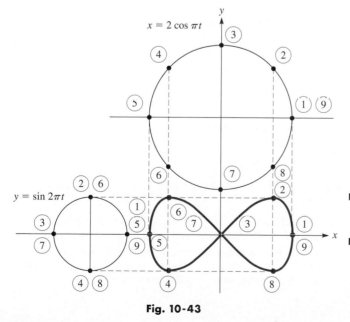

Fig. 10-43

EXAMPLE 6   If we place a circle on the x-axis and another on the y-axis, we may represent the coordinates (x, y) for the curve of Example 5 by the lengths of the projections (see Example 1 of Section 10-5) of a point moving around each circle. A careful study of Fig. 10-43 will clarify this. We note that the radius of the circle giving the x-values is 2 and that the radius of the circle giving the y-values is 1. This is due to the way in which x and y are defined. Also, due to these definitions, the point revolves around the y-circle twice as fast as the corresponding point around the x-circle.

See the chapter introduction.

On an oscilloscope, the curve would result when two electric signals are used. The first would have twice the amplitude and one-half the frequency of the other.

Most graphing calculators can be used to display a curve defined by parametric equations. To do this it is necessary to use the *mode* feature and make the selection for *parametric equations*. Use the manual for the calculator, as there are some differences in how this is done on the various calculators. In the example that follows, we display the graphs of parametric equations on a graphing calculator.

■EXAMPLE 7   Use a graphing calculator to display the graph defined by the parametric equations $x = 2 \cos \pi t$ and $y = \sin 2\pi t$. These are the same equations as those used in Examples 5 and 6.

First we select the parametric equation option from the *mode* feature, and enter the parametric equations $x_{1T} = 2 \cos \pi t$ and $y_{1T} = \sin 2\pi t$. Then we make the following *window* settings:

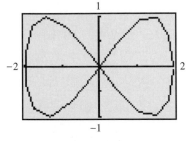

Fig. 10-44

Tmin = 0   (standard default settings, and the usual choice)

Tmax = 2   (the periods are 2 and 1; the longer period is 2)

Tstep = .1047   (standard default setting; curve is smoother with 0.01)

Xmin = −2, Xmax = 2   (smallest and largest possible values of x), Xscl = 1

Ymin = −1, Ymax = 1   (smallest and largest possible values of y), Yscl = 0.5

The calculator graph is shown in Fig. 10-44.   ■

EXERCISES *10-6*

In Exercises 1–8, sketch the curves of the given functions by addition of ordinates.

1. $y = 1 + \sin x$

2. $y = 3 - 2 \cos x$

3. $y = \frac{1}{3}x + \sin 2x$

4. $y = x - \sin x$

5. $y = \frac{1}{10}x^2 - \sin \pi x$

6. $y = \frac{1}{4}x^2 + \cos 3x$

7. $y = \sin x + \cos x$

8. $y = \sin x + \sin 2x$

In Exercises 9–20, display the graphs of the given functions on a graphing calculator.

9. $y = x^3 + 10 \sin 2x$

10. $y = \dfrac{1}{x^2 + 1} - \cos \pi x$

11. $y = \sin x - \sin 2x$

12. $y = \cos 3x - \sin x$

13. $y = 20 \cos 2x + 30 \sin x$

14. $y = \frac{1}{2} \sin 4x + \cos 2x$

15. $y = 2 \sin x - \cos x$

16. $y = 8 \sin 0.5x - 12 \sin x$

17. $y = \sin \pi x - \cos 2x$

18. $y = 2 \cos 4x - \cos\left(x - \dfrac{\pi}{4}\right)$

19. $y = 2 \sin\left(2x - \dfrac{\pi}{6}\right) + \cos\left(2x + \dfrac{\pi}{3}\right)$

20. $y = 3 \cos 2\pi x + \sin \frac{\pi}{2}x$

In Exercises 21–24, plot the Lissajous figures.

21. $x = \sin t, y = \sin t$

22. $x = 2 \cos t, y = \cos(t + 4)$

23. $x = \cos \pi t, y = \sin \pi t$

24. $x = \cos\left(t + \dfrac{\pi}{4}\right), y = \sin 2t$

In Exercises 25–32, use a graphing calculator to display the Lissajous figures.

25. $x = \cos \pi\left(t + \dfrac{1}{6}\right), y = 2 \sin \pi t$

26. $x = \sin^2 \pi t, y = \cos \pi t$

27. $x = 2 \cos 3t, y = \cos 2t$

28. $x = 2 \sin \pi t, y = 3 \sin 3\pi t$

29. $x = \sin t, y = \sin 5t$

30. $x = 2 \cos t, y = \sin 5t$

31. $x = 2 \cos \pi t, y = 3 \sin(2\pi t - \frac{\pi}{4})$

32. $x = 2 \cos 3\pi t, y = \cos 5\pi t$

In Exercises 33–40, sketch the appropriate curves. A graphing calculator may be used.

33. An analysis of the temperature records for Louisville, Kentucky, indicate that the average daily temperature T (in °F) during the year is approximately $T = 56 - 22 \cos[\frac{\pi}{6}(x - 0.5)]$, where x is measured in months ($x = 0.5$ is Jan 15, etc.). Sketch the graph of T vs. x for one year.

34. An analysis of data shows that the mean density d (in mg/cm³) of a calcium compound in the bones of women is given by $d = 139.3 + 48.6 \sin(0.0674x - 0.210)$, where x represents the ages of women ($20 \le x \le 80$ years). (A woman is considered to be osteoporotic if $d < 115$ mg/cm³.) Sketch the graph.

35. The vertical displacement y (in ft) of a buoy floating in water is given by $y = 3.0 \cos 0.2t + 1.0 \sin 0.4t$, where t is in seconds. Sketch the graph of y as a function of t for the first 40 s.

36. The strain e (dimensionless) on a cable caused by vibration is $e = 0.0080 - 0.0020 \sin 30t + 0.0040 \cos 10t$, where t is measured in seconds. Sketch two cycles of e as a function of t.

37. The electric current i (in mA) in a certain circuit is given by $i = 0.32 + 0.50 \sin t - 0.20 \cos 2t$, where t is in milliseconds. Sketch two cycles of i as a function of t.

38. The available solar energy depends on the amount of sunlight, and the available time in a day for sunlight depends on the time of the year. An approximate correction factor (in min) to standard time is $C = 10 \sin \frac{1}{29}(n - 80) - 7.5 \cos \frac{1}{58}(n - 80)$, where n is the number of the day of the year. Sketch C as a function of n.

39. Two signals are seen on an oscilloscope as being at right angles. The equations for the displacements of these signals are $x = 4 \cos \pi t$ and $y = 2 \sin 3\pi t$. Sketch the figure that appears on the oscilloscope.

40. In the study of optics, light is said to be *elliptically polarized* if certain optic vibrations are out of phase. These may be represented by Lissajous figures. Determine the Lissajous figure for two light waves given by $w_1 = \sin \omega t$ and $w_2 = \sin(\omega t + \frac{\pi}{4})$.

CHAPTER EQUATIONS

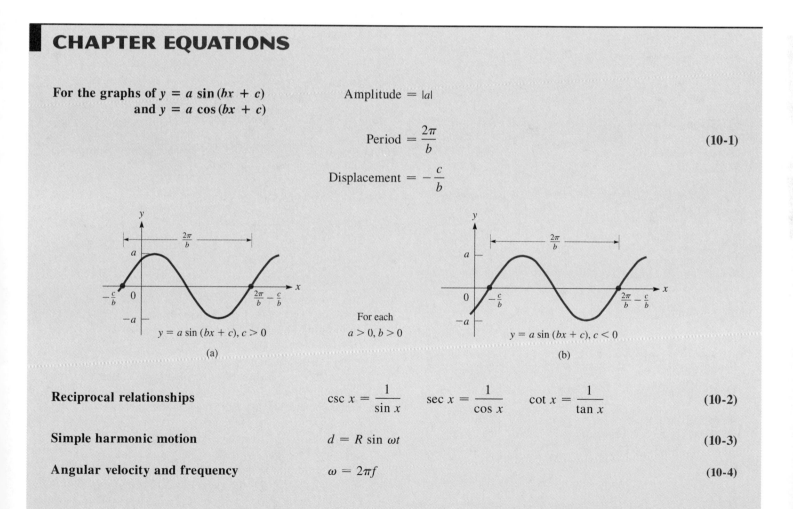

For the graphs of $y = a \sin(bx + c)$
and $y = a \cos(bx + c)$

$$\text{Amplitude} = |a|$$

$$\text{Period} = \frac{2\pi}{b}$$

$$\text{Displacement} = -\frac{c}{b}$$

$(10\text{-}1)$

For each
$a > 0, b > 0$

$y = a \sin(bx + c), c > 0$

(a)

$y = a \sin(bx + c), c < 0$

(b)

| **Reciprocal relationships** | $\csc x = \dfrac{1}{\sin x}$ $\qquad \sec x = \dfrac{1}{\cos x} \qquad \cot x = \dfrac{1}{\tan x}$ | $(10\text{-}2)$ |

Simple harmonic motion $\qquad d = R \sin \omega t$ $\qquad (10\text{-}3)$

Angular velocity and frequency $\qquad \omega = 2\pi f$ $\qquad (10\text{-}4)$

REVIEW EXERCISES

In Exercises 1–28, sketch the curves of the given trigonometric functions. Check each using a graphing calculator.

1. $y = \frac{2}{3} \sin x$ **2.** $y = -4 \sin x$

3. $y = -2 \cos x$ **4.** $y = 2.3 \cos x$

5. $y = 2 \sin 3x$ **6.** $y = 4.5 \sin 12x$

7. $y = 2 \cos 2x$ **8.** $y = 24 \cos 6x$

9. $y = 3 \cos \frac{1}{3}x$ **10.** $y = 3 \sin \frac{1}{2}x$

11. $y = \sin \pi x$ **12.** $y = 3 \sin 4\pi x$

13. $y = 5 \cos 2\pi x$ **14.** $y = -\cos 6\pi x$

15. $y = -0.5 \sin \frac{\pi}{6}x$ **16.** $y = 8 \sin \frac{\pi}{4}x$

17. $y = 2 \sin\left(3x - \frac{\pi}{2}\right)$ **18.** $y = 3 \sin\left(\frac{x}{2} + \frac{\pi}{2}\right)$

19. $y = -2 \cos(4x + \pi)$ **20.** $y = 0.8 \cos\left(\frac{x}{6} - \frac{\pi}{2}\right)$

21. $y = -\sin\left(\pi x + \frac{\pi}{6}\right)$ **22.** $y = 2 \sin(3\pi x - \pi)$

23. $y = 8 \cos\left(4\pi x - \frac{\pi}{2}\right)$ **24.** $y = 3 \cos(2\pi x + \pi)$

25. $y = 3 \tan x$ **26.** $y = \frac{1}{4} \sec x$

27. $y = -\frac{1}{3} \csc x$ **28.** $y = -5 \cot x$

In Exercises 29–32, sketch the curves of the given functions by addition of ordinates.

29. $y = 2 + \frac{1}{2} \sin 2x$ **30.** $y = \frac{1}{2}x - \cos \frac{1}{3}x$

31. $y = \sin 2x + 3 \cos x$ **32.** $y = \sin 3x + 2 \cos 2x$

In Exercises 33–40, display the curves of the given functions on a graphing calculator.

33. $y = 2 \sin x - \cos 2x$ **34.** $y = \sin 3x - 2 \cos x$

35. $y = \cos\left(x + \frac{\pi}{4}\right) - 2 \sin 2x$

36. $y = 2 \cos \pi x + \cos(2\pi x - \pi)$

37. $y = \frac{\sin x}{x}$ **38.** $y = \sqrt{x} \sin 0.5x$

(W) 39. $y = \sin^2 x + \cos^2 x$ ($\sin^2 x = (\sin x)^2$)
What conclusion can be drawn from the graph?

(W) 40. $y = \sin\left(x + \frac{\pi}{4}\right) - \cos\left(x - \frac{\pi}{4}\right) + 1$
What conclusion can be drawn from the graph?

In Exercises 41–44, give the specific form of the indicated equation by evaluating a, b, and c through an inspection of the given curve.

41. $y = a \sin(bx + c)$ **42.** $y = a \cos(bx + c)$
(Figure 10-45)   (Figure 10-45)

43. $y = a \cos(bx + c)$ **44.** $y = a \sin(bx + c)$
(Figure 10-46)   (Figure 10-46)

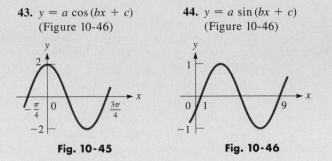

Fig. 10-45   **Fig. 10-46**

In Exercises 45–48, display the Lissajous figures on a graphing calculator.

45. $x = -\cos 2\pi t$, $y = 2 \sin \pi t$

46. $x = \sin\left(t + \frac{\pi}{6}\right)$, $y = \sin t$

47. $x = \cos\left(2\pi t + \frac{\pi}{4}\right)$, $y = \cos \pi t$

48. $x = \cos\left(t - \frac{\pi}{6}\right)$, $y = \cos\left(2t + \frac{\pi}{3}\right)$

In Exercises 49–64, sketch the appropriate curves. A graphing calculator may be used.

49. The range R of a rocket is given by $R = \dfrac{v_0^2 \sin 2\theta}{g}$. Sketch R as a function of θ for $v_0 = 1000$ m/s and $g = 9.8$ m/s^2. See Fig. 10-47.

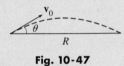

Fig. 10-47

50. The blade of a saber saw moves vertically up and down at 18 strokes per second. The vertical displacement y (in cm) is given by $y = 1.2 \sin 36\pi t$, where t is in seconds. Sketch at least two cycles of the graph of y vs. t.

51. The velocity v (in cm/s) of a piston in a certain engine is given by $v = \omega D \cos \omega t$, where ω is the angular velocity of the crankshaft in radians per second and t is the time in seconds. Sketch the graph of v vs. t if the engine is at 3000 r/min and $D = 3.6$ cm.

52. A light wave for the color yellow can be represented by the equation $y = A \sin 3.4 \times 10^{15} t$. With A as a constant, sketch two cycles of y as a function of t (in s).

53. The electric current i (in A) in a circuit in which there is a *full-wave rectifier* is $i = 10 |\sin 120\pi t|$. Sketch the graph of $i = f(t)$ for $0 \leq t \leq 0.05$ s. What is the period of the current?

54. A circular disk suspended by a thin wire attached to the center of one of its flat faces is twisted through an angle θ. Torsion in the wire tends to turn the disk back in the opposite direction (thus, the name *torsion pendulum* is given to this device). The angular displacement θ (in rad) as a function of time t (in s) is $\theta = \theta_0 \cos(\omega t + \alpha)$, where θ_0 is the maximum angular displacement, ω is a constant that depends on the properties of the disk and wire, and α is the phase angle. Sketch the graph of θ vs. t if $\theta_0 = 0.100$ rad, $\omega = 2.50$ rad/s, and $\alpha = \pi/4$. See Fig. 10-48.

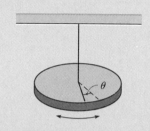

Fig. 10-48

55. At 40° N latitude the number of hours h of daylight each day during the year is approximately
$h = 12.2 + 2.8 \sin[\frac{\pi}{6}(x - 2.7)]$, where x is measured in months ($x = 0.5$ is Jan. 15, etc.) Sketch the graph of h vs. x for one year. (Some of the cities near 40° N are Philadelphia, Madrid, Naples, Ankara, and Beijing.)

W **56.** The equation in Exercise 55 can be used for the number of hours of daylight at 40° S latitude with the appropriate change. Explain what change is necessary and determine the proper equation. Sketch the graph. (This would be appropriate for southern Argentina and Wellington, New Zealand.)

57. If the upper end of a spring is not fixed and is being moved with a sinusoidal motion, the motion of the bob at the end of the spring is affected. Sketch the curve if the motion of the upper end of a spring is being moved by an external force and the bob moves according to the equation $y = 4 \sin 2t - 2 \cos 2t$.

58. The loudness L (in decibels) of a fire siren as a function of the time t (in s) is approximately $L = 40 - 35 \cos 2t + 60 \sin t$. Sketch this function for $0 \le t \le 10$ s.

59. The path of a roller mechanism used in an assembly line process is given by $x = \theta - \sin \theta$ and $y = 1 - \cos \theta$. Sketch the path for $0 \le \theta \le 2\pi$.

60. The equations for two voltage signals that give a resulting curve on an oscilloscope are $x = 6 \sin \pi t$ and $y = 4 \cos 4\pi t$. Sketch the graph of the curve displayed on the oscilloscope.

61. The impedance Z (in Ω) and resistance R (in Ω) for an alternating-current circuit are related by $Z = R \sec \theta$, where θ is called the *phase angle*. Sketch the graph for Z as a function of θ for $-\pi/2 < \theta < \pi/2$.

62. For an object sliding down an inclined plane at constant speed, the coefficient of friction μ between the object and the plane is given by $\mu = \tan \theta$, where θ is the angle between the plane and the horizontal. Sketch the graph of μ vs. θ.

63. The charge q (in C) on a certain capacitor as a function of the time t (in s) is given by $q = 0.0003(3 - 2 \sin 100t \cos 100t)$. Sketch two cycles of q vs. t.

64. The instantaneous power p (in W) in an electric circuit is defined as the product of the instantaneous voltage e and the instantaneous current i (in A). If we have $e = 100 \cos 200t$ and $i = 2 \cos(200t + \frac{\pi}{4})$, plot the graph e vs. t and the graph of i vs. t on the same coordinate system. Then sketch the graph of p vs. t by multiplying appropriate values of e and i.

Writing Exercise

65. A wave passing through a string can be described at any instant by the equation $y = a \sin(bx + c)$. Write one or two paragraphs explaining the change in the wave (a) if a is doubled, (b) if b is doubled, and (c) if c is doubled.

PRACTICE TEST

In Problems 1–4, sketch the graphs of the given functions.

1. $y = 0.5 \cos \frac{\pi}{2} x$

2. $y = 2 + 3 \sin x$

3. $y = 3 \sec x$

4. $y = 2 \sin(2x - \frac{\pi}{3})$

5. A wave is traveling in a string. The displacement y (in in.) as a function of the time t (in s) from its equilibrium position is given by $y = A \cos(2\pi/T)t$. T is the period (in s) of the motion. If $A = 0.200$ in. and $T = 0.100$ s, sketch two cycles of y vs t.

6. Sketch the graph of $y = 2 \sin x + \cos 2x$ by addition of ordinates.

7. Use a graphing calculator to display the Lissajous figure for which $x = \sin \pi t$ and $y = 2 \cos 2\pi t$.

8. Sketch two cycles of the curve of a projection on the end of a radius on the y-axis. The radius is of length R and it is rotating counterclockwise about the origin at 2.00 rad/s. It starts at an angle of $\pi/6$ with the positive x-axis.

CHAPTER *11* — EXPONENTS AND RADICALS

In finding the rate at which solar radiation changes at a solar energy collector, the following expression is found.

$$\frac{(t^4 + 100)^{1/2} - 2t^3(t + 6)(t^4 + 100)^{-1/2}}{[(t^4 + 100)^{1/2}]^2}$$

In Section 11-2 we show that this can be written in a much simpler form.

In Chapter 1 we introduced exponents and radicals. To this point, only a basic understanding of the meaning and the elementary operations with them has been necessary. However, in the later chapters a more detailed understanding of exponents and radicals, and the operations with them, will be required. Therefore, in this chapter we shall introduce and develop the operations which are necessary.

We will show the relationship between exponents and radicals when we introduce fractional exponents. In applications, it is common and generally more convenient to use fractional exponents rather than radicals.

As we develop these necessary algebraic operations, we will show some of their uses in various technical areas of application. They are used to help develop a number of formulas in areas such as electronics, hydrodynamics, optics, solar energy, and machine design.

11-1 SIMPLIFYING EXPRESSIONS WITH INTEGRAL EXPONENTS

The laws of exponents were given in Section 1-4. For reference, they are:

$$a^m \times a^n = a^{m+n} \tag{11-1}$$

$$\frac{a^m}{a^n} = a^{m-n} \quad \text{or} \quad \frac{a^m}{a^n} = \frac{1}{a^{n-m}}, \qquad a \neq 0 \tag{11-2}$$

$$(a^m)^n = a^{mn} \tag{11-3}$$

$$(ab)^n = a^n b^n, \qquad \left(\frac{a}{b}\right)^n = \frac{a^n}{b^n}, \qquad b \neq 0 \tag{11-4}$$

$$a^0 = 1, \qquad a \neq 0 \tag{11-5}$$

$$a^{-n} = \frac{1}{a^n}, \qquad a \neq 0 \tag{11-6}$$

Although Eqs. (11-1) to (11-4) were originally defined for positive integers as exponents, we showed in Section 1-4 that with the definitions given in Eqs. (11-5) and (11-6), they are valid for all integral exponents. Later in this chapter we shall show how fractions may be used as exponents. Since these equations are very important to the development of the topics in this chapter, they should again be reviewed and learned thoroughly.

In this section we review the use of exponents while using Eqs. (11-1) to (11-6). Then we show how exponents are used and handled in more involved expressions.

EXAMPLE 1 Applying Eq. (11-1), we have

$$a^5 \times a^{-3} = a^{5+(-3)} = a^{5-3} = a^2$$

Applying Eq. (11-1) and then Eq. (11-6), we have

$$a^3 \times a^{-5} = a^{3-5} = a^{-2} = \frac{1}{a^2}$$

NOTE ▸ *Negative exponents are generally not used in the expression of a final result,* unless specified otherwise. However, they are often used in intermediate steps. ∎

EXAMPLE 2 Applying Eq. (11-1), then (11-6), and then (11-4), we have

$$(10^3 \times 10^{-4})^2 = (10^{3-4})^2 = (10^{-1})^2 = \left(\frac{1}{10}\right)^2 = \frac{1}{10^2} = \frac{1}{100}$$

Often, several combinations of the laws can be used to simplify an expression. For example, this expression can be simplified by using Eq. (11-1), then (11-3), and then (11-6), as follows:

$$(10^3 \times 10^{-4})^2 = (10^{3-4})^2 = (10^{-1})^2 = 10^{-2} = \frac{1}{10^2} = \frac{1}{100}$$

The result is in a proper form as either $1/10^2$ or $1/100$. If the exponent is large, then it is common to leave the exponent in the answer. ∎

EXAMPLE 3 Applying Eqs. (11-2) and (11-5), we have

$$\frac{a^2 b^3 c^0}{ab^7} = \frac{a^{2-1}(1)}{b^{7-3}} = \frac{a}{b^4}$$

Applying Eqs. (11-4) and (11-3), we have

$$(x^{-2}y)^3 = (x^{-2})^3(y^3) = x^{-6}y^3 = \frac{y^3}{x^6}$$

Here, the simplification was completed by the use of Eq. (11-6). ∎

EXAMPLE 4 $\left(\dfrac{4}{a^2}\right)^{-3} = \dfrac{1}{\left(\dfrac{4}{a^2}\right)^3} = \dfrac{1}{\dfrac{4^3}{a^6}} = \dfrac{a^6}{4^3}$ or $\left(\dfrac{4}{a^2}\right)^{-3} = \dfrac{4^{-3}}{a^{-6}} = \dfrac{a^6}{4^3}$

In the first expression we used Eq. (11-6) first, then Eq. (11-4), and then we inverted the divisor. In the second expression we used Eq. (11-4) and then Eq. (11-6). ∎

EXAMPLE 5 $(x^2y)^2\left(\dfrac{2}{x}\right)^{-2} = \dfrac{(x^4y^2)}{\left(\dfrac{2}{x}\right)^2} = \dfrac{x^4y^2}{\dfrac{4}{x^2}} = \dfrac{x^4y^2}{1} \times \dfrac{x^2}{4} = \dfrac{x^6y^2}{4}$

or

$$(x^2y)^2\left(\frac{2}{x}\right)^{-2} = (x^4y^2)\left(\frac{2^{-2}}{x^{-2}}\right) = (x^4y^2)\left(\frac{x^2}{2^2}\right) = \frac{x^6y^2}{4}$$

The physical units in the denominator of a denominate number can be expressed in terms of negative exponents. This is illustrated in the following example.

Named for the French mathematician and scientist Blaise Pascal (1623–1662).

EXAMPLE 6 The metric unit for pressure is the *pascal,* where 1 Pa = 1 N/m^2. Using a negative exponent, this can be expressed as

$$1\text{ Pa} = 1\text{ N/m}^2 = 1\text{ N}\cdot\text{m}^{-2}$$

where $1/\text{m}^2 = \text{m}^{-2}$.

Named for the English physicist James Prescott Joule (1818–1899).

The metric unit for energy is the *joule,* where 1 J = 1 kg · (m · s^{-1})^2, or

$$1\text{ J} = 1\text{ kg}\cdot\text{m}^2\cdot\text{s}^{-2} = 1\text{ kg}\cdot\text{m}^2/\text{s}^2$$

Care must be taken to apply the laws of exponents properly. Certain common problems are pointed out in the following examples.

EXAMPLE 7 The expression $(-5x)^0$ equals 1, whereas the expression $-5x^0$ equals -5. For $(-5x)^0$ the parentheses show that the expression $-5x$ is raised to the zero power, whereas for $-5x^0$ only x is raised to the zero power and we have

$$-5x^0 = -5(1) = -5$$

CAUTION ▶ Also, $(-5)^0 = 1$, but $-5^0 = -1$. Again, for $(-5)^0$ parentheses show -5 raised to the zero power, whereas for -5^0 only 5 is raised to the zero power.

Similarly, $(-2)^2 = 4$ and $-2^2 = -4$

CAUTION ▶ For the same reasons, $2x^{-1} = \dfrac{2}{x}$ whereas $(2x)^{-1} = \dfrac{1}{2x}$

EXAMPLE 8 $(2a + b^{-1})^{-2} = \dfrac{1}{(2a+b^{-1})^2} = \dfrac{1}{\left(2a+\dfrac{1}{b}\right)^2} = \dfrac{1}{\left(\dfrac{2ab+1}{b}\right)^2}$

$= \dfrac{1}{\dfrac{(2ab+1)^2}{b^2}} = \dfrac{b^2}{(2ab+1)^2}$ not necessary to expand the denominator

Another order of operations for simplifying this expression is

$$(2a+b^{-1})^{-2} = \left(2a+\frac{1}{b}\right)^{-2} = \left(\frac{2ab+1}{b}\right)^{-2}$$

$$= \frac{(2ab+1)^{-2}}{b^{-2}} = \frac{b^2}{(2ab+1)^2}$$ positive exponents used in the final result

EXAMPLE 9   There is an error that is commonly made in simplifying the type of expression in Example 8. We must be careful to see that

CAUTION ▶

$$(2a + b^{-1})^{-2} \quad \text{is } \textbf{\textit{not}} \text{ equal to} \quad (2a)^{-2} + (b^{-1})^{-2}, \quad \text{or} \quad \frac{1}{4a^2} + b^2$$

Remember: As noted in Section 6-1, when raising a binomial (or any multinomial) to a power, we cannot simply raise each term to the power to obtain the result.

For reference, Eq. (11-4) is $(ab)^n = a^n b^n$.

However, when raising a product of factors to a power, we use Eq. (11-4). Thus,

$$(2ab^{-1})^{-2} = (2a)^{-2}(b^{-1})^{-2} = \frac{b^2}{(2a)^2} = \frac{b^2}{4a^2}$$

We see that we must be careful to distinguish between the power of a sum of terms and the power of a product of factors. ■

CAUTION ▶

From the preceding examples, we see that *when a factor is moved from the denominator to the numerator of a fraction, or conversely, the* **sign** *of the exponent is changed.* We should carefully note the word *factor*; this rule does not apply to moving *terms* in the numerator or the denominator.

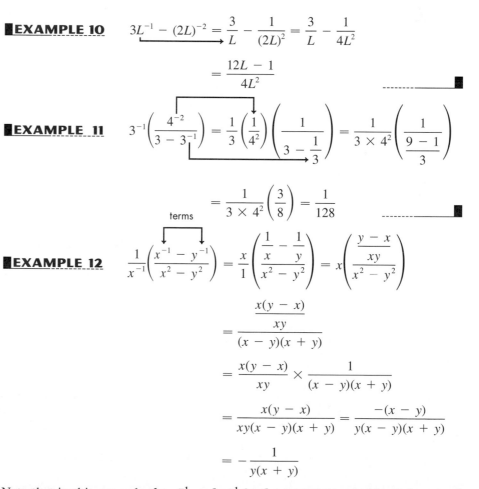

EXAMPLE 10 $3L^{-1} - (2L)^{-2} = \dfrac{3}{L} - \dfrac{1}{(2L)^2} = \dfrac{3}{L} - \dfrac{1}{4L^2}$

$$= \frac{12L - 1}{4L^2}$$ ■

EXAMPLE 11 $3^{-1}\left(\dfrac{4^{-2}}{3 - 3^{-1}}\right) = \dfrac{1}{3}\left(\dfrac{1}{4^2}\right)\left(\dfrac{1}{3 - \dfrac{1}{3}}\right) = \dfrac{1}{3 \times 4^2}\left(\dfrac{1}{\dfrac{9 - 1}{3}}\right)$

$$= \frac{1}{3 \times 4^2}\left(\frac{3}{8}\right) = \frac{1}{128}$$ ■

terms

EXAMPLE 12 $\dfrac{1}{x^{-1}}\left(\dfrac{x^{-1} - y^{-1}}{x^2 - y^2}\right) = \dfrac{x}{1}\left(\dfrac{\dfrac{1}{x} - \dfrac{1}{y}}{x^2 - y^2}\right) = x\left(\dfrac{\dfrac{y - x}{xy}}{x^2 - y^2}\right)$

$$= \frac{\dfrac{x(y - x)}{xy}}{(x - y)(x + y)}$$

$$= \frac{x(y - x)}{xy} \times \frac{1}{(x - y)(x + y)}$$

$$= \frac{x(y - x)}{xy(x - y)(x + y)} = \frac{-(x - y)}{y(x - y)(x + y)}$$

$$= -\frac{1}{y(x + y)}$$

CAUTION ▶   Note that in this example *the x^{-1} and y^{-1} in the numerator could not be moved directly to the denominator with positive exponents* because they are only terms of the original numerator. ■

EXAMPLE 13 $3(x + 4)^2(x - 3)^{-2} - 2(x - 3)^{-3}(x + 4)^3$

$$= \frac{3(x + 4)^2}{(x - 3)^2} - \frac{2(x + 4)^3}{(x - 3)^3} = \frac{3(x - 3)(x + 4)^2 - 2(x + 4)^3}{(x - 3)^3}$$

$$= \frac{(x + 4)^2[3(x - 3) - 2(x + 4)]}{(x - 3)^3} = \frac{(x + 4)^2(x - 17)}{(x - 3)^3}$$

Expressions such as the one in this example are commonly found in problems in calculus.

EXERCISES *11-1*

In Exercises 1–48, express each of the given expressions in simplest form with only positive exponents.

1. $x^7 x^{-4}$

2. $y^9 y^{-2}$

3. $a^2 a^{-6}$

4. ss^{-5}

5. 5×5^{-3}

6. $(3^2 \times 4^{-3})^3$

7. $(2ax^{-1})^2$

8. $(3xy^{-2})^3$

9. $(5an^{-2})^{-1}$

10. $(6s^2 t^{-1})^{-2}$

11. $(-4)^0$

12. -4^0

13. $-7x^0$

14. $(-7x)^0$

15. $3x^{-2}$

16. $(3x)^{-2}$

17. $(7ax)^{-3}$

18. $7ax^{-3}$

19. $\left(\dfrac{2}{n^3}\right)^{-3}$

20. $\left(\dfrac{3}{x^3}\right)^{-2}$

21. $\left(\dfrac{a}{b^{-2}}\right)^{-3}$

22. $\left(\dfrac{2n^{-2}}{D^{-1}}\right)^{-2}$

23. $(a + b)^{-1}$

24. $a^{-1} + b^{-1}$

25. $3x^{-2} + 2y^{-2}$

26. $(3x + 2y)^{-2}$

27. $(2 \times 3^{-2})^2 \left(\dfrac{3}{2}\right)^{-1}$

28. $(3^{-1} \times 7^2)\left(\dfrac{3}{7}\right)^2$

29. $\left(\dfrac{3a^2}{4b}\right)^{-3}\left(\dfrac{4}{a}\right)^{-5}$

30. $(2np^{-2})^{-2}(4^{-1}p^2)^{-1}$

31. $\left(\dfrac{V^{-1}}{2t}\right)^{-2}\left(\dfrac{t^2}{V^{-2}}\right)^{-3}$

32. $\left(\dfrac{a^{-2}}{b^2}\right)^{-3}\left(\dfrac{a^{-3}}{b^5}\right)^2$

33. $2a^{-2} + (2a^{-2})^4$

34. $3(a^{-1}z^2)^{-3} + c^{-2}z^{-1}$

35. $2 \times 3^{-1} + 4 \times 3^{-2}$

36. $5 \times 2^{-2} - 3^{-1} \times 2^3$

37. $(R_1^{-1} + R_2^{-1})^{-1}$

38. $(2a - b^{-2})^{-1}$

39. $(n^{-2} - 2n^{-1})^2$

40. $(2^{-3} - 4^{-1})^{-2}$

41. $\dfrac{6^{-1}}{4^{-2} + 2}$

42. $\dfrac{x - y^{-1}}{x^{-1} - y}$

43. $\dfrac{x^{-2} - y^{-2}}{x^{-1} - y^{-1}}$

44. $\dfrac{ax^{-2} + a^{-2}x}{a^{-1} + x^{-1}}$

45. $2t^{-2} + t^{-1}(t + 1)$

46. $3x^{-1} - x^{-3}(y + 2)$

47. $(x - 1)^{-1} + (x + 1)^{-1}$

48. $4(2x - 1)(x + 2)^{-1} - (2x - 1)^2(x + 2)^{-2}$

In Exercises 49–60 perform the indicated operations.

49. Express $4^2 \times 64$ (a) as a power of 4 and (b) as a power of 2.

50. Express $1/81$ (a) as a power of 9 and (b) as a power of 3.

51. (a) By use of Eqs. (11-4) and (11-6), show that
$$\left(\dfrac{a}{b}\right)^{-n} = \left(\dfrac{b}{a}\right)^n$$
(b) Verify the equation in part (a) by evaluating each side with $a = 3.576$, $b = 8.091$, and $n = 7$.

(W) 52. For what integral values of n is $(-3)^{-n} = -3^{-n}$? Explain.

53. The metric unit of energy, the *joule* (J), can be expressed as $kg \cdot s^{-2} \cdot m^2$. Simplify these units and include *newtons* (see Appendix B) and only positive exponents in the final result.

54. The units for the electric quantity called *permittivity* are $C^2 \cdot N^{-1} \cdot m^{-2}$. Given that $1 \text{ F} = 1 \text{ C}^2 \cdot J^{-1}$, show that the units of permittivity are F/m. See Appendix B.

55. When studying a solar energy system, the units encountered are $kg \cdot s^{-1}(m \cdot s^{-2})^2$. Simplify these units and include *joules* (see Example 6) and only positive exponents in the final result.

56. The metric units for the velocity v of an object are $m \cdot s^{-1}$, and the units for the acceleration a of the object are $m \cdot s^{-2}$. What are the units for v/a?

57. Given that $v = a^p t^r$, where v is the velocity of an object, a is its acceleration, and t is the time, use the metric units given in Exercise 56 to show that $p = r = 1$.

58. An expression encountered in finance is
$$\frac{p(1 + i)^{-1}[(1 + i)^{-n} - 1]}{(1 + i)^{-1} - 1}$$
where n is an integer. Simplify this expression.

59. In analyzing the tuning of an electronic circuit, the expression $[\omega\omega_0^{-1} - \omega_0\omega^{-1}]^2$ is used. Expand and simplify this expression.

60. In optics, the combined focal length F of two lenses is given by $F = [f_1^{-1} + f_2^{-1} + d(f_1 f_2)^{-1}]^{-1}$, where f_1 and f_2 are the focal lengths of the lenses and d is the distance between them. Simplify the right side of this equation.

11-2 FRACTIONAL EXPONENTS

In Section 11-1 we reviewed the use of integral exponents, including exponents that are negative integers and zero. We now show how rational numbers may be used as exponents. With the appropriate definitions, all the laws of exponents are valid for all rational numbers as exponents.

Equation (11-3) states that $(a^m)^n = a^{mn}$. If we were to let $m = \frac{1}{2}$ and $n = 2$, we would have $(a^{1/2})^2 = a^1$. However, we already have a way of writing a quantity which when squared equals a. This is written as $\sqrt{a}$. To be consistent with previous definitions and to allow the laws of exponents to hold, we define

The radical in Eq. (11-7) is the symbol for the *n*th root of a number. Do not confuse it with $\sqrt{a}$, the square root of *a*.

$$a^{1/n} = \sqrt[n]{a}$$

(11-7)

In order that Eqs. (11-3) and (11-7) may hold at the same time, we define

$$a^{m/n} = \sqrt[n]{a^m} = (\sqrt[n]{a})^m$$

(11-8)

These definitions are valid for all the laws of exponents. We must note that Eqs. (11-7) and (11-8) are valid as long as $\sqrt[n]{a}$ does not involve the even root of a negative number. Such numbers are imaginary and are considered in Chapter 12.

For reference, Eq. (11-1) is $a^m \times a^n = a^{m+n}$.

EXAMPLE 1   We now verify that Eq. (11-1) holds for the above definitions:

$$a^{1/4}a^{1/4}a^{1/4}a^{1/4} = a^{(1/4)+(1/4)+(1/4)+(1/4)} = a^1$$

Now, $a^{1/4} = \sqrt[4]{a}$ by definition. Also, by definition $\sqrt[4]{a}\sqrt[4]{a}\sqrt[4]{a}\sqrt[4]{a} = a$. Equation (11-1) is thereby verified for $n = 4$ in Eq. (11-7).

Equation (11-3) is verified by the following:

Eq. (11-3) is $(a^m)^n = a^{mn}$.

$$(a^{1/4})(a^{1/4})(a^{1/4})(a^{1/4}) = (a^{1/4})^4 = a^1 = (\sqrt[4]{a})^4 \qquad \blacksquare$$

We may interpret $a^{m/n}$ in Eq. (11-8) as the mth power of the nth root of a, as well as the nth root of the mth power of a. This is illustrated in the following example.

EXAMPLE 2

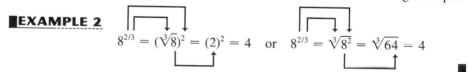

$$8^{2/3} = (\sqrt[3]{8})^2 = (2)^2 = 4 \quad \text{or} \quad 8^{2/3} = \sqrt[3]{8^2} = \sqrt[3]{64} = 4$$

■

Although both interpretations of Eq. (11-8) are possible, as indicated in Example 2, in evaluating numerical expressions involving fractional exponents without a calculator, it is almost always best to *find the root first, as indicated by the denominator* of the fractional exponent. This allows us to find the root of the smaller number, which is normally easier to find.

EXAMPLE 3   To evaluate $(64)^{5/2}$, we should proceed as follows:

$$(64)^{5/2} = [(64)^{1/2}]^5 = 8^5 = 32{,}768$$

If we raised 64 to the fifth power first, we would have

$$(64)^{5/2} = (64^5)^{1/2} = (1{,}073{,}741{,}824)^{1/2}$$

We would now have to evaluate the indicated square root. This demonstrates why it is preferable to find the indicated root first. ■

For reference, basic forms of Eqs. (11-1) to (11-7) are as follows:

$$a^m \times a^n = a^{m+n} \qquad \text{(11-1)}$$

$$\frac{a^m}{a^n} = a^{m-n} \qquad \text{(11-2)}$$

$$(a^m)^n = a^{mn} \qquad \text{(11-3)}$$

$$(ab)^n = a^n b^n \qquad \text{(11-4)}$$

$$a^0 = 1 \qquad \text{(11-5)}$$

$$a^{-n} = \frac{1}{a^n} \qquad \text{(11-6)}$$

$$a^{1/n} = \sqrt[n]{a} \qquad \text{(11-7)}$$

Named for the Scottish physicist Lord Kelvin (1824–1907).

Fig. 11-1

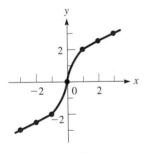

Fig. 11-2

■**EXAMPLE 4**   (a) $(16)^{3/4} = (16^{1/4})^3 = 2^3 = 8$

(b) $4^{-1/2} = \dfrac{1}{4^{1/2}} = \dfrac{1}{2}$ (c) $9^{3/2} = (9^{1/2})^3 = 3^3 = 27$

We note in (b) that Eq. (11-6) must also hold for negative rational exponents. In writing $4^{-1/2}$ as $1/4^{1/2}$, the *sign* of the exponent is changed. ---------■

Fractional exponents allow us to find roots of numbers on a calculator. By use of the appropriate key (on most calculators $\boxed{x^y}$ or $\boxed{\wedge}$), we may raise any positive number to any power. For roots, we use the equivalent fractional exponent. Powers that are fractions or decimal in form are entered directly.

■**EXAMPLE 5**   The thermodynamic temperature T (in kelvins (K)) is related to the pressure P (in kPa) of a gas by the equation $T = 80.5P^{2/7}$. Find the value of T for $P = 750$ kPa.
  Substituting, we have

$$T = 80.5(750)^{2/7}$$

The calculator display for this calculation is shown in Fig. 11-1. ---------■

When finding powers of negative numbers, some calculators will show an error. If this is the case, enter the positive value of the number and then enter a negative sign for the result when appropriate. From Section 1-6 we recall that *an even root of a negative number is imaginary, and an odd root of a negative number is negative*. If it is an integral power, the basic laws of signs are used.

■**EXAMPLE 6**   Plot the graph of the function $y = 2x^{1/3}$.
  In obtaining points for the graph, we use $(1/3)$ as the power when using the calculator. If your calculator does not evaluate powers of negative numbers, enter positive values for the negative values of x, and then make the results negative. Since $x^{1/3} = \sqrt[3]{x}$, we know that $x^{1/3}$ is negative for negative values of x because we have an odd root of a negative number. We get the following table of values.

x	-3	-2	-1	0	1	2	3
y	-2.9	-2.5	-2.0	0	2.0	2.5	2.9

The graph is shown in Fig. 11-2. Of course, this curve can easily be shown on a graphing calculator. ---------■

Another reason for developing fractional exponents is that they are often easier to use in more complex expressions involving roots. This is true in algebra and in topics from more advanced mathematics. Any expression with radicals can also be expressed with fractional exponents and then simplified. We now show some additional examples with fractional exponents.

■**EXAMPLE 7**   (a) $(8a^2b^4)^{1/3} = [(8^{1/3})(a^2)^{1/3}(b^4)^{1/3}]$ using Eq. (11-4)
$$= 2a^{2/3}b^{4/3} \qquad \text{using Eqs. (11-7) and (11-3)}$$

(b) $a^{3/4}a^{4/5} = a^{3/4+4/5} = a^{31/20}$ using Eq. (11-1) ---------■

EXAMPLE 8 $\left(\dfrac{4^{-3/2}x^{2/3}y^{-7/4}}{2^{3/2}x^{-1/3}y^{3/4}}\right)^{2/3} = \left(\dfrac{x^{2/3}x^{1/3}}{2^{3/2}4^{3/2}y^{3/4}y^{7/4}}\right)^{2/3}$ using Eq. (11-6)

$$= \left(\dfrac{x^{2/3+1/3}}{2^{3/2}4^{3/2}y^{3/4+7/4}}\right)^{2/3} \quad \text{using Eq. (11-1)}$$

$$= \dfrac{x^{(1)(2/3)}}{2^{(3/2)(2/3)}4^{(3/2)(2/3)}y^{(10/4)(2/3)}} \quad \text{using Eq. (11-4)}$$

$$= \dfrac{x^{2/3}}{8y^{5/3}} \quad\quad\quad\text{—————} \blacksquare$$

EXAMPLE 9 $(4x^4)^{-1/2} - 3x^{-3} = \dfrac{1}{(4x^4)^{1/2}} - \dfrac{3}{x^3}$ using Eq. (11-6)

$$= \dfrac{1}{2x^2} - \dfrac{3}{x^3} \quad \text{using Eq. (11-7)}$$

$$= \dfrac{x-6}{2x^3} \quad \text{common denominator} \quad\text{—————} \blacksquare$$

See the chapter introduction.

EXAMPLE 10   The rate R at which solar radiation changes at a solar energy collector during a day is given by the equation

$$R = \dfrac{(t^4 + 100)^{1/2} - 2t^3(t + 6)(t^4 + 100)^{-1/2}}{[(t^4 + 100)^{1/2}]^2}$$

Solar collectors supply the power for most space satellites.

Here, R is measured in kW/(m$^2 \cdot$ h), t is the number of hours from noon, and -6 h $\le t \le 8$ h. Express the right side of this equation in simpler form and find R for $t = 0$ (noon) and for $t = 4$ h (4 P.M.)

Performing the simplification, we have the following steps:

$$R = \dfrac{(t^4 + 100)^{1/2} - \dfrac{2t^3(t + 6)}{(t^4 + 100)^{1/2}}}{(t^4 + 100)} \quad\quad \begin{array}{l}\text{using Eq. (11-6)}\\[4pt]\text{using Eq. (11-3)}\end{array}$$

$$= \dfrac{\dfrac{(t^4 + 100)^{1/2}(t^4 + 100)^{1/2} - 2t^3(t + 6)}{(t^4 + 100)^{1/2}}}{(t^4 + 100)} \quad \text{common denominator}$$

$$= \dfrac{(t^4 + 100) - 2t^3(t + 6)}{(t^4 + 100)^{1/2}} \times \dfrac{1}{t^4 + 100} \quad \text{invert divisor and multiply}$$

$$= \dfrac{100 - 12t^3 - t^4}{(t^4 + 100)^{1/2}(t^4 + 100)} = \dfrac{100 - 12t^3 - t^4}{(t^4 + 100)^{3/2}} \quad \text{using Eq. (11-1)}$$

For $t = 0$: $R = \dfrac{100 - 12(0^3) - 0^4}{(0^4 + 100)^{3/2}} = 0.10$ kW/(m$^2 \cdot$ h)

For $t = 4$ h: $R = \dfrac{100 - 12(4^3) - 4^4}{(4^4 + 100)^{3/2}} = -0.14$ kW/(m$^2 \cdot$ h)

We see that the radiation is increasing at noon, and the negative sign tells us that it is decreasing at 4 P.M.   ————— ■

EXERCISES *11-2*

In Exercises 1–24, evaluate the given expressions.

1. $25^{1/2}$

2. $27^{1/3}$

3. $81^{1/4}$

4. $125^{2/3}$

5. $100^{25/2}$

6. $16^{5/4}$

7. $8^{-1/3}$

8. $16^{-1/4}$

9. $64^{-2/3}$

10. $32^{-4/5}$

11. $5^{1/2}5^{3/2}$

12. $(4^4)^{3/2}$

13. $(3^6)^{2/3}$

14. $\dfrac{121^{-1/2}}{100^{1/2}}$

15. $\dfrac{1000^{1/3}}{400^{-1/2}}$

16. $\dfrac{7^{-1/2}}{6^{-1}7^{1/2}}$

17. $\dfrac{15^{2/3}}{5^2 15^{-1/3}}$

18. $\dfrac{(-27)^{1/3}}{6}$

19. $\dfrac{(-8)^{2/3}}{-2}$

20. $\dfrac{-4}{(-64)^{-2/3}}$

21. $125^{-2/3} - 100^{-3/2}$

22. $32^{0.4} + 25^{-0.5}$

23. $\dfrac{16^{-0.25}}{5} + \dfrac{2^{-0.6}}{2^{0.4}}$

24. $\dfrac{4^{-1}}{36^{-1/2}} - \dfrac{5^{-1/2}}{5^{1/2}}$

In Exercises 25–28, use a calculator to evaluate each expression.

25. $17.98^{1/4}$

26. $750.81^{2/3}$

27. $4.0187^{-4/9}$

28. $0.1863^{-1/6}$

In Exercises 29–52, simplify the given expressions. Express all answers with positive exponents.

29. $a^{2/3}a^{1/2}$

30. $x^{5/6}x^{-1/3}$

31. $\dfrac{y^{-1/2}}{y^{2/5}}$

32. $\dfrac{s^{1/4}s^{2/3}}{s^{-1}}$

33. $\dfrac{x^{3/10}}{x^{-1/5}x^2}$

34. $\dfrac{a^{-2/5}a^2}{a^{-3/10}}$

35. $(8a^3b^6)^{1/3}$

36. $(8b^{-4}c^2)^{2/3}$

37. $(16a^4b^3)^{-3/4}$

38. $(32C^5D^4)^{-2/5}$

39. $\left(\dfrac{a^{5/7}}{a^{2/3}}\right)^{7/4}$

40. $\left(\dfrac{4a^{5/6}b^{-1/5}}{a^{2/3}b^2}\right)^{-1/2}$

41. $\dfrac{1}{2}(4x^2 + 1)^{-1/2}(8x)$

42. $\dfrac{2}{3}(x^3 + 1)^{-1/3}(3x^2)$

43. $\left(\dfrac{6x^{-1/2}y^{2/3}}{18x^{-1}}\right)\left(\dfrac{2y^{1/4}}{x^{1/3}}\right)$

44. $\dfrac{3^{-1}a^{1/2}}{4^{-1/2}b} \div \dfrac{9^{1/2}a^{-1/3}}{2b^{-1/4}}$

45. $(T^{-1} + 2T^{-2})^{-1/2}$

46. $(a^{-2} - a^{-4})^{-1/4}$

47. $(a^3)^{-4/3} + a^{-2}$

48. $(4x^6)^{-1/2} - 2x^{-1}$

49. $[(a^{1/2} - a^{-1/2})^2 + 4]^{1/2}$

50. $4x^{1/2} + \frac{1}{2}x^{-1/2}(4x + 1)$

51. $x^2(2x - 1)^{-1/2} + 2x(2x - 1)^{1/2}$

52. $(3x - 1)^{-2/3}(1 - x) - (3x - 1)^{1/3}$

In Exercises 53–56, graph the given functions.

53. $f(x) = 3x^{1/2}$

54. $f(x) = 2x^{2/3}$

55. $f(x) = x^{4/5}$

56. $f(x) = 4x^{3/2}$

In Exercises 57–60, perform the indicated operations.

(W) **57.** A factor used in determining the performance of a solar-energy storage system is $(A/S)^{-1/4}$, where A is the actual storage capacity and S is a standard storage capacity. If this factor is 0.5, explain how to find the ratio A/S.

58. A factor used in measuring the loudness sensed by the human ear is $(I/I_0)^{0.3}$, where I is the intensity of the sound and I_0 is a reference intensity. Evaluate this factor for $I = 3.2 \times 10^{-6}$ W/m^2 (ordinary conversation) and $I_0 = 10^{-12}$ W/m^2.

59. The electric current i (in A) in a circuit with a battery of voltage E, a resistance R, and an inductance L, is

$i = \dfrac{E}{R}(1 - e^{-Rt/L})$, where t is the time after the circuit is closed. See Fig. 11-3. Find i for $E = 6.20$ V, $R = 1.20\ \Omega$, $L = 3.24$ H, and $t = 0.00100$ s. (The number e is irrational and can be found from the calculator by using the [e^x] key with $x = 1$.)

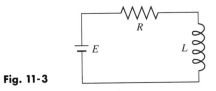

Fig. 11-3

60. For a heat-seeking rocket in pursuit of an aircraft, the distance d (in km) from the rocket to the aircraft is

$d = \dfrac{500(\sin \theta)^{1/2}}{(1 - \cos \theta)^{3/2}}$, where θ is shown in Fig. 11-4. Find d for $\theta = 125.0°$.

Fig. 11-4

11-3 SIMPLEST RADICAL FORM

Roots and radicals were first introduced in Section 1-6 and were used again when we developed the concept of a fractional exponent. As mentioned in the previous section, it is possible to use fractional exponents for any operation required with radicals. For operations involving multiplication and division, using the fractional exponent form has certain advantages. However, for the addition and subtraction of radicals, there is normally little advantage in changing form.

We shall now define the operations with radicals so that these definitions are consistent with the laws of exponents. This will enable us from now on to use either fractional exponents or radicals, whichever is more convenient for the operation being performed.

$$\sqrt[n]{a^n} = (\sqrt[n]{a})^n = a \tag{11-9}$$

$$\sqrt[n]{a}\,\sqrt[n]{b} = \sqrt[n]{ab} \tag{11-10}$$

$$\sqrt[m]{\sqrt[n]{a}} = \sqrt[mn]{a} \tag{11-11}$$

$$\frac{\sqrt[n]{a}}{\sqrt[n]{b}} = \sqrt[n]{\frac{a}{b}} \quad (b \neq 0) \tag{11-12}$$

NOTE▶ *The number under the radical is called the* **radicand,** *and the number indicating the root being taken is called the* **order** *(or* **index***) of the radical.* To avoid difficulties with imaginary numbers (which are considered in the next chapter), *we shall assume that all letters represent positive numbers.*

▌EXAMPLE 1 Following are illustrations of the use of each of Eqs. (11-9) to (11-12).

(a) $\sqrt[5]{4^5} = (\sqrt[5]{4})^5 = 4$ using Eq. (11-9)

(b) $\sqrt[3]{2}\,\sqrt[3]{3} = \sqrt[3]{2 \times 3} = \sqrt[3]{6}$ using Eq. (11-10)

(c) $\sqrt[3]{\sqrt{5}} = \sqrt[3 \times 2]{5} = \sqrt[6]{5}$ using Eq. (11-11)

(d) $\dfrac{\sqrt{7}}{\sqrt{3}} = \sqrt{\dfrac{7}{3}}$ using Eq. (11-12)

▌EXAMPLE 2 In Example 5 of Section 1-6, we saw that

$$\sqrt{16 + 9} \quad \text{is } \textit{not} \text{ equal to} \quad \sqrt{16} + \sqrt{9}$$

However, using Eq. (11-10),

$$\sqrt{16 \times 9} = \sqrt{16} \times \sqrt{9}$$
$$= 4 \times 3 = 12$$

NOTE▶ Therefore, we must *be careful to distinguish between the root of a sum of terms and the root of a product of factors.* This is the same as with powers of sums and powers of products, as shown in Example 9 of Section 11-1. It should be the same, as a root can be interpreted as a fractional exponent.

There are certain operations that are performed on radicals in order to put them in their simplest form. The following two examples illustrate one of these operations.

■EXAMPLE 3   To simplify $\sqrt{75}$, we know that $75 = (25)(3)$ and that $\sqrt{25} = 5$. As in Section 1-6 and now using Eq. (11-10), we write

$$\sqrt{75} = \sqrt{(25)(3)} = \sqrt{25}\,\sqrt{3} = 5\sqrt{3}$$
$$\quad\quad\quad\;\; \underset{\text{perfect square}}{\big\lfloor}$$

This illustrates one step that should always be carried out in simplifying radicals: **NOTE ▶** *Always remove all perfect nth-power factors from the radicand of a radical of order n.*   ---------■

■EXAMPLE 4   (a) $\sqrt{72} = \sqrt{(36)(2)} = \sqrt{36}\,\sqrt{2} = 6\sqrt{2}$
$$\quad\quad\quad\quad\quad \underset{\text{perfect square}}{\big\lfloor}$$

(b) $\sqrt{a^3 b^2} = \sqrt{(a^2)(a)(b^2)} = \sqrt{a^2}\,\sqrt{a}\,\sqrt{b^2} = ab\sqrt{a}$
$$\quad\quad\quad \underset{\text{perfect squares}}{\big\lfloor}$$

(c) cube root $\longrightarrow$ $\sqrt[3]{40} = \sqrt[3]{(8)(5)} = \sqrt[3]{8}\,\sqrt[3]{5} = 2\sqrt[3]{5}$
$$\quad\quad\quad\quad \underset{\text{perfect cube}}{\big\lfloor}$$

(d) fifth root $\longrightarrow$ $\sqrt[5]{64 x^8 y^{12}} = \sqrt[5]{(32)(2)(x^5)(x^3)(y^{10})(y^2)}$
$$\quad\quad\quad\quad\quad\quad\quad\quad\quad \underset{\text{perfect fifth powers}}{\big\lfloor}$$

$$= \sqrt[5]{(32)(x^5)(y^{10})}\,\sqrt[5]{2x^3 y^2}$$
$$= 2xy^2 \sqrt[5]{2x^3 y^2}\quad\quad\quad ---------■$$

NOTE ▶   The next two examples illustrate another procedure that is used to simplify certain radicals. This procedure is to *reduce the order of the radical,* when it is possible to do so.

■EXAMPLE 5 $\sqrt[6]{8} = \sqrt[6]{2^3} = 2^{3/6} = 2^{1/2} = \sqrt{2}$

In this example we started with a sixth root and ended with a square root. Thus, the order of the radical was reduced. Fractional exponents are often helpful when we perform this operation.   ---------■

■EXAMPLE 6   (a) $\sqrt[8]{16} = \sqrt[8]{2^4} = 2^{4/8} = 2^{1/2} = \sqrt{2}$

(b) $\dfrac{\sqrt[4]{9}}{\sqrt{3}} = \dfrac{\sqrt[4]{3^2}}{\sqrt{3}} = \dfrac{3^{2/4}}{3^{1/2}} = 1$

(c) $\dfrac{\sqrt[6]{8}}{\sqrt{7}} = \dfrac{\sqrt[6]{2^3}}{\sqrt{7}} = \dfrac{2^{1/2}}{7^{1/2}} = \sqrt{\dfrac{2}{7}}$

(d) $\sqrt[9]{27 x^6 y^{12}} = \sqrt[9]{3^3 x^6 y^9 y^3} = 3^{3/9} x^{6/9} y^{9/9} y^{3/9} = 3^{1/3} x^{2/3} y y^{1/3}$
$$= y\sqrt[3]{3x^2 y}\quad\quad\quad ---------■$$

If a radical is to be written in its *simplest form,* the two operations illustrated in the last four examples must be performed. Therefore, we have the following:

> **Steps to Reduce a Radical to Simplest Form**
> 1. *Remove all perfect nth-power factors from a radical of order n.*
> 2. *If possible, reduce the order of the radical.*

When working with fractions, it has traditionally been the practice to write a fraction with radicals in a form in which the denominator contains no radicals. Such a fraction was not considered to be in simplest form unless this was done. This step of simplification was performed primarily for ease of calculation, but with a calculator it does not matter to any extent that there is a radical in the denominator. However, the procedure of writing a radical in this form, called **rationalizing the denominator,** is at times useful for other purposes. Therefore, the following examples show how the process of rationalizing the denominator is carried out.

EXAMPLE 7   To write $\sqrt{\frac{2}{5}}$ in an equivalent form in which the denominator is not included under the radical sign, we *create a perfect square in the denominator* by multiplying the numerator and the denominator under the radical by 5. This gives us $\sqrt{\frac{10}{25}}$, which may be written as $\frac{1}{5}\sqrt{10}$ or $\frac{\sqrt{10}}{5}$. These steps are written as follows:

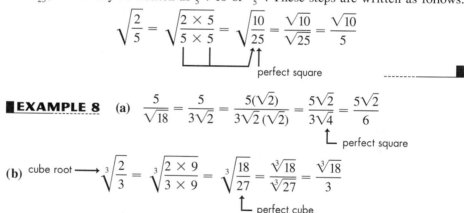

$$\sqrt{\frac{2}{5}} = \sqrt{\frac{2 \times 5}{5 \times 5}} = \sqrt{\frac{10}{25}} = \frac{\sqrt{10}}{\sqrt{25}} = \frac{\sqrt{10}}{5}$$

perfect square

EXAMPLE 8   (a) $\dfrac{5}{\sqrt{18}} = \dfrac{5}{3\sqrt{2}} = \dfrac{5(\sqrt{2})}{3\sqrt{2}\,(\sqrt{2})} = \dfrac{5\sqrt{2}}{3\sqrt{4}} = \dfrac{5\sqrt{2}}{6}$

perfect square

(b) cube root $\longrightarrow$ $\sqrt[3]{\dfrac{2}{3}} = \sqrt[3]{\dfrac{2 \times 9}{3 \times 9}} = \sqrt[3]{\dfrac{18}{27}} = \dfrac{\sqrt[3]{18}}{\sqrt[3]{27}} = \dfrac{\sqrt[3]{18}}{3}$

perfect cube

In (a), a perfect square was made by multiplying by $\sqrt{2}$. We can indicate this as we have shown or by multiplying the numerator by $\sqrt{2}$ and multiplying the 2 under the radical in the denominator by 2, in which case the denominator would be $3\sqrt{2 \times 2}$. In (b), we want a perfect cube, since a cube root is being found.

EXAMPLE 9   The period T (in s) for one cycle of a simple pendulum is given by $T = 2\pi\sqrt{L/g}$, where L is the length of the pendulum and g is the acceleration due to gravity. Rationalize the denominator on the right side of this equation if $L = 3.0$ ft and $g = 32$ ft/s^2.

Substituting, and then rationalizing, we have

$$T = 2\pi\sqrt{\frac{3.0}{32}} = 2\pi\sqrt{\frac{3.0 \times 2}{32 \times 2}} = 2\pi\sqrt{\frac{6.0}{64}} = \frac{2\pi}{8}\sqrt{6.0} = \frac{\pi}{4}\sqrt{6.0} = 1.9 \text{ s}$$

See Appendix C for a graphing calculator program TBLROOTS. It displays a table of square roots and cube roots.

EXAMPLE 10 Simplify $\sqrt{\dfrac{1}{2a^2} + \dfrac{2}{b^2}}$ and rationalize the denominator.

first combine fractions over
lowest common denominator $\longrightarrow$
$$\sqrt{\frac{1}{2a^2} + \frac{2}{b^2}} = \sqrt{\frac{b^2 + 4a^2}{2a^2b^2}} = \frac{\sqrt{b^2 + 4a^2}}{ab\sqrt{2}} \longleftarrow \text{sum of squares—radical cannot be simplified}$$

$$= \frac{\sqrt{b^2 + 4a^2}\sqrt{2}}{ab\sqrt{2 \times 2}} = \frac{\sqrt{2(b^2 + 4a^2)}}{2ab}$$

EXERCISES *11-3*

In Exercises 1–56, write each expression in simplest radical form. If a radical appears in the denominator, rationalize the denominator.

1. $\sqrt{24}$
2. $\sqrt{150}$
3. $\sqrt{45}$
4. $\sqrt{98}$
5. $\sqrt{x^2y^5}$
6. $\sqrt{pq^2r^7}$
7. $\sqrt{x^2y^4z^3}$
8. $\sqrt{12ab^2}$
9. $\sqrt{18a^3bc^4}$
10. $\sqrt{54m^5n^3}$
11. $\sqrt[3]{16}$
12. $\sqrt[4]{48}$
13. $\sqrt[5]{96}$
14. $\sqrt[3]{-16}$
15. $\sqrt[3]{8a^2}$
16. $\sqrt[3]{5a^4b^2}$
17. $\sqrt[4]{64r^3s^4t^5}$
18. $\sqrt[5]{16x^5y^3z^{11}}$
19. $\sqrt[3]{8}\sqrt[3]{4}$
20. $\sqrt[3]{4}\sqrt[3]{64}$
21. $\sqrt[3]{ab^4}\sqrt[3]{a^2b}$
22. $\sqrt[4]{3m^5n^8}\sqrt[4]{9mn}$
23. $\sqrt{\dfrac{3}{2}}$
24. $\sqrt{\dfrac{a}{b^3}}$
25. $\sqrt[3]{\dfrac{3}{4}}$
26. $\sqrt[4]{\dfrac{2}{5}}$
27. $\sqrt[5]{\dfrac{1}{9}}$
28. $\sqrt[6]{\dfrac{5}{4}}$
29. $\sqrt[4]{400}$
30. $\sqrt[8]{81}$
31. $\sqrt[6]{64}$
32. $\sqrt[9]{27}$
33. $\sqrt{4 \times 10^4}$
34. $\sqrt{4 \times 10^5}$
35. $\sqrt{4 \times 10^6}$
36. $\sqrt[3]{16 \times 10^5}$
37. $\sqrt[4]{4a^2}$
38. $\sqrt[6]{b^2c^4}$
39. $\sqrt[4]{\dfrac{1}{4}}$
40. $\dfrac{\sqrt[4]{320}}{\sqrt[4]{5}}$
41. $\sqrt[4]{\sqrt[3]{16}}$
42. $\sqrt[5]{\sqrt[4]{9}}$
43. $\sqrt{\sqrt{\sqrt{2}}}$
44. $\sqrt{b^4\sqrt{a}}$
45. $\sqrt{\dfrac{1}{2} - \dfrac{1}{3}}$
46. $\sqrt{\dfrac{5}{4} - \dfrac{1}{8}}$
47. $\sqrt{\dfrac{1}{a^2} + \dfrac{1}{b}}$
48. $\sqrt{\dfrac{x}{y} + \dfrac{y}{x}}$
49. $\sqrt{\dfrac{x}{x^2 + 1}}$
50. $\sqrt{\dfrac{x - 2}{x + 2}}$
51. $\sqrt{a^2 + 2ab + b^2}$
52. $\sqrt{a^2 + b^2}$
53. $\sqrt{4x^2 - 1}$
54. $\sqrt{9x^2 - 6x + 1}$
55. $\sqrt{x^2 + \dfrac{1}{4}}$
56. $\sqrt{\dfrac{1}{2} + 2r + 2r^2}$

In Exercises 57–60, perform the required operation.

(W) 57. An approximate equation for the efficiency E (in percent) of an engine is $E = 100(1 - 1/\sqrt[5]{R^2})$, where R is the compression ratio. Explain how this equation can be written with fractional exponents and then find E for $R = 7.35$.

58. When analyzing the velocity of an object that falls through a very great distance, the expression $a\sqrt{2g/a}$ is derived. Show by rationalizing the denominator that this expression takes on a simpler form.

59. A formula for the angular velocity ω of a disc is
$$\omega = \sqrt{\frac{576EIg}{WL^3}}.$$ Rationalize the denominator on the right side.

60. In analyzing an electronic filter circuit, the expression
$$\frac{8A}{\pi^2\sqrt{1 + (f_0/f)^2}}$$ is used. Rationalize the denominator, expressing the answer without the fraction f_0/f.

11-4 ADDITION AND SUBTRACTION OF RADICALS

When we add or subtract algebraic expressions, we combine similar terms, those that differ only in numerical coefficients. This is true when adding or subtracting radicals. *The radicals must be similar to perform the addition,* rather than simply indicate the addition. *Radicals are* **similar** *if they differ only in their numerical coefficients.* This means they must have the same order and have the same radicand.

In order to add radicals, we first express each radical in its simplest form, rationalize any denominators, and then combine those which are similar. For those that are not similar, we can only indicate the addition.

■**EXAMPLE 1**   **(a)** $2\sqrt{7} - 5\sqrt{7} + \sqrt{7} = -2\sqrt{7}$ all similar radicals

This result follows the distributive law, as it should. We can write

$$2\sqrt{7} - 5\sqrt{7} + \sqrt{7} = (2 - 5 + 1)\sqrt{7} = -2\sqrt{7}$$

We can also see that the terms combine just as

$$2x - 5x + x = -2x$$

(b) $\sqrt[5]{6} + 4\sqrt[5]{6} - 2\sqrt[5]{6} = 3\sqrt[5]{6}$ all similar radicals

(c) $\sqrt{5} + 2\sqrt{3} - 5\sqrt{5} = 2\sqrt{3} - 4\sqrt{5}$ answer contains two terms

similar radicals

We note in (c) that we are only able to indicate the final subtraction since the radicals are not similar.

■**EXAMPLE 2**   **(a)** $\sqrt{2} + \sqrt{8} = \sqrt{2} + \sqrt{4 \times 2} = \sqrt{2} + \sqrt{4}\sqrt{2}$
$$= \sqrt{2} + 2\sqrt{2} = 3\sqrt{2}$$

(b) $\sqrt[3]{24} + \sqrt[3]{81} = \sqrt[3]{8 \times 3} + \sqrt[3]{27 \times 3} = \sqrt[3]{8}\sqrt[3]{3} + \sqrt[3]{27}\sqrt[3]{3}$
$$= 2\sqrt[3]{3} + 3\sqrt[3]{3} = 5\sqrt[3]{3}$$

CAUTION ▶ Notice that $\sqrt{8}$, $\sqrt[3]{24}$, and $\sqrt[3]{81}$ were simplified before performing the addition. We also note that $\sqrt{2} + \sqrt{8}$ *is* **not** *equal to* $\sqrt{2 + 8}$.

We note in the illustrations of Example 2 that the radicals do not initially appear to be similar. However, after each is simplified we are able to recognize the similar radicals.

■**EXAMPLE 3**   **(a)** $6\sqrt{7} - \sqrt{28} + 3\sqrt{63} = 6\sqrt{7} - \sqrt{4 \times 7} + 3\sqrt{9 \times 7}$
$$= 6\sqrt{7} - 2\sqrt{7} + 3(3\sqrt{7})$$
$$= 6\sqrt{7} - 2\sqrt{7} + 9\sqrt{7}$$
$$= 13\sqrt{7} \qquad \text{all similar radicals}$$

(b) $3\sqrt{125} - \sqrt{20} + \sqrt{27} = 3\sqrt{25 \times 5} - \sqrt{4 \times 5} + \sqrt{9 \times 3}$
$$= 3(5\sqrt{5}) - 2\sqrt{5} + 3\sqrt{3}$$
$$= 13\sqrt{5} + 3\sqrt{3} \qquad \text{not similar to others}$$

■**EXAMPLE 4** $\sqrt{24} + \sqrt{\dfrac{3}{2}} = \sqrt{4 \times 6} + \sqrt{\dfrac{3 \times 2}{2 \times 2}} = \sqrt{4}\sqrt{6} + \sqrt{\dfrac{6}{4}}$

$$= 2\sqrt{6} + \frac{\sqrt{6}}{2} = \frac{4\sqrt{6} + \sqrt{6}}{2} = \frac{5}{2}\sqrt{6}$$

One radical was simplified by removing the perfect square factor, and in the other we rationalized the denominator. Note that we would not be able to combine the radicals if we did not rationalize the denominator of the second radical.

Our main purpose in this section is to add radicals in radical form. However, a decimal value can be obtained by use of a calculator, and in the following example we use decimal values to verify the result of the addition.

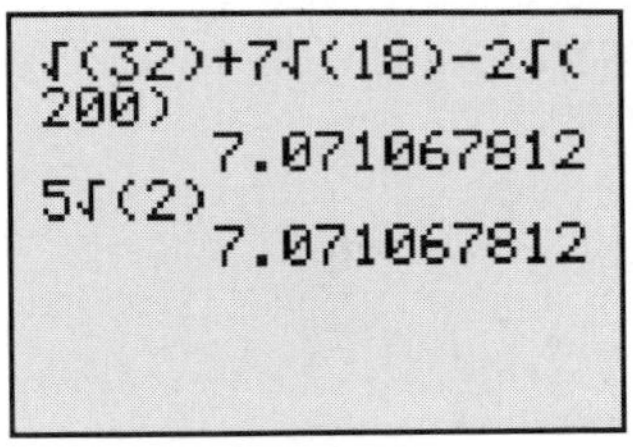

```
√(32)+7√(18)-2√(
200)
        7.071067812
5√(2)
        7.071067812
```

Fig. 11-5

■**EXAMPLE 5**   Use a calculator to verify the result of the following addition.

$$\sqrt{32} + 7\sqrt{18} - 2\sqrt{200} = \sqrt{16 \times 2} + 7\sqrt{9 \times 2} - 2\sqrt{100 \times 2}$$
$$= 4\sqrt{2} + 7(3\sqrt{2}) - 2(10\sqrt{2})$$
$$= 4\sqrt{2} + 21\sqrt{2} - 20\sqrt{2} = 5\sqrt{2}$$

The calculator display that verifies this addition is shown in Fig. 11-5. ■

Next is an example of adding radical expressions that contain literal numbers.

■**EXAMPLE 6**

$$\sqrt{\frac{2}{3a}} - 2\sqrt{\frac{3}{2a}} = \sqrt{\frac{2(3a)}{3a(3a)}} - 2\sqrt{\frac{3(2a)}{2a(2a)}} = \sqrt{\frac{6a}{9a^2}} - 2\sqrt{\frac{6a}{4a^2}}$$
$$= \frac{1}{3a}\sqrt{6a} - \frac{2}{2a}\sqrt{6a} = \frac{1}{3a}\sqrt{6a} - \frac{1}{a}\sqrt{6a}$$
$$= \frac{\sqrt{6a} - 3\sqrt{6a}}{3a} = \frac{-2\sqrt{6a}}{3a} = -\frac{2}{3a}\sqrt{6a}$$ ■

EXERCISES *11-4*

In Exercises 1–36, express each radical in simplest form, rationalize denominators, and perform the indicated operations.

1. $2\sqrt{3} + 5\sqrt{3}$

2. $8\sqrt{11} - 3\sqrt{11}$

3. $2\sqrt{7} + \sqrt{5} - 3\sqrt{7}$

4. $8\sqrt{6} - 2\sqrt{3} - 5\sqrt{6}$

5. $\sqrt{5} + \sqrt{20}$

6. $\sqrt{7} + \sqrt{63}$

7. $2\sqrt{3} - 3\sqrt{12}$

8. $4\sqrt{2} - \sqrt{50}$

9. $\sqrt{8a} - \sqrt{32a}$

10. $\sqrt{27x} + 2\sqrt{18x}$

11. $2\sqrt{28} + 3\sqrt{175}$

12. $5\sqrt{300} - 7\sqrt{48}$

13. $2\sqrt{20} - \sqrt{125} - \sqrt{45}$

14. $2\sqrt{44} - \sqrt{99} + \sqrt{2}\sqrt{88}$

15. $3\sqrt{75R} + 2\sqrt{48R} - 2\sqrt{18R}$

16. $2\sqrt{28} - \sqrt{108} - 2\sqrt{175}$

17. $\sqrt{60} + \sqrt{\frac{5}{3}}$

18. $\sqrt{84} - \sqrt{\frac{3}{7}}$

19. $\sqrt{\frac{1}{2}} + \sqrt{\frac{25}{2}} - \sqrt{18}$

20. $\sqrt{6} - \sqrt{\frac{2}{3}} - \sqrt{18}$

21. $\sqrt[3]{81} + \sqrt[3]{3000}$

22. $\sqrt[3]{-16} + \sqrt[3]{54}$

23. $\sqrt[4]{32} - \sqrt[8]{4}$

24. $\sqrt[6]{\sqrt{2}} - \sqrt[12]{2^{13}}$

25. $\sqrt{a^3b} - \sqrt{4ab^5}$

26. $\sqrt{2x^2y} + \sqrt[3]{8}\sqrt{y^5}$

27. $\sqrt{6}\sqrt{5}\sqrt{3} - \sqrt{40a^2}$

28. $\sqrt{60b^2n} - b\sqrt{135n}$

29. $\sqrt[3]{24a^2b^4} - \sqrt[3]{3a^5b}$

30. $\sqrt[3]{32a^6b^4} + 3a\sqrt[5]{243ab^9}$

31. $\sqrt{\dfrac{a}{c^5}} - \sqrt{\dfrac{c}{a^3}}$

32. $\sqrt{\dfrac{2x}{3y}} + \sqrt{\dfrac{27y}{8x}}$

33. $\sqrt[3]{\dfrac{a}{b}} - \sqrt[3]{\dfrac{8b^2}{a^2}}$

34. $\sqrt[4]{\dfrac{c}{b}} - \sqrt[4]{bc}$

35. $\sqrt{\dfrac{a-b}{a+b}} - \sqrt{\dfrac{a+b}{a-b}}$

36. $\sqrt{\dfrac{16}{x} + 8 + x} - \sqrt{1 - \dfrac{1}{x}}$

In Exercises 37–40, express each radical in simplest form, rationalize denominators, and perform the indicated operations. Then use a calculator to verify the result.

37. $3\sqrt{45} + 3\sqrt{75} - 2\sqrt{500}$

38. $2\sqrt{40} + 3\sqrt{90} - 5\sqrt{250}$

39. $2\sqrt{\frac{2}{3}} + \sqrt{24} - 5\sqrt{\frac{3}{2}}$

40. $\sqrt{\frac{2}{7}} - 2\sqrt{\frac{7}{2}} + 5\sqrt{56}$

In Exercises 41–44, solve the given problems.

41. Find the exact sum of the positive roots of $x^2 - 2x - 2 = 0$ and $x^2 + 2x - 11 = 0$.

(W) 42. For the quadratic equation $ax^2 + bx + c = 0$, if a, b, and c are integers, the sum of the roots is a rational number. Explain.

43. A rectangular piece of plywood 4 ft by 8 ft has corners cut from it as shown in Fig. 11-6. Find the perimeter of the remaining piece in exact form and in decimal form.

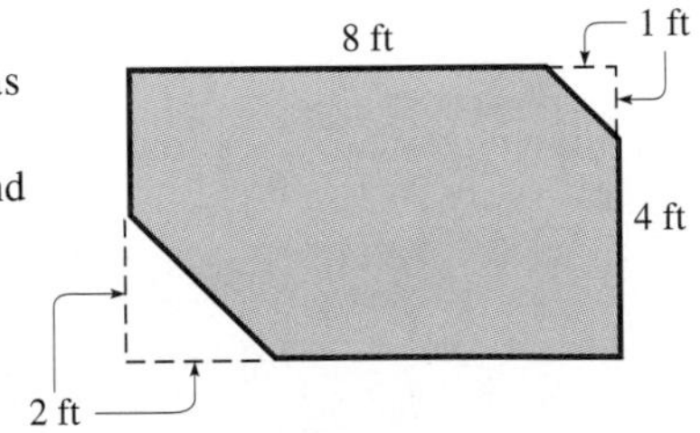

Fig. 11-6

44. Three squares with areas of 150 cm², 54 cm², and 24 cm² are displayed on a computer monitor. What is the sum (in radical form) of the perimeters of these squares?

11-5 MULTIPLICATION AND DIVISION OF RADICALS

For reference, Eq. (11-10) is
$\sqrt[n]{a}\sqrt[n]{b} = \sqrt[n]{ab}$.

When multiplying expressions containing radicals, we use Eq. (11-10), along with the normal procedures of algebraic multiplication. *Note that the orders of the radicals being multiplied in Eq. (11-10) are the same.* The following examples illustrate the method.

EXAMPLE 1 (a) $\sqrt{5}\sqrt{2} = \sqrt{5 \times 2} = \sqrt{10}$

(b) $\sqrt{33}\sqrt{3} = \sqrt{33 \times 3} = \sqrt{99} = \sqrt{9 \times 11} = \sqrt{9}\sqrt{11}$
$= 3\sqrt{11}$ ⌞ perfect square

or $\sqrt{33}\sqrt{3} = \sqrt{33 \times 3} = \sqrt{11 \times 3 \times 3}$
$= 3\sqrt{11}$

Note that we express the resulting radical in simplest form. ∎

EXAMPLE 2 (a) $\sqrt[3]{6}\sqrt[3]{4} = \sqrt[3]{6(4)} = \sqrt[3]{24} = \sqrt[3]{8}\sqrt[3]{3}$
$= 2\sqrt[3]{3}$ ⌞ perfect cube

(b) $\sqrt[5]{8a^3b^4}\sqrt[5]{8a^2b^3} = \sqrt[5]{(8a^3b^4)(8a^2b^3)} = \sqrt[5]{64a^5b^7} = \sqrt[5]{32a^5b^5}\sqrt[5]{2b^2}$
$= 2ab\sqrt[5]{2b^2}$ perfect fifth power ⌝ ∎

EXAMPLE 3 $\sqrt{2}(3\sqrt{5} - 4\sqrt{2}) = 3\sqrt{2}\sqrt{5} - 4\sqrt{2}\sqrt{2} = 3\sqrt{10} - 4\sqrt{4}$
$= 3\sqrt{10} - 4(2) = 3\sqrt{10} - 8$ ∎

EXAMPLE 4

$(5\sqrt{7} - 2\sqrt{3})(4\sqrt{7} + 3\sqrt{3}) = (5\sqrt{7})(4\sqrt{7}) + (5\sqrt{7})(3\sqrt{3}) - (2\sqrt{3})(4\sqrt{7}) - (2\sqrt{3})(3\sqrt{3})$
$= (5)(4)\sqrt{7}\sqrt{7} + (5)(3)\sqrt{7}\sqrt{3} - (2)(4)\sqrt{3}\sqrt{7} - (2)(3)\sqrt{3}\sqrt{3}$
$= 20(7) + 15\sqrt{21} - 8\sqrt{21} - 6(3)$
$= 140 + 7\sqrt{21} - 18$
$= 122 + 7\sqrt{21}$

```
(5√(7)-2√(3))(4√
(7)+3√(3))
        154.0780299
122+7√(21)
        154.0780299
```

Fig. 11-7

The calculator display that verifies this multiplication is shown in Fig. 11-7. ∎

When raising a single-term radical expression to a power, we use the basic meaning of the power. When raising a binomial to a power, we proceed as with any binomial. We use Eqs. (6-3) and (6-4) along with Eq. (11-10) if the binomial is squared. These are illustrated in the next two examples.

For reference, Eqs. (6-3) and (6-4) are
$(x + y)^2 = x^2 + 2xy + y^2$
$(x - y)^2 = x^2 - 2xy + y^2$

EXAMPLE 5 (a) $(2\sqrt{7})^2 = 2^2(\sqrt{7})^2 = 4(7) = 28$
(b) $(2\sqrt{7})^3 = 2^3(\sqrt{7})^3 = 8(\sqrt{7})^2(\sqrt{7}) = 8(7)\sqrt{7}$
$= 56\sqrt{7}$ ∎

EXAMPLE 6 (a) $(3 + \sqrt{5})^2 = 3^2 + 2(3)\sqrt{5} + (\sqrt{5})^2 = 9 + 6\sqrt{5} + 5$
$= 14 + 6\sqrt{5}$
(b) $(\sqrt{a} - \sqrt{b})^2 = (\sqrt{a})^2 - 2\sqrt{a}\sqrt{b} + (\sqrt{b})^2$
$= a + b - 2\sqrt{ab}$ ∎

CAUTION ▶

Again, we note that *to multiply radicals and combine them under one radical sign, it is necessary that the order of the radicals be the same.* If necessary we can make the order of each radical the same by appropriate operations on each radical separately. Fractional exponents are frequently useful for this purpose.

EXAMPLE 7 (a) $\sqrt[3]{2}\sqrt{5} = 2^{1/3}5^{1/2} = 2^{2/6}5^{3/6} = (2^2 5^3)^{1/6} = \sqrt[6]{500}$

(b) $\sqrt[3]{4a^2b}\,\sqrt[4]{8a^3b^2} = (2^2 a^2 b)^{1/3}(2^3 a^3 b^2)^{1/4} = (2^2 a^2 b)^{4/12}(2^3 a^3 b^2)^{3/12}$

$\qquad = (2^8 a^8 b^4)^{1/12}(2^9 a^9 b^6)^{1/12} = (2^{17} a^{17} b^{10})^{1/12}$

$\qquad = 2a(2^5 a^5 b^{10})^{1/12}$

$\qquad = 2a\sqrt[12]{32a^5 b^{10}}$

Division of Radicals

We already have dealt with some cases of division of radicals in the previous sections. When we have had the indicated division of one radical by another, we generally have expressed the answer with no radicals in the denominator by rationalizing the denominator. Although generally no longer important for calculations, the process of rationalizing can make the form of some radical expressions much simpler. Rationalizing *numerators* is also an important step in developing certain expressions in more advanced mathematics. Therefore, if a change is to be made in an expression involving division by a radical, rationalizing the denominator or the numerator is the principal step to be carried out.

When we rationalize the denominator, we change the fraction to an equivalent form in which the denominator is free of radicals. In doing so, multiplication of the numerator and the denominator by the proper quantity is the first important step. We already have dealt with the simpler forms like $a/\sqrt{b}$ in Section 11-3. We now consider fractions in which the denominator is the sum or difference of terms.

If the denominator is the sum (or difference) of two terms, at least one of which is a radical, the fraction can be rationalized by multiplying both the numerator and the denominator by the difference (or sum) of the same two terms, if the radicals are square roots.

EXAMPLE 8 The fraction $\dfrac{1}{\sqrt{3} - \sqrt{2}}$

can be rationalized by multiplying the numerator and the denominator by $\sqrt{3} + \sqrt{2}$. In this way the radicals will be removed from the denominator.

$$\frac{1}{\sqrt{3} - \sqrt{2}} \times \frac{\sqrt{3} + \sqrt{2}}{\sqrt{3} + \sqrt{2}} = \frac{\sqrt{3} + \sqrt{2}}{(\sqrt{3})^2 - (\sqrt{2})^2} = \frac{\sqrt{3} + \sqrt{2}}{3 - 2} = \sqrt{3} + \sqrt{2}$$

change sign

The reason this technique works is that an expression of the form $a^2 - b^2$ is created in the denominator, where a or b (or both) is a radical. We see that the result is a denominator free of radicals.

EXAMPLE 9 Rationalize the denominator of $\dfrac{3\sqrt{6} - \sqrt{2}}{\sqrt{6} + \sqrt{2}}$ and simplify the result.

$$\frac{3\sqrt{6} - \sqrt{2}}{\sqrt{6} + \sqrt{2}}\left(\frac{\sqrt{6} - \sqrt{2}}{\sqrt{6} - \sqrt{2}}\right) = \frac{18 - 4\sqrt{12} + 2}{4}$$

change sign

$$= \frac{20 - 4(2\sqrt{3})}{4} = 5 - 2\sqrt{3}$$

We note that, after rationalizing the denominator, the result has a much simpler form than the original expression.

EXAMPLE 10 In studying the properties of a semiconductor, the expression

$$\frac{\sqrt{1 + 2V}}{C_1 + C_2\sqrt{1 + 2V}}$$

is used. Here, C_1 and C_2 are constants and V is the voltage across a junction of the semiconductor. Rationalize the denominator of this expression.

Multiplying numerator and denominator by $C_1 - C_2\sqrt{1 + 2V}$, we have

$$\frac{\sqrt{1 + 2V}}{C_1 + C_2\sqrt{1 + 2V}} = \frac{\sqrt{1 + 2V}(C_1 - C_2\sqrt{1 + 2V})}{(C_1 + C_2\sqrt{1 + 2V})(C_1 - C_2\sqrt{1 + 2V})}$$

$$= \frac{C_1\sqrt{1 + 2V} - C_2\sqrt{1 + 2V}\,\sqrt{1 + 2V}}{C_1^2 - C_2^2(\sqrt{1 + 2V})^2}$$

$$= \frac{C_1\sqrt{1 + 2V} - C_2(1 + 2V)}{C_1^2 - C_2^2(1 + 2V)}$$

As we noted earlier, in certain types of algebraic operations it may be necessary to rationalize the numerator of an expression. This procedure is illustrated in the following example.

EXAMPLE 11 Rationalize the numerator of the expression $\dfrac{\sqrt{2x + 3} - \sqrt{2x}}{6}$.

We follow the same basic procedure as in rationalizing the denominator of an expression. In this case we multiply the numerator and the denominator of this fraction by $\sqrt{2x + 3} + \sqrt{2x}$. This gives us the following solution:

$$\frac{\sqrt{2x + 3} - \sqrt{2x}}{6} = \frac{(\sqrt{2x + 3} - \sqrt{2x})(\sqrt{2x + 3} + \sqrt{2x})}{6(\sqrt{2x + 3} + \sqrt{2x})}$$

$$= \frac{(\sqrt{2x + 3})^2 - (\sqrt{2x})^2}{6(\sqrt{2x + 3} + \sqrt{2x})} = \frac{(2x + 3) - 2x}{6(\sqrt{2x + 3} + \sqrt{2x})}$$

$$= \frac{3}{6(\sqrt{2x + 3} + \sqrt{2x})} = \frac{1}{2(\sqrt{2x + 3} + \sqrt{2x})}$$

Note that the resulting *numerator* does not contain any radicals.

EXERCISES *11-5*

In Exercises 1–48, perform the indicated operations, expressing answers in simplest form with rationalized denominators.

1. $\sqrt{3}\sqrt{10}$
2. $\sqrt{2}\sqrt{51}$
3. $\sqrt{6}\sqrt{2}$
4. $\sqrt{7}\sqrt{14}$

5. $\sqrt[3]{4}\sqrt[3]{2}$
6. $\sqrt[5]{4}\sqrt[5]{16}$
7. $(5\sqrt{2})^2$
8. $(3\sqrt{5})^3$

9. $\sqrt{8}\sqrt{\frac{5}{2}}$
10. $\sqrt[6]{\frac{6}{7}}\sqrt[6]{\frac{2}{3}}$

11. $\sqrt{3}(\sqrt{2} - \sqrt{5})$
12. $3\sqrt{5}(\sqrt{15} - 2\sqrt{5})$

13. $(2 - \sqrt{5})(2 + \sqrt{5})$
14. $(2 - \sqrt{5})^2$

15. $(3\sqrt{5} - 2\sqrt{3})(6\sqrt{5} + 7\sqrt{3})$

16. $(3\sqrt{7a} - \sqrt{8})(\sqrt{7a} + \sqrt{2})$

17. $(3\sqrt{11} - \sqrt{x})(2\sqrt{11} + 5\sqrt{x})$

18. $(2\sqrt{10} + 3\sqrt{15})(\sqrt{10} - 7\sqrt{15})$

19. $\sqrt{a}(\sqrt{ab} + \sqrt{c^3})$
20. $\sqrt{3x}(\sqrt{3x} - \sqrt{xy})$

21. $\dfrac{\sqrt{6} - 3}{\sqrt{6}}$
22. $\dfrac{5 - \sqrt{10}}{\sqrt{10}}$

23. $\dfrac{\sqrt{2a} - b}{\sqrt{a}}$
24. $\dfrac{\sqrt{7x} - \sqrt{14}}{\sqrt{7}}$

25. $(\sqrt{2a} - \sqrt{b})(\sqrt{2a} + 3\sqrt{b})$ 26. $(2\sqrt{mn} + 3\sqrt{n})^2$

27. $\sqrt{2}\sqrt[3]{3}$
28. $\sqrt[5]{16}\sqrt[3]{8}$

29. $\dfrac{1}{\sqrt{7} + \sqrt{3}}$
30. $\dfrac{\sqrt{8}}{2\sqrt{3} - \sqrt{5}}$

31. $\dfrac{3}{2\sqrt{5} - 6}$
32. $\dfrac{6\sqrt{5}}{5 - 2\sqrt{5}}$

33. $\dfrac{\sqrt{2} - 1}{\sqrt{7} - 3\sqrt{2}}$
34. $\dfrac{2\sqrt{15} - 3}{\sqrt{15} + 4}$

35. $\dfrac{2\sqrt{3} - 5\sqrt{5}}{\sqrt{3} + 2\sqrt{5}}$
36. $\dfrac{\sqrt{15} - 3\sqrt{5}}{2\sqrt{15} - \sqrt{5}}$

37. $\dfrac{2\sqrt{x}}{\sqrt{x} - \sqrt{y}}$
38. $\dfrac{6\sqrt{a}}{2\sqrt{a} - b}$

39. $\dfrac{8}{3\sqrt{a} - 2\sqrt{b}}$
40. $\dfrac{\sqrt{2c} + 3d}{\sqrt{2c} - d}$

41. $(\sqrt[5]{\sqrt{6}} - \sqrt{5})(\sqrt[5]{\sqrt{6}} + \sqrt{5})$

42. $\sqrt[3]{5} - \sqrt{17}\sqrt[3]{5} + \sqrt{17}$

43. $\left(\sqrt{\dfrac{2}{a}} + \sqrt{\dfrac{a}{2}}\right)\left(\sqrt{\dfrac{2}{a}} - 2\sqrt{\dfrac{a}{2}}\right)$

44. $(3 + \sqrt{6 - 2a})(2 - \sqrt{6 - 2a})$

45. $\dfrac{\sqrt{x + y}}{\sqrt{x - y} - \sqrt{x}}$
46. $\dfrac{\sqrt{1 + a}}{a - \sqrt{1 - a}}$

47. $\dfrac{\sqrt{x^2 + y^2}}{\sqrt{x^{-2} + y^{-2}}}$
48. $\dfrac{\sqrt{x^4 - y^4}}{\sqrt{x^{-2} - y^{-2}}}$

In Exercises 49–52, perform the indicated operations, expressing answers in simplest form with rationalized denominators. Then verify the result with a calculator.

49. $(\sqrt{11} + \sqrt{6})(\sqrt{11} - 2\sqrt{6})$

50. $(2\sqrt{5} - \sqrt{7})(3\sqrt{5} + \sqrt{7})$

51. $\dfrac{2\sqrt{6} - \sqrt{5}}{3\sqrt{6} - 4\sqrt{5}}$
52. $\dfrac{\sqrt{7} - 4\sqrt{2}}{5\sqrt{7} - 4\sqrt{2}}$

In Exercises 53–56, combine the terms into a single fraction, but do not rationalize the denominators.

53. $2\sqrt{x} + \dfrac{1}{\sqrt{x}}$
54. $\dfrac{3}{2\sqrt{3x - 4}} - \sqrt{3x - 4}$

55. $\dfrac{x^2}{\sqrt{2x + 1}} + 2x\sqrt{2x + 1}$
56. $4\sqrt{x^2 + 1} - \dfrac{4x}{\sqrt{x^2 + 1}}$

In Exercises 57–60, rationalize the numerator of each fraction.

57. $\dfrac{\sqrt{5} + \sqrt{2}}{3\sqrt{6}}$
58. $\dfrac{\sqrt{19} - 3}{5}$

59. $\dfrac{\sqrt{x + h} - \sqrt{x}}{h}$
60. $\dfrac{\sqrt{3x + 4} + \sqrt{3x}}{8}$

In Exercises 61–68, solve the given problems.

61. By substitution, show that $x = 1 - \sqrt{2}$ is a solution to the equation $x^2 - 2x - 1 = 0$.

(W) 62. For the quadratic equation $ax^2 + bx + c = 0$, if a, b, and c are integers, the product of the roots is a rational number. Explain.

63. For an object oscillating at the end of a spring and on which there is a force that retards the motion, the equation $m^2 + bm + k^2 = 0$ must be solved. Here, b is a constant related to the retarding force, and k is the spring constant. By substitution, show that $m = \frac{1}{2}(\sqrt{b^2 - 4k^2} - b)$ is a solution.

64. A square plastic sheet of side x is stretched by an amount equal to $\sqrt{x}$ in each direction. Find the expression for the percent increase in area of the sheet.

65. Among the products of a specialty furniture company are tables with tops in the shape of a regular octagon (8 sides). Express the area A of a table top as a function of the side s of the octagon.

66. An expression used in determining the characteristics of a spur gear is $\dfrac{50}{50 + \sqrt{V}}$. Rationalize the denominator.

67. In finding the time to drain a tank, the expression $\dfrac{h_2 - h_1}{\sqrt{2g}(\sqrt{h_2} - \sqrt{h_1})}$ is used. Rationalize the denominator.

68. In analyzing a tuned amplifier circuit, the expression $\dfrac{2Q}{\sqrt{\sqrt{2} - 1}}$ is used. Rationalize the denominator.

CHAPTER EQUATIONS

Exponents	$a^m \times a^n = a^{m+n}$	**(11-1)**
	$\dfrac{a^m}{a^n} = a^{m-n}$ or $\dfrac{a^m}{a^n} = \dfrac{1}{a^{n-m}}$ $(a \neq 0)$	**(11-2)**
	$(a^m)^n = a^{mn}$	**(11-3)**
	$(ab)^n = a^n b^n, \quad \left(\dfrac{a}{b}\right)^n = \dfrac{a^n}{b^n}$ $(b \neq 0)$	**(11-4)**
	$a^0 = 1 \quad (a \neq 0)$	**(11-5)**
	$a^{-n} = \dfrac{1}{a^n} \quad (a \neq 0)$	**(11-6)**
Fractional exponents	$a^{1/n} = \sqrt[n]{a}$	**(11-7)**
	$a^{m/n} = \sqrt[n]{a^m} = (\sqrt[n]{a})^m$	**(11-8)**
Radicals	$\sqrt[n]{a^n} = (\sqrt[n]{a})^n = a$	**(11-9)**
	$\sqrt[n]{a}\,\sqrt[n]{b} = \sqrt[n]{ab}$	**(11-10)**
	$\sqrt[m]{\sqrt[n]{a}} = \sqrt[mn]{a}$	**(11-11)**
	$\dfrac{\sqrt[n]{a}}{\sqrt[n]{b}} = \sqrt[n]{\dfrac{a}{b}} \quad (b \neq 0)$	**(11-12)**

REVIEW EXERCISES

In Exercises 1–28, express each expression in simplest form with only positive exponents.

1. $2a^{-2}b^0$ **2.** $(2c)^{-1}z^{-2}$ **3.** $\dfrac{2c^{-1}}{d^{-3}}$ **4.** $\dfrac{-5x^0}{3y^{-1}}$

5. $3(25)^{3/2}$ **6.** $32^{2/5}$ **7.** $400^{-3/2}$ **8.** $1000^{-2/3}$

9. $\left(\dfrac{3}{t^7}\right)^{-2}$ **10.** $\left(\dfrac{2x^3}{3}\right)^{-3}$ **11.** $\dfrac{-8^{2/3}}{49^{-1/2}}$ **12.** $\dfrac{81^{-0.75}}{6^{-3}}$

13. $(2a^{1/3}b^{5/6})^6$ **14.** $(ax^{-1/2}y^{1/4})^8$

15. $(-32m^{15}n^{10})^{3/5}$ **16.** $(27x^{-6}y^9)^{2/3}$

17. $2L^{-2} - 4C^{-1}$ **18.** $a^4a^{-1} + (a^4)^{-1}$

19. $\dfrac{2x^{-1}}{2x^{-1} + y^{-1}}$ **20.** $\dfrac{3a}{(2a)^{-1} - a}$

21. $(a - 3b^{-1})^{-1}$ **22.** $(2s^{-2} + t)^{-2}$

23. $(x^3 - y^{-3})^{1/3}$ **24.** $(W^2 + 2WH + H^2)^{-1/2}$

25. $(8a^3)^{2/3}(4a^{-2} + 1)^{1/2}$ **26.** $\left[\dfrac{(9a)^0(4x^2)^{1/3}(3b^{1/2})}{(2b^0)^2}\right]^{-6}$

27. $2x(x - 1)^{-2} - 2(x^2 + 1)(x - 1)^{-3}$

28. $4(1 - x^2)^{1/2} - (1 - x^2)^{-1/2}$

In Exercises 29–72, perform the indicated operations and express the result in simplest radical form with rationalized denominators.

29. $\sqrt{68}$ **30.** $\sqrt{96}$ **31.** $\sqrt{ab^5c^2}$ **32.** $\sqrt{x^3y^4z^6}$

33. $\sqrt{9a^3b^4}$ **34.** $\sqrt{8x^5y^2}$ **35.** $\sqrt{84st^3u^2}$ **36.** $\sqrt{52x^2y^5}$

37. $\dfrac{5}{\sqrt{2s}}$ **38.** $\dfrac{3a}{\sqrt{5x}}$ **39.** $\sqrt{\dfrac{11}{27}}$ **40.** $\sqrt{\dfrac{7}{8}}$

41. $\sqrt[4]{8m^6n^9}$ **42.** $\sqrt[3]{9a^7b^{-3}}$ **43.** $\sqrt[4]{\sqrt[3]{64}}$

44. $\sqrt{a^{-3}\sqrt[5]{b^{12}}}$ **45.** $\sqrt{200} + \sqrt{32}$ **46.** $2\sqrt{68x} - \sqrt{153x}$

47. $\sqrt{63} - 2\sqrt{112} - \sqrt{28}$ **48.** $2\sqrt{20} - \sqrt{80} - 2\sqrt{125}$

49. $a\sqrt{2x^3} + \sqrt{8a^2x^3}$ **50.** $2\sqrt{m^2n^3} - \sqrt{n^5}$

51. $\sqrt[3]{8a^4} + b\sqrt[3]{a}$ **52.** $\sqrt[4]{2xy^5} - \sqrt[4]{32xy}$

53. $\sqrt{5}(2\sqrt{5} - \sqrt{11})$ **54.** $2\sqrt{8}(5\sqrt{2} - \sqrt{6})$

55. $2\sqrt{2}(\sqrt{6} - \sqrt{10})$ **56.** $3\sqrt{5}(\sqrt{15} + 2\sqrt{35})$

57. $(2 - 3\sqrt{17})(3 + \sqrt{17})$ **58.** $(5\sqrt{6} - 4)(3\sqrt{6} + 5)$

59. $(2\sqrt{7} - 3\sqrt{a})(3\sqrt{7} + \sqrt{a})$

60. $(3\sqrt{2} - \sqrt{13})(5\sqrt{2} + 3\sqrt{13})$

61. $\dfrac{\sqrt{3x}}{2\sqrt{3x} - \sqrt{y}}$ **62.** $\dfrac{5\sqrt{a}}{2\sqrt{a} - c}$ **63.** $\dfrac{\sqrt{2}}{\sqrt{3} - 4\sqrt{2}}$

64. $\dfrac{4}{3 - 2\sqrt{7}}$

65. $\dfrac{\sqrt{7} - \sqrt{5}}{\sqrt{5} + 3\sqrt{7}}$

66. $\dfrac{4 - 2\sqrt{6}}{3 + 2\sqrt{6}}$

67. $\dfrac{2\sqrt{x} - a}{3\sqrt{x} + 5a}$

68. $\dfrac{3\sqrt{y} - \sqrt{z}}{2\sqrt{y} + 5\sqrt{z}}$

69. $\sqrt{4b^2 + 1}$

70. $\sqrt{a^{-2} + \dfrac{1}{b^2}}$

71. $\left(\dfrac{2 - \sqrt{15}}{2}\right)^2 - \left(\dfrac{2 - \sqrt{15}}{2}\right)$

72. $\sqrt{2 + \dfrac{b}{a} + \dfrac{a}{b}} + \sqrt{a^4b^2 + 2a^3b^2 + a^2b^2}$

In Exercises 73 and 74, perform the indicated operations and express the result in simplified radical form with rationalized denominators. In Exercises 75 and 76, without a calculator show that the given equations are true. Then verify each result with a calculator.

73. $(\sqrt{7} - 2\sqrt{15})(3\sqrt{7} - \sqrt{15})$

74. $\dfrac{2\sqrt{3} - 7\sqrt{14}}{3\sqrt{3} + 2\sqrt{14}}$

75. $\sqrt{\sqrt{2} - 1}\,(\sqrt{2} + 1) = \sqrt{\sqrt{2} + 1}$

76. $\sqrt{\sqrt{3} + 1}\,(\sqrt{3} - 1) = \sqrt{2}(\sqrt{3} - 1)$

In Exercises 77–88, perform the indicated operations.

77. The average annual increase i (in %) of the cost of living over n years is given by $i = 100[(C_2/C_1)^{1/n} - 1]$, where C_1 is the cost of living index for the first year and C_2 is the cost of living index for the last year of the period. Evaluate i if $C_1 = 130.7$ and $C_2 = 172.0$ are the values for 1990 and 2000, respectively.

78. Kepler's third law of planetary motion may be given as $T = kr^{3/2}$, where T is the time for one revolution of a planet around the sun, r is its mean radius from the sun, and $k = 1.115 \times 10^{-12}$ year/mi$^{3/2}$. Find the time for one revolution of Venus about the sun if $r = 6.73 \times 10^7$ mi.

79. The speed v of a ship of weight W whose engines produce power P is $v = k\sqrt[3]{P/W}$. Express this equation (a) with a fractional exponent and (b) as a radical with the denominator rationalized.

80. In analyzing the orbit of an earth satellite, the expression $\sqrt{1 + \dfrac{2E}{m}\left(\dfrac{h}{GM}\right)^2}$ is used. Combine terms under the radical and express the result with the denominator rationalized.

81. When studying atomic structure, the expression $\dfrac{v}{n_2^{-2} - n_1^{-2}}$ is used. Write this in simplest form with only positive exponents.

82. In hydrodynamics, the expression $v(4 + 3ar^{-1} - a^3r^{-3})^{-1}$ arises. Express it in simplest form with positive exponents.

83. A square is decreasing in size on a computer screen. To find how fast the side of the square changes, we must rationalize the numerator of $\dfrac{\sqrt{A + h} - \sqrt{A}}{h}$. Perform this operation.

84. A plane takes off 200 mi east of its destination, and a 50 mi/h wind is from the south. If the plane's velocity is 250 mi/h and its heading is always toward its destination, its path can be described as $y = 100[(0.005x)^{0.8} - (0.005x)^{1.2}]$, $0 \le x \le 200$ mi. Sketch the path and check using a graphing calculator.

85. In an experiment, a laser beam follows the path shown in Fig. 11-8. Express the length of the path in simplest radical form.

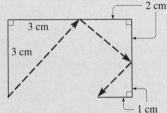

Fig. 11-8

86. The flow rate V (in ft/s) through a storm drain pipe is found from $V = 110(0.55)^{2/3}(0.0018)^{1/2}$. Find the value of V.

87. The frequency of a certain electric circuit is given by $\dfrac{1}{2\pi\sqrt{\dfrac{LC_1C_2}{C_1 + C_2}}}$ Express this in simplest rationalized radical form.

88. A computer analysis of an experiment showed that the fraction f of viruses surviving X-ray dosages was given by $f = \dfrac{20}{d + \sqrt{3d + 400}}$, where d is the dosage. Express this with the denominator rationalized.

Writing Exercise

89. In calculating the forces on a tower by the wind, it is necessary to evaluate $0.018^{0.13}$. Write a paragraph explaining why this form is preferable to the equivalent radical form.

PRACTICE TEST

In Problems 1–12, simplify the given expressions. For those with exponents, express each result with only positive exponents. For the radicals, rationalize the denominator where applicable.

1. $2\sqrt{20} - \sqrt{125}$

2. $\dfrac{100^{3/2}}{8^{-2/3}}$

3. $(as^{-1/3}t^{3/4})^{12}$

4. $(2x^{-1} + y^{-2})^{-1}$

5. $(\sqrt{2x} - 3\sqrt{y})^2$

6. $\sqrt[3]{\sqrt[4]{4}}$

7. $\dfrac{3 - 2\sqrt{2}}{2\sqrt{x}}$

8. $\sqrt{27a^4b^3}$

9. $2\sqrt{2}\,(3\sqrt{10} - \sqrt{6})$

10. $(2x + 3)^{1/2} + (x + 1)(2x + 3)^{-1/2}$

11. $\left(\dfrac{4a^{-1/2}b^{3/4}}{b^{-2}}\right)\left(\dfrac{b^{-1}}{2a}\right)$

12. $\dfrac{2\sqrt{15} + \sqrt{3}}{\sqrt{15} - 2\sqrt{3}}$

13. Express $\dfrac{3^{-1/2}}{2}$ in simplest radical form with a rationalized denominator.

14. In the study of fluid flow in pipes, the expression $0.220N^{-1/6}$ is found. Evaluate this expression for $N = 64 \times 10^6$.

CHAPTER 12 — COMPLEX NUMBERS

In Chapter 1, when we were introducing the topic of numbers, imaginary numbers were mentioned. Again, in Chapter 7 when we were considering quadratic equations and the character of their roots, we briefly came across this type of number. However, until this point we have purposely avoided any extended discussion of imaginary numbers. In this chapter we will discuss the properties and basic forms of *complex numbers,* which include the real numbers as well as the imaginary numbers.

Despite their names, complex numbers and imaginary numbers have very real and useful applications in mathematics and many technical areas. One of the most important areas of application occurs in electricity and electronics where these numbers are used extensively. Other applications are found in mechanical vibrations, optics, and acoustics.

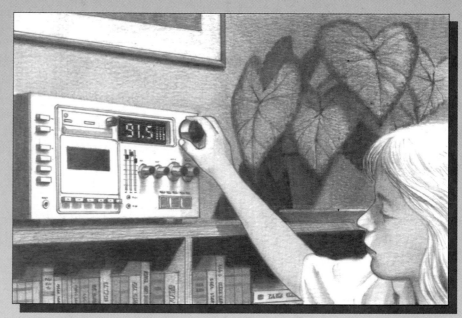

In Section 12-7 we see that tuning in a radio station involves a basic application of complex numbers to electricity.

12-1 BASIC DEFINITIONS

Since the square of any positive number or any negative number results in a positive number, we see that it is not possible to square any real number and have the result be a negative number. Therefore, we must define a new number system in order to include square roots of negative numbers. With the proper definitions, we shall find that these numbers can be used to great advantage in certain applications.

If the radicand in a square root is negative, we can express the indicated root as the product of $\sqrt{-1}$ and the square root of a positive number. *The symbol $\sqrt{-1}$ is defined as the* **imaginary unit** *and is denoted by the symbol j.* Therefore, we have

$$j = \sqrt{-1} \quad \text{and} \quad j^2 = -1 \qquad (12\text{-}1)$$

Generally, mathematicians use the symbol i for $\sqrt{-1}$, and therefore most nontechnical textbooks use i. However, one of the major technical applications of complex numbers is in electronics, where i represents electric current. Therefore, we shall use j for $\sqrt{-1}$, which is also the standard symbol in electronics textbooks.

The following examples illustrate the use of Eq. (12-1), in which we defined $j = \sqrt{-1}$ and $j^2 = -1$.

EXAMPLE 1 Express the following square roots in terms of j.

(a) $\sqrt{-9} = \sqrt{(9)(-1)} = \sqrt{9}\sqrt{-1} = 3j$

(b) $\sqrt{-0.25} = \sqrt{0.25}\sqrt{-1} = 0.5j$

(c) $\sqrt{-5} = \sqrt{5}\sqrt{-1} = \sqrt{5}j = j\sqrt{5}$ the form $j\sqrt{5}$ is better than $\sqrt{5}j$

NOTE▶ When a radical appears, we write the result in a form with the j before the radical as in the last illustration. This clearly shows that the j is not *under* the radical. ∎

EXAMPLE 2 $(\sqrt{-4})^2 = (\sqrt{4}\sqrt{-1})^2 = (j\sqrt{4})^2 = 4j^2 = -4$

The simplification of this expression does not follow Eq. (11-10), which states that $\sqrt{ab} = \sqrt{a}\sqrt{b}$ for square roots. This is the reason it was noted as being valid only if a and b are not negative. Therefore, we see that *Eq. (11-10) does not always hold for negative values of a and b.*

In simplifying $(\sqrt{-4})^2$, we can write it as $(\sqrt{-4})(\sqrt{-4})$. However, *we cannot now write this product as* $\sqrt{(-4)(-4)}$, for this leads to an incorrect result of $+4$. In fact *we cannot use Eq. (11-10) if both a and b are negative.* It cannot be overemphasized that

CAUTION▶ $(\sqrt{-4})^2$ *is not equal to* $\sqrt{(-4)(-4)}$

Each $\sqrt{-4}$ must be written as $j\sqrt{4}$ before continuing the simplification. ∎

EXAMPLE 3 To further illustrate the method of handling square roots of negative numbers, consider the difference between $\sqrt{-3}\sqrt{-12}$ and $\sqrt{(-3)(-12)}$. For these expressions we have

$$\sqrt{-3}\sqrt{-12} = (j\sqrt{3})(j\sqrt{12}) = (\sqrt{3}\sqrt{12})j^2 = (\sqrt{36})j^2$$
$$= 6(-1) = -6$$
$$\sqrt{(-3)(-12)} = \sqrt{36} = 6$$

For $\sqrt{-3}\sqrt{-12}$ we have the product of square roots of negative numbers, whereas for $\sqrt{(-3)(-12)}$ we have the product of negative numbers under the radical. We must be careful to note the difference. ∎

NOTE▶ From Examples 2 and 3 we see that *when we are dealing with square roots of negative numbers, **each should be expressed in terms of j before proceeding.*** To do this, for any positive real number a we write

$$\boxed{\sqrt{-a} = j\sqrt{a} \quad (a > 0)}$$ **(12-2)**

EXAMPLE 4 (a) $\sqrt{-6} = \sqrt{(6)(-1)} = \sqrt{6}\sqrt{-1} = j\sqrt{6}$

this step is correct if only one is negative, as in this case

(b) $-\sqrt{-75} = -\sqrt{(25)(3)(-1)} = -\sqrt{(25)(3)}\sqrt{-1} = -5j\sqrt{3}$

CAUTION▶ We note that $-\sqrt{-75}$ *is not equal to* $\sqrt{75}$. ∎

At times we need to raise imaginary numbers to some power. Using the definitions of exponents and of j, we have the following results:

$$j = j \qquad\qquad j^5 = j^4 j = j$$
$$j^2 = -1 \qquad\qquad j^6 = j^4 j^2 = (1)(-1) = -1$$
$$j^3 = j^2 j = -j \qquad\qquad j^7 = j^4 j^3 = (1)(-j) = -j$$
$$j^4 = j^2 j^2 = (-1)(-1) = 1 \qquad j^8 = j^4 j^4 = (1)(1) = 1$$

The powers of j go through the cycle of j, -1, $-j$, 1, j, -1, $-j$, 1, and so forth. Noting this and the fact that j raised to a power which is a multiple of 4 equals 1 allows us to raise j to any integral power almost on sight.

■ EXAMPLE 5 (a) $j^{10} = j^8 j^2 = (1)(-1) = -1$

(b) $j^{45} = j^{44} j = (1)(j) = j$

(c) $j^{531} = j^{528} j^3 = (1)(-j) = -j$

exponents 8, 44, 528 are multiples of 4 ■

Rectangular Form of a Complex Number

Using real numbers and the imaginary unit j, we define a new kind of number. *A* **complex number** *is any number that can be written in the form* $a + bj$, *where a and b are real numbers. If* $a = 0$ *and* $b \neq 0$, *we have a number of the form* bj, *which is a* **pure imaginary number.** *If* $b = 0$, *then* $a + bj$ *is a real number. The form* $a + bj$ *is known as the* **rectangular form** *of a complex number, where a is known as the* **real part** *and b is known as the* **imaginary part.** We see that complex numbers include all real numbers and all pure imaginary numbers.

A comment here about the words *imaginary* and *complex* is in order. The choice of the names of these numbers is historical in nature, and unfortunately it leads to some misconceptions about the numbers. The use of *imaginary* does not imply that the numbers do not exist. Imaginary numbers do in fact exist, as they are defined above. In the same way, the use of *complex* does not imply that the numbers are complicated and therefore difficult to understand. With the proper definitions and operations, we can work with complex numbers, just as with any type of number.

In the following section we will define the basic operations for complex numbers. Before doing this it is necessary to define the equality of complex numbers. Since a complex number is the sum of a real number and an imaginary number, it is not positive or negative in the usual sense, but each real part and each imaginary part are positive or negative. Therefore, *we define two complex numbers to be equal if the real parts are equal and the imaginary parts are equal.* That is,

NOTE ▶ *two complex numbers,* $a + bj$ *and* $x + yj$, *are equal if* $a = x$ *and* $b = y$.

This is illustrated in the following examples.

■ EXAMPLE 6 (a) $a + bj = 3 + 4j$ if $a = 3$ and $b = 4$

(b) $x + yj = 5 - 3j$ if $x = 5$ and $y = -3$

real part   imaginary part ■

EXAMPLE 7 What values of x and y satisfy the equation $4 - 6j - x = j + jy$?

One way to solve this equation is to arrange the terms so that all the known terms are on the right and all terms containing x and y are on the left. This leads to

$$-x - jy = -4 + 7j$$

From the definition of equality of complex numbers,

$$-x = -4 \quad \text{and} \quad -y = 7$$
$$x = 4 \quad \text{and} \quad y = -7$$

EXAMPLE 8 What values of x and y satisfy the equation

$$x + 3(xj + y) = 5 - j - jy$$

Rearranging the terms so that the known terms are on the right and the terms containing x and y are on the left, we have

$$x + 3y + 3jx + jy = 5 - j$$

Next, factoring j from the two terms on the left will put the expression on the left into proper form. This leads to

$$(x + 3y) + (3x + y)j = 5 - j$$

Using the definition of equality, we have

$$x + 3y = 5 \quad \text{and} \quad 3x + y = -1$$

We now solve this system of equations. The solution is $x = -1$ and $y = 2$. Actually, the solution can be obtained at any point by writing each side of the equation in the form $a + bj$ and then equating first the real parts and then the imaginary parts.

CONJUGATE

The **conjugate** *of the complex number $a + bj$ is the complex number $a - bj$.* We see that the sign of the imaginary part is changed to obtain the conjugate.

EXAMPLE 9 **(a)** $3 - 2j$ is the conjugate of $3 + 2j$. We may also say that $3 + 2j$ is the conjugate of $3 - 2j$. Thus each is the conjugate of the other.

(b) $-2 - 5j$ and $-2 + 5j$ are conjugates.

(c) $6j$ and $-6j$ are conjugates.

(d) 3 is the conjugate of 3 (imaginary part is zero).

EXERCISES *12-1*

In Exercises 1–12, express each number in terms of j.

1. $\sqrt{-81}$ **2.** $\sqrt{-121}$ **3.** $-\sqrt{-4}$ **4.** $-\sqrt{-49}$

5. $\sqrt{-0.36}$ **6.** $-\sqrt{-0.01}$ **7.** $\sqrt{-8}$ **8.** $\sqrt{-48}$

9. $\sqrt{-\frac{7}{4}}$ **10.** $-\sqrt{-\frac{5}{9}}$ **11.** $-\sqrt{-4e^2}$ **12.** $\sqrt{-\pi^4}$

In Exercises 13–24, simplify each of the given expressions.

13. (a) $(\sqrt{-7})^2$ (b) $\sqrt{(-7)^2}$

14. (a) $\sqrt{(-15)^2}$ (b) $(\sqrt{-15})^2$

15. (a) $\sqrt{(-2)(-8)}$ (b) $\sqrt{-2}\sqrt{-8}$

16. (a) $\sqrt{-9}\sqrt{-16}$ (b) $\sqrt{(-9)(-16)}$

17. (a) j^7 (b) j^{49} **18.** (a) $-j^{22}$ (b) j^{408}

19. $j^2 - j^6$ **20.** $2j^5 - j^2$

21. $j^{15} - j^{13}$ **22.** $3j^{48} + j^{200}$ **23.** $-\sqrt{j^2}$ **24.** $-\sqrt{-j^2}$

In Exercises 25–36, perform the indicated operations and simplify each complex number to its rectangular form.

25. $2 + \sqrt{-9}$ **26.** $-6 + \sqrt{-64}$ **27.** $3j - \sqrt{-100}$

28. $-\sqrt{1} - \sqrt{-400}$ **29.** $\sqrt{-4j^2} + \sqrt{-4}$ **30.** $5 - 2\sqrt{25j^2}$

31. $2j^2 + 3j$ **32.** $j^3 - 6$ **33.** $\sqrt{18} - \sqrt{-8}$

34. $\sqrt{-27} + \sqrt{12}$ **35.** $(\sqrt{-2})^2 + j^4$ **36.** $(2\sqrt{2})^2 - j^6$

In Exercises 37–40, find the conjugate of each complex number.

37. (a) $6 - 7j$ (b) $8 + j$ **38.** (a) $-3 + 2j$ (b) $-9 - j$

39. (a) $2j$ (b) -4 **40.** (a) 6 (b) $-5j$

In Exercises 41–48, find the values of x and y that satisfy the given equations.

41. $7x - 2yj = 14 + 4j$ **42.** $2x + 3jy = -6 + 12j$

43. $6j - 7 = 3 - x - yj$ **44.** $9 - j = xj + 1 - y$

45. $x - y = 1 - xj - yj - j$

46. $2x - 2j = 4 - 2xj - yj$

47. $x + 2 + 7j = yj - 2xj$

48. $2x + 6xj + 3 = yj - y + 7j$

In Exercises 49–56, answer the given questions.

49. Are $8j$ and $-8j$ the solutions to the equation $x^2 + 64 = 0$?

50. Are $2j\sqrt{5}$ and $-2j\sqrt{5}$ the solutions to the equation $x^2 + 20 = 0$?

51. Are $2j$ and $-2j$ solutions to the equation $x^4 + 16 = 0$?

52. Are $3j$ and $-3j$ solutions to the equation $x^3 + 27j = 0$?

53. Evaluate $j + j^2 + j^3 + j^4 + j^5 + j^6 + j^7 + j^8$.

54. What is the smallest positive value of n for which $j^{-1} = j^n$?

(W) **55.** Is it possible that a given complex number and its conjugate are equal? Explain.

(W) **56.** Is it possible that a given complex number and the negative of its conjugate are equal? Explain.

$12\text{-}2$ BASIC OPERATIONS WITH COMPLEX NUMBERS

The addition, subtraction, multiplication, and division of complex numbers are based on the operations for binomials with real coefficients (see Chapters 1 and 6). These operations are performed without regard for the fact that j has a special meaning. **NOTE▶** However, *we must be careful to express all complex numbers in terms of j before performing these operations.* Once this is done, we may proceed as with real numbers. We have the following definitions for these operations on complex numbers.

Addition:
$$(a + bj) + (c + dj) = (a + c) + (b + d)j \qquad \textbf{(12-3)}$$

Subtraction:
$$(a + bj) - (c + dj) = (a - c) + (b - d)j \qquad \textbf{(12-4)}$$

Multiplication:
$$(a + bj)(c + dj) = (ac - bd) + (ad + bc)j \qquad \textbf{(12-5)}$$

Division:
$$\frac{a + bj}{c + dj} = \frac{(a + bj)(c - dj)}{(c + dj)(c - dj)} = \frac{(ac + bd) + (bc - ad)j}{c^2 + d^2} \qquad \textbf{(12-6)}$$

Compare Eq. (12-5) with Eq. (6-6).

Eqs. (12-3) and (12-4) show that we *add and subtract complex numbers by combining the real parts and combining the imaginary parts.* Consider these examples.

■EXAMPLE 1   (a) $(3 - 2j) + (-5 + 7j) = (3 - 5) + (-2 + 7)j$
$$= -2 + 5j$$

(b) $(7 + 9j) - (6 - 4j) = (7 - 6) + (9 - (-4))j$
$$= 1 + 13j$$

EXAMPLE 2 $(3\sqrt{-4} - 4) - (6 - 2\sqrt{-25}) - \sqrt{-81}$

$$= [3(2j) - 4] - [6 - 2(5j)] - 9j \qquad \text{write in terms of } j$$

$$= [6j - 4] - [6 - 10j] - 9j$$

$$= 6j - 4 - 6 + 10j - 9j$$

$$= -10 + 7j$$

When complex numbers are multiplied, Eq. (12-5) indicates that *we proceed as in any algebraic multiplication,* properly expressing numbers in terms of j and evaluating the power of j. This is illustrated in the next two examples.

EXAMPLE 3 $(6 - \sqrt{-4})(\sqrt{-9}) = (6 - 2j)(3j) \qquad \text{write in terms of } j$

$$= 18j - 6j^2 = 18j - 6(-1)$$

$$= 6 + 18j$$

EXAMPLE 4

$$(-9.4 - 6.2j)(2.5 + 1.5j) = (-9.4)(2.5) + (-9.4)(1.5j) + (-6.2j)(2.5) + (-6.2j)(1.5j)$$

$$= -23.5 - 14.1j - 15.5j - 9.3j^2$$

$$= -23.5 - 29.6j - 9.3(-1)$$

$$= -14.2 - 29.6j$$

Some calculators are programmed to perform operations with complex numbers. In Fig. 12-1 the display of the operations of Examples 2 and 5 are shown for such a calculator. Note the use of *i* rather than *j*.

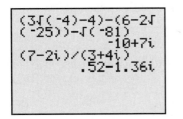

Fig. 12-1

The procedure shown by Eq. (12-6) for dividing by a complex number is the same procedure we used for rationalizing the denominator of a fraction with a radical in the denominator. We use this procedure so that we can express any result in the proper form of a complex number. Therefore, *to divide by a complex number, multiply the numerator and the denominator by the conjugate of the denominator.* This is illustrated in the following examples.

EXAMPLE 5 $\dfrac{7 - 2j}{3 + 4j} = \dfrac{(7 - 2j)(3 - 4j)}{(3 + 4j)(3 - 4j)} \qquad \text{multiply by conjugate of denominator}$

$$= \dfrac{21 - 28j - 6j + 8j^2}{9 - 16j^2} = \dfrac{21 - 34j + 8(-1)}{9 - 16(-1)}$$

$$= \dfrac{13 - 34j}{25}$$

This could be written in the form $a + bj$ as $\frac{13}{25} - \frac{34}{25}j$, but this type of result is generally left as a single fraction. In decimal form, the result would be expressed as $0.52 - 1.36j$.

EXAMPLE 6

(a) $\dfrac{6 - j^3}{2j} = \dfrac{6 - (-j)}{2j} = \dfrac{6 + j}{2j} = \dfrac{6 + j}{2j}\left(\dfrac{-2j}{-2j}\right)$

$$= \dfrac{-12j - 2j^2}{-4j^2} = \dfrac{2 - 12j}{4} = \dfrac{1 - 6j}{2}$$

(b) $\dfrac{1}{j} + \dfrac{2}{3 + j} = \dfrac{(3 + j) + 2j}{j(3 + j)} = \dfrac{3 + 3j}{3j - 1} = \dfrac{3 + 3j}{-1 + 3j} \cdot \dfrac{-1 - 3j}{-1 - 3j}$

$$= \dfrac{-3 - 12j - 9j^2}{1 - 9j^2} = \dfrac{6 - 12j}{10} = \dfrac{3 - 6j}{5}$$

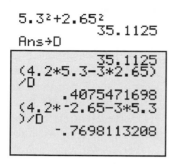

Fig. 12-2

EXAMPLE 7 In an alternating-current circuit, the voltage E is given by $E = IZ$, where I is the current in amperes and Z is the impedance in ohms. Each of these can be represented by complex numbers. Find the complex number representation for I if $E = 4.20 - 3.00j$ volts and $Z = 5.30 + 2.65j$ ohms. (This type of circuit is discussed in more detail in Section 12-7.)

Since $I = E/Z$, we have

$$I = \frac{4.20 - 3.00j}{5.30 + 2.65j} = \frac{(4.20 - 3.00j)(5.30 - 2.65j)}{(5.30 + 2.65j)(5.30 - 2.65j)} \quad \substack{\text{multiply by conjugate} \\ \text{of denominator}}$$

$$= \frac{22.26 - 11.13j - 15.90j + 7.95j^2}{5.30^2 - 2.65^2 j^2} = \frac{22.26 - 7.95 - 27.03j}{5.30^2 + 2.65^2}$$

$$= \frac{14.31 - 27.03j}{35.11} = 0.408 - 0.770j \text{ amperes}$$

The display for the solution using a calculator that does not operate with complex numbers is shown in Fig. 12-2. Note that the first step was to store the value of the denominator, and that we use a negative sign for j^2.

━━━━━━━━━━━━━━━ **EXERCISES** *12-2* ━━━━━━━━━━━━━━━

In Exercises 1–36, perform the indicated operations, expressing all answers in the form $a + bj$.

1. $(3 - 7j) + (2 - j)$
2. $(-4 - j) + (-7 - 4j)$
3. $(7j - 6) - (3 + j)$
4. $(5.4 - 3.4j) - (2.9j + 5.5)$
5. $0.23 - (0.46 - 0.19j) + 0.67j$
6. $(7 - j) - (4 - 4j) + (6 - j)$
7. $(12j - 21) - (15 - 18j) - 9j$
8. $(0.062j - 0.073) - 0.030j - (0.121 - 0.051j)$
9. $(7 - j)(7j)$
10. $(-2.2j)(1.5j - 4.0)$
11. $(4 - j)(5 + 2j)$
12. $(3 - 5j)(6 + 7j)$
13. $(20j - 30)(30j + 10)$
14. $(8j - 5)(7 + 4j)$
15. $(\sqrt{-18}\sqrt{-4})(3j)$
16. $\sqrt{-6}\sqrt{-12}\sqrt{3}$
17. $7j^3 - 7\sqrt{-9}$
18. $6j - 5j^2\sqrt{-63}$
19. $j\sqrt{-7} - j^6\sqrt{112} + 3j$
20. $j^2\sqrt{-7} - \sqrt{-28} + 8$
21. $(3 - 7j)^2$
22. $(4j + 5)^2$
23. $(1 - j)^3$
24. $(1 + j)(1 - j)^2$
25. $\dfrac{6j}{2 - 5j}$
26. $\dfrac{4}{3 + 7j}$
27. $\dfrac{0.25}{3.0 - \sqrt{-1.0}}$
28. $\dfrac{1 - j}{3j}$
29. $\dfrac{6 + 5j}{3 - 4j}$
30. $\dfrac{j\sqrt{2} - 5}{j\sqrt{2} + 3}$
31. $\dfrac{2 + 3j\sqrt{3}}{5 - j\sqrt{3}}$
32. $\dfrac{j^5 - j^3}{3 + j}$
33. $\dfrac{j^2 - j}{2j - j^8}$
34. $\dfrac{3}{2j} + \dfrac{5}{6 - j}$
35. $\dfrac{4j}{1 - j} - \dfrac{3 + j}{2 + 3j}$
36. $\dfrac{(6j + 5)(2 - 4j)}{(5 - j)(4j + 1)}$

In Exercises 37–40, solve the given problems.

37. Show that $-1 - j$ is a solution to the equation $x^2 + 2x + 2 = 0$.
38. Show that $1 - j\sqrt{3}$ is a solution to the equation $x^2 + 4 = 2x$.
39. Multiply $-3 + j$ by its conjugate.
40. Divide $2 - 3j$ by its conjugate.

In Exercises 41–44, solve the given problems. Refer to Example 7.

41. If $I = 0.835 - 0.427j$ amperes and $Z = 250 + 170j$ ohms, find the complex number representation for E.
42. If $E = 5.70 - 3.65j$ volts and $I = 0.360 - 0.525j$ amperes, find the complex number representation for Z.
43. If $E = 85 + 74j$ volts and $Z = 2500 - 1200j$ ohms, find the complex number representation for I.
44. In an alternating-current circuit, two impedances Z_1 and Z_2 have a total impedance Z_T of $Z_T = \dfrac{Z_1 Z_2}{Z_1 + Z_2}$. Find Z_T for $Z_1 = 2 + 3j$ ohms and $Z_2 = 3 - 4j$ ohms.

In Exercises 45–48, demonstrate the indicated properties.

45. Show that the sum of a complex number and its conjugate is a real number.
46. Show that the difference between a complex number and its conjugate is an imaginary number ($b \neq 0$).
47. Explain why the product of a complex number and its conjugate is real and nonnegative.
48. Explain how to show that the reciprocal of the imaginary unit is the negative of the imaginary unit.

$12\text{-}3$ GRAPHICAL REPRESENTATION OF COMPLEX NUMBERS

We showed in Section 1-1 how we could represent real numbers as points on a line. Because complex numbers include all real numbers as well as imaginary numbers, it is necessary to represent them graphically in a different way. Since there are two numbers associated with each complex number (the real part and the imaginary part), we find that we can represent complex numbers by representing the real parts by the *x*-values of the rectangular coordinate system, and the imaginary parts by the *y*-values. In this way, *each complex number is represented as a point in the plane,* the point being designated as *a + bj. When the rectangular coordinate system is used in this manner, it is called the* **complex plane.** *The horizontal axis is called the* **real axis,** *and the vertical axis is called the* **imaginary axis.**

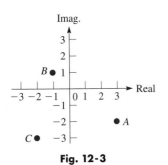

Fig. 12-3

■EXAMPLE 1   In Fig. 12-3, point *A* represents the complex number $3 - 2j$; point *B* represents $-1 + j$; point *C* represents $-2 - 3j$. We note that these complex numbers are represented by the points $(3, -2)$, $(-1, 1)$, and $(-2, -3)$, respectively, of the standard rectangular coordinate system.

We must keep in mind that the meaning given to the points representing complex numbers in the complex plane is different from the meaning given to the points in the standard rectangular coordinate system. A point in the complex plane represents a single complex number, whereas a point in the rectangular coordinate system represents a pair of real numbers.   ■

Let us represent two complex numbers and their sum in the complex plane. Consider, for example, the two complex numbers $1 + 2j$ and $3 + j$. By algebraic addition the sum is $4 + 3j$. When we draw lines from the origin to these points (see Fig. 12-4), we note that if we think of the complex numbers as being vectors, their sum is the vector sum. Because complex numbers can be used to represent vectors, these numbers are particularly important. *Any complex number can be thought of as representing a vector from the origin to its point in the complex plane.* This leads us to the method used to add complex numbers graphically.

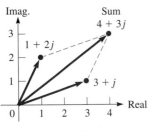

Fig. 12-4

Steps to Add Complex Numbers Graphically

1. *Find the point corresponding to one of the numbers and draw a line from the origin to this point.*
2. *Repeat step 1 for the second number.*
3. *Complete a parallelogram with the lines drawn as adjacent sides. The resulting fourth vertex is the point representing the sum.*

Note that *this is equivalent to adding vectors by graphical means.*

■EXAMPLE 2   Add the complex numbers $5 - 2j$ and $-2 - j$ graphically.

The solution is indicated in Fig. 12-5. We can see that the fourth vertex of the parallelogram is at $3 - 3j$. This is, of course, the algebraic sum of these two complex numbers.   ■

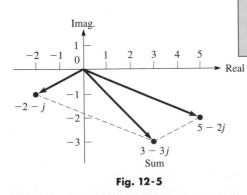

Fig. 12-5

EXAMPLE 3 Add the complex numbers -3 and $1 + 4j$ graphically.

First we note that $-3 = -3 + 0j$, which means that the point representing -3 is on the negative real axis. In Fig. 12-6, we show the numbers -3 and $1 + 4j$ on the graph and complete the parallelogram. From the graph we see that the sum is $-2 + 4j$.

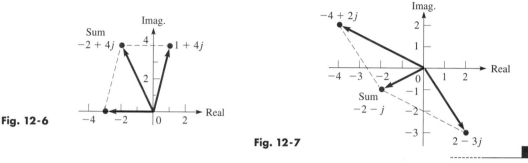

Fig. 12-6

Fig. 12-7

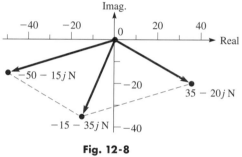

Fig. 12-8

EXAMPLE 4 Subtract $4 - 2j$ from $2 - 3j$ graphically.

Subtracting $4 - 2j$ is equivalent to adding $-4 + 2j$. Therefore, we complete the solution by adding $-4 + 2j$ and $2 - 3j$, as shown in Fig. 12-7. The result is $-2 - j$.

EXAMPLE 5 Two forces acting on the overhead bolt can be represented by $35 - 20j$ N and $-50 - 15j$ N. Find the resultant force graphically.

The forces are shown in Fig. 12-8. From the graph we can see that the sum of the forces, which is the resultant force, is $-15 - 35j$ N.

EXERCISES *12-3*

In Exercises 1–4, locate the given numbers in the complex plane.

1. $2 + 6j$ **2.** $-5 + j$

3. $-4 - 3j$ **4.** $3 - 4j$

In Exercises 5–24, perform the indicated operations graphically. Check them algebraically.

5. $2 + (3 + 4j)$ **6.** $2j + (-2 + 3j)$

7. $(5 - j) + (3 + 2j)$ **8.** $(3 - 2j) + (-1 - j)$

9. $5 - (1 - 4j)$ **10.** $(2 - j) - j$

11. $(2 - 4j) + (-2 + j)$ **12.** $(-1 - 2j) + (6 - j)$

13. $(3 - 2j) - (4 - 6j)$ **14.** $(-2 - 4j) - (2 - 5j)$

15. $(1 + 4j) - (3 + j)$ **16.** $(-j - 2) - (-1 - 3j)$

17. $(1.5 - 0.5j) + (3.0 + 2.5j)$

18. $(3.5 + 2.0j) - (-4.0 - 1.5j)$

19. $(3 - 6j) - (-1 + 5j)$ **20.** $(-6 - 3j) + (2 - 7j)$

21. $(2j + 1) - 3j - (j + 1)$ **22.** $(6 - j) - 9 - (2j - 3)$

23. $(j - 6) - j + (j - 7)$ **24.** $j - (1 - j) + (3 + 2j)$

In Exercises 25–28, show the given number, its negative, and its conjugate on the same coordinate system.

25. $3 + 2j$ **26.** $-2 + 4j$

27. $-3 - 5j$ **28.** $5 - j$

In Exercises 29 and 30, show the number $a + bj$, $3(a + bj)$, and $-3(a + bj)$ on the same coordinate system. The multiplication of a complex number by a real number is called **scalar multiplication** *of the complex number.*

29. $3 - j$ **30.** $-1 - 3j$

In Exercises 31 and 32, perform the indicated vector additions graphically. Check them algebraically.

31. Two ropes hold a boat at a dock. The tensions in the ropes can be represented by $40 + 10j$ lb and $50 - 25j$ lb. Find the resultant force.

32. Relative to the air, a plane heads north of west with a velocity that can be represented by $-320 + 140j$ mi/h. The wind is blowing from south of west with a velocity that can be represented by $40 + 140j$ mi/h. Find the resultant velocity of the plane.

12-4 POLAR FORM OF A COMPLEX NUMBER

We have just seen the relationship between complex numbers and vectors. Since we can use one to represent the other, we shall use this fact to write complex numbers in another way. The new form that we develop has certain advantages when the basic operations of multiplication and division are performed on complex numbers. We will discuss these operations later in the chapter.

By drawing a vector from the origin to the point in the complex plane that represents the number $x + yj$, we see the relation between vectors and complex numbers. Further observation indicates that an angle in standard position has been formed. Also, the point $x + yj$ is r units from the origin. In fact, *we can find any point in the complex plane by knowing the angle θ and the value of r.* The necessary equations relating x, y, r, and θ are similar to those we developed for vectors (Eqs. (9-1) to (9-3)). By referring to Fig. 12-9, we can see that

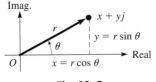

Fig. 12-9

$$x = r \cos \theta \qquad\qquad y = r \sin \theta \tag{12-7}$$

$$r^2 = x^2 + y^2 \qquad\qquad \tan \theta = \frac{y}{x} \tag{12-8}$$

Substituting Eq. (12-7) into the rectangular form $x + yj$ of a complex number, we have

$$x + yj = r \cos \theta + j(r \sin \theta)$$

or

$$x + yj = r(\cos \theta + j \sin \theta) \tag{12-9}$$

The right side of Eq. (12-9) is called the **polar form** *of a complex number. Sometimes it is referred to as the* **trigonometric form.** An abbreviated form of writing the polar form that is sometimes used is r cis θ. *The length r is called the* **absolute value,** *or the* **modulus,** *and the angle θ is called the* **argument** *of the complex number.* Therefore, Eq. (12-9), along with Eqs. (12-8), defines the polar form of a complex number.

■**EXAMPLE 1** Represent the complex number $3 + 4j$ graphically and give its polar form.

From the rectangular form $3 + 4j$, we see that $x = 3$ and $y = 4$. Using Eqs. (12-8), we have

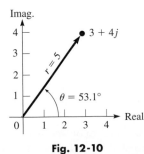

Fig. 12-10

$$r = \sqrt{3^2 + 4^2} = 5 \qquad \theta = \tan^{-1}\frac{4}{3} = 53.1°$$

Thus, the polar form is

$$5(\cos 53.1° + j \sin 53.1°)$$

The graphical representation is shown in Fig. 12-10.

A note on significant digits is in order here. In writing a complex number as $3 + 4j$ in Example 1, no approximate values are intended. However, in expressing the polar form as $5(\cos 53.1° + j \sin 53.1°)$, we rounded off the angle to the nearest $0.1°$, as it is not possible to express the result exactly in degrees. Therefore, in dealing with nonexact numbers, we shall express angles to the nearest $0.1°$. Other results, when approximate, will be expressed to three significant digits, unless a different accuracy is given in the problem. Of course, in applied situations most numbers used are approximate, as they are derived through measurement.

Another convenient and widely used notation for the polar form is r/θ. We must remember in using this form that it represents a complex number and is simply a shorthand way of writing $r(\cos \theta + j \sin \theta)$. Therefore,

$$\boxed{r/\theta = r(\cos \theta + j \sin \theta)} \qquad \text{(12-10)}$$

■**EXAMPLE 2** (a) $3(\cos 40° + j \sin 40°) = 3/40°$

(b) $6.26(\cos 217.3° + j \sin 217.3°) = 6.26/217.3°$

(c) $5/120° = 5(\cos 120° + j \sin 120°)$

(d) $14.5/306.2° = 14.5(\cos 306.2° + j \sin 306.2°)$ ■

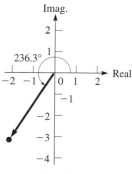

Fig. 12-11

■**EXAMPLE 3** Represent the complex number $-2.08 - 3.12j$ graphically and give its polar forms.

The graphical representation is shown in Fig. 12-11. From Eqs. (12-8), we have

$$r = \sqrt{(-2.08)^2 + (-3.12)^2} = 3.75$$

$$\theta_{\text{ref}} = \tan^{-1}\frac{3.12}{2.08} = 56.3° \qquad \theta = 180° + 56.3° = 236.3°$$

Since both the real and imaginary parts are negative, we know that θ is a third-quadrant angle. Therefore, we found the reference angle before finding θ. This means the polar forms are

$$3.75(\cos 236.3° + j \sin 236.3°) = 3.75/236.3° \qquad ■$$

■**EXAMPLE 4** The impedance Z (in Ω) in an alternating-current circuit is given by $Z = 3560/-32.4°$. Express this in rectangular form.

From the polar form we have $r = 3560 \ \Omega$ and $\theta = -32.4°$ (it is common to use negative angles in this type of application). This means that we can also write

$$Z = 3560(\cos(-32.4°) + j \sin(-32.4°))$$

See Fig. 12-12. This means that

$$x = 3560 \cos(-32.4°) = 3010$$
$$y = 3560 \sin(-32.4°) = -1910$$

Therefore, the rectangular form is

$$Z = 3010 - 1910j \ \Omega \qquad ■$$

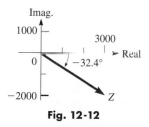

Fig. 12-12

Complex numbers can be converted between rectangular and polar forms on most calculators. This will be illustrated in the next section.

EXAMPLE 5   Represent the numbers 5, −5, 7*j*, and −7*j* in polar form.

Since any positive real number lies on the positive real axis in the complex plane, it is expressed in polar form by

$$a = a(\cos 0° + j \sin 0°) = a\underline{/0°}$$

Negative real numbers, being on the negative real axis, are written as

$$a = |a| (\cos 180° + j \sin 180°) = |a| \underline{/180°}$$

Thus,

$$5 = 5(\cos 0° + j \sin 0°) = 5\underline{/0°}$$
$$-5 = 5(\cos 180° + j \sin 180°) = 5\underline{/180°}$$

Positive pure imaginary numbers lie on the positive imaginary axis and are expressed in polar form by

$$bj = b(\cos 90° + j \sin 90°) = b\underline{/90°}$$

Similarly, negative pure imaginary numbers, being on the negative imaginary axis, are written as

$$bj = |b| (\cos 270° + j \sin 270°) = |b| \underline{/270°}$$

Thus,

$$7j = 7(\cos 90° + j \sin 90°) = 7\underline{/90°}$$
$$-7j = 7(\cos 270° + j \sin 270°) = 7\underline{/270°}$$

The graphical representations of the *complex numbers* 5, −5, 7*j*, and −7*j* are shown in Fig. 12-13.

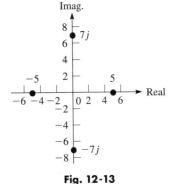

Fig. 12-13

EXERCISES *12-4*

In Exercises 1–16, represent each complex number graphically and give the polar form of each.

1. 8 + 6*j*   **2.** −8 − 15*j*   **3.** 3 − 4*j*

4. −5 + 12*j*   **5.** −2.00 + 3.00*j*   **6.** 7.00 − 5.00*j*

7. −5.50 − 2.40*j*   **8.** 4.60 − 4.60*j*   **9.** 1 + *j*√3

10. √2 − *j*√2   **11.** 3.514 − 7.256*j*   **12.** 6.231 + 9.527*j*

13. −3   **14.** 6   **15.** 9*j*   **16.** −2*j*

In Exercises 17–36, represent each complex number graphically and give the rectangular form of each.

17. 5.00(cos 54.0° + *j* sin 54.0°)

18. 3.00(cos 232.0° + *j* sin 232.0°)

19. 1.60(cos 150.0° + *j* sin 150.0°)

20. 2.50(cos 315.0° + *j* sin 315.0°)

21. 6(cos 180° + *j* sin 180°)

22. 12(cos 270° + *j* sin 270°)

23. 8(cos 360° + *j* sin 360°)

24. 15(cos 0° + *j* sin 0°)

25. 12.36(cos 345.56° + *j* sin 345.56°)

26. 220.8(cos 155.13° + *j* sin 155.13°)

27. cos 240.0° + *j* sin 240.0°

28. cos 299.0° + *j* sin 299.0°

29. 4.75$\underline{/172.8°}$ **30.** 1.50$\underline{/62.3°}$

31. 0.9326$\underline{/229.54°}$ **32.** 277.8$\underline{/-342.63°}$

33. 7.32$\underline{/-270°}$ **34.** 18.3$\underline{/180.0°}$

35. 86.42$\underline{/94.62°}$ **36.** 4629$\underline{/182.44°}$

In Exercises 37–40, solve the given problems.

37. The voltage of a certain generator is represented by 2.84 − 1.06*j* kV. Write this voltage in polar form.

38. Find the magnitude and direction of a force on a bolt that is represented by 40.5 + 24.5*j* N.

39. The electric field intensity of a light wave can be described by 12.4$\underline{/78.3°}$ V/m. Write this in rectangular form.

40. The current in a certain microprocessor circuit is represented by 3.75$\underline{/15.0°}$ μA. Write this in rectangular form.

12-5 EXPONENTIAL FORM OF A COMPLEX NUMBER

Another important form of a complex number is the *exponential form*. It is commonly used in electronics, engineering, and physics applications. As we will see in the next section, it is also convenient for multiplication and division of complex numbers, as the rectangular form is for addition and subtraction.

The **exponential form** *of a complex number is written as* $re^{j\theta}$, where r and θ have the same meanings as given in the previous section, although θ is expressed in radians. *The number e is a special irrational number and has an approximate value*

> The number e is named for the Swiss mathematician Leonhard Euler (1707–1783). His works in mathematics and other fields filled about 70 volumes.

$$e = 2.7182818284590452$$

This number e is very important in mathematics, and we will see it again in the next chapter. For now it is necessary to accept the value for e, although in calculus its meaning is shown along with the reason it has the above value. We can find its value on a calculator by using the e^x key, with $x = 1$.

We now define

$$\boxed{re^{j\theta} = r(\cos\theta + j\sin\theta)} \tag{12-11}$$

CAUTION▶

By expressing θ in radians, the expression $j\theta$ is an exponent, and it can be shown to obey all the laws of exponents as discussed in Chapter 11. Therefore, we shall always *express θ in radians when using exponential form.*

█EXAMPLE 1 Express the number $3 + 4j$ in exponential form.

From Example 1 of Section 12-4, we know that this complex number may be written in polar form as $5(\cos 53.1° + j\sin 53.1°)$. Therefore, we know that $r = 5$. We now express $53.1°$ in terms of radians as

$$\frac{53.1\pi}{180} = 0.927 \text{ rad}$$

Thus, the exponential form is $5e^{0.927j}$. This means that

$$3 + 4j = 5(\cos 53.1° + j\sin 53.1°) = 5e^{0.927j}$$

degrees to radians

value of r

█EXAMPLE 2 Express the number $8.50\underline{/136.3°}$ in exponential form.

Since this complex number is in polar form, we note that $r = 8.50$ and that we must express $136.3°$ in radians. Changing $136.3°$ to radians, we have

$$\frac{136.3\pi}{180} = 2.38 \text{ rad}$$

Therefore, the required exponential form is $8.50e^{2.38j}$. This means that

$$8.50\underline{/136.3°} = 8.50e^{2.38j}$$

We see that the principal step in changing from polar form to exponential form is to change θ from degrees to radians.

EXAMPLE 3 Express the number $3.07 - 7.43j$ in exponential form.

From the rectangular form, we have $x = 3.07$ and $y = -7.43$. Therefore,

$$r = \sqrt{(3.07)^2 + (-7.43)^2} = 8.04$$

$$\theta_{\text{ref}} = \tan^{-1}\frac{7.43}{3.07} = 67.6° \qquad \theta = 360° - 67.6° = 292.4°$$

For reference, Eq. (8-8) is π rad = 180°.

Since $292.4° = 5.10$ rad, the exponential form is $8.04e^{5.10j}$. (Using radian mode, we could show that $\theta_{\text{ref}} = \tan^{-1}(7.43/3.07) = 5.10$). This means that

$$3.07 - 7.43j = 8.04e^{5.10j}$$

EXAMPLE 4 Express the complex number $2.00e^{4.80j}$ in polar and rectangular forms.

We first express 4.80 rad as 275.0°. From the exponential form we know that $r = 2.00$. Thus, the polar form is

$$2.00(\cos 275.0° + j \sin 275.0°)$$

Using the distributive law we rewrite the polar form and then evaluate. Thus,

$$2.00e^{4.80j} = 2.00(\cos 275.0° + j \sin 275.0°)$$
$$= 2.00 \cos 275.0° + (2.00 \sin 275.0°)j$$
$$= 0.174 - 1.99j$$

Calculators that operate with complex numbers treat the exponential and polar forms as being the same, since each uses r and θ. Figures 12-14 and 12-15 show the conversions of Examples 5 and 6 on such a calculator.

EXAMPLE 5 Express $(1.846e^{1.229j})^2$ in polar and rectangular forms.

First, we note that $(1.846e^{1.229j})^2 = 1.846^2 e^{2.458j}$. Since 2.458 rad = 140.83°, the polar form is $1.846^2 \underline{/140.83°} = 3.408 \underline{/140.83°}$.

For the rectangular form, using radian mode we have

$$(1.846e^{1.229j})^2 = 1.846^2 e^{2.458j} = 1.846^2(\cos 2.458 + j \sin 2.458)$$
$$= -2.642 + 2.152j$$

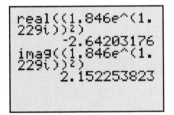

```
real((1.846e^(1.
229i))²)
            -2.64203176
imag((1.846e^(1.
229i))²)
            2.152253823
```

Fig. 12-14

As we noted earlier, an important application of the use of complex numbers is in alternating-current analysis. When an alternating current flows through a given circuit, usually the current and voltage have different phases. That is, they do not reach their peak values at the same time. Therefore, one way of accounting for the magnitude as well as the phase of an electric current or voltage is to write it as a complex number. Here the modulus is the actual magnitude of the current or voltage, and the argument θ is a measure of the phase.

EXAMPLE 6 A current of $2.00 - 4.00j$ amperes flows in a given circuit. Express this current in exponential form and find the magnitude of the current.

From the rectangular form, we have $x = 2.00$ and $y = -4.00$. Therefore,

$$r = \sqrt{(2.00)^2 + (-4.00)^2} = 4.47 \text{ A}$$

$$\theta = \tan^{-1}\frac{-4.00}{2.00} = -63.4°$$

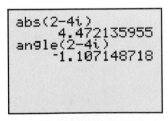

```
abs(2-4i)
            4.472135955
angle(2-4i)
            -1.107148718
```

Fig. 12-15

It is normal to express the phase in terms of negative angles when the imaginary part is negative. Changing 63.4° to radians, we have $63.4° = 1.11$ rad. This means the exponential form is $4.47e^{-1.11j}$. The modulus is 4.47, which means the magnitude of the current is 4.47 A. Note in the calculator display in Fig. 12-15 that the angle is shown in radians and is negative.

At this point we shall summarize the three important forms of a complex number. See Fig. 12-16 for the graphical representation.

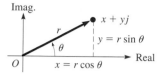

Fig. 12-16

> Rectangular: $x + yj$
>
> Polar: $r(\cos \theta + j \sin \theta) = r\underline{/\theta}$
>
> Exponential: $re^{j\theta}$

It follows that

$$x + yj = r(\cos \theta + j \sin \theta) = r\underline{/\theta} = re^{j\theta} \qquad \textbf{(12-12)}$$

where

$$r^2 = x^2 + y^2 \qquad \tan \theta = \frac{y}{x} \qquad \textbf{(12-8)}$$

In Eq. (12-12) the argument θ is the same for the exponential and polar forms, although it is expressed in radians in exponential form and is usually expressed in degrees in polar form.

EXERCISES *12-5*

In Exercises 1–20, express the given numbers in exponential form.

1. 3.00(cos 60.0° + j sin 60.0°)

2. 5.00(cos 135.0° + j sin 135.0°)

3. 4.50(cos 282.3° + j sin 282.3°)

4. 2.10(cos 228.7° + j sin 228.7°)

5. 375.5(cos 95.46° + j sin 95.46°)

6. 16.72[cos(−7.14°) + j sin (−7.14°)]

7. 0.515$\underline{/198.3°}$ **8.** 4650$\underline{/326.5°}$

9. 4.06$\underline{/-61.4°}$ **10.** 0.0192$\underline{/76.7°}$

11. 9245$\underline{/296.32°}$ **12.** 827.6$\underline{/110.09°}$

13. 3 − 4j **14.** −1 − 5j

15. −3 + 2j **16.** 6 + j

17. 5.90 + 2.40j **18.** 47.3 − 10.9j

19. −634.6 − 528.2j **20.** −8573 + 5477j

In Exercises 21–28, express the given complex numbers in polar and rectangular forms.

21. $3.00e^{0.500j}$ **22.** $2.00e^{1.00j}$

23. $4.64e^{1.85j}$ **24.** $2.50e^{3.84j}$

25. $3.20e^{5.41j}$ **26.** $0.800e^{3.00j}$

27. $0.1724e^{2.391j}$ **28.** $820.7e^{3.492j}$

In Exercises 29–32, perform the indicated operations and express results in rectangular and polar forms.

29. $(4.55e^{1.32j})^2$ **30.** $(0.926e^{0.253j})^3$

31. $(625e^{3.46j})(4.40e^{1.22j})$ **32.** $(18.0e^{5.13j})(25.5e^{0.77j})$

In Exercises 33–36, perform the indicated operations.

33. The impedance in an antenna circuit is 375 + 110j ohms. Write this in exponential form and find the magnitude of the impedance.

34. The intensity of the signal from a radar microwave signal is 37.0[cos (−65.3°) + j sin (−65.3°)] V/m. Write this in exponential form.

35. The displacement of a sound wave is $3.50e^{-1.35j}$ μm. Write this in rectangular form.

36. In an electric circuit, the *admittance* is the reciprocal of the impedance. In a transistor circuit, the impedance is 2800 − 1450j ohms. Find the exponential form of the admittance.

12-6 PRODUCTS, QUOTIENTS, POWERS, AND ROOTS OF COMPLEX NUMBERS

We have previously found products and quotients using the rectangular forms of the given complex numbers. However, these operations can also be performed with complex numbers in polar and exponential forms. These operations not only are convenient but also are useful for purposes of finding powers and roots of complex numbers.

We may find the product of two complex numbers by using the exponential form and the laws of exponents. Multiplying $r_1 e^{j\theta_1}$ by $r_2 e^{j\theta_2}$, we have

$$[r_1 e^{j\theta_1}] \times [r_2 e^{j\theta_2}] = r_1 r_2 e^{j\theta_1 + j\theta_2} = r_1 r_2^{j(\theta_1 + \theta_2)}$$

We use this equation to express the product of two complex numbers in polar form:

$$[r_1 e^{j\theta_1}] \times [r_2 e^{j\theta_2}] = [r_1(\cos \theta_1 + j \sin \theta_1)] \times [r_2(\cos \theta_2 + j \sin \theta_2)]$$

and

$$r_1 r_2 e^{j(\theta_1 + \theta_2)} = r_1 r_2 [\cos(\theta_1 + \theta_2) + j \sin(\theta_1 + \theta_2)]$$

Therefore, the polar expressions are equal, which means that *the product of two complex numbers is*

$$
\begin{aligned}
r_1(\cos \theta_1 &+ j \sin \theta_1) r_2(\cos \theta_2 + j \sin \theta_2) \\
&= r_1 r_2 [\cos(\theta_1 + \theta_2) + j \sin(\theta_1 + \theta_2)] \\
(r_1\underline{/\theta_1})(r_2\underline{/\theta_2}) &= r_1 r_2 \underline{/\theta_1 + \theta_2}
\end{aligned}
\qquad \text{(12-13)}
$$

NOTE▶   The *magnitudes are multiplied,* and the *angles are added.*

■EXAMPLE 1   Multiply the complex numbers $2 + 3j$ and $1 - j$ by using the polar form of each.

For $2 + 3j$: $r_1 = \sqrt{2^2 + 3^2} = 3.61$ $\tan \theta_1 = \dfrac{3}{2}$ $\theta_1 = 56.3°$

For $1 - j$: $r_2 = \sqrt{1^2 + (-1)^2} = 1.41$ $\tan \theta_2 = \dfrac{-1}{1}$ $\theta_2 = 315.0°$

$$(3.61)(\cos 56.3° + j \sin 56.3°)(1.41)(\cos 315.0° + j \sin 315.0°)$$

sum

$$
\begin{aligned}
&= (3.61)(1.41)[\cos(56.3° + 315.0°) + j \sin(56.3° + 315.0°)] \\
&= 5.09(\cos 371.3° + j \sin 371.3°) \\
&= 5.09(\cos 11.3° + j \sin 11.3°)
\end{aligned}
$$

Note that in the final result the angle is expressed as $11.3°$. The angle is usually expressed between $0°$ and $360°$, unless specified otherwise. As we have seen, in some applications it is common to use negative angles.   ■

■**EXAMPLE 2**   When we use the $r\underline{/\theta}$ polar form to multiply the two complex numbers in Example 1, we have

$$r_1 = 3.61 \qquad \theta_1 = 56.3°$$
$$r_2 = 1.41 \qquad \theta_2 = 315.0°$$

$$(3.61\underline{/56.3°})(1.41\underline{/315.0°}) = (3.61)(1.41)\underline{/56.3° + 315.0°}$$

$$= 5.09\underline{/371.3°}$$
$$= 5.09\underline{/11.3°}$$

Again note that the angle in the final result is between 0° and 360°. ----------■

If we wish to *divide* one complex number in exponential form by another, we arrive at the following result:

$$r_1 e^{j\theta_1} \div r_2 e^{j\theta_2} = \frac{r_1}{r_2} e^{j(\theta_1 - \theta_2)} \qquad \textbf{(12-14)}$$

Therefore, *the result of dividing one complex number in polar form by another is given by*

$$\frac{r_1(\cos\theta_1 + j\sin\theta_1)}{r_2(\cos\theta_2 + j\sin\theta_2)} = \frac{r_1}{r_2}[\cos(\theta_1 - \theta_2) + j\sin(\theta_1 - \theta_2)]$$

$$\frac{r_1\underline{/\theta_1}}{r_2\underline{/\theta_2}} = \frac{r_1}{r_2}\underline{/\theta_1 - \theta_2} \qquad \textbf{(12-15)}$$

NOTE ▶   The *magnitudes are divided,* and the *angles are subtracted.*

■**EXAMPLE 3**   Divide the first complex number of Example 1 by the second. Using the polar forms, we have

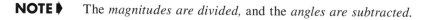

$$\frac{3.61(\cos 56.3° + j\sin 56.3°)}{1.41(\cos 315.0° + j\sin 315.0°)} = \frac{3.61}{1.41}[\cos(56.3° - 315.0°) + j\sin(56.3° - 315.0°)]$$

$$= 2.56[\cos(-258.7°) + j\sin(-258.7°)]$$
$$= 2.56(\cos 101.3° + j\sin 101.3°) \qquad ----------■$$

■**EXAMPLE 4**   Repeating Example 3 using $r\underline{/\theta}$ polar form, we have

$$\frac{3.61\underline{/56.3°}}{1.41\underline{/315.0°}} = \frac{3.61}{1.41}\underline{/56.3° - 315.0°} = 2.56\underline{/-258.7°}$$

$$= 2.56\underline{/101.3°} \qquad ----------■$$

CAUTION ▶

We have just seen that multiplying and dividing numbers in polar form can be easily performed. However, if we are to add or subtract numbers in polar form, *we must do the addition or subtraction by using rectangular form.*

■ EXAMPLE 5 Perform the addition $1.563 \underline{/37.56°} + 3.827 \underline{/146.23°}$.

In order to do this addition, we must change each number to rectangular form.

$$1.563 \underline{/37.56°} + 3.827 \underline{/146.23°}$$
$$= 1.563(\cos 37.56° + j \sin 37.56°) + 3.827(\cos 146.23° + j \sin 146.23°)$$
$$= 1.2390 + 0.9528j - 3.1813 + 2.1273j$$
$$= -1.9423 + 3.0801j$$

Now we change this to polar form.

$$r = \sqrt{(-1.9423)^2 + (3.0801)^2} = 3.641$$

$$\tan \theta = \frac{3.0801}{-1.9423} \qquad \theta = 122.24°$$

Therefore,

$$1.563 \underline{/37.56°} + 3.827 \underline{/146.23°} = 3.641 \underline{/122.24°} \quad \text{------} \blacksquare$$

DeMoivre's Theorem

For reference, Eq. (11-3) is $(a^m)^n = a^{mn}$.

To raise a complex number to a power, we may use the exponential form of the number in Eq.(11-3). This leads to

$$(re^{j\theta})^n = r^n e^{jn\theta} \tag{12-16}$$

Extending this to polar form, we have

$$[r(\cos \theta + j \sin \theta)]^n = r^n(\cos n\theta + j \sin n\theta)$$
$$(r \underline{/\theta})^n = r^n \underline{/n\theta} \tag{12-17}$$

Named for the mathematician Abraham DeMoivre (1667–1754).

Equation (12-17) is known as **DeMoivre's theorem** *and is valid for all real values of n. It is also used for finding roots of complex numbers if n is a fractional exponent.* We note that *the magnitude is raised to the power,* and *the angle is multiplied by the power.*

■ EXAMPLE 6 Using DeMoivre's theorem, find $(2 + 3j)^3$.

From Example 1 of this section, we know $r = 3.61$ and $\theta = 56.3°$. Therefore,

$$[3.61(\cos 56.3° + j \sin 56.3°)]^3 = (3.61)^3[\cos (3 \times 56.3°) + j \sin (3 \times 56.3°)]$$
$$= 47.0(\cos 168.9° + j \sin 168.9°) = 47.0 \underline{/168.9°}$$

Expressing θ in radians, we have $\theta = 56.3° = 0.983$ rad. Therefore,

$$(3.61e^{0.983j})^3 = (3.61)^3 e^{3 \times 0.983j} = 47.0e^{2.95j}$$
$$(2 + 3j)^3 = 47.0(\cos 168.9° + j \sin 168.9°)$$
$$= 47.0 \underline{/168.9°} = 47.0e^{2.95j} = -46 + 9j \quad \text{------} \blacksquare$$

■EXAMPLE 7   Find the cube root of -1.

Since we know that -1 is a real number, we can find its cube root by means of the definition. That is, $(-1)^3 = -1$. We shall check this by DeMoivre's theorem. Writing -1 in polar form, we have

$$-1 = 1(\cos 180° + j \sin 180°)$$

Applying DeMoivre's theorem, with $n = \frac{1}{3}$, we obtain

$$(-1)^{1/3} = 1^{1/3}(\cos \tfrac{1}{3} 180° + j \sin \tfrac{1}{3} 180°) = \cos 60° + j \sin 60°$$

$$= \frac{1}{2} + j\frac{\sqrt{3}}{2} \qquad \text{exact answer}$$

$$= 0.5000 + 0.8660j \qquad \text{decimal approximation}$$

Observe that we did not obtain -1 as the answer. If we check the answer, in the form $\frac{1}{2} + j\frac{\sqrt{3}}{2}$, by actually cubing it, we obtain -1! Therefore, it is a correct answer.

We should note that it is possible to take $\frac{1}{3}$ of any angle up to $1080°$ and still have an angle less than $360°$. Since $180°$ and $540°$ have the same terminal side, let us try writing -1 as $1(\cos 540° + j \sin 540°)$. Using DeMoivre's theorem, we have

$$(-1)^{1/3} = 1^{1/3}(\cos \tfrac{1}{3} 540° + j \sin \tfrac{1}{3} 540°) = \cos 180° + j \sin 180° = -1$$

We have found the answer we originally anticipated.

Angles of $180°$ and $900°$ also have the same terminal side, so we try

$$(-1)^{1/3} = 1^{1/3}(\cos \tfrac{1}{3} 900° + j \sin \tfrac{1}{3} 900°) = \cos 300° + j \sin 300°$$

$$= \frac{1}{2} - j\frac{\sqrt{3}}{2} \qquad \text{exact answer}$$

$$= 0.5000 - 0.8660j \qquad \text{decimal approximation}$$

Checking this, we find that it is also a correct root. We may try $1260°$, but $\frac{1}{3}(1260°) = 420°$, which has the same functional values as $60°$, and would give us the answer $0.5000 + 0.8660j$ again.

We have found, therefore, *three cube roots* of -1. They are

$$-1, \frac{1}{2} + j\frac{\sqrt{3}}{2}, \frac{1}{2} - j\frac{\sqrt{3}}{2}$$

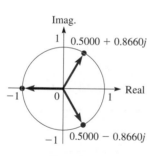

Fig. 12-17

These roots are graphed in Fig. 12-17. Note that they are equally spaced on the circumference of a circle of radius 1. ┄┄┄┄■

When the results of Example 7 are generalized, it can be proved that *there are n nth roots of a complex number*. When graphed, these roots are on a circle of radius $r^{1/n}$ and are equally spaced $360°/n$ apart. Following is the method for finding these n roots of a complex number.

Using DeMoivre's Theorem to Find the _n nth_ Roots of a Complex Number

  1. *Express the number in polar form.*
  2. *Express the root as a fractional exponent.*
  3. *Use Eq. (12-17) with θ to find one root.*
  4. *Use Eq. (12-17) and add 360° to θ, n − 1 times, to find the other roots.*

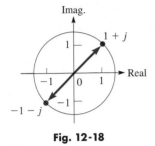

Fig. 12-18

EXAMPLE 8   Find the two square roots of $2j$.

We first write $2j$ in polar form as

$$2j = 2(\cos 90° + j \sin 90°)$$

To find square roots, we use the exponent $n = 1/2$. The first square root is

$$(2j)^{1/2} = 2^{1/2}\left(\cos \frac{90°}{2} + j \sin \frac{90°}{2}\right) = \sqrt{2}(\cos 45° + j \sin 45°) = 1 + j$$

To find the other square root we add 360° to 90°. This gives us

$$(2j)^{1/2} = 2^{1/2}\left(\cos \frac{450°}{2} + j \sin \frac{450°}{2}\right) = \sqrt{2}(\cos 225° + j \sin 225°) = -1 - j$$

Therefore, the two square roots of $2j$ are $1 + j$ and $-1 - j$. We see in Fig. 12-18 that they are on a circle of radius $\sqrt{2}$ and are 180° apart.

EXAMPLE 9   Find the six sixth roots of 64.

First we write 64 in polar form as $64 = 64(\cos 0° + j \sin 0°)$. We then note that we use the exponent 1/6 for the sixth root and that $64^{1/6} = \sqrt[6]{64} = 2$.

See Appendix C for a graphing calculator program DEMOIVRE. It can be used to find the roots of a complex number.

Calculators that operate with complex numbers generally find only the first of the n roots.

First root: $64^{1/6} = 64^{1/6}\left(\cos \frac{0°}{6} + j \sin \frac{0°}{6}\right) = 2(\cos 0° + j \sin 0°) = 2$

add 360°

Second root: $64^{1/6} = 64^{1/6}\left(\cos \frac{0° + 360°}{6} + j \sin \frac{0° + 360°}{6}\right)$

$= 2(\cos 60° + j \sin 60°) = 1 + j\sqrt{3}$

add $2 \times 360°$

Third root: $64^{1/6} = 64^{1/6}\left(\cos \frac{0° + 720°}{6} + j \sin \frac{0° + 720°}{6}\right)$

$= 2(\cos 120° + j \sin 120°) = -1 + j\sqrt{3}$

add $3 \times 360°$

Fourth root: $64^{1/6} = 64^{1/6}\left(\cos \frac{0° + 1080°}{6} + j \sin \frac{0° + 1080°}{6}\right)$

$= 2(\cos 180° + j \sin 180°) = -2$

add $4 \times 360°$

Fifth root: $64^{1/6} = 64^{1/6}\left(\cos \frac{0° + 1440°}{6} + j \sin \frac{0° + 1440°}{6}\right)$

$= 2(\cos 240° + j \sin 240°) = -1 - j\sqrt{3}$

add $5 \times 360°$

Sixth root: $64^{1/6} = 64^{1/6}\left(\cos \frac{0° + 1800°}{6} + j \sin \frac{0° + 1800°}{6}\right)$

$= 2(\cos 300° + j \sin 300°) = 1 - j\sqrt{3}$

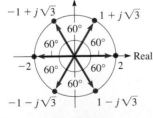

Fig. 12-19

These roots are graphed in Fig. 12-19. Note that they are equally spaced 60° apart on the circumference of a circle of radius 2.

From the text and examples of this and previous sections we see the advantages for the various forms of writing complex numbers. Rectangular form can be used for all of the basic operations, but lends itself best to addition and subtraction. Polar form is used for multiplication, division, raising to powers, and finding roots. Exponential form can be used for multiplication, division, and powers, and also for theoretical purposes (e.g. deriving DeMoivre's theorem).

EXERCISES 12-6

In Exercises 1–16, perform the indicated operations. Leave the result in polar form.

1. $[4(\cos 60° + j \sin 60°)] [2(\cos 20° + j \sin 20°)]$

2. $[3(\cos 120° + j \sin 120°)] [5(\cos 45° + j \sin 45°)]$

3. $(0.5\underline{/140°})(6\underline{/110°})$

4. $(0.4\underline{/320°})(5.5\underline{/150°})$

5. $\dfrac{8(\cos 100° + j \sin 100°)}{4(\cos 65° + j \sin 65°)}$

6. $\dfrac{9(\cos 230° + j \sin 230°)}{3(\cos 80° + j \sin 80°)}$

7. $\dfrac{12\underline{/320°}}{5\underline{/210°}}$

8. $\dfrac{2\underline{/90°}}{4\underline{/75°}}$

9. $[2(\cos 35° + j \sin 35°)]^3$

10. $[3(\cos 120° + j \sin 120°)]^4$

11. $(2\underline{/135°})^8$

12. $(1\underline{/142°})^{10}$

13. $\dfrac{(50\underline{/236°})(2\underline{/84°})}{25\underline{/47°}}$

14. $\dfrac{36\underline{/274°}}{(2\underline{/141°})(6\underline{/195°})}$

15. $\dfrac{(4\underline{/24°})(10\underline{/326°})}{(1\underline{/186°})(8\underline{/77°})}$

16. $\dfrac{(25\underline{/194°})(6\underline{/239°})}{(3\underline{/17°})(10\underline{/29°})}$

In Exercises 17–20, perform the indicated operations. Express results in polar form. See Example 5.

17. $2.78\underline{/56.8°} + 1.37\underline{/207.3°}$

18. $15.9\underline{/142.6°} - 18.5\underline{/71.4°}$

19. $7085\underline{/115.62°} - 4667\underline{/296.34°}$

20. $307.5\underline{/326.54°} + 726.3\underline{/96.41°}$

In Exercises 21–32, change each number to polar form and then perform the indicated operations. Express the result in rectangular and polar forms. Check by performing the same operation in rectangular form.

21. $(3 + 4j)(5 - 12j)$

22. $(-2 + 5j)(-1 - j)$

23. $(7 - 3j)(8 + j)$

24. $(1 + 5j)(4 + 2j)$

25. $\dfrac{7}{1 - 3j}$

26. $\dfrac{8j}{7 + 2j}$

27. $\dfrac{3 + 4j}{5 - 12j}$

28. $\dfrac{-2 + 5j}{-1 - j}$

29. $(3 + 4j)^4$

30. $(-1 - j)^8$

31. $(2 + 3j)^5$

32. $(1 - 2j)^6$

In Exercises 33–44, use DeMoivre's theorem to find all the indicated roots. Be sure to find all roots.

33. The two square roots of $4(\cos 60° + j \sin 60°)$

34. The three cube roots of $27(\cos 120° + j \sin 120°)$

35. The three cube roots of $3 - 4j$

36. The two square roots of $-5 + 12j$

37. The square roots of $1 + j$

38. The cube roots of $\sqrt{3} + j$

39. The fourth roots of 1

40. The cube roots of 8

41. The cube roots of $-27j$

42. The fourth roots of j

43. The fifth roots of -32

44. The sixth roots of -8

In Exercises 45–48, perform the indicated operations.

45. In Example 7, we showed that one cube root of -1 is $\frac{1}{2} - \frac{1}{2}j\sqrt{3}$. Cube this number in rectangular form and show that the result is -1.

(W) 46. Explain why the two square roots of a complex number are negatives of each other.

47. The electric power p (in W) supplied to an element in a circuit is the product of the voltage e and the current i (in A). Find the expression for the power supplied if $e = 6.80\underline{/56.3°}$ volts and $i = 7.05\underline{/-15.8°}$ amperes.

48. The displacement d (in in.) of a weight suspended on a system of two springs is $d = 6.03\underline{/22.5°} + 3.26\underline{/76.0°}$ in. Perform the addition and express the answer in polar form.

$12\text{-}7$ AN APPLICATION TO ALTERNATING-CURRENT (ac) CIRCUITS

In the 1880s it was decided that alternating current (favored by George Westinghouse) would be used to distribute electric power. Thomas Edison had argued for the use of direct current.

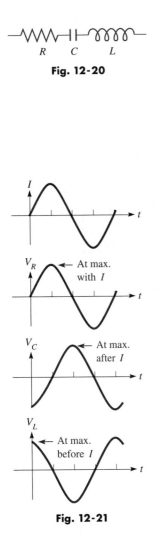

Fig. 12-20

Fig. 12-21

Eqs. (12-18) is based on Ohm's law that states that the current is proportional to the voltage for a constant resistance. It is named for the German physicist Georg Ohm (1787–1854). The ohm (Ω) is named for him.

We shall complete our study of complex numbers by showing their use in one aspect of alternating-current circuit theory. This application will be made to measuring voltage between any two points in a simple ac circuit, similar to the application mentioned in some of the examples and exercises of earlier sections of this chapter. We shall consider a circuit containing a resistance, a capacitance, and an inductance.

A *resistance* is any part of a circuit that tends to obstruct the flow of electric current through the circuit. It is denoted by R (units in ohms, Ω) and in diagrams by —WWW— , as shown in Fig. 12-20. A *capacitance* is two nonconnected plates in a circuit; no current actually flows across the gap between them. In an ac circuit, an electric charge is continually going to and from each plate and, therefore, the current in the circuit is not effectively stopped. It is denoted by C (units in farads, F) and in diagrams by —||— (see Fig. 12-20). An *inductance* is basically a coil of wire in which current is induced because the current is continually changing in the circuit. It is denoted by L (units in henrys, H) and in diagrams by ᴕᴕᴕᴕ (see Fig. 12-20). All of these elements affect the voltage in an alternating-current circuit. We shall state here the relation each has to the voltage and current in the circuit.

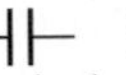

In Chapter 10, when we were discussing the graphs of the trigonometric functions, we noted that the current and voltage in an ac circuit could be represented by a sine or cosine curve. Therefore, each reaches peak values periodically. *If they reach their respective peak values at the same time, we say they are **in phase**. If the voltage reaches its peak before the current, we say that the voltage **leads** the current. If the voltage reaches its peak after the current, we say that the voltage **lags** the current.*

In the study of electricity, it is shown that the voltage across a resistance is in phase with the current. The voltage across a capacitor lags the current by 90°, and the voltage across an inductance leads the current by 90°. This is shown in Fig. 12-21, where, in a given circuit, I represents the current, V_R is the voltage across a resistor, V_C is the voltage across a capacitor, V_L is the voltage across an inductor, and t represents time.

Each element in an ac circuit tends to offer a type of resistance to the flow of current. *The effective resistance of any part of the circuit is called the **reactance**,* and it is denoted by X. The voltage across any part of the circuit whose reactance is X is given by $V = IX$, where I is the current (in amperes) and V is the voltage (in volts). Therefore,

the voltage V_R across a resistor with resistance R,

the voltage V_C across a capacitor with reactance X_C, and

the voltage V_L across an inductor with reactance X_L

are, respectively,

$$V_R = IR \qquad V_C = IX_C \qquad V_L = IX_L \qquad \textbf{(12-18)}$$

To determine the voltage across a combination of these elements of a circuit, we must account for the reactance, as well as the phase of the voltage across the individual elements. Since the voltage across a resistor is in phase with the current, we represent V_R along the positive real axis as a real number. Since the voltage across an inductance leads the current by 90°, we represent this voltage as a positive, pure imaginary number. In the same way, by representing the voltage across a capacitor as a negative, pure imaginary number, we show that the voltage *lags* the current by 90°. These representations are meaningful since the positive imaginary axis is +90° from the positive real axis and the negative imaginary axis is −90° from the positive real axis. See Fig. 12-22.

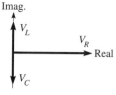

Fig. 12-22

The circuit elements shown in Fig. 12-20 are in *series,* and all circuits we shall consider are series circuits. The total voltage across a series of all three elements is given by $V_R + V_L + V_C$, which we shall represent by V_{RLC}. Therefore,

$$V_{RLC} = IR + IX_Lj - IX_Cj = I[R + j(X_L - X_C)]$$

This expression is also written as

$$V_{RLC} = IZ \tag{12-19}$$

where the symbol Z is called the **impedance** *of the circuit. It is the total effective resistance to the flow of current by a combination of the elements in the circuit,* taking into account the phase of the voltage in each element. From its definition, we see that Z is a complex number.

$$Z = R + j(X_L - X_C) \tag{12-20}$$

with a magnitude

$$|Z| = \sqrt{R^2 + (X_L - X_C)^2} \tag{12-21}$$

Also, as a complex number, it makes an angle θ with the x-axis, given by

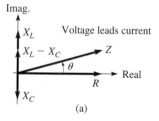

(a)

$$\theta = \tan^{-1} \frac{X_L - X_C}{R} \tag{12-22}$$

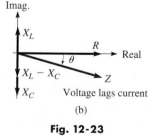

(b)

Fig. 12-23

All of these equations are based on phase relations of voltages with respect to the current. Therefore, *the angle θ represents the phase angle between the current and the voltage.* The standard way of expressing θ is to *use a positive angle if the voltage leads the current* and *use a negative angle if the voltage lags the current.* Using Eq. (12-22), a calculator will give the correct angle even when tan $\theta < 0$.

If the voltage leads the current, then $X_L > X_C$ as shown in Fig. 12-23(a). If the voltage lags the current, then $X_L < X_C$ as shown in Fig. 12-23(b).

In the examples and exercises of this section, the commonly used units and symbols for them are used. For a summary of these units and symbols, including prefixes, see Appendix B.

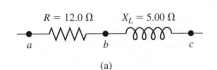

(a)

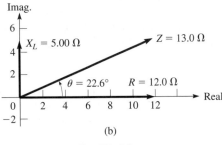

(b)

Fig. 12-24

See Appendix C for a graphing calculator program IMPEDANC. It can be used to calculate the impedance in a circuit.

■**EXAMPLE 1** In the series circuit shown in Fig. 12-24(a), $R = 12.0$ Ω and $X_L = 5.00$ Ω. A current of 2.00 A is in the circuit. Find the voltage across each element, the impedance, the voltage across the combination, and the phase angle between the current and the voltage.

The voltage across the resistor (between points a and b) is the product of the current and the resistance ($V = IR$). This means $V_R = (2.00)(12.0) = 24.0$ V. The voltage across the inductor (between points b and c) is the product of the current and the reactance, or $V_L = (2.00)(5.00) = 10.0$ V.

To find the voltage across the combination, between points a and c, we must first find the magnitude of the impedance. Note that ***the voltage is not the arithmetic sum of V_R and V_L***, as we must account for the phase.

By Eq. (12-20), the impedance is (there is no capacitor)

$$Z = 12.0 + 5.00j$$

with magnitude

$$|Z| = \sqrt{R^2 + X_L^2} = \sqrt{(12.0)^2 + (5.00)^2} = 13.0 \text{ Ω}$$

Thus, the magnitude of the voltage across the combination of the resistor and the inductance is

$$|V_{RL}| = (2.00)(13.0) = 26.0 \text{ V}$$

The phase angle between the voltage and the current is found by Eq. (12-22). This gives

$$\theta = \tan^{-1}\frac{5.00}{12.0} = 22.6°$$

The voltage *leads* the current by 22.6°, and this is shown in Fig. 12-24(b). ■

■**EXAMPLE 2** For a circuit in which $R = 8.00$ Ω, $X_L = 7.00$ Ω, and $X_C = 13.0$ Ω, find the impedance and the phase angle between the current and the voltage.

By the definition of impedance, Eq. (12-20), we have

$$Z = 8.00 + (7.00 - 13.0)j = 8.00 - 6.00j$$

where the magnitude of the impedance is

$$|Z| = \sqrt{(8.00)^2 + (-6.00)^2} = 10.0 \text{ Ω}$$

The phase angle is found by

$$\theta = \tan^{-1}\frac{-6.00}{8.00} = -36.9°$$

The angle $\theta = -36.9°$ is given directly by the calculator, and it is the angle we want. As we noted after Eq. (12-22), we express θ as a negative angle if the voltage lags the current, as it does in this example. See Fig. 12-25.

From the values above, we write the impedance in polar form as $Z = 10.0\underline{/-36.9°}$ ohms. ■

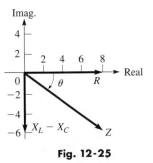

Fig. 12-25

The ampere (A) is named for the French physicist Andre Ampere (1775–1836).

The volt (V) is named for the Italian physicist Alessandro Volta (1745–1827).

The farad (F) is named for the British physicist Michael Faraday (1791–1867).

The henry (H) is named for the U.S. physicist Joseph Henry (1797–1878).

The coulomb (C) is named for the French physicist Charles Coulomb (1736–1806).

Note that the resistance is represented in the same way as a vector along the positive *x*-axis. Actually, resistance is not a vector quantity but is represented in this manner in order to assign an angle as the phase of the current. The important concept in this analysis is that *the phase **difference** between the current and voltage is constant,* and therefore any direction may be chosen arbitrarily for one of them. Once this choice is made, other phase angles are measured with respect to this direction. A common choice, as above, is to make the phase angle of the current zero. If an arbitrary angle is chosen, it is necessary to treat the current, voltage, and impedance as complex numbers.

■**EXAMPLE 3**   In a particular circuit, the current is $2.00 - 3.00j$ amperes and the impedance is $6.00 + 2.00j$ ohms. The voltage across this part of the circuit is

$$V = (2.00 - 3.00j)(6.00 + 2.00j) = 12.0 - 14.0j - 6.00j^2$$
$$= 12.0 - 14.0j + 6.00$$
$$= 18.0 - 14.0j \text{ volts}$$

The magnitude of the voltage is

$$|V| = \sqrt{(18.0)^2 + (-14.0)^2} = 22.8 \text{ V}$$

Since the voltage across a resistor is in phase with the current, this voltage can be represented as having a phase difference of zero with respect to the current. Therefore, the resistance is indicated as an arrow in the positive real direction, denoting the fact that the current and the voltage are in phase. *Such a representation is called a* **phasor.** The arrow denoted by R, as in Fig. 12-24, is actually the phasor representing the voltage across the resistor. Remember, the positive real axis is arbitrarily chosen as the direction of the phase of the current.

To show properly that the voltage across an inductance leads the current by 90°, its reactance (effective resistance) is multiplied by j. We know that there is a positive 90° angle between a positive real number and a positive imaginary number. In the same way, by multiplying the capacitive reactance by $-j$, we show the 90° difference in phase between the voltage and the current in a capacitor, with the current leading. Therefore, jX_L represents the phasor for the voltage across an inductor and $-jX_C$ is the phasor for the voltage across the capacitor. The phasor for the voltage across the combination of the resistance, inductance, and capacitance is Z, where the phase difference between the voltage and the current for the combination is the angle θ.

From this we see that *multiplying a phasor by j means to perform the operation of rotating it through* 90°. For this reason, j is also called the *j-operator.*

■**EXAMPLE 4**   Multiplying a positive real number A by j, we have $A \times j = Aj$, which is a positive imaginary number. In the complex plane, Aj is 90° from A, which means that by multiplying A by j we rotated A by 90°. Similarly, we see that $Aj \times j = Aj^2 = -A$, which is a negative real number, rotated 90° from Aj. Therefore, successive multiplications of A by j give us

$$A \times j = Aj \qquad \text{positive imaginary number}$$
$$Aj \times j = Aj^2 = -A \qquad \text{negative real number}$$
$$-A \times j = -Aj \qquad \text{negative imaginary number}$$
$$-Aj \times j = -Aj^2 = A \qquad \text{positive real number}$$

Fig. 12-26

See Fig. 12-26.

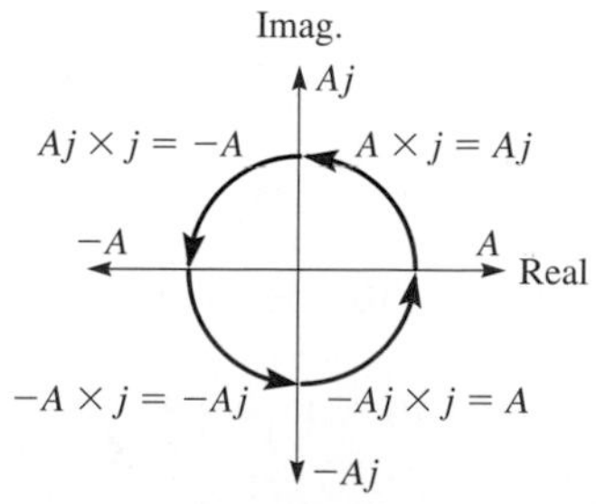

An alternating current is produced by a coil of wire rotating through a magnetic field. If the angular velocity of the wire is ω, the capacitive and inductive reactances are given by

$$X_C = \frac{1}{\omega C} \quad \text{and} \quad X_L = \omega L \tag{12-23}$$

Therefore, if ω, C, and L are known, the reactance of the circuit can be found.

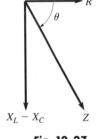

Fig. 12-27

■EXAMPLE 5   If $R = 12.0\ \Omega$, $L = 0.300$ H, $C = 250\ \mu$F, and $\omega = 80.0$ rad/s, find the impedance and the phase difference between the current and the voltage.

$$X_C = \frac{1}{(80.0)(250 \times 10^{-6})} = 50.0\ \Omega$$

$$X_L = (0.300)(80.0) = 24.0\ \Omega$$

$$Z = 12.0 + (24.0 - 50.0)j = 12.0 - 26.0j$$

$$|Z| = \sqrt{(12.0)^2 + (-26.0)^2} = 28.6\ \Omega$$

$$\theta = \tan^{-1}\frac{-26.0}{12.0} = -65.2°$$

$$Z = 28.6\underline{/-65.2°}\ \text{ohms}$$

The voltage *lags* the current (see Fig. 12-27).   --------■

We recall from Section 10-5 that the angular velocity ω is related to the frequency f by the relation $\omega = 2\pi f$. It is very common to use frequency when discussing alternating current.

An important concept in the application of this theory is that of **resonance.** *For resonance, the impedance of any circuit is a minimum, or the total impedance is R.* Thus, $X_L - X_C = 0$. Also, it can be seen that the current and the voltage are in phase under these conditions. Resonance is required for the tuning of radio and television receivers.

See the chapter introduction.

■EXAMPLE 6   In the antenna circuit of a radio, the inductance is 4.20 mH, and the capacitance is variable. What range of values of capacitance is necessary for the radio to receive the AM band of radio stations, with frequencies from 530 kHz to 1600 kHz?

For proper tuning, the circuit should be in resonance, or $X_L = X_C$. This means that

$$2\pi f L = \frac{1}{2\pi f C} \quad \text{or} \quad C = \frac{1}{(2\pi f)^2 L}$$

For $f_1 = 530$ kHz $= 5.30 \times 10^5$ Hz and $L = 4.20$ mH $= 4.20 \times 10^{-3}$ H,

$$C_1 = \frac{1}{(2\pi)^2(5.30 \times 10^5)^2(4.20 \times 10^{-3})} = 2.15 \times 10^{-11}\ \text{F} = 21.5\ \text{pF}$$

and for $f_2 = 1600$ kHz $= 1.60 \times 10^6$ Hz and $L = 4.20 \times 10^{-3}$ H, we have

$$C_2 = \frac{1}{(2\pi)^2(1.60 \times 10^6)^2(4.20 \times 10^{-3})} = 2.36 \times 10^{-12}\ \text{F} = 2.36\ \text{pF}$$

The capacitance should be capable of varying from 2.36 pF to 21.6 pF.   --------■

From Appendix B, the following prefixes are defined as follows:

Prefix	Factor	Symbol
pico	10^{-12}	p
milli	10^{-3}	m
kilo	10^6	k

EXERCISES *12-7*

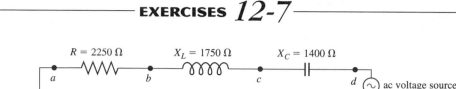

$R = 2250\ \Omega$ $X_L = 1750\ \Omega$ $X_C = 1400\ \Omega$

Fig. 12-28

In Exercises 1–4, use the circuit shown in Fig. 12-28. The current in the circuit is 5.75 mA. Determine the indicated quantities.

1. The voltage across the resistor (between points *a* and *b*).

2. The voltage across the inductor (between points *b* and *c*).

3. (a) The magnitude of the impedance across the resistor and the inductor (between points *a* and *c*).
   (b) The phase angle between the current and the voltage for this combination.
   (c) The voltage across this combination.

4. (a) The magnitude of the impedance across the resistor, inductor, and capacitor (between points *a* and *d*).
   (b) The phase angle between the current and the voltage for this combination.
   (c) The voltage across this combination.

In Exercises 5–8, an ac circuit contains the given combination of circuit elements from among a resistor ($R = 45.0\ \Omega$), a capacitor ($C = 86.2\ \mu F$), and an inductor ($L = 42.9$ mH). If the frequency in the circuit is $f = 60.0$ Hz, find (a) the magnitude of the impedance and (b) the phase angle between the current and voltage.

5. The circuit has the inductor and the capacitor (an *LC* circuit).

6. The circuit has the resistor and the capacitor (an *RC* circuit).

7. The circuit has the resistor and the inductor (an *RL* circuit).

8. The circuit has the resistor, the inductor, and the capacitor (an *RLC* circuit).

In Exercises 9–20, solve the given problems.

9. Given that the current in a given circuit is $3.90 - 6.04j$ mA and the impedance is $5.16 + 1.14j$ kΩ, find the magnitude of the voltage.

10. Given that the voltage in a given circuit is $8.375 - 3.140j$ V and the impedance is $2.146 - 1.114j$ Ω, find the magnitude of the current.

11. A resistance ($R = 25.3\ \Omega$) and a capacitance ($C = 2.75$ nF) are in an AM radio circuit. If $f = 1200$ kHz, find the impedance across the resistor and the capacitor.

12. A resistance ($R = 64.5\ \Omega$) and an inductance ($L = 1.08$ mH) are in a telephone circuit. If $f = 8.53$ kHz, find the impedance across the resistor and inductor.

13. The reactance of an inductor is 1200 Ω for $f = 280$ Hz. What is the inductance?

14. A resistor, an inductor, and a capacitor are connected in series across an ac voltage source. A voltmeter measures 12.0 V, 15.5 V, and 10.5 V, respectively, when placed across each element separately. What is the voltage of the source?

15. An inductance of 12.5 μH and a capacitance of 47.0 nF are in series in an amplifier circuit. Find the frequency for resonance.

16. A capacitance ($C = 95.2$ nF) and an inductance are in series in the circuit of a receiver for navigation signals. Find the inductance if the frequency for resonance is 50.0 kHz.

17. In Example 6, what should be the capacitance in order to receive a 680-kHz radio signal?

18. A 220-V source with $f = 60.0$ Hz is connected in series to an inductance ($L = 2.05$ H) and a resistance R in an electric motor circuit. Find R if the current is 0.250 A.

19. The power P (in W) supplied to a series combination of elements in an ac circuit is $P = VI \cos \theta$, where V is the effective voltage, I is the effective current, and θ is the phase angle between the current and voltage. If $V = 225$ mV across the resistor, capacitor, and inductor combination in Exercise 8, determine the power supplied to these elements.

(W) 20. Explain why the multiplication of a complex number by -1 may be shown as a rotation of the graph of the number about the origin through 180°.

In the development of electromagnetic theory, the existence of radio waves was predicted *mathematically* in the 1860s by the work of James Clerk Maxwell, a Scottish physicist. They were first produced in the laboratory in 1887 by Heinrich Hertz, a German physicist.

CHAPTER EQUATIONS

Chapter Equations for Complex Numbers

Imaginary unit

$$j = \sqrt{-1} \quad \text{and} \quad j^2 = -1 \tag{12-1}$$

$$\sqrt{-a} = j\sqrt{a} \quad (a > 0) \tag{12-2}$$

Basic operations

$$(a + bj) + (c + dj) = (a + c) + (b + d)j \tag{12-3}$$

$$(a + bj) - (c + dj) = (a - c) + (b - d)j \tag{12-4}$$

$$(a + bj)(c + dj) = (ac - bd) + (ad + bc)j \tag{12-5}$$

$$\frac{a + bj}{c + dj} = \frac{(a + bj)(c - dj)}{(c + dj)(c - dj)} = \frac{(ac + bd) + (bc - ad)j}{c^2 + d^2} \tag{12-6}$$

Complex number forms

Rectangular: $x + yj$

Polar: $r(\cos\theta + j\sin\theta) = r\underline{/\theta}$

Exponential: $re^{j\theta}$

$$x = r\cos\theta \qquad y = r\sin\theta \tag{12-7}$$

$$r^2 = x^2 + y^2 \qquad \tan\theta = \frac{y}{x} \tag{12-8}$$

$$x + yj = r(\cos\theta + j\sin\theta) = r\underline{/\theta} = re^{j\theta} \tag{12-12}$$

Product in polar form

$$r_1(\cos\theta_1 + j\sin\theta_1)r_2(\cos\theta_2 + j\sin\theta_2) = r_1 r_2[\cos(\theta_1 + \theta_2) + j\sin(\theta_1 + \theta_2)] \tag{12-13}$$

$$(r_1\underline{/\theta_1})(r_2\underline{/\theta_2}) = r_1 r_2\underline{/\theta_1 + \theta_2}$$

Quotient in polar form

$$\frac{r_1(\cos\theta_1 + j\sin\theta_1)}{r_2(\cos\theta_2 + j\sin\theta_2)} = \frac{r_1}{r_2}[\cos(\theta_1 - \theta_2) + j\sin(\theta_1 - \theta_2)] \tag{12-15}$$

$$\frac{r_1\underline{/\theta_1}}{r_2\underline{/\theta_2}} = \frac{r_1}{r_2}\underline{/\theta_1 - \theta_2}$$

DeMoivre's theorem

$$[r(\cos\theta + j\sin\theta)]^n = r^n(\cos n\theta + j\sin n\theta) \tag{12-17}$$

$$(r\underline{/\theta})^n = r^n\underline{/n\theta}$$

Chapter Equations for Alternating-Current Circuits

Voltage, current, reactance

$$V_R = IX_R \qquad V_C = IX_C \qquad V_L = IX_L \tag{12-18}$$

Impedance

$$V_{RLC} = IZ \tag{12-19}$$

$$Z = R + j(X_L - X_C) \tag{12-20}$$

$$|Z| = \sqrt{R^2 + (X_L - X_C)^2} \tag{12-21}$$

Phase angle

$$\theta = \tan^{-1}\frac{X_L - X_C}{R} \tag{12-22}$$

Capacitive reactance and inductive reactance

$$X_C = \frac{1}{\omega C} \quad \text{and} \quad X_L = \omega L \tag{12-23}$$

REVIEW EXERCISES

In Exercises 1–16, perform the indicated operations, expressing all answers in simplest rectangular form.

1. $(6 - 2j) + (4 + j)$

2. $(12 + 7j) + (-8 + 6j)$

3. $(18 - 3j) - (12 - 5j)$

4. $(-4 - 2j) - (-6 - \sqrt{-49})$

5. $(2 + j)(4 - j)$

6. $(-5 + 3j)(8 - 4j)$

7. $(2j^5)(6 - 3j)(4 + 3j)$

8. $j(3 - 2j) - (j^3)(5 + j)$

9. $\dfrac{3}{7 - 6j}$

10. $\dfrac{4j}{2 + 9j}$

11. $\dfrac{6 - \sqrt{-16}}{\sqrt{-4}}$

12. $\dfrac{3 + \sqrt{-4}}{4 - j}$

13. $\dfrac{5j - (3 - j)}{4 - 2j}$

14. $\dfrac{2 + (j - 6)}{1 - 2j}$

15. $\dfrac{j(7 - 3j)}{2 + j}$

16. $\dfrac{(2 - j)(3 + 2j)}{4 - 3j}$

In Exercises 17–20, find the values of x and y for which the equations are valid.

17. $3x - 2j = yj - 2$

18. $2xj - 2y = (y + 3)j - 3$

19. $2x - j + 4 = 6y + 2xj$

20. $3yj + xj = 6 + 3x + y$

In Exercises 21–24, perform the indicated operations graphically. Check them algebraically.

21. $(-1 + 5j) + (4 + 6j)$

22. $(7 - 2j) + (-5 + 4j)$

23. $(9 + 2j) - (5 - 6j)$

24. $(1 + 4j) - (-3 - 3j)$

In Exercises 25–32, give the polar and exponential forms of each of the complex numbers.

25. $1 - j$

26. $4 + 3j$

27. $-2 - 7j$

28. $6 - 2j$

29. $1.07 + 4.55j$

30. $-327 + 158j$

31. 10

32. $-4j$

In Exercises 33–44, give the rectangular form of each number.

33. $2(\cos 225° + j \sin 225°)$

34. $4(\cos 60° + j \sin 60°)$

35. $5.011(\cos 123.82° + j \sin 123.82°)$

36. $2.417(\cos 296.26° + j \sin 296.26°)$

37. $0.62\underline{/-72°}$

38. $20\underline{/160°}$

39. $27.08\underline{/346.27°}$

40. $1.689\underline{/194.36°}$

41. $2.00e^{0.25j}$

42. $e^{3.62j}$

43. $(35.37e^{1.096j})^2$

44. $(13.6e^{2.158j})(3.27e^{3.888j})$

In Exercises 45–60, perform the indicated operations. Leave the result in polar form.

45. $[3(\cos 32° + j \sin 32°)][5(\cos 52° + j \sin 52°)]$

46. $[2.5(\cos 162° + j \sin 162°)][8(\cos 115° + j \sin 115°)]$

47. $(40\underline{/18°})(0.5\underline{/245°})$

48. $(0.1254\underline{/172.38°})(27.17\underline{/204.34°})$

49. $\dfrac{24(\cos 165° + j \sin 165°)}{3(\cos 106° + j \sin 106°)}$

50. $\dfrac{18(\cos 403° + j \sin 403°)}{4(\cos 192° + j \sin 192°)}$

51. $\dfrac{245.6\underline{/326.44°}}{17.19\underline{/192.83°}}$

52. $\dfrac{100\underline{/206°}}{4\underline{/320°}}$

53. $0.983\underline{/47.2°} + 0.366\underline{/95.1°}$

54. $17.8\underline{/110.4°} - 14.9\underline{/226.3°}$

55. $7644\underline{/294.36°} - 6871\underline{/17.86°}$

56. $4.944\underline{/327.49°} + 8.009\underline{/7.37°}$

57. $[2(\cos 16° + j \sin 16°)]^{10}$

58. $[3(\cos 36° + j \sin 36°)]^6$

59. $(3\underline{/110.5°})^3$

60. $(5.36\underline{/220.3°})^4$

In Exercises 61–64, change each number to polar form and then perform the indicated operations. Express the final result in rectangular and polar forms. Check by performing the same operation in rectangular form.

61. $(1 - j)^{10}$

62. $(\sqrt{3} + j)^8(1 + j)^5$

63. $\dfrac{(5 + 5j)^4}{(-1 - j)^6}$

64. $(\sqrt{3} - j)^{-8}$

In Exercises 65–68, use DeMoivre's theorem to find the indicated roots. Be sure to find all roots.

65. The cube roots of -8

66. The cube roots of 1

67. The fourth roots of $-j$

68. The fifth roots of $32j$

In Exercises 69–72, determine the rectangular form and the polar form of the complex number for which the graphical representation is shown in the given figure.

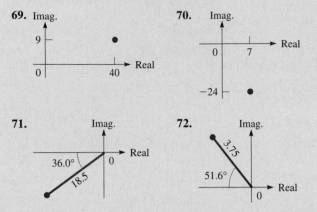

69.   **70.**

71.   **72.**

In Exercises 73–84, find the required quantities.

73. A 60-V ac voltage source is connected in series across a resistor, an inductor, and a capacitor. The voltage across the inductor is 60 V, and the voltage across the capacitor is 60 V. What is the voltage across the resistor?

74. In a series ac circuit with a resistor, an inductor, and a capacitor, $R = 6.50\ \Omega$, $X_C = 3.74\ \Omega$, and $Z = 7.50\ \Omega$. Find X_L.

75. In a series ac circuit with a resistor, an inductor, and a capacitor, $R = 6250\ \Omega$, $Z = 6720\ \Omega$, and $X_L = 1320\ \Omega$. Find the phase angle θ.

76. A coil of wire rotates at 120.0 r/s. If the coil generates a current in a circuit containing a resistance of 12.07 Ω, an inductance of 0.1405 H, and an impedance of 22.35 Ω, what must be the value of a capacitor (in F) in the circuit?

77. What is the frequency f for resonance in a circuit for which $L = 2.65$ H and $C = 18.3\ \mu$F?

78. The displacement of an electromagnetic wave is given by $d = A(\cos \omega t + j \sin \omega t) + B(\cos \omega t - j \sin \omega t)$. Find the expressions for the magnitude and phase angle of d.

79. Two cables lift a crate. The tensions in the cables can be represented by $210 - 120j$ N and $120 + 560j$ N. Express the resultant tension in polar form.

80. A boat is headed across a river with a velocity (relative to the water) that can be represented as $6.5 + 1.7j$ mi/h. The velocity of the river current can be represented as $-1.1 - 4.3j$ mi/h. Express the resultant velocity of the boat in polar form.

81. In the study of shearing effects in the spinal column, the expression $\dfrac{1}{u + j\omega n}$ is found. Express this in rectangular form.

82. In the theory of light reflection on metals, the expression
$$\frac{\mu(1 - kj) - 1}{\mu(1 - kj) + 1}$$
is encountered. Simplify this expression.

83. Show that $e^{j\pi} = -1$.

84. Show that $(e^{j\pi})^{1/2} = j$.

Writing Exercise

85. A computer programmer is writing a program to determine the n nth roots of a real number. Part of the program is to show the number of real roots and the number of pure imaginary roots. Write one or two paragraphs explaining how these numbers of roots can be determined without actually finding the roots.

PRACTICE TEST

1. Add, expressing the result in rectangular form: $(3 - \sqrt{-4}) + (5\sqrt{-9} - 1)$.

2. Multiply, expressing the result in polar form: $(2\underline{/130°})(3\underline{/45°})$.

3. Express $2 - 7j$ in polar form.

4. Express in terms of j: (a) $-\sqrt{-64}$, (b) $-j^{15}$.

5. Add graphically: $(4 - 3j) + (-1 + 4j)$.

6. Simplify, expressing the result in rectangular form: $\dfrac{2 - 4j}{5 + 3j}$.

7. Express $2.56(\cos 125.2° + j \sin 125.2°)$ in exponential form.

8. For an ac circuit in which $R = 3.50\ \Omega$, $X_L = 6.20\ \Omega$, and $X_C = 7.35\ \Omega$, find the impedance and the phase angle between the current and the voltage.

9. Express $3.47 - 2.81j$ in exponential form.

10. Find the values of x and y: $x + 2j - y = yj - 3xj$.

11. What is the capacitance of the circuit in a radio that has an inductance of 8.75 mH if it is to receive a station with frequency 600 kHz?

12. Find the cube roots of j.

13 EXPONENTIAL AND LOGARITHMIC FUNCTIONS

In this chapter we introduce two more important types of functions, the *exponential function* and the *logarithmic function*.

Historically, logarithms were first developed (in the early seventeenth century) to help with the lengthy calculations that were needed in areas such as navigation and astronomy. With the extensive use of calculators and computers, they are not used for such purposes today. However, these functions are very important in many scientific and technical applications, as well as in areas of advanced mathematics.

To illustrate the importance of exponential functions, many applications may be cited. These functions are used extensively in electronics, mechanical systems, thermodynamics, and nuclear physics. They are also used in biology in studying population growth and in business to calculate compound interest.

The use of the logarithmic function, which is closely related to the exponential function, is also extensive. The basic units used to measure intensity of sound and those used to measure the intensity of earthquakes are defined in terms of logarithms. In chemistry, the distinction between a base and an acid is defined in terms of logarithms. In electrical transmission lines, power gains and losses are measured in terms of logarithmic units. Many of these applications are illustrated throughout the chapter.

The growth of population can often be measured by use of an exponential function. In Section 13-6 this is illustrated.

13-1 THE EXPONENTIAL AND LOGARITHMIC FUNCTIONS

To this point we have considered exponents as specific rational numbers. In this section we introduce the *exponential function,* in which the exponent is a variable. This new function is defined on the following page.

Chapter 11 dealt with exponents in the form x^n, where n is a rational number and a constant. We shall now deal with expressions of the form b^x, where b is a constant and x is any real number. The main difference is that in the second expression, *the exponent is a variable.* Therefore, we define the **exponential function** *to be*

$$y = b^x \tag{13-1}$$

In Eq. (13-1), *x is called the* **logarithm** *of the number y to the base b.* In our work with the exponential function we shall restrict all numbers to the real number system. *This leads us to choose the base as a positive number other than* 1. We know that 1 raised to any power will result in 1, which would make y a constant regardless of the value of x. Negative numbers for b would result in imaginary values for y if x were any fractional exponent with an even integer for its denominator.

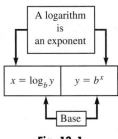

Fig. 13-1

EXAMPLE 1 $y = 2^x$ is an exponential function, where x is the logarithm of y to the base 2. This means that 2 raised to a given power gives the corresponding value of y.

If $x = 3$, $y = 2^3 = 8$; this means that 3 is the logarithm of 8 to the base 2. If $x = 4$, $y = 2^4 = 16$; this means that 4 is the logarithm of 16 to the base 2.

If $x = \frac{1}{2}$, $y = 2^{1/2} \approx 1.41$. Here $\frac{1}{2}$ is the logarithm of 1.41 to the base 2. ∎

Using the definition of a logarithm, we may express x in terms of y in the form

$$x = \log_b y \tag{13-2}$$

This equation is read in accordance with the definition of x in Eq. (13-1), that is, *x equals the logarithm of y to the base b.* This means that x is the power to which the base b must be raised in order to equal the number y; that is, x is a logarithm, and

CAUTION ▶ *a logarithm is an exponent.* Note that Eqs. (13-1) and (13-2) state the same relationship but in a different way. Equation (13-1) is the **exponential form,** and Eq. (13-2) is the **logarithmic form.** See Fig. 13-1.

EXAMPLE 2 The equation $y = 2^x$ would be written as $x = \log_2 y$ if we put it in logarithmic form. When we choose values of y to find the corresponding values of x from this equation, we ask ourselves "2 raised to what power x gives y?"

This means that if $y = 8$, we know that $2^3 = 8$, and therefore $x = 3$. ∎

EXAMPLE 3 (a) $3^2 = 9$ in logarithmic form is $2 = \log_3 9$.

(b) $4^{-1} = 1/4$ in logarithmic form is $-1 = \log_4(1/4)$.

CAUTION ▶ Remember, *the exponent may be negative.* The *base* must be positive. ∎

EXAMPLE 4 (a) $(64)^{1/3} = 4$ in logarithmic form is $\frac{1}{3} = \log_{64} 4$.

(b) $(32)^{3/5} = 8$ in logarithmic form is $\frac{3}{5} = \log_{32} 8$.

(c) $\log_2 32 = 5$ in exponential form is $32 = 2^5$.

(d) $\log_6 \left(\frac{1}{36}\right) = -2$ in exponential form is $\frac{1}{36} = 6^{-2}$. ∎

EXAMPLE 5   (a) Find b, given that $-4 = \log_b\left(\frac{1}{81}\right)$.

Writing this in exponential form, we have $\frac{1}{81} = b^{-4}$. Thus, $\frac{1}{81} = \frac{1}{b^4}$ or $\frac{1}{3^4} = \frac{1}{b^4}$. Therefore, $b = 3$.

(b) Find y, given that $\log_4 y = \frac{1}{2}$.

In exponential form we have $y = 4^{1/2}$, or $y = 2$.

We see that exponential form is very useful for determining values written in logarithmic form. For this reason, it is important that you learn to transform readily from one form to the other.

In order to change a function of the form $y = ab^x$ into logarithmic form, we must **CAUTION▶** first write it as $y/a = b^x$. ***The coefficient of b^x must be equal to 1,*** which is the form of Eq. (13-1). In the same way, the coefficient of $\log_b y$ must be 1 in order to change it into exponential form.

EXAMPLE 6   The power supply P (in W) of a certain satellite is given by $P = 75e^{-0.005t}$, where t is the time (in days) after launch. By writing this equation in logarithmic form, solve for t.

In order to have the equation in the exponential form of Eq. (13-1), we must have only $e^{-0.005t}$ on the right. Therefore, by dividing by 75, we have

$$\frac{P}{75} = e^{-0.005t}$$

Writing this in logarithmic form, we have

$$\log_e\left(\frac{P}{75}\right) = -0.005t$$

or

$$t = \frac{\log_e\left(\dfrac{P}{75}\right)}{-0.005} = -200\log_e\left(\frac{P}{75}\right) \qquad \frac{1}{-0.005} = -200$$

We recall from Section 12-5 that e is a special irrational number equal to about 2.718. It is an important number as a base of logarithms. This is discussed in more detail in Section 13-5.

When we are working with functions, we must keep in mind that a function is defined by the operation being performed on the independent variable, and not by the letter chosen to represent it. However, for consistency, it is standard practice to let y represent the dependent variable and x represent the independent variable. Therefore, *the **logarithmic function** is*

$$\boxed{y = \log_b x} \qquad \text{(13-3)}$$

As with the exponential function, $b > 0$ and $b \neq 1$.

Equations (13-2) and (13-3) do not represent different *functions,* due to the differ-**NOTE▶** ence in location of the variables, since they represent the *same operation* on the independent variable that appears in each. However, Eq. (13-3) expresses the function with the standard dependent and independent variables.

EXAMPLE 7 For the logarithmic function $y = \log_2 x$, we have the standard independent variable x and the standard dependent variable y.

If $x = 16$, $y = \log_2 16$, which means that $y = 4$, since $2^4 = 16$.

If $x = \frac{1}{16}$, $y = \log_2(\frac{1}{16})$, which means that $y = -4$, since $2^{-4} = \frac{1}{16}$. ▪

EXERCISES *13-1*

In Exercises 1–4, evaluate the exponential function $y = 9^x$ for the given values of x.

1. $x = 0.5$ **2.** $x = 4$

3. $x = -2$ **4.** $x = -0.5$

In Exercises 5–16, express the given equations in logarithmic form.

5. $3^3 = 27$ **6.** $5^2 = 25$

7. $4^4 = 256$ **8.** $8^2 = 64$

9. $4^{-2} = \frac{1}{16}$ **10.** $3^{-2} = \frac{1}{9}$

11. $2^{-6} = \frac{1}{64}$ **12.** $(12)^0 = 1$

13. $8^{1/3} = 2$ **14.** $(81)^{3/4} = 27$

15. $(\frac{1}{4})^2 = \frac{1}{16}$ **16.** $(\frac{1}{2})^{-2} = 4$

In Exercises 17–28, express the given equations in exponential form.

17. $\log_3 81 = 4$ **18.** $\log_{11} 121 = 2$

19. $\log_9 9 = 1$ **20.** $\log_{15} 1 = 0$

21. $\log_{25} 5 = \frac{1}{2}$ **22.** $\log_8 16 = \frac{4}{3}$

23. $\log_{243} 3 = \frac{1}{5}$ **24.** $\log_{32}(\frac{1}{8}) = -\frac{3}{5}$

25. $\log_{10} 0.1 = -1$ **26.** $\log_7(\frac{1}{49}) = -2$

27. $\log_{0.5} 16 = -4$ **28.** $\log_{1/3} 3 = -1$

In Exercises 29–44, determine the value of the unknown.

29. $\log_4 16 = x$ **30.** $\log_5 125 = x$

31. $\log_{10} 0.01 = x$ **32.** $\log_{16}(\frac{1}{4}) = x$

33. $\log_7 y = 3$ **34.** $\log_8 N = 3$

35. $\log_8(A - 2) = -\frac{2}{3}$ **36.** $\log_7 y = -2$

37. $\log_b 81 = 2$ **38.** $\log_b 625 = 4$

39. $\log_b 4 = -\frac{1}{3}$ **40.** $\log_b 4 = \frac{2}{3}$

41. $\log_{10} 10^{0.2} = x$ **42.** $\log_5 5^{2.3} = R + 1$

43. $\log_3 27^{-1} = x$ **44.** $\log_b(\frac{1}{4}) = -\frac{1}{2}$

In Exercises 45–48, evaluate the logarithmic function $y = \log_4 x$ for the given values of x.

45. $x = 64$ **46.** $x = \frac{1}{64}$

47. $x = \frac{1}{2}$ **48.** $x = 2$

In Exercises 49–56, perform the indicated operations.

49. The value V of a bank account in which A dollars is invested at 5% interest, compounded annually, is given by $V = A(1.05)^t$, where t is the time in years. Solve for t.

50. The intensity I of an earthquake is given by $I = I_0(10)^R$, where I_0 is a minimum intensity for comparison and R is the Richter scale magnitude of the earthquake. Solve for R.

51. The magnitudes (visual brightnesses), m_1 and m_2, of two stars are related to their (actual) brightnesses, b_1 and b_2, by the equation $m_1 - m_2 = 2.5 \log_{10}(b_2/b_1)$. Solve for b_2.

52. The velocity v of a rocket at the point at which its fuel is completely burned is given by $v = u \log_e(w_0/w)$, where u is the exhaust velocity, w_0 is the lift-off weight, and w is the burnout weight. Solve for w.

53. An equation relating the number N of atoms of radium at any time t in terms of the number N_0 of atoms at $t = 0$ is $\log_e(N/N_0) = -kt$, where k is a constant. Solve for N.

54. The charge q on a capacitor is given by $q = q_0(1 - e^{-at})$, where q_0 is the initial charge, a is a constant, and t is the time. Solve for t.

55. An oil company's records show that the production P (in barrels) of one of its wells had decreased according to the equation $P = 9600(2^{-0.12t})$, where t is measured in years. Solve for t.

56. An equation used in measuring the flow of water in a channel is $C = -a \log_{10}(b/R)$. Solve for R.

13-2 GRAPHS OF $y = b^x$ AND $y = \log_b x$

Graphical representation of functions is often valuable when we wish to show their properties. We shall now show representative graphs of the exponential function $y = b^x$ and the logarithmic function $y = \log_b x$.

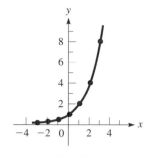

Fig. 13-2

EXAMPLE 1   Plot the graph of $y = 2^x$.

Assuming values for x and then finding the corresponding values for y, we obtain the following table.

x	-3	-2	-1	0	1	2	3
y	$\frac{1}{8}$	$\frac{1}{4}$	$\frac{1}{2}$	1	2	4	8

$\llcorner\, 2^{-3} = \frac{1}{8}$ $\llcorner\, 2^0 = 1$ $\llcorner\, 2^3 = 8$

From these values we plot the curve, as shown in Fig. 13-2. We note that the x-axis is an asymptote of the curve. ----------∎

EXAMPLE 2   Plot the graph of $y = \log_2 x$.

We can find the points for this graph more easily if we first put the equation in exponential form: $x = 2^y$. By assuming values for y, we can find the corresponding values for x.

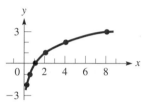

Fig. 13-3

$\overset{\displaystyle 2^{-2} = \frac{1}{4}}{\downarrow}$ $\overset{\displaystyle 2^2 = 4}{\downarrow}$

x	$\frac{1}{4}$	$\frac{1}{2}$	1	2	4	8
y	-2	-1	0	1	2	3

Using these values, we construct the graph seen in Fig. 13-3. ----------∎

Any exponential or logarithmic curve, where $b > 1$, will be similar in shape to those shown in Examples 1 and 2. From these examples we can see that these curves have the following basic properties.

Basic Properties of Exponential and Logarithmic Curves

1. *If $0 < x < 1$, $\log_b x < 0$; if $x = 1$, $\log_b 1 = 0$; if $x > 1$, $\log_b x > 0$.*

2. *If $x > 1$, x increases more rapidly than $\log_b x$.*

3. *For all values of x, $b^x > 0$.*

4. *If $x > 1$, b^x increases more rapidly than x.*

From the graphs in Figs. 13-2 and 13-3 and the above analysis, we note that the domain of the exponential function $y = b^x$ is all real numbers and that its range is $y > 0$. For the logarithmic function $y = \log_b x$, the domain is $x > 0$ and the range is all real numbers.

NOTE▶

We noted above that if $x > 1$, x increases more rapidly than $\log_b x$ and b^x increases more rapidly than x. Actually, as x becomes larger, $\log_b x$ increases very slowly, but b^x increases very rapidly. Using $\log_2 x$ and 2^x and a calculator, we have the following table of values.

Figures 13-2 and 13-3 give a graphical view of the basic properties, and this table gives a numerical view of the basic properties for $x \geq 1$.

x	1	4	16	64
$\log_2 x$	0	2	4	6
2^x	2	16	$65{,}536$	1.8×10^{19}

This shows that we must select values of x carefully when graphing these functions.

The graphing calculator also can be used to graph exponential and logarithmic functions. We can use the information about the domain and range and how the function increases in choosing values for the *window* settings on the calculator.

Although the bases most important to applications are greater than 1, to understand how the curve of the exponential function differs somewhat if $b < 1$, let us consider the following example.

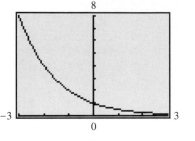

Fig. 13-4

■ EXAMPLE 3   Display the graph of $y = (\frac{1}{2})^x$ on a graphing calculator.

We could simply try some values for the *window* settings to get the graph. However, if we look at the function, we see that

$$\left(\frac{1}{2}\right)^x = \frac{1}{2^x} = 2^{-x}$$

Since x may be any real number, we note that *as x becomes more negative, y will increase rapidly.* Therefore, on a calculator we let $y_1 = (1/2)^x$ (or $y_1 = 0.5^x$) and have the display in Fig. 13-4 with the *window* settings shown.   ------- ■

The log keys on the calculator are $\boxed{\log}$ (for $\log_{10} x$) and $\boxed{\ln}$ (for $\log_e x$). Therefore, for now we restrict graphing logarithmic functions to bases 10 and e on the graphing calculator. In Section 13-5 we will see how to use the calculator to graph a logarithmic function with any positive real number base.

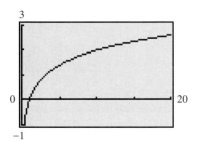

Fig. 13-5

■ EXAMPLE 4   Display the graph of $y = 2 \log_{10} x$ on a graphing calculator.

On a graphing calculator, we set $y_1 = 2 \log x$, and the curve is displayed as shown in Fig. 13-5. The settings for Xmin and Xmax were selected since the domain is $x > 0$ and $\log_{10} x$ increases very slowly. The settings for Ymin and Ymax were selected since $\log_{10} 0.1 = -1$, and y does not reach 4 until $x = 100$. ------ ■

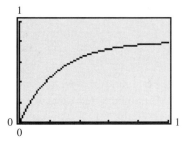

Fig. 13-6

■ EXAMPLE 5   In an electric circuit in which there is a battery, an inductor, and a resistor, the current i (in A) as a function of the time t (in s) is $i = 0.8(1 - e^{-4t})$. Display the graph of this function on a graphing calculator.

Here we note that $e^{-4t} = 1$ for $t = 0$, and this means that $i = 0$ for $t = 0$. Also, e^{-4t} becomes very small in a very short time, and this means that i cannot be greater than 0.8 A. With these considerations, we have the settings and curve as shown in Fig. 13-6. Here we used y for i, x for t, and $y_1 = 0.8(1 - e^{-4x})$.   ------- ■

Inverse Functions

For the exponential function $y = b^x$ and the logarithmic function $y = \log_b x$, if we solve for the independent variable in one of the functions by changing the form, then interchange the variables, we obtain the other function. *Such functions are called* **inverse functions.**

This means that the x- and y-coordinates of inverse functions are interchanged. As a result, the graphs of inverse functions are mirror images of each other across the line $y = x$. This is illustrated in the following example.

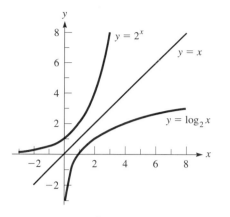

Fig. 13-7

EXAMPLE 6   The functions $y = 2^x$ and $y = \log_2 x$ are inverse functions. We show this by solving $y = 2^x$ for x, and then interchange x and y.

Writing $y = 2^x$ in logarithmic form gives us $x = \log_2 y$. Then interchanging x and y, we have $y = \log_2 x$, which is the inverse function.

Making a table of values for each function we have

$y = 2^x$

x	-3	-2	-1	0	1	2	3
y	$\frac{1}{8}$	$\frac{1}{4}$	$\frac{1}{2}$	1	2	4	8

$y = \log_2 x$

x	$\frac{1}{8}$	$\frac{1}{4}$	$\frac{1}{2}$	1	2	4	8
y	-3	-2	-1	0	1	2	3

We see that the coordinates are interchanged. In Fig. 13-7, note that the graphs of these two functions reflect each other across the line $y = x$.

For a function, there is exactly one value of y in the range for each value of x in the domain. This must also hold for the inverse function. Thus, for a function to have an inverse function, there must be only one x for each y. This is true for $y = b^x$ and $y = \log_b x$, as we have seen earlier in this section.

EXERCISES 13-2

In Exercises 1–16, plot the graphs of the given functions.

1. $y = 3^x$ **2.** $y = 4^x$ **3.** $y = 0.5\pi^x$ **4.** $y = 2e^x$

5. $y = (\frac{1}{3})^x$ **6.** $y = (\frac{1}{4})^x$

7. $y = 0.5(3.06)^{-x}$ **8.** $y = 0.1(10^{-x})$

9. $y = \log_3 x$ **10.** $y = \log_4 x$

11. $y = \log_{32} x$ **12.** $y = \log_{0.5} x$

13. $N = 2.5 \log_3 v$ **14.** $y = 3 \log_2 x$

15. $y = 0.2 \log_4 x$ **16.** $A = 2.4 \log_{10} 2r$

In Exercises 17–24, display the graphs of the given functions on a graphing calculator.

17. $y = 0.3(2.55)^x$ **18.** $y = 1.5(4.15)^x$

19. $y = 0.1(0.25)^x$ **20.** $y = 0.4(0.95)^x$

21. $i = 1.2(2 + 6^{-t})$ **22.** $y = 0.5e^{-x}$

23. $y = 3 \log_e x$ **24.** $y = 5 \log_{10}|x|$

In Exercises 25–32, sketch the indicated graphs. A graphing calculator may be used (except for 29).

25. If an amount of P dollars is invested at an annual interest rate r (expressed as a decimal), the value V of the investment after t years is $V = P(1 + r/n)^{nt}$, if interest is compounded n times a year. If \$1000 is invested at an annual interest rate of 6%, compounded semiannually, express V as a function of t and sketch the graph for $0 \le t \le 8$ years.

26. A certain flexible cable between two towers 300 ft apart can be represented by $y = 40(e^{0.005x} + e^{-0.005x})$. Sketch the graph representing the cable for $-150 \le x \le 150$ ft.

27. A projection of the annual growth rate p (in %) of the number of users of the Internet is $p = 10(1.2^{-t} + 1)$, where t is the number of years after 2000. Sketch the graph of this function from 2000 to 2010.

28. The current i (in A) in a certain electric circuit is given by $i = 16(1 - e^{-250t})$, where t is the time in seconds. Using appropriate values of t, sketch the graph of this function.

29. The time t (in ps) required for N calculations by a certain computer design is $t = N + \log_2 N$. Sketch the graph of this function.

30. An original amount of 100 mg of radium radioactively decomposes such that N mg remain after t years. The function relating t and N is $t = 2350(\log_e 100 - \log_e N)$. Sketch the graph.

W **31.** Sketch the graphs of $y = (1/3)^x$ and $y = 3^{-x}$ and then compare them. Explain what your comparison shows.

W **32.** In Exercise 12, the graph of $y = \log_{0.5} x$ is plotted. By inspecting the graph and noting the properties of $\log_{0.5} x$, describe some of the differences of logarithms to a base less than 1 from those to a base greater than 1.

In Exercises 33–36, show that the given functions are inverse functions of each other. Then display the graphs of each function and the line $y = x$ on a graphing calculator, and note that each is the mirror image of the other across $y = x$.

33. $y = 10^{x/2}$ and $y = 2 \log_{10} x$

34. $y = e^x$ and $y = \log_e x$

35. $y = 3x$ and $y = x/3$

36. $y = 2x + 4$ and $y = 0.5x - 2$

13-3 PROPERTIES OF LOGARITHMS

A logarithm is an exponent. However, a historical curiosity is that logarithms were developed before exponents were used.

Since a logarithm is an exponent, it must follow the laws of exponents. Those laws we will find of the greatest importance at this time are now listed here for reference.

$$b^u b^v = b^{u+v} \tag{13-4}$$

$$\frac{b^u}{b^v} = b^{u-v} \tag{13-5}$$

$$(b^u)^n = b^{nu} \tag{13-6}$$

We will use these laws of exponents to derive certain useful properties of logarithms. The following example illustrates the reasoning used in deriving these properties.

EXAMPLE 1 We know that $8 \times 16 = 128$. Writing these numbers as powers of 2, we have

$$8 = 2^3 \qquad 16 = 2^4 \qquad 128 = 2^7 = 2^{3+4}$$

The logarithmic forms can be written as

$$3 = \log_2 8 \qquad 4 = \log_2 16 \qquad 3 + 4 = \log_2 128$$

This means that

$$\log_2 8 + \log_2 16 = \log_2 128$$

where

$$8 \times 16 = 128$$

The *sum of the logarithms* of 8 and 16 equals the logarithm of 128, where the *product* of 8 and 16 equals 128. ─────────■

Following Example 1, if we let $u = \log_b x$ and $v = \log_b y$ and write these equations in exponential form, we have $x = b^u$ and $y = b^v$. Therefore, forming the product of x and y, we obtain

$$xy = b^u b^v = b^{u+v} \quad \text{or} \quad xy = b^{u+v}$$

Writing this last equation in logarithmic form yields

$$u + v = \log_b xy$$

or

LOGARITHM OF A PRODUCT

$$\log_b xy = \log_b x + \log_b y \tag{13-7}$$

Equation (13-7) states the property that *the logarithm of the product of two numbers is equal to the sum of the logarithms of the numbers.*

Using the same definitions of u and v to form the quotient of x and y, we then have

$$\frac{x}{y} = \frac{b^u}{b^v} = b^{u-v} \quad \text{or} \quad \frac{x}{y} = b^{u-v}$$

Writing this last equation in logarithmic form, we have

$$u - v = \log_b\left(\frac{x}{y}\right)$$

or

LOGARITHM OF A QUOTIENT

$$\log_b\left(\frac{x}{y}\right) = \log_b x - \log_b y \tag{13-8}$$

Equation (13-8) states the property that *the logarithm of the quotient of two numbers is equal to the logarithm of the numerator minus the logarithm of the denominator.*

If we again let $u = \log_b x$ and write this in exponential form, we have $x = b^u$. To find the nth power of x, we write

$$x^n = (b^u)^n = b^{nu}$$

Expressing this equation in logarithmic form yields

$$nu = \log_b(x^n)$$

or

LOGARITHM OF A POWER

$$\log_b(x^n) = n \log_b x \tag{13-9}$$

Equation (13-9) states that *the logarithm of the nth power of a number is equal to n times the logarithm of the number.* The exponent n may be integral or fractional.

In Section 13-1 we showed that the base b of logarithms must be a positive number. Since $x = b^u$ and $y = b^v$, this means that x and y are also positive numbers. Therefore, *the properties of logarithms that have just been derived are valid only for positive values of x and y.*

In advanced mathematics the logarithms of negative and imaginary numbers are defined.

■**EXAMPLE 2**   (a) Using Eq. (13-7), we may express $\log_4 15$ as a sum of logarithms.

$$\log_4 15 = \log_4(3 \times 5) = \log_4 3 + \log_4 5 \quad \begin{array}{l} \text{logarithm of product} \\ \text{sum of logarithms} \end{array}$$

(b) Using Eq. (13-8), we may express $\log_4\left(\frac{5}{3}\right)$ as the difference of logarithms.

$$\log_4\left(\frac{5}{3}\right) = \log_4 5 - \log_4 3 \quad \begin{array}{l} \text{logarithm of quotient} \\ \text{difference of logarithms} \end{array}$$

(c) Using Eq. (13-9), we may express $\log_4(t^2)$ as twice $\log_4 t$.

$$\log_4(t^2) = 2 \log_4 t \quad \begin{array}{l} \text{logarithm of power} \\ \text{multiple of logarithm} \end{array}$$

(d) Using Eq. (13-8) and then Eq. (13-7), we have

$$\log_4\left(\frac{xy}{z}\right) = \log_4(xy) - \log_4 z = \log_4 x + \log_4 y - \log_4 z \quad ■$$

■**EXAMPLE 3** We may also express a sum or difference of logarithms as the logarithm of a single quantity.

(a) $\log_4 3 + \log_4 x = \log_4 (3 \times x) = \log_4 3x$ using Eq. (13-7)

(b) $\log_4 3 - \log_4 x = \log_4 \left(\dfrac{3}{x}\right)$ using Eq. (13-8)

(c) $\log_4 3 + 2 \log_4 x = \log_4 3 + \log_4 (x^2) = \log_4 3x^2$ using Eqs. (13-7) and (13-9)

(d) $\log_4 3 + 2 \log_4 x - \log_4 y = \log_4 \left(\dfrac{3x^2}{y}\right)$ using Eqs. (13-7), (13-8), and (13-9)   ▪

In Section 13-2 we noted that $\log_b 1 = 0$. Also, since $b = b^1$ in logarithmic form is $\log_b b = 1$, we have $\log_b (b^n) = n \log_b b = n(1) = n$.

Summarizing these properties, we have

$$\log_b 1 = 0 \qquad \log_b b = 1 \qquad \textbf{(13-10)}$$

$$\log_b (b^n) = n \qquad \textbf{(13-11)}$$

These equations may be used to find exact values of certain logarithms.

■**EXAMPLE 4**   **(a)** We may evaluate $\log_3 9$ using Eq. (13-11).

$$\log_3 9 = \log_3 (3^2) = 2$$

We can establish the exact value since the base of logarithms and the number being raised to the power are the same. Of course, this could have been evaluated directly from the definition of a logarithm.

(b) We can evaluate $\log_3 (3^{0.4})$ using Eq. (13-11).

$$\log_3 (3^{0.4}) = 0.4$$

Although we did not evaluate $3^{0.4}$, we were able to evaluate $\log_3 3^{0.4}$.   ▪

■**EXAMPLE 5**   **(a)** $\log_2 6 = \log_2 (2 \times 3) = \log_2 2 + \log_2 3 = 1 + \log_2 3$

(b) $\log_5 \frac{1}{5} = \log_5 1 - \log_5 5 = 0 - 1 = -1$

(c) $\log_7 \sqrt{7} = \log_7 (7^{1/2}) = \frac{1}{2} \log_7 7 = \frac{1}{2}$ ▪

■**EXAMPLE 6**   The following illustration shows the evaluation of a logarithm in two different ways. Either method is appropriate.

(a) $\log_5 \left(\frac{1}{25}\right) = \log_5 1 - \log_5 25 = 0 - \log_5 (5^2) = -2$

(b) $\log_5 \left(\frac{1}{25}\right) = \log_5 (5^{-2}) = -2$ ▪

■**EXAMPLE 7**   Use the basic properties of logarithms to solve the following equation for y in terms of x: $\log_b y = 2 \log_b x + \log_b a$.

Using Eq. (13-9) and then Eq. (13-7), we have

$$\log_b y = \log_b (x^2) + \log_b a = \log_b (ax^2)$$

Since we have the logarithm to the base b of different expressions on each side of the resulting equation, the expressions must be equal. Therefore,

$$y = ax^2$$ ▪

EXAMPLE 8 An equation for the current i and the time t in an electric circuit containing a resistance R and a capacitance C is $\log_e i - \log_e I = -t/RC$ (where R and C are both factors of the denominator). Here, I is the current for $t = 0$. Solve for i as a function of t.

Using Eq. (13-8), we rewrite the left side of this equation, obtaining

$$\log_e\left(\frac{i}{I}\right) = -\frac{t}{RC}$$

Rewriting this in exponential form, we have

$$\frac{i}{I} = e^{-t/RC} \qquad \text{or} \qquad i = Ie^{-t/RC}$$

EXERCISES 13-3

In Exercises 1–12, express each as a sum, difference, or multiple of logarithms. See Example 2.

1. $\log_5 33$
2. $\log_3 14$
3. $\log_7\left(\frac{5}{3}\right)$
4. $\log_3\left(\frac{2}{11}\right)$
5. $\log_2(a^3)$
6. $\log_8(n^5)$
7. $\log_6 abc$
8. $\log_2\left(\frac{xy}{z^2}\right)$
9. $\log_5\sqrt[4]{y}$
10. $\log_4\sqrt[7]{x}$
11. $\log_2\left(\frac{\sqrt{x}}{a^2}\right)$
12. $\log_3\left(\frac{\sqrt[3]{y}}{7}\right)$

In Exercises 13–20, express each as the logarithm of a single quantity. See Example 3.

13. $\log_b a + \log_b c$
14. $\log_2 3 + \log_2 x$
15. $\log_5 9 - \log_5 3$
16. $\log_8 V - \log_8 R$
17. $\log_b x^2 - \log_b \sqrt{x}$
18. $\log_4 3^3 + \log_4 9$
19. $2\log_e 2 + 3\log_e n$
20. $\frac{1}{2}\log_b a - 2\log_b 5$

In Exercises 21–28, determine the exact value of each of the given logarithms.

21. $\log_2\left(\frac{1}{32}\right)$
22. $\log_3\left(\frac{1}{81}\right)$
23. $\log_2(2^{2.5})$
24. $\log_5(5^{0.1})$
25. $\log_7\sqrt{7}$
26. $\log_6\sqrt[3]{6}$
27. $\log_3\sqrt[4]{27}$
28. $\log_5\sqrt[3]{25}$

In Exercises 29–36, express each as a sum, difference, or multiple of logarithms. In each case, part of the logarithm may be determined exactly.

29. $\log_3 18$
30. $\log_5 75$
31. $\log_2\left(\frac{1}{6}\right)$
32. $\log_{10}(0.05)$
33. $\log_3\sqrt{6}$
34. $\log_2\sqrt[3]{24}$
35. $\log_{10} 3000$
36. $\log_{10}(40^2)$

In Exercises 37–48, solve for y in terms of x.

37. $\log_b y = \log_b 2 + \log_b x$
38. $\log_b y = \log_b 6 + \log_b x$
39. $\log_4 y = \log_4 x - \log_4 5 + \log_4 3$
40. $\log_3 y = \log_3 7 - 2\log_3(x + 1)$
41. $\log_{10} y = 2\log_{10} 7 - 3\log_{10} x$
42. $\log_b y = 3\log_b\sqrt{x} + 2\log_b 10$
43. $5\log_2 y - \log_2 x = 3\log_2 4 + \log_2 a$
44. $4\log_2 x - 3\log_2 y = \log_2 27$
45. $\log_2 x + \log_2 y = 1$
46. $3\log_4 x + \log_4 y = 1$
47. $\frac{2\log_5 x}{\log_5 3} - \log_5 y = 2$
48. $\log_8 x = 2\log_8 y + 4$

In Exercises 49–52, solve the given problems.

(W) 49. Explain why $\log_{10}(x + 3)$ is not equal to $\log_{10} x + \log_{10} 3$.

(W) 50. Display the graphs of $y = \log_e(e^2 x)$ and $y = 2 + \log_e x$ on a graphing calculator and explain why they are the same.

51. Under certain conditions, the temperature T (in °C) of a cooling object is related to the time t (in min) by the equation $\log_e T = \log_e 65.0 - 0.41t$. Solve for T as a function of t.

52. In analyzing the power gain in an electric circuit, the equation $N = 10(2\log_{10} I_1 - 2\log_{10} I_2 + \log_{10} R_1 - \log_{10} R_2)$ is used. Express this with a single logarithm on the right side.

13-4 LOGARITHMS TO THE BASE 10

In Section 13-1 we stated that a base of logarithms must be a positive number, not equal to one. In the examples and exercises of the previous sections, we used a number of different bases. There are, however, only two bases that are generally used. They are 10 and *e*, where *e* is the irrational number approximately equal to 2.718 that we introduced in Section 12-5 and have used in the previous sections of this chapter.

Base 10 logarithms were developed for calculational purposes and were used a great deal for making calculations until the 1970s, when the modern scientific calculator became widely available. Base 10 logarithms are still used in several scientific measurements, and therefore a need still exists for them. Base *e* logarithms are used extensively in technical and scientific work; we consider them in detail in the next section.

Logarithms to the base 10 are called **common logarithms.** They may be found directly by use of a calculator, and the $\boxed{\log}$ key is used for this purpose. This, of course, is the same key we have used with logarithmic functions to the base 10 on the graphing calculator. This calculator key indicates the common notation. *When no base is shown, it is assumed to be the base 10.*

NOTE ▶

EXAMPLE 1   Using a calculator, as shown in the first two lines of the display in Fig. 13-8, we find that

$$\log 426 = 2.629$$

↑── no base shown means base is 10

```
log(426)
      2.629409599
log(.03654)
      -1.437231457
```

Fig. 13-8

when the result is rounded off. The decimal part of a logarithm is normally expressed to the same accuracy as that of the number of which it is the logarithm, although showing one additional digit in the logarithm is generally acceptable.

Since $10^2 = 100$ and $10^3 = 1000$, and in this case

$$10^{2.629} = 426$$

we see that the 2.629 power of 10 gives a number between 100 and 1000. ■

EXAMPLE 2   Finding log 0.03654, as shown in the third and fourth lines of Fig. 13-8, we obtain

$$\log 0.03654 = -1.4372$$

We note that the logarithm here is negative. This should be the case when we recall the meaning of a logarithm. Raising 10 to a negative power gives us a number between 0 and 1, and here we have

$$10^{-1.4372} = 0.03654$$ ■

We may also use a calculator to find a number *N* if we know log *N*. *In this case we refer to N as the* **antilogarithm** *of log N.* On the calculator we use the $\boxed{10^x}$ key. We note that it shows the basic definiton of a logarithm. (On many scientific calculators the key sequence $\boxed{\text{inv}}$ $\boxed{\log}$ is used. Note that this sequence shows that the exponential and logarithmic functions are inverse functions.)

EXAMPLE 3   Given log $N = 1.1854$, as shown in the first two lines of the display in Fig. 13-9, we find that

$$N = 15.32$$

where the result has been rounded off. Since $10^1 = 10$ and $10^2 = 100$, we see that $10^{1.1854}$ is a number between 10 and 100. ∎

The following example illustrates an application in which a measurement requires the direct use of the value of a logarithm.

EXAMPLE 4   The power gain G (in decibels (dB)) of an electronic device is given by $G = 10 \log (P_0/P_i)$, where P_0 is the output power (in W) and P_i is the input power. Determine the power gain for an amplifier for which $P_0 = 15.8$ W and $P_i = 0.625$ W.

Substituting the given values, we have

$$G = 10 \log \frac{15.8}{0.625} = 14.0 \text{ dB}$$

where lines 3 and 4 of the display in Fig. 13-9 show the evaluation on a calclator. ∎

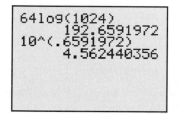

Fig. 13-9

The unit of sound intensity level (used for power gain), the bel (B), is named for the U.S. inventor Alexander Graham Bell (1847–1922). The decibel is the commonly used unit.

As noted in the chapter introduction, logarithms were developed for calculational purposes. They were first used in the seventeenth century for making tedious and complicated calculations that arose in astronomy and navigation. These complicated calculations were greatly simplified, since logarithms allowed them to be performed by means of basic additions, subtractions, multiplications, and divisions. Performing calculations in this way provides an opportunity to understand better the meaning and properties of logarithms. Also, certain calculations cannot be done directly on a calculator but can be done by logarithms.

EXAMPLE 5   A certain computer design has 64 different sequences of 10 binary digits so that the total number of possible states is $(2^{10})^{64} = 1024^{64}$. Evaluate 1024^{64} using logarithms.

Since $\log x^n = n \log x$, we know that $\log 1024^{64} = 64 \log 1024$. While most calculators will not directly evaluate 1024^{64}, we can use one to find the value of $64 \log 1024$. Since 1024^{64} is *exact*, we will show ten calculator digits until we round off the result. We therefore evaluate 1024^{64} as follows:

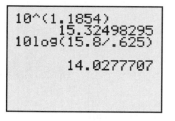

Fig. 13-10

Let $N = 1024^{64}$

$\log N = \log 1024^{64} = 64 \log 1024$ using Eq. (13-9): $\log_b x^n = n \log_b x$

   $= 192.6591972$

$N = 10^{192.6591972}$ meaning of logarithm

   $= 10^{192} \times 10^{0.6591972}$ using Eq. (13-4): $b^u b^v = b^{u+v}$

   $= (10^{192}) \times (4.5624)$ antilogarithm of 0.6591972 is 4.5624 (rounded off)

   $= 4.5624 \times 10^{192}$

By using Eq. (13-4), $10^{0.6591972}$ represents a number between 1 and 10 ($10^0 = 1$ and $10^1 = 10$), and we can write the result immediately in scientific notation.

Although we used a calculator to find 192.6591972 and 4.5624 as shown in Fig. 13-10, the calculation was done essentially by logarithms. ∎

Example 5 shows that calculations using logarithms are based on Eqs. (13-7), (13-8), and (13-9). Multiplication is performed by the addition of logarithms, division is performed by the subtraction of logarithms, and a power is found by a multiple of a logarithm. A root of a number is found by using the fractional exponent form of the power.

EXERCISES *13-4*

In Exercises 1–12, find the common logarithm of each of the given numbers by using a calculator.

1. 567

2. 60.5

3. 0.0640

4. 0.000566

5. 9.24×10^6

6. 3.19^3

7. 1.172×10^{-4}

8. 8.043×10^{-8}

9. $\cos 12.5°$

10. $\tan 50.8°$

11. $\sqrt{274}$

12. $\log_2 16$

In Exercises 13–20, find the antilogarithm of each of the given logarithms by using a calculator.

13. 4.437   **14.** 0.929   **15.** −1.3045   **16.** −6.9788

17. 3.30112   **18.** 8.82436   **19.** −2.23746   **20.** −10.336

In Exercises 21–24, use logarithms to evaluate the given expressions.

21. $(5.98)(14.3)$

22. $\dfrac{895}{73.4^{86}}$

23. $(\sqrt[10]{7.32})(2470)^{30}$

24. $\dfrac{126{,}000^{20}}{2.63^{2.5}}$

In Exercises 25–28, find the logarithms of the given numbers.

25. The signal used by some cordless phones is 9.00×10^8 Hz.

26. The bending moment of a particular concrete column is 4.60×10^6 N·m.

27. About 1.3×10^{-14}% of carbon atoms are carbon-14, the radioactive isotope used in determining the age of sample from ancient sites.

28. In an air sample taken in an urban area, $5/10^6$ of the air was carbon monoxide.

In Exercises 29 and 30, solve the given problems by finding the appropriate logarithms.

29. A stereo amplifier has an input power of 0.750 W and an output power of 25.0 W. What is the power gain? (See Example 4.)

30. Measured on the Richter scale, the magnitude of an earthquake of intensity I is defined as $R = \log (I/I_0)$, where I_0 is a minimum level for comparison. What is the Richter scale reading for the 1995 Philippine earthquake for which $I = 20{,}000{,}000 I_0$?

In Exercises 31 and 32, use logarithms to perform the indicated calculations.

31. A certain type of optical switch in a fiber-optic system allows a light signal to continue in either of two fibers. How many possible paths could a light signal follow if it passes through 400 such switches?

32. The peak current I_m (in A) in an alternating-current circuit is given by $I_m = \sqrt{\dfrac{2P}{Z \cos \theta}}$, where P is the power developed, Z is the magnitude of the impedance, and θ is the phase angle between the current and voltage. Evaluate I_m for $P = 5.25$ W, $Z = 320\ \Omega$, and $\theta = 35.4°$.

Logarithms were invented by the Scottish mathematician John Napier (1550–1617), who published a table of logarithms to the base e. The English mathematician Henry Briggs (1561–1631) realized that logarithms to the base 10 would make calculations much easier. He spent many years laboriously developing a table of base 10 logarithms, which was not completed until after his death.

The invention of logarithms was enthusiastically received by mathematicians and scientists as a long-needed tool for lengthy calculations. The French mathematician Pierre Laplace (1749–1827) stated that logarithms "by shortening the labors doubled the life of the astronomer."

Logarithms were commonly used for calculations until the 1970s when the modern scientific calculator came into use.

13-5 NATURAL LOGARITHMS

As we have noted, another number important as a base of logarithms is the number *e*. *Logarithms to the base e are called* **natural logarithms.** Since *e* is an irrational number equal to about 2.718, it may appear to be a very unnatural choice as a base of logarithms. However, in calculus the reason for its choice and the fact that it is a very natural number for a base of logarithms are shown.

Just as log *x* refers to logarithms to the base 10, the notation ln *x* is used to denote logarithms to the base *e*. We briefly noted this in Section 13-2 in discussing the graphing calculator. Due to the extensive use of natural logarithms, the notation **ln *x*** is more convenient than $\log_e x$, although they mean the same thing.

Since more than one base is important, at times it is useful to change a logarithm from one base to another. If $u = \log_b x$, then $b^u = x$. Taking logarithms of both sides of this last expression to the base *a*, we have

$$\log_a b^u = \log_a x$$

$$u \log_a b = \log_a x$$

$$u = \frac{\log_a x}{\log_a b}$$

However, $u = \log_b x$, which means that

$$\boxed{\log_b x = \frac{\log_a x}{\log_a b}} \tag{13-12}$$

Equation (13-12) allows us to change a logarithm in one base to a logarithm in another base. The following examples illustrate the method of performing this operation.

■**EXAMPLE 1**   Change log 20 to a logarithm with base *e;* that is, find ln 20.
Using Eq. (13-12) with $a = 10$, $b = e$, and $x = 20$, we have

$$\log_e 20 = \frac{\log_{10} 20}{\log_{10} e}$$

or $\ln 20 = \dfrac{\log 20}{\log e} = 2.996$ see lines 1 to 3 of calculator display in Fig. 13-11

This means that $e^{2.996} = 20$.   --------------■

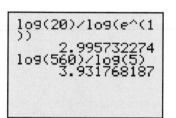

Fig. 13-11

■**EXAMPLE 2**   Find $\log_5 560$.
In Eq. (13-12), if we let $a = 10$ and $b = 5$, we have

$$\log_5 x = \frac{\log x}{\log 5}$$

In this example, $x = 560$. Therefore, we have

$$\log_5 560 = \frac{\log 560}{\log 5} = 3.932$$ see lines 4 to 5 of calculator display in Fig. 13-11

From the definition of a logarithm, this means that

$$5^{3.932} = 560$$ --------------■

Since natural logarithms are used extensively, it is often convenient to have Eq. (13-12) written specifically for use with logarithms to the base 10 and natural logarithms. First, using $a = 10$ and $b = e$, we have

$$\ln x = \frac{\log x}{\log e} \tag{13-13}$$

Then, with $a = e$ and $b = 10$, we have

$$\log x = \frac{\ln x}{\ln 10} \tag{13-14}$$

Note that we really found ln 20 in Example 1 by using Eq. (13-13).

Values of natural logarithms can be found directly on a calculator. The $\boxed{\ln}$ key is used for this purpose. In order to find the antilogarithm of a natural logarithm we use the $\boxed{e^x}$ key. The following example illustrates finding a natural logarithm and an antilogarithm on a graphing calculator.

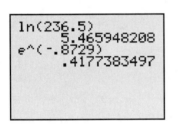

Fig. 13-12

■**EXAMPLE 3**   (a) From the first two lines of the calculator display in Fig. 13-12, we find that

$$\ln 236.5 = 5.4659$$

which means that $e^{5.4659} = 236.5$.

(b) Given that $\ln N = -0.8729$, we determine N by finding $e^{-0.8729}$ on the calculator. This gives us (see lines 3 and 4 of the display in Fig. 13-12)

$$N = 0.4177$$

Using Eq. (13-12), we can display the graph of a logarithmic function with any base on a graphing calculator, as we show in the next example.

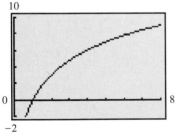

Fig. 13-13

■**EXAMPLE 4**   Display the graph of $y = 3 \log_2 x$ on a graphing calculator.

To display this graph we use the fact that

$$\log_2 x = \frac{\log x}{\log 2} \quad \text{or} \quad \log_2 x = \frac{\ln x}{\ln 2}$$

Therefore, we enter the function

$$y = \frac{3 \log x}{\log 2} \quad \left(\text{or } y = \frac{3 \ln x}{\ln 2}\right)$$

in the calculator. We can select the *window* settings by considering the domain and range of the logarithmic function, and by the fact that $3/\log 2 \approx 10$. This tells us that the range is about 10 times that of the curve $y = \log x$. The curve is shown in the calculator display shown in Fig. 13-13.

Applications of natural logarithms are found in many fields of technology. One such application is shown in the next example, and others are found in the exercises.

EXAMPLE 5 Under certain conditions, the electric current i in a circuit containing a resistance and an inductance (see Section 12-7) is given by

$$\ln \frac{i}{I} = -\frac{Rt}{L}$$

where I is the current at $t = 0$, R is the resistance, t is the time, and L is the inductance. Calculate how long (in seconds) it takes i to reach 0.430 A, if $I = 0.750$ A, $R = 7.50\ \Omega$, and $L = 1.25$ H.

Solving for t, we have

$$t = -\frac{L \ln (i/I)}{R} = -\frac{L(\ln i - \ln I)}{R} \qquad \text{either form can be used}$$

Thus, for the given values, we have

$$t = -\frac{1.25(\ln 0.430 - \ln 0.750)}{7.50} = 0.0927\ \text{s}$$

Therefore, the current changes from 0.750 A to 0.430 A in 0.0927 s.

EXERCISES *13-5*

In Exercises 1–6, use logarithms to the base 10 to find the natural logarithms of the given numbers.

1. 26.0   **2.** 631   **3.** 1.562

4. 0.5017   **5.** 0.0073267   **6.** 0.00044348

In Exercises 7–12, use logarithms to the base 10 to find the indicated logarithms.

7. $\log_7 42$ **8.** $\log_2 86$ **9.** $\log_5 245$

10. $\log_{12} 122$ **11.** $\log_{40} 750$ **12.** $\log_{100} 3720$

In Exercises 13–20, find the natural logarithms of the given numbers.

13. 51.4   **14.** 293   **15.** 1.394

16. 65.62   **17.** 0.9917   **18.** 0.002086

19. $(0.012937)^4$ **20.** $\sqrt{0.000060808}$

In Exercises 21–24, use Eq. (13-14) to find the common logarithms of the given numbers.

21. 45.17   **22.** 8765

23. 0.68528   **24.** 0.0014298

In Exercises 25–32, find the natural antilogarithms of the given logarithms.

25. 2.190   **26.** 3.420   **27.** 0.0084210

28. 0.632   **29.** −0.7429   **30.** −2.94218

31. −23.504   **32.** −0.00804

In Exercises 33–36, use a graphing calculator to display the indicated graphs.

33. The graph of $y = \log_5 x$ **34.** The graph of $y = 2 \log_8 x$

35. Graphically show that $y = 2^x$ and $y = \log_2 x$ are inverse functions. (See Example 6 of Section 13-2).

36. Explain what is meant by the expression $\ln \ln x$. Display the graph of $y = \ln \ln x$ on a calculator.

In Exercises 37–44, solve the given problems.

37. Solve for y in terms of x: $\ln y - \ln x = 1.0986$.

38. Solve for y in terms of x: $\ln y + 2 \ln x = 1 + \ln 5$.

39. If interest is compounded continuously (daily compounded interest closely approximates this), with an interest rate i, a bank account will double in t years according to $i = (\ln 2)/t$. Find i if the account is to double in 8.5 years.

40. One approximate formula for world population growth is $T = 50.0 \ln 2$, where T is the number of years for the population to double. According to this formula, how long does it take for the population to double?

41. For the electric circuit of Example 5, find how long it takes the current to reach 0.1 of the initial value of 0.750 A.

42. The intensity I of light decreases from its value I_0 as it passes a distance x through a medium. Given that $x = k(\ln I_0 - \ln I)$, where k is a constant depending on the medium, find x for $I = 0.850 I_0$ and $k = 5.00$ cm.

43. The distance x traveled by a motorboat in t seconds after the engine is cut off is given by $x = k^{-1} \ln (kv_0 t + 1)$, where v_0 is the velocity of the boat at the time the engine is cut and k is a constant. Find how long it takes a boat to go 150 m if $v_0 = 12.0$ m/s and $k = 6.80 \times 10^{-3}$/m.

44. The electric current i (in A) in a circuit containing a 1-H inductor, a 10-Ω resistor, and a 6-V battery is a function of the time t (in s) given by $i = 0.6(1 - e^{-10t})$. Solve for t as a function of i.

13-6 EXPONENTIAL AND LOGARITHMIC EQUATIONS

Exponential Equations

An equation in which the variable occurs in an exponent is called an **exponential equation.** Although some exponential equations may be solved by changing to logarithmic form, they are more generally solved by *taking the logarithm of each side* and then using the basic properties of logarithms.

▌**EXAMPLE 1**   (a) We can solve the exponential equation $2^x = 8$ for x by writing it in logarithmic form. This gives us

$$x = \log_2 8 = 3 \qquad 2^3 = 8$$

This method is good if we can directly evaluate the resulting logarithm.

   (b) Since 2^x and 8 are equal, the logarithms of 2^x and 8 are also equal. Therefore, we can also solve $2^x = 8$ in a more general way by taking logarithms (to any proper base) of both sides and equating these logarithms. This gives us

$$\log 2^x = \log 8 \qquad\qquad \text{or} \qquad \ln 2^x = \ln 8$$
$$x \log 2 = \log 8 \qquad\qquad\qquad x \ln 2 = \ln 8 \qquad \text{using Eq. (13-9)}$$
$$x = \frac{\log 8}{\log 2} = 3 \qquad\qquad\qquad x = \frac{\ln 8}{\ln 2} = 3 \qquad \text{using a calculator}$$

▌**EXAMPLE 2**   Solve the equation $3^{x-2} = 5$.
   Taking logarithms of each side and equating them, we have

$$\log 3^{x-2} = \log 5$$

or

$$(x - 2) \log 3 = \log 5 \qquad \text{using Eq. (13-9)}$$

Solving this last equation for x, we have

$$x = 2 + \frac{\log 5}{\log 3} = 3.465$$

This solution means that

$$3^{3.465-2} = 3^{1.465} = 5$$

which can be checked by a calculator.

▌**EXAMPLE 3**   Solve the equation $2(4^{x-1}) = 17^x$.
   By taking logarithms of each side, we have the following:

$$\log 2 + (x - 1) \log 4 = x \log 17 \qquad \text{using Eqs. (13-7) and (13-9)}$$
$$x \log 4 - x \log 17 = \log 4 - \log 2$$
$$x(\log 4 - \log 17) = \log 4 - \log 2$$
$$x = \frac{\log 4 - \log 2}{\log 4 - \log 17} = \frac{\log (4/2)}{\log 4 - \log 17} \qquad \text{using Eq. (13-8)}$$
$$= \frac{\log 2}{\log 4 - \log 17}$$
$$= -0.479$$

For reference, Eqs. (13-7), (13-8), and (13-9) are

$$\log_b xy = \log_b x + \log_b y$$
$$\log_b \left(\frac{x}{y}\right) = \log_b x - \log_b y$$
$$\log_b(x^n) = n \log_b x$$

EXAMPLE 4   At constant temperature, the atmospheric pressure p (in Pa) at an altitude h (in m) is given by $p = p_0 e^{kt}$, where p_0 is the pressure where $h = 0$ (usually taken as sea level). Given that $p_0 = 101.3$ kPa (atmospheric pressure at sea level) and $p = 68.9$ kPa for $h = 3050$ m, find the value of k.

Since the equation is defined in terms of e, we can solve it most easily by taking natural logarithms of *each side.* By doing this we have the following solution.

$$\ln p = \ln(p_0 e^{kh}) = \ln p_0 + \ln e^{kh} \qquad \text{using Eq. (13-7)}$$
$$= \ln p_0 + kh \ln e = \ln p_0 + kh \qquad \text{using Eq. (13-9), } \ln e = 1$$
$$\ln p - \ln p_0 = kh$$
$$k = \frac{\ln p - \ln p_0}{h}$$

Substituting the given values, we have

$$k = \frac{\ln(68.9 \times 10^3) - \ln(101.3 \times 10^3)}{3050} = -0.000126/\text{m}$$

Logarithmic Equations

Some of the important measurements in scientific and technical work are defined in terms of logarithms. Using these formulas can lead to solving a **logarithmic equation,** *which is an equation with the logarithm of an expression involving the variable.* In solving logarithmic equations, we use the basic properties of logarithms to help change them into a usable form. There is, however, no general algebraic method for solving such equations, and we shall consider only some special cases.

EXAMPLE 5   The human ear responds to sound on a scale which is approximately proportional to the intensity of the sound. Therefore, the loudness of sound (measured in dB) is defined by the equation $b = 10 \log(I/I_0)$, where I is the intensity of the sound and I_0 is the minimum intensity detectable.

A busy street has a loudness of 70 dB, and riveting has a loudness of 100 dB. To find how many times greater the intensity I_r of the sound of riveting is than the itensity I_c of the sound of the city street, we substitute in the above equation. This gives

$$70 = 10 \log\left(\frac{I_c}{I_0}\right) \quad \text{and} \quad 100 = 10 \log\left(\frac{I_r}{I_0}\right)$$

To solve for I_c and I_r, we divide each side by 10 and then use exponential form.

$$7.0 = \log\left(\frac{I_c}{I_0}\right) \quad \text{and} \quad 10 = \log\left(\frac{I_r}{I_0}\right)$$
$$\frac{I_c}{I_0} = 10^{7.0} \qquad\qquad \frac{I_r}{I_0} = 10^{10}$$
$$I_c = I_0(10^{7.0}) \qquad\quad I_r = I_0(10^{10})$$

This demonstrates that sound intensity levels change much more than loudness levels. (See Exercise 42.)

Since we want the number of times I_r is greater than I_c, we divide I_r by I_c.

$$\frac{I_r}{I_c} = \frac{I_0(10^{10})}{I_0(10^{7.0})} = \frac{10^{10}}{10^{7.0}} = 10^{3.0} \quad \text{or} \quad I_r = 10^{3.0}I_c = 1000I_c$$

Thus, the sound of riveting is 1000 times as intense as the sound of the city street.

See the chapter introduction.

EXAMPLE 6   An analysis of the population of Canada from 1980 to 2000 led to the equation $\log_2 P = \log_2 23.9 + 0.450$, where P is the projected population (in millions) in 2010. Determine this population.

The solution is as follows:

$$\log_2 P - \log_2 23.9 = 0.450$$

$$\log_2 (P/23.9) = 0.450 \qquad \text{using Eq. (13-8)}$$

$$P/23.9 = 2^{0.450} \qquad \text{exponential form}$$

$$P = 23.9(2^{0.450})$$

$$= 32.6 \text{ million} \qquad \text{projected 2010 population} \qquad ■$$

For reference, Eqs. (13-7), (13-8), and (13-9) are

$$\log_b xy = \log_b x + \log_b y$$

$$\log_b \left(\frac{x}{y}\right) = \log_b x - \log_b y$$

$$\log_b(x^n) = n \log_b x$$

EXAMPLE 7   Solve the logarithmic equation $2 \ln 2 + \ln x = \ln 3$.

Using the properties of logarithms, we have the following solution.

$$2 \ln 2 + \ln x = \ln 3$$

$$\ln 2^2 + \ln x - \ln 3 = 0 \qquad \text{using Eq. (13-9)}$$

$$\ln \frac{4x}{3} = 0 \qquad \text{using Eqs. (13-7) and (13-8)}$$

$$\frac{4x}{3} = e^0 = 1 \qquad \text{exponential form}$$

$$4x = 3$$

$$x = \frac{3}{4}$$

Since $\ln (3/4) = \ln 3 - \ln 4$, this solution checks in the *original equation.* $\qquad ■$

EXAMPLE 8   Solve the logarithmic equation $2 \log x - 1 = \log (1 - 2x)$.

$$\log x^2 - \log (1 - 2x) = 1$$

$$\log \frac{x^2}{1 - 2x} = 1 \qquad \text{using Eq. (13-8)}$$

$$\frac{x^2}{1 - 2x} = 10^1 \qquad \text{exponential form}$$

$$x^2 = 10 - 20x$$

$$x^2 + 20x - 10 = 0$$

$$x = \frac{-20 \pm \sqrt{400 + 40}}{2} = -10 \pm \sqrt{110}$$

Since logarithms of negative numbers are not defined and $-10 - \sqrt{110}$ is negative and cannot be used in the first term of the original equation, we have

$$x = -10 + \sqrt{110} = 0.488 \qquad \text{Use a calculator to check this result.} \qquad ■$$

EXERCISES *13-6*

In Exercises 1–36, solve the given equations.

1. $2^x = 16$

2. $3^x = \frac{1}{81}$

3. $5^x = 0.3$

4. $\pi^x = 15$

5. $3^{-x} = 0.525$

6. $15^{-x} = 1.326$

7. $e^{2x} = 3.625$

8. $e^{-x} = 17.54$

9. $6^{x+1} = 10$

10. $5^{x-1} = 2$

11. $4(3^x) = 5$

12. $3(14^x) = 40$

13. $0.8^x = 0.4$

14. $0.6^x = 2^{x^2}$

15. $(15.6)^{x+2} = 23^x$

16. $5^{x+2} = e^{2x}$

17. $3 \log_8 x = -2$

18. $5 \log_{32} x = -3$

19. $2 \ln x = 1$

20. $3 \ln 2x = 2$

21. $\log_2 x + \log_2 7 = \log_2 21$

22. $2 \log_2 3 - \log_2 x = \log_2 45$

23. $2 \log (3 - x) = 1$

24. $3 \log (2x - 1) = 1$

25. $\log 4x + \log x = 2$

26. $\log 12x^2 - \log 3x = 3$

27. $\ln x - \ln \frac{1}{3} = 1$

28. $2 \ln 2 - \ln x = -1$

29. $3 \ln 2 + \ln (x - 1) = \ln 24$

30. $\ln (2x - 1) - 2 \ln 4 = 3 \ln 2$

31. $\frac{1}{2} \log (x + 2) + \log 5 = 1$

32. $\frac{1}{2} \log (x - 1) + \log \frac{1}{x} = 0$

33. $\log_5 (x - 3) + \log_5 x = \log_5 4$

34. $\log_7 x + \log_7 (2x - 5) = \log_7 3$

35. $\log (2x - 1) + \log (x + 4) = 1$

36. $\log_2 x + \log_2 (x + 2) = 3$

In Exercises 37–48, determine the indicated quantities.

37. In computer design, the number N of bits of memory is often expressed as a power of 2, or $N = 2^x$. Find x if $N = 2.68 \times 10^8$ bits.

38. Assume the U.S. national debt in 2000 was 5.7 trillion dollars and increases from then at an annual rate of 5.0%. In what year will it have doubled? (Use the formula in Exercise 49, p. 354.)

39. The temperature T (in °C) of a cooling gold ingot is given by $T = 22 + 98(0.30)^{0.20t}$, where t is the time in minutes. How long will it take the ingot to cool to 35°C?

40. The electric current i (in A) in a circuit containing a resistance R, an inductor L, and a voltage source E is given by $i = \frac{E}{R}(1 - e^{-Rt/L})$, where t is the time in seconds. Find t if $i = 0.750$ A, $E = 6.00$ V, $R = 4.50$ Ω, and $L = 2.50$ H.

41. In chemistry, the pH value of a solution is a measure of its acidity. The pH value is defined by $pH = -\log (H^+)$, where H^+ is the hydrogen ion concentration. If the pH of a sample of rainwater is 4.764, find the hydrogen ion concentration. (If pH < 7, the solution is acid. If pH > 7, the solution is basic.) Acid rain has a pH between 4 and 5, and normal rain is slightly acidic with a pH of about 5.6.

42. Referring to Example 5, show that if the difference in loudness of two sounds is d decibels, the louder sound is $10^{d/10}$ more intense than the quieter sound.

43. Measured on the Richter scale, the magnitude of an earthquake of intensity I is defined as $R = \log (I/I_0)$, where I_0 is a minimum level for comparison. How many times I_0 was the 1906 San Francisco earthquake whose magnitude was 8.3 on the Richter scale?

44. How many times more intense was the 1906 San Francisco earthquake than the 1994 Northridge (southern California) earthquake, $R = 6.8$? (See Exercise 43.)

45. Pure water is running into a certain brine solution, and the same amount of solution is running out. The number n of kilograms of salt in the solution after t minutes is found by solving the equation $\ln n = -0.04t + \ln 20$. Solve for n as a function of t.

46. In an electric circuit containing a resistor and a capacitor with an initial charge q_0, the charge q on the capacitor at any time t after closing the switch can be found by solving the equation $\ln q = -\dfrac{t}{RC} + \ln q_0$. Here, R is the resistance and C is the capacitance. Solve for q as a function of t.

47. An earth satellite loses 0.1% of its remaining power each week. An equation relating the power P, the initial power P_0, and the time t (in weeks) is $\ln P = t \ln 0.999 + \ln P_0$. Solve for P as a function of t.

48. In finding the path of a certain plane, the equation $\ln r = \ln a - \ln \cos \theta - (v/w) \ln (\sec \theta + \tan \theta)$ is used. Solve for r. (The conditions of the plane's flight are similar to those described in Exercise 84, p. 320.)

To solve more complicated problems, we may use graphical methods. For example, to solve the equation $2^x + 3^x = 50$, we can set up the function $y = 2^x + 3^x - 50$ and then find its zeros graphically. Note that the given equation can be written as $2^x + 3^x - 50 = 0$, and therefore the zeros of the function that has been set up will give the desired solution. In Exercises 49–52, solve the given equations in this way by displaying the appropriate graphs on a calculator.

49. $2^x + 3^x = 50$

50. $4^x + x^2 = 25$

51. The curve in which a uniform wire or rope hangs under its own weight is called a *catenary*. An example of a catenary that we see every day is a wire strung between utility poles as shown in Fig. 13-14. For a particular wire, the equation of the catenary it forms is $y = 2(e^{x/4} + e^{-x/4})$, where (x, y) is a point on the curve. Find x for $y = 5.8$ m.

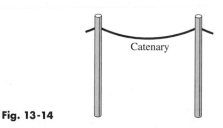

Catenary

Fig. 13-14

52. In finding the current i in a certain electric circuit, the equation $i = 2 \ln (t + 2) - t$ relates the current and the time t (in s). Find t for $i = 0.05$ A.

13-7 GRAPHS ON LOGARITHMIC AND SEMILOGARITHMIC PAPER

When constructing the graphs of some functions, one of the variables changes much more rapidly than the other. We saw this in graphing the exponential and logarithmic functions in Section 13-2. The following example illustrates this point.

■ **EXAMPLE 1** Plot the graph of $y = 4(3^x)$.

Constructing the following table of values,

x	−1	0	1	2	3	4	5
y	1.3	4	12	36	108	324	972

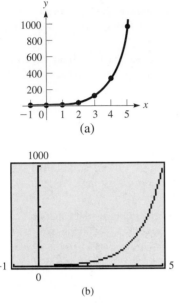

(a)

(b)

Fig. 13-15

we then plot these values as shown in Fig. 13-15(a).

We see that as x changes from −1 to 5, y changes much more rapidly, from about 1 to nearly 1000. Also, because of the scale that must be used, we see that it is not possible to show accurately the differences in the y-values on the graph.

Even a graphing calculator cannot show the graph accurately for the values near $x = 0$, if we wish to view all of this part of the curve. In fact, the graphing calculator view shows the curve as being on the axis for these values, as we see in Fig. 13-15(b). Note in this figure that we have shown the same values of the range of the function as we did in Fig. 13-15(a). ▬

It is possible to graph a function with a large change in values, for one or both variables, more accurately than can be done on the standard rectangular coordinate system. This is done by using a scale marked off in distances proportional to the logarithms of the values being represented. *Such a scale is called a* **logarithmic scale.** For example, log 1 = 0, log 2 = 0.301, and log 10 = 1. Thus, on a logarithmic scale the 2 is placed 0.301 unit of distance from the 1 to the 10. Figure 13-16 shows a logarithmic scale with the numbers represented and the distance used for each.

On a logarithmic scale the distances between the integers are not equal, but this scale does allow for a much greater range of values and much greater accuracy for many of the values. There is another advantage to using logarithmic scales. Many equations that would have more complex curves when graphed on the standard rectangular coordinate system will have simpler curves, often straight lines, when graphed using logarithmic scales. In many cases this makes the analysis of the curve much easier.

Zero and negative numbers do not appear on the logarithmic scale. In fact, all numbers used on the logarithmic scale must be positive, since the domain of the logarithmic function includes only positive real numbers. Thus, the logarithmic scale must start at some number greater than zero. This number is a power of 10 and can be very small, say, $10^{-6} = 0.000001$, but it is positive.

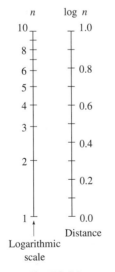

Fig. 13-16

If we wish to use a large range of values for only one of the variables, we use what is known as **semilogarithmic,** or **semilog,** graph paper. On this graph paper, only one axis (usually the y-axis) uses a logarithmic scale. If we wish to use a large range of values for both variables, we use **logarithmic,** or **log-log,** graph paper. Both axes are marked with logarithmic scales.

The following examples illustrate the use of semilog and log-log graph paper.

■**EXAMPLE 2** Construct the graph of $y = 4(3^x)$ on semilogarithmic graph paper.

This is the same function as in Example 1, and we repeat the table of values.

x	-1	0	1	2	3	4	5
y	1.3	4	12	36	108	324	972

Again, we see that the range of y-values is large. When we plotted this curve on the rectangular coordinate system in Example 1, we had to use large units along the y-axis. This made the values of 1.3, 4, 12, and 36 appear at practically the same level. However, when we use semilog graph paper, we can label each axis such that all y-values are accurately plotted as well as the x-values.

The logarithmic scale is shown in **cycles,** and we must label the base line of the first cycle as 1 times a power of 10 (0.01, 0.1, 1, 10, 100, and so on) with the following cycle labeled with the next power of 10. The lines between are labeled with 2, 3, 4, and so on, times the proper power of 10. See the vertical scale in Fig. 13-17. We now plot the points in the table on the graph. The resulting graph is a straight line, as we see in Fig. 13-17. Taking logarithms of each side of the equation, we have

$$\log y = \log [4(3^x)] = \log 4 + \log 3^x \qquad \text{using Eq. (13-7)}$$
$$= \log 4 + x \log 3 \qquad \text{using Eq. (13-9)}$$

However, since $\log y$ was plotted automatically (because we used semilogarithmic paper), the graph really represents

$$u = \log 4 + x \log 3$$

where $u = \log y$; $\log 3$ and $\log 4$ are constants, and therefore this equation is of the form $u = mx + b$, which is a straight line (see Section 5-2).

The logarithmic scale in Fig. 13-12 has *three cycles,* since all values of three powers of 10 are represented. --------▪

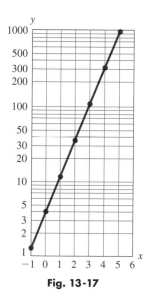

Fig. 13-17

■**EXAMPLE 3** Construct the graph of $x^4y^2 = 1$ on logarithmic paper.

First, we solve for y and make a table of values. Considering positive values of x and y, we have

$$y = \sqrt{\frac{1}{x^4}} = \frac{1}{x^2}$$

x	0.5	1	2	8	20
y	4	1	0.25	0.0156	0.0025

We plot these values on log-log paper on which both scales are logarithmic, as shown in Fig. 13-18. We again see that we have a straight line. Taking logarithms of both sides of the equation, we have

$$\log (x^4y^2) = \log 1$$
$$\log x^4 + \log y^2 = 0 \qquad \text{using Eq. (13-7)}$$
$$4 \log x + 2 \log y = 0 \qquad \text{using Eq. (13-9)}$$

If we let $u = \log y$ and $v = \log x$, we then have

$$4v + 2u = 0 \quad \text{or} \quad u = -2v$$

which is the equation of a straight line, as shown in Fig. 13-18. Note, however, that not all graphs on logarithmic paper are straight lines. --------▪

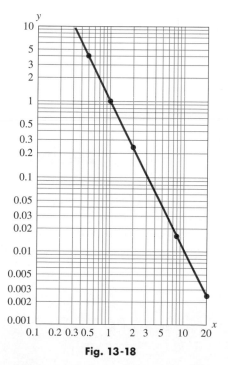

Fig. 13-18

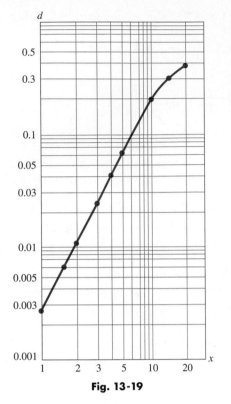

Fig. 13-19

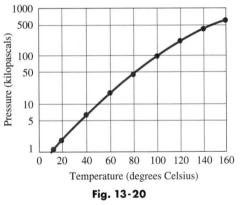

Fig. 13-20

EXAMPLE 4   The deflection (in ft) of a certain cantilever beam as a function of the distance x (in ft) from one end is

$$d = 0.0001(30x^2 - x^3)$$

If the beam is 20.0 ft long, plot a graph of d as a function of x on log-log paper. Constructing a table of values, we have

x (ft)	1.00	1.50	2.00	3.00	4.00
d (ft)	0.00290	0.00641	0.0112	0.0243	0.0416

x (ft)	5.00	10.0	15.0	20.0
d (ft)	0.0625	0.200	0.338	0.400

Since the beam is 20.0 ft long, there is no meaning to values of x greater than 20.0 ft. The graph is shown in Fig. 13-19.

Logarithmic and semilogarithmic paper may be useful for plotting data derived from experimentation. Often the data cover too large a range of values to be plotted on ordinary graph paper. The next example illustrates the use of semilogarithmic paper to plot data.

EXAMPLE 5   The vapor pressure of water depends on the temperature. The following table gives the vapor pressure (in kPa) for the corresponding values of temperature (in °C).

T (°C)	10	20	40	60	80	100	120	140	160
P (kPa)	1.19	2.33	7.34	19.9	47.3	101	199	361	617

These data are then plotted on semilogarithmic paper, as shown in Fig. 13-20. Intermediate values of temperature and pressure can then be read directly from the graph.

EXERCISES *13-7*

In Exercises 1–12, plot the graphs of the given functions on semilogarithmic paper.

1. $y = 2^x$
2. $y = 5^x$
3. $y = 2(4^x)$
4. $y = 5(10^x)$
5. $y = 3^{-x}$
6. $y = 2^{-x}$
7. $y = x^3$
8. $y = x^5$
9. $y = 3x^6$
10. $y = 2x^4$
11. $y = 2x^3 + 4x$
12. $y = 4x^3 + 2x^2$

In Exercises 13–24, plot the graphs of the given functions on log-log paper.

13. $y = 0.01x^4$
14. $y = 0.02x^3$
15. $y = \sqrt{x}$
16. $y = x^{2/3}$
17. $y = x^2 + 2x$
18. $y = x + \sqrt{x}$
19. $xy = 4$
20. $xy^3 = 10$
21. $y^2x = 1$
22. $x^2y^3 = 1$
23. $x^2y^2 = 25$
24. $x^3y = 8$

In Exercises 25–34, plot the indicated graphs.

25. On the moon, the distance s (in ft) a rock will fall due to gravity is $s = 2.66t^2$, where t is the time (in s) of fall. Plot the graph of s as a function of t for $0 \leq t \leq 10$ s on (a) a regular rectangular coordinate system and (b) a semilogarithmic coordinate system.

26. By pumping, the air pressure in a tank is reduced by 18% each second. Thus, the pressure p (in kPa) in the tank is given by $p = 101(0.82)^t$, where t is the time (in s). Plot the graph of p as a function of t for $0 \leq t \leq 30$ s on (a) a regular rectangular coordinate system and (b) a semilogarithmic coordinate system.

27. Strontium 90 decays according to the equation $N = N_0 e^{-0.028t}$, where N is the amount present after t years and N_0 is the original amount. Plot N as a function of t on semilog paper if $N_0 = 1000$ g.

28. The electric power P (in W) in a certain battery as a function of the resistance R (in Ω) in the circuit is given by $P = \dfrac{100R}{(0.50 + R)^2}$. Plot P as a function of R on semilog paper, using the logarithmic scale for R and values of R from 0.01 Ω to 10 Ω. Compare the graph with that in Fig. 3-20 on page 90.

29. The acceleration g (in m/s^2) produced by the gravitational force of the earth on a spacecraft is given by $g = 3.99 \times 10^{14}/r^2$, where r is the distance from the center of the earth to the spacecraft. On log-log paper, graph g as a function of r from $r = 6.37 \times 10^6$ m (the earth's surface) to $r = 3.91 \times 10^8$ m (the distance to the moon).

30. In undergoing an adiabatic (no *heat* gained or lost) expansion of a gas, the relation between the pressure p (in kPa) and the volume v (in m^3) is $p^2 v^3 = 850$. On log-log paper, graph p as a function of v from $v = 0.10$ m^3 to $v = 10$ m^3.

31. The intensity level B (in dB) and the frequency f (in Hz) for a sound of constant loudness were measured as follows:

f (Hz)	100	200	500	1000	2000	5000	10,000
B (dB)	40	30	22	20	18	24	30

Plot the data for B as a function of f on semilog paper, using the logarithmic scale for f.

32. The atmospheric pressure p (in kPa) at a given altitude h (in km) is given in the following table.

h (km)	0	10	20	30	40
p (kPa)	101	25	6.3	2.0	0.53

On semilog paper, plot p as a function of h.

33. One end of a very hot steel bar is sprayed with a stream of cool water. The rate of cooling R (in °F/s) as a function of the distance d (in in.) from the end of the bar is then measured. The following results are obtained:

d (in.)	0.063	0.13	0.19	0.25
R (°F/s)	600	190	100	72

d (in.)	0.38	0.50	0.75	1.0	1.5
R (°F/s)	46	29	17	10	6.0

On log-log paper, plot R as a function of d. Such experiments are made to determine the hardenability of steel.

34. The magnetic intensity H (in A/m) and flux density B (in teslas) of annealed iron are given in the following table.

B (T)	0.0042	0.043	0.67	1.01
H (A/m)	10	50	100	150

B (T)	1.18	1.44	1.58	1.72
H (A/m)	200	500	1000	10,000

Plot H as a function of B on logarithmic paper.

In Exercises 35 and 36, plot the indicated semilogarithmic graphs for the following application.

In a particular electric circuit, called a low-pass filter, the input voltage V_i is across a resistor and a capacitor, and the output voltage V_0 is across the capacitor (see Fig. 13-21). The voltage gain G (in dB) in such a circuit is given by

$$G = 20 \log \frac{1}{\sqrt{1 + (\omega T)^2}}$$

where $\tan \phi = -\omega T$.

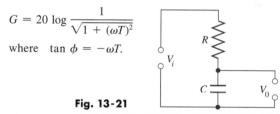

Fig. 13-21

Here, ϕ is the phase angle of V_0/V_i. For values of ωT of 0.01, 0.1, 0.3, 1.0, 3.0, 10.0, 30.0, and 100, plot the indicated graphs. These graphs are called a Bode diagram for the circuit.

35. Calculate values of G for the given values of ωT and plot a semilogarithmic graph of G vs. ωT.

36. Calculate values of ϕ (as negative angles) for the given values of ωT and plot a semilogarithmic graph of ϕ vs. ωT.

CHAPTER EQUATIONS

Exponential function	$y = b^x$	(13-1)
Logarithmic form	$x = \log_b y$	(13-2)
Logarithmic function	$y = \log_b x$	(13-3)
Laws of exponents	$b^u b^v = b^{u+v}$	(13-4)
	$\dfrac{b^u}{b^v} = b^{u-v}$	(13-5)
	$(b^u)^n = b^{nu}$	(13-6)
Properties of logarithms	$\log_b xy = \log_b x + \log_b y$	(13-7)
	$\log_b \left(\dfrac{x}{y}\right) = \log_b x - \log_b y$	(13-8)
	$\log_b (x^n) = n \log_b x$	(13-9)
	$\log_b 1 = 0 \qquad \log_b b = 1$	(13-10)
	$\log_b (b^n) = n$	(13-11)
Changing base of logarithms	$\log_b x = \dfrac{\log_a x}{\log_a b}$	(13-12)
	$\ln x = \dfrac{\log x}{\log e}$	(13-13)
	$\log x = \dfrac{\ln x}{\ln 10}$	(13-14)

REVIEW EXERCISES

In Exercises 1–12, determine the value of x.

1. $\log_{10} x = 4$ **2.** $\log_9 x = 3$

3. $\log_5 x = -1$ **4.** $\log_4 x = -\frac{1}{2}$

5. $\log_2 64 = x$ **6.** $\log_{12} 144 = x - 3$

7. $\log_8 32 = x$ **8.** $\log_9 27 = x$

9. $\log_x 36 = 2$ **10.** $\log_x 243 = 5$

11. $\log_x 10 = \frac{1}{2}$ **12.** $\log_x 8 = \frac{3}{4}$

In Exercises 13–24, express each as a sum, difference, or multiple of logarithms. Wherever possible, evaluate logarithms of the result.

13. $\log_3 2x$ **14.** $\log_5 \left(\dfrac{7}{a}\right)$

15. $\log_3 (t^2)$ **16.** $\log_6 \sqrt{5}$

17. $\log_2 28$ **18.** $\log_7 98$

19. $\log_3 \left(\dfrac{9}{x}\right)$ **20.** $\log_6 \left(\dfrac{5}{36}\right)$

21. $\log_4 \sqrt{48}$ **22.** $\log_6 \sqrt{72y}$

23. $\log_{10} (1000x^4)$ **24.** $\log_3 (9^2 \times 6^3)$

In Exercises 25–36, solve for y in terms of x.

25. $\log_6 y = \log_6 4 - \log_6 x$

26. $\log_3 y = \frac{1}{2} \log_3 7 + \frac{1}{2} \log_3 x$

27. $(\log_2 3)(\log_2 y) - \log_2 x = 3$

28. $6 \log_4 y = 8 \log_4 4 - 3 \log_4 x$

29. $\log_5 x + \log_5 y = \log_5 3 + 1$

30. $\log_7 y = 2 \log_7 5 + \log_7 x + 2$

31. $3(\log_8 y - \log_8 x) = 1$

32. $2(\log_9 y + 2 \log_9 x) = 1$

33. $2(\log_4 y - 3 \log_4 x) = 3$ **34.** $\dfrac{\log_7 x}{\log_7 4} - \log_7 y = 1$

35. $2^y = e^x$ **36.** $10^y = 3^{x+1}$

In Exercises 37–44, graph the given functions. A graphing calculator may be used.

37. $y = 0.5(5^x)$ **38.** $y = 3(2^{-x})$

39. $R = 0.2 \log_4 r$ **40.** $y = 10 \log_{16} x$

41. $y = \log_{3.15} x$ **42.** $y = 0.1 \log_{4.05} x$

43. $y = 1 - e^{-|x|}$ **44.** $s = 2(1 - e^{-0.2t})$

In Exercises 45–48, use logarithms to the base 10 to find the natural logarithms of the given numbers.

45. 8.86   **46.** 33.0

47. $\sin 2.07$ **48.** $\sqrt{0.542}$

In Exercises 49–52, use natural logarithms to find logarithms to the base 10 of the given numbers.

49. 65.89   **50.** 0.0781

51. 0.1197^3 **52.** $\log 1000$

In Exercises 53–60, solve the given equations.

53. $e^{2x} = 5$ **54.** $2(5^x) = 15$

55. $3^{x+2} = 5^x$ **56.** $6^{x+2} = 12^{x-1}$

57. $\log_4 x + \log_4 6 = \log_4 12$

58. $2 \log_3 2 - \log_3 (x + 1) = \log_3 5$

59. $\log_8(x + 2) = 2 - \log_8 2$

60. $\log (x + 2) + \log x = 0.4771$

In Exercises 61 and 62, plot the graphs of the given functions on semilogarithmic paper. In Exercises 63 and 64, plot the graphs of the given functions on log-log paper.

61. $y = 6^x$ **62.** $y = 5x^3$

63. $y = \sqrt[3]{x}$ **64.** $xy^4 = 16$

If x is eliminated between Eqs. (13-1) and (13-2), we have

$$y = b^{\log_b y} \tag{13-15}$$

In Exercises 65–68, evaluate the given expressions using Eq. (13-15).

65. $10^{\log 4}$ **66.** $2e^{\ln 7.5}$

67. $3e^{2 \ln 2}$ **68.** $5(10^{2 \log 3})$

In Exercises 69–84, solve the given problems.

69. If A dollars are invested in a certificate of deposit at i percent annual interest, compounded continuously, its value after t years is $V = Ae^{it}$. Solve for t.

70. For a radioactive substance with half-life h (the time for half the original amount to decay), an equation relating the time t for N_0 atoms to decay is $h \ln (N_0/N) = t \ln 2$. Solve for N.

71. In a certain electric circuit, the current i is given by $i = i_0 e^{-5t}$, where i_0 is the current for $t = 0$ and t is the time. Solve for t.

72. A state lottery pays $500 for a $1 ticket if a person picks the correct three-digit number determined by the random draw of three numbered balls. The probability of a 50% chance of winning in x drawings is $1 - 0.999^x = 0.5$. For how many drawings does a person have to buy a ticket to have a 50% chance of winning?

73. An approximate formula for the population (in millions) of the state of Kansas since 1970 is $P = 2.25e^{0.00486t}$, where t is the number of years since 1970. Sketch the graph of P vs. t for 1970 to 2000.

74. A computer analysis of the luminous efficiency E (in lumens/W) of a tungsten lamp as a function of its input power P (in W) is given by $E = 22.0(1 - 0.65e^{-0.008P})$. Sketch the graph of E as a function of P for $0 \le P \le 1000$ W.

75. An equation that may be used for the angular velocity ω of the slider mechanism in Fig. 13-22 is $2 \ln \omega = \ln 3g + \ln \sin \theta - \ln l$, where g is the acceleration due to gravity. Solve for $\sin \theta$.

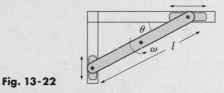

Fig. 13-22

76. Taking into account the weight loss of fuel, the maximum velocity v_m of a rocket is $v_m = u(\ln m_0 - \ln m_s) - gt_f$, where m_0 is the initial mass of the rocket and fuel, m_s is the mass of the rocket shell, t_f is the time during which fuel is expended, u is the velocity of the expelled fuel, and g is the acceleration due to gravity. Solve for m_0.

77. An equation used to calculate the capacity C (in bits/s) of a telephone channel is $C = B \log_2 (1 + R)$. Solve for R.

78. An equation used in studying the action of a protein molecule is $\ln A = \ln \theta - \ln (1 - \theta)$. Solve for θ.

79. The magnitudes (visual brightnesses), m_1 and m_2, of two stars are related to their (actual) brightnesses, b_1 and b_2, by the equation $m_1 - m_2 = 2.5 \log (b_2/b_1)$. As a result of this definition, magnitudes may be negative, and *magnitudes decrease as brightnesses increase*. The magnitude of the brightest star, Sirius, is -1.4, and the magnitudes of the faintest stars observable with the naked eye are about 6.0. How much brighter is Sirius than these faintest stars?

80. The power gain of an electronic device such as an amplifier is defined as $n = 10 \log (P_0/P_i)$, where n is measured in decibels, P_0 (in W) is the power output, and P_i (in W) is the power input. If $P_0 = 0.125$ W, calculate the power gain. (See Example 5 on page 369.)

81. In studying the frictional effects on a flywheel, the revolutions per minute R that it makes as a function of the time t (in min) is given by $R = 4520(0.750)^{2.50t}$. Find t for $R = 1950$ r/min.

82. The efficiency e of a gasoline engine as a function of its compression ratio r is given by $e = 1 - r^{1-\gamma}$, where γ is a constant. Find γ for $e = 0.55$ and $r = 7.5$.

83. For a particular solar energy system, the collector area A required to supply a fraction F of the total energy is given by $A = 480 \, F^{2.2}$. Plot A (in m^2) as a function of F, from $F = 0.1$ to $F = 0.9$, on semilog paper.

84. The current I (in μA) and resistance R (in Ω) were measured as follows in a certain microcomputer circuit.

R (Ω)	100	200	500	1000	2000	5000	10,000
I (μA)	81	41	16	8.2	4.0	1.6	0.8

Plot I as a function of R on log-log paper.

Writing Exercise

85. A machine design student noted that the edge of a robotic link was shaped like a logarithmic curve. Using a graphing calculator, the student viewed various logarithmic curves, including $y = \log x^2$ and $y = 2 \log x$, for which the student thought the graphs would be identical, but a difference was observed. Write a paragraph explaining what the difference is and why it occurs.

PRACTICE TEST

In Problems 1–4, determine the value of x.

1. $\log_9 x = -\frac{1}{2}$

2. $\log_3 x - \log_3 2 = 2$

3. $\log_x 64 = 3$

4. $3^{3x+1} = 8$

5. Graph the function $y = 2 \log_4 x$.

6. Graph the function $y = 2(3^x)$ on semilog paper.

7. Express $\log_5 \left(\dfrac{4a^3}{7}\right)$ as a combination of a sum, difference, and multiple of logarithms, including $\log_5 2$.

8. Solve for y in terms of x: $3 \log_7 x - \log_7 y = 2$.

9. An equation used for a certain electric circuit is $\ln i - \ln I = -t/RC$. Solve for i.

10. Evaluate: $\dfrac{2 \ln 0.9523}{\log 6066}$.

11. Evaluate: $\log_5 732$.

12. If A_0 dollars are invested at 8%, compounded continuously for t years, the value A of the investment is given by $A = A_0 e^{0.08t}$. Determine how long it takes for the investment to double in value.

14 ADDITIONAL TYPES OF EQUATIONS AND SYSTEMS OF EQUATIONS

In Chapter 3 we determined how to graph a function, as well as how to solve an equation graphically. Since then we have introduced a number of important types of functions and the graphs of these functions. We have also discussed the solutions of linear and quadratic equations, as well as systems of linear equations.

In this chapter we will discuss graphical and algebraic solutions of systems of equations. In these systems at least one of the equations will not be linear. We also consider solutions of two special types of equations. The first type can be solved by methods developed by quadratic equations, and the second type involves radicals.

Applications of these types of equations and systems of equations are found in many fields of science and technology. These include physics, electricity, business, and structural design.

The dimensions of a heating system vent are important to its design. In Section 14-1 we discuss a problem in vent design.

14-1 GRAPHICAL SOLUTION OF SYSTEMS OF EQUATIONS

In this section we first discuss the graphs of a special group of nonlinear equations. Then we will find graphical solutions of systems of equations that involve these equations and other nonlinear equations.

The equations we shall now consider have graphs that are known as the **conic sections.** These curves are the *circle, parabola, ellipse,* and *hyperbola.* We previously introduced the parabola when we discussed the quadratic function in Chapter 7. The following examples illustrate these curves. A more complete discussion of these equations is found in Chapter 21.

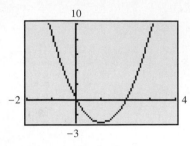

Fig. 14-1

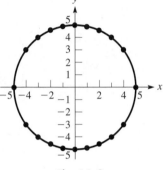

Fig. 14-2

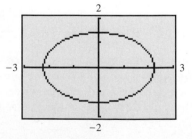

Fig. 14-3

■ EXAMPLE 1   Graph the equation $y = 3x^2 - 6x$.

We graphed equations of this form in Section 7-4. Since the general quadratic function is $y = ax^2 + bx + c$, for the function $y = 3x^2 - 6x$, we see that $a = 3$, $b = -6$, and $c = 0$. Therefore, $-b/(2a) = 1$, which tells us that the x-coordinate of the vertex is 1. For $x = 1$, $y = -3$, which means the vertex is $(1, -3)$. Also, we know that it is a minimum point since $a > 0$.

With this information, and the fact that the y-intercept of the curve is $(0, 0)$, we display the graph on a graphing calculator as shown in Fig. 14-1, with the *window* settings as shown.

As we showed in Section 7-4, the curve is a **parabola,** and a parabola always results if the equation is of the form of the quadratic function $y = ax^2 + bx + c$. ■

■ EXAMPLE 2   Plot the graph of the equation $x^2 + y^2 = 25$.

We first solve this equation for y, and we obtain $y = \sqrt{25 - x^2}$, or $y = -\sqrt{25 - x^2}$, which we write as $y = \pm\sqrt{25 - x^2}$. We now assume values for x and find the corresponding values for y.

x	0	±1	±2	±3	±4	±5
y	±5	±4.9	±4.6	±4	±3	0

If we try values greater than 5, we have imaginary numbers. These cannot be plotted, for we assume that both x and y are real. (The complex plane is only for *numbers* of the form $a + bj$ and does not represent pairs of numbers representing two variables.) When we give the value $x = \pm4$ when $y = \pm3$, this is simply a short way of representing four points. These points are $(4, 3)$, $(4, -3)$, $(-4, 3)$, $(-4, -3)$.

In Fig. 14-2 *the resulting curve is a* **circle.** A circle with its center at the origin results from an equation of the form $x^2 + y^2 = r^2$, where r is the radius. ■

From the graph of the circle in Fig. 14-2, we see that *the equation of a circle does not represent a function.* There are *two* values of y for most of the values of x in the domain. We must take this into account when displaying the graph of such an equation on a graphing calculator. This is illustrated in the next example.

■ EXAMPLE 3   Display the graph of the equation $2x^2 + 5y^2 = 10$ on a graphing calculator.

First, solving for y, we get $y = \pm\sqrt{\dfrac{10 - 2x^2}{5}}$.

NOTE ▶   To display the graph of this equation on a calculator, *we must enter both functions.* We enter one with the plus sign as

$$y_1 = \sqrt{(10 - 2x^2)/5}$$

and the other with the minus sign as

$$y_2 = -\sqrt{(10 - 2x^2)/5}$$

Trying some *window* settings (or noting that the domain is from $-\sqrt{5}$ to $\sqrt{5}$ and the range is from $-\sqrt{2}$ to $\sqrt{2}$), we get the graphing calculator display that is shown in Fig. 14-3.

The curve is an **ellipse.** An ellipse will be the resulting curve if the equation is of the form $ax^2 + by^2 = c$, where the constants a, b, and c have the same sign, and for which $a \neq b$. ■

When these types of curves are displayed on the graphing calculator, it is possible that there are small gaps in the curve. The appearance of such gaps depends on the pixel width and the functional values used by the calculator.

EXAMPLE 4 Display the graph of $2x^2 - y^2 = 4$ on a graphing calculator. Solving for y, we get

$$y = \pm\sqrt{2x^2 - 4}$$

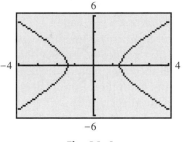

As in Example 3, we enter two functions in the calculator, one with the plus sign and the other with the minus sign, and we have the display shown in Fig. 14-4. The *window* settings are chosen by noting that the values $-\sqrt{2} < x < \sqrt{2}$ are not in the domain of either $y_1 = \sqrt{2x^2 - 4}$ or $y_2 = -\sqrt{2x^2 - 4}$. These values of x would lead to imaginary values of y.

The curve is a **hyperbola,** which results when we have an equation of the form $ax^2 + by^2 = c$, if a and b have *different* signs.

Fig. 14-4

Solving Systems of Equations

As in solving systems of linear equations, we solve any system by finding the values of x and y that satisfy both equations at the same time. To solve a system graphically, we graph the equations and find the coordinates of all points of intersection. If the curves do not intersect, the system has no real solutions.

SOLVING A WORD PROBLEM

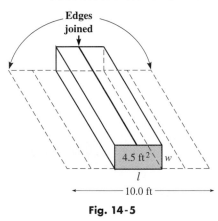

Fig. 14-5

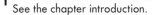

See the chapter introduction.

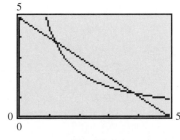

Fig. 14-6

EXAMPLE 5 For proper ventilation, the vent for a hot-air heating system is to have a rectangular cross-sectional area of 4.5 ft² and is to be made from sheet metal 10.0 ft wide. Find the dimensions of this cross-sectional area of the vent.

In Fig. 14-5, we have let l = the length and w = the width of the area. Since the area is 4.5 ft², we have $lw = 4.5$. Also, since the sheet metal is 10.0 ft wide, this is the perimeter of the area. This gives us $2l + 2w = 10.0$, or $l + w = 5.0$. This means that the system of equations to be solved is

$$lw = 4.5$$
$$l + w = 5.0$$

Solving each equation for l, we have

$$l = 4.5/w \quad \text{and} \quad l = 5.0 - w$$

We now display the graphs of these two equations on a graphing calculator, using x for w and y for l. Since negative values of l and w have no meaning to the solution and since the straight line $l = 5.0 - w$ has intercepts of $(5.0, 0)$ and $(0, 5.0)$, we choose the *window* settings as shown in Fig. 14-6.

Using the *trace* and *zoom* features (or the *intersect* feature) with the graph in Fig. 14-6, we find that the solutions are approximately $(1.2, 3.8)$ and $(3.8, 1.2)$. Using the length as the longer dimension, we have the solution of

$$l = 3.8 \text{ ft} \quad \text{and} \quad w = 1.2 \text{ ft}$$

We see that this checks with the statement of the problem.

In Example 5 we graphed the equation $xy = 4.5$ (having used x for l and y for w). The graph of this equation is also a *hyperbola,* another form of which is $xy = c$. We now show the solutions of two more systems of equations.

■**EXAMPLE 6**   Graphically solve the system of equations:

$$9x^2 + 4y^2 = 36$$
$$y = 3^x$$

The first equation is of the form represented by an ellipse, as shown in Example 3. The second equation is an exponential function, as discussed in Chapter 13. Solving the first equation for y, we have $y = \pm\frac{1}{2}\sqrt{36 - 9x^2}$. Since this is an ellipse, we note that its domain extends from $x = -2$ to $x = 2$ ($9x^2$ cannot be greater than 36). Since the exponential curve cannot be negative and increases rapidly, we need only graph the upper part of the ellipse (using the $+$ sign). With these considerations, we choose the *window* settings as shown in the calculator display in Fig. 14-7.

The points of intersection give the approximate solutions of $x = -2.0$, $y = 0.1$ and $x = 0.9$, $y = 2.7$.

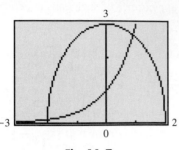

Fig. 14-7

■**EXAMPLE 7**   Graphically solve the system of equations:

$$x^2 = 2y$$
$$3x - y = 5$$

We note that the two curves in this system are a parabola and a straight line. Solving the equation of the parabola for y, we get $y = \frac{1}{2}x^2$. This parabola has its vertex at the origin, and it opens upward since $a > 0$. This means that it is not possible to have a point of intersection for $y < 0$.

The straight line has intercepts of $(0, -5)$ and $(5/3, 0)$. This means that any possible point of intersection must be in the first quadrant. With these considerations, we have the *window* settings as shown in the calculator display in Fig. 14-8.

The curves do not intersect, which means that there are no real solutions to the system of equations.

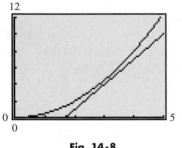

Fig. 14-8

EXERCISES *14-1*

In Exercises 1–24, solve the given systems of equations graphically by using a graphing calculator. Find all values to at least the nearest 0.1.

1. $y = 2x$
   $x^2 + y^2 = 16$

2. $3x - y = 4$
   $y = 6 - 2x^2$

3. $x^2 + 2y^2 = 8$
   $x - 2y = 4$

4. $y = 3x - 6$
   $xy = 6$

5. $y = x^2 - 2$
   $4y = 12x - 17$

6. $x^2 + 4y^2 = 4$
   $2y = 12 - x$

7. $y = x^2$
   $xy = 4$

8. $y = -2x^2$
   $y = x^2 - 6$

9. $y = -x^2 + 4$
   $x^2 + y^2 = 9$

10. $y = 2x^2 - 1$
   $x^2 + 2y^2 = 16$

11. $x^2 - 4y^2 = 16$
   $x^2 + y^2 = 1$

12. $y = 2x^2 - 4x$
   $xy = -4$

13. $2x^2 + 3y^2 = 19$
   $x^2 + y^2 = 9$

14. $x^2 - y^2 = 4$
   $2x^2 + y^2 = 16$

15. $x^2 + y^2 = 1$
   $xy = \frac{1}{2}$

16. $x^2 + y^2 = 25$
   $x^2 - y^2 = 7$

17. $y = x^2$
   $y = \sin x$

18. $y = 4x - x^2$
   $y = 2\cos x$

19. $y = e^{-x}$
   $x + y = 2$

20. $y = 2^x$
   $x^2 + y^2 = 4$

21. $x^2 - y^2 = 1$
   $y = \log_2 x$

22. $x^2 + 4y^2 = 16$
   $y = 2\ln x$

23. $y = \ln(x - 1)$
   $y = \sin\frac{1}{2}x$

24. $y = \cos x$
   $y = \log_3 x$

In Exercises 25–28, set up the indicated systems of equations and solve them graphically.

25. A helicopter is located 5.2 mi north of east of a radio tower such that it is three times as far north as it is east from the tower. Find the northern and eastern components of the displacement from the tower.

26. A 4.60-m insulating strip is placed completely around a rectangular solar panel with an area of 1.20 m². What are the dimensions of the panel?

27. The power developed in an electric resistor is i^2R, where i is the current. If a first current passes through a 2.0-Ω resistor and a second current passes through a 3.0-Ω resistor, the total power produced is 12 W. If the resistors are reversed, the total power produced is 16 W. Find the currents (in A) if ($i > 0$).

(W) 28. A circular hot tub is located on the square deck of a home. The side of the deck is 24 ft more than the radius of the hot tub, and there are 780 ft² of deck around the tub. Find the radius of the hot tub and the length of the side of the deck. Explain your answer.

14-2 ALGEBRAIC SOLUTION OF SYSTEMS OF EQUATIONS

Often the graphical method is the easiest way to solve a system of equations. With a graphing calculator it is possible to find the result with good accuracy. However, the graphical method does not usually give the *exact* answer. Using algebraic methods to find exact solutions for some systems of equations is either not possible or quite involved. There are systems, however, for which there are relatively simple algebraic solutions. In this section we consider two useful methods, both of which we discussed before when we were studying systems of linear equations.

Solution by Substitution

The first method is *substitution*. If we can solve one of the equations for one of its variables, we can substitute this solution into the other equation. We then have only one unknown in the resulting equation, and we can then solve this equation by methods discussed in earlier chapters.

EXAMPLE 1 By substitution, solve the system of equations

$$2x - y = 4$$
$$x^2 - y^2 = 4$$

We solve the first equation for y, obtaining $y = 2x - 4$. We now substitute $2x - 4$ for y in the second equation, getting

— in second equation, y replaced by $2x - 4$

$$x^2 - (2x - 4)^2 = 4$$

When simplified, this gives a quadratic equation.

$$x^2 - (4x^2 - 16x + 16) = 4$$
$$-3x^2 + 16x - 20 = 0$$
$$x = \frac{-16 \pm \sqrt{256 - 4(-3)(-20)}}{-6} = \frac{-16 \pm \sqrt{16}}{-6} = \frac{-16 \pm 4}{-6} = \frac{10}{3}, 2$$

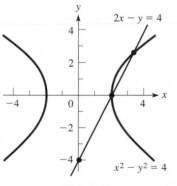

Fig. 14-9

We now find the corresponding values of y by substituting into $y = 2x - 4$. Thus, we have the solutions $x = \frac{10}{3}$, $y = \frac{8}{3}$, and $x = 2$, $y = 0$. As a check, we find that these values also satisfy the equation $x^2 - y^2 = 4$. Compare these solutions with those that would be obtained from Fig. 14-9. ----------∎

EXAMPLE 2   By substitution, solve the system of equations

$$xy = -2$$
$$2x + y = 2$$

From the first equation we have $y = -2/x$. Substituting this into the second equation, we have

in second equation, y replaced by $-\dfrac{2}{x}$

$$2x + \left(-\frac{2}{x}\right) = 2$$
$$2x^2 - 2 = 2x$$
$$x^2 - x - 1 = 0$$
$$x = \frac{1 \pm \sqrt{1 + 4}}{2} = \frac{1 \pm \sqrt{5}}{2}$$

By substituting these values for x into either of the original equations, we find the corresponding values of y, and we have the solutions

$$x = \frac{1 + \sqrt{5}}{2}, y = 1 - \sqrt{5} \quad \text{and} \quad x = \frac{1 - \sqrt{5}}{2}, y = 1 + \sqrt{5}$$

These can be checked by substituting in the original equations. In decimal form they are

$$x \approx 1.618, y \approx -1.236 \quad \text{and} \quad x \approx -0.618, y \approx 3.236$$

The graphical solutions are shown in Fig. 14-10.

The solutions can also be found by first solving the second equation for y, or either equation for x, and then substituting in the other equation.

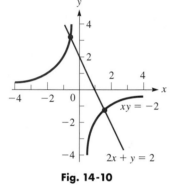

Fig. 14-10

Solution by Addition or Subtraction

The other algebraic method is that of elimination by *addition or subtraction*. This method is most useful if both equations have only squared terms and constants.

EXAMPLE 3   By addition or subtraction, solve the system of equations

$$2x^2 + y^2 = 9$$
$$x^2 - y^2 = 3$$

We note that if we add the corresponding sides of each equation, y^2 is eliminated. This leads to the solution

$$
\begin{aligned}
2x^2 + y^2 &= 9 \\
\underline{x^2 - y^2 = 3} & \\
3x^2 \qquad\; &= 12 \quad \text{add} \\
x^2 &= 4 \\
x &= \pm 2
\end{aligned}
$$

For $x = 2$, we have two corresponding y-values, $y = \pm 1$. Also, for $x = -2$, we have two corresponding y-values, $y = \pm 1$. Thus we have four solutions:

$$x = 2, y = 1; \quad x = 2, y = -1; \quad x = -2, y = 1; \quad x = -2, y = -1$$

Each of these solutions checks in the original equations. The graphical solutions are shown in Fig. 14-11.

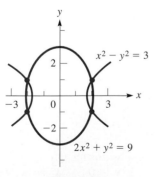

Fig. 14-11

EXAMPLE 4   By addition or subtraction, solve the system of equations

$$3x^2 - 2y^2 = 5$$
$$x^2 + y^2 = 5$$

If we multiply the second equation by 2 and then add the two resulting equations, we get

$$
\begin{array}{ll}
3x^2 - 2y^2 = 5 & \\
\underline{2x^2 + 2y^2 = 10} & \text{each term of second equation multiplied by 2} \\
5x^2 = 15 & \text{add} \\
x^2 = 3 \quad x = \pm\sqrt{3} &
\end{array}
$$

The corresponding values of y for each value of x are $y = \pm\sqrt{2}$. Again we have four solutions:

$$x = \sqrt{3}, y = \sqrt{2}; \quad x = \sqrt{3}, y = -\sqrt{2};$$
$$x = -\sqrt{3}, y = \sqrt{2}; \quad x = -\sqrt{3}, y = -\sqrt{2}$$

Each of these solutions checks when substituted in the original equations. The graphical solutions are shown in Fig. 14-12.   ∎

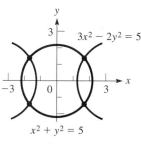

Fig. 14-12

SOLVING A WORD PROBLEM

EXAMPLE 5   A certain number of machine parts cost $1000. If they cost $5 less per part, 10 additional parts could be purchased for the same amount of money. What is the cost of each part?

Since the cost of each part is required, we let c = the cost per part. Also, we let n = the number of parts. From the first statement of the problem, we see that $cn = 1000$. Also, from the second statement, we have $(c - 5)(n + 10) = 1000$. Therefore, we are to solve the system of equations

$$cn = 1000$$
$$(c - 5)(n + 10) = 1000$$

Solving the first equation for n, and multiplying out the second equation, we have $n = \dfrac{1000}{c}$ and $cn + 10c - 5n - 50 = 1000$. Now, substituting the expression for n into the second equation, we solve for c.

$$c\left(\frac{1000}{c}\right) + 10c - 5\left(\frac{1000}{c}\right) - 50 = 1000$$

$$1000 + 10c - \frac{5000}{c} - 50 = 1000$$

$$10c - \frac{5000}{c} - 50 = 0$$

$$c^2 - 5c - 500 = 0$$

$$(c + 20)(c - 25) = 0$$

$$c = -20, 25$$

Since a negative answer has no significance in this particular situation, we see that the solution is $c =$ $25 per part. Checking with the original statement of the problem, we see that this is correct.   ∎

EXERCISES $14\text{-}2$

In Exercises 1–24, solve the given systems of equations
algebraically.

1. $y = x + 1$
$y = x^2 + 1$

2. $y = 2x - 1$
$y = 2x^2 + 2x - 3$

3. $x + 2y = 3$
$x^2 + y^2 = 26$

4. $s = t + 1$
$t^2 + s^2 = 25$

5. $x + y = 1$
$x^2 - y^2 = 1$

6. $x + y = 2$
$2x^2 - y^2 = 1$

7. $2x - y = 2$
$2x^2 + 3y^2 = 4$

8. $6y - x = 6$
$x^2 + 3y^2 = 36$

9. $wh = 1$
$w + h = 2$

10. $xy = 2$
$x + y = 3$

11. $xy = 3$
$3x - 2y = -7$

12. $xy = -4$
$2x + y = -2$

13. $y = x^2$
$y = 3x^2 - 8$

14. $M = L^2 - 1$
$2L^2 - M^2 = 2$

15. $x^2 - y = -1$
$x^2 + y^2 = 5$

16. $x^2 + y = 5$
$x^2 + y^2 = 25$

17. $D^2 - 1 = R$
$D^2 - 2R^2 = 1$

18. $2y^2 - 4x = 7$
$y^2 + 2x^2 = 3$

19. $x^2 + y^2 = 25$
$x^2 - 2y^2 = 7$

20. $3x^2 - y^2 = 4$
$x^2 + 4y^2 = 10$

21. $x^2 + 3y^2 = 37$
$2x^2 - 9y^2 = 14$

22. $5x^2 - 4y^2 = 15$
$3y^2 + 4x^2 = 12$

23. $x^2 + y^2 + 4x = 1$
$x^2 + y^2 - 2y = 9$

24. $x^2 + y^2 - 4x - 2y + 4 = 0$
$x^2 + y^2 - 2x - 4y + 4 = 0$

(Hint for Exercises 23 and 24: First subtract one equation from
the other to get an equation relating x and y. Then substitute this
equation in either given equation.)

In Exercises 25–32, solve the indicated systems of equations alge-
braically. In Exercises 27–32, it is necessary to set up the systems
of equations properly.

25. A rocket is fired from behind a ship and follows the path given
by $h = 3x - 0.05x^2$, where h is its altitude (in mi) and x is the
horizontal distance traveled (in mi). A missile fired from the
ship follows the path given by $h = 0.8x - 15$. For $h > 0$ and
$x > 0$, find where the paths of the rocket and missile cross.

26. A 2-kg block collides with an 8-kg block. Using the physical
laws of conservation of energy and conservation of momentum, along with given conditions, the following equations involving the velocities are established.

$v_1^2 + 4v_2^2 = 41$
$2v_1 + 8v_2 = 12$

Find these velocities (in m/s) if $v_2 > 0$.

27. A rectangular computer chip has a surface area of 2.1 cm² and
a perimeter of 5.8 cm. Find the length and the width of the
chip.

28. The impedance Z in an alternating-current circuit is 2.00 Ω.
If the resistance R is numerically equal to the square of the reactance X, find R and X. See Section 12-7.

29. A roof truss is in the shape of a right triangle. If there are
4.60 m of lumber in the truss and the longest side is 2.20 m
long, what are the lengths of the other two sides of the truss?

30. In a certain roller mechanism, the radius of one steel ball is
2.00 cm greater than the radius of a second steel ball. If the
difference in their masses is 7100 g, find the radii of the balls.
The density of steel is 7.70 g/cm³.

31. A jet travels at 610 mi/h relative to the air. It takes the jet 1.6 h
longer to travel the 3660 mi from London to Washington, D.C.,
against the wind than it takes from Washington to London
with the wind. Find the velocity of the wind. See Fig. 14-13.

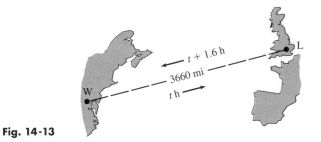

Fig. 14-13

32. In a marketing survey, a company found that the total gross
income for selling t tables at a price of p dollars each was
$35,000. It then increased the price of each table by $100 and
found that the total income was only $27,000 because 40 fewer
tables were sold. Find p and t.

$14\text{-}3$ EQUATIONS IN QUADRATIC FORM

Often we encounter equations that can be solved by methods applicable to quadratic
equations, even though these equations are not actually quadratic. They do have the
property, however, that *with a proper substitution* **they may be written in the form**
of a quadratic equation. All that is necessary is that the equation have terms including some variable quantity, its square, and perhaps a constant term. The following example illustrates these types of equations.

NOTE▶

EXAMPLE 1   **(a)** The equation $x - 2\sqrt{x} - 5 = 0$ is an equation in quadratic form, because if we let $y = \sqrt{x}$ we have $x = (\sqrt{x})^2 = y^2$, and the resulting equation is $y^2 - 2y - 5 = 0$.

(b) $t^{-4} - 5t^{-2} + 3 = 0$

By letting $y = t^{-2}$, we have $\overset{\overset{\displaystyle (t^{-2})^2}{\big\downarrow}}{y^2} - 5y + 3 = 0$.

(c) $t^3 - 3t^{3/2} - 7 = 0$

By letting $y = t^{3/2}$, we have $\overset{\overset{\displaystyle (t^{3/2})^2}{\big\downarrow}}{y^2} - 3y - 7 = 0$.

(d) $(x + 1)^4 - (x + 1)^2 - 1 = 0$

By letting $y = (x + 1)^2$, we have $\overset{\overset{\displaystyle [(x + 1)^2]^2}{\big\downarrow}}{y^2} - y - 1 = 0$.

(e) $x^{10} - 2x^5 + 1 = 0$

By letting $y = x^5$, we have $\overset{\overset{\displaystyle (x^5)^2}{\big\downarrow}}{y^2} - 2y + 1 = 0$.

The following examples illustrate the method of solving equations in quadratic form.

EXAMPLE 2   Solve the equation $x^4 - 5x^2 + 4 = 0$.

We first let $y = x^2$ and obtain the resulting equivalent equation:

$$y^2 - 5y + 4 = 0$$

This may be factored and solved as

$$(y - 4)(y - 1) = 0$$
$$y = 4 \quad \text{or} \quad y = 1$$

Since we want values for x and since $y = x^2$, we have

$$x^2 = 4 \quad \text{or} \quad x^2 = 1$$
$$x = \pm 2 \quad \text{or} \quad x = \pm 1$$

Substitution in the original equation shows each value to be a solution.

EXAMPLE 3   Solve the equation $2x^4 + 7x^2 = 4$.

As in Example 2, we let $y = x^2$ to write the resulting equation in quadratic form.

$$2y^2 + 7y - 4 = 0$$
$$(2y - 1)(y + 4) = 0$$
$$y = \tfrac{1}{2} \quad \text{or} \quad y = -4$$
$$x^2 = \tfrac{1}{2} \quad \text{or} \quad x^2 = -4 \qquad y = x^2$$
$$x = \pm \frac{1}{\sqrt{2}} \quad \text{or} \quad x = \pm 2j$$

Substitution in the original equation shows each value to be a solution.

Two of the solutions in Example 3 are complex numbers. We were able to find these solutions directly from the definition of the square root of a negative number. In some cases (see Exercises 23 and 24 of this section), it is necessary to use the method of Section 12-6 to find such complex number solutions.

■EXAMPLE 4 Solve the equation $x - \sqrt{x} - 2 = 0$.
By letting $y = \sqrt{x}$, we have

$$y^2 - y - 2 = 0$$
$$(y - 2)(y + 1) = 0$$
$$y = 2 \quad \text{or} \quad y = -1$$

Since $y = \sqrt{x}$, we note that y cannot be negative, and this means $y = -1$ cannot lead to a solution. For $y = 2$, we have $x = 4$. Checking, we find that $x = 4$ satisfies the original equation. Therefore, the only solution is $x = 4$. ■

EXTRANEOUS ROOTS

CAUTION ▶

Example 4 illustrates a very important point. *Whenever an operation involving the unknown is performed on an equation, this operation may introduce roots into a subsequent equation that are not roots of the original equation. Therefore, we must check all answers in the original equation.* Only operations involving constants— that is, adding, subtracting, multiplying by, or dividing by constants—are certain not to introduce the **extraneous roots**. We first encountered the concept of an extraneous root in Section 6-8, when we discussed equations involving fractions.

■EXAMPLE 5 Solve the equation $x^{-2} + 3x^{-1} + 1 = 0$.
By substituting $y = x^{-1}$, we have $y^2 + 3y + 1 = 0$. To solve this equation we may use the quadratic formula:

$$y = \frac{-3 \pm \sqrt{9 - 4}}{2} = \frac{-3 \pm \sqrt{5}}{2}$$

Since $x = 1/y$, we have

$$x = \frac{2}{-3 + \sqrt{5}} \quad \text{or} \quad x = \frac{2}{-3 - \sqrt{5}}$$

These answers in decimal form are

$$x \approx -2.618 \quad \text{or} \quad x \approx -0.382$$

These results check when substituted in the original equation. In checking these decimal answers, it is more accurate to use the calculator values, before rounding them off. This can be done by storing the calculator values in memory. ■

■EXAMPLE 6 Solve the equation $(x^2 - x)^2 - 8(x^2 - x) + 12 = 0$.
By substituting $y = x^2 - x$, we have

$$y^2 - 8y + 12 = 0$$
$$(y - 2)(y - 6) = 0$$
$$y = 2 \quad \text{or} \quad y = 6$$

This means that

$$x^2 - x = 2 \quad \text{or} \quad x^2 - x = 6$$

Solving each of these equations, we have

$$x^2 - x - 2 = 0 \qquad x^2 - x - 6 = 0$$
$$(x - 2)(x + 1) = 0 \qquad (x - 3)(x + 2) = 0$$
$$x = 2 \quad \text{or} \quad x = -1 \qquad x = 3 \quad \text{or} \quad x = -2$$

Each of these values checks when substituted in the original equation. ■

SOLVING A WORD PROBLEM

Fig. 14-14

The first permanent photograph was taken in 1826 by the French inventor Joseph Niepce (1765–1833).

▎**EXAMPLE 7** A rectangular photograph has an area of 120 cm². The diagonal of the photograph is 17 cm. Find its length and width. See Fig. 14-14.

Since the required quantities are the length and width, let l = the length of the photograph and let w = its width. Since the area is 120 cm², $lw = 120$. Also, using the Pythagorean theorem and the fact that the diagonal is 17 cm, we have the equation $l^2 + w^2 = 17^2 = 289$. Therefore, we are to solve the system of equations

$$lw = 120 \qquad l^2 + w^2 = 289$$

Solving the first equation for l, we have $l = 120/w$. Substituting this into the second equation, we have

$$\left(\frac{120}{w}\right)^2 + w^2 = 289$$

$$\frac{14{,}400}{w^2} + w^2 = 289$$

$$14{,}400 + w^4 = 289w^2$$

Let $x = w^2$.

$$x^2 - 289x + 14{,}400 = 0$$

$$x = \frac{-(-289) \pm \sqrt{(-289)^2 - 4(1)(14{,}400)}}{2(1)}$$

$$x = 225 \qquad \text{or} \qquad x = 64$$

Therefore, $w^2 = 225$ or $w^2 = 64$.

Solving for w, we get $w = \pm 15$ or $w = \pm 8.0$. Only the positive values are meaningful in this problem, which means if $w = 8.0$ cm, the $l = 15$ cm. Normally, we designate the longer dimension as the length (although by letting $w = 15$ cm, we get $l = 8.0$ cm). Checking with the statement of the problem, we see that these dimensions for the photograph give an area of 120 cm² and a diagonal of 17 cm. ▮

━━━━━ **EXERCISES** *14-3* ━━━━━

In Exercises 1–24, solve the given equations algebraically.

1. $x^4 - 13x^2 + 36 = 0$
2. $x^4 - 20x^2 + 64 = 0$
3. $4x^4 - 5x^2 = 9$
4. $4R^4 + 15R^2 = 4$
5. $x^{-2} - 2x^{-1} - 8 = 0$
6. $10x^{-2} + 3x^{-1} - 1 = 0$
7. $x^{-4} + 2x^{-2} = 24$
8. $x^{-4} + 1 = 2x^{-2}$
9. $2x - 7\sqrt{x} + 5 = 0$
10. $4x + 3\sqrt{x} = 1$
11. $3\sqrt[3]{x} - 5\sqrt[6]{x} + 2 = 0$
12. $\sqrt{x} + 3\sqrt[4]{x} = 28$
13. $x^{2/3} - 2x^{1/3} - 15 = 0$
14. $x^3 + 2x^{3/2} - 80 = 0$
15. $2n^{1/2} - 5n^{1/4} = 3$
16. $4x^{4/3} + 9 = 13x^{2/3}$
17. $(x - 1) - \sqrt{x - 1} - 2 = 0$
18. $(C + 1)^{-2/3} + 5(C + 1)^{-1/3} - 6 = 0$
19. $(x^2 - 2x)^2 - 11(x^2 - 2x) + 24 = 0$
20. $(x^2 - 1)^2 + (x^2 - 1)^{-2} = 2$
21. $x - 3\sqrt{x - 2} = 6$ (Let $y = \sqrt{x - 2}$.)

22. $x^6 + 7x^3 - 8 = 0$

23. $\dfrac{1}{s^2 + 1} + \dfrac{2}{s^2 + 3} = 1$
24. $\dfrac{2}{x^2 + 2} + \dfrac{1}{2x^2 - 7} = \dfrac{4}{3}$

In Exercises 25–28, solve the given problems algebraically.

25. The equivalent resistance R_T of two resistors R_1 and R_2 in parallel is given by $R_T^{-1} = R_1^{-1} + R_2^{-1}$. If $R_T = 1.00\ \Omega$ and $R_2 = \sqrt{R_1}$, find R_1 and R_2.

26. An equation used in the study of the dispersion of light is $\mu = A + B\lambda^{-2} + C\lambda^{-4}$. Solve for λ.

27. A rectangular TV screen has an area of 347 in.² and a diagonal of 27.0 in. Find the dimensions of the screen.

28. A roof truss in the shape of a right triangle has a perimeter of 90 ft. If the hypotenuse is 1 ft longer than one of the other sides, what are the sides of the truss?

14-4 EQUATIONS WITH RADICALS

Equations with radicals in them are normally solved by squaring both sides of the equation if the radical represents a square root or by a similar operation for the other roots. However, when we do this, we often introduce *extraneous roots*. Thus, it is very important that all solutions be checked in the original equation.

■**EXAMPLE 1**   Solve the equation $\sqrt{x-4} = 2$.

By squaring both sides of the equation, we have

$$(\sqrt{x-4})^2 = 2^2$$
$$x - 4 = 4$$
$$x = 8$$

This solution checks when put into the original equation.

■**EXAMPLE 2**   Solve the equation $2\sqrt{3x-1} = 3x$.

Squaring both sides of the equation gives us

$$(2\sqrt{3x-1})^2 = (3x)^2 \longleftarrow \text{don't forget to square the 2}$$
$$4(3x-1) = 9x^2$$
$$12x - 4 = 9x^2$$
$$9x^2 - 12x + 4 = 0$$
$$(3x-2)^2 = 0$$
$$x = \frac{2}{3} \quad \text{(double root)}$$

Checking this solution in the original equation, we have

$$2\sqrt{3(\tfrac{2}{3})-1} \overset{?}{=} 3(\tfrac{2}{3}), \qquad 2\sqrt{2-1} \overset{?}{=} 2, \qquad 2 = 2$$

Therefore, the solution $x = \frac{2}{3}$ checks.

■**EXAMPLE 3**   Solve the equation $\sqrt[3]{x-8} = 2$.

Cubing both sides of the equation, we have

$$x - 8 = 8$$
$$x = 16$$

Checking this solution in the original equation, we get

$$\sqrt[3]{16-8} \overset{?}{=} 2, \quad \text{or} \quad 2 = 2$$

Therefore, the solution checks.

In the above examples, the radical is on the left side of the equation and the other terms of the equation are on the right side. However, if one side of the equation contains a radical as well as other terms, we *first isolate the radical.* That is, we rewrite the equation with the radical on one side and collect all other terms on the other side. The examples on the following page illustrate the use of this procedure.

EXAMPLE 4 Solve the equation $\sqrt{x-1} + 3 = x$.

We first isolate the radical by subtracting 3 from each side. This gives us

$$\sqrt{x-1} = x - 3$$

We now square both sides and proceed with the solution.

CAUTION▶

$$(\sqrt{x-1})^2 = (x-3)^2$$
$$x - 1 = x^2 - 6x + 9$$

square the expression on each side, not just the terms separately

$$x^2 - 7x + 10 = 0$$
$$(x-5)(x-2) = 0$$
$$x = 5 \quad \text{or} \quad x = 2$$

The solution $x = 5$ checks, but the solution $x = 2$ gives $4 = 2$. Thus, the solution is $x = 5$. The value $x = 2$ is an extraneous root. ◼

EXAMPLE 5 Solve the equation $\sqrt{x+1} + \sqrt{x-4} = 5$.

This is most easily solved by first isolating one of the radicals by placing the other radical on the right side of the equation. We then square both sides of the resulting equation.

CAUTION▶

$$\sqrt{x+1} = 5 - \sqrt{x-4} \longleftarrow \text{two terms}$$
$$(\sqrt{x+1})^2 = (5 - \sqrt{x-4})^2$$
$$x + 1 = 25 - 10\sqrt{x-4} + (\sqrt{x-4})^2 \longleftarrow \text{be careful!}$$
$$= 25 - 10\sqrt{x-4} + x - 4$$

Now, isolating the radical on one side of the equation and squaring again, we have

$$10\sqrt{x-4} = 20$$
$$\sqrt{x-4} = 2 \qquad \text{divide by 10}$$
$$x - 4 = 4 \qquad \text{square both sides}$$
$$x = 8$$

This solution checks.

We note again that in squaring $5 - \sqrt{x-4}$, we do not simply square 5 and $\sqrt{x-4}$. This is similar to $(5a-2)^2$ in Example 2 of Section 6-1. ◼

EXAMPLE 6 Solve the equation $\sqrt{x} - \sqrt[4]{x} = 2$.

We can solve this most easily by handling it as an equation in quadratic form. By letting $y = \sqrt[4]{x}$, we have

$$y^2 - y - 2 = 0$$
$$(y-2)(y+1) = 0$$
$$y = 2 \quad \text{or} \quad y = -1$$

Since $y = \sqrt[4]{x}$, we know that a negative value of y does not lead to a solution of the original equation. Therefore, we see that $y = -1$ cannot give us a solution. For the other value, $y = 2$, we have

$$\sqrt[4]{x} = 2, \quad \text{or} \quad x = 16$$

This checks, because $\sqrt{16} - \sqrt[4]{16} = 4 - 2 = 2$. Therefore, the only solution of the original equation is $x = 16$. ◼

**SOLVING A
WORD PROBLEM**

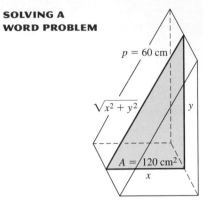

Fig. 14-15

Holography is a method of producing a three-dimensional image without the use of a lens. The theory of holography was developed in the late 1940s by the British engineer Dennis Gabor (1900–1979). After the invention of lasers, the first holographs were produced in the early 1960s.

■**EXAMPLE 7**   Each cross section of a holographic image is in the shape of a right triangle. The perimeter of the cross section is 60 cm, and its area is 120 cm². Find the length of each of the three sides.

If we let the two legs of the triangle be *x* and *y*, as shown in Fig. 14-15, from the formulas for the perimeter *p* and the area *A* of a triangle, we have

$$p = x + y + \sqrt{x^2 + y^2} \quad \text{and} \quad A = \tfrac{1}{2}xy$$

where the hypotenuse was found by use of the Pythagorean theorem. Using the information given in the statement of the problem, we arrive at the equations

$$x + y + \sqrt{x^2 + y^2} = 60 \quad \text{and} \quad xy = 240$$

Isolating the radical in the first equation and then squaring both sides, we have

$$\sqrt{x^2 + y^2} = 60 - x - y$$
$$x^2 + y^2 = 3600 - 120x - 120y + x^2 + 2xy + y^2$$
$$0 = 3600 - 120x - 120y + 2xy$$

Solving the second of the original equations for *y*, we have $y = 240/x$. Substituting, we have

$$0 = 3600 - 120x - 120\left(\frac{240}{x}\right) + 2x\left(\frac{240}{x}\right)$$

$$0 = 3600x - 120x^2 - 120(240) + 480x \qquad \text{multiply by } x$$
$$0 = 30x - x^2 - 240 + 4x \qquad \text{divide by 120}$$
$$x^2 - 34x + 240 = 0 \qquad \text{collect terms on left}$$
$$(x - 10)(x - 24) = 0$$
$$x = 10 \text{ cm} \quad \text{or} \quad x = 24 \text{ cm}$$

If *x* = 10 cm, then *y* = 24 cm, or if *x* = 24 cm, then *y* = 10 cm. Therefore, the legs of the holographic cross section are 10 cm and 24 cm, and the hypotenuse is 26 cm. For these sides, *p* = 60 cm and *A* = 120 cm². We see that these values check with the statement of the problem.   ■

= EXERCISES *14-4*

In Exercises 1–32, solve the given equations. In Exercises 17 and 20, explain how the extraneous root is introduced.

1. $\sqrt{x - 8} = 2$

2. $\sqrt{x + 4} = 3$

3. $\sqrt{8 - 2x} = x$

4. $\sqrt{3x + 4} = x$

5. $\sqrt{3x + 2} = 3x$

6. $2\sqrt{2P + 5} = P$

7. $\sqrt{x - 2} + 3 = x$

8. $\sqrt{5x - 1} + 3 = x$

9. $2\sqrt{3 - x} - x = 5$

10. $x - 3\sqrt{2x + 1} = -5$

11. $\sqrt[3]{y - 5} = 3$

12. $\sqrt[4]{5 - x} = 2$

13. $5\sqrt{s - 6} = s$

14. $\sqrt{x + 3} = 4x$

15. $5\sqrt{x + 3} = 2x$

16. $4\sqrt{x} = x + 3$

Ⓦ 17. $\sqrt{x + 4} + 8 = x$

18. $\sqrt{x + 15} + 5 = x$

19. $\sqrt{5 + \sqrt{x}} = \sqrt{x} - 1$

Ⓦ 20. $\sqrt{13 + \sqrt{x}} = \sqrt{x} + 1$

21. $3\sqrt{1 - 2t} + 1 = 2t$

22. $1 - 2\sqrt{y + 4} = y$

23. $2\sqrt{x + 2} - \sqrt{3x + 4} = 1$ **24.** $\sqrt{x - 1} + \sqrt{x + 2} = 3$

25. $\sqrt{5x + 1} - 1 = 3\sqrt{x}$

26. $\sqrt{2x + 1} + \sqrt{3x} = 11$

27. $\sqrt{2x - 1} - \sqrt{x + 11} = -1$

28. $\sqrt{5x - 4} - \sqrt{x} = 2$

29. $\sqrt[3]{2x - 1} = \sqrt[3]{x + 5}$

30. $\sqrt[4]{x + 10} = \sqrt{x - 2}$

31. $\sqrt{x - 2} = \sqrt[4]{x - 2} + 12$ **32.** $\sqrt{3x + \sqrt{3x + 4}} = 4$

In Exercises 33–36, solve for the indicated letter.

33. The resonant frequency f in an electric circuit with an inductance L and a capacitance C is given by $f = \dfrac{1}{2\pi\sqrt{LC}}$. Solve for L.

34. A formula used in calculating the range R for radio communication is $R = \sqrt{2rh + h^2}$. Solve for h.

35. An equation used in analyzing a certain type of concrete beam is $k = \sqrt{2np + (np)^2} - np$. Solve for p.

36. In the study of spur gears in contact, the equation $kC = \sqrt{R_1^2 - R_2^2} + \sqrt{r_1^2 - r_2^2} - A$ is used. Solve for r_1^2.

In Exercises 37–40, set up the proper equations and solve them.

37. A freighter is 5.2 km farther from a Coast Guard station on a straight coast than from the closest point A on the coast. If the station is 8.3 km from A, how far is it from the freighter?

38. The velocity v of an object that falls through a distance h is given by $v = \sqrt{2gh}$, where g is the acceleration due to gravity. Two objects are dropped from heights that differ by 10.0 m such that the sum of their velocities when they strike the ground is 20.0 m/s. Find the heights from which they are dropped if $g = 9.80$ m/s^2.

39. A point D on Denmark's largest island is 2.4 mi from the nearest point S on the coast of Sweden (assume the coast is straight, which is nearly the case). A person in a motorboat travels straight from D to a point on the beach x km from S and then travels x km farther along the beach away from S. Find x if the person traveled a total of 4.5 mi. See Fig. 14-16.

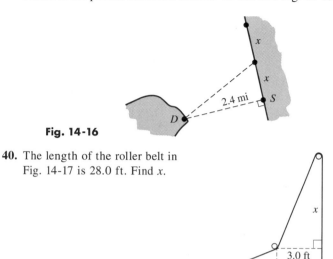

Fig. 14-16

40. The length of the roller belt in Fig. 14-17 is 28.0 ft. Find x.

Fig. 14-17

REVIEW EXERCISES

In Exercises 1–10, solve the given systems of equations by use of a graphing calculator.

1. $x + 2y = 6$
$y = 4x^2$

2. $x + y = 3$
$x^2 + y^2 = 25$

3. $3x + 2y = 6$
$x^2 + 4y^2 = 4$

4. $x^2 - 2y = 0$
$y = 3x - 5$

5. $y = x^2 + 1$
$2x^2 + y^2 = 4$

6. $\dfrac{x^2}{4} + y^2 = 1$
$x^2 - y^2 = 1$

7. $y = 4 - x^2$
$y = 2x^2$

8. $xy = -2$
$y = 1 - 2x^2$

9. $y = x^2 - 2x$
$y = 1 - e^{-x}$

10. $y = \ln x$
$y = \sin x$

In Exercises 11–20, solve each of the given systems of equations algebraically.

11. $y = 4x^2$
$y = 8x$

12. $x + y = 2$
$xy = 1$

13. $2y = x^2$
$x^2 + y^2 = 3$

14. $y = x^2$
$2x^2 - y^2 = 1$

15. $4x^2 + y = 3$
$2x + 3y = 1$

16. $2x^2 + y^2 = 3$
$x + 2y = 1$

17. $4x^2 - 7y^2 = 21$
$x^2 + 2y^2 = 99$

18. $s - t = 6$
$\sqrt{s} - \sqrt{t} = 1$

19. $4x^2 + 3xy = 4$
$x + 3y = 4$

20. $\dfrac{6}{x} + \dfrac{3}{y} = 4$
$\dfrac{36}{x^2} + \dfrac{36}{y^2} = 13$

In Exercises 21–38, solve the given equations.

21. $x^4 - 20x^2 + 64 = 0$

22. $x^6 - 26x^3 - 27 = 0$

23. $x^{3/2} - 9x^{3/4} + 8 = 0$

24. $x^{1/2} + 3x^{1/4} - 28 = 0$

25. $D^{-2} + 4D^{-1} - 21 = 0$

26. $4x^{-4} + 35x^{-2} = 9$

27. $2x - 3\sqrt{x} - 5 = 0$

28. $e^x + e^{-x} = 2$

29. $\dfrac{4}{r^2 + 1} + \dfrac{7}{2r^2 + 1} = 2$

30. $(x^2 + 5x)^2 - 5(x^2 + 5x) = 6$

31. $3\sqrt{2Z + 4} = 2Z$

32. $\sqrt[3]{x - 2} = 3$

33. $\sqrt{5x + 9} + 1 = x$

34. $2\sqrt{5x - 3} - 1 = 2x$

35. $\sqrt{x + 1} + \sqrt{x} = 2$

36. $\sqrt{3 + x} + \sqrt{3x - 2} = 1$

37. $\sqrt{x + 4} + 2\sqrt{x + 2} = 3$

38. $\sqrt{3x - 2} - \sqrt{x + 7} = 1$

(W) *In Exercises 39 and 40, find the value of x. (In each, the expression on the right is called a* **continued radical.** *Also, ... means that the pattern continues indefinitely). (Hint: Square both sides.) Noting the result, complete the solution and explain your method.*

39. $x = \sqrt{2 + \sqrt{2 + \sqrt{2 + \ldots}}}$

40. $x = \sqrt{6 - \sqrt{6 - \sqrt{6 - \ldots}}}$

In Exercises 41–46, solve for the indicated quantities.

41. In the study of atomic structure, the equation
$L = \dfrac{h}{2\pi}\sqrt{l(l + 1)}$ is used. Solve for $l\,(l > 0)$.

42. The frequency ω of a certain *RLC* circuit is given by
$\omega = \dfrac{\sqrt{R^2 + 4(L/C)} + R}{2L}$. Solve for C.

43. In the theory dealing with a suspended cable, the equation
$y = \sqrt{s^2 - m^2} - m$ is used. Solve for m.

44. The equation $V = e^2cr^{-2} - e^2Zr^{-1}$ is used in spectroscopy. Solve for r.

45. In an experiment, an object is allowed to fall, stops, and then falls for twice the initial time. The total distance the object falls is 45 ft. The equations relating the times t_1 and t_2 (in s) of fall are $16t_1^2 + 16t_2^2 = 45$ and $t_2 = 2t_1$. Find the times of fall.

46. If two objects collide and the kinetic energy remains constant, the collision is termed perfectly elastic. Under these conditions, if an object of mass m_1 and initial velocity u_1 strikes a second object (initally at rest) of mass m_2, such that the velocities after collision are v_1 and v_2, the following equations are found:

$m_1u_1 = m_1v_1 + m_2v_2$
$\frac{1}{2}m_1u_1^2 = \frac{1}{2}m_1v_1^2 + \frac{1}{2}m_2v_2^2$

Solve these equations for m_2 in terms of u_1, v_1, and m_1.

In Exercises 47–56, set up the appropriate equations and solve them.

47. A wrench is dropped by a worker at a construction site. Four seconds later the worker hears it hit the ground below. How high is the worker above the ground? (The velocity of sound is 1100 ft/s, and the distance the wrench falls as a function of time is $s = 16t^2$.)

48. A rectangular field is enclosed by fencing and a wall along one side and half of an adjacent side. See Fig. 14-18. If the area of the field is 9000 ft² and 240 ft of fencing are used, what are the dimensions of the field?

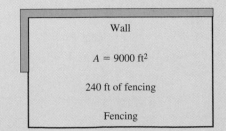

Wall

$A = 9000\ \text{ft}^2$

240 ft of fencing

Fencing

Fig. 14-18

49. In a certain electric circuit the impedance Z is twice the square of the reactance X, and the resistance R is 0.800 Ω. Find the impedance and the reactance. See Section 12-7.

50. For the plywood piece shown in Fig. 14-19, find x and y.

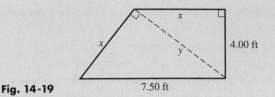

Fig. 14-19   7.50 ft

51. The viewing window on a graphing calculator has an area of 1770 mm² and a diagonal of 62 mm. What are the length and width of the rectangle?

52. The circular solar cell and square solar cell shown in Fig. 14-20 have a combined surface area of 40.0 cm². Find the radius of the circular cell and the side of the square cell.

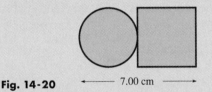

Fig. 14-20   ◄——— 7.00 cm ———►

53. A trough is made from a piece of sheet metal 12.0 in. wide. The cross section of the trough is shown in Fig. 14-21. Find x.

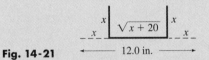

Fig. 14-21   ◄——— 12.0 in. ———►

54. A plastic band 19.0 cm long is bent into the shape of a triangle with sides $\sqrt{x - 1}$, $\sqrt{5x - 1}$, and 9. Find x.

55. A Coast Guard ship travels from Houston to Mobile, and later it returns to Houston at a speed that is 6.0 mi/h faster. If Houston is 510 mi from Mobile and the total travel time is 35 h, find the speed of the ship in each direction.

56. Two trains are approaching the same crossing on tracks that are at right angles to each other. Each is traveling at 60.0 km/h. If one is 6.00 km from the crossing when the other is 3.00 km from it, how much later will they be 4.00 km apart (on a direct line)?

Writing Exercise

57. Using a computer, an engineer designs a triangular support structure with sides (in m) of x, $\sqrt{x - 1}$, and 4.00 m. If the perimeter is to be 9.00 m, the equation to be solved is $x + \sqrt{x - 1} + 4.00 = 9.00$. This equation can be solved by either of two methods used in this chapter. Write one or two paragraphs identifying the methods and explaining how they are used to solve the equation.

PRACTICE TEST

1. Solve for x: $x^{1/2} - 2x^{1/4} = 3$.

2. Solve for x: $3\sqrt{x-2} - \sqrt{x+1} = 1$.

3. Solve for x: $x^4 - 17x^2 + 16 = 0$.

4. Solve for x and y algebraically:

$x^2 - 2y = 5$
$2x + 6y = 1$

5. The velocity v of an object falling under the influence of gravity in terms of its initial velocity v_0, the acceleration due to gravity g, and the height h fallen is given by $v = \sqrt{v_0^2 + 2gh}$. Solve for h.

6. Solve for x and y graphically: $x^2 - y^2 = 4$
$xy = 2$

7. A rectangular desktop has a perimeter of 14.0 ft and an area of 10.0 ft^2. Find the length and the width of the desktop.

CHAPTER 15

EQUATIONS OF HIGHER DEGREE

In Section 15-4 we see how the design of a box to hold a product involves the solution of a higher-degree equation.

In the previous chapters we have discussed methods of solving many types of equations. Except for special cases, however, we have not solved polynomial equations of degree higher than two (a linear equation is a first-degree polynomial equation, and a quadratic equation is a second-degree polynomial equation).

In this chapter we will develop certain methods that are useful for solving higher-degree polynomial equations. Since we will discuss only equations involving polynomials, in this chapter $f(x)$ will be assumed to be a polynomial.

Applications of higher-degree equations arise in a number of technical areas. Included in these are finding the resistance in an electric circuit, determining the dimensions of a container or a structure, robotics, and calculating various business production costs.

15-1 THE REMAINDER THEOREM AND THE FACTOR THEOREM

As stated above, in this chapter we will develop methods of solving higher-degree polynomial equations. Second-degree equations can be solved by use of the quadratic formula. For equations above the second degree, there is no such basic formula that can be used to get the solution. Although methods have been found to solve certain third- and fourth-degree equations, it can be proven that polynomial equations of degree higher than four cannot in general be solved algebraically.

In this section we present two theorems that help identify factors and zeros of a polynomial. These theorems, in turn, will help us solve polynomial equations later in the chapter.

Any function of the form

$$f(x) = a_0 x^n + a_1 x^{n-1} + \cdots + a_n \qquad (15\text{-}1)$$

where $a_0 \neq 0$ and n is a positive integer or zero is called a **polynomial function.** We will be considering only polynomials in which the coefficients $a_0, a_1, \ldots, a_n$ are real numbers.

If we divide a polynomial by $x - r$, we find a result of the form

$$f(x) = (x - r)q(x) + R \qquad (15\text{-}2)$$

where $q(x)$ is the quotient and R is the remainder.

EXAMPLE 1   Divide $f(x) = 3x^2 + 5x - 8$ by $x - 2$.

$$
\begin{array}{r}
3x + 11 \\
x - 2 \overline{\smash{)}\ 3x^2 + 5x - 8} \\
\underline{3x^2 - 6x} \\
11x - 8 \\
\underline{11x - 22} \\
14
\end{array}
$$

Thus,

$$3x^2 + 5x - 8 = (x - 2)(3x + 11) + 14$$

where, for this function $f(x)$ with $r = 2$, we identify $q(x)$ and R as

$$q(x) = 3x + 11 \qquad R = 14$$

The Remainder Theorem

If we now set $x = r$ in Eq. (15-2), we have $f(r) = q(r)(r - r) + R = f(r)(0) + R$, or

$$f(r) = R \qquad (15\text{-}3)$$

This leads us to the **remainder theorem,** which states that *if a polynomial $f(x)$ is divided by $x - r$ until a constant remainder R is obtained, then $f(r) = R$.* As we can see, this means the remainder equals the value of the function of x at $x = r$.

EXAMPLE 2   In Example 1, $f(x) = 3x^2 + 5x - 8$, $R = 14$, and $r = 2$. We find that

$$
\begin{aligned}
f(2) &= 3(2^2) + 5(2) - 8 \\
&= 12 + 10 - 8 \\
&= 14
\end{aligned}
$$

Therefore, $f(2) = 14$ verifies that $f(r) = R$ for this example.

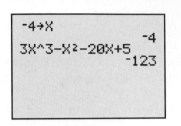

Fig. 15-1

EXAMPLE 3   By using the remainder theorem, determine the remainder when $3x^3 - x^2 - 20x + 5$ is divided by $x + 4$.

In using the remainder theorem, we determine the remainder when the function is divided by $x - r$ by evaluating the function for $x = r$. To have $x + 4$ in the proper form to identify r, we write it as $x - (-4)$. *This means that $r = -4$, and we therefore evaluate* the function $f(x) = 3x^3 - x^2 - 20x + 5$ for $x = -4$, or find $f(-4)$. Therefore,

$$f(-4) = 3(-4)^3 - (-4)^2 - 20(-4) + 5 = -192 - 16 + 80 + 5$$
$$= -123$$

The remainder is -123 when $3x^3 - x^2 - 20x + 5$ is divided by $x + 4$.

NOTE ▶   To evaluate the function with a calculator, *first store in memory the value to be substituted.* In Fig. 15-1, a calculator display of the substitution of -4 in $f(x)$ for this example is shown.

The Factor Theorem

The remainder theorem leads to another important theorem known as the **factor theorem.** It states that *if $f(r) = R = 0$, then $x - r$ is a factor of $f(x)$.* We see in Eq. (15-2) that if the remainder $R = 0$, then $f(x) = (x - r)q(x)$, and this shows that $x - r$ is a factor of $f(x)$. Therefore, we have the following meanings for $f(r) = 0$.

> **Zero, Factor, and Root of the Function $f(x)$**
>
> *If $f(r) = 0$, then $x = r$ is a **zero** of $f(x)$,*
>
> $\qquad x - r$ is a **factor** of $f(x)$, and
>
> $\qquad x = r$ is a **root** of the equation $f(x) = 0$

EXAMPLE 4   Is $t + 1$ a factor of $f(t) = t^3 + 2t^2 - 5t - 6$?
Here $r = -1$, and thus

$$f(-1) = -1 + 2 + 5 - 6 = 0$$

Therefore, since $f(-1) = 0$, $t + 1$ is a factor of $f(t)$.   ▪

EXAMPLE 5   Note that $x + 2$ is not a factor of $f(x)$ in Example 4, since

$$f(-2) = -8 + 8 + 10 - 6 = 4$$

But $x - 2$ is a factor, since $f(2) = 8 + 8 - 10 - 6 = 0$.   ▪

EXAMPLE 6   Determine if $\frac{2}{3}$ is a zero of $f(x) = 3x^3 + 4x^2 - 16x + 8$.
Evaluating $f(\frac{2}{3})$ we have

$$f\left(\frac{2}{3}\right) = 3\left(\frac{2}{3}\right)^3 + 4\left(\frac{2}{3}\right)^2 - 16\left(\frac{2}{3}\right) + 8$$

$$= \frac{8}{9} + \frac{16}{9} - \frac{32}{3} + 8 = \frac{8 + 16 - 96 + 72}{9} = 0$$

Since $f(\frac{2}{3}) = 0$, $\frac{2}{3}$ is a zero of the function.   ▪

We now have one way of determining whether or not an expression of the form $x - r$ is a factor of a function $f(x)$. By finding $f(r)$, we can determine whether or not $x - r$ is a factor and whether or not r is a zero of the function.

— EXERCISES *15-1* —

In Exercises 1–8, find the remainder R by long division and by the remainder theorem.

1. $(x^3 + 2x^2 - x - 2) \div (x - 1)$

2. $(x^3 - 3x^2 - x + 2) \div (x - 2)$

3. $(x^3 + 2x + 3) \div (x + 1)$

4. $(x^4 - 4x^3 - x^2 + x - 100) \div (x + 3)$

5. $(2x^5 - x^2 + 8x + 44) \div (x + 2)$

6. $(4s^3 - 9s^2 - 24s - 17) \div (s - 5)$

7. $(3x^4 - 9x^3 - x^2 + 5x - 10) \div (x - 3)$

8. $(2x^4 - 10x^2 + 30x - 60) \div (x + 4)$

In Exercises 9–16, find the remainder using the remainder theorem.

9. $(x^3 + 2x^2 - 3x + 4) \div (x + 1)$

10. $(2x^3 - 4x^2 + x - 1) \div (x + 2)$

11. $(R^4 + R^3 - 9R^2 + 3) \div (R + 4)$

12. $(4x^4 - x^2 + 5x - 7) \div (x - 3)$

13. $(2x^4 - 7x^3 - x^2 + 8) \div (x - 3)$

14. $(3n^4 - 13n^2 + 10n - 10) \div (n + 4)$

15. $(x^5 - 3x^3 + 5x^2 - 10x + 6) \div (x - 2)$

16. $(3x^4 - 12x^3 - 60x + 4) \div (x - 5)$

In Exercises 17–24, use the factor theorem to determine whether or not the second expression is a factor of the first expression.

17. $y^3 - 8y - 3, y - 3$

18. $3x^3 + 2x^2 - 3x - 2, x + 2$

19. $4x^3 + x^2 - 16x - 4, x - 2$

20. $3x^3 + 14x^2 + 7x - 4, x + 4$

21. $3V^4 - 7V^3 + V + 8, V - 2$

22. $x^5 - 2x^4 + 3x^3 - 6x^2 - 4x + 8, x - 2$

23. $x^6 + 1, x + 1$

24. $x^7 - 128, x + 2$

In Exercises 25–28, determine whether or not the given numbers are zeros of the given functions.

25. $f(x) = x^3 - 2x^2 - 9x + 18$;   2

26. $f(t) = 2t^3 + 7t^2 - 9$; $-\frac{3}{2}$

27. $f(x) = 4x^4 - 4x^3 + 23x^2 + x - 6$; $\frac{1}{2}$

28. $f(x) = 2x^4 + 3x^3 - 12x^2 - 7x + 6$; -3

In Exercises 29–32, answer the given questions.

(W) 29. By division, show that $2x - 1$ is a factor of $f(x) = 4x^3 + 8x^2 - x - 2$. May we therefore conclude that $f(1) = 0$? Explain.

(W) 30. By division, show that $x^2 + 2$ is a factor of $f(x) = 3x^3 - x^2 + 6x - 2$. May we therefore conclude that $f(-2) = 0$? Explain.

31. For what value of k is $x - 2$ a factor of $f(x) = 2x^3 + kx^2 - x + 14$?

32. For what value of k is $x + 1$ a factor of $f(x) = 3x^4 + 3x^3 + 2x^2 + kx - 4$?

15-2 SYNTHETIC DIVISION

In the sections that follow, we will find that division of a polynomial by the factor $x - r$ is also useful in solving polynomial equations. Therefore, we will now develop a method known as **synthetic division,** which greatly simplifies the process of division. It is an abbreviated form of long division by which we can find the coefficients of the quotient and the remainder. This means that synthetic division allows us to find $f(r)$ by finding the remainder. If the degree of the equation is high, using synthetic division is easier than calculating $f(r)$ directly. The method is developed in the following example.

EXAMPLE 1 Divide $x^4 + 4x^3 - x^2 - 16x - 14$ by $x - 2$.

We shall first perform this division in the usual manner.

$$
\begin{array}{r}
x^3 + 6x^2 + 11x + 6 \\
x - 2\,\overline{\smash{\big)}\,x^4 + 4x^3 - x^2 - 16x - 14} \\
\underline{x^4 - 2x^3} \\
6x^3 - x^2 \\
\underline{6x^3 - 12x^2} \\
11x^2 - 16x \\
\underline{11x^2 - 22x} \\
6x - 14 \\
\underline{6x - 12} \\
-2
\end{array}
$$

In performing this division we repeated many terms. We should also note that the only quantities of importance in the function being divided are the numerical coefficients. There is no real need to put in the powers of x all of the time. Therefore, at the left we now write the above example without any x's and also eliminate identical terms.

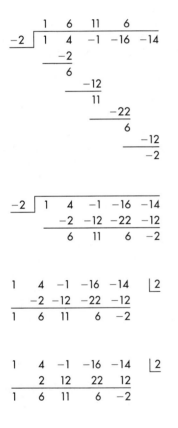

All but the first of the numbers that represent numerical coefficients of the quotients are repeated in the form at the left. Also, all numbers below the dividend may be written in two lines.

$$
\begin{array}{r}
-2\,\lfloor\; 1 \quad 4 \quad -1 \quad -16 \quad -14 \\
-2 \quad -12 \quad -22 \quad -12 \\
\hline
6 \quad 11 \quad 6 \quad -2
\end{array}
$$

$$
\begin{array}{r}
1 \quad 4 \quad -1 \quad -16 \quad -14 \;\lfloor 2 \\
-2 \quad -12 \quad -22 \quad -12 \\
\hline
1 \quad 6 \quad 11 \quad 6 \quad -2
\end{array}
$$

We now write the first coefficient (in this case, 1) in the bottom line. Also, we change the -2 to 2, which is the actual value of r. Then, in the form at the left, we write the 2 on the right. *In this form the 1, 6, 11, and 6 are the coefficients of the x^3, x^2, x, and constant term of the quotient. The -2 is the remainder.*

$$
\begin{array}{r}
1 \quad 4 \quad -1 \quad -16 \quad -14 \;\lfloor 2 \\
2 \quad 12 \quad 22 \quad 12 \\
\hline
1 \quad 6 \quad 11 \quad 6 \quad -2
\end{array}
$$

Finally, it is easier to use addition rather than subtraction in the process, so we change the signs of the numbers in the middle row. Remember that originally the bottom line was found by subtraction. Therefore, we have the last form on the left.

In the last form we note the following: The 1 multiplied by the 2(r) gives 2, the first number in the middle row. Adding the 4 and 2 (of the second column) gives 6, the second number in the bottom row. This 6 multiplied by 2 (r) is 12, the second number in the middle row. This 12 and -1 give 11. The 11 multiplied by 2 is 22. The 22 added to -16 is 6. This 6 multiplied by 2 gives 12. This 12 added to -14 is -2. In general, *the method is called synthetic division.*

We read the bottom line of the last form as

$$1x^3 + 6x^2 + 11x + 6 \text{ with a remainder of } -2$$

and the last form is the one we shall use in performing synthetic division.

Generalizing on Example 1, we have the following steps used in the process of synthetic division.

> ### Procedure for Synthetic Division
>
> 1. *Write the coefficients of $f(x)$. Be certain that the powers are in descending order and that zeros are inserted for missing powers.*
> 2. *Carry down the left coefficient, then multiply it by r, and place this product under the second coefficient of the top line.*
> 3. *Add the two numbers in the second column and place the result below. Multiply this sum by r and place the product under the third coefficient of the top line.*
> 4. *Continue this process until the bottom row has as many numbers as the top row.*

See Appendix C for a graphing calculator program SYNTHDIV. It can be used to find the coefficients and remainder if a polynomial is divided by $x - R$.

The last number in the bottom row is the remainder, and the other numbers are the respective coefficients of the quotient. The first term of the quotient is of degree one less than the dividend.

■EXAMPLE 2   Divide $x^5 + 2x^4 - 4x^2 + 3x - 4$ by $x + 3$ using synthetic division.

Since the powers of x are in descending order, we write down the coefficients of $f(x)$. In doing so we must be certain to include a zero for the missing x^3 term. Next we note that the divisor is $x + 3$, which means that $r = -3$. The -3 is placed to the right. This gives us a top line of

$$\text{coefficients} \longrightarrow 1 \quad 2 \quad 0 \quad -4 \quad 3 \quad -4 \quad \underline{|-3} \longleftarrow r$$

Next we carry the left coefficient, 1, to the bottom line and multiply it by r, -3, placing the product, -3, in the middle line under the second coefficient, 2. We then add the 2 and the -3 and place the result, -1, below. This gives

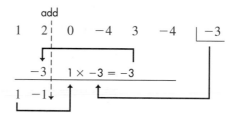

Now we multiply the -1 by r, -3, and place the result, 3, in the middle line under the zero. We now add, and continue the process, obtaining the following result:

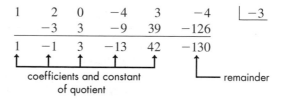

NOTE ▶

Since the degree of the dividend is 5, the degree of the quotient is 4. This means the quotient is $x^4 - x^3 + 3x^2 - 13x + 42$ and the remainder is -130. In turn, this means that for $f(x) = x^5 + 2x^4 - 4x^2 + 3x - 4$, we have $f(-3) = -130$. ■

EXAMPLE 3   By synthetic division, divide $3x^4 - 5x + 6$ by $x - 4$.

$$
\begin{array}{rrrrr|r}
3 & 0 & 0 & -5 & 6 & \underline{4} \\
 & 12 & 48 & 192 & 748 & \\
\hline
3 & 12 & 48 & 187 & 754 &
\end{array}
$$

The quotient is $3x^3 + 12x^2 + 48x + 187$, and the remainder is 754.

EXAMPLE 4   By synthetic division, determine whether or not $t - 4$ is a factor of $t^4 + 2t^3 - 15t^2 - 32t - 16$.

$$
\begin{array}{rrrrr|r}
1 & 2 & -15 & -32 & -16 & \underline{4} \\
 & 4 & 24 & 36 & 16 & \\
\hline
1 & 6 & 9 & 4 & 0 &
\end{array}
$$

Since the remainder is zero, $t - 4$ is a factor. We may also conclude that

$$f(t) = (t - 4)(t^3 + 6t^2 + 9t + 4)$$

since the bottom line gives us the coefficients in the quotient.

EXAMPLE 5   By using synthetic division, determine whether $2x - 3$ is a factor of $2x^3 - 3x^2 + 8x - 12$.

CAUTION▶   We first note that the coefficient of x in the possible factor is not 1. Thus, *we cannot use $r = 3$, since the factor is not of the form $x - r$.* However, $2x - 3 = 2(x - \frac{3}{2})$, which means that if $2(x - \frac{3}{2})$ is a factor of the function, $2x - 3$ is a factor. If we use $r = \frac{3}{2}$ and find that the remainder is zero, then $x - \frac{3}{2}$ is a factor.

$$
\begin{array}{rrrr|r}
2 & -3 & 8 & -12 & \underline{\frac{3}{2}} \\
 & 3 & 0 & 12 & \\
\hline
2 & 0 & 8 & 0 &
\end{array}
$$

Since the remainder is zero, $x - \frac{3}{2}$ is a factor. Also, the quotient is $2x^2 + 8$, which may be factored into $2(x^2 + 4)$. Thus, 2 is also a factor of the function. This means that $2(x - \frac{3}{2})$ is a factor of the function, and this in turn means that $2x - 3$ is a factor. This tells us that

$$2x^3 - 3x^2 + 8x - 12 = (2x - 3)(x^2 + 4)$$

EXAMPLE 6   By synthetic division, determine whether or not $\frac{1}{3}$ is a zero of the function $3x^3 + 2x^2 - 4x + 1$.

This problem is equivalent to dividing the function by $x - \frac{1}{3}$. If the remainder is zero, $\frac{1}{3}$ is a zero of the function.

$$
\begin{array}{rrrr|r}
3 & 2 & -4 & 1 & \underline{\frac{1}{3}} \\
 & 1 & 1 & -1 & \\
\hline
3 & 3 & -3 & 0 &
\end{array}
$$

Since the remainder is zero, we conclude that $\frac{1}{3}$ is a zero of the function.

$$3x^3 + 2x^2 - 4x + 1 = (x - \tfrac{1}{3})(3x^2 + 3x - 3)$$
$$= 3(x - \tfrac{1}{3})(x^2 + x - 1)$$

EXAMPLE 7 Determine whether or not -12.5 is a zero of the function $f(x) = 6x^3 + 61x^2 - 171x + 100$.

If -12.5 is a zero of $f(x)$, then $x - (-12.5)$, or $x + 12.5$, is a factor of $f(x)$, and $f(-12.5) = 0$. We can find the remainder by direct use of the remainder theorem or by synthetic division.

Using synthetic division and a calculator to make the calculations, we have the following setup and calculator sequence.

$$
\begin{array}{rrrr|l}
6 & 61 & -171 & 100 & \quad -12.5 \\
 & -75 & 175 & -50 & \\
\hline
6 & -14 & 4 & 50 &
\end{array}
$$

Since the remainder is 50, and not zero, -12.5 is not a zero of $f(x)$.

EXERCISES *15-2*

In Exercises 1–20, perform the indicated divisions by synthetic division. Exercises 1–16 are the same as Exercises 1–16 of Section 15-1.

1. $(x^3 + 2x^2 - x - 2) \div (x - 1)$

2. $(x^3 - 3x^2 - x + 2) \div (x - 2)$

3. $(x^3 + 2x + 3) \div (x + 1)$

4. $(x^4 - 4x^3 - x^2 + x - 100) \div (x + 3)$

5. $(2x^5 - x^2 + 8x + 44) \div (x + 2)$

6. $(4s^3 - 9s^2 - 24s - 17) \div (s - 5)$

7. $(3x^4 - 9x^3 - x^2 + 5x - 10) \div (x - 3)$

8. $(2x^4 - 10x^2 + 30x - 60) \div (x + 4)$

9. $(x^3 + 2x^2 - 3x + 4) \div (x + 1)$

10. $(2x^3 - 4x^2 + x - 1) \div (x + 2)$

11. $(R^4 + R^3 - 9R^2 + 3) \div (R + 4)$

12. $(4x^4 - x^2 + 5x - 7) \div (x - 3)$

13. $(2x^4 - 7x^3 - x^2 + 8) \div (x - 3)$

14. $(3n^4 - 13n^2 + 10n - 10) \div (n + 4)$

15. $(x^5 - 3x^3 + 5x^2 - 10x + 6) \div (x - 2)$

16. $(3x^4 - 12x^3 - 60x + 4) \div (x - 5)$

17. $(p^6 - 6p^3 - 2p^2 - 6) \div (p - 2)$

18. $(x^5 + 4x^4 - 8) \div (x + 1)$

19. $(x^7 - 128) \div (x - 2)$

20. $(20x^4 + 11x^3 - 89x^2 + 60x - 77) \div (x + 2.75)$

In Exercises 21–32, use the factor theorem and synthetic division to determine whether or not the second expression is a factor of the first.

21. $x^3 + x^2 - x + 2; \quad x + 2$

22. $x^3 + 6x^2 + 10x + 6; \quad x + 3$

23. $x^4 - 6x^2 - 3x - 2; \quad x - 3$

24. $2t^4 - 5t^3 - 24t^2 + 5; \quad t - 5$

25. $2x^5 - x^3 + 3x^2 - 4; \quad x + 1$

26. $x^5 - 3x^4 - x^2 - 6; \quad x - 3$

27. $4x^3 - 6x^2 + 2x - 2; \quad x - \frac{1}{2}$

28. $3x^3 - 5x^2 + x + 1; \quad x + \frac{1}{3}$

29. $2Z^4 - Z^3 - 4Z^2 + 1; \quad 2Z - 1$

30. $6x^4 + 5x^3 - x^2 + 6x - 2; \quad 3x - 1$

31. $4x^4 + 2x^3 - 8x^2 + 3x + 12; \quad 2x + 3$

32. $3x^4 - 2x^3 + x^2 + 15x + 4; \quad 3x + 4$

In Exercises 33–36, use synthetic division to determine whether or not the given numbers are zeros of the given functions.

33. $x^4 - 5x^3 - 15x^2 + 5x + 14; \quad 7$

34. $r^4 + 5r^3 - 18r - 8; \quad -4$

35. $85x^3 + 348x^2 - 263x + 120; \quad -4.8$

36. $2x^3 + 13x^2 + 10x - 4; \quad \frac{1}{2}$

15-3 THE ROOTS OF AN EQUATION

In this section we present certain theorems that are useful in determining the number of roots in the equation $f(x) = 0$ and the nature of some of these roots. In dealing with polynomial equations of higher degree, it is helpful to have as much of this kind of information as is readily obtainable before solving for the roots.

The first of these theorems is so important that it is called the **fundamental theorem of algebra.** It states that

every polynomial equation has at least one (real or complex) root.

The proof of this theorem is of an advanced nature, and therefore we must accept its validity at this time. However, using the fundamental theorem, we can show the validity of other theorems that are useful in solving equations.

Let us now assume that we have a polynomial equation $f(x) = 0$ and that we are looking for its roots. By the fundamental theorem, we know that it has at least one root. Assuming that we can find this root by some means (the factor theorem, for example), we shall call this root r_1. Thus,

$$f(x) = (x - r_1)f_1(x)$$

where $f_1(x)$ is the polynomial quotient found by dividing $f(x)$ by $(x - r_1)$. However, since the fundamental theorem states that any polynomial equation has at least one root, this must apply to $f_1(x) = 0$ as well. Let us assume that $f_1(x) = 0$ has the root r_2. Therefore, this means that $f(x) = (x - r_1)(x - r_2)f_2(x)$. Continuing this process until one of the quotients is a constant a, we have

$$f(x) = a(x - r_1)(x - r_2)\cdots(x - r_n)$$

Note that one linear factor appears each time a root is found and that the degree of the quotient is one less each time. Thus there are n factors, if the degree of $f(x)$ is n.

Therefore, based on the fundamental theorem of algebra, we have the following two related theorems.

1. *A polynomial of the nth degree can be factored into n linear factors.*

2. *A polynomial equation of degree n has exactly n roots.*

These theorems are illustrated in the following example.

■EXAMPLE 1   For the equation $f(x) = 2x^4 - 3x^3 - 12x^2 + 7x + 6 = 0$, we are given the factors that we show. In the next section we will see how to find these factors.

For the function $f(x)$, we have

$$2x^4 - 3x^3 - 12x^2 + 7x + 6 = (x - 3)(2x^3 + 3x^2 - 3x - 2)$$
$$2x^3 + 3x^2 - 3x - 2 = (x + 2)(2x^2 - x - 1)$$
$$2x^2 - x - 1 = (x - 1)(2x + 1)$$
$$2x + 1 = 2(x + \tfrac{1}{2})$$

Therefore,

$$2x^4 - 3x^3 - 12x^2 + 7x + 6 = 2(x - 3)(x + 2)(x - 1)(x + \tfrac{1}{2}) = 0$$

The degree of $f(x)$ is 4. There are four linear factors: $(x - 3)$, $(x + 2)$, $(x - 1)$, and $(x + \tfrac{1}{2})$. There are four roots of the equation: 3, -2, 1, and $-\tfrac{1}{2}$. Thus, we have verified each of the theorems above for this example. ---------■

It is not necessary for each root of an equation to be different from the other roots. For example, the equation $(x - 1)^2 = 0$ has two roots, both of which are 1. Such roots are referred to as *multiple* (or *repeated*) *roots.*

This theorem was first proved in 1799 by the German mathematician Karl Gauss (1777–1855) for his doctoral thesis.

When we solve the equation $x^2 + 1 = 0$, the roots are j and $-j$. In fact, for any equation (with real coefficients) that has a root of the form $a + bj$ ($b \neq 0$), there is also a root of the form $a - bj$. This is so because we can find the solutions of an equation of the form $ax^2 + bx + c = 0$ from the quadratic formula as

$$\frac{-b + \sqrt{b^2 - 4ac}}{2a} \quad \text{and} \quad \frac{-b - \sqrt{b^2 - 4ac}}{2a}$$

and the only difference between these roots is the sign before the radical. Thus, we have the following theorem.

NOTE ▶   *If the coefficients of the equation $f(x) = 0$ are real and $a + bj$ ($b \neq 0$) is a complex root, then its conjugate, $a - bj$, is also a root.*

■**EXAMPLE 2**   Consider the equation $f(x) = (x - 1)^3(x^2 + x + 1) = 0$.

The factor $(x - 1)^3$ shows that there is a triple root of 1, and there is a total of five roots, since the highest-power term would be x^5 if we were to multiply out the function. To find the other two roots, we use the quadratic formula on the *factor* $(x^2 + x + 1)$. This is permissible, since we are finding the values of x for

$$x^2 + x + 1 = 0$$

For this we have

$$x = \frac{-1 \pm \sqrt{1 - 4}}{2}$$

Thus,

$$x = \frac{-1 + j\sqrt{3}}{2} \quad \text{and} \quad x = \frac{-1 - j\sqrt{3}}{2}$$

Therefore, the roots of $f(x) = 0$ are

$$1, \quad 1, \quad 1, \quad \frac{-1 + j\sqrt{3}}{2}, \quad \frac{-1 - j\sqrt{3}}{2} \qquad \blacksquare$$

NOTE ▶   From Example 2 we can see that *whenever enough roots are known so that the remaining factor is quadratic, it is possible to find the remaining roots from the quadratic formula.* This is true for finding real or complex roots.

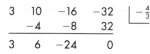

■**EXAMPLE 3**   Solve the equation $3x^3 + 10x^2 - 16x - 32 = 0$; $-\frac{4}{3}$ is a root.

Using synthetic division and the given root we have the setup shown at the left. From this we see that

$$3x^3 + 10x^2 - 16x - 32 = (x + \tfrac{4}{3})(3x^2 + 6x - 24)$$

We know that $x + \frac{4}{3}$ is a factor from the given root and that $3x^2 + 6x - 24$ is a factor found from synthetic division. This second factor can be factored as

$$3x^2 + 6x - 24 = 3(x^2 + 2x - 8) = 3(x + 4)(x - 2)$$

Therefore, we have

$$3x^3 + 10x^2 - 16x - 32 = 3(x + \tfrac{4}{3})(x + 4)(x - 2)$$

This means the roots are $-\frac{4}{3}$, -4, and 2.   ■

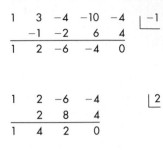

$$
\begin{array}{rrrrr|r}
1 & 3 & -4 & -10 & -4 & \underline{-1} \\
  & -1 & -2 & 6 & 4 & \\
\hline
1 & 2 & -6 & -4 & 0 &
\end{array}
$$

$$
\begin{array}{rrrr|r}
1 & 2 & -6 & -4 & \underline{2} \\
  & 2 & 8 & 4 & \\
\hline
1 & 4 & 2 & 0 &
\end{array}
$$

EXAMPLE 4 Solve $x^4 + 3x^3 - 4x^2 - 10x - 4 = 0$; -1 and 2 are roots.

Using synthetic division and the root -1, we have the first setup shown at the left. This tells us that

$$x^4 + 3x^3 - 4x^2 - 10x - 4 = (x + 1)(x^3 + 2x^2 - 6x - 4)$$

We now know that $x - 2$ must be a factor of $x^3 + 2x^2 - 6x - 4$, since it is a factor of the original function. Again, using synthetic division and this time the root 2, we have the second setup at the left. Thus,

$$x^4 + 3x^3 - 4x^2 - 10x - 4 = (x + 1)(x - 2)(x^2 + 4x + 2)$$

Since the original equation can now be written as

$$(x + 1)(x - 2)(x^2 + 4x + 2) = 0$$

the remaining two roots are found by solving

$$x^2 + 4x + 2 = 0$$

by the quadratic formula. This gives us

$$x = \frac{-4 \pm \sqrt{16 - 8}}{2} = \frac{-4 \pm 2\sqrt{2}}{2} = -2 \pm \sqrt{2}$$

Therefore, the roots are -1, 2, $-2 + \sqrt{2}$, and $-2 - \sqrt{2}$. _____■

EXAMPLE 5 Solve the equation $3x^4 - 26x^3 + 63x^2 - 36x - 20 = 0$, given that 2 is a double root.

Using synthetic division, we have the first setup at the left. It tells us that

$$3x^4 - 26x^3 + 63x^2 - 36x - 20 = (x - 2)(3x^3 - 20x^2 + 23x + 10)$$

$$
\begin{array}{rrrrr|r}
3 & -26 & 63 & -36 & -20 & \underline{2} \\
  & 6 & -40 & 46 & 20 & \\
\hline
3 & -20 & 23 & 10 & 0 &
\end{array}
$$

$$
\begin{array}{rrrr|r}
3 & -20 & 23 & 10 & \underline{2} \\
  & 6 & -28 & -10 & \\
\hline
3 & -14 & -5 & 0 &
\end{array}
$$

Also, since 2 is a double root, it must be a root of $3x^3 - 20x^2 + 23x + 10 = 0$. Using synthetic division again, we have the second setup at the left. This second quotient $3x^2 - 14x - 5$ factors into $(3x + 1)(x - 5)$. The roots are 2, 2, $-\frac{1}{3}$, and 5.

Since the quotient of the first division is the dividend for the second division, both divisions can be done without rewriting the first quotient as follows:

NOTE ▶

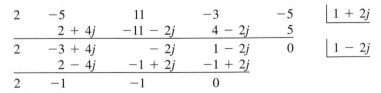

$$
\begin{array}{rrrrr|r}
3 & -26 & 63 & -36 & -20 & \underline{2} \\
  & 6 & -40 & 46 & 20 & \\
\hline
3 & -20 & 23 & 10 & 0 & \underline{2} \\
  & 6 & -28 & -10 & & \\
\hline
3 & -14 & -5 & 0 & &
\end{array}
$$

first division →
second division →

_____■

EXAMPLE 6 Solve the equation $2x^4 - 5x^3 + 11x^2 - 3x - 5 = 0$, given that $1 + 2j$ is a root.

Since $1 + 2j$ is a root, we know that $1 - 2j$ is also a root. Using synthetic division twice, we can then reduce the remaining factor to a quadratic function.

$$
\begin{array}{rrrrr|r}
2 & -5 & 11 & -3 & -5 & \underline{1 + 2j} \\
  & 2 + 4j & -11 - 2j & 4 - 2j & 5 & \\
\hline
2 & -3 + 4j & -2j & 1 - 2j & 0 & \underline{1 - 2j} \\
  & 2 - 4j & -1 + 2j & -1 + 2j & & \\
\hline
2 & -1 & -1 & 0 & &
\end{array}
$$

The quadratic factor $2x^2 - x - 1$ factors into $(2x + 1)(x - 1)$. Therefore, the roots of the equation are $1 + 2j$, $1 - 2j$, 1, and $-\frac{1}{2}$. _____■

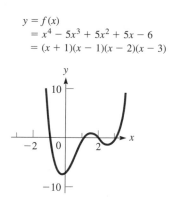

$y = f(x)$
$= x^4 - 5x^3 + 5x^2 + 5x - 6$
$= (x + 1)(x - 1)(x - 2)(x - 3)$

Fig. 15-2

Considering the graphical meaning of the roots of a polynomial equation, we recall that a polynomial equation $f(x) = 0$ of degree n has exactly n roots. If the roots are all real and different, the graph of $f(x)$ crosses the x-axis n times, since $f(x) = 0$ for each value of x where there is a root, and $f(x)$ must change signs as it crosses. See Fig. 15-2, where $n = 4$ and the curve crosses the x-axis four times.

Looking at Fig. 15-2, we see that *the curve must cross the x-axis an odd number of times between two points on the curve and on opposite sides of the x-axis.* Also, *for two points on the curve and on the same side of the x-axis, the curve must cross the x-axis an even number of times between these points, if it crosses at all.*

Considering complex roots and repeated roots, the number of times the curve crosses the x-axis is reduced by 2 for each pair of complex roots (see Fig. 15-3). For a multiple root, the curve crosses the x-axis only once if the multiple is odd, or the curve is tangent to the x-axis if the multiple is even (see Fig. 15-4).

The curve must cross the x-axis at least once if the degree of $f(x)$ is odd, because the range of $f(x)$ includes all real numbers (see Figs. 15-3 and 15-4). The curve may not cross the x-axis at all if the degree of $f(x)$ is even, because the range of $f(x)$ is bounded at a minimum point or a maximum point (as we have seen for the quadratic function) (see Fig. 15-5).

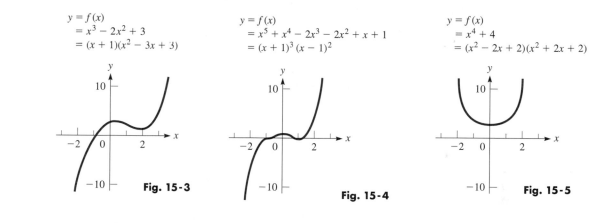

$y = f(x)$
$= x^3 - 2x^2 + 3$
$= (x + 1)(x^2 - 3x + 3)$

Fig. 15-3

$y = f(x)$
$= x^5 + x^4 - 2x^3 - 2x^2 + x + 1$
$= (x + 1)^3 (x - 1)^2$

Fig. 15-4

$y = f(x)$
$= x^4 + 4$
$= (x^2 - 2x + 2)(x^2 + 2x + 2)$

Fig. 15-5

EXERCISES *15-3*

In the following exercises, solve the given equations using synthetic division, given the roots indicated.

1. $x^3 + 2x^2 - x - 2 = 0$ $(r_1 = 1)$
2. $x^3 + 2x^2 + x + 2 = 0$ $(r_1 = -2)$
3. $x^3 + x^2 - 8x - 12 = 0$ $(r_1 = -2)$
4. $R^3 + 1 = 0$ $(r_1 = -1)$
5. $2x^3 + 11x^2 + 20x + 12 = 0$ $(r_1 = -\frac{3}{2})$
6. $4x^3 - 20x^2 - x + 5 = 0$ $(r_1 = \frac{1}{2})$
7. $3x^3 + 2x^2 + 3x + 2 = 0$ $(r_1 = j)$
8. $x^3 + 5x^2 + 9x + 5 = 0$ $(r_1 = -2 + j)$
9. $t^4 + t^3 - 2t^2 + 4t - 24 = 0$ $(r_1 = 2, r_2 = -3)$
10. $x^4 + 2x^3 - 4x^2 - 5x + 6 = 0$ $(r_1 = 1, r_2 = -2)$
11. $x^4 - 9x^2 + 4x + 12 = 0$ (2 is a double root)
12. $4x^4 + 28x^3 + 61x^2 + 42x + 9 = 0$ $(-3$ is a double root)
13. $6x^4 + 5x^3 - 15x^2 + 4 = 0$ $(r_1 = -\frac{1}{2}, r_2 = \frac{2}{3})$

14. $6x^4 - 5x^3 - 14x^2 + 14x - 3 = 0$ $(r_1 = \frac{1}{3}, r_2 = \frac{3}{2})$
15. $2x^4 - x^3 - 4x^2 + 10x - 4 = 0$ $(r_1 = 1 + j)$
16. $s^4 - 8s^3 - 72s - 81 = 0$ $(r_1 = 3j)$
17. $2x^5 + 11x^4 + 16x^3 - 8x^2 - 32x - 16 = 0$
    $(-2$ is a triple root)
18. $x^5 - 3x^4 + 4x^3 - 4x^2 + 3x - 1 = 0$ (1 is a triple root)
19. $2x^5 + x^4 - 15x^3 + 5x^2 + 13x - 6 = 0$
    $(r_1 = 1, r_2 = -1, r_3 = \frac{1}{2})$
20. $12x^5 - 7x^4 + 41x^3 - 26x^2 - 28x + 8 = 0$
    $(r_1 = 1, r_2 = \frac{1}{4}, r_3 = -\frac{2}{3})$
21. $P^5 - 3P^4 - P + 3 = 0$ $(r_1 = 3, r_2 = j)$
22. $4x^5 + x^3 - 4x^2 - 1 = 0$ $(r_1 = 1, r_2 = \frac{1}{2}j)$
23. $x^6 + 2x^5 - 4x^4 - 10x^3 - 41x^2 - 72x - 36 = 0$
    $(-1$ is a double root; $2j$ is a root)
24. $x^6 - x^5 - 2x^3 - 3x^2 - x - 2 = 0$ $(j$ is a double root)

15-4 RATIONAL AND IRRATIONAL ROOTS

The product of the factors $(x + 2)(x - 4)(x + 3)$ is $x^3 + x^2 - 14x - 24$. In forming this product, we find that the constant 24 is determined only by the numbers 2, 4, and 3. We see that these numbers represent the roots of the equation if the given function is set equal to zero. In fact, if we find all the integral roots of an equation with integral coefficients and represent the equation in the form

$$f(x) = (x - r_1)(x - r_2)\cdots(x - r_k)f_{k+1}(x) = 0$$

where all the roots indicated are integers, the constant term of $f(x)$ must have factors of $r_1, r_2, \ldots, r_k$. This leads us to the theorem which states that

in a polynomial equation $f(x) = 0$, if the coefficient of the highest power is 1, then any integral roots are factors of the constant term of $f(x)$.

▪**EXAMPLE 1**   The equation $x^5 - 4x^4 - 7x^3 + 14x^2 - 44x + 120 = 0$ can be written as

$$(x - 5)(x + 3)(x - 2)(x^2 + 4) = 0$$

We now note that $5(3)(2)(4) = 120$. Thus, the roots 5, -3, and 2 are numerical factors of $|120|$. The theorem states nothing about the signs involved.   ▪

If the coefficient a_0 of the highest-power term of $f(x)$ is an integer not equal to 1, the polynomial equation $f(x) = 0$ may have rational roots that are not integers. This coefficient a_0 can be factored from every term of $f(x)$. Thus, any polynomial equation $f(x) = a_0x^n + a_1x^{n-1} + \cdots + a_n = 0$ with integral coefficients can be written in the form

$$f(x) = a_0\left(x^n + \frac{a_1}{a_0}x^{n-1} + \cdots + \frac{a_n}{a_0}\right) = 0$$

Since a_n and a_0 are integers, a_n/a_0 is a rational number. Using the same reasoning as with integral roots applied to the polynomial within the parentheses, we see that any rational roots are factors of a_n/a_0. This leads to the following theorem:

Any rational root of a polynomial equation (with integral coefficients)

$$f(x) = a_0x^n + a_1x^{n-1} + \cdots + a_n = 0$$

is an integral factor of a_n divided by an integral factor of a_0.

We may show this rational root r_r as

$$r_r = \frac{\text{integral factor of } a_n}{\text{integral factor of } a_0} \tag{15-4}$$

▪**EXAMPLE 2**   If $f(x) = 4x^3 - 3x^2 - 25x - 6 = 0$, any rational roots, if they exist, must be integral factors of 6 divided by integral factors of 4. The integral factors of 6 are 1, 2, 3, and 6, and the integral factors of 4 are 1, 2, and 4. Forming all possible positive and negative quotients, any rational roots that exist will be found in the following list: ±1, $\pm\frac{1}{2}$, $\pm\frac{1}{4}$, ±2, ±3, $\pm\frac{3}{2}$, $\pm\frac{3}{4}$, ±6.
   The roots of this equation are -2, 3, and $-\frac{1}{4}$.   ▪

There are 16 different possible rational roots in Example 2, but we cannot tell which of these are the actual roots. Therefore we now present a rule, known as *Descartes' rule of signs,* which will help us to find these roots.

Named for the French mathematician René Descartes (1596–1650). (See page 86.)

Descartes' Rule of Signs

1. *The number of positive roots of a polynomial equation $f(x) = 0$ cannot exceed the number of changes in sign in $f(x)$ in going from one term to the next in $f(x)$.*

2. *The number of negative roots cannot exceed the number of sign changes in $f(-x)$.*

We can reason this way: If $f(x)$ has all positive terms, then any positive number substituted in $f(x)$ must give a positive value for $f(x)$. This indicates that the number substituted in the function is not a root. Thus, there must be at least one negative and one positive term in the function for any positive number to be a root. This is not a proof, but it does indicate the type of reasoning used in developing the theorem.

EXAMPLE 3   By Descartes' rule of signs, determine the maximum number of positive and negative roots of $3x^3 - x^2 - x + 4 = 0$.

Here, $f(x) = 3x^3 - x^2 - x + 4$. The first term is positive, and the second is negative, which indicates a change of sign. The third term is also negative; there is no change of sign from the second to the third term. The fourth term is positive, thus giving us a second change of sign, from the third to the fourth term. Hence, there are two changes in sign, which we can show as follows:

$$f(x) = 3x^3 - x^2 - x + 4$$

1   2   ←——— two sign changes

Since there are *two* changes of sign in $f(x)$, there are *no more than two* positive roots of $f(x) = 0$.

To find the maximum possible number of negative roots, we must find the number of sign changes in $f(-x)$. Thus,

$$f(-x) = 3(-x)^3 - (-x)^2 - (-x) + 4$$
$$= -3x^3 - x^2 + x + 4$$

←——— one sign change

There is only one change of sign in $f(-x)$; therefore, there is one negative root.

NOTE ▶ *When there is just one change of sign in $f(x)$, there is a positive root, and when there is just one change of sign in $f(-x)$, there is a negative root.* ───────■

EXAMPLE 4   For the equation $4x^5 - x^4 - 4x^3 + x^2 - 5x - 6 = 0$, we write

$$f(x) = 4x^5 - x^4 - 4x^3 + x^2 - 5x - 6$$

←——— three sign changes

$$f(-x) = -4x^5 - x^4 + 4x^3 + x^2 + 5x - 6$$

←——— two sign changes

Thus, there are no more than three positive and two negative roots. ───────■

At this point let us summarize the information we can determine about the roots of a polynomial equation $f(x) = 0$ of degree n and with real coefficients.

> ### Roots of a Polynomial Equation of Degree *n*
>
> 1. *There are n roots.*
> 2. *Complex roots appear in conjugate pairs.*
> 3. *Any rational roots must be factors of the constant term divided by factors of the coefficient of the highest-power term.*
> 4. *The maximum number of positive roots is the number of sign changes in f(x), and the maximum number of negative roots is the number of sign changes in f(−x).*
> 5. *Once we find n − 2 of the roots, we can find the remaining roots by the quadratic formula.*

Some calculators have a specific feature for solving higher-degree equations.

Since synthetic division is easy to perform, it is usually used to try possible roots. When a root is found, the quotient is of one degree less than the degree of the dividend. Each root found makes the ensuing work easier. The following examples show the complete method, as well as two other helpful rules.

■**EXAMPLE 5**　Find the roots of the equation $2x^3 + x^2 + 5x - 3 = 0$.

Since $n = 3$, there are three roots. If we can find one of these roots, we can use the quadratic formula to find the other two. We have

$$f(x) = 2x^3 + x^2 + 5x - 3 \quad \text{and} \quad f(-x) = -2x^3 + x^2 - 5x - 3$$

which shows there is one positive root and no more than two negative roots, which may or may not be rational. The *possible* rational roots are ± 1, $\pm\frac{1}{2}$, $\pm\frac{3}{2}$, ± 3.

First, trying the root 1 (always a possibility if there are positive roots), we have the synthetic division shown at the left. The remainder of 5 tells us that 1 is not a root, but we have gained some additional information, if we observe closely. If we try any positive number larger than 1, the results in the last row will be larger positive numbers than we now have. The products will be larger, and therefore the sums will also be larger. Thus, there is no positive root larger than 1. This leads to the following rule: *When we are trying a positive root, if the bottom row contains all positive numbers, then there are no roots larger than the value tried.* This rule tells us that there is no reason to try $+\frac{3}{2}$ and $+3$ as roots.

Now let us try $+\frac{1}{2}$, as shown at the left. The zero remainder tells us that $+\frac{1}{2}$ is a root, and the remaining factor is $2x^2 + 2x + 6$, which itself factors to $2(x^2 + x + 3)$. By the quadratic formula we find the remaining roots by solving the equation $x^2 + x + 3 = 0$. This gives us

$$x = \frac{-1 \pm \sqrt{1 - 12}}{2} = \frac{-1 \pm j\sqrt{11}}{2}$$

The three roots are

$$\frac{1}{2}, \quad \frac{-1 + j\sqrt{11}}{2}, \quad \frac{-1 - j\sqrt{11}}{2}$$

There are no negative roots, because the nonpositive roots are complex. Proceeding in this way, we did not have to try any negative roots.   --------■

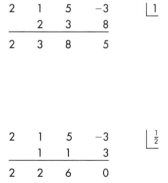

```
2   1   5  -3  |1
    2   3   8
2   3   8   5
```

```
2   1   5  -3  |½
    1   1   3
2   2   6   0
```

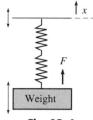

Fig. 15-6

■EXAMPLE 6 During a cycle of the movement of the weight on the double spring shown in Fig. 15-6, the force F (in N) on the weight by the spring is

$$F = x^4 - 7x^3 + 12x^2 + 4x$$

where x is the displacement (in cm) of the top of the double spring. For what values of x is $F = 16$ N?

Substituting 16 for F, we see that we are to solve the equation

$$x^4 - 7x^3 + 12x^2 + 4x - 16 = 0$$

To solve this equation, we write

$$f(x) = x^4 - 7x^3 + 12x^2 + 4x - 16$$
$$f(-x) = x^4 + 7x^3 + 12x^2 - 4x - 16$$

We see that there are four roots; there are no more than three positive roots, and there is one negative root. Since the coefficient of x^4 is 1, any possible rational roots must be integers. These possible rational roots are ± 1, ± 2, ± 4, ± 8, and ± 16. Since there is only one negative root, we shall look for this one first. Trying -2, we have

```
1   -7   +12    +4    -16    |-2
     -2   +18   -60   +112
1   -9   +30   -56    +96
```

NOTE▶

If we were to try any negative roots less than -2 (remember, -3 is less than -2), we would find that the numbers would still alternate from term to term in the quotient. Thus, we have this rule: *When we are trying a negative root, if the signs alternate in the bottom row, then there are no roots less than the value tried.* In this case, we now know that -4, -8, *and* -16 cannot be roots.

Next we try -1, as shown at the left. The remainder of zero tells us that -1 is the negative root.

```
1  -7  +12   +4   -16   |-1
    -1    8  -20    16
1  -8   20  -16     0
```

Now that we have found the one negative root, we look for the positive roots. Trying 1, we have the setup at the left. The remainder of -3 tells us that 1 is not a root. Next we try 2, as shown at the left. We see that 2 is a root.

```
1   -8   20   -16   |1
     1   -7    13
1   -7   13    -3
```

It is not necessary to find any more roots by trial and error. We may now use the quadratic formula or factoring on the equation $x^2 - 6x + 8 = 0$. The remaining roots are 2 and 4. Thus, the roots are -1, 2, 2, and 4. (Note that 2 is a double root.)

These roots now indicate that $F = 16$ N for displacements of -1 cm, 2 cm, and 4 cm.

```
1   -8   20   -16   |2
     2  -12    16
1   -6    8     0
```

By the methods we have presented, we can look for *all roots* of a polynomial equation. These include any possible *complex roots* and *exact values of the rational and irrational roots,* if they exist. These methods allow us to solve a great many polynomial equations for these roots, but there are numerous other equations for which these methods are not sufficient.

When a polynomial equation has more than two irrational roots, we cannot generally find these roots by the methods we have developed. Approximate values can be found on a calculator either graphically or by evaluating the function. These methods are illustrated in the following two examples.

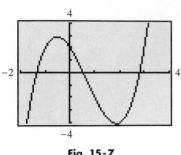

Fig. 15-7

■EXAMPLE 7   Find the roots of the equation $x^3 - 2x^2 - 3x + 2 = 0$, using a graphing calculator.

Since the degree of the equation is 3, we know there are three roots, and since the degree is odd there is at least one real root. Using Descartes' rule of signs we have

$$f(x) = x^3 - 2x^2 - 3x + 2 \qquad \text{two changes of sign}$$
$$f(-x) = -x^3 - 2x^2 + 3x + 2 \qquad \text{one change of sign}$$

This means there is one negative root and no more than two positive roots, if any. Therefore, setting

$$y = x^3 - 2x^2 - 3x + 2$$

and using the *window* settings (after two or three trials) shown in the calculator display in Fig. 15-7, the view shows one negative and two positive roots. Using the *trace* and *zoom* features (or *zero* feature), these roots are

$$-1.34, \; 0.53, \; 2.81 \qquad \text{to the nearest 0.01}$$

Of course, we can use a graphing calculator to help locate and determine the number of real roots of any equation.

SOLVING A WORD PROBLEM

See the chapter introduction.

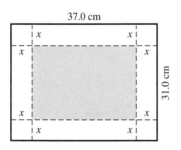

Fig. 15-8

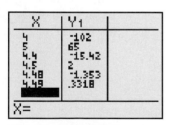

Fig. 15-9

Methods of evaluating functions on a graphing calculator are shown on page 79.

■EXAMPLE 8   The bottom box of a box to hold a jigsaw puzzle is to be made from a rectangular piece of cardboard 37.0 cm by 31.0 cm by cutting out equal squares from the corners, bending up the sides, and taping the corners. See Fig. 15-8. If the volume of the bottom box is to be 2770 cm³, find the side of the square that is to be cut out.

Let x = the side of the square to be cut out. This means

$$2770 = x(37.0 - 2x)(31.0 - 2x) \qquad \text{area} = 2770 \text{ cm}^2$$
$$= 4x^3 - 136x^2 + 1147x$$
$$4x^3 - 136x^2 + 1147x - 2770 = 0 \qquad \text{simplify with terms on the left}$$

We evaluate $f(x) = 4x^3 - 136x^2 + 1147x - 2770$ for integral values of x and find that $f(4) = -102$ and $f(5) = 65$. This means that $f(x) = 0$ between $x = 4$ and $x = 5$. By entering $f(x)$ as Y_1 in a graphing calculator, we can evaluate the function as shown in the calculator display in Fig. 15-9. This shows the following values:

$$f(4) = -102$$
$$f(5) = 65 \qquad \text{different signs means at least one root between 4 and 5}$$
$$f(4.4) = -15.42 \qquad \text{one root between 4.4 and 4.5}$$
$$f(4.5) = 2$$
$$f(4.48) = -1.353 \qquad \text{root between 4.48 and 4.49, closer to 4.49}$$
$$f(4.49) = 0.3318$$

From these values, $x = 4.49$ cm, to three significant digits. This value gives a volume

$$V = 4.49(37.0 - 2(4.49))(31.0 - 2(4.49)) = 2270 \text{ cm}^3$$

which means that the solution checks.

Checking values greater than $x = 5$, we find that there is another root between 6 and 7. Using the same procedure as above, we find that $x = 6.80$ cm is also a solution. Therefore, there are two possible squares that can be cut from each corner to get a bottom box with a volume of 2770 cm³.

EXERCISES *15-4*

In Exercises 1–20, solve the given equations without using a graphing calculator.

1. $x^3 + 2x^2 - x - 2 = 0$ **2.** $x^3 + x^2 - 5x + 3 = 0$

3. $x^3 + 2x^2 - 5x - 6 = 0$ **4.** $t^3 - 12t - 16 = 0$

5. $2x^3 - 5x^2 - 28x + 15 = 0$

6. $2x^3 - x^2 - 3x - 1 = 0$

7. $3x^3 + 11x^2 + 5x - 3 = 0$

8. $4x^3 - 5x^2 - 23x + 6 = 0$

9. $x^4 - 11x^2 - 12x + 4 = 0$

10. $x^4 + x^3 - 2x^2 - 4x - 8 = 0$

11. $x^4 - 2x^3 - 13x^2 + 14x + 24 = 0$

12. $x^4 - x^3 + 2x^2 - 4x - 8 = 0$

13. $2x^4 - 5x^3 - 3x^2 + 4x + 2 = 0$

14. $4n^4 - 17n^2 + 14n - 3 = 0$

15. $12x^4 + 44x^3 + 21x^2 - 11x - 6 = 0$

16. $9x^4 - 3x^3 + 34x^2 - 12x - 8 = 0$

17. $D^5 + D^4 - 9D^3 - 5D^2 + 16D + 12 = 0$

18. $x^6 - x^4 - 14x^2 + 24 = 0$

19. $2x^5 - 5x^4 + 6x^3 - 6x^2 + 4x - 1 = 0$

20. $2x^5 + 5x^4 - 4x^3 - 19x^2 - 16x - 4 = 0$

In Exercises 21–24, use a graphing calculator to solve the given equations to the nearest 0.01.

21. $x^3 - 2x^2 - 5x + 4 = 0$

22. $x^4 - x^3 - 2x^2 - x - 3 = 0$

23. $x^4 + 2x^3 - 3x^2 + 2x - 4 = 0$

24. $2x^5 - 3x^4 + 8x^3 - 4x^2 - 4x + 2 = 0$

In Exercises 25–28, use a calculator to find the irrational root (to the nearest 0.01) that lies between the given values. (See Example 8.)

25. $x^3 - 6x^2 + 10x - 4 = 0$ (0 and 1)

26. $r^4 - r^3 - 3r^2 - r - 4 = 0$ (2 and 3)

27. $3x^3 + 13x^2 + 3x - 4 = 0$ (−1 and 0)

28. $3x^4 - 3x^3 - 11x^2 - x - 4 = 0$ (−2 and −1)

In Exercises 29–44, solve the given problems. Use a graphing calculator in Exercises 41–43.

29. Where does the graph of the function
$f(x) = 4x^3 + 3x^2 - 20x - 15$ cross the x-axis?

30. Where does the graph of the function
$f(s) = 2s^4 - s^3 - 5s^2 + 7s - 6$ cross the s-axis?

31. The angular acceleration α (in rad/s^2) of the wheel of a car is given by $\alpha = -0.2t^3 + t^2$, where t is the time (in s). For what values of t is $\alpha = 2.0$ rad/s^2?

32. In finding one of the dimensions d (in in.) of the support columns of a building, the equation
$3d^3 + 5d^2 - 400d - 18{,}000 = 0$ is found. What is this dimension?

33. The deflection y of a beam at a horizontal distance x from one end is given by $y = k(x^4 - 2Lx^3 + L^3x)$, where L is the length of the beam and k is a constant. For what values of x is the deflection zero?

34. The specific gravity s of a sphere of radius r that sinks to a depth h in water is given by $s = \dfrac{3rh^2 - h^3}{4r^3}$. Find the depth to which a spherical buoy of radius 4.0 cm sinks if $s = 0.50$.

35. A company found that the cost (in $) of producing x lb of a sealant for a space vehicle is given by $C = 8x^3 - 36x^2 + 90$. Find x for $C = \$36$.

36. The angle θ of a robot arm with the horizontal as a function of time t (in s) is given by $\theta = 15 + 20t^2 - 4t^3$ for $0 \le t \le 5$ s. Find t for $\theta = 40°$.

37. The radii of four different-sized ball bearings differ by 1.00 mm in radius from one size to the next. If the volume of the largest equals the volumes of the other three combined, find the radii.

38. The edge of one cubical block of steel is 1.0 in. longer than the edge of another block. Find the length of the edge of each if the sum of their volumes is 91.0 in^3.

39. For electrical resistors connected in parallel, the reciprocal of the combined resistance equals the sum of the reciprocals of the individual resistances. If three resistors are connected in parallel such that the second resistance is 1 Ω more than the first and the third is 4 Ω more than the first, find the resistances for a combined resistance of 1 Ω.

40. Each of three revolving doors has a perimeter of 6.60 m and revolves through a volume of 9.50 m^3 in one revolution about their common vertical side. What are the door's dimensions?

41. A rectangular tray is made from a square piece of sheet metal 10.0 cm on a side by cutting equal squares from each corner, bending up the sides, and then welding them together. How long is the side of the square that must be cut out if the volume of the tray is 70.0 cm^3?

42. A variable electric voltage in a circuit is given by
$V = 0.1t^4 - 1.0t^3 + 3.5t^2 - 5.0t + 2.3$, where t is the time (in s). If the voltage is on for 5.0 s, when is $V = 0$?

43. The pressure difference p (in kPa) at a distance x (in km) from one end of an oil pipeline is given by $p = x^5 - 3x^4 - x^2 + 7x$. If the pipeline is 4 km long, where is $p = 0$?

Ⓦ **44.** An equation $f(x) = 0$ involves only odd powers of x with positive coefficients. Explain why this equation has no real root except $x = 0$.

CHAPTER EQUATIONS

Polynomial function	$f(x) = a_0 x^n + a_1 x^{n-1} + \cdots + a_n$	(15-1)
Remainder theorem	$f(x) = (x - r)q(x) + R$	(15-2)
	$f(r) = R$	(15-3)
Rational roots	$r_r = \dfrac{\text{integral factor of } a_n}{\text{integral factor of } a_0}$	(15-4)

REVIEW EXERCISES

In Exercises 1–4, find the remainder of the indicated division by the remainder theorem.

1. $(2x^3 - 4x^2 - x + 4) \div (x - 1)$

2. $(x^3 - 2x^2 + 9) \div (x + 2)$

3. $(4x^3 + x + 4) \div (x + 3)$

4. $(x^4 - 5x^3 + 8x^2 + 15x - 2) \div (x - 3)$

In Exercises 5–8, use the factor theorem to determine whether or not the second expression is a factor of the first.

5. $x^4 + x^3 + x^2 - 2x - 3; \quad x + 1$

6. $2s^3 - 6s - 4; \quad s - 2$

7. $x^4 + 4x^3 + 5x^2 + 5x - 6; \quad x + 3$

8. $9x^3 + 6x^2 + 4x + 2; \quad 3x + 1$

In Exercises 9–16, use synthetic division to perform the indicated divisions.

9. $(x^3 + 3x^2 + 6x + 1) \div (x - 1)$

10. $(3x^3 - 2x^2 + 7) \div (x - 3)$

11. $(2x^3 - 3x^2 - 4x + 3) \div (x + 2)$

12. $(3D^3 + 8D^2 - 16) \div (D + 4)$

13. $(x^4 - 2x^3 - 3x^2 - 4x - 8) \div (x + 1)$

14. $(x^4 - 6x^3 + x - 8) \div (x - 3)$

15. $(2x^5 - 46x^3 + x^2 - 9) \div (x - 5)$

16. $(x^6 + 63x^3 + 5x^2 - 9x - 8) \div (x + 4)$

In Exercises 17–20, use synthetic division to determine whether or not the given numbers are zeros of the given functions.

17. $y^3 + 5y^2 - 6 = 0; \quad -3$

18. $2x^3 + x^2 - 4x + 4; \quad -2$

19. $2x^4 - x^3 + 2x^2 + x - 1; \quad \frac{1}{2}$

20. $6x^4 - 7x^3 + 2x^2 - 9x - 6; \quad -\frac{2}{3}$

In Exercises 21–32, find all the roots of the given equations, using synthetic division and the given roots.

21. $x^3 + 8x^2 + 17x + 6 = 0 \quad (r_1 = -3)$

22. $2x^3 + 7x^2 - 6x - 8 = 0 \quad (r_1 = -4)$

23. $3x^4 + 5x^3 + x^2 + x - 10 = 0 \quad (r_1 = 1, r_2 = -2)$

24. $x^4 - x^3 - 5x^2 - x - 6 = 0 \quad (r_1 = 3, r_2 = -2)$

25. $4p^4 - p^2 - 18p + 9 = 0 \quad (r_1 = \frac{1}{2}, r_2 = \frac{3}{2})$

26. $x^4 + x^3 - 11x^2 - 9x + 18 = 0 \quad (r_1 = -3, r_2 = 1)$

27. $4x^4 + 4x^3 + x^2 + 4x - 3 = 0 \quad (r_1 = j)$

28. $x^4 + 2x^3 - 4x - 4 = 0 \quad (r_1 = -1 + j)$

29. $x^5 + 3x^4 - x^3 - 11x^2 - 12x - 4 = 0 \quad (-1 \text{ is a triple root})$

30. $24x^5 + 10x^4 + 7x^2 - 6x + 1 = 0 \quad (r_1 = -1, r_2 = \frac{1}{4}, r_3 = \frac{1}{3})$

31. $x^5 + 4x^4 + 5x^3 - x^2 - 4x - 5 = 0 \quad (r_1 = 1,$
$\quad r_2 = -2 + j)$

32. $2x^5 - x^4 + 8x - 4 = 0 \quad (r_1 = \frac{1}{2}, r_2 = 1 + j)$

In Exercises 33–40, solve the given equations.

33. $x^3 + x^2 - 10x + 8 = 0$

34. $x^3 - 8x^2 + 20x - 16 = 0$

35. $2R^3 - 3R^2 + 1 = 0$

36. $2x^3 - 3x^2 - 11x + 6 = 0$

37. $6x^3 - x^2 - 12x - 5 = 0$

38. $6x^3 + 19x^2 + 2x - 3 = 0$

39. $4t^4 - 17t^2 + 14t - 3 = 0$

40. $2x^4 + 5x^3 - 14x^2 - 23x + 30 = 0$

In Exercises 41–56, solve the given problems. Where appropriate, set up the required equations.

W **41.** Explain how to find k if $x + 2$ is a factor of $f(x) = 3x^3 + kx^2 - 8x - 8$. What is k?

W **42.** Explain how to find k if $x - 3$ is a factor of $f(x) = kx^4 - 15x^2 - 5x - 12$. What is k?

43. Where does the graph of the function $f(x) = 6x^4 - 14x^3 + 5x^2 + 5x - 2$ cross the x-axis?

44. Where does the graph of the function $f(x) = 2x^4 - 7x^3 + 11x^2 - 28x + 12$ cross the x-axis?

45. Find the irrational root of the equation $3x^3 - x^2 - 8x - 2 = 0$ that lies between 1 and 2.

46. Find the irrational root of the equation $x^4 + 3x^3 + 6x + 4 = 0$ that lies between -1 and 0.

47. A computer analysis of the number of crimes committed each month in a certain city for the first 10 months of a year showed that $n = x^3 - 9x^2 + 15x + 600$. Here n is the number of monthly crimes and x is the number of the month (as of the last day). In what month were 580 crimes committed?

48. A company determined that the number s (in thousands) of computer chips that it could supply at a price p of less than \$5 is given by $s = 4p^2 - 25$, whereas the demand d (in thousands) for the chips is given by $d = p^3 - 22p + 50$. For what price is the supply equal to the demand?

49. In order to find the diameter d (in cm) of a helical spring subject to given forces, it is necessary to solve the equation $64d^3 - 144d^2 + 108d - 27 = 0$. Solve for d.

50. A cubical tablet for purifying water is wrapped in a sheet of foil 0.500 mm thick. The total volume of tablet and foil is 33.1% greater than the volume of the tablet alone. Find the length of the edge of the tablet.

51. For the mirror shown in Fig. 15-10, the reciprocal of the focal distance f equals the sum of the reciprocals of the object dis-

tance p (in in.) and image distance q (in in.). If $q = p + 4$ and $f = \dfrac{p + 1}{p}$, find p.

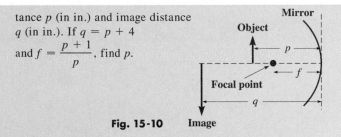

Fig. 15-10

52. Three electric capacitors are connected in series. The capacitance of the second is 1 μF more than that of the first, and the third is 2 μF more than the second. The capacitance of the combination is 1.33 μF. The equation used to determine C, the capacitance of the first capacitor, is

$$\frac{1}{C} + \frac{1}{C + 1} + \frac{1}{C + 3} = \frac{3}{4}$$

Find the values of the capacitances.

53. The height of a cylindrical oil tank is 3.2 m more than the radius. If the volume of the tank is 680 m^3, what are the radius and the height of the tank?

54. A grain storage bin has a square base, each side of which is 5.5 m longer than the height of the bin. If the bin holds 160 m^3 of grain, find its dimensions.

55. A rectangular door has a diagonal brace that is 0.900 ft longer than the height of the door. If the area of the door is 24.3 ft^2, find its dimensions.

56. The radius of one ball bearing is 1.0 mm greater than the radius of a second ball bearing. If the sum of their volumes is 100 mm^3, find the radius of each.

Writing Exercise

57. A computer science student is to write a computer program that will print out the values of n for which $x + r$ is a factor of $x^n + r^n$. Write a paragraph that states which are the values of n and explains how they are found.

PRACTICE TEST

1. Is -3 a zero for the function $2x^3 + 3x^2 + 7x - 6$?

2. Find the remaining roots of the equation $x^4 - 2x^3 - 7x^2 + 20x - 12 = 0$; 2 is a double root.

3. Use synthetic division to perform the division $(x^3 - 5x^2 + 4x - 9) \div (x - 3)$.

4. Use the factor theorem and synthetic division to determine whether or not $2x + 1$ is a factor of $2x^4 + 15x^3 + 23x^2 - 16$.

5. Use the remainder theorem to find the remainder of the division $(x^3 + 4x^2 + 7x - 9) \div (x + 4)$.

6. Solve for x: $2x^4 - x^3 + 5x^2 - 4x - 12 = 0$.

7. The ends of a 10-ft beam are supported at different levels. The deflection y of the beam is given by $y = kx^2(x^3 + 436x - 4000)$, where x is the horizontal distance from one end and k is a constant. Find the values of x for which the deflection is zero.

8. A cubical metal block is heated such that its edge increases by 1.0 mm and its volume is doubled. Find the edge of the cube to tenths.

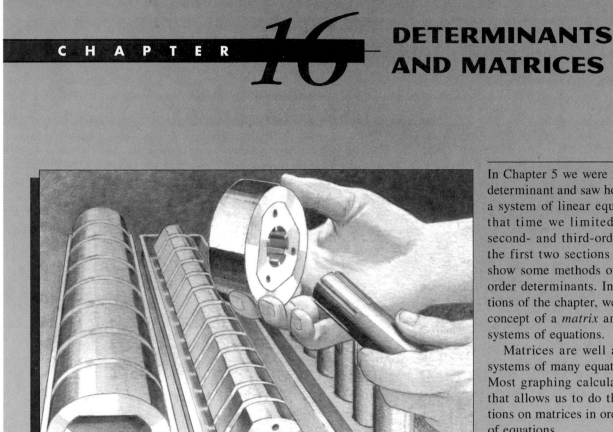 DETERMINANTS AND MATRICES

In Chapter 5 we were first introduced to a determinant and saw how it is used to solve a system of linear equations. However, at that time we limited our discussion to second- and third-order determinants. In the first two sections of this chapter, we show some methods of evaluating higher-order determinants. In the remaining sections of the chapter, we develop the related concept of a *matrix* and its use in solving systems of equations.

Matrices are well adapted for solving systems of many equations on a computer. Most graphing calculators have a feature that allows us to do the necessary operations on matrices in order to solve a system of equations.

Methods developed in this chapter are used in solving applied problems in areas such as electric circuits and analysis of forces. Also, these methods are used extensively in business and industry in making appropriate decisions for research, development, and production.

In Section 16-4 we see how to determine the amount of material and worker time required for the production of machine parts.

16-1 DETERMINANTS; EXPANSION BY MINORS

From Section 5-7 we recall that a *third-order* **determinant** *is defined by the equation*

$$\begin{vmatrix} a_1 & b_1 & c_1 \\ a_2 & b_2 & c_2 \\ a_3 & b_3 & c_3 \end{vmatrix} = a_1b_2c_3 + a_3b_1c_2 + a_2b_3c_1 - a_3b_2c_1 - a_1b_3c_2 - a_2b_1c_3 \qquad \textbf{(16-1)}$$

In Eq. (16-1), if we rearrange the terms on the right and factor a_1, $-a_2$, and a_3 from the terms in which they are contained, we have

$$\begin{vmatrix} a_1 & b_1 & c_1 \\ a_2 & b_2 & c_2 \\ a_3 & b_3 & c_3 \end{vmatrix} = a_1(b_2c_3 - b_3c_2) - a_2(b_1c_3 - b_3c_1) + a_3(b_1c_2 - b_2c_1) \qquad \textbf{(16-2)}$$

Recalling the definition of a second-order determinant, we have

$$\begin{vmatrix} a_1 & b_1 & c_1 \\ a_2 & b_2 & c_2 \\ a_3 & b_3 & c_3 \end{vmatrix} = a_1 \begin{vmatrix} b_2 & c_2 \\ b_3 & c_3 \end{vmatrix} - a_2 \begin{vmatrix} b_1 & c_1 \\ b_3 & c_3 \end{vmatrix} + a_3 \begin{vmatrix} b_1 & c_1 \\ b_2 & c_2 \end{vmatrix} \qquad \textbf{(16-3)}$$

In Eq. (16-3) we note that the third-order determinant is expanded with the terms of the expansion as products of the elements of the first column and specific second-order determinants. In each case the elements of the second-order determinant are those elements that are in neither the same row nor the same column as the element from the first column. *These determinants are called* **minors.**

In general, *the minor of a given element of a determinant is the determinant that results by deleting the row and the column in which the element lies.* Consider the following example.

■EXAMPLE 1

Consider the determinant $\begin{vmatrix} 1 & 2 & 3 \\ 4 & 5 & 6 \\ 7 & 8 & 9 \end{vmatrix}$.

We find the minor of the element 1 by deleting the elements in the first row and first column because the element 1 is located in the first row and in the first column. The minor for the element 6 is formed by deleting the elements in the second row and in the third column, for this is the location of the 6. These minors are shown below.

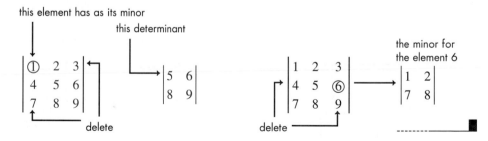

We now see that Eq. (16-3) expresses the expansion of a third-order determinant as the sum of the products of the elements of the first column and their minors, with the second term assigned a minus sign. Actually this is only one of several ways of expressing the expansion. However, it does lead to a general theorem regarding the expansion of a determinant of any order. The foregoing provides a basis for this theorem, although it cannot be considered as a proof. The theorem is given on the next page.

Expansion of a Determinant by Minors

The value of a determinant of order n may be found by forming the n products of the elements of any column (or row) and their minors. A product is given a plus sign if the sum of the number of the column and the number of the row in which the element lies is even, and a minus sign if this sum is odd. The algebraic sum of the terms thus obtained is the value of the determinant.

EXAMPLE 2 Evaluate $\begin{vmatrix} 1 & -3 & -2 \\ 4 & -1 & 0 \\ 4 & 3 & -5 \end{vmatrix}$ by expansion by minors.

Since we may expand by any column or row, using the first row, we have:

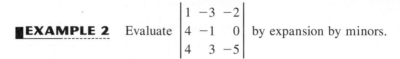

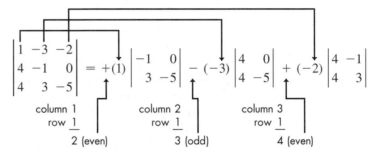

We note that once the first sign has been determined, the others are known since the signs alternate from term to term. Now, using Eq. (5-9), we have

$$\begin{vmatrix} 1 & -3 & -2 \\ 4 & -1 & 0 \\ 4 & 3 & -5 \end{vmatrix} = +(1)(5 - 0) - (-3)(-20 - 0) + (-2)[12 - (-4)]$$

$$= 1(5) + 3(-20) - 2(16) = 5 - 60 - 32 = -87$$ ▪

For reference, Eq. (5-9) is $\begin{vmatrix} a_1 & b_1 \\ a_2 & b_2 \end{vmatrix} = a_1 b_2 - a_2 b_1$.

EXAMPLE 3 Evaluate $\begin{vmatrix} 3 & -2 & 0 & 2 \\ 1 & 0 & -1 & 4 \\ -3 & 1 & 2 & -2 \\ 2 & -1 & 0 & -1 \end{vmatrix}$.

Expanding by the third column (it has two zeros), we have

$$\begin{vmatrix} 3 & -2 & 0 & 2 \\ 1 & 0 & -1 & 4 \\ -3 & 1 & 2 & -2 \\ 2 & -1 & 0 & -1 \end{vmatrix} = +(0)\begin{vmatrix} 1 & 0 & 4 \\ -3 & 1 & -2 \\ 2 & -1 & -1 \end{vmatrix} - (-1)\begin{vmatrix} 3 & -2 & 2 \\ -3 & 1 & -2 \\ 2 & -1 & -1 \end{vmatrix} + (2)\begin{vmatrix} 3 & -2 & 2 \\ 1 & 0 & 4 \\ 2 & -1 & -1 \end{vmatrix} - (0)\begin{vmatrix} 3 & -2 & 2 \\ 1 & 0 & 4 \\ -3 & 1 & -2 \end{vmatrix}$$

$$= \begin{vmatrix} 3 & -2 & 2 \\ -3 & 1 & -2 \\ 2 & -1 & -1 \end{vmatrix} + 2\begin{vmatrix} 3 & -2 & 2 \\ 1 & 0 & 4 \\ 2 & -1 & -1 \end{vmatrix}$$

$$= \left[3\begin{vmatrix} 1 & -2 \\ -1 & -1 \end{vmatrix} - (-3)\begin{vmatrix} -2 & 2 \\ -1 & -1 \end{vmatrix} + 2\begin{vmatrix} -2 & 2 \\ 1 & -2 \end{vmatrix} \right] + 2\left[-(-2)\begin{vmatrix} 1 & 4 \\ 2 & -1 \end{vmatrix} + 0\begin{vmatrix} 3 & 2 \\ 2 & -1 \end{vmatrix} - (-1)\begin{vmatrix} 3 & 2 \\ 1 & 4 \end{vmatrix} \right]$$

expanding first determinant by first column   expanding second determinant by second column

$$= [3(-1 - 2) + 3(2 + 2) + 2(4 - 2)] + 2[2(-1 - 8) + (12 - 2)] = [-9 + 12 + 4] + 2[-18 + 10]$$

$$= 7 + 2(-8) = -9$$ ▪

We see in Examples 2 and 3 that minors can be expanded as determinants or that they can be expanded by minors. These examples also illustrate that expansion by minors effectively reduces by one the order of the determinant being evaluated.

Solving Systems of Linear Equations by Determinants

We can use the expansion of determinants by minors to solve systems of linear equations. **Cramer's rule** *for solving systems of linear equations, as stated in Section 5-7, is valid for any system of n equations in n unknowns.*

SOLVING A WORD PROBLEM

▌**EXAMPLE 4**   Production of a certain computer component is done in three stages, taking a total of 7 h. The second stage is 1 h less than the first, and the third stage is twice as long as the second. How long is each stage of production?

First, we let a = the number of hours of the first production stage, b = the number of hours of the second stage, and c = the number of hours of the third stage.

The total production time of 7 h gives us $a + b + c = 7$. Since the second stage is 1 h less than the first, we then have $b = a - 1$. The fact that the third stage is twice as long as the second gives us $c = 2b$. Stating these equations in standard form for solution, we have

$$a + b + c = 7$$
$$a - b \phantom{{}+c} = 1$$
$$\phantom{a+{}}2b - c = 0$$

Using Cramer's rule, we have

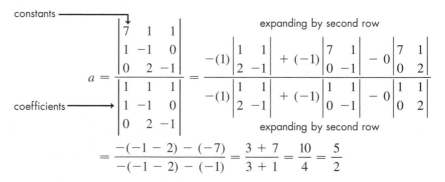

$$= \frac{-(-1 - 2) - (-7)}{-(-1 - 2) - (-1)} = \frac{3 + 7}{3 + 1} = \frac{10}{4} = \frac{5}{2}$$

Here we expanded each determinant by minors of the second row. Now, using the second equation, we have $\frac{5}{2} - b = 1$, or $b = \frac{3}{2}$. Using the third equation, we have $2(\frac{3}{2}) - c = 0$, or $c = 3$. Thus,

$$a = \frac{5}{2}\,\text{h} \qquad b = \frac{3}{2}\,\text{h} \qquad c = 3\,\text{h}$$

Therefore, the first stage takes 2.5 h, the second 1.5 h, and the third 3.0 h. (We are using two significant digits here, although for convenience we used only one in the equations.) These times agree with the given information.   ▬

NOTE▶   In Example 4 we found a by determinants and then found the other values by substituting into the equations. Normally, *as soon as we find the values of all but one of the unknowns in any equation, we then find the value of this unknown by substitution.* This is usually easier than evaluating additional determinants.

EXAMPLE 5   Solve the following system of equations.

$$
\begin{aligned}
x + 2y + z \quad\quad &= 5 \\
2x \quad\quad + z + 2t &= 1 \\
x - y + 3z + 4t &= -6 \\
4x - y \quad\quad - 2t &= 0
\end{aligned}
$$

constants

$$
x = \frac{\begin{vmatrix} 5 & 2 & 1 & 0 \\ 1 & 0 & 1 & 2 \\ -6 & -1 & 3 & 4 \\ 0 & -1 & 0 & -2 \end{vmatrix}}{\begin{vmatrix} 1 & 2 & 1 & 0 \\ 2 & 0 & 1 & 2 \\ 1 & -1 & 3 & 4 \\ 4 & -1 & 0 & -2 \end{vmatrix}}
$$

expanding by fourth row

$$
= \frac{-(0)\begin{vmatrix} 2 & 1 & 0 \\ 0 & 1 & 2 \\ -1 & 3 & 4 \end{vmatrix} + (-1)\begin{vmatrix} 5 & 1 & 0 \\ 1 & 1 & 2 \\ -6 & 3 & 4 \end{vmatrix} - (0)\begin{vmatrix} 5 & 2 & 0 \\ 1 & 0 & 2 \\ -6 & -1 & 4 \end{vmatrix} + (-2)\begin{vmatrix} 5 & 2 & 1 \\ 1 & 0 & 1 \\ -6 & -1 & 3 \end{vmatrix}}{(1)\begin{vmatrix} 0 & 1 & 2 \\ -1 & 3 & 4 \\ -1 & 0 & -2 \end{vmatrix} - 2\begin{vmatrix} 2 & 1 & 2 \\ 1 & 3 & 4 \\ 4 & 0 & -2 \end{vmatrix} + (1)\begin{vmatrix} 2 & 0 & 2 \\ 1 & -1 & 4 \\ 4 & -1 & -2 \end{vmatrix} - (0)\begin{vmatrix} 2 & 0 & 1 \\ 1 & -1 & 3 \\ 4 & -1 & 0 \end{vmatrix}}
$$

expanding by first row

$$
= \frac{-(-26) - 2(-14)}{1(0) - 2(-18) + 1(18)} = \frac{26 + 28}{36 + 18} = \frac{54}{54} = 1
$$

In solving for *x,* we evaluated the determinant in the numerator by expanding by the minors of the fourth row, since it contained two zeros. We evaluated the determinant in the denominator by expanding by the minors of the first row. Now we solve for *y* and again note two zeros in the fourth row of the determinant of the numerator.

$$
y = \frac{\begin{vmatrix} 1 & 5 & 1 & 0 \\ 2 & 1 & 1 & 2 \\ 1 & -6 & 3 & 4 \\ 4 & 0 & 0 & -2 \end{vmatrix}}{54}
$$

expanding by fourth row

$$
= \frac{-4\begin{vmatrix} 5 & 1 & 0 \\ 1 & 1 & 2 \\ -6 & 3 & 4 \end{vmatrix} + (-2)\begin{vmatrix} 1 & 5 & 1 \\ 2 & 1 & 1 \\ 1 & -6 & 3 \end{vmatrix}}{54}
$$

$$
= \frac{-4(-26) - 2(-29)}{54} = \frac{104 + 58}{54} = \frac{162}{54} = 3
$$

Substituting these values for *x* and *y* into the first equation, we can solve for *z.* This gives

$$
(1) + 2(3) + z = 5, \quad\quad z = -2
$$

Again, substituting the values for *x* and *y* into the fourth equation, we can find the value of *t.* This gives

$$
4(1) - (3) - 2t = 0, \quad\quad t = \tfrac{1}{2}
$$

Therefore, the required solution is

$$
x = 1 \quad\quad y = 3 \quad\quad z = -2 \quad\quad t = \tfrac{1}{2}
$$

We can check this solution by substituting these values into either the second or third equation (we used the first and fourth to *find* values of *z* and *t*). ■

—— EXERCISES *16-1* ——

In Exercises 1–14, evaluate the given determinants by expansion by minors.

1. $\begin{vmatrix} 3 & 0 & 0 \\ -2 & 1 & 4 \\ 4 & -2 & 5 \end{vmatrix}$

2. $\begin{vmatrix} 10 & 0 & -3 \\ -2 & -4 & 1 \\ 3 & 0 & 2 \end{vmatrix}$

3. $\begin{vmatrix} -2 & -4 & 2 \\ 1 & 3 & 0 \\ -4 & 5 & 2 \end{vmatrix}$

4. $\begin{vmatrix} 5 & -1 & 2 \\ 8 & 3 & -4 \\ 0 & 2 & -6 \end{vmatrix}$

5. $\begin{vmatrix} -6 & -1 & 3 \\ 2 & -2 & -3 \\ 10 & 1 & -2 \end{vmatrix}$

6. $\begin{vmatrix} 9 & -3 & 1 \\ -1 & 2 & -1 \\ 2 & -1 & 3 \end{vmatrix}$

7. $\begin{vmatrix} -30 & -25 & 54 \\ 12 & 21 & -14 \\ 37 & -46 & 24 \end{vmatrix}$

8. $\begin{vmatrix} 4.8 & -3.7 & 3.6 \\ -3.8 & 5.7 & 6.5 \\ 2.1 & -1.8 & 2.3 \end{vmatrix}$

9. $\begin{vmatrix} 1 & 0 & 1 & 0 \\ 2 & 4 & -3 & 1 \\ 1 & 1 & 1 & 1 \\ 3 & 5 & 0 & 2 \end{vmatrix}$

10. $\begin{vmatrix} 2 & 0 & 3 & 1 \\ -1 & -1 & 4 & 0 \\ 1 & 2 & 1 & 2 \\ 3 & 3 & -2 & -1 \end{vmatrix}$

11. $\begin{vmatrix} 1 & 2 & -1 & -2 \\ 3 & 1 & 2 & 1 \\ -1 & 3 & -1 & 2 \\ 2 & 1 & 3 & -3 \end{vmatrix}$

12. $\begin{vmatrix} 3 & -1 & 2 & -5 \\ 1 & 4 & 2 & 5 \\ -1 & 1 & 1 & 3 \\ 1 & 2 & -1 & -2 \end{vmatrix}$

13. $\begin{vmatrix} 1 & 2 & 1 & 2 & 1 \\ 1 & 0 & 0 & 1 & 0 \\ 0 & 1 & 1 & 0 & 1 \\ 1 & 1 & 2 & 2 & 1 \\ 0 & 1 & 1 & 0 & 2 \end{vmatrix}$

14. $\begin{vmatrix} 3 & 1 & 1 & 1 & 2 \\ 1 & 1 & 0 & 0 & 1 \\ 1 & 1 & 2 & 2 & 3 \\ 0 & 2 & 1 & 0 & 3 \\ 1 & 1 & 0 & 1 & 0 \end{vmatrix}$

Ⓦ *In Exercises 15 and 16, determine the number of terms in the complete expansion of a determinant of the given order. Explain how this number was found.*

15. Fourth order

16. Sixth order

In Exercises 17–24, solve the given systems of equations by determinants. Evaluate the determinants by expansion by minors.

17. $2x + y + z = 6$
    $x - 2y + 2z = 10$
    $3x - y - z = 4$

18. $2x + y = -1$
    $4x - 2y - z = 5$
    $2x + 3y + 3z = -2$

19. $3u + 6v + 2w = -2$
    $u + 3v - 4w = 2$
    $2u - 3v - 2w = -2$

20. $x + 3y + z = 4$
    $2x - 6y - 3z = 10$
    $4x - 9y + 3z = 4$

21. $x + t = 0$
    $3x + y + z = -1$
    $2y - z + 3t = 1$
    $2z - 3t = 1$

22. $2x + y + z = 4$
    $2y - 2z - t = 3$
    $3y - 3z + 2t = 1$
    $6x - y + t = 0$

23. $x + 2y - z = 6$
    $y - 2z - 3t = -5$
    $3x - 2y + t = 2$
    $2x + y + z - t = 0$

24. $2p + 3r + s = 4$
    $p - 2r - 3s + 4t = -1$
    $3p + r + s - 5t = 3$
    $-p + 2r + s + 3t = 2$

In Exercises 25–28, solve the indicated systems by determinants, using methods of this section.

25. In applying Kirchhoff's laws (see Exercise 20 on page 156) to the electric circuit in Fig. 16-1, the following equations are found. Determine the indicated currents (in A).

    $I_A + I_B + I_C + I_D = 0$
    $2I_A - I_B = -2$
    $3I_C - 2I_D = 0$
    $I_B - 3I_C = 6$

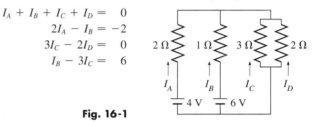

Fig. 16-1

26. In analyzing the motion of four equal particles that are equally spaced along a string, the equation shown below is found. Here, C depends on the string and the mass of each object. Solve for C ($C > 0$).

$$\begin{vmatrix} C & -1 & 0 & 0 \\ -1 & C & -1 & 0 \\ 0 & -1 & C & -1 \\ 0 & 0 & -1 & C \end{vmatrix} = 0$$

27. A firm plans to produce three different appliances, A, B, and C. Total production is to be 1500 appliances each week. A total of 3500 worker-hours are available for production, and 950 worker-hours are available for inspection. Each type A appliance requires 3 worker-hours for production and 1 worker-hour for inspection. Each type B appliance requires 2 worker-hours for production and 20 worker-minutes for inspection, and each type C appliance requires 2 worker-hours for production and 30 worker-minutes for inspection. How many of each are to be produced each week?

28. An alloy is to be made from four other alloys containing copper (Cu), nickel (Ni), zinc (Zn), and iron (Fe). The first is 80% Cu and 20% Ni. The second is 60% Cu, 20% Ni, and 20% Zn. The third is 30% Cu, 60% Ni, and 10% Fe. The fourth is 20% Ni, 40% Zn, and 40% Fe. How much is needed such that the final alloy has 56 g of Cu, 28 g of Ni, 10 g of Zn, and 6 g of Fe?

16-2 SOME PROPERTIES OF DETERMINANTS

Expansion of determinants by minors allows us to evaluate a determinant of any order. However, even a fourth-order determinant usually requires a great deal of calculational work to evaluate.

In this section we briefly present some basic properties of determinants with which they can be evaluated, usually with much less work than using minors. These properties illustrate how determinants can be evaluated without a special calculator or computer program. Following are these properties and an illustration of each.

1. *If each element above or each element below the principal diagonal of a determinant is zero, then the product of the elements of the principal diagonal is the value of the determinant.*

EXAMPLE 1   Following property 1, we have

$$\begin{vmatrix} 2 & 1 & 5 & 8 \\ 0 & -5 & 7 & 9 \\ 0 & 0 & 4 & -6 \\ 0 & 0 & 0 & 3 \end{vmatrix} = 2(-5)(4)(3) = -120$$

Since all the elements below the principal diagonal are zero, there is no need to expand the determinant. Note, however, that if the determinant is expanded by the first column, and successive determinants are expanded by their first columns, the same value is found.

2. *If **all** corresponding rows and columns of a determinant are interchanged, the value of the determinant is unchanged.*

EXAMPLE 2   Interchanging the first row and first column, the second row and second column, and the third row and third column of the determinant,

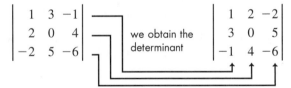

By expanding, we can show that the value of each determinant is −18. We obtain very similar expansions with equal values if we expand the first by the first column and the second by the first row.

3. *If two columns (or rows) of a determinant are identical, the value of the determinant is zero.*

EXAMPLE 3   Property 3 tells us that

identical $\begin{vmatrix} 3 & 5 & 2 \\ -4 & 6 & 9 \\ -4 & 6 & 9 \end{vmatrix} = 0$

since the second and third rows are identical. This is verified by expanding by minors of the first row. All of these minors have a value of zero.

4. *If two columns (or rows) of a determinant are interchanged, the value of the determinant is changed in sign.*

■**EXAMPLE 4**   The values of the determinants

$$\begin{vmatrix} 3 & 0 & 2 \\ 1 & 1 & 5 \\ 2 & 1 & 3 \end{vmatrix} \quad \text{and} \quad \begin{vmatrix} 2 & 0 & 3 \\ 5 & 1 & 1 \\ 3 & 1 & 2 \end{vmatrix}$$

differ in sign, since the first and third columns are interchanged. By expanding, we find that the value of the first determinant is -8 and the value of the second determinant is 8.   ■

5. *If all elements of a column (or row) are multiplied by the same number k, the value of the determinant is multiplied by k.*

■**EXAMPLE 5**   From Property 5 we know that

$$\begin{vmatrix} -1 & 0 & 6 \\ 6 & 3 & -6 \\ 0 & 5 & 3 \end{vmatrix} = 3 \begin{vmatrix} -1 & 0 & 6 \\ 2 & 1 & -2 \\ 0 & 5 & 3 \end{vmatrix}$$

since each element of the second row in the left determinant is three times the corresponding element of the second row in the right determinant and *the other rows are unchanged.*

By expansion, we can show that

$$\begin{vmatrix} -1 & 0 & 6 \\ 6 & 3 & -6 \\ 0 & 5 & 3 \end{vmatrix} = 141 \quad \text{and} \quad \begin{vmatrix} -1 & 0 & 6 \\ 2 & 1 & -2 \\ 0 & 5 & 3 \end{vmatrix} = 47$$

which verifies the original determinant equation, since $141 = 3(47)$.   ■

6. *If all the elements of any column (or row) are multiplied by the same number k, and the resulting numbers are added to the corresponding elements of another column (or row), the value of the determinant is unchanged.*

■**EXAMPLE 6**   The value of the following determinant is unchanged if we multiply each element of the first row by 2 and add these numbers to the corresponding elements of the second row. That is,

$$4 \times 2 = 8 \qquad -1 \times 2 = -2 \qquad 3 \times 2 = 6$$

$$\begin{vmatrix} 4 & -1 & 3 \\ 2 & 2 & 1 \\ 1 & 0 & -3 \end{vmatrix} = \begin{vmatrix} 4 & -1 & 3 \\ 2+8 & 2+(-2) & 1+6 \\ 1 & 0 & -3 \end{vmatrix} = \begin{vmatrix} 4 & -1 & 3 \\ 10 & 0 & 7 \\ 1 & 0 & -3 \end{vmatrix} \quad \text{or} \quad \begin{vmatrix} 4 & -1 & 3 \\ 10 & 0 & 7 \\ 1 & 0 & -3 \end{vmatrix} = \begin{vmatrix} 4 & -1 & 3 \\ 2 & 2 & 1 \\ 1 & 0 & -3 \end{vmatrix}$$

When each determinant is expanded, the value -37 is obtained. The great value in Property 6 is that by its use we can purposely place zeros in the resulting determinant.   ■

With the use of these six properties, determinants of higher order can be evaluated much more easily. The technique is to

obtain zeros in a given column (or row) in all positions except one.

We can then expand by this column (or row), thereby reducing the order of the determinant. Property 6 is probably the most valuable for obtaining the zeros.

█EXAMPLE 7   Using these properties, we evaluate the following determinant.

$$\begin{vmatrix} 3 & 2 & -1 & 1 \\ -1 & 1 & 2 & 3 \\ 2 & 2 & 1 & 4 \\ 0 & -1 & -2 & 2 \end{vmatrix} = \begin{vmatrix} 0 & 5 & 5 & 10 \\ -1 & 1 & 2 & 3 \\ 2 & 2 & 1 & 4 \\ 0 & -1 & -2 & 2 \end{vmatrix}$$

Each element of the second row is multiplied by 3, and the resulting numbers are added to the corresponding elements of the first row. Here we have used Property 6. In this way a zero has been placed in column 1, row 1.

$$= \begin{vmatrix} 0 & 5 & 5 & 10 \\ -1 & 1 & 2 & 3 \\ 0 & 4 & 5 & 10 \\ 0 & -1 & -2 & 2 \end{vmatrix}$$

Each element of the second row is multiplied by 2, and the resulting numbers are added to the corresponding elements of the third row. Again, we have used Property 6. Also, a zero has been placed in the first column, third row. We now have three zeros in the first column.

$$= -(-1) \begin{vmatrix} 5 & 5 & 10 \\ 4 & 5 & 10 \\ -1 & -2 & 2 \end{vmatrix}$$

Expand the determinant by the first column. We have now reduced the determinant to a third-order determinant.

$$= 5 \begin{vmatrix} 1 & 1 & 2 \\ 4 & 5 & 10 \\ -1 & -2 & 2 \end{vmatrix}$$

Factor 5 from each element of the first row. Here we are using Property 5.

$$= 5(2) \begin{vmatrix} 1 & 1 & 1 \\ 4 & 5 & 5 \\ -1 & -2 & 1 \end{vmatrix}$$

Factor 2 from each element of the third column. Again we are using Property 5. Also, by doing this we have reduced the size of the numbers, and the resulting numbers are somewhat easier to work with.

$$= 10 \begin{vmatrix} 1 & 1 & 1 \\ 0 & 1 & 1 \\ -1 & -2 & 1 \end{vmatrix}$$

Each element of the first row is multiplied by -4, and the resulting numbers are added to the corresponding elements of the second row. Here we are using Property 6. We have placed a zero in the first column, second row.

$$= 10 \begin{vmatrix} 1 & 1 & 1 \\ 0 & 1 & 1 \\ 0 & -1 & 2 \end{vmatrix}$$

Each element of the first row is added to the corresponding element of the third row. Again, we have used Property 6. A zero has been placed in the first column, third row. We now have two zeros in the first column.

$$= 10(1) \begin{vmatrix} 1 & 1 \\ -1 & 2 \end{vmatrix}$$

Expand the determinant by the first column.

$$= 10(2 + 1) = 30$$

Expand the second-order determinant.

A somewhat more systematic method is to place zeros below the principal diagonal and then use Property 1.

--------█

The methods of this section and that of expansion by minors illustrate the ways in which determinants were evaluated before the extensive use of computers and calculators. As we noted in Chapter 5, most graphing calculators can be used to quickly evaluate determinants. In fact, most can evaluate determinants up to the sixth- (or possibly higher) order determinants. Figure 16-2 shows the calculator evaluation of the fourth-order determinant of Example 7.

To display and evaluate a determinant on a calculator, see page 158.

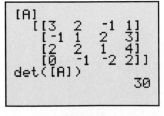

Fig. 16-2

EXERCISES $16\text{-}2$

In Exercises 1–4, evaluate each determinant by inspection. Observation will allow evaluation by using the properties of this section.

1. $\begin{vmatrix} 4 & -5 & 8 \\ 0 & 3 & -8 \\ 0 & 0 & -5 \end{vmatrix}$

2. $\begin{vmatrix} 3 & 0 & 0 \\ 0 & 10 & 0 \\ -9 & -1 & -5 \end{vmatrix}$

3. $\begin{vmatrix} 3 & -2 & 4 & 2 \\ 5 & -1 & 2 & -1 \\ 3 & -2 & 4 & 2 \\ 0 & 3 & -6 & 0 \end{vmatrix}$

4. $\begin{vmatrix} -12 & -24 & -24 & 15 \\ 12 & 32 & 32 & -35 \\ -22 & 18 & 18 & 18 \\ 44 & 0 & 0 & -26 \end{vmatrix}$

In Exercises 5–8, use the given value of the determinant at the right and the properties of this section to evaluate the following determinants.

$\begin{vmatrix} 2 & -3 & 1 \\ -4 & 1 & 3 \\ 1 & -3 & -2 \end{vmatrix} = 40$

5. $\begin{vmatrix} 2 & 1 & -3 \\ -4 & 3 & 1 \\ 1 & -2 & -3 \end{vmatrix}$

6. $\begin{vmatrix} 2 & -3 & 1 \\ -4 & 1 & 3 \\ 2 & -6 & -4 \end{vmatrix}$

7. $\begin{vmatrix} 2 & -3 & -1 \\ -4 & 1 & -3 \\ 1 & -3 & 2 \end{vmatrix}$

8. $\begin{vmatrix} 2 & -4 & 1 \\ -3 & 1 & -3 \\ 1 & 3 & -2 \end{vmatrix}$

In Exercises 9–16, evaluate the determinants, using the properties of this section. Do not evaluate directly more than one second-order determinant for each.

9. $\begin{vmatrix} 3 & 1 & 0 \\ -2 & 3 & -1 \\ 4 & 2 & 5 \end{vmatrix}$

10. $\begin{vmatrix} -4 & 3 & -2 \\ -2 & 2 & 4 \\ -1 & 5 & -3 \end{vmatrix}$

11. $\begin{vmatrix} 4 & 3 & 6 & 0 \\ 3 & 0 & 0 & 4 \\ 5 & 0 & 1 & 2 \\ 2 & 1 & 1 & 7 \end{vmatrix}$

12. $\begin{vmatrix} 6 & -3 & -6 & 3 \\ -2 & 1 & 2 & -1 \\ 18 & 7 & -1 & 5 \\ 0 & -1 & 10 & 10 \end{vmatrix}$

13. $\begin{vmatrix} 1 & 3 & -3 & 5 \\ 4 & 2 & 1 & 2 \\ 3 & 2 & -2 & 2 \\ 0 & 1 & 2 & -1 \end{vmatrix}$

14. $\begin{vmatrix} -2 & 2 & 1 & 3 \\ 1 & 4 & 3 & 1 \\ 4 & 3 & -2 & -2 \\ 3 & -2 & 1 & 5 \end{vmatrix}$

15. $\begin{vmatrix} 1 & 2 & 0 & 1 & 0 \\ 0 & 2 & 1 & 0 & 1 \\ 1 & 0 & -1 & 1 & -1 \\ -2 & 0 & -1 & 2 & 1 \\ 1 & 0 & 2 & -1 & -2 \end{vmatrix}$

16. $\begin{vmatrix} -1 & 3 & 5 & 0 & -5 \\ 0 & 1 & 7 & 3 & -2 \\ 5 & -2 & -1 & 0 & 3 \\ -3 & 0 & 2 & -1 & 3 \\ 6 & 2 & 1 & -4 & 2 \end{vmatrix}$

In Exercises 17–24, solve the given systems of equations by determinants. Evaluate each by the properties of this section.

17. $2x - y + z = 5$
$x + 2y + 3z = 10$
$3x + 3y + 2z = 5$

18. $2x + y + z = 5$
$x + 3y - 3z = -13$
$3x + 2y - z = -1$

19. $3x + 2y + z = 1$
$9x + 2z = 5$
$6x - 4y - z = 3$

20. $3a + b + 2c = 4$
$a - b + 4c = 2$
$6a + 3b - 2c = 10$

21. $2x + y + z = 2$
$3y - z + 2t = 4$
$y + 2z + t = 0$
$3x + 2z = 4$

22. $2x + y + z = 0$
$x - y + 2t = 2$
$2y + z + 4t = 2$
$5x + 2z + 2t = 4$

23. $D + E + 2F = 1$
$2D - E + G = -2$
$D - E - F - 2G = 4$
$2D - E + 2F - G = 0$

24. $3x + y + t = 0$
$3z + 2t = 8$
$6x + 2y + 2z + t = 3$
$3x - y - z - t = 0$

In Exercises 25–28, solve the given problems by using determinants.

25. In applying Kirchhoff's laws (see Exercise 20 on page 156) to the circuit shown in Fig. 16-3, the following equations are found. Determine the indicated currents (in A).

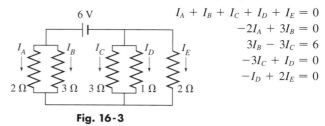

$I_A + I_B + I_C + I_D + I_E = 0$
$-2I_A + 3I_B = 0$
$3I_B - 3I_C = 6$
$-3I_C + I_D = 0$
$-I_D + 2I_E = 0$

Fig. 16-3

26. In analyzing the forces A, B, C, and D shown on the beam in Fig. 16-4, the following equations are used. Find these forces.

$A + B = 850$
$A + B + 400 = 0.8C + 0.6D$
$0.6C = 0.8D$
$5A - 5B + 4C - 3D = 0$

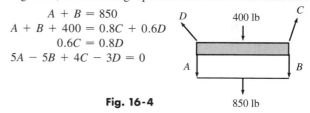

Fig. 16-4

27. In testing for air pollution, a given air sample contained 6.0 parts per million (ppm) of four pollutants, sulfur dioxide (SO_2), nitric oxide (NO), nitrogen dioxide (NO_2), and carbon monoxide (CO). The ppm of CO was 10 times that of SO_2, which in turn equaled those of NO and NO_2. There was a total of 0.8 ppm of SO_2 and NO. How many ppm of each were present in the air sample?

28. Three computer programs A, B, and C, use 15% of a computer's 6.0 GB (gigabyte) hard-drive memory. If two additional programs are added, one requiring the same memory as A and the other half the memory of C, 22% of the memory will be used. However, if two other programs are added to A, B, and C, one requiring half the memory of B and the other the same memory as C, 25% of the memory will be used. How many megabytes of memory are required for each of A, B, and C?

16-3 MATRICES: DEFINITIONS AND BASIC OPERATIONS

Systems of linear equations occur in several areas of important technical and scientific applications. We indicated a few of these in Chapter 5 and in the first two sections of this chapter. The importance of linear systems has led to the development of numerous methods for their solution.

As the use of computers has increased, another mathematical concept that we can use to solve systems of linear equations is now used much more widely than in the past. It is also used in numerous applications other than with systems of equations, in fields such as business, economics, and psychology, as well as the scientific and technical areas. Since it is readily adaptable to use on a computer and on many graphing calculators, its use will continue to increase. At this point, however, we shall only be able to introduce its definitions and basic operations.

Matrices were first used by the British mathematician Arthur Cayley (1821–1895).

We first introduced the term *matrix* on page 158 when we showed the evaluation of a determinant on a calculator.

A **matrix** *is an ordered rectangular array of numbers.* To distinguish such an array from a determinant, we shall enclose it within brackets. As with a determinant, *the individual members are called* **elements** *of the matrix.*

■**EXAMPLE 1**   Some examples of matrices are shown here.

$$\begin{bmatrix} 2 & 8 \\ 1 & 0 \end{bmatrix} \qquad \begin{bmatrix} 2 & -4 & 6 \\ -1 & 0 & 5 \end{bmatrix} \qquad \begin{bmatrix} 4 & 6 \\ 0 & -1 \\ -2 & 5 \\ 3 & 0 \end{bmatrix}$$

$$\begin{bmatrix} -1 & 8 & 6 & 7 & 9 \\ 2 & 6 & 0 & 4 & 3 \\ 5 & -1 & 8 & 10 & 2 \end{bmatrix} \qquad \begin{bmatrix} -1 & 2 & 0 & 9 \end{bmatrix}$$

As we can see, it is not necessary for the number of columns and number of rows to be the same, although such is the case for a determinant. However, *if the number of rows does equal the number of columns, the matrix is called a* **square matrix.** We shall find that square matrices are of special importance. *If all the elements of a matrix are zero, the matrix is called a* **zero matrix.** It is convenient to designate a given matrix by a capital letter.

CAUTION▶   We must be careful to distinguish between a matrix and a determinant. *A matrix is simply any* **rectangular array** *of numbers, whereas a determinant is a specific value associated with a* **square matrix.**

■**EXAMPLE 2**   Consider the following matrices:

$$A = \begin{bmatrix} 5 & 0 & -1 \\ 1 & 2 & 6 \\ 0 & -4 & -5 \end{bmatrix} \qquad B = \begin{bmatrix} 9 \\ 8 \\ 1 \\ 5 \end{bmatrix} \qquad C = \begin{bmatrix} -1 & 6 & 8 & 9 \end{bmatrix} \qquad O = \begin{bmatrix} 0 & 0 \\ 0 & 0 \end{bmatrix}$$

Matrix A is an example of a square matrix, matrix B is an example of a matrix with four rows and one column, matrix C is an example of a matrix with one row and four columns, and matrix O is an example of a zero matrix.

To be able to refer to specific elements of a matrix and to give a general representation, a double-subscript notation is usually employed. That is,

$$A = \begin{bmatrix} a_{11} & a_{12} & a_{13} \\ a_{21} & a_{22} & a_{23} \\ a_{31} & a_{32} & a_{33} \end{bmatrix}$$

row $\underset{\text{row}}{\overset{}{\frown}}$ column

We see that the first subscript refers to the row in which the element lies and the second subscript refers to the column in which the element lies.

NOTE ▶   *Two matrices are said to be* **equal** *if and only if they are identical.* That is, they must have the same number of columns, the same number of rows, and the elements must respectively be equal.

EXAMPLE 3   **(a)** $\begin{bmatrix} a_{11} & a_{12} & a_{13} \\ a_{21} & a_{22} & a_{23} \end{bmatrix} = \begin{bmatrix} 1 & -5 & 0 \\ 4 & 6 & -3 \end{bmatrix}$

if and only if $a_{11} = 1$, $a_{12} = -5$, $a_{13} = 0$, $a_{21} = 4$, $a_{22} = 6$, and $a_{23} = -3$.

(b) The matrices

$$\begin{bmatrix} 1 & 2 & 3 \\ -1 & -2 & -5 \end{bmatrix} \quad \text{and} \quad \begin{bmatrix} 1 & 2 & -5 \\ -1 & -2 & 3 \end{bmatrix}$$

are not equal, since the elements in the third column are reversed.

(c) The matrices

$$\begin{bmatrix} 2 & 3 \\ -1 & 5 \end{bmatrix} \quad \text{and} \quad \begin{bmatrix} 2 & 3 & 0 \\ -1 & 5 & 0 \end{bmatrix}$$

are not equal, since the number of columns is different. This is true despite the fact that both elements of the third column are zeros.   -------- ∎

EXAMPLE 4   The forces acting on a bolt are in equilibrium, as shown in Fig. 16-5. Analyzing the horizontal and vertical components as in Section 9-4, we find the following matrix equation. Find forces F_1 and F_2.

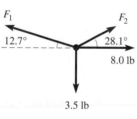

Fig. 16-5

$$\begin{bmatrix} 0.98F_1 - 0.88F_2 \\ 0.22F_1 + 0.47F_2 \end{bmatrix} = \begin{bmatrix} 8.0 \\ 3.5 \end{bmatrix}$$

From the equality of matrices, we know that $0.98F_1 - 0.88F_2 = 8.0$ and $0.22F_1 + 0.47F_2 = 3.5$. Therefore, to find the forces F_1 and F_2, we must solve the system of equations

$$0.98F_1 - 0.88F_2 = 8.0$$
$$0.22F_1 + 0.47F_2 = 3.5$$

Using determinants, we have

$$F_1 = \frac{\begin{vmatrix} 8.0 & -0.88 \\ 3.5 & 0.47 \end{vmatrix}}{\begin{vmatrix} 0.98 & -0.88 \\ 0.22 & 0.47 \end{vmatrix}} = \frac{8.0(0.47) - 3.5(-0.88)}{0.98(0.47) - 0.22(-0.88)} = 10.5 \text{ lb}$$

Using determinants again, or by substituting this value into either equation, we find that $F_2 = 2.6$ lb. These values check when substituted into the original matrix equation.   -------- ∎

Matrix Addition and Subtraction

*If two matrices have the same number of rows and the same number of columns, their **sum** is defined as the matrix consisting of the sums of the corresponding elements.* If the number of rows or the number of columns of the two matrices is not equal, they cannot be added.

EXAMPLE 5 (a)

$$\begin{bmatrix} 8 & 1 & -5 & 9 \\ 0 & -2 & 3 & 7 \end{bmatrix} + \begin{bmatrix} -3 & 4 & 6 & 0 \\ 6 & -2 & 6 & 5 \end{bmatrix} = \begin{bmatrix} 8 + (-3) & 1 + 4 & -5 + 6 & 9 + 0 \\ 0 + 6 & -2 + (-2) & 3 + 6 & 7 + 5 \end{bmatrix}$$

$$= \begin{bmatrix} 5 & 5 & 1 & 9 \\ 6 & -4 & 9 & 12 \end{bmatrix}$$

(b) The matrices

$$\begin{bmatrix} 3 & -5 & 8 \\ 2 & 9 & 0 \\ 4 & -2 & 3 \end{bmatrix} \text{ and } \begin{bmatrix} 3 & -5 & 8 & 0 \\ 2 & 9 & 0 & 0 \\ 4 & -2 & 3 & 0 \end{bmatrix}$$

cannot be added since the second matrix has one more column than the first matrix. This is true even though the extra column contains only zeros. ━━━━■

*The product of a number and a matrix (known as **scalar multiplication** of a matrix) is defined as the matrix whose elements are obtained by multiplying each element of the given matrix by the given number.* Thus, we obtain matrix kA by multiplying the elements of matrix A by k. In this way, $A + A$ and $2A$ are the same matrix.

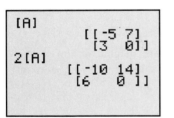

Fig. 16-6

EXAMPLE 6 For the matrix A, where

$$A = \begin{bmatrix} -5 & 7 \\ 3 & 0 \end{bmatrix} \text{ we have } 2A = \begin{bmatrix} 2(-5) & 2(7) \\ 2(3) & 2(0) \end{bmatrix} = \begin{bmatrix} -10 & 14 \\ 6 & 0 \end{bmatrix}$$

Figure 16-6 shows a calculator display for A and $2A$. ━━━━■

By combining the definitions for the addition of matrices and for the scalar multiplication of a matrix, we can define the subtraction of matrices. That is, *the **difference** of matrices A and B is given by $A - B = A + (-B)$.* Therefore, we change the sign of each element of B, and proceed as in addition.

The operations of addition, subtraction, and multiplication of a matrix by a number are like those for real numbers. For these operations, the algebra of matrices is like the algebra of real numbers. We see that the following laws hold for matrices.

$A + B = B + A$	(commutative law)	**(16-4)**
$A + (B + C) = (A + B) + C$	(associative law)	**(16-5)**
$k(A + B) = kA + kB$		**(16-6)**
$A + O = A$		**(16-7)**

Here, we have let O represent the zero matrix. We shall find in the next section that not all laws for matrix operations are like those for real numbers.

EXERCISES *16-3*

In Exercises 1–8, determine the value of the literal numbers in each of the given matrix equalities.

1. $\begin{bmatrix} a & b \\ c & d \end{bmatrix} = \begin{bmatrix} 1 & -3 \\ 4 & 7 \end{bmatrix}$

2. $\begin{bmatrix} x & y & z \\ r & -s & -t \end{bmatrix} = \begin{bmatrix} -2 & 7 & -9 \\ 4 & -4 & 5 \end{bmatrix}$

3. $\begin{bmatrix} x \\ x+y \end{bmatrix} = \begin{bmatrix} 2 \\ 5 \end{bmatrix}$

4. $[a + bj \quad 2c - dj \quad 3e + fj] = [5j \quad a+6 \quad 3b+c]$
$(j = \sqrt{-1})$

5. $\begin{bmatrix} x & x+y \\ x-z & y+z \\ x+t & y-t \end{bmatrix} = \begin{bmatrix} 2 & 3 \\ 4 & -1 \end{bmatrix}$

6. $\begin{bmatrix} 2x - 3y \\ x + 4y \end{bmatrix} = \begin{bmatrix} 13 \\ 1 \end{bmatrix}$

7. $\begin{bmatrix} C + D \\ 2C - D \\ D - 2E \end{bmatrix} = \begin{bmatrix} 5 \\ 4 \\ 6 \end{bmatrix}$

8. $\begin{bmatrix} x & y & z \\ x+y & 2x-y & x+2 \end{bmatrix} = \begin{bmatrix} 2 & -3 \\ z & t \end{bmatrix}$

In Exercises 9–12, find the indicated sums of matrices.

9. $\begin{bmatrix} 2 & 3 \\ -5 & 4 \end{bmatrix} + \begin{bmatrix} -1 & 7 \\ 5 & -2 \end{bmatrix}$

10. $\begin{bmatrix} 1 & 0 & 9 \\ 3 & -5 & -2 \end{bmatrix} + \begin{bmatrix} 4 & -1 & 7 \\ 2 & 0 & -3 \end{bmatrix}$

11. $\begin{bmatrix} 50 & -82 \\ -34 & 57 \\ -15 & 62 \end{bmatrix} + \begin{bmatrix} -55 & 82 \\ 45 & 14 \\ 26 & -67 \end{bmatrix}$

12. $\begin{bmatrix} 4.7 & 2.1 & -9.6 \\ -6.8 & 4.8 & 7.4 \\ -1.9 & 0.7 & 5.9 \end{bmatrix} + \begin{bmatrix} -4.9 & -9.6 & -2.1 \\ 3.4 & 0.7 & 0.0 \\ 5.6 & 10.1 & -1.6 \end{bmatrix}$

In Exercises 13–20, use the following matrices to find the indicated matrices.

$A = \begin{bmatrix} -1 & 4 & -7 & 0 \\ 2 & -6 & -1 & 2 \end{bmatrix} \quad B = \begin{bmatrix} 1 & 5 & -6 & 3 \\ 4 & -1 & 8 & -2 \end{bmatrix},$

$C = \begin{bmatrix} 3 & -6 & 9 \\ -4 & 1 & 2 \end{bmatrix}$

13. $A + B$

14. $A - B$

15. $A + C$

16. $B + C$

17. $2A + B$

18. $2B + A$

19. $A - 2B$

20. $3A - B$

In Exercises 21–24, use matrices A and B to show that the indicated laws hold for these matrices.

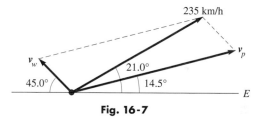

$A = \begin{bmatrix} -1 & 2 & 3 & 7 \\ 0 & -3 & -1 & 4 \\ 9 & -1 & 0 & -2 \end{bmatrix} \quad B = \begin{bmatrix} 4 & -1 & -3 & 0 \\ 5 & 0 & -1 & 1 \\ 1 & 11 & 8 & 2 \end{bmatrix}$

21. $A + B = B + A$

22. $A + O = A$

23. $-(A - B) = B - A$

24. $3(A + B) = 3A + 3B$

In Exercises 25 and 26, find the unknown quantities in the given matrix equations.

25. An airplane is flying in a direction 21.0° north of east at 235 km/h but is headed 14.5° north of east. The wind is from the southeast. Find the speed of the wind v_w and the speed of the plane v_p relative to the wind from the given matrix equation. See Fig. 16-7.

$$\begin{bmatrix} v_p \cos 14.5° - v_w \cos 45.0° \\ v_p \sin 14.5° + v_w \sin 45.0° \end{bmatrix} = \begin{bmatrix} 235 \cos 21.0° \\ 235 \sin 21.0° \end{bmatrix}$$

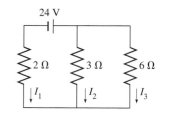

Fig. 16-7

26. Find the electric currents shown in Fig. 16-8 by solving the following matrix equation.

$$\begin{bmatrix} I_1 + I_2 + I_3 \\ -2I_1 + 3I_2 \\ -3I_2 + 6I_3 \end{bmatrix} = \begin{bmatrix} 0 \\ 24 \\ 0 \end{bmatrix}$$

Fig. 16-8

In Exercises 27 and 28, perform the indicated matrix operations.

27. The contractor of a housing development constructs four different types of houses, with either a carport, a one-car garage, or a two-car garage. The following matrix shows the number of houses of each type and the type of garage.

	Type A	Type B	Type C	Type D
Carport	8	6	0	0
1-car garage	5	4	3	0
2-car garage	0	3	5	6

If the contractor builds two additional identical developments, find the matrix showing the total number of each house–garage type built in the three developments.

28. The inventory of a drug supply company shows that the following numbers of cases of bottles of vitamins C and E are in stock: Vitamin C—25 cases of 100-mg bottles, 10 cases of 250-mg bottles, and 32 cases of 500-mg bottles; vitamin E—30 cases of 100-mg bottles, 18 cases of 250-mg bottles, and 40 cases of 500-mg bottles. This is represented by matrix A below. After two shipments are sent out, each of which can be represented by matrix B below, find the matrix that represents the remaining inventory.

$$A = \begin{bmatrix} 25 & 10 & 32 \\ 30 & 18 & 40 \end{bmatrix} \qquad B = \begin{bmatrix} 10 & 5 & 6 \\ 12 & 4 & 8 \end{bmatrix}$$

16-4 MULTIPLICATION OF MATRICES

The definition for the multiplication of matrices does not have an intuitive basis. However, through the solution of a system of linear equations we can, at least in part, show why multiplication is defined as it is. Consider Example 1.

■EXAMPLE 1   If we solve the system of equations

$$2x + y = 1$$
$$7x + 3y = 5$$

we get $x = 2$, $y = -3$. Checking this solution in each of the equations, we get

$$2(2) + 1(-3) = 1$$
$$7(2) + 3(-3) = 5$$

Let us represent the coefficients of the equations by the matrix $\begin{bmatrix} 2 & 1 \\ 7 & 3 \end{bmatrix}$ and the solutions by the matrix $\begin{bmatrix} 2 \\ -3 \end{bmatrix}$. If we now indicate the multiplication of these matrices and perform it as shown

$$\begin{bmatrix} 2 & 1 \\ 7 & 3 \end{bmatrix}\begin{bmatrix} 2 \\ -3 \end{bmatrix} = \begin{bmatrix} 2(2) + 1(-3) \\ 7(2) + 3(-3) \end{bmatrix} = \begin{bmatrix} 1 \\ 5 \end{bmatrix}$$

we note that we obtain a matrix that properly represents the right-side values of the equations. (Note the products and sums in the resulting matrix.)   -------━■

Following reasons along the lines indicated in Example 1, we shall now define the **multiplication of matrices.** If the number of columns in a first matrix equals the number of rows in a second matrix, the product of these matrices is formed as follows: *The element in a specified row and a specified column of the product matrix is the sum of the products formed by multiplying each element in the specified row of the first matrix by the corresponding element in the specific column of the second matrix.* The product matrix will have the same number of rows as the first matrix and the same number of columns as the second matrix. Consider the following examples.

■**EXAMPLE 2** Find the product AB, where

$$A = \begin{bmatrix} 2 & 1 \\ -3 & 0 \\ 1 & 2 \end{bmatrix} \qquad B = \begin{bmatrix} -1 & 6 & 5 & -2 \\ 3 & 0 & 1 & -4 \end{bmatrix}$$

Since there are two columns in matrix A and two rows in matrix B, the product can be formed. To find the element in the first row and first column of the product, we find the sum of the products of corresponding elements of the first row of A and first column of B. To find the element in the first row and second column of the product, we find the sum of the products of corresponding elements in the first row of A and the second column of B. We continue this process until we have found the three rows (the number of rows in A) and the four columns (the number of columns in B) of the product. The product is formed as follows:

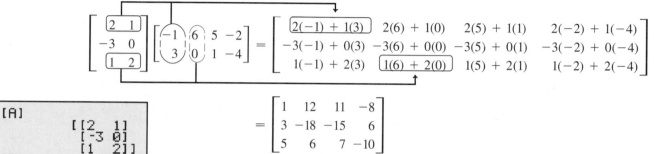

$$= \begin{bmatrix} 1 & 12 & 11 & -8 \\ 3 & -18 & -15 & 6 \\ 5 & 6 & 7 & -10 \end{bmatrix}$$

The specific combination of elements used to form the element in the first row and first column and the element in the third row and second column of the product are outlined in color.

Figure 16-9 shows the calculator display for matrices A and B and the matrix product AB.

If we attempt to form the product BA, we find that B has four columns and A has three rows. Since the number of columns in B does not equal the number of rows in A, the product BA cannot be formed. In this way we see that $AB \neq BA$, which means that *matrix multiplication is not commutative (except in special cases)*. Therefore, we see that matrix multiplication differs from the multiplication of real numbers.

```
[A]
       [[2   1]
        [-3  0]
        [1   2]]
[B]
   [[-1  6  5  -2]
    [3  0  1  -4]]
```

```
[A][B]
[[1  12  11   -8 ...
 [3  -18 -15  6  ...
 [5  6   7    -10...
```

Fig. 16-9

■**EXAMPLE 3** The product of two matrices below may be formed because the first matrix has four columns and the second matrix has four rows. The matrix is formed as shown.

$$\begin{bmatrix} -1 & 9 & 3 & -2 \\ 2 & 0 & -7 & 1 \end{bmatrix} \begin{bmatrix} 6 & -2 \\ 1 & 0 \\ 3 & -5 \\ 3 & 9 \end{bmatrix} = \begin{bmatrix} -1(6) + 9(1) + 3(3) + (-2)(3) & -1(-2) + 9(0) + 3(-5) + (-2)(9) \\ 2(6) + 0(1) + (-7)(3) + 1(3) & 2(-2) + 0(0) + (-7)(-5) + 1(9) \end{bmatrix}$$

$$= \begin{bmatrix} -6 + 9 + 9 - 6 & 2 + 0 - 15 - 18 \\ 12 + 0 - 21 + 3 & -4 + 0 + 35 + 9 \end{bmatrix}$$

$$= \begin{bmatrix} 6 & -31 \\ -6 & 40 \end{bmatrix}$$

Identity Matrix

There are two special matrices of particular importance in the multiplication of matrices. The first of these is the **identity matrix I,** *which is a square matrix with 1's for elements of the principal diagonal with all other elements zero.* (The principal diagonal starts with the element a_{11}.) It has the property that if it is multiplied by another square matrix with the same number of rows and columns, then the second matrix equals the product matrix.

■**EXAMPLE 4**   Show that $AI = IA = A$ for the matrix

$$A = \begin{bmatrix} 2 & -3 \\ 4 & 1 \end{bmatrix}$$

Since A has two rows and two columns, we choose I with two rows and two columns. Therefore, for this case

$$I = \begin{bmatrix} 1 & 0 \\ 0 & 1 \end{bmatrix} \longleftarrow \text{elements of principal diagonal are 1's}$$

Forming the indicated products, we have results as follows:

$$AI = \begin{bmatrix} 2 & -3 \\ 4 & 1 \end{bmatrix}\begin{bmatrix} 1 & 0 \\ 0 & 1 \end{bmatrix}$$

$$= \begin{bmatrix} 2(1) + (-3)(0) & 2(0) + (-3)(1) \\ 4(1) + 1(0) & 4(0) + 1(1) \end{bmatrix} = \begin{bmatrix} 2 & -3 \\ 4 & 1 \end{bmatrix}$$

$$IA = \begin{bmatrix} 1 & 0 \\ 0 & 1 \end{bmatrix}\begin{bmatrix} 2 & -3 \\ 4 & 1 \end{bmatrix}$$

$$= \begin{bmatrix} 1(2) + 0(4) & 1(-3) + 0(1) \\ 0(2) + 1(4) & 0(-3) + 1(1) \end{bmatrix} = \begin{bmatrix} 2 & -3 \\ 4 & 1 \end{bmatrix}$$

Therefore, we see that $AI = IA = A$. ----------■

Inverse of a Matrix

For a given square matrix A, its **inverse** A^{-1} is the other important special matrix. *The matrix A and its inverse A^{-1} have the property that*

$$AA^{-1} = A^{-1}A = I \qquad \text{(16-8)}$$

If the product of two square matrices equals the identity matrix, the matrices are called inverses of each other. Under certain conditions the inverse of a given square matrix may not exist, although for most square matrices the inverse does exist. In the next section we shall develop the procedure for finding the inverse of a square matrix, and the section that follows shows how the inverse is used in the solution of systems of equations. At this point we simply show that the product of certain matrices equals the identity matrix and that therefore these matrices are inverses of each other.

EXAMPLE 5   For the given matrices A and B, show that $AB = BA = I$, and therefore that $B = A^{-1}$.

$$A = \begin{bmatrix} 1 & -3 \\ -2 & 7 \end{bmatrix} \qquad B = \begin{bmatrix} 7 & 3 \\ 2 & 1 \end{bmatrix}$$

Forming the products AB and BA, we have the following:

$$AB = \begin{bmatrix} 1 & -3 \\ -2 & 7 \end{bmatrix}\begin{bmatrix} 7 & 3 \\ 2 & 1 \end{bmatrix} = \begin{bmatrix} 7-6 & 3-3 \\ -14+14 & -6+7 \end{bmatrix} = \begin{bmatrix} 1 & 0 \\ 0 & 1 \end{bmatrix}$$

$$BA = \begin{bmatrix} 7 & 3 \\ 2 & 1 \end{bmatrix}\begin{bmatrix} 1 & -3 \\ -2 & 7 \end{bmatrix} = \begin{bmatrix} 7-6 & -21+21 \\ 2-2 & -6+7 \end{bmatrix} = \begin{bmatrix} 1 & 0 \\ 0 & 1 \end{bmatrix}$$

Since $AB = I$ and $BA = I$, $B = A^{-1}$ and $A = B^{-1}$. ----------- ■

The following example illustrates one kind of application of the multiplication of matrices.

SOLVING A WORD PROBLEM

See the chapter introduction.

EXAMPLE 6   A company makes three types of machine parts. In one day it produces 40 of type X, 50 of type Y, and 80 of type Z. Each of type X requires 4 units of material and 1 worker-hour to produce; each of type Y requires 5 units of material and 2 worker-hours to produce; each of type Z requires 3 units of material and 2 worker-hours to produce. By representing the number of each type produced as matrix A and the material and time requirements as matrix B, we have

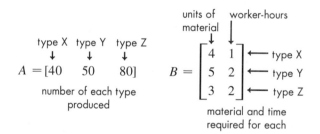

The product AB gives the total number of units of material and the total number of worker-hours needed for the day's production in a one-row, two-column matrix.

$$AB = \begin{bmatrix} 40 & 50 & 80 \end{bmatrix}\begin{bmatrix} 4 & 1 \\ 5 & 2 \\ 3 & 2 \end{bmatrix}$$

$$= \begin{bmatrix} 160+250+240 & 40+100+160 \end{bmatrix} = \begin{bmatrix} 650 & 300 \end{bmatrix}$$

Therefore, 650 units of material and 300 worker-hours are required. ----------- ■

We now have seen how multiplication is defined for matrices. We see that *matrix multiplication is not commutative;* that is, $AB \neq BA$ in general. This is a major difference from the multiplication of real numbers. Another difference is that it is possible that $AB = O$, even though neither A nor B is O. There are, however, some similarities. We have seen, for example, that $AI = A$, where we make I and the number 1 equivalent for the two types of multiplication. Also, *the distributive property $A(B + C) = AB + AC$ holds for matrix multiplication.* This points out additional properties of the algebra of matrices.

EXERCISES *16-4*

In Exercises 1–12, perform the indicated multiplications.

1. $[4 \ -2] \begin{bmatrix} -1 & 0 \\ 2 & 6 \end{bmatrix}$

2. $[-1 \ \ 5 \ -2] \begin{bmatrix} 6 & 3 \\ 2 & -1 \\ 0 & 2 \end{bmatrix}$

3. $\begin{bmatrix} 2 & -3 \\ 5 & -1 \end{bmatrix} \begin{bmatrix} 3 & 0 & -1 \\ 7 & -5 & 8 \end{bmatrix}$

4. $\begin{bmatrix} -7 & 8 \\ 5 & 0 \end{bmatrix} \begin{bmatrix} -9 & 10 \\ 1 & 4 \end{bmatrix}$

5. $\begin{bmatrix} 2 & -3 & 1 \\ 0 & 7 & -3 \end{bmatrix} \begin{bmatrix} 9 \\ -2 \\ 5 \end{bmatrix}$

6. $\begin{bmatrix} 0 & -1 & 2 \\ 4 & 11 & 2 \end{bmatrix} \begin{bmatrix} 3 & -1 \\ 1 & 2 \\ 6 & 1 \end{bmatrix}$

7. $\begin{bmatrix} -8 & 9 \\ 7 & -8 \\ -6 & 4 \end{bmatrix} \begin{bmatrix} -5 & 8 \\ -7 & 5 \end{bmatrix}$

8. $\begin{bmatrix} 12 & -47 \\ 43 & -18 \\ 36 & -22 \end{bmatrix} \begin{bmatrix} 25 \\ 66 \end{bmatrix}$

9. $\begin{bmatrix} -1 & 7 \\ 3 & 5 \\ 10 & -1 \\ -5 & 12 \end{bmatrix} \begin{bmatrix} 2 & 1 \\ 5 & -3 \end{bmatrix}$

10. $[5 \ \ 4] \begin{bmatrix} 4 & -4 \\ -5 & 5 \end{bmatrix}$

11. $\begin{bmatrix} -9.2 & 2.3 & 0.5 \\ -3.8 & -2.4 & 9.2 \end{bmatrix} \begin{bmatrix} 6.5 & -5.2 \\ 4.9 & 1.7 \\ -1.8 & 6.9 \end{bmatrix}$

12. $\begin{bmatrix} 1 & 2 & -6 & 6 & 1 \\ -2 & 4 & 0 & 1 & 2 \end{bmatrix} \begin{bmatrix} 1 \\ -1 \\ 0 \\ 5 \\ 2 \end{bmatrix}$

In Exercises 13–16, find, if possible, AB and BA.

13. $A = [1 \ -3 \ \ 8] \qquad B = \begin{bmatrix} -1 \\ 5 \\ 7 \end{bmatrix}$

14. $A = \begin{bmatrix} -3 & 2 & 0 \\ 1 & -4 & 5 \end{bmatrix} \qquad B = \begin{bmatrix} -2 & 0 \\ 4 & -6 \\ 5 & 1 \end{bmatrix}$

15. $A = \begin{bmatrix} -1 & 2 & 3 \\ 5 & -1 & 0 \end{bmatrix} \qquad B = \begin{bmatrix} 1 \\ -5 \\ 2 \end{bmatrix}$

16. $A = \begin{bmatrix} -2 & 1 & 7 \\ 3 & -1 & 0 \\ 0 & 2 & -1 \end{bmatrix} \qquad B = [4 \ -1 \ \ 5]$

In Exercises 17–20, show that AI = IA = A.

17. $A = \begin{bmatrix} 1 & 8 \\ -2 & 2 \end{bmatrix}$

18. $A = \begin{bmatrix} -3 & 4 \\ 1 & 2 \end{bmatrix}$

19. $A = \begin{bmatrix} 1 & 3 & -5 \\ 2 & 0 & 1 \\ 1 & -2 & 4 \end{bmatrix}$

20. $A = \begin{bmatrix} -1 & 2 & 0 \\ 4 & -3 & 1 \\ 2 & 1 & 3 \end{bmatrix}$

In Exercises 21–24, determine whether or not $B = A^{-1}$.

21. $A = \begin{bmatrix} 5 & -2 \\ -2 & 1 \end{bmatrix} \qquad B = \begin{bmatrix} 1 & 2 \\ 2 & 5 \end{bmatrix}$

22. $A = \begin{bmatrix} 3 & -4 \\ 5 & -7 \end{bmatrix} \qquad B = \begin{bmatrix} 7 & -4 \\ 5 & -2 \end{bmatrix}$

23. $A = \begin{bmatrix} 1 & -2 & 3 \\ 2 & -5 & 7 \\ -1 & 3 & -5 \end{bmatrix} \qquad B = \begin{bmatrix} 4 & -1 & 1 \\ 3 & -2 & -1 \\ 1 & -1 & -1 \end{bmatrix}$

24. $A = \begin{bmatrix} 1 & -1 & 3 \\ 3 & -4 & 8 \\ -2 & 3 & -4 \end{bmatrix} \qquad B = \begin{bmatrix} 8 & -5 & -4 \\ 4 & -2 & -1 \\ -1 & 1 & 1 \end{bmatrix}$

In Exercises 25–28, determine by matrix multiplication whether or not A is the proper matrix of solution values.

25. $\begin{aligned} 3x - 2y &= -1 \\ 4x + y &= 6 \end{aligned} \qquad A = \begin{bmatrix} 1 \\ 2 \end{bmatrix}$

26. $\begin{aligned} 4x + y &= -5 \\ 3x + 4y &= 6 \end{aligned} \qquad A = \begin{bmatrix} -2 \\ 3 \end{bmatrix}$

27. $\begin{aligned} 3x + y + 2z &= 1 \\ x - 3y + 4z &= -3 \\ 2x + 2y + z &= 1 \end{aligned} \qquad A = \begin{bmatrix} -1 \\ 2 \\ 1 \end{bmatrix}$

28. $\begin{aligned} 2x - y + z &= 7 \\ x - 3y + 2z &= 6 \\ 3x + y - z &= 8 \end{aligned} \qquad A = \begin{bmatrix} 3 \\ -2 \\ -1 \end{bmatrix}$

In Exercises 29–36, perform the indicated matrix multiplications.

29. Using two rows and columns, show that $(-I)^2 = I$.

(W) 30. For $J = \begin{bmatrix} j & 0 \\ 0 & j \end{bmatrix}$, where $j = \sqrt{-1}$, show that $J^2 = -I$, $J^3 = -J$, and $J^4 = I$. Explain the similarity with j^2, j^3, and j^4.

31. Show that $A^2 - I = (A + I)(A - I)$ for $A = \begin{bmatrix} 2 & 4 \\ 3 & 5 \end{bmatrix}$.

32. In the study of polarized light, the matrix product
$$\begin{bmatrix} 1 & 0 \\ 0 & -j \end{bmatrix} \begin{bmatrix} 1 & 0 \\ 1 & -j \end{bmatrix} \begin{bmatrix} 1 \\ 1 \end{bmatrix} \text{ occurs } (j = \sqrt{-1}). \text{ Find this product.}$$

33. In studying the motion of electrons, one of the Pauli spin matrices used is $s_y = \begin{bmatrix} 0 & -j \\ j & 0 \end{bmatrix}$, where $j = \sqrt{-1}$. Show that $s_y^2 = I$.

34. In analyzing the motion of a robotic mechanism, the following matrix multiplication is used. Perform the multiplication and evaluate each element of the result.
$$\begin{bmatrix} \cos 60° & -\sin 60° & 0 \\ \sin 60° & \cos 60° & 0 \\ 0 & 0 & 1 \end{bmatrix} \begin{bmatrix} 2 \\ 4 \\ 0 \end{bmatrix}$$

35. In an *ammeter,* nearly all the electric current flows through a *shunt,* and the remaining known fraction of current is measured by the meter. See Fig. 16-10. From the given matrix equation find voltage v_2 and current i_2 in terms of v_1, i_1, and resistance R, whichever may be applicable.

$$\begin{bmatrix} v_2 \\ i_2 \end{bmatrix} = \begin{bmatrix} 1 & 0 \\ -\dfrac{1}{R} & 1 \end{bmatrix} \begin{bmatrix} v_1 \\ i_1 \end{bmatrix}$$

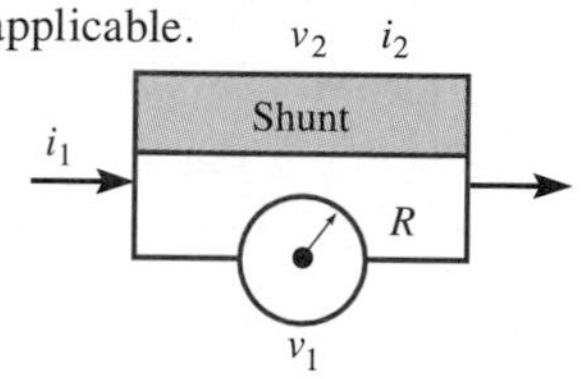

Fig. 16-10

36. In the theory related to the reproduction of color photography, the equations

$$\begin{bmatrix} X \\ Y \\ Z \end{bmatrix} = \begin{bmatrix} 1.0 & 0.1 & 0 \\ 0.5 & 1.0 & 0.1 \\ 0.3 & 0.4 & 1.0 \end{bmatrix} \begin{bmatrix} x \\ y \\ z \end{bmatrix}$$

are found. The X, Y, and Z represent the red, green, and blue densities of the reproductions, respectively, and the x, y, and z represent the red, green, and blue densities, respectively, of the subject. Give the equations relating X, Y, and Z and x, y, and z.

16-5 FINDING THE INVERSE OF A MATRIX

In this section we show how to find the inverse of a matrix, and in the following section we show how the inverse is used in solving a system of linear equations.

We shall first show two methods of finding the inverse of a two-row, two-column (2×2) matrix. The first method is as follows:

INVERSE OF A 2 × 2 MATRIX

1. *Interchange the elements on the principal diagonal.*

2. *Change the signs of the off-diagonal elements.*

3. *Divide each resulting element by the determinant of the given matrix.*

This method, which can be used with second-order square matrices *but not with higher-order matrices,* is illustrated in the following example.

■EXAMPLE 1   Find the inverse of the matrix

$$A = \begin{bmatrix} 2 & -3 \\ 4 & -7 \end{bmatrix}$$

First we interchange the elements on the principal diagonal and change the signs of the off-diagonal elements. This gives us the matrix

$$\begin{bmatrix} -7 & 3 \\ -4 & 2 \end{bmatrix} \begin{array}{l} \longleftarrow \text{signs changed} \\ \longleftarrow \text{elements interchanged} \end{array}$$

Now we find the determinant of the original matrix, which means we evaluate

$$\begin{vmatrix} 2 & -3 \\ 4 & -7 \end{vmatrix} = -14 - (-12) = -2$$

We now divide each element of the second matrix by -2. This gives

$$A^{-1} = \frac{1}{-2}\begin{bmatrix} -7 & 3 \\ -4 & 2 \end{bmatrix} = \begin{bmatrix} \dfrac{-7}{-2} & \dfrac{3}{-2} \\ \dfrac{-4}{-2} & \dfrac{2}{-2} \end{bmatrix} = \begin{bmatrix} \dfrac{7}{2} & -\dfrac{3}{2} \\ 2 & -1 \end{bmatrix} \longleftarrow \text{inverse}$$

Checking by multiplication gives

$$AA^{-1} = \begin{bmatrix} 2 & -3 \\ 4 & -7 \end{bmatrix}\begin{bmatrix} \frac{7}{2} & -\frac{3}{2} \\ 2 & -1 \end{bmatrix} = \begin{bmatrix} 7 - 6 & -3 + 3 \\ 14 - 14 & -6 + 7 \end{bmatrix} = \begin{bmatrix} 1 & 0 \\ 0 & 1 \end{bmatrix} = I$$

Since $AA^{-1} = I$, the matrix A^{-1} is the proper inverse matrix.

GAUSS–JORDAN METHOD

Named for the German mathematician Karl Gauss (1777–1855) and the German geodesist Wilhelm Jordan (1842–1899).

The second method, called the *Gauss–Jordan method,* involves *transforming the given matrix into the identity matrix while* **transforming the identity matrix into the inverse.** There are three types of steps allowable in making these transformations:

1. *Any two rows may be interchanged.*
2. *Every element in any row may be multiplied by any number other than zero.*
3. *Any row may be replaced by a row whose elements are the sum of a nonzero multiple of itself and a nonzero multiple of another row.*

NOTE ▶

Note that these are **row operations,** not column operations, and are the operations used in solving a system of equations by addition or subtraction.

■ EXAMPLE 2 Find the inverse of the matrix

$$A = \begin{bmatrix} 2 & -3 \\ 4 & -7 \end{bmatrix}$$ this is the same matrix as in Example 1

First, we set up the given matrix with the identity matrix as follows:

$$\begin{bmatrix} 2 & -3 & | & 1 & 0 \\ 4 & -7 & | & 0 & 1 \end{bmatrix}$$

The vertical line simply shows the separation of the two matrices.

We wish to transform the left matrix into the identity matrix. Therefore, the first requirement is a 1 for element a_{11}. Therefore, we divide all elements of the first row by 2. This gives the following setup:

$$\begin{bmatrix} 1 & -\frac{3}{2} & | & \frac{1}{2} & 0 \\ 4 & -7 & | & 0 & 1 \end{bmatrix}$$

Next we want to have a zero for element a_{21}. Therefore, we shall subtract 4 times each element of row 1 from the corresponding element in row 2, replacing the elements of row 2. This gives us the following setup:

$$\begin{bmatrix} 1 & -\frac{3}{2} & | & \frac{1}{2} & 0 \\ 4-4(1) & -7-4(-\frac{3}{2}) & | & 0-4(\frac{1}{2}) & 1-4(0) \end{bmatrix} \text{ or } \begin{bmatrix} 1 & -\frac{3}{2} & | & \frac{1}{2} & 0 \\ 0 & -1 & | & -2 & 1 \end{bmatrix}$$

Next, we want to have 1, not -1, for element a_{22}. Therefore, we multiply each element of row 2 by -1. This gives

$$\begin{bmatrix} 1 & -\frac{3}{2} & | & \frac{1}{2} & 0 \\ 0 & 1 & | & 2 & -1 \end{bmatrix}$$

Finally, we want zero for element a_{12}. Therefore, we add $\frac{3}{2}$ times each element of row 2 to the corresponding elements of row 1, replacing row 1. This gives

$$\begin{bmatrix} 1+\frac{3}{2}(0) & -\frac{3}{2}+\frac{3}{2}(1) & | & \frac{1}{2}+\frac{3}{2}(2) & 0+\frac{3}{2}(-1) \\ 0 & 1 & | & 2 & -1 \end{bmatrix} \text{ or } \begin{bmatrix} 1 & 0 & | & \frac{7}{2} & -\frac{3}{2} \\ 0 & 1 & | & 2 & -1 \end{bmatrix}$$

At this point, we have transformed the given matrix into the identity matrix, and the identity matrix into the inverse. Therefore, the matrix to the right of the vertical bar in the last setup is the required inverse. Thus,

$$A^{-1} = \begin{bmatrix} \frac{7}{2} & -\frac{3}{2} \\ 2 & -1 \end{bmatrix}$$

See Fig. 16-11 for a calculator window showing matrix A and its inverse A^{-1}. ■

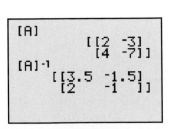

[A]
                [[2 -3]
                 [4 -7]]
[A]⁻¹
                [[3.5 -1.5]
                 [2   -1]]

Fig. 16-11

In transforming a matrix into the identity matrix, we work on one column at a time, transforming the columns in order from left to right. It is generally best to make the element on the principal diagonal for the column 1 first and then make all other elements in the column 0. This was done in Example 2, and we now illustrate it with another 2×2 matrix, and then we find the inverse of a 3×3 matrix. The method is applicable for any square matrix.

EXAMPLE 3   Find the inverse of the matrix $\begin{bmatrix} -3 & 6 \\ 4 & 5 \end{bmatrix}$.

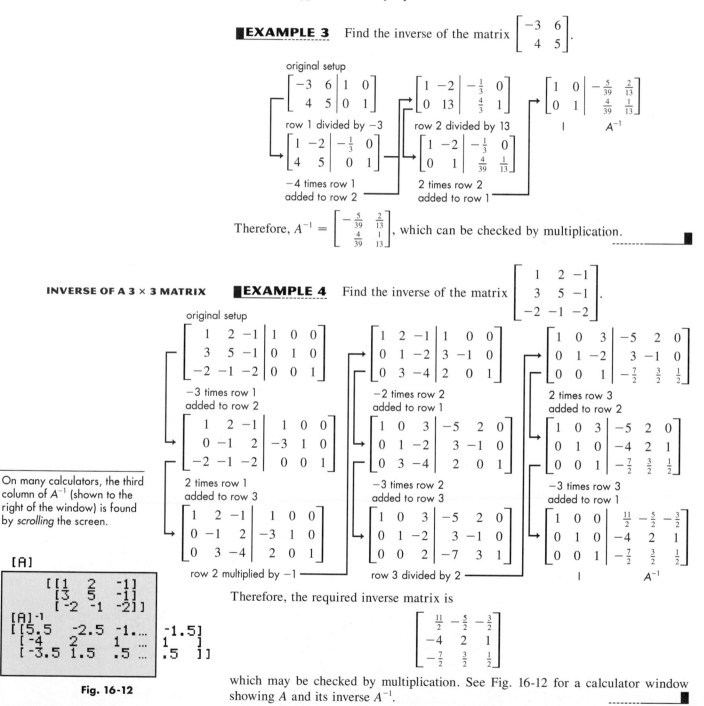

Therefore, $A^{-1} = \begin{bmatrix} -\frac{5}{39} & \frac{2}{13} \\ \frac{4}{39} & \frac{1}{13} \end{bmatrix}$, which can be checked by multiplication.

INVERSE OF A 3 × 3 MATRIX

EXAMPLE 4   Find the inverse of the matrix $\begin{bmatrix} 1 & 2 & -1 \\ 3 & 5 & -1 \\ -2 & -1 & -2 \end{bmatrix}$.

On many calculators, the third column of A^{-1} (shown to the right of the window) is found by *scrolling* the screen.

```
[A]
      [[1    2   -1]
       [3    5   -1]
       [-2  -1   -2]]
[A]⁻¹
  [[5.5   -2.5  -1.…   -1.5]
   [-4      2     1 …    1   ]
   [-3.5  1.5    .5 …   .5  ]]
```

Fig. 16-12

Therefore, the required inverse matrix is

$$\begin{bmatrix} \frac{11}{2} & -\frac{5}{2} & -\frac{3}{2} \\ -4 & 2 & 1 \\ -\frac{7}{2} & \frac{3}{2} & \frac{1}{2} \end{bmatrix}$$

which may be checked by multiplication. See Fig. 16-12 for a calculator window showing A and its inverse A^{-1}.

We have shown calculator displays for many of the matrices and operations on them. All the matrix operations presented in this text can be done on most graphing calculators. There are many different models of calculators, and the manual for any given model should be used to see how the various operations are performed on it. In the following example we review the procedure for entering a matrix, displaying it, and then finding its inverse on a typical graphing calculator by showing the windows that are used.

■EXAMPLE 5   By using a graphing calculator, find the inverse of the matrix

$$A = \begin{bmatrix} 2 & -2 & 3 & 2 \\ 3 & 1 & 5 & 2 \\ -2 & 5 & 2 & -3 \\ 4 & -5 & -1 & 4 \end{bmatrix}$$

Using the *matrix* feature, in Fig. 16-13(a) we show the calculator window used when the elements of the matrix are entered. Scrolling by use of the arrow key will probably be required to enter all the necessary values. We then display the matrix A as shown in Fig. 16-13(b).

In Fig. 16-13(c) we show the display for the inverse of matrix A. To get this display, we enter [A] and then use the $\boxed{x^{-1}}$ key (to show the inverse) as shown. To avoid lengthy decimal approximations in the display, we have used the *mode* feature to limit the number of decimal places to three. On most calculators this still requires scrolling to see all of the values, but the values are easier to read. We have shown those that need scrolling to the right of the window.

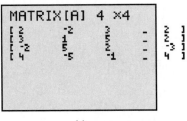

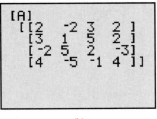

(a)

[A]
[[2 -2 3   2]
 [3   1 5   2]
 [-2 5 2 -3]
 [4 -5 -1   4]]

(b)

Fig. 16-13

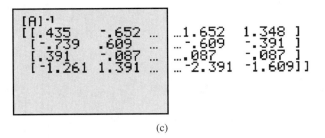

(c)

Therefore, from these calculator windows, we have found that the value of the inverse matrix is

$$A^{-1} = \begin{bmatrix} 0.435 & -0.652 & 1.652 & 1.348 \\ -0.739 & 0.609 & -0.609 & -0.391 \\ .391 & -0.087 & 0.087 & -0.087 \\ -1.261 & 1.391 & -2.391 & -1.609 \end{bmatrix}$$

where values have been rounded off.

The elements of A^{-1} can be checked directly on the calculator by showing that the product $AA^{-1} = I$ (if the rounded values are used, the values of I will be approximate, but should be sufficiently close to the necessary 1's and 0's). ■

EXERCISES $16\text{-}5$

In Exercises 1–8, find the inverse of each of the given matrices by the method of Example 1 of this section.

1. $\begin{bmatrix} 2 & -5 \\ -2 & 4 \end{bmatrix}$

2. $\begin{bmatrix} -6 & 3 \\ 3 & -2 \end{bmatrix}$

3. $\begin{bmatrix} -1 & 5 \\ 4 & 10 \end{bmatrix}$

4. $\begin{bmatrix} 8 & -1 \\ -4 & -5 \end{bmatrix}$

5. $\begin{bmatrix} 0 & -4 \\ 2 & 6 \end{bmatrix}$

6. $\begin{bmatrix} 7 & -2 \\ -6 & 2 \end{bmatrix}$

7. $\begin{bmatrix} -50 & -45 \\ 26 & 80 \end{bmatrix}$

8. $\begin{bmatrix} 7.2 & -3.6 \\ -1.3 & -5.7 \end{bmatrix}$

In Exercises 9–24, find the inverse of each of the given matrices by transforming the identity matrix, as in Examples 2–4.

9. $\begin{bmatrix} 1 & 2 \\ 2 & 3 \end{bmatrix}$

10. $\begin{bmatrix} 1 & 5 \\ -1 & -4 \end{bmatrix}$

11. $\begin{bmatrix} 2 & 4 \\ -1 & -1 \end{bmatrix}$

12. $\begin{bmatrix} -2 & 6 \\ 3 & -4 \end{bmatrix}$

13. $\begin{bmatrix} 2 & 5 \\ -1 & 2 \end{bmatrix}$

14. $\begin{bmatrix} -2 & 3 \\ -3 & 5 \end{bmatrix}$

15. $\begin{bmatrix} 2 & -1 \\ 4 & 6 \end{bmatrix}$

16. $\begin{bmatrix} 1 & -3 \\ 7 & -5 \end{bmatrix}$

17. $\begin{bmatrix} 1 & -3 & -2 \\ -2 & 7 & 3 \\ 1 & -1 & -3 \end{bmatrix}$

18. $\begin{bmatrix} 1 & 2 & -1 \\ 3 & 7 & -5 \\ -1 & -2 & 0 \end{bmatrix}$

19. $\begin{bmatrix} 1 & -1 & -3 \\ 0 & -1 & -2 \\ 2 & 1 & -1 \end{bmatrix}$

20. $\begin{bmatrix} 1 & 4 & 1 \\ -3 & -13 & -1 \\ 0 & -2 & 5 \end{bmatrix}$

21. $\begin{bmatrix} 1 & 3 & 2 \\ -2 & -5 & -1 \\ 2 & 4 & 0 \end{bmatrix}$

22. $\begin{bmatrix} 1 & 3 & 4 \\ -1 & -4 & -2 \\ 4 & 9 & 20 \end{bmatrix}$

23. $\begin{bmatrix} 2 & 4 & 0 \\ 3 & 4 & -2 \\ -1 & 1 & 2 \end{bmatrix}$

24. $\begin{bmatrix} -2 & 6 & 1 \\ 0 & 3 & -3 \\ 4 & -7 & 3 \end{bmatrix}$

In Exercises 25–32, find the inverse of each of the given matrices by using a graphing calculator, as in Example 5. The matrices in Exercises 27–30 are the same as those in Exercises 21–24.

25. $\begin{bmatrix} 2 & 8 \\ -1 & 6 \end{bmatrix}$

26. $\begin{bmatrix} 7 & -3 \\ 6 & -2 \end{bmatrix}$

27. $\begin{bmatrix} 1 & 3 & 2 \\ -2 & -5 & -1 \\ 2 & 4 & 0 \end{bmatrix}$

28. $\begin{bmatrix} 1 & 3 & 4 \\ -1 & -4 & -2 \\ 4 & 9 & 20 \end{bmatrix}$

29. $\begin{bmatrix} 2 & 4 & 0 \\ 3 & 4 & -2 \\ -1 & 1 & 2 \end{bmatrix}$

30. $\begin{bmatrix} -2 & 6 & 1 \\ 0 & 3 & -3 \\ 4 & -7 & 3 \end{bmatrix}$

31. $\begin{bmatrix} 1 & -2 & 1 & 0 \\ 1 & -2 & 2 & -3 \\ 0 & 1 & -1 & 1 \\ -2 & 3 & -2 & 3 \end{bmatrix}$

32. $\begin{bmatrix} 3 & -2 & -1 & 4 \\ 2 & 0 & 5 & 1 \\ -1 & 2 & 1 & -2 \\ 4 & 1 & 3 & 5 \end{bmatrix}$

In Exercises 33–36, solve the given problems.

33. For the matrix $A = \begin{bmatrix} a & b \\ c & d \end{bmatrix}$, show that

$$\frac{1}{ad - bc}\begin{bmatrix} a & b \\ c & d \end{bmatrix}\begin{bmatrix} d & -b \\ -c & a \end{bmatrix} = \begin{bmatrix} 1 & 0 \\ 0 & 1 \end{bmatrix}$$

This verifies the method of Example 1.

(W) 34. Describe the relationship between the elements of the matrix $\begin{bmatrix} a & 0 & 0 \\ 0 & b & 0 \\ 0 & 0 & c \end{bmatrix}$ and the elements of its inverse.

35. For the *four-terminal network* shown in Fig. 16-14, it can be shown that the voltage matrix V is related to the coefficient matrix A and the current matrix I by $V = A^{-1}I$, where

$$V = \begin{bmatrix} v_1 \\ v_2 \end{bmatrix} \qquad A = \begin{bmatrix} a_{11} & a_{12} \\ a_{21} & a_{22} \end{bmatrix} \qquad I = \begin{bmatrix} i_1 \\ i_2 \end{bmatrix}$$

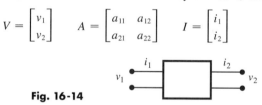

Fig. 16-14

Find the individual equations for v_1 and v_2 that give each in terms of i_1 and i_2.

36. The rotations of a robot arm such as that shown in Fig. 16-15 are often represented by matrices. The values represent trigonometric functions of the angles of rotation. For the following rotation matrix R, find R^{-1}.

$$R = \begin{bmatrix} 0.8 & 0.0 & -0.6 \\ 0.0 & 1.0 & 0.0 \\ 0.6 & 0.0 & 0.8 \end{bmatrix}$$

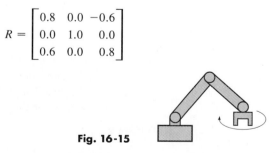

Fig. 16-15

16-6 MATRICES AND LINEAR EQUATIONS

As we stated at the beginning of Section 16-3, matrices can be used to solve systems of equations, and in this section we show one method by which this is done. As we develop this method, it will be apparent that there is a great deal of numerical work involved. However, methods such as this one are easily programmed for use on a computer, which can do the arithmetic work very rapidly. Also, most graphing calculators can perform these operations, and we will show an example at the end of the section in which a calculator is used to solve a system of four equations more readily than with earlier methods. It is the *method* of solving the system of equations that is of primary importance here.

Let us consider the system of equations

$$a_1 x + b_1 y = c_1$$
$$a_2 x + b_2 y = c_2$$

Recalling the definition of equality of matrices, we can write this system as

$$\begin{bmatrix} a_1 x + b_1 y \\ a_2 x + b_2 y \end{bmatrix} = \begin{bmatrix} c_1 \\ c_2 \end{bmatrix} \tag{16-9}$$

If we let

$$A = \begin{bmatrix} a_1 & b_1 \\ a_2 & b_2 \end{bmatrix} \qquad X = \begin{bmatrix} x \\ y \end{bmatrix} \qquad C = \begin{bmatrix} c_1 \\ c_2 \end{bmatrix} \tag{16-10}$$

the left side of Eq. (16-9) can be written as the product of matrices A and X,

$$AX = C \tag{16-11}$$

If we now multiply (on the left) each side of this matrix equation by A^{-1}, we have

$$A^{-1}AX = A^{-1}C$$

Since $A^{-1}A = I$, we have

$$IX = A^{-1}C$$

However, $IX = X$. Therefore,

$$X = A^{-1}C \tag{16-12}$$

NOTE ▶ Equation (16-12) states that *we can solve a system of linear equations by multiplying the one-column matrix of the constants on the right by the inverse of the matrix of the coefficients.* The result is a one-column matrix whose elements are the required values for the solution. Also, note that

CAUTION ▶
$$X = A^{-1}C \qquad \text{and } \textit{\textbf{not}} \qquad CA^{-1}$$

as the order of matrix multiplication must be carefully followed.

EXAMPLE 1   Use matrices to solve the system of equations

$$2x - y = 7$$
$$5x - 3y = 18$$

We set up the matrix of coefficients and the matrix of constants as

$$A = \begin{bmatrix} 2 & -1 \\ 5 & -3 \end{bmatrix} \quad \text{and} \quad C = \begin{bmatrix} 7 \\ 18 \end{bmatrix}$$

By either of the methods of the previous section, we can determine the inverse of matrix A to be

$$A^{-1} = \begin{bmatrix} 3 & -1 \\ 5 & -2 \end{bmatrix}$$

We now form the matrix product $A^{-1}C$.

$$A^{-1}C = \begin{bmatrix} 3 & -1 \\ 5 & -2 \end{bmatrix} \begin{bmatrix} 7 \\ 18 \end{bmatrix} = \begin{bmatrix} 21 - 18 \\ 35 - 36 \end{bmatrix} = \begin{bmatrix} 3 \\ -1 \end{bmatrix}$$

Since $X = A^{-1}C$, this means that

$$\begin{bmatrix} x \\ y \end{bmatrix} = \begin{bmatrix} 3 \\ -1 \end{bmatrix}$$

Therefore, the required solution is $x = 3$ and $y = -1$, which checks when these values are substituted into the original equations.   --------∎

EXAMPLE 2   For the electric circuit shown in Fig. 16-16, the equations used to find the currents (in amperes) i_1 and i_2 are

$$2.30i_1 + 6.45(i_1 + i_2) = 15.0 \qquad 8.75i_1 + 6.45i_2 = 15.0$$
$$1.25i_2 + 6.45(i_1 + i_2) = 12.5 \quad \text{or} \quad 6.45i_1 + 7.70i_2 = 12.5$$

Using matrices to solve this system of equations, we set up the matrix A of coefficients, the matrix C of constants, and the matrix X of currents as

$$A = \begin{bmatrix} 8.75 & 6.45 \\ 6.45 & 7.70 \end{bmatrix} \qquad C = \begin{bmatrix} 15.0 \\ 12.5 \end{bmatrix} \qquad X = \begin{bmatrix} i_1 \\ i_2 \end{bmatrix}$$

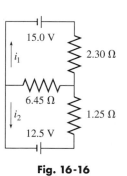

Fig. 16-16

We now find the inverse of A as

$$A^{-1} = \frac{1}{8.75(7.70) - 6.45(6.45)} \begin{bmatrix} 7.70 & -6.45 \\ -6.45 & 8.75 \end{bmatrix} = \begin{bmatrix} 0.2988 & -0.2503 \\ -0.2503 & 0.3395 \end{bmatrix}$$

Therefore,

$$X = A^{-1}C = \begin{bmatrix} 0.2988 & -0.2503 \\ -0.2503 & 0.3395 \end{bmatrix} \begin{bmatrix} 15.0 \\ 12.5 \end{bmatrix}$$

$$= \begin{bmatrix} 0.2988(15.0) - 0.2503(12.5) \\ -0.2503(15.0) + 0.3395(12.5) \end{bmatrix} = \begin{bmatrix} 1.35 \\ 0.49 \end{bmatrix}$$

Therefore, the required currents are $i_1 = 1.35$ A and $i_2 = 0.49$ A. These values check when substituted into the original equations.   --------∎

EXAMPLE 3 Use matrices to solve the system of equations

$$x + 4y - z = 4$$
$$x + 3y + z = 8$$
$$2x + 6y + z = 13$$

Setting up matrices A, C, and X, we have

$$A = \begin{bmatrix} 1 & 4 & -1 \\ 1 & 3 & 1 \\ 2 & 6 & 1 \end{bmatrix} \qquad C = \begin{bmatrix} 4 \\ 8 \\ 13 \end{bmatrix} \qquad X = \begin{bmatrix} x \\ y \\ z \end{bmatrix}$$

To give another example of finding the inverse of a 3 × 3 matrix, we shall briefly show the steps for finding A^{-1}.

$$\begin{bmatrix} 1 & 4 & -1 & 1 & 0 & 0 \\ 1 & 3 & 1 & 0 & 1 & 0 \\ 2 & 6 & 1 & 0 & 0 & 1 \end{bmatrix} \rightarrow \begin{bmatrix} 1 & 4 & -1 & 1 & 0 & 0 \\ 0 & 1 & -2 & 1 & -1 & 0 \\ 0 & -2 & 3 & -2 & 0 & 1 \end{bmatrix} \rightarrow \begin{bmatrix} 1 & 0 & 7 & -3 & 4 & 0 \\ 0 & 1 & -2 & 1 & -1 & 0 \\ 0 & 0 & 1 & 0 & 2 & -1 \end{bmatrix}$$

$$\begin{bmatrix} 1 & 4 & -1 & 1 & 0 & 0 \\ 0 & -1 & 2 & -1 & 1 & 0 \\ 2 & 6 & 1 & 0 & 0 & 1 \end{bmatrix} \quad \begin{bmatrix} 1 & 0 & 7 & -3 & 4 & 0 \\ 0 & 1 & -2 & 1 & -1 & 0 \\ 0 & -2 & 3 & -2 & 0 & 1 \end{bmatrix} \quad \begin{bmatrix} 1 & 0 & 7 & -3 & 4 & 0 \\ 0 & 1 & 0 & 1 & 3 & -2 \\ 0 & 0 & 1 & 0 & 2 & -1 \end{bmatrix}$$

$$\begin{bmatrix} 1 & 4 & -1 & 1 & 0 & 0 \\ 0 & -1 & 2 & -1 & 1 & 0 \\ 0 & -2 & 3 & -2 & 0 & 1 \end{bmatrix} \quad \begin{bmatrix} 1 & 0 & 7 & -3 & 4 & 0 \\ 0 & 1 & -2 & 1 & -1 & 0 \\ 0 & 0 & -1 & 0 & -2 & 1 \end{bmatrix} \quad \begin{bmatrix} 1 & 0 & 0 & -3 & -10 & 7 \\ 0 & 1 & 0 & 1 & 3 & -2 \\ 0 & 0 & 1 & 0 & 2 & -1 \end{bmatrix}$$

Thus, $A^{-1} = \begin{bmatrix} -3 & -10 & 7 \\ 1 & 3 & -2 \\ 0 & 2 & -1 \end{bmatrix}$ and

$$X = A^{-1}C = \begin{bmatrix} -3 & -10 & 7 \\ 1 & 3 & -2 \\ 0 & 2 & -1 \end{bmatrix} \begin{bmatrix} 4 \\ 8 \\ 13 \end{bmatrix} = \begin{bmatrix} -12 - 80 + 91 \\ 4 + 24 - 26 \\ 0 + 16 - 13 \end{bmatrix} = \begin{bmatrix} -1 \\ 2 \\ 3 \end{bmatrix}$$

This means that $x = -1$, $y = 2$, and $z = 3$.

EXAMPLE 4 Use matrices to solve the system of equations:

$$x + 2y - z = -4$$
$$3x + 5y - z = -5$$
$$-2x - y - 2z = -5$$

Setting up matrices A, C, and X, we have

$$A = \begin{bmatrix} 1 & 2 & -1 \\ 3 & 5 & -1 \\ -2 & -1 & -2 \end{bmatrix} \qquad C = \begin{bmatrix} -4 \\ -5 \\ -5 \end{bmatrix} \qquad X = \begin{bmatrix} x \\ y \\ z \end{bmatrix}$$

Finding A^{-1} (see Example 4 of Section 16-5) and solving for X, we have

$$A^{-1} = \begin{bmatrix} \frac{11}{2} & -\frac{5}{2} & -\frac{3}{2} \\ -4 & 2 & 1 \\ -\frac{7}{2} & \frac{3}{2} & \frac{1}{2} \end{bmatrix}$$

$$X = A^{-1}C = \begin{bmatrix} \frac{11}{2} & -\frac{5}{2} & -\frac{3}{2} \\ -4 & 2 & 1 \\ -\frac{7}{2} & \frac{3}{2} & \frac{1}{2} \end{bmatrix} \begin{bmatrix} -4 \\ -5 \\ -5 \end{bmatrix} = \begin{bmatrix} -2 \\ 1 \\ 4 \end{bmatrix}$$

This means that the solution is $x = -2$, $y = 1$, $z = 4$.

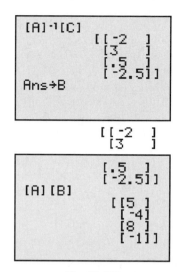

Fig. 16-17

EXAMPLE 5   Use a calculator to perform the necessary matrix operations in solving the following system of equations.

$$2r + 4s - t + u = 5$$
$$r - 2s + 3t - u = -4$$
$$3r + s + 2t - 4u = 8$$
$$4r + 5s - t + 3u = -1$$

First we set up matrices A, X, and C.

$$A = \begin{bmatrix} 2 & 4 & -1 & 1 \\ 1 & -2 & 3 & -1 \\ 3 & 1 & 2 & -4 \\ 4 & 5 & -1 & 3 \end{bmatrix} \qquad X = \begin{bmatrix} r \\ s \\ t \\ u \end{bmatrix} \qquad C = \begin{bmatrix} 5 \\ -4 \\ 8 \\ -1 \end{bmatrix}$$

It is now necessary only to enter matrices A and C in the calculator and find the matrix product $A^{-1}C$, as shown in the upper window in Fig. 16-17. (There is no need to record or display A^{-1}.) This shows that the solution is

$$r = -2 \qquad s = 3 \qquad t = 0.5 \qquad u = -2.5$$

This solution can be checked on the calculator by storing the resulting matrix X as matrix B and finding the matrix product AB, which should equal matrix C as shown in Fig. 16-17. The product AB is equivalent to substituting each value into the original equations, as shown in Eq. (16-11). ----------▪

EXERCISES 16-6

In Exercises 1–8, solve the given systems of equations by using the inverse of the coefficient matrix. The numbers in parentheses refer to exercises from Section 16-5, where the inverses may be checked.

1. $2x - 5y = -14$ (1)
  $-2x + 4y = 11$

2. $-x + 5y = 4$ (3)
  $4x + 10y = -4$

3. $2x + 4y = -9$ (11)
  $-x - y = 2$

4. $2x + 5y = -6$ (13)
  $-x + 2y = -6$

5. $x - 3y - 2z = -8$ (17)
  $-2x + 7y + 3z = 19$
  $x - y - 3z = -3$

6. $x - y - 3z = -1$ (19)
  $-y - 2z = -2$
  $2x + y - z = 2$

7. $x + 3y + 2z = 5$ (21)
  $-2x - 5y - z = -1$
  $2x + 4y = -2$

8. $2x + 4y = -2$ (23)
  $3x + 4y - 2z = -6$
  $-x + y + 2z = 5$

In Exercises 9–16, solve the given systems of equations by using the inverse of the coefficient matrix.

9. $2x - 3y = 3$
  $4x - 5y = 4$

10. $x + 2y = 3$
  $3x + 4y = 11$

11. $2.5x + 2.8y = -3.0$
  $3.5x - 1.6y = 9.6$

12. $12x - 5y = -400$
  $31x + 25y = 180$

13. $x + 2y + 2z = -4$
  $4x + 9y + 10z = -18$
  $-x + 3y + 7z = -7$

14. $x - 4y - 2z = -7$
  $-x + 5y + 5z = 18$
  $3x - 7y + 10z = 38$

15. $2x + 4y + z = 5$
  $-2x - 2y - z = -6$
  $-x + 2y + z = 0$

16. $4x + y = 2$
  $-2x - y + 3z = -18$
  $2x + y - z = 8$

In Exercises 17–24, solve the given systems of equations by using the inverse of the coefficient matrix. Use a calculator to perform the necessary matrix operations and display the results and the check. See Example 5.

17. $2x - y - z = 7$
  $4x - 3y + 2z = 4$
  $3x + 5y + z = -10$

18. $6x + 2y + 9z = 13$
  $7x + 6y - 6z = 6$
  $5x - 4y + 3z = 15$

19. $u - 3v - 2w = 9$
  $3u + 2v + 6w = 20$
  $4u - v + 3w = 25$

20. $2x + y - z = 1$
  $3x - 2y - 8z = -3$
  $x + 3y + z = 10$

21. $x - 5y + 2z - t = -18$
  $3x + y - 3z + 2t = 17$
  $4x - 2y + z - t = -1$
  $-2x + 3y - z + 4t = 11$

22. $2p + q + 5r + s = 5$
  $p + q - 3r - 4s = -1$
  $3p + 6q - 2r + s = 8$
  $2p + 2q + 2r - 3s = 2$

23. $2v + 3w + x - y - 2z = 6$
  $6v - 2w - x + 3y - z = 21$
  $v + 3w - 4x + 2y + 3z = -9$
  $3v - w - x + 7y + 4z = 5$
  $v + 6w + 6x - 4y - z = -4$

24. $4x - y + 2z - 2t + u = -15$
  $8x + y - z + 4t - 2u = 26$
  $2x - 6y - 2z + t - u = 10$
  $2x + 5y + z - 3t + 8u = -22$
  $4x - 3y + 2z + 4t + 2u = -4$

In Exercises 25–28, solve the indicated systems of equations using the inverse of the coefficient matrix. In Exercises 27 and 28, it is necessary to set up the appropriate equations.

25. Forces *A* and *B* hold up a beam that weighs 254 N, as shown in Fig. 16-18. The equations used to find the forces are

$$A \sin 47.2° + B \sin 64.4° = 254$$
$$A \cos 47.2° - B \cos 64.4° = 0$$

Find the forces.

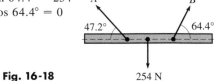

Fig. 16-18   254 N

26. In applying Kirchhoff's laws (see Exercise 20 on page 156) to the circuit shown in Fig. 16-19, the following equations are found. Determine the indicated currents (in A).

$$I_A + I_B + I_C = 0$$
$$2I_A - 5I_B = 6$$
$$5I_B - I_C = -3$$

27. A research chemist wants to make 10.0 L of gasoline containing 2.0% of a new experimental additive. Gasoline without additive and two mixtures of gasoline with additive, one with 5.0% and the other with 6.0%, are to be used. If four times as much gasoline without additive than the 5.0% mixture is to be used, how much of each is needed?

28. A river tour boat takes 5.0 h to cruise downstream and 7.0 h for the return upstream. If the river flows at 4.0 mi/h, how fast does the boat travel in still water, and how far downstream does the boat go before starting the return trip?

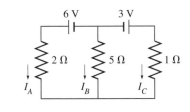

Fig. 16-19

CHAPTER EQUATIONS

Determinants

$$\begin{vmatrix} a_1 & b_1 & c_1 \\ a_2 & b_2 & c_2 \\ a_3 & b_3 & c_3 \end{vmatrix} = a_1b_2c_3 + a_3b_1c_2 + a_2b_3c_1 - a_3b_2c_1 - a_1b_3c_2 - a_2b_1c_3 \qquad (16\text{-}1)$$

$$\begin{vmatrix} a_1 & b_1 & c_1 \\ a_2 & b_2 & c_2 \\ a_3 & b_3 & c_3 \end{vmatrix} = a_1(b_2c_3 - b_3c_2) - a_2(b_1c_3 - b_3c_1) + a_3(b_1c_2 - b_2c_1) \qquad (16\text{-}2)$$

Minors

$$\begin{vmatrix} a_1 & b_1 & c_1 \\ a_2 & b_2 & c_2 \\ a_3 & b_3 & c_3 \end{vmatrix} = a_1 \begin{vmatrix} b_2 & c_2 \\ b_3 & c_3 \end{vmatrix} - a_2 \begin{vmatrix} b_1 & c_1 \\ b_3 & c_3 \end{vmatrix} + a_3 \begin{vmatrix} b_1 & c_1 \\ b_2 & c_2 \end{vmatrix} \qquad (16\text{-}3)$$

Basic laws for matrices

$$A + B = B + A \qquad \text{(commutative law)} \qquad (16\text{-}4)$$
$$A + (B + C) = (A + B) + C \qquad \text{(associative law)} \qquad (16\text{-}5)$$
$$k(A + B) = kA + kB \qquad (16\text{-}6)$$
$$A + O = A \qquad (16\text{-}7)$$

Inverse matrix

$$AA^{-1} = A^{-1}A = I \qquad (16\text{-}8)$$

Solving systems of equations by matrices

$$\begin{bmatrix} a_1x + b_1y \\ a_2x + b_2y \end{bmatrix} = \begin{bmatrix} c_1 \\ c_2 \end{bmatrix} \qquad (16\text{-}9)$$

$$A = \begin{bmatrix} a_1 & b_1 \\ a_2 & b_2 \end{bmatrix}, \qquad X = \begin{bmatrix} x \\ y \end{bmatrix}, \qquad C = \begin{bmatrix} c_1 \\ c_2 \end{bmatrix} \qquad (16\text{-}10)$$

$$AX = C \qquad (16\text{-}11)$$
$$X = A^{-1}C \qquad (16\text{-}12)$$

REVIEW EXERCISES

In Exercises 1–8, evaluate the given determinants by expansion by minors.

1. $\begin{vmatrix} 1 & 2 & -1 \\ 4 & 1 & -3 \\ -3 & -5 & 2 \end{vmatrix}$

2. $\begin{vmatrix} 3 & -1 & 2 \\ 7 & -1 & 4 \\ 2 & 1 & -3 \end{vmatrix}$

3. $\begin{vmatrix} -1 & 3 & -7 \\ 0 & 5 & 4 \\ 4 & -3 & -2 \end{vmatrix}$

4. $\begin{vmatrix} 60 & -54 & -76 \\ -10 & 24 & 40 \\ 25 & -37 & 18 \end{vmatrix}$

5. $\begin{vmatrix} 2 & 6 & 2 & 5 \\ 2 & 0 & 4 & -1 \\ 4 & -3 & 6 & 1 \\ 3 & -1 & 0 & -2 \end{vmatrix}$

6. $\begin{vmatrix} 1 & -2 & 2 & 4 \\ 0 & 1 & 2 & 3 \\ 3 & 2 & 2 & 5 \\ 2 & 1 & -2 & 0 \end{vmatrix}$

7. $\begin{vmatrix} 1 & 3 & -2 & 4 \\ 2 & 0 & 3 & -2 \\ 5 & -1 & 5 & -3 \\ -6 & 4 & -1 & 2 \end{vmatrix}$

8. $\begin{vmatrix} 2 & 3 & -1 & -1 \\ -3 & -2 & 5 & -6 \\ 2 & 1 & -3 & 2 \\ 4 & 0 & -2 & 1 \end{vmatrix}$

In Exercises 9–14, evaluate the determinants of Exercises 1–6 by using the basic properties of determinants.

In Exercises 15–20, evaluate the determinants of Exercises 1–6 by using a graphing calculator.

In Exercises 21–24, determine the values of the literal numbers.

21. $\begin{bmatrix} 2a \\ a - b \end{bmatrix} = \begin{bmatrix} 8 \\ 5 \end{bmatrix}$

22. $\begin{bmatrix} x - y \\ 2x + 2z \\ 4y + z \end{bmatrix} = \begin{bmatrix} 1 \\ 3 \\ -1 \end{bmatrix}$

23. $\begin{bmatrix} 2x & 3y & 2z \\ x + y & 2y + z & z - x \end{bmatrix} = \begin{bmatrix} 4 & -9 & 5 \\ a & b & c \end{bmatrix}$

24. $\begin{bmatrix} a + bj & b \\ aj & b - aj \end{bmatrix} = \begin{bmatrix} 6j & 2d \\ 2cj & ej^2 \end{bmatrix}$ $(j = \sqrt{-1})$

In Exercises 25–32, use the given matrices and perform the indicated operations.

$A = \begin{bmatrix} 2 & -3 \\ 4 & 1 \\ -5 & 0 \\ 2 & -3 \end{bmatrix}$ $B = \begin{bmatrix} -1 & 0 \\ 4 & -6 \\ -3 & -2 \\ 1 & -7 \end{bmatrix}$ $C = \begin{bmatrix} 5 & -6 \\ 2 & 8 \\ 0 & -2 \end{bmatrix}$

25. $A + B$

26. $2C$

27. $-3B$

28. $B - A$

29. $A - C$

30. $2C - B$

31. $2A - 3B$

32. $2(A - B)$

In Exercises 33–36, perform the indicated matrix multiplications.

33. $\begin{bmatrix} 2 & -1 \\ -2 & 1 \end{bmatrix}\begin{bmatrix} 1 & -1 \\ 2 & -2 \end{bmatrix}$

34. $\begin{bmatrix} 6 & -4 & 1 & 0 \\ 2 & 0 & -4 & 3 \end{bmatrix}\begin{bmatrix} 7 & -1 & 6 \\ 4 & 0 & 1 \\ 3 & -2 & 5 \\ 9 & 1 & 0 \end{bmatrix}$

35. $\begin{bmatrix} -1 & 7 \\ 2 & 0 \\ 4 & -1 \end{bmatrix}\begin{bmatrix} 1 & -4 & 5 \\ 5 & 1 & 0 \end{bmatrix}$

36. $\begin{bmatrix} 0 & -1 & 6 \\ 8 & 1 & 4 \\ 7 & -2 & -1 \end{bmatrix}\begin{bmatrix} 5 & -1 & 7 & 1 & 5 \\ 0 & 1 & 0 & 4 & 1 \\ 1 & -2 & 3 & 0 & 1 \end{bmatrix}$

In Exercises 37–44, find the inverses of the given matrices.

37. $\begin{bmatrix} 2 & -5 \\ 2 & -4 \end{bmatrix}$

38. $\begin{bmatrix} -1 & -6 \\ 2 & 10 \end{bmatrix}$

39. $\begin{bmatrix} 7 & -1 \\ 4 & 8 \end{bmatrix}$

40. $\begin{bmatrix} 50 & -12 \\ 42 & -80 \end{bmatrix}$

41. $\begin{bmatrix} 1 & 1 & -2 \\ -1 & -2 & 1 \\ 0 & 3 & 4 \end{bmatrix}$

42. $\begin{bmatrix} -1 & -1 & 2 \\ 2 & 3 & 0 \\ 1 & 4 & 1 \end{bmatrix}$

43. $\begin{bmatrix} 2 & -4 & 3 \\ 4 & -6 & 5 \\ -2 & 1 & -1 \end{bmatrix}$

44. $\begin{bmatrix} 3 & 1 & -4 \\ -3 & 1 & -2 \\ -6 & 0 & 3 \end{bmatrix}$

In Exercises 45–52, solve the given systems of equations using the inverse of the coefficient matrix.

45. $2x - 3y = -9$
    $4x - y = -13$

46. $5A - 7B = 62$
    $6A + 5B = -6$

47. $33x + 52y = -450$
    $45x - 62y = 1380$

48. $0.24x - 0.26y = -3.1$
    $0.40x + 0.34y = -1.3$

49. $2u - 3v + 2w = 7$
    $3u + v - 3w = -6$
    $u + 4v + w = -13$

50. $2x + 2y - z = 8$
    $x + 4y + 2z = 5$
    $3x - 2y + z = 17$

51. $x + 2y + 3z = 1$
    $3x - 4y - 3z = 2$
    $7x - 6y + 6z = 2$

52. $3x + 2y + z = 2$
    $2x + 3y - 6z = 3$
    $x + 3y + 3z = 1$

In Exercises 53–56, solve the given systems of equations by determinants.

53. $3x - 2y + z = 6$
    $2x + 3z = 3$
    $4x - y + 5z = 6$

54. $7n + p + 2r = 3$
    $4n - 2p + 4r = -2$
    $2n + 3p - 6r = 3$

55. $2x - 3y + z - t = -8$
    $4x + 3z + 2t = -3$
    $2y - 3z - t = 12$
    $x - y - z + t = 3$

56. $3x + 2y - 2z - 2t = 0$
    $5y + 3z + 4t = 3$
    $6y - 3z + 4t = 9$
    $6x - y + 2z - 2t = -3$

In Exercises 57–60, solve the given systems of equations by using the inverse of the coefficient matrix. Use a calculator to perform the necessary matrix operations and display the results and the check.

57. $3x - y + 6z - 2t = 8$
$2x + 5y + z + 2t = 7$
$4x - 3y + 8z + 3t = -17$
$3x + 5y - 3z + t = 8$

58. $A + B + 2C - 3D = 15$
$3A + 3B - 8C - 2D = 9$
$6A - 4B + 6C + D = -6$
$2A + 2B - 4C - 2D = 8$

59. $4r - s + 8t - 2u + 4v = -1$
$3r + 2s - 4t + 3u - v = 4$
$3r + 3s + 2t + 5u + 6v = 13$
$6r - s + 2t - 2u + v = 0$
$r - 2s + 4t - 3u + 3v = 1$

60. $2v + 3w + 2x - 2y + 5z = -1$
$7v + 8w + 3x + y - 4z = 3$
$v - 2w - 4x - 4y - 8z = -9$
$3v - w + 7x + 5y - 3z = -18$
$4v + 5w + x + 3y - 6z = 7$

In Exercises 61 and 62, evaluate the given determinants by minors.

61. $\begin{vmatrix} 1 + \sqrt{2} & 2 - \sqrt{3} & 0 \\ 3 + \sqrt{5} & 7 + \sqrt{6} & \sqrt{2} \\ 2 + \sqrt{3} & 1 - \sqrt{2} & 0 \end{vmatrix}$

62. $\begin{vmatrix} \cos \frac{\pi}{3} & \sin \frac{\pi}{6} & \cos 0 \\ \cos \pi & \sin \pi & \tan \frac{\pi}{4} \\ \sin \frac{\pi}{2} & \cos \frac{\pi}{2} & \sin 0 \end{vmatrix}$

In Exercises 63 and 64, evaluate the given determinants by use of the basic properties of determinants.

63. $\begin{vmatrix} \ln e & \log_3 1 & \frac{1}{2} \log 100 \\ \ln \frac{1}{e} & \log_2 8 & \log 0.1 \\ \ln \sqrt{e} & \log 10 & \log_4 2 \end{vmatrix}$

64. $\begin{vmatrix} j & 1 & 1 + j \\ -j & -1 + j & -j \\ 2j & 2 & 2 + 3j \end{vmatrix}$ $(j = \sqrt{-1})$

In Exercises 65–68, use matrices A and B.

$A = \begin{bmatrix} 1 & 0 \\ 3 & 4 \end{bmatrix}$, $B = \begin{bmatrix} 0 & 1 & 0 \\ 0 & 0 & 1 \\ 1 & 0 & 0 \end{bmatrix}$

65. Find A^2, A^3, and A^4.
66. Show that $(A^2)^2 = A^4$.
67. Show that $B^3 = I$.
68. Show that $B^4 = B$.

In Exercises 69 and 70, use the matrix N.

$N = \begin{bmatrix} 0 & -1 \\ 1 & 0 \end{bmatrix}$

69. Show that $N^{-1} = -N$.
70. Show that $N^2 = -I$.

In Exercises 71–74, use matrices A and B.

$A = \begin{bmatrix} 1 & -2 \\ 0 & 3 \end{bmatrix}$ $B = \begin{bmatrix} -3 & 1 \\ 2 & -1 \end{bmatrix}$

71. Show that $(A + B)(A - B) \neq A^2 - B^2$.
72. Show that $(A + B)^2 \neq A^2 + 2AB + B^2$.
73. Show that the inverse of $2A$ is $A^{-1}/2$.
W 74. Without evaluating the determinant of B, explain why the determinant of $2B$ equals four times the determinant of B.

In Exercises 75–78, solve the given systems of equations by any appropriate method of this chapter.

75. Two electric resistors, R_1 and R_2, are tested with currents and voltages such that the following equations are found.

$2R_1 + 3R_2 = 26$
$3R_1 + 2R_2 = 24$

Find the resistances R_1 and R_2 (in Ω).

76. A company produces two products, each of which is processed in two departments. Considering the worker time available, the numbers x and y of each product produced each week can be found by solving the system of equations

$4.0x + 2.5y = 1200$
$3.2x + 4.0y = 1200$

Find x and y.

77. A beam is supported as shown in Fig. 16-20. Find the tension T by solving the system of equations.

$0.500F = 0.866T$
$0.866F + 0.500T = 350$

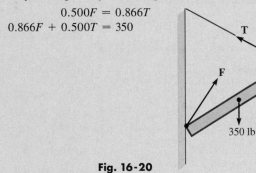

Fig. 16-20

78. To find the electric currents (in A) indicated in Fig. 16-21, it is necessary to solve the following equations.

$I_A + I_B + I_C = 0$
$5I_A - 2I_B = -4$
$2I_B - I_C = 0$

Find I_A, I_B, and I_C.

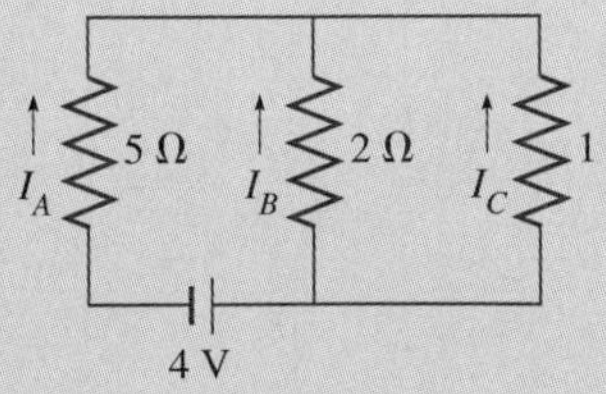

Fig. 16-21

In Exercises 79–84, solve the given problems by using any appropriate method of this chapter.

79. A crime suspect passes an intersection in a car traveling at 110 mi/h. The police pass the intersection 3.0 min later in a car traveling at 135 mi/h. How long is it before the police overtake the suspect?

80. A contractor needs a backhoe and a generator for two different jobs. Renting the backhoe for 5.0 h and the generator for 6.0 h costs $425 for one job. On the other job, renting the backhoe for 2.0 h and the generator for 8.0 h costs $310. What are the hourly charges for the backhoe and the generator?

81. By mass, three alloys have the following percentages of lead, zinc, and copper.

	Lead	Zinc	Copper
Alloy A	60%	30%	10%
Alloy B	40%	30%	30%
Alloy C	30%	70%	

How many grams of each of alloys A, B, and C must be mixed to get 100 g of an alloy that is 44% lead, 38% zinc, and 18% copper?

82. Three computer printers can print a total of 8200 lines/min when printing at the same time. With the first printing for 2 min and the second printing for 3 min, a total of 12,200 lines can be printed. With the first printing for 1 min, the second for 2 min, and the third for 3 min, a total of 17,600 lines can be printed. How many lines per minute can each print?

83. On a 750-mi trip from Salt Lake City to San Francisco that took a total of 5.5 h, a person took a limousine to the airport, then a plane, and finally a car to reach the final destination. The limousine took as long as the final car trip and the time for connections. The limousine averaged 55 mi/h, the plane averaged 400 mi/h, and the car averaged 40 mi/h. The plane traveled four times as far as the limousine and car combined. How long did each part of the trip and the connections take?

84. The area of a quadrilateral is

$$A = \frac{1}{2}\left(\begin{vmatrix} x_0 & x_1 \\ y_0 & y_1 \end{vmatrix} + \begin{vmatrix} x_1 & x_2 \\ y_1 & y_2 \end{vmatrix} + \begin{vmatrix} x_2 & x_3 \\ y_2 & y_3 \end{vmatrix} + \begin{vmatrix} x_3 & x_0 \\ y_3 & y_0 \end{vmatrix} \right)$$

where $(x_0, y_0), (x_1, y_1), (x_2, y_2),$ and (x_3, y_3) are the rectangular coordinates of the vertices of the quadrilateral, listed counterclockwise.

This *surveyor's formula* (see Exercise 34, page 151) can be generalized to find the area of any polygon. By carefully noting its form for a quadrilateral, write it for a triangle. Then express the form for a triangle with one third-order determinant.

In Exercises 85–88, perform the indicated matrix operations.

85. An automobile maker has two assembly plants at which cars with either 4, 6, or 8 cylinders and with either standard or automatic transmission are assembled. The annual production at the first plant of cars with the number of cylinders–transmission type (standard, automatic) is as follows:
4: 12,000, 15,000; 6: 24,000, 8000; 8: 4000, 30,000
At the second plant the annual production is
4: 15,000, 20,000; 6: 12,000, 3000; 8: 2000, 22,000
Set up matrices for this production and by matrix addition find the matrix for the total production by the number of cylinders and type of transmission.

86. Set up a matrix representing the information given in Exercise 81. A given shipment contains 500 g of alloy A, 800 g of alloy B, and 700 g of alloy C. Set up a matrix for this information. By multiplying these matrices, obtain a matrix that gives the total weight of lead, zinc, and copper in the shipment.

87. The matrix equation

$$\left[\begin{bmatrix} R_1 & -R_2 \\ -R_2 & R_1 \end{bmatrix} + R_2 \begin{bmatrix} 1 & 0 \\ 0 & 1 \end{bmatrix} \right] \begin{bmatrix} i_1 \\ i_2 \end{bmatrix} = \begin{bmatrix} 6 \\ 0 \end{bmatrix}$$

may be used to represent the system of equations relating the currents and resistances of the circuit in Fig. 16-22. Find this system of equations by performing the indicated matrix operations.

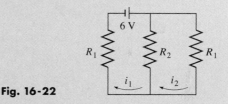

Fig. 16-22

88. A person prepared a meal of the following items, each having the given number of grams of protein, carbohydrates, and fat, respectively. Beef stew: 25, 21, 22; coleslaw: 3, 10, 10; (light) ice cream: 7, 25, 6. If the calorie count of each gram of protein, carbohydrate, and fat is 4.1 Cal/g, 3.9 Cal/g, and 8.9 Cal/g, respectively, find the total Calorie count of each item by matrix multiplication. (1 Cal = 1 kcal.)

Writing Exercise

89. A hardware company has 60 different retail stores in which 3500 different products are sold. Write a paragraph explaining why matrices provide an efficient method of inventory control for this company.

PRACTICE TEST

1. For matrices A and B, find $A - 2B$.

$$A = \begin{bmatrix} 3 & -1 & 4 \\ 2 & 0 & -2 \end{bmatrix} \qquad B = \begin{bmatrix} 1 & 4 & 5 \\ -1 & -2 & 3 \end{bmatrix}$$

2. Evaluate using minors:

$$\begin{vmatrix} 4 & 0 & -2 \\ 3 & -3 & 2 \\ -4 & 1 & -1 \end{vmatrix}$$

3. For matrices C and D, find CD and DC.

$$C = \begin{bmatrix} 1 & 0 & 4 \\ 2 & -2 & 1 \\ -1 & 3 & 2 \end{bmatrix} \qquad D = \begin{bmatrix} 2 & -2 \\ 4 & -5 \\ 6 & 1 \end{bmatrix}$$

4. Evaluate using the properties of determinants:

$$\begin{vmatrix} 1 & 0 & 4 & -2 \\ -2 & -1 & 3 & 0 \\ 3 & 2 & -1 & 2 \\ 1 & 1 & -1 & -2 \end{vmatrix}$$

5. For matrix C of Problem 3, find C^{-1}.

6. Solve by using the inverse of the coefficient matrix.

$$2x - 3y = 11$$
$$x + 2y = 2$$

7. Solve the following system of equations by using the inverse of the coefficient matrix. Use a calculator to perform the necessary matrix operations and display the result and check. Write down the display shown in the window with the solutions.

$$7x - 2y + z = 6$$
$$2x + 3y - 4z = 6$$
$$4x - 5y + 2z = 10$$

8. Fifty shares of stock A and 30 shares of stock B cost $2600. Thirty shares of stock A and 40 shares of stock B cost $2000. What is the price per share of each stock? Solve by setting up the appropriate equations and then using the inverse of the coefficient matrix.

17 INEQUALITIES

Until now we have devoted a great deal of time to the solution of equations. Solving equations is very important in mathematics, but there are also many situations in mathematics and its applications when the solution of an inequality is required.

In mathematics, for example, we state the domain of a function as being *all values* that give real numbers for the function. Inequalities are also very useful in later topics in mathematics, such as calculus.

In electricity it may be necessary to find the values of a current that are *less than* a specified value. In designing a link in a robotic mechanism, it might be necessary to determine the forces that are *greater than* a specified value. Computers are often programmed to switch from one part of a program to another based on a result that is *greater than* (or *less than*) some given value.

An important application occurs in business, where inequalities are used to set production levels for maximizing profits or minimizing costs.

In our discussion of inequalities, we will find use of many of the functions and methods developed in earlier chapters.

In Section 17-5 we use inequalities to show how a company can maximize profit in making products such as speaker systems.

17-1 PROPERTIES OF INEQUALITIES

In Chapter 1 we first introduced the signs of inequality. To this point only a basic understanding of their meanings has been necessary to show certain intervals associated with a variable. In this section we review the meanings and develop certain basic properties of inequalities. We also show the meaning of the solution of an inequality and how it is shown on the number line.

The expression $a < b$ is read as "*a* is less than *b*," and the expression $a > b$ is read as "*a* is greater than *b*." *These signs define what is known as the* **sense** *(indicated by the direction of the sign) of the inequality.* Two inequalities are said to have the same sense if the signs of inequality point in the same direction. They are said to have the opposite sense if the signs of inequality point in opposite directions. *The two sides of the inequality are called* **members** *of the inequality.*

EXAMPLE 1   The inequalities $x + 3 > 2$ and $x + 1 > 0$ have the same sense, as do the inequalities $3x - 1 < 4$ and $x^2 - 1 < 3$.

The inequalities $x - 4 < 0$ and $x > -4$ have the opposite sense, as do the inequalities $2x + 4 > 1$ and $3x^2 - 7 < 1$.   ∎

The **solution** *of an inequality consists of all values of the variable that make the inequality a true statement.* Most inequalities with which we shall deal are **conditional inequalities,** *which are true for some, but not all, values of the variable.* Also, *some inequalities are true for all values of the variable, and they are called* **absolute inequalities.** A solution of an inequality consists of only real numbers, as the terms *greater than* or *less than* have not been defined for complex numbers.

EXAMPLE 2   The inequality $x + 1 > 0$ is true for all values of x greater than -1. Therefore, the values of x that satisfy this inequality are written as $x > -1$. This illustrates the difference between the solution of an equation and the solution of an inequality. The solution of an equation normally consists of a few specific numbers, whereas *the solution to an inequality normally consists of an interval of values of the variable.* Any and all values within this interval are termed solutions of the inequality. Since the inequality $x + 1 > 0$ is satisfied only by the values of x in the interval $x > -1$, it is a *conditional inequality*.

NOTE ▸

The inequality $x^2 + 1 > 0$ is true for all real values of x, since x^2 is never negative. It is an *absolute inequality*.   ∎

There are occasions when it is convenient to combine an inequality with an equality. For such purposes, the symbols $\leq$, meaning *less than or equal to,* and $\geq$, meaning *greater than or equal to,* are used.

EXAMPLE 3   If we wish to state that x is positive, we can write $x > 0$. However, the value zero is not included in the solution. If we wish to state that x is not negative, we write $x \geq 0$. Here, zero is part of the solution.

In order to state that x is less than or equal to -5, we write $x \leq -5$.   ∎

In the sections that follow, we will solve inequalities. It is often useful to show the solution on the number line. The next example shows how this is done.

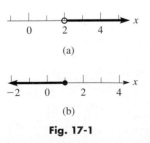

(a)

(b)

Fig. 17-1

EXAMPLE 4   **(a)** To graph $x > 2$, we draw a small open circle at 2 on the number line (which is equivalent to the *x*-axis). Then we draw a solid line to the right of this point with an arrowhead pointing to the right, indicating all values greater than 2. See Fig. 17-1(a). The *open circle* shows that the point is not part of the indicated solution.

(b) To graph $x \leq 1$, we follow the same basic procedure as in part (a), except that we use a solid circle and the arrowhead points to the left. See Fig. 17-1(b). The *solid circle* shows that the point is part of the indicated solution.   ∎

PROPERTIES OF INEQUALITIES

We shall now show the basic operations performed on inequalities. These are the same operations as those performed on equations, but in certain cases the results take on a different form. *The following are the* **properties of inequalities.**

1. *The sense of an inequality is not changed when the same number is added to—or subtracted from—both members of the inequality.* Symbolically, this may be stated as "if $a > b$, then $a + c > b + c$ and $a - c > b - c$."

■**EXAMPLE 5**   Using Property 1 on the inequality $9 > 6$, we have the following results.

Fig. 17-2

$9 > 6$	$9 > 6$
add 4 to each member	subtract 12 from each member
$9 + 4 > 6 + 4$	$9 - 12 > 6 - 12$
$13 > 10$	$-3 > -6$

In Fig. 17-2 we see that 9 is to the right of 6, 13 is to the right of 10, and -3 is to the right of -6.

2. *The sense of an inequality is not changed if both members are multiplied or divided by the same positive number.* Symbolically, this is stated as "if $a > b$, then $ac > bc$, and $a/c > b/c$, provided that $c > 0$."

■**EXAMPLE 6**   Using Property 2 on the inequality $8 < 15$, we have the following results.

$8 < 15$	$8 < 15$
multiply both members by 2	divide both members by 2
$2(8) < 2(15)$	$\dfrac{8}{2} < \dfrac{15}{2}$
$16 < 30$	$4 < \dfrac{15}{2}$

CAUTION ▶

3. *The sense of an inequality is* **reversed** *if both members are multiplied or divided by the same negative number.* Symbolically, this is stated as "if $a > b$, then $ac < bc$, and $a/c < b/c$, provided that $c < 0$." Be very careful to note that *we obtain different results depending on whether both members are multiplied by a positive number or by a negative number.*

■**EXAMPLE 7**   Using Property 3 on the inequality $4 > -2$, we have the following results.

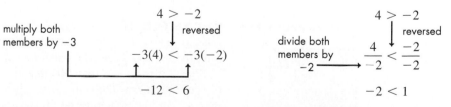

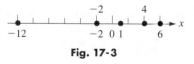

Fig. 17-3

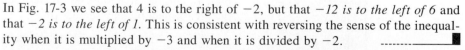

In Fig. 17-3 we see that 4 is to the right of -2, but that *-12 is to the left of 6* and that *-2 is to the left of 1.* This is consistent with reversing the sense of the inequality when it is multiplied by -3 and when it is divided by -2.

4. *If both members of an inequality are positive numbers and n is a positive integer, then the inequality formed by taking the nth power of each member, or the nth root of each member, is in the same sense as the given inequality. Symbolically, this is stated as "if $a > b$, then $a^n > b^n$, and $\sqrt[n]{a} > \sqrt[n]{b}$, provided that $n > 0$, $a > 0$, $b > 0$."*

EXAMPLE 8   Using Property 4 on the inequality $16 > 9$, we have:

$$16 > 9 \qquad\qquad\qquad 16 > 9$$

square both members $\qquad$ take square root of both members

$$16^2 > 9^2 \qquad\qquad \sqrt{16} > \sqrt{9}$$

$$256 > 81 \qquad\qquad\qquad 4 > 3$$

Many inequalities have more than two members. In fact, inequalities with three members are common. All the basic properties hold for inequalities with more than two members. Some care must be used, however, in stating these inequalities.

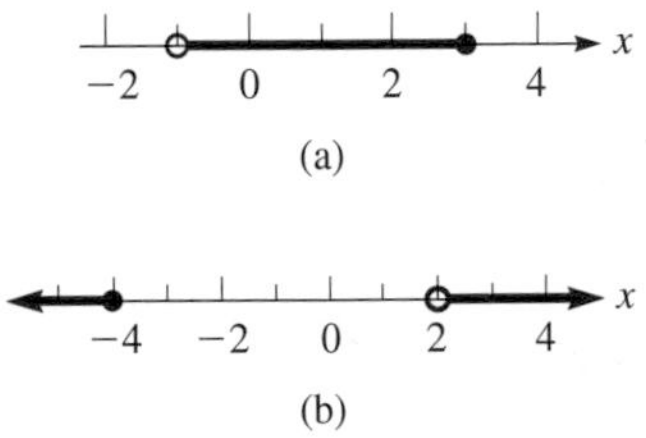

EXAMPLE 9   **(a)** To state that 5 is less than 6, and also greater than 2, we may write $2 < 5 < 6$, or $6 > 5 > 2$. (Generally the *less than* form is preferred.)

(b) To state that a number x may be greater than -1 *and* also less than or equal to 3, we write $-1 < x \le 3$. (It can also be written as $x > -1$ *and* $x \le 3$.) This is shown in Fig. 17-4(a). Note the use of the open circle and the solid circle.

CAUTION▶   **(c)** By writing $x \le -4$ *or* $x > 2$, we state that x is less than or equal to -4, *or* greater than 2. ***It may not be stated as $2 < x \le -4$,*** for this shows x as being less than -4, and also greater than 2, and *no such numbers exist.* See Fig. 17-4(b).

NOTE▶   Notice the use of the words *and* and *or* in Example 9. In stating inequalities, ***and*** is used when the solution consists of values that satisfy ***both*** statements. The word ***or*** is used when the solution consists of values that make ***either*** statement true. (In everyday speech, *or* can sometimes also mean that either one statement is true or another statement is true, but *not* that both statements are true.)

EXAMPLE 10   The inequality $x^2 - 3x + 2 > 0$ is satisfied if x is either greater than 2 *or* less than 1. This is written as $x > 2$ *or* $x < 1$, but it is incorrect to state it as $1 > x > 2$. (If we wrote it this way, we would be saying that the same value of x is less than 1 *and* at the same time greater than 2. Of course, as we noted for this type of situation in Example 9, no such number exists.) Any inequality must be valid for all values satisfying it. However, we could say that the inequality is not satisfied for $1 \le x \le 2$, which means those values of x greater than *or* equal to 1 *and* less than *or* equal to 2 (between or equal to 1 and 2).

NOTE▶

SOLVING A WORD PROBLEM   **EXAMPLE 11**   The design of a rectangular solar panel shows that the length l is between 80 cm and 90 cm and the width w between 40 cm and 80 cm. See Fig. 17-5. Find the values of area the panel may have.

Since l is to be less than 90 cm and w less than 80 cm, the area must be less than $(90 \text{ cm})(80 \text{ cm}) = 7200 \text{ cm}^2$. Also, since l is to be greater than 80 cm and w greater than 40 cm, the area must be greater than $(80 \text{ cm})(40 \text{ cm}) = 3200 \text{ cm}^2$. Therefore, the area A may be represented as

$$3200 \text{ cm}^2 < A < 7200 \text{ cm}^2$$

This means the area is greater than 3200 cm^2 *and* less than 7200 cm^2.

(a)

(b)

Fig. 17-4

Fig. 17-5 $40 \text{ cm} < w < 80 \text{ cm}$

$80 \text{ cm} < l < 90 \text{ cm}$

A

EXAMPLE 12   A semiconductor *diode* has the property that an electric current flows through it in only one direction. If it is an alternating-current circuit, the current in the circuit flows only during the half-cycle when the diode allows it to flow. If a source of current given by $i = 2 \sin \pi t$ (i in mA, t in seconds) is connected in series with a diode, write the inequalities for the current and the time. Assume the source is on for 3.0 s and a positive current passes through the diode.

We are to find the values of t that correspond to $i > 0$. From the properties of the sine function, we know that $2 \sin \pi t$ has a period of $2\pi/\pi = 2.0$ s. Therefore, the current is zero for $t = 0$, 1.0 s, 2.0 s, and 3.0 s.

The source current is positive for $0 < t < 1.0$ s and for 2.0 s $< t < 3.0$ s.

The source current is negative for 1.0 s $< t < 2.0$ s.

Therefore, in the circuit

$$i > 0 \quad \text{for} \quad 0 < t < 1.0 \text{ s} \quad \text{and} \quad 2.0 \text{ s} < t < 3.0 \text{ s, and}$$
$$i = 0 \quad \text{for} \quad t = 0, \, 1.0 \text{ s} \leq t \leq 2.0 \text{ s}.$$

A graph of the current in the circuit as a function of time is shown in Fig. 17-6(a). In Fig. 17-6(b) the values of t for which $i > 0$ are shown.

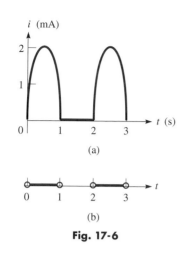

(a)

(b)

Fig. 17-6

EXERCISES *17-1*

In Exercises 1–8, for the inequality $4 < 9$, *state the inequality that results when the given operations are performed on both members.*

1. Add 3.
2. Subtract 6.
3. Multiply by 5.
4. Multiply by -2.
5. Divide by -1.
6. Divide by 2.
7. Square both.
8. Take square roots.

In Exercises 9–20, give the inequalities equivalent to the following statements about the number x.

9. Greater than -2
10. Less than 7
11. Less than or equal to 4
12. Greater than or equal to -6
13. Greater than 1 and less than 7
14. Greater than or equal to -2 and less than 6
15. Less than -9, or greater than or equal to -4
16. Less than or equal to 8, or greater than or equal to 12
17. Less than 1, or greater than 3 and less than or equal to 5
18. Greater than or equal to 0 and less than or equal to 2, or greater than 5
19. Greater than -2 and less than 2, or greater than or equal to 3 and less than 4
20. Less than -4, or greater than or equal to 0 and less than or equal to 1, or greater than or equal to 5

Ⓦ *In Exercises 21–24, give verbal statements equivalent to the given inequalities involving the number x.*

21. $0 < x \leq 2$
22. $x < 5$ or $x > 7$

23. $x < -1$ or $1 \leq x < 2$
24. $-1 \leq x < 3$ or $5 < x < 7$

In Exercises 25–36, graph the given inequalities on the number line.

25. $x < 3$
26. $x \geq -1$
27. $x \leq 1$ or $x > 3$
28. $x < -3$ or $x \geq 0$
29. $0 \leq x < 5$
30. $-4 < y < -2$
31. $x < -1$ or $1 \leq x < 4$
32. $-1 < x < 2$ or $x > 3$
33. $-3 < x < -1$ or $1 < x \leq 3$
34. $1 < x \leq 2$ or $3 \leq x < 4$
35. $t < -3$ or $t > -3$
36. $x < 1$ or $1 < x \leq 4$

In Exercises 37–44, some applications of inequalities are shown.

37. When *Pioneer II* left the solar system in 1989, it was 3×10^{12} mi from the sun. Assuming it never returns, express its distance d from the sun in the future as an inequality. Graph these distances of d.

38. A busy person glances at a digital clock that shows 9:36. Another glance a short time later shows the clock at 9:44. Express the amount of time t (in min) that could have elapsed between glances by use of inequalities. Graph these values of t.

39. An earth satellite put into orbit near the earth's surface will have an elliptic orbit if its velocity v is between 18,000 mi/h and 25,000 mi/h. Write this as an inequality and graph these values of v.

40. Fossils found in Jurassic rocks indicate that dinosaurs flourished during the Jurassic geological period, 140 MY (million years ago) to 200 MY. Write this as an inequality, with *t* representing past time. Graph the values of *t*.

41. In executing a program, a computer must perform a set of calculations. Any one of the calculations takes no more than 2565 steps. Express the number *n* of steps required for a given calculation by an inequality. (Note that *n* is a positive *integer*.)

42. The velocity *v* of an ultrasound wave in soft human tissue may be represented as 1550 ± 60 m/s, where the ± 60 m/s gives the possible variation in the velocity. Express the possible velocities by an inequality.

43. The electric intensity *E* within a charged spherical conductor is zero. The intensity on the surface and outside the sphere equals a constant *k* divided by the square of the distance *r* from the center of the sphere. State these relations for a sphere of radius *a* by using inequalities and graph *E* as a function of *r*.

44. If the current from the source in Example 12 is $i = 5 \cos 4\pi t$ and the diode allows only negative current to flow, write the inequalities and draw the graph for the current in the circuit as a function of time for $0 \le t \le 1$ s.

17-2 SOLVING LINEAR INEQUALITIES

Using the properties and definitions discussed in Section 17-1, we can now proceed to solve inequalities. In this section we solve linear inequalities in one variable. Similar to linear functions as defined in Chapter 5, *a* **linear inequality** *is one in which each term contains only one variable and the exponent of each variable is 1.* We will consider linear inequalities in two variables in Section 17-5.

The procedure for solving a linear inequality in one variable is like that we used in solving basic equations in Chapter 1. We solve the inequality by isolating the variable, and to do this we perform the same operations on each member of the inequality. The operations are based on the properties given in Section 17-1.

■EXAMPLE 1   In each of the following inequalities, by performing the indicated operation we isolate *x* and thereby solve the inequality.

$x + 2 < 4$	$\dfrac{x}{2} > 4$	$2x \le 4$
Subtract 2 from each member.	Multiply each member by 2.	Divide each member by 2.
$x < 2$	$x > 8$	$x \le 2$

Each solution can be checked by substituting any number in the indicated interval into the original inequality. For example, any value less than 2 will satisfy the first inequality, whereas 2 or any number less than 2 will satisfy the third inequality. ■

■EXAMPLE 2   Solve the following inequality: $3 - 2x \ge 15$.
We have the following solution:

$$3 - 2x \ge 15 \qquad \text{original inequality}$$
$$-2x \ge 12 \qquad \text{subtract 3 from each member}$$

inequality reversed ⟶

$$x \le -6 \qquad \text{divide each member by } -2$$

CAUTION▶   Again, carefully note that *the sign of inequality was reversed when each number was divided by* -2. We check the solution by substituting -7 in the original inequality, obtaining $17 \ge 15$. ■

EXAMPLE 3   Solve the inequality $2x \leq 3 - x$.

The solution proceeds as follows:

$$2x \leq 3 - x \qquad \text{original inequality}$$
$$3x \leq 3 \qquad \text{add } x \text{ to each member}$$
$$x \leq 1 \qquad \text{divide each member by 3}$$

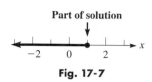

Part of solution

Fig. 17-7

This solution checks and is represented in Fig. 17-7, as we showed in Section 17-1.

This inequality could have been solved by combining x-terms on the right. In doing so, we would obtain $1 \geq x$. Since this might be misread, it is best to combine the variable terms on the left, as we did above.

EXAMPLE 4   Solve the inequality $\frac{3}{2}(1 - x) > \frac{1}{4} - x$.

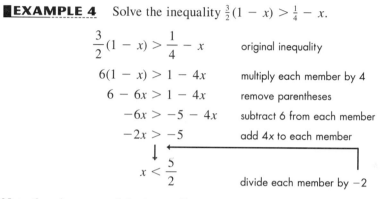

$$\frac{3}{2}(1 - x) > \frac{1}{4} - x \qquad \text{original inequality}$$
$$6(1 - x) > 1 - 4x \qquad \text{multiply each member by 4}$$
$$6 - 6x > 1 - 4x \qquad \text{remove parentheses}$$
$$-6x > -5 - 4x \qquad \text{subtract 6 from each member}$$
$$-2x > -5 \qquad \text{add } 4x \text{ to each member}$$
$$x < \frac{5}{2} \qquad \text{divide each member by } -2$$

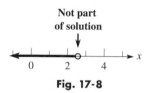

Not part of solution

Fig. 17-8

Note that the sense of the inequality was reversed when we divided by -2. This solution is shown in Fig. 17-8. Any value of $x < 5/2$ checks when substituted into the original inequality.

The following example illustrates an application that involves the solution of an inequality.

EXAMPLE 5   The velocity v (in ft/s) of a missile in terms of the time t (in s) is given by $v = 960 - 32t$. For how long is the velocity positive? (Since velocity is a vector, this can also be interpreted as asking "how long is the missile moving upward?")

In terms of inequalities, we are asked to find the values of t for which $v > 0$. This means that we must solve the incquality $960 - 32t > 0$. The solution is as follows:

$$960 - 32t > 0 \qquad \text{original inequality}$$
$$-32t > -960 \qquad \text{subtract 960 from each member}$$
$$t < 30 \text{ s} \qquad \text{divide each member by } -32$$

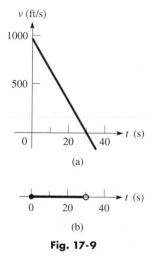

(a)

(b)

Fig. 17-9

Negative values of t have no meaning in this problem. Checking $t = 0$, we find that $v = 960$ ft/s. Therefore, the complete solution is $0 \leq t < 30$ s.

In Fig. 17-9(a) we show the graph of $v = 960 - 32t$, and in Fig. 17-9(b) we show the solution $0 \leq t < 30$ s on the number line (which is really the t-axis in this case). Note that the values of v are above the t-axis for those values of t that are part of the solution. This shows the relationship of the graph of v as a function of t, and the solution as graphed on the number line (the t-axis).

Inequalities with Three Members

■**EXAMPLE 6**   Solve: $-1 < 2x + 3 < 6$.
We have the following solution.

$$-1 < 2x + 3 < 6 \qquad \text{original inequality}$$

$$-4 < 2x < 3 \qquad \text{subtract 3 from each member}$$

$$-2 < x < \frac{3}{2} \qquad \text{divide each member by 2}$$

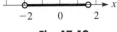
Fig. 17-10

The solution is shown in Fig. 17-10.

■**EXAMPLE 7**   Solve the inequality $2x < x - 4 \le 3x + 8$.
Since we cannot isolate x in the middle member (or in any member), we rewrite the inequality as

$$2x < x - 4 \quad \text{and} \quad x - 4 \le 3x + 8$$

We then solve each of these inequalities, keeping in mind that the solution must satisfy both of them. Therefore, we have

$$2x < x - 4 \quad \text{and} \quad x - 4 \le 3x + 8$$
$$-2x \le 12$$
$$x < -4 \quad \text{and} \quad x \ge -6$$

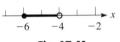
Fig. 17-11

The solution can be written as $-6 \le x < -4$, and this solution is shown in Fig. 17-11.

SOLVING A WORD PROBLEM

■**EXAMPLE 8**   In emptying a wastewater tank, one pump can remove no more than 40 L/min. If it operates for 8.0 min and a second pump operates for 5.0 min, what must be the pumping rate of the second pump if 480 L are to be removed?
Let $x =$ the pumping rate of the first pump and $y =$ the pumping rate of the second pump. Since the first operates for 8.0 min and the second for 5.0 min to remove 480 L, we have

$$\underset{\text{pump}}{\overset{\text{first}}{}} \quad \underset{\text{pump}}{\overset{\text{second}}{}} \quad \text{total} \longleftarrow \text{amounts pumped}$$
$$8.0x + 5.0y = 480$$

Since we know that the first pump can remove no more than 40 L/min, which means that $0 \le x \le 40$ L/min, we solve for x, then substitute in this inequality.

$$x = 60 - 0.625y \qquad \text{solve for } x$$
$$0 \le 60 - 0.625y \le 40 \qquad \text{substitute in inequality}$$
$$-60 \le -0.625y \le -20 \qquad \text{subtract 60 from each member}$$
$$96 \ge y \ge 32 \qquad \text{divide each member by } -0.625$$
$$32 \le y \le 96 \text{ L/min} \qquad \text{use } \le \text{ symbol (optional step)}$$

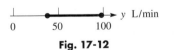
Fig. 17-12

This means that the second pump must be able to pump at least 32 L/min and no more than 96 L/min. See Fig. 17-12.
We note that although this was a three-member inequality and it was combined with equalities, the solution was performed in the same way as with a two-member inequality.

Solving an Inequality with a Calculator

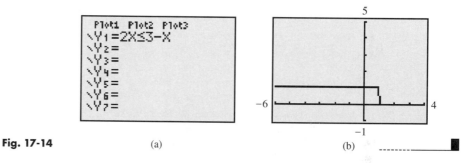

Most graphing calculators can be used to show the graphical solution of an inequality. We will show this method for linear inequalities in the examples that follow. We will also use it with nonlinear inequalities in the next section.

On most calculators, a value of 1 is shown if an inequality is satisfied, and a value of 0 is shown if the inequality is not satisfied. Thus, if we enter $8 > 3$ (consult the manual to determine how $>$ is entered), the calculator displays 1, and if we then enter $3 > 8$ it shows 0, as seen in the display in Fig. 17-13. We can also use an inequality feature to graphically show the solution of an inequality. This is illustrated in the following examples.

EXAMPLE 9 Display the solution of the inequality $2x \leq 3 - x$ (see Example 3) on a graphing calculator.

We set y_1 *equal to the inequality,* as shown in Fig. 17-14(a), and then graph y_1. From Fig. 17-14(b) we see that $y_1 = 1$ up to $x = 1$. Thus, the solution that is shown **NOTE ▶** is $x < 1$. We should note that *this method does not distinguish between $<$ and $\leq$ (or $>$ and $\geq$)*, so the same solution would be displayed for $2x < 3 - x$. Substituting $x = 1$ into the inequality, we find $2 = 2$, and therefore the complete solution is $x \leq 1$. Compare Fig. 17-14(b) with Fig. 17-7.

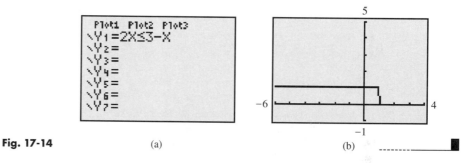

Fig. 17-14   (a)   (b)

EXAMPLE 10 Display the solution of the inequality $-1 < 2x + 3 < 6$ (see Example 6) on a graphing calculator.

In order to display the solution, we must write the inequality as

$$-1 < 2x + 3 \qquad \text{and} \qquad 2x + 3 < 6 \qquad \text{see Example 7}$$

We then enter $y_1 = -1 < 2x + 3$ and $2x + 3 < 6$ (consult the manual to determine how "and" is entered) in the calculator as shown in Fig. 17-15(a). From Fig. 17-15(b) we see that the solution is $-2 < x < 1.5$. The *trace* and *zoom* features can be used to get accurate values. Compare Fig. 17-15(b) with Fig. 17-10.

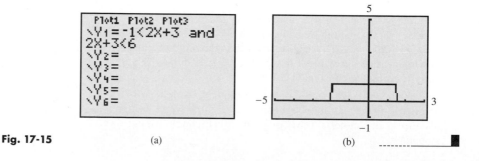

Fig. 17-15   (a)   (b)

EXERCISES *17-2*

In Exercises 1–24, solve the given inequalities. Graph each solution.

1. $x - 3 > -4$

2. $x + 2 \leq 6$

3. $\frac{1}{2}x < 3$

4. $-4t > 12$

5. $3x - 5 \leq -11$

6. $\frac{1}{3}x + 2 \geq 1$

7. $6 - y > 8$

8. $3 - 3x < -1$

9. $\frac{4x - 5}{2} \leq x$

10. $1.50 - 5.24x > 3.75 + 2.25x$

11. $2 - (x + 1) > x + 3$

12. $-2(x + 4) > \frac{1 - 5x}{3}$

13. $2.50(1.50 - 3.40x) < 3.84 - 8.45x$

14. $2x - 7 \leq 4 - (x + 2)$

15. $\frac{1}{3} - \frac{x}{2} < x + \frac{3}{2}$

16. $\frac{x}{5} - 2 > \frac{2}{3}(x + 3)$

17. $-1 < 2x + 1 < 3$

18. $2 < 3R + 1 \leq 8$

19. $-4 \leq 1 - x < -1$

20. $0 \leq 3 - 2x \leq 6$

21. $2x < x - 1 \leq 3x + 5$

22. $x + 1 \leq 7 - x < 2x$

23. $2s - 3 < s - 5 < 3s - 3$

24. $x - 1 < 2x + 2 < 3x + 1$

In Exercises 25–32, solve the inequalities by displaying the solutions on a graphing calculator. See Examples 9 and 10.

25. $3x - 2 < 8 - x$

26. $4(x - 2) > x + 6$

27. $\frac{1}{2}(x + 15) \geq 5 - 2x$

28. $\frac{1}{3}x - 2 \leq \frac{1}{2}x + 1$

29. $1 < 5 - 2t < 9$

30. $-3 < 2 - \frac{s}{3} \leq -1$

31. $x - 3 < 2x + 5 < 6x + 7$

32. $x - 3 < 2x + 4 \leq 1 - x$

In Exercises 33–44, solve the given problems by setting up and solving appropriate inequalities. Graph each solution.

33. Determine the values of x that are in the domain of the function $f(x) = \sqrt{2x - 10}$.

34. Determine the values of x that are in the domain of the function $f(x) = 1/\sqrt{3 - 0.5x}$.

35. The voltage drop V across a resistor is the product of the current i (in A) and the resistance R (in Ω). Find the possible voltage drops across a variable resistor R, if the minimum and maximum resistances are 1.6 kΩ and 3.6 kΩ, respectively, and the current is constant at 2.5 mA.

36. The minimum legal speed on a certain interstate highway is 45 mi/h, and the maximum legal speed is 65 mi/h. What legal distances can a motorist travel in 4 h on this highway without stopping?

37. The value V (in $) of each building lot in a development is estimated as $V = 40{,}000 + 4000t$, where t is the time in years. For how long is the value of each lot no more than $64,000?

38. A beam is supported at each end as shown in Fig. 17-16. Analyzing the forces on the beam leads to the equation $F_1 = 13 - 3d$. For what values of d is F_1 more than 6 N?

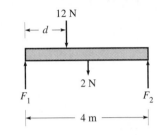

Fig. 17-16

39. The mass m (in g) of silver plate on a dish is increased by electroplating. The mass of silver on the plate is given by $m = 125 + 15.0t$, where t is the time (in h) of electroplating. For what values of t is m between 131 g and 164 g?

40. The relationship between Fahrenheit degrees F and Celsius degrees C is $5F = 9C + 160$. For what values of C is $F \geq 98.6°$? (98.6° is normal body temperature.)

41. During a given rush hour, the numbers of vehicles shown in Fig. 17-17 go in the indicated directions in a one-way street section of a city. By finding the possible values of x and y and the equation relating x and y find the possible values of y.

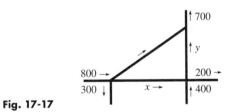

Fig. 17-17

42. One computer printer can print no more than 2500 lines/min. If it operates for 3.0 min and a second printer operates for 2.0 min, what printing rates must the second printer have in order to print 9000 lines?

43. In obtaining a solution of hydrochloric acid containing 2 L of pure acid, a chemist mixes x liters of a 5% solution with y liters of a 10% solution. If y is no more than 8 L, what are the possible values of x?

44. An oil company plans to install eight storage tanks, each with a capacity of x liters, and five additional tanks, each with a capacity of y liters, such that the total capacity of all tanks is 440,000 L. If capacity y will be at least 40,000 L, what are the possible values of capacity x?

17-3 SOLVING NONLINEAR INEQUALITIES

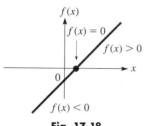

Fig. 17-18

In this section we develop methods of solving inequalities that involve polynomial and fractional expressions. Among these methods are a graphical method and a review of the calculator method, both of which can be used with inequalities that include nonfactorable and nonalgebraic expressions. In order to develop the method for inequalities with polynomial expressions, we shall first take another look at a linear inequality.

If we graph the linear function $f(x) = ax + b$ ($a \neq 0$), we see that all values of $f(x)$ are positive for all values of x on one side of the point where $f(x) = 0$, and all values of $f(x)$ are negative on the other side of this point. See Fig. 17-18. This leads us to another method of solving a linear inequality. This method is to *express the given inequality with **zero** on the right side and then determine the **sign** of the resulting function on either side of the zero of the function.*

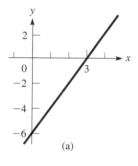

(a)

Zero of
$f(x) = 2x - 6$

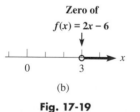

(b)

Fig. 17-19

EXAMPLE 1   Solve the inequality $2x - 5 > 1$.

We first find the equivalent inequality with zero on the right. This is done by subtracting 1 from each member. Thus, we have $2x - 6 > 0$. We now set the left member equal to zero. Thus,

$$2x - 6 = 0 \quad \text{for} \quad x = 3$$

which means that 3 is the zero of the function $f(x) = 2x - 6$. We know that the function $f(x)$ has one sign for $x < 3$ and has the opposite sign for $x > 3$. Testing values in these intervals, we find, for example, that

$$f(x) = -2 \quad \text{for} \quad x = 2 \quad \text{and} \quad f(x) = +2 \quad \text{for} \quad x = 4$$

Thus, for $x > 3$, $2x > 6$, and this means the solution to the original inequality is $x > 3$. The solution shown in Fig. 17-19(b) corresponds to the positive values of the function $f(x) = 2x - 6$, shown in Fig. 17-19(a).

We could have solved this inequality by any of the methods developed in the previous section, but the important idea here is to use the *sign* of the function, with *zero* on the right.

We can extend this method to solving inequalities with polynomials of higher degree. We first find the equivalent inequality with zero on the right and then *factor the function on the left into linear factors and any quadratic factors that lead to complex roots.* As all values of x are considered, each linear factor can change sign at the value for which it is zero. The quadratic factors do not change sign.

This method is especially useful in solving inequalities involving fractions. If a linear factor occurs in the numerator, the function is zero at the value of x for which the linear factor is zero. If such a factor appears in the denominator, the function is undefined where the factor is zero. *The values of x for which a function is zero or undefined are called the* **critical values** *of the function.* As all negative and positive values of x are considered (starting with negative numbers of large absolute value, proceeding through zero, and ending with large positive numbers), a function can change sign only at a critical value.

CRITICAL VALUES

On the next page we outline the method of solving an inequality by using the critical values of the function. Then several examples of the method are shown.

> ### Using Critical Values to Solve an Inequality
>
> 1. *Determine the equivalent inequality with zero on the right.*
> 2. *Find all linear factors of the function.*
> 3. *To find the critical values, set each linear factor equal to zero and solve for x.*
> 4. *Determine the sign of the function to the left of the leftmost critical value, between critical values, and to the right of the rightmost critical value.*
> 5. *Those intervals in which the function has the proper sign satisfy the inequality.*

EXAMPLE 2   Solve the inequality $x^2 - 3 > 2x$.

We first find the equivalent inequality with zero on the right. Therefore, we have $x^2 - 2x - 3 > 0$. We then factor the left member and have

$$(x + 1)(x - 3) > 0$$

Setting each factor equal to zero, we find the left critical value is -1 and the right critical value is 3. All values of x to the left of -1 give the same sign for the function. All values between -1 and 3 give the function the same sign. All values of x to the right of 3 give the same sign to the function. Therefore, ***we must determine the sign of $f(x)$ for each of the intervals $x < -1$, $-1 < x < 3$, and $x > 3$.***

For $x < -1$, both factors are negative, which means their product is positive, or $(x + 1)(x - 3) > 0$. For $-1 < x < 3$, the left factor is positive and the right factor is negative, which means their product is negative, or $(x + 1)(x - 3) < 0$. For $x > 3$, both factors are positive, which means their product is positive, or $(x + 1)(x - 3) > 0$.

Summarizing these results for $f(x) = (x + 1)(x - 3)$, we have

If $x < -1, f(x) > 0$;   if $-1 < x < 3, f(x) < 0$;   if $x > 3, f(x) > 0$

Therefore, the solution to the inequality is $x < -1$ or $x > 3$.

The solution that is shown in Fig. 17-20(b) corresponds to the positive values of the function $f(x) = x^2 - 2x - 3$, shown in Fig. 17-20(a).   ■

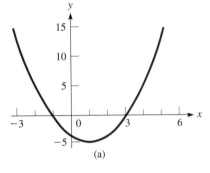

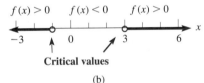

Critical values

(b)

Fig. 17-20

EXAMPLE 3   Solve the inequality $x^3 - 4x^2 + x + 6 < 0$.

By methods developed in Chapter 15, we factor the function on the left and obtain $(x + 1)(x - 2)(x - 3) < 0$. The critical values are $-1, 2, 3$. We wish to determine the sign of the left member for the intervals $x < -1$, $-1 < x < 2$, $2 < x < 3$, and $x > 3$. The following table shows each interval, the sign of each factor in each interval, and the resulting sign of the function

$$f(x) = (x + 1)(x - 2)(x - 3)$$

Interval	$(x + 1)(x - 2)(x - 3)$			Sign of $f(x)$
$x < -1$	$-$	$-$	$-$	$-$
$-1 < x < 2$	$+$	$-$	$-$	$+$
$2 < x < 3$	$+$	$+$	$-$	$-$
$x > 3$	$+$	$+$	$+$	$+$

Since we want values for $f(x) < 0$, the solution is $x < -1$ or $2 < x < 3$. The solution shown in Fig. 17-21(b) corresponds to values of $f(x) < 0$ in Fig. 17-21(a).   ■

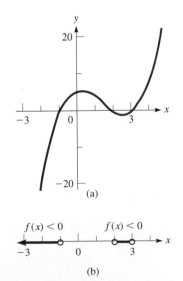

(b)

Fig. 17-21

EXAMPLE 4 Solve the inequality $x^3 - x^2 + x - 1 > 0$.

In order to factor this function, we can use the methods of Chapter 15, or we might note that it is factorable by grouping (see Section 6-2). Either method leads to the inequality

$$(x^2 + 1)(x - 1) > 0$$

In this case we have a quadratic factor $x^2 + 1$ that leads to imaginary roots (j and $-j$) and is never negative. This means we have only one linear factor and therefore only one critical value.

Setting $x - 1 = 0$, we get the critical value $x = 1$. Since $x - 1$ is positive for $x > 1$ and negative for $x < 1$, the solution to the inequality is $x > 1$.

The solution is shown in Fig. 17-22(b), and we see that this solution corresponds to the positive values of $f(x) = x^3 - x^2 + x - 1$ shown in Fig. 17-22(a).

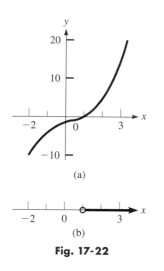

(a)

(b)

Fig. 17-22

The following example illustrates an applied situation that involves the solution of an inequality.

EXAMPLE 5 The force F (in N) acting on a cam varies according to the time t (in s) and it is given by the function $F = 2t^2 - 12t + 20$. For what values of t, $0 \le t \le 6$ s, is the force at least 4 N?

For a force of at least 4 N, we know that $F \ge 4$ N, or $2t^2 - 12t + 20 \ge 4$. This means we are to solve the inequality $2t^2 - 12t + 16 \ge 0$, and the solution is as follows:

$$2t^2 - 12t + 16 \ge 0$$
$$t^2 - 6t + 8 \ge 0$$
$$(t - 2)(t - 4) \ge 0$$

The critical values are $t = 2$ and $t = 4$, which lead to the following table.

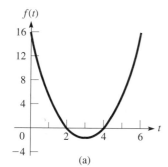

(a)

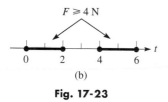

(b)

Fig. 17-23

Interval	$(t - 2)(t - 4)$		Sign of $(t - 2)(t - 4)$
$0 \le t < 2$	−	−	+
$2 < t < 4$	+	−	−
$4 < t \le 6$	+	+	+

We see that the values of t that satisfy the *greater than* part of the problem are $0 \le t < 2$ and $4 < t \le 6$. Since we know that $(t - 2)(t - 4) = 0$ for $t = 2$ and $t = 4$, the solution is

$$0 \le t \le 2 \text{ s} \qquad \text{or} \qquad 4 \text{ s} \le t \le 6 \text{ s}$$

The graph of $f(t) = 2t^2 - 12t + 16$ is shown in Fig. 17-23(a). The solution, shown in Fig. 17-23(b), corresponds to the values of $f(t)$ that are zero or positive or for which $F \ge 4$ N.

We note here that if the cam rotates in 6-s intervals, the force on the cam is periodic, varying from 2 N to 20 N.

EXAMPLE 6 Find the values of x for which $\sqrt{\dfrac{x-3}{x+4}}$ represents a real number.

For the expression to represent a real number, the fraction under the radical must be greater than or equal to zero. This means we must solve the inequality

$$\frac{x-3}{x+4} \geq 0$$

The critical values are found from the factors that are in the numerator or in the denominator. Thus, the critical values are -4 and 3. Considering now the *greater than* part of the $\geq$ sign, we set up the following table.

Interval	$\dfrac{x-3}{x+4}$	*Sign of* $\dfrac{x-3}{x+4}$
$x < -4$	$\dfrac{-}{-}$	$+$
$-4 < x < 3$	$\dfrac{-}{+}$	$-$
$x > 3$	$\dfrac{+}{+}$	$+$

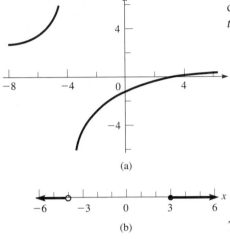

(a)

(b)

Fig. 17-24

Thus, the values that satisfy the *greater than* part of the problem are those for which $x < -4$ or for which $x > 3$. Now, considering the equality part of the $\geq$ sign, we note that $x = 3$ is valid, for the fraction is zero. However, *if $x = -4$, we have division by zero, and thus x may not equal -4.* Therefore, the inequality is satisfied for $x < -4$ or $x \geq 3$, and these are the values for which the original expression represents a real number. The graph of $f(x) = (x-3)/(x+4)$ is shown in Fig. 17-24(a) and the graph of the solution is shown in Fig. 17-24(b).

EXAMPLE 7 Solve the inequality $\dfrac{(x-2)^2(x+3)}{4-x} < 0$.

The critical values are -3, 2, and 4. Thus, we have the following table.

Interval	$\dfrac{(x-2)^2(x+3)}{4-x}$	*Sign of* $\dfrac{(x-2)^2(x+3)}{4-x}$
$x < -3$	$\dfrac{+\quad -}{+}$	$-$
$-3 < x < 2$	$\dfrac{+\quad +}{+}$	$+$
$2 < x < 4$	$\dfrac{+\quad +}{+}$	$+$
$x > 4$	$\dfrac{+\quad +}{-}$	$-$

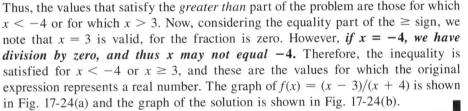

(a)

(b)

Fig. 17-25

Thus, the solution is $x < -3$ or $x > 4$. The graph of the function is shown in Fig. 17-25(a), and the graph of the solution is shown in Fig. 17-25(b).

EXAMPLE 8   Solve the inequality $\dfrac{x+3}{x-1} > 2$.

CAUTION

NOTE

It seems that all we have to do is multiply both members by $x - 1$ and solve the resulting linear inequality. However, *this will lead to an incorrect result.* The reason is that we assume $x - 1$ is positive when we multiply, but $x - 1$ is negative for $x < 1$. To avoid this problem, *first subtract 2 from each member* and combine terms on the left. *This procedure should be used with any inequality involving fractions with the variable in the denominator.* Therefore, the solution is

$$\frac{x+3}{x-1} - 2 > 0 \qquad \text{subtract 2 from both members}$$

$$\frac{x+3-2(x-1)}{x-1} > 0 \qquad \text{combine over the common denominator}$$

$$\frac{5-x}{x-1} > 0 \qquad \text{this is the form to use}$$

The critical values are 1 and 5, and we have the following table of signs.

Interval	$(5-x)/(x-1)$	Sign of $(5-x)/(x-1)$
$x < 1$	$+ \;/\; -$	$-$
$1 < x < 5$	$+ \;/\; +$	$+$
$x > 5$	$- \;/\; +$	$-$

Thus, the solution is $1 < x < 5$. The graph of $f(x) = (5-x)/(x-1)$ is shown in Fig. 17-26(a), and the graph of the solution is shown in Fig. 17-26(b). ∎

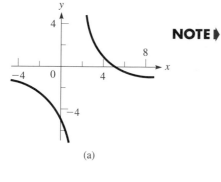

(a)

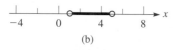

(b)

Fig. 17-26

CALCULATOR SOLUTION

We can use the graphing calculator method of displaying the solution that was developed in Examples 9 and 10 on page 457. This method can be used for any of the inequalities of this section, and we illustrate it again in the following example.

EXAMPLE 9   Solve the inequality of Example 8, $\dfrac{x+3}{x-1} > 2$.

On the calculator, we set $y_1 = (x+3)/(x-1) > 2$, and we then have the display shown in Fig. 17-27. This confirms the solution of $1 < x < 5$. ∎

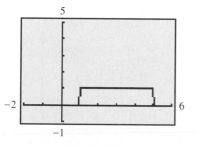

Fig. 17-27

Solving Inequalities Graphically

We can get an approximate solution of an inequality from the graph of a function, including functions that are not factorable or not algebraic. The method follows several of the earlier examples. First, write the equivalent inequality with zero on the right and then graph this function. Those values of x corresponding to the proper values of y (either above or below the x-axis) are those that satisfy the inequality.

EXAMPLE 10   Use a graphing calculator to solve the inequality $x^3 > x^2 - 3$.

Finding the equivalent inequality with zero on the right, we have $x^3 - x^2 + 3 > 0$. On the calculator we then let $y_1 = x^3 - x^2 + 3$ and graph this function. This gives us the calculator display shown in Fig. 17-28.

We can see that the curve crosses the x-axis only once, between $x = -2$ and $x = -1$. Using the *trace* and *zoom* features (or *zero* feature), we find that this value is about $x = -1.17$. Since we want values of x that correspond to positive values of y, the solution is $x > -1.17$. ∎

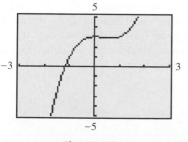

Fig. 17-28

EXERCISES *17-3*

In Exercises 1–20, solve the given inequalities. Graph each solution. It is suggested that you also graph the function on a graphing calculator as a check.

1. $x^2 - 1 < 0$

2. $x^2 + 3x \geq 0$

3. $2x^2 \leq 4x$

4. $x^2 - 4x > 5$

5. $2x^2 - 12 \leq -5x$

6. $9t^2 + 6t > -1$

7. $x^2 + 4x \leq -4$

8. $6x^2 + 1 < 5x$

9. $x^2 + 4 > 0$

10. $x^4 + 2 < 1$

11. $x^3 + x^2 - 2x > 0$

12. $x^3 - 2x^2 + x \geq 0$

13. $s^3 + 2s^2 - s \geq 2$

14. $n^4 - 2n^3 + 8n + 12 \leq 7n^2$

15. $\dfrac{2x - 3}{x + 6} \leq 0$

16. $\dfrac{x + 5}{x - 1} > 0$

17. $\dfrac{x^2 - 6x - 7}{x + 5} > 0$

18. $\dfrac{(x - 2)^2(5 - x)}{(4 - x)^3} \leq 0$

19. $\dfrac{x}{x + 1} > 1$

20. $\dfrac{2p}{p - 1} > 3$

In Exercises 21–28, solve the inequalities by displaying the solutions on a graphing calculator. See Example 9.

21. $3x^2 + 5x \geq 2$

22. $12x^2 + x > 1$

23. $\dfrac{x - 8}{3 - x} < 0$

24. $\dfrac{3x + 1}{x + 3} \geq 0$

25. $\dfrac{6 - x}{3 - x - 4x^2} \geq 0$

26. $\dfrac{4 - x}{3 + 2x - x^2} > 0$

27. $\dfrac{x^4(9 - x)(x - 5)(2 - x)}{(4 - x)^5} > 0$

28. $\dfrac{2}{x - 3} < 4$

In Exercises 29–32, determine the values for x for which the radicals represent real numbers.

29. $\sqrt{(x - 1)(x + 2)}$

30. $\sqrt{x^2 - 3x}$

31. $\sqrt{-x - x^2}$

32. $\sqrt{\dfrac{x^3 + 6x^2 + 8x}{3 - x}}$

In Exercises 33–40, solve the given inequalities graphically by using a graphing calculator. See Example 10.

33. $x^3 - x > 2$

34. $0.5x^3 < 3 - 2x^2$

35. $x^4 < x^2 - 2x - 1$

36. $3x^4 + x + 1 > 5x^2$

37. $2^x > x + 2$

38. $\log x < 1 - 2x^2$

39. $\sin x < 0.1x^2 - 1$

40. $4 \cos 2x > 2x - 3$

In Exercises 41–48, answer the given questions by solving the appropriate inequalities.

41. The electric power p (in W) delivered to part of a circuit is given by $p = 6i - 4i^2$, where i is the current (in A). For what positive values of i is the power greater than 2 W?

42. The weight w (in tons) of fuel in a rocket after launch is $w = 2000 - t^2 - 140t$, where t is the time (in min). During what period of time is the weight of fuel greater than 500 tons?

43. In programming a computer, the formula $63n = 2^x - 1$ may be used. Here, n is the number of items to be added and x is the number of bits needed to represent the sum. Find x if $n < 100$.

44. The object distance p (in cm) and image distance q (in cm) for a camera of focal length 3.00 cm is given by $p = 3.00q/(q - 3.00)$. For what values of q is $p > 12.0$ cm?

45. The length of a rectangular microprocessor chip is 2.0 mm more than its width. If its area is less than 35 mm^2, what values are possible for the width if it must be at least 3.0 mm?

46. A laser source is 2.0 in. from the nearest point P on a flat mirror, and the laser beam is directed at a point Q that is on the mirror and is x in. from P. The beam is then reflected to the receiver, which is x in. from Q. What is x if the total length of the beam is greater than 6.5 in.? See Fig. 17-29.

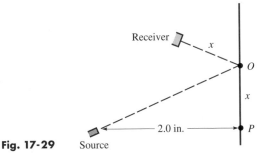

Fig. 17-29   Source

47. A plane takes off from Winnipeg and flies due east at 620 km/h. At the same time, a second plane takes off from the surface of Lake Winnipeg 310 km due north of Winnipeg and flies due north at 560 km/h. For how many hours are the planes less than 1000 km apart?

48. An open box (no top) is formed from a piece of cardboard 8.00 in. square by cutting equal squares from the corners, turning up the resulting sides, and taping the edges together. Find the edges of the squares that are cut out in order that the volume of the box is greater than 32.0 in.^{3}.

17-4 INEQUALITIES INVOLVING ABSOLUTE VALUES

Inequalities involving absolute values are often useful in later topics in mathematics such as calculus, and in applications such as the accuracy of measurements. In this section we show the meaning of such inequalities and how they are solved.

If we wish to write the inequality $|x| > 1$ without absolute-value signs, we must note that we are considering values of x that are *numerically* larger than 1. Thus, we may write this inequality in the equivalent form $x < -1$ or $x > 1$. We now note that *the original inequality, with an absolute-value sign, can be written in terms of two equivalent inequalities, neither involving absolute values.* If we are asked to write the inequality $|x| < 1$ without absolute-value signs, we write $-1 < x < 1$, since we are considering values of x numerically less than 1.

Following reasoning similar to this, whenever absolute values are involved in inequalities, the following two relations allow us to write equivalent inequalities without absolute values.

> If $|f(x)| > n$, then $f(x) < -n$ or $f(x) > n$. (17-1)
>
> If $|f(x)| < n$, then $-n < f(x) < n$. (17-2)

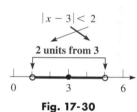

$|x - 3| < 2$

2 units from 3

Fig. 17-30

■EXAMPLE 1   Solve the inequality $|x - 3| < 2$.

Here, we want values of x such that $x - 3$ is numerically smaller than 2, or the values of x within 2 units of $x = 3$. These are given by the inequality $1 < x < 5$. Now, using Eq. (17-2), we have

$$-2 < x - 3 < 2$$

By adding 3 to all three members of this inequality, we have

$$1 < x < 5$$

which is the proper interval. See Fig. 17-30. ■

■EXAMPLE 2   Solve the inequality $|2x - 1| > 5$.

By using Eq. (17-1), we have

$$2x - 1 < -5 \quad \text{or} \quad 2x - 1 > 5$$

Completing the solution, we have

$$2x < -4 \quad \text{or} \quad 2x > 6 \qquad \text{add 1 to each member}$$
$$x < -2 \quad \text{or} \quad x > 3 \qquad \text{divide each member by 2}$$

Fig. 17-31

This means that the given inequality is satisfied for $x < -2$ or for $x > 3$. We must be very careful to remember that *we cannot write this as $3 < x < -2$*. The solution is shown in Fig. 17-31.

The meaning of this inequality is that the numerical value of $2x - 1$ is greater than 5. By considering values in these intervals, we can see that this is true for values of x less than -2 or greater than 3. ■

EXAMPLE 3   Solve the inequality $2\left|\dfrac{2x}{3} + 1\right| \geq 4$.

The solution is as follows:

$$2\left|\frac{2x}{3} + 1\right| \geq 4 \qquad \text{original inequality}$$

$$\left|\frac{2x}{3} + 1\right| \geq 2 \qquad \text{divide each member by 2}$$

$$\frac{2x}{3} + 1 \leq -2 \quad \text{or} \quad \frac{2x}{3} + 1 \geq 2 \qquad \text{using Eq. (17-1)}$$

$$2x + 3 \leq -6 \quad \text{or} \quad 2x + 3 \geq 6$$

$$2x \leq -9 \quad \text{or} \quad 2x \geq 3$$

$$x \leq -\frac{9}{2} \quad \text{or} \quad x \geq \frac{3}{2} \qquad \text{solution}$$

Fig. 17-32

This solution is shown in Fig. 17-32. Note that the sign of equality does not change the method of solution. It simply indicates that $-\frac{9}{2}$ and $\frac{3}{2}$ are included in the solution.

EXAMPLE 4   Solve the inequality $|3 - 2x| < 3$.
We have the following solution.

$$|3 - 2x| < 3 \qquad \text{original inequality}$$

$$-3 < 3 - 2x < 3 \qquad \text{using Eq. (17-2)}$$

$$-6 < -2x < 0$$

$$3 > x > 0 \qquad \text{divide by } -2 \text{ and reverse signs of inequality}$$

$$0 < x < 3 \qquad \text{solution}$$

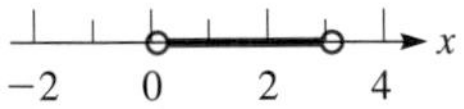

Fig. 17-33

The meaning of the inequality is that the numerical value of $3 - 2x$ is less than 3. This is true for values of x between 0 and 3. The solution is shown in Fig. 17-33.

EXAMPLE 5   A technician measures an electric current and reports that it is 0.036 A with a possible error of ± 0.002 A. Write this result for the current i using an inequality with absolute values.

The statement of the problem tells us that the current is no less than 0.034 A and no more than 0.038 A. Another way of stating this is that the numerical difference between the true value of i (unknown exactly) and the measured value, 0.036 A, is less than or equal to 0.002 A. Using an absolute-value inequality, this is written as

$$|i - 0.036| \leq 0.002$$

where values are in amperes.
We can see that this inequality is correct by using (Eq. 17-2).

$$-0.002 \leq i - 0.036 \leq 0.002$$

$$0.034 \leq i \leq 0.038 \qquad \text{add 0.036 to each member}$$

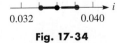

Fig. 17-34

This verifies that i should not be less than 0.034 A or no more than 0.038 A. The solution is shown in Fig. 17-34.

CALCULATOR SOLUTION

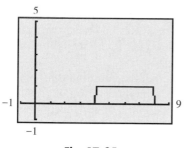

Fig. 17-35

Most graphing calculators can be used to display the solution of an inequality that involves absolute values as they can display the solutions to other inequalities (see Examples 9 and 10 on page 457 and Example 9 on page 463.) Since calculators differ in their operation, check the manual of your calculator to see how an absolute value is entered and displayed. The following example shows the display of the solution of an inequality on a graphing calculator.

■**EXAMPLE 6** Display the solution to the inequality $\left| \dfrac{x}{2} - 3 \right| < 1$ on a graphing calculator.

On the calculator we set $y_1 = \text{abs}(x/2 - 3) < 1$ and obtain the display shown in Fig. 17-35. From this display we see that the solution is $4 < x < 8$. It is possible that it will take two or three window settings to get an accurate solution. ■

EXERCISES *17-4*

In Exercises 1–20, solve the given inequalities. Graph each solution.

1. $|x - 4| < 1$

2. $|x + 1| < 3$

3. $|5x + 4| > 6$

4. $\left| \dfrac{1}{2}N - 1 \right| > 1$

5. $|6x - 5| \le 4$

6. $|5 - x| \le 2$

7. $|3 - 4x| > 3$

8. $|3x + 1| \ge 2$

9. $\left| \dfrac{t + 1}{5} \right| < 5$

10. $\left| \dfrac{2x - 9}{4} \right| < 1$

11. $|20x + 85| \le 43$

12. $|2.6x - 9.1| > 10.4$

13. $2|x - 4| > 8$

14. $3|4 - 3x| \le 10$

15. $4|2 - 5x| \ge 6$

16. $2.5|7.1 - 2.0x| \le 6.5$

17. $\left| \dfrac{3R}{5} + 1 \right| < 8$

18. $\left| \dfrac{4x}{3} - 5 \right| \ge 7$

19. $\left| 6.5 - \dfrac{x}{2} \right| \ge 2.3$

20. $\left| 27 - \dfrac{2x}{3} \right| > 17$

In Exercises 21–24, solve the given inequalities by displaying the solutions on a graphing calculator. See Example 6..

21. $|2x - 5| < 3$

22. $2|6 - x| > 5$

23. $\left| 4 - \dfrac{x}{2} \right| \ge 1$

24. $\left| \dfrac{x}{3} + 2 \right| \le 2$

In Exercises 25–28, solve the given quadratic inequalities. Check each by displaying the solution on a graphing calculator.

25. $|x^2 + x - 4| > 2$
(After using Eq. (17-1), you will have two inequalities. The solution includes the values of x that satisfy *either* of the inequalities.)

26. $|x^2 + 3x - 1| > 3$ (See Exercise 25.)

27. $|x^2 + x - 4| < 2$
(Use Eq. (17-2), then treat the resulting inequality as two inequalities of the form $f(x) > -n$ and $f(x) < n$. The solution includes the values of x that satisfy *both* of the inequalities.)

28. $|x^2 + 3x - 1| < 3$ (See Exercise 27.)

In Exercises 29–32, use inequalities involving absolute values to solve the given problems.

(W) **29.** The production p (in barrels) of oil at a refinery is estimated at $2{,}000{,}000 \pm 200{,}000$. Express p using an inequality with absolute values, and describe the production in a verbal statement.

30. The *Mach number M* of a moving object is the ratio of its velocity v to the velocity of sound v_s, and v_s varies with temperature. A jet traveling at 1650 km/h changes its altitude from 500 m to 5500 m. At 500 m (with the temperature at 27°C), $v_s = 1250$ km/h, and at 5500 m (-3°C), $v_s = 1180$ km/h. Express the range of M using an inequality with absolute values.

31. The voltage V in a certain circuit is given by $V = 6.0 - 200i$, where i is the current (in A). For what values of the current is the absolute value of the voltage less than 2.0 V?

32. A rocket is fired from a plane flying horizontally at 9000 ft. The height h (in ft) of the rocket above the plane is given by $h = 560t - 16t^2$, where t is the time (in s) of flight of the rocket. When is the rocket more than 4000 ft above or below the plane? See Fig. 17-36.

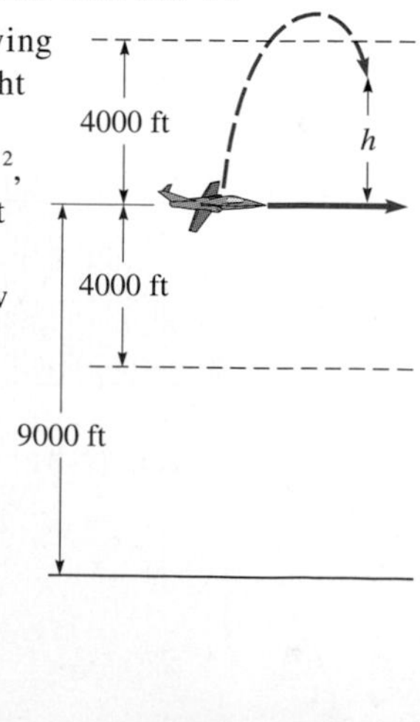

4000 ft

h

4000 ft

9000 ft

Fig. 17-36

$17\text{-}5$ GRAPHICAL SOLUTION OF INEQUALITIES WITH TWO VARIABLES

To this point we have considered inequalities with one variable and certain methods of solving them. We may also graphically solve inequalities involving two variables, such as x and y. In this section we consider the solution of such inequalities, as well as one important type of application.

Let us consider the function $y = f(x)$. We know that the coordinates of points on the graph satisfy the equation $y = f(x)$. However, for points above the graph of the function, we have $y > f(x)$, and for points below the graph of the function we have $y < f(x)$. Consider the following example.

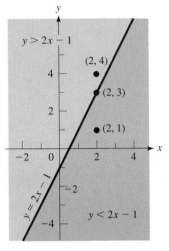

Fig. 17-37

EXAMPLE 1   Consider the linear function $y = 2x - 1$, the graph of which is shown in Fig. 17-37. This equation is satisfied for points on the line. For example, the point $(2, 3)$ is on the line, and we have $3 = 2(2) - 1 = 3$. Therefore, for points on the line, we have $y = 2x - 1$, or $y - 2x + 1 = 0$.

The point $(2, 4)$ is above the line, since we have $4 > 2(2) - 1$, or $4 > 3$. Therefore, for points above the line, we have $y > 2x - 1$, or $y - 2x + 1 > 0$. In the same way, for points below the line, $y < 2x - 1$ or $y - 2x + 1 < 0$. We note this is true for the point $(2, 1)$, since $1 < 2(2) - 1$, or $1 < 3$.

The line for which $y = 2x - 1$ and the regions for which $y > 2x - 1$ and for which $y < 2x - 1$ are shown in Fig. 17-37.

Summarizing,

$$y > 2x - 1 \qquad \text{for points } \textit{above} \text{ the line}$$
$$y = 2x - 1 \qquad \text{for points } \textit{on} \text{ the line}$$
$$y < 2x - 1 \qquad \text{for points } \textit{below} \text{ the line} \qquad \text{--------}\blacksquare$$

The illustration in Example 1 leads us to the graphical method of indicating the points that satisfy an inequality with two variables. First, we solve the inequality for y and then determine the graph of the function $y = f(x)$. *If we wish to solve the inequality $y > f(x)$, we indicate the appropriate points by shading in the region above the curve. For the inequality $y < f(x)$, we indicate the appropriate points by shading in the region below the curve.* We note that the complete solution to the inequality consists of all points in an entire region of the plane.

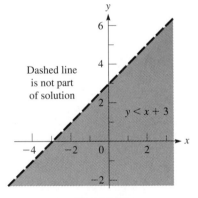

Dashed line is not part of solution

Fig. 17-38

EXAMPLE 2   Draw a sketch of the graph of the inequality $y < x + 3$.

First, we draw the function $y = x + 3$, as shown by the dashed line in Fig. 17-38. Since we wish to find all the points that satisfy the inequality $y < x + 3$, we show these points by shading in the region below the line. The line is shown as a *dashed line* to indicate that points on it do not satisfy the inequality.

Most graphing calculators can be used to display the solution of an inequality involving two variables by shading in an area above a curve, below a curve, or between curves. The manner in which this is done varies according to the model of the calculator. Therefore, the manual should be used to determine how this is done on any particular model of calculator. In Fig. 17-39, such a graphing calculator display is shown for this inequality. $\text{--------}\blacksquare$

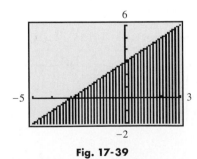

Fig. 17-39

EXAMPLE 3   After a snowstorm, it is estimated that it will take 30 min to plow each mile of Route 15 and 45 min to plow each mile of Route 80. If no more than 60 plowing-hours are available, what combinations of Route 15 and Route 80 can be plowed?

Let x = miles of Route 15 that can be plowed and y = miles of Route 80 that can be plowed. The time to plow along each route is the product of the time for each mile and the number of miles to be plowed. This gives us

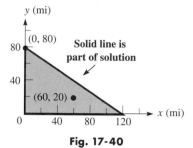

Fig. 17-40

$$\underset{\underset{\text{30 min} \rightharpoondown}{\uparrow}}{(0.50 \text{ h/mi})(x \text{ mi})} + \underset{\underset{\text{45 min}}{\uparrow}}{(0.75 \text{ h/mi})(y \text{ mi})} \overset{\underset{\text{max. available time}}{}}{\leq} \quad 60 \text{ h}$$

with labels: time to plow Rt. 15, time to plow Rt. 80, max. available time

$$0.50x + 0.75y \leq 60$$
$$y \leq 80 - 0.67x$$

Noting that negative values of x and y do not have meaning, we have the graph in Fig. 17-40, shading in the region below the line since we have $y < 80 - 0.67x$ for that region. Any point in the shaded region, or on the axes or the line around the shaded region, gives a solution. The *solid line* indicates that points on it are part of the solution.

The point (0, 80), for example, is a solution and tells us that 80 mi of Route 80 can be plowed if none of Route 15 is plowed. In this case, all 60 h of plowing time are used for Route 80. Another possibility is shown by the point (60, 20), which indicates that 60 mi of Route 15 and 20 mi of Route 80 can be plowed. In this case, not all of the 60 plowing hours are used.

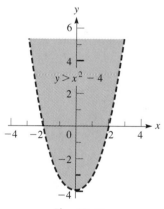

Fig. 17-41

EXAMPLE 4   Draw a sketch of the graph of the inequality $y > x^2 - 4$.

Although the graph of $y = x^2 - 4$ is not a straight line, the method of solution is the same. We graph the function $y = x^2 - 4$ as a dashed curve, since it is not part of the solution, as shown in Fig. 17-41. We then shade in the region above the curve to indicate the points that satisfy the inequality.

EXAMPLE 5   Draw a sketch of the region that is defined by the system of inequalities $y \geq -x - 2$ and $y + x^2 < 0$.

Similar to the solution of a system of equations, *the solution of a system of inequalities is any pair of values (x, y) that satisfies both inequalities.* This means we want the region common to both inequalities. In Fig. 17-42, we first shade in the region above the line $y = -x - 2$ and then shade in the region below the parabola $y = -x^2$. The region defined by this system is the darkly shaded region below the parabola that is also above and on the line. The calculator display for this region is shown in Fig. 17-43.

If we are asked to find the region defined by $y \geq -x - 2$ *or* $y + x^2 < 0$, it consists of both shaded regions and all points on the line.

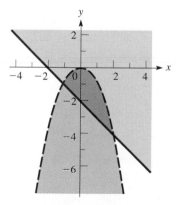

Fig. 17-42

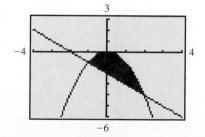

Fig. 17-43

Linear Programming

An important area in which graphs of inequalities with two or more variables are used is the branch of mathematics known as **linear programming** (in this context, "programming" does not mean computer programming). This subject is widely applied in industry, business, economics, and technology. The analysis of many social problems can also be made by the use of linear programming.

Linear programming is used to analyze problems such as those related to maximizing profits, minimizing costs, or the use of materials, with some specified constraints of production. The following is an example of the use of linear programming.

SOLVING A WORD PROBLEM

See the chapter introduction.

The loudspeaker was developed in the early 1920s by the U.S. inventor Kellogg Rice.

▌EXAMPLE 6   A company makes two types of stereo speaker systems, their good-quality system and their highest-quality system. The production of these systems requires assembly of the speaker system itself and the production of the cabinets in which they are installed. The good-quality system requires 3 worker-hours for speaker assembly and 2 worker-hours for cabinet production for each complete system. The highest-quality system requires 4 worker-hours for speaker assembly and 6 worker-hours for cabinet production for each complete system. Available skilled labor allows for a maximum of 480 worker-hours per week for speaker assembly and a maximum of 540 worker-hours per week for cabinet production. It is anticipated that all systems will be sold and that the profit will be $30 for each good-quality system and $75 for each highest-quality system. How many of each system should be produced to provide the greatest profit?

First, let $x =$ the number of good-quality systems and $y =$ the number of highest-quality systems made in one week. Thus, the profit p is given by

$$p = 30x + 75y$$

We know that negative numbers are not valid for either x or y, and therefore we have $x \geq 0$ and $y \geq 0$. Also, the number of available worker-hours per week for each part of the production restricts the number of systems that can be made. Both speaker assembly and cabinet production are required for all systems. The number of worker-hours needed to produce the x good-quality systems is $3x$ in the speaker assembly shop. Also, $4y$ worker-hours are required in the speaker assembly shop for the highest-quality systems. Thus,

$$3x + 4y \leq 480$$

since no more than 480 worker-hours are available in the speaker assembly shop. In the cabinet shop, we have

$$2x + 6y \leq 540$$

since no more than 540 worker-hours are available in the cabinet shop.

Therefore, we wish to maximize the profit p under the **constraints**

$x \geq 0$ $y \geq 0$	number of systems produced cannot be negative
$3x + 4y \leq 480$	worker-hours for speaker assembly
$2x + 6y \leq 540$	worker-hours for cabinet production

In order to do this we sketch the region of points that satisfy this system of inequalities. From the previous examples, we see that the appropriate region is in the first quadrant (since $x \geq 0$ and $y \geq 0$) and under both lines. See Fig. 17-44.

Any point in the shaded region defined by the preceding system of inequalities is *known as a* **feasible point.** In this case it means that it is possible to produce the number of systems of each type according to the coordinates of the point. For example, the point (50, 25) is in the region, which means that it is possible to produce 50 good-quality systems and 25 highest-quality systems under the given constraints of available skilled labor. However, we wish to find the point that indicates the number of each kind of system that produces the greatest profit.

If we assume values for the profit, the resulting equations are straight lines. Thus, by finding the greatest value of p for which the line passes through a feasible point, we may solve the given problem. If $p = \$3000$, or $p = \$6000$, we have the lines shown. Both are possible with various combinations of speaker systems being produced. However, we note the line for $p = \$6000$ passes through feasible points farther from the origin. It is also clear, since the lines are parallel, that *the greatest profit attainable is given by the line passing through P,* where $3x + 4y = 480$ and $2x + 6y = 540$ intersect. The coordinates of P are (72, 66). Thus, the production should be 72 good-quality systems and 66 highest-quality systems to produce a weekly profit of $p = 30(72) + 75(66) = \$7110$.

For this type of problem, *the solution will be given by one of the vertices of the region.* However, it could be any one of them, which means it is possible that only one type of product should be produced (see Exercise 38 of this section). Therefore, *we can solve the problem by finding the appropriate region and then testing the coordinates of the vertex points.* Here the vertex points are (160, 0), which indicates a profit of \$4800; (0, 90), which indicates a profit of \$6750; and (72, 66), which indicates a profit of \$7110.

NOTE ◗

See Appendix C for a graphing calculator program LINPROG. It can be used to shade the area for two linear constraints.

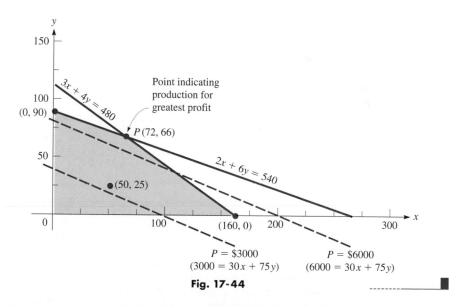

Fig. 17-44

For a problem involving minimum values, such as minimum costs, the feasible points will be *above* some or all of the constraint lines. However, the solution will still be at one of the vertices of the region. See Exercises 39 and 40.

EXERCISES *17-5*

In Exercises 1–16, draw a sketch of the graph of the given inequality.

1. $y > x - 1$
2. $y < 3x - 2$
3. $y \geq 2x + 5$
4. $y \leq 3 - x$
5. $3x + 2y + 6 > 0$
6. $x + 4y - 8 < 0$
7. $y < x^2$
8. $y \leq 2x^2 - 3$
9. $2x^2 - 4x - y > 0$
10. $y \leq x^3$
11. $y < 32x - x^4$
12. $y \leq \sqrt{2x + 5}$
13. $y > \dfrac{1}{x^2 + 1}$
14. $y < \ln x$
15. $y > \sin 2x$
16. $y > |x| - 3$

In Exercises 17–24, draw a sketch of the graph of the region in which the points satisfy the given system of inequalities.

17. $y > x$
    $y > 1 - x$
18. $y \leq 2x$
    $y \geq x - 1$
19. $y \leq 2x^2$
    $y > x - 2$
20. $y > x^2$
    $y < x + 4$
21. $y > \frac{1}{2}x^2$
    $y \leq 4x - x^2$
22. $y > 4 - x$
    $y < \sqrt{16 - x^2}$
23. $y \geq 0$
    $y \leq \sin x$
    $0 \leq x \leq 3\pi$
24. $y > 0$
    $y > 1 - x$
    $y < e^x$

In Exercises 25–32, use a graphing calculator to display the solution of the given inequality or system of inequalities.

25. $2x + y < 5$
26. $4x - y > 1$
27. $y \geq 1 - x^2$
28. $y < |4 - 2x|$
29. $y > 2x - 1$
    $y < x^4 - 8$
30. $y < 3 - x$
    $y > 3x - x^3$
31. $y > x^2 + 2x - 8$
    $y < \dfrac{1}{x} - 2$
32. $y > -2x^2$
    $y < 1 - e^{-x}$

In Exercises 33–36, set up the necessary inequalities and sketch the graph of the region in which the points satisfy the indicated system of inequalities.

33. A telephone company is installing two types of fiber-optic cable in an area. It is estimated that no more than 300 m of type A cable, and at least 200 m but no more than 400 m of type B cable, are needed. Graph the possible lengths of cable that are needed.

34. A refinery can produce gasoline and diesel fuel, in amounts of any combination, except that equipment restricts total production to 2500 gal/day. Graph the different possible production combinations of the two fuels.

35. The elements of an electric circuit dissipate p watts of power. The power p_R dissipated by a resistor in the circuit is given by $p_R = Ri^2$, where R is the resistance (in Ω) and i is the current (in A). Graph the possible values of p and i for $p > p_R$ and $R = 0.5\ \Omega$.

36. One pump can remove wastewater at the rate of 75 gal/min, and a second pump works at the rate of 45 gal/min. Graph the possible values of the time (in min) that each of these pumps operates such that together they pump more than 4500 gal.

In Exercises 37–40, solve the given linear programming problems.

37. A manufacturer produces a business calculator and a graphing calculator. Each calculator is assembled in two sets of operations, where each operation is in production 8 h during each day. The average time required for a business calculator in the first operation is 3 min, and 6 min is required in the second operation. The graphing calculator averages 6 min in the first operation and 4 min in the second operation. All of the calculators can be sold; the profit for a business calculator is $8, and the profit for a graphing calculator is $10. How many of each type of calculator should be made each day in order to maximize profit?

(W) 38. Using the information given in Example 6, with the one change that the profit on each good-quality speaker system is $60 (instead of $30), how many of each system should be made? Explain why this one change in data makes such a change in the solution.

39. Brands A and B of breakfast cereal are both enriched with vitamins P and Q. The necessary information about these cereals is as follows:

	Cereal A	Cereal B	RDA
Vitamin P	1 unit/oz	2 units/oz	10 units
Vitamin Q	5 units/oz	3 units/oz	30 units
Cost	12¢/oz	18¢/oz	

(RDA is the *Recommended Daily Allowance*.) Find the amount of each cereal that together satisfies the RDA of vitamins P and Q at the lowest cost. (Be careful: In this case we wish to *minimize* cost.)

40. A computer company makes parts A and B in each of two different plants. It costs $4000 per day to operate the first plant and $5000 per day to operate the second plant. Each day the first plant produces 100 of part A and 200 of part B, while at the second plant 250 of part A and 100 of part B are produced. How many days should each plant operate to produce 2000 of each part and keep operating costs at a minimum? (Be careful: In this case we wish to *minimize* cost.)

CHAPTER EQUATIONS

If $|f(x)| > n$, then $f(x) < -n$ or $f(x) > n$.   (17-1)

If $|f(x)| < n$, then $-n < f(x) < n$.   (17-2)

REVIEW EXERCISES

In Exercises 1–16, solve each of the given inequalities algebraically. Graph each solution.

1. $2x - 12 > 0$

2. $2.4(T - 4.0) \geq 5.5 - 2.4T$

3. $4 < 2x - 1 < 11$

4. $2x < x + 1 < 4x + 7$

5. $5x^2 + 9x < 2$

6. $x^2 + 2x > 63$

7. $6x^2 - x > 35$

8. $2x^3 + 4 \leq x^2 + 8x$

9. $\dfrac{(2x - 1)(3 - x)}{x + 4} > 0$

10. $x^4 + x^2 \leq 0$

11. $\dfrac{1}{x} < 2$

12. $\dfrac{1}{x - 2} < \dfrac{1}{4}$

13. $|3x + 2| \leq 4$

14. $|4 - 3x| \geq 1$

15. $|3 - 5x| > 7$

16. $\left|2 - \dfrac{y}{2}\right| \leq 5$

In Exercises 17–24, solve the given inequalities on a graphing calculator such that the display is the graph of the solution.

17. $5 - 3x < 0$

18. $6 \leq 4x - 2 < 9$

19. $2 \leq \dfrac{4n - 2}{3} < 3$

20. $3x < 2x + 1 < x - 5$

21. $\dfrac{8 - R}{2R + 1} \leq 0$

22. $\dfrac{(3 - x)^2}{2x + 7} \leq 0$

23. $|x - 2| > 3$

24. $2|2x - 9| < 8$

In Exercises 25–28, use a graphing calculator to solve the given inequalities. Graph the appropriate function and from the graph determine the solution.

25. $x^3 + x + 1 < 0$

26. $\dfrac{2}{x + 2} > 3$

27. $e^{-t} > 0.5$

28. $\sin 2x < 0.8$ $(0 < x < 4)$

In Exercises 29–40, draw a sketch of the region in which the points satisfy the given inequality or system of inequalities.

29. $y > 4 - x$

30. $y < \dfrac{1}{2}x + 2$

31. $2y - 3x - 4 \leq 0$

32. $3y - x + 6 \geq 0$

33. $y > x^2 + 1$

34. $y \leq \dfrac{1}{x^2 - 4}$

35. $y - |x + 1| < 0$

36. $2y + 2x^3 + 6x > 3$

37. $y > x + 1$
$y < 4 - x^2$

38. $y > 2x - x^2$
$y \geq -2$

39. $y \leq \dfrac{1}{x^2 + 1}$
$y < x - 1$

40. $y < \cos \dfrac{1}{2}x$
$y > \dfrac{1}{2}e^x$
$-\pi < x < \pi$

In Exercises 41–48, use a graphing calculator to display the region in which the points satisfy the given inequality or system of inequalities.

41. $y < 3x + 5$

42. $y > 2 - \frac{1}{4}x$

43. $y > 8 + 7x - x^2$

44. $y < x^3 + 4x^2 - x - 4$

45. $y < 32x - x^4$

46. $y > 2x - 1$
$y < 6 - 3x^2$

47. $y > 1 - x \sin 2x$
$y < 5 - x^2$

48. $y > |x - 1|$
$y < 4 + \ln x$

In Exercises 49–52, determine the values of x for which the given radicals represent real numbers.

49. $\sqrt{3 - x}$

50. $\sqrt{x + 5}$

51. $\sqrt{x^2 + 4x}$

52. $\sqrt{\dfrac{x - 1}{x + 2}}$

In Exercises 53–64, solve the given problems using inequalities. (All data are accurate to at least two significant digits.)

53. The pressure p (in kPa) at a depth d (in m) in the ocean is given by $p = 101 + 10.1d$. For what values of d is $p > 500$ kPa?

54. After conducting tests, it was determined that the stopping distance x (in ft) of a car traveling 60 mi/h was $|x - 290| \leq 35$. Express this inequality without absolute values and find the interval of stopping distances that were found in the tests.

55. A heating unit with 80% efficiency and a second unit with 90% efficiency deliver 360,000 Btu of heat to an office complex. If the first unit consumes an amount of fuel that contains no more than 261,000 Btu, what is the Btu content of the fuel consumed by the second unit?

56. A rectangular parking lot is to have a perimeter of 100 m and an area no greater than 600 m². What are the possible dimensions of the lot?

57. The electric power p (in W) dissipated in a resistor is given by $p = Ri^2$, where R is the resistance (in Ω) and i is the current (in A). For a given resistor, $R = 12.0\ \Omega$, and the power varies between 2.50 W and 8.00 W. Find the values of the current.

58. The reciprocal of the total resistance of two electric resistances in parallel equals the sum of the reciprocals of the resistances. If a 2.0 Ω resistance is in parallel with a resistance R, with a total resistance greater than 0.5 Ω, find R.

59. The efficiency e (in %) of a certain gasoline engine is given by $e = 100(1 - r^{-0.4})$, where r is the *compression ratio* for the engine. For what values of r is $e > 50\%$?

60. A rocket is fired such that its height h (in mi) is given by $h = 41t - t^2$. For what values of t (in min) is the height greater than 400 mi?

61. In developing a new product, a company estimates that it will take no more than 1200 min of computer time for research and no more than 1000 min of computer time for development. Graph the possible combinations of the computer times that are needed.

62. A natural-gas supplier has a maximum of 120 worker-hours per week for delivery and for customer service. Graph the possible combinations of times available for these two services.

63. A company produces two types of cameras, the regular model and the deluxe model. For each regular model produced there is a profit of $8, and for each deluxe model the profit is $15. The same amount of materials is used to make each model, but the supply is sufficient only for 450 cameras per day. The deluxe model requires twice the time to produce as the regular model. If only regular models were made, there would be time enough to produce 600 per day. Assuming all cameras will be sold, how many of each model should be produced if the profit is to be a maximum?

64. A company that manufactures compact disc players gets two different parts, A and B, from two different suppliers. Each package of parts from the first supplier costs $2.00, and contains 6 of each type of part. Each package of parts from the second supplier costs $1.50 and contains 4 of A and 8 of B. How many packages should be bought from each supplier to keep the total cost to a minimum, if production requirements are 600 of A and 900 of B?

Writing Exercise

65. In planning a new city development, an engineer uses a rectangular coordinate system to locate points within the development. A park in the shape of a quadrilateral has corners at $(0,0)$, $(0,20)$, $(40,20)$, and $(20,40)$ (measurements in meters). Write two or three paragraphs explaining how to describe the park region with inequalities and find these inequalities.

PRACTICE TEST

1. State conditions on x and y in terms of inequalities if the point (x, y) is in the second quadrant.

In Problems 2–6, solve the given inequalities algebraically and graph each solution.

2. $\dfrac{-x}{2} \geq 3$

3. $3x + 1 < -5$

4. $-1 < 1 - 2x < 5$

5. $\dfrac{x^2 + x}{x - 2} \leq 0$

6. $|2x + 1| \geq 3$

7. Sketch the region in which the points satisfy the following system of inequalities
$y < x^2$
$y \geq x + 1$

8. Determine the values of x for which $\sqrt{x^2 - x - 6}$ represents a real number.

9. The length of a rectangular lot is 20 m more than its width. If the area is to be at least 4800 m², what values may the width be?

10. Type A wire costs $0.10 per foot, and type B wire costs $0.20 per foot. Show the possible combinations of lengths of wire that can be purchased for less than $5.00.

11. The range of the visible spectrum in terms of the wavelength λ of light ranges from about $\lambda = 400$ nm (violet) to about $\lambda = 700$ nm (red). Express these values using an inequality with absolute values.

12. Solve the inequality $x^2 > 12 - x$ on a graphing calculator such that the display is the graph of the solution.

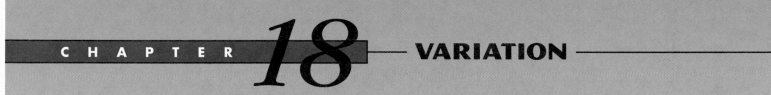

CHAPTER *18* — VARIATION

Experimentation and observation often lead to the discovery of important relationships among related variables. In studying the measured values, it is often possible to see how one changes as the other also changes.

In this chapter we see how such information can be used to set up functional relationships among these variables. Then, using known values, we can set up specific functions that relate the variables for specific situations.

We begin this chapter by reviewing the meanings of *ratio* and *proportion,* which were first introduced in Chapter 1. Then we will see how ratio and proportion lead to *variation* and setting up numerous functional relationships.

Applications of variation are found in all areas of science and technology. It is often used in acoustics, biology, chemistry, computer technology, economics, electronics, environmental technology, hydrodynamics, mechanics, navigation, optics, physics, space technology, thermodynamics, and other fields.

Newton's universal law of gravitation is expressed in the language of variation. In Section 18-2 we discuss a space-age application.

18-1 RATIO AND PROPORTION

In Chapter 1 we introduced the terms *ratio* and *proportion* when we first solved equations. Since then we have seen how they are used in the definitions of the trigonometric functions in Chapter 4 and in a few other specific cases. However, only a basic understanding of their meanings has been necessary to this point. In order to develop the meaning of *variation,* we now review and expand our discussion of these important terms.

From Chapter 1 we recall that *the quotient a/b is called the* **ratio** *of a to b.* Therefore, a fraction is a ratio.

Any measurement made is the ratio of the measured magnitude to an accepted unit of measurement. For example, when we say that an object is 5 ft long, we are saying that the length of that object is five times as long as an accepted unit of length, the foot. Other examples of ratios are density (weight/volume), relative density (density of object/density of water), and pressure (force/area). As these examples illustrate, ratios may compare quantities of the same kind, or they may express a division of magnitudes of different quantities (such a ratio is also called a **rate**).

EXAMPLE 1   The approximate airline distance from Chicago to Dallas is 800 mi, and the approximate airline distance from Chicago to Cleveland is 300 mi. The ratio of these distances is

$$\frac{800 \text{ mi}}{300 \text{ mi}} = \frac{8}{3}$$

Since both units are in miles, the resulting ratio is a dimensionless number.

If a jet travels from Chicago to Dallas in 2 h, its average speed is

$$\frac{800 \text{ mi}}{2 \text{ h}} = 400 \text{ mi/h}$$

In this case we must attach the proper units to the resulting ratio.   --------▪

The first jet-propelled airplane was flown in Germany in 1928.

As we noted in Example 1, we must be careful to attach the proper units to the resulting ratio. Generally, the ratio of measurements of the same kind should be expressed as a dimensionless number. Consider the following example.

EXAMPLE 2   The length of a certain room is 24 ft, and the width of the room is 18 ft. Therefore, the ratio of the length to the width is $\frac{24}{18}$, or $\frac{4}{3}$.

If the width of the room is expressed as 6 yd, we have the ratio 24 ft/6 yd = 4 ft/1 yd. However, this does not clearly show the ratio. It is better and more meaningful first to change the units of one of the measurements to the units of the other measurement. Changing the length from 6 yd to 18 ft, we express the ratio as $\frac{4}{3}$, as we saw above. From this ratio we can easily see that the length is $\frac{4}{3}$ as long as the width.   --------▪

Dimensionless ratios are often used in definitions in mathematics and in technology. For example, the irrational number π is the dimensionless ratio of the circumference of a circle to its diameter. The specific gravity of a substance is the ratio of its density to the density of water. Other illustrations are found in the exercises for this section.

From Chapter 1 we also recall that *an equation stating that two ratios are equal is called a* **proportion.** By this definition, a proportion is

$$\boxed{\frac{a}{b} = \frac{c}{d}}$$ **(18-1)**

Consider the following example.

EXAMPLE 3 On a certain map, 1 in. represents 10 mi. Thus, on this map we have a ratio of 1 in./10 mi. To find the distance represented by 3.5 in., we can set up the proportion

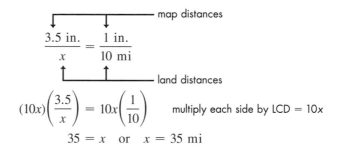

$$\frac{3.5 \text{ in.}}{x} = \frac{1 \text{ in.}}{10 \text{ mi}}$$

$$(10x)\left(\frac{3.5}{x}\right) = 10x\left(\frac{1}{10}\right) \qquad \text{multiply each side by LCD} = 10x$$

$$35 = x \quad \text{or} \quad x = 35 \text{ mi}$$

The ratio 1 in./10 mi is the *scale* of the map and has a special meaning, relating map distances in inches to land distances in miles. In a case like this we should not change either unit to the other, even though they are both units of length. ∎

SOLVING A WORD PROBLEM

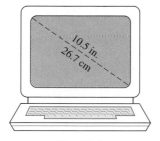

Fig. 18-1

EXAMPLE 4 Given that 1 in. = 2.54 cm, what is the length in centimeters of the diagonal of a rectangular computer screen that is 10.5 in. long? See Fig. 18-1.

If we equate the ratio of known lengths to the ratio of the given length to the required length, we can find the required length by solving the resulting proportion (which is an equation). This gives us

$$\frac{1 \text{ in.}}{2.54 \text{ cm}} = \frac{10.5 \text{ in.}}{x \text{ cm}}$$

$$x = (10.5)(2.54)$$

$$= 26.7 \text{ cm} \qquad \text{rounded off}$$

Therefore, the diagonal of the computer screen is 10.5 in., or 26.7 cm. ∎

In Appendix B there are additional illustrations of changing units. Also, a listing of all the various units used in the text is included.

EXAMPLE 5 The magnitude of an electric field E is the ratio of the force F on a charge q to the magnitude of q. We can write this as $E = F/q$. If we know the force exerted on a particular charge at some point in the field, we can determine the force that would be exerted on another charge placed at the same point. For example, if we know that a force of 10 nN is exerted on a charge of 4.0 nC, we can then determine the force that would be exerted on a charge of 6.0 nC by the proportion

$$\frac{10 \times 10^{-9}}{4.0 \times 10^{-9}} = \frac{F}{6.0 \times 10^{-9}}$$

$$F = \frac{(6.0 \times 10^{-9})(10 \times 10^{-9})}{4.0 \times 10^{-9}}$$

$$= 15 \times 10^{-9} = 15 \text{ nN} \qquad \qquad \text{∎}$$

SOLVING A WORD PROBLEM

▮**EXAMPLE 6** A certain alloy is 5 parts tin and 3 parts lead. How many grams of each are there in 40 g of the alloy?

First, we let $x =$ the number of grams of tin in the given amount of the alloy. Next, we note that there are 8 total parts of alloy, of which 5 are tin. Thus, 5 is to 8 as x is to 40. This gives the equation

$$\text{parts tin} \longrightarrow \frac{5}{8} = \frac{x}{40} \longleftarrow \text{grams of tin}$$
$$\text{total parts} \qquad\qquad\qquad\quad \longleftarrow \text{total grams}$$

$$x = 40\left(\frac{5}{8}\right) = 25 \text{ g}$$

Therefore, there are 25 g of tin and 15 g of lead. The ratio 25 to 15 is the same as 5 to 3. ▬

═══════════════ EXERCISES *18-1* ═══════════════

In Exercises 1–8, express the ratios in the simplest form.

1. 18 V to 3 V
2. 27 ft to 18 ft
3. 96 h to 3 days
4. 120 s to 4 min
5. 20 qt to 2.5 gal
6. 6500 cL to 2.6 L
7. 0.14 kg to 3500 mg
8. 2000 μm to 6 mm

In Exercises 9–20, find the required ratios.

9. The *efficiency* of a power amplifier is defined as the ratio of the power output to the power input. Find the efficiency of an amplifier for which the power output is 2.6 W and the power input is 9.6 W.

10. A virus 3.0×10^{-5} cm long appears to be 1.2 cm long through a microscope. What is the *magnification* (ratio of image length to object length) of the microscope?

11. The *coefficient of friction* for two contacting surfaces is the ratio of the frictional force between them to the perpendicular force that presses them together. If it takes 45 N to overcome friction to move a 110-N crate along the floor, what is the coefficient of friction between the crate and the floor? See Fig. 18-2.

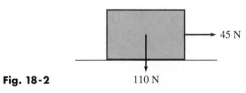

Fig. 18-2

12. The *atomic mass* of an atom of carbon is defined to be 12 u. The ratio of the atomic mass of an atom of oxygen to that of an atom of carbon is $\frac{4}{3}$. What is the atomic mass of an atom of oxygen? (The symbol u represents the *unified atomic mass unit,* where 1 u = 1.66×10^{-27} kg.)

13. An important design feature of an aircraft wing is its *aspect ratio*. It is defined as the ratio of the square of the span of the wing (wingtip to wingtip) to the total area of the wing. If the span of the wing for a certain aircraft is 32.0 ft, and the area is 195 ft^2, find the aspect ratio.

14. For an automobile engine, the ratio of the cylinder volume to compressed volume is the *compression ratio.* If the cylinder volume of 820 cm^3 is compressed to 110 cm^3, find the compression ratio.

15. The *specific gravity* of a substance is the ratio of its density to the density of water. If the density of steel is 487 lb/ft^3 and that of water is 62.4 lb/ft^3, what is the specific gravity of steel?

16. The *percent grade* of a road is the ratio of vertical rise to the horizontal change in distance (expressed in percent). If a highway rises 75 m for each 1200 m along the horizontal, what is the percent grade?

17. The *percent error* in a measurement is the ratio of the error in the measurement to the measurement itself, expressed as a percent. When writing a computer program, the memory remaining is determined as 2450 bytes and then it is correctly found to be 2540 bytes. What is the percent error in the first reading?

18. The electric *current* in a given circuit is the ratio of the voltage to the resistance. What is the current (1 V/1 Ω = 1 A) for a circuit where the voltage is 24.0 V and the resistance is 10.0 Ω?

19. The *mass* of an object is the ratio of its weight to the acceleration g due to gravity. If a space probe weighs 8460 N on earth, where $g = 9.80$ m/s^2, find its mass. (See Appendix B.)

20. *Power* is defined as the ratio of work done to the time required to do the work. If an engine performs 3650 J of work in 15.0 s, find the power developed by the engine. (See Appendix B.)

In Exercises 21–24, find the required quantities from the given proportions.

21. In an electric instrument called a "Wheatstone bridge," electric resistances are related by

$$\frac{R_1}{R_2} = \frac{R_3}{R_4}$$

Find R_2 if $R_1 = 6.00\ \Omega$, $R_3 = 62.5\ \Omega$, and $R_4 = 15.0\ \Omega$. See Fig. 18-3.

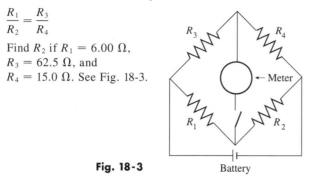

Fig. 18-3

22. For two connected gears, the relation

$$\frac{d_1}{d_2} = \frac{N_1}{N_2}$$

holds, where d is the diameter of the gear and N is the number of teeth. Find N_1 if $d_1 = 2.60$ in., $d_2 = 11.7$ in., and $N_2 = 45$. The ratio N_2/N_1 is called the *gear ratio*. See Fig. 18-4.

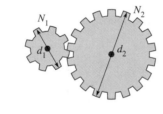

Fig. 18-4

23. According to Boyle's law, the relation

$$\frac{p_1}{p_2} = \frac{V_2}{V_1}$$

holds for pressures p_1 and p_2 and volumes V_1 and V_2 of a gas at constant temperature. Find V_1 if $p_1 = 36.6$ kPa, $p_2 = 84.4$ kPa, and $V_2 = 0.0447\ \text{m}^3$.

24. In a transformer, an electric current in one coil of wire induces a current in a second coil. For a transformer

$$\frac{i_1}{i_2} = \frac{t_2}{t_1}$$

where i is the current and t is the number of windings in each coil. In a neon sign amplifier $i_1 = 1.2$ A, and the *turns ratio* $t_2/t_1 = 160$. Find i_2. See Fig. 18-5.

In Exercises 25–40, answer the given questions by setting up and solving the appropriate proportions.

25. Given that $1.00\ \text{in.}^2 = 6.45\ \text{cm}^2$, what area in square inches is $36.3\ \text{cm}^2$?

26. Given that 1.000 kg $= 2.205$ lb, what mass in kilograms is equivalent to 175.5 lb?

27. Given that 1.00 hp $= 746$ W, what power in horsepower is 250 W?

28. Given that 1.50 L $= 1.59$ qt, what capacity in quarts is 2.75 L?

29. Given that 2.00 km $= 1.24$ mi, what distance in kilometers is 5.00 mi?

30. Given that $10^4\ \text{cm}^2 = 10^6\ \text{mm}^2$, what area in square centimeters is $2.50 \times 10^5\ \text{mm}^2$?

31. How many meters per second are equivalent to 45.0 km/h?

32. How many gallons per hour are equivalent to 540 L/min?

33. A particular type of automobile engine produces $62{,}500\ \text{cm}^3$ of carbon monoxide in 2.00 min. How much carbon monoxide is produced in 45.0 s?

34. An airplane consumes 36.0 gal of gasoline in flying 425 mi. Under similar conditions, how far can it fly on 52.5 gal?

35. By weight, the ratio of chlorine to sodium in table salt is 35.46 to 23.00. How much sodium is contained in 50.00 kg of salt?

36. Ten clicks on an adjustment screw cause an inlet valve opening to change by 0.035 cm. How many clicks are required for a valve adjustment of 0.049 cm?

37. In testing for quality control, it was found that 17 of every 500 computer chips produced by a company in a day were defective. If a total of 595 defective parts were found, what was the total number of chips produced during that day?

38. An electric current of 0.772 mA passes into two wires in which it is divided into currents in the ratio of 2.83 to 1.09. What are the currents in the two wires?

39. One computer line printer can print 2400 lines/min, and a second can print 2800 lines/min. If they print a total of 9100 lines while printing together, how many lines does each print?

40. Of the earth's water area, the Pacific Ocean covers 46.0% and the Atlantic Ocean covers 23.9%. Together they cover a total of $2.53 \times 10^8\ \text{km}^2$. What is the area of each?

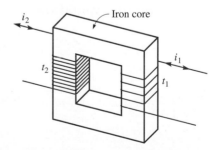

Fig. 18-5

18-2 VARIATION

Named for the French physicist, Jacques Charles (1746–1823).

Scientific laws are often stated in terms of ratios and proportions. For example, Charles' law can be stated as "for a perfect gas under constant pressure, the ratio of any two volumes this gas may occupy equals the ratio of the absolute temperatures." Symbolically, this could be stated as $V_1/V_2 = T_1/T_2$. Thus, if the ratio of the volumes and one of the values of the temperature are known, we can easily find the other temperature.

By multiplying both sides of the proportion of Charles' law by V_2/T_1, we can change the form of the proportion to $V_1/T_1 = V_2/T_2$. This statement says that the ratio of the volume to the temperature (for constant pressure) is constant. Thus, if any pair of values of volume and temperature is known, this ratio of V_1/T_1 can be calculated. This ratio of V_1/T_1 can be called a constant k, which means that Charles' law can be written as $V/T = k$. We now have the statement that the ratio of the volume to temperature is always constant; or, as it is normally stated, "The volume is proportional to the temperature." Therefore, we write $V = kT$, the clearest and most informative statement of Charles' law.

Thus, *for any two quantities always in the same proportion, we say that one is* **proportional to** (*or* **varies directly as**) *the second. To show that y is proportional to x (or varies directly as x), we write*

DIRECT VARIATION

$$y = kx$$ (18-2)

where k is the **constant of proportionality.** This type of relationship is known as **direct variation.**

◼**EXAMPLE 1**   The circumference of a circle is proportional to (varies directly as) the radius r. We write this as $c = 2\pi r$. Since we know that $c = 2\pi r$ for a circle, we know in this case that $k = 2\pi$. ------------◼

◼**EXAMPLE 2**   The fact that the electric resistance R of a wire varies directly as (is proportional to) its length l is written as $R = kl$. As the length of the wire increases (or decreases), this equation tells us that the resistance increases (or decreases) proportionally. ------------◼

It is very common that, when two quantities are related, the product of the two quantities remains constant. In such a case $yx = k$, or

INVERSE VARIATION

$$y = \frac{k}{x}$$ (18-3)

This is read as "y **varies inversely as** *x" or "y* **is inversely proportional to** *x."* This type of relationship is known as **inverse variation.**

Named for the English physicist, Robert Boyle (1627–1691).

◼**EXAMPLE 3**   Boyle's law states that "at a given temperature, the pressure p of an ideal gas varies inversely as the volume V." We write this as $p = k/V$. In this case, as the volume of the gas increases, the pressure decreases. ------------◼

In Fig. 18-6(a) the graph of the equation for direct variation $y = kx$ ($x \geq 0$) is shown. It is a straight line, with slope of k ($k > 0$) and y-intercept of 0. We see that y increases as x increases. In Fig. 18-6(b) the graph of the equation for inverse variation $y = k/x$ ($k > 0, x > 0$) is shown. It is a *hyperbola* (a different form of the equation from that of Example 4 of Section 14-1). As x increases, y decreases.

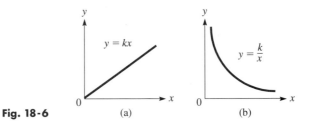

Fig. 18-6   (a)   (b)

For many relationships, one quantity varies as a specified power of another quantity. The terms *varies directly* and *varies inversely* are used in the following examples with a specified power of the independent variable.

■**EXAMPLE 4**   The statement that the volume V of a sphere varies directly as the cube of its radius r is written as $V = kr^3$. In this case we know that $k = 4\pi/3$. We see that as the radius increases, the volume increases much more rapidly. For example, if $r = 2.00$ cm, $V = 33.5$ cm^3, and if $r = 3.00$ cm, $V = 113$ cm^3.   ■

■**EXAMPLE 5**   A company finds that the number n of units of a product that are sold is inversely proportional to the square of the price p of the product. This is written as $n = k/p^2$. As the price of the product is raised, the number of units that are sold decreases much more rapidly.   ■

One quantity may vary as the product of two or more other quantities. Such variation is called **joint variation.** *We write*

JOINT VARIATION

$$\boxed{y = kxz} \qquad \text{(18-4)}$$

to show that y varies jointly as x and z.

■**EXAMPLE 6**   The cost C of a piece of sheet metal varies jointly as the area A of the piece and the cost c per unit area. This we write as $C = kAc$. Here, C increases if the *product Ac* increases.   ■

Direct, inverse, and joint variations may be combined. A given relationship may be a combination of two or all three of these types of variation.

Formulated by the great English mathematician and physicist, Isaac Newton (1642–1727).

■**EXAMPLE 7**   Newton's *universal law of gravitation* can be stated: "The force F of gravitation between two objects varies jointly as the masses m_1 and m_2 of the objects and inversely as the square of the distance r between their centers." We write this as

$$F = \frac{Gm_1m_2}{r^2} \quad \begin{matrix} \longleftarrow \text{ force varies jointly as masses} \\ \text{and} \\ \longleftarrow \text{ inversely as the square of the distance} \end{matrix}$$

where G is the constant of proportionality.

Note the use of the word *and* in this example. It is used to indicate that F varies

CAUTION ▶ in more than one way, but *it is **not** interpreted as addition.*   ■

Calculating the Constant of Proportionality

Once we have used the given statement to set up a general equation in terms of the variables and the constant of proportionality, we may calculate the value of the constant of proportionality if one *complete set of values* of the variables is known. *This value can then be substituted into the general equation to find the specific equation* relating the variables. We can then find the value of any one of the variables for any set of the others.

▌EXAMPLE 8 If y varies inversely as x, and $x = 15$ when $y = 4$, find the value of y when $x = 12$.

First, we write

$$y = \frac{k}{x} \qquad \text{general equation from statement}$$

to show that y varies inversely as x. Next, we substitute $x = 15$ and $y = 4$ into the equation. This leads to

$$4 = \frac{k}{15} \quad \text{or} \quad k = 60 \qquad \text{evaluate } k$$

Thus, for this problem the constant of proportionality is 60, and this may be substituted into $y = k/x$, giving

$$y = \frac{60}{x} \qquad \text{specific equation relating } y \text{ and } x$$

as the equation between y and x. Now, for any given value of x, we may find the value of y. For $x = 12$, we have

$$y = \frac{60}{12} = 5 \qquad \text{evaluating } y \text{ for } x = 12$$

SOLVING A WORD PROBLEM

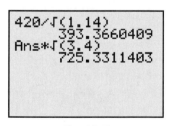

Fig. 18-7

▌EXAMPLE 9 The frequency f of vibration of a wire varies directly as the square root of the tension T of the wire. If $f = 420$ Hz when $T = 1.14$ N, find f when $T = 3.40$ N.

The steps in making this evaluation are outlined below.

$$f = k\sqrt{T} \qquad \text{set up general equation: } f \text{ varies directly as } \sqrt{T}$$
$$420 \text{ Hz} = k\sqrt{1.14 \text{ N}} \qquad \text{substitute given set of values and evaluate } k$$
$$k = 393 \text{ Hz/N}^{1/2}$$
$$f = 393\sqrt{T} \qquad \text{substitute value of } k \text{ to get specific equation}$$
$$f = 393\sqrt{3.40} \qquad \text{evaluate } f \text{ for } T = 3.40 \text{ N}$$
$$= 725 \text{ Hz}$$

We note that k has a set of units associated with it, and this usually will be the case in applied situations. As long as we do not change the units that are used for any of the variables, the units for the final variable that is evaluated will remain the same. The calculator screen for this calculation is shown in Fig. 18-7.

SOLVING A WORD PROBLEM

EXAMPLE 10   The heat H developed in an electric resistor varies jointly as the time t and the square of the current i in the resistor. If H_0 joules of heat are developed in t_0 seconds with i_0 amperes passing through the resistor, how much heat is developed if both the time and the current are doubled?

$$H = kti^2 \qquad \text{set up general equation}$$

$$H_0\,\text{J} = k(t_0\ \text{s})(i_0\ \text{A})^2 \qquad \text{substitute given values and}$$

$$k = \frac{H_0}{t_0 i_0^2}\,\text{J}/(\text{s}\cdot\text{A}^2) \qquad \text{evaluate } k$$

$$H = \frac{H_0 t i^2}{t_0 i_0^2} \qquad \text{substitute for } k \text{ to get specific equation}$$

We are asked to determine H when both the time and the current are doubled. This means we are to substitute $t = 2t_0$ and $i = 2i_0$. Making this substitution,

$$H = \frac{H_0(2t_0)(2i_0)^2}{t_0 i_0^2} = \frac{8H_0 t_0 i_0^2}{t_0 i_0^2} = 8H_0$$

The heat developed is eight times that for the original values of i and t.

SOLVING A WORD PROBLEM

See the chapter introduction.

The first landing on the moon was by the crew of the U.S. spacecraft *Apollo 11* in July 1969.

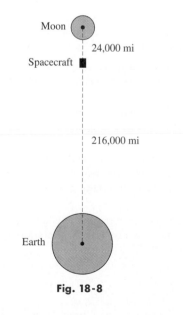

Fig. 18-8

EXAMPLE 11   In Example 7 we stated Newton's universal law of gravitation. This law was formulated in the late 17th century, but it has numerous modern space-age applications. Use this law to solve the following problem.

A spacecraft is traveling from the earth to the moon, which are 240,000 mi apart. The mass of the moon is 0.0123 that of the earth. How far from the earth is the gravitational force of the earth on the spacecraft equal to the gravitational force of the moon on the spacecraft?

From Example 7, we have the gravitational force between two objects as

$$F = \frac{Gm_1 m_2}{r^2}$$

where the constant of proportionality G is the same for any two objects. Since we want the force between the earth and the spacecraft to equal the force between the moon and the spacecraft, we have

$$\frac{Gm_s m_e}{r^2} = \frac{Gm_s m_m}{(240{,}000 - r)^2}$$

where m_s, m_e, and m_m are the masses of the spacecraft, the earth, and the moon, respectively; r is the distance from the earth to the spacecraft; and $240{,}000 - r$ is the distance from the moon to the spacecraft. Since $m_m = 0.0123m_e$, we have

$$\frac{Gm_s m_e}{r^2} = \frac{Gm_s(0.0123m_e)}{(240{,}000 - r)^2}$$

$$\frac{1}{r^2} = \frac{0.0123}{(240{,}000 - r)^2} \qquad \text{divide each side by } Gm_s m_e$$

$$(240{,}000 - r)^2 = 0.0123r^2 \qquad \text{multiply each side by LCD}$$

$$240{,}000 - r = 0.111r \qquad \text{take square roots}$$

$$1.111r = 240{,}000$$

$$r = 216{,}000\ \text{mi}$$

Therefore, the spacecraft is 216,000 mi from the earth and 24,000 mi from the moon when the gravitational forces are equal. See Fig. 18-8.

In Exercises 1–8, set up the general equations from the given statements.

1. y varies directly as z.

2. p varies inversely as q.

3. s varies inversely as the square of t.

4. w is proportional to the cube of L.

5. f is proportional to the square root of x.

6. n is inversely proportional to the $\frac{3}{2}$ power of s.

7. w varies jointly as x and the cube of y.

8. q varies as the square of r and inversely as the fourth power of t.

(W) *In Exercises 9–12, express the meaning of the given equation in a verbal statement, using the language of variation. (k and π are constants.)*

9. $A = \pi r^2$

10. $s = \dfrac{k}{t^{1.2}}$

11. $n = \dfrac{k\sqrt{t}}{u}$

12. $V = \pi r^2 h$

In Exercises 13–16, give the specific equation relating the variables after evaluating the constant of proportionality for the given set of values.

13. V varies directly as the square of H, and $V = 2$ when $H = 64$.

14. n is inversely proportional to the square of p, and $n = \frac{1}{27}$ when $p = 3$.

15. p is proportional to q and inversely proportional to the cube of r, and $p = 6$ when $q = 3$ and $r = 2$.

16. v is proportional to t and the square of s, and $v = 80$ when $s = 2$ and $t = 5$.

In Exercises 17–24, find the required value by setting up the general equation and then evaluating.

17. Find y when $x = 10$ if y varies directly as x, and $y = 20$ when $x = 8$.

18. Find y when $x = 5$ if y varies directly as the square of x, and $y = 6$ when $x = 8$.

19. Find s when $t = 10$ if s is inversely proportional to t, and $s = 100$ when $t = 5$.

20. Find p for $q = 0.8$ if p is inversely proportional to the square of q, and $p = 18$ when $q = 0.2$.

21. Find y for $x = 6$ and $z = 5$ if y varies directly as x and inversely as z, and $y = 60$ when $x = 4$ and $z = 10$.

22. Find r when $n = 16$ if r varies directly as the square root of n and $r = 4$ when $n = 25$.

23. Find f when $p = 2$ and $c = 4$ if f varies jointly as p and the cube of c, and $f = 8$ when $p = 4$ and $c = 0.1$.

24. Find v when $r = 2$, $s = 3$, and $t = 4$ if v varies jointly as r and s and inversely as the square of t, and $v = 8$ when $r = 2$, $s = 6$, and $t = 6$.

In Exercises 25–48, solve the given applied problems involving variation.

25. The volume V of carbon dioxide (CO_2) that is exhausted from a room in a given time varies directly as the initial volume V_0 that is present. If 75 ft^3 of CO_2 are removed in an hour from a room with an initial volume of 160 ft^3, how much is removed in an hour if the initial volume is 130 ft^3?

26. The amount of heat H required to melt ice is proportional to the mass m of ice that is melted. If it takes 2.93×10^5 J to melt 875 g of ice, how much heat is required to melt 625 g?

27. In electroplating, the mass m of the material deposited varies directly as the time t during which the electric current is on. Set up the equation for this relationship if 2.50 g are deposited in 5.25 h.

28. Hooke's law states that the force needed to stretch a spring is proportional to the amount the spring is stretched. If 10.0 lb stretches a certain spring 4.00 in, how much will the spring be stretched by a force of 6.00 lb?

29. The rate H of heat removal by an air conditioner is proportional to the electric power input P. The constant of proportionality is the *performance coefficient*. Find the performance coefficient of an air conditioner for which $H = 1800$ W and $P = 720$ W.

30. The energy E available daily from a solar collector varies directly as the percent p that the sun shines during the day. If a collector provides 1200 kJ for 75% sunshine, how much does it provide for a day during which there is 35% sunshine?

31. The time t required to empty a wastewater holding tank is inversely proportional to the cross-sectional area A of the drainage pipe. If it takes 2.0 h to empty a tank with a drainage pipe for which $A = 48$ in.^{2}, how long will it take to empty the tank if $A = 68$ in.^{2}?

(W) 32. The time t required to make a particular trip is inversely proportional to the average speed v. If a jet takes 2.75 h at an average speed of 520 km/h, how long will it take at an average speed of 620 km/h? Explain the meaning of the constant of proportionality.

(W) 33. In a physics experiment a given force was applied to three objects. The mass m and the resulting acceleration a were recorded as follows:

m (g)	2.0	3.0	4.0
a (cm/s^2)	30	20	15

(a) Is the relationship $a = f(m)$ one of direct or inverse variation? Explain. (b) Find $a = f(m)$.

W **34.** The lift L of each of three model airplane wings of width w was measured and recorded as follows:

w (cm)	20	40	60
L (N)	10	40	90

(a) Is the relationship $L = f(w)$ one of direct or inverse variation? Explain. (b) Find $L = f(w)$.

35. The power P required to propel a ship varies directly as the cube of the speed s of the ship. If 5200 hp will propel a ship at 12.0 mi/h, what power is required to propel it at 15.0 mi/h?

36. The f-number lens setting of a camera varies directly as the square root of the time t that the film is exposed. If the f-number is 8 (written as $f/8$) for $t = 0.0200$ s, find the f-number for $t = 0.0098$ s.

37. The force F on the blade of a wind generator varies jointly as the blade area A and the square of the wind velocity v. Find the equation relating F, A, and v if $F = 19.2$ lb when $A = 3.72$ ft^2 and $v = 31.4$ ft/s.

38. The escape velocity v a spacecraft needs to leave the gravitational field of a planet varies directly as the square root of the product of the planet's radius R and its acceleration due to gravity g. For Mars and earth, $R_M = 0.533R_e$ and $g_M = 0.400g_e$. Find v_M for Mars if $v_e = 11.2$ km/s.

39. The average speed s of oxygen molecules in the air is directly proportional to the square root of the absolute temperature T. If the speed of the molecules is 460 m/s at 273 K, what is the speed at 300 K?

40. The time t required to test a computer memory unit varies directly as the square of the number n of memory cells in the unit. If a unit with 4800 memory cells can be tested in 15.0 s, how long does it take to test a unit with 8400 memory cells?

41. The electric resistance R of a wire varies directly as its length l and inversely as its cross-sectional area A. Find the relation between resistance, length, and area for a wire that has a resistance of 0.200 Ω for a length of 225 ft and cross-sectional area of 0.0500 in.^{2}.

42. The general gas law states that the pressure P of an ideal gas varies directly as the thermodynamic temperature T and inversely as the volume V. If $P = 610$ kPa for $V = 10.0$ cm^3 and $T = 290$ K, find V for $P = 400$ kPa and $T = 400$ K.

43. The power P in an electric circuit varies jointly as the resistance R and the square of the current I. If the power is 10.0 W when the current is 0.500 A and the resistance is 40.0 Ω, find the power if the current is 2.00 A and the resistance is 20.0 Ω.

44. The difference $m_1 - m_2$ in magnitudes (visual brightnesses) of two stars varies directly as the base 10 logarithm of the ratio b_2/b_1 of their actual brightnesses. For two particular stars, if $b_2 = 100b_1$ for $m_1 = 7$ and $m_2 = 2$, find the equation relating m_1, m_2, b_1, and b_2.

45. The power gain G by a parabolic microwave dish varies directly as the square of the diameter d of the opening and inversely as the square of the wavelength λ of the wave carrier. Find the equation relating G, d, and λ if $G = 5.5 \times 10^4$ for $d = 2.9$ m and $\lambda = 0.030$ m.

46. The intensity I of sound varies directly as the power P of the source and inversely as the square of the distance r from the source. Two sound sources are separated by a distance d, and one has twice the power output of the other. Where should an observer be located on a line between them such that the intensity of each sound is the same?

47. The x-component of the acceleration of an object moving around a circle with constant angular velocity ω varies jointly as cos ωt and the square of ω. If the x-component of the acceleration is -11.4 ft/s^2 when $t = 1.00$ s for $\omega = 0.524$ rad/s, find the x-component of the acceleration when $t = 2.00$ s.

48. The tangent of the proper banking angle θ of the road for a car making a turn is directly proportional to the square of the car's velocity v and inversely proportional to the radius r of the turn. If 7.75° is the proper banking angle for a car traveling at 20.0 m/s around a turn of radius 300 m, what is the proper banking angle for a car traveling at 30.0 m/s around a turn of radius 250 m? See Fig. 18-9.

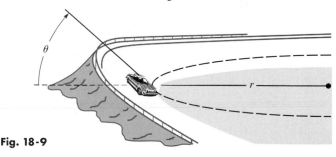

Fig. 18-9

CHAPTER EQUATIONS

Proportion	$\dfrac{a}{b} = \dfrac{c}{d}$	**(18-1)**
Direct variation	$y = kx$	**(18-2)**
Inverse variation	$y = \dfrac{k}{x}$	**(18-3)**
Joint variation	$y = kxz$	**(18-4)**

REVIEW EXERCISES

In Exercises 1–8, find the indicated ratios.

1. 4 Mg to 20 kg

2. 300 nm to 6 μm

3. 20 mL to 5 cL

4. 12 ks to 2 h

5. The mechanical advantage of a lever is the ratio of the output force F_0 to the input force F_i. Find the mechanical advantage if $F_0 = 28$ kN and $F_i = 5000$ N.

6. For an automobile, the ratio of the number n_1 of teeth on the ring gear to the number n_2 of teeth on the pinion gear is the *rear axle ratio* of the car. Find this ratio if $n_1 = 64$ and $n_2 = 20$.

7. The pressure p exerted on a surface is the ratio of the force F on the surface to its area A. Find the pressure on a square patch, 2.25 in. on a side, on a tank if the force on the patch is 37.4 lb.

8. The electric resistance R of a resistor is the ratio of the voltage V across the resistor to the current i in the resistor. Find R if $V = 0.632$ V and $i = 2.03$ mA.

In Exercises 9–20, answer the given questions by setting up and solving the appropriate proportions.

9. On a map of Hawaii 1.3 in. represents 20.0 mi. If the distance on the map between Honolulu and Kahului on Maui is 6.0 in., how far is Kahului from Honolulu?

10. Given that 1.000 lb = 453.6 g, what is the weight in pounds of a 14.0-g computer disk?

11. Given that 1.00 Btu = 1060 J, how much heat in joules is produced by a heating element that produces 2660 Btu?

12. Given that 1.00 L = 61.0 in.^{3}, what capacity in liters has a cubical box that is 3.23 in. along an edge?

13. A computer printer can print 3600 characters in 30 s. How many characters can it print in 5.0 min?

14. A solar heater with a collector area of 58.0 m^2 is required to heat 2560 kg of water. Under the same conditions, how much water can be heated by a rectangular solar collector 9.50 m by 8.75 m?

15. The dosage of a certain medicine is 25 mL for each 10 lb of the patient's weight. What is the dosage for a person weighing 56 kg?

16. A woman invests $50,000 and a man invests $20,000 in a partnership. If profits are to be shared in the ratio that each invested in the partnership, how much does each receive from $10,500 in profits?

17. On a certain blueprint, a measurement of 25.0 ft is represented by 2.00 in. What is the actual distance between two points if they are 5.75 in. apart on the blueprint?

18. The chlorine concentration in a water supply is 0.12 part per million. How much chlorine is there in a cylindrical holding tank 4.22 m in radius and 5.82 m high filled from the water supply?

19. One fiber-optic cable carries 60.0% as many messages as another fiber-optic cable. Together they carry 12,000 messages. How many does each carry?

20. Two types of roadbed material, one 50% rock and the other 100% rock, are used in the ratio of 4 to 1 to form a roadbed. If a total of 150 tons are used, how much rock is in the roadbed?

In Exercises 21–24, give the specific equation relating the variables after evaluating the constant of proportionality for the given set of values.

21. y varies directly as the square of x, and $y = 27$ when $x = 3$.

22. f varies inversely as l, and $f = 5$ when $l = 8$.

23. v is directly proportional to x and inversely proportional to the cube of y, and $v = 10$ when $x = 5$ and $y = 4$.

24. r varies jointly as u, v, and the square of w, and $r = 8$ when $u = 2$, $v = 4$, and $w = 3$.

In Exercises 25–52, solve the given applied problems.

25. For a lever balanced at the fulcrum, the relation

$$\frac{F_1}{F_2} = \frac{L_2}{L_1}$$

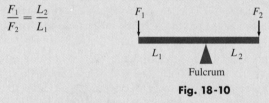

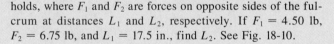
Fulcrum

Fig. 18-10

holds, where F_1 and F_2 are forces on opposite sides of the fulcrum at distances L_1 and L_2, respectively. If $F_1 = 4.50$ lb, $F_2 = 6.75$ lb, and $L_1 = 17.5$ in., find L_2. See Fig. 18-10.

26. A company finds that the volume V of sales of a certain item and the price P of the item are related by

$$\frac{P_1}{P_2} = \frac{V_2}{V_1}$$

Find V_2 if $P_1 = \$8.00$, $P_2 = \$6.00$, and $V_1 = 3000$ per week.

27. The charge C on a capacitor varies directly as the voltage V across it. If the charge is 6.3 μC with a voltage of 220 V across a capacitor, what is the charge on it with a voltage of 150 V across it?

28. The amount of natural gas burned is proportional to the amount of oxygen consumed. If 24.0 lb of oxygen is consumed in burning 15.0 lb of natural gas, how much air, which is 23.2% oxygen by weight, is consumed to burn 50.0 lb of natural gas?

29. The power P of a gas engine is proportional to the area A of the piston. If an engine with a piston area of 8.00 in.^{2} can develop 30.0 hp, what power is developed by an engine with a piston area of 6.00 in.^{2}?

30. The decrease in temperature above a region is directly proportional to the altitude above the region. If the temperature *T* at the base of the rock of Gibraltar is 22.0° and a plane 3.50 km above notes that the temperature is 1.0°C, what is the temperature at the top of Gibraltar, the altitude of which is 430 m? (Assume there are no other temperature effects.)

31. The distance *d* an object falls under the influence of gravity varies directly as the square of the time *t* of fall. If an object falls 64.0 ft in 2.00 s, how far will it fall in 3.00 s?

32. The kinetic energy *E* of a moving object varies jointly as the mass *m* of the object and the square of its velocity *v*. If a 5.00-kg object, traveling at 10.0 m/s, has a kinetic energy of 250 J, find the kinetic energy of an 8.00-kg object moving at 50.0 m/s.

33. In a particular computer design, *N* numbers can be sorted in a time proportional to the square of log *N*. How many times longer does it take to sort 8000 numbers than to sort 2000 numbers?

34. The velocity *v* of a jet of fluid flowing from an opening in the side of a container is proportional to the square root of the depth *d* of the opening. If the velocity of the jet from an opening at a depth of 1.22 m is 4.88 m/s, what is the velocity of a jet from an opening at a depth of 7.62 m? See Fig. 18-11.

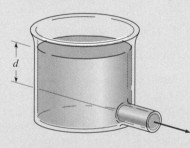

Fig. 18-11

35. In any given electric circuit containing an inductance *L* and a capacitance *C*, the resonant frequency *f* is inversely proportional to the square root of the capacitance. If the resonant frequency in a circuit is 25.0 Hz and the capacitance is 95.0 μF, what is the resonant frequency of this circuit if the capacitance is 25.0 μF?

36. The rate of emission *R* of radiant energy from the surface of a body is proportional to the fourth power of the thermodynamic temperature *T*. Given that a 25.0-W (the rate of emission) lamp has an operating temperature of 2500 K, what is the operating temperature of a similar 40.0-W lamp?

37. The frequency *f* of a radio wave is inversely proportional to its wavelength λ. The constant of proportionality is the velocity of the wave, which equals the speed of light. Find this velocity if an FM radio wave has a frequency of 90.9 MHz and a wavelength of 3.29 m.

38. The acceleration of gravity *g* on a satellite in orbit around the earth varies inversely as the square of its distance *r* from the center of the earth. If *g* = 8.7 m/s^2 for a satellite at an altitude of 400 km above the surface of the earth, find *g* if it is 1000 km above the surface. The radius of the earth is 6.4 × 10^6 m.

39. Using *holography* (a method of producing an image without using a lens), an image of concentric circles is formed. The radius *r* of each circle varies directly as the square root of the wavelength λ of the light used. If *r* = 3.56 cm for λ = 575 nm, find *r* if λ = 483 nm.

40. A metal circular ring has a circular cross section of radius *r*. If *R* is the radius of the ring (measured to the middle of the cross section), the volume *V* of metal in the ring varies directly as *R* and the square of *r*. If *V* = 2550 mm^3 for *r* = 2.32 mm and *R* = 24.0 mm, find *V* for *r* = 3.50 mm and *R* = 32.0 mm. See Fig. 18-12.

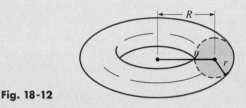

Fig. 18-12

41. The stopping distance *d* of a car varies directly as the square of the velocity *v* of the car when the brakes are applied. A car moving at 32 mi/h can stop in 52 ft. What is the stopping distance for the car if it is moving at 55 mi/h?

42. Kepler's third law of planetary motion states that the square of the period of any planet is proportional to the cube of the mean radius (about the sun) of that planet, with the constant of proportionality being the same for all planets. Using the fact that the period of the earth is one year and its mean radius is 93.0 million miles, calculate the mean radius for Venus, given that its period is 7.38 months.

43. The range *R* of a projectile varies jointly as the square of its initial velocity v_0 and the sine of twice the angle θ from the horizontal at which it is fired. See Fig. 18-13. A bullet for which v_0 = 850 m/s and θ = 22.0° has a range of 5.12 × 10^4 m. Find the range if v_0 = 750 m/s and θ = 43.2°.

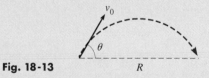

Fig. 18-13

44. The load *L* that a helical spring can support varies directly as the cube of its wire diameter *d* and inversely as its coil diameter *D*. A spring for which *d* = 0.120 in. and *D* = 0.953 in. can support 45.0 lb. What is the coil diameter of a similar spring that supports 78.5 lb and for which *d* = 0.156 in.?

45. The volume rate of flow *R* of blood through an artery varies directly as the fourth power of the radius *r* of the artery and inversely as the distance *d* along the artery. If an operation is successful in effectively increasing the radius of an artery by 25% and decreasing its length by 2%, by how much is the volume rate of flow increased?

46. The safe, uniformly distributed load L on a horizontal beam, supported at both ends, varies jointly as the width w and the square of the depth d and inversely as the distance D between supports. Given that one beam has double the dimensions of another, how many times heavier is the safe load it can support than the first can support?

47. A bank statement exactly 30 years old is discovered. It states, "This 10-year-old account is now worth $185.03 and pays 4% interest compounded annually." An investment with annual compound interest varies directly as $1 + r$ to the power n, where r is the interest rate expressed as a decimal and n is the number of years of compounding. What was the value of the original investment, and what is it worth now?

48. The distance s that an object falls due to gravity varies jointly as the acceleration g due to gravity and the square of the time t of fall. The acceleration due to gravity on the moon is 0.172 of that on earth. If a rock falls for t_0 seconds on earth, how many times farther would the rock fall on the moon in $3t_0$ seconds?

49. The heat loss L through fiberglass insulation varies directly as the time t and inversely as the thickness d of the fiberglass. If the loss through 8.0 in. of fiberglass is 1200 Btu in 30 min, what is the loss through 6.0 in. in 1 h 30 min?

50. A quantity important in analyzing the rotation of an object is its *moment of inertia I*. For a ball bearing, the moment of inertia varies directly as its mass m and the square of its radius r. Find the general expression for I if $I = 39.9$ g·cm^2 for $m = 63.8$ g and $r = 1.25$ cm.

51. In the study of polarized light, the intensity I is proportional to the square of the cosine of the angle θ of transmission. If $I = 0.025$ W/m^2 for $\theta = 12.0°$, find I for $\theta = 20.0°$.

52. The force F that acts on a pendulum bob is proportional to the mass m of the bob and the sine of the angle θ the pendulum makes with the vertical. If $F = 0.120$ N for $m = 0.350$ kg and $\theta = 2.00°$, find F for $m = 0.750$ kg and $\theta = 3.50°$.

Writing Exercise

53. A fruit packing company plans to reduce the size of its fruit juice can (a right circular cylinder) by 10% and keep the price of each can the same (effectively raising the price). The radius and the height of the new can are to be equally proportional to those of the old can. Write one or two paragraphs explaining how to determine the percent decrease in the radius and the height of the old can that is required to make the new can.

PRACTICE TEST

1. Express the ratio of 180 s to 4 min in simplest form.

2. The force F between two parallel wires carrying electric currents is inversely proportional to the distance d between the wires. If a force of 0.750 N exists between wires that are 1.25 cm apart, what is the force between them if they are separated by 1.75 cm?

3. Given that 1.00 in. = 2.54 cm, what length in inches is 7.24 cm?

4. The difference p in pressure in a fluid between that at the surface and that at a point below varies jointly as the density d of the fluid and the depth h of the point. The density of water is 1000 kg/m^3, and the density of alcohol is 800 kg/m^3. This difference in pressure at a point 0.200 m below the surface of water is 1.96 kPa. What is the difference in pressure at a point 0.300 m below the surface of alcohol? (All data are accurate to three significant digits.)

5. The perimeter of a rectangular solar panel is 210.0 in. The ratio of the length to the width is 7 to 3. What are the dimensions of the panel?

6. The crushing load L of a pillar varies directly as the fourth power of its radius r and inversely as the square of its length l. If one pillar has twice the radius and three times the length of a second pillar, what is the ratio of the crushing load of the first pillar to that of the second pillar?

19 SEQUENCES AND THE BINOMIAL THEOREM

In mathematics and its applications, numerical results will often follow certain patterns. In this chapter we study briefly certain types of sequences of numbers. A *sequence* is a set of numbers arranged in some specified manner and usually follows a pattern. Of the many types of sequences, we will develop certain basic ones, including those used in the expansion of a binomial to a power.

Applications of sequences can be found in many scientific and technical areas. These include physics and chemistry in studying radioactivity, and biology in studying population growth. One of the more important applications comes in business when calculating payments and interest.

Sequences and binomial expansions are also of importance in developing mathematical topics that themselves have a number of technical applications.

Sequences are basic to many calculations in business, including compound interest. In Section 19-2 we demonstrate such a calculation.

19-1 ARITHMETIC SEQUENCES

A *sequence* of numbers may consist of numbers chosen in any way that we may wish to select. This could include a random selection of numbers, although the sequences that are useful in technical applications and later topics of mathematics follow some type of pattern. In this section we consider the first of these sequences, an *arithmetic sequence,* which is defined on the following page. In our development of sequences, we shall consider only those that consist of real numbers or literal numbers that represent real numbers.

An **arithmetic sequence** (*or* **arithmetic progression**) *is a set of numbers in which each number after the first can be obtained from the preceding one by adding to it a fixed number called the* **common difference.** This definition can be expressed in terms of the *recursion formula*

$$a_n = a_{n-1} + d \qquad \text{(19-1)}$$

where a_n is any term, a_{n-1} is the preceding term, and d is the common difference.

EXAMPLE 1 **(a)** The sequence 2, 5, 8, 11, 14, ..., is an arithmetic sequence with a common difference $d = 3$. We can obtain any term by adding 3 to the previous term. We see that the fifth term is $a_5 = a_4 + d$, or $14 = 11 + 3$.

(b) The sequence 7, 2, -3, -8, ..., is an arithmetic sequence with a common difference of $d = -5$. We can obtain any term after the first by adding -5 to the previous term.

The three dots after the 14 in part (a) and after the -8 in part (b) mean that the sequences continue. ◼

If we know the first term of an arithmetic sequence, we can find any other term in the sequence by successively adding the common difference enough times for the desired term to be obtained. This, however, is a very inefficient method, and we can learn more about the sequence if we establish a general way of finding any particular term.

In general, if a_1 is the first term and d is the common difference, the second term is $a_1 + d$, the third term is $a_1 + 2d$, and so forth. If we are looking for the nth term, we note that we need only add d to the first term $n - 1$ times. Thus, *the nth term, a_n, of an arithmetic sequence is given by*

***n*TH TERM**

$$a_n = a_1 + (n - 1)d \qquad \text{(19-2)}$$

Equation (19-2) can be used to find any given term in any arithmetic sequence. We can refer to a_n as the *last term* of an arithmetic sequence if no terms beyond it are included in the sequence. Such a sequence is called a *finite* sequence. If the terms in a sequence continue without end, the sequence is called an *infinite* sequence.

EXAMPLE 2 Find the tenth term of the arithmetic sequence 2, 5, 8, ...

By subtracting any given term from the following term, we find that the common difference is $d = 3$. From the terms given, we know that the first term is $a_1 = 2$. From the statement of the problem, the desired term is the tenth, or $n = 10$. Thus, we may find the tenth term, a_{10}, by

$$\begin{array}{ccc} a_1 & n & d \\ \downarrow & \downarrow & \downarrow \end{array}$$
$$a_{10} = 2 + (10 - 1)3 = 2 + (9)(3)$$
$$= 29$$

Therefore, the tenth term is 29.

The three dots written after the 8 mean that the sequence continues. Since there is no additional information, this would indicate that it is an infinite arithmetic sequence. ◼

EXAMPLE 3   Find the common difference between successive terms of the arithmetic sequence for which the first term is 5 and the 32nd term is -119.

We are to find d given that $a_1 = 5$, $a_{32} = -119$, and $n = 32$. Substitution in Eq. (19-2) gives

$$-119 = 5 + (32 - 1)d$$
$$31d = -124$$
$$d = -4$$

There is no information as to whether this is a finite or an infinite sequence. The solution is the same in either case.   --------■

SOLVING A WORD PROBLEM

EXAMPLE 4   How many numbers between 10 and 1000 are divisible by 6?

We must first find the smallest and the largest numbers in this range that are divisible by 6. These numbers are 12 and 996. Obviously, the common difference between one multiple of 6 and the next is 6. Thus, we can solve this as an arithmetic sequence with $a_1 = 12$, $a_n = 996$, and $d = 6$. Substituting these values in Eq. (19-2), we have

$$996 = 12 + (n - 1)6$$
$$6n = 990$$
$$n = 165$$

Thus, 165 numbers between 10 and 1000 are divisible by 6.

All the positive multiples of 6 are included in the infinite arithmetic sequence 6, 12, 18, ..., whereas those between 10 and 1000 are included in the finite arithmetic sequence 12, 18, 24, ..., 996.   --------■

SOLVING A WORD PROBLEM

EXAMPLE 5   A package delivery company uses a metal (low-friction) ramp to slide packages from the sorting area to the loading area. If a package is pushed to start it down the ramp at 25 cm/s and the package accelerates as it slides such that it gains 35 cm/s during each second, after how many seconds is the velocity 305 cm/s? See Fig. 19-1.

Here we see that the velocity (in cm/s) of the package after each second is

$$60, 95, 130, \ldots, 305, \ldots$$

Therefore, $a_1 = 60$ (the 25 cm/s was at the beginning, that is, after 0 s), $d = 35$, $a_n = 305$, and we are to find n.

$$305 = 60 + (n - 1)(35)$$
$$245 = 35n - 35$$
$$35n = 280$$
$$n = 8.0$$

This means that the velocity of a package sliding down the ramp is 305 cm/s after 8.0 s.   --------■

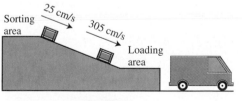

Fig. 19-1

Sum of *n* Terms

Another important quantity related to an arithmetic sequence is the sum of the first *n* terms. We can indicate this sum by starting the sum with either the first term or with the last term, as shown by these two equations:

$$S_n = a_1 + (a_1 + d) + (a_1 + 2d) + \cdots + (a_n - d) + a_n$$

or

$$S_n = a_n + (a_n - d) + (a_n - 2d) + \cdots + (a_1 + d) + a_1$$

If we now add the corresponding members of these two equations, we obtain the result

$$2S_n = (a_1 + a_n) + (a_1 + a_n) + (a_1 + a_n) + \cdots + (a_1 + a_n) + (a_1 + a_n)$$

Each term on the right in parentheses has the same expression $(a_1 + a_n)$, and there are *n* such terms. This tells us that *the sum of the first n terms is given by*

$$S_n = \frac{n}{2}(a_1 + a_n) \qquad \text{(19-3)}$$

The use of Eq. (19-3) is illustrated in the following examples.

Karl Friedrich Gauss (1777–1856) is considered as one of the greatest mathematicians of all time. A popular story about him is that at the age of 10 his schoolmaster asked his class to add all the numbers from 1 to 100, assuming it would take some time to do. He had barely finished stating the problem when Gauss handed in the correct answer of 5050. He had done it essentially as we did Example 6.

EXAMPLE 6   Find the sum of the first 1000 positive integers.

The first 1000 integers form a finite arithmetic sequence for which $a_1 = 1$, $a_{1000} = 1000$, $n = 1000$, and $d = 1$. Substituting into Eq. (19-3) (in which we do not use the value of *d*), we have

$$S_{1000} = \frac{1000}{2}(1 + 1000) = 500(1001)$$

$$= 500{,}500 \qquad \blacksquare$$

EXAMPLE 7   Find the sum of the first 10 terms of the arithmetic sequence in which the first term is 4 and the common difference is -5.

We are to find S_n, given that $n = 10$, $a_1 = 4$, and $d = -5$. Since Eq. (19-3) uses the value of a_n but not the value of *d*, we first find a_{10} by using Eq. (19-2). This gives us

$$a_{10} = 4 + (10 - 1)(-5) = 4 - 45$$

$$= -41$$

Now we can solve for S_{10} by using Eq. (19-3)

$$S_{10} = \frac{10}{2}(4 - 41) = 5(-37)$$

$$= -185 \qquad \blacksquare$$

If any three of the five values a_1, a_n, *n*, *d*, and S_n are given for a particular arithmetic sequence, the other two may be found from Eqs. (19-2) and (19-3). Consider the following example.

EXAMPLE 8   For an arithmetic sequence, given that $a_1 = 2$, $d = \frac{3}{2}$, and $S_n = 72$, find n and a_n.

First we substitute the given values in Eqs. (19-2) and (19-3) in order to identify what is known and how we may proceed. Substituting $a_1 = 2$ and $d = \frac{3}{2}$ in Eq. (19-2), we obtain

$$a_n = 2 + (n - 1)\left(\frac{3}{2}\right)$$

Substituting $S_n = 72$ and $a_1 = 2$ in Eq. (19-3), we obtain

$$72 = \frac{n}{2}(2 + a_n)$$

We note that n and a_n appear in both equations, which means that we must solve them simultaneously. Substituting the expression for a_n from the first equation into the second equation, we proceed with the solution.

$$72 = \frac{n}{2}\left[2 + 2 + (n - 1)\left(\frac{3}{2}\right)\right]$$

$$72 = 2n + \frac{3n(n - 1)}{4}$$

$$288 = 8n + 3n^2 - 3n$$

$$3n^2 + 5n - 288 = 0$$

$$n = \frac{-5 \pm \sqrt{25 - 4(3)(-288)}}{6} = \frac{-5 \pm \sqrt{3481}}{6} = \frac{-5 \pm 59}{6}$$

Since n must be a positive integer, we find that $n = \dfrac{-5 + 59}{6} = 9$. Using this value in the expression for a_n, we find

$$a_9 = 2 + (9 - 1)\left(\frac{3}{2}\right) = 14$$

Therefore, $n = 9$ and $a_9 = 14$.

SOLVING A WORD PROBLEM

EXAMPLE 9   The voltage across a resistor increases such that during each second the increase is 0.002 mV less than during the previous second. Given that the increase during the first second is 0.350 mV, what is the total voltage increase during the first 10.0 s?

We are asked to find the sum of the voltage increases 0.350 mV, 0.348 mV, 0.346 mV, ..., so as to include 10 increases. This means we want the sum of an arithmetic sequence for which $a_1 = 0.350$, $d = -0.002$, and $n = 10$. Since we need a_n to use Eq. (19-3), we first calculate it using Eq. (19-2).

$$a_{10} = 0.350 + (10 - 1)(-0.002) = 0.332 \text{ mV}$$

Now we use Eq. (19-3) to find the sum with $a_1 = 0.350$, $a_{10} = 0.332$, and $n = 10$.

$$S_{10} = \frac{10}{2}(0.350 + 0.332) = 3.410 \text{ mV}$$

Thus, the total voltage increase is 3.410 mV.

EXERCISES *19-1*

In Exercises 1–4, write the first five terms of the arithmetic sequence with the given values.

1. $a_1 = 4, d = 2$

2. $a_1 = 6, d = -\frac{1}{2}$

3. $a_3 = \frac{5}{2}, a_5 = -\frac{3}{2}$

4. $a_2 = -2, a_5 = 43$

In Exercises 5–12, find the nth term of the arithmetic sequence with the given values.

5. $1, 4, 7, \ldots, n = 8$

6. $-6, -4, -2, \ldots, n = 10$

7. $\frac{7\pi}{4}, \frac{3\pi}{2}, \frac{5\pi}{4}, \ldots, n = 17$

8. $2, \frac{1}{2}, -1, \ldots, n = 25$

9. $a_1 = -7, d = 4, n = 80$

10. $a_1 = \frac{3}{2}, d = \frac{1}{6}, n = 601$

11. $a_1 = b, d = 2b, n = 25$

12. $a_1 = -c, d = 3c, n = 30$

In Exercises 13–16, find the sum of the n terms of the indicated arithmetic sequence.

13. $n = 20, a_1 = 4, a_{20} = 40$

14. $n = 8, a_1 = -12, d = -2$

15. $-2, -\frac{5}{2}, -3, \ldots, n = 10$

16. $3k, \frac{10}{3}k, \frac{11}{3}k, \ldots, n = 40$

In Exercises 17–28, find any of the values of a_1, d, a_n, n, or S_n that are missing for an arithmetic sequence.

17. $a_1 = 5, d = 8, a_n = 45$

18. $a_1 = -2, n = 60, a_n = 28$

19. $a_1 = \frac{5}{3}, n = 20, S_{20} = \frac{40}{3}$

20. $a_1 = 0.1, a_n = -5.9, S_n = -8.7$

21. $d = 3, n = 30, S_{30} = 1875$

22. $d = 9, a_n = 86, S_n = 455$

23. $a_1 = 74, d = -5, a_n = -231$

24. $a_1 = -\frac{9}{7}, n = 19, a_{19} = -\frac{36}{7}$

25. $a_1 = -5k, d = \frac{1}{2}k, S_n = \frac{23}{2}k$

26. $d = -2c, n = 50, S_{50} = 0$

27. $a_1 = -c, a_n = \dfrac{b}{2}, S_n = 2b - 4c$

28. $a_1 = 3b, n = 7, d = \dfrac{b}{3}$

In Exercises 29–48, find the indicated quantities for the appropriate arithmetic sequence.

29. $a_6 = 56, a_{10} = 72$ (find a_1, d, S_n for $n = 10$)

30. $a_{17} = -91, a_2 = -73$ (find a_1, d, S_n for $n = 40$)

(W) 31. Is ln 3, ln 6, ln 12, . . . an arithmetic sequence? Explain. If it is, what is the fifth term?

(W) 32. Is sin 2°, sin 4°, sin 6°, . . . an arithmetic sequence? Explain. If it is, what is the fifth term?

33. Find the sum of the first 100 positive integers. (See the margin note on page 492.)

34. Find the sum of the first 100 positive odd integers.

35. Find the number of multiples of 8 between 99 and 999.

36. Is $x, x + 2y, 2x + 3y, \ldots$ an arithmetic sequence? If it is, find the sum of the first 100 terms.

37. A beach now has an area of 9500 m² but is eroding such that it loses 100 m² more of its area each year than during the previous year. If it lost 400 m² during the last year, what will its area be 8 years from now?

38. During a period of heavy rains, on a given day 110,000 ft³/s of water was being released from a dam. In order to minimize downstream flooding, engineers then reduced the releases by 10,000 ft³/s each day thereafter. How much water was released during the first week of these releases?

39. There are 12 seats in the first row around a semicircular stage. Each row behind the first has 4 more seats than the row in front of it. How many rows of seats are there if there is a total of 300 seats?

40. A bank loan of $8000 is repaid in annual payments of $1000 plus 10% interest on the unpaid balance. What is the total amount of interest paid?

41. A car depreciates $1800 during the first year after it is bought. Each year thereafter it depreciates $150 less than the year before. After how many years will it be considered to have no value, and what was the original cost?

42. The sequence of ships bells is as follows: at 12:30 A.M. 1 bell is rung, and each half hour later 1 more bell is rung than the previous time until 8 bells are rung. The sequence is then repeated starting at 4:30 A.M., again until 8 bells are rung. This pattern is followed throughout the day. How many bells are rung in one day?

43. If a tool dropped from a helicopter falls 4.9 m during the first second, 14.7 m during the second second, 24.5 m during the third second, and so on, how high was the helicopter if the tool takes 10.0 s to reach the ground?

44. In preparing a bid for constructing a new building, a contractor determines that the foundation and basement will cost $605,000 and the first floor will cost $360,000. Each floor above the first will cost $15,000 more than the one below it. How much will the building cost if it is to be 18 floors high?

45. Derive a formula for S_n in terms of n, a_1, and d.

(W) 46. A *harmonic sequence* is a sequence of numbers whose reciprocals form an arithmetic sequence. Is a harmonic sequence also an arithmetic sequence? Explain.

47. Show that the sum of the first n positive integers is $\frac{1}{2}n(n + 1)$.

48. Show that the sum of the first n positive odd integers is n^2.

19-2 GEOMETRIC SEQUENCES

A second type of important sequence of numbers is the **geometric sequence** (or **geometric progression**). *In a geometric sequence each number after the first can be obtained from the preceding one by multiplying it by a fixed number, called the* **common ratio.** We can express this definition in terms of the *recursion formula*

$$a_n = r a_{n-1}$$ (19-4)

where a_n is any term, a_{n-1} is the preceding term, and r is the common ratio. One important application of geometric sequences is in computing compound interest on savings accounts. Other applications are found in areas such as biology and physics.

EXAMPLE 1 (a) The sequence 2, 4, 8, 16, ..., is a geometric sequence with a common ratio of 2. Any term after the first can be obtained by multiplying the previous term by 2. We see that the fourth term $a_4 = r a_3$, or $16 = 2(8)$.

  (b) The sequence 9, -3, 1, $-1/3$, ..., is a geometric sequence with a common ratio of $-1/3$. We can obtain any term after the first by multiplying the previous term by $-1/3$.

If we know the first term, we can then find any other desired term by multiplying by the common ratio a sufficient number of times. When we do this for a general geometric sequence, we can find the *n*th term in terms of the first term a_1, the common ratio *r*, and *n*. Thus, the second term is $a_1 r$, the third term is $a_1 r^2$, and so forth. In general, the expression for the *n*th term is

nTH TERM

$$a_n = a_1 r^{n-1}$$ (19-5)

EXAMPLE 2 Find the eighth term of the geometric sequence 8, 4, 2, ...
  By dividing any given term by the previous term, we find the common ratio to be $\frac{1}{2}$. From the terms given, we see that $a_1 = 8$. From the statement of the problem, we know that $n = 8$. Thus, we substitute into Eq. (19-5) to find a_8.

$$a_8 = 8 \left(\frac{1}{2} \right)^{8-1} = \frac{8}{2^7} = \frac{1}{16}$$

EXAMPLE 3 Find the tenth term of the geometric sequence for which $a_1 = \frac{8}{625}$ and $r = -\frac{5}{2}$.
  Using Eq. (19-5) to find a_{10}, we have

$$a_{10} = \frac{8}{625} \left(-\frac{5}{2} \right)^{10-1} = \frac{8}{625} \left(-\frac{5^9}{2^9} \right) = -\left(\frac{2^3}{5^4} \right) \left(\frac{5^9}{2^9} \right) = -\frac{5^5}{2^6}$$

$$= -\frac{3125}{64}$$

■EXAMPLE 4   Find the seventh term of a geometric sequence for which the second term is 3, the fourth term is 9, and $r > 0$.

We can find r if we let $a_1 = 3$, $a_3 = 9$, and $n = 3$. (At this point we are considering a sequence made up of 3, the next number, and 9. These are the second, third, and fourth terms of the original sequence.) Thus

$$9 = 3r^2, \qquad r = \sqrt{3} \qquad \text{(since } r > 0\text{)}$$

We can now find a_1 of the original sequence by considering just the first two terms of the sequence, a_1 and $a_2 = 3$.

$$3 = a_1(\sqrt{3})^{2-1}, \qquad a_1 = \sqrt{3}$$

We can now find the seventh term, using $a_1 = \sqrt{3}$, $r = \sqrt{3}$, and $n = 7$.

$$a_7 = \sqrt{3}(\sqrt{3})^{7-1} = \sqrt{3}(3^3) = 27\sqrt{3}$$

We could have shortened this procedure one step by letting the second term be the first term of a new sequence of six terms. If the first term is of no importance in itself, this is acceptable.   --------■

SOLVING A WORD PROBLEM

■EXAMPLE 5   In an experiment, 22.0% of a substance changes chemically each 10.0 min. If there is originally 120 g of the substance, how much of it will remain after 45.0 min?

Let $P =$ the portion of the substance remaining after each minute. From the statement of the problem we know that $r = 0.780$, since 78.0% remains after each 10.0-min period. We also know that $a_1 = 120$ g, and we let n represent the number of minutes of elapsed time. This means that $P = 120(0.780)^{n/10.0}$. It is necessary to divide by 10.0 because the ratio is given for a 10.0-min period. In order to find P when $n = 45.0$ min, we write

$$P = 120(0.780)^{45.0/10.0} = 120(0.780)^{4.50}$$
$$= 39.2 \text{ g}$$

This means that 39.2 g remain after 45.0 min. Note that the power 4.50 represents 4.50 ten-minute periods.   --------■

Sum of *n* Terms

A general expression for the sum S_n of the first n terms of a geometric sequence may be found by directly forming the sum and multiplying this equation by r.

$$S_n = a_1 + a_1r + a_1r^2 + \cdots + a_1r^{n-1}$$
$$rS_n = a_1r + a_1r^2 + a_1r^3 + \cdots + a_1r^n$$

If we now subtract the second of these equations from the first, we obtain $S_n - rS_n = a_1 - a_1r^n$. All other terms cancel by subtraction. Now, factoring S_n from the terms on the left and a_1 from the terms on the right, we solve for S_n. Thus, *the sum S_n of the first n terms of a geometric sequence is*

$$S_n = \frac{a_1(1 - r^n)}{1 - r} \qquad (r \neq 1) \tag{19-6}$$

EXAMPLE 6   Find the sum of the first seven terms of the geometric sequence in which the first term is 2 and the common ratio is $\frac{1}{2}$.

We are to find S_n given that $a_1 = 2$, $r = \frac{1}{2}$, and $n = 7$. Using Eq. (19-6), we have

$$S_7 = \frac{2(1 - (\frac{1}{2})^7)}{1 - \frac{1}{2}} = \frac{2(1 - \frac{1}{128})}{\frac{1}{2}} = 4\left(\frac{127}{128}\right) = \frac{127}{32}$$

SOLVING A WORD PROBLEM

See the chapter introduction.

There is an unconfirmed story that someone once asked Einstein what was the greatest invention ever made, and his reply was "compound interest."

Fig. 19-2

EXAMPLE 7   If \$100 is invested each year at 5% interest compounded annually, what would be the total amount of the investment after 10 years (before the 11th deposit is made)?

After 1 year the amount invested will have added to it the interest for the year. Therefore, for the last (10th) \$100 invested, its value will become

$$\$100(1 + 0.05) = \$100(1.05) = \$105$$

The next to last \$100 will have interest added twice. After 1 year its value becomes \$100(1.05), and after 2 years it is $\$100(1.05)(1.05) = \$100(1.05)^2$. In the same way, the value of the first \$100 becomes $\$100(1.05)^{10}$, since it will have interest added 10 times. This means that we are to find the sum of the sequence

$$100(1.05) + 100(1.05)^2 + 100(1.05)^3 + \cdots + 100(1.05)^{10}$$

| 1 year in account | 2 years in account | 3 years in account | 10 years in account |

or

$$100[1.05 + (1.05)^2 + (1.05)^3 + \cdots + (1.05)^{10}]$$

For the sequence in the brackets, we have $a_1 = 1.05$, $r = 1.05$, and $n = 10$. Thus,

$$S_{10} = \frac{1.05[1 - (1.05)^{10}]}{1 - 1.05} = 13.2068$$

The total value of the \$100 investments is $100(13.2068) = \$1320.68$. We see that \$320.68 in interest has been earned. See Fig. 19-2.

EXERCISES *19-2*

In Exercises 1–4, write down the first five terms of the geometric sequence with the given values.

1. $a_1 = 45$, $r = \frac{1}{3}$ 　　　**2.** $a_1 = 9$, $r = -\frac{2}{3}$
3. $a_1 = \frac{1}{6}$, $r = 3$ 　　　**4.** $a_1 = -3$, $r = 2$

In Exercises 5–12, find the nth term of the geometric sequence with the given values.

5. $\frac{1}{2}$, 1, 2, ... $(n = 6)$ 　　**6.** 10, 1, 0.1, ... $(n = 8)$
7. 125, −25, 5, ... $(n = 7)$ 　**8.** 0.1, 0.3, 0.9, ... $(n = 5)$
9. $a_1 = -27$, $r = -\frac{1}{3}$, $n = 10$
10. $a_1 = 48$, $r = \frac{1}{2}$, $n = 12$
11. 10^{100}, -10^{98}, 10^{96}, ... $(n = 51)$
12. $-2, 4k, -8k^2, ...$ $(n = 6)$

In Exercises 13–16, find the sum of the first n terms of the geometric sequence with the given values.

13. $a_1 = \frac{1}{8}$, $r = 4$, $n = 5$ 　**14.** 162, −54, 18, ... $(n = 6)$
15. 192, 96, 48, ... $(n = 6)$ 　**16.** $a_1 = 9$, $a_n = -243$, $n = 4$

In Exercises 17–24, find any of the values of a_1, r, a_n, n, or S_n that are missing.

17. $a_1 = \frac{1}{16}$, $r = 4$, $n = 6$
18. $r = 0.2$, $a_n = 0.00032$, $n = 7$
19. $r = \frac{3}{2}$, $n = 5$, $S_5 = 211$ 　**20.** $r = -\frac{1}{2}$, $a_n = \frac{1}{8}$, $n = 7$
21. $a_n = 27$, $n = 4$, $S_4 = 40$ 　**22.** $a_1 = 3$, $n = 7$, $a_7 = 192$
23. $a_1 = 75$, $r = \frac{1}{5}$, $a_n = \frac{3}{625}$ 　**24.** $r = -2$, $n = 6$, $S_6 = 42$

In Exercises 25–40, find the indicated quantities.

(W) **25.** Is 3, 3^{x+1}, 3^{2x+1}, ... a geometric sequence? Explain. If it is, find a_{20}.

26. Find the sum of the first eight terms of the geometric sequence for which the fifth term is 5, the seventh term is 10, and $r > 0$.

27. Each stroke of a pump removes 8.2% of the remaining air from a container. What percent of the air remains after 50 strokes?

28. From 1980 to 1990, Charlotte, North Carolina, had a 2.34% average growth rate of population, compounded annually. If the population in 1980 was 314,000, what was the 1990 population?

29. An electric current decreases by 12.5% each 1.00 μs. If the initial current is 3.27 mA, what is the current after 8.20 μs?

30. A copying machine is set to reduce the dimensions of material copied by 10%. A drawing 12.0 cm wide is reduced, and then the copies are in turn reduced. What is the width of the drawing on the sixth reduction?

31. How much is an investment of $250 worth after 8 years if it earns annual interest of 7.2% compounded monthly? (7.2% annual interest compounded monthly means that 0.6% (7.2%/12) interest is added each month.)

32. A chemical spill pollutes a stream. A monitoring device finds 620 ppm (parts per million) of the chemical 1.0 mi below the spill, and the readings decrease by 12.5% for each mile farther downstream. How far downstream is the reading 100 ppm?

33. During each oscillation, a pendulum swings through 85% of the distance of the previous oscillation. If the pendulum swings through 80.8 cm in the first oscillation, through what total distance does it move in 12 oscillations?

34. A series of deposits, each of value A and made at equal time intervals, earns an interest rate of i for the time interval. The deposits have a total value of

$$A(1 + i) + A(1 + i)^2 + A(1 + i)^3 + \cdots + A(1 + i)^n$$

after n time intervals (just before the next deposit). Find a formula for this sum.

35. A thermometer is removed from hot water at 100.0°C into a room at 20.0°C. The temperature difference D between the thermometer and the air decreases by 35.0% each minute. What is the temperature reading on the thermometer 10.0 min later?

36. The power on a space satellite is supplied by a radioactive isotope. On a given satellite the power decreases by 0.2% each day. What percent of the initial power remains after one year?

37. If you decided to save money by putting away 1¢ on a given day, 2¢ one week later, 4¢ a week later, and so on, how much would you have to put away 6 months (26 weeks) after putting away the 1¢?

38. How many direct ancestors (parents, grandparents, and so on) does a person have in the 10 generations that preceded him or her (assuming that no ancestor appears in more than one line of descent)?

39. Derive a formula for S_n in terms of a_1, r, and a_n.

(W) **40.** Write down several terms of a general geometric sequence. Then take the logarithm of each term. Explain why the resulting sequence is an arithmetic sequence.

$19\text{-}3$ INFINITE GEOMETRIC SERIES

In the previous sections we developed formulas for the sum of the first n terms of an arithmetic sequence and of a geometric sequence. *The indicated sum of the terms of a sequence is called a* **series.**

EXAMPLE 1 (a) The infinite arithmetic sequence 2, 5, 8, 11, 14, ... has the series

$$2 + 5 + 8 + 11 + 14 + \cdots$$

associated with it.

(b) The finite geometric sequence 1, $\frac{1}{2}$, $\frac{1}{4}$, $\frac{1}{8}$ has the series

$$1 + \frac{1}{2} + \frac{1}{4} + \frac{1}{8}$$

associated with it.

The series associated with a finite sequence will sum up to a real number. The series associated with an infinite arithmetic sequence will not sum up to a real number, as the terms being added become larger and larger numerically. The sum is unbounded, as we can see in Example 1(a). The series associated with an infinite geometric sequence may or may not sum up to a real number, as we shall now show.

Let us now consider the sum of the first n terms of the infinite geometric sequence $1, \frac{1}{2}, \frac{1}{4}, \ldots$. This is the sum of the n terms of the associated geometric series

$$1 + \frac{1}{2} + \frac{1}{4} + \cdots + \frac{1}{2^{n-1}}$$

Here $a_1 = 1$ and $r = \frac{1}{2}$, and we find that we get the values of S_n for the given values of n in the following table:

n	2	3	4	5	6	7	8	9	10
S_n	$\frac{3}{2}$	$\frac{7}{4}$	$\frac{15}{8}$	$\frac{31}{16}$	$\frac{63}{32}$	$\frac{127}{64}$	$\frac{255}{128}$	$\frac{511}{256}$	$\frac{1023}{512}$

$1 + \frac{1}{2} + \frac{1}{4} + \frac{1}{8}$ ⟶   ⟵ $1 + \frac{1}{2} + \frac{1}{4} + \frac{1}{8} + \frac{1}{16} + \frac{1}{32} + \frac{1}{64}$

The series for $n = 4$ and $n = 7$ are shown. We see that as n gets larger, the numerator of each fraction becomes more nearly twice the denominator. In fact, we find that if we continue to compute S_n as n becomes larger, S_n can be found as close to the value 2 as desired, although it will never actually reach the value 2. For example, if $n = 100$, $S_{100} = 2 - 1.6 \times 10^{-30}$, which could be written as

1.9999999999999999999999999999984

to 32 significant digits. In the formula for the sum of the first n terms of a geometric sequence

$$S_n = a_1 \frac{1 - r^n}{1 - r}$$

the term r^n becomes exceedingly small, and if we consider n as being sufficiently large, we can see that this term is effectively zero. *If this term were exactly zero,* then the sum would be

$$S_n = 1 \frac{1 - 0}{1 - \frac{1}{2}} = 2$$

The symbol ∞ for infinity was first used by the English mathematician, John Wallis (1616–1703).

The only problem is that we cannot find any number large enough for n to make $\left(\frac{1}{2}\right)^n$ zero. There is, however, an accepted notation for this. This notation is

$$\lim_{n \to \infty} r^n = 0 \qquad (\text{if } |r| < 1)$$

and it is read as "the limit, as n *approaches* infinity, of r to the nth power is zero."

CAUTION ▶

The symbol ∞ is read as **infinity,** *but it must not be thought of as a number.* It is simply a symbol that stands for a *process* of considering numbers which become large without bound. The number called the **limit** of the sums is simply the number the sums get closer and closer to, as n is considered to approach infinity. This notation and terminology are of particular importance in the calculus.

If we consider values of r such that $|r| < 1$ and let the values of n become unbounded, we find that $\lim_{n\to\infty} r^n = 0$. The formula for *the sum of the terms of an infinite geometric series then becomes*

$$S = \frac{a_1}{1 - r} \qquad (|r| < 1) \tag{19-7}$$

where a_1 is the first term and r is the common ratio. If $|r| \geq 1$, S is unbounded in value.

EXAMPLE 2   Find the sum of the infinite geometric series

$$4 - \frac{1}{2} + \frac{1}{16} - \frac{1}{128} + \cdots$$

Here we see that $a_1 = 4$. We find r by dividing any term by the previous term, and we find that $r = -\frac{1}{8}$. We then find the sum by substituting in Eq. (19-7). This gives us

$$S = \frac{4}{1 - (-\frac{1}{8})} = \frac{4}{1 + \frac{1}{8}}$$

$$= \frac{4}{1} \times \frac{8}{9} = \frac{32}{9}$$

EXAMPLE 3   Find the fraction that has as its decimal form $0.121212\ldots$.
This decimal form can be considered as being

$$0.12 + 0.0012 + 0.000012 + \cdots$$

which means that we have an infinite geometric series in which $a_1 = 0.12$ and $r = 0.01$. Thus,

$$S = \frac{0.12}{1 - 0.01} = \frac{0.12}{0.99}$$

$$= \frac{4}{33}$$

Therefore, the decimal $0.121212\ldots$ and the fraction $\frac{4}{33}$ represent the same number.

The decimal in Example 3 is called a **repeating decimal,** because *a particular sequence of digits in the decimal form repeats endlessly.* This example verifies the theorem that any repeating decimal represents a rational number. However, not all repeating decimals start repeating immediately. If the numbers never do repeat, the decimal represents an irrational number. For example, there are no repeating decimals that represent π, $\sqrt{2}$, or e. The decimal form of the number e does repeat at one point, but the repetition stops. As we noted in Section 12-5, the decimal form of e to 16 decimal places is 2.7 1828 1828 4590 452. We see that the sequence of digits 1828 repeats only once.

EXAMPLE 4 Find the fraction that has as its decimal form the repeating decimal 0.50345345345....

We first separate the decimal into the beginning, nonrepeating part, and the infinite repeating decimal, which follows. Thus, we have

$$0.50345345345\ldots = 0.50 + 0.00345345345\ldots$$

This means that we are to add $\frac{50}{100}$ to the fraction that represents the sum of the terms of the infinite geometric series $0.00345 + 0.00000345 + \cdots$. For this series, $a_1 = 0.00345$ and $r = 0.001$. We find this sum to be

$$S = \frac{0.00345}{1 - 0.001} = \frac{0.00345}{0.999} = \frac{115}{33,300} = \frac{23}{6660}$$

Therefore,

$$0.50345345\ldots = \frac{5}{10} + \frac{23}{6660} = \frac{5(666) + 23}{6660} = \frac{3353}{6660}$$

SOLVING A WORD PROBLEM

EXAMPLE 5 Each swing of a certain pendulum bob is 95% as long as the preceding swing. How far does the bob travel in coming to rest if the first swing is 40.0 in. long?

We are to find the sum of the terms of an infinite geometric series for which $a_1 = 40.0$ and $r = 95\% = \frac{19}{20}$. Substituting these values into Eq. (19-7), we obtain

$$S = \frac{40.0}{1 - \frac{19}{20}} = \frac{40.0}{\frac{1}{20}} = (40.0)(20) = 800 \text{ in.}$$

The pendulum bob travels 800 in. (about 67 ft) in coming to rest.

EXERCISES *19-3*

In Exercises 1–4, find the indicated quantity for an infinite geometric series.

1. $a_1 = 4, r = \frac{1}{2}, S = ?$ **2.** $a_1 = 6, r = -\frac{1}{3}, S = ?$

3. $a_1 = 5, S = \frac{25}{4}, r = ?$

4. $S = 4 + 2\sqrt{2}, r = \frac{1}{\sqrt{2}}, a_1 = ?$

In Exercises 5–12, find the sums of the given infinite geometric series.

5. $20 - 1 + 0.05 - \cdots$ **6.** $9 + 8.1 + 7.29 + \cdots$

7. $1 + \frac{7}{8} + \frac{49}{64} + \cdots$ **8.** $6 - 4 + \frac{8}{3} - \cdots$

9. $1 + 0.0001 + 0.00000001 + \cdots$

10. $30 - 9 + 2.7 - \cdots$

11. $(2 + \sqrt{3}) + 1 + (2 - \sqrt{3}) + \cdots$

12. $(1 + \sqrt{2}) - 1 + (\sqrt{2} - 1) - \cdots$

In Exercises 13–24, find the fractions equal to the given decimals.

13. $0.33333\ldots$ **14.** $0.55555\ldots$

15. $0.404040\ldots$ **16.** $0.070707\ldots$

17. $0.181818\ldots$ **18.** $0.336336336\ldots$

19. $0.273273273\ldots$ **20.** $0.82222\ldots$

21. $0.366666\ldots$ **22.** $0.66424242\ldots$

23. $0.100841841841\ldots$

24. $0.184561845618456\ldots$

In Exercises 25–28, solve the given problems by use of the sum of an infinite geometric series.

25. Liquid is continuously collected in a wastewater holding tank such that during a given hour only 92.0% as much liquid is collected as in the previous hour. If 28.0 gal are collected in the first hour, what must be the minimum capacity of the tank?

26. If 75% of all aluminum cans are recycled, what is the total number of recycled cans that can be made from 400,000 cans that are recycled over and over until all the aluminum from these cans is used up?

27. The amounts of plutonium 237 that decay each day because of radioactivity form a geometric sequence. Given that the amounts that decay during each of the first four days are 5.882 g, 5.782 g, 5.684 g, and 5.587 g, respectively, what total amount will decay?

(W) 28. A square has sides of 20 cm. Another square is inscribed in the original square by joining the midpoints of the sides. Assuming that such inscribed squares can be formed endlessly, find the sum of the areas of all the squares and explain how this sum is found. See Fig. 19-3.

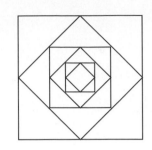

Fig. 19-3

19-4 THE BINOMIAL THEOREM

If we wish to find the roots of the equation $(x + 2)^5 = 0$, we note that there are five factors of $x + 2$, which in turn tells us that the only distinct root is $x = -2$. However, if we wish to expand the expression $(x + 2)^5$, a number of repeated multiplications would be needed. This would be a relatively tedious operation. In this section we develop *the* **binomial theorem,** *by which it is possible to expand binomials to any given power without direct multiplication.* This type of expansion can be helpful and laborsaving in developing certain mathematical topics. We may also expand certain expressions where direct multiplication is not actually possible. Also, the binomial theorem is used to develop the necessary expressions for use in certain technical applications.

By direct multiplication, we may obtain the following expansions of the binomial $a + b$.

$$(a + b)^0 = 1$$
$$(a + b)^1 = a + b$$
$$(a + b)^2 = a^2 + 2ab + b^2$$
$$(a + b)^3 = a^3 + 3a^2b + 3ab^2 + b^3$$
$$(a + b)^4 = a^4 + 4a^3b + 6a^2b^2 + 4ab^3 + b^4$$
$$(a + b)^5 = a^5 + 5a^4b + 10a^3b^2 + 10a^2b^3 + 5ab^4 + b^5$$

An inspection indicates certain properties of these expansions, and we shall assume that these properties are valid for the expansion of $(a + b)^n$, where n is any positive integer.

Properties of the Binomial $(a + b)^n$

  1. *There are $n + 1$ terms.*

  2. *The first term is a^n, and the final term is b^n.*

  3. *Progressing from the first term to the last, the exponent of a decreases by 1 from term to term, the exponent of b increases by 1 from term to term, and the sum of the exponents of a and b in each term is n.*

  4. *If the coefficient of any term is multiplied by the exponent of a in that term, and this product is divided by the number of that term, we obtain the coefficient of the next term.*

  5. *The coefficients of terms equidistant from the ends are equal.*

■EXAMPLE 1 Using the basic properties, develop the expansion for $(a + b)^5$.

Since the exponent of the binomial is 5, we have $n = 5$.

From Property 1, we know that there are six terms.

From Property 2, we know that the first term is a^5 and the final term is b^5.

From Property 3, we know that the factors of a and b in terms 2, 3, 4, and 5 are a^4b, a^3b^2, a^2b^3, and ab^4, respectively.

From Property 4, we obtain the coefficients of terms 2, 3, 4, and 5. In the first term, a^5, the coefficient is 1. Multiplying by 5, the power of a, and dividing by 1, the number of the term, we obtain 5, which is the coefficient of the second term. Thus, the second term is $5a^4b$. Again using Property 4, we obtain the coefficient of the third term. The coefficient of the second term is 5. Multiplying by 4, and dividing by 2, we obtain 10. This means that the third term is $10a^3b^2$.

From Property 5, we know that the coefficient of the fifth term is the same as the second and that the coefficient of the fourth term is the same as the third. These properties are illustrated in the following diagram.

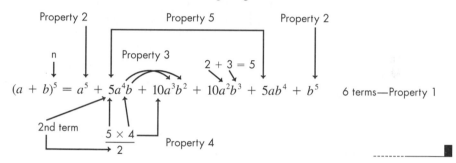

It is not necessary to use the above properties directly to expand a given binomial. If they are applied to $(a + b)^n$, a general formula for the expansion of a binomial may be obtained. In developing and stating the general formula, it is convenient to use the **factorial notation $n!$,** where

$$n! = n(n - 1)(n - 2)\cdots(2)(1) \qquad (19\text{-}8)$$

We see that $n!$, read "n factorial," represents the product of the first n positive integers.

■EXAMPLE 2 **(a)** $3! = (3)(2)(1) = 6$

(b) $5! = (5)(4)(3)(2)(1) = 120$

(c) $8! = (8)(7)(6)(5)(4)(3)(2)(1) = 40,320$

(d) $3! + 5! = 6 + 120 = 126$

(e) $\dfrac{4!}{2!} = \dfrac{(4)(3)(2)(1)}{(2)(1)} = 12$

CAUTION ▶ In evaluating factorials we must remember that they represent products of numbers. In part (d) above we can see that $3! + 5!$ is **not** $(3 + 5)!$. Also, in part (e) we see that $4!/2!$ is **not** $2!$.

The evaluation of factorials can be done on a calculator. The manual should be consulted as to how to evaluate a factorial on any particular calculator. The evaluations for parts (c) and (d) of Example 2 are shown in Fig. 19-4.

```
8!
              40320
3!+5!
                126
```

Fig. 19-4

The Binomial Formula

See Appendix C for a graphing calculator program BINEXPAN. It gives the first K coefficients of $(AX + B)^N$.

Based on the binomial properties, *the* **binomial theorem** *states that the following* **binomial formula** *is valid for all positive integer values of n* (the binomial theorem is proven through advanced methods).

$$(a + b)^n = a^n + na^{n-1}b + \frac{n(n-1)}{2!}a^{n-2}b^2 + \frac{n(n-1)(n-2)}{3!}a^{n-3}b^3 + \cdots + b^n \qquad \textbf{(19-9)}$$

■**EXAMPLE 3** Using the binomial formula, expand $(2x + 3)^6$.

In using the binomial formula for $(2x + 3)^6$, we use $2x$ for a, 3 for b, and 6 for n. Thus,

$$(2x + 3)^6 = (2x)^6 + 6(2x)^5(3) + \frac{(6)(5)}{2}(2x)^4(3^2) + \frac{(6)(5)(4)}{(2)(3)}(2x)^3(3^3) + \frac{(6)(5)(4)(3)}{(2)(3)(4)}(2x)^2(3^4) + \frac{(6)(5)(4)(3)(2)}{(2)(3)(4)(5)}(2x)(3^5) + 3^6$$

$$= 64x^6 + 576x^5 + 2160x^4 + 4320x^3 + 4860x^2 + 2916x + 729 \qquad \text{---------}■$$

For the first few integral powers of a binomial $a + b$, the coefficients can be obtained by setting them up in the following pattern, known as **Pascal's triangle.**

PASCAL'S TRIANGLE

Named for the French scientist and mathematician Blaise Pascal (1623–1662).

$n = 0$							1						
$n = 1$						1		1					
$n = 2$					1		2		1				
$n = 3$				1		3		3		1			
$n = 4$			1		4		6		4		1		
$n = 5$		1		5		10		10		5		1	
$n = 6$	1		6		15		20		15		6		1

see expansions on page 502

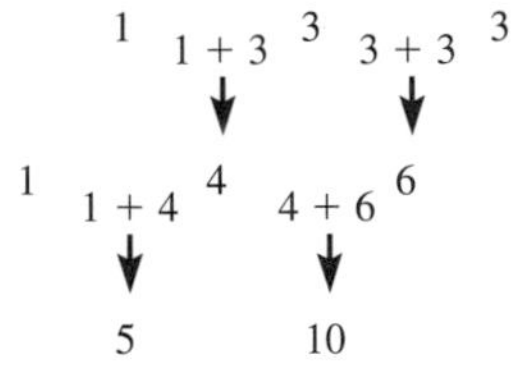

Fig. 19-5

We note that the first and last coefficients shown in each row are 1, and the second and next-to-last coefficients are equal to n. Other coefficients are obtained by adding the two nearest coefficients in the row above, as illustrated in Fig. 19-5 for the indicated section of Pascal's triangle. This pattern may be continued indefinitely, although use of Pascal's triangle is cumbersome for high values of n.

■**EXAMPLE 4** Using Pascal's triangle, expand $(5s - 2t)^4$.

Here we note that $n = 4$. Thus, the coefficients of the five terms are 1, 4, 6, 4, and 1, respectively. Also, here we use $5s$ for a and $-2t$ for b. We are expanding this expression as $[(5s) + (-2t)]^4$. Therefore,

from Pascal's triangle for $n = 4$

$$(5s - 2t)^4 = (5s)^4 + 4(5s)^3(-2t) + 6(5s)^2(-2t)^2 + 4(5s)(-2t)^3 + (-2t)^4$$

$$= 625s^4 - 1000s^3t + 600s^2t^2 - 160st^3 + 16t^4 \qquad \text{---------}■$$

In certain uses of a binomial expansion, it is not necessary to obtain all terms. Only the first few terms are required. The following example illustrates finding the first four terms of an expansion.

EXAMPLE 5   Find the first four terms of the expansion of $(x + 7)^{12}$.
Here we use x for a, 7 for b, and 12 for n. Thus, from the binomial formula we have

$$(x + 7)^{12} = x^{12} + 12x^{11}(7) + \frac{(12)(11)}{2}x^{10}(7^2) + \frac{(12)(11)(10)}{(2)(3)}x^9(7^3) + \cdots$$

$$= x^{12} + 84x^{11} + 3234x^{10} + 75,460x^9 + \cdots$$

If we let $a = 1$ and $b = x$ in the binomial formula, we obtain the **binomial series**

BINOMIAL SERIES

$$(1 + x)^n = 1 + nx + \frac{n(n - 1)}{2!}x^2 + \frac{n(n - 1)(n - 2)}{3!}x^3 + \cdots \qquad \text{(19-10)}$$

which through advanced methods can be shown to be valid for any real number n if $|x| < 1$. When n is either negative or a fraction, we obtain an infinite series. In such a case, we calculate as many terms as may be needed although such a series is not obtainable through direct multiplication. The binomial series may be used to develop important expressions that are used in applications and more advanced mathematics topics.

EXAMPLE 6   In the analysis of forces on beams, the expression $1/(1 + m^2)^{3/2}$ is used. Use the binomial series to find the first four terms of the expansion.
Using negative exponents, we have

$$1/(1 + m^2)^{3/2} = (1 + m^2)^{-3/2}$$

Now, in using Eq. (19-10), we have $n = -3/2$ and $x = m^2$.

$$(1 + m^2)^{-3/2} = 1 + \left(-\frac{3}{2}\right)(m^2) + \frac{(-\frac{3}{2})(-\frac{3}{2} - 1)}{2!}(m^2)^2 + \frac{(-\frac{3}{2})(-\frac{3}{2} - 1)(-\frac{3}{2} - 2)}{3!}(m^2)^3 + \cdots$$

Therefore,

$$\frac{1}{(1 + m^2)^{3/2}} = 1 - \frac{3}{2}m^2 + \frac{15}{8}m^4 - \frac{35}{16}m^6 + \cdots$$

EXERCISES *19-4*

In Exercises 1–8, expand and simplify the given expressions by use of the binomial formula.

1. $(t + 1)^3$

2. $(x - 2)^3$

3. $(2x - 1)^4$

4. $(x^2 + 3)^4$

5. $(2 + 0.1)^5$

6. $(xy - z)^5$

7. $(2a - b^2)^6$

8. $\left(\dfrac{a}{x} + x\right)^6$

In Exercises 9–12, expand and simplify the given expressions by use of Pascal's triangle.

9. $(5x - 3)^4$

10. $(b + 4)^5$

11. $(2a + 1)^6$

12. $(x - 3)^7$

In Exercises 13–20, find the first four terms of the indicated expansions.

13. $(x + 2)^{10}$

14. $(x - 3)^8$

15. $(2a - 1)^7$

16. $(3b + 2)^9$

17. $(x^{1/2} - y)^{12}$

18. $(2a - x^{-1})^{11}$

19. $\left(b^2 + \dfrac{1}{2b}\right)^{20}$

20. $\left(2x^2 + \dfrac{y}{3}\right)^{15}$

In Exercises 21–28, find the first four terms of the indicated expansions by use of the binomial series.

21. $(1 + x)^8$

22. $(1 + x)^{-1/3}$

23. $(1 - x)^{-2}$

24. $(1 - 2\sqrt{x})^9$

25. $\sqrt{1 + x}$

26. $\dfrac{1}{\sqrt{1 + x}}$

27. $\dfrac{1}{\sqrt{9 - 9x}}$

28. $\sqrt{4 + x^2}$

In Exercises 29–32, solve the given problems involving factorial notation.

29. Using a calculator, evaluate (a) 17! + 4!, (b) 21!, (c) 17! × 4!, (d) 68!.

30. Using a calculator, evaluate (a) 8! − 7!, (b) 8!/7!, (c) 8! × 7!, (d) 56!.

(W) **31.** Show that $n! = n \times (n - 1)!$ for $n \geq 2$. To use this equation for $n = 1$, explain why it is necessary to define $0! = 1$ (this is a standard definition of 0!).

32. Show that $\dfrac{(n + 1)!}{(n - 2)!} = n^3 - n$ for $n \geq 2$. See Exercise 31.

In Exercises 33–36, find the indicated terms by use of the following information. The $r + 1$ term of the expansion of $(a + b)^n$ is given by

$$\frac{n(n - 1)(n - 2) \cdots (n - r + 1)}{r!} a^{n-r} b^r$$

33. The term involving b^5 in $(a + b)^8$

34. The term involving y^6 in $(x + y)^{10}$

35. The fifth term of $(2x - 3b)^{12}$

36. The sixth term of $(\sqrt{a} - \sqrt{b})^{14}$

In Exercises 37–40, find the indicated expansions.

37. A company purchases a piece of equipment for A dollars, and the equipment depreciates at a rate of r each year. Its value V after n years is by $V = A(1 - r)^n$. Expand this expression for $n = 5$.

38. In designing a type of tubing, the expression $(D + 0.1)^4$ is used. Expand this expression.

39. In the theory associated with the magnetic field due to an electric current, the expression $1 - \dfrac{x}{\sqrt{a^2 + x^2}}$ is found. By expanding $(a^2 + x^2)^{-1/2}$, find the first three nonzero terms that could be used to approximate the given expression.

40. In the theory related to the dispersion of light, the expression $1 + \dfrac{A}{1 - \lambda_0^2/\lambda^2}$ arises. (a) Let $x = \lambda_0^2/\lambda^2$ and find the first four terms of the expansion of $(1 - x)^{-1}$. (b) Find the same expansion by using long division. (c) Write the original expression in expanded form using the results of (a) and (b).

CHAPTER EQUATIONS

Arithmetic sequences	Recursion formula	$a_n = a_{n-1} + d$	(19-1)
	nth term	$a_n = a_1 + (n - 1)d$	(19-2)
	Sum of n terms	$S_n = \dfrac{n}{2}(a_1 + a_n)$	(19-3)
Geometric sequences	Recursion formula	$a_n = ra_{n-1}$	(19-4)
	nth term	$a_n = a_1 r^{n-1}$	(19-5)
	Sum of n terms	$S_n = \dfrac{a_1(1 - r^n)}{1 - r} \quad (r \neq 1)$	(19-6)
Sum of geometric series		$S = \dfrac{a_1}{1 - r} \quad (|r| < 1)$	(19-7)
Factorial notation		$n! = n(n - 1)(n - 2) \cdots (2)(1)$	(19-8)
Binomial formula		$(a + b)^n = a^n + na^{n-1}b + \dfrac{n(n - 1)}{2!}a^{n-2}b^2 + \dfrac{n(n - 1)(n - 2)}{3!}a^{n-3}b^3 + \cdots + b^n$	(19-9)
Binomial series		$(1 + x)^n = 1 + nx + \dfrac{n(n - 1)}{2!}x^2 + \dfrac{n(n - 1)(n - 2)}{3!}x^3 + \cdots$	(19-10)

REVIEW EXERCISES

In Exercises 1–8, find the indicated term of each sequence.

1. 1, 6, 11, ...   (17th)   **2.** 1, −3, −7, ...   (21st)

3. 0.5, 0.1, 0.02, ...   (9th)

4. 0.025, 0.01, 0.004, ...   (7th)

5. $8, \frac{7}{2}, -1, \ldots$ (16th)

6. $-1, -\frac{5}{3}, -\frac{7}{3}, \ldots$ (25th)

7. $\frac{3}{4}, \frac{1}{2}, \frac{1}{3}, \ldots$ (7th)   **8.** $5^{-2}, 5^0, 5^2, \ldots$ (7th)

In Exercises 9–12, find the sum of each sequence with the indicated values.

9. $a_1 = -4, n = 15, a_{15} = 17$ (arith.)

10. $a_1 = 3, d = -\frac{2}{3}, n = 10$

11. $a_1 = 16, r = -\frac{1}{2}, n = 14$

12. $a_1 = 64, a_n = 729, n = 7$ (geom., $r > 0$)

In Exercises 13–24, find the indicated quantities for the appropriate sequences.

13. $a_1 = 17, d = -2, n = 9, S_9 = ?$

14. $d = \frac{4}{3}, a_1 = -3, a_n = 17, n = ?$

15. $a_1 = 4, r = \sqrt{2}, n = 7, a_7 = ?$

16. $a_n = \frac{49}{8}, r = -\frac{2}{7}, S_n = \frac{1911}{32}, a_1 = ?$

17. $a_1 = 8, a_n = -2.5, S_n = 22, d = ?$

18. $a_1 = 2, d = 0.2, n = 11, S_{11} = ?$

19. $n = 6, r = -0.25, S_6 = 204.75, a_6 = ?$

20. $a_1 = 10, r = 0.1, S_n = 11.111, n = ?$

21. $a_1 = -1, a_n = 32, n = 12, S_{12} = ?$ (arith.)

22. $a_1 = 1, a_n = 64, S_n = 325, n = ?$ (arith.)

23. $a_1 = 1, n = 7, a_7 = 64, S_7 = ?$

24. $a_1 = \frac{1}{4}, n = 6, a_6 = 8, S_6 = ?$

In Exercises 25–28, find the sums of the given infinite geometric series.

25. $9 + 6 + 4 + \cdots$

26. $80 - 20 + 5 - \cdots$

27. $1 + 1.02^{-1} + 1.02^{-2} + \cdots$

28. $3 - \sqrt{3} + 1 - \cdots$

In Exercises 29–32, find the fractions equal to the given decimals.

29. $0.030303\ldots$ **30.** $0.363363\ldots$

31. $0.0727272\ldots$ **32.** $0.25399399399\ldots$

In Exercises 33–36, expand and simplify the given expression. In Exercises 37–40, find the first four terms of the appropriate expansion.

33. $(x - 2)^4$ **34.** $(3 + 0.1)^4$

35. $(x^2 + 1)^5$ **36.** $(3n^{1/2} - a)^6$

37. $(a + 2b^2)^{10}$ **38.** $\left(\dfrac{x}{4} - y\right)^{12}$

39. $\left(p^2 - \dfrac{q}{6}\right)^9$ **40.** $(2s^2 - \frac{3}{2}t^{-1})^{14}$

In Exercises 41–48, find the first four terms of the indicated expansions by use of the binomial series.

41. $(1 + x)^{12}$ **42.** $(1 - x)^{10}$

43. $\sqrt{1 + x^2}$ **44.** $(4 - 4\sqrt{x})^{-1}$

45. $\sqrt{1 - a^2}$ **46.** $\sqrt{1 + b^4}$

47. $(2 - 4x)^{-3}$ **48.** $(1 + 4x)^{-1/4}$

In Exercises 49–72, solve the given problems by use of an appropriate sequence or expansion. All numbers are accurate to at least two significant digits.

49. Find the sum of the first 1000 positive even integers.

50. How many integers divisible by 4 lie between 23 and 121?

51. Each stroke of a pile driver moves a post 2 in. less than the previous stroke. If the first stroke moves the post 24 in., which stroke moves the post 4 in.?

52. During each hour, an exhaust fan removes 15.0% of the carbon dioxide present in the air in a room at the beginning of the hour. What percent of the carbon dioxide remains after 10.0 h?

53. Each 1.0 mm of a filter through which light passes reduces the intensity of the light by 12%. How thick should the filter be to reduce the intensity of the light to 20%?

54. A pile of dirt and 10 holes are in a straight line. It is 20 ft from the dirt pile to the nearest hole, and the holes are 8 ft apart. If a backhoe takes two trips to fill each hole, how far must it travel in filling all the holes if it starts and ends at the dirt pile?

55. A roof support with equally spaced vertical pieces is shown in Fig. 19-6. Find the total length of the vertical pieces if the shortest one is 10.0 in. long.

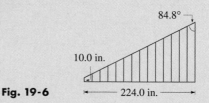

Fig. 19-6

56. During each microsecond the current in an electric circuit decreases by 9.3%. If the initial current is 2.45 mA, how long does it take to reach 0.50 mA?

57. A machine that costs $8600 depreciates 1.0% in value each month. What is its value 5 years after it was purchased?

58. The level of chemical pollution in a lake is 4.50 ppb (parts per billion). If the level increases by 0.20 ppb in the following month and by 5.0% less each month thereafter, what will be the maximum level?

59. A piece of paper 0.0040 in. thick is cut in half. These two pieces are then placed one on the other and cut in half. If this is repeated such that the paper is cut in half 40 times, how high will the pile be?

60. After the power is turned off, an object on a nearly frictionless surface slows down such that it travels 99.9% as far during one second as during the previous second. If it travels 100 cm during the first second after the power is turned off, how far does it travel while stopping?

61. A person invests $1000 each year at the beginning of the year. What is the total value of these investments after 20 years if they earn 7.5% annual interest compounded semiannually?

62. In testing a type of insulation, the temperature in a room was made to fall to 2/3 of the initial temperature after 1.0 h, to 2/5 of the initial temperature after 2.0 h, to 2/7 of the initial temperature after 3.0 h, and so on. If the initial temperature was 50.0°C, what was the temperature after 12.0 h? (This is an illustration of a *harmonic sequence.*)

63. Two competing businesses make the same item and sell it initially for $100. One increases the price by $8 each year for five years, and the other increases the price by 8% each year for five years. What is the difference in price after five years?

64. A well driller charges $10.00 for drilling the first meter of a well and for each meter thereafter charges 0.20% more than for the preceding meter. How much is charged for drilling a 150-m well?

65. In hydrodynamics, while studying compressible fluid flow, the expression $\left(1 + \dfrac{a-1}{2}m^2\right)^{a/(a-1)}$ arises. Find the first three terms of the expansion of this expression.

66. In finding the partial pressure P_F of fluorine gas under certain conditions, the equation
$$P_F = \frac{(1 + 2 \times 10^{-10}) - \sqrt{1 + 4 \times 10^{-10}}}{2} \text{ atm}$$
is found. By using three terms of the expansion for $\sqrt{1+x}$, approximate the value of this expression.

67. During one year a beach eroded 1.2 m to a line 48.3 m from the wall of a building. If the erosion is 0.1 m more each year than the previous year, when will the waterline reach the wall?

68. On a highway with a steep incline, a runaway truck ramp is constructed so that a vehicle which has lost its brakes can stop. The ramp is designed to slow a truck in succeeding 20-m distances by 10 km/h, 12 km/h, 14 km/h, ... If the ramp is 160 m long, will it stop a truck moving at 120 km/h when it reaches the ramp?

69. Each application of an insecticide destroys 75% of a certain insect. How many applications are needed to destroy at least 99.9% of the insects?

70. A wire hung between two poles is parabolic in shape. To find the length of wire between two points on the wire, the expression $\sqrt{1 + 0.08x^2}$ is used. Find the first three terms of the binomial expansion of this expression.

(W) 71. Do the reciprocals of the terms of a geometric sequence form a geometric sequence? Explain.

72. The terms a, $a + 12$, $a + 24$ form an arithmetic sequence, and the terms a, $a + 24$, $a + 12$ form a geometric sequence. Find these sequences.

Writing Exercise

73. Derive a formula for the value V after 1 year of an amount A invested at $r\%$ (as a decimal) annual interest, compounded n times during the year. If $A = \$1000$ and $r = 0.10$ (10%), write two or three paragraphs explaining why the amount of interest increases as n increases and stating your approach to finding the maximum possible amount of interest.

PRACTICE TEST

1. Find the sum of the first seven terms of the sequence $6, -2, \frac{2}{3}, \ldots$.

2. For a given sequence $a_1 = 6$, $d = 4$, and $S_n = 126$. Find n.

3. Find the fraction equal to the decimal $0.454545\ldots$.

4. Find the first three terms of the expansion of $\sqrt{1 - 4x}$.

5. Expand and simplify the expression $(2x - y)^5$.

6. What is the value after 20 years of an investment of $2500 if it draws 5% annual interest compounded annually?

7. Find the sum of the first 100 even integers.

8. A ball is dropped from a height of 8.00 ft, and on each rebound it rises to 1/2 of the height it last fell. If it bounces indefinitely, through what total distance will it move?

20

ADDITIONAL TOPICS IN TRIGONOMETRY

The definitions of the trigonometric functions were first introduced in Section 4-2 and were again summarized in Section 8-1. If we take a close look at these definitions, we find that there are many relationships among the various functions. In this chapter we study some of these basic relationships, as well as develop others.

We also will develop the concept of the inverse trigonometric functions. This was introduced in Chapter 4 when we discussed the use of a calculator in finding angles from given values of the trigonometric functions.

The trigonometric relationships that we shall develop in this chapter are important for a number of reasons. We have already made limited use of some of them in Section 10-4 when we graphed certain trigonometric functions. We also used an important *trigonometric identity* when we derived the law of cosines in Chapter 9.

Later in this chapter we take up the solutions of equations with trigonometric functions. At that time we will find that the solution of many of these equations depends on the proper use of the various identities we will have developed.

In the study of calculus, there are certain types of problems that require the use of trigonometric identities for their solution. This even includes some problems in which trigonometric functions do not appear.

These basic trigonometric relationships are used to develop expressions and solve equations that are used in a number of technical areas. These include electricity and electronics, optics, solar energy, and robotics.

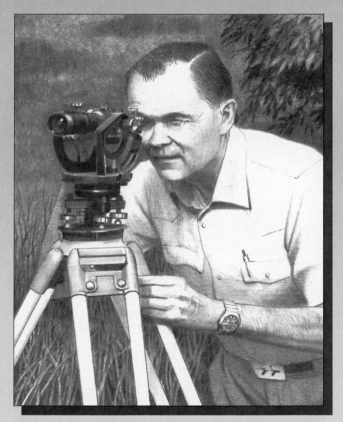

Basic trigonometric relationships can be used to develop useful formulas. In Section 20-3 we derive the formula $A = \frac{1}{2}c^2 \sin 2\theta$, where A is the area of a right triangle, c is the hypotenuse, and θ is either acute angle. It is used in surveying.

20-1 FUNDAMENTAL TRIGONOMETRIC IDENTITIES

From Chapters 4 and 8, we recall that the definition of the sine of an angle θ is $\sin\theta = y/r$ and that the definition of the cosecant of an angle θ is $\csc\theta = r/y$ (see Fig. 20-1). Since $y/r = 1/(r/y)$, we see that $\sin\theta = 1/\csc\theta$. The definitions hold true for *any* angle, which means this relation between $\sin\theta$ and $\csc\theta$ is true for *any* angle. *This type of relation, which is true for any value of the variable, is called an* **identity.** Of course, values where division by zero would be indicated are excluded.

In this section we develop several important identities among the trigonometric functions. We also show how the basic identities are used to verify other identities.

From the definitions, we have

$$\sin\theta\,\csc\theta = \frac{y}{r}\times\frac{r}{y} = 1 \quad\text{or}\quad \sin\theta = \frac{1}{\csc\theta} \quad\text{or}\quad \csc\theta = \frac{1}{\sin\theta}$$

$$\cos\theta\,\sec\theta = \frac{x}{r}\times\frac{r}{x} = 1 \quad\text{or}\quad \cos\theta = \frac{1}{\sec\theta} \quad\text{or}\quad \sec\theta = \frac{1}{\cos\theta}$$

$$\tan\theta\,\cot\theta = \frac{y}{x}\times\frac{x}{y} = 1 \quad\text{or}\quad \tan\theta = \frac{1}{\cot\theta} \quad\text{or}\quad \cot\theta = \frac{1}{\tan\theta}$$

$$\frac{\sin\theta}{\cos\theta} = \frac{y/r}{x/r} = \frac{y}{x} = \tan\theta; \qquad \frac{\cos\theta}{\sin\theta} = \frac{x/r}{y/r} = \frac{x}{y} = \cot\theta$$

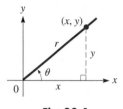

Fig. 20-1

Also, from the definitions and the Pythagorean theorem in the form of $x^2 + y^2 = r^2$, we arrive at the following identities.

By dividing the Pythagorean relation through by r^2, we have

$$\left(\frac{x}{r}\right)^2 + \left(\frac{y}{r}\right)^2 = 1 \quad\text{which leads us to}\quad \cos^2\theta + \sin^2\theta = 1$$

By dividing the Pythagorean relation by x^2, we have

$$1 + \left(\frac{y}{x}\right)^2 = \left(\frac{r}{x}\right)^2 \quad\text{which leads us to}\quad 1 + \tan^2\theta = \sec^2\theta$$

By dividing the Pythagorean relation by y^2, we have

$$\left(\frac{x}{y}\right)^2 + 1 = \left(\frac{r}{y}\right)^2 \quad\text{which leads us to}\quad \cot^2\theta + 1 = \csc^2\theta$$

BASIC IDENTITIES

The term $\cos^2\theta$ is the common way of writing $(\cos\theta)^2$, and it means to square the value of the cosine of the angle. Obviously, the same holds true for the other functions.

Summarizing these results, we have the following important identities.

$$\sin\theta = \frac{1}{\csc\theta} \qquad \textbf{(20-1)}$$

$$\cos\theta = \frac{1}{\sec\theta} \qquad \textbf{(20-2)}$$

$$\tan\theta = \frac{1}{\cot\theta} \qquad \textbf{(20-3)}$$

$$\tan\theta = \frac{\sin\theta}{\cos\theta} \qquad \textbf{(20-4)}$$

$$\cot\theta = \frac{\cos\theta}{\sin\theta} \qquad \textbf{(20-5)}$$

$$\sin^2\theta + \cos^2\theta = 1 \qquad \textbf{(20-6)}$$

$$1 + \tan^2\theta = \sec^2\theta \qquad \textbf{(20-7)}$$

$$1 + \cot^2\theta = \csc^2\theta \qquad \textbf{(20-8)}$$

In using these basic identities, θ may stand for any angle or number or expression representing an angle or a number.

▌EXAMPLE 1   (a) $\sin(x + 1) = \dfrac{1}{\csc(x + 1)}$ using Eq. (20-1)

(b) $\tan 157° = \dfrac{\sin 157°}{\cos 157°}$ using Eq. (20-4)

(c) $\sin^2\left(\dfrac{\pi}{4}\right) + \cos^2\left(\dfrac{\pi}{4}\right) = 1$ using Eq. (20-6)

▌EXAMPLE 2   We shall check the last two illustrations of Example 1 for the particular values of θ that are used.

(a) Using a calculator, we find that

$$\sin 157° = 0.3907311285 \quad \text{and} \quad \cos 157° = -0.9205048535$$

Considering Eq. (20-4) and illustration (b) in Example 1, by dividing we find that

$$\frac{\sin 157°}{\cos 157°} = \frac{0.3907311285}{-0.9205048535} = -0.4244748162$$

We also find that $\tan 157° = -0.4244748162$, which shows that

$$\tan 157° = \frac{\sin 157°}{\cos 157°}$$

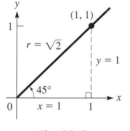

Fig. 20-2

(b) To check illustration (c) of Example 1, we refer to the values found in Example 2 of Section 4-3, or to Fig. 20-2. These tell us that

$$\sin 45° = \frac{1}{\sqrt{2}} = \frac{\sqrt{2}}{2} \quad \text{and} \quad \cos 45° = \frac{\sqrt{2}}{2}$$

Since $\frac{\pi}{4} = 45°$, by adding the squares of $\sin 45°$ and $\cos 45°$, we have

$$\sin^2\left(\frac{\pi}{4}\right) + \cos^2\left(\frac{\pi}{4}\right) = \left(\frac{\sqrt{2}}{2}\right)^2 + \left(\frac{\sqrt{2}}{2}\right)^2 = \frac{1}{2} + \frac{1}{2} = 1$$

We see that this checks with Eq. (20-6) for these values.

Proving Trigonometric Identities

A great many identities exist among the trigonometric functions. We are going to use the basic identities that have been developed in Eqs. (20-1) through (20-8), along with a few additional ones developed in later sections to prove the validity of still other identities.

CAUTION▶   *The ability to prove trigonometric identities depends to a large extent on being very familiar with the basic identities* so that you can *recognize them in somewhat **different forms.***

If you do not learn these basic identities and learn them well, you will have difficulty in following the examples and doing the exercises. The more readily you recognize these forms, the more easily you will be able to prove such identities.

In proving identities, we should look for combinations that appear in, or are very similar to, those in the basic identities. This is illustrated in the following examples.

EXAMPLE 3 In proving the identity

$$\sin x = \frac{\cos x}{\cot x}$$

we know that $\cot x = \dfrac{\cos x}{\sin x}$. Since $\sin x$ appears on the left, substituting for $\cot x$ on the right will eliminate $\cot x$ and introduce $\sin x$. This should help us proceed in proving the identity. Thus,

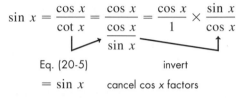

$$\sin x = \frac{\cos x}{\cot x} = \frac{\cos x}{\dfrac{\cos x}{\sin x}} = \frac{\cos x}{1} \times \frac{\sin x}{\cos x}$$

$$\qquad\qquad\quad \text{Eq. (20-5)} \qquad\qquad\qquad \text{invert}$$

$$= \sin x \qquad \text{cancel } \cos x \text{ factors}$$

By showing that the right side may be changed exactly to $\sin x$, the expression on the left side, we have proved the identity. $\blacksquare$

Some important points should be made in relation to the proof of the identity of Example 3. We must recognize what basic identities may be useful. The proof of an identity requires the use of basic algebraic operations, and these must be done carefully and correctly. Although in Example 3 we changed the right side to the form on the left, we could have changed the left to the form on the right. From this and the fact that various substitutions are possible, we see that a variety of procedures can be used to prove any given identity.

As we point out here, performing the *algebraic operations* carefully and correctly is very important when working with trigonometric expressions. Operations such as substituting, factoring, and simplifying fractions are frequently used.

EXAMPLE 4 Prove that $\tan \theta \csc \theta = \sec \theta$.

In proving this identity, we know that $\tan \theta = \dfrac{\sin \theta}{\cos \theta}$ and also that $\dfrac{1}{\cos \theta} = \sec \theta$. Thus, by substituting for $\tan \theta$, we introduce $\cos \theta$ in the denominator, which is equivalent to introducing $\sec \theta$ in the numerator. Therefore, changing only the left side, we have

$$\tan \theta \csc \theta = \frac{\sin \theta}{\cos \theta} \csc \theta = \frac{\sin \theta}{\cos \theta} \frac{1}{\sin \theta}$$

$$\qquad \text{Eq. (20-4)} \qquad\qquad\qquad \text{Eq. (20-1)}$$

$$= \frac{1}{\cos \theta} \qquad \text{cancel } \sin \theta \text{ factors}$$

$$= \sec \theta \qquad \text{using Eq. (20-2)}$$

Having changed the left side into the form on the right side, we have proven the identity.

NOTE ▶ Many variations of the preceding steps are possible. Also, we could have changed the right side to obtain the form on the left. For example,

$$\tan \theta \csc \theta = \sec \theta = \frac{1}{\cos \theta} \qquad\qquad \text{using Eq. (20-2)}$$

$$= \frac{\sin \theta}{\cos \theta \sin \theta} = \frac{\sin \theta}{\cos \theta} \frac{1}{\sin \theta} \qquad \begin{array}{l}\text{multiply numerator and} \\ \text{denominator by } \sin \theta \text{ and rewrite}\end{array}$$

$$= \tan \theta \csc \theta \qquad\qquad \text{using Eqs. (20-4) and (20-1)} \qquad \blacksquare$$

In proving the identities of Examples 3 and 4, we have shown that the expression on one side of the equal sign can be changed into the expression on the other side. Although making the restriction that we change only one side is not entirely necessary, *we shall restrict the method of proof to changing only one side into the same form as the other side.* In this way we know the form we are to obtain, and by looking ahead we are better able to make the proper changes.

CAUTION ▶

There is no set procedure for working with identities. The most important factors are to (1) *recognize the proper forms,* (2) *see what effect a change may have* before performing it, and (3) *perform it correctly.* Normally, *it is easier to change the form of the more complicated side to the same form as the less complicated side.* If the forms are about the same, a close look often suggests possible steps to use.

■EXAMPLE 5 Prove the identity $\dfrac{\cos x \csc x}{\cot^2 x} = \tan x$.

First, we note that the left-hand side has several factors and the right-hand side has only one. Therefore, let us transform the left-hand side. Next, we note that we want $\tan x$ as the final result. We know that $\cot x = 1/\tan x$. Thus,

$$\frac{\cos x \csc x}{\cot^2 x} = \frac{\cos x \csc x}{\dfrac{1}{\tan^2 x}} = \cos x \csc x \tan^2 x$$

At this point, we have two factors of $\tan x$ on the left. Since we want only one, let us factor out one. Therefore,

$$\cos x \csc x \tan^2 x = \tan x(\cos x \csc x \tan x)$$

Now, replacing $\tan x$ within the parentheses by $\sin x/\cos x$, we have

$$\tan x(\cos x \csc x \tan x) = \frac{\tan x(\cos x \csc x \sin x)}{\cos x}$$

Now we may cancel $\cos x$. Also, $\csc x \sin x = 1$ from Eq. (19-1). Finally,

$$\frac{\tan x (\cos x \csc x \sin x)}{\cos x} = \tan x\left(\frac{\cos x}{\cos x}\right)(\csc x \sin x)$$

$$= \tan x (1)(1) = \tan x$$

Since we have transformed the left-hand side into $\tan x$, we have proven the identity. Of course, it is not necessary to rewrite expressions as we did in this example. This was done here only to include the explanations. ▬▬▬▬**■**

■EXAMPLE 6 In finding the radiation rate of an accelerated electric charge, it is necessary to show that $\sin^3 \theta = \sin \theta - \sin \theta \cos^2 \theta$. Show this by changing the left side.

Since each term on the right has a factor of $\sin \theta$, we see that we can proceed by writing $\sin^3 \theta$ as $\sin \theta (\sin^2 \theta)$. Then the factor $\sin^2 \theta$ and the $\cos^2 \theta$ on the right suggest the use of Eq. (20-6). Thus we have

$$\sin^3 \theta = \sin \theta(\sin^2 \theta) = \sin \theta(1 - \cos^2 \theta)$$

$$= \sin \theta - \sin \theta \cos^2 \theta \qquad \text{multiplying}$$

Since we wanted to substitute for $\sin^2 \theta$, we used Eq. (20-6) in the form

$$\sin^2 \theta = 1 - \cos^2 \theta \qquad \text{▬▬▬▬■}$$

■EXAMPLE 7   Prove the identity $\dfrac{\sec^2 y}{\cot y} - \tan^3 y = \tan y$.

Here we shall simplify the left side. We can remove cot y from the denominator, since cot $y = 1/\tan y$. Also, the presence of $\sec^2 y$ suggests the use of Eq. (20-7). Therefore, we have

$$\frac{\sec^2 y}{\cot y} - \tan^3 y = \frac{\sec^2 y}{\dfrac{1}{\tan y}} - \tan^3 y = \sec^2 y \tan y - \tan^3 y$$

$$= \tan y\,(\sec^2 y - \tan^2 y) = \tan y (1)$$

$$= \tan y$$

Here we have used Eq. (20-7) in the form $\sec^2 y - \tan^2 y = 1$.   ⸻ ∎

■EXAMPLE 8   Prove the identity $\dfrac{1 - \sin x}{\sin x \cot x} = \dfrac{\cos x}{1 + \sin x}$.

The combination $1 - \sin x$ also suggests $1 - \sin^2 x$, since multiplying $(1 - \sin x)$ by $(1 + \sin x)$ gives $1 - \sin^2 x$, which can then be replaced by $\cos^2 x$. Thus, changing only the left side, we have

$$\frac{1 - \sin x}{\sin x \cot x} = \frac{(1 - \sin x)(1 + \sin x)}{\sin x \cot x (1 + \sin x)} \qquad \text{multiply numerator and}$$
$$\text{denominator by } 1 + \sin x$$

$$= \frac{1 - \sin^2 x}{\sin x \left(\dfrac{\cos x}{\sin x}\right)(1 + \sin x)} = \frac{\cos^2 x}{\cos x\,(1 + \sin x)} \longleftarrow \text{Eq. (20-6)}$$

cancel sin x

$$= \frac{\cos x}{1 + \sin x} \qquad \text{cancel cos } x$$

⸻ ∎

■EXAMPLE 9   Prove the identity $\sec^2 x + \csc^2 x = \sec^2 x \csc^2 x$.

Here we note the presence of $\sec^2 x$ and $\csc^2 x$ on each side. This suggests the possible use of the square relationships. By replacing the $\sec^2 x$ on the right-hand side by $1 + \tan^2 x$, we can create $\csc^2 x$ plus another term. The left-hand side is the $\csc^2 x$ plus another term, so this procedure should help. Thus, changing only the right side,

$$\sec^2 x + \csc^2 x = \sec^2 x \csc^2 x$$

$$= (1 + \tan^2 x)(\csc^2 x) \qquad \text{using Eq. (20-7)}$$

$$= \csc^2 x + \tan^2 x \csc^2 x \qquad \text{multiplying}$$

$$= \csc^2 x + \left(\frac{\sin^2 x}{\cos^2 x}\right)\left(\frac{1}{\sin^2 x}\right) \qquad \text{using Eqs. (20-4) and (20-1)}$$

$$= \csc^2 x + \frac{1}{\cos^2 x} \qquad \text{cancel sin}^2 x$$

$$= \csc^2 x + \sec^2 x \qquad \text{using Eq. (20-2)}$$

We could have used many other variations of this procedure, and they would have been perfectly valid.   ⸻ ∎

EXAMPLE 10 Simplify the expression $\dfrac{\csc x}{\tan x + \cot x}$.

We proceed with a simplification such as this in a manner similar to proving an identity, although we do not know just what the result should be. One procedure for this simplification is shown as follows.

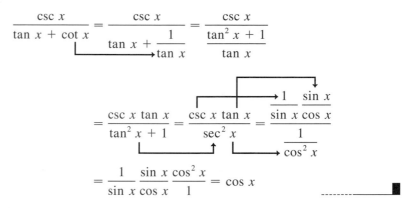

$$= \frac{1}{\sin x \cos x} \frac{\sin x \cos^2 x}{1} = \cos x$$

A graphing calculator can be used to check an identity or a simplification. This is done by graphing the function on each side of an identity, or the initial expression and the final expression for a simplification. If the two graphs are the same, the identity or simplification is probably shown to be correct, although this is not strictly a proof.

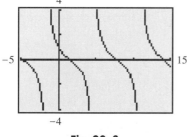

Fig. 20-3

EXAMPLE 11 **(a)** Use a graphing calculator to verify the identity of Example 8.

Noting this identity as $\dfrac{1 - \sin x}{\sin x \cot x} = \dfrac{\cos x}{1 + \sin x}$, on a graphing calculator, we let

$$y_1 = (1 - \sin x)/(\sin x/\tan x) \qquad (\text{noting that } \cot x = 1/\tan x)$$
$$y_2 = \cos x/(1 + \sin x)$$

We then graph these two functions as shown in Fig. 20-3. We choose a domain that obviously includes more than one period of each function. Also, after the first curve is plotted, we must watch the screen carefully to see if any new points are plotted for the second curve (use a heavier curve if possible). Since these curves are the same, the identity appears to be verified (although it has not been *proven*).

--------- EXERCISES *20-1* ---------

In Exercises 1–4, use a calculator to check the indicated basic identities for the given angles.

1. Eq. (20-3) for $\theta = 56°$ **2.** Eq. (20-5) for $\theta = 280°$

3. Eq. (20-6) for $\theta = \dfrac{4\pi}{3}$ **4.** Eq. (20-7) for $\theta = \dfrac{5\pi}{6}$

In Exercises 5–40, prove the given identities.

5. $\dfrac{\cot \theta}{\cos \theta} = \csc \theta$ **6.** $\dfrac{\tan y}{\sin y} = \sec y$

7. $\dfrac{\sin x}{\tan x} = \cos x$ **8.** $\dfrac{\csc \theta}{\sec \theta} = \cot \theta$

9. $\sin y \cot y = \cos y$ **10.** $\cos x \tan x = \sin x$

11. $\sin x \sec x = \tan x$ **12.** $\cot \theta \sec \theta = \csc \theta$

13. $\csc^2 x(1 - \cos^2 x) = 1$ **14.** $\cos^2 x(1 + \tan^2 x) = 1$

15. $\sin x(1 + \cot^2 x) = \csc x$ **16.** $\sec \theta(1 - \sin^2 \theta) = \cos \theta$

17. $\tan y(\cot y + \tan y) = \sec^2 y$

18. $\csc x(\csc x - \sin x) = \cot^2 x$

19. $\sin x \tan x + \cos x = \sec x$

20. $\sec x \csc x - \cot x = \tan x$

21. $\cos \theta \cot \theta + \sin \theta = \csc \theta$

22. $\csc x \sec x - \tan x = \cot x$

23. $\cot\theta\sec^2\theta - \cot\theta = \tan\theta$

24. $\sin y + \sin y\cot^2 y = \csc y$

25. $\tan x + \cot x = \sec x\csc x$

26. $\tan x + \cot x = \tan x\csc^2 x$

27. $\cos^2 x - \sin^2 x = 1 - 2\sin^2 x$

28. $\tan^2 y\sec^2 y - \tan^4 y = \tan^2 y$

29. $\dfrac{\sin x}{1 - \cos x} = \csc x + \cot x$

30. $\dfrac{1 + \cos x}{\sin x} = \dfrac{\sin x}{1 - \cos x}$

31. $\tan^2 x\cos^2 x + \cot^2 x\sin^2 x = 1$

32. $\dfrac{\sin\theta}{\csc\theta} + \dfrac{\cos\theta}{\sec\theta} = 1$

33. $\dfrac{\sec\theta}{\cos\theta} - \dfrac{\tan\theta}{\cot\theta} = 1$

34. $\dfrac{\csc\theta}{\sin\theta} - \dfrac{\cot\theta}{\tan\theta} = 1$

35. $2\sin^4 x - 3\sin^2 x + 1 = \cos^2 x(1 - 2\sin^2 x)$

36. $\dfrac{\sin^4 x - \cos^4 x}{1 - \cot^4 x} = \sin^4 x$

37. $\dfrac{1}{2}\sin\pi t\left(\dfrac{\sin\pi t}{1 - \cos\pi t} + \dfrac{1 - \cos\pi t}{\sin\pi t}\right) = 1$

38. $\dfrac{\cot\omega t}{\sec\omega t - \tan\omega t} - \dfrac{\cos\omega t}{\sec\omega t + \tan\omega t} = \sin\omega t + \csc\omega t$

39. $1 + \sin^2 x + \sin^4 x + \cdots = \sec^2 x$

40. $1 - \tan^2 x + \tan^4 x - \cdots = \cos^2 x \quad (-\tfrac{\pi}{4} < x < \tfrac{\pi}{4})$

In Exercises 41–48, simplify the given expressions. The result will be one of $\sin x$, $\cos x$, $\tan x$, $\cot x$, $\sec x$, or $\csc x$.

41. $\dfrac{\tan x\csc^2 x}{1 + \tan^2 x}$

42. $\dfrac{\cos x - \cos^3 x}{\sin x - \sin^3 x}$

43. $\cot x(\sec x - \cos x)$

44. $\sin x(\tan x + \cot x)$

45. $\dfrac{\tan x + \cot x}{\csc x}$

46. $\dfrac{1 + \tan x}{\sin x} - \sec x$

47. $\dfrac{\cos x + \sin x}{1 + \tan x}$

48. $\dfrac{\sec x - \cos x}{\tan x}$

In Exercises 49–56, use a graphing calculator to verify the given identities by comparing the graphs of each side.

49. $\sin x(\csc x - \sin x) = \cos^2 x$

50. $\cos y(\sec y - \cos y) = \sin^2 y$

51. $\sec\theta\tan\theta\csc\theta = \tan^2\theta + 1$

52. $\sin x\cos x\tan x = 1 - \cos^2 x$

53. $\dfrac{\sec x + \csc x}{1 + \tan x} = \csc x$

54. $\dfrac{\cot x + 1}{\cot x} = 1 + \tan x$

55. $\dfrac{1 - 2\cos^2 x}{\sin x\cos x} = \tan x - \cot x$

56. $\cos^3 x\csc^3 x\tan^3 x = \csc^2 x - \cot^2 x$

In Exercises 57–60, solve the given problems involving trigonometric identities.

57. When designing a solar energy collector, it is necessary to account for the latitude and longitude of the location, the angle of the sun, and the angle of the collector. In doing this, the equation

$\cos\theta = \cos A\cos B\cos C + \sin A\sin B$

is used. If $\theta = 90°$, show that $\cos C = -\tan A\tan B$.

58. In studying the gravitational force between two objects, the expression $(r - R\cos\theta)^2 + (R\sin\theta)^2$ occurs. Show that this expression can be written as $r^2 - 2rR\cos\theta + R^2$.

59. Show that the length l of the straight brace shown in Fig. 20-4 can be found from the equation

$l = \dfrac{a(1 + \tan\theta)}{\sin\theta}$

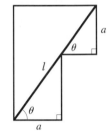

Fig. 20-4

60. In determining the path of least time between two points under certain conditions, it is necessary to show that

$\sqrt{\dfrac{1 + \cos\theta}{1 - \cos\theta}}\,\sin\theta = 1 + \cos\theta$

Show this by transforming the left-hand side.

In Exercises 61–64, solve the given problems.

61. Show that $\sin^2 x(1 - \sec^2 x) + \cos^2 x(1 + \sec^4 x)$ has a constant value.

62. Show that $\cot y\csc y\sec y - \csc y\cos y\cot y$ has a constant value.

63. Prove that $\sec^2\theta + \csc^2\theta = \sec^2\theta\csc^2\theta$ by expressing each function in terms of its x, y, and r definition. See Example 9.

64. Prove that $\dfrac{\csc\theta}{\tan\theta + \cot\theta} = \cos\theta$ by expressing each function in terms of its x, y, and r definition. See Example 10.

In Exercises 65–68, use the given substitutions to show that the given equations are valid. In each, $0 < \theta < \pi/2$.

65. If $x = \cos\theta$, show that $\sqrt{1 - x^2} = \sin\theta$.

66. If $x = 3\sin\theta$, show that $\sqrt{9 - x^2} = 3\cos\theta$.

67. If $x = 2\tan\theta$, show that $\sqrt{4 + x^2} = 2\sec\theta$.

68. If $x = 4\sec\theta$, show that $\sqrt{x^2 - 16} = 4\tan\theta$.

20-2 THE SUM AND DIFFERENCE FORMULAS

For reference, Eq. (12-13) is
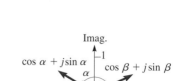

There are other important relations among the trigonometric functions. The most important and useful relations are those that involve twice an angle and half an angle. To obtain these relations, we first derive the expressions for the sine and cosine of the sum and difference of two angles. These expressions will lead directly to the desired relations of double and half angles.

Equation (12-13) gives the polar (or trigonometric) form of the product of two complex numbers. We can use this formula to derive the expression for the sine and cosine of the sum and difference of two angles. These expressions will lead directly to the desired relations of double and half angles in the next section.

Using Eq. (12-13) to find the product of the complex numbers $\cos \alpha + j \sin \alpha$ and $\cos \beta + j \sin \beta$, which are represented in Fig. 20-5, we have

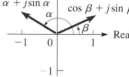

Fig. 20-5

$$(\cos \alpha + j \sin \alpha)(\cos \beta + j \sin \beta) = \cos(\alpha + \beta) + j \sin(\alpha + \beta)$$

Expanding the left side, and then switching sides, we have

$$\cos(\alpha + \beta) + j \sin(\alpha + \beta) = (\cos \alpha \cos \beta - \sin \alpha \sin \beta) + j(\sin \alpha \cos \beta + \cos \alpha \sin \beta)$$

Since two complex numbers are equal if their real parts are equal and their imaginary parts are equal, we have

$$\sin(\alpha + \beta) = \sin \alpha \cos \beta + \cos \alpha \sin \beta \qquad \textbf{(20-9)}$$

and

$$\cos(\alpha + \beta) = \cos \alpha \cos \beta - \sin \alpha \sin \beta \qquad \textbf{(20-10)}$$

■EXAMPLE 1   Verify that $\sin 90° = 1$, by finding $\sin(60° + 30°)$.

$$\sin 90° = \sin(60° + 30°) = \sin 60° \cos 30° + \cos 60° \sin 30° \qquad \text{using Eq. (20-9)}$$

$$= \frac{\sqrt{3}}{2} \times \frac{\sqrt{3}}{2} + \frac{1}{2} \times \frac{1}{2} \qquad \text{for values, see Section 4-3}$$

$$= \frac{3}{4} + \frac{1}{4} = 1$$

It should be obvious from this example that

CAUTION ▸ $\sin(\alpha + \beta)$ *is not equal to* $\sin \alpha + \sin \beta$

which is something that many students simply assume before they become familiar with the formulas and ideas of this section. If we used such a formula, we would get $\sin 90° = \frac{1}{2}\sqrt{3} + \frac{1}{2} = 1.366$ for the combination $(60° + 30°)$. This is not possible, since the values of the sine never exceed 1 in value. Also, if we used the combination $(45° + 45°)$, we would get 1.414, a different value for the same number, $\sin 90°$.   --------■

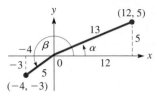

Fig. 20-6

EXAMPLE 2   Given that $\sin \alpha = \frac{5}{13}$ (α in the first quadrant) and $\sin \beta = -\frac{3}{5}$ (for β in the third quadrant), find $\cos(\alpha + \beta)$.

Since $\sin \alpha = \frac{5}{13}$ for α in the first quadrant, from Fig. 20-6 we have $\cos \alpha = \frac{12}{13}$. Also, since $\sin \beta = -\frac{3}{5}$ for β in the third quadrant, from Fig. 20-6 we also have $\cos \beta = -\frac{4}{5}$.

Then, by using Eq. (20-10), we have

$$\cos(\alpha + \beta) = \cos \alpha \cos \beta - \sin \alpha \sin \beta$$

$$= \frac{12}{13}\left(-\frac{4}{5}\right) - \frac{5}{13}\left(-\frac{3}{5}\right)$$

$$= -\frac{48}{65} + \frac{15}{65} = -\frac{33}{65}$$

From Eqs. (20-9) and (20-10) we can easily find expressions for $\sin(\alpha - \beta)$ and $\cos(\alpha - \beta)$. This is done by finding $\sin(\alpha + (-\beta))$ and $\cos(\alpha + (-\beta))$. Thus, we have

$$\sin(\alpha - \beta) = \sin(\alpha + (-\beta)) = \sin \alpha \cos(-\beta) + \cos \alpha \sin(-\beta)$$

Since $\cos(-\beta) = \cos \beta$ and $\sin(-\beta) = -\sin \beta$ (see Exercise 45 of Section 8-2), we have

$$\boxed{\sin(\alpha - \beta) = \sin \alpha \cos \beta - \cos \alpha \sin \beta} \qquad \textbf{(20-11)}$$

In the same manner, we find that

$$\boxed{\cos(\alpha - \beta) = \cos \alpha \cos \beta + \sin \alpha \sin \beta} \qquad \textbf{(20-12)}$$

EXAMPLE 3   Find $\cos 15°$ from $\cos(45° - 30°)$.

$$\cos 15° = \cos(45° - 30°) = \cos 45° \cos 30° + \sin 45° \sin 30° \qquad \text{using Eq. (20-12)}$$

$$= \frac{\sqrt{2}}{2} \times \frac{\sqrt{3}}{2} + \frac{\sqrt{2}}{2} \times \frac{1}{2} = \frac{\sqrt{6} + \sqrt{2}}{4} \qquad \text{(exact)}$$

$$= 0.9659$$

EXAMPLE 4   Reduce $\sin x \cos(x - y) - \cos x \sin(x - y)$ to a single term.

We could expand $\cos(x - y)$ and $\sin(x - y)$ and simplify. This would be tedious, and it is unnecessary. This expression fits the form of the right side of Eq. (20-11) with $\alpha = x$ and $\beta = x - y$. Thus,

$$\sin x \cos(x - y) - \cos x \sin(x - y) = \sin[x - (x - y)]$$

$$= \sin y$$

EXAMPLE 5   Evaluate $\cos 23° \cos 67° - \sin 23° \sin 67°$.

We note that this expression fits the form of the right side of Eq. (20-10), so

$$\cos 23° \cos 67° - \sin 23° \sin 67° = \cos(23° + 67°)$$

$$= \cos 90°$$

$$= 0$$

NOTE ▸   Again, we are able to evaluate this expression by ***recognizing the form*** of the given expression. Evaluation by a calculator will verify the result.

By dividing the right side of Eq. (20-9) by that of Eq. (20-10), we can determine expressions for $\tan(\alpha + \beta)$, and by dividing the right side of Eq. (20-11) by that of Eq. (20-12), we can determine an expression for $\tan(\alpha - \beta)$. The derivation of these formulas is Exercise 37 of this section. These formulas can be written together, as

$$\tan(\alpha \pm \beta) = \frac{\tan \alpha \pm \tan \beta}{1 \mp \tan \alpha \tan \beta} \qquad \textbf{(20-13)}$$

The formula for $\tan(\alpha + \beta)$ uses the upper signs, and the formula for $\tan(\alpha - \beta)$ uses the lower signs.

Certain trigonometric identities can be proven by the formulas derived in this section. The following examples illustrate this use of these formulas.

EXAMPLE 6   Show that $\tan(\alpha + \beta)\tan(\alpha - \beta) = \dfrac{\tan^2 \alpha - \tan^2 \beta}{1 - \tan^2 \alpha \tan^2 \beta}$.

Using both of Eqs. (20-13), we have

$$\tan(\alpha + \beta)\tan(\alpha - \beta) = \left(\frac{\tan \alpha + \tan \beta}{1 - \tan \alpha \tan \beta}\right)\left(\frac{\tan \alpha - \tan \beta}{1 + \tan \alpha \tan \beta}\right)$$

$$= \frac{\tan^2 \alpha - \tan^2 \beta}{1 - \tan^2 \alpha \tan^2 \beta}$$

EXAMPLE 7   Prove that $\sin(180° + x) = -\sin x$.

By using Eq. (20-9) we have

$$\sin(180° + x) = \sin 180° \cos x + \cos 180° \sin x$$

Since $\sin 180° = 0$ and $\cos 180° = -1$, we have

$$\sin 180° \cos x + \cos 180° \sin x = (0)\cos x + (-1)\sin x$$

or

$$\sin(180° + x) = -\sin x$$

Although x may or may not be an acute angle, we see that this agrees with the results in Section 8-2 for the sine of a third-quadrant angle if x is acute. See Fig. 20-7.

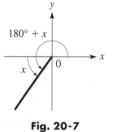

Fig. 20-7

EXAMPLE 8   Show that $\dfrac{\sin(\alpha - \beta)}{\sin \alpha \sin \beta} = \cot \beta - \cot \alpha$.

$$\frac{\sin(\alpha - \beta)}{\sin \alpha \sin \beta} = \frac{\sin \alpha \cos \beta - \cos \alpha \sin \beta}{\sin \alpha \sin \beta} \qquad \text{using Eq. (20-11)}$$

$$= \frac{\sin \alpha \cos \beta}{\sin \alpha \sin \beta} - \frac{\cos \alpha \sin \beta}{\sin \alpha \sin \beta}$$

$$= \frac{\cos \beta}{\sin \beta} - \frac{\cos \alpha}{\sin \alpha}$$

$$= \cot \beta - \cot \alpha \qquad \text{using Eq. (20-5)}$$

EXAMPLE 9   Show that $\sin\left(\dfrac{\pi}{4} + x\right) \cos\left(\dfrac{\pi}{4} + x\right) = \dfrac{1}{2}(\cos^2 x - \sin^2 x)$.

The solution is as follows:

$$\sin\left(\frac{\pi}{4} + x\right)\cos\left(\frac{\pi}{4} + x\right) = \left(\sin\frac{\pi}{4}\cos x + \cos\frac{\pi}{4}\sin x\right)\left(\cos\frac{\pi}{4}\cos x - \sin\frac{\pi}{4}\sin x\right) \qquad \text{using Eqs. (20-9) and (20-10)}$$

$$= \sin\frac{\pi}{4}\cos\frac{\pi}{4}\cos^2 x - \sin^2\frac{\pi}{4}\sin x \cos x + \cos^2\frac{\pi}{4}\sin x \cos x - \sin^2 x \sin\frac{\pi}{4}\cos\frac{\pi}{4} \qquad \text{expanding}$$

$$= \frac{\sqrt{2}}{2}\frac{\sqrt{2}}{2}\cos^2 x - \left(\frac{\sqrt{2}}{2}\right)^2 \sin x \cos x + \left(\frac{\sqrt{2}}{2}\right)^2 \sin x \cos x - \frac{\sqrt{2}}{2}\frac{\sqrt{2}}{2}\sin^2 x \qquad \text{evaluating}$$

$$= \frac{1}{2}\cos^2 x - \frac{1}{2}\sin^2 x = \frac{1}{2}(\cos^2 x - \sin^2 x) \qquad\qquad\qquad \blacksquare$$

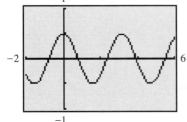

Fig. 20-8

If we check this identity by comparing the graphs of

$$y_1 = \sin(\pi/4 + x)\cos(\pi/4 + x) \qquad \text{and} \qquad y_2 = (1/2)((\cos x)^2 - (\sin x)^2)$$

on a graphing calculator, both curves are shown by the graph in Fig. 20-8.   ∎

EXAMPLE 10   In analyzing the motion of an object at the end of a spring, it is stated that

$$\sin \omega t = \sin(\omega t + \alpha)\cos \alpha - \cos(\omega t + \alpha)\sin \alpha$$

Show that this is correct.

If we let $x = \omega t + \alpha$, we note that the right side of the equation becomes $\sin x \cos \alpha - \cos x \sin \alpha$, which is the form for $\sin(x - \alpha)$. By replacing x with $\omega t + \alpha$, we obtain $\sin(\omega t + \alpha - \alpha)$, which is $\sin \omega t$. These steps (with $x = \omega t + \alpha$) are shown below.

$$\sin \omega t = \sin x \cos \alpha - \cos x \sin \alpha = \sin(x - \alpha)$$
$$= \sin(\omega t + \alpha - \alpha) = \sin \omega t$$

Therefore, the original equation has been shown to be true.

Again, the proper recognition of a basic form leads to the solution.   ∎

—— **EXERCISES** *20-2* ——

In Exercises 1–4, determine the values of the given functions as indicated.

1. Find $\sin 105°$ by using $105° = 60° + 45°$.

2. Find $\tan 75°$ by using $75° = 30° + 45°$.

3. Find $\cos 15°$ by using $15° = 60° - 45°$.

4. Find $\sin 15°$ by using $15° = 45° - 30°$.

In Exercises 5–8, evaluate the given functions with the following information: $\sin \alpha = 4/5$ (α in first quadrant) and $\cos \beta = -12/13$ (β in second quadrant).

5. $\sin(\alpha + \beta)$ 6. $\tan(\beta - \alpha)$

7. $\cos(\alpha + \beta)$ 8. $\sin(\alpha - \beta)$

In Exercises 9–16, reduce each of the given expressions to a single term. Expansion of any term is not necessary; proper identification of the form of the expression leads to the proper result.

9. $\sin x \cos 2x + \sin 2x \cos x$

10. $\sin 3x \cos x - \sin x \cos 3x$

11. $\cos(x + y)\cos y + \sin(x + y) \sin y$

12. $\cos(2x - y)\cos y - \sin(2x - y) \sin y$

13. $\dfrac{\tan(x - y) + \tan y}{1 - \tan(x - y)\tan y}$

14. $\sin x \cos(x + 1) + \cos x \sin(x + 1)$

15. $\sin 3x \cos(3x - \pi) - \cos 3x \sin(3x - \pi)$

16. $\cos(x + \pi)\cos(x - \pi) + \sin(x + \pi)\sin(x - \pi)$

In Exercises 17–20, evaluate each of the given expressions. Proper recognition of the given form leads to the result. Verify each result by use of a calculator.

17. $\sin 122° \cos 32° - \cos 122° \sin 32°$

18. $\cos 250° \cos 70° + \sin 250° \sin 70°$

19. $\cos 312° \cos 48° - \sin 312° \sin 48°$

20. $\dfrac{\tan 18° + \tan 27°}{1 - \tan 18° \tan 27°}$

In Exercises 21–32, prove the given identities.

21. $\sin(180° - x) = \sin x$ **22.** $\cos(180° - x) = -\cos x$

23. $\cos(-x) = \cos x$ (*Hint:* $-x = 0 - x$.)

24. $\sin(-x) = -\sin x$

25. $\tan(180° + x) = \tan x$

26. $\sin(90° + x) = \cos x$

27. $\cos(\frac{\pi}{3} + x) = \dfrac{\cos x - \sqrt{3} \sin x}{2}$

W **28.** $\tan(90° + x) = -\cot x$ (Explain why Eq. (20-13) cannot be used for this, but Eqs. (20-9) and (20-10) can be used.)

29. $\sin(x + y)\sin(x - y) = \sin^2 x - \sin^2 y$

30. $\cos(x + y)\cos(x - y) = \cos^2 x - \sin^2 y$

31. $\cos(\alpha + \beta) + \cos(\alpha - \beta) = 2 \cos \alpha \cos \beta$

32. $\cos(x - y) + \sin(x + y) = (\cos x + \sin x)(\cos y + \sin y)$

In Exercises 33–36, verify each identity by comparing the graph of the left side with the graph of the right side on a graphing calculator.

33. $\cos(30° + x) = \dfrac{\sqrt{3} \cos x - \sin x}{2}$

34. $\sin(120° - x) = \dfrac{\sqrt{3} \cos x + \sin x}{2}$

35. $\tan(\frac{\pi}{4} + x) = \dfrac{1 + \tan x}{1 - \tan x}$ **36.** $\cos(\frac{\pi}{2} - x) = \sin x$

In Exercises 37–40, derive the given equations in the indicated manner. Equations (20-14), (20-15), and (20-16) are known as the product formulas.

37. By dividing the right side of Eq. (20-9) by that of Eq (20-10), and dividing the right side of Eq. (20-11) by that of Eq. (20-12), derive Eq. (20-13).

$$\tan(\alpha \pm \beta) = \dfrac{\tan \alpha \pm \tan \beta}{1 \mp \tan \alpha \tan \beta} \qquad \textbf{(20-13)}$$

(*Hint:* Divide numerator and denominator by $\cos \alpha \cos \beta$.)

38. By adding Eqs. (20-9) and (20-11), derive the equation

$$\sin \alpha \cos \beta = \tfrac{1}{2}[\sin(\alpha + \beta) + \sin(\alpha - \beta)] \qquad \textbf{(20-14)}$$

39. By adding Eqs. (20-10) and (20-12), derive the equation

$$\cos \alpha \cos \beta = \tfrac{1}{2}[\cos(\alpha + \beta) + \cos(\alpha - \beta)] \qquad \textbf{(20-15)}$$

40. By subtracting Eq. (20-10) from Eq. (20-12), derive

$$\sin \alpha \sin \beta = \tfrac{1}{2}[\cos(\alpha - \beta) - \cos(\alpha + \beta)] \qquad \textbf{(20-16)}$$

In Exercises 41–44, additional trigonometric identities are shown. Derive them by letting $\alpha + \beta = x$ and $\alpha - \beta = y$, which leads to $\alpha = \frac{1}{2}(x + y)$ and $\beta = \frac{1}{2}(x - y)$. The resulting equations are known as the factor formulas.

41. Use Eq. (20-14) and the substitutions above to derive the equation

$$\sin x + \sin y = 2 \sin \tfrac{1}{2}(x + y) \cos \tfrac{1}{2}(x - y) \qquad \textbf{(20-17)}$$

42. Use Eqs. (20-9) and (20-11) and the substitutions above to derive the equation

$$\sin x - \sin y = 2 \sin \tfrac{1}{2}(x - y) \cos \tfrac{1}{2}(x + y) \qquad \textbf{(20-18)}$$

43. Use Eq. (20-15) and the substitutions above to derive the equation

$$\cos x + \cos y = 2 \cos \tfrac{1}{2}(x + y) \cos \tfrac{1}{2}(x - y) \qquad \textbf{(20-19)}$$

44. Use Eq. (20-16) and the substitutions above to derive the equation

$$\cos x - \cos y = -2 \sin \tfrac{1}{2}(x + y) \sin \tfrac{1}{2}(x - y) \qquad \textbf{(20-20)}$$

In Exercises 45–48, use the equations of this section to solve the given problems.

45. An alternating electric current i is given by the equation $i = i_0 \sin(\omega t + \alpha)$. Show that this can be written as $i = i_1 \sin \omega t + i_2 \cos \omega t$, where $i_1 = i_0 \cos \alpha$ and $i_2 = i_0 \sin \alpha$.

46. A weight w is held in equilibrium by forces F and T as shown in Fig. 20-9. Equations relating w, F, and T are

$$F \cos \theta = T \sin \alpha$$
$$w + F \sin \theta = T \cos \alpha$$

Show that $w = \dfrac{T \cos(\theta + \alpha)}{\cos \theta}$.

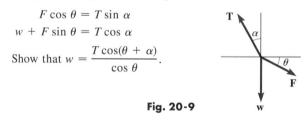

Fig. 20-9

47. For the two bevel gears shown in Fig. 20-10, the equation $\tan \alpha = \dfrac{\sin \beta}{R + \cos \beta}$ is used. Here R is the ratio of gear 1 to gear 2. Show that $R = \dfrac{\sin(\beta - \alpha)}{\sin \alpha}$.

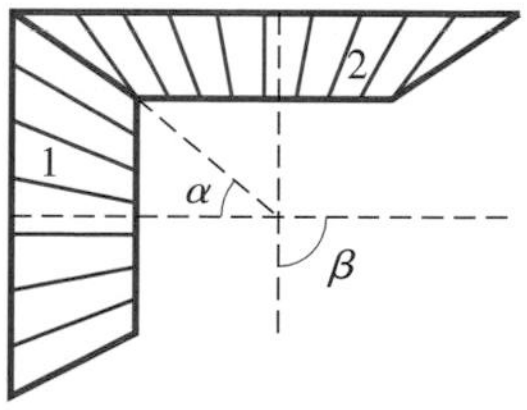

Fig. 20-10

48. In the analysis of the angles of incidence i and reflection r of a light ray subject to certain conditions, the following expression is found:

$$E_2\left(\dfrac{\tan r}{\tan i} + 1\right) = E_1\left(\dfrac{\tan r}{\tan i} - 1\right). \text{ Show that }$$

$$E_2 = E_1 \dfrac{\sin(r - i)}{\sin(r + i)}.$$

20-3 DOUBLE-ANGLE FORMULAS

If we let $\beta = \alpha$ in Eqs. (20-9) and (20-10), we can derive the important double-angle formulas. By making this substitution in Eq. (20-9), we have

$$\sin(\alpha + \alpha) = \sin(2\alpha) = \sin\alpha\cos\alpha + \cos\alpha\sin\alpha = 2\sin\alpha\cos\alpha$$

Using the same substitution in Eq. (20-10), we have

$$\cos(\alpha + \alpha) = \cos\alpha\cos\alpha - \sin\alpha\sin\alpha = \cos^2\alpha - \sin^2\alpha$$

Again using this substitution in the $\tan(\alpha + \beta)$ form of Eq. (20-13), we have

$$\tan(\alpha + \alpha) = \frac{\tan\alpha + \tan\alpha}{1 - \tan\alpha\tan\alpha} = \frac{2\tan\alpha}{1 - \tan^2\alpha}$$

Then using the basic identity Eq. (20-6), other forms of the equation for $\cos 2\alpha$ may be derived. Summarizing these forms, we have

$$\sin 2\alpha = 2\sin\alpha\cos\alpha \tag{20-21}$$
$$\cos 2\alpha = \cos^2\alpha - \sin^2\alpha \tag{20-22}$$
$$= 2\cos^2\alpha - 1 \tag{20-23}$$
$$= 1 - 2\sin^2\alpha \tag{20-24}$$
$$\tan 2\alpha = \frac{2\tan\alpha}{1 - \tan^2\alpha} \tag{20-25}$$

These double-angle formulas are widely used in applications of trigonometry, especially in calculus. They should be recognized quickly in any of the above forms.

EXAMPLE 1 (a) If $\alpha = 30°$, we have

$$\cos 60° = \cos 2(30°) = \cos^2 30° - \sin^2 30° = \left(\frac{\sqrt{3}}{2}\right)^2 - \left(\frac{1}{2}\right)^2 = \frac{1}{2} \quad \text{using Eq. (20-22)}$$

(b) If $\alpha = 3x$, we have

$$\sin 6x = \sin 2(3x) = 2\sin 3x\cos 3x \quad \text{using Eq. (20-21)}$$

(c) If $2\alpha = x$, we may write $\alpha = x/2$, which means that

$$\sin x = \sin 2\left(\frac{x}{2}\right) = 2\sin\frac{x}{2}\cos\frac{x}{2} \quad \text{using Eq. (20-21)}$$

(d) If $\alpha = \frac{\pi}{6}$, we have

$$\tan\frac{\pi}{3} = \tan 2(\tfrac{\pi}{6}) = \frac{2\tan\frac{\pi}{6}}{1 - \tan^2(\frac{\pi}{6})} = \frac{2(\sqrt{3}/3)}{1 - (\sqrt{3}/3)^2} = \sqrt{3} \quad \text{using Eq. (20-25)} \blacksquare$$

EXAMPLE 2 Simplify the expression $\cos^2 2x - \sin^2 2x$.

Since this is the difference of the square of the cosine of an angle and the square of the sine of the same angle, it fits the right side of Eq. (20-22). Therefore, letting $\alpha = 2x$, we have

$$\cos^2 2x - \sin^2 2x = \cos 2(2x) = \cos 4x \qquad \blacksquare$$

See the chapter introduction.

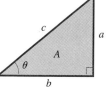

Fig. 20-11

EXAMPLE 3 To find the area A of a right triangular piece of land, a surveyor may use the formula $A = \frac{1}{4}c^2 \sin 2\theta$, where c is the hypotenuse and θ is *either* of the acute angles. Derive this formula.

In Fig. 20-11 we see that $\sin \theta = a/c$ and $\cos \theta = b/c$, which gives us

$$a = c \sin \theta \quad \text{and} \quad b = c \cos \theta$$

The area is given by $A = \frac{1}{2}ab$, which leads to the solution

$$A = \frac{1}{2}ab = \frac{1}{2}(c \sin \theta)(c \cos \theta)$$

$$= \frac{1}{2}c^2 \sin \theta \cos \theta = \frac{1}{2}c^2\left(\frac{1}{2} \sin 2\theta\right) \quad \text{using Eq. (20-21)}$$

$$= \frac{1}{4}c^2 \sin 2\theta$$

In using Eq. (20-21), we divided both sides by 2 to get $\sin \theta \cos \theta = \frac{1}{2} \sin 2\theta$.

If we had labeled the upper acute angle in Fig. 20-11 as θ, we would then have $a = c \cos \theta$ and $b = c \sin \theta$. Substituting these values in the formula for the area gives the same solution.

EXAMPLE 4 (a) Verifying the values of $\sin 90°$ using the functions of $45°$, we have

$$\sin 90° = \sin 2(45°) = 2 \sin 45° \cos 45° = 2\left(\frac{\sqrt{2}}{2}\right)\left(\frac{\sqrt{2}}{2}\right) = 1 \quad \text{using Eq. (20-21)}$$

(b) Using Eq. (20-25), $\tan 142° = \dfrac{2 \tan 71°}{1 - \tan^2 71°}$. Using a calculator to verify this, we have

$$\tan 142° = -0.7812856265 \quad \text{and} \quad \frac{2 \tan 71°}{1 - \tan^2 71°} = -0.7812856265$$

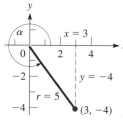

(a)

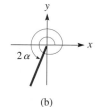

(b)

Fig. 20-12

EXAMPLE 5 Knowing that $\cos \alpha = 3/5$ for an angle in the fourth quadrant, we then determine from Fig. 20-12(a) that

$$\sin \alpha = -4/5$$

Therefore, we have

$$\sin 2\alpha = 2 \sin \alpha \cos \alpha \quad \text{Eq. (20-21)}$$

$$= 2\left(-\frac{4}{5}\right)\left(\frac{3}{5}\right) = -\frac{24}{25}$$

In Fig. 20-12(b) the angle 2α is shown. It is a third-quadrant angle, which verifies the sign of the result. (Since $\cos \alpha = 3/5$, $\alpha \approx 307°$ and $2\alpha \approx 614°$, which is a third-quadrant angle.)

EXAMPLE 6 Prove the identity $\dfrac{2}{1 + \cos 2x} = \sec^2 x$.

$$\frac{2}{1 + \cos 2x} = \frac{2}{1 + (2 \cos^2 x - 1)} \quad \text{using Eq. (20-23)}$$

$$= \frac{2}{2 \cos^2 x} = \sec^2 x \quad \text{using Eq. (20-2)}$$

EXAMPLE 7   Show that $\dfrac{\sin 3x}{\sin x} + \dfrac{\cos 3x}{\cos x} = 4\cos 2x$.

Since the left side is the more complex side, we will change it to the form on the right.

$$\frac{\sin 3x}{\sin x} + \frac{\cos 3x}{\cos x} = \frac{\sin 3x \cos x + \cos 3x \sin x}{\sin x \cos x} \qquad \text{combining fractions}$$

$$= \frac{\sin(3x + x)}{\frac{1}{2}\sin 2x} \quad \longleftarrow \text{using Eq. (20-9)}$$
$$\phantom{= \frac{\sin(3x + x)}{\frac{1}{2}\sin 2x}} \quad \longleftarrow \text{using Eq. (20-21)}$$

$$= \frac{2\sin 4x}{\sin 2x} = \frac{2(2\sin 2x \cos 2x)}{\sin 2x} \qquad \text{using Eq. (20-21)}$$

$$= 4\cos 2x$$

Checking this identity by comparing the graphs of

$$y_1 = (\sin 3x)/(\sin x) + (\cos 3x)/(\cos x) \qquad \text{and} \qquad y_2 = 4\cos 2x$$

on a graphing calculator, both curves are shown by the graph in Fig. 20-13. ∎

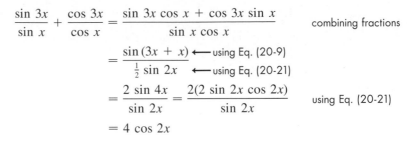

Fig. 20-13

EXERCISES *20-3*

In Exercises 1–4, determine the values of the indicated functions in the given manner.

1. Find sin 60° by using the functions of 30°.
2. Find sin 120° by using the functions of 60°.
3. Find tan 120° by using the functions of 60°.
4. Find cos 60° by using the functions of 30°.

In Exercises 5–8, use a calculator to verify the values found by using the double-angle formulas.

5. Find sin 258° directly and by using functions of 129°.
6. Find tan 84° directly and by using functions of 42°.
7. Find cos 96° directly and by using functions of 48°.
8. Find cos 276° directly and by using functions of 138°.

In Exercises 9–12, evaluate the indicated functions with the given information.

9. Find $\sin 2x$ if $\cos x = \frac{4}{5}$ (in first quadrant).
10. Find $\cos 2x$ if $\sin x = -\frac{12}{13}$ (in third quadrant).
11. Find $\tan 2x$ if $\sin x = 0.5$ (in second quadrant).
12. Find $\sin 4x$ if $\sin x = 0.6$ (in first quadrant).

In Exercises 13–20, simplify the given expressions. Expansion of any term is not necessary; proper recognition of the form of the expression leads to the proper result.

13. $4\sin 4x \cos 4x$

14. $4\sin^2 x \cos^2 x$

15. $1 - 2\sin^2 4x$

16. $\dfrac{4\tan 4\theta}{1 - \tan^2 4\theta}$

17. $2\cos^2 \frac{1}{2}x - 1$

18. $2\sin \frac{1}{2}x \cos \frac{1}{2}x$

19. $4\sin^2 2x - 2$

20. $\cos 3x \sin 3x$

In Exercises 21–32, prove the given identities.

21. $\cos^2 \alpha - \sin^2 \alpha = 2\cos^2 \alpha - 1$

22. $\cos^2 \alpha - \sin^2 \alpha = 1 - 2\sin^2 \alpha$

23. $\dfrac{\cos x - \tan x \sin x}{\sec x} = \cos 2x$

24. $\cos^4 x - \sin^4 x = \cos 2x$

25. $\dfrac{\sin 4\theta}{\sin 2\theta} = 2\cos 2\theta$

26. $2 + \dfrac{\cos 2\theta}{\sin^2 \theta} = \csc^2 \theta$

27. $\dfrac{\sin 2\theta}{1 + \cos 2\theta} = \tan \theta$

28. $\dfrac{2\tan \alpha}{1 + \tan^2 \alpha} = \sin 2\alpha$

29. $1 - \cos 2\theta = \dfrac{2}{1 + \cot^2 \theta}$

30. $\dfrac{\cos^3 \theta + \sin^3 \theta}{\cos \theta + \sin \theta} = 1 - \dfrac{1}{2}\sin 2\theta$

31. $\dfrac{\sin 3x}{\sin x} - \dfrac{\cos 3x}{\cos x} = 2$

32. $\dfrac{\cos 3x}{\sin x} + \dfrac{\sin 3x}{\cos x} = 2\cot 2x$

In Exercises 33–36, verify each identity by comparing the graph of the left side with the graph of the right side on a graphing calculator.

33. $\tan 2\theta = \dfrac{2}{\cot \theta - \tan \theta}$

34. $\dfrac{1 - \tan^2 x}{\sec^2 x} = \cos 2x$

35. $(\sin x + \cos x)^2 = 1 + \sin 2x$

36. $2\csc 2x \tan x = \sec^2 x$

(W) *In Exercises 37–38, prove the given identities. Explain your general approach to the proof.*

37. $\sin 3x = 3 \cos^2 x \sin x - \sin^3 x$

38. $\cos 3x = \cos^3 x - 3 \sin^2 x \cos x$

In Exercises 39–44, solve the given problems.

39. The CN Tower in Toronto is 553 m high, and it has an observation deck at the 335 m level. How far from the top of the CN Tower must a 553-m-high helicopter be in order that the angle subtended at the helicopter by the part of the tower above the deck equals the angle subtended by the part of the tower below the deck?

40. The cross section of a radio wave reflector is defined by $x = \cos 2\theta$, $y = \sin \theta$. Find the relation between x and y by eliminating θ.

41. To find the horizontal range R of a projectile, the equation $R = vt \cos \alpha$ is used, where α is the angle between the line of fire and the horizontal, v is the initial velocity of the projectile, and t is the time of flight. It can be shown that $t = (2v \sin \alpha)/g$, where g is the acceleration due to gravity. Show that $R = (v^2 \sin 2\alpha)/g$.

42. In analyzing light reflection from a cylinder onto a flat surface, the expression $3 \cos \theta - \cos 3\theta$ arises. Show that this equals $2 \cos \theta \cos 2\theta + 4 \sin \theta \sin 2\theta$.

43. The instantaneous electric power p in an inductor is given by the equation $p = vi \sin \omega t \sin(\omega t - \pi/2)$. Show that this equation can be written as $p = -\frac{1}{2} vi \sin 2\omega t$.

44. In the study of the stress at a point in a bar, the equation $s = a \cos^2 \theta + b \sin^2 \theta - 2t \sin \theta \cos \theta$ arises. Show that this equation can be written as $s = \frac{1}{2}(a + b) + \frac{1}{2}(a - b)\cos 2\theta - t \sin 2\theta$.

20-4 HALF-ANGLE FORMULAS

If we let $\theta = \alpha/2$ in the identity $\cos 2\theta = 1 - 2 \sin^2 \theta$ and then solve for $\sin(\alpha/2)$,

$$\sin \frac{\alpha}{2} = \pm \sqrt{\frac{1 - \cos \alpha}{2}}$$

(20-26)

Also, with the same substitution in the identity $\cos 2\theta = 2 \cos^2 \theta - 1$, which is then solved for $\cos(\alpha/2)$, we have

$$\cos \frac{\alpha}{2} = \pm \sqrt{\frac{1 + \cos \alpha}{2}}$$

(20-27)

CAUTION▸ In each of Eqs. (20-26) and (20-27), ***the sign chosen depends on the quadrant in which $\frac{\alpha}{2}$ lies.***

We can use these half-angle formulas to find values of the functions of angles that are half of those for which the functions are known. The following examples illustrate how these identities are used in evaluations and in identities.

■**EXAMPLE 1**   We can find $\sin 15°$ by using the relation

$$\sin 15° = \sqrt{\frac{1 - \cos 30°}{2}} \qquad \text{using Eq. (20-26)}$$

$$= \sqrt{\frac{1 - 0.8660}{2}} = 0.2588$$

Here the plus sign is used, since $15°$ is in the first quadrant. ■

EXAMPLE 2 We can find cos 165° by use of the relation

$$\cos 165° = -\sqrt{\frac{1 + \cos 330°}{2}} \quad \text{using Eq. (20-27)}$$

$$= -\sqrt{\frac{1 + 0.8660}{2}} = -0.9659$$

Here the minus sign is used, since 165° is in the second quadrant, and the cosine of a second-quadrant angle is negative. ━━━■

EXAMPLE 3 Simplify $\sqrt{\dfrac{1 - \cos 114°}{2}}$ by expressing the result in terms of one-half the given angle. Then, using a calculator, show that the values are equal.

We note that the given expression fits the form of the right side of Eq. (20-26), which means that

$$\sqrt{\frac{1 - \cos 114°}{2}} = \sin \tfrac{1}{2}(114°) = \sin 57°$$

Using a calculator shows that

$$\sqrt{\frac{1 - \cos 114°}{2}} = 0.8386705679 \quad \text{and} \quad \sin 57° = 0.8386705679$$

which verifies the equation for these values. ━━━■

EXAMPLE 4 Simplify the expression $\sqrt{\dfrac{9 + 9 \cos 6x}{2}}$.

$$\sqrt{\frac{9 + 9 \cos 6x}{2}} = \sqrt{\frac{9(1 + \cos 6x)}{2}} = 3\sqrt{\frac{1 + \cos 6x}{2}}$$

$$= 3 \cos \tfrac{1}{2}(6x) \quad \text{using Eq. (20-27) with } \alpha = 6x$$

$$= 3 \cos 3x$$

Noting the original expression, we see that cos 3x cannot be negative. ━━━■

EXAMPLE 5 In the kinetic theory of gases, the expression $\sqrt{(1 - \cos \alpha)^2 + \sin^2 \alpha}$ is found. Show that this expression equals $2 \sin \tfrac{1}{2}\alpha$.

$$\sqrt{(1 - \cos \alpha)^2 + \sin^2 \alpha} = \sqrt{1 - 2 \cos \alpha + \cos^2 \alpha + \sin^2 \alpha} \quad \text{expanding}$$

$$= \sqrt{1 - 2 \cos \alpha + 1} \quad \text{using Eq. (20-6)}$$

$$= \sqrt{2 - 2 \cos \alpha}$$

$$= \sqrt{2(1 - \cos \alpha)} \quad \text{factoring}$$

This last expression is very similar to that for $\sin \tfrac{1}{2}\alpha$, except that no 2 appears in the denominator. Therefore, multiplying the numerator and the denominator under the radical by 2 leads to the solution.

$$\sqrt{2(1 - \cos \alpha)} = \sqrt{\frac{4(1 - \cos \alpha)}{2}} = 2\sqrt{\frac{1 - \cos \alpha}{2}}$$

$$= 2 \sin \tfrac{1}{2}\alpha \quad \text{using Eq. (20-26)}$$

Noting the original expression, we see that $\sin \tfrac{1}{2}\alpha$ cannot be negative. ━━━■

■EXAMPLE 6   Given that $\tan \alpha = \frac{8}{15}$ $(180° < \alpha < 270°)$, find $\cos(\frac{\alpha}{2})$.

Knowing that $\tan \alpha = \frac{8}{15}$ for a third-quadrant angle, we determine from Fig. 20-14 that $\cos \alpha = -\frac{15}{17}$. This means

$$\cos \frac{\alpha}{2} = -\sqrt{\frac{1 + (-15/17)}{2}} = -\sqrt{\frac{2}{34}} \qquad \text{using Eq. (20-27)}$$

$$= -\frac{1}{17}\sqrt{17} = -0.2425$$

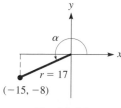

Fig. 20-14

Since $180° < \alpha < 270°$, we know that $90° < \frac{\alpha}{2} < 135°$, and therefore $\frac{\alpha}{2}$ is in the second quadrant. Since the cosine is negative for second-quadrant angles, we use the negative value of the radical.

■EXAMPLE 7   Prove the identity $\sec \frac{\alpha}{2} + \csc \frac{\alpha}{2} = \dfrac{2\left(\sin \dfrac{\alpha}{2} + \cos \dfrac{\alpha}{2}\right)}{\sin \alpha}$.

By expressing $\sin \alpha$ as $2 \sin \frac{\alpha}{2} \cos \frac{\alpha}{2}$, we have

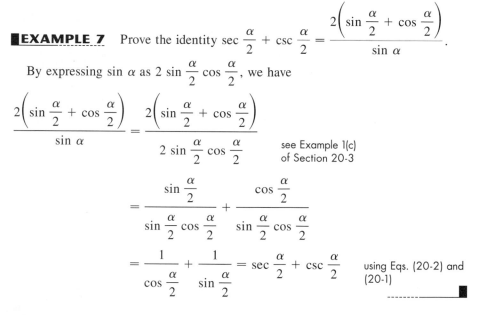

$$\frac{2\left(\sin \dfrac{\alpha}{2} + \cos \dfrac{\alpha}{2}\right)}{\sin \alpha} = \frac{2\left(\sin \dfrac{\alpha}{2} + \cos \dfrac{\alpha}{2}\right)}{2 \sin \dfrac{\alpha}{2} \cos \dfrac{\alpha}{2}} \qquad \text{see Example 1(c) of Section 20-3}$$

$$= \frac{\sin \dfrac{\alpha}{2}}{\sin \dfrac{\alpha}{2} \cos \dfrac{\alpha}{2}} + \frac{\cos \dfrac{\alpha}{2}}{\sin \dfrac{\alpha}{2} \cos \dfrac{\alpha}{2}}$$

$$= \frac{1}{\cos \dfrac{\alpha}{2}} + \frac{1}{\sin \dfrac{\alpha}{2}} = \sec \frac{\alpha}{2} + \csc \frac{\alpha}{2} \qquad \text{using Eqs. (20-2) and (20-1)}$$

■EXAMPLE 8   We can find relations for the other functions of $\frac{\alpha}{2}$ by expressing these functions in terms of $\sin(\frac{\alpha}{2})$ and $\cos(\frac{\alpha}{2})$. For example,

$$\sec \frac{\alpha}{2} = \frac{1}{\cos \dfrac{\alpha}{2}} = \pm \frac{1}{\sqrt{\dfrac{1 + \cos \alpha}{2}}} \qquad \text{using Eq. (20-27)}$$

$$= \pm \sqrt{\frac{2}{1 + \cos \alpha}}$$

■EXAMPLE 9   Show that $2 \cos^2 \dfrac{x}{2} - \cos x = 1$.

The first step is to substitute for $\cos \frac{x}{2}$, which will result in each term on the left being in terms of x, and no $\frac{x}{2}$ terms will exist. This might allow us to combine terms. So we perform this operation, and we have for the left side

$$2 \cos^2 \frac{x}{2} - \cos x = 2\left(\frac{1 + \cos x}{2}\right) - \cos x \qquad \text{using Eq. (20-27) with both sides squared}$$

$$= 1 + \cos x - \cos x = 1$$

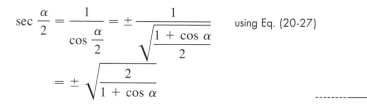

Fig. 20-15

From Fig. 20-15, we verify that the graph of $y_1 = 2(\cos(x/2))^2 - \cos x$ is the same as the graph of $y_2 = 1$.

EXERCISES *20-4*

In Exercises 1–4, use the half-angle formulas to evaluate the given functions.

1. $\cos 15°$

2. $\sin 22.5°$

3. $\sin 75°$

4. $\cos 112.5°$

In Exercises 5–8, simplify the given expressions by giving the results in terms of one-half the given angle. Then use a calculator to verify the result.

5. $\sqrt{\dfrac{1 - \cos 236°}{2}}$

6. $\sqrt{\dfrac{1 + \cos 98°}{2}}$

7. $\sqrt{1 + \cos 164°}$

8. $\sqrt{2 - 2\cos 328°}$

In Exercises 9–12, use the half-angle formulas to simplify the given expressions.

9. $\sqrt{\dfrac{1 - \cos 6x}{2}}$

10. $\sqrt{\dfrac{4 + 4\cos 8\beta}{2}}$

11. $\sqrt{8 + 8\cos 4x}$

12. $\sqrt{2 - 2\cos 16x}$

In Exercises 13–16, evaluate the indicated functions with the given information.

13. Find the value of $\sin(\frac{\alpha}{2})$ if $\cos \alpha = \frac{12}{13}$ $(0° < \alpha < 90°)$.

14. Find the value of $\cos(\frac{\alpha}{2})$ if $\sin \alpha = -\frac{4}{5}$ $(180° < \alpha < 270°)$.

15. Find the value of $\cos(\frac{\alpha}{2})$ if $\tan \alpha = -0.2917$ $(90° < \alpha < 180°)$.

16. Find the value of $\sin(\frac{\alpha}{2})$ if $\cos \alpha = 0.4706$ $(270° < \alpha < 360°)$.

In Exercises 17–20, derive the required expressions.

17. Derive an expression for $\csc(\frac{\alpha}{2})$ in terms of $\cos \alpha$.

18. Derive an expression for $\sec(\frac{\alpha}{2})$ in terms of $\sec \alpha$.

19. Derive an expression for $\tan(\frac{\alpha}{2})$ in terms of $\sin \alpha$ and $\cos \alpha$.

20. Derive an expression for $\cot(\frac{\alpha}{2})$ in terms of $\sin \alpha$ and $\cos \alpha$.

In Exercises 21–26, prove the given identities.

21. $\sin \dfrac{\alpha}{2} = \dfrac{1 - \cos \alpha}{2 \sin \dfrac{\alpha}{2}}$

22. $2 \cos \dfrac{x}{2} = (1 + \cos x)\sec \dfrac{x}{2}$

23. $2 \sin^2 \dfrac{x}{2} + \cos x = 1$

24. $2 \cos^2 \dfrac{\theta}{2} \sec \theta = \sec \theta + 1$

25. $\cos \dfrac{\theta}{2} = \dfrac{\sin \theta}{2 \sin \dfrac{\theta}{2}}$

26. $\cos^2 \dfrac{x}{2}\left[1 + \left(\dfrac{\sin x}{1 + \cos x}\right)^2\right] = 1$

In Exercises 27 and 28, verify each identity by comparing the graph of the left side with the graph of the right side on a graphing calculator.

27. $2 \sin^2 \dfrac{\alpha}{2} - \cos^2 \dfrac{\alpha}{2} = \dfrac{1 - 3 \cos \alpha}{2}$

28. $\tan \dfrac{\alpha}{2} = \dfrac{\sin \alpha}{1 + \cos \alpha}$

In Exercises 29–32, use the half-angle formulas to solve the given problems.

29. In electronics, in order to find the *root-mean-square current* in a circuit, it is necessary to express $\sin^2 \omega t$ in terms of $\cos 2\omega t$. Show how this is done.

30. In studying interference patterns of radio signals, the expression $2E^2 - 2E^2 \cos(\pi - \theta)$ arises. Show that this can be written as $4E^2 \cos^2(\theta/2)$.

31. The index of refraction n, the angle A of a prism, and the minimum angle of deflection ϕ are related by

$$n = \dfrac{\sin \frac{1}{2}(A + \phi)}{\sin \frac{1}{2}A}$$

See Fig. 20-16. Show that an equivalent expression is

$$n = \sqrt{\dfrac{1 - \cos A \cos \phi + \sin A \sin \phi}{1 - \cos A}}$$

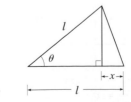

Fig. 20-16

32. For the structure shown in Fig. 20-17, show that $x = 2l \sin^2 \frac{1}{2}\theta$.

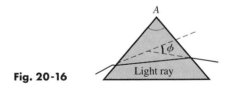

Fig. 20-17

20-5 SOLVING TRIGONOMETRIC EQUATIONS

One of the most important uses of the trigonometric identities is in the solution of equations involving trigonometric functions. The solution of this type of equation consists of the angles that satisfy the equation. When solving for the angle, we generally first solve for a value of a function of the angle and then find the angle from this value of the function.

When equations are written in terms of more than one function, the identities provide a way of changing many of them to equations or factors involving only one function of the same angle. Thus, *the solution is found by using algebraic methods and trigonometric identities and values.* From Chapter 8 we recall that we must be careful regarding the sign of the value of a trigonometric function in finding the angle. Figure 20-18 shows again the quadrants in which the functions are positive. Functions not listed are negative.

NOTE ▶

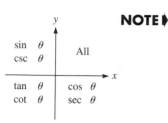

Positive functions

Fig. 20-18

■EXAMPLE 1   Solve the equation $2 \cos \theta - 1 = 0$ for all values of θ such that $0 \le \theta < 2\pi$.

Solving the equation for $\cos \theta$, we obtain $\cos \theta = \frac{1}{2}$. The problem asks for all values of θ from 0 to 2π that satisfy the equation. We know that the cosines of angles in the first and fourth quadrants are positive. Also, we know that $\cos \frac{\pi}{3} = \frac{1}{2}$, which means that $\frac{\pi}{3}$ is the reference angle. Therefore, the solution proceeds as follows:

$$2 \cos \theta - 1 = 0$$
$$2 \cos \theta = 1 \qquad \text{solve for } \cos \theta$$
$$\cos \theta = \frac{1}{2}$$
$$\theta = \frac{\pi}{3}, \frac{5\pi}{3} \qquad \theta \text{ in quadrants I and IV}$$

■EXAMPLE 2   Solve the equation $2 \cos^2 x - \sin x - 1 = 0 \ (0 \le x < 2\pi)$.

By use of the identity $\sin^2 x + \cos^2 x = 1$, this equation may be put in terms of $\sin x$ only. Thus, we have

$$2(1 - \sin^2 x) - \sin x - 1 = 0 \qquad \text{use identity}$$
$$-2 \sin^2 x - \sin x + 1 = 0 \qquad \text{solve for } \sin x$$
$$2 \sin^2 x + \sin x - 1 = 0$$

or

$$(2 \sin x - 1)(\sin x + 1) = 0$$

Just as in solving algebraic equations, we can set each factor equal to zero to find valid solutions. Thus, $\sin x = \frac{1}{2}$ or $\sin x = -1$. For the domain from 0 to 2π, the value $\sin x = \frac{1}{2}$ gives values of x as $\frac{\pi}{6}$ and $\frac{5\pi}{6}$, and $\sin x = -1$ gives the value $x = \frac{3\pi}{2}$. Thus, the complete solution is

$$x = \frac{\pi}{6}, \frac{5\pi}{6}, \frac{3\pi}{2}$$

These values check when substituted in the original equation.

GRAPHICAL SOLUTIONS

We can also graphically solve equations involving trigonometric functions. As with algebraic equations, graphical solutions are approximate, whereas with analytical solutions we can often get exact solutions. In finding graphical solutions we will follow the same procedure we introduced in Section 3-5. That is, we *collect all terms on the left of the equal sign, with zero on the right.* We then *graph the function on the left and determine the zeros of this function by finding the values of x where the curve crosses the x-axis.* We also used this method in Section 7-4 in solving quadratic equations and in Section 15-4 in solving higher-degree equations.

■**EXAMPLE 3**   Graphically solve the equation $2 \cos^2 x - \sin x - 1 = 0$ such that $0 \le x < 2\pi$ by using a graphing calculator. (This is the same equation as in Example 2.)

Since all the terms of the equation are on the left, with zero on the right, we now set $y = 2 \cos^2 x - \sin x - 1$. We then enter this function in the graphing calculator as Y_1, and the graph is displayed in Fig. 20-19. Since angles are expressed in radians and $2\pi = 6.3$, Xmax was chosen as 6.4. Usng the *trace* and *zoom* features (or the *zero* feature, or by letting $y_1 = 2 \cos^2 x - \sin x - 1$, $y_2 = 0$ and using the *intersect* feature) of a graphing calculator, we find that $y = 0$ for

$$x = 0.52, \ 2.62, \ 4.71$$

These values agree with those found in Example 2. We note that for $x = 4.71$ the curve *touches* the x-axis but does not cross it. This means that it is *tangent* to the x-axis.

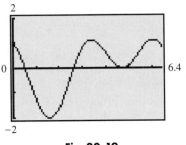

Fig. 20-19

In the examples that follow, we show the analytical solution to each equation and a graph that could be used to determine the graphical solution. The graph also can be used as an approximate check on the analytical solution.

■**EXAMPLE 4**   Solve the equation $\sec^2 x + 2 \tan x - 6 = 0$ $(0 \le x < 2\pi)$.

By use of the identity $1 + \tan^2 x = \sec^2 x$, we may express this equation in terms of $\tan x$ only. Therefore,

$$1 + \tan^2 x + 2 \tan x - 6 = 0 \qquad \text{use identity}$$
$$\tan^2 x + 2 \tan x - 5 = 0$$

We note that this expression is not factorable. Therefore, using the quadratic formula, we solve for $\tan x$.

$$\tan x = \frac{-2 \pm \sqrt{4 + 20}}{2} = -1 \pm 2.4495$$

Therefore, $\tan x = 1.4495$ and $\tan x = -3.4495$. In radians, we find that $\tan x = 1.4495$ for $x = 0.9669$. Since $\tan x$ is also positive in the third quadrant, we have $x = 4.108$ as well. In the same way, using $\tan x = -3.4495$, we obtain $x = 1.853$ and $x = 4.995$. Therefore, the correct solutions are

$$x = 0.9669, \ 1.853, \ 4.108, \ 4.995$$

These values check in the original equation. Also, they agree with the values of x for which the graph of $y = \sec^2 x + 2 \tan x - 6$ crosses the x-axis in Fig. 20-20 (which can be viewed on a graphing calculator).

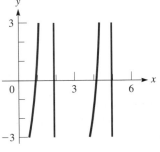

Fig. 20-20

EXAMPLE 5 The vertical displacement y of an object at the end of a spring, which itself is being moved up and down, is given by $y = 3.50 \sin t + 1.20 \sin 2t$. Find the first two values of t (in seconds) for which $y = 0$.

Using the double-angle formula for $\sin 2t$ leads to the solution.

$$3.50 \sin t + 1.20 \sin 2t = 0 \qquad \text{setting } y = 0$$
$$3.50 \sin t + 2.40 \sin t \cos t = 0 \qquad \text{using identities}$$
$$\sin t(3.50 + 2.40 \cos t) = 0 \qquad \text{factoring}$$
$$\sin t = 0 \quad \text{or} \quad \cos t = -1.46$$
$$t = 0.00, 3.14, \ldots$$

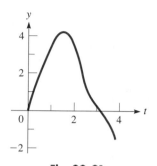

Fig. 20-21

Since $\cos t$ cannot be numerically larger than 1, there are no values of t for which $\cos t = -1.46$. Thus, the required times are $t = 0.00$ s, 3.14 s.

We can see that these values agree with the values of t for which the graph of $y = 3.50 \sin t + 1.20 \sin 2t$ crosses the t-axis in Fig. 20-21. (In using a graphing calculator, use x for t.) --------_____∎

EXAMPLE 6 Solve the equation $\cos \frac{x}{2} = 1 + \cos x \, (0 \le x < 2\pi)$.

By using the half-angle formula for $\cos(x/2)$ and then squaring both sides of the resulting equation, this equation can be solved.

$$\pm\sqrt{\frac{1 + \cos x}{2}} = 1 + \cos x \qquad \text{using identity}$$
$$\frac{1 + \cos x}{2} = 1 + 2 \cos x + \cos^2 x \qquad \text{squaring both sides}$$
$$2 \cos^2 x + 3 \cos x + 1 = 0 \qquad \text{simplifying}$$
$$(2 \cos x + 1)(\cos x + 1) = 0 \qquad \text{factoring}$$
$$\cos x = -\frac{1}{2}, -1$$
$$x = \frac{2\pi}{3}, \frac{4\pi}{3}, \pi$$

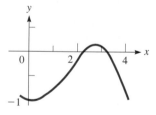

Fig. 20-22

In finding this solution, we squared both sides of the original equation. In doing this we may have introduced extraneous solutions (see Section 14-3). Thus, we must check each solution in the original equation to see if it is valid. Hence

$$\cos \frac{\pi}{3} \overset{?}{=} 1 + \cos \frac{2\pi}{3} \quad \text{or} \quad \frac{1}{2} \overset{?}{=} 1 + \left(-\frac{1}{2}\right) \quad \text{or} \quad \frac{1}{2} = \frac{1}{2}$$

$$\cos \frac{2\pi}{3} \overset{?}{=} 1 + \cos \frac{4\pi}{3} \quad \text{or} \quad -\frac{1}{2} \overset{?}{=} 1 + \left(-\frac{1}{2}\right) \quad \text{or} \quad -\frac{1}{2} \ne \frac{1}{2}$$

$$\cos \frac{\pi}{2} \overset{?}{=} 1 + \cos \pi \quad \text{or} \quad 0 \overset{?}{=} 1 - 1 \quad \text{or} \quad 0 = 0$$

Thus, the apparent solution $x = \frac{4\pi}{3}$ is not a solution of the original equation. The correct solutions are $x = \frac{2\pi}{3}$ and $x = \pi$.

We can see that these values agree with the values of x for which the graph of $y = \cos(x/2) - 1 - \cos x$ crosses the x-axis in Fig. 20-22. In using a graphing calculator, be sure to enter the function as we have shown it here. --------_____∎

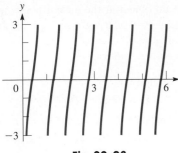

Fig. 20-23

■EXAMPLE 7   Solve the equation $\tan 2\theta - \cot 2\theta = 0$ $(0 \le \theta < 2\pi)$.

$$\tan 2\theta - \frac{1}{\tan 2\theta} = 0 \qquad \text{using } \cot 2\theta = \frac{1}{\tan 2\theta}$$

$$\tan^2 2\theta = 1 \qquad \text{multiplying by } \tan 2\theta \text{ and adding 1 to each side}$$

$$\tan 2\theta = \pm 1 \qquad \text{taking square roots}$$

Since we require values of θ such that $0 \le \theta < 2\pi$, we must have values of 2θ such that $0 \le 2\theta < 4\pi$. Therefore,

$$2\theta = \frac{\pi}{4}, \frac{3\pi}{4}, \frac{5\pi}{4}, \frac{7\pi}{4}, \frac{9\pi}{4}, \frac{11\pi}{4}, \frac{13\pi}{4}, \frac{15\pi}{4}$$

This means that the solutions are

$$\theta = \frac{\pi}{8}, \frac{3\pi}{8}, \frac{5\pi}{8}, \frac{7\pi}{8}, \frac{9\pi}{8}, \frac{11\pi}{8}, \frac{13\pi}{8}, \frac{15\pi}{8}$$

These values satisfy the original equation. Since we multiplied through by $\tan 2\theta$ in the solution, any value of θ that leads to $\tan 2\theta = 0$ would not be valid, since this would indicate division by zero in the original equation.

We see that these solutions agree with the values of θ for which the graph of $y = \tan 2\theta - \cot 2\theta$ crosses the θ axis in Fig. 20-23. (In using a graphing calculator, use x for θ.) ■

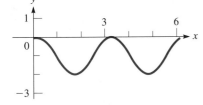

Fig. 20-24

■EXAMPLE 8   Solve the equation $\cos 3x \cos x + \sin 3x \sin x = 1$ $(0 \le x < 2\pi)$.

The left side of this equation is of the general form $\cos(A - x)$, where $A = 3x$. Therefore,

$$\cos 3x \cos x + \sin 3x \sin x = \cos(3x - x) = \cos 2x$$

The original equation becomes

$$\cos 2x = 1$$

This equation is satisfied if $2x = 0$ or $2x = 2\pi$. The solutions are $x = 0$ and $x = \pi$. Only through recognition of the proper trigonometric form can we readily solve this equation.

We see that these solutions agree with the two values of θ for which the graph of $y = \cos 3x \cos x + \sin 3x \sin x - 1$ touches the x-axis in Fig. 20-24. ■

■EXAMPLE 9   Solve the equation $\sin 2x + 3 = x^2$.

Although we can substitute $2 \sin x \cos x$ for $\sin 2x$, we cannot express x^2 in terms of a trigonometric function. However, we can find an approximate graphical solution to this equation.

First, we write the equation with all terms on the left and zero on the right. This gives us

$$y = \sin 2x + 3 - x^2$$

We then enter this function in a graphing calculator as Y_1, and the graph is displayed in Fig. 20-25. Using the *trace* and *zoom* (or *zero* feature) features, we find that

$$x = -1.90, 1.67$$

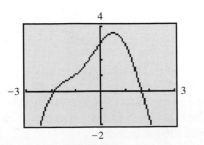

Fig. 20-25

These are the only values for which $y = 0$. (The solution can also be found by letting $y_1 = 2x + 3$, $y_2 = x^2$, and using the *intersect* feature of a graphing calculator.) ■

EXERCISES $20\text{-}5$

In Exercises 1–16, solve the given trigonometric equations analytically (using identities when necessary for exact values when possible) for values of x for $0 \le x < 2\pi$.

1. $\sin x - 1 = 0$ **2.** $2 \cos x + 1 = 0$

3. $2(2 + \cos x) = 3 + \cos x$

4. $4 \tan x + 2 = 3(1 + \tan x)$

5. $4 \cos^2 x - 1 = 0$ **6.** $3 \tan^2 x - 1 = 0$

7. $2 \sin^2 x - \sin x = 0$ **8.** $\sin 4x - \sin 2x = 0$

9. $\sin 2x \sin x + \cos x = 0$

10. $\sin x - \sin \dfrac{x}{2} = 0$

11. $2 \cos^2 x - 2 \cos 2x - 1 = 0$

12. $\tan^2 x + 6 = 5 \tan x$

13. $4 \tan x - \sec^2 x = 0$ **14.** $\sin x \sin \frac{1}{2}x = 1 - \cos x$

15. $\sin 2x \cos x - \cos 2x \sin x = 0$

16. $\cos 3x \cos x - \sin 3x \sin x = 0$

In Exercises 17–32, solve the given trigonometric equations analytically and by use of a graphing calculator. Compare results. Use values of x for $0 \le x < 2\pi$.

17. $\tan x + 1 = 0$ **18.** $2 \sin x + 1 = 0$

19. $3 - 4 \cos x = 7 - (2 - \cos x)$

20. $7 \sin x - 2 = 3(2 - \sin x)$

21. $4 \sin^2 x - 3 = 0$ **22.** $\sin^2 x - 1 = 0$

23. $\sin 4x - \cos 2x = 0$ **24.** $3 \cos x - 4 \cos^2 x = 0$

25. $2 \sin x = \tan x$ **26.** $\cos 2x + \sin^2 x = 0$

27. $\sin^2 x - 2 \sin x = 1$

28. $2 \cos^2 2x + 1 = 3 \cos 2x$

29. $\tan x + 3 \cot x = 4$

30. $\tan^2 x + 4 = 2 \sec^2 x$

31. $\sin 2x + \cos 2x = 0$ **32.** $2 \sin 4x + \csc 4x = 3$

In Exercises 33–36, solve the indicated equations analytically.

33. To find the angle θ subtended by a certain object on a camera film, it is necessary to solve the equation

$$\frac{p^2 \tan \theta}{0.0063 + p \tan \theta} = 1.6$$

where p is the distance from the camera to the object. Find θ if $p = 4.8$ m.

34. In finding the maximum illuminance from a point source of light, it is necessary to solve the equation $\cos \theta \sin 2\theta - \sin^3 \theta = 0$. Find θ if $0 < \theta < 90°$.

35. The vertical displacement y (in m) of the end of a robot arm is given by $y = 2.30 \cos 0.1t - 1.35 \sin 0.2t$. Find the first four values of t (in s) for which $y = 0$.

36. Looking for a lost ship in the North Atlantic Ocean, a plane flew from Reykjavik, Iceland, 160 km west. It then turned and flew due north and then made a final turn to fly directly back to Reykjavik. If the total distance flown was 480 km, how long were the final two legs of the flight? Solve by setting up and solving an appropriate trigonometric equation. (The Pythagorean theorem may be used only as a check.)

In Exercises 37–44, solve the given equations graphically.

37. $3 \sin x - x = 0$ **38.** $4 \cos x + 3x = 0$

39. $2 \sin 2x = x^2 + 1$ **40.** $\sqrt{x} - \sin 3x = 1$

41. $2 \ln x = 1 - \cos 2x$ **42.** $e^x = 1 + \sin x$

43. In finding the frequencies of vibration of a vibrating wire, the equation $x \tan x = 2.00$ occurs. Find x if $0 < x < \pi/2$.

44. An equation used in astronomy is $\theta - e \sin \theta = M$. Solve for θ for $e = 0.25$ and $M = 0.75$.

$20\text{-}6$ THE INVERSE TRIGONOMETRIC FUNCTIONS

When we studied the exponential and logarithmic functions, we often found it useful to change an expression from exponential to logarithmic form or from logarithmic to exponential form. Each of these forms has its advantages for particular purposes. We also showed that the exponential function $y = b^x$ can be written in logarithmic form with x as a function of y, or $x = \log_b y$. We then represented both functions with y in terms of x, saying that the letters used for the dependent and independent variables did not matter when we wished to express a functional relationship. Since it is standard practice to use y as the dependent variable and x as the independent variable, we wrote the logarithmic function as $y = \log_b x$.

These two functions, the exponential function $y = b^x$ and the logarithmic function $y = \log_b x$, are called **inverse functions.** *This means that if we solve for the independent variable in terms of the dependent variable in one function, we will arrive at the functional relationship expressed by the other.* It also means that, for every value of x, there is only one corresponding value of y.

Just as we are able to solve $y = b^x$ for the exponent by writing it in logarithmic form, at times it is necessary to solve for the independent variable (the angle) in trigonometric functions. Therefore, we define the **inverse sine function**

$$y = \sin^{-1} x \qquad \left(-\frac{\pi}{2} \le y \le \frac{\pi}{2} \right) \tag{20-28}$$

where *y is the angle whose sine is x.* This means that x is the value of the sine of the angle y, or $x = \sin y$. (It is necessary to show the range as $-\pi/2 \le y \le \pi/2$, as we will see shortly.)

CAUTION ▶   *In Eq. (20-28), the -1 is* **not** *an exponent.*

The -1 in $\sin^{-1} x$ is the notation showing the inverse function. We first introduced this notation in Chapter 4 when we were finding the angle with a known value of one of the functions.

The notations Arcsin x, arcsin x, Sin$^{-1} x$ are also used to designate the inverse sine. Some calculators use the $\boxed{\text{inv}}$ key and then the $\boxed{\text{sin}}$ key to find values of the inverse sine. However, since most calculators use the notation $\sin^{-1}$ as a second function on the $\boxed{\text{sin}}$ key, we will continue to use $\sin^{-1} x$ for the inverse sine.

Similar definitions are used for the other inverse trigonometric functions. They also have meanings similar to that of Eq. (20-28).

■EXAMPLE 1   **(a)** $y = \cos^{-1} x$ is read as "y is the angle whose cosine is x." In this case, $x = \cos y$.

(b) $y = \tan^{-1} 2x$ is read as "y is the angle whose tangent is $2x$." In this case, $2x = \tan y$.

(c) $y = \csc^{-1}(1 - x)$ is read as "y is the angle whose cosecant is $1 - x$." In this case, $1 - x = \csc y$, or $x = 1 - \csc y$.   ■

We have seen that $y = \sin^{-1} x$ means that $x = \sin y$. From our previous work with the trigonometric functions, we know that there is an unlimited number of possible values of y for a given value of x in $x = \sin y$. Consider the following example.

■EXAMPLE 2   **(a)** For $x = \sin y$, we know that

$$\sin \frac{\pi}{6} = \frac{1}{2} \quad \text{and} \quad \sin \frac{5\pi}{6} = \frac{1}{2}$$

In fact, $x = \frac{1}{2}$ also for values of y of $-\frac{7\pi}{6}, \frac{13\pi}{6}, \frac{17\pi}{6}$, and so on.

(b) For $x = \cos y$, we know that

$$\cos 0 = 1 \quad \text{and} \quad \cos 2\pi = 1$$

In fact, the $\cos y = 1$ for y equal to any even multiple of π.   ■

From Chapter 3, we know that *to have a properly defined **function,** there must be only one value of the dependent variable for a given value of the independent variable.* (A *relation,* on the other hand, may have more than one such value.) Therefore, as in Eq. (20-28), in order to have only one value of y for each value of x in the domain of the inverse trigonometric functions, it is not possible to include all values of y in the range. For this reason, *the range of each of the **inverse trigonometric functions*** is defined as follows:

$$-\frac{\pi}{2} \leq \sin^{-1} x \leq \frac{\pi}{2} \qquad 0 \leq \cos^{-1} x \leq \pi \qquad -\frac{\pi}{2} < \tan^{-1} x < \frac{\pi}{2} \qquad 0 < \cot^{-1} x < \pi$$

$$0 \leq \sec^{-1} x \leq \pi \quad \left(\sec^{-1} x \neq \frac{\pi}{2}\right) \qquad -\frac{\pi}{2} \leq \csc^{-1} x \leq \frac{\pi}{2} \quad (\csc^{-1} x \neq 0)$$

$$(20\text{-}29)$$

For any of the inverse trigonometric functions $y = f(x)$, we must use a value of y in the range as defined in Eqs. (20-29) that corresponds to a given value of x in the domain. We will discuss the domains and the reasons for these definitions, along with the graphs of the inverse trigonometric functions, following the next two examples.

■**EXAMPLE 3** (a) $\sin^{-1}\left(\dfrac{1}{2}\right) = \dfrac{\pi}{6}$ first-quadrant angle

This is the only value of the function that lies within the defined range. The value $\frac{5\pi}{6}$ is not correct, even though $\sin(\frac{5\pi}{6}) = \frac{1}{2}$, since $\frac{5\pi}{6}$ lies outside the defined range.

(b) $\cos^{-1}\left(-\dfrac{1}{2}\right) = \dfrac{2\pi}{3}$ second-quadrant angle

Other values such as $\frac{4\pi}{3}$ and $-\frac{2\pi}{3}$ are not correct, since they are not within the defined range for the function $\cos^{-1} x$.   ■

■**EXAMPLE 4** $\tan^{-1}(-1) = -\dfrac{\pi}{4}$ fourth-quadrant angle

CAUTION▶

This is the only value within the defined range for the function $\tan^{-1} x$. We must remember that *when x is negative for $\sin^{-1} x$ and $\tan^{-1} x$, the value of y is a fourth-quadrant angle, expressed as a **negative angle.*** This is a direct result of the definition. (The single exception is $\sin^{-1}(-1) = -\pi/2$, which is a quadrantal angle and is not *in* the fourth quadrant.)   ■

In choosing these values to be the ranges of the inverse trigonometric functions, we first note that the *domain* of $y = \sin^{-1} x$ and $y = \cos^{-1} x$ are each $-1 \leq x \leq 1$, since the sine and cosine functions take on only these values. Therefore, for each value in this domain we use only one value of y in the range of the function. Although the domain of $y = \tan^{-1} x$ is all real numbers, we still use only one value of y in the range.

The ranges of the inverse trigonometric functions are chosen so that if x is positive, the resulting value is an angle in the first quadrant. However, care must be taken in choosing the range for negative values of x.

We cannot choose second-quadrant angles for $\sin^{-1} x$. Since the sine of a second-quadrant angle is also positive, this would lead to ambiguity. The sine is negative for fourth-quadrant angles, and to have a continuous range of values we express the fourth-quadrant angles in the form of negative angles. This range is also chosen for $\tan^{-1} x$, for similar reasons. However, the range for $\cos^{-1} x$ cannot be chosen in this way, since the cosine of a fourth-quadrant angle is also positive. Thus, to keep a continuous range of values for $\cos^{-1} x$, the second-quadrant angles are chosen for negative values of x.

As for the values for the other functions, we chose values such that if x is positive, the result is also an angle in the first quadrant. As for negative values of x, it rarely makes any difference, since either positive values of x arise, or we can use one of the other functions. Our definitions, however, are those which are generally used.

The graphs of the inverse trigonometric functions can be used to show the domains and ranges. We can obtain the graph of the inverse sine function by first sketching the sine curve $x = \sin y$ *along the y-axis*. We then mark the specific part of this curve for which $-\frac{\pi}{2} \le y \le \frac{\pi}{2}$ as the graph of the inverse sine function. The graphs of the other inverse trigonometric functions are found in the same manner. In Figs. 20-26, 20-27, and 20-28, the graphs of $x = \sin y$, $x = \cos y$, and $x = \tan y$, respectively, are shown. The heavier, colored portions indicate the graphs of the respective inverse trigonometric functions.

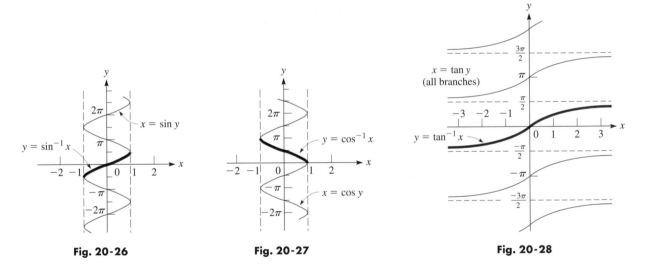

Fig. 20-26   Fig. 20-27   Fig. 20-28

The following examples further illustrate the values and meanings of the inverse trigonometric functions.

■**EXAMPLE 5**   **(a)** $\sin^{-1}(-\sqrt{3}/2) = -\pi/3$ **(b)** $\cos^{-1}(-1) = \pi$
(c) $\tan^{-1} 0 = 0$ **(d)** $\tan^{-1}(\sqrt{3}) = \pi/3$

Using a calculator in radian mode, we find the following values:

(e) $\sin^{-1} 0.6294 = 0.6808$ **(f)** $\sin^{-1}(-0.1568) = -0.1574$
(g) $\cos^{-1}(-0.8026) = 2.5024$ **(h)** $\tan^{-1}(-1.9268) = -1.0921$

We note that the calculator gives values that are in the defined range for each function.

EXAMPLE 6   Given that $y = \pi - \sec^{-1} 2x$, solve for x.

We first find the expression for $\sec^{-1} 2x$ and then use the meaning of the inverse secant. The solution follows.

$$y = \pi - \sec^{-1} 2x$$

$$\sec^{-1} 2x = \pi - y \qquad \text{solve for } \sec^{-1} 2x$$

$$2x = \sec(\pi - y) \qquad \text{use meaning of inverse secant}$$

$$x = -\frac{1}{2} \sec y \qquad \sec(\pi - y) = -\sec y$$

As $\sec 2x$ and $2 \sec x$ are different functions, $\sec^{-1} 2x$ and $2 \sec^{-1} x$ are also different functions. Since the values of $\sec^{-1} 2x$ are restricted, so are the resulting values of y.

EXAMPLE 7   The instantaneous power p in an electric inductor is given by the equation $p = vi \sin \omega t \cos \omega t$. Solve for t.

Noting the product $\sin \omega t \cos \omega t$ suggests using $\sin 2\alpha = 2 \sin \alpha \cos \alpha$. Then, using the meaning of the inverse sine, we can complete the solution.

$$p = vi \sin \omega t \cos \omega t$$

$$= \frac{1}{2} vi \sin 2\omega t \qquad \text{using double-angle formula}$$

$$\sin 2\omega t = \frac{2p}{vi}$$

$$2\omega t = \sin^{-1}\left(\frac{2p}{vi}\right) \qquad \text{using meaning of inverse sine}$$

$$t = \frac{1}{2\omega} \sin^{-1}\left(\frac{2p}{vi}\right)$$

If we know the value of one of the inverse functions, we can find the trigonometric functions of the angle. If general relations are desired, a representative triangle is very useful. The following examples illustrate these methods.

EXAMPLE 8   Find $\cos(\sin^{-1} 0.5)$.

Knowing that the values of inverse trigonometric functions are *angles,* we see that $\sin^{-1} 0.5$ is a first-quadrant angle. Thus, we find $\sin^{-1} 0.5 = \pi/6$. The problem is now to find $\cos(\pi/6)$. This is, of course, $\sqrt{3}/2$, or 0.8660. Thus,

$$\cos(\sin^{-1} 0.5) = \cos(\pi/6) = 0.8660.$$

EXAMPLE 9   (a) $\sin(\cot^{-1} 1) = \sin(\pi/4) \qquad$ first-quadrant angle

$$= \frac{\sqrt{2}}{2} = 0.7071$$

(b) $\tan[\cos^{-1}(-1)] = \tan \pi \qquad$ quadrantal angle

$$= 0$$

(c) $\cos[\sin^{-1}(-0.2395)] = 0.9709 \qquad$ using a calculator

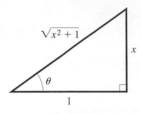

Fig. 20-29

■EXAMPLE 10   Find $\sin(\tan^{-1} x)$.

We know that $\tan^{-1} x$ is another way of stating "the angle whose tangent is x." Thus, let us draw a right triangle (as in Fig. 20-29) and label one of the acute angles as θ, the side opposite θ as x, and the side adjacent to θ as 1. In this way we see that, by definition, $\tan \theta = \frac{x}{1}$, or $\theta = \tan^{-1} x$, which means θ is the desired angle. By the Pythagorean theorem, the hypotenuse of this triangle is $\sqrt{x^2 + 1}$. Now we find that $\sin \theta$, which is the same as $\sin(\tan^{-1} x)$, is $x/\sqrt{x^2 + 1}$. Thus,

$$\sin(\tan^{-1} x) = \frac{x}{\sqrt{x^2 + 1}}$$

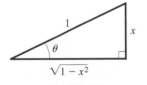

Fig. 20-30

■EXAMPLE 11   Find $\cos(2 \sin^{-1} x)$.

From Fig. 20-30, we see that $\theta = \sin^{-1} x$. From the double-angle formulas, we have

$$\cos 2\theta = 1 - 2 \sin^2 \theta$$

Thus, since $\sin \theta = x$, we have

$$\cos(2 \sin^{-1} x) = 1 - 2x^2$$

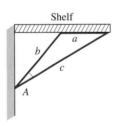

Fig. 20-31

■EXAMPLE 12   A triangular brace of sides a, b, and c supports a shelf, as shown in Fig. 20-31. Find the expression for the angle between sides b and c.

The law of cosines leads to the solution, which follows.

$$a^2 = b^2 + c^2 - 2bc \cos A \qquad \text{law of cosines}$$

$$2bc \cos A = b^2 + c^2 - a^2 \qquad \text{solving for cos } A$$

$$\cos A = \frac{b^2 + c^2 - a^2}{2bc}$$

$$A = \cos^{-1}\left(\frac{b^2 + c^2 - a^2}{2bc}\right) \qquad \text{using meaning of inverse cosine}$$

EXERCISES *20-6*

In Exercises 1–8, write down the meaning of each of the given equations. See Example 1.

1. $y = \tan^{-1} x$

2. $y = \sec^{-1} x$

3. $y = \cot^{-1} 3x$

4. $y = \csc^{-1} 4x$

5. $y = 2 \sin^{-1} x$

6. $y = 3 \tan^{-1} x$

7. $y = 5 \cos^{-1}(2x - 1)$

8. $y = 4 \sin^{-1}(3x + 2)$

In Exercises 9–28, evaluate the given expressions.

9. $\cos^{-1} 0.5$

10. $\sin^{-1} 1$

11. $\tan^{-1} 1$

12. $\cos^{-1} 0$

13. $\tan^{-1}(-\sqrt{3})$

14. $\sin^{-1}(-0.5)$

15. $\sec^{-1} 2$

16. $\cot^{-1} \sqrt{3}$

17. $\sin^{-1}(-\sqrt{2}/2)$

18. $\cos^{-1}(-\sqrt{3}/2)$

19. $\csc^{-1}\sqrt{2}$

20. $\cot^{-1}(-\sqrt{3})$

21. $\tan(\sin^{-1} 0)$

22. $\csc(\tan^{-1} 1)$

23. $\sin(\tan^{-1} \sqrt{3})$

24. $\tan[\sin^{-1}(\sqrt{2}/2)]$

25. $\cos[\tan^{-1}(-1)]$

26. $\sec[\cos^{-1}(-0.5)]$

27. $\cos(2 \sin^{-1} 1)$

28. $\sin(2 \tan^{-1} 2)$

In Exercises 29–32, find the exact value of x.

29. $\tan^{-1} x = \sin^{-1} \frac{2}{5}$

30. $\cot^{-1} x = \cos^{-1} \frac{1}{3}$

31. $\sec^{-1} x = -\sin^{-1}(-\frac{1}{2})$

32. $\sin^{-1} x = -\tan^{-1}(-1)$

In Exercises 33–44, use a calculator to evaluate the given expressions.

33. $\tan^{-1}(-3.7321)$

34. $\cos^{-1}(-0.6561)$

35. $\sin^{-1}(-0.8326)$

36. $\tan^{-1} 0.2846$

37. $\cos^{-1} 0.1291$

38. $\sin^{-1} 0.2119$

39. $\tan^{-1} 8.2614$

40. $\sin^{-1}(-0.8881)$

41. $\tan[\cos^{-1}(-0.6281)]$

42. $\cos[\tan^{-1}(-1.2256)]$

43. $\sin[\tan^{-1}(-0.2297)]$

44. $\tan[\sin^{-1}(-0.3019)]$

In Exercises 45–52, solve the given equations for x.

45. $y = \sin 3x$

46. $y = \cos(x - \pi)$

47. $y = \tan^{-1}(x/4)$

48. $y = 2 \sin^{-1}(x/6)$

49. $y = 1 + 3 \sec 3x$

50. $4y = 5 - 2 \csc 8x$

51. $1 - y = \cos^{-1}(1 - x)$

52. $2y = \cot^{-1} 3x - 5$

In Exercises 53–60, find an algebraic expression for each of the given expressions.

53. $\tan(\sin^{-1} x)$

54. $\sin(\cos^{-1} x)$

55. $\cos(\sec^{-1} x)$

56. $\cot(\cot^{-1} x)$

57. $\sec(\csc^{-1} 3x)$

58. $\tan(\sin^{-1} 2x)$

59. $\sin(2 \sin^{-1} x)$

60. $\cos(2 \tan^{-1} x)$

In Exercises 61–64, solve the given problems with the use of the inverse trigonometric functions.

61. In the analysis of ocean tides, the equation $y = A \cos 2(\omega t + \phi)$ is used. Solve for t.

62. For an object of weight w on an inclined plane that is at an angle θ to the horizontal, the equation relating w and θ is $\mu w \cos \theta = w \sin \theta$, where μ is the coefficient of friction between the surfaces in contact. Solve for θ.

63. The electric current in a certain circuit is given by $i = I_m[\sin(\omega t + \alpha)\cos \phi + \cos(\omega t + \alpha)\sin \phi]$. Solve for t.

64. The time t as a function of the displacement d of a piston is given by $t = \dfrac{1}{2\pi f} \cos^{-1} \dfrac{d}{A}$. Solve for d.

In Exercises 65 and 66, prove that the given expressions are equal. Use the relation for $\sin(\alpha + \beta)$ and show that the sine of the sum of the angles on the left equals the sine of the angle on the right.

65. $\sin^{-1} \dfrac{3}{5} + \sin^{-1} \dfrac{5}{13} = \sin^{-1} \dfrac{56}{65}$

66. $\tan^{-1} \dfrac{1}{3} + \tan^{-1} \dfrac{1}{2} = \dfrac{\pi}{4}$

In Exercises 67 and 68, verify the given expressions.

67. $\sin^{-1} 0.5 + \cos^{-1} 0.5 = \pi/2$

68. $\tan^{-1} \sqrt{3} + \cot^{-1} \sqrt{3} = \pi/2$

(W) *In Exercises 69 and 70, solve for the angle A for the triangles in the given figures in terms of the given sides and angles. Explain your method.*

69. Fig. 20-32

70. Fig. 20-33

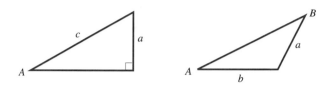

In Exercises 71 and 72, derive the given expressions.

71. The height of the Statue of Liberty is 151 ft. See Fig. 20-34. From the deck of a boat at a horizontal distance d from the statue, the angles of elevation of the top of the statue and the top of its pedestal are α and β, respectively. Show that
$$\alpha = \tan^{-1}\left(\frac{151}{d} + \tan \beta\right).$$

151 ft

Fig. 20-34

72. Show that the length L of the pulley belt shown in Fig. 20-35 is $L = 24 + 11\pi + 10 \sin^{-1} \dfrac{5}{13}$.

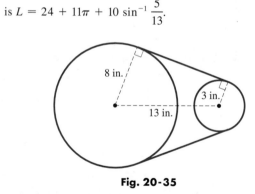

8 in.

3 in.

13 in.

Fig. 20-35

CHAPTER EQUATIONS

Basic trigonometric identities	$\sin \theta = \dfrac{1}{\csc \theta}$	**(20-1)**	$\tan \theta = \dfrac{\sin \theta}{\cos \theta}$	**(20-4)**	$\sin^2 \theta + \cos^2 \theta = 1$	**(20-6)**
	$\cos \theta = \dfrac{1}{\sec \theta}$	**(20-2)**	$\cot \theta = \dfrac{\cos \theta}{\sin \theta}$	**(20-5)**	$1 + \tan^2 \theta = \sec^2 \theta$	**(20-7)**
	$\tan \theta = \dfrac{1}{\cot \theta}$	**(20-3)**			$1 + \cot^2 \theta = \csc^2 \theta$	**(20-8)**

Sum and difference identities

$$\sin(\alpha + \beta) = \sin \alpha \cos \beta + \cos \alpha \sin \beta \qquad (20\text{-}9)$$

$$\cos(\alpha + \beta) = \cos \alpha \cos \beta - \sin \alpha \sin \beta \qquad (20\text{-}10)$$

$$\sin(\alpha - \beta) = \sin \alpha \cos \beta - \cos \alpha \sin \beta \qquad (20\text{-}11)$$

$$\cos(\alpha - \beta) = \cos \alpha \cos \beta + \sin \alpha \sin \beta \qquad (20\text{-}12)$$

$$\tan(\alpha \pm \beta) = \frac{\tan \alpha \pm \tan \beta}{1 \mp \tan \alpha \tan \beta} \qquad (20\text{-}13)$$

Double-angle formulas

$$\sin 2\alpha = 2 \sin \alpha \cos \alpha \qquad (20\text{-}21)$$

$$\cos 2\alpha = \cos^2 \alpha - \sin^2 \alpha \qquad (20\text{-}22)$$

$$= 2 \cos^2 \alpha - 1 \qquad (20\text{-}23)$$

$$= 1 - 2 \sin^2 \alpha \qquad (20\text{-}24)$$

$$\tan 2\alpha = \frac{2 \tan \alpha}{1 - \tan^2 \alpha} \qquad (20\text{-}25)$$

Half-angle formulas

$$\sin \frac{\alpha}{2} = \pm \sqrt{\frac{1 - \cos \alpha}{2}} \qquad (20\text{-}26)$$

$$\cos \frac{\alpha}{2} = \pm \sqrt{\frac{1 + \cos \alpha}{2}} \qquad (20\text{-}27)$$

Inverse trigonometric functions

$$y = \sin^{-1} x \quad \left(-\frac{\pi}{2} \le y \le \frac{\pi}{2} \right) \qquad (20\text{-}28)$$

$$-\frac{\pi}{2} \le \sin^{-1} x \le \frac{\pi}{2} \quad 0 \le \cos^{-1} x \le \pi \quad -\frac{\pi}{2} < \tan^{-1} x < \frac{\pi}{2} \quad 0 < \cot^{-1} x < \pi \qquad (20\text{-}29)$$

$$0 \le \sec^{-1} x \le \pi \ \left(\sec^{-1} x \ne \frac{\pi}{2} \right) \quad -\frac{\pi}{2} \le \csc^{-1} x \le \frac{\pi}{2} \ (\csc^{-1} x \ne 0)$$

▌ REVIEW EXERCISES

In Exercises 1–8, determine the values of the indicated functions in the given manner.

1. Find $\sin 120°$ by using $120° = 90° + 30°$.

2. Find $\cos 30°$ by using $30° = 90° - 60°$.

3. Find $\sin 135°$ by using $135° = 180° - 45°$.

4. Find $\tan 225°$ by using $225° = 180° + 45°$.

5. Find $\cos 180°$ by using $180° = 2(90°)$.

6. Find $\sin 180°$ by using $180° = 2(90°)$.

7. Find $\tan 60°$ by using $60° = 2(30°)$.

8. Find $\cos 45°$ by using $45° = \frac{1}{2}(90°)$.

In Exercises 9–16, simplify the given expressions by using one of the basic formulas of the chapter. Then use a calculator to verify the result by finding the value of the original expression and the value of the simplified expression.

9. $\sin 14° \cos 38° + \cos 14° \sin 38°$

10. $\cos^2 148° - \sin^2 148°$

11. $2 \sin 46° \cos 46°$

12. $1 - 2 \sin^2 82°$

13. $\cos 73° \cos 142° + \sin 73° \sin 142°$

14. $\cos 3° \cos 215° - \sin 3° \sin 215°$

15. $\dfrac{4 \tan 12°}{1 - \tan^2 12°}$

16. $\sqrt{\dfrac{1 - \cos 166°}{2}}$

In Exercises 17–24, simplify each of the given expressions. Expansion of any term is not necessary; recognition of the proper form leads to the proper result.

17. $\sin 2x \cos 3x + \cos 2x \sin 3x$

18. $\cos 7x \cos 3x + \sin 7x \sin 3x$

19. $8 \sin 6x \cos 6x$

20. $\dfrac{\tan x + \tan 2x}{1 - \tan x \tan 2x}$

21. $2 - 4 \sin^2 6x$

22. $\cos^2 2x - \sin^2 2x$

23. $\sqrt{2 + 2 \cos 2x}$

24. $\sqrt{32 - 32 \cos 4x}$

In Exercises 25–32, evaluate the given expressions.

25. $\sin^{-1}(-1)$ **26.** $\sec^{-1}\sqrt{2}$

27. $\cos^{-1} 0.9659$ **28.** $\tan^{-1}(-0.6249)$

29. $\tan[\sin^{-1}(-0.5)]$ **30.** $\cos[\tan^{-1}(-\sqrt{3})]$

31. $\sin^{-1}(\tan \pi)$ **32.** $\cos^{-1}[\tan(-\pi/4)]$

In Exercises 33–52, prove the given identities.

33. $\dfrac{\sec y}{\csc y} = \tan y$ **34.** $\cos \theta \csc \theta = \cot \theta$

35. $\sin x(\csc x - \sin x) = \cos^2 x$

36. $\cos y(\sec y - \cos y) = \sin^2 y$

37. $\dfrac{1}{\sin \theta} - \sin \theta = \cot \theta \cos \theta$

38. $\sin \theta \sec \theta \csc \theta \cos \theta = 1$

39. $\cos \theta \cot \theta + \sin \theta = \csc \theta$

40. $\dfrac{\sin x \cot x + \cos x}{\cot x} = 2 \sin x$

41. $\dfrac{\sec^4 x - 1}{\tan^2 x} = 2 + \tan^2 x$

42. $\cos^2 y - \sin^2 y = \dfrac{1 - \tan^2 y}{1 + \tan^2 y}$

43. $2 \csc 2x \cot x = 1 + \cot^2 x$

44. $\sin x \cot^2 x = \csc x - \sin x$

45. $\dfrac{1 - \sin^2 \theta}{1 - \cos^2 \theta} = \cot^2 \theta$ **46.** $\dfrac{1 + \cos 2\theta}{\cos^2 \theta} = 2$

47. $\dfrac{\cos 2\theta}{\cos^2 \theta} = 1 - \tan^2 \theta$ **48.** $\dfrac{\sin 2\theta \sec \theta}{2} = \sin \theta$

49. $\sin \dfrac{\theta}{2} \cos \dfrac{\theta}{2} = \dfrac{\sin \theta}{2}$ **50.** $\sin \dfrac{x}{2} = \dfrac{\sec x - 1}{2 \sec x \sin\left(\dfrac{x}{2}\right)}$

51. $\sec x + \tan x = \dfrac{\cos x}{1 - \sin x}$

52. $\cos(x - y)\cos y - \sin(x - y)\sin y = \cos x$

In Exercises 53–60, verify each identity by comparing the graph of the left side with the graph of the right side on a graphing calculator.

53. $\dfrac{\cos \theta - \sin \theta}{\cos \theta + \sin \theta} = \dfrac{\cot \theta - 1}{\cot \theta + 1}$

54. $\sin 3y \cos 2y - \cos 3y \sin 2y = \sin y$

55. $\sin 4x(\cos^2 2x - \sin^2 2x) = \dfrac{\sin 8x}{2}$

56. $\csc 2x + \cot 2x = \cot x$

57. $\dfrac{\sin x}{\csc x - \cot x} = 1 + \cos x$

58. $\cos x - \sin \dfrac{x}{2} = \left(1 - 2 \sin \dfrac{x}{2}\right)\left(1 + \sin \dfrac{x}{2}\right)$

59. $\tan \dfrac{\alpha}{2} = \csc \alpha - \cot \alpha$

60. $\sec \dfrac{x}{2} + \csc \dfrac{x}{2} = \dfrac{2\left(\sin \dfrac{x}{2} + \cos \dfrac{x}{2}\right)}{\sin x}$

In Exercises 61–64, solve for x.

61. $y = 2 \cos 2x$ **62.** $y - 2 = 2 \tan\left(x - \dfrac{\pi}{2}\right)$

63. $y = \dfrac{\pi}{4} - 3 \sin^{-1} 5x$ **64.** $2y = \sec^{-1} 4x - 2$

In Exercises 65–76, solve the given equations for x ($0 \le x < 2\pi$).

65. $3(\tan x - 2) = 1 + \tan x$

66. $5 \sin x = 3 - (\sin x + 2)$

67. $2(1 - 2 \sin^2 x) = 1$ **68.** $\sec^2 x = 2 \tan x$

69. $\cos^2 2x - 1 = 0$ **70.** $2 \sin 2x + 1 = 0$

71. $4 \cos^2 x = 3$ **72.** $\cos 2x = \sin x$

73. $\sin 2x = \cos 3x$

74. $\cos 3x \cos x + \sin 3x \sin x = 0$

75. $\sin^2\left(\dfrac{x}{2}\right) - \cos x + 1 = 0$

76. $\sin x + \cos x = 1$

In Exercises 77–80, solve the given equations graphically.

77. $x + \ln x - 3 \cos^2 x = 2$

78. $e^{\sin x} - 2 = x \cos^2 x$

79. $2 \tan^{-1} x + x^2 = 3$

80. $3 \sin^{-1} x = 6 \sin x + 1$

In Exercises 81–84, find an algebraic expression for each of the given expressions.

81. $\tan(\cot^{-1} x)$ **82.** $\cos(\csc^{-1} x)$

83. $\sin(2 \cos^{-1} x)$ **84.** $\cos(\pi - \tan^{-1} x)$

In Exercises 85–88, use the given substitutions to show that the equations are valid for $0 \le \theta < \pi/2$.

85. If $x = 2 \cos \theta$, show that $\sqrt{4 - x^2} = 2 \sin \theta$.

86. If $x = 2 \sec \theta$, show that $\sqrt{x^2 - 4} = 2 \tan \theta$.

87. If $x = \tan \theta$, show that $\dfrac{x}{\sqrt{1 + x^2}} = \sin \theta$.

88. If $x = \cos \theta$, show that $\dfrac{\sqrt{1 - x^2}}{x} = \tan \theta$.

In Exercises 89–104, use the methods and formulas of this chapter to solve the given problems.

89. Solve the inequality $\sin 2x > 2 \sin x$ for $0 \le x < 2\pi$.

90. Show that $(\cos 2\alpha + \sin^2 \alpha)\sec^2 \alpha$ has a constant value.

91. Show that $y = A \sin 2t + B \cos 2t$ may be written as $y = C \sin(2t + \alpha)$, where $C = \sqrt{A^2 + B^2}$ and $\tan \alpha = B/A$. (*Hint:* Let $A/C = \cos \alpha$ and $B/C = \sin \alpha$.)

92. In a right triangle with sides a and b and hypotenuse c, show that $\sin(A/2) = \sqrt{(c - b)/2c}$, where angle A is opposite a.

93. Forces **A** and **B** act on a bolt such that **A** makes an angle θ with the x-axis and **B** makes an angle θ with the y-axis as shown in Fig. 20-36. The resultant **R** has components $R_x = A \cos \theta - B \sin \theta$ and $R_y = A \sin \theta + B \cos \theta$. Using these components, show that $R = \sqrt{A^2 + B^2}$.

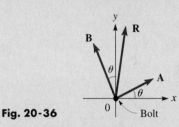

Fig. 20-36

94. For a certain alternating-current generator, the expression $I \cos \theta + I \cos(\theta + 2\pi/3) + I \cos(\theta + 4\pi/3)$ arises. Simplify this expression.

95. Some comets follow a parabolic path that can be described by the equation $r = (k/2)\csc^2(\theta/2)$, where r is the distance to the sun and k is a constant. Show that this equation can be written as $r = k/(1 - \cos \theta)$.

96. In studying the interference of light waves, the identity
$$\frac{\sin \tfrac{3}{2}x}{\sin \tfrac{1}{2}x} \sin x = \sin x + \sin 2x$$
is used. Prove this identity.
(*Hint:* $\sin \tfrac{3}{2}x = \sin(x + \tfrac{1}{2}x)$.)

97. In the study of chemical spectroscopy, the equation
$$\omega t = \sin^{-1} \frac{\theta - \alpha}{R}$$
arises. Solve for θ.

98. The power p in a certain electric circuit is given by $p = 2.5[\cos \alpha \sin(\omega t + \phi) - \sin \alpha \cos(\omega t + \phi)]$. Solve for t.

99. In surveying, when determining an azimuth (a measure used for reference purposes), it might be necessary to simplify the expression $(2 \cos \alpha \cos \beta)^{-1} - \tan \alpha \tan \beta$. Perform this operation by expressing it in the simplest possible form when $\alpha = \beta$.

100. In analyzing the motion of an automobile universal joint, the equation $\sec^2 A - \sin^2 B \tan^2 A = \sec^2 C$ is used. Show that this equation is true if $\tan A \cos B = \tan C$.

101. A roof truss is in the shape of an isosceles triangle of height 6.4 ft. If the total length of the three members is 51.2 ft, what is the length of a rafter? Solve by setting up and solving an appropriate trigonometric equation. (The Pythagorean theorem may be used only as a check.)

102. The angle of elevation of the top of the Washington Monument from a point on level ground 250 ft from the center of its base is twice the angle of elevation of the top from a point 610 ft farther away. Find the height of the monument.

103. If a plane surface inclined at angle θ moves horizontally, the angle for which the lifting force of the air is a maximum is found by solving the equation $2 \sin \theta \cos^2 \theta - \sin^3 \theta = 0$, where $0 < \theta < 90°$. Solve for θ.

104. To determine the angle between two sections of a certain robot arm, the equation $1.20 \cos \theta + 0.135 \cos 2\theta = 0$ is to be solved. Find the required angle θ if $0° < \theta < 180°$.

Writing Exercise

105. Analyzing the forces on a bridge support, an engineer finds the equation $\tan \theta = 0.4250$. Write a paragraph explaining the difference in solving this equation and evaluating the expression $\tan^{-1} 0.4250$.

PRACTICE TEST

1. Prove that $\sec \theta - \dfrac{\tan \theta}{\csc \theta} = \cos \theta$.

2. Solve for x $(0 \le x < 2\pi)$ analytically, using trigonometric relations where necessary: $\sin 2x + \sin x = 0$.

3. Find an algebraic expression for $\cos(\sin^{-1} x)$.

4. The electric current as a function of the time for a particular circuit is given by $i = 8.00e^{-20t}(1.73 \cos 10.0t - \sin 10.0t)$. Find the time (in s) when the current is first zero.

5. Prove that $\dfrac{\tan \alpha + \tan \beta}{\tan \alpha - \tan \beta} = \dfrac{\sin(\alpha + \beta)}{\sin(\alpha - \beta)}$.

6. Prove that $\cot^2 x - \cos^2 x = \cot^2 x \cos^2 x$.

7. Find $\cos \tfrac{1}{2}x$ if $\sin x = -\tfrac{3}{5}$ and $270° < x < 360°$.

8. The intensity of a certain type of polarized light is given by $I = I_0 \sin 2\theta \cos 2\theta$. Solve for θ.

9. Solve graphically: $x - 2 \cos x = 5$.

CHAPTER 21

PLANE ANALYTIC GEOMETRY

We first introduced the graph of a function in Chapter 3. Since then we have seen the graphs of certain types of functions that fit particular forms. For example, the graph of a linear function is a straight line, the graph of the quadratic function is a parabola, and the graph of the sine function is a periodic waveform.

The underlying principle of *analytic geometry* is the relationship of an algebraic equation and the geometric properties of the curve that represents the equation. In this chapter we develop equations for a number of important geometric curves and show how to find the properties of these curves through an analysis of their equations. Most important among these curves are the *conic sections,* which we briefly introduced in Chapter 14.

The concepts developed in analytic geometry are very useful in the study of calculus. Also, there are a great many technical and scientific applications. These include projectile motion, planetary orbits, and fluid motion. Other important applications range from the design of gears, airplane wings, and automobile headlights to the construction of bridges and nuclear cooling towers.

In Section 21-4 we show an important application of analytic geometry in the design of automobile headlights.

The development of geometry made little progress from the time of the ancient Greeks until the 1600s, when algebra was systematically applied to geometric problems. The French mathematician René Descartes (1596–1650) is recognized as the founder of analytic geometry.

21-1 BASIC DEFINITIONS

As we noted above, analytic geometry deals with the relationship between an algebraic equation and the geometric curve it represents. In this section we develop certain basic concepts that will be needed for future use in establishing the proper relationships between an equation and a curve.

The Distance Formula

The first of these concepts involves the distance between any two points in the coordinate plane. If these points lie on a line parallel to the x-axis, the **directed distance** from the first point $A(x_1, y)$ to the second point $B(x_2, y)$ is denoted as $\overline{AB}$ and is defined as $x_2 - x_1$. We see that $\overline{AB}$ *is positive if B is to the right of A, and it is negative if B is to the left of A.* Similarly, the directed distance between two points $C(x, y_1)$ and $D(x, y_2)$ on a line parallel to the y-axis is $y_2 - y_1$. We see that $\overline{CD}$ *is positive if D is above C, and it is negative if D is below C.*

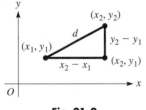

Fig. 21-1

■**EXAMPLE 1**   The line segment joining $A(-1, 5)$ and $B(-4, 5)$ in Fig. 21-1 is parallel to the x-axis. Therefore, the directed distance

$$\overline{AB} = -4 - (-1) = -3$$

and the directed distance

$$\overline{BA} = -1 - (-4) = 3$$

Also in Fig. 21-1, the line segment joining $C(2, -3)$ and $D(2, 6)$ is parallel to the y-axis. The directed distance

$$\overline{CD} = 6 - (-3) = 9$$

and the directed distance

$$\overline{DC} = -3 - 6 = -9 \qquad \text{------------} ■$$

We now wish to find the length of a line segment joining any two points in the plane. If these points are on a line that is not parallel to either axis (see Fig. 21-2), we use the Pythagorean theorem to find the distance between them. By making a right triangle with the line segment joining the points as the hypotenuse and line segments parallel to the axes as legs, we have *the **distance formula,** which gives the distance between any two points in the plane.* This formula is

$$\boxed{d = \sqrt{(x_2 - x_1)^2 + (y_2 - y_1)^2}} \qquad \text{(21-1)}$$

Here we choose the positive square root since we are concerned only with the magnitude of the length of the line segment.

■**EXAMPLE 2**   The distance between $(3, -1)$ and $(-2, -5)$ is given by

$$d = \sqrt{[(-2) - 3]^2 + [(-5) - (-1)]^2}$$
$$= \sqrt{(-5)^2 + (-4)^2} = \sqrt{25 + 16}$$
$$= \sqrt{41} = 6.403$$

See Fig. 21-3.

NOTE ▶   *It makes no difference which point is chosen as (x_1, y_1) and which is chosen as (x_2, y_2),* since the differences in the x-coordinates and the y-coordinates are squared. We obtain the same value for the distance when we calculate it as

$$d = \sqrt{[3 - (-2)]^2 + [(-1) - (-5)]^2}$$
$$= \sqrt{5^2 + 4^2} = \sqrt{41} = 6.403 \qquad \text{------------} ■$$

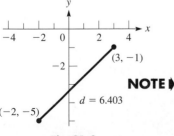

Fig. 21-3

(The figure 21-2 caption: **Fig. 21-2**)

The Slope of a Line

Another important quantity for a line is its *slope,* which we defined in Chapter 5. Here we give it a somewhat more general definition and develop its meaning in more detail, as well as review the basic meaning.

The **slope** *gives a measure of the direction of a line and is defined as the vertical directed distance from one point to another on the same straight line, divided by the horizontal directed distance from the first point to the second.* Thus, the slope, m, is given by

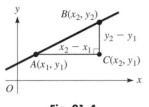

Fig. 21-4

$$m = \frac{y_2 - y_1}{x_2 - x_1} \qquad \textbf{(21-2)}$$

See Fig. 21-4. When the line is horizontal, $y_2 = y_1$ and $m = 0$. When the line is vertical, $x_2 = x_1$ and the slope is undefined.

■ **EXAMPLE 3**   The slope of a line joining $(3, -5)$ and $(-2, -6)$ is

$$m = \frac{-6 - (-5)}{-2 - 3} = \frac{-6 + 5}{-5} = \frac{1}{5}$$

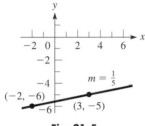

Fig. 21-5

See Fig. 21-5. Again we may interpret either of the points as (x_1, y_1) and the other as (x_2, y_2). We can also obtain the slope of this same line from

$$m = \frac{-5 - (-6)}{3 - (-2)} = \frac{1}{5} \qquad \text{-------}■$$

The larger the numerical value of the slope of a line, the more nearly vertical is the line. Also, *a line rising to the right has a positive slope, and a line falling to the right has a negative slope.*

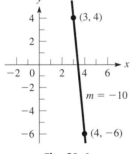

Fig. 21-6

■ **EXAMPLE 4**   (a) The line in Example 3 has a positive slope, which is numerically small. From Fig. 21-5 it can be seen that the line rises slightly to the right.
   (b) The line joining $(3, 4)$ and $(4, -6)$ has a slope of

$$m = \frac{4 - (-6)}{3 - 4} = -10 \qquad \text{or} \qquad m = \frac{-6 - 4}{4 - 3} = -10$$

This line falls sharply to the right, as shown in Fig. 21-6.   -------■

If a given line is extended indefinitely in either direction, it must cross the x-axis at some point unless it is parallel to the x-axis. *The angle measured from the x-axis in a positive direction to the line is called the* **inclination** *of the line* (see Fig. 21-7). The inclination of a line parallel to the x-axis is defined to be zero. *An alternate definition of slope, in terms of the inclination α, is*

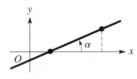

Fig. 21-7

$$m = \tan \alpha \quad (0° \leq \alpha < 180°) \qquad \textbf{(21-3)}$$

Since the slope can be defined in terms of any two points on the line, we can choose the x-intercept and any other point. Therefore, from the definition of the tangent of an angle, we see that Eq. (21-3) is in agreement with Eq. (21-2).

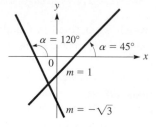

Fig. 21-8

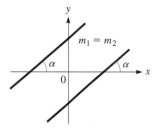

Fig. 21-9

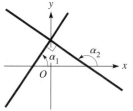

Fig. 21-10

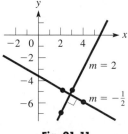

Fig. 21-11

■EXAMPLE 5 **(a)** The slope of a line with an inclination of 45° is

$$m = \tan 45° = 1.000$$

(b) The slope of a line having an inclination of 120° is

$$m = \tan 120° = -\sqrt{3} = -1.732$$

See Fig. 21-8.

We see that if the inclination is an acute angle, the slope is positive and the line rises to the right. If the inclination is obtuse, the slope is negative and the line falls to the right.

Any two parallel lines crossing the x-axis have the same inclination. Therefore, as shown in Fig. 21-9, the *slopes of parallel lines are equal*. This can be stated as

$$m_1 = m_2 \qquad \text{(for } \| \text{ lines)} \qquad \textbf{(21-4)}$$

If two lines are perpendicular, this means that there must be 90° between their inclinations (Fig. 21-10). The relation between their inclinations is

$$\alpha_2 = \alpha_1 + 90°$$

which can be written as

$$90° - \alpha_2 = -\alpha_1$$

If neither line is vertical (the slope of a vertical line is undefined) and we take the tangent in this last relation, we have

$$\tan(90° - \alpha_2) = \tan(-\alpha_1)$$

or

$$\cot \alpha_2 = -\tan \alpha_1$$

since a function of the complement of an angle equals the cofunction of that angle (see Section 4-4) and since $\tan(-\alpha) = -\tan \alpha$ (see Exercise 45 of Section 8-2). But $\cot \alpha = 1/\tan \alpha$, which means $1/\tan \alpha_2 = -\tan \alpha_1$. Using the inclination definition of slope, we have as *the relation between slopes of perpendicular lines*,

$$m_2 = -\frac{1}{m_1} \quad \text{or} \quad m_1 m_2 = -1 \qquad \text{(for } \perp \text{ lines)} \qquad \textbf{(21-5)}$$

■EXAMPLE 6 The line through $(3, -5)$ and $(2, -7)$ has a slope of

$$m_1 = \frac{-5 + 7}{3 - 2} = 2$$

The line through $(4, -6)$ and $(2, -5)$ has a slope of

$$m_2 = \frac{-6 - (-5)}{4 - 2} = -\frac{1}{2}$$

Since the slopes of the two lines are negative reciprocals, we know that the lines are perpendicular. See Fig. 21-11.

See Appendix C for a graphing calculator program SLOPEDIS. It calculates the slope of a line through two points and the distance between the points.

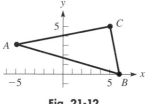

Fig. 21-12

Using the formulas for distance and slope, we can show certain basic geometric relationships. The following examples illustrate the use of the formulas and thereby show the use of algebra in solving problems that are basically geometric. This illustrates the methods of analytic geometry.

EXAMPLE 7   Show that the line segments joining $A(-5, 3)$, $B(6, 0)$, and $C(5, 5)$ form a right triangle. See Fig. 21-12.

If these points are vertices of a right triangle, the slopes of two of the sides must be negative reciprocals. This would show perpendicularity. Thus, we find the slopes of the three lines to be

$$m_{AB} = \frac{3 - 0}{-5 - 6} = -\frac{3}{11}, \qquad m_{AC} = \frac{3 - 5}{-5 - 5} = \frac{1}{5}, \qquad m_{BC} = \frac{0 - 5}{6 - 5} = -5$$

We see that the slopes of AC and BC are negative reciprocals, which means that $AC \perp BC$. From this we can conclude that the triangle is a right triangle. ∎

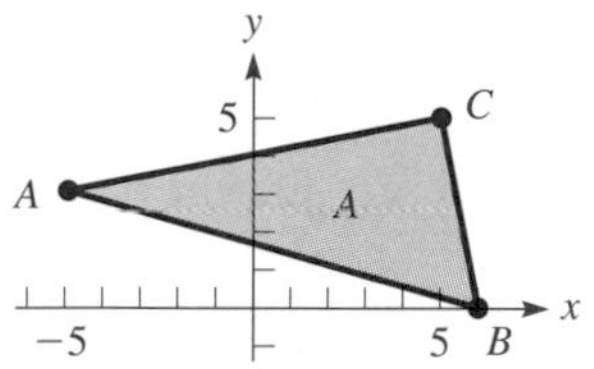

EXAMPLE 8   Find the area of the triangle in Example 7. See Fig. 21-13.

Since the right angle is at C, the legs of the triangle are AC and BC. The area is one-half the product of the lengths of the legs of a right triangle. The lengths of the legs are

$$d_{AC} = \sqrt{(-5 - 5)^2 + (3 - 5)^2} = \sqrt{104} = 2\sqrt{26}$$
$$d_{BC} = \sqrt{(6 - 5)^2 + (0 - 5)^2} = \sqrt{26}$$

Therefore, the area is

$$A = \frac{1}{2}(2\sqrt{26})(\sqrt{26}) = 26$$
∎

Fig. 21-13

EXERCISES *21-1*

In Exercises 1–10, find the distance between the given pairs of points.

1. $(3, 8)$ and $(-1, -2)$ **2.** $(-1, 3)$ and $(-8, -4)$

3. $(4, -5)$ and $(4, -8)$ **4.** $(-3, 7)$ and $(2, 10)$

5. $(-12, 20)$ and $(32, -13)$ **6.** $(23, -9)$ and $(-25, 11)$

7. $(\sqrt{32}, -\sqrt{18})$ and $(-\sqrt{50}, \sqrt{8})$

8. $(e, -\pi)$ and $(-2e, -\pi)$

9. $(1.22, -3.45)$ and $(-1.07, -5.16)$

10. $(-5.6, 2.3)$ and $(8.2, -7.5)$

In Exercises 11–20, find the slopes of the lines through the points in Exercises 1–10.

In Exercises 21–24, find the slopes of the lines with the given inclinations.

21. $30°$ **22.** $62.5°$ **23.** $132.7°$ **24.** $135°$

In Exercises 25–28, find the inclinations of the lines with the given slopes.

25. 0.364 **26.** 0.824 **27.** -6.691 **28.** -1.428

In Exercises 29–32, determine whether or not the lines through the two pairs of points are parallel or perpendicular.

29. $(6, -1)$ and $(4, 3)$; $(-5, 2)$ and $(-7, 6)$

30. $(-3, 9)$ and $(4, 4)$; $(9, -1)$ and $(4, -8)$

31. $(-1, -4)$ and $(2, 3)$; $(-5, 2)$ and $(-19, 8)$

32. $(-1, -2)$ and $(3, 6)$; $(2, -6)$ and $(5, 0)$

In Exercises 33–36, determine the value of k.

33. The distance between $(-1, 3)$ and $(11, k)$ is 13.

34. The distance between $(k, 0)$ and $(0, 2k)$ is 10.

35. Points $(6, -1)$, $(3, k)$, and $(-3, -7)$ are on the same line.

36. The points in Exercise 35 are the vertices of a right triangle, with the right angle at $(3, k)$.

In Exercises 37–40, show that the given points are vertices of the given geometric figures.

37. $(2, 3)$, $(4, 9)$, and $(-2, 7)$ are vertices of an isosceles triangle.

38. $(-1, 3)$, $(3, 5)$, and $(5, 1)$ are the vertices of a right triangle.

39. $(-5, -4)$, $(7, 1)$, $(10, 5)$ and $(-2, 0)$ are the vertices of a parallelogram.

40. $(-5, 6)$, $(0, 8)$, $(-3, 1)$, and $(2, 3)$ are the vertices of a square.

In Exercises 41–44, find the indicated areas and perimeters.

41. Find the area of the triangle in Exercise 38.

42. Find the area of the square in Exercise 40.

43. Find the perimeter of the triangle in Exercise 37.

44. Find the perimeter of the parallelogram in Exercise 39.

In Exercises 45–48, use the following definition to find the midpoints between the given points on a straight line.

The *midpoint* between points (x_1, y_1) and (x_2, y_2) on a straight line is the point

$$\left(\frac{x_1 + x_2}{2}, \frac{y_1 + y_2}{2} \right)$$

45. $(-4, 9)$ and $(6, 1)$ **46.** $(-1, 6)$ and $(-13, -8)$

47. $(-12.4, 25.7)$ and $(6.8, -17.3)$

48. $(2.6, 5.3)$ and $(-4.2, -2.7)$

21-2 THE STRAIGHT LINE

In Chapter 5 we derived the *slope-intercept form* of the equation of the straight line. Here we extend the development to include other forms of the equation of a straight line. Also, other methods of finding and applying these equations are shown. For completeness, we review some of the material in Chapter 5.

Using the definition of slope, we can derive the general type of equation that represents a straight line. This is another basic method of analytic geometry. That is, equations of a particular form can be shown to represent a particular type of curve. When we recognize the form of the equation, we know the kind of curve it represents. As we have seen, this is of great assistance in sketching the graph.

A straight line can be defined as a *curve* with a constant slope. This means that the value for the slope is the same for any two different points on the line that might be chosen. Thus, considering point (x_1, y_1) on a line to be fixed (Fig. 21-14), and another point $P(x, y)$ that *represents* any other point on the line, we have

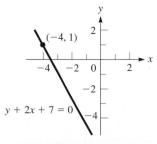

Fig. 21-14

$$m = \frac{y - y_1}{x - x_1}$$

which can be written as

$$\boxed{y - y_1 = m(x - x_1)} \qquad \textbf{(21-6)}$$

Equation (21-6) is the **point-slope form** *of the equation of a straight line.* It is useful when we know the slope of a line and some point through which the line passes.

■**EXAMPLE 1**   Find the equation of the line that passes through $(-4, 1)$ with a slope of -2. See Fig. 21-15.

Substituting in Eq. (21-6), we find that

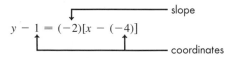

which can be simplified to

$$y + 2x + 7 = 0$$

Fig. 21-15

See Appendix C for a graphing calculator program GRAPHLIN. It displays the graph of a line through two given points.

█EXAMPLE 2   Find the equation of the line through $(2, -1)$ and $(6, 2)$.

We first find the slope of the line through these points:

$$m = \frac{2 + 1}{6 - 2} = \frac{3}{4}$$

Then, by using either of the two known points and Eq. (21-6), we can find the equation of the line:

$$y - (-1) = \frac{3}{4}(x - 2)$$

or

$$4y + 4 = 3x - 6$$

or

$$4y - 3x + 10 = 0$$

This line is shown in Fig. 21-16.

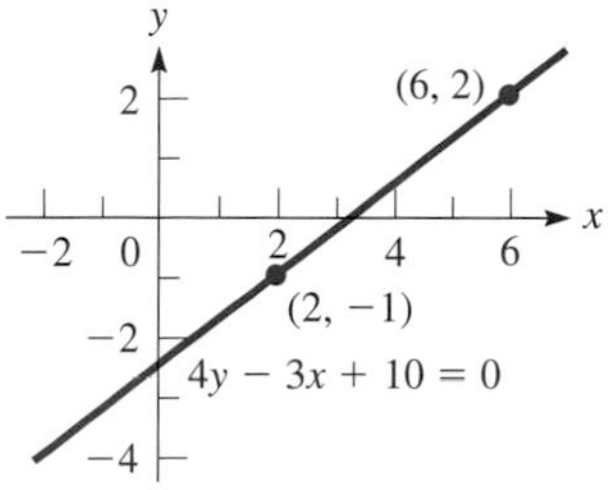

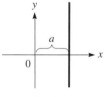

Fig. 21-16

Equation (21-6) can be used for any line except for one parallel to the y-axis. Such a line has an undefined slope. However, it does have the property that all points have the same x-coordinate, regardless of the y-coordinate. *We represent a line parallel to the y-axis (see Fig. 21-17) as*

$$\boxed{x = a} \tag{21-7}$$

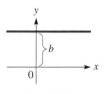

Fig. 21-17

A line parallel to the x-axis has a slope of zero. From Eq. (21-6), we can find its equation to be $y = y_1$. To keep the same form as Eq. (21-7), we normally write this as

$$\boxed{y = b} \tag{21-8}$$

Fig. 21-18

See Fig. 21-18.

█EXAMPLE 3   **(a)** The line $x = 2$ is a line parallel to the y-axis and 2 units to the right of it. This line is shown in Fig. 21-19.

Fig. 21-19   **Fig. 21-20**

(b) The line $y = -4$ is a line parallel to the x-axis and 4 units below it. This line is shown in Fig. 21-20.

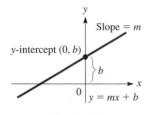

Fig. 21-21

If we choose the special point $(0, b)$, which is the y-intercept of the line, as the point to use in Eq. (21-6), we have $y - b = m(x - 0)$, or

$$\boxed{y = mx + b} \qquad (21\text{-}9)$$

Equation (21-9) is the **slope-intercept form** *of the equation of a straight line,* and we first derived it in Chapter 5. Its primary usefulness lies in the fact that once we find the equation of a line and then write it in slope-intercept form, we know that the slope of the line is the coefficient of the x-term and that it crosses the y-axis at the coordinate indicated by the constant term. See Fig. 21-21.

▌**EXAMPLE 4**   Find the slope and the y-intercept of the straight line whose equation is $2y + 4x - 5 = 0$.

We write this equation in slope-intercept form:

$$2y = -4x + 5$$

slope ⌐   ⌐ y-coordinate of intercept

$$y = -2x + \frac{5}{2}$$

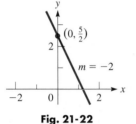

Fig. 21-22

Since the coefficient of x in this form is -2, the slope is -2. The constant on the right is $5/2$, which means that the y-intercept is $(0, 5/2)$. See Fig. 21-22.   ▪

SOLVING A WORD PROBLEM

▌**EXAMPLE 5**   The pressure p_0 at the surface of a body of water (due to the atmosphere) is 101 kPa. The pressure p at a depth of 10.0 m is 199 kPa. In general, the pressure difference $p - p_0$ varies directly as the depth h. Sketch a graph of p as a function of h.

The solution is as follows:

$$
\begin{array}{ll}
p - p_0 = kh & \text{direct variation} \\
199 - 101 = k(10.0) & \text{substitute given values} \\
k = 9.80 \text{ kPa/m} & \\
p - 101 = 9.80h & \text{substitute in first equation} \\
p = 9.80h + 101 &
\end{array}
$$

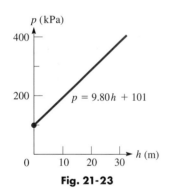

Fig. 21-23

We see that this is the equation of a straight line. The slope is 9.80, and the p-intercept is $(0, 101)$. Negative values do not have any physical meaning. The graph is shown in Fig. 21-23.   ▪

From Eqs. (21-6) and (21-9) and from the examples of this section, we see that the equation of the straight line has certain characteristics: We have a term in y, a term in x, and a constant term if we simplify as much as possible. *This form is represented by the equation*

$$\boxed{Ax + By + C = 0} \qquad (21\text{-}10)$$

which is known as the **general form** *of the equation of the straight line*. We saw this form before in Chapter 5. Now we have shown why it represents a straight line.

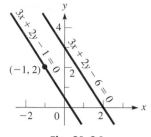

Fig. 21-24

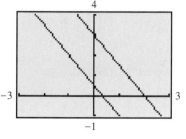

Fig. 21-25

SOLVING A WORD PROBLEM

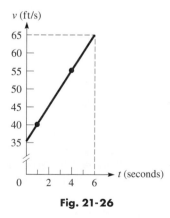

Fig. 21-26

█**EXAMPLE 6** Find the general form of the equation of the line parallel to the line $3x + 2y - 6 = 0$ and that passes through the point $(-1, 2)$.

Since the line whose equation we want is parallel to the line $3x + 2y - 6 = 0$, it has the same slope. Thus, writing $3x + 2y - 6 = 0$ in slope-intercept form,

$$2y = -3x + 6 \qquad \text{solving for } y$$

$$y = -\frac{3}{2}x + 3$$

Since the slope of $3x + 2y - 6 = 0$ is $-3/2$, the slope of the required line is also $-3/2$. Using $m = -3/2$, the point $(-1, 2)$, and the point-slope form, we have

$$y - 2 = -\frac{3}{2}(x + 1)$$

$$2y - 4 = -3(x + 1)$$

$$3x + 2y - 1 = 0$$

This is the general form of the equation. Both lines are shown in Fig. 21-24.

With $y_1 = -3x/2 + 3$ and $y_2 = -3x/2 + 1/2$, these lines are shown in the calculator display in Fig. 21-25. --------━

In many physical situations, a linear relationship exists between variables. A few examples of this are (1) the distance traveled by an object and the elapsed time, when the velocity is constant, (2) the amount a spring stretches and the force applied, (3) the change in electric resistance and the change in temperature, (4) the force applied to an object and the resulting acceleration, and (5) the pressure at a certain point within a liquid and the depth of the point.

█**EXAMPLE 7** For a period of 6.0 s, the velocity v of a car varies linearly with the elapsed time t. If $v = 40$ ft/s when $t = 1.0$ s and $v = 55$ ft/s when $t = 4.0$ s, find the equation relating v and t and graph the function. From the graph, find the initial velocity and the velocity after 6.0 s. What is the meaning of the slope of the line?

With v as the dependent variable and t as the independent variable, the slope is

$$m = \frac{v_2 - v_1}{t_2 - t_1}$$

Using the information given in the statement of the problem, we have

$$m = \frac{55 - 40}{4.0 - 1.0} = 5.0$$

Then, using the point-slope form of the equation of a straight line, we have

$$v - 40 = 5.0(t - 1.0)$$

$$v = 5.0t + 35$$

The given values are sufficient to graph the line in Fig. 21-26. There is no need to include negative values of t, since they have no physical meaning. We see that the line crosses the v-axis at 35. This means that the initial velocity (for $t = 0$) is 35 ft/s. Also, when $t = 6.0$ s, we see that $v = 65$ ft/s.

The slope is the ratio of the change in velocity to the change in time. This is the car's *acceleration*. Here the speed of the car increases 5.0 ft/s each second. We can express this acceleration as 5.0 (ft/s)/s = 5.0 ft/s^2. --------━

─── **EXERCISES** *21-2* ───

In Exercises 1–20, find the equation of each of the lines with the given properties. Sketch the graph of each line.

1. Passes through $(-3, 8)$ with a slope of 4.

2. Passes through $(-2, -1)$ with a slope of -2.

3. Passes through $(2, -5)$ and $(4, 2)$.

4. Passes through $(-3, 5)$ and $(-2, 3)$.

5. Passes through $(1, 3)$ and has an inclination of $45°$.

6. Has a y-intercept $(0, -2)$ and an inclination of $120°$.

7. Passes through $(5.3, -2.7)$ and is parallel to the x-axis.

8. Passes through $(-4, -2)$ and is perpendicular to x-axis.

9. Is parallel to the y-axis and is 3 units to the left of it.

10. Is parallel to the x-axis and is 4.1 units below it.

11. Has an x-intercept $(4, 0)$ and a y-intercept of $(0, -6)$.

12. Has an x-intercept of $(-3, 0)$ and a slope of 2.

13. Is perpendicular to a line with a slope of 3 and passes through $(1, -2)$.

14. Is perpendicular to a line with a slope of -4 and has a y-intercept of $(0, 3)$.

15. Is parallel to a line through $(-1, 2)$ and $(3, 1)$ and passes through $(1, 2)$.

16. Is parallel to a line through $(7, -1)$ and $(4, 3)$ and has a y-intercept of $(0, -2)$.

17. Is perpendicular to the line $6.0x - 2.4y - 3.9 = 0$ and passes through $(7.5, -4.7)$.

18. Is parallel to the line $2y - 6x - 5 = 0$ and passes through $(-4, -5)$.

19. Has a slope of -3 and passes through the intersection of the lines $5x - y = 6$ and $x + y = 12$.

20. Passes through the point of intersection of $2x + y - 3 = 0$ and $x - y - 3 = 0$ and through the point $(4, -3)$.

In Exercises 21–28, reduce the equations to slope-intercept form and find the slope and the y-intercept. Sketch each line.

21. $4x - y = 8$

22. $2x - 3y - 6 = 0$

23. $3x + 5y - 10 = 0$

24. $4y = 6x - 9$

25. $3x - 2y - 1 = 0$

26. $4x + 2y - 5 = 0$

27. $11.2x + 1.6 = 3.2y$

28. $11.5x + 4.60y = 5.98$

In Exercises 29–32, determine the value of k.

29. Find k if the lines $4x - ky = 6$ and $6x + 3y + 2 = 0$ are parallel.

30. Find k if the lines given in Exercise 29 are perpendicular.

(W) 31. Find k if the lines $3x - y = 9$ and $kx + 3y = 5$ are perpendicular. Explain how this value is found.

(W) 32. Find k if the lines given in Exercise 31 are parallel. Explain how this value is found.

In Exercises 33–40, determine whether the given lines are parallel, perpendicular, or neither.

33. $3x - 2y + 5 = 0$ and $4y = 6x - 1$

34. $8x - 4y + 1 = 0$ and $4x + 2y - 3 = 0$

35. $6x - 3y - 2 = 0$ and $x + 2y - 4 = 0$

36. $3y - 2x = 4$ and $6x - 9y = 5$

37. $5x + 2y - 3 = 0$ and $10y = 7 - 4x$

38. $48y - 36x = 71$ and $52x = 17 - 39y$

39. $4.5x - 1.8y = 1.7$ and $2.4x + 6.0y = 0.3$

40. $3.5y = 4.3 - 1.5x$ and $3.6x + 8.4y = 1.7$

In Exercises 41–52, some applications involving straight lines are shown.

41. The velocity v of a box sliding down a long ramp is given by $v = v_0 + at$, where v_0 is the initial velocity, a is the acceleration, and t is the time. If $v_0 = 12.2$ ft/s and $v = 35.4$ ft/s when $t = 4.50$ s, find v as a function of t. Sketch the graph.

42. The voltage V across part of an electric circuit is given by $V = E - iR$, where E is a battery voltage, i is the current, and R is the resistance. If $E = 6.00$ V and $V = 4.35$ V for $i = 9.17$ mA, find V as a function of i. Sketch the graph (i and V may be negative).

43. The velocity of sound v increases 0.607 m/s for each increase in temperature T of $1.00°C$. If $v = 343$ m/s for $T = 20.0°C$, express v as a function of T.

44. An acid solution is made from x liters of a 20% solution and y liters of a 30% solution. If the final solution contains 20 L of acid, find the equation relating x and y.

45. One computer printer can print x characters per second, and a second printer can print y characters per second. If the first prints for 50 s and the second for 60 s and they print a total of 12,200 characters, find the equation relating x and y.

(W) 46. A wall is 15 cm thick. At the outside, the temperature is $3°C$, and at the inside, it is $23°C$. If the temperature changes at a constant rate through the wall, write an equation of the temperature T in the wall as a function of the distance x from the outside to the inside of the wall. What is the meaning of the slope of the line?

47. One heating unit uses x gallons of fuel at 72% efficiency, and another heating unit uses y gallons at 90% efficiency. If 135,000 Btu of heat is delivered by these units together, express y as a function of x.

(W) 48. The length of a rectangular solar cell is 10 cm more than the width w. Express the perimeter p of the cell as a function of w. What is the meaning of the slope of the line?

49. A light beam is reflected off the edge of an optic fiber at an angle of 0.0032°. The diameter of the fiber is 48 μm. Find the equation of the reflected beam with the x-axis (at the center of the fiber) and the y-axis as shown in Fig. 21-27.

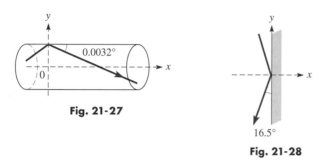

Fig. 21-27

50. A police report stated that a bullet caromed downward off a wall at an angle of 16.5° with the wall, as shown in Fig. 21-28. What is the equation of the path of the bullet after impact?

(W) **51.** A survey of the traffic on a particular highway showed that the number of cars passing a particular point each minute varied linearly from 6:30 A.M. to 8:30 A.M. on workday mornings. The study showed that an average of 45 cars passed the point in one minute at 7 A.M. and that 115 cars passed in one minute at 8 A.M. If n is the number of cars passing the point in one minute and t is the number of minutes after 6:30 A.M., find the equation relating n and t, and graph the equation. From the graph, determine n at 6:30 A.M. and at 8:30 A.M. What is the meaning of the slope of the line?

52. In a research project on cancer, a tumor was determined to weigh 30 mg when first discovered. While being treated, it grew smaller by 2 mg each month. Find the equation relating the weight w of the tumor as a function of the time t in months. Graph the equation.

In Exercises 53–56, treat the given nonlinear functions as linear functions in order to sketch their graphs. At times, this can be useful in showing certain values of a function. For example, $y = 2 + 3x^2$ can be shown as a straight line by graphing y as a function of x^2. A table of values for this graph is shown along with the corresponding graph in Fig. 21-29.

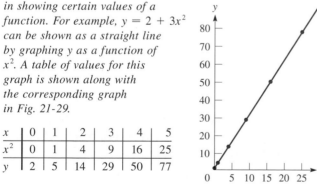

x	0	1	2	3	4	5
x^2	0	1	4	9	16	25
y	2	5	14	29	50	77

Fig. 21-29

53. The number n of memory cells of a certain computer that can be tested in t seconds is given by $n = 1200\sqrt{t}$. Sketch n as a function of $\sqrt{t}$.

54. The force F (in lb) applied to a lever to balance a certain weight on the opposite side of the fulcrum is given by $F = 40/d$, where d is the distance (in ft) of the force from the fulcrum. Sketch F as a function of $1/d$.

55. A spacecraft is launched such that its altitude h (in km) is given by $h = 300 + 2t^{3/2}$ for $0 \le t < 100$ s. Sketch this as a linear function.

56. The current i (in A) in a certain electric circuit is given by $i = 6(1 - e^{-t})$. Sketch this as a linear function.

In Exercises 57–60, show that the given nonlinear functions are linear when plotted on semilogarithmic or logarithmic paper. In Section 13-7, we noted that graphs on this paper often become straight lines.

57. A function of the form $y = ax^n$ is straight when plotted on logarithmic paper, since $\log y = \log a + n \log x$ is in the form of a straight line. The variables are $\log y$ and $\log x$; the slope can be found from $(\log y - \log a)/\log x = n$, and the intercept is a. (To get the slope from the graph, it is necessary to measure vertical and horizontal distances between two points. The log y-intercept is found where $\log x = 0$, and this occurs when $x = 1$.) Plot $y = 3x^4$ on logarithmic paper to verify this analysis.

58. A function of the form $y = a(b^x)$ is a straight line on semilogarithmic paper, since $\log y = \log a + x \log b$ is in the form of a straight line. The variables are $\log y$ and x, the slope is $\log b$, and the intercept is a. (To get the slope from the graph, we calculate $(\log y - \log a)/x$ for some set of values x and y. The intercept is read directly off the graph where $x = 0$.) Plot $y = 3(2^x)$ on semilogarithmic paper to verify this analysis.

59. If experimental data are plotted on logarithmic paper and the points lie on a straight line, it is possible to determine the function (see Exercise 57). The following data come from an experiment to determine the functional relationship between the pressure p and the volume V of a gas undergoing an adiabatic (no heat loss) change. From the graph on logarithmic paper, determine p as a function of V.

V (m³)	0.100	0.500	2.00	5.00	10.0
p (kPa)	20.1	2.11	0.303	0.0840	0.0318

60. If experimental data are plotted on semilogarithmic paper, and the points lie on a straight line, it is possible to determine the function (see Exercise 58). The following data come from an experiment designed to determine the relationship between the voltage across an inductor and the time, after the switch is opened. Determine v as a function of t.

v (V)	40	15	5.6	2.2	0.8
t (ms)	0.0	20	40	60	80

21-3 THE CIRCLE

We have found that we can obtain a general equation that represents a straight line by considering a fixed point on the line and then a general point $P(x, y)$ which can represent any other point on the same line. Mathematically, we can state this as "the line is the **locus** of a point $P(x, y)$ that *moves* from a fixed point with constant slope along the line." That is, the point $P(x, y)$ can be considered as a variable point that moves along the line.

In this way we can define a number of important curves. *A* **circle** *is defined as the locus of a point $P(x, y)$ that moves so that it is always equidistant from a fixed point. We call this fixed distance the* **radius,** *and we call the fixed point the* **center** *of the circle.* Thus, using this definition, calling the fixed point (h, k) and the radius r, we have

$$\sqrt{(x - h)^2 + (y - k)^2} = r$$

or, by squaring both sides, we have

$$(x - h)^2 + (y - k)^2 = r^2$$ **(21-11)**

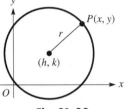

Fig. 21-30

Equation (21-11) is called the **standard equation** *of a circle with center at (h, k) and radius r.* See Fig. 21-30.

EXAMPLE 1   The equation $(x - 1)^2 + (y + 2)^2 = 16$ represents a circle with center at $(1, -2)$ and a radius of 4. We determine these values by considering the equation of the circle to be in the form of Eq. (21-11) as

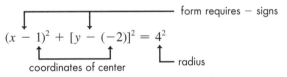

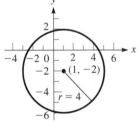

Fig. 21-31

Note carefully the way in which we found the y-coordinate of the center. *We must have a minus sign before each of the coordinates.* Here, to get the y-coordinate, we had to write $+2$ as $-(-2)$. This circle is shown in Fig. 21-31.

EXAMPLE 2   Find the equation of the circle with center at $(2, 1)$ and that passes through $(4, 8)$.

In Eq. (21-11) we can determine the equation if we can find h, k, and r for this circle. From the given information, $h = 2$ and $k = 1$. To find r, we use the fact that *all points on the circle must satisfy the equation of the circle.* The point $(4, 8)$ must satisfy Eq. (21-11), with $h = 2$ and $k = 1$. Thus,

$$(4 - 2)^2 + (8 - 1)^2 = r^2 \quad \text{or} \quad r^2 = 53$$

Therefore, the equation of the circle is

$$(x - 2)^2 + (y - 1)^2 = 53$$

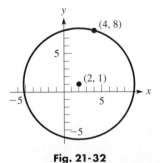

Fig. 21-32

This circle is shown in Fig. 21-32.

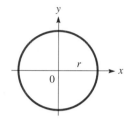

Fig. 21-33

SOLVING A WORD PROBLEM

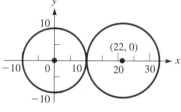

Fig. 21-34

If the center of the circle is at the origin, which means that the coordinates of the center are $(0, 0)$, the equation of the circle (see Fig. 21-33) becomes

$$x^2 + y^2 = r^2$$ **(21-12)**

The following example illustrates an application using this type of circle and one with its center not at the origin.

▌**EXAMPLE 3**   A student is drawing a friction drive in which two circular disks are in contact with each other. They are represented by circles in the drawing. The first has a radius of 10.0 cm, and the second has a radius of 12.0 cm. What is the equation of each circle if the origin is at the center of the first circle and the positive x-axis passes through the center of the second circle? See Fig. 21-34.

Since the center of the smaller circle is at the origin, we can use Eq. (21-12). Given that the radius is 10.0 cm, we have as its equation

$$x^2 + y^2 = 100$$

The fact that the two disks are in contact tells us that they meet at the point $(10.0, 0)$. Knowing that the radius of the larger circle is 12.0 cm tells us that its center is at $(22.0, 0)$. Thus, using Eq. (21-11) with $h = 22.0$, $k = 0$, and $r = 12.0$,

$$(x - 22.0)^2 + (y - 0)^2 = 12.0^2$$

or

$$(x - 22.0)^2 + y^2 = 144$$

as the equation of the larger circle. ▬

SYMMETRY

NOTE▶

A circle with its center at the origin exhibits an important property of the graphs of many equations. *It is* **symmetrical** *to the x-axis and also to the y-axis.* Symmetry to the x-axis can be thought of as meaning that the lower half of the curve is a reflection of the upper half, and conversely. It can be shown that *if $-y$ can replace y in an equation without changing the equation, the graph of the equation is* **symmetrical to the x-axis.** Symmetry to the y-axis is similar. *If $-x$ can replace x in the equation without changing the equation, the graph is symmetrical to the y-axis.*

NOTE▶

This type of circle is also symmetrical to the origin as well as being symmetrical to both axes. The meaning of symmetry to the origin is that the origin is the midpoint of any two points (x, y) and $(-x, -y)$ that are on the curve. Thus, *if $-x$ can replace x and $-y$ can replace y at the same time, without changing the equation, the graph of the equation is symmetrical to the origin.*

▌**EXAMPLE 4**   The equation of the circle with its center at the origin and with a radius of 6 is $x^2 + y^2 = 36$.

The symmetry of this circle can be shown analytically by the substitutions mentioned above. Replacing x by $-x$, we obtain $(-x)^2 + y^2 = 36$. Since $(-x)^2 = x^2$, this equation can be rewritten as $x^2 + y^2 = 36$. Since this substitution did not change the equation, the graph is symmetrical to the y-axis.

Replacing y by $-y$, we obtain $x^2 + (-y)^2 = 36$, which is the same as $x^2 + y^2 = 36$. This means that the curve is symmetrical to the x-axis.

Replacing x by $-x$ and simultaneously replacing y by $-y$, we obtain $(-x)^2 + (-y)^2 = 36$, which is the same as $x^2 + y^2 = 36$. This means that the curve is symmetrical to the origin. This circle is shown in Fig. 21-35. ▬

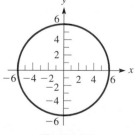

Fig. 21-35

If we multiply out each of the terms in Eq. (21-11), we may combine the resulting terms to obtain

$$x^2 - 2hx + h^2 + y^2 - 2ky + k^2 = r^2$$

$$\boldsymbol{x^2 + y^2 - 2hx - 2ky + (h^2 + k^2 - r^2) = 0} \qquad \textbf{(21-13)}$$

Since each of h, k, and r is constant for any given circle, the coefficients of x and y and the term within parentheses in Eq. (21-13) are constants. Equation (21-13) can then be written as

$$\boxed{x^2 + y^2 + Dx + Ey + F = 0} \qquad \textbf{(21-14)}$$

Equation (21-14) is called the **general equation** *of the circle.* It tells us that any equation which can be written in that form will represent a circle.

■EXAMPLE 5 Find the center and radius of the circle

$$x^2 + y^2 - 6x + 8y - 24 = 0$$

CAUTION ▶

We can find this information if we write the given equation in standard form. To do so, **we must complete the square in the x-terms and also in the y-terms**. This is done by first writing the equation in the form

$$(x^2 - 6x \quad) + (y^2 + 8y \quad) = 24$$

To complete the square of the x-terms, we take half of -6, which is -3, square it, and add the result, 9, to each side of the equation. In the same way, we complete the square of the y-terms by adding 16 to each side of the equation, which gives

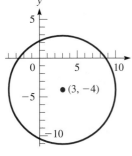

y
5

0   5   10   x
-5   ●(3, −4)
−10

Fig. 21-36

$$\overset{\left(\frac{-6}{2}\right)^2}{} \qquad \overset{\left(\frac{8}{2}\right)^2}{} \qquad \text{add to both sides}$$
$$(x^2 - 6x + 9) + (y^2 + 8y + 16) = 24 + 9 + 16$$
$$(x - 3)^2 + (y + 4)^2 = 49$$
$$(x - 3)^2 + (y - (-4))^2 = 7^2$$

coordinates of center   radius

Thus, the center is $(3, -4)$, and the radius is 7 (see Fig. 21-36). ▬

SOLVING A WORD PROBLEM

■EXAMPLE 6 A certain pendulum is found to swing through an arc of the circle $3x^2 + 3y^2 - 9.60y - 2.80 = 0$. What is the length (in m) of the pendulum, and from what point is it swinging?

We see that this equation represents a circle by dividing through by 3. This gives us $x^2 + y^2 - 3.20y - 2.80/3 = 0$. The length of the pendulum is the radius of the circle, and the point from which it swings is the center. These are found as follows:

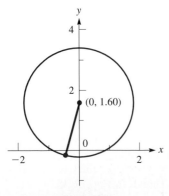

y
4

2
●(0, 1.60)
0
−2   2   x

Fig. 21-37

$$x^2 + (y^2 - 3.20y + 1.60^2) = 1.60^2 + 2.80/3 \qquad \text{complete squares in both } x\text{- and } y\text{-terms}$$
$$x^2 + (y - 1.60)^2 = 3.493 \qquad \text{standard form}$$

Since $\sqrt{3.493} = 1.87$, the length of the pendulum is 1.87 m. The point from which it is swinging is $(0, 1.60)$. See Fig. 21-37.

Replacing x by $-x$, the equation does not change. Replacing y by $-y$, the equation does change (the 3.20y term changes sign). Thus, the circle is symmetric only to the y-axis. ▬

NOTE ▶

In Section 14-1 we noted that the equation of a circle does not represent a *function* since there are two values of *y* for most values of *x* in the domain. In fact, *it might be necessary to use the quadratic formula to find the two functions to enter into a graphing calculator* in order to view the curve. This is illustrated in the following example.

▌EXAMPLE 7   Display the graph of the circle $3x^2 + 3y^2 + 6y - 20 = 0$ on a graphing calculator.

To fit the form of a quadratic equation in *y*, we write

$$3y^2 + 6y + (3x^2 - 20) = 0$$

Now, using the quadratic formula to solve for *y*, we let

$$a = 3 \qquad b = 6 \qquad c = 3x^2 - 20$$

Therefore,

$$y = \frac{-6 \pm \sqrt{6^2 - 4(3)(3x^2 - 20)}}{2(3)}$$

which means we get the *two functions*

$$y_1 = \frac{-6 + \sqrt{276 - 36x^2}}{6} \quad \text{and} \quad y_2 = \frac{-6 - \sqrt{276 - 36x^2}}{6}$$

which are entered into the calculator to get the view shown in Fig. 21-38.

The *window* values were chosen so that the length along the *x*-axis is about 1.5 times that along the *y*-axis so as to have less distortion in the circle. There may be gaps at the left and right sides of the circle.   --------▬▬

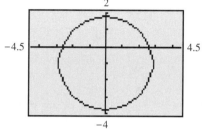

Fig. 21-38

EXERCISES *21-3*

In Exercises 1–4, determine the center and the radius of each circle.

1. $(x - 2)^2 + (y - 1)^2 = 25$

2. $(x - 3)^2 + (y + 4)^2 = 49$

3. $(x + 1)^2 + y^2 = 4$ **4.** $x^2 + (y - 6)^2 = 64$

In Exercises 5–20, find the equation of each of the circles from the given information.

5. Center at $(0, 0)$, radius 3 **6.** Center at $(0, 0)$, radius 1

7. Center at $(2, 2)$, radius 4 **8.** Center at $(0, 2)$, radius 2

9. Center at $(-2, 5)$, radius $\sqrt{5}$

10. Center at $(-3, -5)$, radius $2\sqrt{3}$

11. Center at $(12, -15)$, radius 18

12. Center at $(\frac{3}{2}, -2)$, radius $\frac{5}{2}$

13. Center at $(2, 1)$, passes through $(4, -1)$

14. Center at $(-1, 4)$, passes through $(-2, 3)$

15. Center at $(-3, 5)$, tangent to the *x*-axis

16. Center at $(2, -4)$, tangent to the *y*-axis

17. Tangent to both axes and the lines $y = 4$ and $x = 4$

18. Tangent to both axes, radius 4, in the second quadrant

19. Center on the line $5x = 2y$, radius 5, tangent to the *x*-axis

20. The points $(3, 8)$ and $(-3, 0)$ are the ends of a diameter.

In Exercises 21–32, determine the center and radius of each circle. Sketch each circle.

21. $x^2 + (y - 3)^2 = 4$

22. $(x - 2)^2 + (y + 3)^2 = 49$

23. $4(x + 1)^2 + 4(y - 5)^2 = 81$

24. $2(x + 4)^2 + 2(y + 3)^2 = 25$

25. $x^2 + y^2 - 2x - 8 = 0$

26. $x^2 + y^2 - 4x - 6y - 12 = 0$

27. $x^2 + y^2 + 4.20x - 2.60y = 3.51$

28. $x^2 + y^2 + 22x + 14y = 26$

29. $4x^2 + 4y^2 - 16y = 9$

30. $9x^2 + 9y^2 + 18y = 7$

31. $2x^2 + 2y^2 - 4x - 8y - 1 = 0$

32. $3x^2 + 3y^2 - 12x + 4 = 0$

In Exercises 33–36, determine whether the circles with the given equations are symmetrical to either axis or to the origin.

33. $x^2 + y^2 = 100$

34. $x^2 + y^2 - 4x - 5 = 0$

35. $3x^2 + 3y^2 + 24y = 8$

36. $5x^2 + 5y^2 - 10x + 20y = 3$

In Exercises 37–48, solve the given problems.

37. Determine whether the circle $x^2 - 6x + y^2 - 7 = 0$ crosses the *x*-axis.

38. Find the points of intersection of the circle $x^2 + y^2 - x - 3y = 0$ and the line $y = x - 1$.

(W) 39. Find the locus of a point $P(x, y)$ that moves so that its distance from (2, 4) is twice its distance from (0, 0). Describe the locus.

(W) 40. Find the equation of the locus of a point $P(x, y)$ that moves so that the line joining it and (2, 0) is always perpendicular to the line joining it and $(-2, 0)$. Describe the locus.

41. Use a graphing calculator to view the circle $x^2 + y^2 + 5y - 4 = 0$.

42. Use a graphing calculator to view the circle $2x^2 + 2y^2 + 2y - x - 1 = 0$.

43. In a hoisting device, two of the pulley wheels may be represented by $x^2 + y^2 = 14.5$ and $x^2 + y^2 - 19.6y + 86.0 = 0$. How far apart (in in.) are the wheels?

44. The design of a machine part shows it as a circle represented by the equation $x^2 + y^2 = 42.5$, with a circular hole represented by $x^2 + y^2 + 3.06y - 1.24 = 0$ cut out. What is the least distance (in in.) from the edge of the hole to the edge of the machine part?

45. A wire is rotating in a circular path through a magnetic field to induce an electric current in the wire. The wire is rotating at 60.0 Hz with a constant velocity of 37.7 m/s. Taking the origin at the center of the circle of rotation, find the equation of the path of the wire.

46. A communications satellite remains stationary at an altitude of 22,500 mi over a point on the earth's equator. It therefore rotates once each day about the earth's center. Its velocity is constant, but the horizontal and vertical components, v_H and v_V, of the velocity constantly change. Show that the equation relating v_H and v_V (in mi/h) is that of a circle. The radius of the earth is 3960 mi.

47. In analyzing the stress on a beam, *Mohr's circle* is often used. To form it, normal stress is plotted as the *x*-coordinate and shear stress is plotted as the *y*-coordinate. The center of the circle is midway between the minimum and maximum values of normal stress on the *x*-axis. Find the equation of Mohr's circle if the minimum normal stress is 100×10^{-6} and the maximum normal stress is 900×10^{-6} (stress is unitless). Sketch the graph.

48. A Norman window has the form of a rectangle surmounted by a semicircle. An architect designs a Norman window on a coordinate system as shown in Fig. 21-39. If the circumference of the circular part of the window is on the circle $x^2 + y^2 - 3.00y + 1.25 = 0$, find the area of the window. Measurements are in meters.

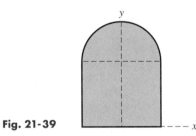

Fig. 21-39

21-4 THE PARABOLA

Another important curve is the parabola. We came across this curve several times in earlier chapters. In Chapter 7 we showed that the graph of a quadratic function is a parabola. In this section we define the parabola more generally and thereby find the general form of its equation.

A **parabola** *is defined as the locus of a point $P(x, y)$ that moves so that it is always equidistant from a given line and a given point. The given line is called the* **directrix,** *and the given point is called the* **focus.** The line through the focus that is perpendicular to the directrix is called the **axis** of the parabola. The point midway between the directrix and focus is the **vertex** of the parabola. Using this definition, we shall find the equation of the parabola for which the focus is the point $(p, 0)$ and the directrix is the line $x = -p$. By choosing the focus and directrix in this manner, we can find a general representation of the equation of a parabola with its vertex at the origin.

According to the definition of the parabola, the distance from a point $P(x, y)$ on the parabola to the focus $(p, 0)$ must equal the distance from $P(x, y)$ to the directrix $x = -p$. The distance from P to the focus can be found by using the distance formula. The distance from P to the directrix is the perpendicular distance, and this can be found as the distance between two points on a line parallel to the x-axis. These distances are indicated in Fig. 21-40.

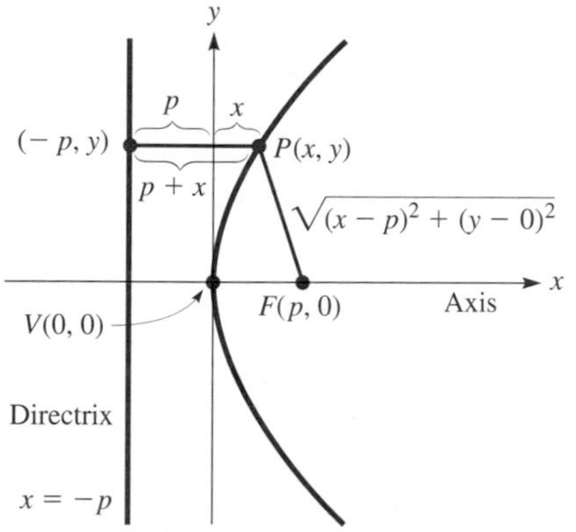

Thus, we have

$$\sqrt{(x - p)^2 + (y - 0)^2} = x + p$$

Squaring both sides of this equation, we have

$$(x - p)^2 + y^2 = (x + p)^2$$

or

$$x^2 - 2px + p^2 + y^2 = x^2 + 2px + p^2$$

Simplifying, we obtain

$$\boxed{y^2 = 4px}$$

(21-15)

Fig. 21-40

Equation (21-15) is called the **standard form** *of the equation of a parabola with its axis along the x-axis and the vertex at the origin.* Its symmetry to the x-axis can be proven since $(-y)^2 = 4px$ is the same as $y^2 = 4px$.

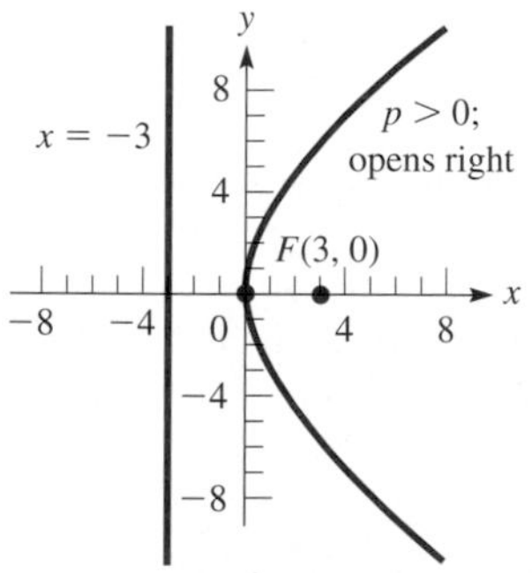

■**EXAMPLE 1** Find the coordinates of the focus and the equation of the directrix and sketch the graph of the parabola $y^2 = 12x$.

Since the equation of this parabola fits the form of Eq. (21-15), we know that the vertex is at the origin. The coefficient of 12 tells us that

$$4p = 12, \qquad p = 3$$

Since $p = 3$, the focus is the point $(3, 0)$, and the directrix is the line $x = -3$, as shown in Fig. 21-41. ------------ ∎

Fig. 21-41

■**EXAMPLE 2** If the focus is to the left of the origin, with the directrix an equal distance to the right, the coefficient of the x-term is negative. This tells us that the parabola opens to the left, rather than to the right, as is the case when the focus is to the right of the origin. For example, the parabola $y^2 = -8x$ has its vertex at the origin, its focus at $(-2, 0)$, and the line $x = 2$ as its directrix. We determine this from the equation as follows:

$$y^2 = -8x \qquad 4p = -8, \qquad p = -2$$

Since $p = -2$, we find

the focus is $(-2, 0)$
the directrix is the line $x = -(-2)$, or $x = 2$

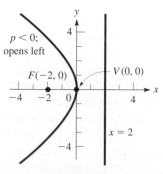

Fig. 21-42

The parabola opens to the left, as shown in Fig. 21-42. ------------ ∎

Fig. 21-43

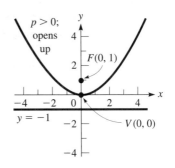

Fig. 21-44

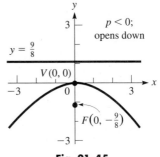

Fig. 21-45

See the chapter introduction.

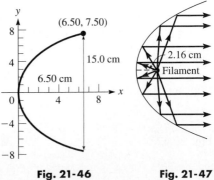

Fig. 21-46   **Fig. 21-47**

If we chose the focus as the point $(0, p)$ and the directrix as the line $y = -p$ (see Fig. 21-43), we would find that the resulting equation is

$$x^2 = 4py \qquad \text{(21-16)}$$

This is the standard form of the equation of a parabola with the y-axis as its axis and the vertex at the origin. Its symmetry to the y-axis can be proved, since $(-x)^2 = 4py$ is the same as $x^2 = 4py$. We note that **the difference between this equation and Eq. (21-15) is that x is squared and y appears to the first power in Eq. (21-16), rather than the reverse, as in Eq. (21-15).**

■**EXAMPLE 3**   The parabola $x^2 = 4y$ fits the form of Eq. (21-16). Therefore, its axis is along the y-axis and its vertex is at the origin. From the equation, we find the value of p, which in turn tells us the location of the vertex and the directrix.

$$x^2 = 4y \qquad 4p = 4, \qquad p = 1$$

Focus $(0, p)$ is $(0, 1)$; directrix $y = -p$ is $y = -1$. The parabola is shown in Fig. 21-44, and we see in this case that it opens upward.   --------■

■**EXAMPLE 4**   The parabola $2x^2 = -9y$ fits the form of Eq. (21-16) if we write it in the form

$$x^2 = -\frac{9}{2}y$$

Here, we see that $4p = -9/2$. Therefore, its axis is along the y-axis, and its vertex is at the origin. Since $4p = -9/2$, we have

$$p = -\frac{9}{8} \qquad \text{focus}\left(0, -\frac{9}{8}\right) \qquad \text{directrix } y = \frac{9}{8}$$

The parabola opens downward, as shown in Fig. 21-45.   --------■

■**EXAMPLE 5**   In calculus it can be shown that a light ray coming from the focus of a parabola will be reflected off the parabolic surface parallel to the axis of the parabola. This property of a parabola involving light reflection has many very useful applications. One of these is in the design of automobile headlights.

An automobile headlight reflector is designed with cross sections of the reflecting surface being equal parabolas. It has an opening of 15.0 cm and is 6.50 cm deep, as shown in Fig. 21-46. Determine where the filament of the bulb should be located so that reflected light rays are parallel in order to create a light beam.

By finding the equation of the parabola we can find the location of the focus, which is the desired location of the filament. Since the parabolic opening is 15.0 cm wide and 6.50 cm deep, the point $(6.50, 7.50)$ is on the parabola.

Since we have placed the parabola with its vertex at the origin and its axis along the x-axis, its general form is given by Eq. (21-15), or $y^2 = 4px$. We can find the value of p by using the fact that $(6.50, 7.50)$ is a point on the parabola and the fact that the coordinates must satisfy the equation. This means that

$$7.50^2 = 4p(6.50) \quad \text{or} \quad p = 2.16$$

The equation of the parabola is $y^2 = 8.64x$. The filament should be at the focus, which means it should be on the axis of the parabola, 2.16 cm from the vertex. In this way the reflected rays are parallel to the axis, as shown in Fig. 21-47.   --------■

Equations (21-15) and (21-16) give us the general form of the equation of a parabola with its vertex at the origin and its focus on one of the coordinate axes. The next example shows the use of the definition to find the equation of a parabola that has its vertex at a point other than at the origin.

EXAMPLE 6   Using the definition of the parabola, find the equation of the parabola with its focus at $(2, 3)$ and its directrix the line $y = -1$. See Fig. 21-48.

Choosing a general point $P(x, y)$ on the parabola and equating the distances from this point to $(2, 3)$ and to the line $y = -1$, we have

$$\sqrt{(x - 2)^2 + (y - 3)^2} = y + 1$$

distance P to F = distance P to $y = -1$

Squaring both sides of this equation and simplifying, we have

$$(x - 2)^2 + (y - 3)^2 = (y + 1)^2$$
$$x^2 - 4x + 4 + y^2 - 6y + 9 = y^2 + 2y + 1$$

or

$$8y = 12 - 4x + x^2$$

We note that this type of equation has appeared frequently in earlier chapters. The x-term and the constant (12 in this case) are characteristic of a parabola that does not have its vertex at the origin if the directrix is parallel to the x-axis. ∎

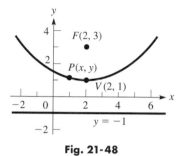

Fig. 21-48

We can readily view a parabola on a graphing calculator. If the axis is along the y-axis, or parallel to it, such as in Examples 3, 4, and 6, we simply solve for y and use this function. However, if the axis is along the x-axis, or parallel to it, as in Examples 1, 2, and 5, we get *two* functions to graph. This is similar to Example 7 on page 557. These cases are shown in the following example.

EXAMPLE 7   To display the graph of the parabola in Example 6 on a graphing calculator, we solve for y and enter this function in the calculator. Therefore, we enter the function

$$y_1 = (12 - 4x + x^2)/8$$

in the calculator, and we get the display shown in Fig. 21-49.

To display the graph of the parabola in Example 1, where $y^2 = 12x$, when we solve for y, we get $y = \pm\sqrt{12x}$. Therefore, we enter the two functions

$$y_1 = \sqrt{12x} \qquad \text{and} \qquad y_2 = -\sqrt{12x}$$

in the calculator, and we get the display shown in Fig. 21-50. ∎

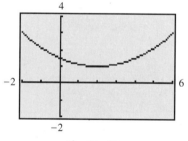

Fig. 21-49

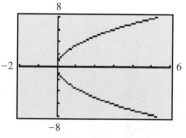

Fig. 21-50

We can conclude that *the equation of a parabola is characterized by the presence of the square of either (but not both) x or y, and a first power term in the other.* We will consider further the equation of the parabola in Sections 21-7 and 21-8.

The parabola has numerous technical applications. The reflection property illustrated in Example 5 has other important applications, such as the design of a radar antenna. The path of a projectile is parabolic. The cables of a suspension bridge are parabolic. These and other applications are illustrated in the exercises.

EXERCISES *21-4*

In Exercises 1–12, determine the coordinates of the focus and the equation of the directrix of the given parabolas. Sketch each curve.

1. $y^2 = 4x$
2. $y^2 = 16x$
3. $y^2 = -4x$
4. $y^2 = -16x$
5. $x^2 = 8y$
6. $x^2 = 10y$
7. $x^2 = -4y$
8. $x^2 = -12y$
9. $2y^2 = 5x$
10. $3x^2 = 8y$
11. $y = 0.48x^2$
12. $x = 7.6y^2$

In Exercises 13–20, find the equations of the parabolas satisfying the given conditions.

13. Focus $(3, 0)$, directrix $x = -3$
14. Focus $(-2, 0)$, directrix $x = 2$
15. Focus $(0, 4)$, vertex $(0, 0)$
16. Focus $(-3, 0)$, vertex $(0, 0)$
17. Vertex $(0, 0)$, directrix $y = -0.16$
18. Vertex $(0, 0)$, directrix $y = 2.3$
19. Vertex $(0, 0)$, axis along the y-axis, passes through $(-1, 8)$
20. Vertex $(0, 0)$, symmetric to the x-axis, passes through $(2, -1)$

In Exercises 21–40, solve the given problems.

21. Find the equation of the parabola with focus $(6, 1)$ and directrix $x = 0$ by use of the definition. Sketch the curve.

22. Find the equation of the parabola with focus $(1, 1)$ and directrix $y = 5$ by use of the definition. Sketch the curve.

23. Use a graphing calculator to view the parabola $y^2 + 2x + 8y + 13 = 0$.

24. Use a graphing calculator to view the parabola $y^2 - 2x - 6y + 19 = 0$.

25. The equation of a parabola with vertex (h, k) and axis parallel to the x-axis is $(y - k)^2 = 4p(x - h)$. (This is shown in Section 21-7.) Sketch the parabola for which (h, k) is $(2, -3)$ and $p = 2$.

26. The equation of a parabola with vertex (h, k) and axis parallel to the y-axis is $(x - h)^2 = 4p(y - k)$. (This is shown in Section 21-7.) Sketch the parabola for which (h, k) is $(-1, 2)$ and $p = -3$.

27. The chord of a parabola that passes through the focus and is parallel to the directrix is called the *latus rectum* of the parabola. Find the length of the latus rectum of the parabola $y^2 = 4px$.

28. Find the equation of the circle that has the focus and the vertex of the parabola $x^2 = 8y$ as the ends of a diameter.

29. The Golden Gate Bridge at San Francisco Bay is a suspension bridge, and its supporting cables are parabolic. See Fig. 21-51. With the origin at the low point of the cable, what equation represents the cable if the towers are 4200 ft apart and the maximum sag is 300 ft?

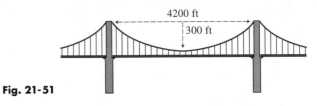

Fig. 21-51

30. The entrance to a building is a parabolic arch 5.6 m high at the center and 7.4 m wide at the base. What equation represents the arch if the vertex is at the top of the arch?

31. The rate of development of heat H (in W) in a resistor of resistance R (in Ω) of an electric circuit is given by $H = Ri^2$, where i is the current (in A) in the resistor. Sketch the graph of H vs. i, if $R = 6.0\ \Omega$.

32. What is the length of the horizontal bar across the parabolically shaped window shown in Fig. 21-52?

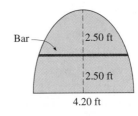

Fig. 21-52

33. The primary mirror in the Hubble space telescope has a parabolic cross section, which is shown in Fig. 21-53. What is the focal length (vertex to focus) of the mirror?

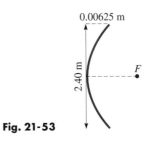

Fig. 21-53

34. A rocket is fired horizontally from a plane. Its horizontal distance x and vertical distance y from the point at which it was fired are given by $x = v_0 t$ and $y = \frac{1}{2}gt^2$, where v_0 is the initial velocity of the rocket, t is the time, and g is the acceleration due to gravity. Express y as a function of x and show that it is the equation of a parabola.

35. A wave entering parallel to the axis of a radio wave antenna with a parabolic cross section is reflected through the focus. What is the equation of the parabola for the antenna with the reflected wave shown in Fig. 21-54?

$y = -12.0x + 3.6$

F

Fig. 21-54

36. A wire is fastened 36.0 ft up on each of two telephone poles that are 200 ft apart. Halfway between the poles the wire is 30.0 ft above the ground. Assuming the wire is parabolic, find the height of the wire 50.0 ft from either pole.

37. The total annual fraction f of energy supplied by solar energy to a home is given by $f = 0.065 \sqrt{A}$, where A is the area of the solar collector. Sketch the graph of f as a function of A $(0 < A \leq 200 \text{ m}^2)$.

38. The velocity v (in ft/s) of a jet of water flowing from an opening in the side of a certain container is given by $v = 8 \sqrt{h}$, where h is the depth (in ft) of the opening. Sketch a graph of v vs. h.

39. A small island is 4 km from a straight shoreline. A ship channel is equidistant between the island and the shoreline. Write an equation for the channel.

40. Under certain circumstances, the maximum power P (in W) in an electric circuit varies as the square of the voltage of the source E_0 and inversely as the internal resistance R_i (in Ω) of the source. If 10 W is the maximum power for a source of 2.0 V and internal resistance of 0.10 Ω, sketch the graph of P vs. E_0 if R_i remains constant.

21-5 THE ELLIPSE

The next important curve is the ellipse. *An **ellipse** is defined as the locus of a point $P(x, y)$ that moves so that the sum of its distances from two fixed points is constant. These fixed points are the **foci** of the ellipse.* Letting this sum of distances be $2a$ and the foci be the points $(-c, 0)$ and $(c, 0)$, we have

$$\sqrt{(x - c)^2 + y^2} + \sqrt{(x + c)^2 + y^2} = 2a$$

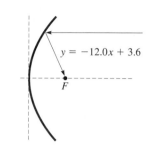

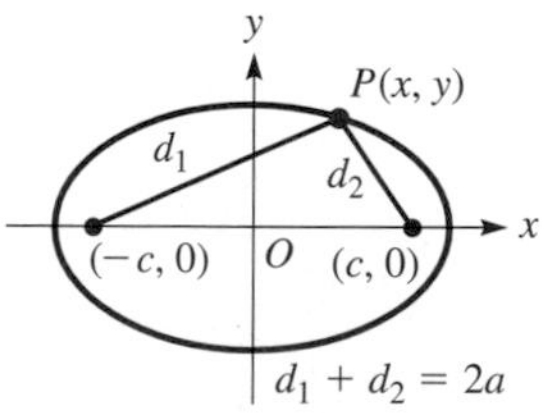

Fig. 21-55

See Fig. 21-55. The ellipse has its center at the origin such that c is the length of the line segment from the center to a focus. We shall also see that a has a special meaning. Now, from Section 14-4, we see that we should move one radical to the right and then square each side. This leads to the following steps.

$$\sqrt{(x + c)^2 + y^2} = 2a - \sqrt{(x - c)^2 + y^2}$$
$$(x + c)^2 + y^2 = 4a^2 - 4a\sqrt{(x - c)^2 + y^2} + (\sqrt{(x - c)^2 + y^2})^2$$
$$x^2 + 2cx + c^2 + y^2 = 4a^2 - 4a\sqrt{(x - c)^2 + y^2} + x^2 - 2cx + c^2 + y^2$$
$$4a\sqrt{(x - c)^2 + y^2} = 4a^2 - 4cx$$
$$a\sqrt{(x - c)^2 + y^2} = a^2 - cx$$
$$a^2(x^2 - 2cx + c^2 + y^2) = a^4 - 2a^2cx + c^2x^2$$
$$(a^2 - c^2)x^2 + a^2y^2 = a^2(a^2 - c^2)$$

We now define $a^2 - c^2 = b^2$ (the reason will be shown presently). Therefore,

$$b^2x^2 + a^2y^2 = a^2b^2$$

Dividing through by a^2b^2, we have

$$\frac{x^2}{a^2} + \frac{y^2}{b^2} = 1$$

(21-17)

A graphical analysis of this equation is found on the next page.

The x-intercepts are $(-a, 0)$ and $(a, 0)$. This means that $2a$ (the sum of distances used in the derivation) is also the distance between the x-intercepts. *The points $(a, 0)$ and $(-a, 0)$ are the* **vertices** *of the ellipse, and the line between them is the* **major axis** [see Fig. 21-56(a)]. *Thus, a is the length of the* **semimajor axis.**

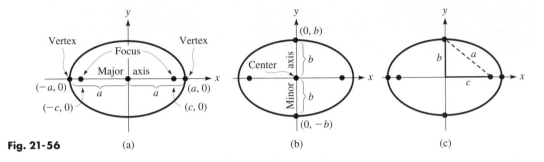

Fig. 21-56   (a)   (b)   (c)

We can now state that *Eq. (21-17) is called the* **standard equation** *of the ellipse with its major axis along the x-axis and its center at the origin.*

The y-intercepts of this ellipse are $(0, -b)$ and $(0, b)$. *The line joining these intercepts is called the* **minor axis** *of the ellipse* (Fig. 21-56(b)), *which means b is the length of the* **semiminor** *axis.* The intercept $(0, b)$ is equidistant from $(-c, 0)$ and $(c, 0)$. Since the sum of the distances from these points to $(0, b)$ is $2a$, the distance from $(c, 0)$ to $(0, b)$ must be a. Thus, we have a right triangle with line segments of lengths a, b, and c, with a as hypotenuse (Fig. 21-56(c)). Therefore,

$$a^2 = b^2 + c^2 \qquad \textbf{(21-18)}$$

is the relation between distances a, b, and c. This also shows why b was defined as it was in the derivation of Eq. (21-17).

If we choose points on the y-axis as the foci, *the standard equation of the ellipse, with its center at the origin and its major axis along the y-axis, is*

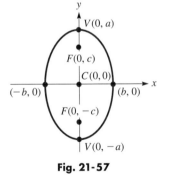

Fig. 21-57

$$\frac{y^2}{a^2} + \frac{x^2}{b^2} = 1 \qquad \textbf{(21-19)}$$

In this case the vertices are $(0, a)$ and $(0, -a)$, the foci are $(0, c)$ and $(0, -c)$, and the ends of the minor axis are $(b, 0)$ and $(-b, 0)$. See Fig. 21-57.

The ellipses represented by Eqs. (21-17) and (21-19) are both symmetrical to both axes and to the origin.

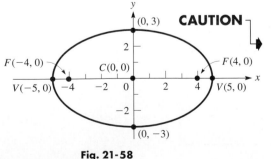

Fig. 21-58

CAUTION

■**EXAMPLE 1**   The ellipse $\dfrac{x^2}{25} + \dfrac{y^2}{9} = 1$ seems to fit the form of either Eq. (21-17) or Eq. (21-19). Since $a^2 = b^2 + c^2$, we know that **a is always larger than b.** Since the square of the larger number appears under x^2, we know the equation is in the form of Eq. (21-17). Therefore, $a^2 = 25$ and $b^2 = 9$, or $a = 5$ and $b = 3$. This means that the vertices are $(5, 0)$ and $(-5, 0)$ and the minor axis extends from $(0, -3)$ to $(0, 3)$. See Fig. 21-58.

We find c from the relation $c^2 = a^2 - b^2$. This means that $c^2 = 16$ and the foci are $(4, 0)$ and $(-4, 0)$.

     ■

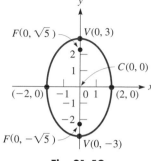

$F(0, \sqrt{5})$ $V(0, 3)$

$C(0, 0)$

$(-2, 0)$ $(2, 0)$

$F(0, -\sqrt{5})$ $V(0, -3)$

Fig. 21-59

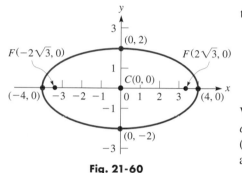

$F(-2\sqrt{3}, 0)$ $(0, 2)$ $F(2\sqrt{3}, 0)$

$C(0, 0)$

$(-4, 0)$ $(4, 0)$

$(0, -2)$

Fig. 21-60

SOLVING A WORD PROBLEM

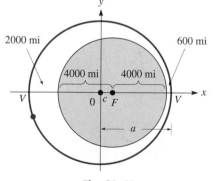

2000 mi   600 mi

4000 mi   4000 mi

V 0 c F V

a

Fig. 21-61

EXAMPLE 2   The ellipse

$$\frac{x^2}{4} + \frac{y^2}{9} = 1$$

$b^2 \uparrow$ $\uparrow a^2$

has vertices $(0, 3)$ and $(0, -3)$. The minor axis extends from $(-2, 0)$ to $(2, 0)$. The equation fits the form of Eq. (21-19) since the larger number appears under y^2. Therefore, $a^2 = 9$, $b^2 = 4$, and $c^2 = 5$. The foci are $(0, \sqrt{5})$ and $(0, -\sqrt{5})$. This ellipse is shown in Fig. 21-59. ∎

EXAMPLE 3   Find the coordinates of the vertices, the ends of the minor axis, and the foci of the ellipse $4x^2 + 16y^2 = 64$.

This equation must be put in standard form first, which we do by dividing through by 64. When this is done, we obtain

$$\frac{x^2}{16} + \frac{y^2}{4} = 1 \leftarrow$$

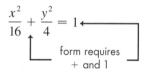

form requires + and 1

We see that $a^2 = 16$ and $b^2 = 4$, which tells us that $a = 4$ and $b = 2$. Then, $c = \sqrt{16 - 4} = \sqrt{12} = 2\sqrt{3}$. Since a^2 appears under x^2, the vertices are $(4, 0)$ and $(-4, 0)$. The ends of the minor axis are $(0, 2)$ and $(0, -2)$, and the foci are $(2\sqrt{3}, 0)$ and $(-2\sqrt{3}, 0)$. See Fig. 21-60. ∎

EXAMPLE 4   A satellite to study the earth's atmosphere has a minimum altitude of 600 mi and a maximum altitude of 2000 mi. If the path of the satellite about the earth is an ellipse with the center of the earth at one focus, what is the equation of its path? Assume the radius of the earth is 4000 mi.

We set up the coordinate system such that the center of the ellipse is at the origin and the center of the earth is at the right focus, as shown in Fig. 21-61. We know that the distance between vertices is

$$2a = 2000 + 4000 + 4000 + 600 = 10{,}600 \text{ mi}$$
$$a = 5300 \text{ mi}$$

From the right focus to the right vertex is 4600 mi. This tells us

$$c = a - 4600 = 5300 - 4600 = 700 \text{ mi}$$

We can now calculate b^2 as

$$b^2 = a^2 - c^2 = 5300^2 - 700^2 = 2.76 \times 10^7 \text{ mi}^2$$

Since $a^2 = 5300^2 = 2.81 \times 10^7 \text{ mi}^2$, the equation is

$$\frac{x^2}{2.81 \times 10^7} + \frac{y^2}{2.76 \times 10^7} = 1$$

or

$$2.76x^2 + 2.81y^2 = 7.76 \times 10^7$$ ∎

EXAMPLE 5   Find the equation of the ellipse with its center at the origin and an end of its minor axis at $(2, 0)$ and which passes through $(-1, \sqrt{6})$.

Since the center is at the origin and an end of the minor axis is at $(2, 0)$, we know that the ellipse is of the form of Eq. (21-19) and that $b = 2$. Thus, we have

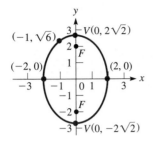

Fig. 21-62

$$\frac{y^2}{a^2} + \frac{x^2}{2^2} = 1$$

In order to find a^2, we use the fact that the ellipse passes through $(-1, \sqrt{6})$. This means that these coordinates satisfy the equation of the ellipse. This gives

$$\frac{(\sqrt{6})^2}{a^2} + \frac{(-1)^2}{4} = 1, \qquad \frac{6}{a^2} = \frac{3}{4}, \qquad a^2 = 8$$

Therefore, the equation of the ellipse, shown in Fig. 21-62, is

$$\frac{y^2}{8} + \frac{x^2}{4} = 1$$

The following example illustrates the use of the definition of the ellipse to find the equation of an ellipse with its center at a point other than the origin.

EXAMPLE 6   Using the definition, find the equation of the ellipse with foci at $(1, 3)$ and $(9, 3)$, with major axis of 10.

Recalling that the sum of distances in the definition equals the length of the major axis, we now use the same method as in the derivation of Eq. (21-17).

$\sqrt{(x-1)^2 + (y-3)^2} + \sqrt{(x-9)^2 + (y-3)^2} = 10$	use definition of ellipse
$\sqrt{(x-1)^2 + (y-3)^2} = 10 - \sqrt{(x-9)^2 + (y-3)^2}$	isolate a radical
$x^2 - 2x + 1 + y^2 - 6y + 9 = 100 - 20\sqrt{(x-9)^2 + (y-3)^2} + x^2 - 18x + 81 + y^2 - 6y + 9$	square both sides and simplify
$20\sqrt{(x-9)^2 + (y-3)^2} = 180 - 16x$	isolate radical
$5\sqrt{(x-9)^2 + (y-3)^2} = 45 - 4x$	divide by 4
$25(x^2 - 18x + 81 + y^2 - 6y + 9) = 2025 - 360x + 16x^2$	square both sides
$9x^2 - 90x + 25y^2 - 150y + 225 = 0$	simplify

The additional x- and y-terms are characteristic of the equation of an ellipse whose center is not at the origin (see Fig. 21-63).

To view this ellipse on a graphing calculator as shown in Fig. 21-64, we solve for y to get the two functions needed. The solutions are $y = \dfrac{15 \pm 3\sqrt{10x - x^2}}{5}$.

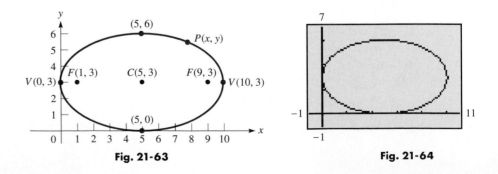

Fig. 21-63   Fig. 21-64

We can conclude that *the equation of an ellipse is characterized by the presence of both an x^2-term and a y^2-term, having different coefficients (in value but not in sign)*. The difference between the equation of an ellipse and that of a circle is that the coefficients of the squared terms in the equation of the circle are the same, whereas those of the ellipse differ. We will consider the equation of the ellipse further in Sections 21-7 and 21-8.

The ellipse has many applications. The orbits of the planets about the sun are elliptical. Gears, cams, and springs are often elliptical in shape. Arches are often constructed in the form of a semiellipse. These and other applications are illustrated in the exercises.

The Polish astronomer Nicolaus Copernicus (1473–1543) is credited as being the first to suggest that the earth revolved about the sun, rather than the previously held belief that the earth was the center of the universe.

EXERCISES *21-5*

In Exercises 1–12, find the coordinates of the vertices and foci of the given ellipses. Sketch each curve.

1. $\dfrac{x^2}{4} + \dfrac{y^2}{1} = 1$

2. $\dfrac{x^2}{100} + \dfrac{y^2}{64} = 1$

3. $\dfrac{x^2}{25} + \dfrac{y^2}{36} = 1$

4. $\dfrac{x^2}{49} + \dfrac{y^2}{81} = 1$

5. $4x^2 + 9y^2 = 36$

6. $x^2 + 36y^2 = 144$

7. $49x^2 + 4y^2 = 196$

8. $25x^2 + y^2 = 25$

9. $8x^2 + y^2 = 16$

10. $2x^2 + 3y^2 = 600$

11. $4x^2 + 25y^2 = 0.25$

12. $9x^2 + 4y^2 = 0.09$

In Exercises 13–20, find the equations of the ellipses satisfying the given conditions. The center of each is at the origin.

13. Vertex $(15, 0)$, focus $(9, 0)$

14. Minor axis 8, vertex $(0, -5)$

15. Focus $(0, 2)$, major axis 6

16. Sum of lengths of major and minor axes 18, focus $(3, 0)$

17. Vertex $(8, 0)$, passes through $(2, 3)$

18. Focus $(0, 2)$, passes through $(-1, \sqrt{3})$

19. Passes through $(2, 2)$ and $(1, 4)$

20. Passes through $(-2, 2)$ and $(1, \sqrt{6})$

In Exercises 21–40, solve the given problems.

21. Find the equation of the ellipse with foci $(-2, 1)$ and $(4, 1)$ and a major axis of 10 by use of the definition. Sketch the curve.

22. Find the equation of the ellipse with foci $(1, 4)$ and $(1, 0)$ that passes through $(4, 4)$ by use of the definition. Sketch the curve.

23. Use a graphing calculator to view the ellipse
$4x^2 + 3y^2 + 16x - 18y + 31 = 0$.

24. Use a graphing calculator to view the ellipse
$4x^2 + 8y^2 + 4x - 24y + 1 = 0$.

25. The equation of an ellipse with center (h, k) and major axis parallel to the x-axis is $\dfrac{(x - h)^2}{a^2} + \dfrac{(y - k)^2}{b^2} = 1$. (This is shown in Section 21-7.) Sketch the ellipse that has a major axis of 6, a minor axis of 4, and for which (h, k) is $(2, -1)$.

26. The equation of an ellipse with center (h, k) and major axis parallel to the y-axis is $\dfrac{(y - k)^2}{a^2} + \dfrac{(x - h)^2}{b^2} = 1$. (This is shown in Section 21-7.) Sketch the ellipse that has a major axis of 8, a minor axis of 6, and for which (h, k) is $(1, 3)$.

(W) 27. For what values of k does the ellipse $x^2 + ky^2 = 1$ have its vertices on the y-axis? Explain how these values are found.

(W) 28. For what value of k does the ellipse $x^2 + k^2y^2 = 25$ have a focus at $(3, 0)$? Explain how this value is found.

29. Show that the ellipse $2x^2 + 3y^2 - 8x - 4 = 0$ is symmetrical to the x-axis.

30. Show that the ellipse $5x^2 + y^2 - 3y - 7 = 0$ is symmetrical to the y-axis.

31. The *eccentricity e* of an ellipse is defined as $e = c/a$. A cam in the shape of an ellipse can be described by the equation $x^2 + 9y^2 = 81$. Find the eccentricity of this elliptical cam.

32. The planet Pluto moves about the sun in an elliptical orbit, with the sun at one focus. The closest that Pluto approaches the sun is 2.8 billion miles, and the farthest it gets from the sun is 4.6 billion miles. Find the eccentricity of Pluto's orbit. (See Exercise 31.)

33. A draftsman draws a series of triangles with a base from $(-3, 0)$ to $(3, 0)$ and a perimeter of 14 cm (all measurements in centimeters). Find the equation of the curve on which all of the third vertices of the triangles are located.

34. The electric power P (in W) dissipated in a resistance R (in Ω) is given by $P = Ri^2$, where i is the current (in A) in the resistor. Find the equation for the total power of 64 W dissipated in two resistors, with resistances 2.0 Ω and 8.0 Ω, respectively, and with currents i_1 and i_2, respectively. Sketch the graph, assuming that negative values of current are meaningful.

35. An ellipse has a focal property such that a light ray or sound wave emanating from one focus will be reflected through the other focus. Many buildings, such as Statuary Hall in the U.S. Capitol and the Taj Mahal, are built with elliptical ceilings with the property that a sound from one focus is easily heard at the other focus. If a building has a ceiling whose cross sections are part of an ellipse that can be described by the equation $36x^2 + 225y^2 = 8100$ (measurements in meters), how far apart must two persons stand in order to whisper to each other using this focal property?

36. An airplane wing is designed such that a certain cross section is an ellipse 8.40 ft wide and 1.20 ft thick. Find an equation that can be used to describe the perimeter of this cross section.

37. A road passes through a tunnel with a semielliptical cross section 64 ft wide and 18 ft high at the center. What is the height of the tallest vehicle that can pass through the tunnel at a point 22 ft from the center? See Fig. 21-65.

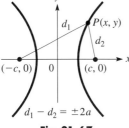

Fig. 21-65

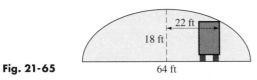

38. An architect designs a window in the shape of an ellipse 4.50 ft wide and 3.20 ft high. Find the perimeter of the window from the formula $p = \pi(a + b)$. This formula gives a good *approximation* for the perimeter when a and b are nearly equal.

39. The ends of a horizontal tank 20.0 ft long are ellipses, which can be described by the equation $9x^2 + 20y^2 = 180$, where x and y are measured in feet. The area of an ellipse is $A = \pi ab$. Find the volume of the tank.

40. A laser beam 6.80 mm in diameter is incident on a plane surface at an angle of 62.0°, as shown in Fig. 21-66. What is the elliptical area that the laser covers on the surface? (See Exercise 39.)

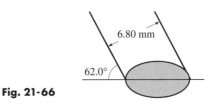

Fig. 21-66

21-6 THE HYPERBOLA

The final curve we shall discuss in detail is the hyperbola. *A* **hyperbola** *is defined as the locus of a point P(x, y) that moves so that the difference of the distances from two fixed points is a constant. These fixed points are the* **foci** *of the hyperbola.* We choose the foci of the hyperbola as the points $(-c, 0)$ and $(c, 0)$ (see Fig. 21-67) and the constant difference to be $2a$. As with the ellipse, these choices make c the length of the line segment from the center to a focus and a (as we will see) the length of the line segment from the center to a vertex. Therefore,

Fig. 21-67

$$\sqrt{(x + c)^2 + y^2} - \sqrt{(x - c)^2 + y^2} = 2a$$

Following the same procedure as in the preceding section, we find the equation of the hyperbola to be

$$\frac{x^2}{a^2} - \frac{y^2}{b^2} = 1 \qquad \textbf{(21-20)}$$

CAUTION ▶   When we derive this equation, *we have a definition of the relation between, a, b, and c that is different from that for the ellipse.* This relation is

$$c^2 = a^2 + b^2 \qquad \textbf{(21-21)}$$

In Eq. (21-20), if we let $y = 0$, we find that the x-intercepts are $(-a, 0)$ and $(a, 0)$, just as they are for the ellipse. *These are the* **vertices** *of the hyperbola*. For $x = 0$, we find that we have imaginary solutions for y, which means there are no points on the curve that correspond to a value of $x = 0$.

To find the meaning of b, we solve Eq. (21-20) for y in a special form:

$$\frac{y^2}{b^2} = \frac{x^2}{a^2} - 1$$

$$= \frac{x^2}{a^2} - \frac{a^2 x^2}{a^2 x^2}$$

$$= \frac{x^2}{a^2}\left(1 - \frac{a^2}{x^2}\right)$$

Multiplying through by b^2 and then taking the square root of each side, we have

$$y^2 = \frac{b^2 x^2}{a^2}\left(1 - \frac{a^2}{x^2}\right)$$

$$y = \pm\frac{bx}{a}\sqrt{1 - \frac{a^2}{x^2}} \qquad \textbf{(21-22)}$$

We note that, if large values of x are assumed in Eq. (21-22), the quantity under the radical becomes approximately 1. In fact, the larger x becomes, the nearer 1 this expression becomes, since the x^2 in the denominator of a^2/x^2 makes this term nearly zero. Thus, for large values of x, Eq. (21-22) is approximately

$$y = \pm\frac{bx}{a} \qquad \textbf{(21-23)}$$

NOTE ▶

Equation (21-23) is seen to represent two straight lines, each of which passes through the origin. One has a slope of b/a, and the other has a slope of $-b/a$. *These lines are called the* **asymptotes** *of the hyperbola*. An **asymptote** *is a line that the curve approaches as one of the variables approaches some particular value*. The graph of the tangent function also has asymptotes, as we saw in Fig. 10-23. We can designate this limiting procedure with notation introduced in Chapter 19 by saying that

$$y \to \frac{bx}{a} \quad \text{as} \quad x \to \pm\infty$$

Since straight lines are easily sketched, the easiest way to sketch a hyperbola is to draw its asymptotes and then to draw the hyperbola out from each vertex so that it comes closer and closer to each of these asymptotes as x becomes numerically larger. To draw in the asymptotes, the usual procedure is to first draw a small rectangle $2a$ by $2b$, with the origin at the center, as shown in Fig. 21-68. Then straight lines are drawn through opposite vertices of the rectangle. These straight lines are the asymptotes of the hyperbola. Therefore, we see that the significance of the value of b lies in the slope of the asymptotes of the hyperbola.

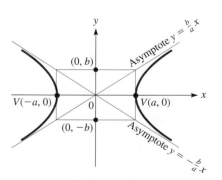

Fig. 21-68

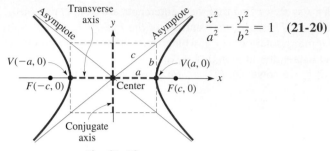

Fig. 21-69

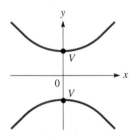

Fig. 21-70

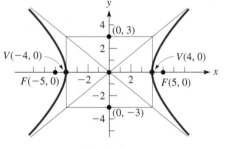

Fig. 21-71

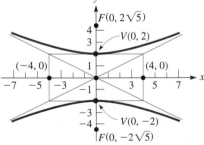

Fig. 21-72

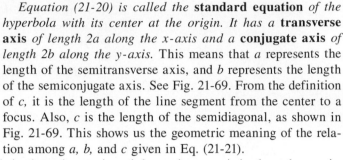

$$\frac{x^2}{a^2} - \frac{y^2}{b^2} = 1 \quad \textbf{(21-20)}$$

Equation (21-20) is called the **standard equation** *of the hyperbola with its center at the origin. It has a* **transverse axis** *of length 2a along the x-axis and a* **conjugate axis** *of length 2b along the y-axis.* This means that a represents the length of the semitransverse axis, and b represents the length of the semiconjugate axis. See Fig. 21-69. From the definition of c, it is the length of the line segment from the center to a focus. Also, c is the length of the semidiagonal, as shown in Fig. 21-69. This shows us the geometric meaning of the relation among a, b, and c given in Eq. (21-21).

If the tranverse axis is along the y-axis and the conjugate axis is along the x-axis, the equation of a hyperbola with its center at the origin (see Fig. 21-70) is

$$\frac{y^2}{a^2} - \frac{x^2}{b^2} = 1 \qquad \textbf{(21-24)}$$

The hyperbolas represented by Eqs. (21-20) and (21-24) are both symmetrical to both axes and to the origin.

■EXAMPLE 1 The hyperbola $\dfrac{x^2}{16} - \dfrac{y^2}{9} = 1$

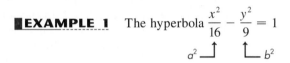

fits the form of Eq. (21-20). We know that it fits Eq. (21-20) and not Eq. (21-24) since the x^2-term is the positive term with 1 on the right. From the equation we see that $a^2 = 16$ and $b^2 = 9$, or $a = 4$ and $b = 3$. In turn this means the vertices are $(4, 0)$ and $(-4, 0)$ and the conjugate axis extends from $(0, -3)$ to $(0, 3)$.

Since $c^2 = a^2 + b^2$, we find that $c^2 = 25$, or $c = 5$. The foci are $(-5, 0)$ and $(5, 0)$.

Drawing the rectangle and the asymptotes in Fig. 21-71, we then sketch in the hyperbola from each vertex toward each asymptote. --------■

■EXAMPLE 2 The hyperbola $\dfrac{y^2}{4} - \dfrac{x^2}{16} = 1$

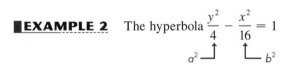

has vertices at $(0, -2)$ and $(0, 2)$. Its conjugate axis extends from $(-4, 0)$ to $(4, 0)$. The foci are $(0, -2\sqrt{5})$ and $(0, 2\sqrt{5})$. We find this directly from the equation since the y^2-term is the positive term with 1 on the right. This means the equation fits the form of Eq. (21-24) with $a^2 = 4$ and $b^2 = 16$. Also, $c^2 = 20$, which means that $c = \sqrt{20} = 2\sqrt{5}$.

Since $2a$ extends along the y-axis, we see that the equations of the asymptotes are $y = \pm(a/b)x$. This is not a contradiction of Eq. (21-23) but an extension of it for a hyperbola with its transverse axis along the y-axis. The ratio a/b gives the slope of the asymptote. The hyperbola is shown in Fig. 21-72. --------■

EXAMPLE 3   Determine the coordinates of the vertices of the hyperbola

$$4x^2 - 9y^2 = 36$$

First, by dividing through by 36, we have

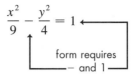

$$\frac{x^2}{9} - \frac{y^2}{4} = 1$$

form requires
— and 1

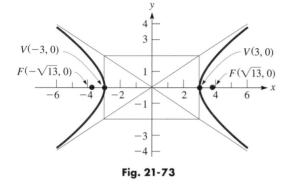

Fig. 21-73

From this form we see that $a^2 = 9$ and $b^2 = 4$. In turn this tells us that $a = 3$, $b = 2$, and $c = \sqrt{9 + 4} = \sqrt{13}$. Since a^2 appears under x^2, the equation fits the form of Eq. (21-20). Therefore, the vertices are $(-3, 0)$ and $(3, 0)$ and the foci are $(-\sqrt{13}, 0)$ and $(\sqrt{13}, 0)$. The hyperbola is shown in Fig. 21-73.

SOLVING A WORD PROBLEM

EXAMPLE 4   In physics it is shown that where the velocity of a fluid is greatest, the pressure is the least. In designing an experiment to study this effect in the flow of water, a pipe is constructed such that its lengthwise cross section is hyperbolic. The pipe is 1.0 m long, 0.2 m in diameter at the narrowest point in the middle, and 0.4 m in diameter at each end. What is the equation that represents the cross section of the pipe as shown in Fig. 21-74?

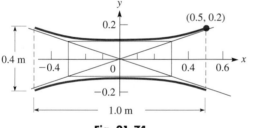

Fig. 21-74

As shown, the hyperbola has its transverse axis along the y-axis and its center at the origin. This means the general equation is given by Eq. (21-24). Since the radius at the middle of the pipe is 0.1 m, we know that $a = 0.1$ m. Also, since it is 1.0 m long and the radius at the end is 0.2 m, we know the point $(0.5, 0.2)$ is on the hyperbola. This point must satisfy the equation.

$$\frac{y^2}{a^2} - \frac{x^2}{b^2} = 1 \qquad \text{Eq. (21-24)}$$

point $(0.5, 0.2)$ satisfies equation

$$a = 0.1 \rightarrow \frac{0.2^2}{0.1^2} - \frac{0.5^2}{b^2} = 1$$

$$4 - \frac{0.25}{b^2} = 1, \qquad 3b^2 = 0.25, \qquad b^2 = 0.083$$

$$\frac{y^2}{0.1^2} - \frac{x^2}{0.083} = 1 \qquad \text{substituting } a = 0.1, b^2 = 0.083 \text{ in Eq. (21-24)}$$

$$100y^2 - 12x^2 = 1 \qquad \text{equation of cross section}$$

If we use a graphing calculator to display a hyperbola represented by either Eq. (21-20) or (21-24) we have two functions when we solve for y. One represents the upper half of the hyperbola, and the other represents the lower half. For the hyperbola in Example 3, the functions are $y_1 = \sqrt{(4x^2 - 36)/9}$ and $y_2 = -\sqrt{(4x^2 - 36)/9}$. For the hyperbola in Example 4, they are $y_1 = \sqrt{(12x^2 + 1)/100}$ and $y_2 = -\sqrt{(12x^2 + 1)/100}$.

Equations (21-20) and (21-24) give us the standard forms of the equation of the hyperbola with its center at the origin and its foci on one of the coordinate axes. There is another important equation form that represents a hyperbola, and it is

$$xy = c \qquad \text{(21-25)}$$

The asymptotes of this hyperbola are the coordinate axes, and the foci are on the line $y = x$ if c is positive or on the line $y = -x$ if c is negative.

The hyperbola represented by Eq. (21-25) is symmetrical to the origin, for if $-x$ replaces x and $-y$ replaces y at the same time, we obtain $(-x)(-y) = c$, or $xy = c$. The equation is unchanged. However, if $-x$ replaces x or if $-y$ replaces y, but not both, the sign on the left is changed. This means it is not symmetrical to either axis. Here, c represents a constant and is not related to the focus.

■EXAMPLE 5   Plot the graph of the equation $xy = 4$.

We find the values in the table below and then plot the appropriate points. Here it is permissible to use a limited number of points, since we know the equation represents a hyperbola. Therefore, using $y = 4/x$, we obtain the values

x	-8	-4	-1	$-\frac{1}{2}$	$\frac{1}{2}$	1	4	8
y	$-\frac{1}{2}$	-1	-4	-8	8	4	1	$\frac{1}{2}$

Note that neither x nor y may equal zero. The hyperbola is shown in Fig. 21-75.

If the constant on the right is negative (for example, if $xy = -4$), then the two branches of the hyperbola are in the second and fourth quadrants. ▪

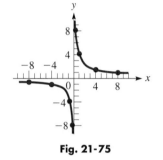

Fig. 21-75

The first reasonable measurement of the speed of light was made by the Danish astronomer Olaf Roemer (1644–1710). He measured the time required for light to come from the moons of Jupiter across the earth's orbit.

■EXAMPLE 6   For a light wave, the product of its frequency f of vibration and its wavelength λ is a constant, and this constant is the speed of light c. For green light, for which $f = 600$ THz, $\lambda = 500$ nm. Graph λ as a function of f for any light wave.

From the statement above, we know that $f\lambda = c$, and from the given values we have

$$(600 \text{ THz})(500 \text{ nm}) = (6.0 \times 10^{14} \text{ Hz})(5.0 \times 10^{-7} \text{ m}) = 3.0 \times 10^{8} \text{ m/s}$$

which means $c = 3.0 \times 10^{8}$ m/s. We are to sketch $f\lambda = 3.0 \times 10^{8}$. Solving for λ as $\lambda = 3.0 \times 10^{8}/f$, we have the following table (only positive values have meaning).

f (THz)	750	600	500	430
λ (nm)	400	500	600	700

See Fig. 21-76. (Violet light has wavelengths of about 400 nm, orange light has wavelengths of about 600 nm, and red light has wavelengths of about 700 nm.) ▪

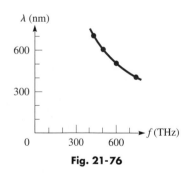

Fig. 21-76

We can conclude that *the equation of a hyperbola is characterized by the presence of both an x^2-term and a y^2-term, having different signs, or by the presence of an xy-term with no squared terms.* We will consider the equation of the hyperbola further in Sections 21-7 and 21-8.

The hyperbola has some very useful applications. The LORAN radio navigation system is based on the use of hyperbolic paths. Some reflecting telescopes use hyperbolic mirrors. The paths of comets that never return to pass by the sun are hyperbolic. Some applications are illustrated in the exercises.

EXERCISES *21-6*

In Exercises 1–12, find the coordinates of the vertices and the foci of the given hyperbolas. Sketch each curve.

1. $\dfrac{x^2}{25} - \dfrac{y^2}{144} = 1$

2. $\dfrac{x^2}{16} - \dfrac{y^2}{4} = 1$

3. $\dfrac{y^2}{9} - \dfrac{x^2}{1} = 1$

4. $\dfrac{y^2}{2} - \dfrac{x^2}{2} = 1$

5. $4x^2 - y^2 = 4$

6. $x^2 - 9y^2 = 81$

7. $2y^2 - 5x^2 = 10$

8. $3y^2 - 2x^2 = 300$

9. $4x^2 - y^2 + 4 = 0$

10. $9x^2 - y^2 - 9 = 0$

11. $4x^2 - y^2 = 0.64$

12. $9y^2 - x^2 = 0.36$

In Exercises 13–20, find the equations of the hyperbolas satisfying the given conditions. The center of each is at the origin.

13. Vertex $(3, 0)$, focus $(5, 0)$

14. Vertex $(0, 1)$, focus $(0, \sqrt{3})$

15. Conjugate axis = 12, vertex $(0, 10)$

16. Sum of lengths of transverse and conjugate axes 28, focus $(10, 0)$

17. Passes through $(2, 3)$, focus $(2, 0)$

18. Passes through $(8, \sqrt{3})$, vertex $(4, 0)$

19. Passes through $(5, 4)$ and $(3, \frac{4}{5}\sqrt{5})$

20. Passes through $(1, 2)$ and $(2, 2\sqrt{2})$

In Exercises 21–36, solve the given problems.

21. Sketch the graph of the hyperbola $xy = 2$.

22. Sketch the graph of the hyperbola $xy = -4$.

23. Find the equation of the hyperbola with foci $(1, 2)$ and $(11, 2)$, and a transverse axis of 8, by use of the definition. Sketch the curve.

24. Find the equation of the hyperbola with vertices $(-2, 4)$ and $(-2, -2)$, and a conjugate axis of 4, by use of the definition. Sketch the curve.

25. Use a graphing calculator to view the hyperbola $x^2 - 4y^2 + 4x + 32y - 64 = 0$.

26. Use a graphing calculator to view the hyperbola $5y^2 - 4x^2 + 8x + 40y + 56 = 0$.

27. The equation of a hyperbola with center (h, k) and transverse axis parallel to the x-axis is $\dfrac{(x - h)^2}{a^2} - \dfrac{(y - k)^2}{b^2} = 1$. (This is shown in Section 21-7.) Sketch the hyperbola that has a transverse axis of 4, a conjugate axis of 6, and for which (h, k) is $(-3, 2)$.

28. The equation of a hyperbola with center (h, k) and transverse axis parallel to the y-axis is $\dfrac{(y - k)^2}{a^2} - \dfrac{(x - h)^2}{b^2} = 1$. (This is shown in Section 21-7.) Sketch the hyperbola that has a transverse axis of 2, a conjugate axis of 8, and for which (h, k) is $(5, 0)$.

29. Two concentric (same center) hyperbolas are called conjugate hyperbolas if the transverse and conjugate axes of one are, respectively, the conjugate and transverse axes of the other. What is the equation of the hyperbola conjugate to the hyperbola in Exercise 14?

30. As with an ellipse, the *eccentricity e* of a hyperbola is defined as $e = c/a$. Find the eccentricity of the hyperbola $2x^2 - 3y^2 = 24$.

31. A plane flying at a constant altitude of 2000 m is observed from the control tower of an airport. Show that the equation relating the horizontal distance x and direct line distance l from the tower to the plane is that of a hyperbola. Sketch the graph of l as a function of x. See Fig. 21-77.

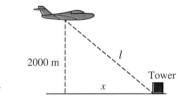

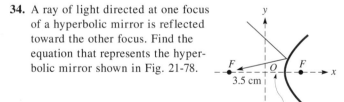

Fig. 21-77

32. Two holes of radius r are drilled from a circular area of radius R such that 24 in.^{2} of material remains. Show that the equation relating R and r is that of a hyperbola.

33. Ohm's law in electricity states that the product of the current i and the resistance R equals the voltage V across the resistance. If a battery of 6.00 V is placed across a variable resistor R, find the equation relating i and R and sketch the graph of i as a function of R.

34. A ray of light directed at one focus of a hyperbolic mirror is reflected toward the other focus. Find the equation that represents the hyperbolic mirror shown in Fig. 21-78.

Fig. 21-78

35. A radio signal is sent simultaneously from stations A and B 600 km apart on the Carolina coast. A ship receives the signal from A 1.20 ms before it receives the signal from B. Given that radio signals travel at 300 km/ms, draw a graph showing the possible locations of the ship. This problem illustrates the basis of LORAN.

36. For monochromatic (single-color) light coming from two point sources, curves of maximum intensity occur where the difference in the distances from the sources is an integral number of wavelengths. If a thin translucent film is placed in the plane of the sources, find the equation of the curves of maximum intensity in the film where the difference in paths is two wavelengths and the sources are separated by four wavelengths. Assume the sources are on the x-axis with the origin midway between and use units of one wavelength for both x and y.

21-7 TRANSLATION OF AXES

The equations we have considered for the parabola, the ellipse, and the hyperbola are those for which the center of the ellipse or hyperbola, or vertex of the parabola, is at the origin. In this section we consider, without specific use of the definition, the equations of these curves for the cases in which the axis of the curve is parallel to one of the coordinate axes. This is done by **translation of axes.**

In Fig. 21-79, we choose a point (h, k) in the xy-coordinate plane as the origin of another coordinate system, the $x'y'$-coordinate system. The x'-axis is parallel to the x-axis, and the y'-axis is parallel to the y-axis. Every point now has two sets of coordinates, (x, y) and (x', y'). We see that

$$x = x' + h \quad \text{and} \quad y = y' + k \qquad \textbf{(21-26)}$$

Equations (21-26) can also be written in the form

$$x' = x - h \quad \text{and} \quad y' = y - k \qquad \textbf{(21-27)}$$

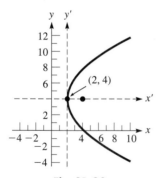

Fig. 21-79

■**EXAMPLE 1**   Find the equation of the parabola with vertex $(2, 4)$ and focus $(4, 4)$.

If we let the origin of the $x'y'$-coordinate system be the point $(2, 4)$, then the point $(4, 4)$ is the point $(2, 0)$ in the $x'y'$-system. This means $p = 2$ and $4p = 8$. See Fig. 21-80. In the $x'y'$-system, the equation is

$$(y')^2 = 8(x')$$

Using Eqs. (21-27), we have

$$(y - 4)^2 = 8(x - 2)$$

coordinates of vertex $(2, 4)$

Fig. 21-80

as the equation of the parabola in the xy-coordinate system.   ■

Following the method of Example 1, by writing the equation of the curve in the $x'y'$-system and then using Eqs. (21-27), we have the following more general forms of the equations of the parabola, ellipse, and hyperbola.

Parabola, vertex (h, k):	$(y - k)^2 = 4p(x - h)$	(axis parallel to x-axis)	**(21-28)**
	$(x - h)^2 = 4p(y - k)$	(axis parallel to y-axis)	**(21-29)**
Ellipse, center (h, k):	$\dfrac{(x - h)^2}{a^2} + \dfrac{(y - k)^2}{b^2} = 1$	(major axis parallel to x-axis)	**(21-30)**
	$\dfrac{(y - k)^2}{a^2} + \dfrac{(x - h)^2}{b^2} = 1$	(major axis parallel to y-axis)	**(21-31)**
Hyperbola, center (h, k):	$\dfrac{(x - h)^2}{a^2} - \dfrac{(y - k)^2}{b^2} = 1$	(transverse axis parallel to x-axis)	**(21-32)**
	$\dfrac{(y - k)^2}{a^2} - \dfrac{(x - h)^2}{b^2} = 1$	(transverse axis parallel to y-axis)	**(21-33)**

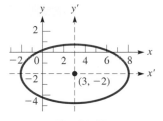

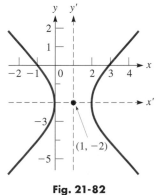

Fig. 21-81

EXAMPLE 2   Describe the curve of the equation

$$\frac{(x-3)^2}{25} + \frac{(y+2)^2}{9} = 1$$

We see that this equation fits the form of Eq. (21-30) with $h = 3$ and $k = -2$. It is the equation of an ellipse with its center at $(3, -2)$ and its major axis parallel to the x-axis. The semimajor axis is $a = 5$, and the semiminor axis is $b = 3$. The ellipse is shown in Fig. 21-81.

EXAMPLE 3   Find the center of the hyperbola
$2x^2 - y^2 - 4x - 4y - 4 = 0$.

To analyze this curve, we first complete the square in the x-terms and in the y-terms. This will allow us to recognize properly the choice of h and k.

$$2x^2 - 4x - y^2 - 4y = 4$$
$$2(x^2 - 2x \quad) - (y^2 + 4y \quad) = 4$$
$$2(x^2 - 2x + 1) - (y^2 + 4y + 4) = 4 + 2 - 4$$

Fig. 21-82

CAUTION ▶

We note here that when we added 1 to complete the square of the x-terms within the parentheses, **we were actually adding 2 to the left side.** Thus, we added 2 to the right side. Similarly, when we added 4 to the y-terms within the parentheses, **we were actually subtracting 4 from the left side.** Continuing, we have

$$2(x - 1)^2 - (y + 2)^2 = 2$$

coordinates of center $(1, -2)$

$$\frac{(x-1)^2}{1} - \frac{(y+2)^2}{2} = 1$$

Therefore, the center of the hyperbola is $(1, -2)$. See Fig. 21-82.

SOLVING A WORD PROBLEM

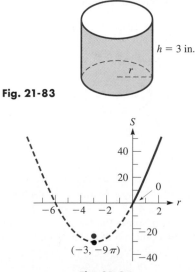

Fig. 21-83

EXAMPLE 4   Cylindrical glass beakers are to be made with a height of 3 in. Express the surface area in terms of the radius of the base, and sketch the curve.

The total surface area S of a beaker is the sum of the area of the base and the lateral surface area of the side. In general, S in terms of the radius r of the base and height h of the side is $S = \pi r^2 + 2\pi rh$. Since $h = 3$ in., we have

$$S = \pi r^2 + 6\pi r$$

which is the desired relationship. See Fig. 21-83.

To get the equation relating S and r, we complete the square of the r terms.

$$S = \pi(r^2 + 6r)$$
$$S + 9\pi = \pi(r^2 + 6r + 9) \qquad \text{complete the square}$$
$$S + 9\pi = \pi(r + 3)^2$$

vertex $(-3, -9\pi)$

$$(r + 3)^2 = \frac{1}{\pi}(S + 9\pi)$$

This represents a parabola with vertex $(-3, -9\pi)$. Since $4p = 1/\pi$, $p = 1/(4\pi)$, the focus is $(-3, \frac{1}{4\pi} - 9\pi)$ as shown in Fig. 21-84. The part of the graph for negative r is dashed since only positive values have meaning.

Fig. 21-84

EXERCISES *21-7*

In Exercises 1–8, describe the curve represented by each equation. Identify the type of curve and its center (or vertex if it is a parabola). Sketch each curve.

1. $(y - 2)^2 = 4(x + 1)$

2. $\dfrac{(x + 4)^2}{4} + \dfrac{(y - 1)^2}{1} = 1$

3. $\dfrac{(x - 1)^2}{4} - \dfrac{(y - 2)^2}{9} = 1$

4. $(y + 5)^2 = -8(x - 2)$

5. $\dfrac{(x + 1)^2}{1} + \dfrac{y^2}{9} = 1$

6. $\dfrac{(y - 4)^2}{16} - \dfrac{(x + 2)^2}{4} = 1$

7. $(x + 3)^2 = -12(y - 1)$

8. $\dfrac{x^2}{0.16} + \dfrac{(y + 1)^2}{0.25} = 1$

In Exercises 9–20, find the equation of each of the curves described by the given information.

9. Parabola: vertex $(-1, 3)$, focus $(3, 3)$

10. Parabola: vertex $(2, -1)$, directrix $y = 3$

11. Parabola: axis and directrix are the coordinate axes; focus $(12, 0)$

12. Parabola: focus $(2, 4)$, directrix $x = 6$

13. Ellipse: center $(-2, 2)$, focus $(-5, 2)$, vertex $(-7, 2)$

14. Ellipse: center $(0, 3)$, focus $(12, 3)$, major axis 26 units

15. Ellipse: vertices $(-2, -3)$ and $(-2, 5)$, end of minor axis $(0, 1)$

16. Ellipse: foci $(1, -2)$ and $(1, 10)$, minor axis 5 units

17. Hyperbola: vertex $(-1, 1)$, focus $(-1, 4)$, center $(-1, 2)$

18. Hyperbola: foci $(2, 1)$ and $(8, 1)$, conjugate axis 6 units

19. Hyperbola: vertices $(2, 1)$ and $(-4, 1)$, focus $(-6, 1)$

20. Hyperbola: center $(1, -4)$, focus $(1, 1)$, transverse axis 8 units

In Exercises 21–36, determine the center (or vertex if the curve is a parabola) of the given curve. Sketch each curve.

21. $x^2 + 2x - 4y - 3 = 0$

22. $y^2 - 2x - 2y - 9 = 0$

23. $4x^2 + 9y^2 + 24x = 0$

24. $2x^2 + 9y^2 + 8x - 72y + 134 = 0$

25. $9x^2 - y^2 + 8y - 7 = 0$

26. $5x^2 - 4y^2 + 20x + 8y = 4$

27. $2x^2 - 4x = 9y - 2$

28. $0.04x^2 + 0.16y^2 = 0.01y$

29. $4x^2 - y^2 + 32x + 10y + 35 = 0$

30. $2x^2 + 2y^2 - 24x + 16y + 95 = 0$

31. $9x^2 + 4y^2 - 12x + 16y + 16 = 0$

32. $5x^2 - 3y^2 - 40x + 95 = 0$

33. $7x^2 - y^2 - 14x - 16y - 64 = 0$

34. $5x^2 - 2y^2 + 12y + 18 = 0$

35. $9x^2 + 9y^2 - 6x - 24y + 14 = 0$

36. $4y^2 - 15x - 12y + 29 = 0$

In Exercises 37–44, solve the given problems.

37. Find the equation of the hyperbola with asymptotes $x - y = -1$ and $x + y = -3$ and vertex $(3, -1)$.

38. The circle $x^2 + y^2 + 4x - 5 = 0$ passes through the foci and the ends of the minor axis of an ellipse that has its major axis along the x-axis. Find the equation of the ellipse.

39. A first parabola has its vertex at the focus of a second parabola and its focus at the vertex of the second parabola. If the equation of the second parabola is $y^2 = 4x$, find the equation of the first parabola.

40. Identify the curve represented by the equation $4y^2 - x^2 - 6x - 2y - 14 = 0$ and then view its graph on a graphing calculator.

41. The stream of water from a fire hose follows a parabolic curve. If the stream from a hose nozzle fastened at the ground reaches a maximum height of 60 ft at a horizontal distance of 95 ft from the nozzle, find the equation that describes the stream. Take the origin at the location of the nozzle. Sketch the graph of the stream.

42. For a constant capacitive reactance and a constant resistance, sketch the graph of the impedance and inductive reactance (as abscissas) for an alternating-current circuit. (See Section 12-7.)

43. Two wheels in a friction drive assembly are equal ellipses, as shown in Fig. 21-85. The wheels are always in contact, with the center of the left wheel fixed in position and with the right wheel able to move horizontally. Find the equation that can be used to describe the circumference of each wheel in the position shown.

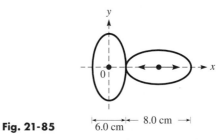

Fig. 21-85

44. An agricultural test station plans to divide a large tract of land into rectangular sections such that the perimeter of each section is 480 m. Express the area A of each section in terms of its width w. Identify the type of curve the equation represents and sketch the graph of A as a function of w. For what value of w is A the greatest?

21-8 THE SECOND-DEGREE EQUATION

The equations of the circle, parabola, ellipse, and hyperbola are all special cases of the same general equation. In this section we discuss this equation and how to identify the particular form it takes when it represents a specific type of curve.

Each of these curves can be represented by a **second-degree equation** *of the form*

$$Ax^2 + Bxy + Cy^2 + Dx + Ey + F = 0 \qquad \textbf{(21-34)}$$

The coefficients of the second-degree equation terms determine the type of curve that results. Recalling the discussions of the general forms of the equations of the circle, parabola, ellipse, and hyperbola from the previous sections of this chapter, Eq. (21-34) represents the indicated curve for given conditions of *A, B,* and *C,* as follows:

1. If $A = C$, $B = 0$, a circle.
2. If $A \neq C$ (but they have the same sign), $B = 0$, an ellipse.
3. If A and C have different signs, $B = 0$, a hyperbola.
4. If $A = 0$, $C = 0$, $B \neq 0$, a hyperbola.
5. If either $A = 0$ or $C = 0$ (but not both), $B = 0$, a parabola.
    (Special cases, such as a single point or no real locus, can also result.)

Another conclusion about Eq. (21-34) is that, if either $D \neq 0$ or $E \neq 0$ (or both), the center of the curve (or the vertex of a parabola) is not at the origin. If $B \neq 0$, the axis of the curve has been rotated. We have considered only one such case (the hyperbola $xy = c$) in this chapter. Rotation of axes is covered as one of the supplementary topics following the final chapter.

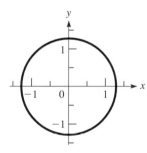

Fig. 21-86

■**EXAMPLE 1**   The equation $2x^2 = 3 - 2y^2$ represents a circle. This can be seen by putting the equation in the form of Eq. (21-34). This form is

$$2x^2 + 2y^2 - 3 = 0$$
$$A = 2 \uparrow \qquad \uparrow C = 2$$

We see that $A = C$. Also, since there is no xy-term, we know that $B = 0$. This means that the equation represents a circle. If we write it as $x^2 + y^2 = \frac{3}{2}$, we see that it fits the form of Eq. (21-12). The circle is shown in Fig. 21-86. ■

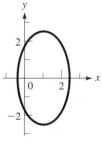

Fig. 21-87

■**EXAMPLE 2**   The equation $3x^2 = 6x - y^2 + 3$ represents an ellipse. Before we analyze the equation, we should put it in the form of Eq. (21-34). For this equation, this form is

$$3x^2 + y^2 - 6x - 3 = 0$$
$$A = 3 \uparrow \qquad \uparrow C = 1$$

Here we see that $B = 0$, A and C have the same sign, and $A \neq C$. Therefore, it is an ellipse. The $-6x$ term indicates that the center of the ellipse is not at the origin. The ellipse is shown in Fig. 21-87. ■

■EXAMPLE 3 Identify the curve represented by $2x^2 + 12x = y^2 - 14$. Determine the appropriate quantities for the curve, and sketch the graph.

Writing this equation in the form of Eq. (21-34), we have

$$2x^2 - y^2 + 12x + 14 = 0$$

$$A = 2 \qquad C = -1$$

We identify this equation as representing a hyperbola, since A and C have different signs and $B = 0$. We now write it in the standard form of a hyperbola.

$$2x^2 + 12x - y^2 = -14$$

$$2(x^2 + 6x \quad) - y^2 = -14 \qquad \text{complete the square}$$

$$2(x^2 + 6x + 9) - y^2 = -14 + 18$$

$$2(x + 3)^2 - y^2 = 4$$

center is $(-3, 0)$

$$\frac{(x + 3)^2}{2} - \frac{y^2}{4} = 1 \qquad \frac{x'^2}{2} - \frac{y'^2}{4} = 1$$

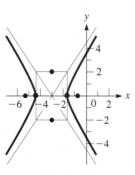

Fig. 21-88

Thus, we see that the center (h, k) of the hyperbola is the point $(-3, 0)$. Also, $a = \sqrt{2}$ and $b = 2$. This means that the vertices are $(-3 + \sqrt{2}, 0)$ and $(-3 - \sqrt{2}, 0)$, and the conjugate axis extends from $(-3, 2)$ to $(-3, -2)$. Also, $c^2 = 2 + 4 = 6$, which means that $c = \sqrt{6}$. The foci are $(-3 + \sqrt{6}, 0)$ and $(-3 - \sqrt{6}, 0)$. The graph is shown in Fig. 21-88. ■

■EXAMPLE 4 Identify the curve represented by $4y^2 - 23 = 4(4x + 3y)$ and find the appropriate important quantities. Then view it on a graphing calculator.

Writing the equation in the form of Eq. (21-34), we have

$$4y^2 - 16x - 12y - 23 = 0$$

Therefore, we recognize the equation as representing a parabola, since $A = 0$ and $B = 0$. Now, writing the equation in the standard form of a parabola, we have

$$4y^2 - 12y = 16x + 23$$

$$4(y^2 - 3y \quad) = 16x + 23 \qquad \text{complete the square}$$

$$4\left(y^2 - 3y + \frac{9}{4}\right) = 16x + 23 + 9$$

$$4\left(y - \frac{3}{2}\right)^2 = 16(x + 2)$$

vertex $\left(-2, \frac{3}{2}\right)$

$$\left(y - \frac{3}{2}\right)^2 = 4(x + 2) \qquad \text{or} \quad y'^2 = 4x'$$

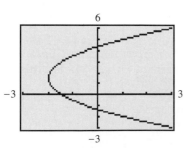

Fig. 21-89

We now note that the vertex is $(-2, 3/2)$ and that $p = 1$. This means that the focus is $(-1, 3/2)$ and the directrix is $x = -3$.

To view the graph of this equation on a graphing calculator, we first solve the equation for y and get $y = (3 \pm 4\sqrt{x + 2})/2$. Entering these two functions in the calculator, we get the view shown in Fig. 21-89. ■

See Appendix C for a graphing calculator program GRAPHCON. It displays the graph of the conic
$Ax^2 + Cy^2 + Dx + Ey + F = 0$.

In Chapter 14, when these curves were first introduced, they were referred to as **conic sections.** If a plane is passed through a cone, the intersection of the plane and the cone results in one of these curves; the curve formed depends on the angle of the plane with respect to the axis of the cone. This is shown in Fig. 21-90.

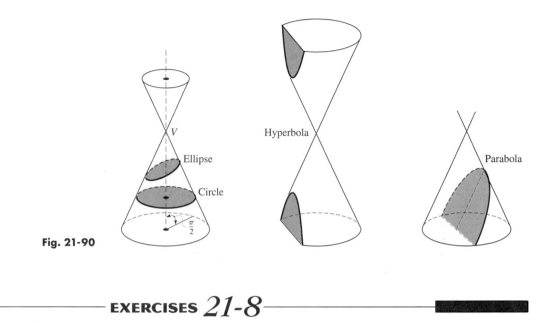

Fig. 21-90

─── EXERCISES *21-8* ───

In Exercises 1–20, identify each of the equations as representing either a circle, a parabola, an ellipse, a hyperbola, or none of these.

1. $x^2 + 2y^2 - 2 = 0$
2. $x^2 - y = 0$
3. $2x^2 - y^2 - 1 = 0$
4. $y(y + x^2) = 4$
5. $2x^2 + 2y^2 - 3y - 1 = 0$
6. $x(x - 3) = y(1 - 2y^2)$
7. $2.2x^2 - x - y = 1.6$
8. $2x^2 + 4y^2 - y - 2x = 4$
9. $x^2 = y^2 - 1$
10. $32x^2 = 21y - 47y^2$
11. $3.6x^2 = 1.1y - 3.6y^2$
12. $y = 3 - 6x^2$
13. $y(3 - 2x) = x(5 - 2y)$
14. $x(13 - 5x) = 5y^2$
15. $2xy + x - 3y = 6$
16. $(y + 1)^2 = x^2 + y^2 - 1$
17. $2x(x - y) = y(3 - y - 2x)$
18. $2x^2 = x(x - 1) + 4y^2$
19. $x(y + 3x) = x^2 + xy - y^2 + 1$
20. $4x(x - 1) = 2x^2 - 2y^2 + 3$

In Exercises 21–28, identify the curve represented by each of the given equations. Determine the appropriate important quantities for the curve and sketch the graph.

21. $x^2 = 8(y - x - 2)$
22. $x^2 = 6x - 4y^2 - 1$
23. $y^2 = 2(x^2 - 2x - 2y)$
24. $4x^2 + 4 = 9 - 8x - 4y^2$
25. $y^2 + 42 = 2x(10 - x)$
26. $x^2 - 4y = y^2 + 4(1 - x)$
27. $4(y^2 - 4x - 2) = 5(4y - 5)$
28. $2(2x^2 - y) = 8 - y^2$

In Exercises 29–32, view the curve for each equation on a graphing calculator. In Exercises 29 and 30, identify the type of curve before viewing it. In Exercises 31 and 32, the axis of each curve has been rotated.

29. $x^2 + 2y^2 - 4x + 12y + 14 = 0$
30. $4y^2 - x^2 + 40y - 4x + 60 = 0$
31. $x^2 + 6xy + 9y^2 - 2x + 14y - 10 = 0$
32. $x^2 - xy + y^2 - 6 = 0$

In Exercises 33–36, use the given values to determine the type of curve represented.

33. For the equation $x^2 + ky^2 = a^2$, what type of curve is represented if (a) $k = 1$, (b) $k < 0$, and (c) if $k > 0$ ($k \neq 1$)?

34. For the equation $\dfrac{x^2}{4 - C} - \dfrac{y^2}{C} = 1$, what type of curve is represented if (a) $C < 0$, (b) $0 < C < 4$? (For $C > 4$, see Exercise 36.)

Ⓦ 35. In Eq. (21-34), if $A = C \neq 0$ and $B = D = E = F = 0$, describe the locus of the equation.

Ⓦ 36. For the equation in Exercise 34, describe the locus of the equation if $C > 4$.

In Exercises 37–40, determine the type of curve from the given information.

37. The diagonal brace in a rectangular metal frame is 3.0 cm longer than the length of one of the sides. Determine the type of curve represented by the equation relating the lengths of the sides of the frame.

38. One circular solar cell has a radius that is 2.0 in. less than the radius r of a second circular solar cell. Determine the type of curve represented by the equation relating the total area A of both cells and r.

39. A flashlight emits a cone of light onto the floor. What type of curve is the perimeter of the lighted area on the floor, if the floor cuts completely through the cone of light?

40. The supersonic jet airliner Concorde creates a conical shock wave behind it. What type of curve is outlined on the surface of a lake by the shock wave if the Concorde is flying horizontally?

21-9 POLAR COORDINATES

The Swiss mathematician Jakob Bernoulli (1654–1705) was among the first to make significant use of polar coordinates.

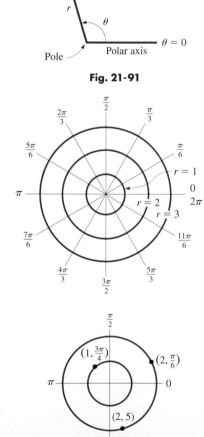

Fig. 21-91

Thus far we have graphed all curves in one coordinate system. This system, the rectangular coordinate system, is probably the most useful and widely applicable system. However, for certain types of curves, other coordinate systems prove to be better adapted. These coordinate systems are widely used, especially when certain applications of higher mathematics are involved. We shall discuss one of these systems here.

Instead of designating a point by its x- and y-coordinates, we can specify its location by its radius vector and the angle the radius vector makes with the x-axis. Thus, the r and θ that are used in the definitions of the trigonometric functions can also be used as the coordinates of points in the plane. The important aspect of choosing coordinates is that, for each set of values, there must be only one point which corresponds to this set. We can see that this condition is satisfied by the use of r and θ as coordinates. *In* **polar coordinates,** *the origin is called the* **pole,** *and the half-line for which the angle is zero (equivalent to the positive x-axis) is called the* **polar axis.** The coordinates of a point are designated as (r, θ). We shall use radians when measuring the value of θ. See Fig. 21-91.

When using polar coordinates, we generally label the lines for some of the values of θ; namely, those for $\theta = 0$ (the polar axis), $\theta = \pi/2$ (equivalent to the positive y-axis), $\theta = \pi$ (equivalent to the negative x-axis), $\theta = 3\pi/2$ (equivalent to the negative y-axis), and possibly others. In Fig. 21-92, these lines and those for multiples of $\pi/6$ are shown. Also, the circles for $r = 1$, $r = 2$, and $r = 3$ are shown in this figure.

Fig. 21-92

■EXAMPLE 1 **(a)** If $r = 2$ and $\theta = \pi/6$, we have the point as shown in Fig. 21-93. The coordinates (r, θ) of this point are written as $(2, \pi/6)$ when polar coordinates are used. This point corresponds to $(\sqrt{3}, 1)$ in rectangular coordinates.

(b) In Fig. 21-93, the polar coordinate point $(1, 3\pi/4)$ is also shown. It is equivalent to the point $(-\sqrt{2}/2, \sqrt{2}/2)$ in rectangular coordinates.

(c) In Fig. 21-93, the polar coordinate point $(2, 5)$ is also shown. It is equivalent approximately to the point $(0.6, -1.9)$ in rectangular coordinates. Remember, the 5 is an angle in radian measure.

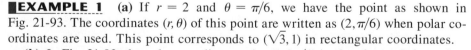

Fig. 21-93

One difference between rectangular coordinates and polar coordinates is that, for each point in the plane, there are limitless possibilities for the polar coordinates of that point. For example, the point $(2, \frac{\pi}{6})$ can also be represented by $(2, \frac{13\pi}{6})$ since the angles $\frac{\pi}{6}$ and $\frac{13\pi}{6}$ are coterminal. We also remove one restriction on r that we imposed in the definition of the trigonometric functions. That is, r is allowed to take on positive and negative values. If r is negative, θ is located as before, but ***the point is found r units from the pole but on the opposite side*** from that on which it is positive.

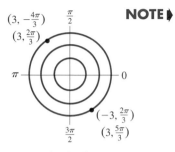

Fig. 21-94

■EXAMPLE 2   The coordinates $(3, 2\pi/3)$ and $(3, -4\pi/3)$ represent the same point. However, the point $(-3, 2\pi/3)$ is on the opposite side of the pole, three units from the pole. Another possible set of coordinates for the point $(-3, 2\pi/3)$ is $(3, 5\pi/3)$. See Fig. 21-94.   ■

When plotting a point in polar coordinates, it is generally easier to *first locate the terminal side of θ and then measure r along this terminal side*. This is illustrated in the following example.

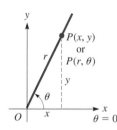

Fig. 21-95

■EXAMPLE 3   Plot the points $A(2, 5\pi/6)$ and $B(-3.2, -2.4)$ in the polar coordinate system.

To locate A we determine the terminal side of $\theta = 5\pi/6$ and then determine $r = 2$. See Fig. 21-95.

To locate B we find the terminal side of $\theta = -2.4$, measuring clockwise from the polar axis (and recalling that $\pi = 3.14 = 180°$). Then we locate $r = -3.2$ on the opposite side of the pole. See Fig. 21-95.

We will find that points with negative values of r occur frequently when plotting curves in polar coordinates.   ■

Polar and Rectangular Coordinates

The relationships between the polar coordinates of a point and the rectangular coordinates of the same point come from the definitions of the trigonometric functions. Those most commonly used are (see Fig. 21-96):

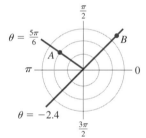

Fig. 21-96

$$x = r \cos \theta \qquad y = r \sin \theta \qquad \text{(21-35)}$$

$$\tan \theta = \frac{y}{x} \qquad r = \sqrt{x^2 + y^2} \qquad \text{(21-36)}$$

The following examples show the use of Eqs. (21-35) and (21-36) in changing coordinates in one system to coordinates in the other system. Also, these equations are used to transform equations from one system to the other.

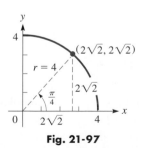

Fig. 21-97

■EXAMPLE 4   Using Eqs. (21-35), we can transform the polar coordinates of $(4, \pi/4)$ into the rectangular coordinates $(2\sqrt{2}, 2\sqrt{2})$, since

$$x = 4 \cos \frac{\pi}{4} = 4\left(\frac{\sqrt{2}}{2}\right) = 2\sqrt{2} \qquad \text{and} \qquad y = 4 \sin \frac{\pi}{4} = 4\left(\frac{\sqrt{2}}{2}\right) = 2\sqrt{2}$$

See Fig. 21-97.   ■

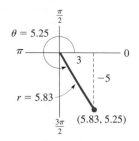

Fig. 21-98

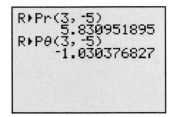

Fig. 21-99

The cyclotron was invented in 1931 at the University of California. It was the first accelerator to deflect particles into circular paths.

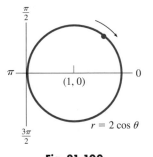

Fig. 21-100

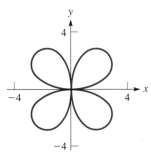

Fig. 21-101

EXAMPLE 5 Using Eqs. (21-36), we can transform the rectangular coordinates $(3, -5)$ into polar coordinates.

$$\tan \theta = -\frac{5}{3}, \qquad \theta = 5.25 \qquad (\text{or } -1.03)$$

$$r = \sqrt{3^2 + (-5)^2} = 5.83$$

We know that θ is a fourth-quadrant angle since x is positive and y is negative. Therefore, the point $(3, -5)$ in rectangular coordinates can be expressed as the point $(5.83, 5.25)$ in polar coordinates (see Fig. 21-98). Other polar coordinates for the point are also possible.

Calculators are programmed to make conversions between rectangular coordinates and polar coordinates. The manual for any model should be consulted to determine how any particular model is used for these conversions. For a calculator that uses the *angle* feature, the display for the conversions of Example 5 is shown in Fig. 21-99.

EXAMPLE 6 If an electrically charged particle enters a magnetic field at right angles to the field, the particle follows a circular path. This fact is used in the design of nuclear particle accelerators.

A proton (positively charged) enters a magnetic field such that its path may be described by the rectangular equation $x^2 + y^2 = 2x$, where measurements are in meters. Find the polar equation of this circle.

We change this equation expressed in the rectangular coordinates x and y into an equation expressed in the polar coordinates r and θ by using the relations $r^2 = x^2 + y^2$ and $x = r \cos \theta$ as follows:

$$
\begin{array}{ll}
x^2 + y^2 = 2x & \text{rectangular equation} \\
r^2 = 2r \cos \theta & \text{substitute} \\
r = 2 \cos \theta & \text{divide by } r
\end{array}
$$

This is the polar equation of the circle, which is shown in Fig. 21-100.

EXAMPLE 7 Find the rectangular equation of the *rose* $r = 4 \sin 2\theta$.

Using the trigonometric identity $\sin 2\theta = 2 \sin \theta \cos \theta$ and Eqs. (21-35) and (21-36) leads to the solution.

$$
\begin{array}{ll}
r = 4 \sin 2\theta & \text{polar equation} \\
= 4(2 \sin \theta \cos \theta) = 8 \sin \theta \cos \theta & \text{using identity} \\
\sqrt{x^2 + y^2} = 8\left(\dfrac{y}{r}\right)\left(\dfrac{x}{r}\right) = \dfrac{8xy}{r^2} = \dfrac{8xy}{x^2 + y^2} & \text{using Eqs. (21-35) and (21-36)} \\
x^2 + y^2 = \dfrac{64x^2y^2}{(x^2 + y^2)^2} & \text{squaring both sides} \\
(x^2 + y^2)^3 = 64x^2y^2 & \text{simplifying}
\end{array}
$$

Plotting the graph of this equation from the rectangular equation would be complicated. However, as we will see in the next section, plotting this graph in polar coordinates is quite simple. The curve is shown in Fig. 21-101.

━━━━━━━━ EXERCISES *21-9* ━━━━━━━━

In Exercises 1–12, plot the given polar coordinate points on polar coordinate paper.

1. $\left(3, \dfrac{\pi}{6}\right)$ **2.** $(2, \pi)$ **3.** $\left(\dfrac{5}{2}, -\dfrac{2\pi}{5}\right)$

4. $\left(5, -\dfrac{\pi}{3}\right)$ **5.** $\left(-2, \dfrac{7\pi}{6}\right)$ **6.** $\left(-5, \dfrac{\pi}{4}\right)$

7. $\left(-3, -\dfrac{5\pi}{4}\right)$ **8.** $\left(-4, -\dfrac{5\pi}{3}\right)$ **9.** $\left(0.5, -\dfrac{8\pi}{3}\right)$

10. $(2.2, -6\pi)$ **11.** $(2, 2)$ **12.** $(-1, -1)$

In Exercises 13–16, find a set of polar coordinates for each of the points for which the rectangular coordinates are given.

13. $(\sqrt{3}, 1)$ **14.** $(-1, -1)$

15. $\left(-\dfrac{\sqrt{3}}{2}, -\dfrac{1}{2}\right)$ **16.** $(-5, 4)$

In Exercises 17–20, find the rectangular coordinates for each of the points for which the polar coordinates are given.

17. $\left(8, \dfrac{4\pi}{3}\right)$ **18.** $(-4, -\pi)$

19. $(3.0, -0.40)$ **20.** $(-1.0, 1.0)$

In Exercises 21–28, find the polar equation of each of the given rectangular equations.

21. $x = 3$ **22.** $y = x$
23. $x + 2y = 3$ **24.** $x^2 + y^2 = 0.81$
25. $x^2 + (y - 2)^2 = 4$ **26.** $x^2 - y^2 = 0.01$
27. $x^2 + 4y^2 = 4$ **28.** $y^2 = 4x$

In Exercises 29–40, find the rectangular equation of each of the given polar equations. In Exercises 29–36, identify the curve that is represented by the equation.

29. $r = \sin \theta$ **30.** $r = 4 \cos \theta$

31. $r \cos \theta = 4$ **32.** $r \sin \theta = -2$

33. $r = \dfrac{2}{\cos \theta - 3 \sin \theta}$ **34.** $r = e^{r \cos \theta} \csc \theta$

35. $r = 4 \cos \theta + 2 \sin \theta$ **36.** $r \sin(\theta + \pi/6) = 3$

37. $r = 2(1 + \cos \theta)$ **38.** $r = 1 - \sin \theta$

39. $r^2 = \sin 2\theta$ **40.** $r^2 = 16 \cos 2\theta$

In Exercises 41–44, find the required equations.

41. Under certain conditions, the *x*- and *y*-components of a magnetic field *B* are given by the equations

$$B_x = \frac{-ky}{x^2 + y^2} \quad \text{and} \quad B_y = \frac{kx}{x^2 + y^2}$$

Write these equations in terms of polar coordinates.

42. In designing a domed roof for a building, an architect uses the equation $x^2 + \dfrac{y^2}{k^2} = 1$, where *k* is a constant. Write this equation in polar form.

43. The shape of a cam can be described by the polar equation $r = 3 - \sin \theta$. Find the rectangular equation for the shape of the cam.

44. The polar equation of the path of a weather satellite of the earth is $r = \dfrac{4800}{1 + 0.14 \cos \theta}$, where *r* is measured in miles. Find the rectangular equation of the path of this satellite. The path is an ellipse, with the earth at one of the foci.

21-10 CURVES IN POLAR COORDINATES

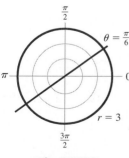

Fig. 21-102

The basic method for finding a curve in polar coordinates is the same as in rectangular coordinates. We assume values of θ and then find the corresponding values of *r*. These points are plotted and joined, thus forming the curve that represents the function. However, there are certain basic curves that can be sketched directly from the equation.

■**EXAMPLE 1**   **(a)** The graph of the polar equation $r = 3$ is a circle of radius 3, with center at the pole. This is the case, since $r = 3$ for all values of θ. It is not necessary to find specific points for this circle, which is shown in Fig. 21-102.

   (b) The graph of $\theta = \pi/6$ is a straight line through the pole. It represents all points for which $\theta = \pi/6$ for all values of *r*. This line is shown in Fig. 21-102. ■

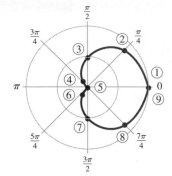

Fig. 21-103

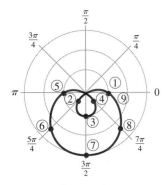

Fig. 21-104

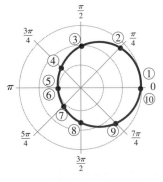

Fig. 21-105

EXAMPLE 2   Plot the graph of $r = 1 + \cos \theta$.

We find the following values of r corresponding to the chosen values of θ.

θ	0	$\frac{\pi}{4}$	$\frac{\pi}{2}$	$\frac{3\pi}{4}$	π	$\frac{5\pi}{4}$	$\frac{3\pi}{2}$	$\frac{7\pi}{4}$	2π
r	2	1.7	1	0.3	0	0.3	1	1.7	2
Point number	1	2	3	4	5	6	7	8	9

We now see that the points start repeating, and it is unnecessary to find additional points. The curve is called a **cardioid** and is shown in Fig. 21-103. ▪

EXAMPLE 3   Plot the graph of $r = 1 - 2 \sin \theta$.

Choosing values of θ, and then finding the corresponding values of r, we find the following table of values.

θ	0	$\frac{\pi}{4}$	$\frac{\pi}{2}$	$\frac{3\pi}{4}$	π	$\frac{5\pi}{4}$	$\frac{3\pi}{2}$	$\frac{7\pi}{4}$	2π
r	1	-0.4	-1	-0.4	1	2.4	3	2.4	1
Point number	1	2	3	4	5	6	7	8	9

Particular care should be taken in plotting the points for which r is negative. This curve is known as a **limaçon** and is shown in Fig. 21-104. ▪

EXAMPLE 4   A cam is shaped such that the edge of the upper "half" is represented by the equation $r = 2.0 + \cos \theta$ and the lower "half" by the equation $r = \dfrac{3.0}{2.0 - \cos \theta}$, where measurements are in inches. Plot the curve that represents the shape of the cam.

We get the points for the edge of the cam by using values of θ from 0 to π for the upper "half" and from π to 2π for the lower half. The table of values follows.

$r = 2.0 + \cos \theta$	θ	0	$\frac{\pi}{4}$	$\frac{\pi}{2}$	$\frac{3\pi}{4}$	π
	r	3.0	2.7	2.0	1.3	1.0
	Point number	1	2	3	4	5

$r = \dfrac{3.0}{2.0 - \cos \theta}$	θ	π	$\frac{5\pi}{4}$	$\frac{3\pi}{2}$	$\frac{7\pi}{4}$	2π
	r	1.0	1.1	1.5	2.3	3.0
	Point number	6	7	8	9	10

The upper "half" is part of a limaçon, and the lower "half" is a semiellipse. The cam is shown in Fig. 21-105. ▪

EXAMPLE 5   Plot the graph of $r = 2 \cos 2\theta$.

In finding values of r we must be careful first to multiply the values of θ by 2 before finding the cosine of the angle. Also, for this reason, we take values of θ as multiples of $\pi/12$, so as to get enough useful points. The table of values follows.

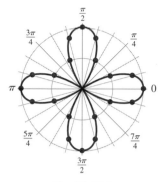

Fig. 21-106

θ	0	$\frac{\pi}{12}$	$\frac{\pi}{6}$	$\frac{\pi}{4}$	$\frac{\pi}{3}$	$\frac{5\pi}{12}$	$\frac{\pi}{2}$
r	2	1.7	1	0	-1	-1.7	-2

θ	$\frac{7\pi}{12}$	$\frac{2\pi}{3}$	$\frac{3\pi}{4}$	$\frac{5\pi}{6}$	$\frac{11\pi}{12}$	π
r	-1.7	-1	0	1	1.7	2

For values of θ starting with π, the values of θ repeat. We have a four-leaf **rose,** as shown in Fig. 21-106. ∎

EXAMPLE 6   Plot the graph of $r^2 = 9 \cos 2\theta$.

Choosing the indicated values of θ, we get the values of r shown in the following table of values.

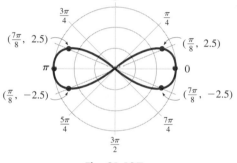

Fig. 21-107

θ	0	$\frac{\pi}{8}$	$\frac{\pi}{4}$	...	$\frac{3\pi}{4}$	$\frac{7\pi}{8}$	π
r	±3	±2.5	0		0	±2.5	±3

There are no values of r corresponding to values of θ in the range $\pi/4 < \theta < 3\pi/4$, since twice these angles are in the second and third quadrants, and the cosine is negative for such angles. The value of r^2 cannot be negative. Also, the values of r repeat for $\theta > \pi$. The figure is called a **lemniscate** and is shown in Fig. 21-107. ∎

EXAMPLE 7   View the graph of $r = 1 - 2 \cos \theta$ on a graphing calculator.

Using the *mode* feature, a polar equation is displayed using the polar graph option or the parametric graph option, depending on the calculator. (Review the manual for the calculator.)

With the polar graph option, the function is entered directly. The values for the viewing window are determined by settings for *x, y,* and the angles θ that will be used. These values are set in a manner similar to those used for parametric equations. (See page 296 for an example of graphing parametric equations.)

With the parametric graph option, to graph $r = f(\theta)$, we note that $x = r \cos \theta$ and $y = r \sin \theta$. This tells us that

$$x = f(\theta)\cos \theta$$
$$y = f(\theta)\sin \theta$$

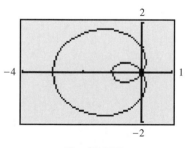

Fig. 21-108

Thus, for $r = 1 - 2 \cos \theta$, by using

$$x = (1 - 2 \cos \theta)\cos \theta$$
$$y = (1 - 2 \cos \theta)\sin \theta$$

the graph can be displayed, as shown in Fig. 21-108. ∎

$$\text{EXERCISES } 21\text{-}10$$

In Exercises 1–28, plot the curves of the given polar equations in polar coordinates.

1. $r = 4$ **2.** $r = 2$ **3.** $\theta = 3\pi/4$

4. $\theta = -1.5$ **5.** $r = 4 \sec \theta$ **6.** $r = 4 \csc \theta$

7. $r = 2 \sin \theta$ **8.** $r = 3 \cos \theta$

9. $r = 1 - \cos \theta$ (cardioid)

10. $r = \sin \theta - 1$ (cardioid)

11. $r = 2 - \cos \theta$ (limaçon)

12. $r = 2 + 3 \sin \theta$ (limaçon)

13. $r = 4 \sin 2\theta$ (rose) **14.** $r = 2 \sin 3\theta$ (rose)

15. $r^2 = 4 \sin 2\theta$ (lemniscate)

16. $r^2 = 2 \sin \theta$

17. $r = 2^{\theta}$ (spiral) **18.** $r = 1.5^{-\theta}$ (spiral)

19. $r = 4|\sin 3\theta|$ **20.** $r = 2 \sin \theta \tan \theta$ (cissoid)

21. $r = \dfrac{1}{2 - \cos \theta}$ (ellipse)

22. $r = \dfrac{1}{1 - \cos \theta}$ (parabola)

23. $r = \dfrac{6}{1 - 2 \cos \theta}$ (hyperbola)

24. $r = \dfrac{6}{3 - 2 \sin \theta}$ (ellipse)

25. $r = 4 \cos \frac{1}{2}\theta$ **26.** $r = 2 + \cos 3\theta$

27. $r = 2(1 - \sin(\theta - \pi/4))$ **28.** $r = 4 \tan \theta$

In Exercises 29–32, view the curves of the given polar equations on a graphing calculator.

29. $r = \theta$ $(-20 \le \theta \le 20)$ **30.** $r = 0.5^{\sin \theta}$

31. $r = 2 \sec \theta + 1$ **32.** $r = 2 \cos(\cos 2\theta)$

In Exercises 33–36, sketch the indicated graphs.

33. An architect designs a patio shaped such that it can be described as the area within the polar curve $r = 4.0 - \sin \theta$, where measurements are in meters. Sketch the curve that represents the perimeter of the patio.

34. A missile is fired at an airplane and is always directed toward the airplane. The missile is traveling at twice the speed of the airplane. An equation that describes the distance r between the missile and the airplane is $r = \dfrac{70 \sin \theta}{(1 - \cos \theta)^2}$, where θ is the angle between their directions at all times. See Fig. 21-109. This is a *relative pursuit curve*. Sketch the graph of this equation for $\pi/4 \le \theta \le \pi$.

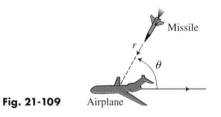

Fig. 21-109 Airplane

35. In studying the photoelectric effect, an equation used for the rate R at which photoelectrons are ejected at various angles θ is $R = \dfrac{\sin^2 \theta}{(1 - 0.5 \cos \theta)^2}$. Sketch the graph.

36. Sketch the graph of the rectangular equation
$4(x^6 + 3x^4y^2 + 3x^2y^4 + y^6 - x^4 - 2x^2y^2 - y^4) + y^2 = 0$
(*Hint:* The equation can be written as
$4(x^2 + y^2)^3 - 4(x^2 + y^2)^2 + y^2 = 0$.) Transform this equation to polar coordinates and then sketch the curve.

CHAPTER EQUATIONS

Distance formula	Fig. 21-2	$d = \sqrt{(x_2 - x_1)^2 + (y_2 - y_1)^2}$	**(21-1)**
Slope	Fig. 21-4	$m = \dfrac{y_2 - y_1}{x_2 - x_1}$	**(21-2)**
	Fig. 21-7	$m = \tan \alpha \quad (0° \le \alpha < 180°)$	**(21-3)**
	Fig. 21-9	$m_1 = m_2 \quad$ (for $\parallel$ lines)	**(21-4)**
	Fig. 21-10	$m_2 = -\dfrac{1}{m_1} \quad$ or $\quad m_1 m_2 = -1 \quad$ (for $\perp$ lines)	**(21-5)**

Straight line	Fig. 21-14	$y - y_1 = m(x - x_1)$	**(21-6)**
	Fig. 21-17	$x = a$	**(21-7)**
	Fig. 21-18	$y = b$	**(21-8)**
	Fig. 21-21	$y = mx + b$	**(21-9)**
		$Ax + By + C = 0$	**(21-10)**
Circle	Fig. 21-30	$(x - h)^2 + (y - k)^2 = r^2$	**(21-11)**
	Fig. 21-33	$x^2 + y^2 = r^2$	**(21-12)**
		$x^2 + y^2 + Dx + Ey + F = 0$	**(21-14)**
Parabola	Fig. 21-40	$y^2 = 4px$	**(21-15)**
	Fig. 21-43	$x^2 = 4py$	**(21-16)**
Ellipse	Fig. 21-56	$\dfrac{x^2}{a^2} + \dfrac{y^2}{b^2} = 1$	**(21-17)**
	Fig. 21-56	$a^2 = b^2 + c^2$	**(21-18)**
	Fig. 21-57	$\dfrac{y^2}{a^2} + \dfrac{x^2}{b^2} = 1$	**(21-19)**
Hyperbola	Fig. 21-69	$\dfrac{x^2}{a^2} - \dfrac{y^2}{b^2} = 1$	**(21-20)**
	Fig. 21-69	$c^2 = a^2 + b^2$	**(21-21)**
	Fig. 21-68	$y = \pm \dfrac{bx}{a}$ (asymptotes)	**(21-23)**
	Fig. 21-70	$\dfrac{y^2}{a^2} - \dfrac{x^2}{b^2} = 1$	**(21-24)**
	Fig. 21-75	$xy = c$	**(21-25)**
Translation of axes	Fig. 21-79	$x = x' + h$ and $y = y' + k$	**(21-26)**
		$x' = x - h$ and $y' = y - k$	**(21-27)**
Parabola, vertex (h, k):		$(y - k)^2 = 4p(x - h)$ (axis parallel to x-axis)	**(21-28)**
		$(x - h)^2 = 4p(y - k)$ (axis parallel to y-axis)	**(21-29)**
Ellipse, center (h, k):		$\dfrac{(x - h)^2}{a^2} + \dfrac{(y - k)^2}{b^2} = 1$ (major axis parallel to x-axis)	**(21-30)**
		$\dfrac{(y - k)^2}{a^2} + \dfrac{(x - h)^2}{b^2} = 1$ (major axis parallel to y-axis)	**(21-31)**
Hyperbola, center (h, k):		$\dfrac{(x - h)^2}{a^2} - \dfrac{(y - k)^2}{b^2} = 1$ (transverse axis parallel to x-axis)	**(21-32)**
		$\dfrac{(y - k)^2}{a^2} - \dfrac{(x - h)^2}{b^2} = 1$ (transverse axis parallel to y-axis)	**(21-33)**
Second-degree equation		$Ax^2 + Bxy + Cy^2 + Dx + Ey + F = 0$	**(21-34)**
Polar coordinates	Fig. 21-96	$x = r \cos \theta$ $y = r \sin \theta$	**(21-35)**
		$\tan \theta = \dfrac{y}{x}$ $r = \sqrt{x^2 + y^2}$	**(21-36)**

REVIEW EXERCISES

In Exercises 1–12, find the equation of the indicated curve subject to the given conditions. Sketch each curve.

1. Straight line: passes through $(1, -7)$ with a slope of 4
2. Straight line: passes through $(-1, 5)$ and $(-2, -3)$
3. Straight line: perpendicular to $3x - 2y + 8 = 0$ and has a y-intercept of $(0, -1)$
4. Straight line: parallel to $2x - 5y + 1 = 0$ and has an x-intercept of $(2, 0)$
5. Circle: center at $(1, -2)$, passes through $(4, -3)$
6. Circle: tangent to the line $x = 3$, center at $(5, 1)$
7. Parabola: focus $(3, 0)$, vertex $(0, 0)$
8. Parabola: directrix $y = -5$, vertex $(0, 0)$
9. Ellipse: vertex $(10, 0)$, focus $(8, 0)$, center $(0, 0)$
10. Ellipse: center $(0, 0)$, passes through $(0, 3)$ and $(2, 1)$
11. Hyperbola: $V(0, 13)$, $C(0, 0)$, conj. axis of 24
12. Hyperbola: $F(0, 10)$, $F(0, -10)$, $V(0, 8)$

In Exercises 13–24, find the indicated quantities for each of the given equations. Sketch each curve.

13. $x^2 + y^2 + 6x - 7 = 0$, center and radius
14. $x^2 + y^2 - 4x + 2y - 20 = 0$, center and radius
15. $x^2 = -20y$, focus and directrix
16. $y^2 = 24x$, focus and directrix
17. $16x^2 + y^2 = 16$, vertices and foci
18. $2y^2 - 9x^2 = 18$, vertices and foci
19. $2x^2 - 5y^2 = 0.25$, vertices and foci
20. $2x^2 + 25y^2 = 800$, vertices and foci
21. $x^2 - 8x - 4y - 16 = 0$, vertex and focus
22. $y^2 - 4x + 4y + 24 = 0$, vertex and directrix
23. $4x^2 + y^2 - 16x + 2y + 13 = 0$, center
24. $x^2 - 2y^2 + 4x + 4y + 6 = 0$, center

In Exercises 25–32, plot the given curves in polar coordinates.

25. $r = 4(1 + \sin \theta)$
26. $r = 1 - 3 \cos \theta$
27. $r = 4 \cos 3\theta$
28. $r = 3 \sin \theta - 4 \cos \theta$
29. $r = \dfrac{3}{\sin \theta + 2 \cos \theta}$
30. $r = \dfrac{1}{2(\sin \theta - 1)}$
31. $r = 2 \sin \left(\dfrac{\theta}{2} \right)$
32. $r = 1 - \cos 2\theta$

In Exercises 33–36, find the polar equation of each of the given rectangular equations.

33. $y = 2x$
34. $2xy = 1$
35. $x^2 + xy + y^2 = 2$
36. $x^2 + (y + 3)^2 = 16$

In Exercises 37–40, find the rectangular equation of each of the given polar equations.

37. $r = 2 \sin 2\theta$
38. $r^2 = \sin \theta$
39. $r = \dfrac{4}{2 - \cos \theta}$
40. $r = 4 \tan \theta \sec \theta$

In Exercises 41–44, determine the number of real solutions of the given systems of equations by sketching the indicated curves. (See Section 14-1.)

41. $x^2 + y^2 = 9$
    $4x^2 + y^2 = 16$
42. $y = e^x$
    $x^2 - y^2 = 1$
43. $x^2 + y^2 - 4y - 5 = 0$
    $y^2 - 4x^2 - 4 = 0$
44. $x^2 - 4y^2 + 2x - 3 = 0$
    $y^2 - 4x - 4 = 0$

In Exercises 45–48, view the curves of the given equations on a graphing calculator.

45. $x^2 - 4y^2 + 4x + 24y - 48 = 0$
46. $x^2 + 2xy + y^2 - 3x + 8y = 0$
47. $r = 3 \cos(3\theta/2)$
48. $r = 5 - 2 \sin 4\theta$

In Exercises 49–80, solve the given problems.

49. In two ways show that the line segments joining $(-3, 11)$, $(2, -1)$, and $(14, 4)$ form a right triangle.
50. Show that the altitudes of the triangle with vertices $(2, -4)$, $(3, -1)$, and $(-2, 5)$ meet at a single point.
51. Find the area of the square that can be inscribed in the ellipse $7x^2 + 2y^2 = 18$.
52. Using a graphing calculator, determine the number of points of intersection of the polar curves $r = 4|\cos 2\theta|$ and $r = 6 \sin(\cos(\cos 3\theta))$.
53. By means of the definition of a parabola, find the equation of the parabola with focus at $(3, 1)$ and directrix the line $y = -3$. Find the same equation by the method of translation of axes.
54. For what value of k does $x^2 - ky^2 = 1$ represent an ellipse with vertices on the y-axis?
55. The total resistance R_T of two resistances in series in an electric circuit is the sum of the resistances. If a variable resistor R is in series with a 2.5-Ω resistor, express R_T as a function of R and sketch the graph.
56. The acceleration of an object is defined as the change in velocity v divided by the corresponding change in time t. Find the equation relating the velocity v and time t for an object for which the acceleration is 20 ft/s^2 and $v = 5.0 \text{ ft/s}$ when $t = 0$ s.
57. One computer printer prints 2500 lines/min for x min, and a second printer prints 1500 lines/min for y min. If they print a total of 37,500 lines together, express y as a function of x and sketch the graph.

58. An airplane touches down when landing at 100 mi/h. Its velocity v while coming to a stop is given by $v = 100 - 20{,}000t$, where t is the time in hours. Sketch the graph of v vs. t.

59. It takes 2.010 kJ of heat to raise the temperature of 1.000 kg of steam by 1.000°C. In a steam generator, a total of y kJ is used to raise the temperature of 50.00 kg of steam from 100°C to T°C. Express y as a function of T and sketch the graph.

60. The temperature in a certain region is 27°C, and at an altitude of 2500 m above the region it is 12°C. If the equation relating the temperature T and the altitude h is linear, find the equation.

61. The radar gun on a police helicopter 490 ft above a multilane highway is directed vertically down onto the highway. If the radar gun signal is cone-shaped with a vertex angle of 14°, what area of the highway is covered by the signal?

62. One of the most famous Ferris wheels is Vienna's (Austria) *Prata*. Find the equation representing its circumference, given that $c = 191$ m. Place the origin of the coordinate system 2.0 m below the bottom of the wheel, and the center of the wheel on the y-axis.

63. The top horizontal cross section of a dam is parabolic. The open area within this cross section is 80 ft across and 50 ft from front to back. Find the equation of the edge of the open area with the vertex at the origin of the coordinate system and the axis along the x-axis.

64. The *quality factor Q* of a series resonant electric circuit with resistance R, inductance L, and capacitance C is given by $Q = \dfrac{1}{R}\sqrt{\dfrac{L}{C}}$. Sketch the graph of Q and L for a circuit in which $R = 1000\ \Omega$ and $C = 4.00\ \mu\text{F}$.

65. A rectangular parking lot is to have a perimeter of 600 m. Express the area A in terms of the width w and sketch the graph.

66. At very low temperatures, certain metals have an electric resistance of zero. This phenomenon is called *superconductivity*. A magnetic field also affects the superconductivity. A certain level of magnetic field H_T, the threshold field, is related to the thermodynamic temperature T by $H_T/H_0 = 1 - (T/T_0)^2$, where H_0 and T_0 are specifically defined values of magnetic field and temperature. Sketch the graph of H_T/H_0 vs. T/T_0.

67. The electric power P (in W) supplied by a battery is given by $P = 12.0i - 0.500i^2$, where i is the current (in A). Sketch the graph of P vs. i.

68. The Colosseum in Rome is in the shape of an ellipse 188 m long and 156 m wide. Find the area of the Colosseum. ($A = \pi ab$ for an ellipse.)

69. A specialty electronics company makes an ultrasonic device to repel animals. It emits a 20–25 kHz sound (above those heard by people), which is unpleasant to animals. The sound covers an elliptical area starting at the device, with the longest dimension extending 120 ft from the device and the focus of the area 15 ft from the device. Find the area covered by the signal. ($A = \pi ab$)

70. A study indicated that the fraction f of cells destroyed by various dosages d of X-rays is given by the graph in Fig. 21-110. Assuming that the curve is a quarter-ellipse, find the equation relating f and d for $0 \le f \le 1$ and $0 < d \le 10$ units.

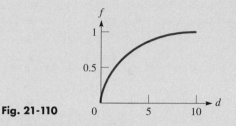

Fig. 21-110

71. A machine-part designer wishes to make a model for an elliptical cam by placing two pins in a design board, putting a loop of string over the pins, and marking off the outline by keeping the string taut. (Note that the definition of the ellipse is being used.) If the cam is to measure 10 cm by 6 cm, how long should the loop of string be and how far apart should the pins be?

72. Soon after reaching the vicinity of the moon, *Apollo 11* (the first spacecraft to land a man on the moon) went into an elliptical lunar orbit. The closest the craft was to the moon in this orbit was 70 mi, and the farthest it was from the moon was 190 mi. What was the equation of the path if the center of the moon was at one of the foci of the ellipse? Assume that the major axis is along the x-axis and that the center of the ellipse is at the origin. The radius of the moon is 1080 mi.

73. The vertical cross section of the cooling tower of a nuclear power plant is hyperbolic, as shown in Fig. 21-111. Find the radius r of the smallest circular horizontal cross section.

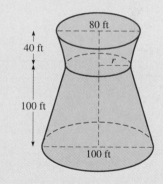

Fig. 21-111

74. Tremors from an earthquake are recorded at the California Institute of Technology (Pasadena, California) 36 s before they are recorded at Stanford University (Palo Alto, California). If the seismographs are 510 km apart and the shock waves from the tremors travel at 5.0 km/s, what is the curve on which lies the point where the earthquake occurred?

75. An electronic instrument located at point P records the sound of a rifle shot and the impact of the bullet striking the target at the same instant. Show that P lies on a branch of a hyperbola.

76. A 60-ft rope passes over a pulley 10 ft above the ground, and a crate on the ground is attached at one end. The other end of the rope is held at a level of 4 ft above the ground and is drawn away from the pulley. Express the height of the crate over the ground in terms of the distance the person is from directly below the crate. Sketch the graph of distance and height. See Fig. 21-112. (Neglect the thickness of the crate.)

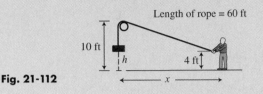

Fig. 21-112

77. A satellite at an altitude proper to make one revolution per day around the center of the earth will have for an excellent approximation of its projection on the earth of its path the curve $r^2 = R^2 \cos 2(\theta + \frac{\pi}{2})$, where R is the radius of the earth. Sketch the path of the projection.

78. The vertical cross sections of two pipes as drawn on a drawing board are shown in Fig. 21-113. Find the polar equation of each.

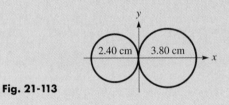

Fig. 21-113

79. The path of a certain plane is $r = 200(\sec \theta + \tan \theta)^{-5}/\cos \theta$, $0 < \theta < \pi/2$. Sketch the path and check it on a graphing calculator. (This path is the same as for the plane in Exercise 84 on page 320.)

80. The sound produced by a jet engine was measured at a distance of 100 m in all directions. The loudness of the sound d (in decibels) was found to be $d = 115 + 10 \cos \theta$, where the 0° line for the angle θ is directed in front of the engine. Sketch the graph of d vs. θ in polar coordinates (use d as r).

Writing Exercise

81. Under a force that varies inversely as the square of the distance from an attracting object (such as the sun exerts on the earth), it can be shown that the equation of the path an object follows is given in general by

$$\frac{1}{r} = a + b \cos \theta$$

where a and b are constants for a particular path. First, transform this equation into rectangular coordinates. Then, write one or two paragraphs explaining why this equation represents one of the conic sections, depending on the values of a and b. It is through this kind of analysis that we know the paths of the planets and comets are conic sections.

PRACTICE TEST

1. Identify the type of curve represented by the equation $2(x^2 + x) = 1 - y^2$.

2. Sketch the graph of the straight line $4x - 2y + 5 = 0$ by finding its slope and y-intercept.

3. Find the polar equation of the curve whose rectangular equation is $x^2 = 2x - y^2$.

4. Find the vertex and the focus of the parabola $x^2 = -12y$. Sketch the graph.

5. Find the equation of the circle with center at $(-1, 2)$ and that passes through $(2, 3)$.

6. Find the equation of the straight line that passes through $(-4, 1)$ and $(2, -2)$.

7. Where is the focus of a parabolic reflector that is 12.0 cm across and 4.00 cm deep?

8. A hallway 16 ft wide has a ceiling whose cross section is a semiellipse. The ceiling is 10 ft high at the walls and 14 ft high at the center. Find the height of the ceiling 4 ft from each wall.

9. Plot the polar curve $r = 3 + \cos \theta$.

10. Find the center and vertices of the conic section $4y^2 - x^2 - 4x - 8y - 4 = 0$. Show completely the sketch of the curve.

22

INTRODUCTION TO STATISTICS

In finding the volume of a cylindrical support column, an engineer would use the formula $V = \pi r^2 h$, which is true for all right circular cylinders. However, to find the safe load a cable can support, it is not possible simply to write down an equation. The safe load depends on the diameter of the cable, the quality of the material from which it is made, and many other possible differences of this cable from other cables.

Thus, to determine the load the cable can support, the engineer may test cables of particular specifications. This testing will show that most cables can support approximately the same load, but that there is some variation and occasionally perhaps a great variation for some reason or other.

The engineer, in testing a number of cables and thereby selecting a certain type of cable, is making use of the basic methods of statistics. That is, the engineer (1) collects the data, (2) then analyzes the data, and (3) interprets the data. In this chapter we will first discuss some of the basic methods of tabulating and analyzing this type of statistical information.

Applications of statistics occur in nearly all fields of study, including science and technology. A section of this chapter on *normal distributions* is devoted to very large sets of statistical data that may be used in analyzing the production of various products. Also, one of the most important applications in industry today is noted when we briefly introduce the topic of *control charts* that are used in quality control. Other applications will be shown in the examples and exercises.

The final sections of this chapter show how to start with a set of points and find an equation that best "fits" this data found through experimentation and observation. Such an equation can show a basic relationship between the variables and can make it possible to analyze the experiment better. This can be particularly useful in research and development leading to new and improved products.

This chapter gives an introduction to some of the basic concepts and uses of statistics. In a more complete coverage a number of other useful methods with important applications would be developed.

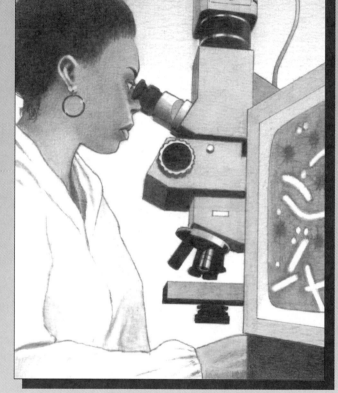

Statistical analysis is used extensively in medical research. In Section 22-4, a use of curve fitting is shown.

The microscope was invented in 1609 by a Dutch spectacle-maker, Zacharias Janssen.

22-1 FREQUENCY DISTRIBUTIONS

In statistics we first collect data and thereby form a set of numbers. These numbers could be measurements of electric current, test scores, lengths of objects, or numerous other possibilities. *Data that have been collected but not yet organized are called* **raw data.** In order to obtain useful information from the data it is necessary to organize it in some way. Normally, a first step in organizing the data is to arrange the numerical values in ascending (or descending) order, called an **array.** An example of raw data and an array is shown in the following example.

■EXAMPLE 1   A survey of 50 users of home computers with Internet access asked each user to estimate carefully the number of hours they spent each week on the Internet. Following are the estimates.

12, 20, 15, 14, 7, 10, 12, 25, 18, 5, 10, 24, 16, 3, 12, 14, 28, 8, 13, 18,
15, 8, 11, 15, 14, 22, 14, 19, 6, 10, 18, 4, 16, 24, 18, 5, 13, 20, 12, 12,
25, 11, 8, 12, 20, 5, 10, 15, 13, 8

As we can see, no clear pattern can be seen from this raw data. Arranging these in numerical order to form an array, we can summarize the array by showing the number of persons reporting each estimate as follows:

(hours-persons) 3-1, 4-1, 5-3, 6-1, 7-1, 8-4, 10-4, 11-2, 12-6
13-3, 14-4, 15-4, 16-2, 18-4, 19-1, 20-3, 22-1, 24-2, 25-2, 28-1   ──────■

In Example 1, although a pattern is somewhat clearer from the array than from the raw data, a still clearer pattern is found by *grouping the data.* In the process of grouping, the detail of the raw data is lost, but the advantage is that a much clearer overall pattern of the data can be obtained.

The grouping of data is done by first defining what values are to be included in each group and then tabulating the number of all values that are within each group. *Each group is called a* **class,** and *the number of values in the class is called the* **frequency.** *The table is called a* **frequency distribution table.** This is illustrated in the following example.

■EXAMPLE 2   In Example 1 we note that the estimates vary from 3 h to 28 h. We can see that if we form *classes* of 0–4 h, 5–9 h, etc., we will have five possible estimates in each class, and that there will be six classes. This gives us the following table of values showing the number of persons (frequency) reporting the indicated estimate of hours on the Internet:

Estimate (hours)	0–4	5–9	10–14	15–19	20–24	25–29
Frequency (persons)	2	9	19	11	6	3

This table shows us the frequency distribution. The 0, 5, 10, and so on are the *lower class limits,* and the 4, 9, 14, and so on are the *upper class limits.* Each class includes five values, which is the *class width.* As with the class limits we have chosen, it is generally preferable to have the same width for each class.

We can see from this frequency distribution table that the pattern of hours on the Internet by the persons responding to the survey is clearer.   ──────■

At times it is also helpful to know the **relative frequency** *of the class, which is the frequency of the class divided by the total frequency of all classes.* The relative frequency can be expressed as a fraction, decimal, or percent.

■EXAMPLE 3   The relative frequency of each class for the data in Example 2 can be shown as in the following table:

Estimated hours on Internet	Frequency	Relative Frequency (%)	
0–4	2	4	2/50 = 0.04 = 4%
5–9	9	18	
10–14	19	38	
15–19	11	22	
20–24	6	12	
25–29	3	6	
Total	50	100	

Just as graphs are useful in representing algebraic functions, so are they a very convenient method of representing frequency distributions. There are several useful types of graphs for such distributions. *Among the most important of which are the* **histogram** *and the* **frequency polygon.** These graphical representations are gener ally constructed to represent the data in a frequency distribution table.

In order to represent *grouped* data, where the raw data values are generally not all the same within a given class, we find it necessary to use a representative value for each class. For this we use the **class mark,** *which is found by dividing the sum of the lower and upper class limits by 2.* The following examples illustrate the use of a class mark with a histogram and a frequency polygon.

■EXAMPLE 4   *A histogram represents a particular set of data by displaying each class of the data as a rectangle.* Each rectangle is labeled at the center of its base by the *class mark.* The width of each rectangle represents the *class width,* and the height of the rectangle represents the *frequency* of the class.

For the data in Example 2 on estimated hours on the Internet, the class marks are $(0 + 4)/2 = 2$, $(5 + 9)/2 = 7$, and so on. Therefore, a histogram representing these data is shown in Fig. 22-1.

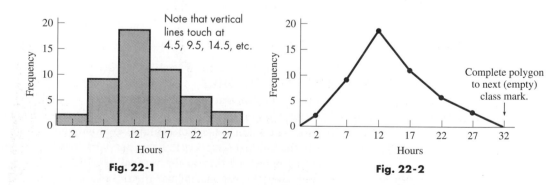

Note that vertical lines touch at 4.5, 9.5, 14.5, etc.

Fig. 22-1

Complete polygon to next (empty) class mark.

Fig. 22-2

■EXAMPLE 5   *A frequency polygon is used to represent a set of data by plotting the class marks as abscissas (x-values) and the frequencies as ordinates (y-values).* The resulting points are joined by straight-line segments.

A frequency polygon representing the data in Example 2 on estimated hours on the Internet is shown in Fig. 22-2.

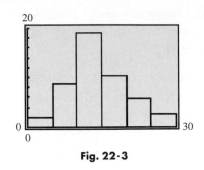

Fig. 22-3

Computer spreadsheets are very useful for this type of analysis.

A graphing calculator may be used to display histograms and frequency polygons. The manual should be reviewed to see how this is done on any particular model. On a calculator it is necessary to enter both sets of numbers, choose proper *window* settings, and select the correct type of statistical graph. A calculator display for the histogram in Fig. 22-1 is shown in Fig. 22-3.

When dividing data into classes, we must decide how many classes to use. A general guideline is that *the number of classes should be no less than 5 nor greater than 15* (this might go to 20 if we have thousands of values to analyze). The principal consideration is that we get a good pattern of the distribution. This is illustrated in the following example.

EXAMPLE 6   If we divided the data of estimated Internet hours of Example 1 into nine classes, the frequency table would be as follows:

Estimate (hours)	3–5	6–8	9–11	12–14	15–17	18–20	21–23	24–26	27–29
Freq. (respondents)	5	6	6	13	6	8	1	4	1

We can see that this does not give nearly as good an idea of the pattern as the six classes shown in Example 1.

Another way of analyzing data is to use *cumulative* totals. The way this is generally done is to change the frequency into a "less than" or a "more than" **cumulative frequency.** To do this we *add the class frequencies,* starting either at the greatest class boundary or the lowest class boundary. In our discussion, we will restrict our attention to a "less than" cumulative frequency. The graphical display that is generally used for cumulative frequency is called an **ogive** (pronounced oh-jive).

EXAMPLE 7   For the data on estimated hours on the Internet in Examples 2 and 3, the cumulative frequency is shown in the following table.

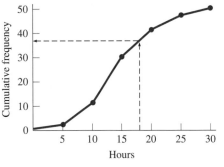

Fig. 22-4

Estimated hours on Internet	Cumulative Frequency
Less than 5	2
Less than 10	11
Less than 15	30
Less than 20	41
Less than 25	47
Less than 30	50

The ogive showing the cumulative frequency for the values in this table is shown in Fig. 22-4. The vertical scale shows the *frequency,* and the horizontal scale shows the *class boundaries.*

One important use of an ogive is to determine the number of values above or below a certain value. For example, to *approximate* the number of respondents that use the Internet *less than* 18 hours per week, we draw a line from the horizontal axis to the ogive and then to the vertical axis as shown in Fig. 22-4. From this we see that *about* 37 respondents use the Internet *less than* 18 hours per week.

If the data with which we are dealing has only a limited number of values and we do not divide it into classes, we can use the methods we have devleoped. We use the specific values rather than *class* values. Consider the following example.

EXAMPLE 8 A test station measured the loudness of the sound of jet aircraft taking off from a certain airport. The decibel readings were measured to the nearest 5 decibels, and the readings for the first 20 jets were as follows: 110, 95, 100, 115, 105, 110, 120, 110, 115, 105, 90, 95, 105, 110, 100, 115, 105, 120, 95, 110.

Since there are only seven different values in the twenty readings, the best idea of the pattern of readings is determined by using these seven values. Using these seven values, we have the following frequency distribution table:

Decibel reading	90	95	100	105	110	115	120
Frequency	1	3	2	4	5	3	2

The histogram for this table is shown in Fig. 22-5, and the frequency polygon is shown in Fig. 22-6.

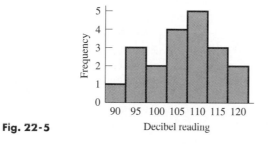

Fig. 22-5

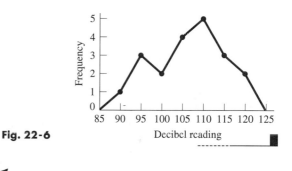

Fig. 22-6

EXERCISES *22-1*

In Exercises 1–8, use the following set of numbers as the raw data.

   103, 108, 106, 104, 109, 104, 110, 108, 108, 104,
   113, 106, 107, 106, 107, 109, 105, 111, 109, 108

1. Form a frequency distribution table for these numbers.

2. Find the relative frequencies of these numbers.

3. Form a frequency distribution table with five classes for these numbers.

4. Find the relative frequencies for the data in the frequency distribution table of Exercise 3.

5. Draw a histogram for the data of Exercise 3.

6. Draw a frequency polygon for the data of Exercise 3.

7. Form a cumulative frequency table for the data of Exercise 3.

8. Draw an ogive for the data of Exercise 3.

In Exercises 9–28, find the indicated quantities.

9. In testing a computer system, the number of instructions it could perform in 1 ns was measured at different points in a program. The numbers of instructions were recorded as follows:

   19, 21, 22, 25, 22, 20, 18, 21, 20, 19, 22, 21, 19, 23, 21

   Form a frequency distribution table for these values.

10. For the data of Exercise 9, draw a histogram.

11. For the data of Exercise 9, draw a frequency polygon.

12. For the data of Exercise 9, form a relative frequency distribution table.

13. A strobe light is designed to flash every 2.25 s at a certain setting. Sample bulbs were tested with the following results:

Time (s) *between* flashes (class mark)	2.21	2.22	2.23	2.24
Number of bulbs	2	7	18	41

Time (s)	2.25	2.26	2.27	2.28	2.29
No. bulbs	56	32	8	3	3

Draw a histogram for these data.

14. For the data of Exercise 13, draw a frequency polygon.

15. For the data of Exercise 13, form a cumulative frequency distribution table.

16. For the data of Exercise 13, draw an ogive.

17. In testing a braking system, the distance required to stop a car from 70 mi/h was measured in 120 trials. The results are shown in the following distribution table.

Stopping distance (ft)	155–159	160–164	165–169	170–174
Times car stopped	2	15	32	36

Stopping distance (ft)	175–179	180–184	185–189
Times car stopped	24	10	1

Form a relative frequency distribution table for these data.

18. For the data in Exercise 17, form a cumulative frequency distribution table.

19. For the data of Exercise 17, draw an ogive.

20. From the ogive in Exercise 19, estimate the number of cars that stopped in less than 176 ft.

21. The dosage, in milliroentgens (mR), given by a particular X-ray machine was measured 20 times, with the following readings:

    4.25, 4.36, 3.96, 4.21, 4.44, 3.83, 4.37, 4.27, 4.33, 4.34, 4.15, 3.90, 4.41, 4.51, 4.18, 4.26, 4.29, 4.09, 4.36, 4.23

    Form a histogram with six classes and the lowest class mark of 3.80 mR.

22. For the data used for the histogram in Exercise 21, draw a frequency polygon.

23. The life of a certain type of battery was measured for a sample of batteries with the following results (in number of hours):

    34, 30, 32, 35, 31, 28, 29, 30, 32, 25, 31, 30, 28, 36, 33, 34, 30, 33, 31, 34, 29, 30, 32

    Draw a frequency polygon using six classes.

24. For the data in Exercise 23, draw a cumulative frequency distribution table using six classes.

25. The diameters of a sample of fiber-optic cables were measured with the following results (diameters are class marks):

Diam. (mm)	0.0055	0.0056	0.0057	0.0058	0.0059	0.0060
No. cables	4	15	32	36	59	64

Diam. (mm)	0.0061	0.0062	0.0063	0.0064	0.0065	0.0066
No. cables	22	18	10	12	4	4

Draw a histogram for these data.

26. For the data of Exercise 25, draw a histogram with six classes. Compare the pattern of distribution with that of the histogram in Exercise 25.

(W) 27. Toss four coins 50 times and tabulate the number of heads that appear for each toss. Draw a frequency polygon showing the number of tosses for which 0, 1, 2, 3, or 4 heads appeared. Describe the distribution (is it about what should be expected?).

(W) 28. Most calculators can generate random numbers (between 0 and 1). On a calculator, display 50 random numbers and record the first digit. Draw a histogram showing the number of times for which each first digit (0, 1, 2, . . . , 9) appeared. Describe the distribution (is it about what should be expected?).

22-2 MEASURES OF CENTRAL TENDENCY

Tables and graphical representations give a general description of data. However, it is often useful and convenient to find representative values for the location of the center of the distribution, and other numbers to give a measure of the deviation from this central value. In this way we can obtain an arithmetical description of the data. We shall now discuss the values commonly used to measure the location of the center of the distribution. These are referred to as *measures of central tendency*.

MEDIAN

Some use of statistics was developed in the 1600s, but very little was done with statistical measures until the early 1800s.

The first of these measures of central tendency is the **median.** *The median is the middle number, that number for which there are as many above it as below it in the distribution.* If there is no middle number, the median is that number halfway between the two numbers nearest the middle of the distribution.

EXAMPLE 1 Given the numbers $5, 2, 6, 4, 7, 4, 7, 2, 8, 9, 4, 11, 9, 1, 3$, we first arrange them in numerical order. This arrangement is

┌ middle number
↓

$1, 2, 2, 3, 4, 4, 4, 5, 6, 7, 7, 8, 9, 9, 11$

Since there are 15 numbers, the middle number is the eighth. Since the eighth number is 5, the median is 5.

If the number 11 is not included in this set of numbers, and there are only 14 numbers in all, the median is that number halfway between the seventh and eighth numbers. Since the seventh is 4 and the eighth is 5, the median is 4.5. ∎

EXAMPLE 2 In the distribution of Internet hours in Example 1 of Section 22-1, the median is 13 h. There are 50 values in all, and when listed in order, the 24th to 26th values are each 13 h. The number halfway between the 25th and 26th value is the median. Since both are 13 h, the median is 13 h. ∎

ARITHMETIC MEAN   Another very widely applied measure of central tendency is the **arithmetic mean.** *The mean is calculated by finding the sum of all the values and then dividing by the number of values.* (The arithmetic mean is the number most people call the "average." However, in statistics the word *average* has the more general meaning of a measure of central tendency.)

EXAMPLE 3   The arithmetic mean of the numbers given in Example 1 is determined by finding the sum of all the numbers and dividing by 15. Therefore, by letting $\bar{x}$ (read as "*x* bar") represent the mean, we have

$$\bar{x} = \frac{5 + 2 + 6 + 4 + 7 + 4 + 7 + 2 + 8 + 9 + 4 + 11 + 9 + 1 + 3}{15}$$

sum of values

$$= \frac{82}{15} = 5.5$$

number of values

Thus, the mean is 5.5. (The mean is usually calculated to one more significant digit than was present in the original data.)

If we wish to find the arithmetic mean of a large number of values and if some of them appear more than once, the calculation can be simplified. The mean can be calculated by multiplying each value by its frequency, adding these results, and then dividing by the total number of values (the sum of the frequencies). Letting $\bar{x}$ represent the mean of the values $x_1, x_2, \ldots, x_n$, which occur with frequencies $f_1, f_2, \ldots, f_n$, respectively, we have

> This is called *a weighted mean* since each value is given a weighting based on the number of times it occurs.

$$\bar{x} = \frac{x_1 f_1 + x_2 f_2 + \cdots + x_n f_n}{f_1 + f_2 + \cdots + f_n} \tag{22-1}$$

EXAMPLE 4   Using Eq. (22-1) to find the arithmetic mean of the numbers of Example 1, we first set up a table of values and their respective frequencies, as follows:

Value	1	2	3	4	5	6	7	8	9	11
Frequency	1	2	1	3	1	1	2	1	2	1

We now calculate the arithmetic mean $\bar{x}$ by using Eq. (22-1):

multiply each value by its frequency and add results

$$\bar{x} = \frac{1(1) + 2(2) + 3(1) + 4(3) + 5(1) + 6(1) + 7(2) + 8(1) + 9(2) + 11(1)}{1 + 2 + 1 + 3 + 1 + 1 + 2 + 1 + 2 + 1}$$

sum of frequencies

$$= \frac{82}{15} = 5.5$$

We see that this agrees with the result of Example 3.

Summations such as those in Eq. (22-1) occur frequently in statistics and other branches of mathematics. In order to simplify writing these sums, the symbol Σ is used to indicate the process of summation. (Σ is the Greek capital letter sigma.) Σx means the sum of the x's.

EXAMPLE 5   We can show the sum of the numbers $x_1, x_2, x_3, \ldots, x_n$ as

$$\Sigma x = x_1 + x_2 + x_3 + \cdots + x_n$$

If these numbers are $3, 7, 2, 6, 8, 4$, and 9, we have

$$\Sigma x = 3 + 7 + 2 + 6 + 8 + 4 + 9 = 39$$

Using the summation symbol Σ, we can write Eq. (22-1) for the arithmetic mean as

$$\bar{x} = \frac{x_1 f_1 + x_2 f_2 + x_3 f_3 + \cdots + x_n f_n}{f_1 + f_2 + f_3 + \cdots + f_n} = \frac{\Sigma x f}{\Sigma f} \qquad \textbf{(22-1)}$$

The summation notation Σx is an abbreviated form of the more general notation $\sum_{i=1}^{n} x_i$. This more general form can be used to indicate the sum of the first n numbers of a sequence or to indicate the sum of a certain set within the sequence. For example, for a set of at least 5 numbers, $\sum_{i=3}^{5} x_i$ indicates the sum of the third through the fifth of these numbers (in Example 5, $\sum_{i=3}^{5} x_i = 16$). We will use the abbreviated form Σx to indicate the sum of all the numbers being considered.

EXAMPLE 6   We find the arithmetic mean of the Internet hours in Example 1 of Section 22-1 (page 592) by

> The arithmetic mean is one of a number of statistical measures that can be found on a calculator. The use of a calculator will be shown in the next section.

$$\bar{x} = \frac{\Sigma x f}{\Sigma f} = \frac{3(1) + 4(1) + 5(3) + \cdots + 12(6) + \cdots + 28(1)}{50}$$

$$= \frac{687}{50} = 13.7 \text{ h} \qquad \text{(rounded off to tenths)}$$

MODE   Another measure of central tendency is *the* **mode,** *which is the value that appears most frequently.* If two or more values appear with the same greatest frequency, each is a mode. If no value is repeated, there is no mode.

EXAMPLE 7   (a) The mode of the numbers in Example 1 is 4, since it appears three times and no other value appears more than twice.
   (b) The modes of the numbers

$$1, 2, 2, 4, 5, 5, 6, 7$$

are 2 and 5, since each appears twice and no other number is repeated.
   (c) There is no mode for the values

$$1, 2, 5, 6, 7, 9$$

since none of the values is repeated.

■**EXAMPLE 8**   To find the frictional force between two specially designed surfaces, the force to move a block with one surface along an inclined plane with the other surface is measured 10 times. The results, with forces in newtons, are

$$2.2, 2.4, 2.1, 2.2, 2.5, 2.2, 2.4, 2.7, 2.1, 2.5$$

Find the mean, median, and mode of these forces.

To find the mean, we sum the values of the forces and divide this total by 10. This gives

$$\bar{F} = \frac{\sum F}{10} = \frac{2.2 + 2.4 + 2.1 + 2.2 + 2.5 + 2.2 + 2.4 + 2.7 + 2.1 + 2.5}{10}$$

$$= \frac{23.3}{10} = 2.33 \text{ N}$$

> The mean is useful with many statistical methods and is used extensively. The median is also commonly used and is a good choice if there are some extreme values. The mode is occasionally used (if there is only one) for an estimate of data that is approximately symmetric.

The median is found by arranging the values in order and finding the middle value. The values in order are

$$2.1, 2.1, 2.2, 2.2, 2.2, 2.4, 2.4, 2.5, 2.5, 2.7$$

Since there are 10 values, we see that the fifth value is 2.2 and the sixth is 2.4. The value midway between these is 2.3, which is the median. Therefore, the median force is 2.3 N.

The mode is 2.2 N, since this value appears three times, which is more than any other value.

EXERCISES *22-2*

In Exercises 1–12, use the following sets of numbers.

A: 3, 6, 4, 2, 5, 4, 7, 6, 3, 4, 6, 4, 5, 7, 3

B: 25, 26, 23, 24, 25, 28, 26, 27, 23, 28, 25

C: 0.48, 0.53, 0.49, 0.45, 0.55, 0.49, 0.47, 0.55, 0.48, 0.57, 0.51, 0.46

D: 105, 108, 103, 108, 106, 104, 109, 104, 110, 108, 108, 104, 113, 106, 107, 106, 107, 109, 105, 111, 109, 108

In Exercises 1–4, determine the median of the numbers of the given set.

1. Set *A*   **2.** Set *B*

3. Set *C*   **4.** Set *D*

In Exercises 5–8, determine the arithmetic mean of the numbers of the given set.

5. Set *A*   **6.** Set *B*

7. Set *C*   **8.** Set *D*

In Exercises 9–12, determine the mode of the numbers of the given set.

9. Set *A*   **10.** Set *B*

11. Set *C*   **12.** Set *D*

In Exercises 13–28, the required sets of numbers are those in Exercises 22-1. Find the indicated measures of central tendency.

13. Median of computer instructions in Exercise 9

14. Mean of computer instructions in Exercise 9

15. Mode of computer instructions in Exercise 9

16. Median of times in Exercise 13

17. Mean of times in Exercise 13

18. Mode of times in Exercise 13

19. Median of stopping distances in Exercise 17. (Use the class mark for each class.)

20. Mean of stopping distances in Exercise 17. (Use the class mark for each class.)

21. Mean of X-ray dosages in Exercise 21

22. Median of X-ray dosages in Exercise 21

23. Mode of X-ray dosages in Exercise 21

24. Mean of battery lives in Exercise 23

25. Median of battery lives in Exercise 23

26. Mode of battery lives in Exercise 23

27. Mean of cable diameters in Exercise 25

28. Median of cable diameters in Exercise 25

In Exercises 29–40, find the indicated measures of central tendency.

29. The weekly salaries (in dollars) for the workers in a small factory are as follows:

450, 550, 475, 425, 375, 500, 400,
550, 475, 600, 500, 425, 450, 500

Find the median and the mode of the salaries.

30. Find the mean salary for the salaries in Exercise 29.

31. In a particular month the electrical usage, rounded to the nearest 100 kWh (kilowatt-hours), of 1000 homes in a certain city was summarized as follows:

Usage	500	600	700	800	900	1000	1100	1200
No. homes	22	80	106	185	380	122	90	15

Find the mean of the electrical usage.

32. Find the median and mode of electrical usage in Exercise 31.

33. A test of air pollution in a city gave the following readings of the concentration of sulfur dioxide (in parts per million) for 18 consecutive days:

0.14, 0.18, 0.27, 0.19, 0.15, 0.22, 0.20, 0.18, 0.15,
0.17, 0.24, 0.23, 0.22, 0.18, 0.32, 0.26, 0.17, 0.23

Find the median and mode of these readings.

34. Find the mean of the readings in Exercise 33.

35. The *midrange,* another measure of central tendency, is found by finding the sum of the lowest and the highest values and dividing this sum by 2. Find the midrange of the salaries in Exercise 29.

36. Find the midrange of the sulfur dioxide readings in Exercise 33. (See Exercise 35.)

(W) 37. Add $100 to each of the salaries in Exercise 29. Then find the median, mean, and mode of the resulting salaries. State any conclusion that might be drawn from the results.

(W) 38. Multiply each of the salaries in Exercise 29 by 2. Then find the median, mean, and mode of the resulting salaries. State any conclusion that might be drawn from the results.

(W) 39. Change the final salary in Exercise 29 to $4000, with all other salaries being the same. Then find the mean of these salaries. State any conclusion that might be drawn from the result. (The $4000 here is called an *outlier,* which is an extreme value.)

(W) 40. Find the median and mode of the salaries indicated in Exercise 39. State any conclusion that might be drawn from the results.

22-3 STANDARD DEVIATION

In using statistics, we generally collect a sample of data and draw certain conclusions about the complete collection of possible values. In statistics, *the complete collection of values (measurements, people, scores, etc.) is called the* **population,** *and a* **sample** *is a subset of the population.* In the applied examples and exercises of the previous two sections we were using samples of larger populations. For example, in Example 1 on page 592 we used estimates of the time on the Internet from a *sample* of 50 persons, as it would be impractical to get estimates from the total *population* of *all* users of the Internet.

In the preceding section we discussed measures of central tendency of the data in the various samples. However, regardless of the measure that may be used, it does not tell us whether the values of the population tend to be grouped closely together or spread out over a large range of values. Therefore, we also need some measure of the deviation, or dispersion, of the values from the median or the mean. If the dispersion is small and the numbers are grouped closely together, the measure of central tendency is more reliable and descriptive of the data than in the case in which the spread is greater.

In statistics, there are several measures of dispersion that may be defined. In this section we discuss one that is very widely used: the *standard deviation.* It is defined on the next page.

The **standard deviation** *of a set of* **sample** *values is defined by the equation*

$$s = \sqrt{\frac{\sum (x - \bar{x})^2}{n - 1}}$$

(22-2)

The definition of *s* shows that the following steps are used in computing its value.

> ## Steps for Calculating Standard Deviation
> 1. *Find the arithmetic mean $\bar{x}$ of the numbers of the set.*
> 2. *Subtract the mean from each number of the set.*
> 3. *Square these differences.*
> 4. *Find the sum of these squares.*
> 5. *Divide this sum by n − 1.*
> 6. *Find the square root of this result.*

NOTE ▶

The standard deviation *s* is a positive number. It is a *deviation from the mean*, regardless of whether or not the individual numbers are greater than or less than the mean. Numbers close together will have a small standard deviation, whereas numbers further apart have a larger standard deviation. Therefore, *the standard deviation becomes larger as the spread of data increases.*

Following the steps shown above, we use Eq. (22-2) for the calculation of standard deviation in the following examples.

EXAMPLE 1 Find the standard deviation of the following numbers: 1, 5, 4, 2, 6, 2, 1, 1, 5, 3.

A table of the necessary values is shown below, and steps 1–6 are indicated.

	step 2	step 3
x	$x - \bar{x}$	$(x - \bar{x})^2$
1	−2	4
5	2	4
4	1	1
2	−1	1
6	3	9
2	−1	1
1	−2	4
1	−2	4
5	2	4
3	0	0
30		32 step 4

$$\bar{x} = \frac{30}{10} = 3 \qquad \text{step 1}$$

$$\frac{\sum (x - \bar{x})^2}{n - 1} = \frac{32}{10 - 1} = \frac{32}{9} \qquad \text{step 5}$$

$$s = \sqrt{\frac{32}{9}} = 1.9 \qquad \text{step 6}$$

NOTE ▶

In calculating the standard deviation, it is usually rounded off to one more significant digit than was present in the original data. ────────■

EXAMPLE 2 Find the standard deviation of the numbers in Example 1 of Section 22-2.

Since several of the numbers appear more than once, it is helpful to use the frequency of each number in the table, as follows:

x	f	fx	$x - \bar{x}$	$(x - \bar{x})^2$	$f(x - \bar{x})^2$
			step 2	step 3	
1	1	1	−4.5	20.25	20.25
2	2	4	−3.5	12.25	24.50
3	1	3	−2.5	6.25	6.25
4	3	12	−1.5	2.25	6.75
5	1	5	−0.5	0.25	0.25
6	1	6	0.5	0.25	0.25
7	2	14	1.5	2.25	4.50
8	1	8	2.5	6.25	6.25
9	2	18	3.5	12.25	24.50
11	1	11	5.5	30.25	30.25
	15	82			123.75

step 1

$$\bar{x} = \frac{82}{15} = 5.5$$

$$\frac{\sum f(x - \bar{x})^2}{n - 1} = \frac{123.75}{15 - 1} = \frac{123.75}{14} \qquad \text{step 5}$$

$$s = \sqrt{\frac{123.75}{14}} = 3.0 \qquad \text{step 6}$$

step 4

In some sources, standard deviation is defined such that n, rather than $n - 1$, is used in the denominator of Eq. (22-2). Using n gives us the *population standard deviation*. However, we generally will be using samples of the population, and the use of $n - 1$ gives better estimates of a population standard deviation when using samples. Therefore, in this text we will use Eq. (22-2) and refer to the sample standard deviation simply as the standard deviation.

It is possible to reduce the computational work required to find the standard deviation. Algebraically, it can be shown (although we will not do so here) that the following equation is another form of Eq. (22-2) and therefore gives the same results.

$$s = \sqrt{\frac{n(\sum x^2) - (\sum x)^2}{n(n - 1)}} \qquad \text{(22-3)}$$

Although the form of this equation appears more involved, it does reduce the amount of calculation which is necessary. Consider the following examples.

EXAMPLE 3 Using Eq. (22-3), find s for the numbers in Example 1.

x	x^2
1	1
5	25
4	16
2	4
6	36
2	4
1	1
1	1
5	25
3	9
30	122

$$n = 10$$

$$\sum x^2 = 122$$

$$(\sum x)^2 = 30^2 = 900$$

$$s = \sqrt{\frac{10(122) - 900}{10(9)}} = 1.9$$

EXAMPLE 4   An ammeter measures the electric current in a circuit. In an ammeter, two resistances are connected in parallel, with most of the current passing through a very low resistance called the *shunt*. The resistance of each shunt in a sample of 100 shunts was measured. The results were grouped, and the class mark and frequency for each class are shown in the following table. Calculate the arithmetic mean and the standard deviation of the resistances of the shunts.

R (ohms)	f	fR	fR²
0.200	1	0.200	0.0400
0.210	3	0.630	0.1323
0.220	5	1.100	0.2420
0.230	10	2.300	0.5290
0.240	17	4.080	0.9792
0.250	40	10.000	2.5000
0.260	13	3.380	0.8788
0.270	6	1.620	0.4374
0.280	3	0.840	0.2352
0.290	2	0.580	0.1682
	100	24.730	6.1421

$$\overline{R} = \frac{24.730}{100} = 0.247 \ \Omega$$

$$n = 100$$

$$\Sigma R^2 = 6.1421$$

$$(\Sigma R)^2 = 24.730^2$$

$$s = \sqrt{\frac{100(6.1421) - 24.730^2}{100(99)}} = 0.016$$

The arithmetic mean of the resistances is 0.247 Ω, with a standard deviation of 0.016 Ω. ■

EXAMPLE 5   Find the standard deviation of the estimated hours on the Internet as grouped in Example 2 of Section 22-1 (page 592). In doing this we assume that each value in the class is the same as the class mark. The method is not exact, but with a large set of numbers it provides a good approximation with less arithmetic work.

The statistical measures x, Σx, Σx^2, s_x, σ_x (the population standard deviation), and n (and possibly others) can be displayed on a graphing calculator as shown in Fig. 22-7 for the data of Example 5.

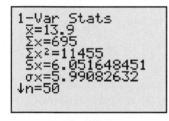

Fig. 22-7

Interval	x	f	fx	fx²
0–4	2	2	4	8
5–9	7	9	63	441
10–14	12	19	228	2736
15–19	17	11	187	3179
20–24	22	6	132	2904
25–29	27	3	81	2187
		50	695	11455

$$n = 50$$

$$\Sigma x^2 = 11455$$

$$(\Sigma x)^2 = 695^2$$

$$s = \sqrt{\frac{50(11455) - 695^2}{50(49)}} = 6.1$$

Thus, $s = 6.1$ h. Compare the mean calculated in Example 6 on page 598 (not using class marks) with that shown in the calculator display to the left (using class marks). ■

In using the statistical measures we have discussed, we must be careful in using and interpreting such measures. Consider the following example.

EXAMPLE 6   **(a)** The numbers 1, 2, 3, 4, 5 have a mean of 3, a median of 3, and a standard deviation is 1.6. These values fairly well describe the center and distribution of the numbers in the set.

(b) The numbers 1, 2, 3, 4, 100 have a mean of 22, a median of 3, and a standard deviation of 44. The large difference between the median and the mean and the very large range of values within one standard deviation of the mean (−22 to 66) indicate that this set of measures does not describe this set of numbers well. In a case like this, the 100 should be checked to see if it is in error. ■

Example 6 illustrates that the statistical measures can be misleading if the numbers in a set are unevenly distributed. Misleading statistics can also come from the source of the data. Consider the probable results of a survey to find the percent of persons in favor of raising income taxes for the wealthy if the survey is taken at the entrance to a welfare office or if it is taken at the entrance to a stock brokerage firm. There are many other considerations in the proper use and interpretation of statistical measures.

EXERCISES *22-3*

In Exercises 1–12, use the following sets of numbers. They are the same as those used in Exercise 22-2.

A: 3, 6, 4, 2, 5, 4, 7, 6, 3, 4, 6, 4, 5, 7, 3

B: 25, 26, 23, 24, 25, 28, 26, 27, 23, 28, 25

C: 0.48, 0.53, 0.49, 0.45, 0.55, 0.49, 0.47, 0.55, 0.48, 0.57, 0.51, 0.46

D: 105, 108, 103, 108, 106, 104, 109, 104, 110, 108, 108, 104, 113, 106, 107, 106, 107, 109, 105, 111, 109, 108

In Exercises 1–4, use Eq. (22-2) to find the standard deviation s for the indicated sets of numbers.

1. Set *A*   **2.** Set *B*   **3.** Set *C*   **4.** Set *D*

In Exercises 5–8, use Eq. (22-3) to find the standard deviation s for the indicated sets of numbers.

5. Set *A*   **6.** Set *B*   **7.** Set *C*   **8.** Set *D*

In Exercises 9–12, use the statistical feature of a calculator to find the arithmetic mean and the standard deviation s for the indicated sets of numbers.

9. Set *A*   **10.** Set *B*   **11.** Set *C*   **12.** Set *D*

In Exercises 13–20, find the standard deviation s for the indicated sets of numbers from Exercises 22-1 and 22-2.

13. The computer instructions in Exercise 9 of Section 22-1

14. The X-ray dosages in Exercise 21 of Section 22-1

15. The battery lives in Exercise 23 of Section 22-1

16. The salaries of Exercise 29 of Section 22-2

17. The strobe light times in Exercise 13 of Section 22-1

18. The stopping distances in Exercise 17 of Section 22-1

19. The fiber-optic cable diameters in Exercise 25 of Section 22-1 (use the statistical feature of a graphing calculator).

20. The electric power usages in Exercise 31 of Section 22-2 (use the statistical feature of a graphing calculator).

22-4 NORMAL DISTRIBUTIONS

The distributions in the previous sections have been for a limited number of values. Let us now consider a very large population, such as the useable lifetime of all of the AA batteries sold in the world in a year. It could have a large number of classes with many values within each class.

We would expect a frequency polygon for this very large population to have its maximum frequency very near the mean and taper off to smaller frequencies on either side. It would probably be shaped very close to the curve shown in Fig. 22-8.

The smooth bell-shaped curve in Fig. 22-8 shows the **normal distribution** of a population large enough that the distribution is considered to be *continuous*. (We can think of this as the points on the frequency distribution curve are so close that they can be considered to be touching.) Using advanced methods, its equation is found to be

Fig. 22-8

$$y = \frac{e^{-(x-\mu)^2/2\sigma^2}}{\sigma\sqrt{2\pi}}$$

(22-4)

Here, μ is the *population mean* and σ is the *population standard deviation,* and π and *e* are the familiar numbers first used in Chapters 2 and 12, respectively.

From Eq. (22-4), we can see that any particular normal distribution for a large population depends on the values of μ and σ. The horizontal location of the curve depends on μ, and the shape (how spread out the curve is) depends on σ, but the bell shape remains. This is illustrated in general in the following example.

■**EXAMPLE 1**   In Fig. 22-9, for the left curve $\mu = 10$ and $\sigma = 5$, whereas for the right curve $\mu = 20$ and $\sigma = 10$.

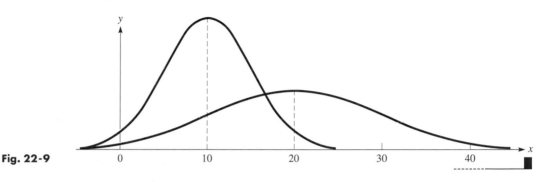

Fig. 22-9

Standard Normal Distribution

As we have just seen, there are innumerable possible normal distributions. However, there is one of particular interest. *The* **standard normal distribution** *is the normal distribution for which the mean is* 0 *and the standard deviation is* 1. Making these substitutions in Eq. (22-4), we have

$$y = \frac{1}{\sqrt{2\pi}} e^{-x^2/2}$$

(22-5)

as the equation of the standard normal distribution curve.

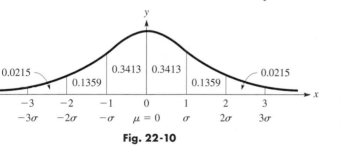

Fig. 22-10

From Eq. (22-5), we see that the standard normal distribution curve is symmetrical to the y-axis (if we replace x by $-x$, the equation is unchanged). Since the mean is zero, the curve is symmetrical about the mean. Using advanced mathematics, it can be shown that the total area under the curve (above the x-axis) is exactly one unit. As shown in Fig. 22-10, it is found that the area between $x = 0$ and $x = 1$ is 0.3413, the area between $x = 1$ and $x = 2$ is 0.1359, and the area between $x = 2$ and $x = 3$ is 0.0215. Since areas for negative values of x are equivalent, the total area between $x = -3$ and $x = 3$ is about 0.9974, which means nearly all of the area is between $x = -3$ and $x = 3$.

From Fig. 22-10, we can see that *about 68% of the area is within one standard deviation of the mean (in the interval $\mu - \sigma$ to $\mu + \sigma$), and about 95% of the area is within two standard deviations of the mean (in the interval $\mu - 2\sigma$ to $\mu + 2\sigma$).* These percentages are often useful in data analysis.

Since a normal distribution curve gives a measure of the frequency of a particular value within the distribution, *the area under any part of the standard normal curve gives the relative frequency of those values of the distribution.* This is illustrated in the following example.

EXAMPLE 2   For a normal distribution of values, 0.3413 (34.13%) of the values are between $x = 0$ and $x = 1$, and 68.26% of the values are between $x = -1$ and $x = 1$. For this distribution, 81.85% of the values lie between $x = -2$ and $x = 1$. $(0.1359 + 0.3413 + 0.3413 = 0.8185)$. ◼

We can find the relative frequency of values for any normal distribution by use of the **standard score** z (or z-*score*), which is defined as

$$z = \frac{x - \mu}{\sigma}$$ **(22-6)**

For the normal standard distribution, where $\mu = 0$, if we let $x = \sigma$, then $z = 1$. If we let $x = 2\sigma$, $z = 2$. Therefore, we can see that *a value of z tells us the number of standard deviations the given value of x is above or below the mean*. From the discussion above, we can see that the value of z can tell us the area under the curve between the mean and the value of x corresponding to that value of z. In turn, *this tells us the relative frequency of all values between the mean and the value of x*.

Table 22-1 gives the area under the standard normal distribution curve for given values of z. Since the curve is symmetric to the y-axis, the values shown are also valid if z is negative.

Table 22-1   Standard Normal (z) Distribution

z	Area	z	Area	z	Area
0.0	0.0000	1.0	0.3413	2.0	0.4772
0.1	0.0398	1.1	0.3643	2.1	0.4821
0.2	0.0793	1.2	0.3849	2.2	0.4861
0.3	0.1179	1.3	0.4032	2.3	0.4893
0.4	0.1554	1.4	0.4192	2.4	0.4918
0.5	0.1915	1.5	0.4332	2.5	0.4938
0.6	0.2257	1.6	0.4452	2.6	0.4953
0.7	0.2580	1.7	0.4554	2.7	0.4965
0.8	0.2881	1.8	0.4641	2.8	0.4974
0.9	0.3159	1.9	0.4713	2.9	0.4981
1.0	0.3413	2.0	0.4772	3.0	0.4987

The following examples illustrate the use of Eq. (22-6) and z-scores.

EXAMPLE 3   For a normal distribution curve based on values of $\mu = 20$ and $\sigma = 5$, find the area between $x = 24$ and $x = 32$. To find this area we use Eq. (22-6) to find the corresponding values of z and then find the difference between these z-scores. These z-scores are

$$z = \frac{24 - 20}{5} = 0.8 \quad \text{and} \quad z = \frac{32 - 20}{5} = 2.4$$

For $z = 0.8$, the area is 0.2881, and for $z = 2.4$, the area is 0.4918. Therefore, the area between $x = 24$ and $x = 32$ (see Fig. 22-11) is

$$0.4918 - 0.2881 = 0.2037$$

This means that the relative frequency of the values between $x = 24$ and $x = 32$ is 20.37%. If we have a large set of measured values with $\mu = 20$ and $\sigma = 5$, we should expect that about 20% of them are between $x = 24$ and $x = 32$. ◼

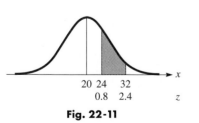

Fig. 22-11

SOLVING A WORD PROBLEM

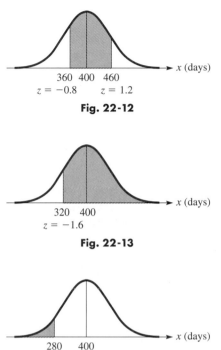

Fig. 22-12

Fig. 22-13

Fig. 22-14

Standard Errors

■**EXAMPLE 4** The lifetimes of a certain type of watch battery are normally distributed. The mean lifetime is 400 days, and the standard deviation is 50 days. For a sample of 5000 new batteries, determine how many batteries will last **(a)** between 360 days and 460 days, **(b)** more than 320 days, and **(c)** less than 280 days.

(a) For this distribution, $\mu = 400$ days and $\sigma = 50$ days. Using Eq. (22-6), we find the z-scores for $x = 360$ days and $x = 460$ days. They are

$$z = \frac{360 - 400}{50} = -0.8 \quad \text{and} \quad z = \frac{460 - 400}{50} = 1.2$$

For $z = -0.8$, the area is to the left of the mean, and since the curve is symmetrical about the mean, we use the $z = 0.8$ value of the area and add it to the area for $z = 1.2$. Therefore, the area is $0.2881 + 0.3849 = 0.6730$. See Fig. 22-12.

This means that 67.30% of the 5000 batteries, or 3365 of the batteries, will last between 360 days and 460 days.

(b) To determine the number of batteries that will last more than 320 days, we first find the z-score for $x = 320$. It is $z = (320 - 400)/50 = -1.6$. This means we want the total area to the right of $z = -1.6$. In this case we add the area for $z = 1.6$ to the total area to the right of the mean. Since the total area under the curve is 1.0000, the total area on either side of the mean is 0.5000. Therefore, the area to the right of $z = -1.6$ is $0.4452 + 0.5000 = 0.9452$. See Fig. 22-13. This means that $0.9452 \times 5000 = 4726$ batteries will last more than 320 days.

(c) To find the number of batteries that will last less than 280 days, we first find that $z = (280 - 400)/50 = -2.4$ for $x = 280$. Since we want the total area to the left of $z = -2.4$, we subtract the area for $z = 2.4$ from 0.5000, the total area to the left of the mean. Since the area for $z = 2.4$ is 0.4918, the total area to the left of $z = -2.4$ is $0.5000 - 0.4918 = 0.0082$. See Fig. 22-14. Therefore, $0.0082 \times 5000 = 41$ batteries will last less than 280 days. ------------■

In Example 4 we assumed that the lifetimes of the batteries were normally distributed. Of course, for any set of 5000 watches, or any number of watches for that matter, the lifetimes that actually occur will not follow normal distribution *exactly*. There will be some variation from normal distribution, but for a large sample this variation should be small. The mean and the standard deviation for any sample would probably vary somewhat from that of the population.

In the study of probability it is shown that if we select all possible samples of size n from a population with a mean μ and standard deviation σ, the mean of the sample means is also μ. Also, *the standard deviation of the sample means,* denoted by $\sigma_{\bar{x}}$, and called **the standard error of the mean,** is

$$\sigma_{\bar{x}} = \frac{\sigma}{\sqrt{n}} \tag{22-7}$$

Also, the *standard error of the standard deviation s of the sample,* denoted by σ_s, and called **the standard error of s,** is

$$\sigma_s = \frac{\sigma}{\sqrt{2n}} \tag{22-8}$$

We can see from these formulas that as the sample size gets larger (*large* is usually considered to be over 30), the less variation there should be in the mean and standard deviation of the sample. This agrees with what we should expect.

EXAMPLE 5 For the sample of 5000 watch batteries in Example 4, we know that $\sigma = 50$ days. Therefore, the standard error in the mean $\bar{x}$ is $50/\sqrt{5000} = 0.7$ days. This means that of all samples of 5000 batteries, about 68% should have a mean lifetime of 400 ± 0.7 days (between 399.3 days and 400.7 days).

The standard error in the standard deviation s is $50/\sqrt{2(5000)} = 0.5$ days. This means that of all samples of 5000 batteries, about 68% should have a standard deviation of 50 ± 0.5 days (between 49.5 days and 50.5 days).

(Considering the significant digits of these values, 68% (within one standard deviation) or even 95% (within two standard deviations) of the sample values of $\bar{x}$ and s should not vary by more than 1 day.)

EXERCISES 22-4

In Exercises 1–4, use a graphing calculator to display the indicated graph of Eq. (22-4).

1. Display the graph of the normal distribution of values for which $\mu = 10$ and $\sigma = 5$. Compare with the graph shown in Fig. 22-9.

2. Display the graph of the normal distribution of values for which $\mu = 20$ and $\sigma = 10$. Compare with the graph shown in Fig. 22-9.

3. Sketch a graph of a normal distribution of values for which $\mu = 100$ and $\sigma = 10$. Then compare with the graph displayed by a graphing calculator.

(W) 4. Sketch a graph of a normal distribution of values for which $\mu = 100$ and $\sigma = 30$. Then compare with the graph displayed by a graphing calculator. How does this graph differ from that of Exercise 3?

In Exercises 5–8, use the following data and refer to Fig. 22-10. A sample of 200 bags of cement are weighed as a quality check. Over a long period it has been found that the mean value and standard deviation for this size bag are known and that the weights are normally distributed. Determine how many bags within this sample should have weights that satisfy the following conditions.

5. Within one standard deviation of the mean

6. Within two standard deviations of the mean

7. Between the mean and two standard deviations above the mean

8. Between one standard deviation below the mean and three standard deviations above the mean

In Exercises 9–12, use the following data. Each AA battery in a sample of 500 batteries is checked for its voltage. It has been previously established for this type of battery (when newly produced) that the voltages are distributed normally with $\mu = 1.50$ V and $\sigma = 0.05$ V.

9. How many batteries have voltages between 1.45 V and 1.55 V?

10. How many batteries have voltages between 1.52 V and 1.58 V?

11. What percent of the batteries have voltages below 1.54 V?

12. What percent of the batteries have voltages above 1.64 V?

In Exercises 13–20, use the following data. The lifetimes of a certain type of automobile tire have been found to be distributed normally with a mean lifetime of 100,000 km and a standard deviation of 10,000 km. Answer the following questions for a sample of 5000 of these tires.

13. How many tires will last between 85,000 km and 100,000 km?

14. How many tires will last between 95,000 km and 115,000 km?

15. How many tires will last more than 118,000 km?

16. If the manufacturer guarantees to replace all tires that do not last 75,000 km, what percent of the tires may have to be replaced under this guarantee?

(W) 17. What is the standard error in the mean for all samples of 5000 of these tires? Explain the meaning of this result.

(W) 18. What is the standard error in the standard deviation of all samples of 5000 of these tires? Explain the meaning of this result.

19. What percent of the samples of 5000 of these tires should have a mean lifetime of more than 10,282 km?

20. What percent of the samples of 5000 of these tires should have a mean standard deviation of between 9900 km and 10,100 km?

In Exercises 21–24, solve the given problems.

21. For the strobe light times in Exercise 13 of Exercises 22-1, find the percent of times within one standard deviation of the mean. From Exercise 17 of Section 22-2 we find that $\bar{x} = 2.248$ s, and from Exercise 17 of Section 22-3 we find that $s = 0.014$ s. Compare the results with that of a normal distribution.

22. Follow the same instructions as in Excercise 21 for the fiber-optic diameters in Exercise 25 of Exercises 22-1. From Exercise 27 of Exercises 22-2, $\bar{x} = 0.00595$ mm, and from Exercise 19 of Section 22-3, $s = 0.00022$ mm.

23. Follow the same instructions as in Exercise 21 for the hours estimated on the Internet in Example 1 of Section 22-1. From Example 6 of Section 22-2 we find that $\bar{x} = 13.7$ h, and from Example 5 of Section 22-3, we find that $s = 6.1$ h.

(W) 24. Discuss the results found in Exercises 21 and 23, considering the methods used to find the mean and the standard deviation.

22-5 STATISTICAL PROCESS CONTROL

One of the most important uses of statistics in industry is Statistical Process Control (SPC), which is used to maintain and improve product quality. Samples are tested during the production at specified intervals to determine whether or not the production process needs adjustment to meet quality requirements.

A particular industrial process is considered to be *in control* if it is stable and predictable, and sample measurements fall within upper and lower control limits. The process is *out of control* if it has an unpredictable amount of variation and there are sample measurements outside the control limits due to special causes.

Minor variations may be expected, for example, from very small fluctuations in voltage, temperature, or material composition. Special causes resulting in an out-of-control process could include line stoppage, material defect, or an incorrect applied pressure.

EXAMPLE 1   The manufacturer of 1.5-V batteries states that the voltage of its batteries is no less than 1.45 V or greater than 1.55 V.

If all samples of batteries that are tested have voltages in the proper range with only expected minor variations, the production process is *in control*.

However, if some samples have batteries with voltages out of the proper range, the process is *out of control*. This would indicate some special cause for the problem, such as an improperly operating machine or an impurity getting into the process. The process would probably be halted until the cause is determined. ------------∎

Control Charts

An important device used in SPC is the *control chart.* It is used to show a trend of a production characteristic over time. Samples are measured at specified intervals of time to see if the measurements are within acceptable limits. The measurements are plotted on a chart to check for trends and abnormalities in the production process.

In making a control chart, we must determine what the mean should be. For a stable process for which previous data is known, it can be based on a production specification or on previous data. For a new or recently modified process, it may be necessary to use present data, although the value may have to be revised for future charts. On a control chart, *this value is used as the population mean,* μ.

It is also necessary to establish the upper and lower control limits. The standard generally used is that 99.7% of the sample measurements should fall within these control limits. This assumes a normal distribution, and we note that this is equal to ± 3 sample standard deviations of the population mean. We will establish these limits by use of a table or a formula that has been made using statistical measures developed in a more complete coverage of quality control. This does follow the normal practice of using a formula or a more complete table in setting up the control limits.

In Fig. 22-15 we show a sample control chart, and on the following pages we illustrate how control charts are made.

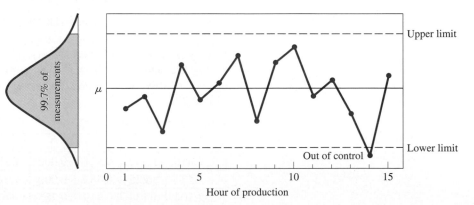

Fig. 22-15

EXAMPLE 2   A pharmaceutical company makes a capsule of a prescription drug that contains 500 mg of the drug, according to the label. In a newly modified process of making the capsule, five capsules are tested every 15 minutes to check the amount of the drug in each capsule. Testing over a 5-hour period gave the following results for the 20 subgroups of samples.

Subgroup	Amount of drug (in mg) of five capsules					Mean x	Range R
1	503	501	498	507	502	502.2	9
2	497	499	500	495	502	498.6	7
3	496	500	507	503	502	501.6	11
4	512	503	488	500	497	500.0	24
5	504	505	500	508	502	503.8	8
6	495	495	501	497	497	497.0	6
7	503	500	507	499	498	501.4	9
8	494	498	497	501	496	497.2	7
9	502	504	505	500	502	502.6	5
10	500	502	500	496	497	499.0	6
11	502	498	510	503	497	502.0	13
12	497	498	496	502	500	498.6	6
13	504	500	495	498	501	499.6	9
14	500	499	498	501	494	498.4	7
15	498	496	502	501	505	500.4	9
16	500	503	504	499	505	502.2	6
17	487	496	499	498	494	494.8	12
18	498	497	497	502	497	498.2	5
19	503	501	500	498	504	501.2	6
20	496	494	503	502	501	499.2	9
					Sums	9998.0	174
					Means	499.9	8.7

As we note from the table, *the **range** R of each sample is the difference between the highest value and the lowest value of the sample.*

From this table of values, we can make an $\bar{x}$ control chart and an R control chart. The $\bar{x}$ chart maintains a check on the average quality level, whereas the R chart maintains a check on the dispersion of the production process. These two control charts are often plotted together and referred to as the $\bar{x}$–R chart.

In order to define the **central line** of the $\bar{x}$ chart, which ideally is equivalent to the value of the population mean μ, we use the mean of the sample means $\bar{\bar{x}}$. For the central line of the R chart we use $\bar{R}$. From the table we see that

$$\bar{\bar{x}} = 499.9 \text{ mg} \quad \text{and} \quad \bar{R} = 8.7 \text{ mg}$$

The **upper control limit** (UCL) and the **lower control limit** (LCL) for each chart are defined in terms of the mean range $\bar{R}$ and an appropriate constant taken from a table of control chart factors. These factors, which are related to the sample size n, are determined by statistical considerations found in a more complete coverage of quality control. We now present a brief table of control chart factors (Table 22-2).

Table 22-2   Control Chart Factors

n	d_2	A	A_2	D_1	D_2	D_3	D_4
5	2.326	1.342	0.577	0.000	4.918	0.000	2.115
6	2.534	1.225	0.483	0.000	5.078	0.000	2.004
7	2.704	1.134	0.419	0.205	5.203	0.076	1.924

The UCL and LCL for the $\bar{x}$ chart are found as follows:

$$\text{UCL}(\bar{x}) = \bar{\bar{x}} + A_2\bar{R} = 499.9 + 0.577(8.7) = 504.9 \text{ mg}$$
$$\text{LCL}(\bar{x}) = \bar{\bar{x}} - A_2\bar{R} = 499.9 - 0.577(8.7) = 494.9 \text{ mg}$$

The UCL and LCL for the $\bar{R}$ chart are found as follows:

$$\text{LCL}(\bar{R}) = D_3\bar{R} = 0.000(8.7) = 0.0 \text{ mg}$$
$$\text{UCL}(\bar{R}) = D_4\bar{R} = 2.115(8.7) = 18.4 \text{ mg}$$

Using these central lines and control limit lines, we now plot the $\bar{x}$ control chart in Fig. 22-16 and the R control chart in Fig. 22-17.

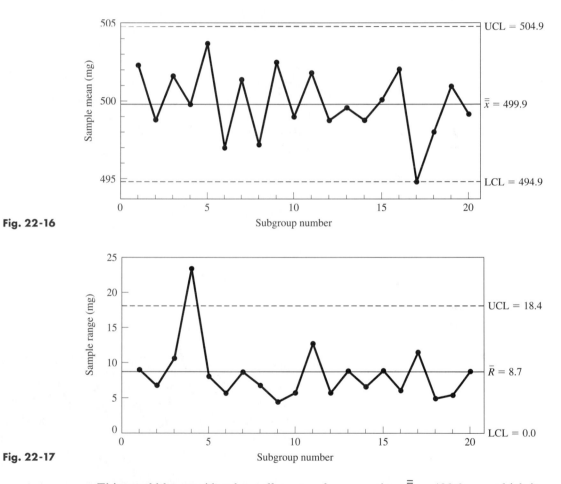

Fig. 22-16

Fig. 22-17

This would be considered a *well-centered process* since $\bar{\bar{x}} = 499.9$ mg, which is very near the target value of 500.0 mg. We do note, however, that subgroup 17 was at a control limit and this might have been due to some special cause, such as the use of a substandard mixture of ingredients. We also note that the process was *out of control* due to some special cause for subgroup 4 since the range was above the upper control limit. We should keep in mind that there are numerous considerations, including human factors, that should be taken into account when making and interpreting control charts and that this is only a very brief introduction to this important industrial use of statistics.

Day	Defective CDs	Proportion defective
1	22	0.022
2	16	0.016
3	14	0.014
4	18	0.018
5	12	0.012
6	25	0.025
7	36	0.036
8	16	0.016
9	14	0.014
10	22	0.022
11	20	0.020
12	17	0.017
13	26	0.026
14	20	0.020
15	22	0.022
16	28	0.028
17	17	0.017
18	15	0.015
19	25	0.025
20	12	0.012
21	16	0.016
22	22	0.022
23	19	0.019
24	16	0.016
25	20	0.020
Sum	490	

In Example 2 the weight (in mg) of a prescription drug was tested. In quality control, weight is a **variable,** *which is a characteristic that can be* **measured.** Other examples of variables are length, voltage, and pressure.

A characteristic that can be **counted** *is an* **attribute,** which is determined to be either acceptable or not acceptable in testing. Examples of attributes are color (acceptable or not), entries on a customer account (correct or incorrect), and defects (not acceptable) in a product. To monitor an attribute in a production process, we get the *proportion* of defective parts by *dividing the number of defective parts in a sample by the total number of parts in the sample,* and then make a *p control chart.* This is illustrated in the following example.

EXAMPLE 3 The manufacturer of compact discs has 1000 CDs checked each day for defects (surface scratches, for example). The data for this procedure for 25 days is shown in the table at the left.

For a *p* control chart, the central line is the value of $\bar{p}$, which in this case is

$$\bar{p} = \frac{490}{25{,}000} = 0.0196$$

The control limits are each three standard deviations from $\bar{p}$. If *n* is the size of the subgroup, the standard deviation σ_p of a proportion is given by

$$\sigma_p = \sqrt{\frac{\bar{p}(1 - \bar{p})}{n}} = \sqrt{\frac{0.0196(1 - 0.0196)}{1000}} = 0.00438$$

Therefore, the control limits are

$$\text{UCL } (p) = 0.0196 + 3(0.00438) = 0.0327$$
$$\text{LCL } (p) = 0.0196 - 3(0.00438) = 0.0065$$

Using this central line and these control limit lines, we now plot the *p* control chart in Fig. 22-18.

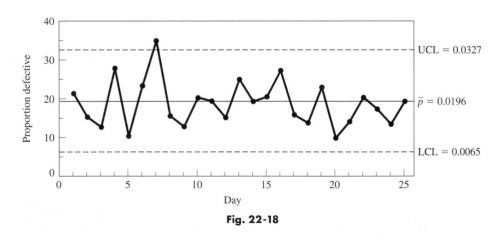

Fig. 22-18

According to the proportion mean of 0.0196, the process produces about 2% defective CDs. We note that the process was out of control on day 7. An adjustment to the production process was probably made to remove the special cause of the additional defective CDs. ---------◼

EXERCISES *22-5*

In Exercises 1–4, use the following data.

Five automobile engines are taken from the production line each hour and tested for their torque (in N · m) when rotating at a constant frequency. The measurements of the sample torques for 20 h of testing are as follows:

Hour	Torques (in N · m) of five engines				
1	366	352	354	360	362
2	370	374	362	366	356
3	358	357	365	372	361
4	360	368	367	359	363
5	352	356	354	348	350
6	366	361	372	370	363
7	365	366	361	370	362
8	354	363	360	361	364
9	361	358	356	364	364
10	368	366	368	358	360
11	355	360	359	362	353
12	365	364	357	367	370
13	360	364	372	358	365
14	348	360	352	360	354
15	358	364	362	372	361
16	360	361	371	366	346
17	354	359	358	366	366
18	362	366	367	361	357
19	363	373	364	360	358
20	372	362	360	365	367

1. Find the central line, upper control limit, and lower control limit for the mean.
2. Find the central line, upper control limit, and lower control limit for the range.
3. Plot an $\bar{x}$ chart.
4. Plot an R chart.

(In order that the table for Exercises 5–8 can be shown without breaking it between columns. Exercises 5–8 are set to the right.)

In Exercises 5–8, use the following data.

Five AC adaptors that are used to charge batteries of a cellular phone are taken from the production line each 15 minutes and tested for their DC output voltage. The output voltages for 24 sample subgroups are as follows:

Subgroup	Output voltages of five adaptors				
1	9.03	9.08	8.85	8.92	8.90
2	9.05	8.98	9.20	9.04	9.12
3	8.93	8.96	9.14	9.06	9.00
4	9.16	9.08	9.04	9.07	8.97
5	9.03	9.08	8.93	8.88	8.95
6	8.92	9.07	8.86	8.96	9.04
7	9.00	9.05	8.90	8.94	8.93
8	8.87	8.99	8.96	9.02	9.03
9	8.89	8.92	9.05	9.10	8.93
10	9.01	9.00	9.09	8.96	8.98
11	8.90	8.97	8.92	8.98	9.03
12	9.04	9.06	8.94	8.93	8.92
13	8.94	8.99	8.93	9.05	9.10
14	9.07	9.01	9.05	8.96	9.02
15	9.01	8.82	8.95	8.99	9.04
16	8.93	8.91	9.04	9.05	8.90
17	9.08	9.03	8.91	8.92	8.96
18	8.94	8.90	9.05	8.93	9.01
19	8.88	8.82	8.89	8.94	8.88
20	9.04	9.00	8.98	8.93	9.05
21	9.00	9.03	8.94	8.92	9.05
22	8.95	8.95	8.91	8.90	9.03
23	9.12	9.04	9.01	8.94	9.02
24	8.94	8.99	8.93	9.05	9.07

5. Find the central line, upper control limit, and lower control limit for the mean.
6. Find the central line, upper control limit, and lower control limit for the range.
7. Plot an $\bar{x}$ chart.
8. Plot an R chart.

For Exercises 9–12, use the following information.

For a production process for which there is a great deal of data since its last modification, the population mean μ and population standard deviation σ are assumed known. For such a process, we have the following values (using additional statistical analysis):

$\bar{x}$ chart: central line = μ, UCL = $\mu + A\sigma$, LCL = $\mu - A\sigma$
R chart: central line = $d_2\sigma$, UCL = $D_2\sigma$, LCL = $D_1\sigma$

The values of A, d_2, D_2, and D_1 are found in the table of control chart factors in Example 2.

9. In the production of robot links and tests for their lengths, it has been found that $\mu = 2.725$ in. and $\sigma = 0.0032$ in. Find the central length, UCL, and LCL for the mean if the sample subgroup size is 5.
10. For the robot link samples of Exercise 9, find the central line, UCL, and LCL for the range.
11. After bottling, the volume of soft drink in six sample bottles is checked each 10 minutes. For this process, $\mu = 750.0$ mL and $\sigma = 2.2$ mL. Find the central line, UCL, and LCL for the range.
12. For the bottling process in Exercise 11, find the central line, UCL, and LCL for the mean.

In Exercises 13 and 14, use the following data.

A telephone company rechecks the entries for 1000 of its new customers each week for name, address, and phone number. The data collected regarding the number of new accounts with errors, along with the proportion of these accounts with errors is given in the following table for a 20-week period.

Week	Accounts with errors	Proportion with errors
1	52	0.052
2	36	0.036
3	27	0.027
4	58	0.058
5	44	0.044
6	21	0.021
7	48	0.048
8	63	0.063
9	32	0.032
10	38	0.038
11	27	0.027
12	43	0.043
13	22	0.022
14	35	0.035
15	41	0.041
16	20	0.020
17	28	0.028
18	37	0.037
19	24	0.024
20	42	0.042
Total	738	

13. For a p chart, find the values for the central line, UCL, and LCL.

14. Plot a p chart.

In Exercises 15 and 16, use the following data.

The maker of electric fuses checks 500 fuses each day for defects. The number of defective fuses, along with the proportion of defective fuses for 24 days, is shown in the following table.

Day	Number defective	Proportion defective
1	26	0.052
2	32	0.064
3	37	0.074
4	16	0.032
5	28	0.056
6	31	0.062
7	42	0.084
8	22	0.044
9	31	0.062
10	28	0.056
11	24	0.048
12	35	0.070
13	30	0.060
14	34	0.068
15	39	0.078
16	26	0.052
17	23	0.046
18	33	0.066
19	25	0.050
20	25	0.050
21	32	0.064
22	23	0.046
23	34	0.068
24	20	0.040
Total	696	

15. For a p chart, find the values for the central line, UCL, and LCL.

16. Plot a p chart.

22-6 LINEAR REGRESSION

We have considered statistical methods for dealing with one variable. We have discussed methods of tabulating, graphing, and measuring the central tendency and the deviations from this value for one variable and have shown an important industrial application in the area of quality control. We now discuss how to find an equation relating two variables for which a set of points is known.

In this section we show a method of finding the equation of a straight line that passes through a set of data points, and in this way we *fit* the line to the points. In general, *the fitting of a curve to a set of points is called* **regression.** Fitting a straight line to a set of points is *linear regression,* and fitting some other type of curve is called *nonlinear regression.* We will consider nonlinear regression in the next section.

Some of the reasons for using regression to find the equation of a curve that passes through a set of points, and thereby "fit" the curve to the points are **(1)** to express a concise relationship between the variables, **(2)** to use the equation to predict certain fundamental results, **(3)** to determine the reliability of certain sets of data, and **(4)** to use the data for testing certain theoretical concepts.

For a given set of several (at least 5 or 6) points for representing pairs of data values, we cannot reasonably expect that the curve of any given equation will pass through all of the points *exactly*. Therefore, when we fit the curve of an equation to such a set of points, we are finding the curve that best approximates passing through the points. It is possible that the curve that best fits the data will not actually pass directly through any of the points, although it should come reasonably close to most of them. Consider the following example.

■**EXAMPLE 1**　All the students enrolled in a mathematics course took an entrance test. To study the reliability of this test as an indicator of future success, an instructor tabulated the test scores of 10 students (selected at random), along with their course averages at the end of the course, and made a graph of the data. See the table below and Fig. 22-19.

Student	Entrance test score, based on 40	Course average, based on 100
A	29	63
B	33	88
C	22	77
D	17	67
E	26	70
F	37	93
G	30	72
H	32	81
I	23	47
J	30	74

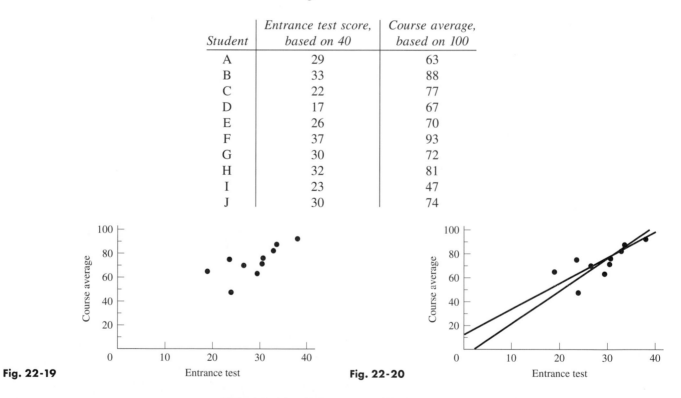

Fig. 22-19

Fig. 22-20

We now ask whether or not there is a functional relationship between the test scores and the course grades. Certainly, no clear-cut relationship exists, but in general we see that the higher the test score, the higher the course grade. This leads to the possibility that there might be some straight line, from which none of the points would vary too significantly. If such a line could be found, then it could be the basis of predictions as to the possible success a student might have in the course, on the basis of his or her grade on the entrance test. Assuming that such a straight line exists, the problem is to find the equation of this line. Figure 22-20 shows two such possible lines.

There are a number of different methods of determining the straight line that best fits the given data points. We shall employ the method that is most widely used: the **method of least squares.** *The basic principle of this method is that the sum of the squares of the deviations of all data points from the best line (in accordance with this method) has the least value possible. By* **deviation** *we mean the difference between the y-value of the line and the y-value for the point (of original data) for a particular value of x.*

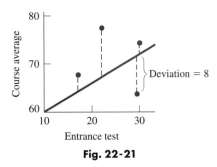

Fig. 22-21

■**EXAMPLE 2**   In Fig. 22-21 the deviations of some of the points of Example 1 are shown. The point (29, 63) (student A of Example 1) has a deviation of 8 from the indicated line in the figure. Thus, we square the value of this deviation to obtain 64. In order to find the equation of the straight line that best fits the given points, the method of least squares requires that the sum of all such squares be a minimum.

Therefore, in applying this method of least squares, it is necessary to use the equation of a straight line and the coordinates of the points of the data. The deviations of all of these data points are determined, and these values are then squared. It is then necessary to determine the constants for the slope m and the y-intercept b in the equation of a straight line $y = mx + b$ for which the sum of the squared values is a minimum. To do this requires certain methods of advanced mathematics. ▪

Using the methods that are required from advanced mathematics, we show that *the equation of the* **least squares line**

$$y = mx + b \qquad \text{(22-9)}$$

can be found by calculating the values of the slope m and the y-intercept b by using the formulas

$$m = \frac{n \sum xy - \left(\sum x \right)\left(\sum y \right)}{n \sum x^2 - \left(\sum x \right)^2} \qquad \text{(22-10)}$$

and

$$b = \frac{\left(\sum x^2 \right)\left(\sum y \right) - \left(\sum xy \right)\left(\sum x \right)}{n \sum x^2 - \left(\sum x \right)^2} \qquad \text{(22-11)}$$

In Eqs. (22-10) and (22-11), *the x's and y's are the values of the coordinates of the points in the given data,* and n is the number of points of data. We can reduce the calculational work in finding the values of m and b by noting that the denominators in Eqs. (22-10) and (22-11) are the same. Therefore, in using a calculator, the value of this denominator can be stored in memory.

EXAMPLE 3 Find the equation of the least-squares line for the points indicated in the following table. Graph the line and data points on the same graph.

x	1	2	3	4	5
y	3	6	6	8	12

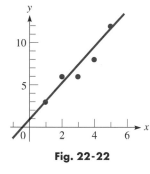

Fig. 22-22

We see from Eqs. (22-10) and (22-11) that we need the sums of x, y, xy, and x^2 in order to find m and b. Thus we set up a table for these values, along with the necessary calculations, as follows:

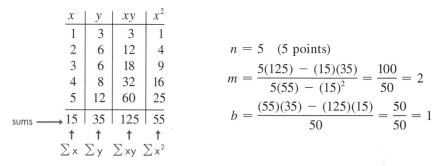

x	y	xy	x^2
1	3	3	1
2	6	12	4
3	6	18	9
4	8	32	16
5	12	60	25
sums ⟶ 15	35	125	55

$$\sum x \quad \sum y \quad \sum xy \quad \sum x^2$$

$n = 5$ (5 points)

$$m = \frac{5(125) - (15)(35)}{5(55) - (15)^2} = \frac{100}{50} = 2$$

$$b = \frac{(55)(35) - (125)(15)}{50} = \frac{50}{50} = 1$$

This means that the equation of the least-squares line is $y = 2x + 1$. This line and the data points are shown in Fig. 22-22.

EXAMPLE 4 Find the least-squares line for the data of Example 1.
Here the x-values will be the entrance-test scores and the y-values are the course averages.

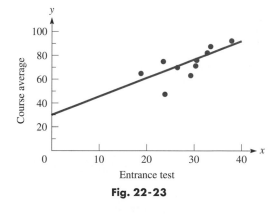

Fig. 22-23

x	y	xy	x^2
29	63	1,827	841
33	88	2,904	1089
22	77	1,694	484
17	67	1,139	289
26	70	1,820	676
37	93	3,441	1369
30	72	2,160	900
32	81	2,592	1024
23	47	1,081	529
30	74	2,220	900
279	732	20,878	8101

$n = 10$

$$m = \frac{10(20,878) - 279(732)}{10(8101) - 279^2} = 1.44$$

$$b = \frac{8101(732) - 20,878(279)}{10(8101) - 279^2} = 33.1$$

Thus the equation of the least-squares line is $y = 1.44x + 33.1$. The line and data points are shown in Fig. 22-23. This line best fits the data, although the fit is obviously approximate. It can be used to predict the approximate course average that a student might be expected to attain, based on the entrance test.

A graphing calculator can be used to find the slope and the intercept, and to display the least-squares line. In the following example we compare the values using Eqs. (22-9), (22-10), and (22-11) with the corresponding calculator displays.

EXAMPLE 5   In a research project to determine the amount of a drug that remains in the bloodstream after a given dosage, the amounts y (in mg of drug/dL of blood) were recorded after t hours.

t (h)	1.0	2.0	4.0	8.0	10.0	12.0
y (mg/dL)	7.6	7.2	6.1	3.8	2.9	2.0

Find the least-squares line for these data, expressing y as a function of t. Sketch the graph of the line and data points.

The table and calculations are shown below.

t	y	ty	t^2
1.0	7.6	7.6	1.0
2.0	7.2	14.4	4.0
4.0	6.1	24.4	16.0
8.0	3.8	30.4	64.0
10.0	2.9	29.0	100
12.0	2.0	24.0	144
37.0	29.6	129.8	329

$n = 6$

$$m = \frac{6(129.8) - 37.0(29.6)}{6(329) - 37.0^2} = -0.523$$

$$b = \frac{(329)(29.6) - (129.8)(37.0)}{6(329) - 37.0^2} = 8.16$$

The equation of the least-squares line is $y = -0.523t + 8.16$. The line and data points are shown in Fig. 22-24. This line is useful in determining the effectiveness of the drug. It can also be used to determine when additional medication may be administered.

Using the linear regression feature of a calculator, Fig. 22-25(a) shows the display of the coefficients of the line $y = ax + b$ (note that the slope is a). The value of r is the *coefficient of correlation* (see Exercises 13–16). Figure 22-25(b) shows a calculator display of the points and the line. We see that they agree with Fig. 22-24.

See the chapter introduction.

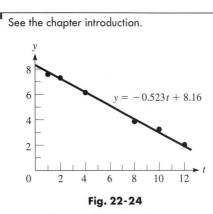

Fig. 22-24

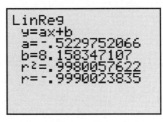

(a)

(b)

Fig. 22-25

EXERCISES 22-6

In Exercises 1–12, find the equation of the least-squares line for the given data. Graph the line and data points on the same graph.

1.

x	4	6	8	10	12
y	1	4	5	8	9

2.

x	1	2	3	4	5	6	7
y	10	17	28	37	49	56	72

3.

x	20	26	30	38	48	60
y	160	145	135	120	100	90

4.

x	1	3	6	5	8	10	4	7	3	8
y	15	12	10	8	9	2	11	9	11	7

5. In an electrical experiment, the following data were found for the values of current and voltage for a particular element of the circuit. Find the voltage V as a function of the current i.

Current (mA)	15.0	10.8	9.30	3.55	4.60
Voltage (V)	3.00	4.10	5.60	8.00	10.50

6. A particular muscle was tested for its speed of shortening as a function of the force applied to it. The results appear below. Find the speed as a function of the force.

Force (N)	60.0	44.2	37.3	24.2	19.5
Speed (m/s)	1.25	1.67	1.96	2.56	3.05

7. The altitude h (in m) of a rocket was measured at several positions at a horizontal distance x (in m) from the launch site, shown in the table. Find the least-squares line for h as a function of x.

x (m)	0	500	1000	1500	2000	2500
h (m)	0	1130	2250	3360	4500	5600

8. In testing an air-conditioning system, the temperature T in a building was measured during the afternoon hours with the results shown in the table. Find the least-squares line for T as a function of the time t from noon.

t (h)	0.0	1.0	2.0	3.0	4.0	5.0
T (°C)	20.5	20.6	20.9	21.3	21.7	22.0

9. The pressure p was measured along an oil pipeline at different distances from a reference point, with results as shown. Find the least-squares line for p as a function of x. Check the values and line with a graphing calculator.

x (ft)	0	100	200	300	400
p (lb/in.^{2})	650	630	605	590	570

10. The heat loss L per hour through various thicknesses of a particular type of insulation was measured as shown in the table. Find the least-squares line for L as a function of t. Check the values and line with a graphing calculator.

t (in.)	3.0	4.0	5.0	6.0	7.0
L (Btu)	5900	4800	3900	3100	2450

11. In an experiment on the photoelectric effect, the frequency of light being used was measured as a function of the stopping potential (the voltage just sufficient to stop the photoelectric effect) with the results given below. Find the least-squares

line for V as a function of f. The frequency for $V = 0$ is known as the *threshold frequency*. From the graph determine the threshold frequency. Check the values and curve with a graphing calculator.

f (PHz)	0.550	0.605	0.660	0.735	0.805	0.880
V (V)	0.350	0.600	0.850	1.10	1.45	1.80

12. If gas is cooled under conditions of constant volume, it is noted that the pressure falls nearly proportionally as the temperature. If this were to happen until there was no pressure, the theoretical temperature for this case is referred to as *absolute zero*. In an elementary experiment, the following data were found for pressure and temperature under constant volume.

T (°C)	0.0	20	40	60	80	100
P (kPa)	133	143	153	162	172	183

Find the least-squares line of P as a function of T, and from the graph determine the value of absolute zero found in this experiment. Check the values and curve with a graphing calculator.

The linear coefficient of correlation, a measure of the relatedness of two variables, is defined by $r = m(s_x/s_y)$, where s_x and s_y are the standard deviations of the x-values and y-values, respectively. Due to its definition, the values of r lie in the range $-1 \leq r \leq 1$. If r is near 1, the correlation is considered good. For values of r between -0.5 and $+0.5$, the correlation is poor. If r is near -1, the variables are said to be negatively correlated; that is, one increases as the other decreases. In Exercises 13–16, compute r for the given sets of data.

13. Exercise 1   14. Exercise 2

15. Exercise 4   16. Example 1

22-7 NONLINEAR REGRESSION

If the experimental points do not appear to be on a straight line, but we recognize them as being approximately on some other type of curve, the method of least squares can be extended to use on these other curves. For example, if the points are apparently on a parabola, we could use the function $y = a + bx^2$. To use the above method, we shall extend the least-squares line to

$$y = m[f(x)] + b \tag{22-12}$$

Here, $f(x)$ must be calculated first, and then the problem can be treated as a least-squares line to find the values of m and b. Some of the functions $f(x)$ that may be considered for use are x^2, $1/x$, and 10^x.

EXAMPLE 1   Find the least-squares curve $y = mx^2 + b$ for the following points.

x	0	1	2	3	4	5
y	1	5	12	24	53	76

In using Eq. (22-12), $f(x) = x^2$. Our first step is to calculate values of x^2, and then we use x^2 as we used x in finding the equation of the least-squares line.

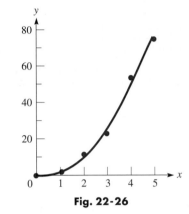

Fig. 22-26

x	$f(x) = x^2$	y	x^2y	$(x^2)^2$
0	0	1	0	0
1	1	5	5	1
2	4	12	48	16
3	9	24	216	81
4	16	53	848	256
5	25	76	1900	625
	55	171	3017	979

$n = 6$

$$m = \frac{6(3017) - 55(171)}{6(979) - 55^2} = 3.05$$

$$b = \frac{(979)(171) - (3017)(55)}{6(979) - 55^2} = 0.52$$

Therefore, the required equation is $y = 3.05x^2 + 0.52$. The graph of this equation and the data points are shown in Fig. 22-26. ▬

EXAMPLE 2   In a physics experiment, the pressure p and volume V of a gas were measured at constant temperature. When the points were plotted, they were seen to approximate the hyperbola $y = c/x$. Find the least-squares approximation to the hyperbola $y = m(1/x) + b$ for the given data. See Fig. 22-27.

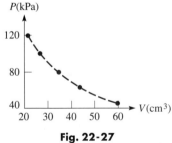

Fig. 22-27

P (kPa)	V (cm^3)	$x (= V)$	$f(x) = \frac{1}{x}$	$y (= P)$	$(\frac{1}{x})y$	$(\frac{1}{x})^2$
120.0	21.0	21.0	0.0476190	120.0	5.7142857	0.0022676
99.2	25.0	25.0	0.0400000	99.2	3.9680000	0.0016000
81.3	31.8	31.8	0.0314465	81.3	2.5566038	0.0009889
60.6	41.1	41.1	0.0243309	60.6	1.4744526	0.0005920
42.7	60.1	60.1	0.0166389	42.7	0.7104825	0.0002769
			0.1600353	403.8	14.4238246	0.0057254

(*Calculator note:* The final digits for the values shown may vary depending on the calculator and how the values are used. Here, all individual values are shown with eight digits (rounded off), although more digits were used. The value of $1/x$ was found from the value of x, with the eight digits shown. However, the values of $(1/x)y$ and $(1/x)^2$ were found from the value of $1/x$, using the extra digits. The sums were found using the rounded-off values shown. However, since the data contains only three digits, any variation in the final digits for $1/x$, $(1/x)y$, or $(1/x)^2$ will not matter.)

$$m = \frac{5(14.4238246) - 0.1600353(403.8)}{5(0.0057254) - 0.1600353^2} = 2490$$

$$b = \frac{(0.0057254)(403.8) - (14.4238246)(0.1600353)}{5(0.0057254) - 0.1600353^2} = 1.2$$

The equation of the hyperbola $y = m(1/x) + b$ is

$$y = \frac{2490}{x} + 1.2$$

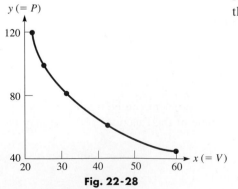

Fig. 22-28

This hyperbola and the data points are shown in Fig. 22-28. ▬

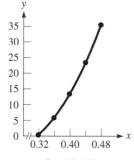

Fig. 22-29

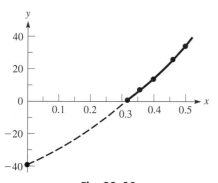

Fig. 22-30

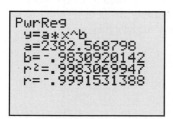

Fig. 22-31

■**EXAMPLE 3** It has been found experimentally that the tensile strength of brass (a copper–zinc alloy) increases (within certain limits) with the percent of zinc. The following table shows the values that have been found. See Fig. 22-29.

Tensile strength (10^5 lb/in.^{2})	0.32	0.36	0.40	0.44	0.48
Percent of zinc	0	5	13	22	34

Fit a curve of the form $y = m(10^x) + b$ to the data. Let x = tensile strength ($\times 10^5$) and y = percent of zinc.

x	$f(x) = 10^x$	y	$(10^x)y$	$(10^x)^2$
0.32	2.0892961	0	0.000000	4.3651583
0.36	2.2908677	5	11.454338	5.2480746
0.40	2.5118864	13	32.654524	6.3095734
0.44	2.7542287	22	60.593031	7.5857758
0.48	3.0199517	34	102.67836	9.1201084
	12.6662306	74	207.38025	32.6286905

(See the note on calculator use in Example 2.)

$$m = \frac{5(207.38025) - 12.6662306(74)}{5(32.6286905) - 12.6662306^2} = 36.8$$

$$b = \frac{32.6286905(74) - 207.38025(12.6662306)}{5(32.6286905) - 12.6662306^2} = -78.3$$

The equation of the curve is $y = 36.8(10^x) - 78.3$. It must be remembered that for practical purposes, y must be positive. The graph of the equation is shown in Fig. 22-30, with the solid portion denoting the meaningful part of the curve. The points of the data are also shown. --------■

As we noted and illustrated in the previous section, a graphing calculator can be used to determine the equation of a regression curve, and to display its graph. A typical calculator can fit a number of different types of equations to data, and the regression equations available on a typical calculator are as follows:

Linear: $y = ax + b$
Quadratic: $y = ax^2 + bx + c$
Cubic: $y = ax^3 + bx^2 + cx + d$
Quartic: $y = ax^4 + bx^3 + cx^2 + dx + e$
Logarithmic: $y = a + b \ln x$
Exponential: $y = ab^x$
Power: $y = ax^b$
Logistic: $y = \dfrac{c}{1 - ae^{-bx}}$
Sinusoidal: $y = a \sin(bx + c) + d$

In comparing an equation obtained using Eq. (22-12) with that obtained using a calculator, we must note the difference in the form of the equation. For example, using the power form $y = ax^b$ on a calculator for Example 2, we get the calculator display shown in Fig. 22-31. The differences are due to (**1**) a power of -1 is assumed in our use of Eq. (22-12), and (**2**) there is a b-term in Eq. (22-12) and no b-term in the power form on the calculator.

EXERCISES *22-7*

In the following exercises, find the equation of the indicated least-squares curve. Sketch the curve and plot the data points on the same graph.

1. For the points in the following table, find the least-squares curve $y = mx^2 + b$.

x	2	4	6	8	10
y	12	38	72	135	200

2. For the points in the following table, find the least-squares curve $y = m\sqrt{x} + b$.

x	0	4	8	12	16
y	1	9	11	14	15

3. For the points in the following table, find the least-squares curve $y = m(1/x) + b$.

x	1.10	2.45	4.04	5.86	6.90	8.54
y	9.85	4.50	2.90	1.75	1.48	1.30

4. For the points in the following table, find the least-squares curve $y = m(10^x) + b$.

x	0.00	0.200	0.500	0.950	1.325
y	6.00	6.60	8.20	14.0	26.0

5. The following data were found for the distance y that an object rolled down an inclined plane in time t. Determine the least-squares curve $y = mt^2 + b$. Compare the equation with that using the *quadratic regression* feature on a graphing calculator.

t (s)	1.0	2.0	3.0	4.0	5.0
y (cm)	6.0	23	55	98	148

6. The increase in length y of a certain metallic rod was measured in relation to particular increases x in temperature. Find the least-squares curve $y = mx^2 + b$. Compare the equation with that using the *quadratic regression* feature of a graphing calculator.

x (°C)	50.0	100	150	200	250
y (cm)	1.00	4.40	9.40	16.4	24.0

7. The pressure p at which Freon, a refrigerant, vaporizes for temperature T is given in the following table. Find the least-squares curve $p = mT^2 + b$. Compare the equation with that using the *quadratic regression* feature of a graphing calculator.

T (°F)	0	20	40	60	80
p (lb/in.²)	23	35	49	68	88

8. A fraction f of annual hot-water loads at a certain facility are heated by solar energy. The fractions f for certain values of the collector area A are given in the following table. Find the least-squares curve $f = m\sqrt{A} + b$. Compare the equation with that using the *power regression* feature of a graphing calculator.

A (m²)	0	12	27	56	90
f	0.0	0.2	0.4	0.6	0.8

9. The makers of a special blend of coffee found that the demand for the coffee depended on the price charged. The price P per pound and the monthly sales S are shown in the following table. Find the least-squares curve $P = m(1/S) + b$.

S (thousands)	240	305	420	480	560
P (dollars)	5.60	4.40	3.20	2.80	2.40

10. The resonant frequency f of an electric circuit containing a 4-μF capacitor was measured as a function of an inductance L in the circuit. The following data were found. Find the least-squares curve $f = m(1/\sqrt{L}) + b$.

L (H)	1.0	2.0	4.0	6.0	9.0
f (Hz)	490	360	250	200	170

11. The displacement y of an object at the end of a spring at given times t is shown in the following table. Find the least-squares curve $y = me^{-t} + b$.

t (s)	0.0	0.5	1.0	1.5	2.0	3.0
y (cm)	6.1	3.8	2.3	1.3	0.7	0.3

(W) 12. The average daily temperatures T (in °F) for each month in Minneapolis (National Weather Service records) are given in the following table.

t	J	F	M	A	M	J	J	A	S	O	N	D
T (°F)	11	18	29	46	57	68	73	71	61	50	33	19

Find the least-squares curve $T = m\cos[\frac{\pi}{6}(t - 0.5)] + b$. Assume the average temperature is for the 15th of each month. Then the values of t (in months) are 0.5, 1.5, ..., 11.5. (The fit is fairly good.) Compare the equation using the *sinusoidal regression* feature of a graphing calculator. What are the main reasons for the differences in the equations?

CHAPTER EQUATIONS

Arithmetic mean	$$\bar{x} = \frac{x_1 f_1 + x_2 f_2 + \cdots + x_n f_n}{f_1 + f_2 + \cdots + f_n} = \frac{\sum xf}{\sum f}$$	(22-1)
Standard deviation	$$s = \sqrt{\frac{\sum (x - \bar{x})^2}{n - 1}}$$	(22-2)
	$$s = \sqrt{\frac{n(\sum x^2) - (\sum x)^2}{n(n - 1)}}$$	(22-3)
Normal distribution	$$y = \frac{e^{-(x-\mu)^2/2\sigma^2}}{\sigma\sqrt{2\pi}}$$	(22-4)
Standard normal distribution	$$y = \frac{1}{\sqrt{2\pi}} e^{-x^2/2}$$	(22-5)
Standard (z) Score	$$z = \frac{x - \mu}{\sigma}$$	(22-6)
Standard error of $\bar{x}$	$$\sigma_{\bar{x}} = \frac{\sigma}{\sqrt{n}}$$	(22-7)
Standard error of s	$$\sigma_s = \frac{\sigma}{\sqrt{2n}}$$	(22-8)
Least-squares line	$$y = mx + b$$	(22-9)
	$$m = \frac{n\sum xy - \left(\sum x\right)\left(\sum y\right)}{n\sum x^2 - \left(\sum x\right)^2}$$	(22-10)
	$$b = \frac{\left(\sum x^2\right)\left(\sum y\right) - \left(\sum xy\right)\left(\sum x\right)}{n\sum x^2 - \left(\sum x\right)^2}$$	(22-11)
Nonlinear curves	$$y = m[f(x)] + b$$	(22-12)

REVIEW EXERCISES

In Exercises 1–10, use the following set of numbers.

1098, 1102, 1101, 1095, 1104, 1097, 1107, 1099, 1104, 1093, 1095, 1102, 1101, 1098, 1106, 1098

1. Determine the median.

2. Determine the mode.

3. Determine the mean.

4. Determine the standard deviation.

5. Construct a frequency distribution table with five classes and a lowest class limit of 1093.

6. Draw a frequency polygon for the data in Exercise 5.

7. Draw a histogram for the data in Exercise 5.

8. Construct a relative frequency table for the data in Exercise 5.

9. Construct a cumulative frequency table for the data of Exercise 5.

10. Draw an ogive for the data of Exercise 5.

In Exercises 11–16, use the following data: An important property of oil is its coefficient of viscosity, which gives a measure of how well it flows. In order to determine the viscosity of a certain motor oil, a refinery took samples from 12 different storage tanks and tested them at 50°C. The results (in pascal-seconds) were 0.24, 0.28, 0.29, 0.26, 0.27, 0.26, 0.25, 0.27, 0.28, 0.26, 0.26, 0.25.

11. Find the mean.　　　　　**12.** Find the median.

13. Find the standard deviation.　**14.** Draw a histogram.

15. Draw a frequency polygon.　**16.** Determine the mode.

In Exercises 17–24, use the following data: A sample of wind generators was tested for power output when the wind speed was 30 km/h. The following table gives the class marks of the powers produced and the number of generators in each class.

Power (W)	650	660	670	680	690
No. generators	3	2	7	12	27

Power (W)	700	710	720	730
No. generators	34	15	16	5

17. Find the median.　　　　**18.** Find the mean.

19. Find the mode.　　　　**20.** Draw a histogram.

21. Find the standard deviation.　**22.** Draw a frequency polygon.

23. Make a cumulative frequency table.

24. Draw an ogive.

In Exercises 25–28, use the following data: A Geiger counter records the presence of high-energy nuclear particles. Even though no apparent radioactive source is present, some particles will be recorded. These are primarily cosmic rays, which are caused by very high energy particles from outer space. In an experiment to measure the amount of cosmic radiation, the number of counts were recorded during 200 5-s intervals. The following table gives the number of counts and the number of 5-s intervals having this number of counts. Draw a frequency curve for these data.

Counts	0	1	2	3	4	5	6	7	8	9	10
Intervals	3	10	25	45	29	39	26	11	7	2	3

25. Find the median.　　　　**26.** Find the mean.

27. Draw a histogram.

28. Make a relative frequency table.

In Exercises 29–32, use the following data: Police radar on Interstate 80 recorded the speeds of 110 cars in a 65 mi/h zone. The following table shows the class marks of the speeds recorded and the number of cars in each class.

Speed (mi/h)	40	45	50	55	60	65	70	75	80	85
No. cars	3	4	4	5	8	22	48	10	4	2

29. Find the mean.　　　　**30.** Find the median.

31. Find the standard deviation.　**32.** Draw an ogive.

In Exercises 33 and 34, use the following information: A company that makes electric light bulbs tests 500 bulbs each day for defects. The number of defective bulbs, along with the proportion of defective bulbs for 20 days, is shown in the following table.

Day	Number defective	Proportion defective
1	23	0.046
2	31	0.062
3	19	0.038
4	27	0.054
5	29	0.058
6	39	0.078
7	26	0.052
8	17	0.034
9	28	0.056
10	33	0.066
11	22	0.044
12	29	0.058
13	20	0.040
14	35	0.070
15	21	0.042
16	32	0.064
17	25	0.050
18	23	0.046
19	29	0.058
20	32	0.064
Total	540	

33. For a p chart, find the values of the central line, UCL, and LCL.

34. Plot a p chart.

In Exercises 35 and 36, use the following information: Five ball bearings are taken from the production line every 15 min and their diameters are measured. The diameters of the sample ball bearings for 16 successive subgroups are given in the following table.

Subgroup	Diameters (mm) of five ball bearings				
1	4.98	4.92	5.02	4.91	4.93
2	5.03	5.01	4.94	5.06	5.07
3	5.05	5.03	5.00	5.02	4.96
4	5.01	4.92	4.91	4.99	5.03
5	4.92	4.97	5.02	4.95	4.94
6	5.02	4.95	5.01	5.07	5.15
7	4.93	5.03	5.02	4.96	4.99
8	4.85	4.91	4.88	4.92	4.90
9	5.02	4.95	5.06	5.04	5.06
10	4.98	4.98	4.93	5.01	5.00
11	4.90	4.97	4.93	5.05	5.02
12	5.03	5.05	4.92	5.03	4.98
13	4.90	4.96	5.00	5.02	4.97
14	5.09	5.04	5.05	5.02	4.97
15	4.88	5.00	5.02	4.97	4.94
16	5.02	5.09	5.03	4.99	5.03

35. Plot an $\bar{x}$ chart.　　　**36.** Plot an R chart.

In Exercises 37–40, use the following data: After analyzing data for a long period of time, it was determined that samples of 500 readings of an organic pollutant for an area are distributed normally. For this pollutant, $\mu = 2.20 \ \mu g/m^3$ and $\sigma = 0.50 \ \mu g/m^3$.

37. In a sample, how many readings are between $1.50 \ \mu g/m^3$ and $2.50 \ \mu g/m^3$?

38. In a sample, how many readings are between $2.50 \ \mu g/m^3$ and $3.50 \ \mu g/m^3$?

39. In a sample, how many readings are above $1.00 \ \mu g/m^3$?

40. In a sample, how many readings are below $2.00 \ \mu g/m^3$?

In Exercises 41–48, find the indicated least-squares curve. Sketch the curve and data points on the same graph.

41. In a certain experiment, the resistance R of a certain resistor was measured as a function of the temperature T. The data found are shown in the following table. Find the least-squares line, expressing R as a function of T.

T (°C)	0.0	20.0	40.0	60.0	80.0	100
R (Ω)	25.0	26.8	28.9	31.2	32.8	34.7

42. An air-pollution monitoring station took samples of air each hour during the later morning hours and tested each sample for the number n of parts per million (ppm) of carbon monoxide. The results are shown in the table, where t is the number of hours after 6 A.M. Find the least-squares line for n as a function of t.

t (h)	0.0	1.0	2.0	3.0	4.0	5.0	6.0
n (ppm)	8.0	8.2	8.8	9.5	9.7	10.0	10.7

43. The *Mach number* of a moving object is the ratio of its speed to the speed of sound (740 mi/h). The following table shows the speed s of a jet aircraft, in terms of Mach numbers, and the time t after it starts to accelerate. Find the least-squares line of s as a function of t. Compare the equation with that using the *linear regression* feature of a graphing calculator.

t (min)	0.00	0.60	1.20	1.80	2.40	3.00
s (Mach number)	0.88	0.97	1.03	1.11	1.19	1.25

44. In an experiment to determine the relation between the load y on a spring and the length x of the spring, the following data were found. Find the least-squares line that expresses y as a function of x. Compare the equation with that using the *linear regression* feature of a graphing calculator.

Load (lb)	0.0	1.0	2.0	3.0	4.0	5.0
Length (in.)	10.0	11.2	12.3	13.4	14.6	15.9

45. The distance s of a missile above the ground at time t after being released from a plane is given by the following table. Find the least-squares curve of the form $s = mt^2 + b$ for these data. Compare the equation with that using the *quadratic regression* feature of a graphing calculator.

t (s)	0.0	3.0	6.0	9.0	12.0	15.0	18.0
s (m)	3000	2960	2820	2600	2290	1900	1410

46. In an elementary experiment that measured the wavelength L of sound as a function of the frequency f, the following results were obtained.

Frequency (Hz)	240	320	400	480	560
Wavelength (cm)	140	107	81.0	70.0	60.0

Find the least-squares curve of the form $L = m(1/f) + b$ for these data.

47. After being heated, the temperature T of an insulated liquid is measured at times t as follows:

t (h)	0	2	4	6	8	10
T (°C)	100	85	72	63	54	48

Plot these points and choose an appropriate function $f(x)$ for $y = m[f(x)] + b$. Then find the equation of the least-squares curve.

48. The vertical distance y of the cable of a suspension bridge above the surface of the bridge is measured at a horizontal distance x along the bridge from its center. See Fig. 22-32. The results are as follows:

x (m)	0	100	200	300	400	500
y (m)	15	17	23	33	47	65

Plot these points and choose an appropriate function $f(x)$ for $y = m[f(x)] + b$. Then find the equation of the least-squares curve.

Fig. 22-32

In Exercises 49–52, solve the given problems.

49. The nth root of the product of n positive numbers is the *geometric mean* of the numbers. Find the geometric mean of the carbon monoxide readings in Exercise 42.

50. One use of the geometric mean (see Exercise 49) is to find an average ratio. By finding the geometric mean, find the average Mach number for the jet in Exercise 43.

51. Show that Eqs. (22-10) and (22-11) satisfy the equation $\bar{y} = m\bar{x} + b$.

52. Given that $\Sigma(x - \bar{x})^2 = \Sigma x^2 - n\bar{x}^2$, derive Eq. (22-3) from Eq. (22-2).

Writing Exercise

53. A study is to be made of the effectiveness of a treatment for glaucoma (a severe eye disorder), depending on the age of the person treated. Write two or three paragraphs explaining what data could be found and how they can be analyzed and then used to predict the effects of the treatment on future patients.

PRACTICE TEST

In Problems 1–3, use the following set of numbers.

5, 6, 1, 4, 9, 5, 7, 3, 8, 10, 5, 8, 4, 9, 6

1. Find the median.

2. Find the mode.

3. Draw a histogram with five classes and the lowest class limit at 1.

In Problems 4–8, use the following data: Two machine parts are considered satisfactorily assembled if their total thickness (to the nearest 0.01 in.) is between or equal to 0.92 in. and 0.94 in. One hundred assemblies are tested, and the class mark of the thicknesses and the number of assemblies in each class are given in the following table.

Total thickness (in.)	0.90	0.91	0.92	0.93	0.94	0.95	0.96
Number	3	9	31	38	12	5	2

4. Find the mean.

5. Find the standard deviation.

6. Draw a frequency polygon.

7. Make a relative frequency table.

8. Draw an ogive (less than).

9. For a set of values that are normally distributed, what percent of them is below the value (greater than the mean) for which the z-score is 0.2257?

10. The machine-part assemblies in Problems 4–8 were tested in groups of five each hour for 20 hours. Explain, in general, how to use the data from the test subgroups to plot an R chart.

11. Find the equation of the least-squares line for the points indicated in the following table. Graph the line and data points on the same graph.

x	1	3	5	7	9
y	5	11	17	20	27

12. The period of a pendulum as a function of its length was measured, giving the following results:

Length (ft)	1.00	3.00	5.00	7.00	9.00
Period (s)	1.10	1.90	2.50	2.90	3.30

Find the equation of the least-squares curve of the form $y = m\sqrt{x} + b$, which expresses the period as a function of the length.

SUPPLEMENTARY TOPICS

$S\text{-}1$ GAUSSIAN ELIMINATION

Named for the German mathematician Karl Gauss (1777–1855).

In Chapters 5 and 16 we developed several methods of solving systems of simultaneous linear equations. These included algebraic methods, determinants, and the use of the inverse of a matrix.

We now show a general method that can be used to solve a system of linear equations. The procedure is similar to that used in finding the inverse of a matrix in Section 16-5 and is known as **Gaussian elimination.** It is commonly used in computer programs, particularly when there are several equations in the system.

Gaussian elimination is based on the use of the following two operations on the equations of a system.

1. *Both sides of an equation may be multiplied by a constant.*

2. *A multiple of one equation may be added to another equation.*

Using three linear equations in three unknowns as an example, by using the above operations we can change the system

$$
\begin{aligned}
a_1x + b_1y + c_1z &= d_1 \\
a_2x + b_2y + c_2z &= d_2 \\
a_3x + b_3y + c_3z &= d_3
\end{aligned}
\tag{S-1}
$$

into the equivalent system

$$
\begin{aligned}
x + b_4y + c_4z &= d_4 \\
y + c_5z &= d_5 \\
z &= d_6
\end{aligned}
\tag{S-2}
$$

The solution is now completed by substituting the known value of z into the second equation and then substituting the known values of y and z into the first equation.

$$
\begin{aligned}
2x + y &= 4 \\
3x - 2y &= 3
\end{aligned}
$$

$$
\begin{aligned}
x + \tfrac{1}{2}y &= 2 \\
3x - 2y &= 3
\end{aligned}
$$

$$
\begin{aligned}
x + \tfrac{1}{2}y &= 2 \\
- \tfrac{7}{2}y &= -3
\end{aligned}
$$

$$
\begin{aligned}
x + \tfrac{1}{2}y &= 2 \\
y &= \tfrac{6}{7}
\end{aligned}
$$

$$
\begin{aligned}
x + \tfrac{1}{2}\!\left(\tfrac{6}{7}\right) &= 2 \\
x &= \tfrac{11}{7}
\end{aligned}
$$

EXAMPLE 1 Solve the given system of equations by Gaussian elimination. Given system of equations (equations and solution at left)

We want the coefficient of x in the first equation to be 1. Therefore, divide the first equation by 2, the coefficient of x.

We next eliminate x in the second equation by subtracting 3 times the first equation from the second equation.

Now we solve the second equation for y by dividing by $-\tfrac{7}{2}$.

To find the value of x, we substitute the value of $y = \tfrac{6}{7}$ into the first equation.

The solution is $x = \tfrac{11}{7}$, $y = \tfrac{6}{7}$, which checks when substituted into the original equations.

627

EXAMPLE 2 Solve the given system of equations by Gaussian elimination.
Given system of equations (equations and solution at left)

$$x + 3y - 2z = -5$$
$$2x - y + 4z = 7$$
$$-3x + 2y - 3z = -1$$

Since the coefficient of x in the first equation is 1, we proceed to the next step.
Subtract 2 times first equation from second equation and add three times first equation to third equation to eliminate x from the second and third equations.
To get the coefficient of y equal to 1 in the second equation, we divide it by -7.

$$x + 3y - 2z = -5$$
$$- 7y + 8z = 17$$
$$11y - 9z = -16$$

$$x + 3y - 2z = -5$$
$$y - \frac{8}{7}z = -\frac{17}{7}$$
$$11y - 9z = -16$$

To eliminate y from the third equation, we subtract 11 times the second equation from the third equation.

$$x + 3y - 2z = -5$$
$$y - \frac{8}{7}z = -\frac{17}{7}$$
$$\frac{25}{7}z = \frac{75}{7}$$

We solve for z by multiplying the third equation by $\frac{7}{25}$ (or dividing by $\frac{25}{7}$).

$$x + 3y - 2z = -5$$
$$y - \frac{8}{7}z = -\frac{17}{7}$$
$$z = 3$$

The value of y is found by substituting $z = 3$ into the second equation.

$$y - \frac{8}{7}(3) = -\frac{17}{7}$$
$$y = 1$$

The value of x is found by substituting $y = 1$ and $z = 3$ back into the first equation.

$$x + 3(1) - 2(3) = -5$$
$$x = -2$$

   The solution is $x = -2$, $y = 1$, $z = 3$, which checks when substituted into the original equations. ---------■

EXAMPLE 3 Solve the given system of equations by Gaussian elimination.
Given system of equations (equations and solution at left)

$$4y + z = 2$$
$$2x + 6y - 2z = 3$$
$$4x + 8y - 5z = 4$$

Since the first equation does not contain x, which means that $a_1 = 0$, we cannot divide by a_1. Therefore, we interchange the first and second equations, then divide the new first equation by 2.

$$x + 3y - z = \frac{3}{2}$$
$$4y + z = 2$$
$$4x + 8y - 5z = 4$$

We eliminate x from the third equation by subtracting 4 times the first equation from the third equation.

$$x + 3y - z = \frac{3}{2}$$
$$4y + z = 2$$
$$- 4y - z = -2$$

Make the coefficient of y in the second equation equal to 1 by dividing the second equation by 4.

$$x + 3y - z = \frac{3}{2}$$
$$y + \frac{1}{4}z = \frac{1}{2}$$
$$- 4y - z = -2$$

Eliminate y from the third equation by adding 4 times the second equation to the third equation.

$$x + 3y - z = \frac{3}{2}$$
$$y + \frac{1}{4}z = \frac{1}{2}$$
$$0 = 0$$

Since the third equation, $0 = 0$, is correct, we can continue. Although there is no specific value for z, it is possible to express both x and y in terms of z.

$$x + 3y - z = \frac{3}{2}$$
$$y = \frac{1}{2} - \frac{1}{4}z$$

Solve the second equation for y. In this case it is expressed in terms of z. We no longer need to include the third equation.

$$x + 3(\frac{1}{2} - \frac{1}{4}z) - z = \frac{3}{2}$$
$$x = \frac{7}{4}z$$

Substitute the solution for y in the first equation, and solve for x in terms of z.

We have expressed the solution as $x = \frac{7}{4}z$ and $y = \frac{1}{2} - \frac{1}{4}z$. Since both x and y are expressed in terms of z and there is no specific value of z, the value of z can be

CAUTION▶ chosen arbitrarily. This means *there is an unlimited number of solutions.* For example, if $z = 4$, $x = 7$ and $y = -\frac{1}{2}$. If $z = -2$, $x = -\frac{7}{2}$ and $y = 1$. ---------■

When we solved systems of linear equations in Chapters 5 and 16, we found that not all systems have unique solutions, as in Examples 1 and 2. In Example 3 we illustrated the use of Gaussian elimination on a system of equations for which the solution is not unique.

In Example 3 one of the equations became $0 = 0$, and there was an unlimited number of solutions. If any of the equations of a system becomes $0 = a$, $a \neq 0$, then the system is inconsistent and there is no solution.

If a system of equations has more unknowns than equations, or if it can be written in this way, as in Example 3, it usually has an unlimited number of solutions. It is possible, however, that such a system is inconsistent.

If a system of equations has more equations than unknowns, it is inconsistent unless enough equations become $0 = 0$ such that at least one solution is found. The following example illustrates two systems of equations in which there are more equations than unknowns.

▌EXAMPLE 4   Solve the following systems of equations by Gaussian elimination.

First system

$$x + 2y = 5$$
$$3x - y = 1$$
$$4x + y = 6$$

$$x + 2y = 5$$
$$-7y = -14$$
$$-7y = -14$$

$$x + 2y = 5$$
$$y = 2$$
$$-7y = -14$$

$$x + 2y = 5$$
$$y = 2$$
$$0 = 0$$
$$x = 1$$

Second system

$$x + 2y = 5$$
$$3x - y = 1$$
$$4x + y = 2$$

$$x + 2y = 5$$
$$-7y = -14$$
$$-7y = -18$$

$$x + 2y = 5$$
$$y = 2$$
$$-7y = -18$$

$$x + 2y = 5$$
$$y = 2$$
$$0 = -4$$

The solutions are shown at the left. We note that each system has three equations and two unknowns.

In the solution of the first system, the third equation becomes $0 = 0$, and only two equations are needed to find the solution $x = 1$, $y = 2$.

In the solution of the second system, the third equation becomes $0 = -4$, which means the system is inconsistent and there is no solution.

The solutions are shown graphically in Figs. S1-1 and S1-2. In Fig. S1-1 each of the three lines passes through the point (1, 2), whereas in Fig. S1-2 there is no point common to the three lines.

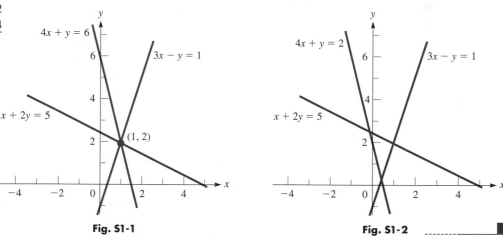

Fig. S1-1   **Fig. S1-2** --------- ▌

EXERCISES S-1

In Exercises 1–20, solve the given systems of equations by Gaussian elimination. If there is an unlimited number of solutions, find two of them.

1. $x + 2y = 4$
$3x - y = 5$

2. $2x + y = 1$
$5x + 2y = 1$

3. $5x - 3y = 2$
$-2x + 4y = 3$

4. $-3x + 2y = 4$
$4x + y = -5$

5. $2x + y - z = 0$
$4x + y + z = 2$
$-2x - 2y + 3z = 0$

6. $2y + 2z = -1$
$3x - 4y + 3z = 1$
$4x + 2y + 5z = 4$

7. $x + 3y + 3z = -3$
$2x + 2y + z = -5$
$-2x - y + 4z = 6$

8. $3x - y + 2z = 3$
$4x - 2y + z = 3$
$6x + 6y + 3z = 4$

9. $w + 2x - y + 3z = 12$
$2w - 2y - z = 3$
$3x - y - z = -1$
$-w + 2x + y + 2z = 3$

10. $2x - 3y + 2z + 2t = 3$
$4x + 2y - 3z = -4$
$2x - y + 3z + 2t = 3$
$6x + 3y - 2z - t = 2$

11. $x - 4y + z = 2$
$3x - y + 4z = -4$

12. $4x + z = 6$
$2x - y - 2z = -2$

13. $2x - y + z = 5$
$3x + 2y - 2z = 4$
$5x + 8y - 8z = 5$

14. $3x + 2y - z = 3$
$2x - y - 3z = 2$
$-x + 4y + 5z = -1$

15. $2x - 4y = 7$
$3x + 5y = -6$
$9x - 7y = 15$

16. $4x - y = 5$
$2x + 2y = 3$
$6x - 4y = 7$
$2x + y = 4$

17. $3x + 5y = -2$
$24x - 18y = 13$
$15x - 33y = 19$
$6x + 68y = -33$

18. $x + 3y - z = 1$
$3x - y + 4z = 4$
$-2x + 2y + 3z = 17$
$3x + 7y + 5z = 23$

19. $x - 2y - 2z = 3$
$2x + y + 3z = 4$
$-2x - y - z = 5$
$3x + 3y - 2z = 2$

20. $2x - y - 2z - t = 4$
$4x + 2y + 3z + 2t = 3$
$-2x - y + 4z = -2$

In Exercises 21–24, set up systems of equations and solve by Gaussian elimination.

21. One personal computer can perform x calculations per second, and a second personal computer can perform y calculations per second. If each operates for two seconds, 25.0 million calculations are performed. If the first operates for four seconds and the second for three seconds, 43.2 million calculations are performed. Find x and y.

22. The voltage across an electric resistor equals the current (in A) times the resistance (in Ω). If a current of 3.00 A passes through each of two resistors, the sum of the voltages is 10.5 V. If 2.00 A passes through the first resistor and 4.00 A passes through the second resistor, the sum of the voltages is 13.0 V. Find the resistances.

23. Three machines together produce 650 parts each hour. Twice the production of the second machine is 10 parts/h more than the sum of the production of the other two machines. If the first operates for three hours and the others operate for two hours, 1550 parts are produced. Find the production rate of each machine.

24. A total of $12,000 is invested, part at 6.5%, part at 6.0%, and part at 5.5%, yielding a total annual interest of $726. The income from the 6.5% part yields $128 less than that for the other two parts combined. How much is invested at each rate?

S-2 ROTATION OF AXES

In Chapter 21 we discussed the circle, parabola, ellipse, and hyperbola and how these curves are represented by the second-degree equation

$$Ax^2 + Bxy + Cy^2 + Dx + Ey + F = 0 \qquad \text{(S2-1)}$$

Our discussion included the properties of the curves and their equations with center (vertex of a parabola) at the origin. However, except for the special case of the hyperbola $xy = c$, we did not cover what happens when the axes are rotated about the origin.

If a set of axes is rotated about the origin through an angle θ, *as shown in Fig. S2-1, we say that there has been a* **rotation of axes.** In this case each point P in the plane has two sets of coordinates, (x, y) in the original system and (x', y') in the rotated system.

If we now let r equal the distance from the origin O to point P and let ϕ be the angle between the x'-axis and the line OP, we have

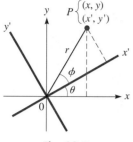

$$x' = r \cos \phi \qquad y' = r \sin \phi \tag{S2-2}$$
$$x = r \cos(\theta + \phi) \qquad y = r \sin(\theta + \phi) \tag{S2-3}$$

Using the cosine and sine of the sum of two angles, we can write Eqs. (S2-3) as

$$x = r \cos \phi \cos \theta - r \sin \phi \sin \theta$$
$$y = r \cos \phi \sin \theta + r \sin \phi \cos \theta \tag{S2-4}$$

Now, using Eqs. (S2-2), we have

$$\boxed{\begin{aligned} x &= x' \cos \theta - y' \sin \theta \\ y &= x' \sin \theta + y' \cos \theta \end{aligned}} \tag{S2-5}$$

Fig. S2-1

In our derivation, we have used the special case when θ is acute and P is in the first quadrant of both sets of axes. When simplifying equations of curves using Eqs. (S2-5), we find that a rotation through a positive acute angle θ is sufficient. It can be shown, however, that Eqs. (S2-5) hold for any θ and position of P.

■EXAMPLE 1   Transform $x^2 - y^2 + 8 = 0$ by rotating the axes through 45°.
When $\theta = 45°$, the rotation equations (S2-5) become

$$x = x' \cos 45° - y' \sin 45° = \frac{x'}{\sqrt{2}} - \frac{y'}{\sqrt{2}}$$

$$y = x' \sin 45° + y' \cos 45° = \frac{x'}{\sqrt{2}} + \frac{y'}{\sqrt{2}}$$

Substituting into the equation $x^2 - y^2 + 8 = 0$ gives

$$\left(\frac{x'}{\sqrt{2}} - \frac{y'}{\sqrt{2}}\right)^2 - \left(\frac{x'}{\sqrt{2}} + \frac{y'}{\sqrt{2}}\right)^2 + 8 = 0$$

$$\frac{1}{2}x'^2 - x'y' + \frac{1}{2}y'^2 - \frac{1}{2}x'^2 - x'y' - \frac{1}{2}y'^2 + 8 = 0$$

$$x'y' = 4$$

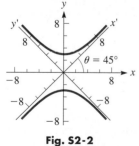

Fig. S2-2

The graph and both sets of axes are shown in Fig. S2-2. The original equation represents a hyperbola. We have the $xy = c$ form with rotation through 45°.   ■

When we showed the type of curve represented by the second-degree equation Eq. (S2-1) in Section 21-8, the standard forms of the parabola, ellipse, and hyperbola required that $B = 0$. This means that there is no xy-term in the equation. In Section 21-8, we considered only one case ($xy = c$) for which $B \neq 0$.

If we can remove the xy-term from a second-degree equation, the analysis of the graph is simplified. By a proper rotation of axes we find that Eq. (S2-1) can be transformed into an equation that has no $x'y'$-term.

By substituting Eqs. (S2-5) into Eq. (S2-1) and then simplifying, we have

$$(A \cos^2 \theta + B \sin \theta \cos \theta + C \sin^2 \theta)x'^2 + [B \cos 2\theta - (A - C)\sin 2\theta]x'y'$$
$$+ (A \sin^2 \theta - B \sin \theta \cos \theta + C \cos^2 \theta)y'^2 + (D \cos \theta + E \sin \theta)x' + (E \cos \theta - D \sin \theta)y' + F = 0$$

If there is to be no $x'y'$-term, its coefficient must be zero. This means that $B \cos 2\theta - (A - C)\sin 2\theta = 0$, or

ANGLE OF ROTATION

$$\tan 2\theta = \frac{B}{A - C} \qquad (A \neq C) \qquad \text{(S2-6)}$$

Eq. (S2-6) gives the angle of rotation except when $A = C$. In this case the coefficient of the $x'y'$-term is $B \cos 2\theta$, which is zero if $2\theta = 90°$. Thus,

$$\theta = 45° \qquad (A = C) \qquad \text{(S2-7)}$$

Consider the following example.

▌**EXAMPLE 2**   By rotation of axes, transform $8x^2 + 4xy + 5y^2 = 9$ into a form without an xy-term. Identify and sketch the curve.

Here, $A = 8$, $B = 4$, and $C = 5$. Therefore, using Eq. (S2-6), we have

$$\tan 2\theta = \frac{4}{8 - 5} = \frac{4}{3}$$

Since $\tan 2\theta$ is positive, we may take 2θ as an acute angle, which means θ is also acute. For the transformation we need $\sin \theta$ and $\cos \theta$. We find these values by first finding the value of $\cos 2\theta$ and then using the half-angle formulas.

$$\cos 2\theta = \frac{1}{\sec 2\theta} = \frac{1}{\sqrt{1 + \tan^2 2\theta}} = \frac{1}{\sqrt{1 + (\frac{4}{3})^2}} = \frac{3}{5} \qquad \begin{array}{l}\text{using Eqs. (20-2)}\\\text{and (20-7)}\end{array}$$

For reference: Eq. (20-2) is
$$\cos \theta = \frac{1}{\sec \theta}$$
Eq. (20-7) is
$$1 + \tan^2 \theta = \sec^2 \theta$$
Eq. (20-26) is
$$\sin \frac{\alpha}{2} = \pm \sqrt{\frac{1 - \cos \alpha}{2}}$$
Eq. (20-27) is
$$\cos \frac{\alpha}{2} = \pm \sqrt{\frac{1 + \cos \alpha}{2}}$$

Now, using the half-angle formulas, Eqs. (20-26) and (20-27), we have

$$\sin \theta = \sqrt{\frac{1 - \cos 2\theta}{2}} = \sqrt{\frac{1 - \frac{3}{5}}{2}} = \frac{1}{\sqrt{5}}, \quad \cos \theta = \sqrt{\frac{1 + \cos 2\theta}{2}} = \sqrt{\frac{1 + \frac{3}{5}}{2}} = \frac{2}{\sqrt{5}}$$

Here, θ is about $26.6°$. Now substituting these values into Eqs. (S2-5), we have

$$x = x'\left(\frac{2}{\sqrt{5}}\right) - y'\left(\frac{1}{\sqrt{5}}\right) = \frac{2x' - y'}{\sqrt{5}}, \quad y = x'\left(\frac{1}{\sqrt{5}}\right) + y'\left(\frac{2}{\sqrt{5}}\right) = \frac{x' + 2y'}{\sqrt{5}}$$

Now, substituting into the equation $8x^2 + 4xy + 5y^2 = 9$ gives

$$8\left(\frac{2x' - y'}{\sqrt{5}}\right)^2 + 4\left(\frac{2x' - y'}{\sqrt{5}}\right)\left(\frac{x' + 2y'}{\sqrt{5}}\right) + 5\left(\frac{x' + 2y'}{\sqrt{5}}\right)^2 = 9$$
$$8(4x'^2 - 4x'y' + y'^2) + 4(2x'^2 + 3x'y' - 2y'^2) + 5(x'^2 + 4x'y' + 4y'^2) = 45$$
$$45x'^2 + 20y'^2 = 45$$
$$\frac{x'^2}{1} + \frac{y'^2}{\frac{9}{4}} = 1$$

This is an ellipse with semimajor axis of 3/2 and semiminor axis of 1. See Fig. S2-3.   ∎

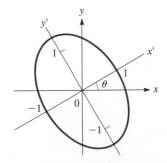

Fig. S2-3

In Example 2, tan 2θ was positive, and we made 2θ and θ positive. If, when using Eq. (S2-6), tan 2θ is negative, we then make 2θ obtuse ($90° < 2\theta < 180°$). In this case cos 2θ will be negative, but θ will be acute ($45° < \theta < 90°$).

In Section 21-7 we showed the use of translation of axes in writing an equation in standard form if $B = 0$. In this section we have seen how rotation of axes is used to eliminate the xy-term. It is possible that both a translation of axes and a rotation of axes are needed to write an equation in standard form.

In Section 21-8 we identified a conic section by inspecting the values of A and C when $B = 0$. If $B \neq 0$, these curves are identified as follows:

1. If $B^2 - 4AC = 0$, a parabola

2. If $B^2 - 4AC < 0$, an ellipse

3. If $B^2 - 4AC > 0$, a hyperbola

Special cases such as a point, parallel or intersecting lines, or no curve may result.

■**EXAMPLE 3**   For the equation $16x^2 - 24xy + 9y^2 + 20x - 140y - 300 = 0$, identify the curve and simplify it to standard form. Sketch the graph and display it on a graphing calculator.

With $A = 16$, $B = -24$, and $C = 9$, using Eq. (S2-6), we have

$$\tan 2\theta = \frac{-24}{16 - 9} = -\frac{24}{7}$$

In this case tan 2θ is negative, and we take 2θ to be an obtuse angle. We then find that cos $2\theta = -7/25$. In turn we find that sin $\theta = 4/5$ and cos $\theta = 3/5$. Here θ is about $53.1°$. Using these values in Eqs. (S2-5), we find that

$$x = \frac{3x' - 4y'}{5} \qquad y = \frac{4x' + 3y'}{5}$$

Substituting these into the original equation and simplifying, we get

$$y'^2 - 4x' - 4y' - 12 = 0$$

This equation represents a parabola with its axis parallel to the x'-axis. The vertex is found by completing the square:

$$(y' - 2)^2 = 4(x' + 4)$$

The vertex is the point $(-4, 2)$ in the $x'y'$-rotated system. Therefore,

$$y''^2 = 4x''$$

is the equation in the $x''y''$-rotated and then translated system. The graph and the coordinate systems are shown in Fig. S2-4.

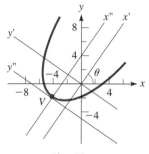

Fig. S2-4

To display the curve on a graphing calculator, we solve for y by using the quadratic formula. Writing the equation as

$$9y^2 + (-24x - 140)y + (16x^2 + 20x - 300) = 0$$

Therefore, we see that in using the quadratic formula, $a = 9$, $b = -24x - 140$, and $c = 16x^2 + 20x - 300$. Now, solving for y, we have

$$y = \frac{24x + 140 \pm \sqrt{(-24x - 140)^2 - 4(9)(16x^2 + 20x - 300)}}{18}$$

We ***enter both functions indicated by the ± sign*** to get the display in Fig. S2-5. ■

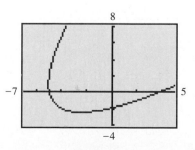

Fig. S2-5

EXERCISES S-2

In Exercises 1–4, transform the given equations by rotating the axes through the given angle. Identify and sketch each curve.

1. $x^2 - y^2 = 25$, $\theta = 45°$

2. $x^2 + y^2 = 16$, $\theta = 60°$

3. $8x^2 - 4xy + 5y^2 = 36$, $\theta = \tan^{-1} 2$

4. $2x^2 + 24xy - 5y^2 = 8$, $\theta = \tan^{-1} \frac{3}{4}$

In Exercises 5–10, transform each equation to a form without an xy-term by a rotation of axes. Identify and sketch each curve. Then display each curve on a graphing calculator.

5. $x^2 + 2xy + y^2 - 2x + 2y = 0$

6. $5x^2 - 6xy + 5y^2 = 32$

7. $3x^2 + 4xy = 4$

8. $9x^2 - 24xy + 16y^2 - 320x - 240y = 0$

9. $11x^2 - 6xy + 19y^2 = 20$

10. $x^2 + 4xy - 2y^2 = 6$

In Exercises 11 and 12, transform each equation to a form without an xy-term by a rotation of axes. Then transform the equation to a standard form by a translation of axes. Identify and sketch each curve. Then display each curve on a graphing calculator.

11. $16x^2 - 24xy + 9y^2 - 60x - 80y + 400 = 0$

12. $73x^2 - 72xy + 52y^2 + 100x - 200y + 100 = 0$

APPENDIX A — STUDY AIDS

A-1 INTRODUCTION

The primary objective of this text is to give you an understanding of mathematics so that you can use it effectively as a tool in your technology. Without an understanding of the basic methods, knowledge is usually short-lived. However, if you do understand, you will find your work much more enjoyable and rewarding. This is true in any course you may take, be it in mathematics or in any other field.

Mathematics is an indispensable tool in almost all scientific fields of study. You will find it used to a greater and greater degree as you work in your chosen field. Generally, in the introductory portions of allied courses, it is enough to have a grasp of elementary concepts in algebra and geometry. However, as you progress in your field, the need for more mathematics will be apparent. This text is designed to develop these necessary tools so that they will be available to you in your technical courses. *You cannot derive the full benefit from your mathematics course unless you devote the necessary amount of time to develop a sound understanding of the subject.*

Your mathematics background probably includes some geometry and algebra. Therefore, some of the topics covered in this book may seem familiar to you, especially in the earlier chapters. However, it is likely that your background in some of these areas is not complete, either because you have not studied mathematics for a while or because you did not fully understand the topics when you first encountered them. *If a topic is familiar, take the opportunity to clarify any points on which you are not certain* and do not reason that there is no sense in studying it again. In almost every topic, you probably will find certain points that can use further study. *If the topic is new to you, use your time effectively to develop an understanding of the methods involved* and do not simply memorize problems of a certain type.

There is only one good way to develop the understanding and working knowledge necessary in any course, and that is to **work with it.** Many students consider mathematics difficult. They say that it is their lack of mathematical ability and the complexity of the material that makes it difficult. Some of the topics in mathematics, especially in the more advanced areas, do require a certain aptitude for full comprehension. However, *a large proportion of poor grades in elementary mathematics courses results from the fact that* **the student is not willing to put in the necessary time to develop a full understanding.** The student takes a quick glance through the material, tries a few exercises, is largely unsuccessful, and then decides that the material is "impossible." A detailed reading of the text, following the illustrative examples carefully, and then solving the exercises would lead to more success and therefore make the work much more enjoyable and rewarding. No matter what text is used, what methods are used in the course, or what other variables may be introduced, *if you do not put in an adequate amount of time studying, you will not derive the proper results.* More detailed suggestions for study are included in the following section.

A-2 SUGGESTIONS FOR STUDY

When you are studying the material presented in this text, the following suggestions may help you to derive full benefit from the time you devote to it.

1. Before attempting to do the exercises, read through the material preceding them.

2. Follow the illustrative examples carefully, being certain that you know how to proceed from step to step. You should then have a good idea of the methods involved.

3. Work through the exercises, spending a reasonable amount of time on each problem. If you cannot solve a certain problem in a reasonable amount of time, leave it and go to the next. Return to this problem later. If you find many problems difficult, you should reread the explanatory material and the examples to determine what point or points you have not understood.

4. When you have completed the exercises, or at least most of them, glance back through the explanatory material to be sure you understand the methods and principles.

5. If you have gone through the first four steps and certain points still elude you, ask to have these points clarified in class. Do not be afraid to ask questions; only be sure that you have made a sincere effort on your own before you ask them.

Some study habits that are useful not only here but in all of your other subjects are the following:

1. Put in the time required to develop the material fully, being certain that you are making effective use of your time. A good place to study helps immeasurably.

2. Learn the *methods and principles* being presented. Memorize as little as possible, for although certain basic facts are more expediently learned by memorization, these should be kept to a minimum.

3. Keep up with the material in all of your courses. Do not let yourself get so behind in your studies that it becomes difficult to make up the time. Usually the time is never really made up. Studying only before tests is a poor way of learning and is usually rather ineffective.

4. When you are taking examinations, always read each question carefully before attempting the solution. Solve those you find easiest first, and do not spend too much time on any one problem. Also, use all the time available for the examination. If you finish early, use the remainder of the time to check your work.

If you consider these suggestions carefully, and follow good study habits, you should enjoy a successful learning experience in this course as well as in other courses.

A-3 SOLVING WORD PROBLEMS

Drill-type problems require a working knowledge of the methods presented, and some algebraic steps to change the algebraic form may be required to complete the solution. *Word problems,* however, require a proper interpretation of the statement of the problem before they can be put in a form for solution.

We have to put word problems in symbolic form in order to solve them, and it is this procedure that most students find difficult. Because such problems require more than going through a certain routine, they demand more analysis and appear to be more difficult. Among the reasons for the student's difficulty at solving word problems are (1) unsuccessful previous attempts at solving word problems, leading the student to believe that all word problems are "impossible", (2) a poorly organized approach to the solution, and (3) failure to read the problem carefully, thereby having an improper and incomplete interpretation of the statement given. These can be overcome with proper attitude and care.

NOTE▶
A specific procedure for solving word problems is shown on page 41, when word problems are first covered in our study of algebra. There are over 70 completely worked examples of word problems (as well as numerous other examples that show a similar analysis) throughout this text, illustrating proper interpretations and approaches to these problems. After Chapter 1, these examples are noted in the margin by the phrase *solving a word problem.*

RISERS

The procedure shown on page 41 is similar to that used by most instructors and texts. One of the variations that a number of instructors use is called *RISERS.* This is a word formed from the first letters (an *acronym*) of the words that outline the procedure. These are *Read, Imagine, Sketch, Equate, Relate,* and *Solve.* We now briefly outline this procedure here.

Read the statement of the problem carefully.
Imagine. Take time to get a mental image of the situation described.
Sketch a figure.
Equate, *on the sketch,* the known and unknown quantities.
Relate the known and unknown quantities with an equation.
Solve the equation.

There are problems where a *sketch* may simply be words and numbers placed so that we may properly *equate* the known and unknown quantities. For example, in Example 2 on page 42, the *sketch, equate,* and *relate* steps might look like this:

sketch	34 resistors		56 Ω
	1.5-Ω resistors	2.0-Ω resistors	
equate	x	$34 - x$	
resistance	$1.5x$	$2.0(34 - x)$	56
relate	$1.5x$	$+\quad 2.0(34 - x)$	$= 56$

NOTE▶
If you follow the method on page 41, or this RISERS variation, or any appropriate step-by-step method, and *write out the solution neatly,* you will find that word problems lend themselves to solution more readily than you had previously found.

B

UNITS OF MEASUREMENT; THE METRIC SYSTEM

B-1 INTRODUCTION

Most scientific and technical calculations involve numbers that represent a measurement or count of a specific physical quantity. *Such numbers are called* **denominate numbers,** *and associated with these denominate numbers are* **units of measurement.** For calculations and results to be meaningful, we must know these units. For example, if we measure the length of an object to be 12, we must know whether it is being measured in feet, yards, or some other specified unit of length.

Certain universally accepted **base units** *are used to measure fundamental quantities.* The units for numerous other quantities are expressed in terms of the base units. Fundamental quantities for which base units are defined are (1) length, (2) mass or force, depending on the system of units being used, (3) time, (4) electric current, (5) temperature, (6) amount of substance, and (7) luminous intensity. *Other units, referred to as* **derived units,** *are expressible in terms of the units for these quantities.*

Even though all other quantities can be expressed in terms of the fundamental ones, many have units that are given a specified name. This is done primarily for those quantities that are used very commonly, although it is not done for all such quantities. For example, the volt is defined as a meter^2-kilogram/second^3-ampere, which is in terms of (a unit of length)^2(a unit of mass)/(a unit of time)^3(a unit of electric current). The unit for acceleration has no special name and is left in terms of the base units, for example, feet/second^2. For convenience, special symbols are usually used to designate units. The units for acceleration would be written as ft/s^2.

Two basic systems of units, the **SI metric system** and the **United States Customary** system, are in use today. The U.S. Customary system traditionally has been known as the *British system.* (The SI metric system is now used in Great Britain, although many measurements in the traditional British system are also still used.) Nearly every country in the world now uses the SI metric system. In the United States both systems are used, and international trade has led most major U.S. industrial firms to convert their products to the metric system. Also, due to world trade, to further promote conversion to the metric system, the U.S. Congress has passed legislation stating that the metric system is the preferred system and requiring Federal agencies to use the metric system in their business-related activities. It should also be noted that the metric system is used worldwide in nearly all scientific work.

Therefore, for the present, both systems are of importance, although the metric system will eventually be used almost universally. For that reason, where units are used, some of the exercises and examples have metric units and others have U.S. Customary units. Technicians and engineers need to have some knowledge of both systems.

SI METRIC SYSTEM

Although more than one system has been developed in which metric units are used, the system now accepted as the metric system is the **International System of Units (SI).** This was established in 1960 and uses some different definitions for base units from the previously developed metric units. However, the measurement of the base units in the SI system is more accessible, and the differences are very slight. *Therefore, when we refer to the metric system, we are using SI units.*

As we have stated, in each system the base units are specified, and all other units are then expressible in terms of these base units. Table B-1 on the next page lists the fundamental quantities, as well as many other commonly used quantities, along with their symbols and the names of units used to represent each quantity shown.

In the U.S. Customary system, the base unit of length is the *foot,* and that of force is the *pound.* In the metric system, the base unit of length is the *meter,* and that of mass is the *kilogram.* Here, we see a difference in the definition of the systems that causes some difficulty when units are converted from one system to the other. That is, a base unit in the U.S. Customary system is a unit of force, and a base unit in the metric system is a unit of mass.

The distinction between mass and force is very significant in physics, and the weight of an object is the force with which it is attracted to the earth. Weight, which is therefore a force, is different from mass, which is a measure of the inertia an object exhibits. Although they are different quantities, mass and weight are, however, very closely related. In fact, the weight of an object equals its mass multiplied by the acceleration due to gravity. Near the surface of the earth, the acceleration due to gravity is nearly constant, although it decreases as the distance from the earth increases. Therefore, near the surface of the earth, the weight of an object is directly proportional to its mass. However, at great distances from the earth, the weight of an object will be zero, whereas its mass does not change.

Since force and mass are different, it is not strictly correct to convert pounds to kilograms. However, since the pound is the base unit in the U.S. Customary system and the kilogram is the base unit in the metric system, at the earth's surface 1 kg corresponds to 2.21 lb, in the sense that the force of gravity on a 1-kg mass is 2.21 lb.

When designating units for weight, we use pounds in the U.S. Customary system. In the metric system, although kilograms are used for weight, it is preferable to specify the mass of an object in kilograms. The force of gravity on an object is designated in newtons, and the use of the term *weight* is avoided unless its meaning is completely clear.

As for the other fundamental quantities, both systems use the *second* as the base unit of time. In the SI system, the *ampere* is defined as the base unit of electric current, and this can also be used in the U.S. Customary system. As for temperature, *degrees Fahrenheit* is used with the U.S. Customary system, and *degrees Celsius* (formerly centigrade) is used with the metric system (actually, the *kelvin* is defined as the base unit, where the temperature in kelvins is the temperature in degrees Celsius plus 273.16). In the SI system, the base unit for the amount of a substance is the *mole,* and the base unit of luminous intensity is the *candela.* These last two are of limited importance to our use in this text.

The failure to convert units caused the $125 000 000 Mars Climate Orbiter to fly too close to Mars and break up in the Martian atmosphere in September 1999. A spacecraft team submitted force data in pounds, but the mission controllers assumed the data were in newtons. The system for checking data did not note the change in units. (1 lb = 4.448 N)

Table B-1 Quantities and Their Associated Units

Quantity	Quantity Symbol	U.S. Customary Name	U.S. Customary Symbol	Metric (SI) Name	Metric (SI) Symbol	In Terms of Other SI Units
Length	s	foot	ft	**meter**	m	
Mass	m	slug		**kilogram**	kg	
Force	F	pound	lb	newton	N	$m \cdot kg/s^2$
Time	t	second	s	**second**	s	
Area	A		ft^2		m^2	
Volume	V		ft^3		m^3	
Capacity	V	gallon	gal	liter	L	$(1 \text{ L} = 1 \text{ dm}^3)$
Velocity	v		ft/s		m/s	
Acceleration	a		ft/s^2		m/s^2	
Density	d, ρ		lb/ft^3		kg/m^3	
Pressure	p		lb/ft^2	pascal	Pa	N/m^2
Energy, work	E, W		$ft \cdot lb$	joule	J	$N \cdot m$
Power	P	horsepower	hp	watt	W	J/s
Period	T		s		s	
Frequency	f		l/s	hertz	Hz	1/s
Angle	θ	radian	rad	radian	rad	
Electric current	I, i	ampere	A	**ampere**	A	
Electric charge	q	coulomb	C	coulomb	C	$A \cdot s$
Electric potential	V, E	volt	V	volt	V	$J/(A \cdot s)$
Capacitance	C	farad	F	farad	F	s/Ω
Inductance	L	henry	H	henry	H	$\Omega \cdot s$
Resistance	R	ohm	Ω	ohm	Ω	V/A
Thermodynamic temperature	T			**kelvin**	K	(temp. interval
Temperature	T	degrees Fahrenheit	°F	degrees Celsius	°C	1 °C = 1 K)
Quantity of heat	Q	British thermal unit	Btu	joule	J	
Amount of substance	n			**mole**	mol	
Luminous intensity	I	candlepower	cp	**candela**	cd	

Special Notes:

1. The SI base units are shown in boldface type.

2. The unit symbols shown above are those that are used in the text. Many of them were adopted with the adoption of the SI system. This means, for example, that we use s rather than sec for seconds and A rather than amp for amperes. Also, other units, such as volt, are not spelled out, a common practice in the past. When a given unit is used with both systems, we use the SI symbol for the unit.

3. The liter and degree Celsius are not actually SI units. However, they are recognized for use with the SI system due to their practical importance. Also, the symbol for liter has several variations. Presently L is recognized for use in the United States and Canada, l is recognized by the International Committee of Weights and Measures, and ℓ is also recognized for use in several countries.

4. Other units of time, along with their symbols, which are recognized for use with the SI system and are used in this text, are minute, min; hour, h; day, d.

5. Many additional specialized units are used with the SI system. However, most of those that appear in this text are shown in the table. A few of the specialized units are noted when used in the text. One which is frequently used is that for revolution, r.

6. Other common U.S. units used in the text are inch, in.; yard, yd; mile, mi; ounce, oz.; ton; quart, qt; acre.

7. There are a number of units that were used with the metric system prior to the development of the SI system. However, many of these are not to be used with the SI system. Among those that were commonly used are the dyne, erg, and calorie.

Due to greatly varying sizes of certain quantities, the metric system employs certain prefixes to units to denote different orders of magnitude. These prefixes, with their meanings and symbols, are shown in Table B-2.

Table B-2 Metric Prefixes

Prefix	Factor	Symbol	Prefix	Factor	Symbol
exa	10^{18}	E	deci	10^{-1}	d
peta	10^{15}	P	centi	10^{-2}	c
tera	10^{12}	T	milli	10^{-3}	m
giga	10^{9}	G	micro	10^{-6}	μ
mega	10^{6}	M	nano	10^{-9}	n
kilo	10^{3}	k	pico	10^{-12}	p
hecto	10^{2}	h	femto	10^{-15}	f
deca	10^{1}	da	atto	10^{-18}	a

■**EXAMPLE 1** Some commonly used units that use prefixes in Table B-2, along with their meanings, are shown below.

Unit	Symbol	Meaning	Unit	Symbol	Meaning
megohm	MΩ	10^{6} ohms	milligram	mg	10^{-3} gram
kilometer	km	10^{3} meters	microfarad	μF	10^{-6} farad
centimeter	cm	10^{-2} meter	nanosecond	ns	10^{-9} second

(Mega is shortened to meg when used with "ohm.")   --------------■

When designating units of area or volume, where square or cubic units are used, we use exponents in the designation. For example, we use m^2 rather than sq m and in.^{3} rather than cu in.

B-2 REDUCTIONS AND CONVERSIONS

When using denominate numbers, it may be necessary to change from one set of units to another. *A change within a given system is called a* **reduction,** *and a change from one system to another is called a* **conversion.** Table B-3 gives some basic conversion factors. Some calculators are programmed to do conversions.

Table B-3 Conversion Factors

1 in. = 2.54 cm (exact)	1 ft^3 = 28.32 L	1 lb = 453.6 g	1 Btu = 778.0 ft·lb
1 km = 0.6214 mi	1 L = 1.057 qt	1 kg = 2.205 lb	1 hp = 550 ft·lb/s (exact)
		1 lb = 4.448 N	1 hp = 746.0 W

The advantages of the metric system are evident. Reductions within the system are made by using powers of 10, which amounts to moving the decimal point. Reductions in the U.S. Customary system are made by using many different multiples. Comparisons and changes within the metric system are much simpler.

NOTE ▶   To change a given number of one set of units into another set of units, *we perform algebraic operations with units in the same manner as we do with any algebraic symbol.* Consider the following example.

EXAMPLE 1   If we had a number representing feet per second to be multiplied by another number representing seconds per minute, as far as the units are concerned, we have

$$\frac{\text{ft}}{\text{s}} \times \frac{\text{s}}{\text{min}} = \frac{\text{ft} \times \cancel{\text{s}}}{\cancel{\text{s}} \times \text{min}} = \frac{\text{ft}}{\text{min}}$$

This means that the final result would be in feet per minute.   ∎

In changing a number of one set of units to another set of units, we use reduction and conversion factors and the principle illustrated in Example 1. The convenient way to use the values in the tables is in the form of fractions. Since the given values are equal to each other, their quotient is 1. For example, since 1 in. = 2.54 cm,

NOTE ▶

$$\frac{1 \text{ in.}}{2.54 \text{ cm}} = 1 \quad \text{or} \quad \frac{2.54 \text{ cm}}{1 \text{ in.}} = 1$$

since each represents the division of a certain length by itself. Multiplying a quantity by 1 does not change its value. The following examples illustrate reduction and conversion of units.

EXAMPLE 2   Reduce 20 kg to milligrams.

$$20 \text{ kg} = 20 \text{ kg}\left(\frac{10^3 \text{ \cancel{g}}}{1 \text{ \cancel{kg}}}\right)\left(\frac{10^3 \text{ mg}}{1 \text{ \cancel{g}}}\right)$$
$$= 20 \times 10^6 \text{ mg} = 2.0 \times 10^7 \text{ mg}$$

We note that this result is found essentially by moving the decimal point three places when changing from kilograms to grams, and another three places when changing from grams to milligrams.   ∎

EXAMPLE 3   Change 30 mi/h to feet per second.

$$30\frac{\text{mi}}{\text{h}} = \left(30\frac{\cancel{\text{mi}}}{\cancel{\text{h}}}\right)\left(\frac{5280 \text{ ft}}{1 \cancel{\text{mi}}}\right)\left(\frac{1 \cancel{\text{h}}}{60 \cancel{\text{min}}}\right)\left(\frac{1 \cancel{\text{min}}}{60 \text{ s}}\right) = \frac{(30)(5280) \text{ ft}}{(60)(60) \text{ s}} = 44\frac{\text{ft}}{\text{s}}$$

The only units remaining after the division are those required.   ∎

EXAMPLE 4   Change 575 g/cm³ to kilograms per cubic meter.

$$575\frac{\text{g}}{\text{cm}^3} = \left(575\frac{\text{g}}{\text{cm}^3}\right)\left(\frac{100 \text{ cm}}{1 \text{ m}}\right)^3\left(\frac{1 \text{ kg}}{1000 \text{ g}}\right)$$
$$= \left(575\frac{\cancel{\text{g}}}{\cancel{\text{cm}^3}}\right)\left(\frac{10^6 \cancel{\text{cm}^3}}{1 \text{ m}^3}\right)\left(\frac{1 \text{ kg}}{10^3 \cancel{\text{g}}}\right)$$
$$= 575 \times 10^3 \frac{\text{kg}}{\text{m}^3} = 5.75 \times 10^5 \frac{\text{kg}}{\text{m}^3}$$ ∎

EXAMPLE 5 Change 62.8 lb/in.^{2} to newtons per square meter.

$$62.8\,\frac{\text{lb}}{\text{in.}^2} = \left(62.8\,\frac{\text{lb}}{\text{in.}^2}\right)\left(\frac{4.448\ \text{N}}{1\ \text{lb}}\right)\left(\frac{1\ \text{in.}}{2.54\ \text{cm}}\right)^2\left(\frac{100\ \text{cm}}{1\ \text{m}}\right)^2$$

$$= \left(62.8\,\frac{\text{lb}}{\text{in.}^2}\right)\left(\frac{4.448\ \text{N}}{1\ \text{lb}}\right)\left(\frac{1\ \text{in.}^2}{2.54^2\ \text{cm}^2}\right)\left(\frac{10^4\ \text{cm}^2}{1\ \text{m}^2}\right)$$

$$= \frac{(62.8)\,(4.448)\,(10^4)}{(2.54)^2}\,\frac{\text{N}}{\text{m}^2} = 4.33 \times 10^5\,\frac{\text{N}}{\text{m}^2}$$

EXERCISES FOR APPENDIX *B*

In Exercises 1–4, give the symbol and the meaning for the given unit.

1. megahertz
2. kilowatt
3. millimeter
4. picosecond

In Exercises 5–8, give the name and the meaning for the units whose symbols are given.

5. kV
6. GΩ
7. mA
8. pF

In Exercises 9–44, make the indicated reductions and conversions.

9. Reduce 1 km to centimeters.
10. Reduce 1 kV to millivolts.
11. Reduce 1 mi to inches.
12. Reduce 1 gal to pints.
13. Convert 5.25 in. to centimeters.
14. Convert 6.50 kg to pounds.
15. Convert 15.7 qt to liters.
16. Convert 185 km to miles.
17. Reduce 1 ft^2 to square inches.
18. Reduce 1 yd^3 to cubic feet.
19. Reduce 250 mm^2 to square meters.
20. Reduce 0.125 MΩ to milliohms.
21. Convert 4.50 lb to grams.
22. Convert 0.360 in. to meters.
23. Convert 829 in.^{3} to liters.
24. Convert 0.0680 kL to cubic feet.
25. Convert 2.25 hp to newton centimeters per second.
26. Convert 8.75 Btu to joules.
27. Reduce 25 kW · h to megajoules.
28. Reduce 30 ns to minutes.
29. Convert 75.0 W to horsepower.
30. Convert 326 mL to quarts.
31. A weather satellite orbiting the earth weighs 6500 lb. How many tons is this?

32. An airplane is flying at 37,000 ft. What is its altitude in miles?
33. A car's gasoline tank holds 56 L. Convert this capacity to gallons.
34. A hockey puck has a mass of about 0.160 kg. What is its weight in pounds?
35. The speed of sound is about 1130 ft/s. Change this speed to kilometers per hour.
36. Water flows from a kitchen faucet at the rate of 8.5 gal/min. What is this rate in liters per second?
37. The speedometer of a car is calibrated in kilometers per hour. If the speed of such a car reads 60, how fast in miles per hour is the car moving?
38. The acceleration due to gravity is about 980 cm/s^2. Convert this to feet per squared second.
39. Fifteen grams of a medication are to be dissolved in 0.060 L of water. Express this concentration in milligrams per deciliter.
40. The earth's surface receives energy from the sun at the rate of 1.35 kW/m^2. Reduce this to joules per second square centimeter.
41. At sea level, atmospheric pressure is about 14.7 lb/in.^{2}. Express this in pascals.
42. The density of water is about 62.4 lb/ft^3. Convert this to kilograms per cubic meter.
43. A typical electric current density in a wire is 1.2×10^6 A/m^2. Express this in milliamperes per square centimeter.
44. A certain automobile engine produces a maximum torque of 110 N · m. Convert this to foot pounds.

In Exercises 45–48, use the following information. A commercial jet with 230 passengers on a 2850-km flight from Vancouver to Chicago averaged 765 km/h and used fuel at the rate of 5650 L/h.

45. How many hours long was the flight?
46. How long in seconds did it take to use 1.0 L of fuel?
47. What was the fuel consumption in km/L?
48. What was the fuel consumption in L/passenger?

THE GRAPHING CALCULATOR

C-1 INTRODUCTION

Until the 1970s the personal calculating device for engineers and scientists was the slide rule. The microprocessor chip became commercially available in 1971, and the scientific calculator became widely used during the 1970s. Then in the 1980s the graphing calculator was developed and is now used extensively.

The scientific calculator is still used by many when only accurate calculations are required. However, the graphing calculator can perform all the operations of a scientific calculator and numerous other operations. It also has the major advantage that entries can be seen in the viewing *window,* and this allows a visual check of entered data.

For these reasons, the graphing calculator is now used much more than the scientific calculator by engineers and scientists and in the more advanced mathematics courses. Therefore, as we stated on page 10, we restrict our coverage of calculator use in this text to graphing calculators.

This appendix includes a brief discussion of graphing calculator features. We then show its use in the listing of the over 120 examples with sample screen displays throughout the text. A set of exercises that use the basic calculational and graphing features is also included. The final section of this appendix gives some graphing calculator programs that can be used to perform more extensive operations with fewer steps.

C-2 THE GRAPHING CALCULATOR

Since their introduction in the 1980s, many types and models of graphing calculators have been developed. All models can do all the calculational operations (and more) and display any of the graphs required in this text. Some models can also do symbolic operations. However, regardless of the model you are using, *to determine how the features of a particular model are used, refer to the manual for that model.*

In using a calculator, we must keep in mind that certain operations do not have defined results. If such an operation is attempted, the calculator will show an error display. Operations that can result in an error display include division by zero, square root of a negative number, logarithm of a negative number, and the inverse trigonometric function of a value outside the defined interval. Also, an error display may result if an improper sequence of keys is used.

On a graphing calculator, negative numbers are entered by using the $(-)$ key, and subtraction is entered by using the $-$ key. There is additional discussion of this on page 11 in Section 1-3. Use of the $-$ key for the $(-)$ key will usually result in an error display.

Most of the keys are used to access two or three different features. These features may also have subroutines that may be used. To activate a second use, the *2nd* (or *shift*) key is used. The *alpha* key is used for entering alphabetical characters when using more than one literal symbol (*x* is usually available directly) or when entering a calculator program.

Graphing Calculator Features

We now list some of the more important features of a graphing calculator with an explanation of their use. (This listing uses the designations on a TI-83 calculator. Your calculator may use a different designation for some features.)

Feature	*Use*
MODE	Determines how numbers and graphs are displayed. A *menu* is displayed that shows settings that may be used. The *mode* can be set, for example, (1) for the number of decimal places displayed in numbers; (2) to display angles in degrees or in radians; (3) to display the graph of a function or of parametric equations; (4) to display a graph in rectangular or polar coordinates. See the *mode* feature for other possible settings.
Y=	Used to enter functions. These functions are to be graphed, or used in some other way.
WINDOW	Used to set the boundaries of the viewed portion of a graph. (On some calculators this is the *range* feature.)
GRAPH	Used to display the graph of a function. (On some models this feature puts the calculator in graphing mode.)
TRACE	Used to move the cursor from pixel to pixel along a displayed graph.
ZOOM	Used to adjust the viewing window, usually to magnify the view of a particular part of a graph.
STAT	Used to access various statistical features.
MATH	Used to access specific mathematical features. These include fractions, *n th* roots, maximum and minimum values, absolute values, and others.
TEST	Used to enter special symbols such as $>$ and $<$, and to enter special logic words such as *and* and *or*.
MATRX	Used to enter matrix values and perform matrix operations such as the evaluation of a determinant.
DRAW	Used to make certain types of drawings, such as shading in an area between curves or a tangent line to a curve.
PRGM	Used to enter and access calculator programs. A set of sample programs is given in the last section of this appendix.

A *Graphing Calculator Lab Manual* supplement for this text is also available. It provides general directions for using all graphing calculator features, and specific key steps for some models.

There are many other features on a graphing calculator that are used to perform specific types of calculations and operations. Again, *to determine how the features of a particular model are used, refer to the manual for that model.*

Examples of the use of these features are found in the text examples in the listing that follows on the next page.

Examples of Calculator Use

The first calculational use of a graphing calculator is shown in Section 1-3 on page 11. The first graphical use is shown in Section 3-5 on page 93.

The following list shows the pages on which examples of the use of a graphing calculator are given in the text. Calculator *window* screens are shown with each example. The list is divided into calculator uses that are primarily calculational and those that are primarily graphical.

EXERCISES *C-2*

The following exercises provide an opportunity for practice in performing calculations and a few basic graphs on a graphing calculator. There are many additional exercises throughout the book that require these types of calculations. Many additional exercises that require graphing of various types of curves are found after graphing is introduced in Chapter 3.

In Exercises 1–92, perform the indicated calculations on a graphing calculator.

1. $47.08 + 8.94$
2. $654.1 + 407.7$
3. $4724 - 561.9$
4. $0.9365 - 8.077$
5. 0.0396×471
6. 26.31×0.9393
7. $76.7 \div 194$
8. $52,060 \div 75.09$
9. 3.76^2
10. 0.986^2
11. $\sqrt{0.2757}$
12. $\sqrt{60.36}$
13. $\dfrac{1}{0.0749}$
14. $\dfrac{1}{607.9}$
15. $(19.66)^{2.3}$
16. $(8.455)^{1.75}$
17. $\sin 47.3°$
18. $\sin 1.15$
19. $\cos 3.85$
20. $\cos 119.1°$
21. $\tan 306.8°$
22. $\tan 0.537$
23. $\sec 6.11$
24. $\csc 242.0°$
25. $\sin^{-1} 0.6607$ (in degrees)
26. $\cos^{-1}(-0.8311)$ (in radians)
27. $\tan^{-1}(-2.441)$ (in radians)
28. $\sin^{-1} 0.0737$ (in degrees)
29. $\log 3.857$
30. $\log 0.9012$
31. $\ln 808$
32. $\ln 70.5$
33. $10^{0.545}$
34. $10^{-0.0915}$
35. $e^{-5.17}$
36. $e^{1.672}$
37. $(4.38 + 9.07) \div 6.55$
38. $(382 + 964) \div 844$
39. $4.38 + (9.07 \div 6.55)$
40. $382 + (964 \div 844)$
41. $\dfrac{5.73 \times 10^{11}}{20.61 - 7.88}$
42. $\dfrac{7.09 \times 10^{23}}{284 + 839}$
43. $50.38\pi^2$
44. $\dfrac{5\pi}{14.6}$
45. $\sqrt{1.65^2 + 6.44^2}$
46. $\sqrt{0.735^2 + 0.409^2}$
47. $3(3.5)^4 - 4(3.5)^2$
48. $\dfrac{3(-1.86)}{(-1.86)^2 + 1}$
49. $29.4 \cos 72.5°$
50. $\dfrac{477}{\sin 58.7°}$

51. $\dfrac{4 + \sqrt{(-4)^2 - 4(3)(-9)}}{2(3)}$
52. $\dfrac{-5 - \sqrt{5^2 - 4(4)(-7)}}{2(4)}$
53. $\dfrac{0.176(180)}{\pi}$
54. $\dfrac{209.6\pi}{180}$
55. $\dfrac{1}{2}\left(\dfrac{51.4\pi}{180}\right)(7.06)^2$
56. $\dfrac{1}{2}\left(\dfrac{148.2\pi}{180}\right)(49.13)^2$
57. $\sin^{-1}\dfrac{27.3 \sin 36.5°}{46.8}$
58. $\dfrac{0.684 \sin 76.1°}{\sin 39.5°}$
59. $\sqrt{3924^2 + 1762^2 - 2(3924)(1762)\cos 106.2°}$
60. $\cos^{-1}\dfrac{8.09^2 + 4.91^2 - 9.81^2}{2(8.09)(4.91)}$
61. $\sqrt{5.81 \times 10^8} + \sqrt[3]{7.06 \times 10^{11}}$
62. $(6.074 \times 10^{-7})^{2/5} - (1.447 \times 10^{-5})^{4/9}$
63. $\dfrac{3}{2\sqrt{7} - \sqrt{6}}$
64. $\dfrac{7\sqrt{5}}{4\sqrt{5} - \sqrt{11}}$
65. $\tan^{-1}\dfrac{7.37}{5.06}$
66. $\tan^{-1}\dfrac{46.3}{-25.5}$
67. $2 + \dfrac{\log 12}{\log 7}$
68. $\dfrac{10^{0.4115}}{\pi}$
69. $\dfrac{26}{2}(-1.450 + 2.075)$
70. $\dfrac{4.55(1 - 1.08^{15})}{1 - 1.08}$
71. $\sin^2\left(\dfrac{\pi}{7}\right) + \cos^2\left(\dfrac{\pi}{7}\right)$
72. $\sec^2\left(\dfrac{2}{9}\pi\right) - \tan^2\left(\dfrac{2}{9}\pi\right)$
73. $\sin 31.6° \cos 58.4° + \sin 58.4° \cos 31.6°$
74. $\cos^2 296.7° - \sin^2 296.7°$
75. $\sqrt{(1.54 - 5.06)^2 + (-4.36 - 8.05)^2}$
76. $\sqrt{(7.03 - 2.94)^2 + (3.51 - 6.44)^2}$
77. $\dfrac{(4.001)^2 - 16}{4.001 - 4}$
78. $\dfrac{(2.001)^2 + 3(2.001) - 10}{2.001 - 2}$
79. $\dfrac{4\pi}{3}(8.01^3 - 8.00^3)$
80. $4\pi(76.3^2 - 76.0^2)$
81. $0.01\left(\dfrac{1}{2}\sqrt{2} + \sqrt{2.01} + \sqrt{2.02} + \dfrac{1}{2}\sqrt{2.03}\right)$
82. $0.2\left[\dfrac{1}{2}(3.5)^2 + 3.7^2 + 3.9^2 + \dfrac{1}{2}(4.1)^2\right]$
83. $\dfrac{e^{0.45} - e^{-0.45}}{e^{0.45} + e^{-0.45}}$
84. $\ln \sin 2e^{-0.055}$

85. $\ln\dfrac{2-\sqrt{2}}{2-\sqrt{3}}$

86. $\sqrt{\dfrac{9}{2}+\dfrac{9\sin 0.2\pi}{8\pi}}$

87. $2+\dfrac{0.3}{4}-\dfrac{(0.3)^2}{64}+\dfrac{(0.3)^2}{512}$

88. $\dfrac{1}{2}+\dfrac{\pi\sqrt{3}}{360}-\dfrac{1}{4}\left(\dfrac{\pi}{180}\right)^2$

89. $160(1-e^{-1.50})$

90. $e^{-3.60}(\cos 1.20+2\sin 1.20)$

91. $\dfrac{(10)(9)(8)(7)(6)}{5!}$

92. $\dfrac{20!-15!}{20!+15!}$

In Exercises 93–108, display the graphs of the given functions on a graphing calculator. Use window settings to properly display the curve.

93. $y=2x$

94. $y=6-x$

95. $y=x^2$

96. $y=4-2x^2$

97. $y=0.5x^3$

98. $y=2x^2-x^4$

99. $y=\sqrt{4-x}$

100. $y=\sqrt{25-x^2}$

101. $y=\log x$

102. $y=2^x$

103. $y=2e^x$

104. $y=3\ln x$

105. $y=\sin 2x$

106. $y=2\cos x$

107. $y=\tan^{-1}x$

108. $y=\dfrac{6}{x}$

C-3 GRAPHING CALCULATOR PROGRAMS

Programs like those used on a computer can be stored in the memory of a graphing calculator. As mentioned earlier, such programs are used to perform more extensive operations with fewer steps, often significantly fewer steps. Generally, it is necessary only to enter certain data to obtain the required results.

Following are 14 programs that were written for use on a TI-83 graphing calculator. For other calculator models, it may be necessary to adapt the steps indicated to the format and designated operations of that model.

The chapter, program title, and page on which the program reference appears are shown. A brief description of each program is also given.

Chapter 2 PYTHAGTH page 55

This program solves for any side of a right triangle, given the other two sides. The input is chosen from the menu.

```
:Menu("PYTHAGTH","FINDLEGA",A,"FINDLEGB",B,
"FINDHYPC",C)
:Lbl A:Prompt B,C
:Disp "A=", √(C²–B²):Stop
:Lbl B:Prompt A,C
:Disp "B=", √(C²–A²):Stop
:Lbl C:Prompt A,B
:Disp "C=", √(A²+B²)
```

Chapter 4 SLVRTTRI page 119

This program solves a right triangle, given both legs.

```
:Prompt A,B
:Disp "C=", √(A²+B²)
:Disp "θA=",tan⁻¹(A/B)
:Disp "θB=",tan⁻¹(B/A)
```

Chapter 7 QUADFORM page 211

This program solves the quadratic equation $Ax^2+Bx+C=0$ for real roots.

```
:Prompt A,B,C
:"B²–4AC→D
:If D<0: Then:Disp "IMAGINARY ROOTS":Stop:End
:Disp "ROOTS ARE",(⁻B+√(D))/(2A),(⁻B–√(D))/(2A)
```

Chapter 9 ADDVCTR page 255

This program adds two vectors, given their magnitudes and angles.

```
:Disp "ENTER MAGNITUDES"
:Input "A=",A:Input "B=",B
:Disp "ENTER ANGLES"
:Input "θA=",P:Input "θB=",Q
:Acos(P)+Bcos(Q)→R
:Asin(P)+Bsin(Q)→S
:tan⁻¹(S/R)→T
:Disp "R=", √(R²+S²)
:If R<0:Goto A:If S<0:Goto B
:Disp "θR=",T:Stop
:Lbl A:Disp "θR=",T+180:Stop
:Lbl B:Disp "θR=",T+360
```

Chapter 10   SINECURV   page 284

This program displays the graphs of $y_1 = \sin x$, $y_2 = 2 \sin x$, $y_3 = \sin 2x$, and $y_4 = 2 \sin(2x - \pi/3)$. There is no input.

```
:¯2→Xmin:7→Xmax
:¯2→Ymin:2→Ymax
:"sin(X)"→Y₁:"2sin(X)"→Y₂
:"sin(2X)"→Y₃:"2sin(2X−π/3)"→Y₄
:Disp "Y₁=sin(X)"
:Disp "Y₂=2sin(X)"
:Disp "Y₃=sin(2X)"
:Disp "Y₄=2sin(2X−π/3)"
:Pause:DispGraph:Pause:ClrHome
```

Chapter 11   TBLROOTS   page 311

This program displays a table of square roots and cube roots. Input to TblSet should be made before running the program.

```
:"√(X)"→Y₁
:"³√(X)"→Y₂
:DispTable:Pause:ClrHome
```

Chapter 12   DEMOIVRE   page 340

This program finds the nth roots of the complex number $a + bj$. Displayed are the real and imaginary parts of each root.

```
:Prompt A,B,N
:√((A²+B²)^(1/N))→R
:If A<0:(tan⁻¹(B/A)+180)/N→θ
:If A>0 and B<0:(tan⁻¹(B/A)+360)/N→θ
:If A=0 and B>0:90/N→θ
:If A=0 and B<0:270/N→θ
:If A>0 and B≥0:(tan⁻¹(B/A))/N→θ
:For(I,1,N)
:Disp "ROOT",I
:Disp "REAL PART",Rcos(θ)
:Disp "IMAG PART",Rsin(θ)
:θ+360/N→θ:Pause:End
```

Chapter 12   IMPEDANC   page 344

This program calculates the impedance Z and phase angle θ, given a resistance R, inductance L, capacitance C, and frequency F. (Any of R, L, or C can be zero.)

```
:Prompt R,L,F,C
:2πFL→I
:If C=0:Goto A
:(2πFC)⁻¹→Q
:If R=0:Goto B
:Disp "Z=",√(R²+(I−Q)²)
:Disp "θ=",tan⁻¹((I−Q)/R):Stop
:Lbl A:Disp "Z=",√(R²+I²):Disp "θ=",tan⁻¹(I/R):Stop
:Lbl B:Disp "Z=",abs(I−Q)
:If I>Q:Disp "θ=90"
:If I<Q:Disp "θ=−90"
```

Chapter 15   SYNTHDIV   page 401

This program gives the coefficients and the remainder if a polynomial is divided by $x - R$. The coefficients are entered as a {LIST}.

```
:Disp "LIST COEFFICIENTS"
:Input L₁
:Input "R=",R
:L₁(1)→F:dim(L₁)→D
:Disp "QUOTIENT="
:For(N,1,D−1)
:Pause F:RF+L₁(N+1)→F:End
:Disp "REMAINDER=",F
```

Chapter 17   LINPROG   page 471

This program double shades the area of feasible points for two linear constraints. *Trace* is used to find the point of maximum value or minimum value.

```
:0→Xmin:0→Ymin
:Input "Xmax=",Xmax
:Input "Xscl=",Xscl
:Input "Ymax=",Ymax
:Input "Yscl=", Yscl
:Input "Y₁=",Y₁
:Input "Y₂=",Y₂
:Disp "FOR MINIMUM"
:Disp "ENTER A=1"
:Disp "FOR MAXIMUM"
:Disp "ENTER A=2"
:Prompt A
:If A=1:Goto A
:Shade(0,Y₁,0,Xmax,1,3)
:Shade(0,Y₂,0,Xmax,2,3)
:Trace:Pause:ClrHome:Stop
:Lbl A
:Shade(Y₁,Ymax,0,Xmax,1,3)
:Shade(Y₂,Ymax,0,Xmax,2,3)
:Trace:Pause:ClrHome
```

Chapter 19   BINEXPAN   page 504

This program gives the first k coefficients of $(ax + b)^n$.

```
:Prompt K,A,B,N
:A^N→T
:Disp "COEFFICIENTS="
:For (C,1,K): Disp T
:T(N−C+1)B/(AC)→T
:Pause:End
```

Chapter 21   SLOPEDIS   page 547

This program calculates the slope m of a line through two points (a, c) and (e, f) and the distance between the points. It also displays the line segment on a split screen. Set proper *window* values before running the program.

```
:Prompt A,C,E,F
:Horiz
:Output (2,1,"M=")
:Output (2,4,(F−C)/(E−A))
:Output (3,1,"D=")
:Output (3,4,√((E−A)²+(F−C)²))
:Line (A,C,E,F)
:Pause:Full
```

Chapter 21   GRAPHLIN   page 549

This program displays the graph of a line through two points, (a, c) and (d, e). The *window* values should be set appropriately before running the program. Default *window* is $(−9, 9)$ by $(−6, 6)$.

```
:Prompt A,C,D,E
:(E−C)/(D−A)→M:C−MA→B
```

```
:"MX+B"→Y₁
:−9→Xmin:9→Xmax:1→Xscl
:−6→Ymin:6→Ymax:1→Yscl
:DispGraph
:Pause:ClrHome
```

Chapter 21   GRAPHCON   page 579

This program displays the graph of a conic of the form $ax^2 + cy^2 + dx + ey + f = 0$. The *window* settings may have to be changed in the program. The default *window* is $(−9, 9)$ by $(−6, 6)$.

```
:Prompt A,C,D,E,F
:If C=0:Goto A
:(−E+√(E²−4C(AX²+DX+F)))/(2C)→Y₁
:(−E−√(E²−4C(AX²+DX+F)))/(2C)→Y₂
:Lbl B
:−9→Xmin:9→Xmax:1→Xscl
:−6→Ymin:6→Ymax:1→Yscl
:DispGraph
:Pause:ClrHome:Stop
:Lbl A:"(AX²+DX+F)/(−E)"→Y₁
:Goto B
```

EXERCISES *C-3*

In each of the following exercises, write and test a graphing calculator program to perform the indicated operations for a topic in the indicated chapter.

1. (Chapter 2) To find the total surface area and volume of a right circular cylinder for given values of the radius r and height h.

2. (Chapter 5) To solve the system of equations $ax + by = c$ and $dx + ey = f$ for given values of the constants.

3. (Chapter 6) To find the coefficients e, f, and g for the product of binomials $(ax + b)(cx + d) = ex^2 + fx + g$ for given values of the constants.

4. (Chapter 8) To find the arc length and area of a circular sector for given values of the radius r and angle θ.

5. (Chapter 9) To solve a triangle given angles A, B, and included side C.

6. (Chapter 13) To find the logarithm to any given base b.

7. (Chapter 18) To find the constant of proportionality k for $y = kr^a s^b / t^c$, given the values of y, r, a, s, b, t, and c.

8. (Chapter 19) For a geometric sequence, to calculate the nth term, sum of n terms, and the sum of the infinite series for given values of the first term a, ratio r, and n.

Since statements will vary for writing exercises Ⓦ, answers here are in abbreviated form. Answers are not included for end-of-chapter writing exercises.

Exercises 1-1, page 5

1. integer, rational, real; irrational, real

3. imaginary; irrational, real

5. $3, \dfrac{7}{2}$ 7. $\dfrac{6}{7}, \sqrt{3}$ 9. $6 < 8$ 11. $\pi > -1$

13. $-4 < -|-3|$ 15. $-\dfrac{1}{3} > -\dfrac{1}{2}$ 17. $\dfrac{1}{3}, -3$

19. $-\dfrac{\pi}{5}, \dfrac{1}{x}$ 21.

23.

25. $-18, -|-3|, -1, \sqrt{5}, \pi, |-8|, 9$

27. (a) positive integer (b) negative integer (c) positive rational number less than 1

29. (a) yes (b) yes   31. (a) to right of origin (b) to left of -4

33. between 0 and 1   35. L, t are variables; a is constant

37. $N = 1000an$ 39. yes; -20 is to right of -30

Exercises 1-2, page 10

1. 4   3. 6   5. -3 7. -24 9. 35   11. 20

13. 40   15. -1 17. 9   19. undefined   21. 20

23. -5 25. -9 27. 24   29. -6 31. 3

33. commutative law of multiplication   35. distributive law

37. associative law of addition

39. associative law of multiplication

41. d   43. b   45. (a) positive (b) negative   47. 2°C

49. 100 m + 200 m = 200 m + 100 m, commutative law of addition

51. 3(50 + 40) cases, distributive law

Exercises 1-3, page 15

1. 8 is exact; 55 is approx.   3. 1 and 19.3 are approx.

5. 3, 4   7. 3, 3   9. 1, 5   11. (a) 0.01 (b) 30.8

13. (a) same (b) 78.0   15. (a) 0.004 (b) same

17. (a) 4.94 (b) 4.9   19. (a) 50,900 (b) 51,000

21. (a) 9550 (b) 9500   23. (a) 0.950 (b) 0.95

25. (a) 51.2 (b) 51   27. (a) 62.1 (b) 68

29. (a) 0.0114 (b) 0.015   31. (a) -0.0022 (b) -0.002

33. (a) 0.1356 (b) 0.1   35. (a) 6.086 (b) 6.5   37. 15.8788

39. 204.2   41. 2.745 MHz, 2.755 MHz

43. Too many sig. digits; time has only 2 sig. digits

45. (a) 19.3 (b) 27

47. (a) $\pi = 3.141592654$ (b) $\dfrac{22}{7} = 3.142857143$

49. (a) 0.2424242424 (b) 3.141592654   51. 196 ft

53. 262,144 bytes   55. 59.14%

Exercises 1-4, page 20

1. x^7 3. $2b^6$ 5. m^2 7. $\dfrac{1}{n^4}$ 9. a^8 11. t^{20}

13. $8n^3$ 15. a^2x^8 17. $\dfrac{8}{b^3}$ 19. $\dfrac{x^8}{16}$ 21. 1

23. -3 25. $\dfrac{1}{6}$ 27. R^2 29. $-t^{14}$ 31. $64x^{12}$

33. 1   35. $-b^2$ 37. $\dfrac{1}{8}$ 39. 1   41. $\dfrac{a}{x^2}$

43. $\dfrac{x^3}{64a^3}$ 45. $64g^2s^6$ 47. $\dfrac{5a}{n}$ 49. -53 51. 253

53. -0.421 55. 9990   57. $\dfrac{r}{6}$ 59. 69 W

Exercises 1-5, page 23

1. 45,000   3. 0.00201   5. 3.23   7. 18.6

9. 4×10^4 11. 8.7×10^{-3} 13. 6×10^0

15. 6.3×10^{-2} 17. 5.6×10^{13} 19. 2.2×10^8

21. 4.85×10^{10} 23. 1.59×10^7 25. 9.965×10^{-3}

27. 3.01×10^{-3} 29. 6.5×10^6 kW   31. 3×10^{-6} W

33. 1,000,000,000,000,000,000,000,000,000°C

35. 0.0000000000016 W   37. 4.2×10^{-8} s

39. 2.46×10^{-1} s   41. 7.3×10^{15} cm^2 43. 3.433 Ω

Exercises 1-6, page 25

1. 5   3. -11 5. -7 7. 20   9. 5   11. -6

13. 5   15. 31   17. 18   19. $2\sqrt{3}$ 21. $4\sqrt{21}$

23. 7   25. 10   27. $3\sqrt{10}$ 29. 9.24   31. 0.6877

33. (a) 60 (b) 84   35. (a) 0.0388 (b) 0.0246

37. 13.3 ft/s   39. 1450 m/s   41. 19.0 in.

43. no, not true if $a < 0$

Exercises 1-7, page 29

1. $8x$ 3. $y + 4x$ 5. $a + c - 2$ 7. $-a^2b - a^2b^2$

9. $4s + 4$ 11. $5x - v - 4$ 13. $5a - 5$

15. $-5a + 2$ **17.** $-2t + 5u$ **19.** $7r + 8s$
21. $-50 + 19j$ **23.** $-9 + 3n$ **25.** $18 - 2t^2$ **27.** $6a$
29. $2a\sqrt{LC} + 1$ **31.** $4c - 6$ **33.** $8p - 5q$
35. $-4x^2 + 22$ **37.** $7V^2 - 3$ **39.** $-6t + 13$
41. $2D + d$ **43.** $-b + 4c - 3a$

Exercises 1-8, page 31

1. $a^3 x$ **3.** $-a^2 c^3 x^3$ **5.** $-8a^3 x^5$ **7.** $2a^8 x^3$
9. $i^2 R + i^2 r$ **11.** $-3s^3 + 15st$ **13.** $5m^3 n + 15m^2 n$
15. $-3x^2 - 3xy + 6x$ **17.** $a^2 b^2 c^5 - ab^3 c^5 - a^2 b^3 c^4$
19. $acx^4 + acx^3 y^3$ **21.** $x^2 + 2x - 15$
23. $2x^2 + 9x - 5$ **25.** $6a^2 - 7ab + 2b^2$
27. $6s^2 + 11st - 35t^2$ **29.** $2x^3 + 5x^2 - 2x - 5$
31. $x^3 + 2x^2 - 8x$ **33.** $x^3 - 2x^2 - x + 2$
35. $x^5 - x^4 - 6x^3 + 4x^2 + 8x$ **37.** $2a^2 - 16a - 18$
39. $2L^3 - 6L^2 - 8L$ **41.** $4x^2 - 20x + 25$
43. $x_1^2 + 6x_1 x_2 + 9x_2^2$ **45.** $x^2 y^2 z^2 - 4xyz + 4$
47. $2x^2 + 32x + 128$ **49.** $-x^3 + 2x^2 + 5x - 6$
51. $6x^4 + 21x^3 + 12x^2 - 12x$
53. $n^2 + 200n + 10{,}000$ **55.** $R_1^2 - R_2^2$

Exercises 1-9, page 34

1. $-4x^2 y$ **3.** $\dfrac{4t^4}{r^2}$ **5.** $4x^2$ **7.** $-6a$ **9.** $a^2 + 4y$
11. $t - 2rt^2$ **13.** $q + 2p - 4q^3$ **15.** $\dfrac{2L}{R} - R$
17. $\dfrac{1}{3a} - \dfrac{2b}{3a} + 1$ **19.** $x^2 + a$ **21.** $2x + 1$ **23.** $x - 1$
25. $4x^2 - x - 1, R = -3$ **27.** $x + 5$ **29.** $x^2 + x - 6$
31. $2x^2 + 4x + 2, R = 4x + 4$ **33.** $x^2 - 2x + 4$
35. $x - y$ **37.** $A + \dfrac{\mu^2 E^2}{2A} - \dfrac{\mu^4 E^4}{8A^3}$ **39.** $3T^2 - 2T - 4$

Exercises 1-10, page 37

1. 9 **3.** -1 **5.** 10 **7.** -5 **9.** -3 **11.** 1
13. $-\dfrac{7}{2}$ **15.** 8 **17.** $\dfrac{10}{3}$ **19.** $-\dfrac{13}{3}$ **21.** 2
23. 0 **25.** 9.5 **27.** -1.5 **29.** 5.7 **31.** 0.85
33. 1.3 mi/h   **35.** 750 gal   **37.** 60 mg
39. true for all x

Exercises 1-11, page 40

1. $\dfrac{b}{a}$ **3.** $\dfrac{4m - 1}{4}$ **5.** $\dfrac{c + 6}{a}$ **7.** $2a + 8$
9. $\theta - kA$ **11.** $\dfrac{E}{I}$ **13.** $\dfrac{P}{2\pi f}$ **15.** $\dfrac{p - p_a}{dg}$ **17.** $\dfrac{APV}{R}$

19. $\dfrac{0.3t - ct^2}{c}$ **21.** $\dfrac{C_0^2 - C_1^2}{2C_1^2}$ **23.** $\dfrac{P + nc}{n}$
25. $\dfrac{Q_1 + PQ_1}{P}$ **27.** $\dfrac{N + N_2 - N_2 T}{T}$
29. $\dfrac{L - \pi r_2 - 2x_1 - x_2}{\pi}$ **31.** $\dfrac{gJP + V_1^2}{V_1}$
33. 10.6 m   **35.** $32.3°C$

Exercises 1-12, page 44

1. $138, $252
3. 1.9 million the first year, 2.6 million the second year
5. 20 acres at $20,000/acre, 50 acres at $10,000/acre
7. 20 girders   **9.** -2.3 μA, -4.6 μA, 6.9 μA
11. 6.9 km, 9.5 km   **13.** $11,000
15. 390 s, first car   **17.** 84.2 km/h, 92.2 km/h
19. 900 m   **21.** 64 mi from A   **23.** 180 g

Review Exercises for Chapter 1, page 45

1. -10 **3.** -20 **5.** -22 **7.** -25 **9.** -4
11. 5 **13.** $4r^2 t^4$ **15.** $-\dfrac{6m^2}{nt^2}$ **17.** $\dfrac{8t^3}{s^2}$ **19.** $3\sqrt{5}$
21. (a) 3   (b) 8800   **23.** (a) 4   (b) 9.0   **25.** 18.0
27. 1.3×10^{-4} **29.** $-a - 2ab$ **31.** $7LC - 3$
33. $2x^2 + 9x - 5$ **35.** $x^2 + 16x + 64$ **37.** $hk - 3h^2 k^4$
39. $7R - 6r$ **41.** $13xy - 10z$ **43.** $2x^3 - x^2 - 7x - 3$
45. $-3x^2 y + 24xy^2 - 48y^3$ **47.** $-9p^2 + 3pq + 18p^2 q$
49. $\dfrac{6q}{p} - 2 + \dfrac{3q^4}{p^3}$ **51.** $2x - 5$ **53.** $x^2 - 2x + 3$
55. $4x^3 - 2x^2 + 6x, R = -1$ **57.** $15r - 3s - 3t$
59. $y^2 + 5y - 1, R = 4$ **61.** $-\dfrac{9}{2}$ **63.** $\dfrac{21}{10}$ **65.** $-\dfrac{7}{3}$
67. 3 **69.** $-\dfrac{19}{5}$ **71.** 1 **73.** 2.5×10^4 mi/h
75. $1{,}020{,}000{,}000$ Hz   **77.** 1.2×10^{-6} cm^2
79. 0.15 Bq/L   **81.** $\dfrac{5a - 2}{3}$ **83.** $\dfrac{4 - 2n}{3}$ **85.** $\dfrac{R}{n^2}$
87. $\dfrac{I - P}{Pr}$ **89.** $\dfrac{dV - m}{dV}$ **91.** $\dfrac{N_1 + TN_3 - N_3}{T}$
93. $\dfrac{RH + AT_1}{A}$ **95.** $\dfrac{d - 3kbx^2 + kx^3}{3kx^2}$ **97.** 110 m
99. 0.0188 Ω **101.** $2rV - aV - bV$
103. $4t + 4h - 2t^2 - 4th - 2h^2$
105. 4.2 gigabytes, 6.3 gigabytes
107. 1900 Ω, 3100 Ω **109.** 9.2 lb
111. 34.6 h after second ship enters Red Sea
113. 400 L, 600 L   **115.** 290 ft^2

Exercises 2-1, page 51

1. $\angle EBD, \angle DBC$　　**3.** $\angle ABC$　　**5.** 25°　　**7.** BD, BC

9. 140°　　**11.** 145°　　**13.** 62°　　**15.** 28°　　**17.** 46°

19. 136°　　**21.** 4.53 in.　　**23.** 3.40 in.

25. 133°　　**27.** 920 m

Exercises 2-2, page 57

1. 56°　　**3.** 48°　　**5.** 9.9 ft　　**7.** 64.5 cm　　**9.** 8.4 ft^2

11. 32,300 cm^2　　**13.** 4.41 ft^2　　**15.** 0.390 m^2

17. 26.6 ft　　**19.** 52.2 cm　　**21.** 67°　　**23.** 227.2 cm

25. $\angle K = \angle N = 90°$; $\angle LMK = \angle OMN$;
　　$\angle KLM = \angle NOM$; $\triangle MKL \sim \triangle MNO$

27. 8　　**29.** 71°　　**31.** 148 ft^2　　**33.** 60 ft^2　　**35.** 930 m

37. 23 ft　　**39.** 7.5 m, 9.0 m

Exercises 2-3, page 61

1. 2.6 m　　**3.** 167.8 in.　　**5.** 12.8 m　　**7.** 214.4 ft

9. 7.3 mm^2　　**11.** 1740 in.^{2}　　**13.** 9.3 m^2　　**15.** 2000 ft^2

17. $p = 4a + 2b$　　**19.** $A = bh + a^2$　　**21.** 12 cm

23. 6.30 ft^2　　**25.** 344 m　　**27.** 2.4 ft, 9.6 ft

Exercises 2-4, page 64

1. (a) AD (b) AF　　**3.** (a) $AF \perp OE$ (b) $\triangle OEC$　　**5.** 17.3 ft

7. 72.6 mm　　**9.** 0.0285 yd^2　　**11.** 4.26 m^2　　**13.** 25°

15. 25°　　**17.** 120°　　**19.** 40°　　**21.** 0.393 rad

23. 2.185 rad　　**25.** $p = \frac{1}{2}\pi r + 2r$　　**27.** $A = \frac{1}{4}\pi r^2 - \frac{1}{2}r^2$

29. 24,900 mi　　**31.** 35.7 in.　　**33.** 52.1 m^2

35. horizontally and opposite to original direction

Exercises 2-5, page 68

1. 84 m^2　　**3.** 4.9 ft^2　　**5.** 9.8 mi^2　　**7.** 19,000 km^2

9. 120,000 ft^2　　**11.** 8100 ft^2

13. 2.73 in.^{2}　　The trapezoids are inside the boundary and do not include some of the area.

15. 2.98 in.^{2} The ends of the areas are curved so that they can get closer to the boundary.

Exercises 2-6, page 71

1. 366 ft^3　　**3.** 399 m^2　　**5.** 2.83 yd^3　　**7.** 20,500 cm^2

9. 1100 in.^{3}　　**11.** 3.358 m^2　　**13.** 0.15 yd^3

15. 0.825 cm^2　　**17.** 604 in.^{2}

19. 0.00034 mi^3 = 5.0 × 10^7 ft^3　　**21.** 3.3 × 10^6 yd^3

23. 2,350,000 ft^3　　**25.** 1560 mm^2　　**27.** 2.5 cm

Review Exercises for Chapter 2, page 74

1. 32°　　**3.** 32°　　**5.** 41　　**7.** 42　　**9.** 7.36　　**11.** 21.1

13. 25.5 mm　　**15.** 3.06 ft^2　　**17.** 309 mm　　**19.** 2580 in.^{2}

21. 6190 cm^3　　**23.** 160,000 ft^3　　**25.** 162 m^2

27. 1230 in.^{2}　　**29.** 25°　　**31.** 65°　　**33.** 53°　　**35.** 2.4

37. $p = \pi a + b + \sqrt{4a^2 + b^2}$　　**39.** $A = ab + \frac{1}{2}\pi a^2$

41. yes　　**43.** n^2; $A = \pi(nr)^2 = n^2(\pi r^2)$　　**45.** 7.9 m

47. 30 m　　**49.** 215 ft^2　　**51.** 3.73 mi　　**53.** 26,200 mi

55. 30 ft^2　　**57.** 1.0 × 10^6 m^2　　**59.** 190 m^3　　**61.** 4.4 mi

63. 233 gal

Exercises 3-1, page 80

1. (a) $A = \pi r^2$ (b) $A = \frac{1}{4}\pi d^2$　　**3.** $V = \frac{1}{6}\pi d^3$　　**5.** $A = 5l$

7. $A = s^2, s = \sqrt{A}$　　**9.** 3, −1　　**11.** 11, 3.8

13. $\dfrac{5}{2}, -\dfrac{1}{2}$　　**15.** $\frac{1}{4}a + \frac{1}{2}a^2$, 0

17. $3s^2 + s + 6, 12s^2 - 2s + 6$　　**19.** −8

21. 62.9, 260　　**23.** −0.2998

25. Square the value of the independent variable and add 2.

27. Cube the value of the independent variable and subtract this from 6 times the value of the independent variable.

29. Multiply by 3 the sum of twice r added to 5 and subtract 1.

31. Subtract 3 from twice t and divide this quantity by the sum of t and 2.

33. $y = x^2, f(x) = x^2$

35. $A = 8430 - 140t, f(t) = 8430 - 140t$

37. 10.4 m　　**39.** 75 ft, $2v + 0.2v^2$, 240 ft, 240 ft

Exercises 3-2, page 84

1. domain: all real numbers; range: all real numbers

3. domain: all real numbers except 0; range: all real numbers except 0

5. domain: all real numbers except 0; range: all real numbers $f(s) > 0$

7. domain: all real numbers $h \geq 0$; range: all real numbers $H(h) \geq 1$; $\sqrt{h}$ cannot be negative

9. all real numbers $y > 2$

11. all real numbers except −4, 2, and 6　　**13.** 2, not defined

15. 2, $\dfrac{3}{4}$　　**17.** $d = 80 + 55t$　　**19.** $w = 5500 - 2t$

21. $m = 0.5h - 390$　　**23.** $C = 5l + 250$

25. (a) $y = 3000 - 0.25x$　　(b) 2900 L

27. $A = \dfrac{1}{16}p^2 + \dfrac{(60 - p)^2}{4\pi}$

29. $A = \pi(6 - x)^2$, domain $0 \leq x \leq 6$, range: $0 \leq A \leq 36\pi$

31. $s = \dfrac{300}{t}$, domain: $t > 0$, range: $s > 0$ (upper limits depend on truck)

33. Domain is all values of $C > 0$, with some upper limit depending on the circuit.

35. $m = \begin{cases} 0.5h - 390 \text{ for } h > 1000 \\ 110 \text{ for } 0 \le h \le 1000 \end{cases}$

Exercises 3-3, page 87

1. $(2, 1), (-1, 2), (-2, -3)$ **3.**

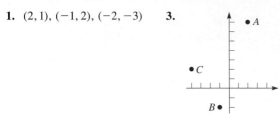

5. isosceles triangle   **7.** rectangle   **9.** $(5, 4)$

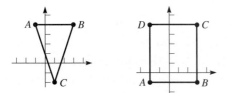

11. $(3, -2)$

13. on a line parallel to the y-axis, 1 unit to the right

15. on a line parallel to the x-axis, 3 units above

17. on a line bisecting the first and third quadrants

19. 0   **21.** to the right of the y-axis   **23.** to the left of a line that is parallel to the y-axis, 1 unit to its left

25. first, third   **27.** (a) 8   (b) 6

Exercises 3-4, page 92

1.
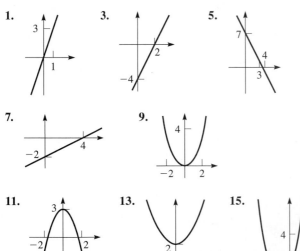

3.

5.

7.

9.

11.

13.

15.

17.
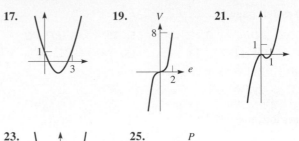

19.

21.

23.
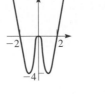

25.
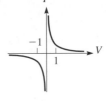

27.
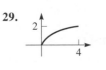

29.
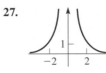

31.

33.
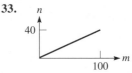

35.
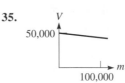

37.
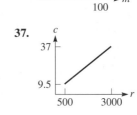

39.
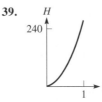

41.
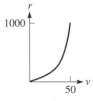

43.
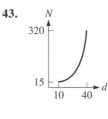

45. $A = 100w - w^2$
$30 \text{ m} \le w \le 70 \text{ m}$

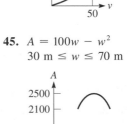

47.

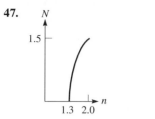

49.

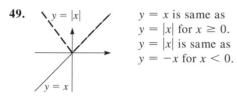

$y = x$ is same as
$y = |x|$ for $x \geq 0$.
$y = |x|$ is same as
$y = -x$ for $x < 0$.

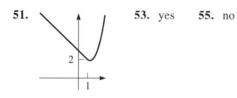

51.

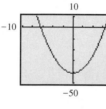

53. yes   **55.** no

Exercises 3-5, page 96

1. 0.7

3. −6.4, 6.4

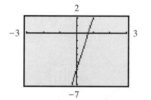

5. −2.8, 1.8

7. −1.6, 2.1

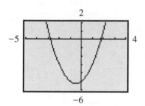

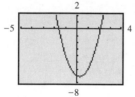

9. −2.0, 0.0, 2.0

11. 0.0, 1.3

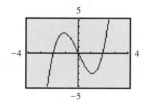

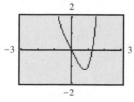

13. 1.4

15. no values

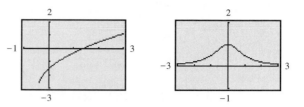

17. all real numbers $y > 0$ or $y \leq -1$

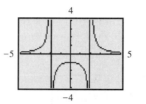

19. all real numbers $y \geq 0$ or $y \leq -4$

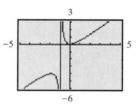

21. all real numbers $Y(y) > 3.46$ (approx.)

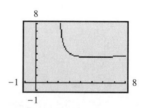

23. all real numbers   **25.** 6.0 V   **27.** 24.4 cm, 29.4 cm

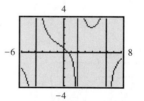

29. 18 cm, 30 cm   **31.** 67 ft   **33.** −0.4142, 2.4142
35. 0.25 ft/min

Exercises 3-6, page 100

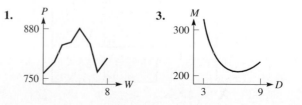

5.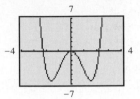

7.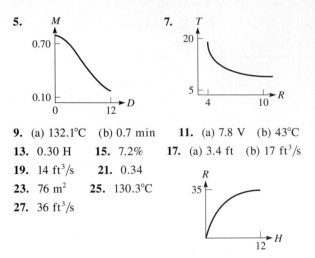

9. (a) 132.1°C   (b) 0.7 min   **11.** (a) 7.8 V   (b) 43°C

13. 0.30 H   **15.** 7.2%   **17.** (a) 3.4 ft   (b) 17 ft³/s

19. 14 ft³/s   **21.** 0.34

23. 76 m²   **25.** 130.3°C

27. 36 ft³/s

Review Exercises for Chapter 3, page 101

1. $A = 4\pi t^2$ **3.** $y = -\dfrac{10}{9}x + \dfrac{250}{9}$ **5.** 16, −47

7. 3, $\sqrt{1 - 4h}$ **9.** $3h^2 + 6hx - 2h$ **11.** −3

13. −3.67, 16.7   **15.** 0.16503, −0.21476

17. domain: all real numbers; range: all real numbers $f(x) \geq 1$

19. domain: all real numbers $t > -4$; range: all real numbers $g(t) > 0$

21. 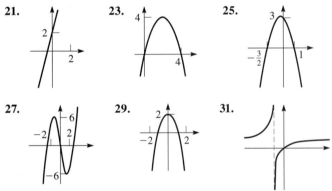 **23.** **25.**

27.  **29.** **31.**

33. 0.4   **35.** 0.2, 5.8

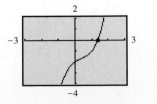 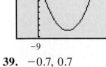

37. 1.4   **39.** −0.7, 0.7

41. all real numbers $y \geq -6.25$

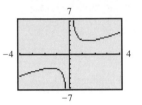

43. all real numbers $y \leq -2.83$ or $y \geq 2.83$

45. either a or b is positive, the other is negative

47. $(1, \sqrt{3})$ or $(1, -\sqrt{3})$ **49.** 13.4   **51.** 72.0°

53. **55.**

57. P **59.** N

61. d **63.** $f(10) = 204$

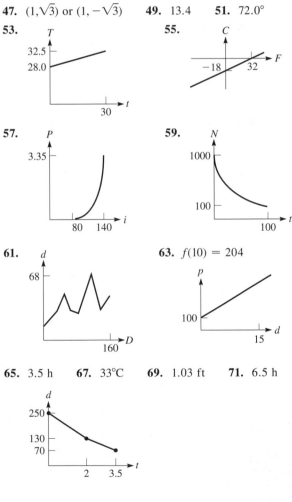

65. 3.5 h   **67.** 33°C   **69.** 1.03 ft   **71.** 6.5 h

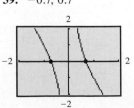

Exercises 4-1, page 107

1. 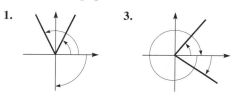   **3.**

5. 405°, −315°   **7.** 210°, −510°   **9.** 430° 30′, −289° 30′

11. 638.1°, −81.9°   **13.** 15.18°   **15.** 82.91°   **17.** 18.85°

19. 235.49°   **21.** 0.977   **23.** 4.97   **25.** 47° 30′

27. −5° 37′   **29.** 15.20°   **31.** 301.27°

33. 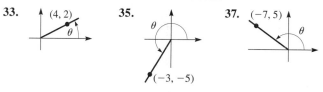   **35.**   **37.** (−7, 5)

39.   **41.** 21.710°

43. 86° 16′ 26″

Exercises 4-2, page 111

1. $\sin \theta = \dfrac{4}{5}$, $\cos \theta = \dfrac{3}{5}$, $\tan \theta = \dfrac{4}{3}$, $\cot \theta = \dfrac{3}{4}$, $\sec \theta = \dfrac{5}{3}$, $\csc \theta = \dfrac{5}{4}$

3. $\sin \theta = \dfrac{8}{17}$, $\cos \theta = \dfrac{15}{17}$, $\tan \theta = \dfrac{8}{15}$, $\cot \theta = \dfrac{15}{8}$, $\sec \theta = \dfrac{17}{15}$, $\csc \theta = \dfrac{17}{8}$

5. $\sin \theta = \dfrac{40}{41}$, $\cos \theta = \dfrac{9}{41}$, $\tan \theta = \dfrac{40}{9}$, $\cot \theta = \dfrac{9}{40}$, $\sec \theta = \dfrac{41}{9}$, $\csc \theta = \dfrac{41}{40}$

7. $\sin \theta = \dfrac{\sqrt{15}}{4}$, $\cos \theta = \dfrac{1}{4}$, $\tan \theta = \sqrt{15}$, $\cot \theta = \dfrac{1}{\sqrt{15}}$, $\sec \theta = 4$, $\csc \theta = \dfrac{4}{\sqrt{15}}$

9. $\sin \theta = \dfrac{1}{\sqrt{2}}$, $\cos \theta = \dfrac{1}{\sqrt{2}}$, $\tan \theta = 1$, $\cot \theta = 1$, $\sec \theta = \sqrt{2}$, $\csc \theta = \sqrt{2}$

11. $\sin \theta = \dfrac{2}{\sqrt{29}}$, $\cos \theta = \dfrac{5}{\sqrt{29}}$, $\tan \theta = \dfrac{2}{5}$, $\cot \theta = \dfrac{5}{2}$, $\sec \theta = \dfrac{\sqrt{29}}{5}$, $\csc \theta = \dfrac{\sqrt{29}}{2}$

13. $\sin \theta = 0.846$, $\cos \theta = 0.534$, $\tan \theta = 1.58$, $\cot \theta = 0.631$, $\sec \theta = 1.87$, $\csc \theta = 1.18$

15. $\sin \theta = 0.1521$, $\cos \theta = 0.9884$, $\tan \theta = 0.1539$, $\cot \theta = 6.498$, $\sec \theta = 1.012$, $\csc \theta = 6.575$

17. $\dfrac{5}{13}, \dfrac{12}{5}$ **19.** $\dfrac{1}{\sqrt{2}}, \sqrt{2}$ **21.** 0.882, 1.33

23. 0.246, 3.94   **25.** $\sin \theta = \dfrac{4}{5}$, $\tan \theta = \dfrac{4}{3}$

27. $\tan \theta = \dfrac{1}{2}$, $\sec \theta = \dfrac{\sqrt{5}}{2}$ **29.** 1   **31.** $\sec \theta$

Exercises 4-3, page 115

1. $\sin 40° = 0.64$, $\cos 40° = 0.77$, $\tan 40° = 0.84$, $\cot 40° = 1.19$, $\sec 40° = 1.31$, $\csc 40° = 1.56$

3. $\sin 15° = 0.26$, $\cos 15° = 0.97$, $\tan 15° = 0.27$, $\cot 15° = 3.73$, $\sec 15° = 1.04$, $\csc 15° = 3.86$

5. 0.381   **7.** 1.58   **9.** 0.9626   **11.** 0.99

13. 0.4085   **15.** 1.57   **17.** 1.32   **19.** 0.07063

21. 70.97°   **23.** 65.70°   **25.** 11.7°   **27.** 49.453°

29. 53.44°   **31.** 81.79°   **33.** 74.1°   **35.** 17.85°

37. $0.9556 = 0.9556$ **39.** y is always less than r.

41. 0.8885   **43.** 0.93614   **45.** 87 dB   **47.** 1.19 A

Exercises 4-4, page 120

1. 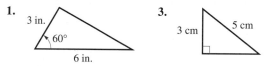   **3.**

5. $B = 12.2°$, $b = 1450$, $c = 6850$

7. $A = 25.8°$, $B = 64.2°$, $b = 311$

9. $A = 57.9°$, $a = 20.2$, $b = 12.6$

11. $A = 21°$, $a = 32$, $B = 69°$

13. $a = 30.21$, $B = 57.90°$, $b = 48.16$

15. $A = 52.15°$, $B = 37.85°$, $c = 71.85$

17. $A = 52.5°$, $b = 0.661$, $c = 1.09$

19. $A = 15.82°$, $a = 0.5239$, $c = 1.922$

21. $A = 65.886°$, $B = 24.114°$, $c = 648.46$

23. $a = 3.3621$, $B = 77.025°$, $c = 14.974$

25. 4.45   **27.** 40.24°

29. $a = c \sin A$, $b = c \cos A$, $B = 90° - A$

31. $A = 90° - B$, $b = a \tan B$, $c = \dfrac{a}{\cos B}$

Exercises 4-5, page 122

1. 97 ft   **3.** 44.0 ft   **5.** 0.4°   **7.** 850.1 cm

9. 25.4 ft   **11.** 26.6°, 63.4°, 90.0°   **13.** 3.4°   **15.** 23.5°

17. 8.1°   **19.** 3.07 cm   **21.** 651 ft   **23.** 30.2°

25. 47.3 m   **27.** 2100 ft

Review Exercises for Chapter 4, page 125

1. $377.0°$, $-343.0°$ **3.** $142.5°$, $-577.5°$ **5.** $31.9°$

7. $38.1°$ **9.** $17° 30'$ **11.** $49° 42'$

13. $\sin \theta = \dfrac{7}{25}$, $\cos \theta = \dfrac{24}{25}$, $\tan \theta = \dfrac{7}{24}$, $\cot \theta = \dfrac{24}{7}$,

$\sec \theta = \dfrac{25}{24}$, $\csc \theta = \dfrac{25}{7}$

15. $\sin \theta = \dfrac{1}{\sqrt{2}}$, $\cos \theta = \dfrac{1}{\sqrt{2}}$, $\tan \theta = 1$, $\cot \theta = 1$,

$\sec \theta = \sqrt{2}$, $\csc \theta = \sqrt{2}$

17. $0.923, 2.40$ **19.** $0.447, 1.12$ **21.** 0.952

23. 1.853 **25.** 1.05 **27.** 0 **29.** $18.2°$ **31.** $57.57°$

33. $12.25°$ **35.** $66.8°$ **37.** $a = 1.83$, $B = 73.0°$, $c = 6.27$

39. $A = 51.5°$, $B = 38.5°$, $c = 104$

41. $B = 52.5°$, $b = 15.6$, $c = 19.7$

43. $A = 31.61°$, $a = 4.006$, $B = 58.39°$

45. $a = 0.6292$, $B = 40.33°$, $b = 0.5341$

47. $A = 48.813°$, $B = 41.187°$, $b = 10.196$

49. $80.3°$ **51.** $12.0°$

53. (a) $A = \frac{1}{2}bh = \frac{1}{2}b(a \sin C) = \frac{1}{2}ab \sin C$ (b) 679.2 m^2

55. 4.92 km   **57.** 56% **59.** 67 ft   **61.** 4.43 m

63. 34 m   **65.** 10.2 in.

67. middle: $\dfrac{580}{\tan 1.1°} = 30{,}200$ m; end: $\dfrac{1160}{\tan 2.2°} = 30{,}200$ m

69. (a) $147{,}000$ mm^2 (b) $155{,}000$ mm^2
(c) Curved surfaces covering each pentagon or hexagon would have greater area than plane area.

71. 1.83 km   **73.** 464 m   **75.** 73.3 cm

Exercises 5-1, page 132

1. yes; no   **3.** yes; yes   **5.** -1; -16 **7.** $\dfrac{1}{4}$, -0.6

9. yes   **11.** no   **13.** no   **15.** yes   **17.** yes   **19.** no

Exercises 5-2, page 136

1. 4 **3.** -5 **5.** $\dfrac{2}{7}$ **7.** $\dfrac{1}{2}$

9.   **11.**

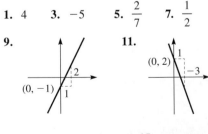

13.   **15.**

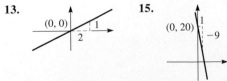

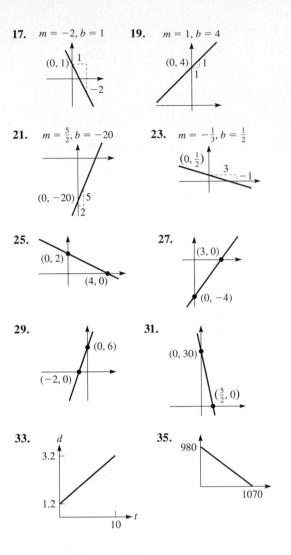

17. $m = -2$, $b = 1$ **19.** $m = 1$, $b = 4$

21. $m = \frac{5}{2}$, $b = -20$ **23.** $m = -\frac{1}{3}$, $b = \frac{1}{2}$

25.   **27.**

29.   **31.**

33.   **35.**

Exercises 5-3, page 140

1. $x = 3.0$, $y = 1.0$ **3.** $x = 3.0$, $y = 0.0$

5. $x = 2.2$, $y = -0.3$ **7.** $x = -0.9$, $y = -2.3$

9. $s = 1.1$, $t = -1.7$ **11.** $x = 0.0$, $y = 3.0$

13. $x = -14.0$, $y = -5.0$ **15.** $r_1 = 4.0$, $r_2 = 7.5$

17. $x = -3.6$, $y = -1.4$ **19.** $x = 1.1$, $y = 0.8$

21. dependent   **23.** $t = -1.9$, $v = -3.2$

25. $x = 1.5$, $y = 4.5$ **27.** inconsistent

29. 50 N, 47 N   **31.** 8.5 ft, 11 ft

Exercises 5-4, page 145

1. $x = 1$, $y = -2$ **3.** $V = 7$, $p = 3$ **5.** $x = -1$, $y = -4$

7. $x = \dfrac{1}{2}$, $y = 2$ **9.** $x = -\dfrac{1}{3}$, $y = 4$

11. $p = \dfrac{9}{22}$, $n = -\dfrac{16}{11}$ **13.** $x = 3$, $y = 1$

15. $x = -1, y = -2$ **17.** $x = 1, y = 2$ **19.** inconsistent

21. $x = -\dfrac{14}{5}, y = -\dfrac{16}{5}$ **23.** $R = 2.38, Z = 0.45$

25. $x = \dfrac{1}{2}, y = -4$ **27.** $x = -\dfrac{2}{3}, y = 0$

29. dependent   **31.** $C = -1, V = -2$

33. $V_1 = 9.0$ V, $V_2 = 6.0$ V   **35.** $x = 6250$ L, $y = 3750$ L

37. $t_1 = 32$ s, $t_2 = 20$ s   **39.** 7.00 kW, 9.39 kW

41. 10 ft, 8.0 ft, 4.0 ft

43. incorrect conclusion or error in sales figures; system of equations is inconsistent

Exercises 5-5, page 151

1. -10 **3.** 29   **5.** 32   **7.** 93   **9.** 0.9   **11.** 96

13. $x = 3, y = 1$ **15.** $x = -1, y = -2$ **17.** $x = 1, y = 2$

19. inconsistent   **21.** $x = -\dfrac{14}{5}, y = -\dfrac{16}{5}$

23. $R = 2.38, Z = 0.45$ **25.** $x = 0.32, y = -4.5$

27. $R = 4.0, t = -3.2$ **29.** $x = -11.2, y = -9.26$

31. $x = -1.0, y = -2.0$ **33.** $F_1 = 15$ lb, $F_2 = 6.0$ lb

35. $x = 18.4$ gal, $y = 17.6$ gal   **37.** 2.5 h, 2.1 h

39. $2000, 6%   **41.** 0.8 L, 1.2 L   **43.** $V = 4.5i - 3.2$

Exercises 5-6, page 155

1. $x = 2, y = -1, z = 1$ **3.** $x = 4, y = -3, z = 3$

5. $l = \dfrac{1}{2}, w = \dfrac{2}{3}, h = \dfrac{1}{6}$ **7.** $x = \dfrac{2}{3}, y = -\dfrac{1}{3}, z = 1$

9. $x = \dfrac{4}{15}, y = -\dfrac{3}{5}, z = \dfrac{1}{3}$ **11.** $p = -2, q = \dfrac{2}{3}, r = \dfrac{1}{3}$

13. $x = \dfrac{3}{4}, y = 1, z = -\dfrac{1}{2}$

15. $r = 0, s = 0, t = 0, u = -1$

17. $P = 800$ h, $M = 125$ h, $I = 225$ h

19. $F_1 = 9.43$ N, $F_2 = 8.33$ N, $F_3 = 1.67$ N

21. $\theta = 0.295t^3 - 5.53t^2 + 24.2t$ **23.** 70 lb, 100 lb, 30 lb

25. unlimited: $x = -10, y = -6, z = 0$ **27.** no solution

Exercises 5-7, page 161

1. 122   **3.** 651   **5.** -439 **7.** 202   **9.** 128

11. 0.128   **13.** $x = -1, y = 2, z = 0$

15. $x = 2, y = -1, z = 1$ **17.** $x = 4, y = -3, z = 3$

19. $l = \dfrac{1}{2}, w = \dfrac{2}{3}, h = \dfrac{1}{6}$ **21.** $x = \dfrac{2}{3}, y = -\dfrac{1}{3}, z = 1$

23. $x = \dfrac{4}{15}, y = -\dfrac{3}{5}, z = \dfrac{1}{3}$ **25.** $p = -2, q = \dfrac{2}{3}, r = \dfrac{1}{3}$

27. $x = \dfrac{3}{4}, y = 1, z = -\dfrac{1}{2}$

29. $A = 125$ N, $B = 60$ N, $F = 75$ N

31. 30.9 mi/h, 45.9 mi/h, 551 mi/h

Review Exercises for Chapter 5, page 162

1. -17 **3.** -1485 **5.** -4 **7.** $\dfrac{2}{7}$

9. $m = -2, b = 4$ **11.** $m = 4, b = -\frac{5}{2}$

13. $x = 2.0, y = 0.0$ **15.** $x = 2.2, y = 2.7$

17. $x = 1.5, y = -1.9$ **19.** $M = 1.5, N = 0.4$

21. $x = 1, y = 2$ **23.** $x = \dfrac{1}{2}, y = -2$

25. $i = -\dfrac{1}{3}, v = \dfrac{7}{4}$ **27.** $x = -\dfrac{6}{19}, y = \dfrac{36}{19}$

29. $x - 1.10, y = 0.54$ **31.** $x = 1, y = 2$

33. $x = \dfrac{1}{2}, y = -2$ **35.** $i = -\dfrac{1}{3}, v = \dfrac{7}{4}$

37. $x = -\dfrac{6}{19}, y = \dfrac{36}{19}$ **39.** $x = 1.10, y = 0.54$

41. best choices: 33, 34   **43.** best choices: 39, 40

45. -115 **47.** 230.08   **49.** $x = 2, y = -1, z = 1$

51. $r = 3, s = -1, t = \dfrac{3}{2}$

53. $x = -0.17, y = 0.16, z = 2.4$

55. $x = 2, y = -1, z = 1$ **57.** $r = 3, s = -1, t = \dfrac{3}{2}$

59. $x = -0.17, y = 0.16, z = 2.4$ **61.** 4   **63.** $-\dfrac{15}{7}$

65. $x = \dfrac{8}{3}, y = -8$ **67.** $x = 1, y = 3$ **69.** -6

71. $-\dfrac{4}{3}$ **73.** $F_1 = 230$ lb, $F_2 = 24$ lb, $F_3 = 200$ lb

75. $a = 440$ m$\cdot°$C, $b = 9.6°$C   **77.** 22,800 km/h, 1400 km/h

79. $R_1 = 0.50\ \Omega, R_2 = 1.5\ \Omega$ **81.** $L = 10$ lb, $w = 40$ lb

83. 68 MB, 24 MB, 48 MB

Exercises 6-1, page 169

1. $40x - 40y$ **3.** $2x^3 - 8x^2$ **5.** $y^2 - 36$ **7.** $9v^2 - 4$

9. $16x^2 - 25y^2$ **11.** $144 - 25a^2b^2$ **13.** $25f^2 + 40f + 16$

15. $4r^2 + 28x + 49$ **17.** $L^4 - 2L^2 + 1$

19. $16a^2 + 56axy + 49x^2y^2$ **21.** $16x^2 - 16xy + 4y^2$

23. $36s^2 - 12st + t^2$ **25.** $x^2 + 6x + 5$

27. $C^4 + 9C^2 + 18$ **29.** $6x^2 + 13x - 5$

31. $20x^2 - 21x - 5$ **33.** $20v^2 + 13v - 15$

35. $6x^2 - 13xy - 63y^2$ **37.** $2x^2 - 8$ **39.** $8a^3 - 2a$

41. $6ax^2 + 24abx + 24ab^2$ **43.** $20n^4 + 100n^3 + 125n^2$

45. $16R^4 - 72R^2r^2 + 81r^4$

47. $x^2 + y^2 + 2xy + 2x + 2y + 1$

49. $x^2 + y^2 + 2xy - 6x - 6y + 9$

51. $125 - 75t + 15t^2 - t^3$

53. $8x^3 + 60x^2t + 150xt^2 + 125t^3$ **55.** $w^2 + 2wh + h^2 - 1$

57. $x^3 + 8$ **59.** $64 - 27x^3$ **61.** $P_0 P_1 c + P_1 G$

63. $4p^2 + 8pDA + 4D^2A^2$ **65.** $\frac{1}{2}\pi R^2 - \frac{1}{2}\pi r^2$

67. $\frac{L}{6}x^3 - \frac{L}{2}ax^2 + \frac{L}{2}a^2x - \frac{L}{6}a^3$ **69.** $4x^2 - 9$

71. $(-2 - 3)^3 = (-2)^3 - 3(-2)^2(3) + 3(-2)(3^2) - 3^3 = -125$

Exercises 6-2, page 173

1. $6(x + y)$ **3.** $5(a - 1)$ **5.** $3x(x - 3)$ **7.** $7b(bh - 4)$

9. $6n(2n + 1)$ **11.** $2(x + 2y - 4z)$

13. $3ab(b - 2 + 4b^2)$ **15.** $4pq(3q - 2 - 7q^2)$

17. $2(a^2 - b^2 + 2c^2 - 3d^2)$ **19.** $(x + 2)(x - 2)$

21. $(10 + 3A)(10 - 3A)$ **23.** $(6a^2 + 1)(6a^2 - 1)$

25. $(9s + 5t)(9s - 5t)$ **27.** $(12n + 13p^2)(12n - 13p^2)$

29. $(x + y + 3)(x + y - 3)$ **31.** $2(x + 2)(x - 2)$

33. $3(x + 3z)(x - 3z)$ **35.** $2(I - 1)(I - 5)$

37. $(x^2 + 4)(x + 2)(x - 2)$

39. $(x^4 + 1)(x^2 + 1)(x + 1)(x - 1)$ **41.** $\frac{3 + b}{2 - b}$

43. $\frac{3}{2(t - 1)}$ **45.** $(3 + b)(x - y)$ **47.** $(a - b)(a + x)$

49. $(x + 2)(x - 2)(x + 3)$ **51.** $(x - y)(x + y + 1)$

53. $2\pi r(h + r)$ **55.** $a(D_1 + D_2)(D_1 - D_2)$ **57.** $\frac{i_3 - i_1}{i_1 - i_2}$

59. $\frac{ER}{A(T_0 - T_1)}$

Exercises 6-3, page 179

1. $(x + 1)(x + 4)$ **3.** $(s - 7)(s + 6)$ **5.** $(t + 8)(t - 3)$

7. $(x + 1)^2$ **9.** $(x - 2y)^2$ **11.** $(3x + 1)(x - 2)$

13. $(3y + 1)(y - 3)$ **15.** $(2s + 11)(s + 1)$

17. $(3f^2 - 1)(f^2 - 5)$ **19.** $(2t - 3)(t + 5)$

21. $(3t - 4u)(t - u)$ **23.** $(4x - 7)(x + 1)$

25. $(9x - 2y)(x + y)$ **27.** $(2m + 5)^2$ **29.** $(2x - 3)^2$

31. $(3t - 4)(3t - 1)$ **33.** $(8b^3 - 1)(b^3 + 4)$

35. $(4p - q)(p - 6q)$ **37.** $(12x - y)(x + 4y)$

39. $2(x - 1)(x - 6)$ **41.** $2(2x - 1)(x + 4)$

43. $ax(x + 6a)(x - 2a)$ **45.** $(a + b + 2)(a + b - 2)$

47. $(5a + 5x + y)(5a - 5x - y)$ **49.** $4(s + 1)(s + 3)$

51. $100(2n + 3)(n - 12)$ **53.** $wx^2(x - 2L)(x - 3L)$

55. $Ad(3u - v)(u - v)$

Exercises 6-4, page 181

1. $(x + 1)(x^2 - x + 1)$ **3.** $(2 - t)(4 + 2t + t^2)$

5. $(3x - 2a)(9x^2 + 6ax + 4a^2)$ **7.** $2(x + 2)(x^2 - 2x + 4)$

9. $6A(A + 1)(A^2 - A + 1)$ **11.** $6x^3y(1 - y)(1 + y + y^2)$

13. $x^3y^3(x + y)(x^2 - xy + y^2)$ **15.** $3a^2(a^2 + 1)(a + 1)(a - 1)$

17. $0.001(R - 4r)(R^2 + 4Rr + 16r^2)$

19. $27L^3(L + 2)(L^2 - 2L + 4)$

21. $(a + b - 4)(a^2 + 2ab + b^2 + 4a + 4b + 16)$

23. $(4 + x^2)(16 - 4x^2 + x^4)$

25. $x^5 - y^5 = (x - y)(x^4 + x^3y + x^2y^2 + xy^3 + y^4)$
$x^7 - y^7 = (x - y)(x^6 + x^5y + x^4y^2 + x^3y^3 + x^2y^4 + xy^5 + y^6)$

27. $2(x + 5)(x^2 - 5x + 25)$ **29.** $D(D - d)(D^2 + Dd + d^2)$

31. $QH(H + Q)(H^2 - HQ + Q^2)$

Exercises 6-5, page 185

1. $\dfrac{14}{21}$ **3.** $\dfrac{2ax^2}{2xy}$ **5.** $\dfrac{2x - 4}{x^2 + x - 6}$ **7.** $\dfrac{ax^2 - ay^2}{x^2 - xy - 2y^2}$

9. $\dfrac{7}{11}$ **11.** $\dfrac{2xy}{4y^2}$ **13.** $\dfrac{2}{x + 1}$ **15.** $\dfrac{s - 5}{2s - 1}$ **17.** $\dfrac{1}{4}$

19. $\dfrac{3x}{4}$ **21.** $\dfrac{1}{5a}$ **23.** $\dfrac{3a - 2b}{2a - b}$

25. $\dfrac{4x^2 + 1}{(2x + 1)(2x - 1)}$ (cannot be reduced) **27.** $3x$

29. $\dfrac{1}{2y^2}$ **31.** $\dfrac{x - 4}{x + 4}$ **33.** $\dfrac{2w^2 - 1}{w^2 + 8}$ **35.** $\dfrac{5x + 4}{x(x + 3)}$

37. $(x^2 + 4)(x - 2)$ **39.** $\dfrac{1}{2t + 1}$ **41.** $\dfrac{x + 3}{x - 3}$

43. $-\dfrac{1}{2}$ **45.** $-\dfrac{2x - 1}{x}$ **47.** $\dfrac{(x + 5)(x - 3)}{(5 - x)(x + 3)}$

49. $\dfrac{x^2 - xy + y^2}{2}$ **51.** $\dfrac{2x}{9x^2 - 3x + 1}$

53. (a) $\dfrac{x^2(x + 2)}{x^2 + 4}$ (b) $\dfrac{x^2}{x^2 - 4}$ Numerator and denominator have no common factor. In each, x^2 is not a factor of the denominator.

55. (a) $\dfrac{(x - 2)(x + 1)}{x(x - 1)}$ (b) $\dfrac{x - 2}{x}$ Numerator and denominator have no common factor. In each, x is not a factor of the numerator.

57. $u + v$ **59.** $\dfrac{E^2(R - r)}{(R + r)^3}$

Exercises 6-6, page 188

1. $\dfrac{3}{28}$ **3.** $6xy$ **5.** $\dfrac{7}{18}$ **7.** $\dfrac{xy^2}{bz^2}$ **9.** $4t$

11. $3(u + v)(u - v)$ **13.** $\dfrac{10}{3(a + 4)}$ **15.** $\dfrac{x^2 - 3}{x^2(x^2 + 3)}$

17. $\dfrac{3x}{5a}$ **19.** $\dfrac{(x + 1)(x - 1)(x - 4)}{4(x + 2)}$ **21.** $\dfrac{x^2}{a + x}$ **23.** $\dfrac{15}{4}$

25. $\dfrac{3}{4x + 3}$ **27.** $\dfrac{7x - 1}{3x + 5}$ **29.** $\dfrac{7x^4}{3a^4}$

31. $\dfrac{4t(2t - 1)(t + 5)}{(2t + 1)^2}$ **33.** $\dfrac{x + y}{2}$ **35.** $(x + y)(3p + 7q)$

37. $\dfrac{2v_1 v_2}{v_1 + v_2}$ **39.** $\dfrac{\pi}{2}$

Exercises 6-7, page 193

1. $\dfrac{9}{5}$ **3.** $\dfrac{8}{x}$ **5.** $\dfrac{5}{4}$ **7.** $\dfrac{3 + 7ax}{4x}$ **9.** $\dfrac{ax - b}{x^2}$

11. $\dfrac{30 + ax^2}{25x^3}$ **13.** $\dfrac{14 - a^2}{10a}$ **15.** $\dfrac{-x^2 + 4x + xy + y - 2}{xy}$

17. $\dfrac{7}{2(2x - 1)}$ **19.** $\dfrac{5 - 3x}{2x(x + 1)}$ **21.** $\dfrac{-3}{4(s - 3)}$

23. $\dfrac{7R + 6}{3(R + 3)(R - 3)}$ **25.** $\dfrac{2x - 5}{(x - 4)^2}$

27. $\dfrac{x + 27}{(x - 5)(x + 5)(x - 6)}$ **29.** $\dfrac{9x^2 + x - 2}{(3x - 1)(x - 4)}$

31. $\dfrac{13t^2 + 27t}{(t - 3)(t + 2)(t + 3)^2}$ **33.** $\dfrac{-2w^3 + w^2 - w}{(w + 1)(w^2 - w + 1)}$

35. $\dfrac{1}{x - 1}$ **37.** $\dfrac{x - y}{y}$ **39.** $-\dfrac{(x + 1)(x - 1)(2x + 1)}{x^2(x + 2)}$

41. $-\dfrac{(3x + 4)(x - 1)}{2x}$ **43.** $\dfrac{2s}{r - s}$

45. $\dfrac{h}{(x + 1)(x + h + 1)}$ **47.** $\dfrac{-2hx - h^2}{x^2(x + h)^2}$

49. $\dfrac{y^2 - rx + r^2}{r^2}$ **51.** $\dfrac{2a - 1}{a^2}$ **53.** $\dfrac{a^2 + 2a - 1}{a + 1}$

55. $\dfrac{a + b}{\dfrac{1}{a} + \dfrac{1}{b}} = ab$ **57.** $\dfrac{3(H - H_0)}{4\pi H}$ **59.** $\dfrac{n(2n + 1)}{2(n + 2)(n - 1)}$

61. $\dfrac{P^2(9x^2 + L^2)}{4L^4}$ **63.** $\dfrac{sL + R}{s^2 LC + sRC + 1}$

Exercises 6-8, page 198

1. 4 **3.** -3 **5.** $\dfrac{7}{2}$ **7.** $\dfrac{16}{21}$ **9.** -9 **11.** $-\dfrac{2}{13}$

13. $\dfrac{5}{3}$ **15.** -2 **17.** $\dfrac{3}{4}$ **19.** $\dfrac{37}{6}$ **21.** -5

23. $\dfrac{63}{8}$ **25.** no solution   **27.** $\dfrac{2}{3}$ **29.** $\dfrac{3b}{1 - 2b}$

31. $\dfrac{(2b - 1)(b + 6)}{2(b - 1)}$ **33.** $\dfrac{2s - 2s_0 - v_0 t}{t}$ **35.** $\dfrac{V_r A - V_0}{V_0 A}$

37. $\dfrac{jX}{1 - g_m z}$ **39.** $\dfrac{PV^3 - bPV^2 + aV - ab}{RV^2}$

41. $\dfrac{N_1^2 R_2 R_3}{N_2^2 R_3 + N_3^2 R_2}$ **43.** $\dfrac{fnR_2 - fR_2}{R_2 + f - fn}$ **45.** 2.4 h

47. 3.2 min   **49.** 80 km/h   **51.** 2.2 V

Review Exercises for Chapter 6, page 199

1. $12ax + 15a^2$ **3.** $4a^2 - 49b^2$ **5.** $4a^2 + 4a + 1$

7. $b^2 + 3b - 28$ **9.** $2x^2 - 13x - 45$ **11.** $4c^{2a} - d^2$

13. $3(s + 3t)$ **15.** $a^2(x^2 + 1)$ **17.** $(W + 12)(W - 12)$

19. $(4x + 8 + t^2)(4x + 8 - t^2)$ **21.** $(3t - 1)^2$

23. $(5t + 1)^2$ **25.** $(x + 8)(x - 7)$

27. $(t + 3)(t - 3)(t^2 + 4)$ **29.** $(2x - 9)(x + 4)$

31. $(2x + 5)(2x - 7)$ **33.** $(5b - 1)(2b + 5)$

35. $4(x + 4y)(x - 4y)$ **37.** $2(5 - 2y^2)(25 + 10y^2 + 4y^4)$

39. $(2x + 3)(4x^2 - 6x + 9)$ **41.** $(a - 3)(b^2 + 1)$

43. $(x + 5)(n - x + 5)$ **45.** $\dfrac{16x^2}{3a^2}$ **47.** $\dfrac{3x + 1}{2x - 1}$

49. $\dfrac{16}{5x(x - y)}$ **51.** $\dfrac{6}{5 - x}$ **53.** $\dfrac{x + 2}{2x(7x - 1)}$

55. $\dfrac{1}{x - 1}$ **57.** $\dfrac{16x - 15}{36x^2}$ **59.** $\dfrac{5y + 6}{2xy}$

61. $\dfrac{-2(2a + 3)}{a(a + 2)}$ **63.** $\dfrac{2x^2 - x + 1}{x(x + 3)(x - 1)}$

65. $\dfrac{12x^2 - 7x - 4}{2(x - 1)(x + 1)(4x - 1)}$ **67.** $\dfrac{x^3 + 6x^2 - 2x + 2}{x(x - 1)(x + 3)}$

69. 2 **71.** $-\dfrac{(a - 1)^2}{2a}$ **73.** 6 **75.** -6

77. $\dfrac{1}{4}[(x + y)^2 - (x - y)^2]$

$\quad = \dfrac{1}{4}(x^2 + 2xy + y^2 - x^2 + 2xy - y^2) = \dfrac{1}{4}(4xy)$

$\quad = xy$

79. $2zS^2 + 2zS$ **81.** $\pi l(r_1 + r_2)(r_1 - r_2)$

83. $4s(s + 2)(s + 12)$ **85.** $R(3R - 4r)$

87. $8n^6 + 36n^5 + 66n^4 + 63n^3 + 33n^2 + 9n + 1$

89. $10aT - 10at + aT^2 - 2aTt + at^2$

91. $4(3x^2 + 12x + 16)$ **93.** $\dfrac{12\pi^2 wv^2 D}{gn^2 t}$ **95.** $\dfrac{R^2 + r^2}{2}$

97. $\dfrac{120 - 60d^2 + 5d^4 - d^6}{120}$

99. $\dfrac{2\pi^2 CN + 2\pi^2 Cn + N^2 - 2nN + n^2}{4\pi^2 C}$

101. $\dfrac{4r^3 - 3ar^2 - a^3}{4r^3}$ **103.** $\dfrac{c^2u^2 - 2gc^2x}{1 - c^2u^2 + 2gc^2x}$

105. $\dfrac{W}{g(h_2 - h_1)}$ **107.** $\dfrac{RHw}{w - RH}$ **109.** $\dfrac{-mb^2s^2 - kL^2}{b^2s}$

111. $\dfrac{i}{sV - V_0}$ **113.** 3.4 h   **115.** 1.5 s   **117.** 11.3

119. 18 Ω

Exercises 7-1, page 206

1. $a = 1, b = -8, c = 5$ **3.** $a = 1, b = -2, c = -4$

5. not quadratic, no x^2-term   **7.** $a = 1, b = -1, c = 0$

9. $2, -2$ **11.** $\frac{3}{2}, -\frac{3}{2}$ **13.** $-1, 9$ **15.** $3, 4$

17. $0, -2$ **19.** $\frac{1}{3}, -\frac{1}{3}$ **21.** $\frac{1}{3}, 4$ **23.** $-4, -4$

25. $\frac{2}{3}, \frac{3}{2}$ **27.** $\frac{1}{2}, -\frac{3}{2}$ **29.** $2, -1$ **31.** $2b, -2b$

33. $0, \frac{5}{2}$ **35.** $\frac{5}{2}, -\frac{9}{2}$ **37.** $0, -2$ **39.** $b - a, -b - a$

41. $0, L$ **43.** 2 A.M., 10 A.M.   **45.** $\frac{3}{2}, 4$ **47.** $-2, \frac{1}{2}$

49. 3 N/cm, 6 N/cm   **51.** 550 mi/h

Exercises 7-2, page 209

1. $-5, 5$ **3.** $-\sqrt{7}, \sqrt{7}$ **5.** $-3, 7$ **7.** $-3 \pm \sqrt{7}$

9. $2, -4$ **11.** $-2, -1$ **13.** $2 \pm \sqrt{2}$ **15.** $-5, 3$

17. $-3, \frac{1}{2}$ **19.** $\frac{1}{6}(3 \pm \sqrt{33})$ **21.** $\frac{1}{4}(1 \pm \sqrt{17})$

23. $\frac{1}{5}(5 \pm \sqrt{5})$ **25.** $-\frac{1}{3}, -\frac{1}{3}$ **27.** $-b \pm \sqrt{b^2 - c}$

Exercises 7-3, page 213

1. $2, -4$ **3.** $-2, -1$ **5.** $2 \pm \sqrt{2}$ **7.** $-5, 3$

9. $-3, \frac{1}{2}$ **11.** $\frac{1}{6}(3 \pm \sqrt{33})$ **13.** $\frac{1}{4}(1 \pm \sqrt{17})$

15. $-\frac{8}{5}, \frac{5}{6}$ **17.** $\frac{1}{2}(-5 \pm \sqrt{-5})$ **19.** $\frac{1}{6}(1 \pm \sqrt{109})$

21. $\frac{3}{2}, -\frac{3}{2}$ **23.** $\frac{3}{4}, -\frac{5}{8}$ **25.** $-0.54, 0.74$

27. $-0.26, 2.43$ **29.** $-c \pm \sqrt{c^2 + 1}$

31. $\dfrac{b + 1 \pm \sqrt{-3b^2 + 2b + 4b^2a + 1}}{2b^2}$ **33.** 4.376 cm

35. $\dfrac{-R \pm \sqrt{R^2 - 4L/C}}{2L}$ **37.** imaginary, unequal

39. real, irrational, unequal   **41.** 11.0 m by 23.8 m

43. 2.6 ft

Exercises 7-4, page 217

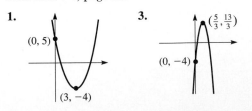

5.

7.

9.

11. $\left(-\frac{3}{2}, \frac{25}{2}\right)$, $(0, 8)$, $(-4, 0)$, $(1, 0)$

13. $(-1, 5)$, $(1, 5)$, $(0, 3)$

15. $\left(-\frac{1}{2}, -\frac{11}{2}\right)$, $(-1, -6)$, $(0, -6)$, $(1, -10)$

17. $-1.2, 1.2$ **19.** $0.5, 3.1$ **21.** no real roots

23. $-2.4, 1.2$

25. The parabola $y = 3x^2$ rises more quickly, and the parabola $y = \frac{1}{3}x^2$ rises more slowly.

27. The parabola $y = -x^2$ is inverted, and the parabola $y = 3x - x^2$ is inverted with a different vertex.

29.   **31.**

33. (a) after 16.2 s
(b) 1130 ft
(c) 3.3 s, 12.3 s

35. 173 ft by 116 ft, or 77 ft by 259 ft

Review Exercises for Chapter 7, page 218

1. $-4, 1$ **3.** $2, 8$ **5.** $\frac{1}{3}, -4$ **7.** $\frac{1}{2}, \frac{5}{3}$ **9.** $0, \frac{25}{6}$

11. $-\frac{3}{2}, \frac{7}{2}$ **13.** $-10, 11$ **15.** $-1 \pm \sqrt{6}$ **17.** $-4, \frac{9}{2}$

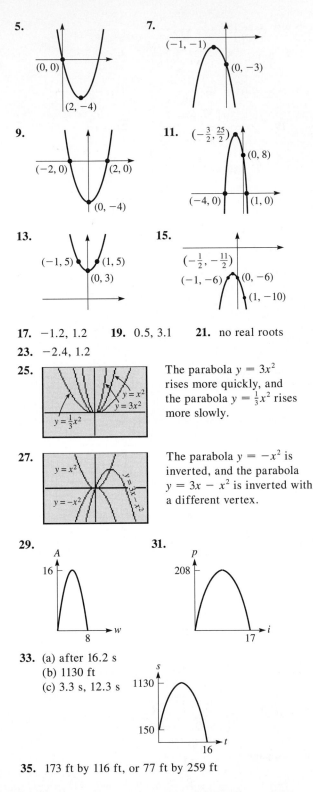

19. $\frac{1}{8}(3 \pm \sqrt{41})$ **21.** $\frac{1}{42}(-23 \pm \sqrt{-4091})$

23. $\frac{1}{6}(-2 \pm \sqrt{58})$ **25.** $-2 \pm 2\sqrt{2}$ **27.** $\frac{1}{3}(-4 \pm \sqrt{10})$

29. $-1, \frac{5}{4}$ **31.** $\frac{1}{4}(-3 \pm \sqrt{-47})$ **33.** $\dfrac{-1 \pm \sqrt{-1}}{a}$

35. $\dfrac{-3 \pm \sqrt{9 + 4a^3}}{2a}$ **37.** $-5, 6$ **39.** $\frac{1}{4}(1 \pm \sqrt{33})$

41. $3 \pm \sqrt{7}$ **43.** 0 (3 is not a solution)

45.

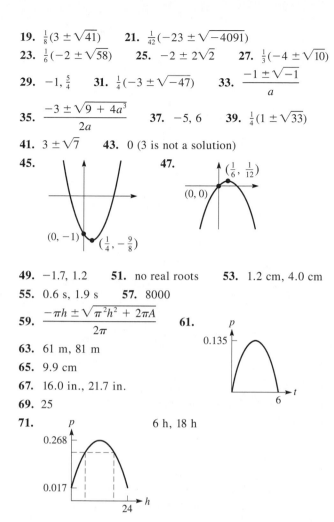

47.

49. $-1.7, 1.2$ **51.** no real roots **53.** 1.2 cm, 4.0 cm

55. 0.6 s, 1.9 s **57.** 8000

59. $\dfrac{-\pi h \pm \sqrt{\pi^2 h^2 + 2\pi A}}{2\pi}$ **61.**

63. 61 m, 81 m

65. 9.9 cm

67. 16.0 in., 21.7 in.

69. 25

71. 6 h, 18 h

Exercises 8-1, page 223

1. $+, -$ **3.** $+, +$ **5.** $-, +$ **7.** $+, -$

9. $-, -$ **11.** $-, +$

13. $\sin \theta = \dfrac{1}{\sqrt{5}}$, $\cos \theta = \dfrac{2}{\sqrt{5}}$, $\tan \theta = \dfrac{1}{2}$, $\cot \theta = 2$,

$\sec \theta = \dfrac{1}{2}\sqrt{5}$, $\csc \theta = \sqrt{5}$

15. $\sin \theta = -\dfrac{3}{\sqrt{13}}$, $\cos \theta = -\dfrac{2}{\sqrt{13}}$, $\tan \theta = \dfrac{3}{2}$,

$\cot \theta = \dfrac{2}{3}$, $\sec \theta = -\dfrac{1}{2}\sqrt{13}$, $\csc \theta = -\dfrac{1}{3}\sqrt{13}$

17. $\sin \theta = \dfrac{12}{13}$, $\cos \theta = -\dfrac{5}{13}$, $\tan \theta = -\dfrac{12}{5}$,

$\cot \theta = -\dfrac{5}{12}$, $\sec \theta = -\dfrac{13}{5}$, $\csc \theta = \dfrac{13}{12}$

19. $\sin \theta = -\dfrac{2}{\sqrt{29}}$, $\cos \theta = \dfrac{5}{\sqrt{29}}$, $\tan \theta = -\dfrac{2}{5}$,

$\cot \theta = -\dfrac{5}{2}$, $\sec \theta = \dfrac{1}{5}\sqrt{29}$, $\csc \theta = -\dfrac{1}{2}\sqrt{29}$

21. II **23.** II **25.** IV **27.** III **29.** IV **31.** II

Exercises 8-2, page 229

1. $\sin 20°$; $-\cos 40°$ **3.** $-\tan 75°$; $-\csc 58°$

5. $\cos 40°$; $-\tan 40°$ **7.** $-\sin 15° = -0.26$

9. $-\cos 73.7° = -0.281$ **11.** $\sec 31.67° = 1.175$

13. -0.523 **15.** -0.7620 **17.** -3.910

19. $237.99°, 302.01°$ **21.** $66.40°, 293.60°$

23. $102.0°, 282.0°$ **25.** $119.5°$ **27.** $263°$ **29.** $306.21°$

31. $299.24°$ **33.** -0.7003 **35.** -0.777 **37.** $<$

39. $=$ **41.** 0.0183 A **43.** 12.6 in.

45. The y-coordinate has the opposite sign for $\sin(-\theta)$ than it

has for $\sin \theta$. $\cos(-\theta) = \dfrac{x}{r}$, $\tan(-\theta) = \dfrac{-y}{x}$, $\cot(-\theta) = \dfrac{x}{-y}$,

$\sec(-\theta) = \dfrac{r}{x}$, $\csc(-\theta) = \dfrac{r}{-y}$

47. (a) 5.7 (b) -1.4

Exercises 8-3, page 234

1. $\dfrac{\pi}{12}, \dfrac{5\pi}{6}$ **3.** $\dfrac{5\pi}{12}, \dfrac{11\pi}{6}$ **5.** $\dfrac{7\pi}{6}, \dfrac{3\pi}{2}$ **7.** $\dfrac{8\pi}{9}, \dfrac{13\pi}{9}$

9. $72°, 270°$ **11.** $10°, 315°$ **13.** $170°, 300°$

15. $15°, 27°$ **17.** 0.401 **19.** 4.40 **21.** 5.821

23. 3.115 **25.** $43.0°$ **27.** $195.2°$ **29.** $140°$

31. $940.8°$ **33.** 0.7071 **35.** 3.732 **37.** -0.8660

39. -8.327 **41.** 0.9056 **43.** -0.89 **45.** -2.1

47. -0.15 **49.** $0.3141, 2.827$ **51.** $2.932, 6.074$

53. $0.8309, 5.452$ **55.** $2.442, 3.841$

57. 0.030 ft·lb **59.** 2900 m

Exercises 8-4, page 238

1. 3.46 in. **3.** 426 mm **5.** $0.4141 = 23.73°$

7. 43 cm^2 **9.** 0.0647 ft **11.** $2.04 = 117.0°$

13. 42.7 cm **15.** 1.81, 14.5 cm^2 **17.** 32.73 min past noon

19. 4240 ft^2 **21.** 0.52 rad/s **23.** 34.73 m^2

25. 2.30 ft **27.** 22.6 m^2 **29.** 369 m^3 **31.** 0.4 km

33. 8.2 r/min **35.** 3960 in./min **37.** 25.7 ft

39. 1400 in./min **41.** 0.433 rad/s **43.** 8800 ft/min

45. 250 rad **47.** 14.9 m^3 **49.** ratios become very close to 1

51. 7.13×10^7 mi

Review Exercises for Chapter 8, page 241

1. $\sin \theta = \dfrac{4}{5}$, $\cos \theta = \dfrac{3}{5}$, $\tan \theta = \dfrac{4}{3}$, $\cot \theta = \dfrac{3}{4}$,

$\sec \theta = \dfrac{5}{3}$, $\csc \theta = \dfrac{5}{4}$

3. $\sin \theta = -\dfrac{2}{\sqrt{53}}$, $\cos \theta = \dfrac{7}{\sqrt{53}}$, $\tan \theta = -\dfrac{2}{7}$,

$\cot \theta = -\dfrac{7}{2}$, $\sec \theta = \dfrac{\sqrt{53}}{7}$, $\csc \theta = -\dfrac{\sqrt{53}}{2}$

5. $-\cos 48°$, $\tan 14°$ **7.** $-\sin 71°$, $\sec 15°$ **9.** $\dfrac{2\pi}{9}$, $\dfrac{17\pi}{20}$

11. $\dfrac{4\pi}{15}$, $\dfrac{9\pi}{8}$ **13.** $252°$, $130°$ **15.** $12°$, $330°$ **17.** $32.1°$

19. $206.7°$ **21.** 1.78 **23.** 0.3534 **25.** 4.5736

27. 2.377 **29.** -0.415 **31.** -0.47 **33.** -1.080

35. -0.4264 **37.** -1.64 **39.** 4.140 **41.** -0.5878

43. -0.8660 **45.** 0.5569 **47.** 1.197

49. $10.30°$, $190.30°$ **51.** $118.23°$, $241.77°$

53. 0.5759, 5.707 **55.** 4.187, 5.238 **57.** $223.76°$

59. $246.78°$ **61.** 10.8 in.   **63.** $3.23 = 185.3°$

65. 0.0562 W   **67.** $0.800 = 45.8°$

69. (a) 4150 mi (b) 6220 mi; great circle route shortest of all routes

71. 4710 cm/s   **73.** 24.4 ft^2 **75.** 1.81×10^6 cm/s

77. 138 m^2 **79.** 3.58×10^5 km

Exercises 9-1, page 247

1. (a) vector: magnitude and direction are specified; (b) scalar: only magnitude is specified

3. (a) vector: magnitude and direction are specified; (b) scalar: only magnitude is specified

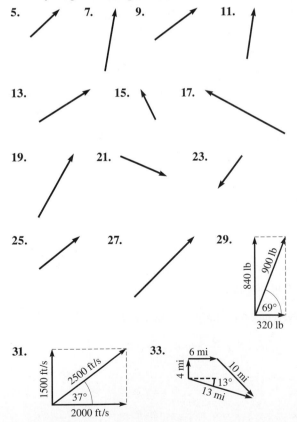

Exercises 9-2, page 251

1. 662, 352 **3.** -349, -664 **5.** 3.22, 7.97

7. -62.9, 44.1 **9.** 2.08, -8.80 **11.** -2.53, -0.788

13. -0.8088, 0.3296 **15.** $88{,}920$, $12{,}240$

17. 23.9 km/h, 7.43 km/h   **19.** 51.9 lb, 18.3 lb

21. 115 km, 88.3 km   **23.** $18{,}530$ mi/h, -825.3 mi/h

Exercises 9-3, page 256

1. $R = 24.2$, $\theta = 52.6°$ with A

3. $R = 7.781$, $\theta = 66.63°$ with A

5. $R = 10.0$, $\theta = 58.8°$ **7.** $R = 2.74$, $\theta = 111.0°$

9. $R = 2130$, $\theta = 107.7°$ **11.** $R = 1.426$, $\theta = 299.12°$

13. $R = 29.2$, $\theta = 10.8°$ **15.** $R = 58.5$, $\theta = 87.8°$

17. $R = 27.27$, $\theta = 33.14°$ **19.** $R = 12.735$, $\theta = 25.216°$

21. $R = 50.2$, $\theta = 50.3°$ **23.** $R = 0.242$, $\theta = 285.9°$

25. $R = 235$, $\theta = 121.7°$ **27.** 2700 lb, $\theta = 107°$

Exercises 9-4, page 259

1. 6.60 lb, $29.5°$ from the 5.75-lb force

3. $11{,}000$ N, $44°$ above horizontal

5. 3070 ft, $17.8°$ S of W   **7.** 229.4 ft, $72.82°$ N of E

9. 25.3 km/h, $29.6°$ S of E   **11.** 175 lb, $9.3°$ above horizontal

13. 540 km/h, $6°$ from direction of plane

15. $18{,}130$ mi/h, $0.03°$ from direction of shuttle

17. $184{,}000$ in./min^2, $\phi = 89.6°$ **19.** 138 mi, $65.0°$ N of E

21. 79.0 m/s, $11.3°$ from direction of plane, $75.6°$ from vertical

23. 4.05 A/m, $11.5°$ with magnet

Exercises 9-5, page 266

1. $b = 38.1$, $C = 66.0°$, $c = 46.1$

3. $a = 2800$, $b = 2620$, $C = 108.0°$

5. $B = 12.20°$, $C = 149.57°$, $c = 7.448$

7. $a = 110.5$, $A = 149.70°$, $C = 9.57°$

9. $A = 125.6°$, $a = 0.0776$, $c = 0.00566$

11. $A = 99.4°$, $b = 55.1$, $c = 24.4$

13. $A = 68.01°$, $a = 5520$, $c = 5376$

15. $A_1 = 61.36°$, $C_1 = 70.51°$, $c_1 = 5.628$; $A_2 = 118.64°$, $C_2 = 13.23°$, $c_2 = 1.366$

35.

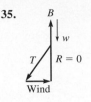

17. $A_1 = 107.3°$, $a_1 = 5280$, $C_1 = 41.3°$; $A_2 = 9.9°$, $a_2 = 952$, $C_2 = 138.7°$

19. no solution   **21.** 1490 ft   **23.** 880 N   **25.** 0.88 mi

27. 13.94 cm   **29.** 27,300 km

31. 77.3° with bank downstream

Exercises 9-6, page 271

1. $A = 50.3°$, $B = 75.7°$, $c = 6.31$

3. $A = 70.9°$, $B = 11.1°$, $c = 4750$

5. $A = 34.72°$, $B = 40.67°$, $C = 104.61°$

7. $A = 18.21°$, $B = 22.28°$, $C = 139.51°$

9. $A = 6.0°$, $B = 16.0°$, $c = 1150$

11. $A = 82.3°$, $b = 21.6$, $C = 11.4°$

13. $A = 36.24°$, $B = 39.09°$, $a = 97.22$

15. $A = 46.94°$, $B = 61.82°$, $C = 71.24°$

17. $A = 137.9°$, $B = 33.7°$, $C = 8.4°$

19. $b = 37$, $C = 25°$, $c = 24$

21. 69.4 mi   **23.** 0.039 km   **25.** 57.3°, 141.7°

27. 16.5 mm   **29.** 5.09 km/h   **31.** 17.8 mi

Review Exercises for Chapter 9, page 273

1. $A_x = 57.4$, $A_y = 30.5$ **3.** $A_x = -0.7485$, $A_y = -0.5357$

5. $R = 602$, $\theta = 57.1°$ with A

7. $R = 5960$, $\theta = 33.60°$ with A   **9.** $R = 965$, $\theta = 8.6°$

11. $R = 26.12$, $\theta = 146.03°$ **13.** $R = 71.93$, $\theta = 336.50°$

15. $R = 99.42$, $\theta = 359.57°$

17. $b = 18.1$, $C = 64.0°$, $c = 17.5$

19. $A = 21.2°$, $b = 34.8$, $c = 51.5$

21. $a = 17{,}340$, $b = 24{,}660$, $C = 7.99°$

23. $A = 39.88°$, $a = 51.94$, $C = 30.03°$

25. $A_1 = 54.8°$, $a_1 = 12.7$, $B_1 = 68.6°$; $A_2 = 12.0°$, $a_2 = 3.24$, $B_2 = 111.4°$

27. $A = 32.3°$, $b = 267$, $C = 17.7°$

29. $A = 148.7°$, $B = 9.3°$, $c = 5.66$

31. $a = 1782$, $b = 1920$, $C = 16.00°$

33. $A = 37°$, $B = 25°$, $C = 118°$

35. $A = 20.6°$, $B = 35.6°$, $C = 123.8°$

37. Add the 3 forms of law of cosines together; simplify; divide by $2abc$.

39. $A_t = \frac{1}{2}ab$; $b = \dfrac{a \sin B}{\sin A}$; substitute

41. -155.7 lb, 81.14 lb   **43.** 2100 ft/s   **45.** 9.8 km

47. 6.1 m/s, 35° with horizontal   **49.** 2.30 m, 2.49 m

51. 32,900 mi   **53.** 2.65 km   **55.** 190 mi

57. 810 N, 36° N of E   **59.** 2510 ft or 3370 ft (ambiguous)

Exercises 10-1, page 279

1. 0, -0.7, -1, -0.7, 0, 0.7, 1, 0.7, 0, -0.7, -1, -0.7, 0, 0.7, 1, 0.7, 0

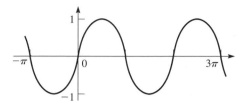

3. -3, -2.1, 0, 2.1, 3, 2.1, 0, -2.1, -3, -2.1, 0, 2.1, 3, 2.1, 0, -2.1, -3

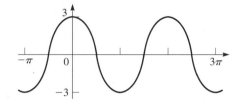

5.
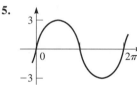

7.
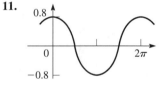

9.
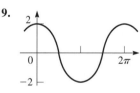

11.
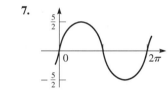

13.
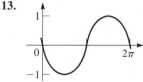

15.
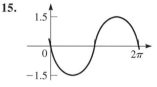

17.
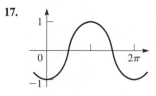

19.
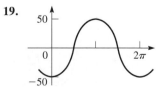

21. 0, 0.84, 0.91, 0.14, -0.76, -0.96, -0.28, 0.66

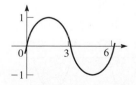

23. 1, 0.54, −0.42, −0.99, −0.65, 0.28, 0.96, 0.75

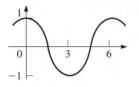

25. $y = 4 \sin x$ **27.** $y = -1.5 \cos x$

Exercises 10-2, page 282

1. $\dfrac{\pi}{3}$ **3.** $\dfrac{\pi}{4}$ **5.** $\dfrac{\pi}{6}$ **7.** $\dfrac{\pi}{8}$ **9.** 1

11. $\dfrac{1}{2}$ **13.** 6π **15.** 3π **17.** 3   **19.** $\dfrac{2}{\pi}$

21. **23.** **25.** **27.** **29.** **31.** **33.** **35.** **37.** **39.**

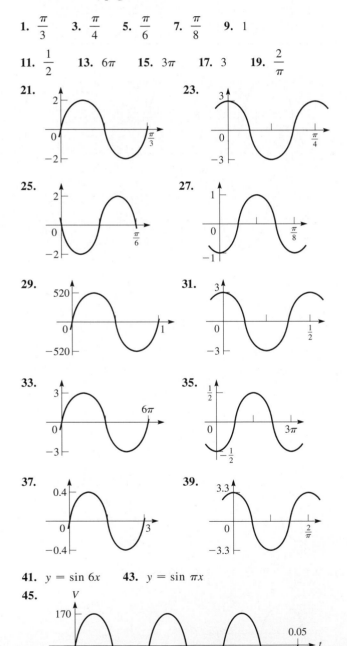

41. $y = \sin 6x$ **43.** $y = \sin \pi x$

45.

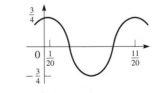

47.

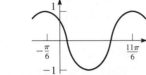

49. $y = \frac{1}{2} \cos 2x$ **51.** $y = -4 \sin \pi x$

Exercises 10-3, page 286

1. $1, 2\pi, \dfrac{\pi}{6}$ **3.** $1, 2\pi, -\dfrac{\pi}{6}$

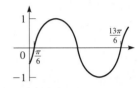

5. $2, \pi, -\dfrac{\pi}{4}$ **7.** $1, \pi, \dfrac{\pi}{2}$

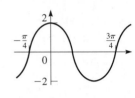
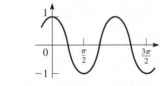

9. $\dfrac{1}{2}, 4\pi, \dfrac{\pi}{2}$ **11.** $3, 6\pi, -\pi$

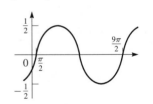
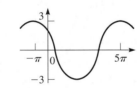

13. $1, 2, -\dfrac{1}{8}$ **15.** $\dfrac{3}{4}, \dfrac{1}{2}, \dfrac{1}{20}$

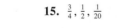
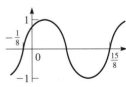

17. $0.6, 1, \dfrac{1}{2\pi}$ **19.** $40, \dfrac{2}{3}, -\dfrac{2}{3\pi}$

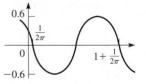
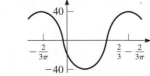

21. $1, \frac{2}{\pi}, \frac{1}{\pi}$

23. $\frac{3}{2}, 2, -\frac{\pi}{6}$

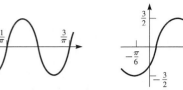

9.

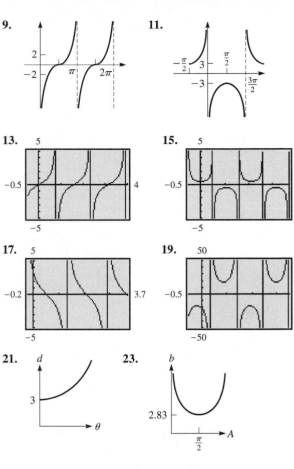

11.

25.

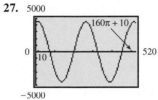

13. 5

15. 5

27. 5000

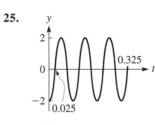

17. 5

19. 50

21. d

23. b

29. $y = 5 \sin\left(\frac{\pi}{8}x + \frac{\pi}{8}\right)$. As shown: amplitude = 5; period = $\frac{2\pi}{b} = 16$, $b = \frac{\pi}{8}$; displacement = $-\frac{c}{b} = -1$, $c = \frac{\pi}{8}$

31. $y = -0.8 \cos 2x$. As shown: amplitude = $|-0.8| = 0.8$; period = $\frac{2\pi}{b} = \pi$, $b = 2$; displacement = $-\frac{c}{b} = 0$, $c = 0$

Exercises 10-4, page 289

1. undef., -1.7, -1, -0.58, 0, 0.58, 1, 1.7, undef., -1.7, -1, -0.58, 0

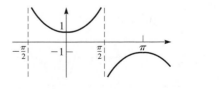

3. undef., 2, 1.4, 1.2, 1, 1.2, 1.4, 2, undef., -2, -1.4, -1.2, -1

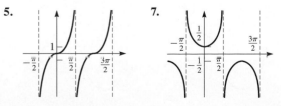

5.

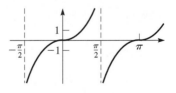

7.

Exercises 10-5, page 292

1. d

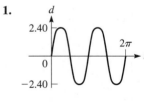

3. d

5. D

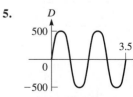

7. e

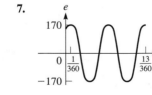

9. y

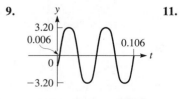

11. p

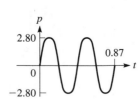

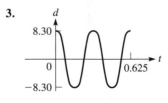

13.
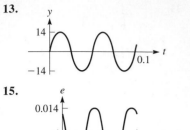

15.

Exercises 10-6, page 296

1.

3.

5.

7.

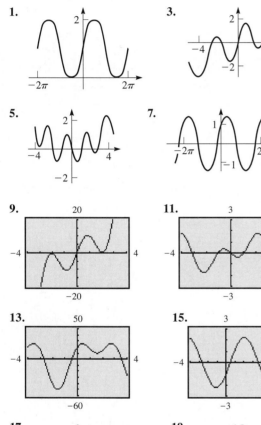

9.

11.

13.

15.

17.

19.

21.
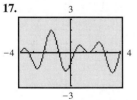

23.
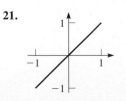

25.

27.

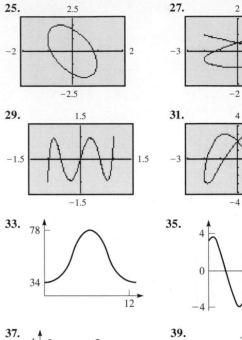

29.

31.

33.

35.
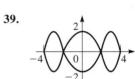

37.

39.
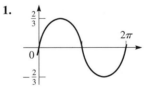

Review Exercises for Chapter 10, page 298

1.

3.
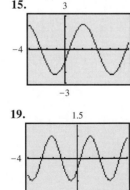

5.

7.
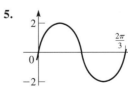

9.

11.
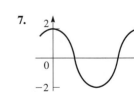

13.

15.
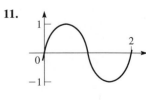

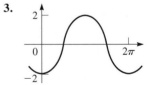
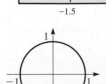
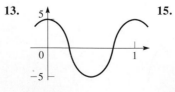
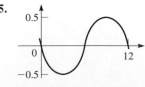
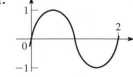

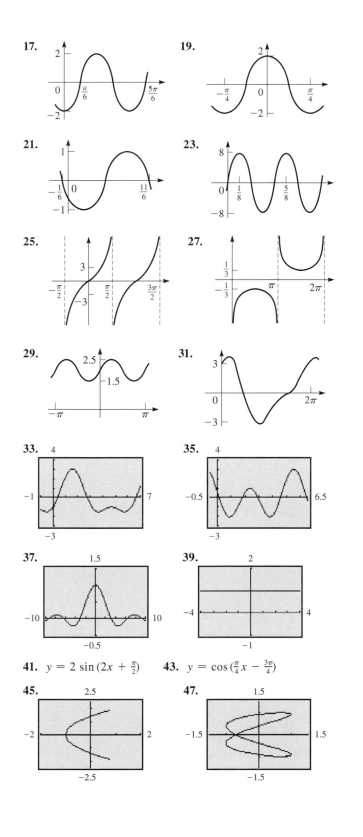

17.

19.

21.

23.

25.

27.

29.

31.

33.

35.

37.

39.

41. $y = 2\sin\left(2x + \frac{\pi}{2}\right)$

43. $y = \cos\left(\frac{\pi}{4}x - \frac{3\pi}{4}\right)$

45.

47.

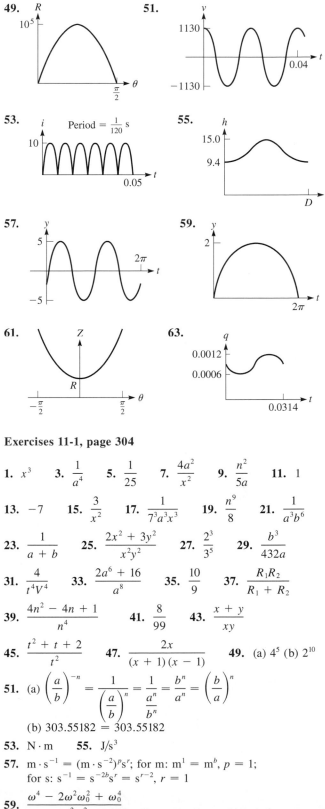

49.

51.

53. Period $= \frac{1}{120}$ s

55.

57.

59.

61.

63.

Exercises 11-1, page 304

1. x^3 **3.** $\dfrac{1}{a^4}$ **5.** $\dfrac{1}{25}$ **7.** $\dfrac{4a^2}{x^2}$ **9.** $\dfrac{n^2}{5a}$ **11.** 1

13. -7 **15.** $\dfrac{3}{x^2}$ **17.** $\dfrac{1}{7^3 a^3 x^3}$ **19.** $\dfrac{n^9}{8}$ **21.** $\dfrac{1}{a^3 b^6}$

23. $\dfrac{1}{a+b}$ **25.** $\dfrac{2x^2 + 3y^2}{x^2 y^2}$ **27.** $\dfrac{2^3}{3^5}$ **29.** $\dfrac{b^3}{432a}$

31. $\dfrac{4}{t^4 V^4}$ **33.** $\dfrac{2a^6 + 16}{a^8}$ **35.** $\dfrac{10}{9}$ **37.** $\dfrac{R_1 R_2}{R_1 + R_2}$

39. $\dfrac{4n^2 - 4n + 1}{n^4}$ **41.** $\dfrac{8}{99}$ **43.** $\dfrac{x+y}{xy}$

45. $\dfrac{t^2 + t + 2}{t^2}$ **47.** $\dfrac{2x}{(x+1)(x-1)}$ **49.** (a) 4^5 (b) 2^{10}

51. (a) $\left(\dfrac{a}{b}\right)^{-n} = \dfrac{1}{\left(\dfrac{a}{b}\right)^n} = \dfrac{1}{\dfrac{a^n}{b^n}} = \dfrac{b^n}{a^n} = \left(\dfrac{b}{a}\right)^n$

(b) $303.55182 = 303.55182$

53. $N \cdot m$ **55.** J/s^3

57. $m \cdot s^{-1} = (m \cdot s^{-2})^p s^r$; for m: $m^1 = m^b$, $p = 1$;
for s: $s^{-1} = s^{-2b} s^r = s^{r-2}$, $r = 1$

59. $\dfrac{\omega^4 - 2\omega^2 \omega_0^2 + \omega_0^4}{\omega^2 \omega_0^2}$

Exercises 11-2, page 308

1. 5   **3.** 3   **5.** 10^{25} **7.** $\dfrac{1}{2}$ **9.** $\dfrac{1}{16}$ **11.** 25

13. 81   **15.** 200   **17.** $\dfrac{3}{5}$ **19.** -2 **21.** $\dfrac{39}{1000}$

23. $\dfrac{3}{5}$ **25.** 2.059   **27.** 0.53891   **29.** $a^{7/6}$ **31.** $\dfrac{1}{y^{9/10}}$

33. $\dfrac{1}{x^{3/2}}$ **35.** $2ab^2$ **37.** $\dfrac{1}{8a^3b^{9/4}}$ **39.** $a^{1/12}$

41. $\dfrac{4x}{(4x^2+1)^{1/2}}$ **43.** $\dfrac{2}{3}x^{1/6}y^{11/12}$ **45.** $\dfrac{T}{(T+2)^{1/2}}$

47. $\dfrac{a^2+1}{a^4}$ **49.** $\dfrac{a+1}{a^{1/2}}$ **51.** $\dfrac{5x^2-2x}{(2x-1)^{1/2}}$

53.   **55.**

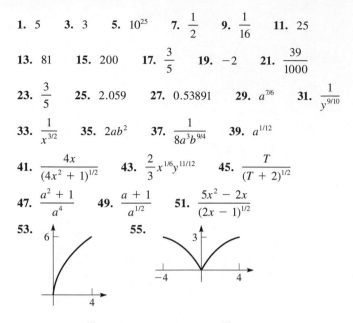

57. If $(A/S)^{-1/4}=0.5=1/2$, then $(A/S)^{1/4}=2$. Raise each to the fourth power and get $A/S=16$.   **59.** 1.91 mA

Exercises 11-3, page 312

1. $2\sqrt{6}$ **3.** $3\sqrt{5}$ **5.** $xy^2\sqrt{y}$ **7.** $xy^2z\sqrt{z}$
9. $3ac^2\sqrt{2ab}$ **11.** $2\sqrt[3]{2}$ **13.** $2\sqrt[5]{3}$ **15.** $2\sqrt[3]{a^2}$

17. $2st\sqrt[4]{4r^3t}$ **19.** 2   **21.** $ab\sqrt[3]{b^2}$ **23.** $\dfrac{1}{2}\sqrt{6}$

25. $\dfrac{1}{2}\sqrt[3]{6}$ **27.** $\dfrac{1}{3}\sqrt[5]{27}$ **29.** $2\sqrt{5}$ **31.** 2   **33.** 200

35. 2000   **37.** $\sqrt{2a}$ **39.** $\dfrac{1}{2}\sqrt{2}$ **41.** $\sqrt[3]{2}$ **43.** $\sqrt[8]{2}$

45. $\dfrac{1}{6}\sqrt{6}$ **47.** $\dfrac{\sqrt{b(a^2+b)}}{ab}$ **49.** $\dfrac{\sqrt{x^3+x}}{x^2+1}$ **51.** $a+b$

53. $\sqrt{4x^2-1}$ **55.** $\dfrac{1}{2}\sqrt{4x^2+1}$

57. Write $1/\sqrt[5]{R^2}$ as $R^{-2/5}$. $E=100(1-R^{-2/5})=55.0\%$

59. $\dfrac{24\sqrt{EIgWL}}{WL^2}$

Exercises 11-4, page 314

1. $7\sqrt{3}$ **3.** $\sqrt{5}-\sqrt{7}$ **5.** $3\sqrt{5}$ **7.** $-4\sqrt{3}$
9. $-2\sqrt{2a}$ **11.** $19\sqrt{7}$ **13.** $-4\sqrt{5}$

15. $23\sqrt{3R}-6\sqrt{2R}$ **17.** $\dfrac{7}{3}\sqrt{15}$ **19.** 0   **21.** $13\sqrt[3]{3}$

23. $\sqrt[4]{2}$ **25.** $(a-2b^2)\sqrt{ab}$ **27.** $(3-2a)\sqrt{10}$

29. $(2b-a)\sqrt[3]{3a^2b}$ **31.** $\dfrac{(a^2-c^3)\sqrt{ac}}{a^2c^3}$

33. $\dfrac{(a-2b)\sqrt[3]{ab^2}}{ab}$ **35.** $\dfrac{2b\sqrt{a^2-b^2}}{b^2-a^2}$

37. $15\sqrt{3}-11\sqrt{5}=1.3840144$ **39.** $\dfrac{1}{6}\sqrt{6}=0.4082483$

41. $3\sqrt{3}$ **43.** $18+3\sqrt{2}=22.2$ ft

Exercises 11-5, page 318

1. $\sqrt{30}$ **3.** $2\sqrt{3}$ **5.** 2   **7.** 50   **9.** $2\sqrt{5}$
11. $\sqrt{6}-\sqrt{15}$ **13.** -1 **15.** $48+9\sqrt{15}$

17. $66+13\sqrt{11x}-5x$ **19.** $a\sqrt{b}+c\sqrt{ac}$ **21.** $\dfrac{2-\sqrt{6}}{2}$

23. $\dfrac{a\sqrt{2}-b\sqrt{a}}{a}$ **25.** $2a-3b+2\sqrt{2ab}$ **27.** $\sqrt[6]{72}$

29. $\dfrac{1}{4}(\sqrt{7}-\sqrt{3})$ **31.** $-\dfrac{3}{8}(\sqrt{5}+3)$

33. $\dfrac{1}{11}(\sqrt{7}+3\sqrt{2}-6-\sqrt{14})$ **35.** $\dfrac{1}{17}(-56+9\sqrt{15})$

37. $\dfrac{2x+2\sqrt{xy}}{x-y}$ **39.** $\dfrac{8(3\sqrt{a}+2\sqrt{b})}{9a-4b}$ **41.** 1

43. $\dfrac{2-a-a^2}{a}$ **45.** $-\dfrac{\sqrt{x^2-y^2}+\sqrt{x^2+xy}}{y}$

47. xy **49.** $-1-\sqrt{66}=-9.1240384$

51. $-\dfrac{16+5\sqrt{30}}{26}=-1.6686972$ **53.** $\dfrac{2x+1}{\sqrt{x}}$

55. $\dfrac{5x^2+2x}{\sqrt{2x+1}}$ **57.** $\dfrac{1}{\sqrt{30}-2\sqrt{3}}$ **59.** $\dfrac{1}{\sqrt{x+h}+\sqrt{x}}$

61. $(1-\sqrt{2})^2-2(1-\sqrt{2})-1=1-2\sqrt{2}+2-2+2\sqrt{2}-1=0$

63. $\left[\dfrac{1}{2}(\sqrt{b^2-4k^2}-b)\right]^2+b\left[\dfrac{1}{2}(\sqrt{b^2-4k^2}-b)\right]$
$+k^2=\dfrac{1}{4}(b^2-4k^2)-\dfrac{b}{2}\sqrt{b^2-4k^2}+\dfrac{1}{4}b^2$
$+\dfrac{b}{2}\sqrt{b^2-4k^2}-\dfrac{1}{2}b^2+k^2=0$

65. $A=2s^2(1+\sqrt{2})$ **67.** $\dfrac{\sqrt{2g}\,(\sqrt{h_2}+\sqrt{h_1})}{2g}$

Review Exercises for Chapter 11, page 319

1. $\dfrac{2}{a^2}$ **3.** $\dfrac{2d^3}{c}$ **5.** 375   **7.** $\dfrac{1}{8000}$ **9.** $\dfrac{t^4}{9}$ **11.** -28

13. $64a^2b^5$ **15.** $-8m^9n^6$ **17.** $\dfrac{2C-4L^2}{CL^2}$ **19.** $\dfrac{2y}{x+2y}$

21. $\dfrac{b}{ab-3}$ **23.** $\dfrac{(x^3y^3-1)^{1/3}}{y}$ **25.** $4a(a^2+4)^{1/2}$

27. $\dfrac{-2(x+1)}{(x-1)^3}$ **29.** $2\sqrt{17}$ **31.** $b^2c\sqrt{ab}$

33. $3ab^2\sqrt{a}$ **35.** $2tu\sqrt{21st}$ **37.** $\dfrac{5\sqrt{2s}}{2s}$ **39.** $\dfrac{1}{9}\sqrt{33}$

41. $mn^2\sqrt[4]{8m^2n}$ **43.** $\sqrt{2}$ **45.** $14\sqrt{2}$ **47.** $-7\sqrt{7}$

49. $3ax\sqrt{2x}$ **51.** $(2a+b)\sqrt[3]{a}$ **53.** $10-\sqrt{55}$

55. $4\sqrt{3}-4\sqrt{5}$ **57.** $-45-7\sqrt{17}$

59. $42-7\sqrt{7a}-3a$ **61.** $\dfrac{6x+\sqrt{3xy}}{12x-y}$ **63.** $-\dfrac{8+\sqrt{6}}{29}$

65. $\dfrac{13-2\sqrt{35}}{29}$ **67.** $\dfrac{6x-13a\sqrt{x}+5a^2}{9x-25a^2}$ **69.** $\sqrt{4b^2+1}$

71. $\dfrac{15-2\sqrt{15}}{4}$ **73.** $51-7\sqrt{105}=-20.728655$

75. $\sqrt{\sqrt{2}-1}(\sqrt{2}+1)=\sqrt{(\sqrt{2}-1)(\sqrt{2}+1)^2}$
$=\sqrt{(\sqrt{2}-1)(3+2\sqrt{2})}=\sqrt{\sqrt{2}+1}$

77. 2.784% **79.** (a) $v=k(P/W)^{1/3}$ (b) $v=\dfrac{k\sqrt[3]{PW^2}}{W}$

81. $\dfrac{n_1^2 n_2^2 v}{n_1^2-n_2^2}$ **83.** $\dfrac{1}{\sqrt{A+h}+\sqrt{A}}$

85. $6\sqrt{2}$ cm **87.** $\dfrac{\sqrt{LC_1C_2(C_1+C_2)}}{2\pi LC_1C_2}$

Exercises 12-1, page 324

1. $9j$ **3.** $-2j$ **5.** $0.6j$ **7.** $2j\sqrt{2}$ **9.** $\dfrac{1}{2}j\sqrt{7}$

11. $-2ej$ **13.** (a) -7 (b) 7 **15.** (a) 4 (b) -4

17. (a) $-j$ (b) j **19.** 0 **21.** $-2j$ **23.** $-j$

25. $2+3j$ **27.** $-7j$ **29.** $2+2j$ **31.** $-2+3j$

33. $3\sqrt{2}-2j\sqrt{2}$ **35.** -1 **37.** (a) $6+7j$ (b) $8-j$

39. (a) $-2j$ (b) -4 **41.** $x=2,\ y=-2$

43. $x=10,\ y=-6$ **45.** $x=0,\ y=-1$

47. $x=-2,\ y=3$ **49.** yes **51.** no **53.** 0

55. Yes; imag. part is zero

Exercises 12-2, page 327

1. $5-8j$ **3.** $-9+6j$ **5.** $-0.23+0.86j$

7. $-36+21j$ **9.** $7+49j$ **11.** $22+3j$

13. $-900-700j$ **15.** $-18j\sqrt{2}$ **17.** $-28j$

19. $3\sqrt{7}+3j$ **21.** $-40-42j$ **23.** $-2-2j$

25. $\dfrac{1}{29}(-30+12j)$ **27.** $0.075+0.025j$

29. $\dfrac{1}{25}(-2+39j)$ **31.** $\dfrac{1}{28}(1+17j\sqrt{3})$

33. $\dfrac{1}{5}(-1+3j)$ **35.** $\dfrac{1}{13}(-35+33j)$

37. $(-1-j)^2+2(-1-j)+2=1+2j-1-2-2j+2$
$=0$

39. 10 **41.** $281+35.2j$ V **43.** $0.016+0.037j$ amperes

45. $(a+bj)+(a-bj)=2a$

47. Product is sum of squares of two real numbers
$(a+bj)(a-bj)=a^2+b^2$

Exercises 12-3, page 329

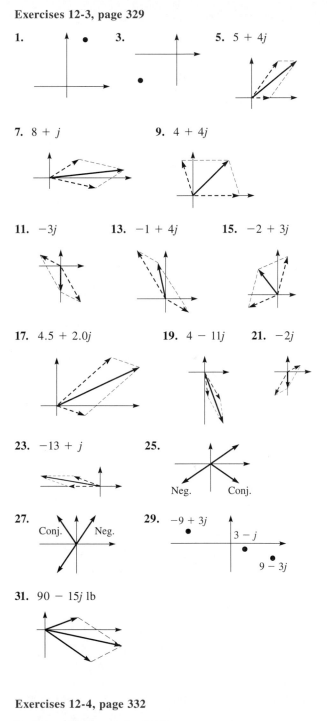

1. **3.** **5.** $5+4j$

7. $8+j$ **9.** $4+4j$

11. $-3j$ **13.** $-1+4j$ **15.** $-2+3j$

17. $4.5+2.0j$ **19.** $4-11j$ **21.** $-2j$

23. $-13+j$ **25.** Neg. Conj.

27. Conj. Neg. **29.** $-9+3j$, $3-j$, $9-3j$

31. $90-15j$ lb

Exercises 12-4, page 332

1. $10(\cos 36.9°+j\sin 36.9°)$

3. $5(\cos 306.9°+j\sin 306.9°)$

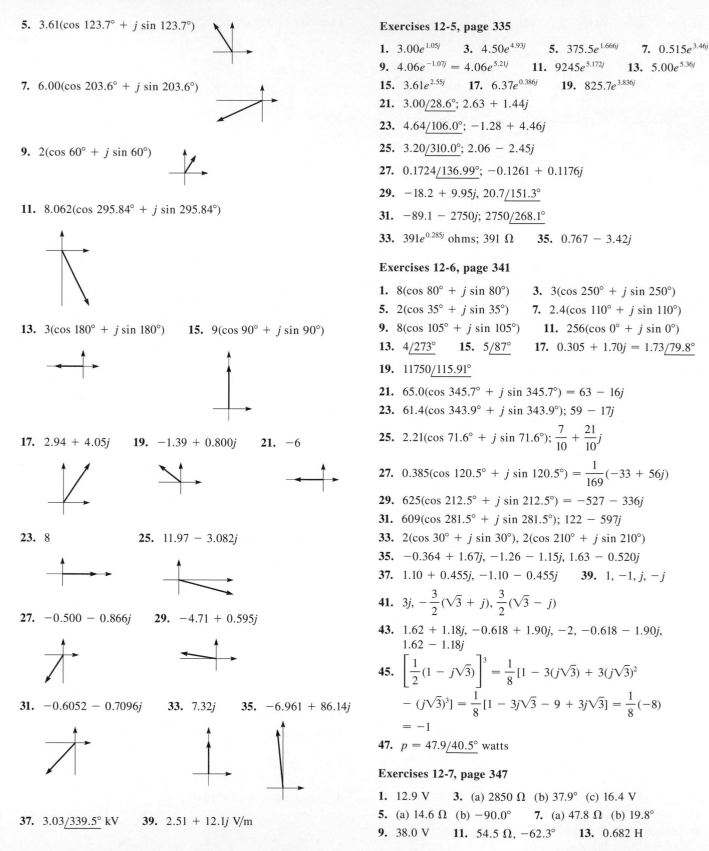

5. $3.61(\cos 123.7° + j \sin 123.7°)$

7. $6.00(\cos 203.6° + j \sin 203.6°)$

9. $2(\cos 60° + j \sin 60°)$

11. $8.062(\cos 295.84° + j \sin 295.84°)$

13. $3(\cos 180° + j \sin 180°)$ **15.** $9(\cos 90° + j \sin 90°)$

17. $2.94 + 4.05j$ **19.** $-1.39 + 0.800j$ **21.** -6

23. 8 **25.** $11.97 - 3.082j$

27. $-0.500 - 0.866j$ **29.** $-4.71 + 0.595j$

31. $-0.6052 - 0.7096j$ **33.** $7.32j$ **35.** $-6.961 + 86.14j$

37. $3.03\underline{/339.5°}$ kV   **39.** $2.51 + 12.1j$ V/m

Exercises 12-5, page 335

1. $3.00e^{1.05j}$ **3.** $4.50e^{4.93j}$ **5.** $375.5e^{1.666j}$ **7.** $0.515e^{3.46j}$
9. $4.06e^{-1.07j} = 4.06e^{5.21j}$ **11.** $9245e^{5.172j}$ **13.** $5.00e^{5.36j}$
15. $3.61e^{2.55j}$ **17.** $6.37e^{0.386j}$ **19.** $825.7e^{3.836j}$
21. $3.00\underline{/28.6°}$; $2.63 + 1.44j$
23. $4.64\underline{/106.0°}$; $-1.28 + 4.46j$
25. $3.20\underline{/310.0°}$; $2.06 - 2.45j$
27. $0.1724\underline{/136.99°}$; $-0.1261 + 0.1176j$
29. $-18.2 + 9.95j$, $20.7\underline{/151.3°}$
31. $-89.1 - 2750j$; $2750\underline{/268.1°}$
33. $391e^{0.285j}$ ohms; $391\ \Omega$ **35.** $0.767 - 3.42j$

Exercises 12-6, page 341

1. $8(\cos 80° + j \sin 80°)$ **3.** $3(\cos 250° + j \sin 250°)$
5. $2(\cos 35° + j \sin 35°)$ **7.** $2.4(\cos 110° + j \sin 110°)$
9. $8(\cos 105° + j \sin 105°)$ **11.** $256(\cos 0° + j \sin 0°)$
13. $4\underline{/273°}$ **15.** $5\underline{/87°}$ **17.** $0.305 + 1.70j = 1.73\underline{/79.8°}$
19. $11750\underline{/115.91°}$
21. $65.0(\cos 345.7° + j \sin 345.7°) = 63 - 16j$
23. $61.4(\cos 343.9° + j \sin 343.9°)$; $59 - 17j$
25. $2.21(\cos 71.6° + j \sin 71.6°)$; $\dfrac{7}{10} + \dfrac{21}{10}j$
27. $0.385(\cos 120.5° + j \sin 120.5°) = \dfrac{1}{169}(-33 + 56j)$
29. $625(\cos 212.5° + j \sin 212.5°) = -527 - 336j$
31. $609(\cos 281.5° + j \sin 281.5°)$; $122 - 597j$
33. $2(\cos 30° + j \sin 30°)$, $2(\cos 210° + j \sin 210°)$
35. $-0.364 + 1.67j$, $-1.26 - 1.15j$, $1.63 - 0.520j$
37. $1.10 + 0.455j$, $-1.10 - 0.455j$ **39.** $1, -1, j, -j$
41. $3j, -\dfrac{3}{2}(\sqrt{3} + j), \dfrac{3}{2}(\sqrt{3} - j)$
43. $1.62 + 1.18j$, $-0.618 + 1.90j$, -2, $-0.618 - 1.90j$, $1.62 - 1.18j$
45. $\left[\dfrac{1}{2}(1 - j\sqrt{3})\right]^3 = \dfrac{1}{8}[1 - 3(j\sqrt{3}) + 3(j\sqrt{3})^2$
$\quad - (j\sqrt{3})^3] = \dfrac{1}{8}[1 - 3j\sqrt{3} - 9 + 3j\sqrt{3}] = \dfrac{1}{8}(-8)$
$\quad = -1$
47. $p = 47.9\underline{/40.5°}$ watts

Exercises 12-7, page 347

1. 12.9 V   **3.** (a) $2850\ \Omega$ (b) $37.9°$ (c) 16.4 V
5. (a) $14.6\ \Omega$ (b) $-90.0°$ **7.** (a) $47.8\ \Omega$ (b) $19.8°$
9. 38.0 V   **11.** $54.5\ \Omega, -62.3°$ **13.** 0.682 H

15. 2.08×10^5 Hz   **17.** 1.30×10^{-11} F = 13.0 pF
19. 1.02 mW

Review Exercises for Chapter 12, page 349

1. $10 - j$ **3.** $6 + 2j$ **5.** $9 + 2j$ **7.** $-12 + 66j$

9. $\dfrac{1}{85}(21 + 18j)$ **11.** $-2 - 3j$ **13.** $\dfrac{1}{10}(-12 + 9j)$

15. $\dfrac{1}{5}(13 + 11j)$ **17.** $x = -\dfrac{2}{3}, y = -2$

19. $x = -\dfrac{1}{2}, y = \dfrac{1}{2}$ **21.** $3 + 11j$ **23.** $4 + 8j$

25. $1.41(\cos 315° + j \sin 315°) = 1.41e^{5.50j}$
27. $7.28(\cos 254.1° + j \sin 254.1°) = 7.28e^{4.43j}$
29. $4.67(\cos 76.8° + j \sin 76.8°); 4.67e^{1.34j}$
31. $10(\cos 0° + j \sin 0°); 10e^{0j}$ **33.** $-1.41 - 1.41j$
35. $-2.789 + 4.163j$ **37.** $0.19 - 0.59j$
39. $26.31 - 6.427j$ **41.** $1.94 + 0.495j$
43. $-728.1 + 1017j$ **45.** $15(\cos 84° + j \sin 84°)$
47. $20\underline{/263°}$ **49.** $8(\cos 59° + j \sin 59°)$
51. $14.29\underline{/133.61°}$ **53.** $1.26\underline{/59.7°}$ **55.** $9682\underline{/249.52°}$
57. $1024(\cos 160° + j \sin 160°)$ **59.** $27\underline{/331.5°}$
61. $32(\cos 270° + j \sin 270°) = -32j$
63. $\dfrac{625}{2}(\cos 270° + j \sin 270°) = -\dfrac{625}{2}j$
65. $1.00 + 1.73j, -2, 1.00 - 1.73j$
67. $\cos 67.5° + j \sin 67.5°, \cos 157.5° + j \sin 157.5°,$
$\cos 247.5° + j \sin 247.5°, \cos 337.5° + j \sin 337.5°$
69. $40 + 9j, 41(\cos 12.7° + j \sin 12.7°)$
71. $-15.0 - 10.9j, 18.5(\cos 216.0° + j \sin 216.0°)$
73. 60 V   **75.** $-21.6°$ **77.** 22.9 Hz   **79.** $550\underline{/53°}$ N
81. $\dfrac{u - j\omega n}{u^2 + \omega^2 n^2}$ **83.** $e^{j\pi} = \cos \pi + j \sin \pi = -1$

Exercises 13-1, page 354

1. 3   **3.** $\dfrac{1}{81}$ **5.** $\log_3 27 = 3$ **7.** $\log_4 256 = 4$

9. $\log_4\left(\dfrac{1}{16}\right) = -2$ **11.** $\log_2\left(\dfrac{1}{64}\right) = -6$

13. $\log_8 2 = \dfrac{1}{3}$ **15.** $\log_{1/4}\left(\dfrac{1}{16}\right) = 2$ **17.** $81 = 3^4$

19. $9 = 9^1$ **21.** $5 = 25^{1/2}$ **23.** $3 = 243^{1/5}$

25. $0.1 = 10^{-1}$ **27.** $16 = (0.5)^{-4}$ **29.** 2   **31.** -2

33. 343   **35.** $\dfrac{9}{4}$ **37.** 9   **39.** $\dfrac{1}{64}$ **41.** 0.2

43. -3 **45.** 3   **47.** $-\dfrac{1}{2}$ **49.** $t = \log_{1.05}(V/A)$

51. $b_2 = b_1\, 10^{0.4(m_1 - m_2)}$ **53.** $N = N_0 e^{-kt}$

55. $t = -8.3 \log_2(P/9600)$

Exercises 13-2, page 357

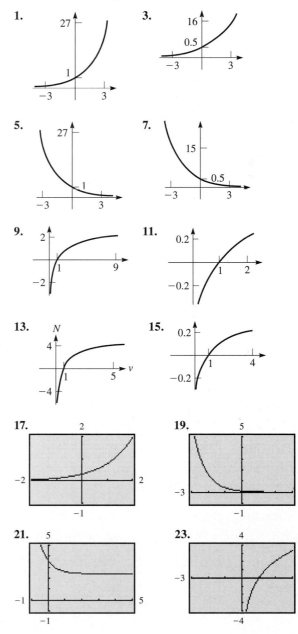

25.

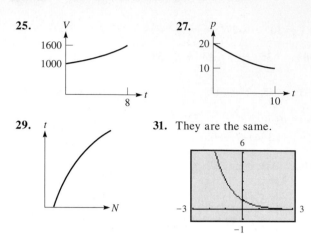

27.

29.

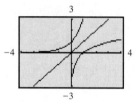

31. They are the same.

33. $x/2 = \log_{10} y$; $x = 2 \log_{10} y$;
$y = 2 \log_{10} x$

35. $x = y/3$; $y = x/3$

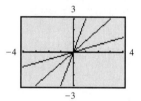

Exercises 13-3, page 361

1. $\log_5 3 + \log_5 11$ **3.** $\log_7 5 - \log_7 3$ **5.** $3 \log_2 a$

7. $\log_6 a + \log_6 b + \log_6 c$ **9.** $\frac{1}{4} \log_5 y$

11. $\frac{1}{2} \log_2 x - 2 \log_2 a$ **13.** $\log_b ac$ **15.** $\log_5 3$

17. $\log_b x^{3/2}$ **19.** $\log_e 4n^3$ **21.** -5 **23.** 2.5 **25.** $\frac{1}{2}$

27. $\frac{3}{4}$ **29.** $2 + \log_3 2$ **31.** $-1 - \log_2 3$

33. $\frac{1}{2}(1 + \log_3 2)$ **35.** $3 + \log_{10} 3$ **37.** $y = 2x$

39. $y = \frac{3x}{5}$ **41.** $y = \frac{49}{x^3}$ **43.** $y = 2(2ax)^{1/5}$

45. $y = \frac{2}{x}$ **47.** $y = \frac{1}{25} x^{2/\log_5 3}$

49. $\log_{10} x + \log_{10} 3 = \log_{10} 3x$ **51.** $T = 65e^{-0.41t}$

Exercises 13-4, page 364

1. 2.754 **3.** -1.194 **5.** 6.966 **7.** -3.9311

9. -0.0104 **11.** 1.219 **13.** $27,400$ **15.** 0.04960

17. 2000.4 **19.** 0.0057882 **21.** 85.5

23. 7.37×10^{101} **25.** 8.9542 **27.** -13.886

29. 15.2 dB   **31.** $2^{400} = 2.58 \times 10^{120}$

Exercises 13-5, page 367

1. 3.258 **3.** 0.4460 **5.** -4.91623 **7.** 1.92

9. 3.418 **11.** 1.795 **13.** 3.940 **15.** 0.3322

17. -0.008335 **19.** -17.39066 **21.** 1.6549

23. -0.16413 **25.** 8.935 **27.** 1.0085 **29.** 0.4757

31. 6.1993×10^{-11}

33.   **35.**

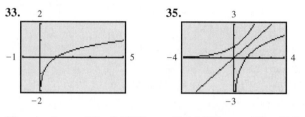

37. $y = 3x$ **39.** 8.155% **41.** 0.384 s   **43.** 21.7 s

Exercises 13-6, page 370

1. 4 **3.** -0.748 **5.** 0.587 **7.** 0.6439 **9.** 0.285

11. 0.203 **13.** 4.11 **15.** 14.2 **17.** $\frac{1}{4}$ **19.** 1.649

21. 3 **23.** -0.162 **25.** 5 **27.** 0.906 **29.** 4

31. 2 **33.** 4 **35.** 1.42 **37.** 28.0 **39.** 8.4 min

41. 1.72×10^{-5} **43.** $10^{8.3} = 2.0 \times 10^8$ **45.** $n = 20e^{-0.04t}$

47. $P = P_0(0.999)^t$ **49.** $x = 3.353$

51. ± 3.7 m

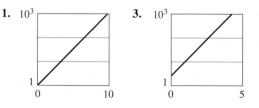

Exercises 13-7, page 374

1.   **3.**

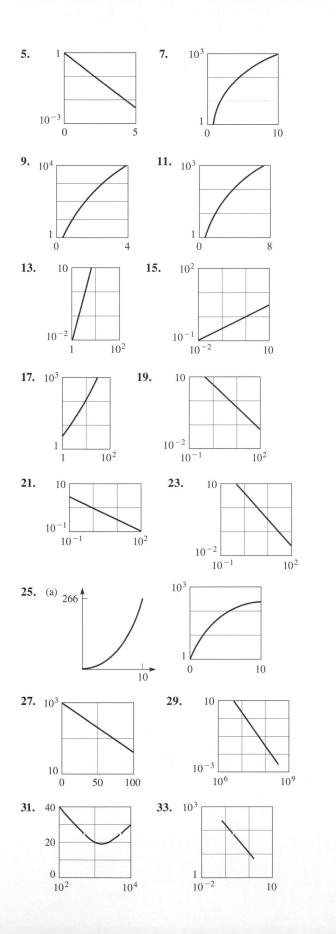

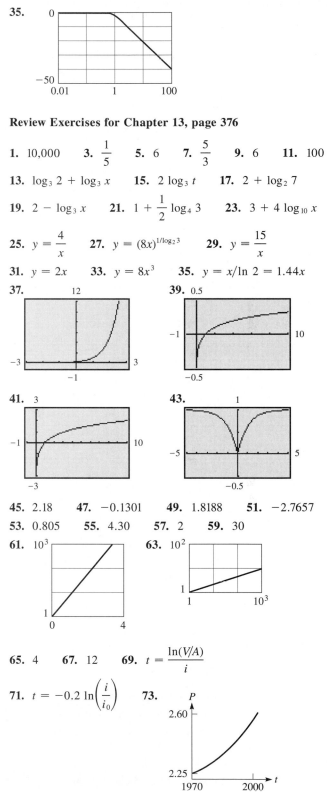

35.

Review Exercises for Chapter 13, page 376

1. 10,000   **3.** $\dfrac{1}{5}$ **5.** 6   **7.** $\dfrac{5}{3}$ **9.** 6   **11.** 100

13. $\log_3 2 + \log_3 x$ **15.** $2 \log_3 t$ **17.** $2 + \log_2 7$

19. $2 - \log_3 x$ **21.** $1 + \dfrac{1}{2} \log_4 3$ **23.** $3 + 4 \log_{10} x$

25. $y = \dfrac{4}{x}$ **27.** $y = (8x)^{1/\log_2 3}$ **29.** $y = \dfrac{15}{x}$

31. $y = 2x$ **33.** $y = 8x^3$ **35.** $y = x/\ln 2 = 1.44x$

37.   **39.**

41.   **43.**

45. 2.18   **47.** −0.1301   **49.** 1.8188   **51.** −2.7657

53. 0.805   **55.** 4.30   **57.** 2   **59.** 30

61.   **63.**

65. 4   **67.** 12   **69.** $t = \dfrac{\ln(V/A)}{i}$

71. $t = -0.2 \ln\left(\dfrac{i}{i_0}\right)$ **73.**

75. $\sin \theta = \dfrac{l\omega^2}{3g}$ **77.** $R = 2^{C/B} - 1$

79. 910 times brighter **81.** 1.17 min

83.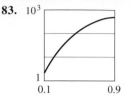

Exercises 14-1, page 382

1. $x = 1.8, y = 3.6; x = -1.8, y = -3.6$

3. $x = 0.0, y = -2.0; x = 2.7, y = -0.7$

5. $x = 1.5, y = 0.2$ **7.** $x = 1.6, y = 2.5$

9. $x = 1.1, y = 2.8; x = -1.1, y = 2.8;$
$x = 2.4, y = -1.8; x = -2.4, y = -1.8$

11. no solution

13. $x = -2.8, y = -1.0; x = 2.8, y = 1.0;$
$x = 2.8, y = -1.0; x = -2.8, y = 1.0$

15. $x = 0.7, y = 0.7; x = -0.7, y = -0.7$

17. $x = 0.0, y = 0.0; x = 0.9, y = 0.8$

19. $x = -1.1, y = 3.0; x = 1.8, y = 0.2$

21. $x = 1.0, y = 0.0$ **23.** $x = 3.6, y = 1.0$

25. 4.9 mi N, 1.6 mi E **27.** 2.2 A, 0.9 A

Exercises 14-2, page 386

1. $x = 0, y = 1; x = 1, y = 2$

3. $x = -\dfrac{19}{5}, y = \dfrac{17}{5}; x = 5, y = -1$ **5.** $x = 1, y = 0$

7. $x = \dfrac{2}{7}(3 + \sqrt{2}), y = \dfrac{2}{7}(-1 + 2\sqrt{2});$

$x = \dfrac{2}{7}(3 - \sqrt{2}), y = \dfrac{2}{7}(-1 - 2\sqrt{2})$

9. $w = 1, h = 1$ **11.** $x = \dfrac{2}{3}, y = \dfrac{9}{2}; x = -3, y = -1$

13. $x = -2, y = 4; x = 2, y = 4$

15. $x = 1, y = 2; x = -1, y = 2$

17. $D = 1, R = 0; D = -1, R = 0; D = \dfrac{1}{2}\sqrt{6}, R = \dfrac{1}{2};$

$D = -\dfrac{1}{2}\sqrt{6}, R = \dfrac{1}{2}$

19. $x = \sqrt{19}, y = \sqrt{6}; x = \sqrt{19}, y = -\sqrt{6};$
$x = -\sqrt{19}, y = \sqrt{6}; x = -\sqrt{19}, y = -\sqrt{6}$

21. $x = -5, y = -2; x = -5, y = 2; x = 5, y = -2;$
$x = 5, y = 2$

23. $x = -3, y = 2; x = -1, y = -2$

25. $x = 50$ mi, $h = 25$ mi **27.** 1.5 cm, 1.4 cm

29. 2.19 m, 0.21 m **31.** 80 mi/h

Exercises 14-3, page 389

1. $-3, -2, 2, 3$ **3.** $\dfrac{3}{2}, -\dfrac{3}{2}, j, -j$ **5.** $-\dfrac{1}{2}, \dfrac{1}{4}$

7. $-\dfrac{1}{2}, \dfrac{1}{2}, \dfrac{1}{6}j\sqrt{6}, -\dfrac{1}{6}j\sqrt{6}$ **9.** $1, \dfrac{25}{4}$ **11.** $\dfrac{64}{729}, 1$

13. $-27, 125$ **15.** 81 **17.** 5 **19.** $-2, -1, 3, 4$

21. 18 **23.** $1, -1, j\sqrt{2}, -j\sqrt{2}$

25. $R_1 = 2.62\ \Omega, R_2 = 1.62\ \Omega$ **27.** 15.9 in., 21.8 in.

Exercises 14-4, page 392

1. 12 **3.** 2 **5.** $\dfrac{2}{3}$ **7.** $\dfrac{1}{2}(7 + \sqrt{5})$ **9.** -1

11. 32 **13.** 4, 9 **15.** 9

17. 12 (Extraneous root introduced in squaring both sides
of $\sqrt{x + 4} = x - 8$.)

19. 16 **21.** $\dfrac{1}{2}$ **23.** $7, -1$ **25.** 0 **27.** 5

29. 6 **31.** 258 **33.** $L = \dfrac{1}{4\pi^2 f^2 C}$

35. $\dfrac{k^2}{2n(1 - k)}$ **37.** 9.2 km **39.** 1.6 mi

Review Exercises for Chapter 14, page 393

1. $x = -0.9, y = 3.5; x = 0.8, y = 2.6$

3. $x = 2.0, y = 0.0; x = 1.6, y = 0.6$

5. $x = 0.8, y = 1.6; x = -0.8, y = 1.6$

7. $x = -1.2, y = 2.7; x = 1.2, y = 2.7$

9. $x = 0.0, y = 0.0; x = 2.4, y = 0.9$

11. $x = 0, y = 0; x = 2, y = 16$

13. $x = \sqrt{2}, y = 1; x = -\sqrt{2}, y = 1$

15. $x = \dfrac{1}{12}(1 + \sqrt{97}), y = \dfrac{1}{18}(5 - \sqrt{97});$

$x = \dfrac{1}{12}(1 - \sqrt{97}), y = \dfrac{1}{18}(5 + \sqrt{97})$

17. $x = 7, y = 5; x = 7, y = -5; x = -7, y = 5;$
$x = -7, y = -5$

19. $x = -2, y = 2; x = \dfrac{2}{3}, y = \dfrac{10}{9}$

21. $-4, -2, 2, 4$ **23.** 1, 16 **25.** $\dfrac{1}{3}, -\dfrac{1}{7}$

27. $\dfrac{25}{4}$ **29.** $\sqrt{3}, -\sqrt{3}, \dfrac{1}{2}j\sqrt{3}, -\dfrac{1}{2}j\sqrt{3}$ **31.** 6

33. 8 **35.** $\dfrac{9}{16}$ **37.** $\dfrac{1}{3}(11 - 4\sqrt{15})$ **39.** 2

41. $l = \dfrac{1}{2}(-1 + \sqrt{1 + 16\pi^2 L^2/h^2})$

43. $m = \dfrac{1}{2}(-y \pm \sqrt{2s^2 - y^2})$ **45.** 0.75 s, 1.5 s

47. 230 ft   **49.** $Z = 1.09\ \Omega,\ X = 0.738\ \Omega$

51. 34 mm, 52 mm   **53.** 3.57 in.   **55.** 26 mi/h, 32 mi/h

Exercises 15-1, page 399

1. 0   **3.** 0   **5.** −40   **7.** −4   **9.** 8   **11.** 51

13. −28   **15.** 14   **17.** yes   **19.** yes   **21.** no

23. no   **25.** yes   **27.** yes

29. $(4x^3 + 8x^2 - x - 2) \div (2x - 1) = 2x^2 + 5x + 2$;
No, because the coefficient of x in $2x - 1$ is 2, not 1.

31. −7

Exercises 15-2, page 403

1. $x^2 + 3x + 2,\ R = 0$ **3.** $x^2 - x + 3,\ R = 0$

5. $2x^4 - 4x^3 + 8x^2 - 17x + 42,\ R = -40$

7. $3x^3 - x + 2,\ R = -4$ **9.** $x^2 + x - 4,\ R = 8$

11. $R^3 - 3R^2 + 3R - 12,\ \text{Rem} = 51$

13. $2x^3 - x^2 - 4x - 12,\ R = -28$

15. $x^4 + 2x^3 + x^2 + 7x + 4,\ R = 14$

17. $p^5 + 2p^4 + 4p^3 + 2p^2 + 2p + 4,\ R = 2$

19. $x^6 + 2x^5 + 4x^4 + 8x^3 + 16x^2 + 32x + 64,\ R = 0$

21. yes   **23.** no   **25.** no   **27.** no   **29.** yes

31. no   **33.** yes   **35.** yes

Exercises 15-3, page 407

(Note: Unknown roots listed)

1. −2, −1   **3.** −2, 3   **5.** −2, −2   **7.** $-j, -\dfrac{2}{3}$

9. $2j, -2j$ **11.** −1, −3   **13.** −2, 1

15. $1 - j, \dfrac{1}{2}, -2$ **17.** $\dfrac{1}{4}(1 + \sqrt{17}), \dfrac{1}{4}(1 - \sqrt{17})$

19. 2, −3   **21.** $-j, -1, 1$ **23.** $-2j, 3, -3$

Exercises 15-4, page 413

1. 1, −1, −2   **3.** 2, −1, −3   **5.** $\dfrac{1}{2}, 5, -3$

7. $\dfrac{1}{3}, -3, -1$ **9.** $-2, -2, 2 \pm \sqrt{3}$

11. 2, 4, −1, −3   **13.** $1, -\dfrac{1}{2}, 1 \pm \sqrt{3}$

15. $\dfrac{1}{2}, -\dfrac{2}{3}, -3, -\dfrac{1}{2}$ **17.** 2, 2, −1, −1, −3

19. $\dfrac{1}{2}, 1, 1, j, -j$ **21.** −1.86, 0.68, 3.18   **23.** −3.24, 1.24

25. 0.59   **27.** −0.77   **29.** $-\dfrac{3}{4}, \sqrt{5}, -\sqrt{5}$

31. 1.8 s, 4.5 s   **33.** 0, L   **35.** 1.5 lb, 4.1 lb

37. 3.0 mm, 4.0 mm; 5.0 mm, 6.0 mm

39. $2\ \Omega, 3\ \Omega, 6\ \Omega$ **41.** 1.23 cm or 2.14 cm

43. 0.0 km, 1.6 km, 2.8 km

Review Exercises for Chapter 15, page 414

1. 1   **3.** −107   **5.** yes   **7.** no

9. $x^2 + 4x + 10,\ R = 11$ **11.** $2x^2 - 7x + 10,\ R = -17$

13. $x^3 - 3x^2 - 4,\ R = -4$

15. $2x^4 + 10x^3 + 4x^2 + 21x + 105,\ R = 516$

17. no   **19.** yes   **21.** (unlisted roots) $\dfrac{1}{2}(-5 \pm \sqrt{17})$

23. (unlisted roots) $\dfrac{1}{3}(-1 \pm j\sqrt{14})$

25. (unlisted roots) $-1 + j\sqrt{2}, -1 - j\sqrt{2}$

27. (unlisted roots) $-j, \dfrac{1}{2}, -\dfrac{3}{2}$ **29.** (unlisted roots) 2, −2

31. (unlisted roots) $-2 - j, \dfrac{1}{2}(-1 \pm j\sqrt{3})$

33. 1, 2, −4   **35.** $1, 1, -\dfrac{1}{2}$ **37.** $\dfrac{5}{3}, -\dfrac{1}{2}, -1$

39. $\dfrac{1}{2}, \dfrac{3}{2}, -1 + \sqrt{2}, -1 - \sqrt{2}$

41. Use k in synthetic division, $k = 4$.

43. 1, 0.4, 1.5, −0.6 (last three are irrational)

45. 1.91   **47.** April   **49.** 0.75 cm   **51.** 2 in.

53. 5.1 m, 8.3 m   **55.** $h = 6.75$ ft, $w = 3.60$ ft

Exercises 16-1, page 421

1. 39   **3.** 30   **5.** 50   **7.** −47,416   **9.** −6

11. 118   **13.** −2

15. 24; Since a complete expansion of a third-order determinant has 6 terms, expanding by a row or column of a fourth-order determinant gives $4 \times 6 = 24$ terms.

17. $x = 2, y = -1, z = 3$ **19.** $u = -1, v = \dfrac{1}{3}, w = -\dfrac{1}{2}$

21. $x = -1, y = 0, z = 2, t = 1$

23. $x = 1, y = 2, z = -1, t = 3$

25. $\dfrac{2}{7}$ A, $\dfrac{18}{7}$ A, $-\dfrac{8}{7}$ A, $-\dfrac{12}{7}$ A   **27.** 500, 300, 700

Exercises 16-2, page 425

1. -60 **3.** 0 **5.** -40 **7.** -40 **9.** 57

11. -13 **13.** -72 **15.** 0 **17.** $x = 0, y = -1, z = 4$

19. $x = \dfrac{1}{3}, y = -\dfrac{1}{2}, z = 1$

21. $x = 2, y = -1, z = -1, t = 3$

23. $D = 1, E = 2, F = -1, G = -2$

25. $\dfrac{33}{16}$ A, $\dfrac{11}{8}$ A, $-\dfrac{5}{8}$ A, $-\dfrac{15}{8}$ A, $-\dfrac{15}{16}$ A

27. ppm of SO_2: 0.5, NO: 0.3, NO_2: 0.2, CO: 5.0

Exercises 16-3, page 429

1. $a = 1, b = -3, c = 4, d = 7$ **3.** $x = 2, y = 3$

5. Elements cannot be equated; different number of rows

7. $C = 3, D = 2, E = -2$

9. $\begin{bmatrix} 1 & 10 \\ 0 & 2 \end{bmatrix}$ **11.** $\begin{bmatrix} -5 & 0 \\ 11 & 71 \\ 11 & -5 \end{bmatrix}$

13. $\begin{bmatrix} 0 & 9 & -13 & 3 \\ 6 & -7 & 7 & 0 \end{bmatrix}$ **15.** cannot be added

17. $\begin{bmatrix} -1 & 13 & -20 & 3 \\ 8 & -13 & 6 & 2 \end{bmatrix}$ **19.** $\begin{bmatrix} -3 & -6 & 5 & -6 \\ -6 & -4 & -17 & 6 \end{bmatrix}$

21. $A + B = B + A = \begin{bmatrix} 3 & 1 & 0 & 7 \\ 5 & -3 & -2 & 5 \\ 10 & 10 & 8 & 0 \end{bmatrix}$

23. $-(A - B) = B - A = \begin{bmatrix} 5 & -3 & -6 & -7 \\ 5 & 3 & 0 & -3 \\ -8 & 12 & 8 & 4 \end{bmatrix}$

25. $v_w = 31.0$ km/h, $v_p = 249$ km/h

27. $\begin{bmatrix} 24 & 18 & 0 & 0 \\ 15 & 12 & 9 & 0 \\ 0 & 9 & 15 & 18 \end{bmatrix}$

Exercises 16-4, page 434

1. $[-8 \ -12]$ **3.** $\begin{bmatrix} -15 & 15 & -26 \\ 8 & 5 & -13 \end{bmatrix}$

5. $\begin{bmatrix} 29 \\ -29 \end{bmatrix}$ **7.** $\begin{bmatrix} -23 & -19 \\ 21 & 16 \\ 2 & -28 \end{bmatrix}$ **9.** $\begin{bmatrix} 33 & -22 \\ 31 & -12 \\ 15 & 13 \\ 50 & -41 \end{bmatrix}$

11. $\begin{bmatrix} -49.43 & 55.2 \\ -53.02 & 79.16 \end{bmatrix}$

13. $AB = [40]$, $BA = \begin{bmatrix} -1 & 3 & -8 \\ 5 & -15 & 40 \\ 7 & -21 & 56 \end{bmatrix}$

15. $AB = \begin{bmatrix} -5 \\ 10 \end{bmatrix}$, BA not defined

17. $AI = IA = A$ **19.** $AI = IA = A$ **21.** $B = A^{-1}$

23. $B = A^{-1}$ **25.** yes **27.** no

29. $\begin{bmatrix} -1 & 0 \\ 0 & -1 \end{bmatrix}\begin{bmatrix} -1 & 0 \\ 0 & -1 \end{bmatrix} = \begin{bmatrix} 1 & 0 \\ 0 & 1 \end{bmatrix}$

31. $A^2 - I = (A + I)(A - I) = \begin{bmatrix} 15 & 28 \\ 21 & 36 \end{bmatrix}$

33. $\begin{bmatrix} 0 & -j \\ j & 0 \end{bmatrix}\begin{bmatrix} 0 & -j \\ j & 0 \end{bmatrix} = \begin{bmatrix} 1 & 0 \\ 0 & 1 \end{bmatrix}$

35. $v_2 = v_1, i_2 = -v_1/R + i_1$

Exercises 16-5, page 439

1. $\begin{bmatrix} -2 & -\dfrac{5}{2} \\ -1 & -1 \end{bmatrix}$ **3.** $\begin{bmatrix} -\dfrac{1}{3} & \dfrac{1}{6} \\ \dfrac{2}{15} & \dfrac{1}{30} \end{bmatrix}$ **5.** $\begin{bmatrix} \dfrac{3}{4} & \dfrac{1}{2} \\ -\dfrac{1}{4} & 0 \end{bmatrix}$

7. $\begin{bmatrix} -\dfrac{8}{283} & -\dfrac{9}{566} \\ \dfrac{13}{1415} & \dfrac{5}{283} \end{bmatrix}$ **9.** $\begin{bmatrix} -3 & 2 \\ 2 & -1 \end{bmatrix}$

11. $\begin{bmatrix} -\dfrac{1}{2} & -2 \\ \dfrac{1}{2} & 1 \end{bmatrix}$ **13.** $\begin{bmatrix} \dfrac{2}{9} & -\dfrac{5}{9} \\ \dfrac{1}{9} & \dfrac{2}{9} \end{bmatrix}$ **15.** $\begin{bmatrix} \dfrac{3}{8} & \dfrac{1}{16} \\ -\dfrac{1}{4} & \dfrac{1}{8} \end{bmatrix}$

17. $\begin{bmatrix} -18 & -7 & 5 \\ -3 & -1 & 1 \\ -5 & -2 & 1 \end{bmatrix}$ **19.** $\begin{bmatrix} 3 & -4 & -1 \\ -4 & 5 & 2 \\ 2 & -3 & -1 \end{bmatrix}$

21. $\begin{bmatrix} 2 & 4 & \dfrac{7}{2} \\ -1 & -2 & -\dfrac{3}{2} \\ 1 & 1 & \dfrac{1}{2} \end{bmatrix}$ **23.** $\begin{bmatrix} \dfrac{5}{2} & -2 & -2 \\ -1 & 1 & 1 \\ \dfrac{7}{4} & -\dfrac{3}{2} & -1 \end{bmatrix}$

25. $\begin{bmatrix} 0.3 & -0.4 \\ 0.05 & 0.1 \end{bmatrix}$ **27.** $\begin{bmatrix} 2 & 4 & 3.5 \\ -1 & -2 & -1.5 \\ 1 & 1 & 0.5 \end{bmatrix}$

29. $\begin{bmatrix} 2.5 & -2 & -2 \\ -1 & 1 & 1 \\ 1.75 & -1.5 & -1 \end{bmatrix}$ **31.** $\begin{bmatrix} 1 & 2 & 3 & 1 \\ 1 & 3 & 3 & 2 \\ 2 & 4 & 3 & 3 \\ 1 & 1 & 1 & 1 \end{bmatrix}$

33. $\dfrac{1}{ad - bc}\begin{bmatrix} ad - bc & -ba + ab \\ cd - dc & -bc + ad \end{bmatrix} = \begin{bmatrix} 1 & 0 \\ 0 & 1 \end{bmatrix}$

35. $v_1 = (a_{22}i_1 - a_{12}i_2)/(a_{11}a_{22} - a_{12}a_{21})$
$v_2 = (-a_{21}i_1 + a_{11}i_2)/(a_{11}a_{22} - a_{12}a_{21})$

Exercises 16-6, page 443

1. $x = \dfrac{1}{2}, y = 3$ **3.** $x = \dfrac{1}{2}, y = -\dfrac{5}{2}$

5. $x = -4, y = 2, z = -1$ **7.** $x = -1, y = 0, z = 3$

9. $x = -\dfrac{3}{2}, y = -2$ **11.** $x = 1.6, y = -2.5$

13. $x = 2, y = -4, z = 1$ **15.** $x = 2, y = -\dfrac{1}{2}, z = 3$

17. $x = 1, y = -2, z = -3$ **19.** $u = 2, v = -5, w = 4$

21. $x = 2, y = 3, z = -2, t = 1$

23. $v = 2, w = -1, x = \dfrac{1}{2}, y = \dfrac{3}{2}, z = -3$

25. $A = 118$ N, $B = 186$ N   **27.** 6.4 L, 1.6 L, 2.0 L

Review Exercises for Chapter 16, page 445

1. 6   **3.** 186   **5.** -438 **7.** 44   **9.** 6

11. 186   **13.** -438 **15.** 6   **17.** 186

19. -438 **21.** $a = 4, b = -1$

23. $x = 2, y = -3, z = \dfrac{5}{2}, a = -1, b = -\dfrac{7}{2}, c = \dfrac{1}{2}$

25. $\begin{bmatrix} 1 & -3 \\ 8 & -5 \\ -8 & -2 \\ 3 & -10 \end{bmatrix}$ **27.** $\begin{bmatrix} 3 & 0 \\ -12 & 18 \\ 9 & 6 \\ -3 & 21 \end{bmatrix}$

29. cannot be subtracted   **31.** $\begin{bmatrix} 7 & -6 \\ -4 & 20 \\ -1 & 6 \\ 1 & 15 \end{bmatrix}$

33. $\begin{bmatrix} 0 & 0 \\ 0 & 0 \end{bmatrix}$ **35.** $\begin{bmatrix} 34 & 11 & -5 \\ 2 & -8 & 10 \\ -1 & -17 & 20 \end{bmatrix}$ **37.** $\begin{bmatrix} -2 & \dfrac{5}{2} \\ -1 & 1 \end{bmatrix}$

39. $\begin{bmatrix} \dfrac{2}{15} & \dfrac{1}{60} \\ -\dfrac{1}{15} & \dfrac{7}{60} \end{bmatrix}$ **41.** $\begin{bmatrix} 11 & 10 & 3 \\ -4 & -4 & -1 \\ 3 & 3 & 1 \end{bmatrix}$

43. $\begin{bmatrix} \dfrac{1}{2} & -\dfrac{1}{2} & -1 \\ -3 & 2 & 1 \\ -4 & 3 & 2 \end{bmatrix}$ **45.** $x = -3, y = 1$

47. $x = 10, y = -15$ **49.** $u = -1, v = -3, w = 0$

51. $x = 1, y = \dfrac{1}{2}, z = -\dfrac{1}{3}$ **53.** $x = 3, y = 1, z = -1$

55. $x = 1, v = 2, z = -3, t = 1$

57. $x = -\dfrac{1}{3}, y = 3, z = \dfrac{2}{3}, t = -4$

59. $r = \dfrac{1}{2}, s = -7, t = -\dfrac{3}{4}, u = 5, v = \dfrac{3}{2}$ **61.** $2\sqrt{2}$

63. 0   **65.** $\begin{bmatrix} 1 & 0 \\ 15 & 16 \end{bmatrix}, \begin{bmatrix} 1 & 0 \\ 63 & 64 \end{bmatrix}, \begin{bmatrix} 1 & 0 \\ 255 & 256 \end{bmatrix}$

67. $B^3 = \begin{bmatrix} 1 & 0 & 0 \\ 0 & 1 & 0 \\ 0 & 0 & 1 \end{bmatrix}$ **69.** $N^{-1} = -N = \begin{bmatrix} 0 & 1 \\ -1 & 0 \end{bmatrix}$

71. $(A + B)(A - B) = \begin{bmatrix} -6 & 2 \\ 4 & 2 \end{bmatrix}, A^2 - B^2 = \begin{bmatrix} -10 & -4 \\ 8 & 6 \end{bmatrix}$

73. $\begin{bmatrix} \dfrac{1}{2} & \dfrac{1}{3} \\ 0 & \dfrac{1}{6} \end{bmatrix} = \dfrac{1}{2} \begin{bmatrix} 1 & \dfrac{2}{3} \\ 0 & \dfrac{1}{3} \end{bmatrix}$

75. $R_1 = 4\ \Omega, R_2 = 6\ \Omega$ **77.** $F = 303$ lb, $T = 175$ lb

79. 0.22 h after police pass intersection

81. 30 g, 50 g, 20 g   **83.** 2.0 h, 1.5 h, 1.0 h, 1.0 h

85. $\begin{bmatrix} 12{,}000 & 24{,}000 & 4{,}000 \\ 15{,}000 & 8{,}000 & 30{,}000 \end{bmatrix}$

$+ \begin{bmatrix} 15{,}000 & 12{,}000 & 2{,}000 \\ 20{,}000 & 3{,}000 & 22{,}000 \end{bmatrix}$

$= \begin{bmatrix} 27{,}000 & 36{,}000 & 6{,}000 \\ 35{,}000 & 11{,}000 & 52{,}000 \end{bmatrix}$

87. $(R_1 + R_2)i_1 - R_2 i_2 = 6$
$-R_2 i_1 + (R_1 + R_2)i_2 = 0$

Exercises 17-1, page 453

1. $7 < 12$ **3.** $20 < 45$ **5.** $-4 > -9$ **7.** $16 < 81$

9. $x > -2$ **11.** $x \leq 4$ **13.** $1 < x < 7$

15. $x < -9$ or $x \geq -4$ **17.** $x < 1$ or $3 < x \leq 5$

19. $-2 < x < 2$ or $3 \leq x < 4$

21. x is greater than 0 and less than or equal to 2.

23. x is less than -1, or greater than or equal to 1 and less than 2.

25. (number line, open circle at 3, shaded to left)

27. (number line, closed circle at 1, open circle at 3)

29. (number line, closed circle at 0, open circle at 5)

31. (number line, -1, 1, 4)

33. (number line, -3, -1, 1, 3)

35. (number line, -3)

37. $d > 3 \times 10^{12}$ mi   (number line, 0 to 3×10^{12})

39. $18{,}000 < v < 25{,}000$ mi/h   (number line, 0, 18,000, 25,000)

41. $0 < n \leq 2565$ steps

43. $E = 0$ for $0 \leq r < a$
$E = k/r^2$ for $r \geq a$

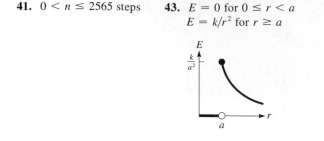

Exercises 17-2, page 458

1. $x > -1$

3. $x < 6$

5. $x \leq -2$

7. $y < -2$

9. $x \leq \dfrac{5}{2}$

11. $x < -1$

13. $x > 1.80$

15. $x > -\dfrac{7}{9}$

17. $-1 < x < 1$

19. $2 < x \leq 5$

21. $-3 \leq x < -1$

23. no values

25. $x < \dfrac{5}{2}$

27. $x \geq -1$

29. $-2 < t < 2$

31. $x > -\dfrac{1}{2}$

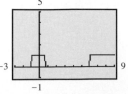

33. $x \geq 5$

35. $4.0 < V < 9.0$ V

37. $0 \leq t \leq 6$ years

39. $0.4 < t < 2.6$ h

41. $0 \leq x \leq 500$
$200 \leq y \leq 700$

43. $24 \leq x \leq 40$ L

Exercises 17-3, page 464

1. $-1 < x < 1$

3. $0 \leq x \leq 2$

5. $-4 \leq x \leq \dfrac{3}{2}$

7. $x = -2$

9. all x

11. $-2 < x < 0, x > 1$

13. $-2 \leq s \leq -1, s \geq 1$

15. $-6 < x \leq \dfrac{3}{2}$

17. $-5 < x < -1, x > 7$

19. $x < -1$

21. $x \leq -2, x \geq \dfrac{1}{3}$

23. $x < 3, x > 8$

25. $-1 < x < \dfrac{3}{4}, x \geq 6$

27. $2 < x < 4, 5 < x < 9$

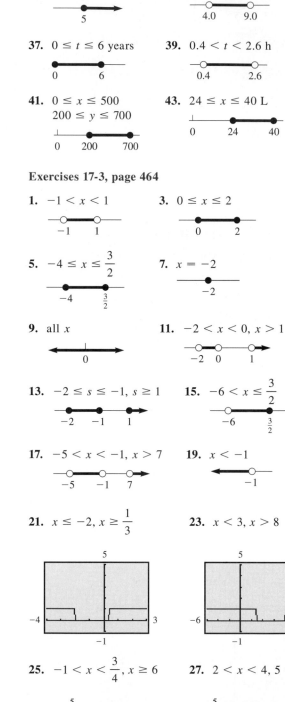

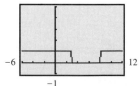

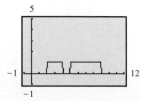

29. $x \leq -2$, $x \geq 1$ **31.** $-1 \leq x \leq 0$

33. $x > 1.52$ **35.** $-1.39 < x < -0.43$

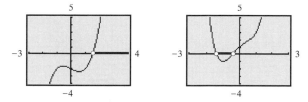

37. $x < -1.69$, $x > 2.00$

39. $x < -4.43$, $-3.11 < x < -1.08$, $x > 3.15$

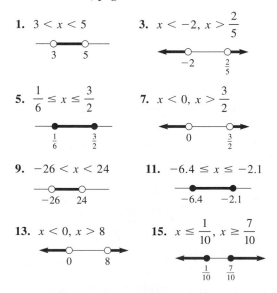

41. $0.5 < i < 1$ A   **43.** $x \leq 12$

45. $3.0 \leq w < 5.0$ mm   **47.** $0 \leq t < 0.92$ h

Exercises 17-4, page 467

1. $3 < x < 5$ **3.** $x < -2$, $x > \dfrac{2}{5}$

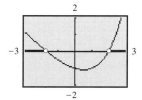

5. $\dfrac{1}{6} \leq x \leq \dfrac{3}{2}$ **7.** $x < 0$, $x > \dfrac{3}{2}$

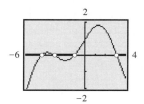

9. $-26 < x < 24$ **11.** $-6.4 \leq x \leq -2.1$

13. $x < 0$, $x > 8$ **15.** $x \leq \dfrac{1}{10}$, $x \geq \dfrac{7}{10}$

17. $-15 < R < \dfrac{35}{3}$ **19.** $x \leq 8.4$, $x \geq 17.6$

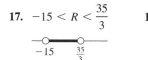

21. $1 < x < 4$ **23.** $x \leq 6$, $x \geq 10$

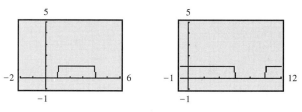

25. $x < -3$, $-2 < x < 1$, $x > 2$

27. $-3 < x < -2$, $1 < x < 2$

29. $|p - 2{,}000{,}000| \leq 200{,}000$; production is at least 1,800,000 barrels, but not greater than 2,200,000 barrels.

31. $0.020 < i < 0.040$ A

Exercises 17-5, page 472

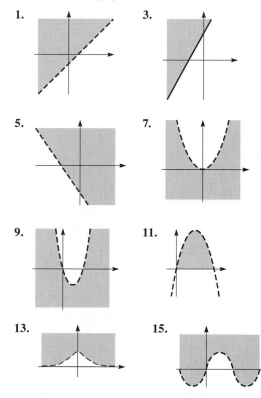

17. **19.**

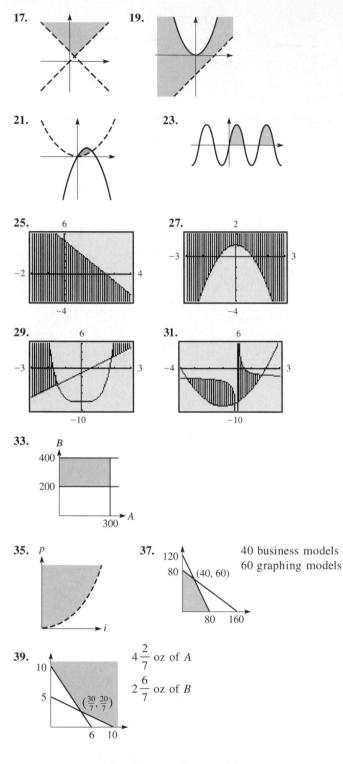

21. **23.**

25. **27.**

29. **31.**

33.

35. *p* **37.** 40 business models
60 graphing models

39. $4\frac{2}{7}$ oz of A

$2\frac{6}{7}$ oz of B

Review Exercises for Chapter 17, page 473

1. $x > 6$ **3.** $\frac{5}{2} < x < 6$

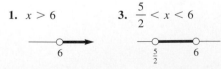

5. $-2 < x < \frac{1}{5}$ **7.** $x < -\frac{7}{3}, x > \frac{5}{2}$

9. $x < -4, \frac{1}{2} < x < 3$ **11.** $x < 0, x > \frac{1}{2}$

13. $-2 \le x \le \frac{2}{3}$ **15.** $x < -\frac{4}{5}, x > 2$

17. $x > \frac{5}{3}$ **19.** $2 \le n < \frac{11}{4}$

21. $R < -\frac{1}{2}, R \ge 8$ **23.** $x < -1, x > 5$

25. $x < -0.68$ **27.** $x < 0.69$

29. **31.**

33. **35.**

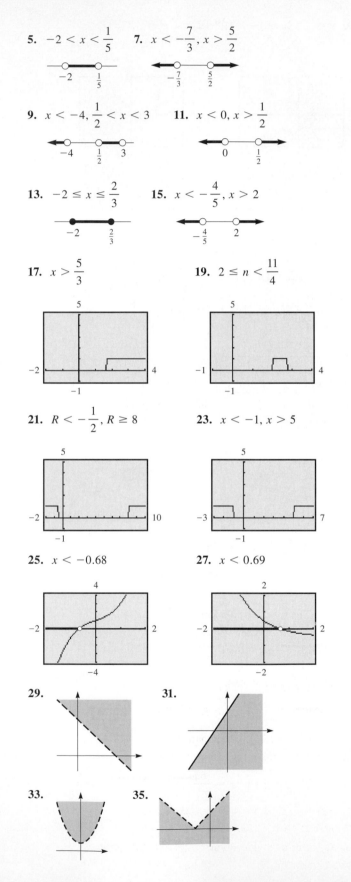

37. **39.**

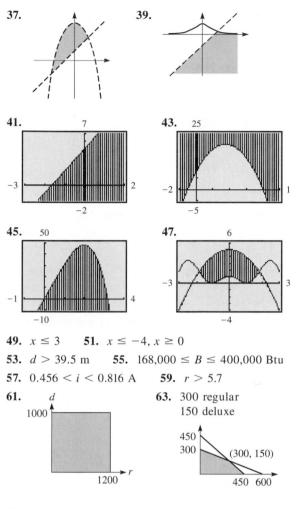

41. 7 **43.** 25

45. 50 **47.** 6

49. $x \leq 3$ **51.** $x \leq -4, x \geq 0$
53. $d > 39.5$ m **55.** $168{,}000 \leq B \leq 400{,}000$ Btu
57. $0.456 < i < 0.816$ A **59.** $r > 5.7$
61. d **63.** 300 regular
150 deluxe

(300, 150)

Exercises 18-1, page 478

1. 6 **3.** $\dfrac{4}{3}$ **5.** 2 **7.** 40 **9.** $0.27 = 27\%$
11. 0.41 **13.** 5.25 **15.** 7.80 **17.** 3.5%
19. 863 kg **21.** 1.44 Ω **23.** 0.103 m³ **25.** 5.63 in.²
27. 0.335 hp **29.** 8.06 km **31.** 12.5 m/s
33. 23,400 cm³ **35.** 19.67 kg **37.** 17,500 chips
39. 4200 lines, 4900 lines

Exercises 18-2, page 484

1. $y = kz$ **3.** $s = \dfrac{k}{t^2}$ **5.** $f = k\sqrt{x}$ **7.** $w = kxy^3$
9. A varies directly as the square of r.
11. n varies as the square root of t and inversely as u.
13. $V = \dfrac{H^2}{2048}$ **15.** $p = \dfrac{16q}{r^3}$ **17.** 25 **19.** 50
21. 180 **23.** 2.56×10^5 **25.** 61 ft³
27. $m = 0.476t$ **29.** 2.5 **31.** 1.4 h
33. (a) inverse (b) $a = 60/m$ **35.** 10,200 hp

37. $F = 0.00523Av^2$ **39.** 480 m/s
41. $R = \dfrac{4.44 \times 10^{-5}l}{A}$ **43.** 80.0 W **45.** $G = \dfrac{5.9d^2}{\lambda^2}$
47. -6.57 ft/s²

Review Exercises for Chapter 18, page 486

1. 200 **3.** $\dfrac{2}{5}$ **5.** 5.6 **7.** 7.39 lb/in.² **9.** 92 mi
11. 2.82×10^6 J = 2.82 MJ **13.** 36,000 characters
15. 310 mL **17.** 71.9 ft **19.** 4500, 7500 **21.** $y = 3x^2$
23. $v = \dfrac{128x}{y^3}$ **25.** 11.7 in. **27.** 4.3 μC **29.** 22.5 hp
31. 144 ft **33.** 1.4 **35.** 48.7 Hz **37.** 2.99×10^8 m/s
39. 3.26 cm **41.** 150 ft **43.** 5.73×10^4 m
45. 150% **47.** $125.00, $600.13 **49.** 4800 Btu
51. 0.023 W/m²

Exercises 19-1, page 494

1. 4, 6, 8, 10, 12 **3.** $\dfrac{13}{2}, \dfrac{9}{2}, \dfrac{5}{2}, \dfrac{1}{2}, -\dfrac{3}{2}$ **5.** 22
7. $-\dfrac{9\pi}{4}$ **9.** 309 **11.** $49b$ **13.** 440 **15.** $-\dfrac{85}{2}$
17. $n = 6, S_6 = 150$ **19.** $d = -\dfrac{2}{19}, a_{20} = -\dfrac{1}{3}$
21. $a_1 = 19, a_{30} = 106$ **23.** $n = 62, S_{62} = -4867$
25. $n = 23, a_{23} = 6k$ **27.** $n = 8, d = \dfrac{1}{14}(b + 2c)$
29. $a_1 = 36, d = 4, S_{10} = 540$ **31.** yes, $d = \ln 2$
33. 5050 **35.** 112 **37.** 2700 m² **39.** 10 rows
41. 12 years, $11,700 **43.** 490 m
45. $S_n = \dfrac{1}{2}n[2a_1 + (n - 1)d]$ **47.** $a_1 = 1, a_n = n$

Exercises 19-2, page 497

1. $45, 15, 5, \dfrac{5}{3}, \dfrac{5}{9}$ **3.** $\dfrac{1}{6}, \dfrac{1}{2}, \dfrac{3}{2}, \dfrac{9}{2}, \dfrac{27}{2}$ **5.** 16 **7.** $\dfrac{1}{125}$
9. $\dfrac{1}{729}$ **11.** 1 **13.** $\dfrac{341}{8}$ **15.** 378
17. $a_6 = 64, S_6 = \dfrac{1365}{16}$ **19.** $a_1 = 16, a_5 = 81$
21. $a_1 = 1, r = 3$ **23.** $n = 7, S_n = \dfrac{58593}{625}$
25. yes; $r = 3^x; a_{20} = 3^{19x+1}$ **27.** 1.4% **29.** 1.09 mA
31. $443.96 **33.** 462 cm **35.** 21.1°C
37. $671,088.64 **39.** $S_n = \dfrac{a_1 - ra_n}{1 - r}$

Exercises 19-3, page 501

1. 8   **3.** $r = \dfrac{1}{5}$ **5.** $\dfrac{400}{21}$ **7.** 8   **9.** $\dfrac{10,000}{9999}$

11. $\dfrac{1}{2}(5 + 3\sqrt{3})$ **13.** $\dfrac{1}{3}$ **15.** $\dfrac{40}{99}$ **17.** $\dfrac{2}{11}$ **19.** $\dfrac{91}{333}$

21. $\dfrac{11}{30}$ **23.** $\dfrac{100,741}{999,000}$ **25.** 350 gal   **27.** 346 g

Exercises 19-4, page 505

1. $t^3 + 3t^2 + 3t + 1$ **3.** $16x^4 - 32x^3 + 24x^2 - 8x + 1$

5. 40.84101

7. $64a^6 - 192a^5b^2 + 240a^4b^4 - 160a^3b^6 + 60a^2b^8 - 12ab^{10} + b^{12}$

9. $625x^4 - 1500x^3 + 1350x^2 - 540x + 81$

11. $64a^6 + 192a^5 + 240a^4 + 160a^3 + 60a^2 + 12a + 1$

13. $x^{10} + 20x^9 + 180x^8 + 960x^7 + \cdots$

15. $128a^7 - 448a^6 + 672a^5 - 560a^4 + \cdots$

17. $x^6 - 12x^{11/2}y + 66x^5y^2 - 220x^{9/2}y^3 + \cdots$

19. $b^{40} + 10b^{37} + \dfrac{95}{2}b^{34} + \dfrac{285}{2}b^{31} + \cdots$

21. $1 + 8x + 28x^2 + 56x^3 + \cdots$

23. $1 + 2x + 3x^2 + 4x^3 + \cdots$

25. $1 + \dfrac{1}{2}x - \dfrac{1}{8}x^2 + \dfrac{1}{16}x^3 - \cdots$

27. $\dfrac{1}{3}\left[1 + \dfrac{1}{2}x + \dfrac{3}{8}x^2 + \dfrac{5}{16}x^3 + \cdots \right]$

29. (a) 3.557×10^{14} (b) 5.109×10^{19} (c) 8.536×10^{15} (d) 2.480×10^{96}

31. $n! = n(n - 1)(n - 2)(\cdots)(2)(1) = n \times (n - 1)!$; for $n = 1$, $1! = 1 \times 0!$ Since $1! = 1$, $0!$ must $= 1$.

33. $56a^3b^5$ **35.** $10,264,320x^8b^4$

37. $V = A(1 - 5r + 10r^2 - 10r^3 + 5r^4 - r^5)$

39. $1 - \dfrac{x}{a} + \dfrac{x^3}{2a^3} - \cdots$

Review Exercises for Chapter 19, page 507

1. 81   **3.** 1.28×10^{-6} **5.** $-\dfrac{119}{2}$ **7.** $\dfrac{16}{243}$

9. $\dfrac{195}{2}$ **11.** $\dfrac{16383}{1536}$ **13.** 81   **15.** 32

17. -1.5 **19.** -0.25 **21.** 186

23. $\dfrac{455}{2}$ (as), 127 (gs), or 43 (gs)   **25.** 27   **27.** 51

29. $\dfrac{1}{33}$ **31.** $\dfrac{4}{55}$ **33.** $x^4 - 8x^3 + 24x^2 - 32x + 16$

35. $x^{10} + 5x^8 + 10x^6 + 10x^4 + 5x^2 + 1$

37. $a^{10} + 20a^9b^2 + 180a^8b^4 + 960a^7b^6 + \cdots$

39. $p^{18} - \dfrac{3}{2}p^{16}q + p^{14}q^2 - \dfrac{7}{18}p^{12}q^3 + \cdots$

41. $1 + 12x + 66x^2 + 220x^3 + \cdots$

43. $1 + \dfrac{1}{2}x^2 - \dfrac{1}{8}x^4 + \dfrac{1}{16}x^6 - \cdots$

45. $1 - \dfrac{1}{2}a^2 - \dfrac{1}{8}a^4 - \dfrac{1}{16}a^6 - \cdots$

47. $\dfrac{1}{8} + \dfrac{3}{4}x + 3x^2 + 10x^3 + \cdots$ **49.** $1,001,000$

51. 11th   **53.** 12.6 mm   **55.** 302.9 in.   **57.** \$4700

59. 4.4×10^9 in. = 69,000 mi   **61.** \$47,340.80

63. \$6.93   **65.** $1 + \dfrac{1}{2}am^2 + \dfrac{1}{8}am^4$ **67.** 21 years

69. 5 applications   **71.** yes; term to term ratios are equal

Exercises 20-1, page 515

(*Note:* "Answers" to trigonometric identities are intermediate steps of suggested reductions of the left member.)

1. $1.483 = \dfrac{1}{0.6745}$

3. $\left(-\dfrac{1}{2}\sqrt{3}\right)^2 + \left(-\dfrac{1}{2}\right)^2 = \dfrac{3}{4} + \dfrac{1}{4} = 1$

5. $\dfrac{\cos \theta}{\sin \theta}\left(\dfrac{1}{\cos \theta}\right) = \dfrac{1}{\sin \theta}$

7. $\dfrac{\sin x}{\dfrac{\sin x}{\cos x}} = \dfrac{\sin x}{1}\left(\dfrac{\cos x}{\sin x}\right)$ **9.** $\sin y\left(\dfrac{\cos y}{\sin y}\right)$

11. $\sin x\left(\dfrac{1}{\cos x}\right)$ **13.** $\csc^2 x\,(\sin^2 x)$

15. $\sin x\,(\csc^2 x) = (\sin x)(\csc x)(\csc x)$
$= \sin x\left(\dfrac{1}{\sin x}\right)\csc x$

17. $\tan y \cot y + \tan^2 y = 1 + \tan^2 y$

19. $\sin x\left(\dfrac{\sin x}{\cos x}\right) + \cos x = \dfrac{\sin^2 x + \cos^2 x}{\cos x} = \dfrac{1}{\cos x}$

21. $\cos \theta\left(\dfrac{\cos \theta}{\sin \theta}\right) + \sin \theta = \dfrac{\cos^2 \theta + \sin^2 \theta}{\sin \theta} = \dfrac{1}{\sin \theta}$

23. $\cot \theta\,(\sec^2 \theta - 1) = \cot \theta \tan^2 \theta = (\cot \theta \tan \theta)\tan \theta$

25. $\dfrac{\sin x}{\cos x} + \dfrac{\cos x}{\sin x} = \dfrac{\sin^2 x + \cos^2 x}{\cos x \sin x} = \dfrac{1}{\cos x \sin x}$

27. $(1 - \sin^2 x) - \sin^2 x$

29. $\dfrac{\sin x\,(1 + \cos x)}{1 - \cos^2 x} = \dfrac{1 + \cos x}{\sin x}$

31. $\dfrac{\sin^2 x}{\cos^2 x}\cos^2 x + \dfrac{\cos^2 x}{\sin^2 x}\sin^2 x = \sin^2 x + \cos^2 x$

33. $\dfrac{\sec\theta}{\dfrac{1}{\sec\theta}} - \dfrac{\tan\theta}{\dfrac{1}{\tan\theta}} = \sec^2\theta - \tan^2\theta$

35. $(2\sin^2 x - 1)(\sin^2 x - 1)$

37. $\dfrac{\sin\pi t}{2}\left(\dfrac{\sin^2\pi t + (1 - \cos\pi t)^2}{(1 - \cos\pi t)\sin\pi t}\right)$

$= \dfrac{\sin^2\pi t + 1 - 2\cos\pi t + \cos^2\pi t}{2(1 - \cos\pi t)} = \dfrac{2(1 - \cos\pi t)}{2(1 - \cos\pi t)}$

39. Infinite series: $\dfrac{1}{1 - \sin^2 x} = \dfrac{1}{\cos^2 x}$ **41.** $\cot x$

43. $\sin x$ **45.** $\sec x$ **47.** $\cos x$

49.   **51.**   **53.**   **55.**

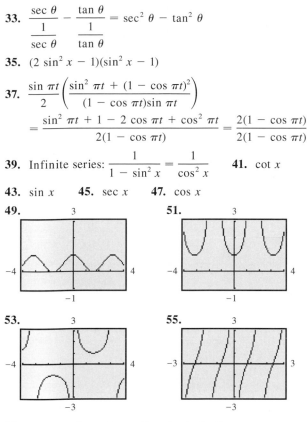

57. $0 = \cos A \cos B \cos C + \sin A \sin B,$

$\cos C = -\dfrac{\sin A \sin B}{\cos A \cos B}$

59. $l = a\csc\theta + a\sec\theta = a\left(\dfrac{1}{\sin\theta} + \dfrac{\tan\theta}{\sin\theta}\right)$

61. $\sin^2 x - \sin^2 x \sec^2 x + \cos^2 x + \cos^2 x \sec^4 x$
$= \sin^2 x - \tan^2 x + \cos^2 x + \sec^2 x = 2$

63. $\left(\dfrac{r}{x}\right)^2 + \left(\dfrac{r}{y}\right)^2 = \dfrac{r^2(x^2 + y^2)}{x^2 y^2}$

65. $\sqrt{1 - \cos^2\theta} = \sqrt{\sin^2\theta}$

67. $\sqrt{4 + 4\tan^2\theta} = 2\sqrt{1 + \tan^2\theta}$

Exercises 20-2, page 520

1. $\sin 105° = \sin 60° \cos 45° + \cos 60° \sin 45°$
$= \dfrac{\sqrt{3}}{2}\dfrac{\sqrt{2}}{2} + \dfrac{1}{2}\dfrac{\sqrt{2}}{2} = 0.9659$

3. $\cos 15° = \cos(60° - 45°)$
$= \cos 60° \cos 45° + \sin 60° \sin 45°$
$= \left(\dfrac{1}{2}\right)\left(\dfrac{1}{2}\sqrt{2}\right) + \left(\dfrac{1}{2}\sqrt{3}\right)\left(\dfrac{1}{2}\sqrt{2}\right)$
$= \dfrac{1}{4}\sqrt{2} + \dfrac{1}{4}\sqrt{6} = \dfrac{1}{4}(\sqrt{2} + \sqrt{6}) = 0.9659$

5. $-\dfrac{33}{65}$ **7.** $-\dfrac{56}{65}$ **9.** $\sin 3x$ **11.** $\cos x$

13. $\tan x$ **15.** 0 **17.** 1 **19.** 1

21. $\sin(180° - x) = \sin 180° \cos x - \cos 180° \sin x$
$= (0)\cos x - (-1)\sin x$

23. $\cos(0° - x) = \cos 0° \cos x + \sin 0° \sin x$
$= (1)\cos x + (0)\sin x$

25. $\tan(180° + x) = \dfrac{\tan 180° + \tan x}{1 - \tan 180° \tan x} = \tan x$

27. $\cos\left(\dfrac{\pi}{3} + x\right) = \cos\dfrac{\pi}{3}\cos x - \sin\dfrac{\pi}{3}\sin x$
$= \dfrac{1}{2}\cos x - \dfrac{1}{2}\sqrt{3}\sin x$

29. $(\sin x \cos y + \cos x \sin y)(\sin x \cos y - \cos x \sin y)$
$= \sin^2 x \cos^2 y - \cos^2 x \sin^2 y$
$= \sin^2 x(1 - \sin^2 y) - (1 - \sin^2 x)\sin^2 y$

31. $(\cos\alpha \cos\beta - \sin\alpha \sin\beta)$
$+ (\cos\alpha \cos\beta + \sin\alpha \sin\beta)$

33.   **35.**

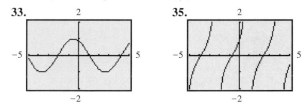

37, 39, 41, 43. Use the indicated method.

45. $i_0 \sin(\omega t + \alpha) = i_0(\sin\omega t \cos\alpha + \cos\omega t \sin\alpha)$

47. $\tan\alpha(R + \cos\beta) = \sin\beta, \ R = \dfrac{\sin\beta - \tan\alpha\cos\beta}{\tan\alpha}$
$= \dfrac{\sin\beta\cos\alpha - \cos\beta\sin\alpha}{\cos\alpha\tan\alpha}$

Exercises 20-3, page 524

1. $\sin 60° = \sin 2(30°) = 2\sin 30° \cos 30°$
$= 2\left(\dfrac{1}{2}\right)\left(\dfrac{1}{2}\sqrt{3}\right) = \dfrac{1}{2}\sqrt{3}$

3. $\tan 120° = \dfrac{2\tan 60°}{1 - \tan^2 60°} = \dfrac{2\sqrt{3}}{1 - (\sqrt{3})^2} = -\sqrt{3}$

5. $\sin 258° = 2\sin 129° \cos 129° = -0.9781476$

7. $\cos 96° = \cos^2 48° - \sin^2 48° = -0.1045285$

9. $\dfrac{24}{25}$ **11.** $-\sqrt{3}$ **13.** $2\sin 8x$ **15.** $\cos 8x$

17. $\cos x$ **19.** $-2\cos 4x$ **21.** $\cos^2\alpha - (1 - \cos^2\alpha)$

23. $\dfrac{\cos x - (\sin x/\cos x)\sin x}{1/\cos x} = \cos^2 x - \sin^2 x$

25. $\dfrac{2\sin 2\theta \cos 2\theta}{\sin 2\theta}$ **27.** $\dfrac{2\sin\theta\cos\theta}{1 + 2\cos^2\theta - 1} = \dfrac{\sin\theta}{\cos\theta}$

29. $1 - (1 - 2\sin^2 2\theta) = \dfrac{2}{\csc^2\theta}$

31. $\dfrac{\sin 3x \cos x - \cos 3x \sin x}{\sin x \cos x} = \dfrac{\sin 2x}{\dfrac{1}{2}\sin 2x}$

33.

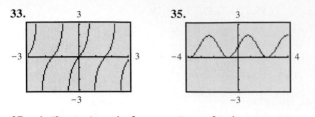

35.

37. $\sin(2x + x) = \sin 2x \cos x + \cos 2x \sin x$
$= (2 \sin x \cos x)(\cos x) + (\cos^2 x - \sin^2 x)\sin x$

39. 474 m   **41.** $R = v\left(\dfrac{2v \sin \alpha}{g}\right)\cos \alpha = \dfrac{v^2(2 \sin \alpha \cos \alpha)}{g}$

43. $vi \sin \omega t \sin\left(\omega t - \dfrac{\pi}{2}\right)$

$= vi \sin \omega t\left(\sin \omega t \cos \dfrac{\pi}{2} - \cos \omega t \sin \dfrac{\pi}{2}\right)$

$= vi \sin \omega t[-(\cos \omega t)(1)] = -\dfrac{1}{2}vi(2 \sin \omega t \cos \omega t)$

Exercises 20-4, page 528

1. $\cos 15° = \cos \dfrac{1}{2}(30°) = \sqrt{\dfrac{1 + \cos 30°}{2}} = \sqrt{\dfrac{1.8660}{2}}$
$= 0.9659$

3. $\sin 75° = \sin \dfrac{1}{2}(150°) = \sqrt{\dfrac{1 - \cos 150°}{2}}$
$= \sqrt{\dfrac{1.8660}{2}} = 0.9659$

5. $\sin 118° = 0.8829476$

7. $\sqrt{2\left(\dfrac{1 + \cos 164°}{2}\right)} = \sqrt{2} \cos 82° = 0.1968205$

9. $\sin 3x$ **11.** $4 \cos 2x$ **13.** $\dfrac{1}{26}\sqrt{26}$

15. 0.1414   **17.** $\pm\sqrt{\dfrac{2}{1 - \cos \alpha}}$

19. $\tan \dfrac{1}{2}\alpha = \dfrac{1 - \cos \alpha}{\sin \alpha} = \dfrac{\sin \alpha}{1 + \cos \alpha}$

21. $\dfrac{1 - \cos \alpha}{2 \sin \dfrac{1}{2}\alpha} = \dfrac{1 - \cos \alpha}{2\sqrt{\dfrac{1}{2}(1 - \cos \alpha)}} = \sqrt{\dfrac{1 - \cos \alpha}{2}}$

23. $2\left(\dfrac{1 - \cos x}{2}\right) + \cos x$

25. $\sqrt{\dfrac{(1 + \cos \theta)(1 - \cos \theta)}{2(1 - \cos \theta)}} = \dfrac{\sin \theta}{\sqrt{4\left(\dfrac{1 - \cos \theta}{2}\right)}}$

27.

29. $\sin \omega t = \pm\sqrt{\dfrac{1 - \cos 2\omega t}{2}},$

$\sin^2 \omega t = \dfrac{1}{2}(1 - \cos 2\omega t)$

31. $\dfrac{\sqrt{\dfrac{1 - \cos(A + \phi)}{2}}}{\sqrt{\dfrac{1 - \cos A}{2}}} = \sqrt{\dfrac{1 - \cos(A + \phi)}{1 - \cos A}}$

Exercises 20-5, page 533

1. $\dfrac{\pi}{2}$ **3.** π **5.** $\dfrac{\pi}{3}, \dfrac{2\pi}{3}, \dfrac{4\pi}{3}, \dfrac{5\pi}{3}$

7. $0, \dfrac{\pi}{6}, \dfrac{5\pi}{6}, \pi$ **9.** $\dfrac{\pi}{2}, \dfrac{3\pi}{2}$

11. $\dfrac{\pi}{4}, \dfrac{3\pi}{4}, \dfrac{5\pi}{4}, \dfrac{7\pi}{4}$ **13.** 0.2618, 1.309, 3.403, 4.451

15. $0, \pi$ **17.** $\dfrac{3\pi}{4} = 2.36, \dfrac{7\pi}{4} = 5.50$

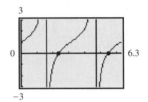

19. 1.9823, 4.3009

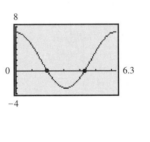

21. $\dfrac{\pi}{3} = 1.05, \dfrac{2\pi}{3} = 2.09,$
$\dfrac{4\pi}{3} = 4.19, \dfrac{5\pi}{3} = 5.24$

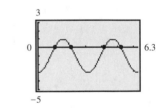

23. $\dfrac{\pi}{12} = 0.26, \dfrac{\pi}{4} = 0.79, \dfrac{5\pi}{12} = 1.31, \dfrac{3\pi}{4} = 2.36,$
$\dfrac{13\pi}{12} = 3.40, \dfrac{5\pi}{4} = 3.93, \dfrac{17\pi}{12} = 4.45, \dfrac{7\pi}{4} = 5.50$

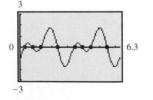

25. $0 = 0.00, \dfrac{\pi}{3} = 1.05, \pi = 3.14, \dfrac{5\pi}{3} = 5.24$

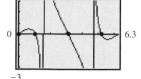

27. 3.569, 5.856

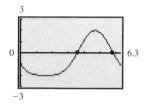

29. 0.7854, 1.249, 3.927, 4.391

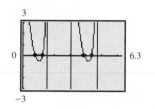

31. $\dfrac{3\pi}{8} = 1.18, \dfrac{7\pi}{8} = 2.75, \dfrac{11\pi}{8} = 4.32, \dfrac{15\pi}{8} = 5.89$

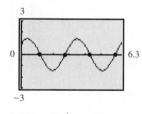

33. 6.56×10^{-4} **35.** 10.2 s, 15.7 s, 21.2 s, 47.1 s
37. −2.28, 0.00, 2.28   **39.** 0.29, 0.95

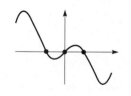

41. 2.10   **43.** 1.08

Exercises 20-6, page 538

1. y is the angle whose tangent is x.

3. y is the angle whose cotangent is $3x$.

5. y is twice the angle whose sine is x.

7. y is 5 times the angle whose cosine is $2x - 1$.

9. $\dfrac{\pi}{3}$ **11.** $\dfrac{\pi}{4}$ **13.** $-\dfrac{\pi}{3}$ **15.** $\dfrac{\pi}{3}$

17. $-\dfrac{\pi}{4}$ **19.** $\dfrac{\pi}{4}$ **21.** 0   **23.** $\dfrac{1}{2}\sqrt{3}$

25. $\dfrac{1}{2}\sqrt{2}$ **27.** −1   **29.** $2/\sqrt{21}$ **31.** $2/\sqrt{3}$

33. −1.3090   **35.** −0.9838   **37.** 1.4413   **39.** 1.4503

41. −1.2389   **43.** −0.2239   **45.** $x = \dfrac{1}{3}\sin^{-1} y$

47. $x = 4\tan y$ **49.** $x = \dfrac{1}{3}\sec^{-1}\!\left(\dfrac{y-1}{3}\right)$

51. $x = 1 - \cos(1 - y)$ **53.** $\dfrac{x}{\sqrt{1-x^2}}$ **55.** $\dfrac{1}{x}$

57. $\dfrac{3x}{\sqrt{9x^2-1}}$ **59.** $2x\sqrt{1-x^2}$

61. $t = \dfrac{1}{2\omega}\cos^{-1}\dfrac{y}{A} - \dfrac{\phi}{\omega}$ **63.** $t = \dfrac{1}{\omega}\!\left(\sin^{-1}\dfrac{i}{I_m} - \alpha - \phi\right)$

65. $\sin\!\left(\sin^{-1}\dfrac{3}{5} + \sin^{-1}\dfrac{5}{13}\right) = \dfrac{3}{5}\cdot\dfrac{12}{13} + \dfrac{4}{5}\cdot\dfrac{5}{13} = \dfrac{56}{65}$

67. $\dfrac{\pi}{6} + \dfrac{\pi}{3} = \dfrac{\pi}{2}$ **69.** $\sin^{-1}\!\left(\dfrac{a}{c}\right)$

71. Let y = height to top of pedestal; $\tan\alpha = \dfrac{151 + y}{d}$,

$\tan\beta = \dfrac{y}{d}$; $\tan\alpha = \dfrac{151 + d\tan\beta}{d}$

Review Exercises for Chapter 20, page 540

1. $\sin(90° + 30°) = \sin 90°\cos 30° + \cos 90°\sin 30°$

$= (1)\left(\dfrac{1}{2}\sqrt{3}\right) + (0)\left(\dfrac{1}{2}\right) = \dfrac{1}{2}\sqrt{3}$

3. $\sin(180° - 45°) = \sin 180°\cos 45° - \cos 180°\sin 45°$

$= 0\left(\dfrac{1}{2}\sqrt{2}\right) - (-1)\left(\dfrac{1}{2}\sqrt{2}\right) = \dfrac{1}{2}\sqrt{2}$

5. $\cos 2(90°) = \cos^2 90° - \sin^2 90° = 0 - 1 = -1$

7. $\tan 2(30°) = \dfrac{2\tan 30°}{1 - \tan^2 30°} = \dfrac{2(\sqrt{3}/3)}{1 - (\sqrt{3}/3)^2} = \sqrt{3}$

9. $\sin 52° = 0.7880108$ **11.** $\sin 92° = 0.9993908$

13. $\cos(-69°) = 0.3583679$ **15.** $2\tan 24° = 0.8904574$

17. $\sin 5x$ **19.** $4\sin 12x$ **21.** $2\cos 12x$ **23.** $2\cos x$

25. $-\dfrac{\pi}{2}$ **27.** 0.2619   **29.** $-\dfrac{1}{3}\sqrt{3}$ **31.** 0

33. $\dfrac{\dfrac{1}{\cos y}}{\dfrac{1}{\sin y}} = \dfrac{1}{\cos y} \cdot \dfrac{\sin y}{1}$

35. $\sin x \csc x - \sin^2 x = 1 - \sin^2 x$

37. $\dfrac{1 - \sin^2 \theta}{\sin \theta} = \dfrac{\cos^2 \theta}{\sin \theta}$

39. $\cos \theta \left(\dfrac{\cos \theta}{\sin \theta} \right) + \sin \theta = \dfrac{\cos^2 \theta + \sin^2 \theta}{\sin \theta}$

41. $\dfrac{(\sec^2 x - 1)(\sec^2 x + 1)}{\tan^2 x} = \sec^2 x + 1$

43. $2\left(\dfrac{1}{\sin 2x} \right)\left(\dfrac{\cos x}{\sin x} \right) = 2\left(\dfrac{1}{2 \sin x \cos x} \right)\left(\dfrac{\cos x}{\sin x} \right)$
$= \dfrac{1}{\sin^2 x}$

45. $\dfrac{\cos^2 \theta}{\sin^2 \theta}$ **47.** $\dfrac{\cos^2 \theta - \sin^2 \theta}{\cos^2 \theta} = 1 - \dfrac{\sin^2 \theta}{\cos^2 \theta}$

49. $\dfrac{1}{2}\left(2 \sin \dfrac{\theta}{2} \cos \dfrac{\theta}{2} \right)$

51. $\dfrac{1}{\cos x} + \dfrac{\sin x}{\cos x} = \dfrac{(1 + \sin x)(1 - \sin x)}{\cos x(1 - \sin x)}$
$= \dfrac{1 - \sin^2 x}{\cos x(1 - \sin x)}$

53.

55.

57.

59.

61. $x = \dfrac{1}{2} \cos^{-1} \dfrac{1}{2} y$ **63.** $x = \dfrac{1}{5} \sin \dfrac{1}{3}\left(\dfrac{1}{4}\pi - y \right)$

65. 1.2925, 4.4341   **67.** $\dfrac{\pi}{6}, \dfrac{5\pi}{6}, \dfrac{7\pi}{6}, \dfrac{11\pi}{6}$

69. $0, \dfrac{\pi}{2}, \pi, \dfrac{3\pi}{2}$ **71.** $\dfrac{\pi}{6}, \dfrac{5\pi}{6}, \dfrac{7\pi}{6}, \dfrac{11\pi}{6}$

73. 0.3142, 1.571, 2.827, 4.084, 4.712, 5.341   **75.** 0

77. 1.56, 2.16, 3.46   **79.** $-2.31, 1.14$

81. $\dfrac{1}{x}$ **83.** $2x\sqrt{1 - x^2}$ **85.** $2\sqrt{1 - \cos^2 \theta}$

87. $\dfrac{\tan \theta}{\sqrt{1 + \tan^2 \theta}} = \dfrac{\tan \theta}{\sec \theta}$ **89.** $\pi < x < 2\pi$

91. $C\left(\dfrac{A}{C} \sin 2t + \dfrac{B}{C} \cos 2t \right)$
$= C(\cos \alpha \sin 2t + \sin \alpha \cos 2t)$

93. $R = \sqrt{(A \cos \theta - B \sin \theta)^2 + (A \sin \theta + B \cos \theta)^2}$
$= \sqrt{A^2(\cos^2 \theta + \sin^2 \theta) + B^2(\sin^2 \theta + \cos^2 \theta)}$

95. $\dfrac{k}{2} \cdot \dfrac{1}{\sin^2\left(\dfrac{\theta}{2} \right)} = \dfrac{k}{2} \cdot \dfrac{1}{\left(\dfrac{1 - \cos \theta}{2} \right)}$

97. $\theta = \alpha + R \sin \omega t$ **99.** $\dfrac{\cos 2\alpha}{2 \cos^2 \alpha}$ **101.** 13.6 ft

103. 54.7°

Exercises 21-1, page 547

1. $2\sqrt{29}$ **3.** 3   **5.** 55   **7.** $2\sqrt{53}$ **9.** 2.86

11. $\dfrac{5}{2}$ **13.** undefined   **15.** $-\dfrac{3}{4}$ **17.** $-\dfrac{5}{9}$

19. 0.747   **21.** $\dfrac{1}{3}\sqrt{3}$ **23.** -1.084 **25.** 20.0°

27. 98.50°   **29.** parallel   **31.** perpendicular

33. $8, -2$ **35.** -3 **37.** two sides equal $2\sqrt{10}$

39. $m_1 = \dfrac{5}{12}, m_2 = \dfrac{4}{3}$ **41.** 10   **43.** $4\sqrt{10} + 4\sqrt{2} = 18.3$

45. $(1, 5)$ **47.** $(-2.8, 4.2)$

Exercises 21-2, page 552

1. $4x - y + 20 = 0$ **3.** $7x - 2y - 24 = 0$

5. $x - y + 2 = 0$ **7.** $y = -2.7$

9. $x = -3$

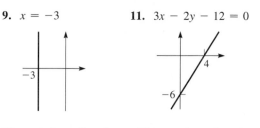

11. $3x - 2y - 12 = 0$

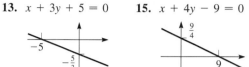

13. $x + 3y + 5 = 0$

15. $x + 4y - 9 = 0$

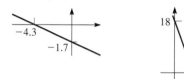

17. $0.4x + y + 1.7 = 0$

19. $3x + y - 18 = 0$

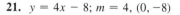

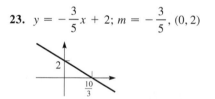

21. $y = 4x - 8$; $m = 4$, $(0, -8)$

23. $y = -\dfrac{3}{5}x + 2$; $m = -\dfrac{3}{5}$, $(0, 2)$

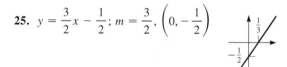

25. $y = \dfrac{3}{2}x - \dfrac{1}{2}$; $m = \dfrac{3}{2}$, $\left(0, -\dfrac{1}{2}\right)$

27. $y = 3.5x + 0.5$; $m = 3.5$, $(0, 0.5)$ **29.** -2

31. The slope of the first line is 3. A line perpendicular to it has a slope of $-\dfrac{1}{3}$. The slope of the second line is $-\dfrac{k}{3}$ so $k = 1$.

33. parallel   **35.** perpendicular   **37.** neither

39. perpendicular

41. $v = 12.2 + 5.16t$ **43.** $v = 0.607T + 331$

45. $5x + 6y = 1220$

47. $y = 150,000 - 0.80x$

49. $y = 10^{-5}(2.4 - 5.6x)$

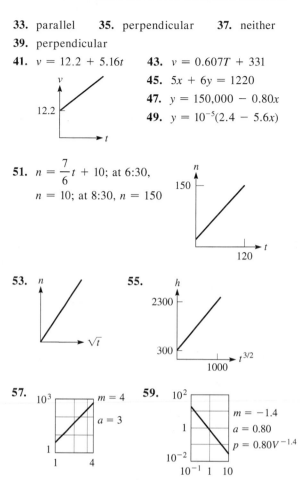

51. $n = \dfrac{7}{6}t + 10$; at 6:30, $n = 10$; at 8:30, $n = 150$

53.

55.

57. $m = 4$ $a = 3$

59. $m = -1.4$ $a = 0.80$ $p = 0.80V^{-1.4}$

Exercises 21-3, page 557

1. $(2, 1)$, $r = 5$ **3.** $(-1, 0)$, $r = 2$ **5.** $x^2 + y^2 = 9$

7. $(x - 2)^2 + (y - 2)^2 = 16$, or $x^2 + y^2 - 4x - 4y - 8 = 0$

9. $(x + 2)^2 + (y - 5)^2 = 5$, or $x^2 + y^2 + 4x - 10y + 24 = 0$

11. $(x - 12)^2 + (y + 15)^2 = 324$, or $x^2 + y^2 - 24x + 30y + 45 = 0$

13. $(x - 2)^2 + (y - 1)^2 = 8$, or $x^2 + y^2 - 4x - 2y - 3 = 0$

15. $(x + 3)^2 + (y - 5)^2 = 25$, or $x^2 + y^2 + 6x - 10y + 9 = 0$

17. $(x - 2)^2 + (y - 2)^2 = 4$, or $x^2 + y^2 - 4x - 4y + 4 = 0$

19. $(x - 2)^2 + (y - 5)^2 = 25$, or $x^2 + y^2 - 4x - 10y + 4 = 0$; and $(x + 2)^2 + (y + 5)^2 = 25$, or $x^2 + y^2 + 4x + 10y + 4 = 0$

21. $(0, 3)$, $r = 2$ **23.** $(-1, 5)$, $r = \dfrac{9}{2}$

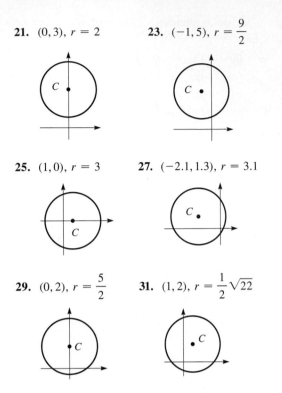

25. $(1, 0)$, $r = 3$ **27.** $(-2.1, 1.3)$, $r = 3.1$

29. $(0, 2)$, $r = \dfrac{5}{2}$ **31.** $(1, 2)$, $r = \dfrac{1}{2}\sqrt{22}$

33. symmetrical to both axes and origin
35. symmetrical to y-axis   **37.** $(7, 0)$, $(-1, 0)$
39. $3x^2 + 3y^2 + 4x + 8y - 20 = 0$, circle

41.

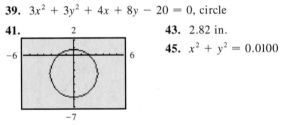

43. 2.82 in.
45. $x^2 + y^2 = 0.0100$

47. $(x - 500 \times 10^{-6})^2 + y^2 = 0.16 \times 10^{-6}$

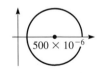

Exercises 21-4, page 562

1. $F(1, 0)$, $x = -1$ **3.** $F(-1, 0)$, $x = 1$

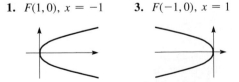

5. $F(0, 2)$, $y = -2$ **7.** $F(0, -1)$, $y = 1$

9. $F\left(\dfrac{5}{8}, 0\right)$, $x = -\dfrac{5}{8}$ **11.** $F\left(0, \dfrac{25}{48}\right)$, $y = -\dfrac{25}{48}$

13. $y^2 = 12x$ **15.** $x^2 = 16y$ **17.** $x^2 = 0.64y$
19. $x^2 = \dfrac{1}{8}y$ **21.** $y^2 - 2y - 12x + 37 = 0$

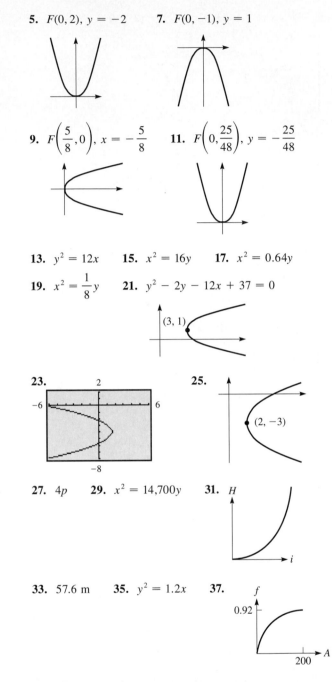

23.   **25.**

27. $4p$ **29.** $x^2 = 14{,}700y$ **31.** H

33. 57.6 m   **35.** $y^2 = 1.2x$ **37.**

39. $y^2 = 8x$ or $x^2 = 8y$ with vertex midway between island and shore

Exercises 21-5, page 567

1. $V(2, 0)$, $V(-2, 0)$,
       $F(\sqrt{3}, 0)$, $F(-\sqrt{3}, 0)$

3. $V(0, 6)$, $V(0, -6)$,
       $F(0, \sqrt{11})$, $F(0, -\sqrt{11})$

5. $V(3,0)$, $V(-3,0)$,
$F(\sqrt{5},0)$, $F(-\sqrt{5},0)$

7. $V(0,7)$, $V(0,-7)$,
$F(0,\sqrt{45})$, $F(0,-\sqrt{45})$

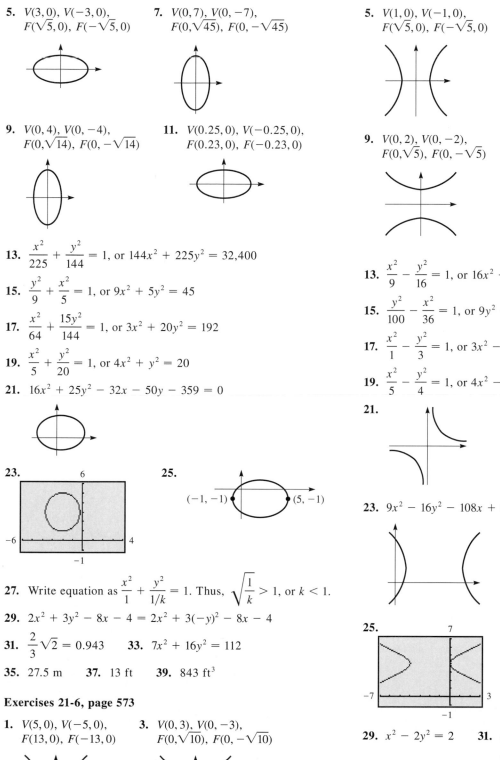

5. $V(1,0)$, $V(-1,0)$,
$F(\sqrt{5},0)$, $F(-\sqrt{5},0)$

7. $V(0,\sqrt{5})$, $V(0,-\sqrt{5})$,
$F(0,\sqrt{7})$, $F(0,-\sqrt{7})$

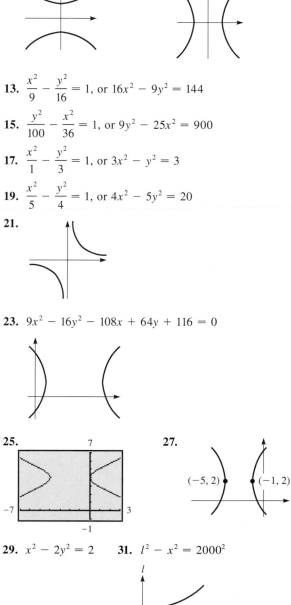

9. $V(0,4)$, $V(0,-4)$,
$F(0,\sqrt{14})$, $F(0,-\sqrt{14})$

11. $V(0.25,0)$, $V(-0.25,0)$,
$F(0.23,0)$, $F(-0.23,0)$

9. $V(0,2)$, $V(0,-2)$,
$F(0,\sqrt{5})$, $F(0,-\sqrt{5})$

11. $V(0.4,0)$, $V(-0.4,0)$,
$F(0.9,0)$, $F(-0.9,0)$

13. $\dfrac{x^2}{225} + \dfrac{y^2}{144} = 1$, or $144x^2 + 225y^2 = 32{,}400$

15. $\dfrac{y^2}{9} + \dfrac{x^2}{5} = 1$, or $9x^2 + 5y^2 = 45$

17. $\dfrac{x^2}{64} + \dfrac{15y^2}{144} = 1$, or $3x^2 + 20y^2 = 192$

19. $\dfrac{x^2}{5} + \dfrac{y^2}{20} = 1$, or $4x^2 + y^2 = 20$

21. $16x^2 + 25y^2 - 32x - 50y - 359 = 0$

23.

25. $(-1,-1)$ $(5,-1)$

27. Write equation as $\dfrac{x^2}{1} + \dfrac{y^2}{1/k} = 1$. Thus, $\sqrt{\dfrac{1}{k}} > 1$, or $k < 1$.

29. $2x^2 + 3y^2 - 8x - 4 = 2x^2 + 3(-y)^2 - 8x - 4$

31. $\dfrac{2}{3}\sqrt{2} = 0.943$ **33.** $7x^2 + 16y^2 = 112$

35. 27.5 m   **37.** 13 ft   **39.** 843 ft^3

Exercises 21-6, page 573

1. $V(5,0)$, $V(-5,0)$,
$F(13,0)$, $F(-13,0)$

3. $V(0,3)$, $V(0,-3)$,
$F(0,\sqrt{10})$, $F(0,-\sqrt{10})$

13. $\dfrac{x^2}{9} - \dfrac{y^2}{16} = 1$, or $16x^2 - 9y^2 = 144$

15. $\dfrac{y^2}{100} - \dfrac{x^2}{36} = 1$, or $9y^2 - 25x^2 = 900$

17. $\dfrac{x^2}{1} - \dfrac{y^2}{3} = 1$, or $3x^2 - y^2 = 3$

19. $\dfrac{x^2}{5} - \dfrac{y^2}{4} = 1$, or $4x^2 - 5y^2 = 20$

21.

23. $9x^2 - 16y^2 - 108x + 64y + 116 = 0$

25.

27. $(-5,2)$ $(-1,2)$

29. $x^2 - 2y^2 = 2$ **31.** $l^2 - x^2 = 2000^2$

33. $i = 6.00/R$

35.

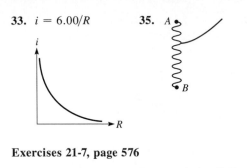

Exercises 21-7, page 576

1. parabola, $(-1, 2)$ **3.** hyperbola, $(1, 2)$

5. ellipse, $(-1, 0)$ **7.** parabola, $(-3, 1)$

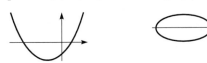

9. $(y - 3)^2 = 16(x + 1)$, or $y^2 - 6y - 16x - 7 = 0$

11. $y^2 = 24(x - 6)$

13. $\dfrac{(x + 2)^2}{25} + \dfrac{(y - 2)^2}{16} = 1$, or
$16x^2 + 25y^2 + 64x - 100y - 236 = 0$

15. $\dfrac{(y - 1)^2}{16} + \dfrac{(x + 2)^2}{4} = 1$, or
$4x^2 + y^2 + 16x - 2y + 1 = 0$

17. $\dfrac{(y - 2)^2}{1} - \dfrac{(x + 1)^2}{3} = 1$, or
$x^2 - 3y^2 + 2x + 12y - 8 = 0$

19. $\dfrac{(x + 1)^2}{9} - \dfrac{(y - 1)^2}{16} = 1$, or
$16x^2 - 9y^2 + 32x + 18y - 137 = 0$

21. parabola, $(-1, -1)$ **23.** ellipse, $(-3, 0)$

25. hyperbola, $(0, 4)$ **27.** parabola, $(1, 0)$

29. hyperbola, $(-4, 5)$ **31.** ellipse, $\left(\dfrac{2}{3}, -2\right)$

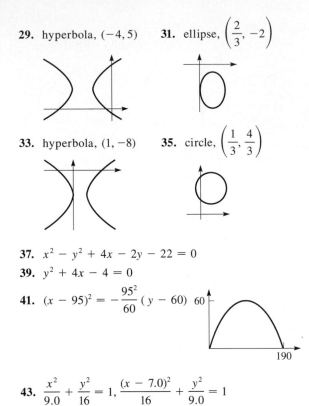

33. hyperbola, $(1, -8)$ **35.** circle, $\left(\dfrac{1}{3}, \dfrac{4}{3}\right)$

37. $x^2 - y^2 + 4x - 2y - 22 = 0$

39. $y^2 + 4x - 4 = 0$

41. $(x - 95)^2 = -\dfrac{95^2}{60}(y - 60)$

43. $\dfrac{x^2}{9.0} + \dfrac{y^2}{16} = 1, \dfrac{(x - 7.0)^2}{16} + \dfrac{y^2}{9.0} = 1$

Exercises 21-8, page 579

1. ellipse   **3.** hyperbola   **5.** circle   **7.** parabola

9. hyperbola   **11.** circle   **13.** none (st. line)

15. hyperbola   **17.** ellipse   **19.** ellipse

21. parabola; $V(-4, 0)$; $F(-4, 2)$

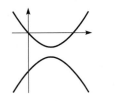

23. hyperbola; $C(1, -2)$;
$V(1, -2 \pm \sqrt{2})$

25. ellipse; $C(5, 0)$;
$V(5, \pm 2\sqrt{2})$

27. parabola; $V\left(-\dfrac{1}{2}, \dfrac{5}{2}\right)$; $F = \left(\dfrac{1}{2}, \dfrac{5}{2}\right)$

29. ellipse

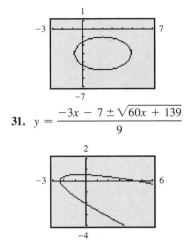

31. $y = \dfrac{-3x - 7 \pm \sqrt{60x + 139}}{9}$

Enter both functions (from quadratic formula).

33. (a) circle (b) hyperbola (c) ellipse

35. a point at the origin   **37.** parabola

39. Circle if light beam is perpendicular to floor; otherwise, an ellipse.

Exercises 21-9, page 583

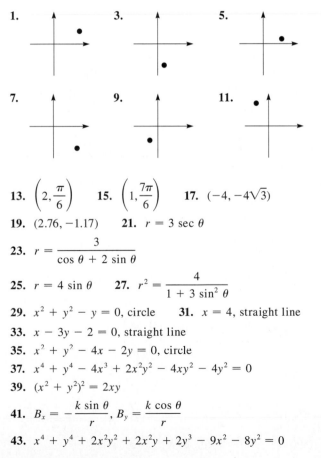

13. $\left(2, \dfrac{\pi}{6}\right)$ **15.** $\left(1, \dfrac{7\pi}{6}\right)$ **17.** $(-4, -4\sqrt{3})$

19. $(2.76, -1.17)$ **21.** $r = 3 \sec \theta$

23. $r = \dfrac{3}{\cos \theta + 2 \sin \theta}$

25. $r = 4 \sin \theta$ **27.** $r^2 = \dfrac{4}{1 + 3 \sin^2 \theta}$

29. $x^2 + y^2 - y = 0$, circle   **31.** $x = 4$, straight line

33. $x - 3y - 2 = 0$, straight line

35. $x^2 + y^2 - 4x - 2y = 0$, circle

37. $x^4 + y^4 - 4x^3 + 2x^2y^2 - 4xy^2 - 4y^2 = 0$

39. $(x^2 + y^2)^2 = 2xy$

41. $B_x = -\dfrac{k \sin \theta}{r}, B_y = \dfrac{k \cos \theta}{r}$

43. $x^4 + y^4 + 2x^2y^2 + 2x^2y + 2y^3 - 9x^2 - 8y^2 = 0$

Exercises 21-10, page 586

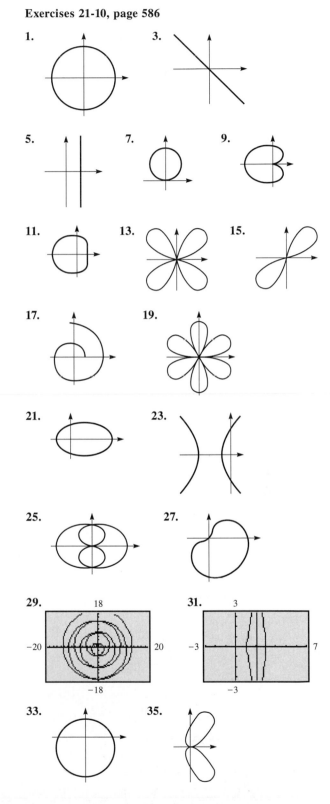

Review Exercises for Chapter 21, page 588

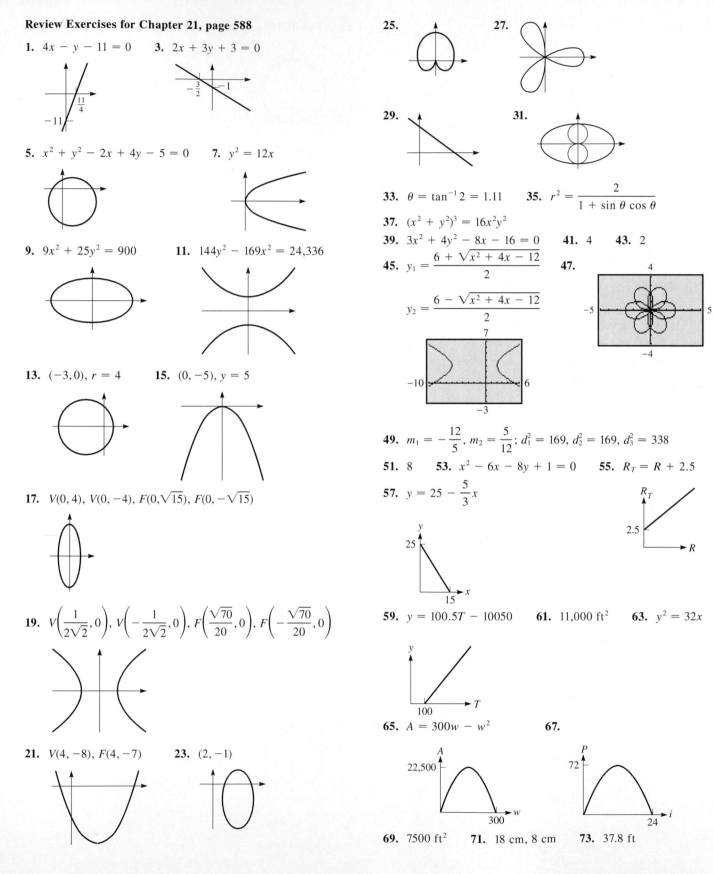

1. $4x - y - 11 = 0$ **3.** $2x + 3y + 3 = 0$

5. $x^2 + y^2 - 2x + 4y - 5 = 0$ **7.** $y^2 = 12x$

9. $9x^2 + 25y^2 = 900$ **11.** $144y^2 - 169x^2 = 24{,}336$

13. $(-3, 0)$, $r = 4$ **15.** $(0, -5)$, $y = 5$

17. $V(0, 4)$, $V(0, -4)$, $F(0, \sqrt{15})$, $F(0, -\sqrt{15})$

19. $V\left(\dfrac{1}{2\sqrt{2}}, 0\right)$, $V\left(-\dfrac{1}{2\sqrt{2}}, 0\right)$, $F\left(\dfrac{\sqrt{70}}{20}, 0\right)$, $F\left(-\dfrac{\sqrt{70}}{20}, 0\right)$

21. $V(4, -8)$, $F(4, -7)$ **23.** $(2, -1)$

33. $\theta = \tan^{-1} 2 = 1.11$ **35.** $r^2 = \dfrac{2}{1 + \sin\theta\cos\theta}$

37. $(x^2 + y^2)^3 = 16x^2y^2$

39. $3x^2 + 4y^2 - 8x - 16 = 0$ **41.** 4   **43.** 2

45. $y_1 = \dfrac{6 + \sqrt{x^2 + 4x - 12}}{2}$ **47.**

$y_2 = \dfrac{6 - \sqrt{x^2 + 4x - 12}}{2}$

49. $m_1 = -\dfrac{12}{5}$, $m_2 = \dfrac{5}{12}$; $d_1^2 = 169$, $d_2^2 = 169$, $d_3^2 = 338$

51. 8   **53.** $x^2 - 6x - 8y + 1 = 0$ **55.** $R_T = R + 2.5$

57. $y = 25 - \dfrac{5}{3}x$

59. $y = 100.5T - 10050$ **61.** 11,000 ft^2 **63.** $y^2 = 32x$

65. $A = 300w - w^2$ **67.**

69. 7500 ft^2 **71.** 18 cm, 8 cm   **73.** 37.8 ft

75. Dist. from rifle to P − dist. from target to P = constant (related to dist. from rifle to target).

77. 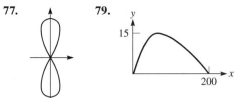   **79.**

27. The greatest frequency should be at 2 and the least at 0 and 4.

(Graphs will generally have approximately the shape shown.)

Exercises 22-1, page 595

1.

No.	103	104	105	106	107	108	109	110	111	112	113
f	1	3	1	3	2	4	3	1	1	0	1

3.

No.	101–103	104–106	107–109	110–112	113–115
f	1	7	9	2	1

5.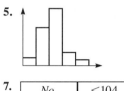

Exercises 22-2, page 599

1. 4   **3.** 0.49   **5.** 4.6   **7.** 0.503   **9.** 4

11. 0.48, 0.49, 0.55   **13.** 21   **15.** 21

17. 2.248 s   **19.** 172 ft   **21.** 4.237 mR

23. 4.36 mR   **25.** 31 h   **27.** 0.00595 mm

29. $475, $500   **31.** 862 kW · h   **33.** 0.195, 0.18

35. $487.50

37. $575, $577, $600; if each value is increased by the same amount, the median, mean, and mode are also increased by this amount.

39. $727; an outlier can make the mean a poor measure of the center of the distribution.

7.

No.	<104	<107	<110	<113	<116
Cum. f	1	8	17	19	20

9.

No. inst.	18	19	20	21	22	23	24	25
f	1	3	2	4	3	1	0	1

11. 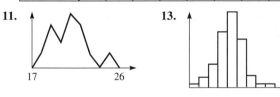   **13.**

15.

Time (s)	<2.22	<2.23	<2.24	<2.25	<2.26	<2.27	<2.28	<2.29	<2.30
Cum. f	2	9	27	68	124	156	164	167	170

17.

Dist. (ft)	155–159	160–164	165–169	170–174	175–179	180–184	185–189
f, (%)	1.7	12.5	26.7	30.0	20.0	8.3	0.8

19.   **21.**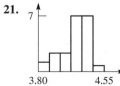

Exercises 22-3, page 604

1. 1.55   **3.** 0.039   **5.** 1.55   **7.** 0.039   **9.** 4.6, 1.55

11. 0.503, 0.039   **13.** 1.8   **15.** 2.6 h

17. 0.014 s   **19.** 0.00022 mm

23. 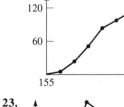   **25.**

Exercises 22-4, page 608

1.
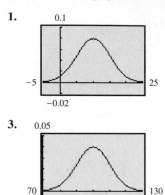

3.

5. 136   **7.** 95   **9.** 341   **11.** 78.81%

13. 2166   **15.** 179

17. 141; about 68% of samples have mean lifetime from 99,859 km to 100,141 km

19. 2.28%   **21.** 76% (normal dist. is 68%)

23. 68% (normal dist. is 68%)

Exercises 22-5, page 613

1. $\bar{\bar{x}} = 361.8$ N · m, UCL = 368.9 N · m, LCL = 354.6 N · m

3.
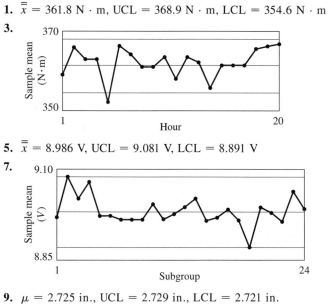

5. $\bar{\bar{x}} = 8.986$ V, UCL = 9.081 V, LCL = 8.891 V

7.
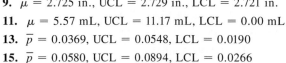

9. $\mu = 2.725$ in., UCL = 2.729 in., LCL = 2.721 in.

11. $\mu = 5.57$ mL, UCL = 11.17 mL, LCL = 0.00 mL

13. $\bar{p} = 0.0369$, UCL = 0.0548, LCL = 0.0190

15. $\bar{p} = 0.0580$, UCL = 0.0894, LCL = 0.0266

Exercises 22-6, page 618

1. $y = 1.0x - 2.6$ **3.** $y = -1.77x + 191$

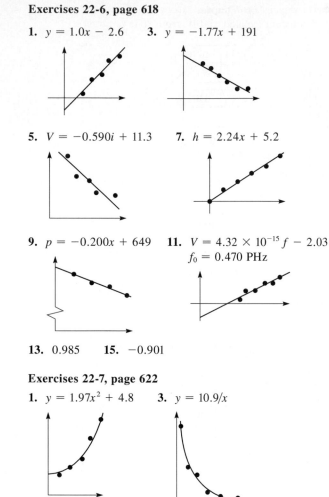

5. $V = -0.590i + 11.3$ **7.** $h = 2.24x + 5.2$

9. $p = -0.200x + 649$ **11.** $V = 4.32 \times 10^{-15} f - 2.03$
                                     $f_0 = 0.470$ PHz

13. 0.985   **15.** −0.901

Exercises 22-7, page 622

1. $y = 1.97x^2 + 4.8$ **3.** $y = 10.9/x$

5. $y = 5.97t^2 + 0.38$ **7.** $p = 0.00967T^2 + 29.4$

9. $p = \dfrac{1343}{S}$ **11.** $y = 6.20e^{-t} - 0.05$

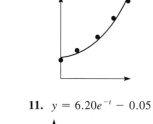

Review Exercises for Chapter 22, page 623

1. 1100   **3.** 1100

5.

No.	1093–1095	1096–1098	1099–1101
f	3	4	3

No.	1102–1104	1105–1107
f	4	2

7.

9.

No.	<1096	<1099	<1102	<1105	<1108
f	3	7	10	14	16

11. 0.264 Pa · s   **13.** 0.014 Pa · s

15.

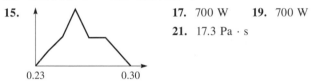

0.23   0.30

17. 700 W   **19.** 700 W
21. 17.3 Pa · s

23.

P(W)	<660	<670	<680	<690
Cum. f	3	5	12	24

P(W)	<700	<710	<720	<730	<740
Cum. f	51	85	100	116	121

25. 4   **27.** 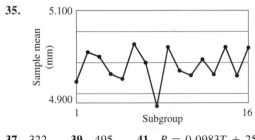   **29.** 66.2 mi/h
31. 9.0 mi/h

33. $\overline{p} = 0.0540$, UCL = 0.0843, LCL = 0.0237

35.

Sample mean (mm), 4.900 to 5.100, Subgroup 1 to 16

37. 322   **39.** 495   **41.** $R = 0.0983T + 25.0$

43. $s = 0.123t + 0.887$ **45.** $s = -4.90t^2 + 3000$

47. $y = -5.2x + 96.2$ (Curve has a good fit to a straight line, but fits $T = 82.6e^{-0.1t} + 17.3$ better.)

49. 9.2 ppm

51. Substitute expressions for m, b, and $\overline{x}$. Simplify to $\overline{y} = \overline{y}$.

Exercises S-1, page 940

1. $x = 2$, $y = 1$ **3.** $x = \dfrac{17}{14}$, $y = \dfrac{19}{14}$

5. $x = -1$, $y = 4$, $z = 2$

7. $x = -2$, $y = -\dfrac{2}{3}$, $z = \dfrac{1}{3}$

9. $w = 1$, $x = 0$, $y = -2$, $z = 3$

11. unlimited: $x = -3$, $y = -1$, $z = 1$; $x = 12$, $y = 0$, $z = -10$

13. inconsistent   **15.** $x = \dfrac{1}{2}$, $y = -\dfrac{3}{2}$

17. $x = \dfrac{1}{6}$, $y = -\dfrac{1}{2}$

19. inconsistent

21. 5.7 calc/s, 6.8 calc/s

23. 250 parts/h, 220 parts/h, 180 parts/h

Exercises S-2, page 944

1. hyperbola;
$2x'y' + 25 = 0$

3. ellipse;
$4x'^2 + 9y'^2 = 36$

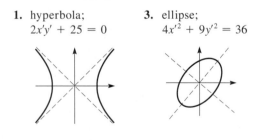

5. parabola;
$x'^2 + \sqrt{2}y' = 0$

7. hyperbola;
$4x'^2 - y'^2 = 4$

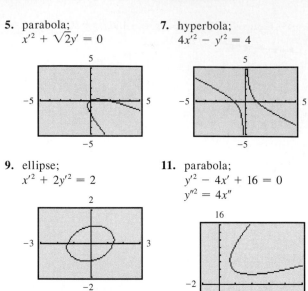

9. ellipse;
$x'^2 + 2y'^2 = 2$

11. parabola;
$y'^2 - 4x' + 16 = 0$
$y''^2 = 4x''$

Exercises for Appendix B, page A-9

1. MHz, 1 MHz = 10^6 Hz   **3.** mm, 1 mm = 10^{-3} m

5. kilovolt, 1 kV = 10^3 V

7. milliampere, 1 mA = 10^{-3} A   **9.** 10^5 cm

11. 63,360 in.   **13.** 13.3 cm   **15.** 14.9 L

17. 144 in.^{2} **19.** 2.50×10^{-4} m^2 **21.** 2040 g

23. 13.6 L   **25.** 1.68×10^5 N · cm/s   **27.** 90 MJ

29. 0.101 hp   **31.** 3.25 tons   **33.** 15 gal

35. 1240 km/h   **37.** 37 mi/h   **39.** 25,000 mg/dL

41. 101,000 Pa   **43.** 1.2×10^5 mA/cm^2

45. 3.73 h   **47.** 0.135 km/L

Exercises C-2, page A-13

(Most answers have been rounded off to four significant digits.)

1. 56.02   **3.** 4162.1   **5.** 18.65   **7.** 0.3954

9. 14.14   **11.** 0.5251   **13.** 13.35   **15.** 944.6

17. 0.7349   **19.** -0.7594 **21.** -1.337

23. 1.015   **25.** 41.35°   **27.** -1.182 **29.** 0.5862

31. 6.695   **33.** 3.508   **35.** 0.005685   **37.** 2.053

39. 5.765   **41.** 4.501×10^{10} **43.** 497.2

45. 6.648   **47.** 401.2   **49.** 8.841   **51.** 2.523

53. 10.08   **55.** 22.36   **57.** 20.3°   **59.** 4729

61. 3.301×10^4 **63.** 1.056   **65.** 55.5°

67. 3.277   **69.** 8.125   **71.** 1.000   **73.** 1.000

75. 12.90   **77.** 8.001   **79.** 8.053   **81.** 0.04259

83. 0.4219   **85.** 0.7822   **87.** 2.0736465

89. 124.3   **91.** 252

93.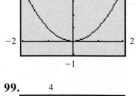

95.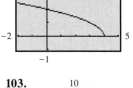

97.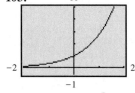

99.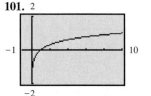

101.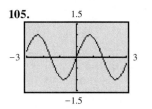

103.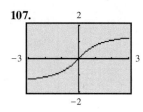

105.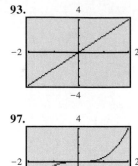

107.

Exercises C-3, page A-17

(Program names are suggested.)

1. CYLINDER
:Prompt R,H
:Disp "TOTAL AREA=",2πR(R+H)
:Disp "VOLUME=",πR^{2}H

3. PRODBINO
:Prompt A,B,C,D
:Disp "E=",A*C
:Disp "F=",A*D+B*C
:Disp "G=",B*D

5. SLVTRASA
:Input "θA=",A
:Input "θB=",B
:Input "θB=",B
:180$-$A$-$B→C
:Disp "θC=",C
:Disp "SA=",Fsin(A)/sin(C)
:Disp "SB=",Fsin(B)/sin(C)

7. CNSTPROP
:Prompt Y,R,A,S,B,T,C
:Disp "K="
:Disp YT^C/((R^A)(S^B))

Chapter 1

1. $\sqrt{9 + 16} = \sqrt{25} = 5$

2. $\dfrac{(7)(-3)(-2)}{(-6)(0)}$ is undefined (division by zero).

3. $\dfrac{3.372 \times 10^{-3}}{7.526 \times 10^{12}} = 4.480 \times 10^{-16}$

```
3.372E-3/7.526E1
2
     4.48046771E-16
```

4. $\dfrac{(+6)(-2) - 3(-1)}{5 - 2} = \dfrac{-12 - (-3)}{3} = \dfrac{-12 + 3}{3}$

$\qquad\qquad = \dfrac{-9}{3} = -3$

5. $\dfrac{346.4 - 23.5}{287.7} - \dfrac{0.944^3}{(3.46)(0.109)} = 1.108$

```
(346.4-23.5)/287
.7-.944^3/(3.46*
.109)
        -1.10820764
```

6. $(2a^0 b^{-2} c^3)^{-3} = 2^{-3} a^{0(-3)} b^{(-2)(-3)} c^{3(-3)} = 2^{-3} a^0 b^6 c^{-9}$

$\qquad = \dfrac{b^6}{8c^9}$

7. $(2x + 3)^2 = (2x + 3)(2x + 3)$
$\qquad = 2x(2x) + 2x(3) + 3(2x) + 3(3)$
$\qquad = 4x^2 + 6x + 6x + 9$
$\qquad = 4x^2 + 12x + 9$

8. $3m^2(am - 2m^3) = 3m^2(am) + 3m^2(-2m^3)$
$\qquad\qquad = 3am^3 - 6m^5$

9. $\dfrac{8a^3 x^2 - 4a^2 x^4}{-2ax^2} = \dfrac{8a^3 x^2}{-2ax^2} - \dfrac{4a^2 x^4}{-2ax^2} = -4a^2 - (-2ax^2)$
$\qquad = -4a^2 + 2ax^2$

10.

$$
\begin{array}{r}
3x \;-\; 5 \qquad \text{(quotient)} \\
2x - 1\,\overline{\big)\; 6x^2 - 13x + 7} \\
\underline{6x^2 \;-\; 3x} \\
-10x + 7 \\
\underline{-10x + 5} \\
2 \quad \text{(remainder)}
\end{array}
$$

11. $(2x - 3)(x + 7)$
$\qquad = 2x(x) + 2x(7) + (-3)(x) + (-3)(7)$
$\qquad = 2x^2 + 14x - 3x - 21 = 2x^2 + 11x - 21$

12. $3x - [4x - (3 - 2x)] = 3x - [4x - 3 + 2x]$
$\qquad\qquad = 3x - [6x - 3]$
$\qquad\qquad = 3x - 6x + 3 = -3x + 3$

13. $5y - 2(y - 4) = 7$
$\qquad 5y - 2y + 8 = 7$
$\qquad 3y + 8 - 8 = 7 - 8$
$\qquad\qquad 3y = -1$
$\qquad\qquad y = -1/3$

14. $\qquad 3(x - 3) = x - (2 - 3d)$
$\qquad\quad 3x - 9 = x - 2 + 3d$
$\quad 3x - 9 - x + 9 = x - 2 + 3d - x + 9$
$\qquad\qquad 2x = 3d + 7$
$\qquad\qquad x = \dfrac{3d + 7}{2}$

15. $0.000\,003\,6 = 3.6 \times 10^{-6}$ (6 places to right)

16.

$-\pi$	-3	0.3	$\sqrt{2}$	$|-4|$	(order)
-3.14	-3	0.3	1.41	4	(value)

17. $3(5 + 8) = 3(5) + 3(8)$ illustrates distributive law.

18. (a) 5, (b) 3.0 (zero is significant)

19. Evaluation:
$1000(1 + 0.05/2)^{2(3)} = 1000(1.025)^6 = \$1159.69$

20. $8(100 - x)^2 + x^2 = 8(100 - x)(100 - x) + x^2$
$\qquad\qquad = 8(10{,}000 - 200x + x^2) + x^2$
$\qquad\qquad = 80{,}000 - 1600x + 8x^2 + x^2$
$\qquad\qquad = 80{,}000 - 1600x + 9x^2$

21. $L = L_0[1 + \alpha(t_2 - t_1)]$
$\qquad = L_0[1 + \alpha t_2 - \alpha t_1]$
$\qquad = L_0 + \alpha L_0 t_2 - \alpha L_0 t_1$
$\qquad L - L_0 + \alpha L_0 t_1 = \alpha L_0 t_2$
$\qquad t_2 = \dfrac{L - L_0 + \alpha L_0 t_1}{\alpha L_0}$

22. Let n = number of pounds of second alloy
$\qquad 0.3(20) + 0.8n = 0.6(n + 20)$
$\qquad\quad 6 + 0.8n = 0.6n + 12$
$\qquad\qquad 0.2n = 6$
$\qquad\qquad\;\; n = 30$ lb

Chapter 2

1. $\angle 1 + \angle 3 + 90° = 180°$ (sum of angles of a triangle)
$\qquad\qquad \angle 3 = 52°$ (vertical angles)
$\qquad\qquad \angle 1 = 180° - 90° - 52° = 38°$

2. $\angle 2 + \angle 4 = 180°$ (straight angle)
$\qquad\quad \angle 4 = 52°$ (corresponding angles)
$\qquad \angle 2 + 52° = 180°$
$\qquad\qquad \angle 2 = 128°$

$AB \parallel CD$

3.

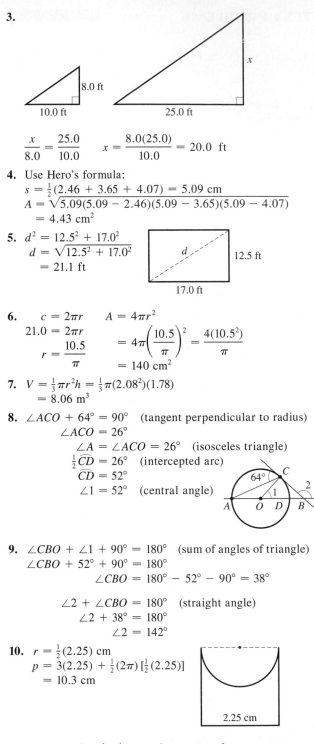

$$\frac{x}{8.0} = \frac{25.0}{10.0} \qquad x = \frac{8.0(25.0)}{10.0} = 20.0 \text{ ft}$$

4. Use Hero's formula:
$$s = \tfrac{1}{2}(2.46 + 3.65 + 4.07) = 5.09 \text{ cm}$$
$$A = \sqrt{5.09(5.09 - 2.46)(5.09 - 3.65)(5.09 - 4.07)}$$
$$= 4.43 \text{ cm}^2$$

5. $d^2 = 12.5^2 + 17.0^2$
$d = \sqrt{12.5^2 + 17.0^2}$
$= 21.1 \text{ ft}$

6. $c = 2\pi r \qquad A = 4\pi r^2$
$21.0 = 2\pi r$
$r = \dfrac{10.5}{\pi}$
$= 4\pi \left(\dfrac{10.5}{\pi}\right)^2 = \dfrac{4(10.5^2)}{\pi}$
$= 140 \text{ cm}^2$

7. $V = \tfrac{1}{3}\pi r^2 h = \tfrac{1}{3}\pi (2.08^2)(1.78)$
$= 8.06 \text{ m}^3$

8. $\angle ACO + 64° = 90°$ (tangent perpendicular to radius)
$\qquad \angle ACO = 26°$
$\qquad\quad \angle A = \angle ACO = 26°$ (isosceles triangle)
$\qquad\quad \tfrac{1}{2}\overarc{CD} = 26°$ (intercepted arc)
$\qquad\qquad \overarc{CD} = 52°$
$\qquad\qquad \angle 1 = 52°$ (central angle)

9. $\angle CBO + \angle 1 + 90° = 180°$ (sum of angles of triangle)
$\angle CBO + 52° + 90° = 180°$
$\qquad\quad \angle CBO = 180° - 52° - 90° = 38°$

$\qquad \angle 2 + \angle CBO = 180°$ (straight angle)
$\qquad\quad \angle 2 + 38° = 180°$
$\qquad\qquad \angle 2 = 142°$

10. $r = \tfrac{1}{2}(2.25) \text{ cm}$
$p = 3(2.25) + \tfrac{1}{2}(2\pi)[\tfrac{1}{2}(2.25)]$
$= 10.3 \text{ cm}$

2.25 cm

11. $A = 2.25^2 - \tfrac{1}{2}\pi[\tfrac{1}{2}(2.25)]^2 = 3.07 \text{ cm}^2$

12. $A = \tfrac{1}{2}(50)\,[0 + 2(90) + 2(145) + 2(260)$
$\qquad + 2(205) + 2(110) + 20]$
$\qquad = 41{,}000 \text{ ft}^2$

Chapter 3

1. $f(x) = 2x - x^2 + \dfrac{8}{x}$

$f(-4) = 2(-4) - (-4)^2 + \dfrac{8}{-4}$
$\qquad = -8 - 16 - 2$
$\qquad = -26$

$f(2.385) = 2(2.385) - (2.385)^2 + \dfrac{8}{2.385} = 2.436$

2. $w = 2000 - 10t$

3. $f(x) = 4 - 2x$

$y = 4 - 2x$	x	y
$y = 4 - 2(-1) = 6$	-1	6
$y = 4 - 2(0) = 4$	0	4
$y = 4 - 2(1) = 2$	1	2
$y = 4 - 2(2) = 0$	2	0
$y = 4 - 2(3) = -2$	3	-2
$y = 4 - 2(4) = -4$	4	-4

4. $y = 2x^2 - 3x - 3$

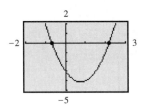

$x = -0.7$ and $x = 2.2$

5. $y = \sqrt{4 + 2x}$

$y = \sqrt{4 + 2x}$	x	y
$y = \sqrt{4 + 2(-2)} = 0$	-2	0
$y = \sqrt{4 + 2(-1)} = 1.4$	-1	1.4
$y = \sqrt{4 + 2(0)} = 2$	0	2
$y = \sqrt{4 + 2(1)} = 2.4$	1	2.4
$y = \sqrt{4 + 2(2)} = 2.8$	2	2.8
$y = \sqrt{4 + 2(4)} = 3.5$	4	3.5

6. On negative x-axis

7. $f(x) = \sqrt{6 - x}$
Domain: $x \leq 6$; x cannot be greater than 6 to have real values of $f(x)$.
Range: $f(x) \geq 0$; $\sqrt{6 - x}$ is the principal square root of $6 - x$ and cannot be negative.

8. Let r = radius of circular part
$\qquad\quad$ square $\qquad$ semicircle
$A = (2r)(2r) + \tfrac{1}{2}(\pi r^2)$
$\quad = 4r^2 + \tfrac{1}{2}\pi r^2$

9.

Voltage V	10.0	20.0	30.0	40.0	50.0	60.0
Current i	145	188	220	255	285	315

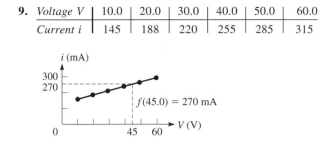

$f(45.0) = 270$ mA

10. $V = 20.0$ V for $i = 188$ mA.
$V = 30.0$ V for $i = 220$ mA.

$$\frac{x}{10} = \frac{12}{32}, \qquad x = 3.8 \quad \text{(rounded off)}$$

$V = 20.0 + 3.8 = 23.8$ V
for $i = 200$ mA

Chapter 4

1. $39' = \left(\frac{39}{60}\right)^\circ = 0.65^\circ$
$37^\circ 39' = 37.65^\circ$

2. $\cos \theta = 0.3726;$ $\theta = 68.12^\circ$

```
cos-1(.3726)
        68.12394435
```

3. Let $x =$ distance from
course to east

$$\frac{x}{22.62} = \sin 4.05^\circ$$

$$x = 22.62 \sin 4.05^\circ$$
$$= 1.598 \text{ km}$$

22.62 km
4.05°
S   x

4. $\sin \theta = \dfrac{2}{3}$

$x = \sqrt{3^2 - 2^2} = \sqrt{5}$

$\tan \theta = \dfrac{2}{\sqrt{5}}$

$r = 3$

5. $\tan \theta = 1.294;$ $\csc \theta = 1.264$

```
(sin(tan-1(1.294)
)-1
        1.263810119
```

6. $B = 90^\circ - 37.4^\circ = 52.6^\circ$

$$\frac{52.8}{c} = \cos 37.4^\circ$$

$$\frac{a}{52.8} = \tan 37.4^\circ$$

$$c = \frac{52.8}{\cos 37.4^\circ}$$

$$a = 52.8 \tan 37.4^\circ$$
$$= 40.4$$

$$= 66.5$$

B
c
a
$A = 37.4^\circ$
$b = 52.8$

7. $2.49^2 + b^2 = 3.88^2$

$b = \sqrt{3.88^2 - 2.49^2} = 2.98$

$$\sin A = \frac{2.49}{3.88}$$

$A = 39.9^\circ \qquad B = 50.1^\circ$

```
√(3.88²-2.49²)
        2.975617583
sin-1(2.49/3.88)
        39.92262926
90-Ans
        50.07737074
```

B
$c = 3.88$
$a = 2.49$
A
b

8. $\lambda = d \sin \theta$
$= 30.05 \sin 1.167^\circ$
$- 0.6120 \ \mu$m

9. $r = \sqrt{5^2 + 2^2} = \sqrt{29}$

$$\sin \theta = \frac{2}{\sqrt{29}} = 0.3714 \qquad \csc \theta = \frac{\sqrt{29}}{2} = 2.693$$

$$\cos \theta = \frac{5}{\sqrt{29}} = 0.9285 \qquad \sec \theta = \frac{\sqrt{29}}{5} = 1.077$$

$$\tan \theta = \frac{2}{5} = 0.4000 \qquad \cot \theta = \frac{5}{2} = 2.500$$

y
(5, 2)
r
$y = 2$
θ
0 $x = 5$ x

10. Distance between points is $x - y$.

$$\frac{18.525}{x} = \tan 13.500^\circ \qquad \frac{18.525}{y} = \tan 21.375^\circ$$

$$x - y = \frac{18.525}{\tan 13.500^\circ} - \frac{18.525}{\tan 21.375^\circ} = 29.831 \text{ m}$$

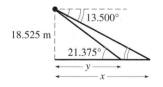

Chapter 5

1. Points $(2, -5)$ and $(-1, 4)$

$$m = \frac{4 - (-5)}{-1 - 2} = \frac{9}{-3} = -3$$

2.

$\begin{aligned} x + 2y &= 5 \\ 4y &= 3 - 2x \\ \hline x &= 5 - 2y \end{aligned}$ $\quad$ $\begin{aligned} 4y &= 3 - 2(5 - 2y) \\ 4y &= 3 - 10 + 4y \\ 0 &= -7 \\ &\text{Inconsistent} \end{aligned}$

3. $3x - 2y = 4$

$2x + 5y = -1$

$$x = \frac{\begin{vmatrix} 4 & -2 \\ -1 & 5 \end{vmatrix}}{\begin{vmatrix} 3 & -2 \\ 2 & 5 \end{vmatrix}} = \frac{20 - 2}{15 - (-4)} = \frac{18}{19}$$

$$y = \frac{\begin{vmatrix} 3 & 4 \\ 2 & -1 \end{vmatrix}}{19} = \frac{-3 - 8}{19} = -\frac{11}{19}$$

4. $2x + y = 4$

$y = -2x + 4$

$m = -2 \qquad b = 4$

5. $2l + 2w = 24$ (perimeter)

$l = w + 6.0$

$\overline{}$

$l + w = 12$

$l - w = 6.0$

$\overline{}$

$2l = 18$

$l = 9$ km

$9 = w + 6$

$w = 3$ km

6.

$\begin{aligned} 2x - 3y &= 6 \\ x = 0&: y = -2 \\ y = 0&: x = 3 \\ \text{Int: } &(0, -2), (3, 0) \\ x = 1.3 &\qquad y = -1.1 \end{aligned}$ $\quad$ $\begin{aligned} 4x + y &= 4 \\ x = 0&: y = 4 \\ y = 0&: x = 1 \\ \text{Int: } &(0, 4), (1, 0) \end{aligned}$

7. Let x = vol. of first alloy

$ y$ = vol. of second alloy

$ z$ = vol. of third alloy

$\begin{aligned} x + y + z &= 100 \quad \text{total vol.} \\ 0.6x + 0.5y + 0.3z &= 40 \quad \text{copper} \\ 0.3x + 0.3y &= 15 \quad \text{zinc} \end{aligned}$

$\overline{}$

$\begin{aligned} x + y + z &= 100 \\ 6x + 5y + 3z &= 400 \\ 3x + 3y &= 150 \end{aligned}$

$\overline{}$

$\begin{aligned} 3x + 2y &= 100 \\ 3x + 3y &= 150 \end{aligned}$

$\overline{}$

$\begin{aligned} y &= 50 \text{ cm}^3 \\ x &= 0 \text{ cm}^3 \\ z &= 50 \text{ cm}^3 \end{aligned}$

8.

$\begin{aligned} 3x + 2y - z &= 4 \\ 2x - y + 3z &= -2 \\ x + 4z &= 5 \end{aligned}$

$\overline{}$

$$y = \frac{\begin{vmatrix} 3 & 4 & -1 \\ 2 & -2 & 3 \\ 1 & 5 & 4 \end{vmatrix}}{\begin{vmatrix} 3 & 2 & -1 \\ 2 & -1 & 3 \\ 1 & 0 & 4 \end{vmatrix}}$$

$$= \frac{-24 + 12 - 10 - 2 - 45 - 32}{-12 + 6 + 0 - 1 - 0 - 16}$$

$$= \frac{-101}{-23} = \frac{101}{23}$$

Chapter 6

1. $2x(2x - 3)^2 = 2x[(2x)^2 - 2(2x)(3) + 3^2]$

$ = 2x(4x^2 - 12x + 9)$

$ = 8x^3 - 24x^2 + 18x$

2. $\dfrac{1}{R} = \dfrac{1}{R_1 + r} + \dfrac{1}{R_2}$

$\dfrac{RR_2(R_1 + r)}{R} = \dfrac{RR_2(R_1 + r)}{R_1 + r} + \dfrac{RR_2(R_1 + r)}{R_2}$

$R_2(R_1 + r) = RR_2 + R(R_1 + r)$

$R_1 R_2 + rR_2 = RR_2 + RR_1 + rR$

$R_1 R_2 - RR_1 = RR_2 + rR - rR_2$

$R_1(R_2 - R) = RR_2 + rR - rR_2$

$R_1 = \dfrac{RR_2 + rR - rR_2}{R_2 - R}$

3. $\dfrac{2x^2 + 5x - 3}{2x^2 + 12x + 18} = \dfrac{(2x - 1)(x + 3)}{2(x + 3)^2} = \dfrac{2x - 1}{2(x + 3)}$

4. $4x^2 - 16y^2 = 4(x^2 - 4y^2) = 4(x + 2y)(x - 2y)$

5. $\dfrac{3}{4x^2} - \dfrac{2}{x^2 - x} - \dfrac{x}{2x - 2} = \dfrac{3}{4x^2} - \dfrac{2}{x(x - 1)} - \dfrac{x}{2(x - 1)}$

$= \dfrac{3(x - 1) - 2(4x) - x(2x^2)}{4x^2(x - 1)}$

$= \dfrac{3x - 3 - 8x - 2x^3}{4x^2(x - 1)}$

$= \dfrac{-2x^3 - 5x - 3}{4x^2(x - 1)}$

6. $\dfrac{x^2 + x}{2 - x} \div \dfrac{x^2}{x^2 - 4x + 4} = \dfrac{x^2 + x}{2 - x} \times \dfrac{x^2 - 4x + 4}{x^2}$

$= \dfrac{x(x + 1)}{2 - x} \times \dfrac{(x - 2)^2}{x^2}$

$= -\dfrac{x(x + 1)(x - 2)^2}{(x - 2)(x^2)}$

$= -\dfrac{(x + 1)(x - 2)}{x}$

7. $\dfrac{1 - \dfrac{3}{2x+2}}{\dfrac{x}{5} - \dfrac{1}{2}} = \dfrac{\dfrac{2(x+1)-3}{2(x+1)}}{\dfrac{2x-5}{10}}$

$= \dfrac{2x+2-3}{2(x+1)} \times \dfrac{10}{2x-5}$

$= \dfrac{5(2x-1)}{(x+1)(2x-5)}$

8. Let $t = $ time working together

$\dfrac{t}{12} + \dfrac{t}{16} = 1$

LCD of 12 and 16 is 48.

$\dfrac{48t}{12} + \dfrac{48t}{16} = 48$

$4t + 3t = 48$

$t = \dfrac{48}{7} = 6.9$ days

Chapter 7

1. $x^2 - 3x - 5 = 0$

Not factorable

$a = 1, \qquad b = -3, \qquad c = -5$

$x = \dfrac{-(-3) \pm \sqrt{(-3)^2 - 4(1)(-5)}}{2(1)}$

$= \dfrac{3 \pm \sqrt{29}}{2}$

2. $2x^2 = 9x - 4$

$2x^2 - 9x + 4 = 0$

$(2x - 1)(x - 4) = 0$

$2x - 1 = 0 \qquad x - 4 = 0$

$x = \frac{1}{2} \quad$ or $\quad x = 4$

3. $\dfrac{3}{x} - \dfrac{2}{x+2} = 1$

$\dfrac{3x(x+2)}{x} - \dfrac{2x(x+2)}{x+2} = x(x+2)$

$3(x+2) - 2x = x^2 + 2x$

$3x + 6 - 2x = x^2 + 2x$

$0 = x^2 + x - 6$

$(x + 3)(x - 2) = 0$

$x = -3, 2$

4. $2x^2 - x = 6 - 2x(3 - x)$

$2x^2 - x = 6 - 6x + 2x^2$

$5x = 6 \quad$ (not quadratic)

$x = \frac{6}{5}$

5. $y = 2x^2 + 8x + 5$

$\dfrac{-b}{2a} = \dfrac{-8}{2(2)} = -2$

$y = 2(-2)^2 + 8(-2) + 5 = -3$

Min. pt. $(a > 0)$ is $(-2, -3)$,

$c = 5$, y-intercept is $(0, 5)$.

6. $P = EI - RI^2$

$RI^2 - EI + P = 0$

$I = \dfrac{-(-E) \pm \sqrt{(-E)^2 - 4RP}}{2R}$

$= \dfrac{E \pm \sqrt{E^2 - 4RP}}{2R}$

7. $x^2 - 6x - 9 = 0$

$x^2 - 6x \qquad = 9$

$x^2 - 6x + 9 = 9 + 9$

$(x - 3)^2 = 18$

$x - 3 = \pm\sqrt{18}$

$x = 3 \pm 3\sqrt{2}$

8. Let $w = $ width of window

$\quad\;\; h = $ height of window

$2w + 2h = 8.4 \quad$ (perimeter)

$w + h = 4.2 \qquad h = 4.2 - w$

$w(4.2 - w) = 3.8 \quad$ (area)

$-w^2 + 4.2w = 3.8$

$w^2 - 4.2w + 3.8 = 0$

$w = \dfrac{-(-4.2) \pm \sqrt{(-4.2)^2 - 4(3.8)}}{2}$

$-2.88, 1.32$

$w = 1.3$ m, $h = 2.9$ m $\quad$ or

$w = 2.9$ m, $h = 1.3$ m

Chapter 8

1. $150° = \left(\frac{\pi}{180}\right)(150) = \frac{5\pi}{6}$

2. $\sin 205° = -\sin(205° - 180°) = -\sin 25°$

3. $x = -9, \quad y = 12$

$r = \sqrt{(-9)^2 + 12^2} = 15$

$\sin \theta = \frac{12}{15} = \frac{4}{5}$

$\sec \theta = \frac{15}{-9} = -\frac{5}{3}$

4. $r = 2.80$ ft

$\omega = 2200$ r/min $= (2200$ r/min$)(2\pi$ rad/r$)$

$\quad = 4400\pi$ rad/min

$v = \omega r = (4400\pi)(2.80) = 39{,}000$ ft/min

5. $3.572 = 3.572\left(\frac{180°}{\pi}\right) = 204.7°$

6. $\tan \theta = 0.2396$

$\theta_{ref} = 13.47°$

$\theta = 13.47° \quad$ or

$\theta = 180° + 13.47° = 193.47°$

7. $\cos \theta = -0.8244$, $\csc \theta < 0$;

$\cos \theta$ negative, $\csc \theta$ negative,

θ in third quadrant

$\theta_{ref} = 0.6017, \qquad \theta = 3.7432$

8. $s = r\theta$, $32.0 = 8.50\theta$

$\theta = \frac{32.0}{8.50} = 3.76$ rad

$A = \frac{1}{2}\theta r^2 = \frac{1}{2}(3.76)(8.50)^2$

$= 136$ ft^2

$s = 32.0$ ft

$r = 8.50$ ft

Chapter 9

1.

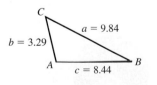

2. $b^2 = a^2 + c^2 - 2ac \cos B$

$b = \sqrt{22.5^2 + 30.9^2 - 2(22.5)(30.9)\cos 78.6°} = 34.4$

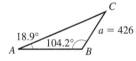

3. $x^2 = 36.50^2 + 21.38^2 - 2(36.50)(21.38)\cos 45.00°$

$x = 26.19$ m $\quad \dfrac{21.38}{\sin \alpha} = \dfrac{26.19}{\sin 45.00°}$

$\sin \alpha = \dfrac{21.38 \sin 45.00°}{26.19}, \quad \alpha = 35.26°$

$\theta = 45.00° - 35.26° = 9.74°$

Displacement is 26.19 m,
9.74° N of E.

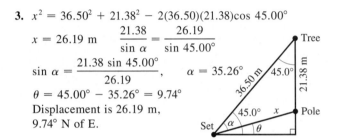

4. $C = 180° - (18.9° + 104.2°) = 56.9°$

$\dfrac{c}{\sin C} = \dfrac{a}{\sin A} \quad c = \dfrac{426 \sin 56.9°}{\sin 18.9°} = 1100$

5. Since a is longest side, find A first.

$a^2 = b^2 + c^2 - 2bc \cos A$

$\cos A = \dfrac{b^2 + c^2 - a^2}{2bc} = \dfrac{3.29^2 + 8.44^2 - 9.84^2}{2(3.29)(8.44)}$

$A = 105.4°$

$\dfrac{b}{\sin B} = \dfrac{a}{\sin A}$

$\sin B = \dfrac{b \sin A}{a} = \dfrac{3.29 \sin 105.4°}{9.84}$

$B = 18.8°$

$C = 180° - (105.4° + 18.8°) = 55.8°$

6. $A_x = 871 \cos 284.3° = 215$
$A_y = 871 \sin 284.3° = -844$

7. $\dfrac{63.0}{\sin 148.5°} = \dfrac{42.0}{\sin A} \qquad \dfrac{x}{\sin 11.1°} = \dfrac{63.0}{\sin 148.5°}$

$\sin A = \dfrac{42.0 \sin 148.5°}{63.0} \qquad x = \dfrac{63.0 \sin 11.1°}{\sin 148.5°}$

$A = 20.4° \qquad\qquad\qquad = 23.2$ mi

$C = 180° - (148.5° + 20.4°) = 11.1°$

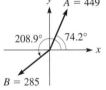

8. $A_x = 449 \cos 74.2° \qquad B_x = 285 \cos 208.9°$
$A_y = 449 \sin 74.2° \qquad B_y = 285 \sin 208.9°$

$R_x = A_x + B_x = 449 \cos 74.2° + 285 \cos 208.9°$
$\quad = -127.3$

$R_y = A_y + B_y = 449 \sin 74.2° + 285 \sin 208.9°$
$\quad = 294.3$

$R = \sqrt{(-127.3)^2 + 294.3^2} = 321$

$\tan \theta_{ref} = \dfrac{294.3}{127.3}, \qquad \theta_{ref} = 66.6°, \qquad \theta = 113.4°$

θ is in second quadrant, since R_x
is negative and R_y is positive.

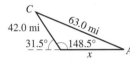

9. $29.6 \sin 36.5° = 17.6$

$17.6 < 22.3 < 29.6$ means two solutions.

$\dfrac{29.6}{\sin B} = \dfrac{22.3}{\sin 36.5°}, \qquad \sin B = \dfrac{29.6 \sin 36.5°}{22.3}$

$B_1 = 52.1° \qquad C_1 = 180° - 36.5° - 52.1° = 91.4°$
$B_2 = 180° - 52.1° = 127.9°,$
$C_2 = 180° - 36.5° - 127.9° = 15.6°$

$\dfrac{c_1}{\sin 91.4°} = \dfrac{22.3}{\sin 36.5°}, \qquad c_1 = \dfrac{22.3 \sin 91.4°}{\sin 36.5°} = 37.5$

$\dfrac{c_2}{\sin 15.6°} = \dfrac{22.3}{\sin 36.5°}, \qquad c_2 = \dfrac{22.3 \sin 15.6°}{\sin 36.5°} = 10.1$

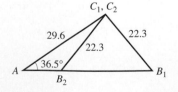

Chapter 10

1. $y = 0.5 \cos \frac{\pi}{2} x$
Amp. = 0.5, disp. = 0,

per. $= \dfrac{2\pi}{\pi/2} = 4$

x	0	1	2	3	4
y	0.5	0	−0.5	0	0.5

2. $y = 2 + 3 \sin x$
For $y_1 = 3 \sin x$,
amp. = 3, per. = 2π,
disp. = 0

x	0	$\frac{\pi}{2}$	π	$\frac{3\pi}{2}$	2π
$y_1 = 3 \sin x$	0	3	0	−3	0
$y = 2 + 3 \sin x$	2	5	2	−1	2

3. $y = 3 \sec x$

$\sec x = \dfrac{1}{\cos x}$

4. $y = 2 \sin(2x - \frac{\pi}{3})$
Amp. = 2, per. $= \frac{2\pi}{2} = \pi$

disp. $= -\dfrac{-\pi/3}{2} = \dfrac{\pi}{6}$

x	$\frac{\pi}{6}$	$\frac{5\pi}{12}$	$\frac{2\pi}{3}$	$\frac{11\pi}{12}$	$\frac{7\pi}{6}$
y	0	2	0	−2	0

$\frac{\pi}{6} + \frac{\pi}{4} = \frac{5\pi}{12}, \ \frac{\pi}{6} + \frac{\pi}{2} = \frac{2\pi}{3}, \ \frac{\pi}{6} + \frac{3\pi}{4} = \frac{11\pi}{12}$

5. $y = A \cos \frac{2\pi}{T} t$
$A = 0.200$ in., $T = 0.100$ s,
$y = 0.200 \cos 20\pi t$,
amp. = 0.200 in., per. = 0.100 s

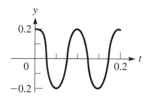

6. $y = 2 \sin x + \cos 2x$
For $y_1 = 2 \sin x$,
amp. = 2, per. = 2π, disp. = 0
For $y_2 = \cos 2x$,
amp. = 1, per. $= \frac{2\pi}{2} = \pi$, disp. = 0

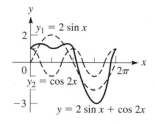

7. $x = \sin \pi t$, $y = 2 \cos 2\pi t$

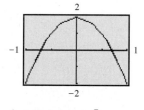

8. $d = R \sin(\omega t + \frac{\pi}{6})$
$\omega = 2.00$ rad/s
$d = R \sin(2.00t + \frac{\pi}{6})$

Amp. = R, per. $= \dfrac{2\pi}{2.00} = \pi$ s = 3.14 s,

disp. $= \dfrac{-\pi/6}{2.00} = -\dfrac{\pi}{12}$ s $= -0.26$ s

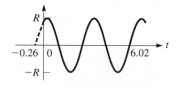

Chapter 11

1. $2\sqrt{20} - \sqrt{125} = 2\sqrt{4 \times 5} - \sqrt{25 \times 5}$
$$= 2(2\sqrt{5}) - 5\sqrt{5}$$
$$= 4\sqrt{5} - 5\sqrt{5} = -\sqrt{5}$$

2. $\dfrac{100^{3/2}}{8^{-2/3}} = (100^{3/2})(8^{2/3}) = [(100^{1/2})^3][(8^{1/3})^2]$
$$= (10^3)(2^2) = 4000$$

3. $(as^{-1/3}t^{3/4})^{12} = a^{12}s^{(-1/3)(12)}t^{(3/4)(12)} = a^{12}s^{-4}t^9 = \dfrac{a^{12}t^9}{s^4}$

4. $(2x^{-1} + y^{-2})^{-1} = \dfrac{1}{2x^{-1} + y^{-2}} = \dfrac{1}{\dfrac{2}{x} + \dfrac{1}{y^2}} = \dfrac{1}{\dfrac{2y^2 + x}{xy^2}}$
$$= \dfrac{xy^2}{2y^2 + x}$$

5. $(\sqrt{2x} - 3\sqrt{y})^2 = (\sqrt{2x})^2 - 2\sqrt{2x}(3\sqrt{y}) + (3\sqrt{y})^2$
$$= 2x - 6\sqrt{2xy} + 9y$$

6. $\sqrt[3]{\sqrt[4]{4}} = \sqrt[12]{4} = \sqrt[12]{2^2} = 2^{2/12} = 2^{1/6} = \sqrt[6]{2}$

7. $\dfrac{3 - 2\sqrt{2}}{2\sqrt{x}} = \dfrac{3 - 2\sqrt{2}}{2\sqrt{x}} \times \dfrac{\sqrt{x}}{\sqrt{x}} = \dfrac{3\sqrt{x} - 2\sqrt{2x}}{2x}$

8. $\sqrt{27a^4b^3} = \sqrt{9 \times 3 \times (a^2)^2(b^2)(b)} = 3a^2b\sqrt{3b}$

9. $2\sqrt{2}(3\sqrt{10} - \sqrt{6}) = 2\sqrt{2}(3\sqrt{10}) - 2\sqrt{2}(\sqrt{6})$
$$= 6\sqrt{20} - 2\sqrt{12}$$
$$= 6(2\sqrt{5}) - 2(2\sqrt{3})$$
$$= 12\sqrt{5} - 4\sqrt{3}$$

10. $(2x + 3)^{1/2} + (x + 1)(2x + 3)^{-1/2}$
$$= (2x + 3)^{1/2} + \dfrac{x + 1}{(2x + 3)^{1/2}}$$
$$= \dfrac{(2x + 3)^{1/2}(2x + 3)^{1/2} + x + 1}{(2x + 3)^{1/2}} = \dfrac{2x + 3 + x + 1}{(2x + 3)^{1/2}}$$
$$= \dfrac{3x + 4}{(2x + 3)^{1/2}}$$

11. $\left(\dfrac{4a^{-1/2}b^{3/4}}{b^{-2}}\right)\left(\dfrac{b^{-1}}{2a}\right) = \dfrac{4a^{-1/2}b^{3/4 - 1}}{2ab^{-2}} = \dfrac{2b^{2 - 1/4}}{a^{1 + 1/2}} = \dfrac{2b^{7/4}}{a^{3/2}}$

12. $\dfrac{2\sqrt{15} + \sqrt{3}}{\sqrt{15} - 2\sqrt{3}} = \dfrac{2\sqrt{15} + \sqrt{3}}{\sqrt{15} - 2\sqrt{3}} \times \dfrac{\sqrt{15} + 2\sqrt{3}}{\sqrt{15} + 2\sqrt{3}}$
$$= \dfrac{2(15) + 4\sqrt{45} + \sqrt{45} + 2(3)}{15 - 4(3)}$$
$$= \dfrac{36 + 15\sqrt{5}}{3}$$
$$= 12 + 5\sqrt{5}$$

13. $\dfrac{3^{-1/2}}{2} = \dfrac{1}{2 \times 3^{1/2}} = \dfrac{1}{2\sqrt{3}}$
$$= \dfrac{\sqrt{3}}{2\sqrt{3}\sqrt{3}} = \dfrac{\sqrt{3}}{6}$$

14. $0.220N^{-1/6} \qquad N = 64 \times 10^6$
$$0.220(64 \times 10^6)^{-1/6} = \dfrac{0.220}{(64 \times 10^6)^{1/6}}$$
$$= \dfrac{0.220}{2 \times 10} = 0.011$$

Chapter 12

1. $(3 - \sqrt{-4}) + (5\sqrt{-9} - 1) = (3 - 2j) + [5(3j) - 1]$
$$= 3 - 2j + 15j - 1$$
$$= 2 + 13j$$

2. $(2\underline{/130°})(3\underline{/45°}) = (2)(3)\underline{/130° + 45°} = 6\underline{/175°}$

3. $2 - 7j: \quad r = \sqrt{2^2 + (-7)^2} = \sqrt{53} = 7.28$
$$\tan\theta = \dfrac{-7}{2} = -3.500, \qquad \theta_{\text{ref}} = 74.1°, \qquad \theta = 285.9°$$
$$2 - 7j = 7.28(\cos 285.9° + j\sin 285.9°)$$
$$= 7.28\underline{/285.9°}$$

4. (a) $-\sqrt{-64} = -(8j) = -8j$
(b) $-j^{15} = -j^{12}j^3 = (-1)(-j) = j$

5. $(4 - 3j) + (-1 + 4j) = 3 + j$

6. $\dfrac{2 - 4j}{5 + 3j} = \dfrac{(2 - 4j)(5 - 3j)}{(5 + 3j)(5 - 3j)} = \dfrac{10 - 26j + 12j^2}{25 - 9j^2}$
$$= \dfrac{10 - 26j - 12}{25 - 9(-1)} = \dfrac{-2 - 26j}{34} = -\dfrac{1 + 13j}{17}$$

7. $2.56(\cos 125.2° + j\sin 125.2°) = 2.56e^{2.185j}$
$$125.2° = \dfrac{125.2\pi}{180} = 2.185 \text{ rad}$$

8. $R = 3.50\ \Omega,\ X_L = 6.20\ \Omega,\ X_C = 7.35\ \Omega$
$$|Z| = \sqrt{R^2 + (X_L - X_C)^2}$$
$$= \sqrt{3.50^2 + (6.20 - 7.35)^2} = 3.68\ \Omega$$
$$\tan\theta = \dfrac{X_L - X_C}{R} = \dfrac{6.20 - 7.35}{3.50} = \dfrac{-1.15}{3.50}$$
$$\theta = -18.2°$$

9. $3.47 - 2.81j = 4.47e^{5.60j}$
$$R = \sqrt{3.47^2 + (-2.81)^2} = 4.47$$
$$\tan\theta = \dfrac{-2.81}{3.47}, \qquad \theta_{\text{ref}} = 39.0°$$
$$\theta = 321.0° = 5.60 \text{ rad}$$

10. $x + 2j - y = yj - 3xj$
$$x - y + 3xj - yj = -2j$$
$$(x - y) + (3x - y)j = 0 - 2j$$
$$x - y = \quad 0$$
$$\underline{3x - y = -2}$$
$$2x = -2$$
$$x = -1, \qquad y = -1$$

11. $L = 8.75 \text{ mH} = 8.75 \times 10^{-3} \text{ H}$
$f = 600 \text{ kHz} = 6.00 \times 10^{5} \text{ Hz}$

$$2\pi f L = \frac{1}{2\pi f C}$$

$$C = \frac{1}{(2\pi f)^2 L} = \frac{1}{(2\pi)^2 (6.00 \times 10^5)^2 (8.75 \times 10^{-3})}$$
$$= 8.04 \times 10^{-12} = 8.04 \text{ pF}$$

12. $j = 1(\cos 90° + j \sin 90°)$

$$j^{1/3} = 1^{1/3}\left(\cos \frac{90°}{3} + j \sin \frac{90°}{3}\right)$$
$$= \cos 30° + j \sin 30° = 0.8660 + 0.5000j$$
$$= 1^{1/3}\left(\cos \frac{90° + 360°}{3} + j \sin \frac{90° + 360°}{3}\right)$$
$$= \cos 150° + j \sin 150° = -0.8660 + 0.5000j$$
$$= 1^{1/3}\left(\cos \frac{90° + 720°}{3} + j \sin \frac{90° + 720°}{3}\right)$$
$$= \cos 270° + j \sin 270° = -j$$

Cube roots of j: $0.8660 + 0.5000j$
$-0.8660 + 0.5000j$
$-j$

Chapter 13

1. $\log_9 x = -\dfrac{1}{2}$

$x = 9^{-1/2}$

$= \dfrac{1}{9^{1/2}} = \dfrac{1}{3}$

2. $\log_3 x - \log_3 2 = 2$

$\log_3 \dfrac{x}{2} = 2$

$\dfrac{x}{2} = 3^2$

$x = 2(3^2) = 18$

3. $\log_x 64 = 3$

$64 = x^3$

$4^3 = x^3$

$x = 4$

4. $3^{3x+1} = 8$

$(3x + 1)\log 3 = \log 8$

$3x + 1 = \dfrac{\log 8}{\log 3}$

$x = \dfrac{1}{3}\left(\dfrac{\log 8}{\log 3} - 1\right) = 0.298$

5. $y = 2 \log_4 x$

x	$\frac{1}{4}$	1	4	16
y	-2	0	2	4

$\log_4 \frac{1}{4} = -1$, $\log_4 16 = 2$

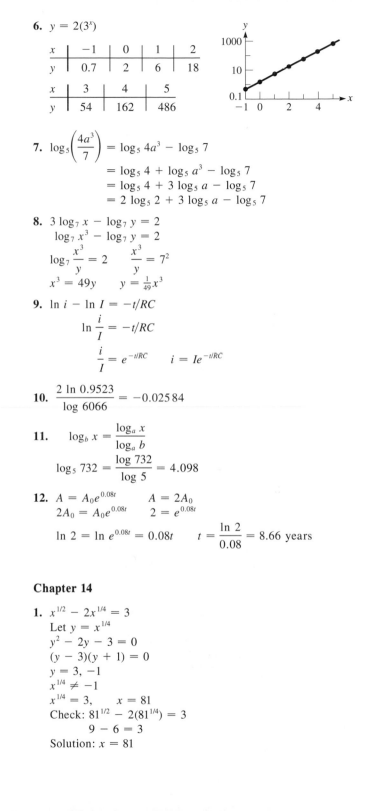

6. $y = 2(3^x)$

x	-1	0	1	2
y	0.7	2	6	18

x	3	4	5
y	54	162	486

7. $\log_5\left(\dfrac{4a^3}{7}\right) = \log_5 4a^3 - \log_5 7$

$= \log_5 4 + \log_5 a^3 - \log_5 7$

$= \log_5 4 + 3 \log_5 a - \log_5 7$

$= 2 \log_5 2 + 3 \log_5 a - \log_5 7$

8. $3 \log_7 x - \log_7 y = 2$

$\log_7 x^3 - \log_7 y = 2$

$\log_7 \dfrac{x^3}{y} = 2 \qquad \dfrac{x^3}{y} = 7^2$

$x^3 = 49y \qquad y = \frac{1}{49}x^3$

9. $\ln i - \ln I = -t/RC$

$\ln \dfrac{i}{I} = -t/RC$

$\dfrac{i}{I} = e^{-t/RC} \qquad i = Ie^{-t/RC}$

10. $\dfrac{2 \ln 0.9523}{\log 6066} = -0.025\,84$

11. $\log_b x = \dfrac{\log_a x}{\log_a b}$

$\log_5 732 = \dfrac{\log 732}{\log 5} = 4.098$

12. $A = A_0 e^{0.08t} \qquad A = 2A_0$
$2A_0 = A_0 e^{0.08t} \qquad 2 = e^{0.08t}$

$\ln 2 = \ln e^{0.08t} = 0.08t \qquad t = \dfrac{\ln 2}{0.08} = 8.66 \text{ years}$

Chapter 14

1. $x^{1/2} - 2x^{1/4} = 3$

Let $y = x^{1/4}$

$y^2 - 2y - 3 = 0$

$(y - 3)(y + 1) = 0$

$y = 3, -1$

$x^{1/4} \neq -1$

$x^{1/4} = 3, \qquad x = 81$

Check: $81^{1/2} - 2(81^{1/4}) = 3$

$9 - 6 = 3$

Solution: $x = 81$

2. $3\sqrt{x-2} - \sqrt{x+1} = 1$

$$3\sqrt{x-2} = 1 + \sqrt{x+1}$$
$$9(x-2) = 1 + 2\sqrt{x+1} + (x+1)$$
$$8x - 20 = 2\sqrt{x+1}$$
$$4x - 10 = \sqrt{x+1}$$
$$16x^2 - 80x + 100 = x + 1$$
$$16x^2 - 81x + 99 = 0$$
$$x = \frac{81 \pm \sqrt{81^2 - 4(16)(99)}}{32}$$
$$= \frac{81 \pm 15}{32} = 3, \frac{33}{16}$$

Check: $x = 3$: $3\sqrt{3-2} - \sqrt{3+1} \overset{?}{=} 1$
$$3 - 2 = 1$$
$$x = \tfrac{33}{16}: 3\sqrt{\tfrac{33}{16} - 2} - \sqrt{\tfrac{33}{16} + 1} \overset{?}{=} 1$$
$$\tfrac{3}{4} - \tfrac{7}{4} \neq 1$$
Solution: $x = 3$

3. $x^4 - 17x^2 + 16 = 0$

Let $y = x^2$
$$y^2 - 17y + 16 = 0$$
$$(y - 1)(y - 16) = 0$$
$$y = 1, 16$$
$$x^2 = 1, 16$$
$$x = -1, 1, -4, 4$$
All values check.

4. $x^2 - 2y = 5$

$$\underline{2x + 6y = 1}$$
$$y = \tfrac{1}{2}(x^2 - 5)$$
$$2x + 6(\tfrac{1}{2})(x^2 - 5) = 1$$
$$3x^2 + 2x - 16 = 0$$
$$(3x + 8)(x - 2) = 0$$
$$x = -\tfrac{8}{3}, 2$$
$$x = -\tfrac{8}{3}: y = \tfrac{1}{2}(\tfrac{64}{9} - 5) = \tfrac{19}{18}$$
$$x = 2: y = \tfrac{1}{2}(4 - 5) = -\tfrac{1}{2}$$
$$x = -\tfrac{8}{3}, y = \tfrac{19}{18}$$
or $x = 2, y = -\tfrac{1}{2}$

5. $v = \sqrt{v_0^2 + 2gh}$

$$v^2 = v_0^2 + 2gh$$
$$2gh = v^2 - v_0^2$$
$$h = \frac{v^2 - v_0^2}{2g}$$

6. $x^2 - y^2 = 4 \qquad xy = 2$

$$y = \pm\sqrt{x^2 - 4} \qquad y = \frac{2}{x}$$

x	y
± 2	0
± 3	± 2.2
± 4	± 3.5

x	y
-4	$-\tfrac{1}{2}$
-2	-1
-1	-2
0	—
1	2
2	1
4	$\tfrac{1}{2}$

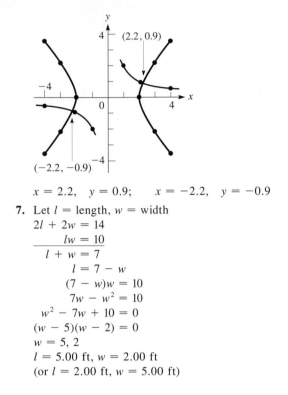

$x = 2.2, \quad y = 0.9; \qquad x = -2.2, \quad y = -0.9$

7. Let $l =$ length, $w =$ width

$$2l + 2w = 14$$
$$\underline{lw = 10}$$
$$l + w = 7$$
$$l = 7 - w$$
$$(7 - w)w = 10$$
$$7w - w^2 = 10$$
$$w^2 - 7w + 10 = 0$$
$$(w - 5)(w - 2) = 0$$
$$w = 5, 2$$
$l = 5.00$ ft, $w = 2.00$ ft
(or $l = 2.00$ ft, $w = 5.00$ ft)

Chapter 15

1. $f(x) = 2x^3 + 3x^2 + 7x - 6$
$$f(-3) = 2(-3)^3 + 3(-3)^2 + 7(-3) - 6$$
$$= -54 + 27 - 21 - 6 = -54$$
$f(-3) \neq 0, \qquad -3$ is not a zero

2. $x^4 - 2x^3 - 7x^2 + 20x - 12 = 0$

$$\begin{array}{rrrrr|l}
1 & -2 & -7 & 20 & -12 & \underline{2} \\
  & 2 & 0 & -14 & 12 & \\
\hline
1 & 0 & -7 & 6 & & \underline{2} \\
  & 2 & 4 & -6 & & \\
\hline
1 & 2 & -3 & & &
\end{array}$$

$x^2 + 2x - 3 = (x + 3)(x - 1)$
Other roots: $x = -3, 1$

3. $(x^3 - 5x^2 + 4x - 9) \div (x - 3)$

$$\begin{array}{rrrr|l}
1 & -5 & 4 & -9 & \underline{3} \\
  & 3 & -6 & -6 & \\
\hline
1 & -2 & -2 & -15 &
\end{array}$$

Quotient: $x^2 - 2x - 2$
Remainder $= -15$

4. $f(x) = 2x^4 + 15x^3 + 23x^2 - 16$
$2x + 1 = 2(x + \tfrac{1}{2})$

$$\begin{array}{rrrrr|l}
2 & 15 & 23 & 0 & -16 & \underline{-\tfrac{1}{2}} \\
  & -1 & -7 & -8 & 4 & \\
\hline
2 & 14 & 16 & -8 & -12 &
\end{array}$$

Remainder is not zero;
$2x + 1$ is not a factor.

5. $(x^3 + 4x^2 + 7x - 9) \div (x + 4)$
$f(x) = x^3 + 4x^2 + 7x - 9$
$f(-4) = (-4)^3 + 4(-4)^2 + 7(-4) - 9$
$\qquad = -64 + 64 - 28 - 9 = -37$
Remainder $= -37$

6. $2x^4 - x^3 + 5x^2 - 4x - 12 = 0$
$\quad f(x) = 2x^4 - x^3 + 5x^2 - 4x - 12$
$\quad f(-x) = 2x^4 + x^3 + 5x^2 + 4x - 12$
$n = 4$; 4 roots
No more than 3 positive roots
One negative root
Rational roots: factors of 12 divided by factors of 2
Possible rational roots: $\pm 1, \pm 2, \pm 3, \pm 4, \pm 6, \pm 12,$
$\quad \pm \frac{1}{2}, \pm \frac{3}{2}$

$$
\begin{array}{rrrrr|l}
2 & -1 & 5 & -4 & -12 & \underline{2} \\
 & 4 & 6 & 22 & 36 & \\
\hline
2 & 3 & 11 & 18 & 24 &
\end{array}
$$

2 is too large.

$$
\begin{array}{rrrrr|l}
2 & -1 & 5 & -4 & -12 & \underline{\frac{3}{2}} \\
 & 3 & 3 & 12 & 12 & \\
\hline
2 & 2 & 8 & 8 & 0 & \underline{-1}
\end{array}
$$

$$
\begin{array}{rrrr}
 & -2 & 0 & -8 \\
\hline
2 & 0 & 8 & 0
\end{array}
$$

$2x^2 + 8 = 0, \qquad x^2 + 4 = 0, \qquad x = \pm 2j$
roots: $\frac{3}{2}, -1, 2j, -2j$

7. $y = kx^2(x^3 + 436x - 4000)$
$y = 0, \ kx^2(x^3 + 436x - 4000) = 0$
$x = 0, \ x^3 + 436x - 4000 = 0$

$$
\begin{array}{rrrr|l}
1 & 0 & 436 & -4000 & \underline{8} \\
 & 8 & 64 & 4000 & \\
\hline
1 & 8 & 500 & 0 &
\end{array}
$$

$y = 0$ for $x = 0$ ft, $\qquad x = 8$ ft

8. Let $x =$ length of edge.
$V = x^3, \qquad 2V = (x + 1)^3$
$2x^3 = x^3 + 3x^2 + 3x + 1$
$x^3 - 3x^2 - 3x - 1 = 0$
$y = x^3 - 3x^2 - 3x - 1$

$x = 3.8$ mm

Chapter 16

1. $A = \begin{bmatrix} 3 & -1 & 4 \\ 2 & 0 & -2 \end{bmatrix} \qquad B = \begin{bmatrix} 1 & 4 & 5 \\ -1 & -2 & 3 \end{bmatrix}$

$2B = \begin{bmatrix} 2 & 8 & 10 \\ -2 & -4 & 6 \end{bmatrix}$

$A - 2B = \begin{bmatrix} 3 - 2 & -1 - 8 & 4 - 10 \\ 2 + 2 & 0 + 4 & -2 - 6 \end{bmatrix}$

$\qquad = \begin{bmatrix} 1 & -9 & -6 \\ 4 & 4 & -8 \end{bmatrix}$

2. $\begin{vmatrix} 4 & 0 & -2 \\ 3 & -3 & 2 \\ -4 & 1 & -1 \end{vmatrix} = 4 \begin{vmatrix} -3 & 2 \\ 1 & -1 \end{vmatrix} - 0 \begin{vmatrix} 3 & 2 \\ -4 & -1 \end{vmatrix}$

$\qquad\qquad + (-2) \begin{vmatrix} 3 & -3 \\ -4 & 1 \end{vmatrix} \quad$ using first row

$\qquad = 4(3 - 2) - 0 - 2(3 - 12)$
$\qquad = 4 - 2(-9) = 4 + 18 = 22$

$= -0 \begin{vmatrix} 3 & 2 \\ -4 & -1 \end{vmatrix} + (-3) \begin{vmatrix} 4 & -2 \\ -4 & -1 \end{vmatrix} - (1) \begin{vmatrix} 4 & -2 \\ 3 & 2 \end{vmatrix}$

$\qquad\qquad\qquad\qquad\qquad\qquad$ using second column

$= 0 - 3(-4 - 8) - (8 + 6) = -3(-12) - 14 = 22$

3. $CD = \begin{bmatrix} 1 & 0 & 4 \\ 2 & -2 & 1 \\ -1 & 3 & 2 \end{bmatrix} \begin{bmatrix} 2 & -2 \\ 4 & -5 \\ 6 & 1 \end{bmatrix}$

$\quad = \begin{bmatrix} 2 + 0 + 24 & -2 + 0 + 4 \\ 4 - 8 + 6 & -4 + 10 + 1 \\ -2 + 12 + 12 & 2 - 15 + 2 \end{bmatrix}$

$\quad = \begin{bmatrix} 26 & 2 \\ 2 & 7 \\ 22 & -11 \end{bmatrix}$

$DC = \begin{bmatrix} 2 & -2 \\ 4 & -5 \\ 6 & 1 \end{bmatrix} \begin{bmatrix} 1 & 0 & 4 \\ 2 & -2 & 1 \\ -1 & 3 & 2 \end{bmatrix} \quad \begin{array}{l} \text{not defined, since} \\ D \text{ has 2 columns} \\ \text{and } C \text{ has 3 rows} \end{array}$

4. $\begin{vmatrix} 1 & 0 & 4 & -2 \\ -2 & -1 & 3 & 0 \\ 3 & 2 & -1 & 2 \\ 1 & 1 & -1 & -2 \end{vmatrix} = \begin{vmatrix} 1 & 0 & 0 & 0 \\ -2 & -1 & 11 & -4 \\ 3 & 2 & -13 & 8 \\ 1 & 1 & -5 & 0 \end{vmatrix}$

$= 1 \begin{vmatrix} -1 & 11 & -4 \\ 2 & -13 & 8 \\ 1 & -5 & 0 \end{vmatrix} = \begin{vmatrix} -1 & 11 & -4 \\ 0 & 9 & 0 \\ 0 & 6 & -4 \end{vmatrix}$

$= -1 \begin{vmatrix} 9 & 0 \\ 6 & -4 \end{vmatrix} = -(-36 - 0) = 36$

5. $\begin{bmatrix} 1 & 0 & 4 & | & 1 & 0 & 0 \\ 2 & -2 & 1 & | & 0 & 1 & 0 \\ -1 & 3 & 2 & | & 0 & 0 & 1 \end{bmatrix} \rightarrow \begin{bmatrix} 1 & 0 & 4 & | & 1 & 0 & 0 \\ 0 & -2 & -7 & | & -2 & 1 & 0 \\ 0 & 3 & 6 & | & 1 & 0 & 1 \end{bmatrix} \rightarrow$

$\begin{bmatrix} 1 & 0 & 4 & | & 1 & 0 & 0 \\ 0 & 1 & \frac{7}{2} & | & 1 & -\frac{1}{2} & 0 \\ 0 & 3 & 6 & | & 1 & 0 & 1 \end{bmatrix} \rightarrow \begin{bmatrix} 1 & 0 & 4 & | & 1 & 0 & 0 \\ 0 & 1 & \frac{7}{2} & | & 1 & -\frac{1}{2} & 0 \\ 0 & 0 & -\frac{9}{2} & | & -2 & \frac{3}{2} & 1 \end{bmatrix} \rightarrow$

$\begin{bmatrix} 1 & 0 & 4 & | & 1 & 0 & 0 \\ 0 & 1 & \frac{7}{2} & | & 1 & -\frac{1}{2} & 0 \\ 0 & 0 & 1 & | & \frac{4}{9} & -\frac{1}{3} & -\frac{2}{9} \end{bmatrix} \rightarrow \begin{bmatrix} 1 & 0 & 0 & | & -\frac{7}{9} & \frac{4}{3} & \frac{8}{9} \\ 0 & 1 & 0 & | & -\frac{5}{9} & \frac{2}{3} & \frac{7}{9} \\ 0 & 0 & 1 & | & \frac{4}{9} & -\frac{1}{3} & -\frac{2}{9} \end{bmatrix}$

$C^{-1} = \begin{bmatrix} -\frac{7}{9} & \frac{4}{3} & \frac{8}{9} \\ -\frac{5}{9} & \frac{2}{3} & \frac{7}{9} \\ \frac{4}{9} & -\frac{1}{3} & -\frac{2}{9} \end{bmatrix}$

6. $2x - 3y = 11$
     $x + 2y = 2$

$$A = \begin{bmatrix} 2 & -3 \\ 1 & 2 \end{bmatrix} \qquad A^{-1} = \begin{bmatrix} \frac{2}{7} & \frac{3}{7} \\ -\frac{1}{7} & \frac{2}{7} \end{bmatrix} \qquad C = \begin{bmatrix} 11 \\ 2 \end{bmatrix}$$

$$A^{-1}C = \begin{bmatrix} \frac{2}{7} & \frac{3}{7} \\ -\frac{1}{7} & \frac{2}{7} \end{bmatrix}\begin{bmatrix} 11 \\ 2 \end{bmatrix} = \begin{bmatrix} \frac{22}{7} + \frac{6}{7} \\ -\frac{11}{7} + \frac{4}{7} \end{bmatrix} = \begin{bmatrix} 4 \\ -1 \end{bmatrix}$$

$x = 4 \qquad y = -1$

7. $A = \begin{bmatrix} 7 & -2 & 1 \\ 2 & 3 & -4 \\ 4 & -5 & 2 \end{bmatrix} \qquad C = \begin{bmatrix} 6 \\ 6 \\ 10 \end{bmatrix}$

```
[A]⁻¹[C]
      [[.5  ]
       [-3  ]
       [-3.5]]
```

$x = 0.5 \qquad y = -3 \qquad z = -3.5$

8. Let A = number of shares of stock A
         B = number of shares of stock B

$50A + 30B = 2600$
$\underline{30A + 40B = 2000}$

$5A + 3B = 260$
$\underline{3A + 4B = 200}$

Let C = coefficient matrix

$$C = \begin{bmatrix} 5 & 3 \\ 3 & 4 \end{bmatrix}, \qquad \begin{vmatrix} 5 & 3 \\ 3 & 4 \end{vmatrix} = 20 - 9 = 11$$

$$C^{-1} = \frac{1}{11}\begin{bmatrix} 4 & -3 \\ -3 & 5 \end{bmatrix} = \begin{bmatrix} \frac{4}{11} & -\frac{3}{11} \\ -\frac{3}{11} & \frac{5}{11} \end{bmatrix}$$

$$\begin{bmatrix} A \\ B \end{bmatrix} = \begin{bmatrix} \frac{4}{11} & -\frac{3}{11} \\ -\frac{3}{11} & \frac{5}{11} \end{bmatrix}\begin{bmatrix} 260 \\ 200 \end{bmatrix} = \begin{bmatrix} \frac{4}{11}(260) - \frac{3}{11}(200) \\ -\frac{3}{11}(260) + \frac{5}{11}(200) \end{bmatrix}$$

$$= \begin{bmatrix} 40 \\ 20 \end{bmatrix}$$

$A = 40$ shares $B = 20$ shares

Chapter 17

1. $x < 0, y > 0$

2. $\dfrac{-x}{2} \geq 3$
     $-x \geq 6$
     $x \leq -6$

   (number line with arrow to left from open circle at -6)

3. $3x + 1 < -5$
     $3x < -6$
     $x < -2$

   (number line with arrow to left from open circle at -2)

4. $-1 < 1 - 2x < 5$
     $-2 < -2x < 4$
     $-2 < x < 1$

   (number line from open circle at -2 to open circle at 1)

5. $\dfrac{x^2 + x}{x - 2} \leq 0, \dfrac{x(x + 1)}{x - 2} \leq 0$

Interval	$\dfrac{x(x + 1)}{x - 2}$	Sign
$x < -1$	$\dfrac{-\ -}{-}$	$-$
$-1 < x < 0$	$\dfrac{-\ +}{-}$	$+$
$0 < x < 2$	$\dfrac{+\ +}{-}$	$-$
$x > 2$	$\dfrac{+\ +}{+}$	$+$

Solution: $x \leq -1$ or $0 \leq x < 2$ (x cannot equal 2)

   (number line at -1, 0, 2)

6. $|2x + 1| \geq 3$
     $2x + 1 \geq 3, \qquad 2x + 1 \leq -3$
     $2x \geq 2 \qquad\qquad 2x \leq -4$
     $x \geq 1$ or $\qquad x \leq -2$

   (number line at -2, 1)

7.

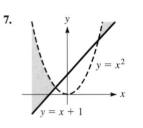

8. If $\sqrt{x^2 - x - 6}$ is real,
     then $x^2 - x - 6 \geq 0$.
     $(x - 3)(x + 2) \geq 0$
     $x \leq -2$ or $x \geq 3$

9. Let w = width, l = length
     $l = w + 20 \qquad w^2 + 20w - 4800 \geq 0$
     $wl \geq 4800 \qquad (w + 80)(w - 60) \geq 0$
     $w(w + 20) \geq 4800 \qquad w \geq 60$ m

10. Let A = length of type A wire
       B = length of type B wire
     $0.10A + 0.20B < 5.00$
     $A + 2B < 50$

   (graph with B axis to 25, A axis to 50, shaded triangle)

11. $|\lambda - 550 \text{ nm}| < 150$ nm   (within 150 nm of
                                              $\lambda = 550$ nm)

12. $x^2 > 12 - x$

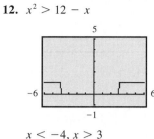

$x < -4, x > 3$

Chapter 18

1. $\dfrac{180 \text{ s}}{4 \text{ min}} = \dfrac{180 \text{ s}}{240 \text{ s}} = \dfrac{3}{4}$

2. $F = \dfrac{k}{d};$ $0.750 = \dfrac{k}{1.25},$ $k = 0.938 \text{ N·cm}$

$F = \dfrac{0.938}{d};$ $F = \dfrac{0.938}{1.75} = 0.536 \text{ N}$

3. $\dfrac{1.00 \text{ in.}}{2.54 \text{ cm}} = \dfrac{x}{7.24 \text{ cm}}$

$x = \dfrac{7.24}{2.54} = 2.85 \text{ in.}$

4. $p = kdh$
Using values for water,
$1.96 = k(1000)(0.200)$
$k = 0.00980 \text{ kPa·m}^2/\text{kg}$
For alcohol,
$p = 0.00980(800)(0.300)$
$\quad = 2.35 \text{ kPa}$

5. $\quad 2l + 2w = 210.0$
$\qquad\qquad l = 105.0 - w$
$\qquad \dfrac{l}{w} = \dfrac{7}{3}$
$\dfrac{105.0 - w}{w} = \dfrac{7}{3}$
$315.0 - 3w = 7w$
$\qquad\quad 10w = 315.0$
$\qquad\qquad w = 31.5 \text{ in.}$
$\qquad\qquad l = 105.0 - 31.5 = 73.5 \text{ in.}$

6. Let L_1 = crushing load of first pillar
$\quad\;\; L_2$ = crushing load of second pillar

$L_2 = \dfrac{kr_2^4}{l_2^2} \qquad L_1 = \dfrac{k(2r_2)^4}{(3l_2)^2}$

$\dfrac{L_1}{L_2} = \dfrac{\dfrac{k(2r_2)^4}{(3l_2)^2}}{\dfrac{kr_2^4}{l_2^2}} = \dfrac{16kr_2^4}{9l_2^2} \times \dfrac{l_2^2}{kr_2^4} = \dfrac{16}{9}$

Chapter 19

1. $6, -2, \frac{2}{3}, \ldots;$
geometric sequence

$a_1 = 6 \qquad r = \dfrac{-2}{6} = -\dfrac{1}{3}$

$S_7 = \dfrac{6[1 - (-\frac{1}{3})^7]}{1 - (-\frac{1}{3})}$

$\quad = \dfrac{6\left(1 + \dfrac{1}{3^7}\right)}{\frac{4}{3}} = \dfrac{1094}{243}$

2. $a_1 = 6, d = 4, s_n = 126;$
arithmetic sequence
$126 = \frac{n}{2}(6 + a_n)$
$\quad a_n = 6 + (n - 1)4 = 2 + 4n$
$126 = \frac{n}{2}[6 + (2 + 4n)]$
$252 = n(8 + 4n) = 4n^2 + 8n$
$\quad n^2 + 2n - 63 = 0$
$(n + 9)(n - 7) = 0$
$\qquad\qquad n = 7$

3. $0.454545\ldots = 0.45 + 0.0045 + 0.000045 + \cdots$
$a = 0.45 \qquad r = 0.01$
$S = \dfrac{0.45}{1 - 0.01} = \dfrac{0.45}{0.99} = \dfrac{5}{11}$

4. $\sqrt{1 - 4x} = (1 - 4x)^{1/2}$

$= 1 + (\tfrac{1}{2})(-4x) + \dfrac{\frac{1}{2}(\frac{1}{2} - 1)}{2}(-4x)^2 + \cdots$

$= 1 - 2x - 2x^2 + \cdots$

5. $(2x - y)^5 = (2x)^5 + 5(2x)^4(-y) + \dfrac{5(4)}{2}(2x)^3(-y)^2$

$+ \dfrac{5(4)(3)}{2(3)}(2x)^2(-y)^3$

$+ \dfrac{5(4)(3)(2)}{2(3)(4)}(2x)(-y)^4 + (-y)^5$

$= 32x^5 - 80x^4y + 80x^3y^2 - 40x^2y^3 + 10xy^4 - y^5$

6. $5\% = 0.05$
Value after 1 year is
$2500 + 2500(0.05) = 2500(1.05)$
$V_{20} = 2500(1.05)^{20}$
$\quad\; = \$6633.24$

7. $2 + 4 + \cdots + 200$
$a_1 = 2, \qquad a_{100} = 200, \qquad n = 100$
$S_{100} = \frac{100}{2}(2 + 200)$
$\quad\;\; = 10{,}100$

8. Ball falls 8.00 ft, rises 4.00 ft, falls 4.00 ft, etc.
Distance $= 8.00 + (4.00 + 4.00)$
$+ (2.00 + 2.00) + \cdots$
$= 8.00 + 8.00 + 4.00 + 2.00 + \cdots$
$= 8.00 + \dfrac{8.00}{1 - 0.5} = 8.00 + 16.0 = 24.0 \text{ ft}$

Chapter 20

1.
$$\sec\theta - \frac{\tan\theta}{\csc\theta} = \frac{1}{\cos\theta} - \frac{\dfrac{\sin\theta}{\cos\theta}}{\dfrac{1}{\sin\theta}} = \frac{1}{\cos\theta} - \frac{\sin^2\theta}{\cos\theta}$$

$$= \frac{1 - \sin^2\theta}{\cos\theta} = \frac{\cos^2\theta}{\cos\theta} = \cos\theta$$

$$\cos\theta = \cos\theta$$

2. $\sin 2x + \sin x = 0$
$2\sin x\cos x + \sin x = 0$
$\sin x(2\cos x + 1) = 0$

$\sin x = 0 \qquad \cos x = -\dfrac{1}{2}$

$x = 0, \pi, \frac{2\pi}{3}, \frac{4\pi}{3}$

3. $\theta = \sin^{-1} x$
$\cos\theta = \cos(\sin^{-1} x)$
$\quad = \sqrt{1 - x^2}$

4. $i = 8.00e^{-20t}(1.73\cos 10.0t - \sin 10.0t)$
Since $8.00e^{-20t}$ will not equal zero, $i = 0$ if
$1.73\cos 10.0t - \sin 10.0t = 0$
$1.73 = \tan 10.0t$
$10.0t = \tan^{-1} 1.73$
$\quad t = 0.105$ s

5.
$$\frac{\tan\alpha + \tan\beta}{\tan\alpha - \tan\beta} = \frac{\dfrac{\sin\alpha}{\cos\alpha} + \dfrac{\sin\beta}{\cos\beta}}{\dfrac{\sin\alpha}{\cos\alpha} - \dfrac{\sin\beta}{\cos\beta}}$$

$$= \frac{\sin\alpha\cos\beta + \sin\beta\cos\alpha}{\sin\alpha\cos\beta - \sin\beta\cos\alpha}$$

$$= \frac{\sin(\alpha + \beta)}{\sin(\alpha - \beta)}$$

6. $\cot^2 x - \cos^2 x = \dfrac{\cos^2 x}{\sin^2 x} - \cos^2 x =$
$\cos^2 x(\csc^2 x - 1) = \cos^2 x\cot^2 x;$
$\cos^2 x\cot^2 x = \cos^2 x\cot^2 x$

7. $\sin x = -\dfrac{3}{5}$
$\cos x = \dfrac{4}{5}$
$$\cos\frac{x}{2} = -\sqrt{\frac{1 + \left(\frac{4}{5}\right)}{2}} = -\sqrt{\frac{9}{10}} = -0.9487$$
since $270° < x < 360°$, $135° < \frac{x}{2} < 180°$;
$\frac{x}{2}$ is in second quadrant, where $\cos\frac{x}{2}$ is negative

$r = 5$
$(4, -3)$

8. $I = I_0\sin 2\theta\cos 2\theta$
$\dfrac{2I}{I_0} = 2\sin 2\theta\cos 2\theta = \sin 4\theta$
$4\theta = \sin^{-1}\dfrac{2I}{I_0}$
$\theta = \dfrac{1}{4}\sin^{-1}\dfrac{2I}{I_0}$

9. $y = x - 2\cos x - 5$
$x = 3.02, 4.42, 6.77$

Chapter 21

1. $\qquad 2(x^2 + x) = 1 - y^2$
$2x^2 + y^2 + 2x - 1 = 0$
$A \neq C$ (same sign)
$B = 0$: ellipse

2. $4x - 2y + 5 = 0$
$y = 2x + \dfrac{5}{2}$
$m = 2 \qquad b = \dfrac{5}{2}$

3. $\qquad x^2 = 2x - y^2$
$x^2 + y^2 = 2x$
$r^2 = 2r\cos\theta$
$r = 2\cos\theta$

4. $x^2 = -12y$
$4p = -12, \qquad p = -3$
$V(0,0) \qquad F(0, -3)$

5. Center $(-1, 2)$; $h = -1, k = 2$
$r = \sqrt{(2 + 1)^2 + (3 - 2)^2}$
$\quad = \sqrt{10}$
$(x + 1)^2 + (y - 2)^2 = 10$
or
$x^2 + y^2 + 2x - 4y - 5 = 0$

6. $m = \dfrac{-2 - 1}{2 + 4} = -\dfrac{1}{2}$
$y - 1 = -\dfrac{1}{2}(x + 4)$
$2y - 2 = -x - 4$
$x + 2y + 2 = 0$

7. $x^2 = 4py$
$(6.00)^2 = 4p(4.00)$
$p = 2.25$
Focus is 2.25 cm
from vertex.

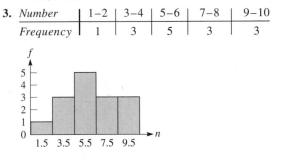

8. $a = 8$ $b = 4$

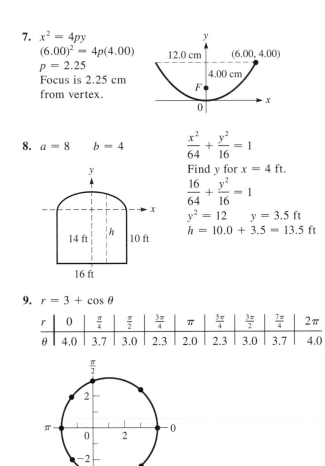

$\dfrac{x^2}{64} + \dfrac{y^2}{16} = 1$

Find y for $x = 4$ ft.

$\dfrac{16}{64} + \dfrac{y^2}{16} = 1$

$y^2 = 12$ $y = 3.5$ ft

$h = 10.0 + 3.5 = 13.5$ ft

9. $r = 3 + \cos\theta$

r	0	$\frac{\pi}{4}$	$\frac{\pi}{2}$	$\frac{3\pi}{4}$	π	$\frac{5\pi}{4}$	$\frac{3\pi}{2}$	$\frac{7\pi}{4}$	2π
θ	4.0	3.7	3.0	2.3	2.0	2.3	3.0	3.7	4.0

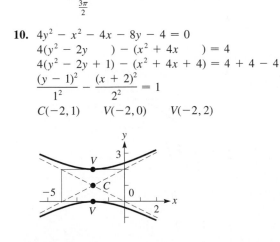

10. $4y^2 - x^2 - 4x - 8y - 4 = 0$
$4(y^2 - 2y\quad) - (x^2 + 4x\quad) = 4$
$4(y^2 - 2y + 1) - (x^2 + 4x + 4) = 4 + 4 - 4$
$\dfrac{(y-1)^2}{1^2} - \dfrac{(x+2)^2}{2^2} = 1$
$C(-2,1)$ $V(-2,0)$ $V(-2,2)$

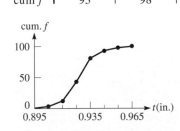

Chapter 22

1.
Number	1	2	3	4	5	6	7	8	9	10
Frequency	1	0	1	2	3	2	1	2	2	1

$\sum f = 15$

Median is eighth number; median = 6.

2. 5 appears 3 times, and no other number appears more than twice; mode = 5.

3.
Number	1–2	3–4	5–6	7–8	9–10
Frequency	1	3	5	3	3

4. Let t = thickness

$$\bar{t} = \frac{3(0.90) + 9(0.91) + 31(0.92) + 38(0.93) + 12(0.94) + 5(0.95) + 2(0.96)}{3 + 9 + 31 + 38 + 12 + 5 + 2}$$

$$= \frac{92.7}{100} = 0.927 \text{ in.}$$

5. $\Sigma x^2 = 3(0.90)^2 + 9(0.91)^2 + 31(0.92)^2 + 38(0.93)^2$
$\qquad\qquad + 12(0.94)^2 + 5(0.95)^2 + 2(0.96)^2 = 85.9464$

$(\Sigma x)^2 = 92.7^2$

$$s = \sqrt{\frac{100(85.9464) - 92.7^2}{100(99)}} = 0.117 \text{ in.}$$

6.

7.
t (in.)	0.90	0.91	0.92	0.93	0.94	0.95	0.96
%	3	9	31	38	12	5	2

8.
t (in.)	<0.905	<0.915	<0.925	<0.935
cum f	3	12	43	81

t (in.)	<0.945	<0.955	<0.965
cum f	93	98	100

9. $0.5000 + 0.2257 = 0.7257$ (total area to the left)
$\qquad\qquad\qquad\qquad\quad = 72.57\%$

10. Find the range of each subgroup (subtract the lowest value from the largest value). Find the mean R of these ranges (divide their sum by 20). This is the value of the central line. Multiply R by the appropriate control chart factors to obtain the upper control limit (UCL) and the lower control limit (LCL). Draw the central line, the UCL line, and the LCL line on a graph. Plot the points for each R corresponding to its subgroup and join successive points by straight-line segments.

11.

x	y	xy	x^2
1	5	5	1
3	11	33	9
5	17	85	25
7	20	140	49
9	27	243	81
25	80	506	165

$n = 5$

$$m = \frac{5(506) - (25)(80)}{5(165) - 25^2}$$
$$= 2.65$$

$$b = \frac{165(80) - (506)(25)}{5(165) - 25^2}$$
$$= 2.75$$

$$y = 2.65x + 2.75$$

12.

x	$\sqrt{x}$	y	$\sqrt{x}\,y$	$(\sqrt{x})^2 = x$
1.00	1.000	1.10	1.1000	1.000
3.00	1.732	1.90	3.2908	3.000
5.00	2.236	2.50	5.5900	5.000
7.00	2.646	2.90	7.6734	7.000
9.00	3.000	3.30	9.9000	9.000
	10.614	11.70	27.5542	25.000

$n = 5$

$$m = \frac{5(27.5542) - (10.614)(11.70)}{5(25.000) - (10.614)^2} = 1.10$$

$$b = \frac{(25.000)(11.70) - (27.5542)(10.614)}{5(25.000) - (10.614)^2} = 0.00$$
(rounded off)

$$y = 1.10\sqrt{x}$$

Acoustics

Acoustical intensity, 292(16), 485(46)
Acoustics, 173(58)
Loudness of siren, 299(58)
Loudness sensed by human ear, 308(58), 371(42), 375(31)
Sound of jet engine, 115(45), 590(80)
Sound reflection in buildings, 568(35)
Sound wave displacement, 335(35)
Velocity of sound, 15(43), 25(39), 146(40), 152(42), 164(78), 552(43), A-9(35)
Ultrasonics, 589(69)
Wavelength of sound, 625(46)

Aeronautics

Aircraft emergency locater transmitter, 200(97)
Aircraft wing design, 68(3, 4), 478(13), 568(36)
Airfoil design, 194(64), 200(94)
Air traffic control, 272(31), 573(31)
Fuel supply, 151(35), 200(81), 479(34), A-9(45, 46, 47, 48)
Glide-slope indicator, 75(46)
Helicopter blade rotation, 239(40)
Helicopter position, 382(25)
Jet speed, 10(52), 16(52), 198(49), 219(66), 239(42), 484(32)
Mach number, 467(30), 625(43, 50)
Plane landing, 267(26), 589(58)
Plane location, 85(34), 251(21), 274(55), 371(48), 464(47)
Plane on radar, 126(55)
Plane route, 239(31), 272(25), 274(56), 590(79)
Plane speed, 132(17), 218(36), 207(51), 248(32), 260(13), 320(84), 329(32), 386(31), 429(25)
Propeller rotation, 242(4)
Radar, 200(85)
Rocket pursuit of aircraft, 308(60), 586(34)
Shock wave of SST Concorde, 580(40)
Speed of SST Concorde, 44(16), 273(45)
Wing lift and drag, 248(29), 485(34), 542(103)

Architecture

Arch design, 213(36), 562(30)
Architecture, 169(65), 198(38), 201(107, 121)
Attic room, 126(61)
Building design, 75(49)
Ceiling panels, 164(80)
CN Tower (Toronto) design, 46(97)
Courtyard design, 61(25)
Deck design, 382(28)
Door design, 413(40), 415(55)
Elliptical window, 568(38)
Fenway Park (Boston) playing area, 69(9, 10)
Floor design, 68(8), 76(5)
Gothic arch, 242(77)
Great Pyramid (Egypt) design, 72(21)
Hallway design, 58(38), 238(23), 590(8)
Kitchen tiling, 47(115)
Norman window, 220(70), 558(48)
Parabolic window, 562(32)
Patio design, 58(33), 65(33), 239(46), 586(33)
Pentagon (U.S. Dept. of Defense) design, 75(65), 267(21)
Ramp for disabled, 75(45), 122(9)
Remodeling, 213(43)
Roof design, 583(42)
Room design, 58(37)
Rotating restaurant, 239(36)
Security fence, 61(28), 123(21), 218(35), 238(27)
Spaceship Earth (Florida) design, 72(23)
Sunroom design, 242(8)
Superdome (New Orleans) design, 75(54)
Support design, 61(23)
Swimming pool design, 68(1, 2), 72(18)
Wall panel, 273(39)
Window design, 61(26), 103(8), 127(75), 220(8)
Window in sun, 126(57)
World Trade Center, 46(76), 124(26)

Astronomy

Astronomy, 533(44)
Black holes, 46(86)
Comet paths, 542(95)
Crater on moon, 274(48)
Diameter of Venus, 242(80)
Distance to Centaurus A, 23(34)
Earth's shadow, 75(56)
Hubble telescope, 75(59), 562(33)
Light from sun, 23(44)
Planetary motion, 320(78), 487(42), 567(32), 590(81)
Star brightness, 354(51), 377(79), 485(44)
Star movement, 240(51)

Automotive Technology

Accident analysis, 260(10)
Auto production, 447(85)
Auto repair, 201(116)
Battery life, 201(113)
Carburetor assembly, 152(40)
Compression ratio, 478(14)
Cost of operating car, 102(58)
Drive shaft rotation, 239(45)
Engine coolant, 16(55)
Engine cylinders, 15(1)
Engine efficiency, 312(57), 378(82), 474(59)
Engine emissions, 479(33)
Engine power, 218(32), 486(29)
Engine torque, 100(8, 16), 613(1, 2, 3, 4), A-9(44)

Electricity and Electronics

The final review exercise in each chapter is a writing exercise. Each of these 22 exercises will require at least a paragraph to provide a good explanation of the problem presented. Also, there are about 150 other exercises throughout the text (at least five in each chapter and marked with Ⓦ before the exercise number or instructions) that require at least one complete sentence (up to a paragraph) to complete the solution.

Following is a listing of these writing exercises. The first number shown is the page number, and the numbers in parentheses are the exercise numbers. The * denotes the final review exercise of the chapter.

BASIC CURVES

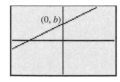

$y = mx + b$

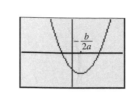

$y = ax^2 + bx + c \quad (a > 0)$

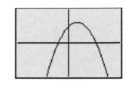

$y = ax^2 + bx + c \quad (a < 0)$

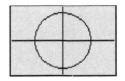

$x^2 + y^2 = a^2$

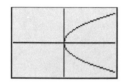

$y^2 = 4px \quad (p > 0)$

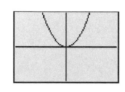

$x^2 = 4py \quad (p > 0)$

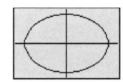

$\dfrac{x^2}{a^2} + \dfrac{y^2}{b^2} = 1$

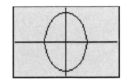

$\dfrac{y^2}{a^2} + \dfrac{x^2}{b^2} = 1$

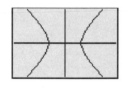

$\dfrac{x^2}{a^2} - \dfrac{y^2}{b^2} = 1$

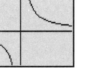

$\dfrac{y^2}{a^2} - \dfrac{x^2}{b^2} = 1$

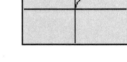

$xy = a \quad (a > 0)$

$y = a\sqrt{x} \quad (a > 0)$

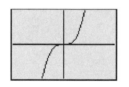

$y = ax^3 \quad (a > 0)$

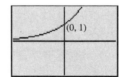

$y = ax^4 \quad (a > 0)$

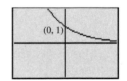

$y = b^x \quad (b > 1)$

$y = b^{-x} \quad (b > 1)$

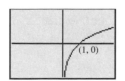

$y = \log_b x$

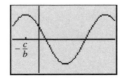

$y = a \sin(bx + c)$
$(a > 0, c > 0)$

$y = a \cos(bx + c)$
$(a > 0, c > 0)$

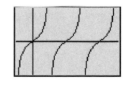

$y = a \tan x \quad (a > 0)$